MICROPROCESSORS AND INTERFACING

By .

NIKHIL MARRIWALA

Department of Electronics & Communication Engg.

University Institute of Engg. and Tech. Kurukshetra University

Kurukshetra

S.K. KATARIA & SONS®

Publisher of Engineering Books

4885/109, Prakash Mahal, Dr. Subhash Bhargav Lane,
Opposite Delhi Medical Association, Daryaganj, New Delhi-110002 (INDIA)
Phone: +91-11-43551243 Mobile: +91-9871775858, 7678425209
e-mail: katariabook@gmail.com, katariabooks@yahoo.com
Website: www.skkatariaandsons.com

(e-books, subscriptions and bulk buying also available on our website)

Published By:

S.K. Kataria & Sons®

4885/109, Prakash Mahal, Dr. Subhash Bhargav Lane,
Opposite Delhi Medical Association, Daryaganj, New Delhi-110002 (INDIA)
Phone: +91-11-43551243, Mobile: +91-9871775858, 7678425209
e-mail: katariabook@gmail.com, katariabooks@yahoo.com
Website: www.skkatariaandsons.com

Price: ₹ 995/-

Laser Typesetting by :

AISA ENTERPRISES, Delhi-110092

Preface

Microprocessor Principles

This book provides students with the fundamentals of microprocessors. Students are introduced to the basic concept of microprocessors and the role that hardware and software play in the functional behavior of a microprocessor using a step-by-step approach of the design of 8-bit and 16-bit microprocessor. Students are then introduced to some of the more common peripheral devices. Special care has been taken to understand the students approach and the book has been designed to introduce microprocessor software development, testing and debugging. Interfacing external memories as well as various input/output devices and sensors form the final part of these modules.

The rapid advancement of microprocessors into our everyday affairs has both simplified and complicated our lives. Whether we use a computer in our job, or come in contact with one elsewhere, most of us have used a computer at one time or another. Most people know that a microprocessor is lurking somewhere inside the machinery, but what a microprocessor is, and what it does remains a mystery.

Purpose of This Book

This book is intended to help remove the mystery concerning the 8-bit and 16-bit microprocessors through detailed coverage of its hardware and software through examples of many different applications.

The book is intended for 2- or- 4 year of Electronics/Electronics and Communication Engineering, Computer Science Engineering, Electrical Engineering and Instrumentation Engineering students. Professional people such as Engineers and technicians will also find it handy. Prior knowledge of digital electronics, including combinational and sequential logic, decoders, memories, Boolean algebra and operations or binary numbers, will be helpful. This presumes knowledge of standard computer-related terms, such as RAM, EPROM, TTL, and so forth. The Appendix in the end provides the user with the knowledge of DOS and BIOS function calls used by the 16-bit microprocessor.

Outline Coverage of This Book

Chapter 1 : Microprocessor and its Overview

For those individuals who have no prior knowledge of microprocessors, , is a good introduction to the microprocessor, highlighting how a microprocessor is used in a small system. We will see how various hardware's are connected to microprocessors as per applications. Topics covered in this chapter are: evolution of microprocessors, register structure of microprocessor, bus organization, arithmetic and logic unit, CPU, memory.

Chapter 2 : 8085 Microprocessor (8-Bit)

This chapter begins by discussing Fundamental concepts of 8085 microprocessor such as its ALU, register units, system buses. The chapter also highlights the working of 8085 microprocessor in detail with its timing diagrams and its pin configuration. Also the interrupt structure and the detailed instruction set of the 8085 microprocessor helps the students in understanding the deeper concepts involved with the processor in an easy and effective manner. A systematic approach has been followed to make the students understand the meaning of the instructions in depth with their respective timing diagrams. In the end the chapter also covers some solved assembly language programming examples.

Chapter 3 : 8086 Microprocessor (16-bit)

This chapter begins by discussing features and its fundamental concepts such as its ALU, register units, system buses and its pin configuration, memory segmentation and physical address computation for 8086 microprocessor. The chapter highlights the different devices used with 8086 microprocessor such as 8284 the clock generator and the wait state generation concept. We also describe the buffering and latching techniques used in the 8086 microprocessor. The chapter also highlights the working of 8086 microprocessor in its different modes, detail with its timing diagrams. The chapter also brings into the picture the various addressing modes used by the 8086 microprocessor with diagrammatic representation of all the modes.

Chapter 4 : 8086 Interrupt Structure

This chapter explains the Interrupt structure for 8086 in detail highlighting the response of 8086 microprocessor when it encounters and interrupts, the 8086 Interrupt vector table and the process to compute Interrupt vector addresses.

Chapter 5 : Memory Interfacing and Address Decoding Techniques for 8086 Microprocessor

This chapter makes the student familiar with the address decoding techniques used for the microprocessor and then interfacing the microprocessor with different memories such as RAM, ROM, and DRAM etc. The students are made familiar with interfacing with the help of several solved examples.

Chapter 6 : 8086 Instruction set

This chapter introduces the user to the 8086 instruction set and there detailed diagrammatic explanation is included to help the students understand how any instruction woks. The instructions include Data transfer instructions, string instructions, logical instructions, arithmetic instructions, transfer of control instructions; process control instructions. The chapter also highlights the instruction format used by 8086 microprocessor with solved examples and easier technique to decode instructions.

Chapter 7 : Techniques for Assembly Language Programming with DOS and Bios Functions for 8086 Microprocessor

This chapter introduces the students to Assembly language programming techniques with various solved examples of all the possible problems that include 8086 microprocessor at this level and to some advanced concepts, such as linking, debugging and execution multiple object files using TASM software. The students are also made familiar with writing procedures; Data tables, modular programming, Macros and assembler directives.

Chapter 8 : I/O Chips and Their Interfacing with 8086 Microprocessor

The students are given an idea for different data transfer schemes used in the microprocessor such as Programmed Data Transfer, Synchronous Data Transfer, Asynchronous Data and Interrupt Driven Data Transfer and 8237 chip in detail and the PIC or the 8259 chip for interrupt processing for the 8086 microprocessor. The interrupt structure of the 8086 microprocessor helps the students in understanding the deeper concepts involved with the processor in an easy and effective manner. This chapter also highlights the most important area for students i.e. the interfacing of different devices with the 8086 microprocessor. Here the students are given detailed knowledge of 8255 0r the PPI chip used with 8086 for interfacing and then many solved examples used with 8255. The students will also be made familiar with 8251 or USART chip. The students are made well versed with the interfacing of different devices with microprocessor such as, interfacing with 8086 with ADCs and DACs, Interfacing Keyboards, Interfacing alphanumeric displays, 7-segment displays, interfacing multiplexed displays, interfacing high power devices with 8086.

Chapter 9 : 80286/287 Microprocessor

In this chapter the students are given an overview of the next 16-bit processor 80286 microprocessor with its Internal Architecture, memory organization and segmentation, General Registers, Memory Segmentation and Segment Registers, physical address computation.

Chapter 10 : 8051 Microcontroller

The chapter highlights the functioning of 8051 microcontroller with detailed description of its block diagram and pin configuration. Different addressing modes and interrupts of microcontroller are also explained. In the end the detailed instruction set and some applications of 8051 microcontroller are explained.

Chapter 11 : Application of Microprocessors

The chapter specifically highlights some working examples of applications of microprocessors. Some of the examples included are smart-scale input circuitary, microcomputer based industrial process control systems and hardware for control sytems and temperature controller.

BOOK AIMS

1. To introduce the microprocessor as a programmable digital system element.
2. To provide an understanding of a microprocessor based system as a combination of hardware and software subsystems and their interaction.
3. To illustrate some basic concepts of microprocessors through the use of assembly language programming.
4. To outline the principles of microprocessor development system.

LEARNING OUTCOMES

On successful reading of this complete book students will be able to:

1. Use a programmers' model to describe the basic operation of a microprocessor
2. Develop, test, debug and document using a typical industry standard microprocessor development system.
3. Write programs for a microprocessor using assembly language, to undertake a simple task incorporating arithmetic, logic and input/output operations.
4. Outline similarities and differences between different types of microprocessor
5. Explain the need for and describe the operation of the additional hardware and software components, such as input/output devices and memory, needed to enable a microprocessor to operate as part of a computer system.

For any further query or suggestion one is most welcome to contact the author at nikhilmarriwala@gmail.com.

AUTHORS

ACKNOWLEDGMENT

A number of individuals have contributed to the preparation of the manuscript of this book. I gratefully acknowledge the contributions of each of them. I would like to express my thanks to the faculty members of the department of Electronics and Communication Engineering of Haryana Engineering College Jagadhri.

In particular I would like to thank, our Director Prof. (Dr.) Dinesh Kumar for his constant support and encouragement. I would like to extend my sincere thanks to Dr. O.P. Sahu Professor NIT Kurukshetra, to Dr. Anil Vohra Professor and Chairman Electronics Science Department Kurukshetra University, Kurukshetra. I would also like to thank Prof. (Dr.) Shakti Kumar Chairman Board of Governors IST, Kalawad and Prof. (Dr.) B.S. Gill Principal IST, Kalawad.

I am thankful to my teachers and professors at M.M.Institute of Technology Mullana, from whom I learnt the importance of self learning and need to understand technology.

I acknowledge the help rendered to me by my fellow colleagues Gaurav Bhasin, Vikas Mittal, Aman Ganesh, Ajay Dhiman, Tarun Gulati, Vijay Garg, Ram Avtar, Monish Gupta and my fellows at UIET, K.U.K.

I thank Mr. S.K Kataria & Sons for reading and thoroughly checking my manuscript and helping me to overcome my initial deficiencies in the book.

Finally I am grateful to my loving parents Mrs. Shobha Marriwala, Sh. Kishori Lal, my loving wife Shalini Marriwala and my son Aryan Marriwala for being a constant source of support.

Nikhil Marriwala

Dedicated to My Loving Wife

"Shalini Marriwala"

Contents

5. MEMORY INTERFACING AND ADDRESS DECODING TECHNIQUES FOR 8086 MICROPROCESSOR 249–308

6. 8086 INSTRUCTION SET 309–408

10. THE 80386/387/486 MICROPROCESSORS 661–710

11. THE PENTIUM PROCESSOR 711–738

12. APPLICATION OF MICROPROCESSORS 739–748

CHAPTER 1

MICROPROCESSORS AND ITS OVERVIEW

1.1. INTRODUCTION TO MICROPROCESSOR

INSIDE THIS CHAPTER

The microprocessor has made a great impact on many aspects of our lives. In future it is going to play a very important role in many activities of mankind. With the advancement in microprocessor technology, it has become possible to solve complex scientific and engineering problems and handle large amount of data. Today microprocessor has made a profound impact on Computer Aided Design (CAD). Even a shortlist of the devices using the microprocessor shows how dependent we have become on it :

1. Smart telephones
2. Automatic teller such as in banks
3. Digital watches
4. Pocket calculators
5. Home security and control devices
6. Talking/singing dolls and other toys
7. Compact disk players.

In this chapter we see how a microprocessor is used in a small system. We will see how various hardware's are connected to microprocessors as per applications. Topics covered in this chapter are: evolution of microprocessors, register structure of microprocessor, bus organization, arithmetic and logic unit, CPU, memory. In addition many other discussions are made related to the microprocessors study.

1.1.1. A Digital System

Microprocessor is the heart of a digital computer or a digital system. Basically, a digital computer or a digital system is a machine which accepts a stream of symbols, stores them, processes them according to precise rules, and produces a stream of symbols at its output. At the simplest level, a digital processor may accept a single number at its input, perform an operation on it, and produce another number at its output. For example, a processor to find the square of a one digit number comes under this category. On the other hand, at a more complex level a large number of symbols may be processed using extensive rules.

A digital system to automatically print a book is a good example of complex system. Such a system should accept a large text or the type written script. Given the number of letters which could be accommodated on a page width line and the rules for hyphenating a word, it should determine the space to be left between the words on a line so that lines are aligned on both the left and right hand sides of a page. The processor should also arrange lines into paragraphs and pages as directed by commands. Also, decisions to leave space figures should be made. A number of such decisions are to be taken before a well laid out book is obtained. Such complex processing would require extensive special facilities such as a large amount of storage, electronic circuits to connect and manipulate characters and printer which has a complete assortment of various sizes and styles of letters.

Digital systems now performes a wide variety of tasks in daily life. In some cases, electronic systems have replaced mechanisms that operated mechanically, hydraulically. Other way, digital systems are usually smaller, more flexible, and easier to use. We can say digital systems have created totally new applications: some of them visible and some are hidden.

1.2. THE BASIC SYSTEM COMPONENTS

The basic operational design of a computer system is called its architecture. John Von Neumann, a pioneer in computer design, is given credit for the architecture of most computers in use today. For example, the 80x86 family uses the Von Neumann architecture (VNA). A typical Von Neumann system has three major components: the Central Processing Unit (or CPU), memory, and input/output (or I/O). The way a system designer combines these components impacts system performance (See Fig. 1.1).

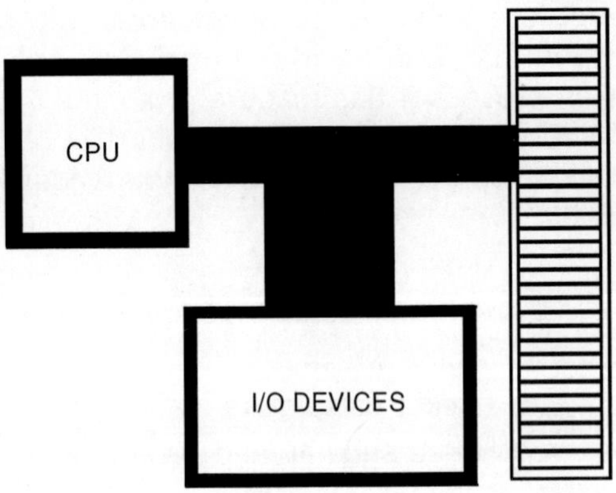

Fig. 1.1.

In VNA machines, like the 80x86 family, the CPU is where all the action takes place. All computations occur inside the CPU. Data and machine instructions reside in memory until required by the CPU. To the CPU, most I/O devices look like memory because the CPU can store data to an output device and read data from an input device. The major difference between memory and I/O locations is the fact that I/O locations are generally associated with external devices in the outside world.

1.2.1. The System Bus

The *system bus* connects the various components of a VNA machine. The 80x86 family has three major buses: the *address* bus, the *data* bus, and the *control* bus. A bus is a collection of wires on which electrical signals pass between components in the system. These buses vary from processor to processor. However, each bus carries comparable information on all processors; e.g., the data bus may have a different implementation on the 80386 than on the 8088, but both carry data between the processor, I/O, and memory.

A typical 80x86 system component uses *standard TTL logic levels.* This means each wire on a bus uses a standard voltage level to represent zero and one. We will always specify zero and one rather than the electrical levels because these levels vary on different processors.

1.2.1.1. The Data Bus

The 80x86 processors use the *data bus* to shuffle data between the various components in a computer system. The size of this bus varies widely in the 80x86 family. Indeed, this bus defines the "size" of the processor.

Every modern x86 CPU from the Pentium on up employs a 64-bit wide data bus. Some of the earlier processors used 8-bit, 16-bit, or 32-bit data buses, but such machines are sufficiently obsolete that we do not need to consider them here.

You'll often hear a processor called an *eight, 16, 32, or 64 bit processor.* While there is a mild controversy concerning the size of a processor, most people now agree that the minimum of either the number of data lines on the processor or the size of the largest general purpose integer register determines the processor size. The modern x86 CPUs all have 64-bit buses, but only provide 32-bit general purpose integer registers,

So most people classify these devices as 32-bit processors.

Although the 80x86 family members with eight, 16, 32, and 64 bit data busses *can* process data up to the width of the bus, they can also access smaller memory units of eight, 16, or 32 bits. Therefore, anything you can do with a small data bus can be done with a larger data bus as well; the larger data bus, however, may access memory faster and can access larger chunks of data in one memory operation.

1.2.1.2. The Address Bus

The data bus on an 80x86 family processor transfers information between a particular memory location or I/O device and the CPU. The only question is, "*Which memory location or I/O device?*" The address bus answers this question. To differentiate memory locations and I/O devices, the system designer assigns a unique memory address to each memory element and I/O device. When the software wants to access some particular memory location or I/O device, it places the corresponding address on the address bus. Circuitry associated with the memory or I/O device recognizes this address and instructs the memory or I/O device to read the data from or place data on to the data bus. In either case, all other memory locations ignore the request. Only the device whose address matches the value on the address bus responds.

With a single address line, a processor could create exactly two unique addresses: zero and one. With *n* address lines, the processor can provide 2^n unique addresses (since there are 2^n unique values in an *n*-bit binary number). Therefore, the number of bits on the address bus will determine the *maximum* number of addressable memory and I/O locations. Early x86 processors, for example, provided only 20 bit address busses. Therefore, they could only access up to 1,048,576 (or 2^{20}) memory locations. Larger address busses can access more memory.

TABLE 1.1. 80x86 Family Address Bus Sizes.

Processor	Address Bus Size	Max Addressable Memory	In English!
8088, 8086, 80186, 80188	20	1,048,576	One Megabyte
80286, 80386sx	24	16,777,216	Sixteen Megabytes
80386dx	32	4,294,976,296	Four Gigabytes
80486, Pentium	32	4,294,976,296	Four Gigabytes
Pentium Pro, II, III, IV	36	68,719,476,736	64 Gigabytes

Future 80x86 processors (e.g., the AMD "Hammer") will probably support 40, 48, and 64-bit address busses. The time is coming when most programmers will consider four gigabytes of storage to be too small, much like they consider one megabyte insufficient today. (There was a time when one megabyte was considered far more than anyone would ever need!).

1.2.1.3. The Control Bus

The control bus is an eclectic collection of signals that control how the processor communicates with the rest of the system. Consider for a moment the data bus. The CPU sends data to memory and receives data from memory on the data bus. This prompts the question, "Is it sending or receiving?" There are two lines on the control bus, *read* and *write*, which specify the direction of data ? How. Other signals include system clocks, interrupt lines, status lines, and so on. The exact make up of the control bus varies among processors in the 80x86 family. However, some control lines are common to all processors and are worth a brief mention.

The *read* and *write* control lines control the direction of data on the data bus. When both contain a logic one, the CPU and memory-I/O are not communicating with one another. If the read line is low (logic zero), the CPU is reading data from memory (that is, the system is transferring data from memory to the CPU). If the write line is low, the system transfers data from the CPU to memory.

The *byte enable lines* are another set of important control lines. These control lines allow 16, 32, and 64 bit processors to deal with smaller chunks of data. Additional details appear in the next section.

The 80x86 family, unlike many other processors, provides two distinct address spaces: one for memory and one for I/O. While the memory address busses on various 80x86 processors vary in size, the I/O address bus on all 80x86 CPUs is 16 bits wide. This allows the processor to address up to 65,536 different I/O *locations.* As it turns out, most devices (like the keyboard, printer, disk drives, etc.) require more than one I/O location. Nonetheless, 65,536 I/O locations are more than sufficient for most applications. The original IBM PC design only allowed the use of 1,024 of these.

Although the 80x86 family supports two address spaces, it does not have two address buses (for I/O and memory). Instead, the system shares the address bus for both I/O and memory addresses. Additional control lines decide whether the address is intended for memory or I/O. When such signals are active, the I/O devices use the address on the L.O. 16 bits of the address bus. When inactive, the I/O devices ignore the signals on the address bus (the memory subsystem takes over at that point).

1.2.2. The Memory Subsystem

A typical 80x86 processor addresses a maximum of 2^n different memory locations, where n is the number of bits on the address bus. As you've seen already, 80x86 processors have 20, 24, 32, and 36 bit address buses (with 64 bits on the way).

Of course, the first question you should ask is, "What exactly is a memory location?" The 80x86 supports *byte addressable memory*. Therefore, the basic memory unit is a byte. So with 20, 24, 32, and 36 address lines, the 80x86 processors can address one megabyte, 16 megabytes, four gigabytes, and 64 gigabytes of memory, respectively.

Think of memory as a linear array of bytes. The address of the first byte is zero and the address of the last byte is $2^n - 1$. For an 8088 with a 20 bit address bus, the following pseudo-Pascal array declaration is a good approximation of memory:

```
Memory: array [0..1048575] of byte;
```

To execute the equivalent of the Pascal statement "Memory [125] := 0;" the CPU places the value zero on the data bus, the address 125 on the address bus, and asserts the write line (since the CPU is writing data to memory) (see Fig. 1.2).

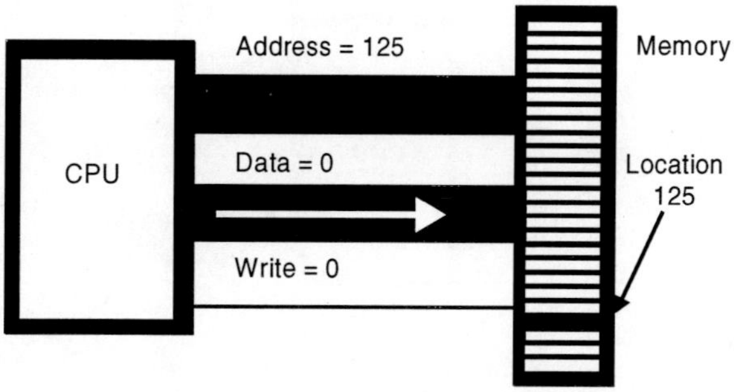

Fig. 1.2. Memory Write Operation.

To execute the equivalent of "CPU := Memory [125];" the CPU places the address 125 on the address bus, asserts the read line (since the CPU is reading data from memory), and then reads the resulting data from the data bus (see Fig. 1.3).

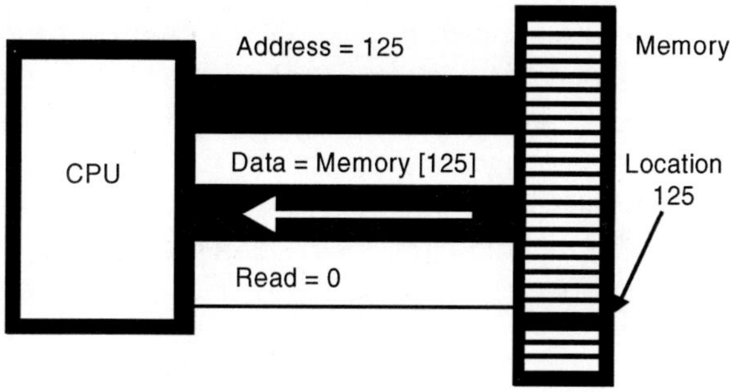

Fig. 1.3. Memory Read Operation.

The above discussion applies *only* when accessing a single byte in memory. So what happens when the processor accesses a word or a double word? Since memory consists of an array of bytes, how can we possibly deal with values larger than eight bits?

Different computer systems have different solutions to this problem. The 80x86 family deals with this problem by storing the L.O. byte of a word at the address specified and the H.O. byte at the next location. Therefore, a word consumes two consecutive memory addresses (as you would expect, since a word consists of two bytes). Similarly, a double word consumes four consecutive memory locations. The address for the double word is the address of its L.O. byte. The remaining three bytes follow this L.O. byte, with the H.O. byte appearing at the address of the double word *plus three* (see Fig. 1.4). Bytes, words, and double words may begin at *any* valid address in memory. We will soon see, however, that starting larger objects at an arbitrary address is not a good idea.

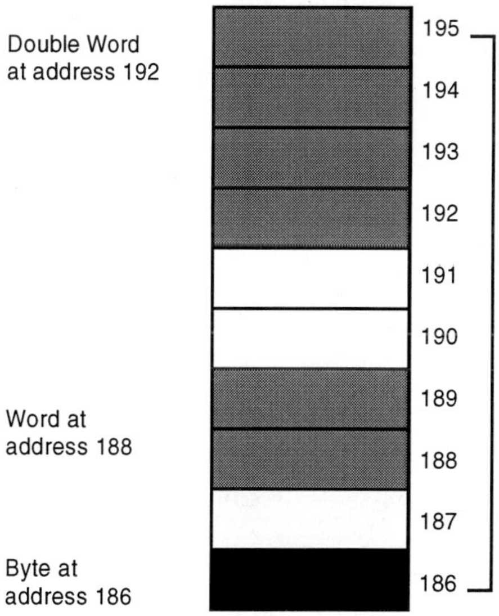

Fig. 1.4. Byte, Word, and DWord Storage in Memory.

Note that it is quite possible for byte, word, and double word values to overlap in memory. For example, in Fig. 1.4 you could have a word variable beginning at address 193, a byte variable at address 194, and a double word value beginning at address 192. These variables would all overlap.

A processor with an eight-bit bus (like the old 8088 CPU) can transfer eight bits of data at a time. Since each memory address corresponds to an eight bit byte, this turns out to be the most convenient arrangement (from the hardware perspective), (see Fig. 1.5).

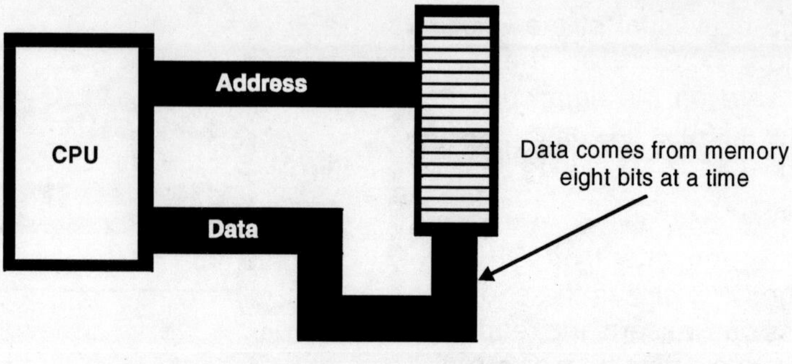

Fig. 1.5. Eight-Bit CPU <-> Memory Interface.

The term "byte addressable memory array" means that the CPU can address memory in chunks as small as a single byte. It also means that this is the *smallest* unit of memory you can access at once with the processor. That is, if the processor wants to access a four bit value, it must read eight bits and then ignore the extra four bits. Also realize that byte addressability does not imply that the CPU can access eight bits on any arbitrary bit boundary. When you specify address 125 in memory, you get the entire eight bits at that address, nothing less, nothing more. Addresses are integers; you cannot, for example, specify address 125.5 to fetch fewer than eight bits.

CPUs with an eight-bit bus can manipulate word and double word values, even through their data bus is only eight bits wide. However, this requires multiple memory operations because these processors can only move eight bits of data at once. To load a word requires two memory operations; to load a double word requires four memory operations.

Some older x86 CPUs (e.g., the 8086 and 80286) have a 16-bit data bus. This allows these processors to access twice as much memory in the same amount of time as their eight bit brethren. These processors organize memory into two *banks*: an "even" bank and an "odd" bank (see Fig. 1.6). Fig. 1.7 illustrates the connection to the CPU (D0-D7 denotes the L.O. byte of the data bus, D8-D15 denotes the H.O. byte of the data bus):

	Even	Odd
Word 3	6	7
Word 2	4	5
Word 1	2	3
Word 0	0	1

Fig. 1.6. Byte Addressing in Word Memory.

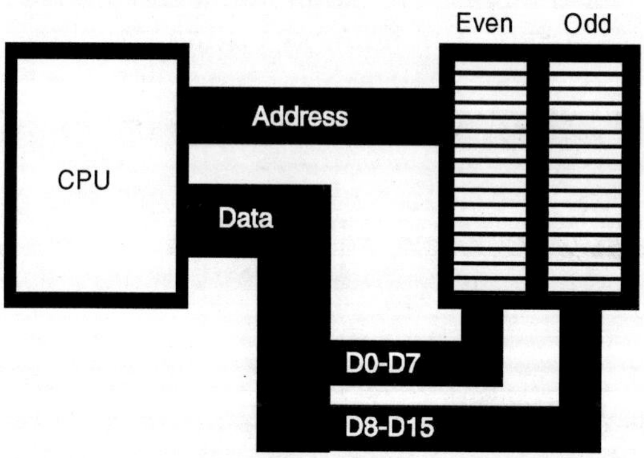

Fig. 1.7. Sixteen-Bit Processor (8086, 80186, 80286, 80386sx) Memory Organization.

The 16 bit members of the 80x86 family can load a word from any arbitrary address. As mentioned earlier, the processor fetches the L.O. byte of the value from the address specified and the H.O. byte from the next consecutive address. This creates a subtle problem if you look closely at the diagram above. What happens when you access a word on an odd address? Suppose you want to read a word from location 125. Okay, the L.O. byte of the word comes from location 125 and the H.O. word comes from location 126. What's the big deal? It turns out that there are two problems with this approach.

First, look again at Fig. 1.7. Data bus lines eight through 15 (the H.O. byte) connect to the odd bank, and data bus lines zero through seven (the L.O. byte) connect to the even bank. Accessing memory location 125 will transfer data to the CPU on the H.O. byte of the data bus; yet we want this data in the L.O. byte! Fortunately, the 80x86 CPUs recognize this situation and automatically transfer the data on D8-D15 to the L.O. byte.

The second problem is even more obscure. When accessing words, we're really accessing two separate bytes, each of which has its own byte address. So the question arises, "What address appears on the address bus?" The 16 bit 80x86 CPUs always place even addresses on the bus. Even bytes always appear on data lines D0-D7 and the odd bytes always appear on data lines D8-D15. If you access a word at an even address, the CPU can bring in the entire 16 bit chunk in one memory operation. Likewise, if you access a single byte, the CPU activates the appropriate bank (using a "byte enable" control line). If the byte appeared at an odd address, the CPU will automatically move it from the H.O. byte on the bus to the L.O. byte.

So what happens when the CPU accesses a *word* at an odd address, like the example given earlier? Well, the CPU cannot place the address 125 onto the address bus and read the 16 bits from memory. There are no odd addresses coming out of a 16 bit 80x86 CPU. The addresses are always even. So if you try to put 125 on the address bus, this will put 124 on to the address bus. Were you to read the 16 bits at this address, you would get the word at addresses 124 (L.O. byte) and 125 (H.O. byte) – not what you'd expect. Accessing a word at an odd address requires two memory operations. First the CPU must read the byte at address 125, then it needs to read the byte at address 126. Finally, it needs to swap the positions of these bytes internally since both entered the CPU on the wrong half of the data bus.

Fortunately, the 16 bit 80x86 CPUs hide these details from you. Your programs can access words at *any* address and the CPU will properly access and swap (if necessary) the data in memory. However, to access a word at an odd address requires two memory operations (just like the 8088/80188). Therefore, accessing words at odd addresses on a 16 bit processor is slower than accessing words at even addresses. **By carefully arranging how you use memory, you can improve the speed of your program on these CPUs.**

Accessing 32 bit quantities always takes at least two memory operations on the 16 bit processors. If you access a 32 bit quantity at an odd address, a 16-bit processor will require three memory operations to access the data.

The 80x86 processors with a 32-bit data bus (e.g., the 80386 and 80486) use four banks of memory connected to the 32 bit data bus (see Fig. 1.8).

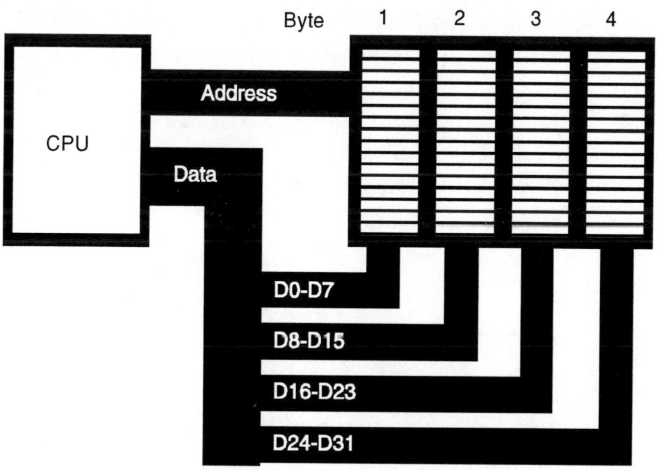

Fig. 1.8. 32-Bit Processor (80386, 80486, Pentium Overdrive) Memory Organization.

The address placed on the address bus is always some multiple of four. Using various "byte enable" lines, the CPU can select which of the four bytes at that address the software wants to access. As with the 16 bit processor, the CPU will automatically rearrange bytes as necessary.

With a 32 bit memory interface, the 80x86 CPU can access any byte with one memory operation. If (address MOD 4) does not equal three, then a 32 bit CPU can access a word at that address using a single memory operation. However, if the remainder is three, then it will take two memory operations to access that word (see Fig. 1.9). This is the same problem encountered with the 16 bit processor, except it occurs half as often.

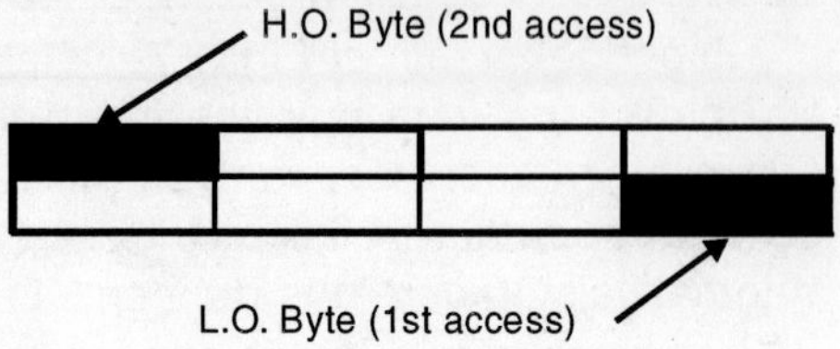

Fig. 1.9. Accessing a Word at (Address mod 4) = 3.

A 32 bit CPU can access a double word in a single memory operation *if* the address of that value is evenly divisible by four. If not, the CPU will require two memory operations.

Once again, the CPU handles all of this automatically. In terms of loading correct data the CPU handles everything for you. However, there is a performance benefit to proper data alignment. As a general rule you should always place word values at even addresses and double word values at addresses which are evenly divisible by four. This will speed up your program.

The Pentium and later processors provide a 64-bit bit data bus and special cache memory that reduces the impact of non-aligned data access. Although there may still be a penalty for accessing data at an inappropriate address, modern x86 CPUs suffer from the problem less frequently than the earlier CPUs.

1.2.3. The I/O Subsystem

Besides the 20, 24, or 32 address lines which access memory, the 80x86 family provides a 16 bit I/O address bus. This gives the 80x86 CPUs two separate address spaces: one for memory and one for I/O operations. Lines on the control bus differentiate between memory and I/O addresses. Other than separate control lines and a smaller bus, I/O addressing behaves exactly like memory addressing. Memory and I/O devices both share the same data bus and the L.O. 16 lines on the address bus.

There are three limitations to the I/O subsystem on the PC: first, the 80x86 CPUs require special instructions to access I/O devices; second, the designers of the PC used the "best" I/O locations for their own purposes, forcing third party developers to use less accessible locations; third, 80x86 systems can address no more than 65,536 (2^{16}) I/O addresses. When you consider that a typical video display card requires over eight megabytes of addressable locations, you can see a problem with the size of I/O bus.

Fortunately, hardware designers can map their I/O devices into the memory address space as easily as they can the I/O address space. So by using the appropriate circuitry, they can make their I/O devices look just like memory. This is how, for example, display adapters on the PC work.

1.3. BASICS

A computer can only do what the programmer has *told* **it to do, in the form of a** *program.* A program is just a sequence of very simple commands that lead the computer to solve some problem. Once the program is written and debugged (you hardly ever get it right the first time), the computer can execute the instructions **very fast**, and always do it the same, every time, without a mistake.

And herein lies the power of a computer. Even though the program consists of very simple instructions, the overall result can be very impressive, due mostly to the speed at which the computer can process the instructions. Even though each step in the program is very simple, the sequence of instructions, executing at *millions* of steps per second, can appear to be very complicated, when taken as a whole. The *trick* is not to think of it as a whole, but as a *series of very simple steps, or commands.*

The microprocessor itself is usually a single Integrated Circuit (IC). Most microprocessors, or very small computers, (here after referred to simply as *micro's*) have much the same commands or instructions that they can perform. They vary mostly in the names used to describe each command. In a typical micro, there are commands to move data around, do simple math (add, subtract, multiply, and divide), bring data into the micro from the outside world, and send data out of the micro to the outside world. Sounds too simple....right?

A typical micro has three basic parts inside. They are the Program Counter (PC), Memory, and Input / Output (I/O). The Program Counter keeps track of which command is to be executed. The Memory contains the commands to be executed. The Input / Output handles the transfer of data to and from the outside world (outside the micro's physical package). The micro we'll be using is housed inside a 40 pin package, or *chip or IC.* There are many other actual parts inside our micro however, we will learn about each and every single one, one step at a time.

1.4. SYSTEM TIMING

Although modern computers are quite fast and getting faster all the time, they still require a finite amount of time to accomplish even the smallest tasks. On Von Neumann machines like the 80x86, most operations are *serialized.* This means that the computer executes commands in a prescribed order. It wouldn't do, for example, to execute the statement I:=I*5+2; before I:=J; in the following sequence:

```
I := J;
I := I * 5 + 2;
```

Clearly we need some way to control which statement executes first and which executes second.

Of course, on real computer systems, operations do not occur instantaneously. Moving a copy of J into I takes a certain amount of time. Likewise, multiplying I by five and then adding two and storing the result back into I takes time. As you might expect, the second Pascal statement above takes quite a bit longer to execute than the first. For those interested in writing fast software, a natural question to ask is, "How does the processor execute statements, and how do we measure how long they take to execute?"

The CPU is a very complex piece of circuitry. Without going into too many details, let us just say that operations inside the CPU must be very carefully coordinated or the CPU will produce erroneous results. To ensure that all operations occur at just the right moment, the 80x86 CPUs use an alternating signal called the *system clock.*

1.4.1. The System Clock

At the most basic level, the *system clock* handles all synchronization within a computer system. The system clock is an electrical signal on the control bus which alternates between zero and one at a periodic rate (see Fig. 1.10). All activity within the CPU is synchronized with the edges (rising or falling) of this clock signal.

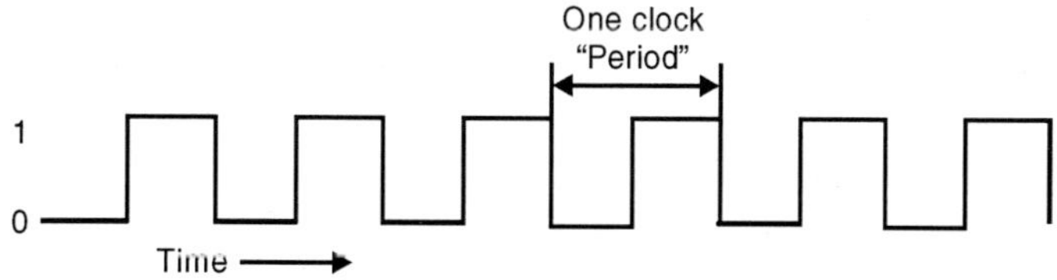

Fig. 1.10. The System Clock.

The frequency with which the system clock alternates between zero and one is the *system clock frequency.* The time it takes for the system clock to switch from zero to one and back to zero is the *clock period.* One full period is also called a *clock cycle.* On most modern systems, the system clock switches between zero and one at rates exceeding several hundred million times per second to several billion times per second. The clock frequency is simply the number of clock cycles which occur each second. A typical Pentium IV chip, circa 2002, runs at speeds of 2 billion cycles per second or faster. "Hertz" (Hz) is the technical term meaning one cycle per second. Therefore, the aforementioned Pentium chip runs at 2000 million hertz, or 2000 megahertz (MHz), also known as two gigahertz. Typical frequencies for 80x86 parts range from 5 MHz up to several Gigahertz (GHz, or billions of cycles per second) and beyond. Note that one clock period (the amount of time for one complete clock cycle) is the reciprocal of the clock frequency. For example, a 1 MHz clock would have a clock period of one microsecond (1/1,000,000th of a second). Likewise, a 10 MHz clock would have a clock period of 100 nanoseconds (100 billionths of a second). A CPU running at 1 GHz would have a clock period of one nanosecond. Note that we usually express clock periods in millionths or billionths of a second.

To ensure synchronization, most CPUs start an operation on either the *falling edge* (when the clock goes from one to zero) or the *rising edge* (when the clock goes from zero to one). The system clock spends most of its time at either zero or one and very little time switching between the two. Therefore clock edge is the perfect synchronization point.

Since all CPU operations are synchronized around the clock, the CPU cannot perform tasks any faster than the clock. However, just because a CPU is running at some clock frequency doesn't mean that it is executing that many operations each second. Many operations take multiple clock cycles to complete so the CPU often performs operations at a significantly lower rate.

1.4.2. Memory Access and the System Clock

Memory access is one of the most common CPU activities. Memory access is definitely an operation synchronized around the system clock or some submultiple of the system clock. That is, reading a value from memory or writing a value to memory occurs no more often than once every clock cycle. Indeed, on many 80x86 processors, it takes several clock cycles to access a memory location. The *memory access time* is the number of clock cycles the system requires to access a memory location; this is an important value since longer memory access times result in lower performance.

Memory access time is the amount of time between a memory operation request (read or write) and the time the memory operation completes. Modern x86 CPUs are so much faster than memory that systems built around these CPUs often use a second clock, the bus clock, that is some submultiple of the CPU speed. For example, typical processors in the 100 MHz to 2 GHz range use 400 MHz, 133 MHz, 100 MHz, or 66 MHz bus clocks (often, the bus speed is selectable on the CPU).

When reading from memory, the memory access time is the amount of time from the point that the CPU places an address on the address bus and the CPU takes the data off the data bus. On typical x86 CPU with a one cycle memory access time, a read looks something like shown in Fig. 1.11. Writing data to memory is similar (see Fig. 1.12).

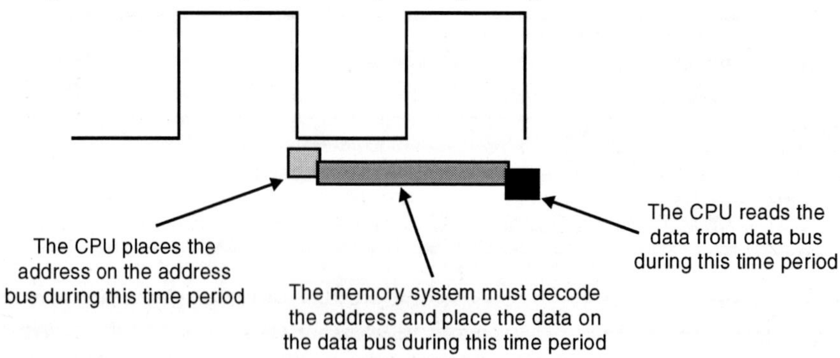

Fig. 1.11. The 80x86 Memory Read Cycle.

Fig. 1.12. The 80x86 Memory Write Cycle.

Note that the CPU doesn't wait for memory. The access time is specified by the bus clock frequency. If the memory subsystem doesn't work fast enough, the CPU will read garbage data on a memory read operation and will not properly store the data on a memory write operation. This will surely cause the system to fail.

Memory devices have various ratings, but the two major ones are capacity and speed (access time). Typical dynamic RAM (random access memory) devices have capacities of 512 (or more) megabytes and speeds of 0.25-100 ns. You can buy bigger or faster devices, but they are much more expensive. A typical 2 GHz Pentium system uses 2.5 ns (400 MHz) memory devices.

Wait just a second here! At 2 GHz the clock period is roughly 0.5 ns. How can a system designer get away with using 2.5 ns memory? The answer is *wait states.*

1.4.3. Wait States

A wait state is nothing more than an extra clock cycle to give some device time to complete an operation. For example, a 100 MHz Pentium system has a 10 ns clock period. This implies that you need 10 ns memory. In fact, the situation is worse than this. In most computer systems there is additional circuitry between the CPU and memory: decoding and buffering logic. This additional circuitry introduces additional delays into the system (see Fig. 1.13). In this diagram, the system loses 10 ns to buffering and decoding. So if the CPU needs the data back in 10 ns, the memory must respond in less than 0 ns (which is impossible).

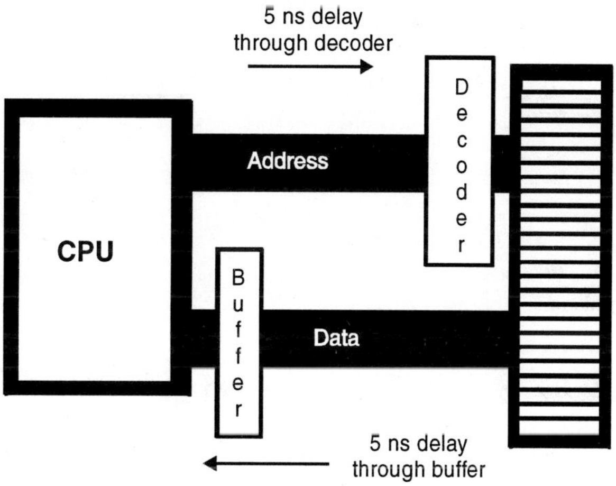

Fig. 1.13. Decoding and Buffer Delays.

If cost-effective memory won't work with a fast processor, how do companies manage to sell fast PCs? One part of the answer is the wait state. For example, if you have a 2 GHz processor with a memory cycle time of 0.5 ns and you lose 2 ns to buffering and decoding, you'll need 2.5 ns memory. What if your system can only support 10 ns memory (i.e., a 100 MHz system bus)? Adding three wait states to extend the memory cycle to 10 ns (one bus clock cycle) will solve this problem.

Almost every general purpose CPU in existence provides a signal on the control bus to allow the insertion of wait states. Generally, the decoding circuitry asserts this line to delay one additional clock period, if necessary. This gives the memory sufficient access time, and the system works properly (see Fig. 1.14).

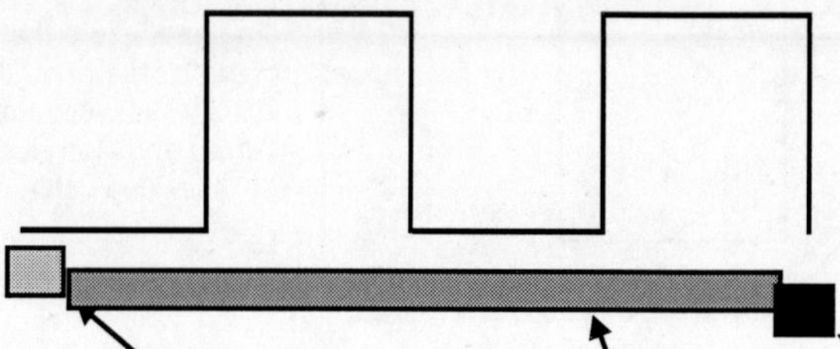

Fig. 1.14. Inserting a Wait State into a Memory Read Operation.

Needless to say, from the system performance point of view, wait states are *not* a good thing. While the CPU is waiting for data from memory it cannot operate on that data. Adding a single wait state to a memory cycle on a typical CPU *doubles* the amount of time required to access the data. This, in turn, *halves* the speed of the memory access. Running with a wait state on every memory access is almost like cutting the processor clock frequency in half. You're going to get a lot less work done in the same amount of time.

However, we're not doomed to slow execution because of added wait states. There are several tricks hardware designers can play to achieve zero wait states *most* of the time. The most common of these is the use of *cache* (pronounced "cash") memory.

1.5. EVOLUTION OF MICROPROCESSORS

The microprocessor technology took birth in 1960. The first generation of computers was a big computing machine called ENIAC (Electronic Numerical Integrator and Computer). That machine was constructed largely of vacuum tubes. It was slow, prone to breakdowns, could perform a limited number of instructions. In 1971-72 Intel manufactured its first microprocessor, the 4004. It was a 4 bit microprocessor.

In 1974 second generation microprocessor became available. The second generation computers used transistors and size of computer was reduced. This generation caused the decrease in computer size and increase in computing power. The 8-bit microprocessors were developed in this generation, such as Intel's 8008, 8080 and 8085.

The third generation of computers came up with the invention of integrated circuits (SSI and MSI). In the late 1970s third generation came-up. In 1979 Intel manufactured its first 16 bit data microprocessor technology made rapid advancement in 32-bit microprocessor and Intel manufactured 80286, 80386, microprocessors. With third generation computers we saw further decrease in size and increase in computing power.

The invention of large scale integration (LSI, thousands of transistors on the same piece of silicon) caused further miniaturization of microprocessor. This generation made another remarkable decrease in digital computer size and increase in computing power and overall improvement in the performance of computer. LSI generation has given birth to fourth generation of computers that we use today. An advanced form of LSI technology, VLSI (Very Large Scale Integration) has further miniaturized the microprocessor and increased processing power of digital computer. Some of the early fourth generation microprocessor technology has improved tremendously. The RISC (Reduced Instruction Set Computer) technology has recently gained acceptance. In this technology microprocessors use only a small portion of their instruction set.

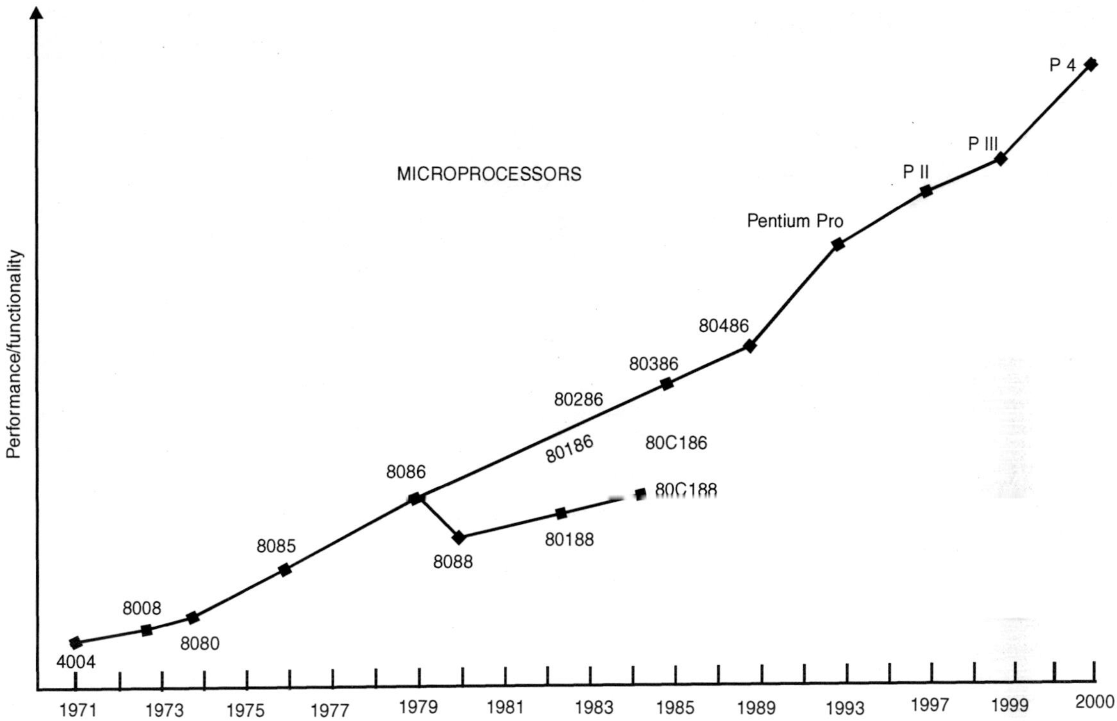

Fig. 1.15. Evolution of Microprocessors by Intel Inc.

In early 1990's fourth generation dominated the microprocessor market. Due to the rapid development in VLSI technology and RISC technology of computers; manufacturing the 32-bit microprocessor technology gained further boost. Some other advanced micro-processors came up, such as Intel 486, Pentium, Pentium II, Pentium III, and recently Intel has manufactured Pentium 4. Other manufacturers have also made a very significant advancement in microprocessor technology such as Advanced Micro Devices (AMD) has manufactured K5, K6 and K7 microprocessors. Similarly, there is a big series of advanced microprocessors from Motorola. RISC technology enhanced the processing speed, without the need for a significant advance in integrated circuit technology. Today, Pentium processors of Intel use many of the architectural techniques employed by RISC machines. In rate 2000 Intel manufactured P4 microprocessor, based on these advanced technologies.

The fifth generation computers are to come that will have intelligence or can think. Figure 1.15 represents the evolution graph of Intel microprocessors. Table 1.2 presents the summary of Intel's Microprocessors.

TABLE 1.2. Intel Microprocessors

Microprocessor	Year of Introduction	Data Bus Width (Bit)	Address Bus Width	Memory Size
4004	1971	4	10	1KB
8085A	1976	8	16	64KB
8086	1978	16	20	1MB
8088	1980	8	20	1MB
80186	1981	16	20	1MB
80188	1982	8	20	1MB
80286	1983	16	24	16MB Real, 4GB Virtual
80386 SX	1984	16	24	16MB
80386EX	1985	16	26	64MB
80386DX	1986	32	32	4GB
80486 DX/SX	1989	32	32	4GB+8KB/16KB Cache
Pentium	1993	64	32	4GB+16KB Cache
Pentium overdrive (replaced 80486)	1994	32	32	4GB+16KB Cache
Pentium Pro	1995	64	36	64GB+32KB L_1 Cache + 256 KB L_2 Cache
Pentium-II	1997	64	36	64GB+32KB L_1 Cache + 512 KB L_2 Cache
Pentium-II Xeon	1998	64	36	64GB+32KB L_1 Cache + 512 KB or 1 MB L_2 Cache
Pentium-III	1999	64	36	64GB+32KB L Cache + 256 KB L_2 Cache
Pentium 4	2000	64	36	64GB+32KB L_1 Cache + 256 KB L_2 Cache (enhanced speed of 1.3, 1.4, and 1.5 GHz

1.5.1. Intel Microprocessors Family

The 4004 Microprocessor

The Intel 4004, a 4-bit microprocessor was the world's first microprocessor. Basically, it was a programmable controller on a chip. It addressed 4096 memory locations of 4-bit width. The instruction set of 4004 merely contained 45 instructions. It was fabricated on P-channel MOSFET technology in early 1971, the then technology allowed it to execute instructions at a very slow rate of 50 kilo-instruction per second. This was much slower than ENIAC computer of 1946. But Intel 4004 weighed much less than ENIAC. Because of this the Intel 4004 microprocessor debuted in early video game systems and small microprocessor-based control system. The speed, word length, and memory size were the main problems of this early microprocessor. Later Intel released the 4040, and updated version of 4004. The 4040 was improved only on the aspect of higher speed. The 4-bit microprocessor still survives in low and applications such as microwave ovens and small control systems. Most of the calculators are still based on 4-bit microprocessors that process 4-bit Binary-Coded-Decimal (BCD) codes.

The 8008 Microprocessor

Realizing that the microprocessor was a commercially viable product, in 1972 Intel released the 8008 microprocessor, an extended 8 bit version of the 4004 microprocessor. An 8-bit word length and 14 address lines of 8008 expanded its memory size to 16 Kbytes (16 KB). A byte is and 8-bit long binary number and kilo (K) is 2^{10} = 1024-1 K.

The 8080 Microprocessor

Increasing demand of 8008 microprocessor, forced to realize its small memory size, slow speed, and limited instruction set. Intel recognized these limitations and introduced the 8080 microprocessor in 1973. The 8080 was the first modern 8-bit microprocessor. Soon other companies (Motorola-MC6800, Zilog-280, etc.) began to introduce their own versions of the 8-bit microprocessor. The specialties of the 8080 were capability to address more memory and execute additional instructions much faster than the 8008. Also, the 8080 was compatible with TTL, whereas the 8008 was not directly compatible. Such improvements are responsible for ushering in the era of the 8080.

The 8085 Microprocessor

Intel Corporation introduced the 8085 microprocessor in 1977—an updated version of the 8080 microprocessor. The 8085 was the last 8-bit, 16 address lines general purpose microprocessor-slightly more advanced than an 8080. The 8085 executed software at an even higher speed. The main advantages of the 8085 microprocessor were its internal clock generator, internal system controller, and higher clock frequency. The higher level of component integration reduced the 8085 cost and increased its applications. Applications that contain the 8085 microprocessor will probably continue to be popular well in to the future.

The 8086 Microprocessor

Intel introduced the 8086 microprocessor-a 16-bit microprocessor, this was to be the **first modern microprocessor.** Around a year later, Intel released the 8088 microprocessor—another 16-bit microprocessor having compatibility with 8085. Both 8086 and 8088 20 address lines-addressed 1 MB of memory, which had 16 times more memory than the 8085. Also, these modern microprocessors executed 2.5 million of instructions per second. This higher execution speed and larger memory size allowed the 8086 and 8088 to replace smaller microcomputers in many applications. Another enhanced feature found in the 8086/8088 was a small 4-byte/6-byte respectively instruction stream queue or Cache that perfected a few instructions before they were executed. The queue speeded up the operation of microprocessor and proved to be the basis for the much larger instruction queues or Caches found in other modern microprocessor. This pre-fetch-and-queue scheme greatly speeds up processing. The process of fetching the next instruction once the current instruction executes is referred as pipelining. The enhanced memory size and additional instructions in the 8086/8088 have led to many sophisticated applications of microprocessors.

In addition, the number of instructions increased from 246 on the 8085, to well over 20,000 variations on the 8086 and 8088. Because of the number and complexity of instructions these microprocessors were referred as **complex-instruction-set-computers** (CISC). The additional instructions caused the task of developing efficient and sophisticated applications. The 16-bit microprocessor also provided more internal register storage space than the 8-bit microprocessor, which allowed the programmers to write software more efficiently. The main reason to evolve 16-bit microprocessor was the need for larger memory system. Applications such as spreadsheets, word processors, spelling checkers, etc. were memory intensive and required more than the 64 kilobytes of memory found in 8-bit microprocessors to execute efficiently. The 16-bit 8086/8088 provided 1 MB memory for such applications.

The 80286 Microprocessors

In 1983 Intel launched 80286 microprocessor-also a 16-bit microprocessor almost identical to 8086 and 8088, except that it addressed a 16 MB memory instead of 1 MB. The instruction set of 80286 was identical to the 8086 and 8088, except for a few additional instructions that managed the extra 15 MB of memory. The clock speed of 80286 was increased, so it executed some instructions in as little as 250 ns.

The 80386 Microprocessor

In 1986 Intel introduced 80386—a 32 bit microprocessor, a faster microprocessor speed, increased memory, and wider data bus. The 80386 was a major overhaul of the 16-bit 8086-80286 architecture. The 80386 was the Intel's first 32-bit microprocessor that contained a 32-bit data bus and a 32-bit memory address. The 32-bit memory address bus increased the memory size to 4GB. The other modified versions of 80386 were 80386-16 bit data, 24 bit address bus and 32 MB memory.

Another version 80386 EX was launched much later called an embedded PC because it contained all the components of AT class personal computer on a single integrated circuit. The 80386 EX contained – 24 lines for input/output data, 16-bit data bus, 26-bit address bus, a DRAM refresh controller, and programmable chip selection logic. The 32-bit microprocessor was needed because of the size of its data bus, which transfers real numbers that require 32-bit microprocessor must efficiently pass them between itself and memory. If the numbers pass through a 16-bit data bus, it takes two read or write cycle is needed. This, significantly, increases the speed of any program that manipulates real numbers. Most of the high level languages, data base management systems, and spread sheets use real numbers for data storage. Real numbers are also used in graphical design packages such as Computer-Aided-Design (CAD) system, as ORCAD, AUTOCAD, etc.

In earlier microprocessors memory management was completely done by software. Whereas, 80386 included a memory management unit that allowed memory resources to be allocated and managed by the operating system, the 80386 included hardware circuitry for memory management and memory assignment that improved its efficiency and reduced software overhead. The instruction set of 80386 was upward compatible with the previous Intel's 8086, 8088 and 80286. Additional instructions were used to reference the 32-bit registers and managed the memory system.

Instructions caused the task of developing efficient and sophisticated applications. The 16-bit microprocessor also provided more internal register storage space than the 8-bit microprocessor, which allowed the programmers to write software more efficiently. The main reason to evolve 16-bit microprocessor was the need for larger memory system. Applications such as spreadsheets, word processors, spelling checkers, etc. were memory intensive and required more than the 64 kilobytes of memory found in 8-bit microprocessors to execute efficiently. The 16-bit 8086/8088 provided 1 MB memory for such applications.

The 80486 Microprocessor

Intel introduced the 80486 microprocessors in 1989, which incorporated an 80386 like microprocessor, an 80387 like numeric coprocessor. The internal structure of the 80486 was modified from the 80386 microprocessor so that about half of its instructions, were executed in one clock instead of two clocks. As the 80486 microprocessor was available in a 50 MHz version, about half of the instructions were executed in 25 ns. The average speed improvement for a typical (8 KB for data and 8 KB for instructions), the Pentium Pro-processor also contains a 256 KD level two (L2) cache. Another significant change is that, this processor uses three execution engines, so it can execute up to three instructions at a time, which can conflict and still execute in parallel. This represents a change from the Pentium, which executes two instructions simultaneously as long as they do I not conflict. The Pentium Pro-processor is optimized to efficiently execute 32 bit code. Because

The Pentium Microprocessor

In 1993, Intel released Pentium processor, which was similar to the 80386 and 80486 processors. This microprocessor was originally named 80586 or P5, but this name was dropped to avoid copyrighting a number. In the beginning, Intel introduced two versions of the Pentium processor operated with a clocking frequency of 60 MHz and 66 MHz. Another higher speed version double clock Pentium with a clocking frequency of 120 MHz and 133 MHz, was available. Another fast version processor produced by Intel was 233 MHz Pentium, which was a three and one-half clocked version. Another major difference was that the Cache size was increased to 16 KB from the 8 KB Cache available in the basic version of the 80486. The Pentium processor contains an 8 KB instruction Cache and an 8 KB data Cache, which allowed a program, disallowed a program that transfers a large amount of memory data to still benefit from a cache. The memory system contained up to 4 GB, with the data bus size increased from the 32 bits found in the 80386 and 80486 to a full 64 bits. Depending on the version of the Pentium, data bus transfer speed was either 60 MHz or 66 MHz. The wider data bus width accommodated double precision floating-point numbers used for modern high-speed, vector generated graphical displays. Recent version of the Pentium included additional instructions, called multimedia extensions, MMX instructions.

In 1994, Intel released the Pentium Overdrive for older 80486 systems that operate at 63 MHz or 83 MHz frequency. The upgraded 83 MHz system performs at a rate somewhere between a 66 MHz Pentium and a 75 MHz Pentium. The Pentium Overdrive represents ideal upgrade path from the 80486 to the Pentium. The most ingenious feature of the Pentium is its dual integer processors. The Pentium executes two instructions, which are not dependent on each other, simultaneously because it contains two independent internal integer processors called superscalar architecture. This allows the Pentium to often execute two instructions per clock period. Another feature which enhances performance of a Pentium is a jump prediction technology that speeds the execution of program's which include loops. As in case of 80486, the Pentium processor also employs an internal floating point co-processor to handle floating point data and speed improvement. These features continued to enhance performance of Intel family of microprocessors.

The Pentium Pro-Microprocessor

In 1995, Intel released the Pentium Pro-processor (code named P6 microprocessor) which contains 21 million transistors, 3 integer units, and a floating point unit to increase the performance of most software. Initially in late 1995, this processor was available with 150 MHz and 166 MHz frequency. In addition to the internal 16 KB level one (L1) cache (8 KB for data and 8 KB for instructions), the Pentium Pro Processor also contains a 256 KB level two L2 cache. Another significant change is that, this processor uses three execution engines, so it can execute up to three instructions at a time, which can conflict and still execute in parallel. This represents a change from the Pentium, which executes two instructions simultaneously as long as they do not conflict. Mainly Pentium Pro-Processor was launched for the server market. Another change in this processor is that the Pentium Pro processor can address either a 4GB memory system or a 64GB memory system. This processor has a 36 bit address bus if configured for a 64 GB memory system.

The Pentium II Microprocessor

In 1997, Intel released Pentium II processor, instead of being an integrated circuit as with previous versions of the microprocessor, Intel placed the Pentium II processor on a small circuit board. The main reason for the change was that the L2 cache found on the main circuit board of the Pentium was not fast enough to justify a new microprocessor. In Pentium II L2 cache and microprocessor are on a circuit board called the Pentium II module. This on board L2 cache operates at a speed of 133 MHz and stores 512 KB of information. The microprocessor on the Pentium II module is actually a Pentium Pro processor with multimedia extensions or MMX extensions, which has no internal L2 cache. In 1998, Intel changed the bus speed of the Pentium II processor. The 266 MHz through 233 MHz Pentium II processor used an external bus speed of 66 MHz.The newer Pentium II processors rated at 350 MHz, 400 MHz and 450 MHz used higher speed 100 MHz memory bus. The higher speed memory bus required the use of, SDRAM in place of the 10 ns SDRAM found in use with the 66 MHz bus speed.

In mid-1998, Intel introduced a new version of Pentium II processors called the Xeon, w: was specifically designed for high-end workstation and server applications. Main differs between the Pentium II and the Pentium II Xeon is that the Xeon is available with an L1 C: size of 32 KB and an L2 cache size of either 512 KB, 1 MB, or 2 MB.

The Pentium III Microprocessor

In 1999, Intel released Pentium III processor, which uses a faster core than the Pentair but it was still a Pentium Pro or P6 processor. It is available in the slot 1 version mounted plastic cartridge and a socket 370 version called a flip-chip, which looks like the older Pen1 package. Another change is that Pentium III is available to clock frequencies of 1GHz. The s. version contains a 512 KB Cache and the flip-chip version contains a 256 KB Cache. The sp are comparable because the Cache in the slot 1 version runs at one-half the clock speed, while (cache in the flip-chip version runs at a clock speed. These two versions use a memory bus s: of 100 MHz, whereas Celeron uses a memory bus clock speed of 66 MHz

The Pentium 4 Microprocessor

The latest release of Intel Corporation, Pentium 4 microprocessor, was initially available late 2000. Like the Pentium Pro through Pentium III, the Pentium 4 uses the Intel P6 architecture, but the main difference is that the Pentium 4 is available in a 1.3, 1.4, and 1.5 GHz speed versions and chipset which supports the Pentium 4 uses the RAMBUS memory technology inset SDRAM technology. These higher microprocessor speeds are made available by an improvement in the size of the internal integration.

It is difficult to make the predictions, but the success in micro-processor technology s1: continue for quite a few years. The microprocessor technology may change to RISC technology) the new technology will embody the CISC instruction set of the 80X86 family of microprobes so that software for the system will survive. The Basic theme behind this technology is that microprocessors will communicate directly with each other, allowing parallel processing wife any change in the instruction set or program. Presently, the superscalar technology uses I microprocessors, but they all share the same register set. The new technology, to be used next version of the Intel microprocessor, will contain many microprocessors, each containing register set that is linked with the other microprocessor's registers.

This technology is expected to offer true parallel processing without writing any special program.

1.5.2. Introduction and Characteristics of Microprocessor System

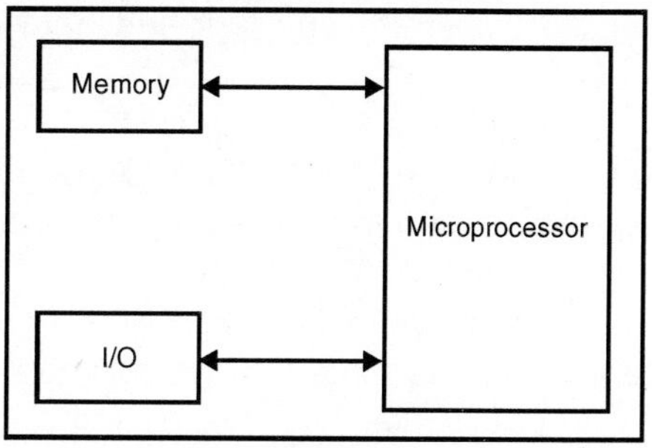

Fig. 1.16. Programmable System Block Diagram.

The microprocessor is a semiconductor device consisting of electronic logic circuits, manufactured by using either a large-scale (LSI) or very large scale integration (VLSI) techniques.

Today microprocessors are manufactured using more complex integration techniques to achieve maximum miniaturization of the device. The microprocessor is capable of performing various computing functions and making decisions to change the sequence of program execution. In broad sense a Microprocessor is a multi-purpose, programmable logic device that reads binary instructions from a storage device called memory, accepts binary data as inputs and processes data according to these instructions, and provides result. as output. A typical programmable system can be represented with three basic

Components : Microprocessor, memory and Input/output (I/O) devices such as shown in Fig. 1.16.

These three components work together or interact with each other to perform a given task; thus they comprise a system. The physical components of this system are called hardware. A set of Instructions written for the microprocessor to perform a task is called a program and a group of programs is called software. The system represented in Fig. 1.16 can be programmed to keep track of a guidance system, turn traffic lights ON and OFF, or compute mathematical functions.

Depending on the applications a programmable system may be simple or sophisticated and the system is recognized by different names depending upon the purpose for which it is designed. The applications of microprocessors are classified basically into two categories, reprogrammable systems and embedded systems. In reprogrammable systems, such as micro-computers, the microprocessor is used for computing and data processing. Reprogrammable systems use general purpose microprocessors capable of handling large data, mass storage devices such as disks, and peripherals such as printers or plotters; a personal computer (PC) is an example of programmable devices.

In embedded system, the microprocessor is a part of final product and is not available for: deprogramming to the end user. A Xerox machine is a typical example of an embedded system. The microprocessors used in embedded system are generally categorized as:

(i) Microcontrollers that include all the three components, microprocessor, memory and Input/output device.

(ii) Integrated microprocessors that include various devices such as timers and various types of I/Os on a chip.

(iii) General purpose microprocessors with three discrete components, microprocessor, Memory and I/Os.

There are some embedded systems such as washing machines, dishwashers, traffic light controllers etc. They use microprocessors to perform their operations.

A block diagram of a digital computer is represented in Fig. 1.17. The memory unit stores programs as well as input, output and intermediate data. The processor unit performs arithmetic's and other data Processing tasks as specified by a program. The control unit supervises the flow of information between the various units. The control unit retrieves the instructions, one by one, from the program that is stored in the memory. For each instruction, the control unit informs the processor to execute the operation specified by the instruction. Both program and data are stored in the memory.

The Control unit supervises the program instructions Fig. 1.17. Block Diagram of a Digital Computer and the processor manipulates the data as specified by the program.

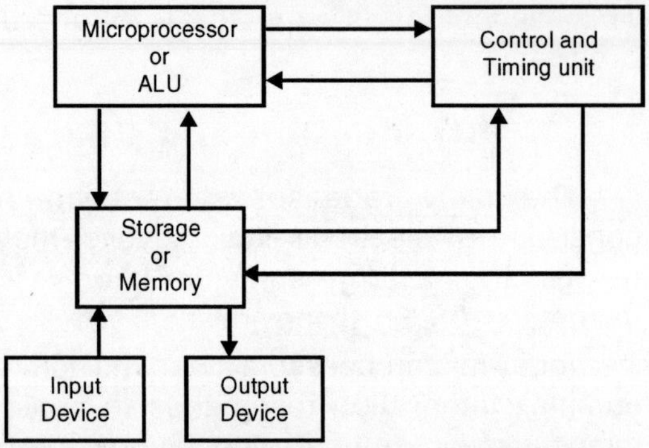

Fig. 1.17. Block Diagram Representation of a Digital Computer.

The program and data prepared by the user are transferred into the memory unit by means of an input device such as a keyboard. An output device, such as a printer receives the result of computations and printed results are presented to the user. The input and output devices are special digital systems driven by electromechanical parts and controlled by electronic digital circuits.

1.5.3. A Microprocessor Based Typical System

A typical microprocessor based system must have some standard elements such as memory, timing! Control unit, input/output (I/O). Other associate circuitry may be necessary depending on the application. Some special purpose devices such as Analog-to-digital (A/D) converters and their counterparts Digital-to-Analog (D/A) Converters, interval timers, co-processor or Numerical Data Processor, interrupt circuitry, speech synthesizers and video display controllers may also be required. A system block diagram containing some standard circuitry and functions normally used is represented in Fig. 1.18.

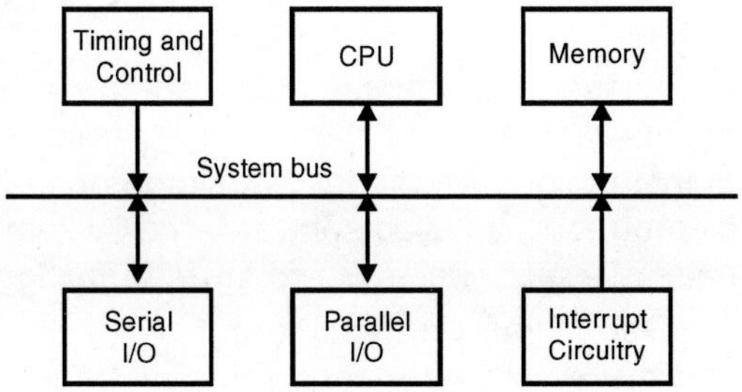

Fig. 1.18. Microprocessor based system.

As represented in the Fig. 1.18, all components communicate via the system bus. The system bus consists of processor address, data, and control signals. The Central Processing Unit (CPU) consists of a microprocessor and the associated logic circuitry required to enable the CPU communication with the system bus. The CPU is the heart of the system and works as the master controller of all operations that can be performed. The CPU executes Instructions that are stored in the System memory. A provision is made for future expansion. For that the system bus is made available to the outside world through a special connector (represented by - - - - in Fig. 1.18). All devices on the system bus must communicate with the processor, usually within a tightly controlled period of time. The timing and control unit govern all systems timings and thus is responsible for the proper operation of all system hardware. The actual microprocessor use depends on the complexity task that will be controlled or performed by the system.

The memory section usually has two components: Random Access Memory (RAM) and Read Only Memory (ROM). Some systems may be able to ivory properly without RAM, but at least a ml all amount of ROM is required by all the systems. The ROM is included to provide the system its intelligence, which is ordinarily needed at start-up (power on) to configure or initialize the peripherals, and sometimes to recover from a catastrophic system failure, such as an unexpected power failure. Some systems use the ROM programs to download the main program into RAM from a larger external system such as a personal computer or mainframe computer. As the need arises, provisions are usually made for adding additional ROM/RAM.

1.5.4. Microprocessor Operation

General operation of a microprocessor does not depend on its complexity but various microprocessors follow the same pattern of operations during program execution: endless fetching, decoding, and execute cycles. During the fetch cycle microprocessor loads an instruction from memory into its internal instruction register. Some advanced microprocessors load more than one instruction into the special buffer to decrease program execution time. This is to facilitate that when microprocessor is decoding the current instruction, other instructions can be read from memory into the instruction stream byte queue or cache, a special type of internal high speed memory. This way, microprocessor saves time by performing two jobs at a time.

During the decode cycle, the microprocessor determines what type of instruction has been fetched. Information from this fetch cycle is then passed to the execute cycle. For the completion of instruction, the execute cycle may need to read more data from memory or write results to memory. During the progress of these cycles, microprocessor also pays attention to other details. If an interrupt signal arrives during execution of an instruction, the processor usually latches on to the request, holding off an interrupt processing until the current instruction finishes execution.

The processor also monitors some other signals such as WAIT, HOLD, or READY inputs. All these are usually included in the microprocessor architecture so that slow devices, such as memories, or peripherals (IIOs), can communicate with the faster processor without loss of data.

Most of the Microprocessors also include a set of control signals that allow the external circuitry to take over the system bus. In a system where multiple processors share the same memory devices, these types of control signals are required to resolve bus contention (two or more processors require the bus system at the same time). As we continue to strive towards faster execution of our programs, multiple processor systems are becoming more popular. Parallel processing is a term often used to describe multiple processor systems and their associated software. Co-processor is an example of multiprocessing.

1.5.5. Microprocessor Manufacturing Technologies

In general there are—two major computer architectures in use : CISC and RISC. CISC and RISC stand for complex instruction set computers and Reduced Instruction Set Computers, respectively. All the 80 x 86 machines can be considered CISC machines. Early series of Pentium is also a mixture of CISC and RISC technologies. The CISC aspect of the Pentium provides for upward compatibility with the other 80 x 86 architectures. The RISC architecture leads to additional performance improvements. Some of these improvements are separate data and instruction caches, dual integer pipelines, and branch prediction. Branch prediction is a technique to maintain a steady flow of instructions into the pipelines.

Microprocessor technology has improved tremendously. A 32-bits microprocessor can now multiply and divide, operate on many different data types (4, 8, 16, 32 bits numbers), and address billions of bytes of information.

Today each new microprocessor hits the market with a fancier instruction set and faster clock speed, and still our needs for faster and better processor keep growing. Now RISC technology has gained acceptance. This technology is based on the fact that most microprocessors use only a small portion of their entire instruction set. By designing a machine that was only the more common types of instructions, processing speed can be increased without the need for a significant advance in IC technology. Today's most of the Intel Pentium processors use RISC architectural machines.

1.5.6. Applications of Microprocessor

The applications of microprocessors have been increasing rapidly since the advent of microprocessor in 1971. The domain of microprocessors application will definitely continue expanding especially in the wake of the fact that new microprocessors and microcomputers with superior performance are now available to the system designer, e.g. Intel's 80486, Pentium II, III, Pentium-IV, Motorola's MC 68030, 80196, microcontrollers etc. The microprocessors are available with enhanced features and functions due to rapid development in electronics, VLSI and ULSI techniques. One of the most emerged application area, of microprocessor is process control. In a typical process control application, the microprocessor continuously monitors one or more process variables and generates outputs to the electro-mechanical elements which, in turn, control the process variables. This kind of control is known as closed loop control. If the microprocessor outputs the control variables to human operators, via displays or line printers, which in turn, apply the necessary control inputs, then the process is called as open loop control.

Here an example is presented to illustrate the use of a microprocessor in open loop process control. For simplicity, the use of microprocessors and microcomputers in automobiles is illustrated here.

Since 1976, engineers have been using microprocessors and microcomputers in automobiles. As the cost of microprocessors and microcomputers has declined, and their power has increased, their use in automobiles has shown tremendous growth in the past decade. Today, most modern cars that need to compete in the international market make expensive use of microprocessor technology for a variety of functions. Several cars use up to ten or more microcomputers responsible for different control functions. As this field is so wide that a complete book may not be sufficient to cover this topic, this section is limited to provide an overview of automotive applications of microprocessors.

Microprocessors in an Automobile

A microprocessor is merely a CPU in a computer system. Many control applications inducing automobiles use microcomputer instead of a microprocessor. In fact, a microcomputer, popularly known as a single chip microcomputer, or a microcontroller, consists of several peripheral devices in addition to a CPU. In automobiles, the tasks to be performed require AID converters, D/A converters and timers as important peripheral devices. Therefore, most microcontrollers designed specifically for use in automobiles, contain all these devices in addition to what a microprocessor contains.

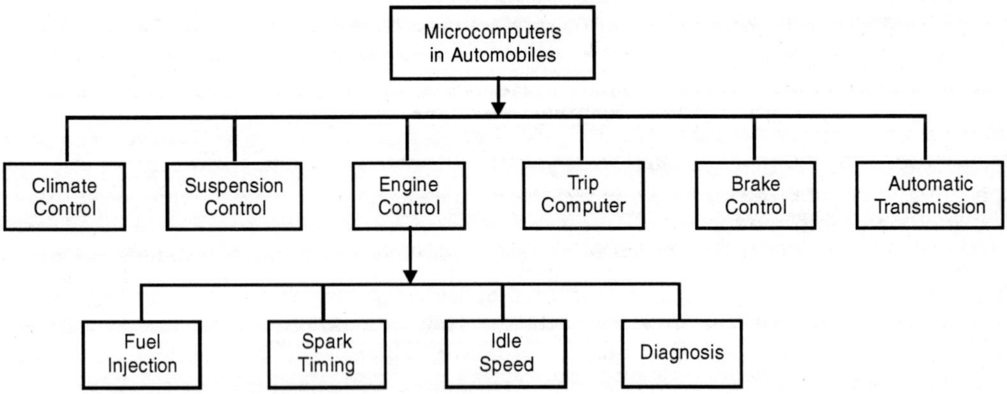

Fig. 1.19. Usage of Microcomputer in Automobiles

Figure 1.19 presents the variety of uses of the microcomputer inside an automobile. Increased competition, stringent federal laws on automobiles, emission controls and improved engine and other automobile technology forced the manufacturers to resort to microcomputer use. Many advantages can be cited in using a microcomputer in an automobile :

- Improved car control for the driver.
- Advanced information system for the driver.
- Smoother ride.
- Improved fuel economy, and.
- Reduced pollution by better emission control.

In future the use of microcomputer in cars will also provide:

- Entertainment center for the passengers,
- Voice control for the elderly and disabled,
- Route management using detailed maps and satellite aided directional information.

Besides the above illustrated applications of microprocessor in automobiles, today

Microprocessor based systems are used in every sphere of life. Besides CPU a microprocessor is also used to control input, output and other devices of a computer. Single chip microcomputers, rolled microcontrollers are also widely used for automatic control in industry for process and machine control, in instrumentation, for commercial and consumer appliances control, etc. Some applications of microprocessors are listed here:

1. A single chip microcomputer *i.e.* microprocessor is used to measure and control the Temperature of furnace and oven.
2. A microprocessor is used to control speed of an electric motor.
3. A microprocessor is used to control the pressure of a boiler.
4. Microprocessors are used in military equipment, radar, tank, fighter planes control, etc.
5. Computers are used for business/office automation for word process, Data Base Management System (DBMS), storing information, sharing information through a network-WAN or internet.
6. Computers are used to interconnect educational institutions, corporations, organizations, in various networks, or internet for data analysis, computation, scientific and technological Calculations, design of machines, preparation of drawings, design of buildings, on line educational tutors etc.
7. Microprocessors are specially used in this modern era for controlling lifts, escalators, and Lights.
8. Microprocessors are used for automatic control of generator's voltage, and fuel control in plants and automobiles.
9. Microcomputers are used for railway ticketing, airways ticketing etc.
10. Microcomputers are used for library accessing and management.
11. Microcomputers are used in film and entertainment industry.
12. Microcomputers are used in printing media.

1.6. MICROPROCESSOR REGISTERS

Generally microprocessors contain general purpose data registers, pointer registers, flag Register and as a special feature segment register.

In the 8085, there is one 8-bit accumulator register, commonly referred as Accumulator (ACC). It is used in various arithmetic and logical operations. There are six general purpose 8-bit registers that can be used by a programmer for a variety of purposes. There is one 16-bit program counter register, commonly called Program Counter (PC). A program counter is used to keep track of the address of the instruction in the memory that has to be executed next. There is another 16-bit register, known as Stack Pointer (SP). It is used by the programmer to maintain a stack in the memory. There is another 8-bit register, known as Flag Register. The five bits of flag register are used to indicate certain conditions that arise during arithmetic and logical operations. Several other registers are used in 8085 for its internal operation; they are discussed later.

In the 80x86 family of microprocessors, there are four 16-bit general purpose data registers referred as *AX, BX,* ex and *DX.* The *AX* register is also known as Accumulator because of its special feature that the other data registers do not possess. These four internal data register are used for temporary storage of data. There are five other 16-bit registers available for use as pointers or index registers. These registers are the Basic Pointer (BP), Source Index (SI), Destination Index (DI), Stack Pointer (SP), and Instruction Pointer (IP). There is one 16-bit flag register used to indicate the results of arithmetic and logical instructions. As a special feature 8086 has four 16-bit segment registers: Data Segment (DS), Code Segment (CS), Stock Segment (SS), and Extra Segment (ES). These four segment registers are used by the processors to control all accesses to memory and I/O.

1.6.1. Bus Organisation of a Microcomputer

Microcomputer is a digital system. The major parts of a microcomputer are Central Processing Unit (CPU), memory unit, and input and output device or I/O. Connecting these three parts are three sets of lines, called buses. These three buses are the address bus, data bus, and the control bus, Figure 1.19 represents a block diagram of a microcomputer.

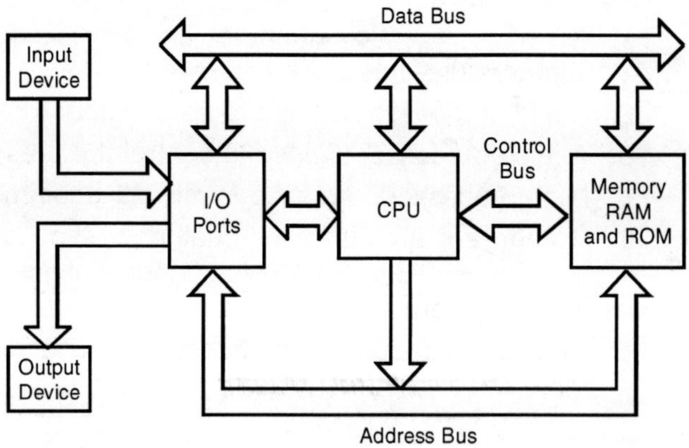

Fig. 1.20. Bus Organization for a Microcomputer.

1.6.2. Address Bus

The address bus consists of 16, 20, 24, or 32 parallel signal lines. The CPU sends out the address of the memory location that is to be written or read to/from these address buses, respectively. The .size of the memory depends on the number of address lines. For N number of address lines CPU can address 2N memory locations. When the CPU reads data from or write data to a port or memory, CPU sends out the address on the address bus.

1.6.3. Control Bus

The CPU sends out signal on the control bus to enable the outputs of addressed memory devices or port devices. The control bus consists of 4 to 10 parallel signal lines. In a microprocessor system the typical control bus signals are Memory Read, Memory Write, I/O Read and I/O Write.

1.6.4. Data Bus

There are different sizes of data bus for various microprocessors. The data bus basically consists of parallel signal lines and is used to transit data. Most of the data buses are bidirectional.

This means that CPU can send data out to memory or to a port, or it can read data in from memory or from a port on these lines. Several devices in a system will have their outputs connected to the data bus, but only one device at a time will have its output enabled.

1.7. ARITHMETIC AND LOGIC UNIT (ALU)

The ALU performs the computing functions. It includes the accumulator, the temporary register, the arithmetic and logic circuits and flags. The temporary data storage registers are used to hold data during an arithmetic and logic operation. The result is stored in the accumulator and flags are set or reset according to the result of the operation. The flags are affected by the arithmetic and logic operations in the ALU. In most of these operations, the result is stored in the accumulator. Therefore, the flags generally reflect data conditions in the accumulator with some exceptions.

In 16-bits microprocessors the flag register used is 16-bit. Only 9 bits are used as flags and 7 bits remains unused. The six flags are conditional flags (CF, AF, PF, SF, ZF and OF) and three flags (IF, TF and DF) are control flags.

NOTE : In brief, ALU is the area of microprocessor where various computing functions are performed on data. The ALU performs all arithmetic operations such as addition, subtraction, multiplication, etc., and logic operations such as AND, OR, and exclusive OR. Results computed by the ALU are stored either in register or in memory.

1.7.1. Timing and Control Unit

This unit synchronizes all the microprocessor operations with the clock and generates the control signal necessary for communication between the microprocessor and peripherals. The control signals are similar to a sync pulse in an oscilloscope. The Read and Write signals are sync pulses indicating the availability of data on the data bus.

This unit supervises the flow of information between the various units. Control unit retrieves the instructions, one by one, from the program that is stored in memory. For each instruction, the control unit informs the processor to execute the operation specified by the instruction. Both program and data are stored in memory. The control unit supervises the program instructions and the processor manipulates the data as specified by the program.

NOTE : It concludes that control and timing unit provides the necessary timing and control signals to all the operations in the microcomputer. It controls the flow of data between the Microprocessor and memory and peripherals.

1.8. CENTRAL PROCESSING UNIT (CPU)

Central processing unit is the heart of computer and it controls the operation of computer. In a microcomputer a microprocessor works as a CPU. The CPU fetches binary coded instructions from memory, decodes the instructions into a series of actions, and carries out these actions in a sequence of steps. CPU contains instruction pointer, general purpose registers and control circuitry.

1.9. MEMORY

Memory is a storage device. It consists of cells; each cell is a flip-flop/or transistor and can store one binary digit. Typically each line/or location is an 8 bit or one byte register that can store 8 binary bits. When several of these registers are arranged in a sequence it is called memory. The user writes the necessary instructions and data in memory through an input device, microprocessor performs the given task and answer/result is generally displayed at an output device or stored in memory. Basically, memory is of two types—main memory and storage memory.

The main memory is basically in two sections: Read Only Memory (ROM) and Write/Read Memory (W/RM), popularly known as Random Access Memory (RAM). Programs stored in ROM can only be read, they can not be altered. The RAM is also known as user memory and is used to store user program and data. The information stored in this memory can be easily read and altered.

The storage memory is used to store programs and results after the completion of program execution. This memory is non-volatile. Most common storage memory devices are—Floppy disk. Hard disk, and Compact disk.

1.10. INPUT/OUTPUT

The input devices are used to enter instructions and data into memory, these devices are keyboard or mouse etc. The microprocessor reads the instructions from memory and processes the data according to those instructions. The result is displayed by an output device such as seven segment display or monitor screen, or printed by a printer.

1.11. HARDWARE, SOFTWARE AND FIRMWARE

The hardware, software, and firmware are computer related terms. All the physical devices and circuitry of the computer are called HARDWARE. All the programs written for a computer are referred as SOFTWARE. FIRMWARE is the programs stored in ROMs and kept permanently.

1.12. TYPES OF COMPUTERS

Computers are classified on the basis of their wide vanities of size and capabilities. These are mainframes, minicomputers and microcomputer types.

1.12.1. Main Frame Computer

These are the largest and most powerful computers; they are implemented using two or more central processing units. These are designed to work at very high speed with large data words, typically 64 bits or more. They have very high data storage capacity. These computers are most often used for complex scientific calculation, large data processing, military defense control and computer graphics displays for science fiction movies. The fastest and most powerful mainframe computers are called supercomputers.

1.12.2. Mini Computer

Mini computers are the scaled down mainframes with slow speed. They work directly with the smaller data words and have smaller memory. They are used for scientific research, data processing applications, etc.

These are the largest and most powerful computers; they are implemented using two or more central processing units. These are designed to work at very high speed with large data words, typically 64 bits or more. They have very high data storage capacity. These computers are most often used for complex scientific calculation, large data processing, military defense control and computer graphics displays for science fiction movies. The fastest and most powerful mainframe computers are called supercomputers.

1.12.3. Microcomputers

Microcomputers are smaller in size. They contain only one CPU, which is usually a single integrated circuit called microprocessor. Once we add ROM, RAM and ports to a microprocessor then it becomes microcomputer. Today microcomputers are used in everything from smart sewing machines to computer aided design systems. They are used, for small industrial control, prices control and where storage and speed requirements are limited.

1.13. INTRODUCTION TO COMPUTER PROGRAMMING LANGUAGES

There are five languages that can be used to write a program for microcomputer. These languages are machine language, assembly language, or low level language and high level language.

1.13.1. Machine Language

Machine language is the simple binary codes and programs are a sequence of the binary codes written for the instructions. Computer/machine understands only these programs in binary codes. It is very difficult to program in machine language because of its several binary codes.

1.13.2. Assembly Language

Assembly language uses mnemonics or op-codes. Basically, op-codes are combination of two to four alphabets. Instructions consist of op-codes and operands and programs are sequence of instructions. Assembly language is a low level language. Because mnemonics are simple words of alphabets, it is user friendly language. Programs that involve, more hardware control and/or must run as quickly as possible are usually best written in assembly language. The most preferred language for microprocessor programming is assembly language. The following' format is used for writing programs in assembly language—example given in format is applicable for 8085 microprocessor.

An assembler program such as TASM/MASM is used to translate 'assembly language to machine language.

1.13.3. High Level Language

High level language is a more English like language, such as FORTRAN, BASIC, C or C++. An interpreter or computer program is used to translate higher level language to machine codes to load into memory and execute. Writing programs in high level language is faster but execution is slower than assembly language and also requires more memory space than the same program written in assembly language. Complex data processing programs that manipulate large amount of data, such as bank records, are usually best written in high level language.

High level language is a more English like language, such as FORTRAN, BASIC, C or C++. An interpreter or computer program is used to translate higher level language to machine codes to load into memory and execute. Writing programs in high level language is faster but execution is slower than assembly language and also requires more memory space than the same program written in assembly language. Complex data processing programs that manipulate large amount of data, such as bank records, are usually best written in high level language.

1.13.4. Memory Hierarchy of a System

The memory system is an essential component of any microprocessor based system—memory is needed for storing programs and data. A small computer with a limited application can fulfill it intended task without needing additional storage capacity. However, the general purpose computers will run more efficiently if they were equipped with external storage beyond the capacity of main memory. One memory may not provide enough space to accommodate all the programs use in a typical microprocessor based system. The *memory* unit of a computer that communicate! Directly with the microprocessor is known as main memory. The devices used to provide secondary storage and back up storage are known as storage memory, also named auxiliary") memory.

1.13.5. Memory Hierarchy of a Common System

Figure 1.21 represents typical memory hierarchy of a microprocessor based system. At the bottom of the figure the relatively slow magnetic tapes (Secondary Memory) are used to store removable files. The magnetic tapes are also used for back up storage. The main memory occupies a central position by being able to communicate directly with the CPU and with the secondary memory devices. When programs not residing in main memory are required by the CPU, they are brought in from secondary memory. Programs not currently needed in main memory are transferred into secondary memory to provide space for currently used programs and data.

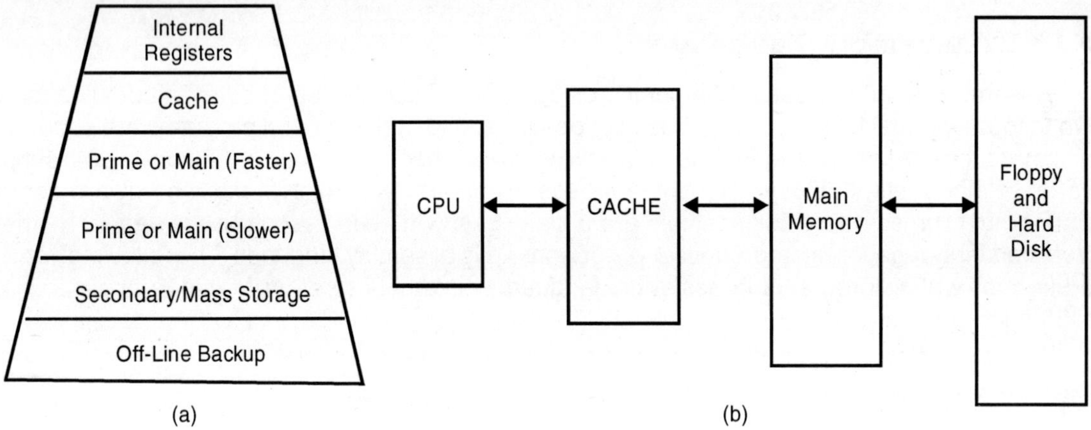

Fig. 1.21. Memory Hierarchy Diagrams.

Sometimes a special very high speed memory known as cache memory is used to increase the speed of processing by making current programs and data available to the CPU at a faster rate. This memory is employed in computer systems to overcome the speed differential between main memory access time and processor logic. The processor logic is usually faster, than main memory access time, with the result that processing speed is limited primarily by the speed of main memory. A technique used to compensate the differential in operating speeds is to employ an extremely fast, small cache between the processor and main memory whose access time is close to processor logic clock cycle time. The purpose of cache is to store segments of programs currently being executed in the CPU and temporary data frequently needed in the present calculations. This process of making data available at a faster rate increases the performance rate/speed of the computer. When the I/O processor manages data transfer between secondary memory and main memory, the cache organization is concerned with the transfer of information between main memory and CPU. Therefore each memory component is involved with a different level in the memory hierarchy system. The use of different levels of memory hierarchy makes the computer system highly economical with increased performance rate. As the storage capacity of the memory less creases, the cost per bit for storing binary information decreases and the access time of the memory more becomes longer. The secondary memory has a large storage capacity, is relatively inexpensive, but has low access speed compared to main memory. The cache memory is very small, relatively expensive and has very high access speed. Therefore, as the memory access speed increases, so does its relative cost. The basic idea of using a memory hierarchy is to obtain the highest possible average access speed while minimizing the total cost of the entire memory system.

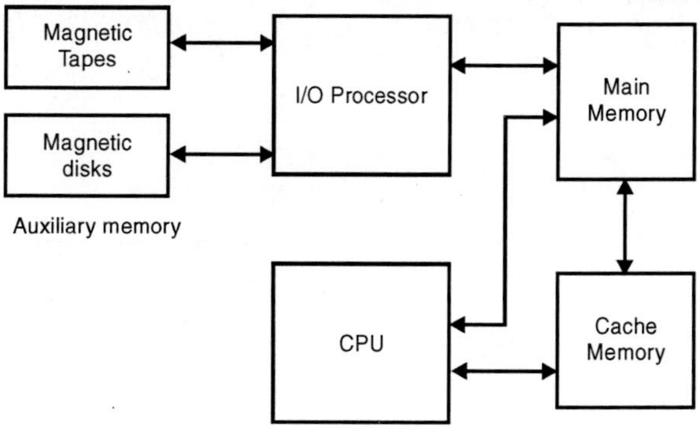

Fig. 1.22. Memory Hierarchy in a Typical Computer System.

The secondary and cache memories are used for different purposes. The cache is used to hold those parts of the program and data that are most heavily used. The secondary memory is used to hold those parts of the program that are not presently used by the CPU. The CPU has direct access to both cache and main memory but not to secondary memory. A transfer from secondary to main memory is usually done by means of direct memory access of large blocks of data. The typical access/time ratio between cache and main memory is about 1 to 7.

Figure 1.22 represents memory hierarchy of a typical computer system. The memory hierarchy system of a computer system enables multiprogramming, refers to the existence of two or more programs in different parts of memory hierarchy at the same time. This enables to keep all the parts of a computer busy by working with several programs in sequence.

Sometimes, computer programs are too long to be accommodated in the total space available m the main memory. Moreover a computer system uses many programs and all the programs cannot reside in main memory all the time. A program with its data normally resides in secondary memory. When the program or a segment of the program is to be executed, it is transferred to main memory to be executed by the CPU. The part of the computer system that supervises flow of information between secondary memory and main memory is called the memory management system.

1.14. TYPES OF MEMORY

Memories can be categorized into two groups i.e. main memory and storage memory. The main memory is also named primary memory and similarly, storage memory is sometimes referred as secondary memory. The main memories are those which can be used for read/write (R/W) and read only operations.

These are named random access memory (RAM) and read only memory (ROM), respectively. The computer uses these memories for executing and storing programs. The main memory should be able to respond fast enough to keep pace with the execution speed of the microprocessor. Therefore, it should be random access memory i.e. the Microprocessor should be able to access information from any register with the same speed. The other group of memory is the storage memory such as magnetic disks and tapes. This memory is used to store programs and results after the completion of program execution. This memory is non-volatile *i.e.,* information remains intact even if the computer system is turned off. Generally these memory devices are not part of any system; they are made part of the system when stored programs need to be accessed. The microprocessor cannot directly execute or process programs stored in these devices; program needs to be copied into the read/write memory first. Therefore, the size of the main memory determines how large a program the computer system can process. The size of the storage memory is unlimited; when one disk or tape is full, the next one can be used. Fig. 1.22 represents the memory types.

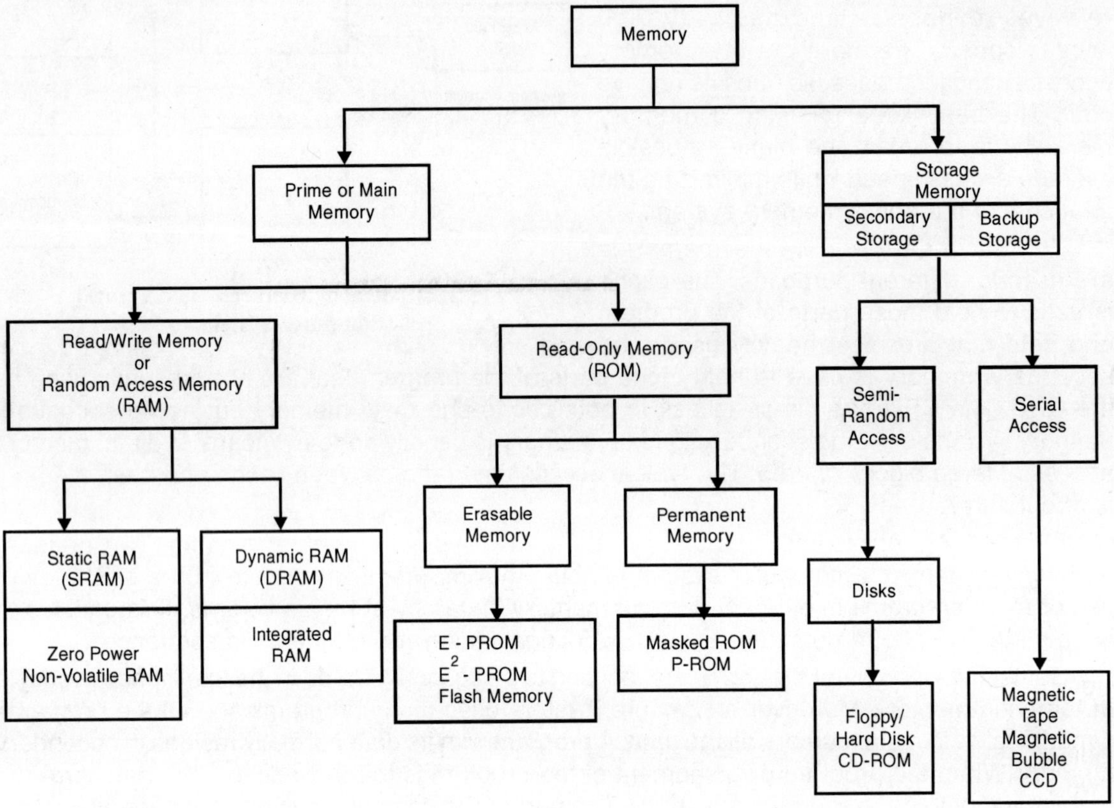

Fig. 1.23. Memory Types

The most common floppy disk is 3.5 inch high density disk. A floppy disk is made of a thin Magnetic material (iron oxide) that can store logic 0's and 1's in the form of magnetic directions. The surface of the disk is divided into a number of concentric tracks, with each track divided into Sectors. Another memory storage device is hard disk. The hard disk is similar to the floppy disk except that the magnetic material is coated on a rigid aluminum base and is enclosed in a sealed -Container. While it is highly precise and reliable, the hard disk requires sophisticated controller: circuitry; thus it is expensive. However, its storage capacity is large. Hard disks are available in various sizes. Storage capacity ranges from several megabytes to several gigabytes. Now-a-days Hard disks of 20 GB, 40 GB etc. are often used.

1.14.1. Floppy Disk

These disks are known as floppy disks due to the nature of their construction. Disk contains: a Round, flexible piece of plastic whose surface is coated with a magnetic oxide. The disk is mounted: side a protective package to allow easy insertion into the floppy drive. The information's are Stored on the disk in data areas called sectors. Generally, a sector contains 512 bytes of Information. The sectors are arranged in groups of nine or more on a single track as represented.

Thus, with the read/write head in a particular position, say over track 34, all sectors in track 34 will pass under the read/write head as the disk rotates. Most floppy drives today contain read/write heads so that both sides of the disk can be used. This doubles the number of sec associated with any given track, and leads to the term double sided use to describe a paretic disk.

After manufacturing a floppy disk, its surface contains a random jumble of magnetic polarity and cannot be used for storage until it is formatted. After formatting a 3.5 inch floppy disk 1; disk will have 80 tracks and 18 sectors per track. The capacity of a disk is found by the follow. Formula:

Capacity sides × tracks × (Sectors/Track) × (Bytes/Sector)

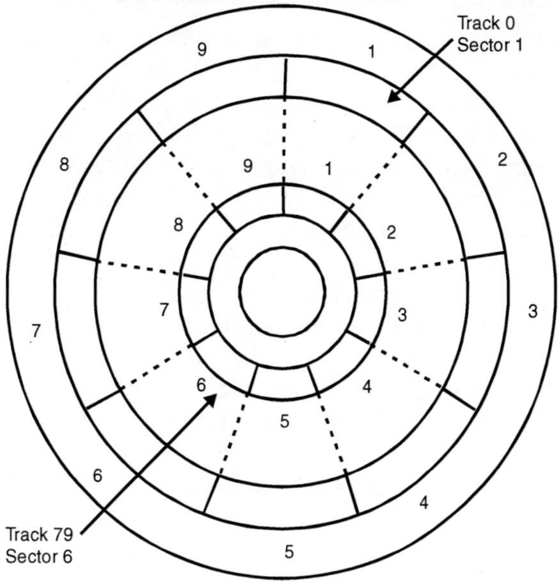

Fig. 1.24. Tracks and Sectors on a 3.5 inch floppy disk.

Therefore, for a 3.5 inch disk with 18 sectors/track, the capacity of the disk:

$$\text{Capacity} = 2 \times 80 \times 18 \times 512$$
$$= 14{,}74{,}560 \text{ bytes}$$

To convert the disk capacity into kilobytes (KB) or megabytes (MB), divide by 1024. Thic views Capacity 1.44 MB.

The 18 sectors/track 3.5 inch disk is referred to as a **high-density** disk. Disk manufacture indicate the storage capabilities of a floppy through the **2DD** (two-sided, double density) and (two-sided, high-density) part numbers.

1.14.2. Hard Disk

The hard disk is a primary intermediate-speed, non-volatile, writable storage device for m computers. The typical hard drive stores information serially on a non-removable disk with m platters. Each platter is magnetically on one or both surfaces. There are one or more read/write heads per recording surface. Each platter is divided into concentric tracks as represented in the Fig. 1.25.

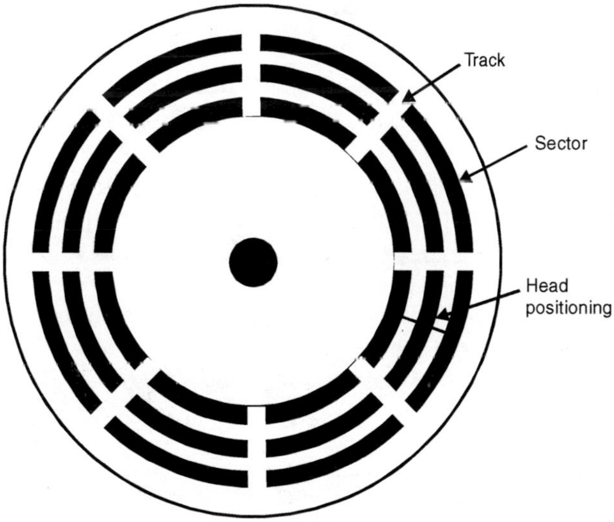

Fig. 1.25. Structure of a Hard Disk

The set of tracks that are at the same distance from the centre of the disk on all platter surfaces are referred to as a cylinder. Each track is divided into sectors containing a fixed number of bytes. The number: of bytes per sector typically ranges from 256 to 4 K the typical byte address includes the cylinder number, head number, sector number and work offset within the sector. The addressing assumes that the number of sectors per track is fixed. In today's high capacity disk, more sectors are included in the longer outer tracks than in the shorter Inner tracks. In addition, a number of spare sectors are reserved to take the place of defective sectors. As a consequence of these design choices, the actual physical address of a sector on the disk is likely to be different from the address of the sector sent to the disk controller. The mapping from this address is typically accomplished in the disk controller or the drive electronics.

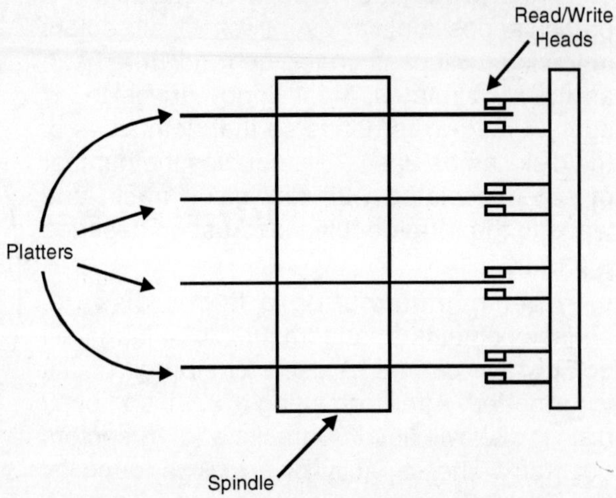

Fig. 1.26. A Hard Disk Construction.

In order to enable information to be accessed, the set of heads is mounted on an actuator that can move the heads ideally over the disk. The time required to move the heads from the current cylinder to the desired cylinder is called the seek time. The time required to rotate the disk from its current position to that having the desired sector under the heads is called the rotational delay. A certain amount of time is required by the disk controller to access and output information. This time is the controller time. The time required to locate a word on the disk is the disk access time, which is the sum of the controller time, the seek time and the rotational delay. Words can be transferred singly; they are often accessed in blocks. The transfer rate for a block of words, once die block has been located, is the disk transfer rate, typically specified in megabyte per second. The transfer rate required by the CPU memory bus to transfer a sector from disk is the number of bytes in the sector divided by the length of time taken to read a sector from the disk. The length of time required to read a sector is equal to the proportion of the cylinder occupied by the sector divided by the rotational speed of the disk.

Hard disks are popular for two major reasons

(1) Their storage capacity is rated in megabyte, even in Gigabytes for modern hard disk. A hard disk stores equivalent of several floppy disks.

(2) Next is that the hard disks transfer's disk data at much faster than floppy disks. The reduction in transfer time is largely due toe rotational speed of the hard disk.

A hard disk contains many read write heads. A hard drive will contain a number of recordings. Surfaces called platters stacked one on top of the other. Each platter is similar to a floppy disk, except that it is made of more rigid material, so that it can spin faster, and can fit more tracks per inch than a floppy disk. Fig. 1.26 represents a hard disk of four platters and eight heads. Because the read/write heads are mounted vertically, they all access the same portion of each platter as the platters spin. Thus, the eight heads define a 3-Dimensional cylinder of tracks. The storage capacity of a hard disk can be calculated using the formula :

$$\text{Capacity} = \text{Heads} \times \text{Cylinders} \times (\text{Sectors/ Tracks})$$

1.14.3. Main Memory

The main memory is the central storage unit in a computer system. The main memory is relatively large and fast memory used to store programs and data during the computer operation.

The technology used for main memory is based on semiconductor integrated circuit. Integral circuit RAM chips are available in two possible operating modes: static and dynamic. The static RAM consists essentially of internal flip-flops that store the binary information. The stored information remains valid as long as power is applied to the unit. The dynamic RAM stores binary information in the form of electric charges that are applied to capacitors. The capacity is provided inside the chip by MOS transistors. The stored charge on the capacitors tends to discharge with time and the capacitors must be periodically recharged by refreshing the dynamic memory. The refreshing is done by cycling through the words every few milliseconds to store decaying charge. The dynamic RAM reduces power consumption and has larger storage capacity in a single memory chip.

The static RAM is easier to use and has shorter read and write cycles. Read Only Memory (ROM) is used for storing programs that are permanent residents in the computer and for tables of constants that do not change in value once the production of computer is completed. Another fact is that the ROM portion of main memory is needed for storage an initial program named boot step loader. The boot step loader is a program whose function is to start the computer software operating when power is turned on. As the RAM is volatile, its control is destroyed once power is turned off. The start up of a computer consists of turning the power and starting the execution of an initial program. Thus, when power is turned on, the hardware of the computer sets the program counter/instruction pointer to the first address of the boot: loader. The boot step program loads a portion of the operating system from disk to main men and control is then transferred to the operating system, which prepares the computer for general use.

1.14.4. Random Access Memory (RAM)

The main memory In a computer is made up of RAM integrated circuit chips, but a portion of the main memory may be constructed with ROM chips. Basically, RAM was used to refer random access memory, but now it is used to designate a read/write memory to distinguish it 1 ROM, although ROM is also a random access. The RAM is used for storing the bulk of the program and data that are subject to change. RAM is of two types i.e. static RAM (SRAM), and dynamic RAM (DRAM pair is such that when one transistor is ON, the other is OFF, and vice-versa. A 1 is assigned to the state that exists when Q2 is ON.

1.14.5. Static Ram Devices

As mentioned earlier a memory consists of cells, each cell stores one bit. For static mer devices, a cell is commonly implemented using six metal oxide semiconductors (MOS) transistor as represented in Fig. 1.27.

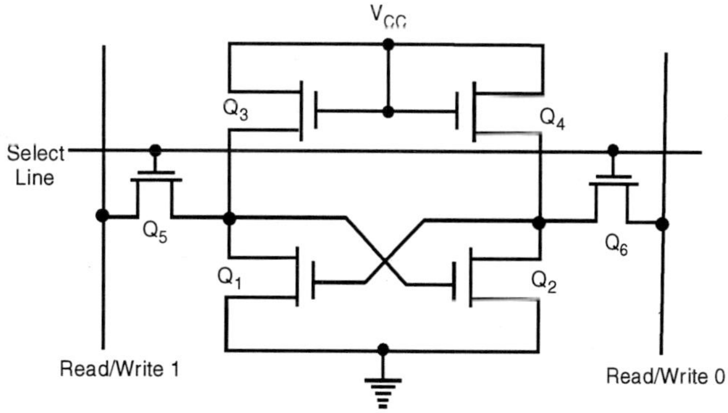

Fig. 1.27. Static Ram Cell.

Information is stored according to the state of transistors QI Q2.The cross-coupled transistor pair is such that when one transistor is ON, the other is OFF, and vice-versa. 1, is assigned to the state that exists when Q2 is ON and Q1 is OFF, and 0 is assigned to the opposite state. The two transistors Q3 and Q4 serve as registers and the other two transistors Q5 and Q6 act as enable gates. While writing, first the cell is selected by raising the voltage level on the select lines. When this is done, transistors Q5 and Q6 act as short circuits, so that the read/write 1 line is applied to the gate of Q2 and the read/write 0 line is applied to gate Q1. To write a 1 into the cell, a 1 is placed on read/write 0; this causes Q2 to be turned ON and Q1 to be turned OFF.

In other words, if a 0 is to be written into the cell, a 1 is placed on read/write 0 and a 0 is placed on read/write 1.In any case, once they are set the states of QI or Q2 will remain unchanged until the next write an operation. A cell can be read A3 simply by applying a voltage to A4 the select line. When this is done, the state of QI is applied to read/write 0 and the state of Q2 is applied to read/write 1.

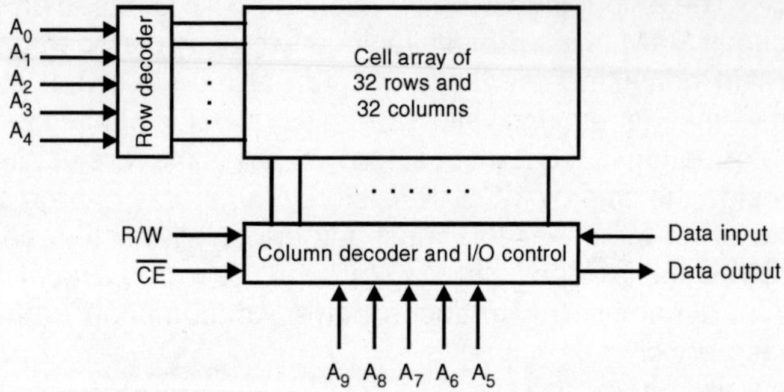

Fig. 1.28. Block Diagram of 1K x 1 Memory Device.

The number of memory cells and their arrangement in a static memory device varies considerably. The general organization of a static RAM is described by the block diagram represented in Fig. 1.28.

The figure illustrates a 1K x 1 RAM. The memory cells are organized in a matrix of 32 rows and 32 columns. Address bits A9 through AO are divided into row and column addresses and together specify one of the 1024 cells. Row address inputs (A4 - AO) are decoded to select one of the 32 rows of storage cells. Column address inputs (A9 - A5) not only select the column but also enable the corresponding 1(0 circuits consisting of drives and sense amplifiers. These circuits permit a stored bit to be output during a read operation and to be hanged during a write operation. The read/write (R/W) control input specifies the type of operation, high for read and low for write. The chip enable input (CE) is used to select the appropriate row of memory devices in the memory model.

Each static RAM device generally contains fewer cells than a dynamic RAM (DRAM) chip of romp arable design because each cell in a static RAM (SRAM) device must contain several transistors. Also, the static RAM consumes more power since one of the two cross coupled transistors Is always on, therefore consuming power continuously. The major advantage of SRAM is that it does not requires to be refreshed.

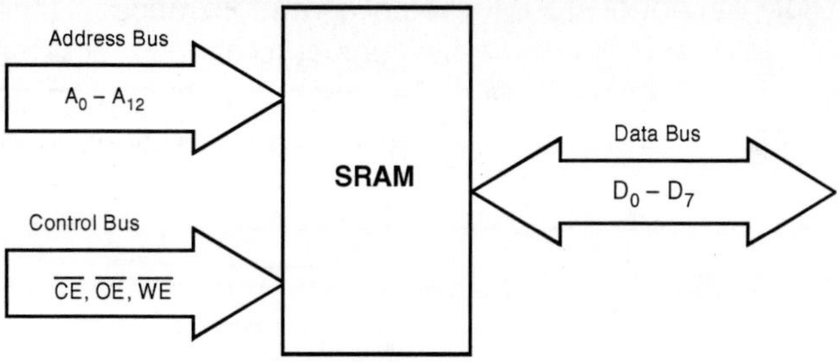

Fig. 1.29. Block Diagram of a Static RAM Cell.

The Fig. 1.29 represents the block diagram of a typical static RAM. It can be observed that the memory device has address lines, data lines and control lines. The data lines of SRAM, however, act as both inputs and outputs. Due to this reason, they are identified as a **Bidirectional Bus.**

1.14.6. Dynamic RAM Device (DRAM)

Although static RAM is the faster memory and the easiest to interface with, but the densest memories are dynamic RAMs (DRAMs). Their high capacity is due to an extremely efficient memory element, the one transistor memory cell consisting of a single access transistor and a capacitor.

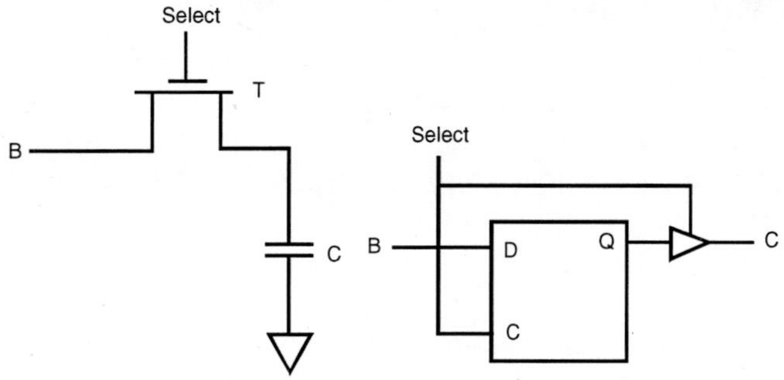

Fig. 1.30. One Transistor Dynamic RAM Cell

The dynamic RAM memory chip is organized in a matrix form by rows and columns of memory cells. The simplex type of dynamic RAM (DRAM) cell contains only one transistor and one capacitor. The Fig. 1.30 represents a typical one transistor dynamic RAM cell. Whether a 1 or o is contained in a cell is determined by whether or not there is a charge on the capacitor. To write the memory cell, the bit line is charged to a logic 1 or 0 voltage while the word line is asserted. This enables the access transistor, making it possible to charge up the storage capacitor with the desired logic voltage. The read operation takes place by asserting the word line. The access transistor is turned ON, sharing the voltage on the capacitor with the bit line. Sensitive amplifier circuits detect small changes on the bit line to determine whether a 1 or 0 was stored in the selected memory element. The destructiveness of the read operation makes DRAMs complex. To read the contents of the storage capacitor, we must discharge it across the bit line. Thus, external circuitry in the DRAM must buffer the values that have been read out and then write them right back.

The second problem with DRAMs, and the most significant one from viewpoint of a system designer is that their contents decay over time. Every once in a while (measured in milliseconds), the charge on the storage capacitor leaks off. To counteract this, the DRAM must be refreshed. Periodically, the memory elements must be read and written back to their storage locations. To make this operation reasonably efficient, the Drams memory array is a two dimensional matrix.

Addresses

(a) Refresh Cycle : The storage capacitor at the heart of a DRAM memory cell is not perfect. Over time, it leaks away the charge it is meant to hold. Thus DRAMs must undergo like *an* abbreviated read cycle; data is extracted from the storage matrix and then immediately written back without appearine: at the output pin is incremented by 1 for each memory refresh cycle so that it sequences through all the row.

(b) During a memory refresh cycle, all memory chips are enabled so that memory refresh is performed on every chip in the memory module simultaneously. This reduces the number of refresh cycles in a regular read cycle, at most one row of memory chips is enabled.

(c) Normally, a dynamic RAM has a data output enabled control. The chip enable control input and data output enable control input are combined internally so that the data output is forced to its high impedance more unless both control inputs are activated. During a memory refresh cycle, the data *output* enable control is deactivated. This is necessary because all the chips in the same column are selected and their data outputs are tied together.

Advantages of Dynamic RAMs

The dynamic RAMs are very attractive to memory designers due to many reasons. These memories are most considered by the designers, especially when the memory size is large. Same of the reasons are :

1. High Density

The DRAMs are highly dense and occupy less space as compared to SRAM. For a SRAM, a typical cell requires six MOS. Transistors, whereas, the structure of a DRAM cell is much simpler and can be implemented with three, or even one MOS transistor. Thus, more cells can be put into a single chip, and the number of memory chips needed to implement memory module is reduced.

2. Low Cost

DRAM is less expensive per bit than SRAM. However, DRAM requires more supporting circuitry, and therefore, there is little or no cost advantage when building a memory system. Therefore, for selecting or making a cost statement, designer has to think of size of memory system.

3. Low Power Consumption

The DRAM consumes considerably less power per bit than SRAM. The power dissipation is typically 0.2 mW per bit for SRAM, whereas the power dissipation is less than 0.05 mW per bit for DRAM. This feature of the DRAM reduces the system power requirements and lowers the cost. Besides this, the power consumption of DRAM is extremely low in stand by mode; this makes it very attractive in the design of memory that is made nonvolatile through use of a back up power source.

There are many more advantages of using Dynamic RAM. The only disadvantage of DRAM is that of continuous need of refreshing the capacitor, which increases the complexity of chip logic.

1.14.7. Read Only Memory

The static and dynamic memories described earlier are volatile memories. Therefore, when electric power is removed from these memory chips, they lose the stored information. Non-volatile memories are very useful in situations where we want the information not to be destroyed when the power to the memory is removed. Although secondary memories such as floppy disk and hard disk are nonvolatile memories too. But, these are very slow to serve as primary memories in microprocessor based systems.

A read only memory (ROM) is essentially a device in which permanent information's are stored. The read only memories that are erasable have been the most popular non-volatile semiconductor memories for programs storage. Designers often use these memories for program storage because data stored in these memory chips can be easily and inexpensively altered. The information must be specified by the designer and then embedded into the ROM to form the required interconnection or electronic device pattern. Once the pattern is established, it stays within the ROM even when power is turned OFF and ON again; that is, ROM is non-volatile. A ROM is most widely used in microcomputer systems for storage of the program that determines overall system operation. For some of the ROM de\ices, information i.e. the microcomputer program must be built in during manufacturing and for others the data must be electrically entered. The process of entering data into a ROM is called programming of a ROM. As the name ROM implies, once entered into the device, the information can be only read.

Types of ROM

Main types of ROM devices available are :

1. Mask-programmable read-only memory (ROM)
2. The one time programmable read-only memory (PROM),

3. The erasable programmable read-only memory (EPROM), and
4. The electrically-erasable programmable read-only memory (E2.PROM). The ROM memory chips that permit erasure of data by exposure to ultraviolet light and can then be reprogrammed, are known as (UV-EPROMs).

The mask programmable ROM requires the manufacturer to first produce a mask according to a given set of contents or program. It is expensive to produce the first ROM, but the subsequent copies are inexpensive. Thus, ROMs are used only after the development stage of a computer system is complete, when it is not likely that the contents of the ROMs will need to be changed. The only pins required on a ROM device are a set of address input pins, a set of datall out pins, and a chip select pin.

The programmable read-only memories (PROM) have nichrome or polysilicon wires arranged in a matrix. These wires can be functionally viewed as diodes or fuses. This memory can be programmed by the user with a special PROM programmer that selectively bums the fuses according to the bit pattern to be stored. This process is known as burning or blowing of the PROM, and the information stored is permanent. The Is are represented by the diode links left intact and the Os by the diode links that are blown. Because the PROMs for all customers are identical when they are shipped, their price is not as dependent on quantity, but their circuitry is more complex and expensive due to their programmability. Therefore, for a few devices PROMs are cheaper than masked ROMs but for large quantity orders masked ROMs are cheaper. As a result, PROMs (or EPROMs) are used during development, but masked ROMs are used in the final mass produced products. PROMs are programmed bit-by-bit by placing the address pins and forcing a current into a bit's data outpin while the voltage supply and appropriate control pins are pulsed. The magnitude of the current, the magnitude and width of the pulses, and the manner In which the pulses are applied varies from one PROM to the next. Generally, the programming cycles are interspersed with verification cycles and the cycles are alternated for approximately twice the time needed to program the bit.

The Erasable Programmable read only memory (EPROM) stores a bit by charging they can be erased, they have the advantage of being reusable. However, unlike Proms, EPROM's may lose their contents over extended periods of time and should not be put in final products that are expected to last for some years. An EPROM is programmed by applying an address to the address pins and high or low voltages to all the data out pins, and then applying the proper voltages and pulses to the supply and control pin.

The electrically erasable read only memory (E^2.PROM) is also a reprogrammable ROM, but offers the added advantage that information can be altered by using electrical signals at the register level rather than erasing all the information's. This has an advantage in field and remote control applications. In microprocessor systems, software update is a common occurrence. If E^2.PROMs are used in the systems, they can be updated from a central computer by using a remote link via telephone lines. Similarly, in a process control where timing information needs to be changed, it can be changed by sending electrical signals from a central place. This memory also includes a chip erase mode, whereby the entire chip can be erased in 10 ms, whereas 15 to 20 minutes are required to erase an EPROM. Besides all the above advantages, the major disadvantage of an **E^2.PROM is that it tends to be more expensive**.

In a microprocessor based product, programs are generally written in ROM, and data that are likely to vary are stored in RAM. For instance, in a microprocessor control oven, programs that run the oven are permanently stored in ROM, and data such as baking period, starting tie and temperature are entered in RAM memory through the keyboard. On the other hand, when microcomputers are used for developing programs (software) or for learning purposes programs are first written in RAM memory, and then stored on a storage memory such as floppy disk or hard disk.

Block Diagram of a ROM

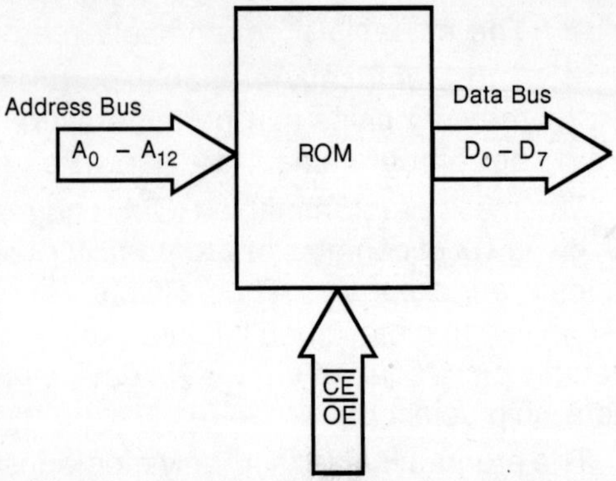

Fig. 1.31. Block Diagram of a ROM.

The Fig. 1.30 represents a block diagram of a typical ROM. It can be observed that the device has three sets of signal lines: the address inputs, data output, and control inputs. This block diagram is valid for a ROM, PROM and EPROM.

The address bus is used to input the signals that select between the data storage locations with in the ROM device. The address bus consists of eleven address lines, Ao through AIO. The bits in the address are arranged so that AIO is the MSB and Ao is the LSB. With an 11-bit address the memory device has 211 = 2048 unique data storage locations. Each bit of data is stored inside a ROM as either a binary 0 or binary 1. Actually, 8 bits of data are stored at every address. Therefore the total storage capacity of the device described here is 2048 x 8 = 16,384 bits, that is, the device described is really a 16 K-bits ROM. By applying the address of a storage location to the address inputs of the ROM, the byte of data held at the address location is read out on to the data lines. In the block diagram of the ROM device the data bus consists of eight lines labeled Do through D7' The control bus represents the control signals that are required to enable or disable the ROM device, two control leads output enable (OE) and chip enable (CE) are identified. For instance, logic 0 at OE enables the three state outputs, Do through D7, of the device. If OE is switched to the 1 logic level, these outputs are disabled, put in the high-impedance state. Moreover, CE must be at logic 0 for the device to be active. Logic 1 at CE puts the device in a low power stand-by mode. When in this state, the data outputs are in the high-impedance of the logic level of OE.

1.15. RISC (REDUCED INSTRUCTION SET COMPUTERS)

The technical term *RISC* has been swamped by the marketing term *RISC* to the point that it's lost almost all meaning as a technical term. Almost everything now is described as RISC, even when it isn't. A historical perspective can help illustrate the point in the development of computer architectures which "RISC" was invented to describe. Register usage evolved from accumulators and index registers down two paths, one leading to general purpose registers and another less popular path to stacks.

1.15.1. Accumulator Architecture

The first computers were accumulator based, performing operations on data stored in a register with data loaded from memory locations. Addresses were coded in program memory, and could only be changed by modifying the program. Initially machines only had one (which all operations implicitly refered to), but later multiple accumulators were sometimes used.

Index Registers

Index registers were used to hold memory addresses, while accumulators were still used for computation. Operations could specify an accumulator for the one operand, and an index register to fetch the second operand from memory.

Memory to Memory Architectures

With a little modification, index and accumulator registers could be used interchangeably—they were now general purpose, meaning that a register could hold either data or an address.

Some designs took this further, allowing the value in memory to be an indirect address itself—sometimes in an unlimited chain of references.

Stack Architectures

General registers can be made even more flexible allowing them to hold multiple values (either data or addresses) in a useful order—in other words, stacks. At the same time, the design can be simplified back to a single stack (as accumulator architecture) with no loss of flexibility, providing swap and duplicate operations are provided.

1.15.2. Register to Register Architectures

Although conceptually powerful, memory to memory operations are limited in speed because memory access is relatively slow, while registers are fast. The obvious improvement is to increase the number of registers, and restrict or simplify memory access (limited to load and store operations) to simplify the overall design. The first major system to follow this idea was the CDC 6600, but the idea was also explored in the IBM 801 project. However it wasn't until the Berkeley RISC design named this type of design a "Reduced Instruction Set Computer". The term *CISC* was also invented at this time solely to give a name to the prior generation (memory-memory designs).

A diagram can show the approximate lineage of register usage:

Accumulator I Accumulator +Index / \ Stack...Memory- (CISC) Data I Load (RISC) Store Program instruction formats changed from simple to complex due to limited memory capacity, then to simple as capacity eased.

1.15.3. Simple Hardware-Interpreted Instructions

Limited hardware meant operations and circuitry had to be simple, hence the use of accumulators. Instruction bits often triggered the operations.

Hardware decoded, then interpreted instructions: More complex instructions were encoded in numeric codes which needed to be decoded to supply the bits to trigger hardware operations.

Multistep instructions, broken down by hardware (microcode):

Commonly used sequences (copy a block of memory, scan a string) could be encoded as a single instruction, which is decoded to several which are executed, one operation at a time.

Instructions broken down before execution, executed as simple instructions:

The Intel P6 series and Motorola 68060 operate like this, to compensate for the limits of complex instructions. This was actually produced in response to single step instructions, but is a logical progression from microcode.

Single-step instructions, no need to breakdown: This is basically the idea presented as RISC. The execution of instructions evolved to perform more operations at a time.

One at a time : One instruction was fetched, decoded if necessary, and executed before the next one. For microcoded processors, all operations finished before the next instruction was fetched.

Independent stages of multiple insturcions at a time (pipeline) : After the decoder was finished, it could be used for the next instruction while the first went to the execute stage. If an instruction took more than one cycle in a stage, the compiler could be used to anticipate this and insert a NOP if the delay could be predicted, or hardware could detect this and stall the pipeline until execution is finished. This is the technique used for initial RISC processors. Multiple parallel instructions, grouped by compiler (VLIW):

The idea was that obviously independent instructions (especially simple ones) could be detected by the compiler, and executed safely without hardware checks.

Parallel instructions grouped partly by compiler and partly by available hardware:

What Intel/HP call EPIC, used in IA-64 and TMS 320C6x.

Parallel instructions grouped by hardware (superscalar) : Self explanatory.

Instructions executed out of order, grouped by hardware : Generally with the use of automated resource renaming.

At the time of introduction, RISC and "load-store" were often synonymous, but RISC usually referred to a list of features:

- All instructions are a single, fixed length.
- Operations are simple enough to execute in a single cycle, using an execution pipeline.
- Operations are performed on registers only, the only memory operations are load and store.
- No microcode is needed, because of the simpler design.
- A large number of registers are available.
- Register windows are used to speed subroutine calls.

Register windows turned out to not be a useful enough idea to catch on, so have been mostly forgotten. RISC is more commonly used to refer to a design philosophy than a list of features (actually implementation techniques), especially since most of them have been applied to CISC designs (pipelines as far back as the Zilog Z8000, register windows in the Hitachi H16 and H32, and microcode eliminated in most modern designs). Basically, RISC asks whether hardware (for complex instructions or memory-to-memory operations) is necessary, or whether it can be replaced by software (simpler instructions or load/store architecture). Higher instruction bandwidth is usually offset by a simpler chip that can run at a higher clock speed, and more available optimisations for the compiler. By contrast, the CISC philosophy has been that if added hardware can result in an overall increase in speed, it's good - the ultimate goal of mapping every high level language statement on to a single CPU instruction. The disadvantage is that it's harder to increase the clock speed of a complex chip. The PowerPC is a good example of this idea applied to a load-store architecture.

1.15.4. Six Principles of RISC Philosophy (Reduced Instruction Set Computer)

(a) Reliance on optimizing Compilers

(b) Few Instructions and Addressing modes

(c) Fixed instruction formats

(d) Instruction execution in one machine cycle

(e) Only call and return instructions access memory

(f) Hardwired control

Characteristics of RISCs

(1) Instructions are simple.

(2) Only a few instructions.

(3) Instructions are of uniform length.

(4) One of few instruction formats.

(5) Few addressing modes.

(6) LOAD and STORE only references the memory.

(7) All operations are done on the registers of the CPU.

(8) Few data types typically integer and floating point.

(9) All machine cycle execute in one clock state, and machine instructions are hardwired.

(10) Large register set.

(11) Strong pipelining.

Benefits

(1) Improved performance due to simple and faster execution of instructions, simple compilers.

(2) Effective pipelining, parallelism in CU.

(3) More responsive to interrupts, since interrupts are checked between elementary operations.

(4) Only 6% of VLSI area is Control Unit, against 50% for CISC

(5) Design implementation time for VLSI is far less in RISCs.

* More recently RISC and CISC approaches are coming together and have features of each other.

Examples of RISC :

(1) Motorola 88000

(2) Sun SPARC

(3) Intel 860/960

RISC typically have been in applications area like work stations, high performance, and low cost lab machine.

1.16. CISC (COMPLEX INSTRUCTION SET COMPUTER)

A Computer with large number of instructions is classified as a Complex Instruction Set Computer.

Design consideration for Instruction set : Machine Language Constructs to simplify the compilation.

The essential goal of CISC Architecture is to attempt to provide a single machine instruction for each statement that is written in high level language.

1.16.1. Characteristics of CISC Architecture

(1) Large number of instructions in the Instruction set; 100 to 250.

(2) Instructions are quite complex and detailed; needs more hardware which may slow down the processing.

(3) Some Instructions perform specialized tasks and are not used frequently.

(4) A large variety of Addressing Modes typically from 5-20.

(5) Variable length Instruction formats.

(6) It has instructions which manipulate operands in memory.

1.16.2. Limitation of CISC

Thus, it is not clear that by increasing the complexity of the instruction set, one have higher overall efficiency.

1. Optimization the generated code and enhancement of pipelining is difficult with complex instructions set computers.
2. Longer opcodes, larger addressing modes, thereby larger size of instruction and complex control unit , and increased execution time of the instruction.
3. Design of the cost is high.

Thus, it is not clear that by increasing the complexity of the instruction set, one have higher overall efficiency.

REVIEW QUESTIONS

1. What is a microprocessor ? What is the difference between microprocessor and microcomputer ?
2. Define a CPU and explain the structure of Von Neumann architecture ?
3. Explain in detail the concept of System bus, Data bus, Conrol bus and Address bus ?
4. Explain with the help of a diagram the memory read and write operatons ?
5. Define memory addressing.
6. Explain the memory structure of 8-bit, 16-bit processors and 32-bit processors ?
7. List some applications of microprocessor ?
8. Distinguish between a microprocessor and a CPU ?
9. What do you mean by bit, byte, word and word length ?
10. Draw the diagram of a microcomputer and explain each unit of it ?
11. What is the role of ALU and CPU ?
12. Explain the term memory organization ?
13. Define the significance of a system clock ?
14. Explain in detail the relationship between memory access and system clock ?
15. Explain with the help of a diagram a general microprocessor configuration with its typical PSW ?
16. Explain in detail the fetch, decode execute cycle in the microprocessor ?
17. Define different memory devices ?
18. Explain RAM with its read and write memory cycle ?
19. Explain ROM with its memory cycle ?
20. Define what is Tristate logic and what do we mean by trisating of buses ?
21. Define memory access time ?
22. What is multimedia computers system ? In what way does it differ from a conventional computer ?
23. What are RISC and CISC processors ?
24. What are the various types of semiconductor memories ? Describe them briefly ?
25. What is the function of an input device ? Give examples of some input devices ?
26. What are Von-Neumann, Harvard, and data flow architecture of computers ? Give examples of each architecture ?
27. What do you understand by single-chip computers ? Where are they used ?
28. What are the various types of memories which are usually used with modern digital computers ?
29. Give the important features of all the memories used in a system ?
30. What is the function of an output device ? Give examples of some output devices ?

❑❑❑

CHAPTER 2

8085 MICROPROCESSOR (8-BIT)

2.1. INTRODUCTION

INSIDE THIS CHAPTER

A microprocessor is the Central Processing Unit (CPU) of a computer. It is the heart of the computer. The Intel corporation have developed a large number of general purpose and special purpose peripheral devices. These devices are very useful for the development of microprocessor based system. "This chapter describes the Intel 8085 microprocessor".

2.2. FUNDAMENTAL CONCEPTS

Before we start with the discussion on 8085 microprocessor we must first try and understand some of the fundamental concepts of "Microprocessors".

2.2.1. Microcomputer

A programmable machine that processes binary data. There are three major parts of a microcomputer system as shown in Fig. 2.1.

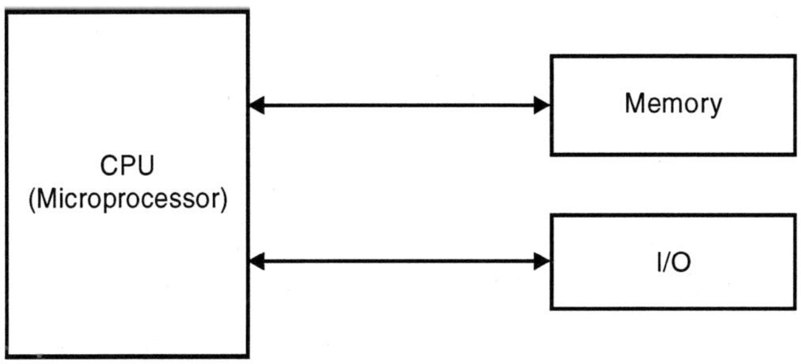

Fig. 2.1. Block diagram of Microcomputer.

1. **The CPU (Central Processing Unit) :** The CPU acts as the brain coordinating all Activities within the computer i.e. The group of circuits that process data and provide control signals and timing.
2. **The Memory :** Memory is the unit where the program instructions and data are temporarily stored.

3. **The I/O (Input/Output) Devices :** The I/O devices allow the computer to input information for processing and then output the result.

Today the CPU circuitry has been reduced to IC's called the microprocessor.

Definition

"Microprocessor" is a multipurpose, programmable, clock-driven, register based electronic device that reads binary instructions from a storage device called memory, accepts binary data as input and processes data according to those instructions, and provides as output.

2.2.2. CPU (Central Processing Unit)

The CPU includes ALU, Registers, Control unit and buses.

1. ALU (Arithmetic Logic Unit) : The group of circuits that perform arithmetic and logical operations is called the ALU. The ALU is a part of the CPU.

2. Control Unit : The group of circuits that provides timing and signals to all operations in the computer and controls data flow. The basic timing of the computer is controlled by a square wave oscillator or a clock generator circuit.

3. Register Unit : The area of CPU consists of various registers. The registers are used primarily to store data temporarily during the execution of a program. Registers can be viewed as a memory location where the memory resides within the CPU.

The block diagram in Fig. 2.2 shows the arrangement of the CPU.

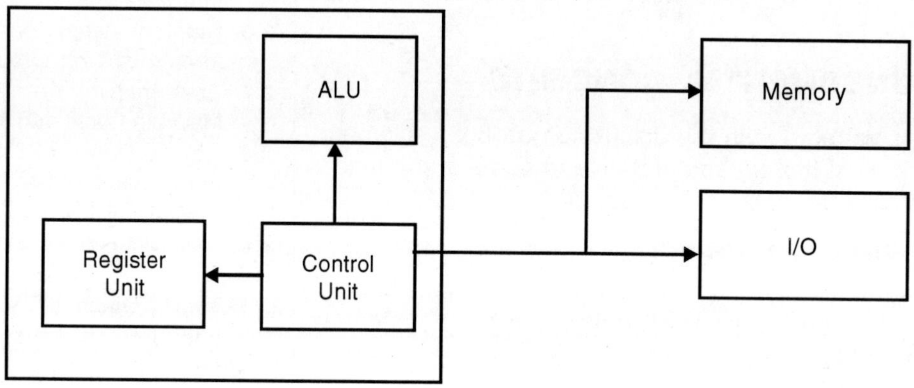

Fig. 2.2.

2.2.3. Scale of Integration

SSI (Small Scale Integration) : The process of designing a few circuits on a single chip. The term refers to the technology used to fabricate discrete logic gates on a chip.

MSI (Medium Scale Integration) : The process of designing more than hundreds of gates on a single chip.

LSI (Large Scale Integration) : The process of designing more than a thousand of gates on a single chip.

VLSI (Very Large Scale Integration) : The process is used to indicate very large scale of integration.

SLSI (Super Scale Integration) : The process is used to indicate the super scale of integration.

2.2.4. Programming a Microprocessor

A microprocessor is a programmable circuit. Microprocessor is programmable because it can be instructed to perform given tasks within its capability. Microprocessor is designed to understand and execute many binary instructions.

Programming a microprocessor means teaching all the steps necessary to complete a job. A microprocessor can be instructed by an user to furnish the following tasks :

1. To take data from the user
2. To modify the above data in a way the user wants
3. To return the modified data back to the user

Instruction Set

The list of all recognizable instructions by the instruction decoder is called the instruction set. Based on Instruction set, two types of computer systems:

CISC (Complex Instruction Set Computers), e.g., 80x86 family has more than 3000 instructions.

RISC (Reduced Instruction Set Computers) - A small number of very fast executing instructions.

2.2.5. Three Bus System Architecture

A collection of electronic signals all dedicated to a particular task is called a bus. The 3-buses used in a microprocessor are :

(i) Address bus

(ii) Data bus

(iii) Control bus

- A bus, as it permits to a microprocessor is a bundle of wires that are grouped together to serve a single purpose.
- The major parts of the microprocessor are connected to each other by 3-sets of parallel lines called buses as shown in Fig. 2.3.

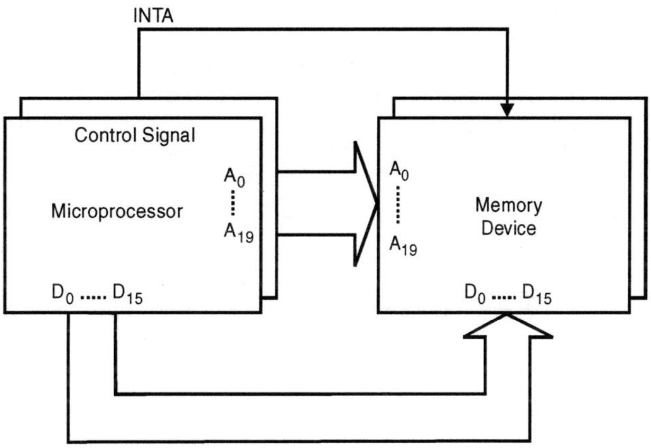

Fig. 2.3. Block diagram for bus organisation.

Some buses pass data, some address and others act as control lines. These lines are daisy chained from one device to the next. This is true for each of the address and data connections.

1. Address Bus

(a) A set of parallel lines connected is called as a bus.

(b) The bus over which the CPU sends out the address of the memory location is called as the address bus.

(c) The address bus carries the address of the memory location to be written to or read from.

(d) The address bus may consis of 16, 20, 24 or 32 parallel signal lines. If there are N address lines, then it can directly address 2^N memory locations.

(e) The address bus is a unidirectional bus as shown in Fig. 2.4.

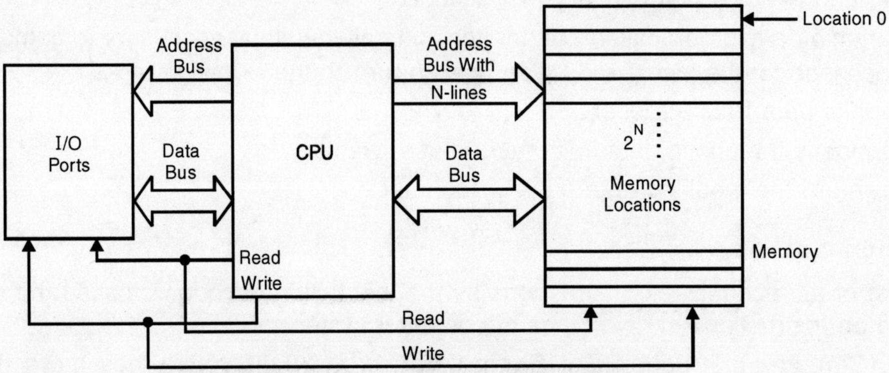

Fig. 2.4. Bus organization.

(f) The address bus is also used to send the port address on the address bus. When the CPU reads data from or writes data to a port, then it sends the port address out on the address bus.

2. Control Bus

(a) The control bus consists of 4 to 10 parallel signal lines. The control bus is used for sending control signals to the memory and input/output devices as shown in Fig. 2.4.

(b) The CPU sends signals on the control bus to enable the outputs of addressed memory devices or I/O port devices.

(c) Some of the control signals generated by the CPU and sends on to the control bus are :

(i) Memory read (ii) I/O read

(iii) Memory write (iv) I/O write.

3. Data Bus

(a) The data bus consists of 8, 16, or 32 parallel lines.

(b) The data bus is a bi-directional bus. That means that the data can be transferred to the CPU and from the CPU to memory or I/O devices.

(c) The data bus also connects the I/O ports and CPU, so that the CPU can write data to or read it from the memory or I/O ports.

(d) Many devices in the system will connect their outputs to the some data bus, but only one device at a time will have its output enabled, so that there is no collision of information.

(e) The number of data lines used in the data bus is equal to the size of data word being written or read.

(f) In order to completely eliminate the possibility of more than one device putting their outputs simultaneously on the data bus, all the devices getting connected to data bus should have "3 state outputs".

This would make it possible to disable the outputs of devices which should be stopped from connecting their outputs to the data bus.

2.2.6. Important Terminology

(i) A bit is a binary digit that can have the value "0 or 1".
(ii) A byte is defined as 8 bits.
(iii) A nibble is half a byte (4-bits).
(iv) A word is two bytes (16-bits).
(v) A double word is four bytes (32-bits).

2.3. 8085 MICROPROCESSOR

Intel 8085 is an 8-bit, NMOS microprocessor. It is a 40 Pin I.C. package fabricated on a single LSI chip.The Intel 8085 uses a single +5 $V_{d.c}$ supply for its operation.

It's clock speed is about 3 MHz. The clock cycle of 8085 is of 320 ns and 246 OP codes. Fig. 2.5 shows the block diagram of Intel 8085.

It consists of 3-main sections :

1. An arithmetic and logic unit.
2. A timing and control unit.
3. A set of registers.

2.3.1. Features of Intel 8085 Microprocessor

1. 8085 is an 8-bit microprocessor i.e. it has an 8-bit ALU.
2. It is a single chip, NMOS device implemented with 6200 transistors.
3. 8085 requires a single +5 $V_{d.c}$ supply.
4. 8085 has an on chip clock generator, hence it does not requires external clock generator, but is requires external tuned circuit like LC, RC or Crystal.
5. The maximum clock frequency is 5 MHz, but normally the 8085 operates at a clock frequency of 3 MHz.
6. 8085 has an 8-bit data bus and 16-bit address bus.
7. It provides 74 instructions with different addressing modes as discussed in section 2.8.
8. It provides 16 address lines, hence it can access 216 = 64 K bytes (65,536) of memory locations.
9. The data bus (D_0–D_7) is multiplexed with lower 8-bit address lines (A_0–A_7), hence it requires external hardware to seperate data lines from address lines.
10. 8085 provides 5 hardware interrupts : TRAP, RST 7.5, RST 6.5, RST 5.5 and INTR.
11. 8085 generates 8-bit I/O address, hence it can access 2^8 = 256 input ports and 256 output ports.
12. 8085 provides control signals $\left(\overline{RD}, \overline{WR}, IO/\overline{M}\right)$ to control the bus cycles hence external bus controller is not required.
13. 8085 provides on 8-bit accumulator, Flag register, 6 general purpose registers (B, C, D, E, H, L) and two special purpose registers (SP and PC).
14. 8085 can be used to implement 3-chip microcomputer using IC 8155 and IC 8355.
15. 8085 supports a mechanism that allows the 8085 to increase its interrupt handling capacity.
16. 8085 provides status for advanced control signals.

17. 8085 provides two serial I/O lines viz SID and SOD, hence serial peripherals can be interfaced with 8085 directly.

2.3.2. 8085 CPU Internal Architecture

The block diagram in Fig. 2.5 shows the internal architecture of 8085 microprocessor. The architecture of 8085 can be divided in different groups such as :

1. Register Set.
2. Address Buffer.
3. Address/Data Buffer.
4. Arithmetic and Logic Unit.
5. Instruction Decoder and Machine Cycle Encoder.
6. Interrupt Control.
7. Incrementer/Decrementer Address Latch.
8. Timing and Control Circuitry.
9. Serial I/O Control.

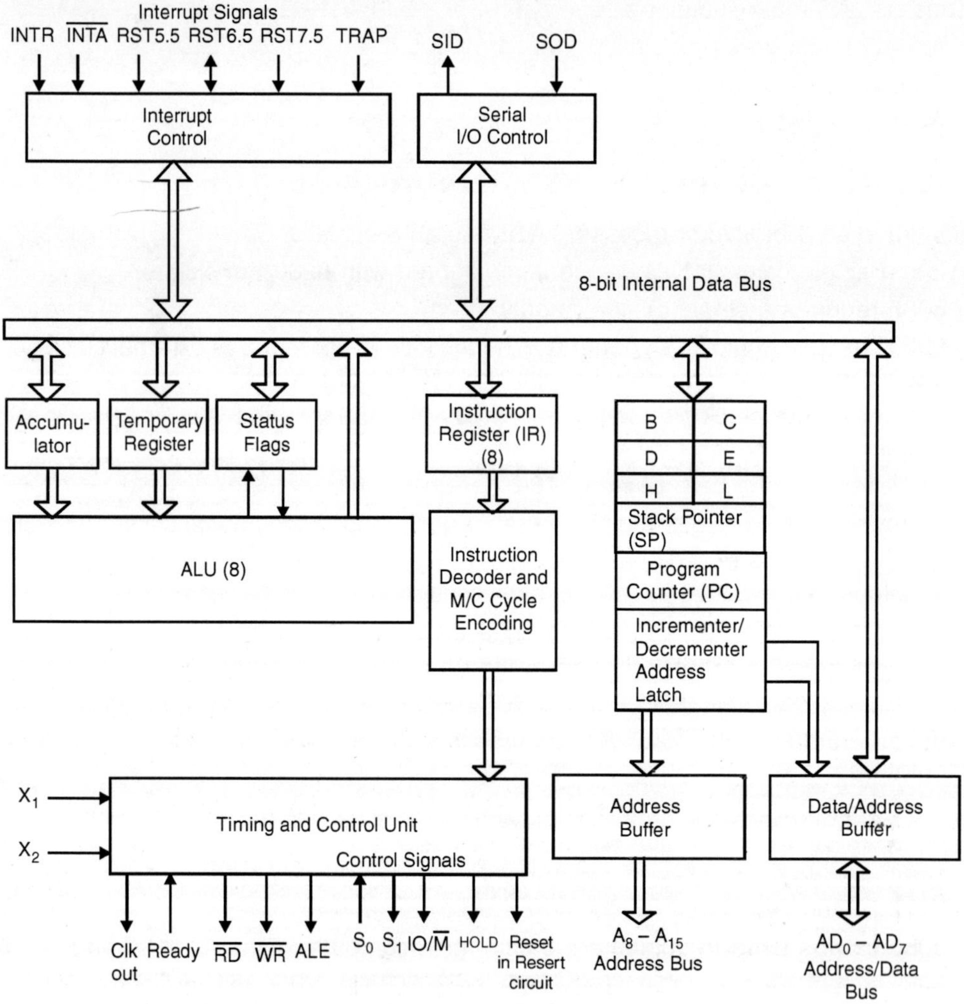

Fig. 2.5. Internal Architecture of 8085 Microprocessor.

2.3.3. Functioning of Different Blocks of 8085 Microprocessor

2.3.3.1. Register Set

(a) 8085 consists of parallel in parallel out (PIPO) registers.

(b) The registers are also called as scratch pad memory or 8085. It stores data and address of memory.

(c) The register organization affects the length of program, the execution time of the program and simplification of the program. To achieve the above parameters, the number of registers should be large.

(d) The architecture of microcomputer depends upon the number and type of the registers used in microprocessor. It consists of 8-bit registers; A, B, C, D, E, H, L and two 16-bit registers PC and SP. These registers are classified as

(i) Temperary registers (w and z).

(ii) General purpose registers.

(iii) Special purpose registers.

2.3.3.2. Temporary Registers

1. Temporary Data Registers : These are also called as operand register (8-bit). These registers provide operands to the ALU.

(i) The ALU can store intermediate result in temporary register. It is not available to the users.

(ii) The temporary data register is also used for other operations such as register to register data transfer, etc.

2. Temporary Registers (w and z)

(i) The temporary registers serve as input to the ALU. The other input to the ALU is from the accumulator.

(ii) These registers hold 8-bit data.

(iii) These registers are not available to the users. They are internally used by the microprocessor.

(iv) These registers are used by control section to hold the data during an arithmetic or logical operations.

3. General Purpose Registers

(i) The 8085 contains 6 general purpose registers of 8-bits each, named as B, C, D, E, H and L.

(ii) B, C, D, E, H and L can be used to store the 8-bits of data or can be used to form a register pair to store 16-bit of data.

The valid register parts available are BC, DE, and HL. The user cannot form a register pair of choice.

(iii) The HL register pair functions as data pointer or memory pointer. If used as memory pointer it holds the 16-bit address of a memory location.

4. Uses of GPRS (General Purpose Register Set) : The main use of these registers is to hold data which is frequently accessed. The GPRS in 8085 helps to increase the speed of operation in the following manner :

(i) It increases the speed of program execution as the data in the microprocessor can be stored in memory or general register. If the data is present in the memory the microprocessor has to perform an operation of memory read. This data is taken by microprocessor, then the required operation is performed and result is stored back in the memory.

To store the result in the memory the microprocessor has to perform one more operation of memory write. Thus there are 2 operations involved in using memory to hold data.

But if the data is present in the GPRS itself there is no such operation required. As the GPRS are part of the microprocessor the microprocessor doesn't have to perform any external memory read and write operation. Thus the time required to execute program using general purpose registers is very less as compared to program using memory.

5. Special Purpose Registers : These registers are used for special use. The different special purpose registers (SPR) in 8085 microprocessor are :

(1) Accumulator
(2) Flag Register
(3) Instruction Register
(4) Program Counter
(5) Stack Pointer.

(1) Accumulator :

(i) Accumulator is the part of general purpose registers of 8085 microprocessor also referred to as A register. The name of this register has been derived from the arithmetic addition process.

(ii) The accumulator is a multi purpose register that can be used for data storage, arithmetic, logical I/O operations.

(iii) Accumulator is used to store result of operation performed by ALU.

(iv) Accumulator is also used as a general purpose register to store 8-bit of data.

(2) Flag Register : A flag is a flip flop. It indicates some condition produced by the execution of an instruction.

Example the zero flag (zf) will set if the result of execution of an instruction is zero.

The flag register of 8085 microprocessor consists of 5 active flags, and 3 inactive flags. The flag register is connected to ALU.

When an operation is performed by ALU the result is transferred on internal data bus and status of the result is stored in the flip flops.

The different types of flags and their respective positions in flag register is as shown below in Fig. 2.6.

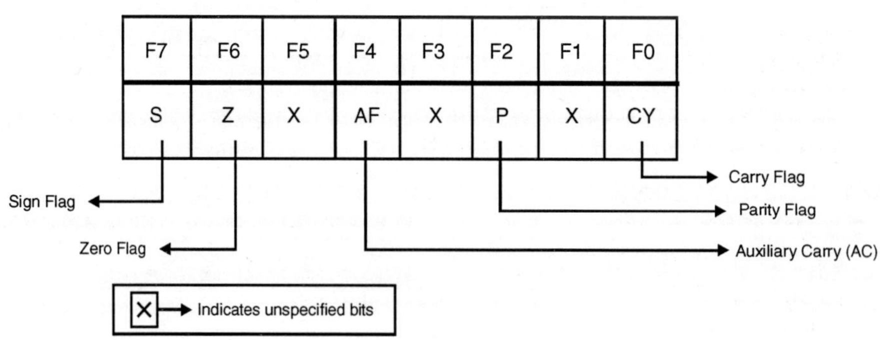

Fig. 2.6. Flag Register of 8085.

(a) Carry Flag (CF)

(i) Carry flag is set whenever there has been a carry out of, or a borrow into, the higher order bit of the result.

(ii) Carry flag is used by the instructions that add and subtract multibyte numbers.

(iii) Rotate instructions can also isolate a bit in memory or a register by placing it in the carry flag.

If there is a '1' in the carry flag bit than there is a carry out from the most significant bit. '0' in the carry flag bit means no carry out from msb.

(b) Parity Flag (PF)

(i) The parity flag is set when the result has even parity, an even number of 1 bits.

(ii) If the parity is odd, parity flag is regested.

(iii) Parity flag is normally used to check for data transmission errors.

'1' bit in the parity flag bit means that lower byte has an even number of 1 bits, and '0' means lower byte has an odd parity.

(c) Auxiliary Carry Flag (ACF) :

(i) The auxiliary carry flag is set, whenever there has been a carry out of the lower nibble into the higher nibble or a borrow from high nibble into the lower nibble of an 8-bit quantity (lower order type of 16-bit quantity) else AF is reset.

(ii) Auxiliary carry flag is used by decimal arithmetic instructions.

'1' in the AF flag bit means that carry out from bit 3 on addition, and a '0' means borrow into bit 3 on addition.

(d) Zero Flag (ZF)

(i) The zero flag is set, when the result of operation is zero, else it is reset.

'1' in the zero flag bit means we have a zero result and '0' means non-zero result.

(e) Sign Flag (SF)

(i) The sign flag is set, when higher order bit of the result is '1'.

(ii) Since negative binary numbers are represented in the 8085 CPU in stand or 2's complement notation, SF indicates sign of the result.

'1' in SF bit means msb is 1 i.e. negative number and '0' means msb is '0' i.e. positive number.

(3) Instruction Register

The instruction register holds the opcode of the instruction that is decoded and executed. This opcode is further sent to the instruction decoder to select one of the 256 alternatives.

(4) Stack Pointer (SP)

(i) Stack is a reserved portion of memory where register pair information can be stored or taken back under software control. This memory area is referred to as STACK Area.

(ii) Stack is used to keep track of data stored on stack.

(iii) Stack is a 16-bit register used to define the stack starting address. It always points forwards the top of the stack.

(iv) The stack pointer is loaded with an initial value by means of a transfer type instruction. This initial value must be the highest address of an assigned stack in memory.

(v) The stack pointer is decremented after each stack write operation and incremented after each stack read operation.

(5) Program Counter (PC)

(i) PC is used to hold the address of program memory.

(ii) PC always points to the next instruction to be fetched i.e. it holds the memory address of the next instruction to be executed.

(iii) The size of the program counter depends upon the number of address bits. For 8085 it is of 16-bits.

(iv) When reset pin is activated, the program counter is set to the address of the first instruction to be fetched and executed.

For 8085 the reset address is 0000H.

(v) In case of jump and call instructions, the address followed by the jump and call instructions is placed in the program counter. If the condition is satisfied then the 8085 fetches the next instruction from the new address specified by Jump or CALL instruction otherwise the 8085 continues with the next instruction after CALL or JUMP instruction.

(vi) During instruction fetch operation, the microprocessor places contents of program counter on the address bus and fetches the first byte of the instruction from that memory location. Then the microprocessor increments program counter to point to the next byte of the instruction.

2.3.5. Address Buffer

(a) This is an 8-bit unidirectional buffer used to address the address lines.

(b) Address buffers are used to drive the high order address bits.

(c) When they are not in use or under certain conditions such as reset, hold, halt, this buffer is used to tri-state to address lines.

2.3.6. Address/Data Buffer

(a) This is an 8-bit bidirectional buffer used for address and data.

(b) It is used to drive the lower order address and data bus.

(c) Under certain conditions such as reset, hold, halt this buffer is used to tri-state to address/data lines.

2.3.7. Arithmetic and Logic Unit

The ALU or arithmetic and logic unit processes data i.e. it performs arithmetical and logical operations.

(a) ALU performs arithmetic operations such as addition, subtraction, division, multiplication.

(b) ALU performs logical operations such as ANDing, ORing, EX-ORing, NOTing etc.

(c) ALU is always controlled by timing and control circuits.

(d) It accepts operands from accumulator and temporary register. It stores result of the arithmetic and logical operations in accumulator or temporary register.
(e) ALU provides status of result to the flag register.
(f) ALU also loops after the branching decisions.

2.3.8. Instruction Decoder and Machine Cycle Encoder

(a) This block accepts a bit pattern (OPCODE) from instruction register, decodes it and gives the decoded information to the control logic.
(b) This information includes what operation is to be performed, who is going to perform it, and how many operand bytes the instruction contains.
(c) The 8085 executes seven types of machine cycles. The machine cycle encoder gives information about which machine cycle is currently being executed.

2.3.9. Interrupt Control

This block in 8085 microprocessor accepts different interrupt requests such as TRAP, RST 7.5, RST 6.5, RST 5.5 and INTR. When a valid interrupt request is present it informs control logic to take action in response to each signal.

2.3.10. Incrementer/Decrementer Address Latch

These are the 16-bit registers used to increment or decrement the contents of PC and SP registers respectively

2.3.11. Timing and Control Unit

(a) The control section of 8085 is made up of synchronous sequential logic circuit.
(b) It controls all internal and external circuits.
(c) It operates with reference to clock signal.
(d) This block accepts information from instruction decoder and generates microsteps to perform it.
(e) This block also accepts clock inputs, performs sequering and synchronising operations. The syncrhronization is required for communication between microprocessor and peripheral devices. To implement this it uses different status and control signals.

2.3.12. Serial I/O Control Groups

(a) The data transferred on to lives is a parallel data, but under certain condition it is advantageous to use serial data transfer.
(b) 8085 implements this by using SID and SOD signals. The data on these lines is accepted or transferred under software control by serial I/O control block.
(c) In 8085 the serial data transfer is performed using two special instruction RIM and SIM.

2.4. PIN DIAGRAM OF 8085

The 8085 is an 8-bit general purpose microprocessor having 40 pins and works on a single power supply as shown in Fig. 2.7.

Figure 2.7 shows the pin diagram of 8085. For convenience to remember the pin diagram various pins have been grouped together as shown in Fig. 2.8 ahead.

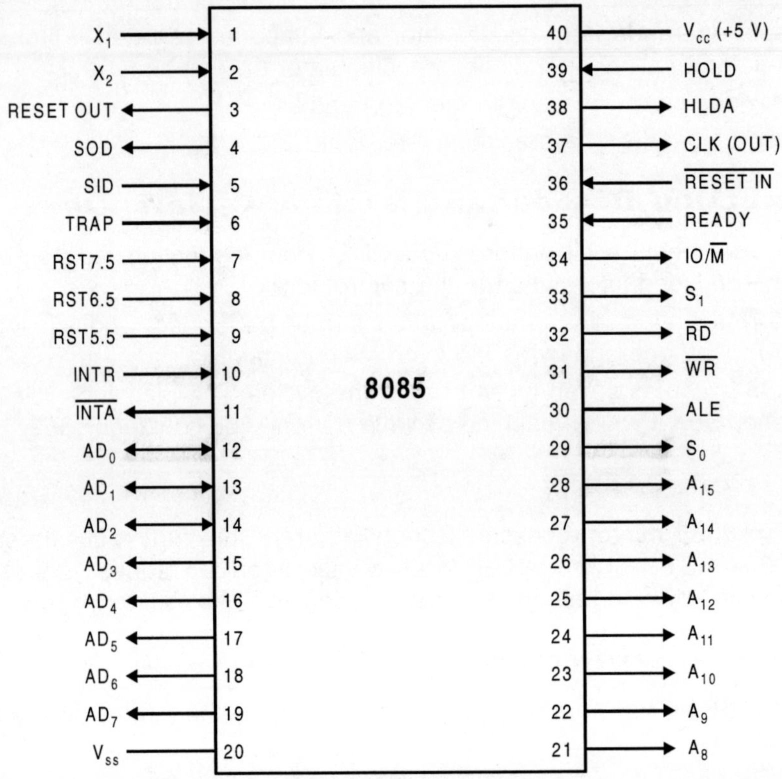

Fig. 2.7. Pin Diagram.

Symbol	Type	Name	Function
$A_8 - A_{16}$	0	Address Bus	The most significant 8 bits of the memory address or the 8 bits of the I/O address, 3 stated during Hold and Halt modes and during Reset.
$AD_0 - AD_7$	I/O	Multiplexed Address/Data Bus	Lower 8 bits of memory address from the first clock cycle (T state) become the data bus during second and third clock cycles.
ALE	0	Address latch enable	1. It occurs during the first clock state of the machine cycle and enables the address to get latched into the on-chip latch of peripherals. 2. The falling edge of ALE is set to guarantee set-up and hold times for the address information. 3. Falling edge be used to strobe the status information. 4. ALE is never 3 states.
S_0, S_1 and IO/M	0	Machine Cycle Status	1. S_1 can be used as an advanced R/W. 2. S_0, S_1 and $\overline{IO}/M$ become valid at the beginning of the machine cycle and remain stable throughout the cycle. 3. The failing edge of ALE may be used to latch the state of these lines.
$\overline{RD}$	0	Read Control	A low on $\overline{RD}$ indicates the selected memory or I/O device is to be read and the Data Bus is available for the data transfer, 3 stated during Hold and Halt modes and during RESET.

Symbol	Type	Name	Function
$\overline{WR}$	0	Write Control	A low level on $\overline{WR}$ indicates the data on the Data Bus is to be written into the selected memory.
Ready	1	Ready	1. If READY is high during a read or writes cycle, it indicates that the memory or peripheral is ready to send and receive data. 2. If READY is low, CPU will wait an integral number of clock cycles for READY to go high before completing the read or write cycle.
Hold	1	Hold	1. HOLD is requesting the use of address and data buses. 2. The CPU, receive HOLD request, it will relinguish the use of the bus as soon as the completing of the current bus transfer. 3. The processor can regain the bus only after the HOLD is removed.
HLDA	0	Hold Acknowledge	CPU has received the HOLD request and relinguishes the bus in the next clock cycle.
INTR	1	Interrupt request	1. Used as general purpose interrupt. 2. If it is active, the Program Counter (PC) will be inhibited from incrementing and $\overline{INTA}$ will be issued. (During this cycle RESTART or CALL instruction can be inserted to jump to the interrupt service routine) 3. INTR is enabled and disabled by software 4. It is disabled by RESET and immediately after an interrupt is accepted.
$\overline{INTA}$	0	Interrupt acknowledge	It can be used to activate an 8259A interrupt is accepted.
RST 5.5	1	Restart interrupts	These interrupt have the same timing as INTR accepted they cause an internal RESTART to be automatically inserted.
RST 6.5 RST 7.5 TRAP	1	TRAP	1. Trap interrupt is non-maskable RESTART interrupt. 2. It has the highest priority of any interrupt.
$\overline{RESET\ IN}$	1	RESET IN	1. Sets the program counter to zero and reset the interrupt enable and HLDA flip-flops. 2. RESET IN is an Schmitt-triggered input, allowing connection to an RC network for power-on RESET delay.
RESET OUT	0	RESET OUT	1. RESET OUT indicates CPU is being reset. 2. Can be used as a system reset.
X1, X2	1	X1, X2	1. Connected to a crystal, LO or RC network to drive the internal clock generator. 2. X1 can be external clock input from logic gate.
CLK	0	CLOCK	1. Clock output for uses as a system clock. 2. The period of clock is twice the X1, X2 are input period
SID	1	Serial Input Data Line	The data on this line is loaded into accumulator bit 7 whenever a RIM instruction is executed.
SOD	0	Serial Output Data Line	The output SOD is set or reset as specified by the SIM instruction.
Vµ		Power	+6 volt supply
Vss		Ground	Reference

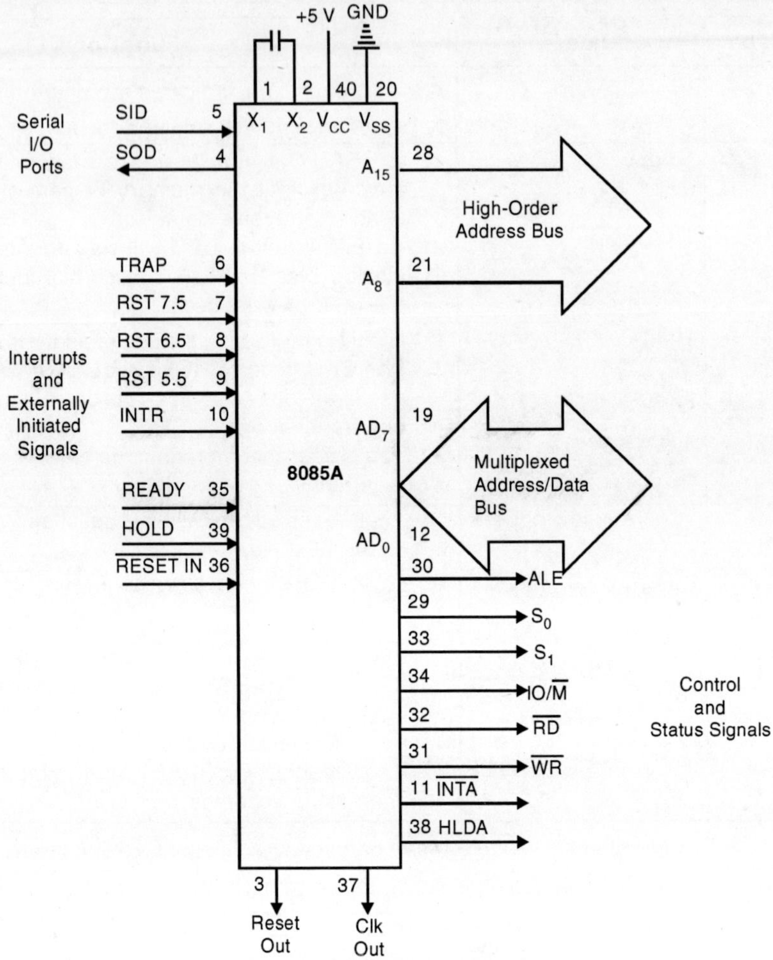

Fig. 2.8. Groups of 8085 Signals.

2.5. DEMULTIPLEXING ADDRESS AND DATA BUS

The address and data lines AD_6 to AD_7 are multiplexed. The lower half of the address bus A_0–A_7 is available only during the T1 state of the machine. This address is also required during the execution of the T2 and T3 states of the machine cycle. Hence it becomes essential to latch the lower half of the address bus so that it remains for the entire machine cycle.

The lower half of the address is latched using an external latch IC 74373, and ALE signal from the 8085 microprocessor.

Fig. 2.9 shows the phenomenon how the address and data are demultiplexed.

The IC 7415373 is an 8-bit latch, consisting of 8-D type flip-flops. The input is transferred to the output when the clock is high.

The ALE signal drive the clock signal. The ALE signal is always high during the T_1 state of the 8085 machine cycle; hence the lower half address (A_0–A_7) bus is latched during the T_1 state.

After that ALE is latch (A_0–A_7) remains unchanged.

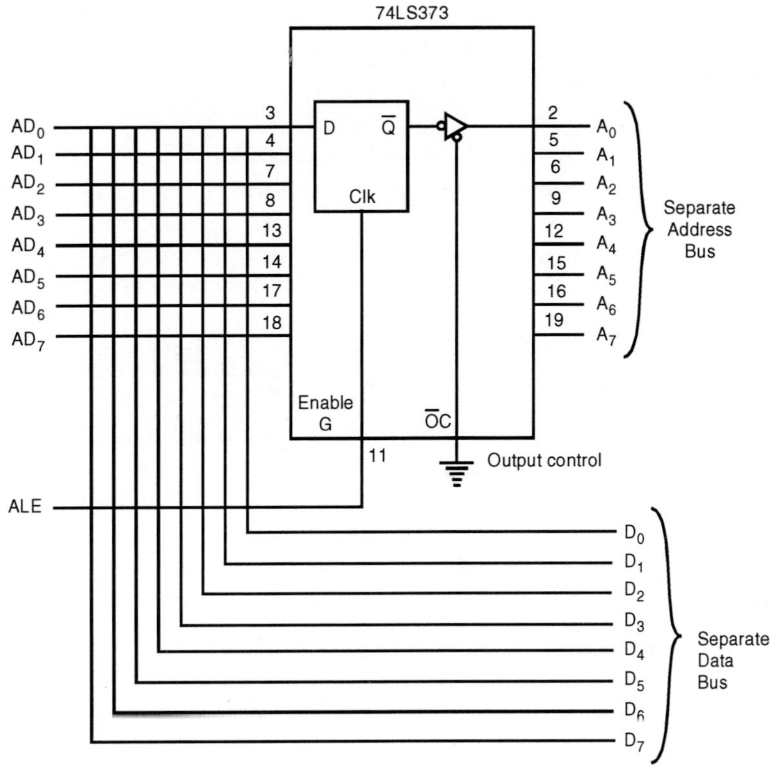

Fig. 2.9. Demultiplexing address and data bus.

2.6. DIFFERENT OPERATIONS PERFORMED BY MICROPROCESSOR

The instruction fetch operation causes an instruction byte that consists the opcode to be fetched from the memory. The fetch operation is completed in 3 steps as stated below :

1. Send the address of the next instruction to the memory.
2. Acquire the instruction code from the memory.
3. Microprocessor then updates the program counter.

Initially the microprocessor sends the address of the memory location from where the opcode is to be fetched. Next the microprocessor reads that instruction byte in the instruction decoder. The instruction is decoded in the instruction decoder where the type of instruction, its size and the task to be completed is specified. Then the microprocessor transfers the opcode, data and address depending on the byte length of the instruction. This process is termed as decoding.

2.6.1. Fetch and Execute Cycle of Microprocessor

1. The fetch and execute cycles of instruction will toggle among one to another. 1st fetch operation is completed then the instruction code is available. The microprocessor then executes this instruction and will go for fetch operation of next instruction.
2. The process explained above is applicable for all instructions except HALT instruction. When HALT instruction is executed, the microprocessor enters HALT loop so will not go for fetch operation. The microprocessor will remain in HALT loop and check reset signal. When a reset is applied, it will return back to fetch operation.

2.7. THE INSTRUCTION CYCLE, MACHINE CYCLE AND T-STATE

For better understanding of each instruction, it is very necessary to understand, how microprocessor executes these instructions ?

Which signals are activated to execute the particular instruction ?

What is the duration of each signal for a particular operation ?

We would try and answer all the above questions in the coming section with the graphical representations of steps with respect to time i.e. the clock and is generally referred to as the "Timing Diagram".

2.7.1. Instruction Cycle

- The CPU fetches one instruction from the memory at a time executes it. The essential steps required the CPU to fetch and execute an instruction is called as "Instruction Cycle".
- An instruction cycle consists of a fetch cycle and an execute cycle. In fetch the CPU fetches the opcodes of the instruction from the memory. The steps required to acquire the data from the memory and complete the operation that is specified in the instruction is constituted by the execute cycle. The time required to execute an instruction is

$$IC = FC + EC$$

IC $\Rightarrow$ Time required for instruction cycle

FC $\Rightarrow$ Time required for fetch cycle

EC $\Rightarrow$ Time required for execution cycle.

The fetch and execute cycles are carried out in synchoronization with the clock.

2.7.2. Machine Cycle

- The time required by the microprocessor to complete the operation of accessing memory or I/O device is called as a "Machine Cycle".
- Operations like opcode fetch, memory read, memory write, I/O read, I/O write are performed in a machine cycle.
- An instruction cycle thus consists of several machine cycles.

2.7.3. T-State

- Microprocessor performs an operation in a specific time period i.e. specific clock cycles. Each clock cycle is called as a "T-state".
- The number of T-states required to perform an operation is called as "Machine cycle".
- Machine cycle consists of 3 to 6 T-states and 1-7 state is of one clock period. An instruction cycle consists of 1 to 5 machine cycles.

2.8. REPRESENTATION OF DIFFERENT SIGNALS USED DURING TIMING DIAGRAM OF 8085

1. Clock Signal : The microprocessor operates with reference to clock signals provided to it At X1 and X2, a clock generator is attached and this frequency is divided by 2 and used by the microprocessor.

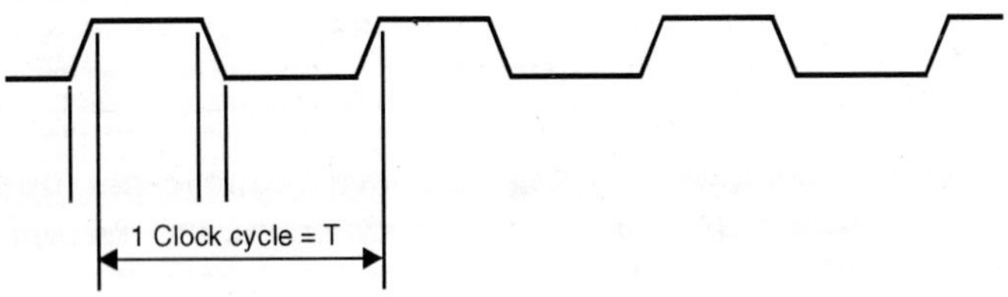

Fig. 2.10. Clock Signal for Microprocessor.

Ideal clock should be exact having very less rise and fall time. But the practical clock is different which contains two components as shown in Fig. 2.10.

TR = Rise time

TF = Fall time

2. Single Line Signal : The status of single line is either low or HIGH i.e. any one state, logic 0 or logic 1. But the change from one state to another is not possible in zero time. It requires some time to go from low to HIGH or from HIGH to LOW as shown in Fig. 2.11. ahead

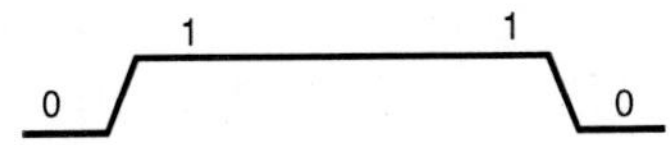

Fig. 2.11. Single Line Signal.

3. Multiple Line Signals : In microprocessor we have muliple lines. Instead of showing each line using single line signals, these are grouped and shown in the form of block as shown in Fig. 2.12.

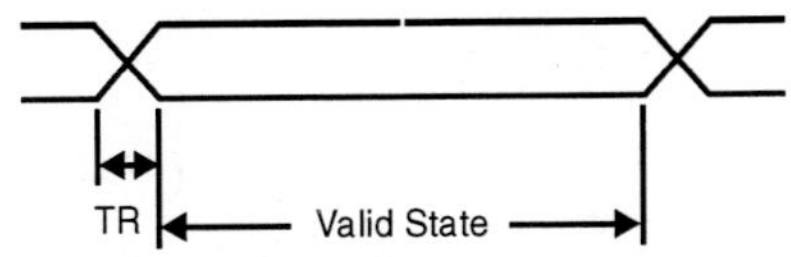

Fig. 2.12. Multiple Line Signals.

As shown in Fig. 2.12. two crossing lines are shown. At that time the signal change occurs and for rest of the time the contents of those line are valid and you can use those signals. If a single line change occurs then also we have to show a crossing to indicate change in contents.

4. Multiple Lines with High Impedance : If multiple lines are having 3 states viz logic 0, logic 1 and tristate, it is as shown in Fig. 2.13.

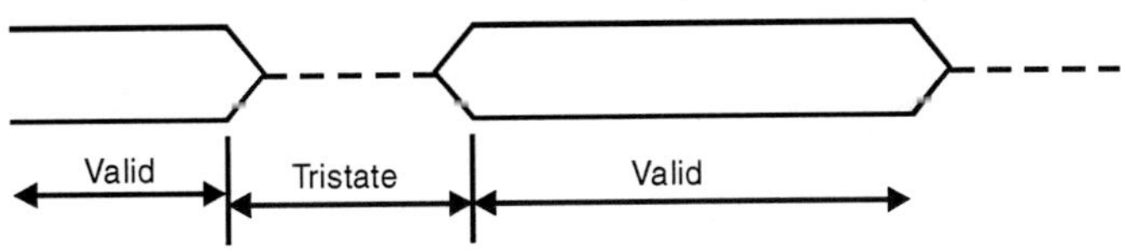

Fig. 2.13. Multiple Lines with High Impedance.

5. Single Line and Single Effect Signal : Because of single line the other signals also changes. A change in one single line causes change in other signal, it is as shown in Fig. 2.14.

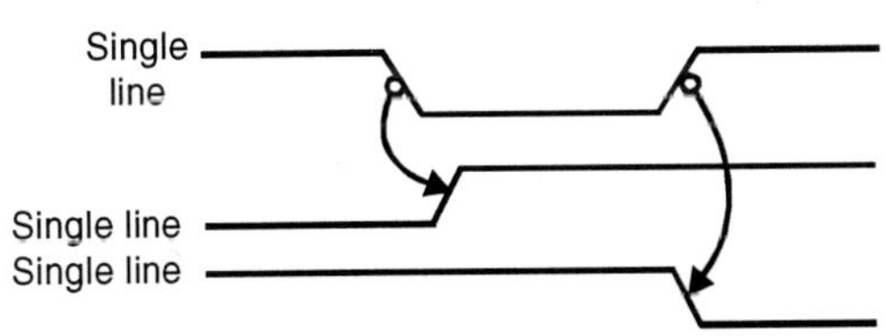

Fig. 2.14. Single Line and Single Effect Signal.

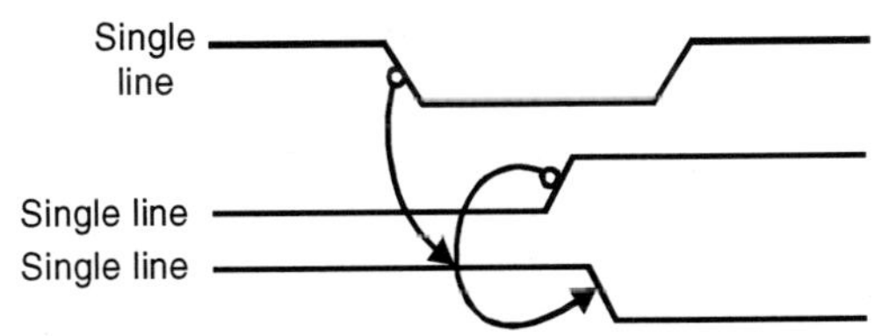

Fig. 2.15. Multiple Lines and Single Effect Signal.

6. Multiple Lines and Single Effect Signal : If a change in multiple lines causes change in single signal, as shown in Fig. 2.15.

7. A_8–A_{15} (Higher Order Address : The higher order address is available on A_8–A_{15}, lines during T_1, T_2 and T_3 state of each machine cycle. During T_4 state or T_4, T_5 and T_6 state of OPCODE fetch the contents of A_8–A_{15} are not specified. The A_8–A_{15} signals with respect to clock signal is as shown in Fig. 2.16.

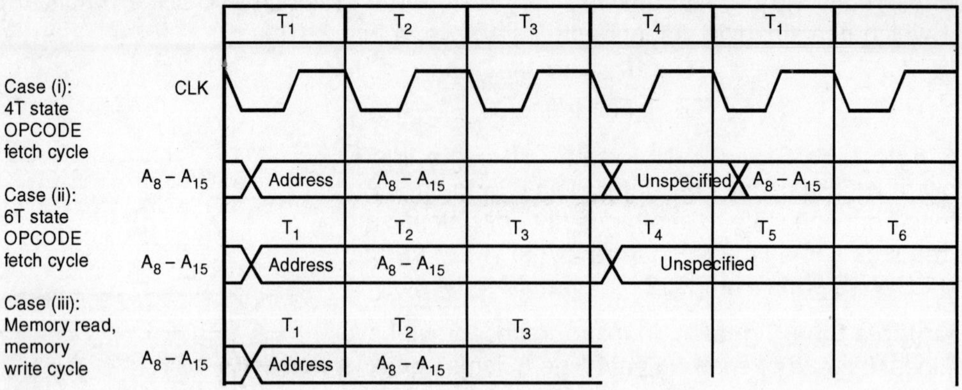

Fig. 2.16. Higher Order Address Signals.

8. Lower Order Address/Data :

➢ The lower order address is available on AD_0–AD_7 lines during T1 state of each machine cycle.

➢ During memory read operation, data from memory or I/O device is transferred on AD_0–AD_7 lines during each state. During memory write operation, data from microprocessor is transferred on AD_0–AD_7 lines as soon as the address AD_0–AD_7 is removed. The AD_0–AD_7 signals with respect to clock signal is as shown in Fig. 2.17.

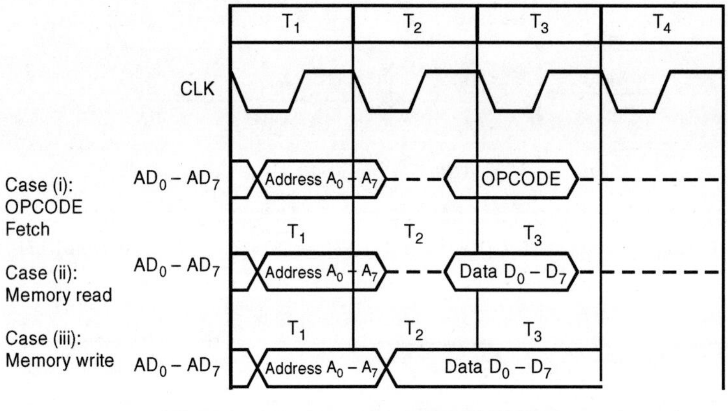

Fig. 2.17. Lower Order Address Signals.

9. ALE (Address Latch Enable) : The ALE signal is available during state of each machine cycle. The ALE is a pulse signal, used to latch the address from AD_0–AD_7. The ALE signal with respect to clock signal is as shown in Fig. 2.18.

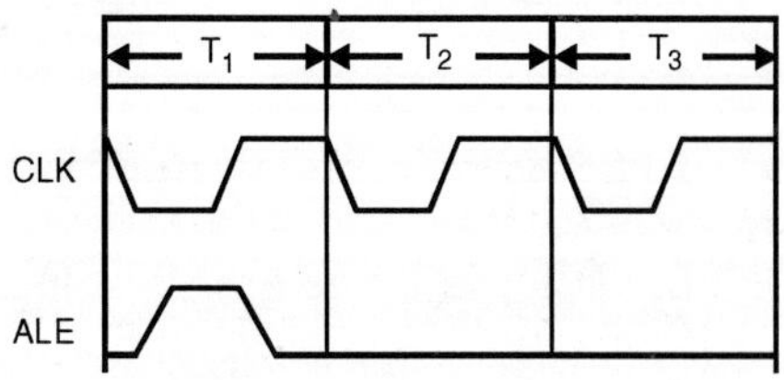

Fig. 2.18. ALE Signal.

10. $IO/\overline{M}$, S_0 S_1 (Status Signals) : The status signals

$IO/\overline{M}$, and S0 and S1 and are available during state to last T state of each machine cycle. The logic levels on these lines are decided by machine cycles as follows :

The status signals with respect to clock signal are as shown in Fig. 2.19.

Machine Cycle	IO/$\overline{M}$	S_1	S_0
OPCODE fetch	0	1	1
Memory read	0	1	0
I/O read	0	0	1
I/O write	1	0	1
Interrupt acknowledge	1	1	1
Bus ideal	0	0	0

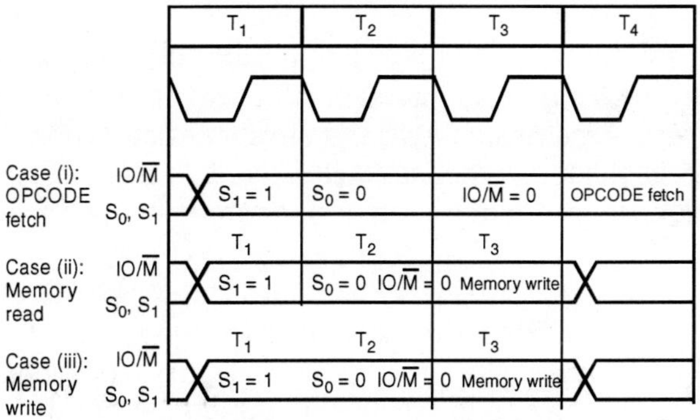

Fig. 2.19. Status Signals.

11. Read and Write Control Signals : The control signals decide the operation to be performed : (i) For memory read, OPCODE fetch, I/O read the $\overline{RD}$ signal is activated during. T_2 and T_3 state of machine cycle. (ii) For memory write I/O write the $\overline{WR}$ signal is activated during T_2 and T_3 state of machine cycle.

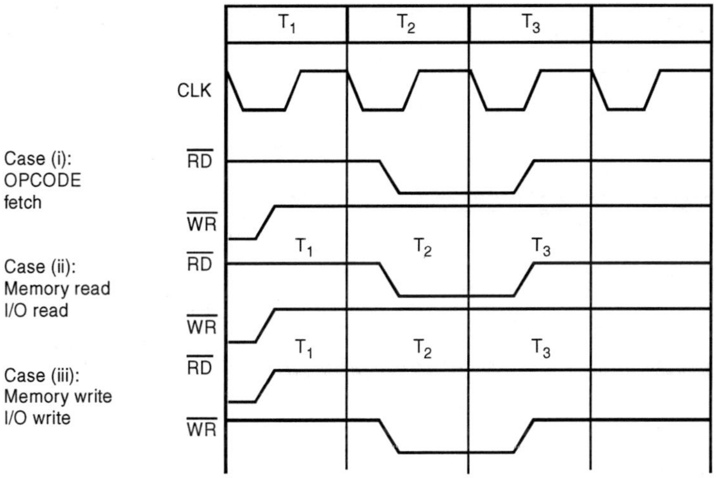

Fig. 2.20. Control Signals.

2.9. INSTRUCTION WORD SIZE

A digital computer understands instructions written in binary codes (machine codes). The machine codes of all instructions are not of the same length. According to the word size the Intel 8085 instructions are classified into the following three types :

(1) 1-byte instruction

(2) 2-byte instruction

(3) 3-byte instruction

One-Byte Instruction. Examples of one-byte instructions are;

MOV A, B : Move the content of register B to register A. 78H is the opcode for MOV A, B. Besides the operation to be performed the opcode also specifies the registers which contain, operands : (data). The opcode 78H can be written in the binary form as 01111000. The 1st two bits, i.e., 01 are for MOV operation; the next three bits 111 are the binary code for register A and the last three bits U(X) are the binary code for register B.

ADD B: Add the content of register B to the content of the accumulator. 80H is the opcode for the instruction ADD B. In this instruction one of the operands is register B (its content) which is indicated in the instruction itself. In this type of instruction (arithmetic group of instruction) it is assumed that the other operand is in the accumulator. The opcode 80H in the binary form is 10000000. The first five hits, *i.e.* 10000 specify the operation to be performed, *i.e.* ADD operation The last three bits 000 are the code for register B for 8085 microprocessor.

RAL: Rotate the content of the accumulator left by one bit. The opcode for RAL is 17H. In this instruction operand is not given, it is implied. The operand is in the accumulator. The instruction has to operate on the content of the accumulator.

All The above three examples are only one byte long. All one byte instructions contain information regarding operands in the opcode itself.

Two-Byte Instruction. In a two-byte instruction the 1st byte of the instruction is its opcode and the 2nd byte is either data or address. Examples are :

(i) MVI B, 05 ; Move 05 to register B,
06, 05 ; MVI B, 05 in the code form.

The 1st byte 06 is the opcode for MVI B and the 2nd byte 05 is the data which is to moved to register B.

(ii) IN 01 ;Read data at port B.
DB/01 ;IN 01 in the code form.

DB is the opcode for the instruction IN and 01 is the address of a port (the port H of Intel 8255.1 of a microprocessor kit). A two-byte instruction is stored in two consecutive memory locations.

Three-Byte Instruction. In a three byte instruction the 1st byte of the instruction is its opcode and the 2nd and 3rd bytes are either 16-bit data or 16-bit address. Examples are :

(i) LXI H,2400H ; Load H-L pair with 2400H.
21, 00, 24 ; LXIH,2400H in the code form

The 1st byte 21 is the opcode for the instruction LXI H. The 2nd byte 00 is 8 LSBs of the data (2400H), which is loaded into register L. The 3rd byte 24 is 8 MPBs of the data (2400H), which is loaded into register H.

(ii) LDA_1500H ; get the content of the memory location 2500H into accumulator 3A, 00, 25; LDA 2500H in the code form.

The 1st byte 3A is the opcode for the instruction LDA. The 2nd byte 00 is 8 LSBs of the address of the memory location 2500H. The 3rd byte 25 is 8 MSBs of the address of the memory location 2500H. In this instruction the data is the content of the memory location 2500H. The data loaded into the accumulator. A 3-byte instruction is stored in three consecutive memory locations.

2.10. INSTRUCTION CYCLE

An instruction is a command given to the computer to perform a specified operation on given data. To perform a particular task a programmer writes a sequence of instructions, called a program. Program and data are stored in the memory. The CPU fetches one instruction from the memory at a time and executes it. It executes all the instructions of a program one by one to produce the final result.

The necessary steps that a CPU carries out to fetch an instruction and necessary data from the memory, and to execute it, constitute an *instruction cycle.* An instruction cycle consists of a *fetch cycle* and *execute cycle.* In fetch cycle a CPU fetches opcode (the machine code of an instruction) from the memory. The necessary steps which are carried out to fetch an opcode from the memory, constitute a fetch cycle. The necessary steps which are carried out to get data, if any, from the memory and to perform the specific operation specified in an instruction, constitute an execute cycle. The time required to fetch an opcode (FC) is a fixed slot of time while the time required to execute an instruction (EC) is variable which depends on the type of instruction to be executed. The total time required to execute an instruction is given by

$$IC = FC + EC$$

2.10.1. Fetch Operation

The 1st byte of an instruction is its opcode. An instruction may be more than one byte long. The other bytes are data or operand address is. The program counter (PC) keeps the memory address of the next instruction to be executed. In the beginning of a fetch cycle the content of the program counter, which is the address of the memory location where opcode is available, is sent to the memory. The memory places the opcode on the data bus so as to transfer it to the CPU. The entire operation of, fetching an opcode takes three clock cycles. A slow memory may take more time. In case of a slow memory the CPU has to wait till the memory sends the opcode. The clock cycle for which the CPU waits is called wait cycle. Most of the CPUs have been designed to introduce wait cycles to cope up with slow memories.

2.10.2. Execute Operation

The opcode fetched from the memory goes to the data register, DR (data/address buffer in Intel 8085) and then to instruction register, IR. From the instruction register it goes to the decoder circuitry which decodes the instruction. The decoder circuitry is within the microprocessor. After the instruc tion is decoded, execution begins. If the operand is in the general purpose registers, execution immediately performed. The time taken in decoding and execution is one clock cycle. If an instruction contains data or operand address which are still in the memory, the CPU has to perform some read operations to get the desired data. After receiving the data it performs execute operation. A read cycle is similar to a fetch cycle. In case of a read cycle the quantity received from the memory are data or operand address instead of an opcode. In some instructions write operation is performed. In write cycle data are sent from the CPU to the memory or an output device. Thus we see that in some cases an execute cycle may involve one or more read or write cycles or both. Figs. 2.21 and 2.22 an instruction and fetch cycle respectively.

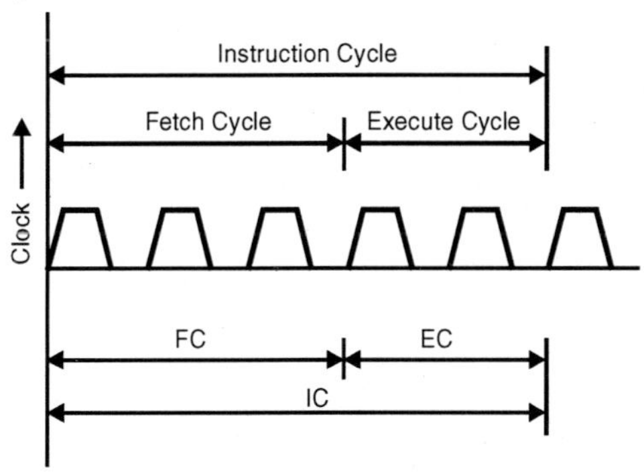

Fig. 2.21. Instruction cycle showin FC, EF, IC.

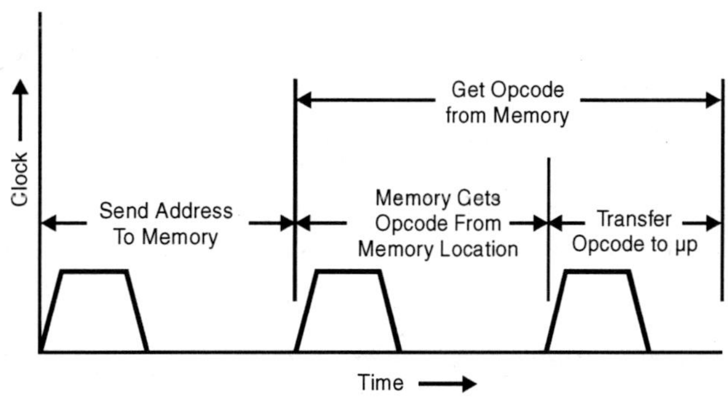

Fig. 2.22. Fetch Cycle.

2.10.3. Machine Cycle and State

The necessary steps carried out to perform the operation of accessing either memory or I/O device, constitute a machine cycle In other words necessary steps carried out to perform a fetch, a read or a write operation constitute a Machine cycle.

In a machine cycle one basic operation such as opcode fetch, memory read, memory write, I /O read or I/O write is performed. An instruction cycle consists of several machine cycles. The opcode of an instruction is fetched in the first machine cycle of an instruction cycle. Most of the single-byte instructions require only one machine cycle to fetch the opcode and execute the instruction. Two-byte and three-byte instructions require more than one machine cycle. Additional machine cycles are needed to read data from or to write data into the memory or I/O devices. Fig. 2.23 shows instruction cycle for MVI r, data. It has two machine cycles: one for fetching opcode, and the other for reading data from the memory and to execute the instruction.

One subdivision of an operation performed in one clock cycle is called a state or T-state. Th subdivisions are internal states synchronised with the system clock. So one clock cycle of the systen clock is referred to as a state.

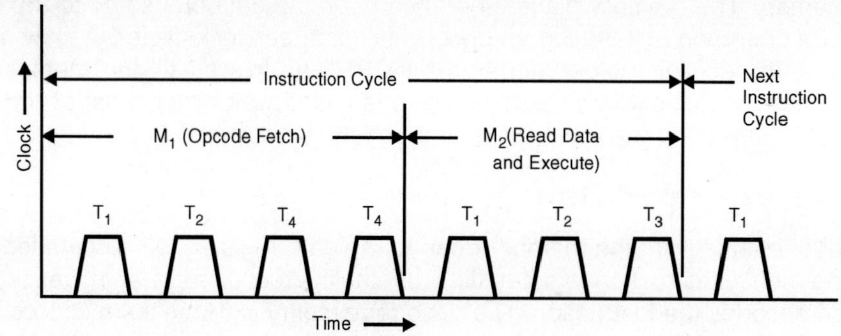

Fig. 2.23. Typical Instruction Cycle.

2.10.4. Instruction and Data Flow

A computer receives instructions and data both in binary form., A group of 8 bits is called a byte Data and instructions for computers are specified in the byte form. The word length of a computer is the number of bits it handles at a time. The word length of an 8-bit microprocessor is of one byte that of a 16-bit microprocessor is of two bytes and so on. Two kinds of words, namely instruction word (instruction code or opcode) and data word are processed during an instruction cycle. In the beginning of a fetch cycle the content of the program counter is transferred to a special register known as memory address register, MAR or simply address register, AR; (address buffer in Intel 8085), The content of MAR is transferred to the memory through the address bus. By sending certain control signals to the memory the microprocessor also indicates that it wants to read the content of the memory. The decoder circuitry in the memory is activated and the memory understand what is to be done. Then the memory sends opcode to the microprocessor through the data bus. The opcode first comes in memory data register (MDR) or simply data register (DR) as shown in Fig. 2.24. In case of Intel 8085, there is a data buffer for this purpose. The operation code is then placed in

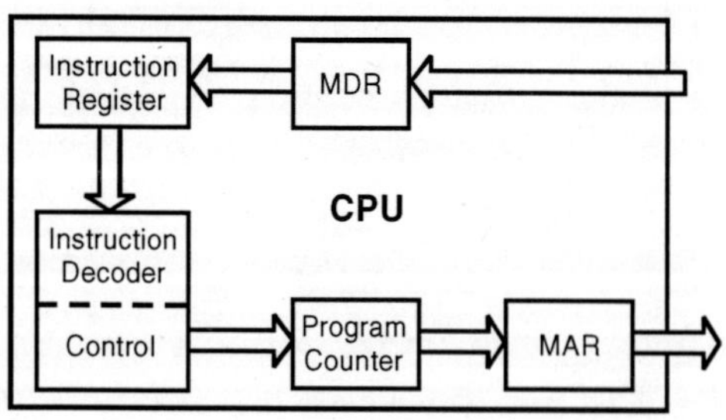

Fig. 2.24. Flow of Instruction word (opcode).

the instruction register (IR). The instruction is decoded by instruction decoder and it is executed. Finally, the content of the program counter is incremented.

The execution of an instruction requires the flow of data word in the most of the instructions flow of a data word is shown in Fig. 2.25. A data word is received either from the memory or input device. The data word flows to the processor through the data bus and is placed in the accumulator or any other general purpose register depending upon the instruction. After the execution of an instruction the data is placed in a register or a memory location. After the execution of an instruction of a program the result (data) is placed in the memory or sent to an output device. When a data word is written into the memory, it is also held in MDR (or DR or Data Buffer) until the wrtie operation is complete.

2.11. TIMING DIAGRAM

The necessary steps which are carried out in a machine cycle can be represented graphically. Such a graphical representation is called *timing diagram.* The timing diagram for opcode fetch memory read, memory write, I/O read and I/O write will be discussed ahead.

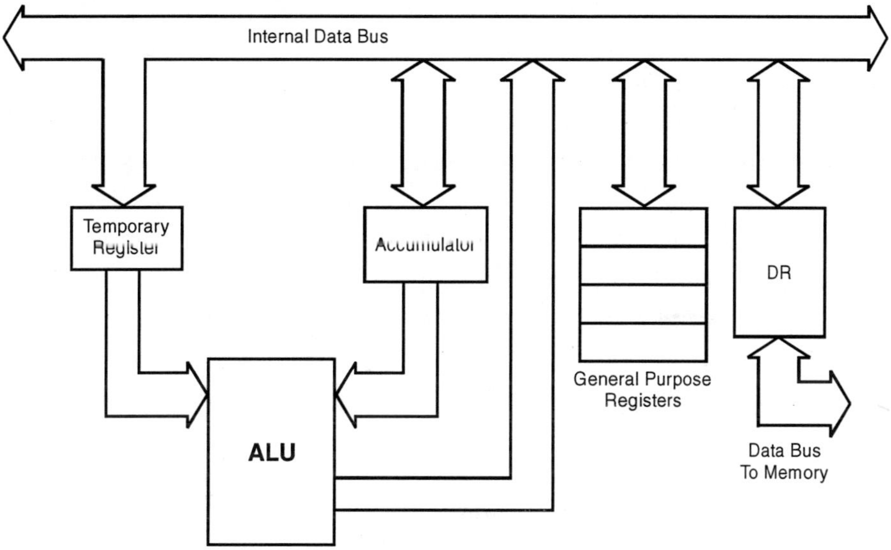

Fig. 2.25. Flow of Data Word.

2.11.1. 8085 Machine Cycles and Their Timings

Machine cycles are operations performed by a microprocessor. There are different operations given below :

1. OPCODE fetch cycle
2. Operand fetch cycle
3. Memory read cycle
4. Memory write cycle
5. I/O read cycle
6. I/O write cycle
7. Interrupt acknowledge cycle.
8. Idle machine cycle.

Now we will see these 8 operations in brief.

1. OPCODE fetch cycle :

➢ "The microprocessor uses this cycle to take the OPCODE of an instruction i.e. fetch OPCODE, In this case, the address of memory location where the OPCODE is stored 'is given by program counter. This is always the first cycle of any instruction cycle. The OPCODE is taken from memory and transferred to instruction register for decoding and execution.

➢ The tune required to complete this cycle is 4 or 6 T states.

2. Operand fetch cycle : In 2 and 3 byte instructions, taking OPCODE, doesn't complete the instruction fetching, so we require operand fetch cycles. For 2 byte instruction we require one operand fetch cycle and for 3 byte instruction we require two operand fetch cycles. In this cycle the address is given by program counter. But the contents are not transferred to instruction register instead they are stored in temporary registers. The time required to complete the operand fetch cycle is 3T states.

3. Memory read cycle : The microprocessor executes these cycles to read data from memory. The address of memory location is given, by instruction. In this case, the program counter is not used, instead, instruction will give address or will specify, where the address is present $IO/\overline{M}$ =0.

4. Memory write cycle : The microprocessor executes these cycles to write data to memory. The address to memory is given by instruction. $IO/\overline{M}$ = 0.

5. I/O read cycle : The microprocessor executes these cycles to read data from I/O device instead of memory. The status of $IO/\overline{M}$ line is HIGH. The address of port is given by instruction.

6. I/O write cycle : The microprocessor executes these cycles to write or send data to I/O device. The status of $IO/\overline{M}$ line is HIGH. The address of port is given by instruction.

7. Interrupt acknowledge cycle : In response to interrupt request input INTR, the microprocessor executes 1 MC cycle to read information from interrupting device.

8. Idle machine cycle :

➢ The microprocessor executes these cycles for internal execution. During this cycle, all data, address and control lines are instated i.e. microprocessor does not perform any external operation.

➢ In 8085 there are different instructions which doesn't complete execution in given time period, these instructions require idle machine cycles to complete the execution of instruction.

Example : DAD instruction.

The above description was in general. Now let's study detailed graphical presentation for machine cycles.

2.11.2. OPCODE Fetch Cycle

The 8085 uses this cycle to fetch instruction OPCODE from memory. In this case, the address of memory is always given by program counter. The program counter is incremented by one to point to the next location. There are 2 types of OPCODE fetch cycles, one using 4 T states to complete the operation and other using 6 T states to complete the operation. Generally instructions uses 4 T states but some instructions performing some additional operation in OPCODE fetch requires 6 T states.

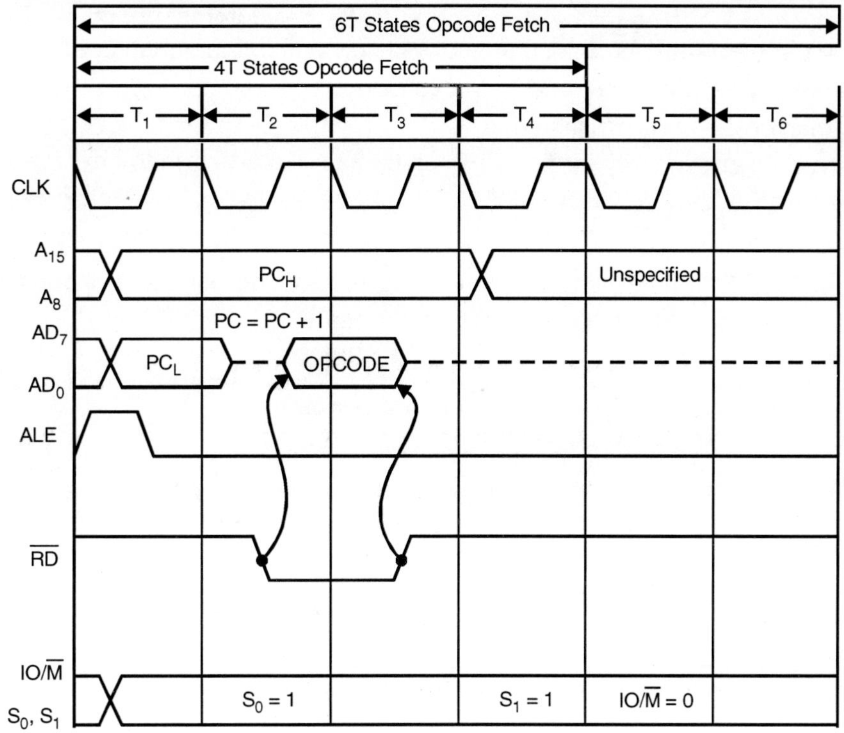

Fig. 2.26. Timing diagram of OPCODE fetch.

Operations :

1. Step 1 (State T_1) : In the state, 8085 sends the status signals, $IO/\overline{M}$ and S_0 $IO/\overline{M} = 0$, $S_0 = 1$, $S_0 = 1 = S_1 = 1$ i.e. OPCODE fetch.

- The 8085 sends a 16 bit address on $A_8 - A_{15}$ and $AD_0 - AD_7$.
- The high order byte of program counter is placed on the $A_8 - A_{15}$ lines and it remains there upto T_3 state. The low order byte of program counter is placed on the $AD_0 - AD_7$ lines which remains there only for T_7.
- During this state, ALE (address latch enable) gives a positive pulse which represents the contents of $AD_0 - AD_7$ as an address. The ALE signal is used to latch the address to $A_0 - A_7$.
- No control signal is generated in state.

2. Step 2 : The contents of PC_L (Lower order address bus) will disappear on $A_0 - A_7$ lines. so that the same lines can be used as data lines. The contents of A_0 to A_7 are still available for memory from external latch.

- The control signal RD is made LOW by the processor which enables the read circuit of addressed memory device. The memory device then sends the contents on data bus i.e. $AD_0 - AD_7$.
- In addition to these operations, microprocessor increments PC contents by 1.

3. Step 3 (State T_3) : During this clock cycle, the data from memory i.e. OPCODE is transferred to instruction register and $\overline{RD}$ control signal is made HIGH. Thus $\overline{RD}$ disables the memory device.

4. Step 4 (State T_4) : The microprocessor performs only internal operation. The OPCODE is decoded by the CPU and upon decoding 8085 knows all the information about :

(i) Whether it should enter T_5 and T_6 states or not.

(ii) How many bytes of instruction it is ?

If the instruction doesn't require T_5 and T_6 states, it will start or go to next machine cycle. If it is multi byte instruction the operand fetch cycles are executed to complete the instruction fetching.

5. **Step 5 (T_5 and T_6) :** T_5 and T_6, states are required to complete decoding and some operations inside the 8085. Operation performed depends on instruction.

Note : Following instructions require T_5 and T_6 states :

Instruction	Operation performed
CALL	Stack pointer is decremented by 1
CALL conditional	Stack pointer is decremented by 1
DCX R_p	Register pair decremented by 1
INX R_p	Register pair incremented by 1
PCHL	HL pair transferred to PC
PUSH RP	Stack pointer is decremented by 1
SPHL	HL pair transferred to SP
RET The conditional	The condition of flags checked

All other instructions except the above instructions require OPCODE fetch of T_1 to T_4 – 4T states only.

The same diagram as shown in Fig. 2.26 is applicable to OPCODE fetch cycle of 6T states. But in case of 4T states; we require T_1 to T_4 only.

2.11.3. Operand Fetch or Memory Read Cycle

- The 8085 uses this cycle : (i) to fetch the operand of an instruction if it is a multi-byte instruction (ii) to read the data from memory.
- The difference between two is : (1) In operand fetch the address of memory location is given by program counter (2) In memory read the address is given by instruction. These operations require 3 T states i.e. T_1 to T_3.

Operations :

1. Step 1 (State T_1) : In T_1 state, 8085 sends appropriate status signals, $IO/\overline{M}$ S_0 and S_1 $IO/\overline{M} = 0$, $S_0 = 0$, $S_1 = 1$ (memory read). The 16 bit address is transferred on AD_0 and AD_7.

- The address is dependent on the operation. In operand fetch cycle, address is given by program counter and then PC is incremented by 1. In memory read cycle, address depends upon instruction; address may be from SP or general purpose register pair.

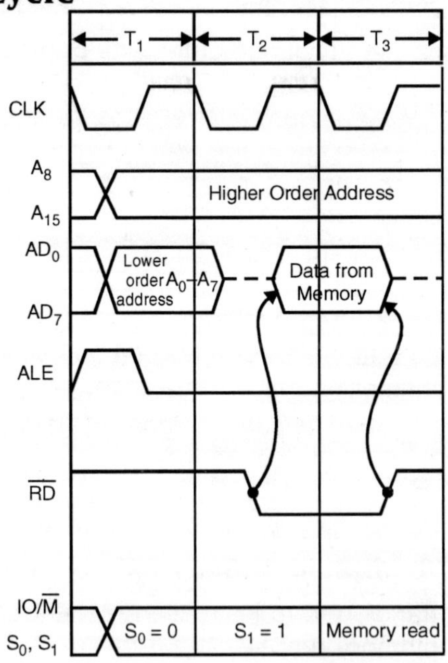

Fig. 2.27. Timing diagram of memory read cycle.

➢ ALE is generated by 8085 to indicate the availability of address $A_0 - A_7$ on $A_0 - A_7$. The ALE is used to latch A_0 to A_7.

2. Step 2 (State T_2) : In T_2 state, the lines $AD_0 - AD_7$ will be used as $D_0 - D_7$ to transfer data. The control signal $\overline{RD}$ is made LOW which will enable the selected memory. The memory will send data on the data bus.

3. Step 3 (State T_3) : In T_3 state, data from memory is transferred to 8085. 8085 accepts this data and transfers on internal data bus, and $\overline{RD}$ is made HIGH. Now where the data is transferred depends on the instruction for which this machine cycle is used. It is generally stored in general purpose registers. If it is operand fetch cycle then it is loaded in temporary registers for further use. In operand fetch cycle the PC contents are incremented by one, to point to next location.

2.11.4. Memory Write Cycle

8085 executes the memory write cycle to store data either in data memory or stack memory. If it is data memory, the address is generated by general purpose register pair. If it is stack memory (area), address is generated by SP.

As shown in Fig. 2.28, it requires total 3T states i.e. T_1 to T_3. Actually, the timing cycle, matches with "memory read cycle", with following few changes :

(1) $\overline{RD} = 1$, $\overline{WR} = 0$

(2) S0 = 1, S1 = 0.

(3) It doesn't increment PC.

Let's study step by step, all the T states.

1. Step 1 (State T_1) : In T_1 state, 8085 sends the appropriate status signals $IO/\overline{M}$ S_0 and S_1 i.e. $IO/\overline{M} = 0$. $S_0 = 1$, $S_1 = 0$, (memory write). The 16 bit address is transferred on $A_8 - A_{15}$ and $AD_0 - AD_7$. ALE is active for short duration.

2. Step 2 (State) : In T_2 state the address on $AD_0 - AD_7$ i.e. $A_0 - A_7$ is removed and the data to be stored IN memory is transferred on these lines. The control signal $\overline{WR}$ is made LOW. The $\overline{WR}$ LOW will make memory active to accept the contents of data bus. The accepted contents are stored at selected location.

3. Step 3 (State T_3) : In state, data from data bus is stored at memory location. WR is made HIGH which deactivates memory and microprocessor completes the machine cycle.

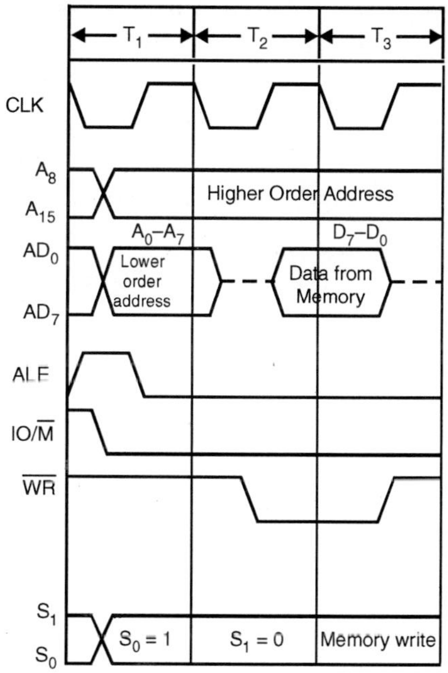

Fig. 2.28. Timing diagram of memory write cycle.

Now at this stage we have built basic platform for drawing timing diagram for a particular instruction.

While drawing tuning diagram of 8085 instructions, we will see some important INSTANTS in the diagram. To make the drawing of tuning diagram very very easy job, you are advised to remember the waste machine cvcles given in next section. Using these machine cycles now we draw the timing diagrams of 8085 instruction. The machine cycles required for instructions is covered in instruction set summary.

2.11.5. I/O Read and I/O Write Machine Cycles

These machine cycles are similar to the memory read and the memory write machine cycles. But in these machine cycles the IO/M signal is high for I/O read and I/O write machine cycles.

Fig. 2.29 and Fig. 2.30 shows the timing diagram for I/O read and I/O write machine cycles.

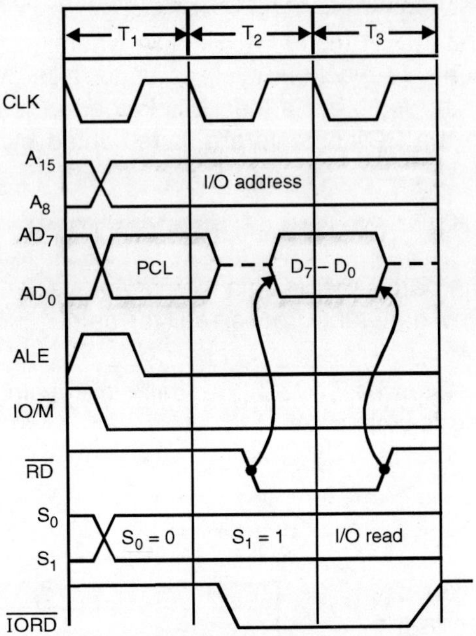

Fig. 2.29. I/O Read Cycle

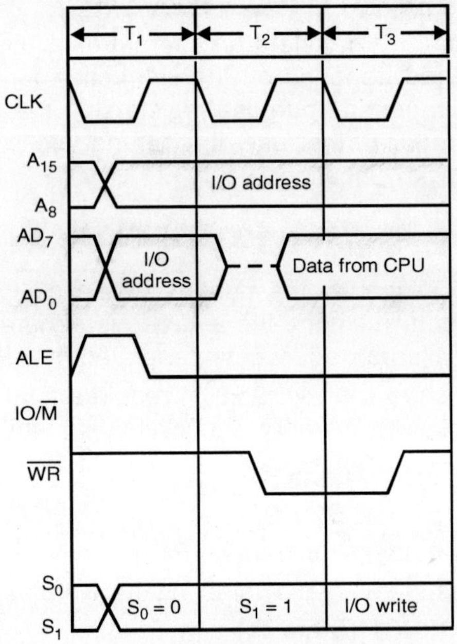

Fig. 2.30. I/O Write Cycle.

2.11.6. Interrupt Acknowledge Machine Cycle

- In response to the INTR signal the 8085 microprocessor executes an interrupt acknowledge machine cycle to read an instruction from the external device.
- In response to the $\overline{INTA}$ signal the external device can place any instruction on the data bus.
- However, only RST and CALL instructions save the contents of PC i.e. the return address before transferring control to the interrupt service routine.

Interrupt Acknowledge Cycle for the RST Instruction

- Figure 2.31 shows the timing diagram of the interrupt acknowledge machine cycle is similar to the opcode fetch machine cycle with two exceptions.
  - (i) The $\overline{INTA}$ signal is activated instead of the RD signal.
  - (ii) The status signals ($IO/\overline{M}$ = 1, S_0 = 1 and S_1 = 1).
- During the interrupt acknowledge machine cycle M_1, the RST is decoded it initiates 1 byte CALL instruction to the specific vector location.
- The machine M_2 and M_3, are memory write machine cycles. They store the contents of program counter on the stack.

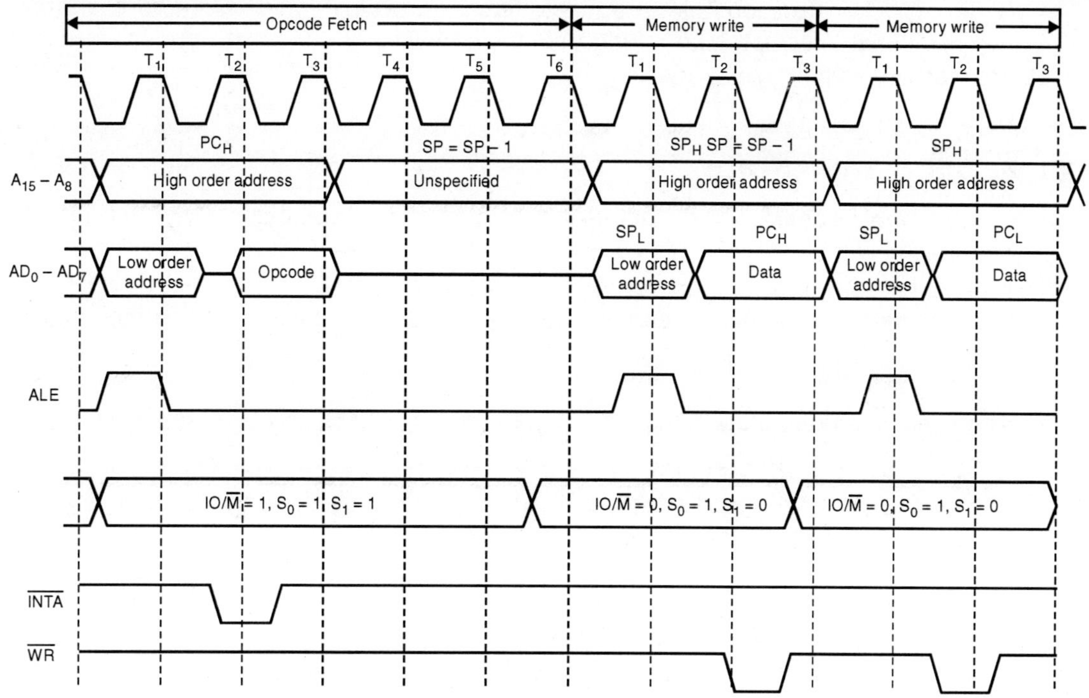

Fig. 2.31. Interrupt acknowledge Machine Cycle for 8085.

Interrupt Acknowledge Cycle for CALL Instruction

➢ Figure 2.32 shows the timing diagram of the interrupt acknowledge machine cycle and the execution of CALL instruction.

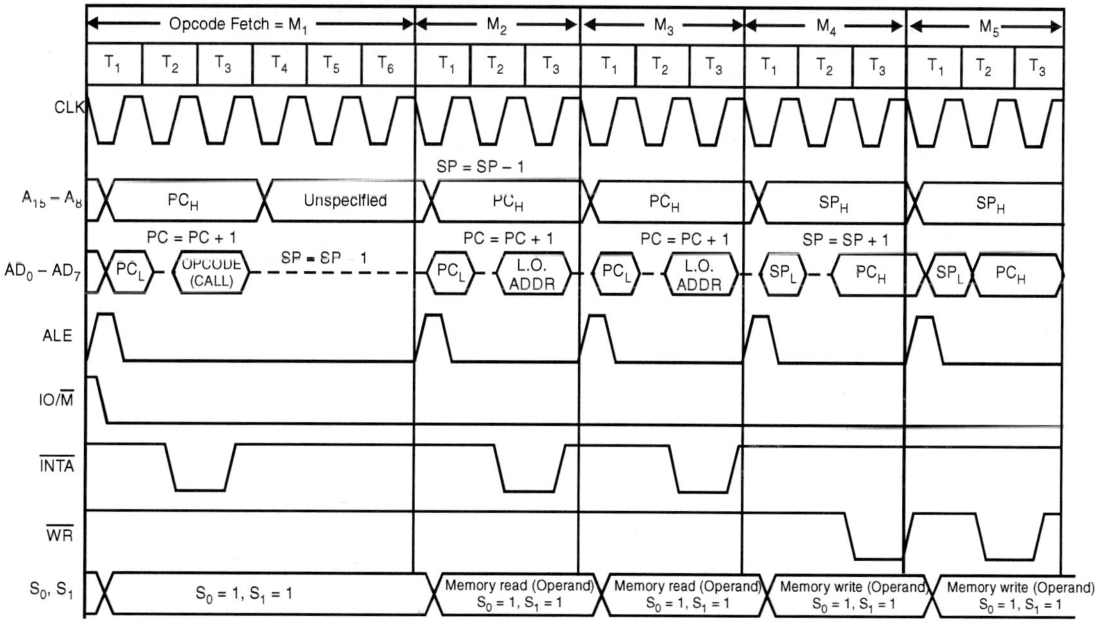

Fig. 2.32. Interrupt acknowledge cycle for call instruction.

- For the CALL instruction it is essential to fetch the two bytes of CALL address through two additional interrupt acknowledge machine cycles. The machine cycles M_4 and M_5 are memory write machine cycles. They store the contents of program counter onto the stack.

2.11.7. 8085 Wait State

- The 8085 microprocessor checks its ready input in T_2 state, if it is LOW, the microprocessor enters wait state T_w after T_2.
- If ready is HIGH, the microprocessor proceeds directly to T_3. If ready goes HIGH during the wait state, then after wait state the microprocessor continues with T3 of the machine cycle. If ready is still LOW, then another wait state is added and it will go on adding Tw till ready is LOW.
- The simplified 8085 state transition diagram with T wait is as shown in Fig. 2.33.

During wait state, the address bus, data bus and control bus are maintained as it was in T_2 state. The main advantage of wait state is, it gives an extra time for selected memory or I/O device. If you are using slower memory or I/O devices then to use it with 8085 it will deactivate ready signal and 8085 will proceed for next states.

The timing diagram of memory read with and without wait state is as shown in Fig. 2.33.

In Fig. 2.33(b) ready pin is LOW in T_2 state so H adds a wait state T_w. Again it will check ready pin, if it is still LOW it will add one more wait state T_w. But if it is HIGH it will proceed for next state T3 and continue next states.

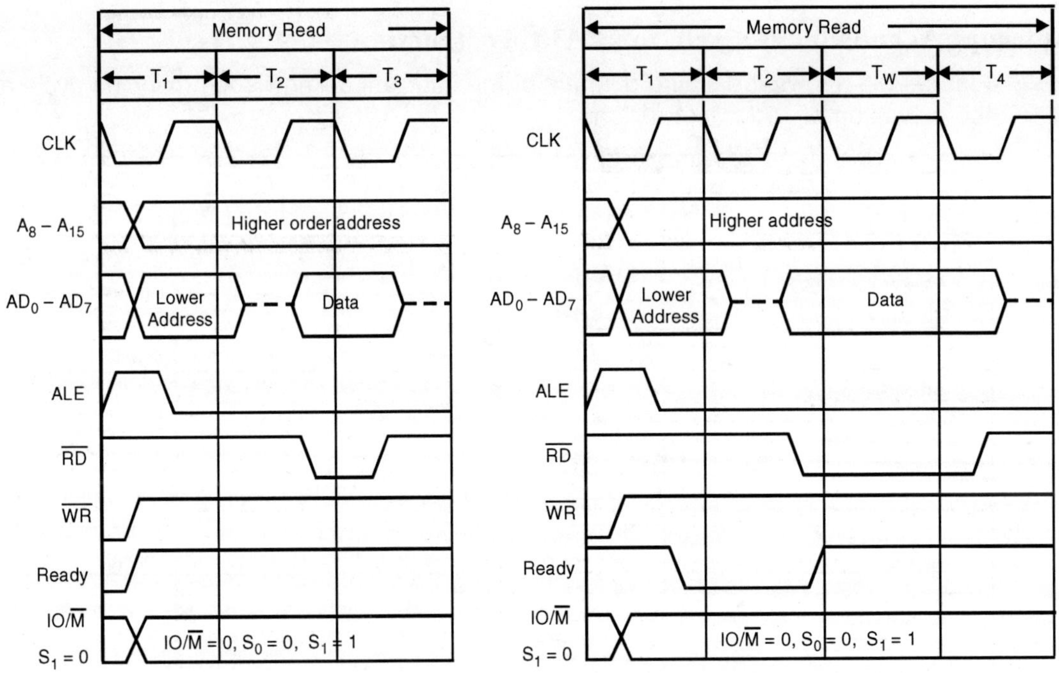

(a) Memory read without wait state *(b) Memory read with one wait state*

Fig. 2.33. Memory read cycle with and without wait state.

2.11.8. Addition of 1 Wait State in Every Machine Cycle

If a memory and I/O device connected wants addition of one wait state in even machine cycle then the circuit shown in Fig. 2.34 (a) will do it.

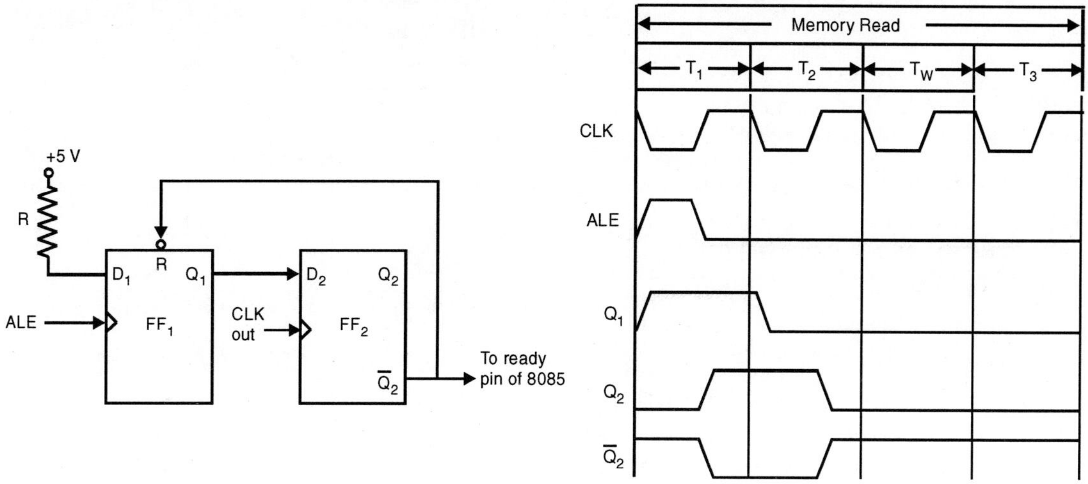

(a) One wait state generator. *(b) Timing Diagram of wait state generator.*

Fig. 2.34.

The circuit of Fig. 2.34 (a) works as follows :

1. The ALE is generated in first clock cycle (T_1 state) of every machine cycle, it is connected as CLK to flip-flop 1 and its D_1 input is HIGH. So Q_1 will be made HIGH by flip-flop 1.
2. The clock out is connected as CLK to flip-flop 2 and during its D input is HIGH because of flip-flop 1. On rising edge of clock but, flip-flop 2 will make T_1 output HIGH and Q_2 output LOW. This $\overline{Q_2}$ is connected to reset of flip-flop 1 and ready input of 8085 so it will reset the flip-flop and introduce wait state after T_2.
3. As ALE is used as CLK. to flip-flop 1, it will be available in next machine cycle, so output of flip-flop 1 will remain LOW for that machine cycle.
4. As input to flip-flop 2 is LOW on next rising edge of CLK, it will make Q_2 output LOW and $\overline{Q_2}$ output HIGH. This removes ready input LOW to HIGH.
5. 8085 will check its ready input in wait state. As it is HIGH, 8085 will go for next states i.e. T_3 onwards.
6. The output of flip-flop 2 will make reset input of flip-flop 1 HIGH. But the output of flip-flop 1 will not change as ALE is not present. During next machine cycle ALE will be generated and will repeat the above steps for addition of one wait state in every machine cycle. The timing diagram for one wait state generator will be as shown in Fig. 2.34 (b).

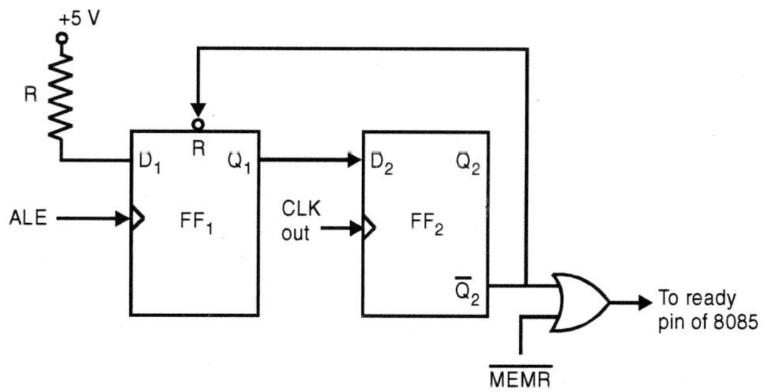

Fig. 2.35. Wait state generator in memory read cycle.

The circuit shown in Fig. 2.35 can be used T_0 add one wait state in every machine cycles. This circuit can be modified to select any machine cycle desired to add wait state.

For example suppose we want to add one wait state in all memory read cycles, in this case combine the $\overline{Q_1}$ output of FF_1 i.e. $\overline{Q_1}$ with memory read signal using OR gate. In this case the ready input will not be LOW by wait state generation circuit because of $\overline{MEMR}$ will not be LOW at any time. But in memory read cycles one of the inputs of OR gate will be LOW because of $\overline{MEMR}$, so the output of OR gate will be controlled by wait state generator circuit to add the wait, states as shown in Fig. 2.35. Similarly the wait state can be selected for any machine cycle by using machine cycle select logic. The status signals and control signals are used as input to machine cycle select logic as shown in Fig. 2.36. In this logic HIGH signals are considered, so NAND gate is used.

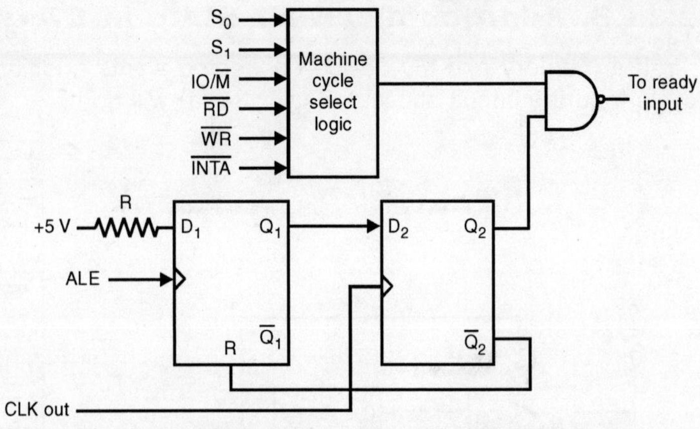

Fig. 2.36. Wait state generator with cycle select logic.

2.11.9. Addition of 2 Wait State in Every Machine Cycle

If a memory and I/O device connected wants addition of two wait state in every machine cycle then the circuit shown in Fig. 2.37 will do it.

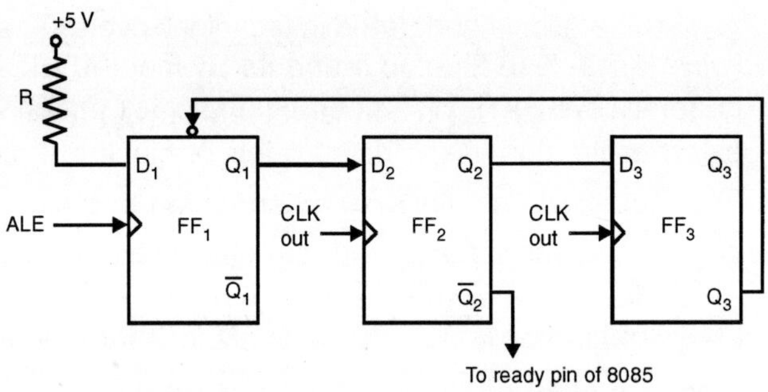

Fig. 2.37. Two wait state generator.

The circuit of Fig. 2.37 works as follows :

1. The ALE is generated in first clock cycle (T_1 state) of every machine cycle, it is connected as CLK to flip-flop 1 and its D_1 input is HIGH. So Q_1 will be made HIGH by flip-flop 1.
2. The clock out is connected as CLK to flip-flop 2 and during T_1, its D_2 input is HIGH because of flip-flop 1. On rising edge of clock out, flip-flop 2 will make $\overline{Q_2}$ output HIGH and output LOW. This $\overline{Q_2}$ is connected to ready input of 8085 so it will introduce one wait state after T_2. (Refer timing diagram).

2.12. INTERRUPT AND SERIAL I/O

The 8085A has 5 interrupt inputs : INTR, RST5.5, RST 6.5, RST 7.5 and TRAP. INTR is identical in function to the 8080 INT. Each of three RESTART inputs, 5.5, 6.5, 7.5, has a programmable mask. TRAP is also a RESTART interrupt except it is nonmaskable and latched into the peripherals by the Address Latch Enable (ALE). During the rest of the machine cycles. Data Bus is used for memory or I/O Data.

The 8085A provides RD, WR, and IO/Memory signals for bus control. An interrupt acknowledge signal (INTA) is also provided. Hold, Ready, and all Interrupts are synchronized. The 8085A also provides serial input data (SID) and serial output data (SOD) lines for simple serial interface.

In addition to these features, the 8085A has three maskable, restart interrupt and one nonmaskable trap interrupt.

8085A vs. 8080A

The 8085A includes the following features on chip in addition to all of the 8080A functions :

(1) Internal clock generator

(2) Clock output

(3) Fully synchronized Ready

(4) Schmitt action on RESET IN

(5) RESET OUT pin

(6) RD, WR, and IO/M Bus Control Signals

(7) Encoded Status information

(8) Multiplexed Address and Data

(9) Direct Restarts and nonmaskable Interrupt

(10) Serial Input/Output lines.

The internal clock generator requires an external crystal or RC network. It will oscillate at twice the basic CPU operating frequency. A 50% duty cycle, two phase, nonoverlapping clock is generated from this oscillator internally and one phase of the clock (¢2) is available as an external clock. The 8085A directly provides the external RDY synchronization previously provided by the 8224. The RESET IN input is provided with a Schmitt action input so that power on reset only requires a resistor and capacitor. RESET OUT is provided for System RESET.

The 8085A provides RD, WR and IO/M signals for Bus control. An INTA which was previously provided by the 8228 in 8080 system is also included in 8085A.

2.13. INTERRUPT FACILITIES

The 8085 contains three levels of interrupt capabilities. The first is essentially identical to that of the 8080 which technique it fully supports. The second technique involves the direct input pins 5.5, 6.5, and 7.5. These are maskable hardware interrupts. The third is Trap, which has its own pin and is non-maskable. Each will be discussed in detail.

8080 Type Interrupts

The 8085 has facilities for servicing interrupts similar to the 8080. The functional items required are an Interrupt Request (INTR) pin, an Interrupt Acknowledge (INTA) pin, an Interrupt Enable (INTE) pin; eight interrupt vectors in low RAM, and the Restart instruction. These perform in the same way as the 8080 interrupt system. Here is a brief review:

1. A program is running normally in the system. The 8214 Priority Interrupt Controller or similar circuit has its compare mask set to some priority level.The Interrupt Enable bit has been set on by some previous routine, enabling interrupts.

2. A device wishes to interrupt the system. It raises its own line, which connects directly to the 8214. The 8214 compare this request with the current status of the system. If the new request is higher in priority than the existing (if any), the interrupt will be allowed. If not, the interrupt will be latched for later use, but no further action is taken.
3. The Interrupts Enabled line existing the 8085 is high, indicating that interrupts are permitted. The 8214 raises the Interrupt line, which causes the MP to finish the current instruction, and then enter an interrupt service cycle. The MP generates the Interrupts Acknowledge line at the beginning of this cycle to permit the 8214 to proceed.
4. Upon receipt of the INTA line, the 8214 along with an 8212 octal latch or similar circuit generates a Restart instruction, which it jams onto the data bus at T3 of the interrupt service cycle. The MP receives this, and removes from it the three-bit modulo-8 vector, which it then multiplies by 8 to find the vector in low RAM.This vector contains one or more instructions, which can service the device causing the interrupt.
5. The execution of the Restart instruction causes the address of the next normal instruction to be executed, obtained from PC, to be placed onto the stack. The next machine cycle will be the M1 of the instruction located in the vector in low RAM.This instruction can now guide the MP to the routine to service the interrupt.
6. At the end of the interrupt service routine, a Return (RTN) instruction will cause the popping of the address off the stack, which was of the next instruction to be serviced if the interrupt had not occurred.

The System Now Finds Itself Back Where It Came From There are three possible variations to the above scenario. First, unlike the 8080, the 8085 will permit the interrupt as described above as long as no other interrupts are pending which are of greater importance. These, of course, are the 5.5, 6.5, 7.5, and Trap. If any of these are pending, they will be serviced first.

Secondly, while the 8214 was the original device to service interrupts on the 8080 system, the 8085 can work with the 8259A Programmable Interrupt Controller as well.This is a more complex device, programmable as to how it handles interrupts, and stackable to two levels, providing as many as 64 levels of interrupt for the '85. The 8259A, moreover, generates Call instructions as well as Restarts. This means that a Call may be jammed onto the data bus during T3 of the interrupt cycle, instead of Restart. While the Restart provides a vector to eight different places in low RAM, depending upon the modulo-8 bits it contains, the Call contains a full two-byte-wide address, which can effectively vector the MP to any-place within the 64K RAM address space. This obviously provides a vastly extended ability to handle interrupts more efficiently.

The third item to be aware of is that the Interrupt Enable flip- flop of the 8080 is now observable as the IE bit #3 of the byte obtained by executing the RIM instruction.

Maskable Interrupts

Three maskable interrupts are provided in the 8085, each with their own pins. They are named RST 5.5, RST 6.5, and RST 7.5, respectively. To see where these names come from, study this chart:

NAME:	ADDRESS:
RST 0	OOH
RST 1	08H
RST 2	10H
RST 3	18H
RST 4	20H
TRAP	24H
RST 5	28H
REST 5.5	2CH
RST 6	30H
RST 6.5	34H
RST 7	38H
RST 7.5	3CH

Note in the chart that the items in lightface are those with which we are already familiar. They are the normal vectors for the Restart instructions 0 through 7, as created by the 8214. They are 8 bytes apart, which is ample room for such jumps as are needed to obtain the interrupt servicing routines.Now look at the bold face items. These items have vector areas, which are between the original vectors in RAM. RST 5.5, for instance, is half way between the RST 5 and the RST 6 vectors, hence the "5". If all the vectors were in use, those located above address 20H would each have only four bytes in which to locate and jump to the interrupt service routine. This should be enough room, however, if used wisely. Note also that the Trap interrupt is located at the 4.5 point in the vectors. The 5.5,6.5, and 7.5 vectors have several items in common. First, they each have their own pin directly into the 8085. These pins, will accept asynchronous interrupt requests without the need for any sort of external priority interrupt device. Secondly, these interrupts are individually maskable. This is accomplished via the Set Interrupt Mask instruction. This instruction allows bits to be set or cleared which will permit or deny an interrupt on one of these lines to force the '85 into an interrupt service cycle. When an input is received on one of these lines and its respective mask bit is enabled (set to 0), the processor will finish the current machine cycle, then enter a interrupt service cycle in which an automatic jam inside the MP will vector it to 2CH, 34H, or 3CH for 5.5,6.5, or 7.5 respectively. Those locations will assumedly have been previously set to contain directions to the interrupt servicing routines.The RST 5.5 and RST 6.5 interrupts are "level sensitive"This means that the device wishing to interrupt will apply a steady high level to the appropriate pin and hold it there until the 8085 gets around to responding. When the '85 recognizes the applied high level, it will permit the interrupt to be serviced in the next machine cycle. The mask bits set by the SIM instruction will directly determine what the RIM instruction sees with respect to the 5.5 and 6.5 interrupt pending bits. If the mask bits are set high (to a 1), these interrupts are masked off. This means that a following RIM will not see them as pending. If the mask bits are set to 0 (enabled), a RIM will see the true condition in bits 4 and 5 of the mask byte.

The RST 7.5 interrupt is "edge sensitive". This means that a pulse applied to this pin, requesting an interrupt, can come and go before the processor gets around to servicing it. This is possible because, unlike the 5.5 and 6.5, the 7.5 has a flip-flop just inside its pin, which instantly registers the fact that an interrupt request was applied to the device. This flip-flop provides a bit which is read in RIM instruction as bit 6. This bit will indicate an interrupt pending if a quick pulse is applied to pin 7.5, even though bit 2 of the SIM instruction, the 7.5 mask bit, is turned on (disabled). Bit 2 of SIM byte, therefore, acts differently as a mask bit than does bits 0 and 1 for 5.5 and 6.5. Whereas bits 0 and 1 will mask off all indication of action on pins 5.5 and 6.5, bit 2 will allow the indication of a 7.5 interrupt pending, but will prevent the actual servicing of the 7.5 vector unless the mask is enabled for it. In this way, even though the mask set by the SIM prevents the MP from servicing a 7.5 interrupt, the fact that such an interrupt did occur, captured by the flip-flop, is indicated to whatever routine next executes a RIM instruction.

While the normal interrupt and 5.5 and 6.5 interrupts' enable bits are reset when these are serviced, the 7.5-interrupt flip-flop must be turned off individually. This may be accomplished by actually.

Status Information

Status information is directly available from the 8085A. ALE serves as a status strobe. The status is partially encoded, and provides the user with advanced timing of the type of bus transfer being done. IO/M cycle status signal is provided directly also. S Carries the following status information:

HALT

WRITE

READ

FETCH

S1 can be interpreted as R/W in all bus transfers.

The three RESTART interrupts cause the internal execution of RST (saving the program counter in the stack and branching to the RESTART address) if the interrupts are enabled and if the interrupt mask is not set. The nonmaskable TRAP causes the internal execution of a RST independent of the state of the interrupt enable or masks.

Name RESTART Address (Box)

TRAP

RST 5.5

RST 6.5

RST 7.5

24_{16}

$2C_{16}$

34_{16}

$3C_{16}$

There are two different types of inputs in the restart interrupts. RST 5.5 and RST 6.5 are high levelsensitive like INTR (and INT on the 8080) and are recognized with the same timing as INTR. RST 7.5 is rising edgesensitive.

For RST 7.5, only a pulse is required to set an internal flip flop which generates the internal interrupt request. The RST 7.5 request flip flop remains set until the request is serviced. Then it is reset automatically. This flip flop may also be reset by using the SIM instruction or by issuing a RESET IN to the 8085A.The RST 7.5 internal flip flop will be set by a pulse on the RST 7.5 pin even when the RST 7.5 interrupt is masked out.

The status of the three RST interrupt masks can only be affected by the SIM instruction and RESET IN.

The interrupts are arranged in a fixed priority that determines which interrupt is to be recognized if more than one is pending as follows: TRAP highest priority, RST 7.5, RST 6.5, RST 5.5, INTR lowest priority This priority scheme does not take into account the priority of a routine that was started by a higher priority interrupt. RST 5.5 can interrupt a RST 7.5 routine if the interrupts were reenabled before the end of the RST 7.5 routine.

The TRAP interrupt is useful for catastrophic errors such as power failure or bus error. The TRAP input is recognized just as any other interrupt but has the highest priority. It is not affected by any flag or mask. The TRAP input is both edge and level sensitive. The TRAP input must go high and remain high to be acknowledged, but will not be recognized again until it goes low, then high again. This avoids any false triggering due to noise or logic glitches. The following diagram illustrates the TRAP interrupt request circuitry within the 8085A.

Note that the servicing of any interrupt (TRAP, RST 7.5, RST 6.5, RST 5.5, INTR) disables all future interrupts (except TRAPs) until an EI instruction is executed.

The TRAP interrupt is special in that it preserves the previous interrupt enable status. Performing the first RIM instruction following a TRAP interrupt allows you to determine whether interrupts were enabled or disabled prior to the TRAP. All subsequent RIM instructions provide current interrupt enable status.

The serial I/O system is also controlled by the RIM and SIM instructions. SID is read by RIM, and SIM sets the SOD data.

2.13.1. 8085 Interrupts

The 8085 microprocessors has 5 interrupt inputs. These interrupts and their call locations are as stated below :

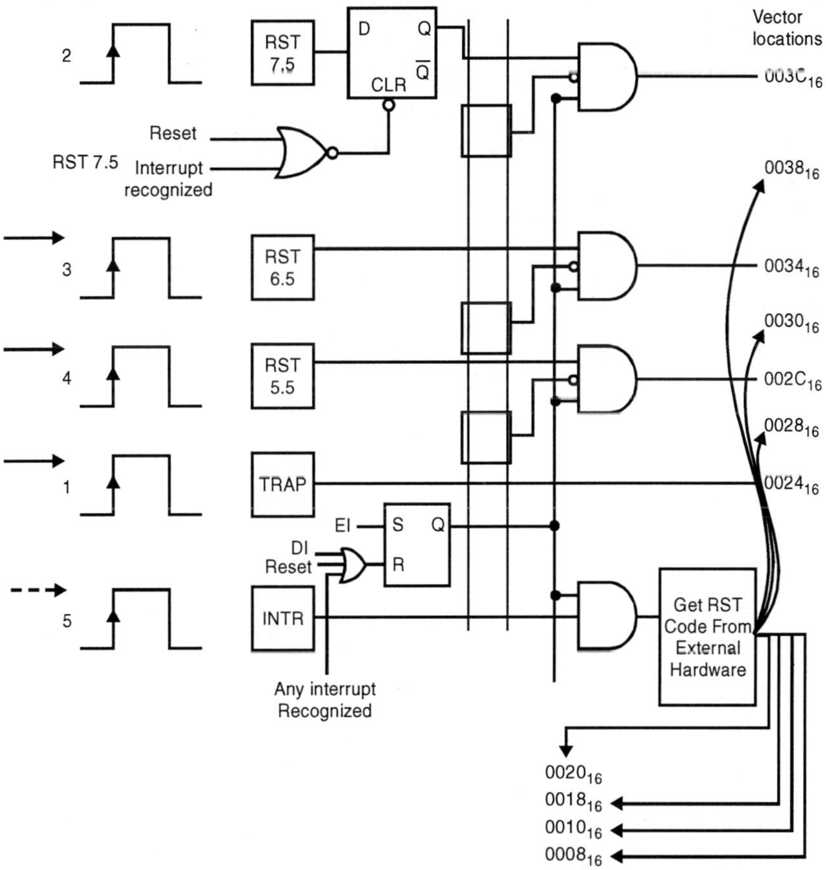

Fig. 2.38. 8085 Interrupt structure.

Location

1. TRAP → 0024H
2. RST 7.5 → 003CH
3. RST 6.5 → 0034H
4. RST 5.5 → 002CH

The TRAP has the highest priority followed by RST 7.5, 6.5, 5.5 and INTR, howover TRAP has a lower priority than HOLD signal used for DMA.

2.14. ADDRESSING MODES

Each instruction requires certain data on which it has to operate. These techniques are called Addressing modes. These are explained below:

(1) Implied Addressing

The addressing mode of certain instructions is implied by the instruction's function. For example, the STC (set carry flag) instruction deals only with the carry flag, the QAA (decimal adjust accumulator) instruction deals with the accumulator.

(2) Register Addressing

Quite a large set of instructions call for register addressing. With these instructions, you must specify one of the registers A through E, H or L as well as the operation code. With these instructions, the accumulator is implied as a second operand.

For example, the instruction CMP E may be interpreted as 'compare the contents of the E register with the contents of the accumulator. Most of the instructions that use register addressing deal with 8-bit values. However, a few of these instructions deal with 16- bit register pairs. For example, the PCHL instruction exchanges the contents of the program counter with the contents of the H and L registers.

(3) Immediate Addressing

Instructions that use immediate addressing have data assembled as a part of the instruction itself. For example, the instruction CPI 'C' may be interpreted as 'compare the contents of the accumulator with the letter C. When assembled, this instruction has the hexadecimal value FE43. Hexadecimal 43 is the internal representation for the letter C. When this instruction is executed, the processor fetches the first instruction byte and determines that it must fetch one more byte.The processor fetches the next byte into one of its internal registers and then performs the compare operation.

Notice that the names of the immediate instructions indicate that they use immediate data. Thus, the name of an add instruction is ADD; the name of an add immediate instruction is AD I.

All but two of the immediate instructions uses the accumulator as an implied operand, as in the CPI instruction shown previously. The MVI (move immediate) instruction can move its immediate data to any of the working registers including the accumulator or to memory. Thus, the instruction MVI D, OFFH moves the hexadecimal value FF to the D register.

The LXI instruction (load register pair immediate) is even more unusual in that its immediate data is a 16-bit value.This instruction is commonly used to load addresses into a register"pair. As mentioned previously, your program must initialize the stack pointer; LXI is the instruction most commonly used for this purpose. For example, the instruction LXI SP, 3OFFH loads the stack pointer with the hexadecimal value 30FF.

(4) Direct Addressing

Jump instructions include a 16-bit address as part of the instruction. For example, the instruction JMP 1000H causes a jump to the hexadecimal address 1000 by replacing the current contents of the program counter with the new value 1000H.

Instructions that include a direct address require three bytes of storage: one for the instruction code, and two for the 16-bit address.

(5) Register Indirect Addressing

Register indirect instructions reference memory via a register pair. Thus, the instruction MOV M, C moves the contents of the C register into the memory address stored in the H and L register pair. The instruction LDAX B loads the accumulator with the byte of data specified by the address in the B and C register pair.

(6) Combined Addressing Modes

Some instructions use a combination of addressing modes. A CALL instruction, for example, combines direct addressing and register indirect addressing.The direct address in a CALL instruction specifies the address of the desired subroutine; the register indirect address is the stack pointer. The CALL instruction pushes the current contents of the program counter into the memory location specified by the stack pointer.

2.14.1. Timing Effects of Addressing Modes

Addressing modes affect both the amount of time required for executing an instruction and the amount of memory required for its storage. For example, instructions that use implied or register addressing, execute very quickly since they deal directly with the processor's hardware or with data already present in hardware registers. Most important, however is that the entire instruction can be fetched with a single memory access. The number of memory accesses required is the single greatest factor in determining execution timing. More memory accesses therefore require more execution time. A CALL instruction for example, requires five memory accesses: three to access the entire instruction and two more to push the contents of the program counter onto the stack.

The processor can access memory once during each processor cycle. Each cycle comprises a variable number of states. (See below and the appendix of "Using The Sdk-85 Microprocessor Trainer"). The length of a state depends on the clock frequency specified for your system, and may range from 480 nanoseconds to 2 microseconds. Thus, the timing for a four state instruction may range from 1.920 microseconds through 8 microseconds. (The 8085 have a maximum clock frequency of 5 MHz and therefore a minimum state length of 200 nanoseconds.)

2.14.2. The Stack

The stack is one of the most important things you must know when programming. Think of the stack as a deck of cards.

When you put a card on the deck, it will be the top card. Then you put another card, then another. When you remove the cards, you remove them backwards, the last card first and so on. The stack works the same way; you put *(push)* words (addresses or register pairs) on the stack and then remove *(pop)* them backwards. That's called LIFO, *Last In First Out.*

The 8085 uses a 16-bit register to know where the stack top is located, and that register is called the SP (Stack Pointer). There are instructions that allow you to modify it's contents but you should NOT change the contents of that register if you don't know what you're doing!

2.14.3. Push & Pop

As you may have guessed, push and pop "pushes" bytes on the stack and then takes them off. When you push something, the stack counter will decrease with 2 (the stack "grows" down, from higher addresses to lower) and then the register pair is loaded onto the stack. When you pop, the register pair is first lifted of the stack, and then SP increases by 2.

N.B.:

Push and Pop only operate on words (2 bytes i.e.: 16 bits). You can push (and pop) all register pairs: BC, DE, HL and PSW (Register A and Flags). When you pop PSW, remember that all flags may be changed. You can't push an immediate value. If you want, you'll have to load a register pair with the value and then push it. Perhaps it's worth noting that when you push something, the contents of the registers will still be the same; they won't be erased or something. Also, if you push

DE, you can pop it back as HL (you don't have to pop it back to the same register where you got it from). The stack is also updated when you CALL and RETurn from subroutines. The PC (program counter which points at the next instruction to be executed) is pushed to the stack and the calling address is loaded into PC. When returning, the PC is loaded with the word popped from the top of the stack (TOS).

So, when is this useful? It's almost always used when you call subroutines. For example, you have an often-used value stored in HL. You have to call a subroutine that you know will destroy HL (with destroy I mean that HL will be changed to another value, which you perhaps don't know). Instead of first saving HL in a memory location and then loading it back after the subroutine, you can push HL before calling and directly after the calling pop it back. Of course, it's often better to use the pushes and pops inside the subroutine. All registers you know will be changed are often pushed in the beginning of a subroutine and then popped at the end, in reverse order! Don't forget - last in first out. If you want to only push one 8-bit register, you still have to push its "friend". Therefore, be aware that if you want to store away D with pushing and popping, remember that E will also be changed back to what it was before. In those cases, if you don't want that to happen, you should try first to change register (try to store the information in E in another register if you can) or else you have to store it in a temporary variable.

Before executing a program, you should keep track of your pushes and pops, since they are responsible for 99% of all computer crashes! For example, if you push HL and then forget to pop it back, the next RET instruction will cause a jump to HL, which can be anywhere in the ROM/RAM and the computer will crash. Note however, it's also a way to jump to the location stored in HL, but then you should really use the JMP instruction, to do the same thing. Push and pop doesn't change any flags, so you can use them between a compare and jump instructions, depending on a condition, which is often very useful.

2.15. MEMORY INTEGRATED CIRCUITS

A typical memory will consist of address lines, data lines, chip enable, read-write line and output enable line, as shown in schematic (Fig. 2.39)

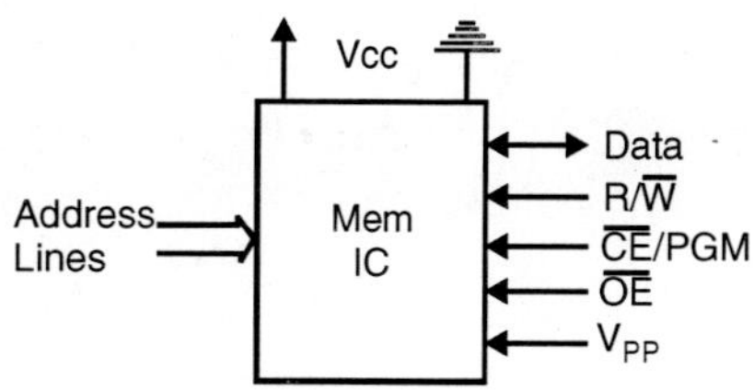

Fig. 2.39

The number of address lines will be determined by "memory capacity". For example, a memory of capacity 1 k × 8 will have 10 address lines from A_0 to A_9. Address lines will be unidirectional. Number of Data lines will also be determined by memory size. For example, 1 k × 8 chip will have 8 data lines and 2k × 4 chip will have 4 data lines. The control signal $R/\overline{W}$ is logic '1' for read and logic '0' for writing to the memory. Chip enable ($\overline{CE}$) signal enables the chip while output enable ($\overline{OE}$) enables the tristate output buffers of data lines. IN program mode, $\overline{CE}$/PGM line is pulsed with VPP as 25 V (in normal Read/Write mode, VPP = 5 V) and data lines acts as input lines for the data to be programmed in case of PROM/EPROM programming.

Common memories available from Intel are of 27 series for EPROM and 23 series for ROMs, 2708, 2716, 2723, 2764 are of 1 k × 8, 4 k × 8, 4k × 8 and 8 k × 8 EPROM's respectively. All these chips have compatible access time to 8 bit microprocessors and 16 bit microprocessors. Another popular device is 6116, 2k RAM, manufactured by Motorola Inc. A more detail study of these IC's and their pin details, reader may refer to the manual of these companies.

Interfacing of 8085 to Memories

There are two different ways by which microprocessor communicates with external devices. Microprocessor issues $IO/\overline{M}$ signal and if this signal is used to distinguish between I/O and memory than that system is said to employ standard I/O or I/O mapped I/O. If $IO/\overline{M}$ signal is not used, naturally CPU does not distinguish between I/O and memory and is said to employ memory mapped I/O.

In the case of Memory mapped I/O, processor does not distinguish I/O from memory 1. Using this scheme, the large instruction set which is available as memory referred instruction can now be used for I/O also whereas one has to use only IN and OUT instruction in standard I/O, In memory mapped I/O MOV r, VI will mean move data from I/O to register.

ADD M will mean add port to ACC. This scheme has limitation that now I/O devices are now addressed as memory, memory space is cut down to half, since the one hit in address is being used to distinguish between memory and I/O, Instruction to load and store data in ACC (LDA and STA) will take 13 clock states whereas IN and OUT instructions are of 10 clock states.

2.15.1. Address Decoding and Memory Addresses

The memory address of the chip ranges from 0000H to 0FFFH, as shown below

	A_{15}	A_{14}	A_{13}	A_{12}	A_{11}	A_{10}	A_9	A_8	A_7	A_6	A_5	A_4	A_3	A_2	A_1	A_0
000H =	0	0	0	0	0	0	0	0	0	0	0	0	0	0	0	0
FFFH =	0	0	0	0	1	1	1	1	1	1	1	1	1	1	1	1
	Chip Enable				Register Select											

The logic level on the address line A_{15}–A_{12} must be 0000 to assert the chip enable and the address lines A_{11}–A_0 can assume any combinations from all 0's to all 1's; as shown above.

$\overline{WR}$ signals of the 8085 are directly connected to the memory chip. The signals $\overline{MEMR}$ and $\overline{MEMW}$ need not be generated seperately, thus this technique saves two gates. The memory address of this chip ranges from 8800H to 8FFFH, as shown below

	A_{15}	A_{14}	A_{13}	A_{12}	A_{11}	A_{10}	A_9	A_8	A_7	A_6	A_5	A_4	A_3	A_2	A_1	A_0
800H =	1	0	0	0	1	0	0	0	0	0	0	0	0	0	0	0
					1	1	1	1	1	1	1	1	1	1	1	1
FFFH =	MSEL$_1$															

The output line 01 of the decodes is connected to $\overline{CE}$ of the memory chip, and it is indentified as MSEL1 because it is asserted only when $IO/\overline{M}$ is low.

The figure shows the block diagram of 8085 memory section. It has 8 address lines, one chip enable ($\overline{CE}$) line, and 5 lines compatible with the control and status signal of the 8085 $IO/\overline{M}$, ALE, $\overline{RD}$, $\overline{WR}$ and RESET. These line eliminate the need for external demultiplexing of the bus AD_7–AD_0 and for generations 8085 places the address 0FFFH on the address bus. The address 0000(0H) goes to the decoder, and the output line 00 of the decoder selects the chip.The remaining address FFFH goes on the address lines of the chip and the internal decoder of the chip decodes the address and selects the register FFFH.

Separate control signals for memory and I/O. The memory section includes 256 × 8 memory locations and an internal latch to demultiplex the bus lines AD_7–AD_0. The memory section also requires a $\overline{CE}$ signal and a $\overline{MEMW}$ and $\overline{MEMR}$ control signals which are generated internally by combining the $IO/\overline{M}$, $\overline{WR}$ and $\overline{RD}$ signals.

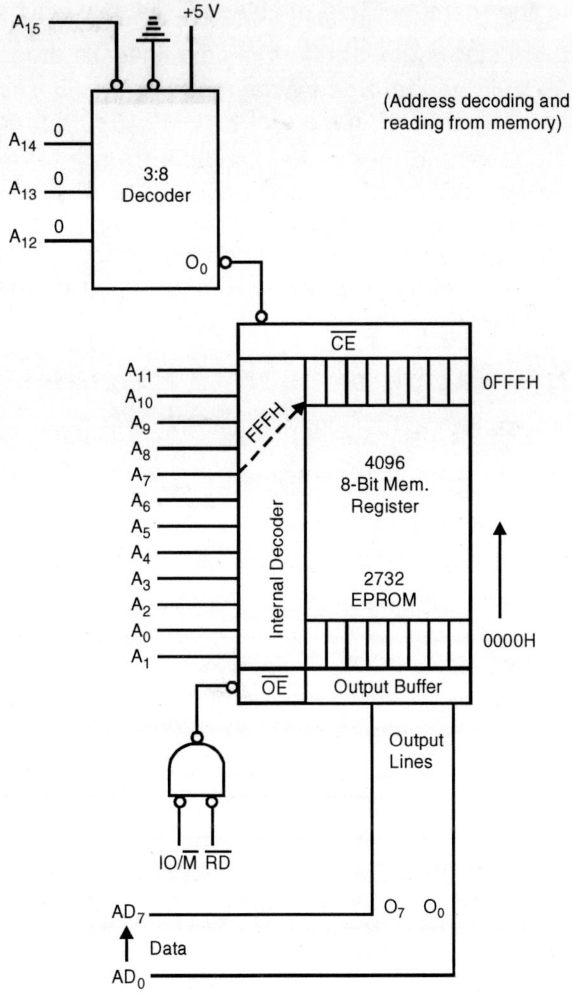

Fig. 2.40

Sample Designs

Example 1 : Design the interface for interfacing single chip of 64 k memory with 8085 microprocessor.

Solution. Memories of 64 k capacity are rarely used by designers of microcomputers, still this exercise will develop the concept of basic design.

Interface Steps :

(i) As the 64k chip will have 16 addressing lines (A_0 to A_{15}). They must be connected to $A_0 - A_7$ obtained from 8212 IC (8212 IC demultiplex $AD_0 - AD_7$ lines and heads A_0 to A_7 lines) and $A_8 - A_{15}$ directly connected to microprocessor $A_8 - A_{15}$ lines.

(ii) Data lines $D_0 - D_7$ of memory will be connected to $AD_0 - AD_7$ lines of microprocessor.

(iii) $\overline{OE}$ is connected to $\overline{RD}$ output of 8085 and $\overline{WR}$ is connected to $\overline{WR}$ output of 8085.

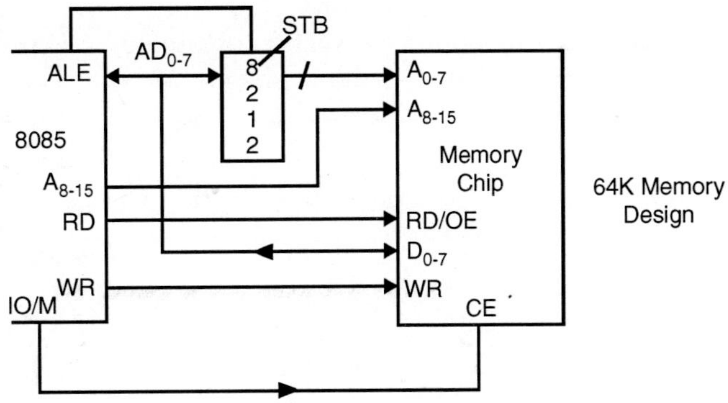

Fig. 2.41. Interfacing single chip of 64K with 8085 microprocessor.

(iv) The chip select input of memory is connected to $IO/\overline{M}$ output of 8085. The chip will be enabled only when $IO/\overline{M}$ goes low.

Example 2. Design the interface to address 64 k of memory using chips of 8 k × 8 each (2764).

Solution. IC memory chip of 8 k × 8 will have 13 address lines $A_0 - A_{12}$. Number of chips needed are :

$$= \frac{64\,k \times 8}{8\,k \times 8} = 8 \text{ chips}$$

These 8 chips will occupy memory space as shown below :

Chip #	0	8K
#	1	8K
#	2	8K
#	3	8K
#	4	8K
#	5	8K
#	6	8K
#	7	8K
#	8	8K
#	9	8K

Memory Chart

(i) Interface Design

Each chip of 8 k will have $A_0 - A_{12}$ lines and all of these will be connected to microprocessor $A_0 - A_{12}$ lines for all 8 chips. Now at a particular moment only one of these chips will be communicating with CPU, hence we need eight different signals to be connected to chip select of each chip. At a particular moment only one out of these eight signal lines will be active.

The free lines are A_{13}, A_{14}, A_{15} which will not be connected to memory chips. Now these free lines can be used to generated eight signals by using a 3 to 8 decoder IC. The eight outputs of IC 3 to 8 decoder will be connected to chip select input of these eight memory chips. The truth table of this scheme is shown below :

A15	A14	A13	Chip Selected
0	0	0	# 0
0	0	1	# 1
0	1	0	# 2
0	1	1	# 3
1	0	0	# 4
1	0	1	# 5
1	1	0	# 6
1	1	1	# 7

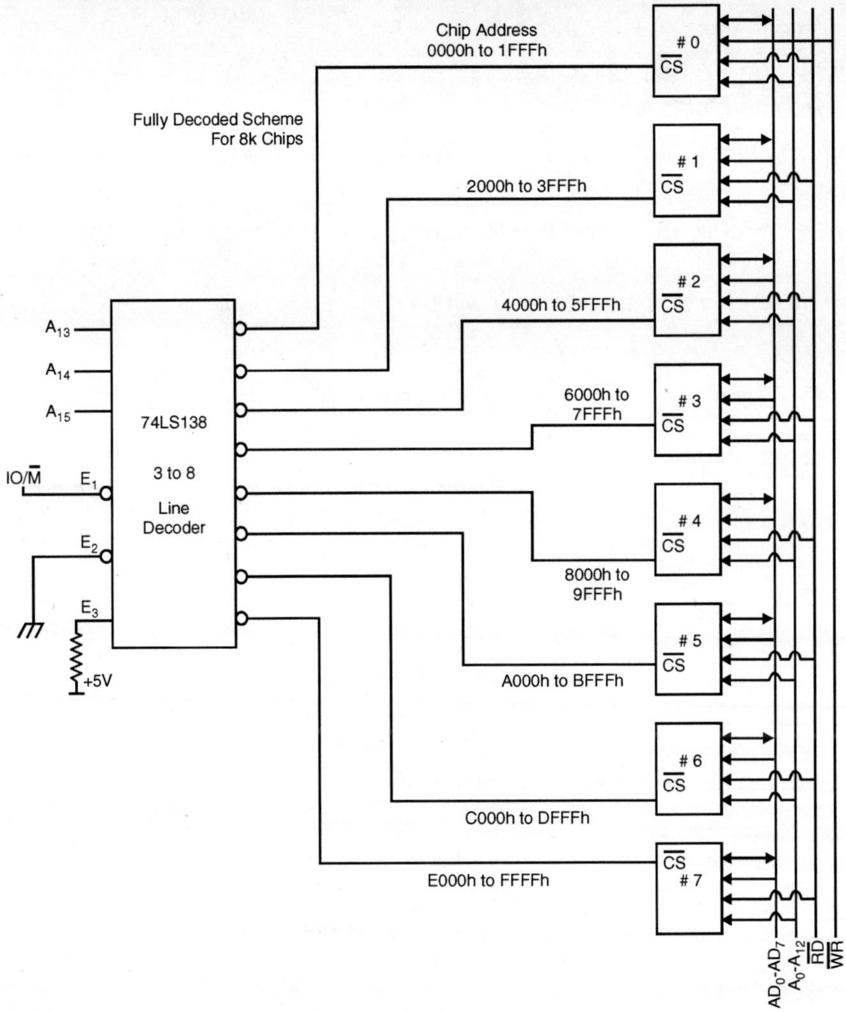

Fig. 2.42.

When A_{13}, A_{14}, A_{15} are all zeros; chip #0 i.e. first chip will be selected and A_{13}, A_{14}, A_{15} with all '1' will select chip 7.

The addresses generated for chip #0 will be

First address	0	0	0	0	0	0	0	0	0	0	0	0	0	0	0	0

0 0 0 0 h

Last address	0	0	0	1	1	1	1	1	1	1	1	1	1	1	1	1

1 F F F h

and for chip #7

First address	1	1	1	0	0	0	0	0	0	0	0	0	0	0	0	0

i.e., E 0 0 0 h

Last address	1	1	1	1	1	1	1	1	1	1	1	1	1	1	1	1

i.e., F F F F h

and similarly it can be derived for other chips also and will be

for	Chip #1	2000h to 3FFFh
	Chip #2	4000h to 5FFFh
	Chip #3	6000h to 7FFFh
	Chip #4	8000h to 9FFFh
	Chip #5	000h to BFFFh
	Chip #6	C000h to DFFFh

(ii) Hardware Implementation

All the address lines from 8 chips i.e. $A_0 - A_{12}$ are connected to microprocessor address lines $A_0 - A_{12}$. Thus each line of $A_0 - A_{12}$ of microprocessor will be driving 8 chips of memory. All the data lines of all memory chips will be connected to CPU's data lines. Thus each data line of CPU will be loaded by eight lines of memories.

$\overline{RD}$ ($\overline{OE}$) of all 8 chips will be tied to $\overline{RD}$ line of CPU and $\overline{WR}$ line of each chip will be tied to CPU $\overline{WR}$ signal. Thus $\overline{RD}$ and $\overline{WR}$ line of CPU will also be loaded by 8 lines load each. The block schematic is shown in Fig. 2.42.

Lines A_{13}, A_{14}, A_{15} will be connected to a 3 to 8 decoder, whose eight active low outputs will be connected to chip select of each memory chip. $IO/\overline{M}$ will be connected to EI (active low) so that 74LS138 will get enabled only when addresses are meant for memory. Additional enable lines for 74LS138 are provided, whose purposes will be clear in later discussion. The 8085, A_{13}, A_{14}, A_{15} lines have to drive single load of TTL low power schotcky (or referred as LS load).

(iii) Loading Considerations

(a) Each line of $AD_0 - AD_7$ drives eight lines of 8 memory chips (one line of $AD_0 - AD_7$ drives one line each of 8 chips) i.e., 8 CMOS loads.

(b) $A_0 - A_{12}$ each of them drives 8 lines or 8 CMOS loads.

(c) Each line of A_{13}, A_{14}, A_{15} and $IO/\overline{M}$ drives 1 LS load.

(d) $\overline{RD}$ drives 8 CMOS loads.

(e) $\overline{WR}$ drives 8 CMOS loads

Microprocessor 8085 has following driving capability. It can drive 30—40 CMOS loads or 34 LS TTL loads or 1 TTL load only.

In design, designer has to lake care that at no time microprocessor is loaded beyond its capability. If the load is beyond its driving capability, additional buffers are needed in between. The scheme described is referred as fully decoded scheme, i.e., even the memory chip is not connected, its provision is made.

Example 3. Design a fully decoded scheme for 64k × 8 lo interface chips of 4k × 8 (2732). Discuss loading considerations also.

Solution. Number of memory chips needed $= \dfrac{64k \times 8}{4k \times 8} = 16$ chips

These 16 chips will occupy memory space as shown below. A single chip of 4k will have 12 address lines *i.e.,* A_0 to A_{11} (from 000h to FFFh).

(i) Interface Design : Each chip of 4k will have $A_0 - A_{11}$ address lines and all of these for all chips will be connected to microprocessor's A_0 to A_{11} address lines. All of the data lines of all chips will be connected to CPU's data lines ($AD_0 - AD_7$ lines). $\overline{OE}$ and $\overline{WR}$ for all chips will be tied to CPU $\overline{RD}$ and $\overline{WR}$ line. Now at a particular moment in a particular machine cycle of memory read/memory write, only one chip will be enabled. Hence 16 different chip select (active low) signals are to be generated, and at a time only one signal will be active.

Memory space chart

	Chip	
0000h 0FFFh	Chip # 0	4K
1000h 1FFFh	# 1	4K
2000h	# 3	4K
2FFFh	# 3	4K
	# 4	4K
	# 5	4K
	# 6	4K
7000h	# 7	4K
7FFFh	# 8	4K
	# 9	4K
	# 10	4K
	# 11	4K
	# 12	4K
	# 13	4K
	# 14	4K
	# 15	4K

The free lines A_{12}, A_{13}, A_{14}, A_{15} will not he connected to memory chips. Therefore these free lines could be used to generate sixteen control signals by using a 4 to 16 line decoder 1C. The sixteen active low outputs of 4 to 16 decoder 1C drives different memory chips will be active depending on A_{15}, A_{14}, A_{13}, A_{12} line. The truth table of this scheme is shown below.

A_{15}	A_{14}	A_{13}	A_{12}	Chip to selected
0	0	0	0	# 0
0	0	0	1	# 1
0	0	1	0	# 2
0	0	1	1	# 3
0	1	0	0	# 4
0	1	0	1	# 5
0	1	1	0	# 6
0	1	1	1	# 7
1	0	0	0	# 8
1	0	0	1	# 9
1	0	1	0	# 10
1	0	1	1	# 11
1	1	0	0	# 12
1	1	0	1	# 13
1	1	1	0	# 14
1	1	1	1	# 15

When A_{12}, A_{13}, A_{14}, A_{15} are all zeroes chip # 0 will be selected and when A_{12}, A_{13}, A_{14}, A_{15} are all '1' then chip #15 will be selected. Addresses for various chips will be as follows :

For chip #0 starting address	0	0	0	0	0	0	0	0	0	0	0	0	0	0	0	0
or		0	0	0	0 h											
Last address	0	0	0	1	1	1	1	1	1	1	1	1	1	1	1	1
or		0	F	F	F h											

and similarly it can he derived for other chips and are as below

Chip #1	1 0 0 0 h to 1 F F F h
Chip #2	2 0 0 0 h to 2 F F F h
Chip #3	3 0 0 0 h to 3 F F F h
Chip #4	4 0 0 0 h to 4 F F F h
Chip #5	5 0 0 0 h to 5 F F F h
Chip #6	6 0 0 0 h to 6 F F F h
Chip #7	7 0 0 0 h to 7 F F F h
Chip #8	8 0 0 0 h to 8 F F F h
Chip #9	9 0 0 0 h to 9 F F F h
Chip #10	A 0 0 0 h to A F F F h
Chip #11	B 0 0 0 h to B F F F h
Chip #12	C 0 0 0 h to C F F F h
Chip #13	D 0 0 0 h to D F F F h
Chip #14	E 0 0 0 h to E F F F h
Chip #15	F 0 0 0 h to F F F F h

(ii) Hardware Implementation : All the address lines $A_0 - A_{11}$ of all chips will he connected to CPU A_0-A_{11} lines and all 8 data lines will be connected to CPU data lines. $\overline{RD}$ ($\overline{OE}$) and $\overline{WR}$ of all chips will be connected to CPU $\overline{RD}$ and $\overline{WR}$ lines. The block schematic is shown in Fig. 2.43.

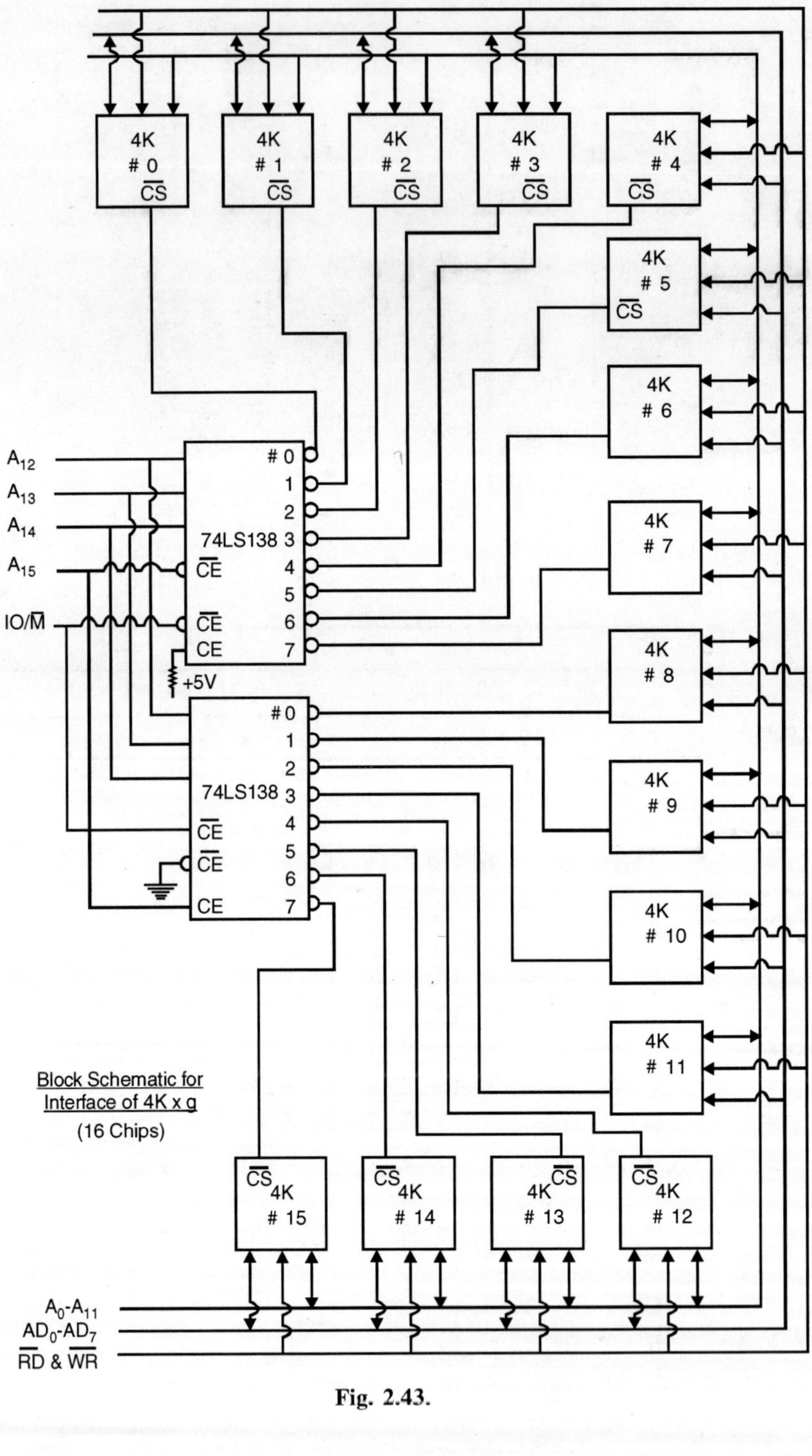

Fig. 2.43.

Lines A_{12}, A_{13}, A_{14}, A_{15} will be connected to a 4 to 16 decoder whose sixteen active low outputs will be connected to chip select of each memory IC. Frequently this 4 to 16 decoder is constructed using 3 to 8 decoder IC's only. Lines A_{12}, A_{13}, A_{14} are connected to both 3 to 8 decoder. A_{15} in conjuction with $IO/\overline{M}$ enables one 3 to 8 decoder. The active output of enabled 3 to 8 decoder IC selects the desired chips.

(iii) **Loading Considerations :**

(a) Each line of $AD_0 - AD_7$ of CPU drives 16 MOS loads

(b) Each line of $A_0 - A_{11}$ drives 16 MOS loads.

(c) Each line of A_{12}, A_{13}, A_{14}, A_{15} drives 2 LS TTL loads.

(d) $\overline{RD}$, $\overline{WR}$ drives 16 MOS loads each.

(e) $IO/\overline{M}$ drives 2 LSTTL loads.

No additional buffers are required as the load is needed within the capability of the microprocessor.

Linear Selection of Memories

If one of the address bit is used to select a chip, it is called as linear selection. In a small microcomputer system this has the advantage of eliminating the extra hardware for decoders. The disadvantages of such scheme is that for each bit used as chip select, the available memory space is cut down to half. For example only five chips, each of 2 k can be interfaced using A_{11}, A_{12}, A_{13}, A_{14}, A_{15} as control signals for chip select. Another disadvantage of linear selection is that the memory address for different chips are not contiguous. This will become clear in the next example.

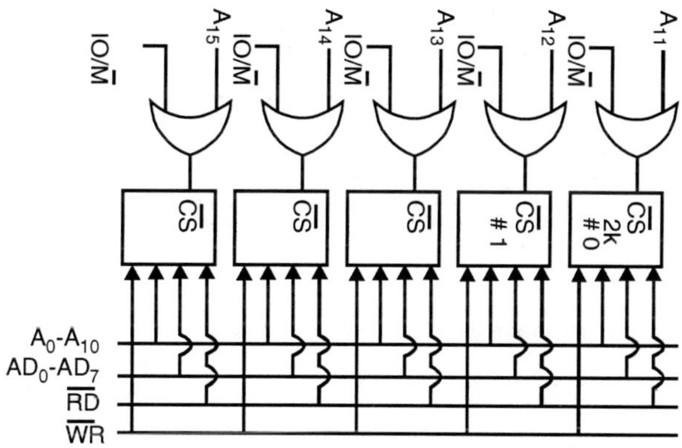

Fig. 2.44. Interfacing 2K × 8 chips using linear select scheme.

Example 4. Design a scheme to address 10 k × 8 of memory using 2k × 8 chips employing linear select scheme. Derive address for each chip.

Solution. Each chip of 2k × 8 will have $A_0 - A_{10}$ lines and for 10 k, five chips will be needed. Now if a single address bit is used for selection of each chip. A_{11}, A_{12}, A_{13}, A_{14}, A_{15} can be used to select five memory chips. $A_0 - A_{10}$ address lines of all memory chips will go to CPU $A_0 - A_{10}$ lines. Similarly Data lines of all chips will go to CPU data lines. $\overline{RD}$ and $\overline{WR}$ will be connected to all memory chips $\overline{OE}$ and $\overline{WR}$ inputs.

OR gates are used to select memories only IO/$\overline{M}$ is active. (OR gate is an AND gate for negative logic). Chip #0 will be selected only when IO/$\overline{M}$ = 0 and A_{11} = 0. To deselect remaining chips A_{12}, A_{13}, A_{14}, A_{15} must be '1'. Hence addresses will be

Chip #0

	A15															A0
Starting Address	1	1	1	1	0	0	0	0	0	0	0	0	0	0	0	0
Last Address	1	1	1	1	0	1	1	1	1	1	1	1	1	1	1	1

Starting Address *i.e.,* F 0 0 0 h

Last Address *i.e.,* F 7 F F h

Chip # 1

Starting Address	1	1	1	0	1	0	0	0	0	0	0	0	0	0	0	0
Last Address	1	1	1	0	1	1	1	1	1	1	1	1	1	1	1	1

Starting Address *i.e.,* E 8 0 0 h

Last Address *i.e.,* E F F F h

Chip # 2

Starting Address	1	1	0	1	1	0	0	0	0	0	0	0	0	0	0	0
Last Address	1	1	1	0	1	1	1	1	1	1	1	1	1	1	1	1

Starting Address *i.e.,* D 8 0 0 h

Last Address *i.e.,* D F F F h

Chip #3

Starting Address	1	0	1	1	1	0	0	0	0	0	0	0	0	0	0	0
Last Address	1	0	1	1	1	1	1	1	1	1	1	1	1	1	1	1

Starting Address *i.e.,* B 8 0 0 h

Last Address *i.e.,* B F F F h

Chip #4

Starting Address	0	1	1	1	0	0	0	0	0	0	0	0	0	0	0	0
First Address	0	1	1	1	1	0	0	0	0	0	0	0	0	0	0	1

Starting Address *i.e.,* 7 8 0 0 h

First Address *i.e.,* 7 F F F h

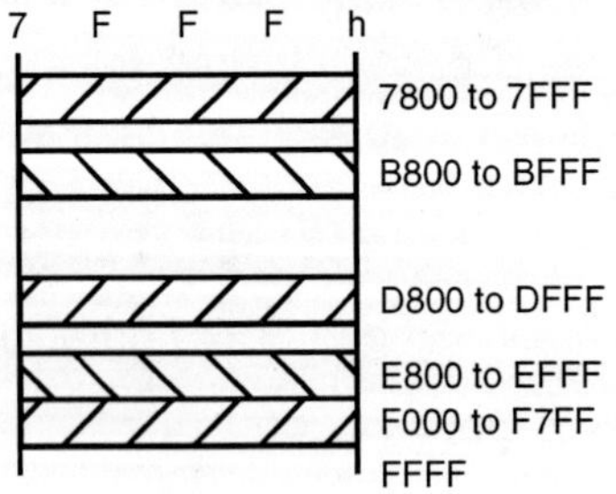

Fig. 2.45. Memory Space Chart (Interfacing circuit is shown in Fig. 2.44)

Interfacing Memories At Fixed Addresses

Many times there is requirement of connecting a memory chip in a particular place of memory map. Suppose it is derived to connect a chip of 4k × 8 from B 0 0 0 h to B F F F h. Then it is seen that this address has constant part of 1011 i.e., B is starting and last address of chip. So, this 1011 should be hardwired as below to generate chip select.

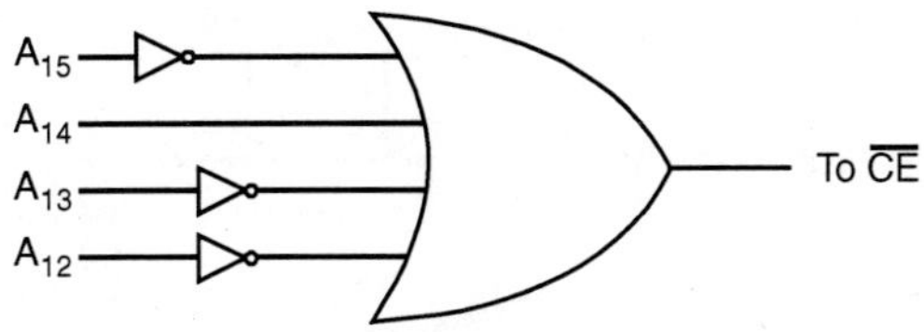

Fig. 2.46. Interfacing 2K × 8 Chips Using Linear Select Scheme.

Such a scheme will have the disadvantage that if at later stages, the space allowed to memory is to be changed, then there will be change in hardware design. So it is desirable to make a scheme in which such changes are possible.

Such a scheme in this case will use a 4 to 16 line decoder.

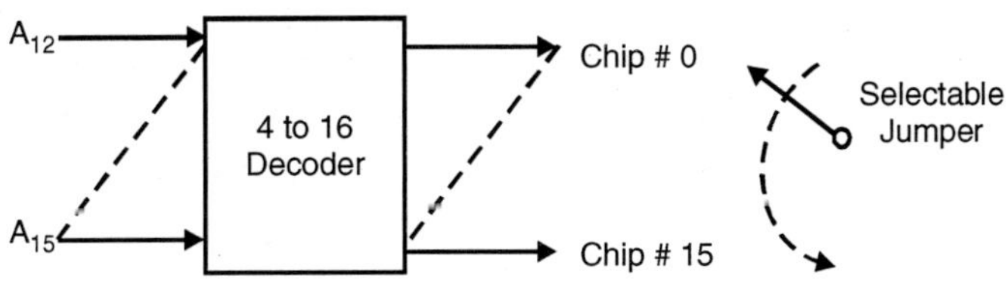

Fig. 2.47.

A selectable jumper can be connected to choose the desired place at which memory is to be located. Such a scheme will need no change in hardware except change in jumper for changing the position of chip address in memory space.

Addressing Memory More Than 64 K

With a microprocessor of 16 address lines (A_0 – A_{15}), 64k of total memory could be addressed. In case the memory to be addressed, is say 12k, so an additional line A_{16} will be needed. This additional line can be obtained from one of the output port of microcomputer. The simple scheme to address 128k is as under :

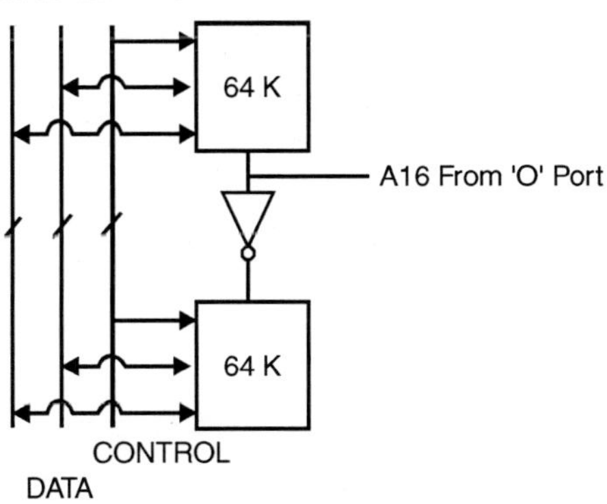

Fig. 2.47. Addressing more than 64K

A_{16} on logic '0' will select upper 64 k bank and A_{16} on logic '1' will select lower 64 k bank.

Using A_{16} and A_{17} from '0' port, and a 2 to 4 line decoder IC, four separate banks, each of 64 k could be addressed, thereby addressing memory of 256 k bytes.

Interfacing Circuit

We will show an interfacing circuit for 2732 EPROM with a 3 : 8 decoder, as shown below in Fig. 2.48.

We assume that the chip has already been programmed. The analysis of the circuit is a stated below :

1. The 8085 address lines A_{11}–A_0 are connected to pins A_{11}–A_0 of the memory chip to add 4096 register.
2. The decoder is used to decode four address lines A_{15}–A_{12}. The output 00 of the decode is connected to chip enable ($\overline{CE}$).

   The $\overline{CE}$ is asserted only when the address on A_{15}–A_{12} is 0000; A_{15} (low) enables the decodes and the input 000 asserts the output 0.
3. For this EPROM, we need one control signal ($\overline{MEMR}$) memory read, active low.

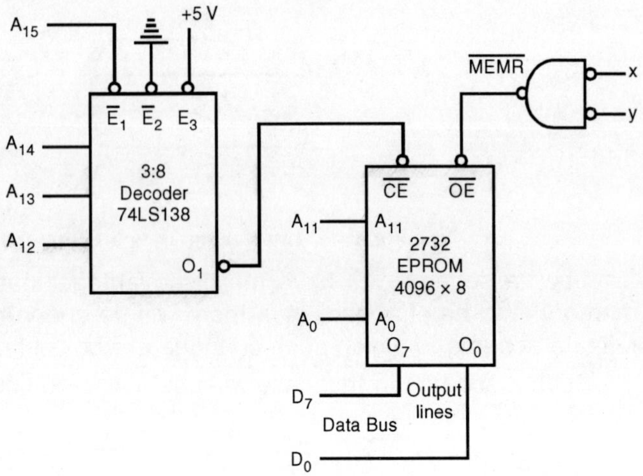

Fig. 2.48. Interfacing circuit for 2732 EPROM with a 3:8 Decoder

2.16. DATA TRANSFER AND DEVICE SELECTION

The process of data transfer is done by using an output device. We interrupt the device to see the result. The OUT instruction solves the purpose. During M3 cycle of the OUT instruction the peripheral places the information on the data bus.

If the data bus is connected to the latch during this period the result can be displayed through the devices such as LED's Printers.

The steps done during this process are :

1. Decode the address bus to generate a address on the bus, this is referred to as I/O address pulse.
2. Combine the device address pulse with the control signal to generate a device select pulse.
3. Use the I/O select pulse to activate the interfacing device.

I/O interfacing can be shown with the help of the following block diagram in Fig. 2.49.

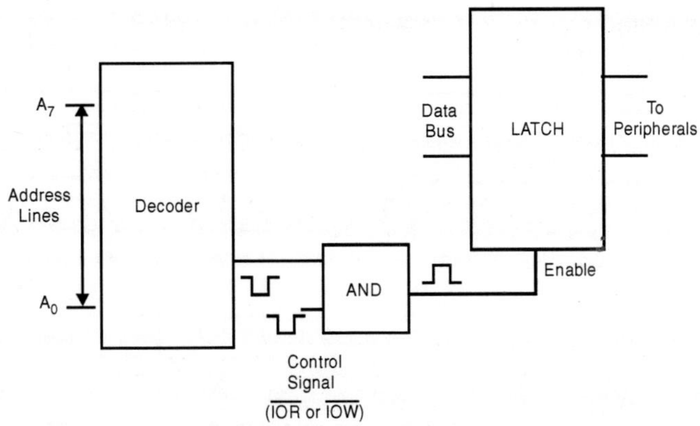

Fig. 2.49. I/O interfacing

The address lines A_7–A_0 are connected to a decoder, which will generate a unique pulse corresponding to each address on the address lines. This pulse is then combined with the control signal as shown above in the Fig. to generate a device select pulse.

Decode Logic for LED Output Port

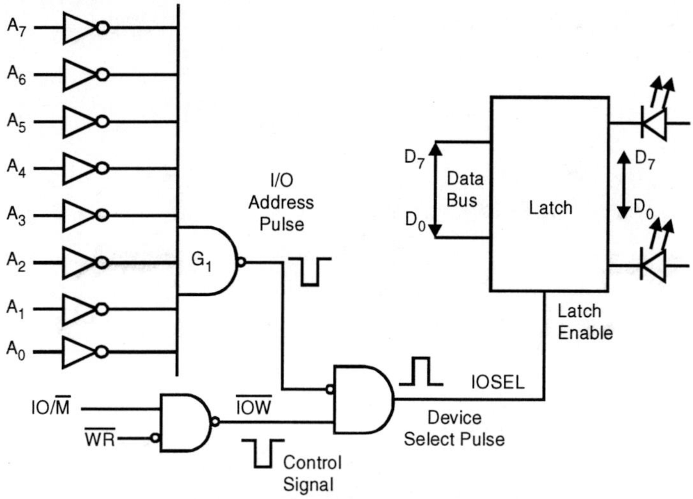

Fig. 2.50. Decoding circuit for the output device.

The above Fig. 2.50 shows the decoding circuit for the output device with address O/H. (A_0–A_7) lines are connected to the 8-input NAND gate that functions as a decoder.

Gate G1 generates a low pulse when the address bus carries address O/H, else the output remains high. Gate G2 combines the O/P of G1 and $\overline{IOW}$ to generate a I/O select pulse. The I/O select pulse clock the data into the latch for display by the LED's.

2.16.1. Input Interfacing

For explaining the term input interfacing we will see the example of 8-key I/P port as shown below in Fig. 2.51.

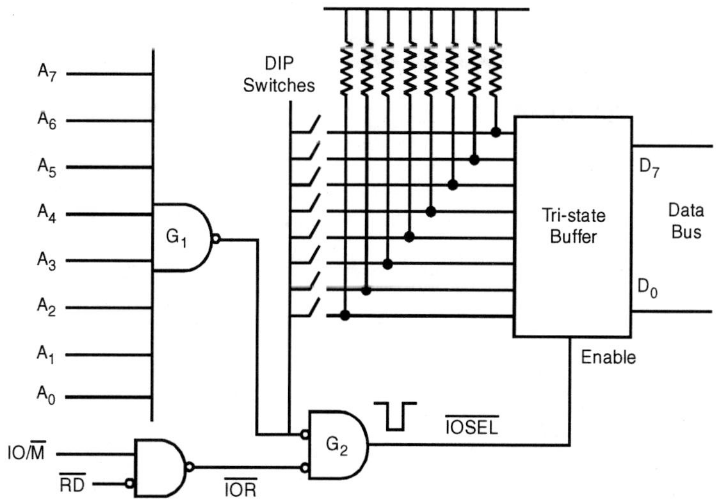

Fig. 2.51. 8-key I/P port interfacing.

We use an 8-input NAND gate to decode the address lines. When A_7–A_0 are high the O/P of NAND gate goes low and is combined with control signal $\overline{IOR}$ in gate G_2. G_2, then generates the device select pulse that is used to enable the tri-state buffer. The data from the keys are put on the data bus D_7–D_0 and loaded into the accumulator.

2.17. MEMORY MAPPED I/O

The input and output devices in memory mapped I/O are identified by 16-bit address. To transfer data between the MPU and IO devices, memory control signals, memory related instructions are used.

The memory mapped I/O technique is similar in many ways to the peripheral I/O technique. This can be shown with the following example :

Address	Code	Mnemonics
2000	32	STA 6000H
2001	00	
2002	60	

The STA is a 3-type instruction, and here the 16-bit address 6000H is entered in the reverse order; the low order-byte 00 is stored in location 2001, and 6000 H to the next location 2002.

Here if we use an output device to display the result i.e. the contents of accumulater register are transffered to the O/P device, this is reffered to as the memory-mapped I/O technique.

2.17.1. Execution of Data Transfer Instruction from the Memory

Here we will see the previous example related in memory mapped I/O and try and see the timing diagram for the instruction.

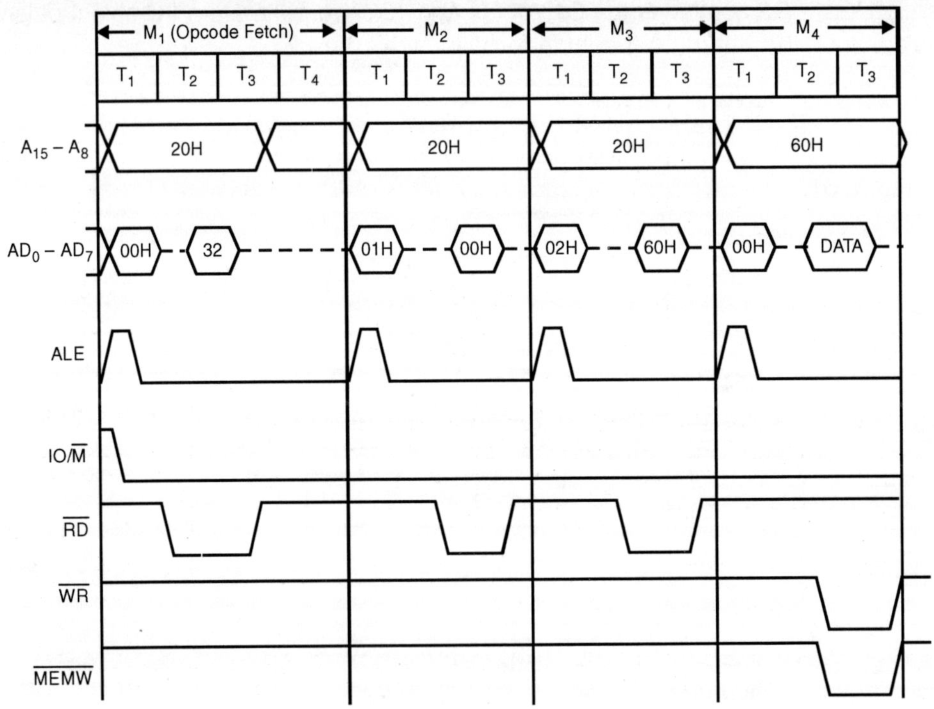

Fig. 2.52. Timing diagram of data transfer instruction from the memory.

As shown in the diagram 8085 requires 4-machine cycles (13T-states) to execute the instruction "STA 6000H"

In the 4th machine cycle (M4), the microprocessor places memory address 6000 on the entire address bus (A_{15}–A_0). The accumulator contents are sent on the date bus, followed by the control signal $\overline{MEMW}$.

2.18. INSTRUCTION FORMAT

An instruction is a binary pattern designed inside a microprocessor to perform a specific function. The entire group of instructions, is called the instruction set. The instruction set determines what functions the microprocessor can perform.

- Each instruction basically consists of 2 parts. The 1st is called as the OPCODE and the 2nd part is called as the OPERAND.
- The OPCODE part of an instruction specifies the operation to be performed. The operands either provides the data or specific data. The operand can be specified in a number of ways. It includes :
  1. 8-bit/16-bit internal general purpose register.
  2. 8-bit/16-bit port address
  3. 8-bit/16-bit memory address
  4. A memory location.
  5. Implicit operand : The operand is not specified, instead it is assumed in register.

Depending on the number of bytes or bit patterns required to specify on operation and the data, the instructions are of 3-type :

1. 1-byte instruction
2. 2-byte instruction
3. 3-byte instruction

1-Byte Instruction : A 1-byte instruction includes the opcode and the operand in the 8-bit only i.e. one byte. The 1-byte instruction specifies the operation to be performed and who is going to perform it. These type of instruction require 1 (single) memory location.

Format :

0 7
OPCODE

For example : MOV, B, C, whose opcode is 41 H which is 1-byte. This instruction copies the contents of register C in B.

2-Byte Instructions : The 2-byte instruction uses 1st byte to specify the operation i.e. opcode and 2nd byte to specify the operand. These instructions require 2-successive locations in the memory. The 1st byte stored is the opcode and the 2nd byte stored is data i.e. operand.

Format :

0 7	15
OPCODE	OPERAND

For example : MVI B, 57H

The 1st byte of this instruction is the opcode for instruction MVIB and second byte is the data that is to be moved to register B.

3-Byte Instructions : The 3-byte instruction use 1st byte to specify the operation i.e. opcode, 2nd and 3rd bytes are used to specify the operand. Generally these instructions are used to specify 16-bit data or memory address with the instruction.

1-byte stored is opcode, 2nd byte stored is lower order 8-bits of 16-bits operand or address and 3rd-byte stored is higher order 8-bits of 16-bits operand or address.

Format :

0 7	15	23
OPCODE	OPERAND	OPERAND

For example : STA 5600H

The 1st byte of this instruction is the opcode for instruction STA 5600H and the second and 3rd bytes are the address 5600H.

OPCODE Formats : The 8085 microprocessor is on 8-bit microprocessor has 8-bit opcodes. Each instruction has a unique opcode. The opcode contains information regarding the operation, type of operation to be performed, registers to be used, flags. The opcode is fixed for each instruction.

Notations used in object code or OPCODES are as shown in the table below :

Notations	Meaning
ddd	Destination Register (S) Source Registers ddd = sss = 111 = A Register 000 = B Register 001 = C Register 010 = D Register 011 = E Register 100 = H Register 101 = L Register
nnn	Restart number 000 to 111
yy	An 8-bit binary data
yyyy	A 16-bit binary data unit
X	Register pair 0 = BC 1 = DE
XX	Register pair 00 = BC 01 = DE 10 = HL 11 = SP (IF Push/Pop) PSW
PPQQ	A, 16-bit memory address

2.19. INSTRUCTION SET

Instruction are generally classified into the functional categories as follows :

1. Data Transfer Group
2. Arithmetic Group
3. Logical Group
4. Branching Group
5. Stack and Machine Control Group

Notation Used in Instructions

Notations used in instructions are :

Notations	Meaning
M	Memory location pointed by HL pair.
r	8-bit register
RS	Source register
RD	16-bit register pair
RP	Destination register
Addr	16-bit address
Data 16-bit	16-bit data
Data	8-bit data

2.19.1. Data Transfer Group

(1) Mov r1, r2 : This group of instructions transfer data to and from register and memory. Conditional flags are not affected by any instruction in this group.

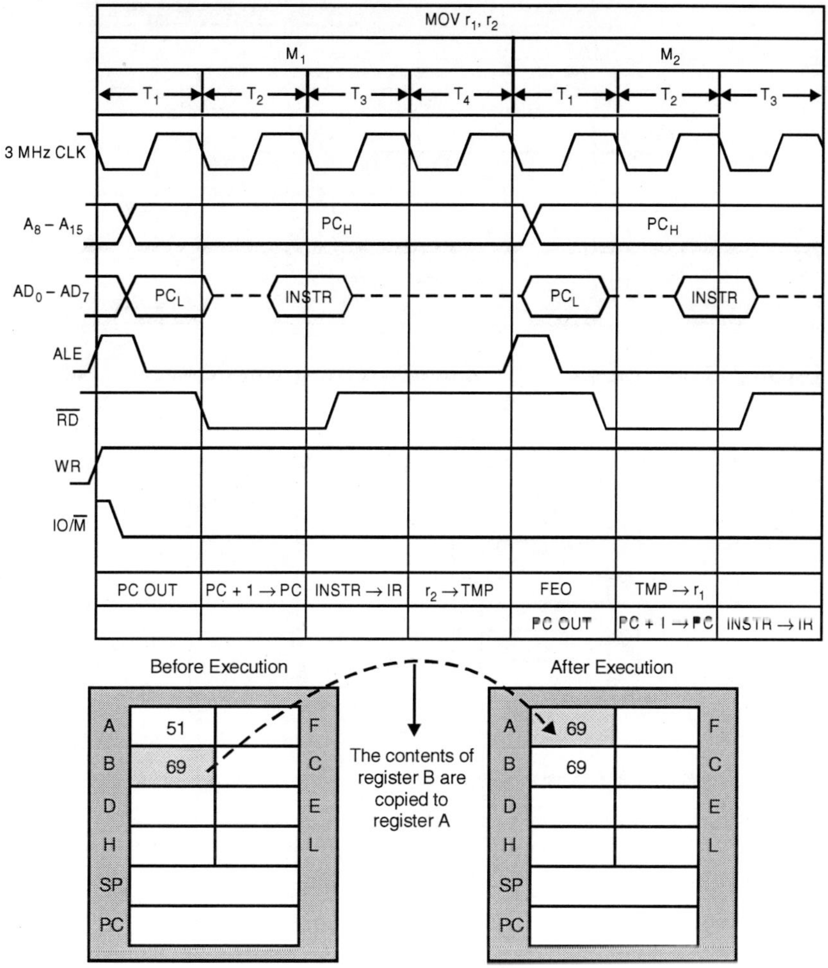

Fig. 2.53. Mov A, B

Mnemonic : Mov R1, R2

This instruction copies data from source to destination register. The source or destination can be any general purpose register such as A, B, C, D, E, H, L. The contents of source register remains unchanged.

Number of Bytes Used : 1-byte

Flags : No flags are modified

Addressing Mode : Register addressing mode.

Example : Mov A, B

The above instruction copies the contents of register B to the register A.

Suppose A = 51H, B = 69H, after execution of the instruction we get

A = 69H and B = 69H as shown in the Fig. 2.53.

0	1	D	D	D	S	S	S

(2) Mov r, M : The content of memory location, whose address is in register H and L, is moved to register r.

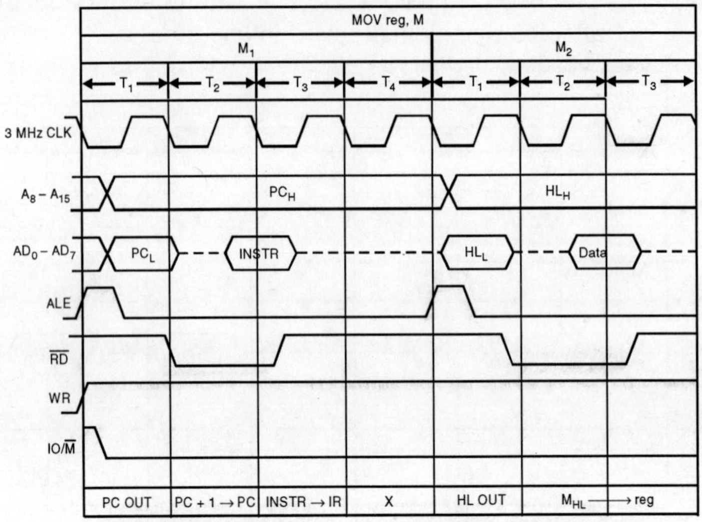

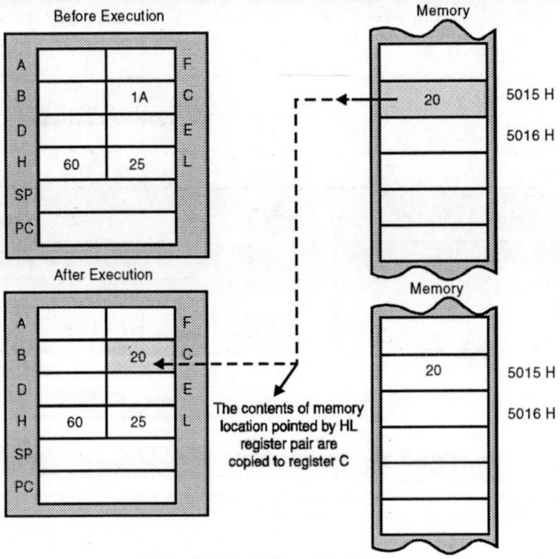

Fig. 2.54. Mov C, M

Mnemonic : Mov R, M

Operation : R = M, or R = (HL). This instruction copies data from memory M to register R. The contents of HL register pair are used as address of the memory location. The contents of the memory location pointed by HL register pair are transferred to specific register R.

Register R can be any general purpose register such as A, B, C, D, E, H, L

Number of Byte Used : 1-byte

Flags : No flags affected

Addressing Mode : Direct addressing mode.

Example : Mov, C, M

Suppose the contents of HL register pair be 6025H the contents of memory location 6025H = 20H. After the execution of the instruction the contents of memory location pointed by the HL register pair i.e. 6025H are copied to the register C as shown in the Fig. 2.54.

0	1	D	D	D	1	1	0

(3) Mov M, r : The content of register r, is moved to the memory location whose address is in register H and L.

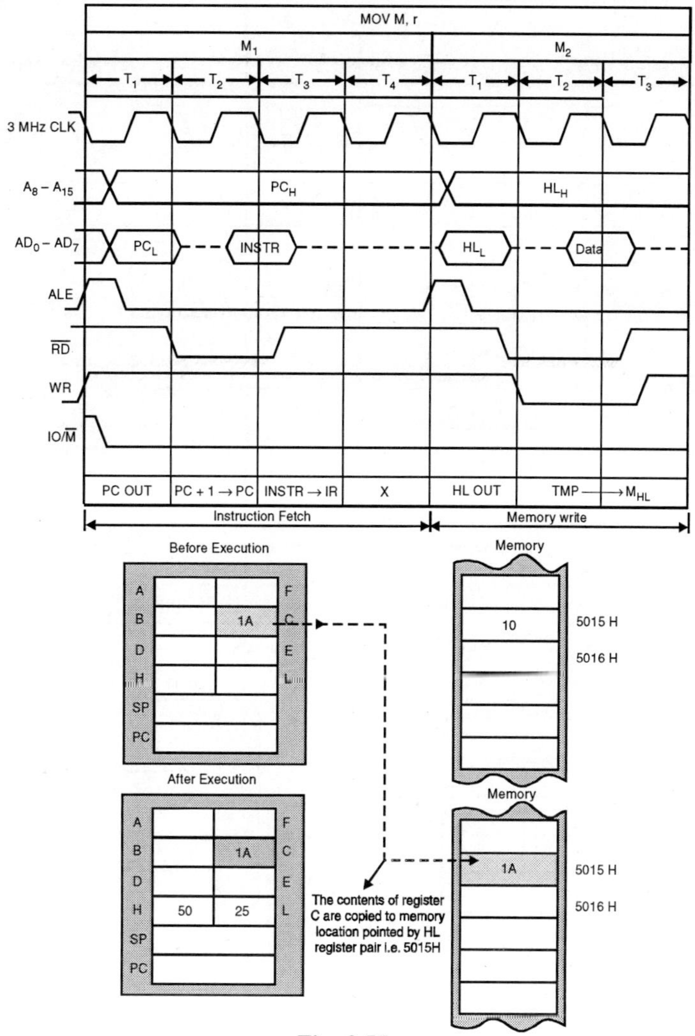

Fig. 2.55.

Mnemonic : Mov M, R

Operation : The above instruction copies data from register R to memory M. The HL register pair is used as memory pointer. The contents at HL register pair are used as memory address and the contents of specified register are transferred to that memory location.

Register R can be any 8-bit G.P.R, A, B, C, D, E, H and L.

Number of bytes Used : 1-byte

Addressing Mode : Indirect addressing mode

Flags Affected : None

Example : Suppose the contents of HL register pair be 5015H, the contents of register C = 1 AH. After execution of the instruction, Mov M, C the contents of register C, i.e. 1AH will be copied to the memory location pointed by the HL register pair i.e. 5015H, as shown in Fig. 2.55.

0	1	1	1	0	S	S	S

(4) MVI r, Data : The content of byte-2 of instruction is moved to register r.

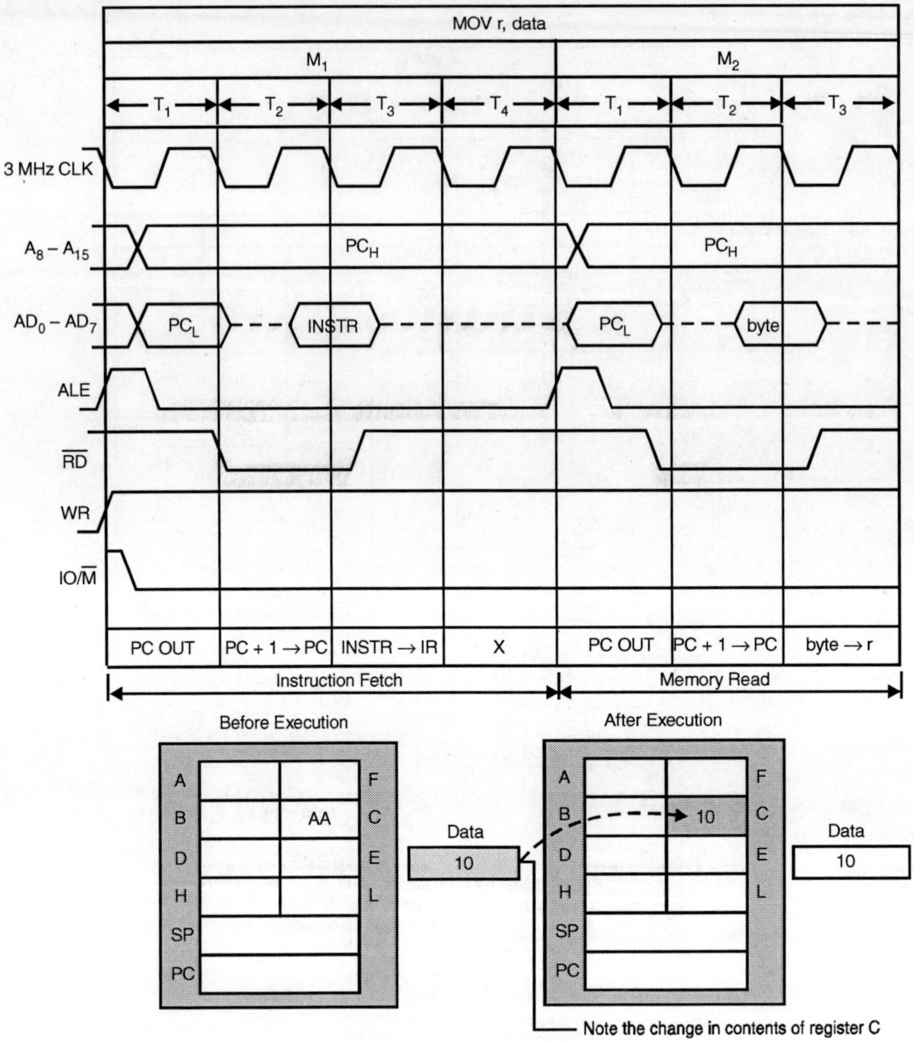

Fig. 2.56. MVI C, 10h

Mnemonic : MVI R, Data

Operation : The above instruction moves the immediate 8-bit data to the specified register in the instruction. It is a 2-byte instruction. The register R can be any 8-bit G.P.R, like A, B, C, D, E, H and L.

Number of Bytes : 2-byte

Addressing Mode : Immediate addressing mode

Flags Affected : None

Example : MVI C, 10H

This instruction will load immediate data 10H in the register C as shown in Fig. 2.56.

0	0	D	D	D	1	1	0
Data							

(5) MVI M, Data : THe content of byte-2 of the instruction is moved to the memory location whose address is in register H and L.

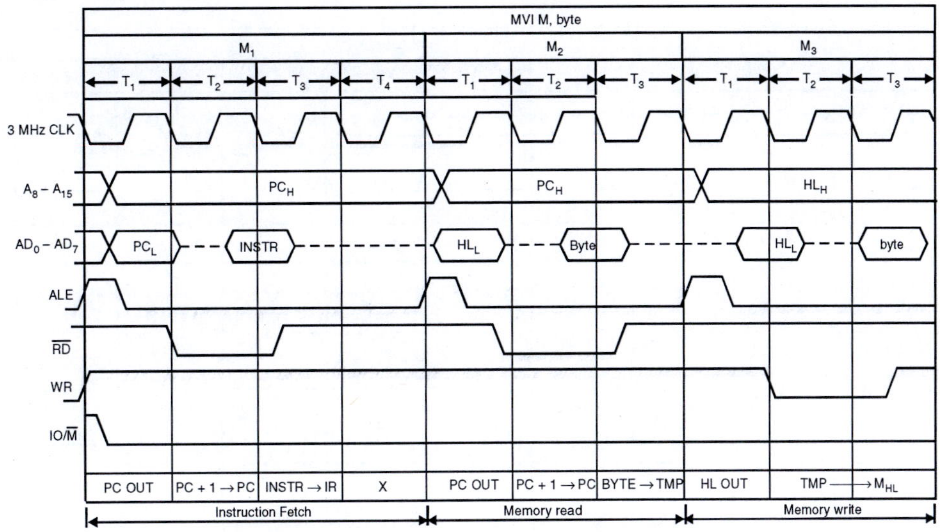

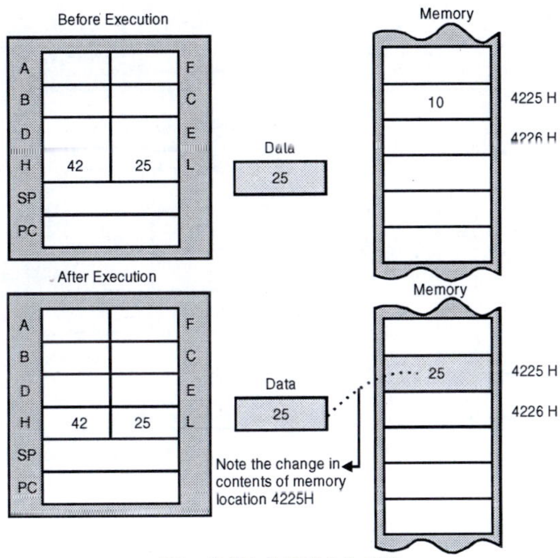

Fig. 2.57. MVI M, 25h

Mneonic : MVI M, Data

Operation : The above instruction move immediate data to memory. The HL register pair used as memory pointer. The contents of HL register pair are used as memory address and the immediate data is transferred to that memory location.

Number of Bytes : 2-byte

Addressing Mode : Immediate addressing mode

Flags Affected : None

Example : MVI M, 25H

Suppose H = 42H, L = 25H, The instruction will store data 25H, in memory location addressed by HL pair i.e. 4225H as shown in Fig. 2.57.

0	0	1	1	0	1	1	0
Data							

(6) LXI, rp, Data-16 : Byte 3 of the instruction is moved into the high order register (rh) of the register pair (rp) byte 2 of the instruction is moved into the low order register (rl) to the register pair (rp).

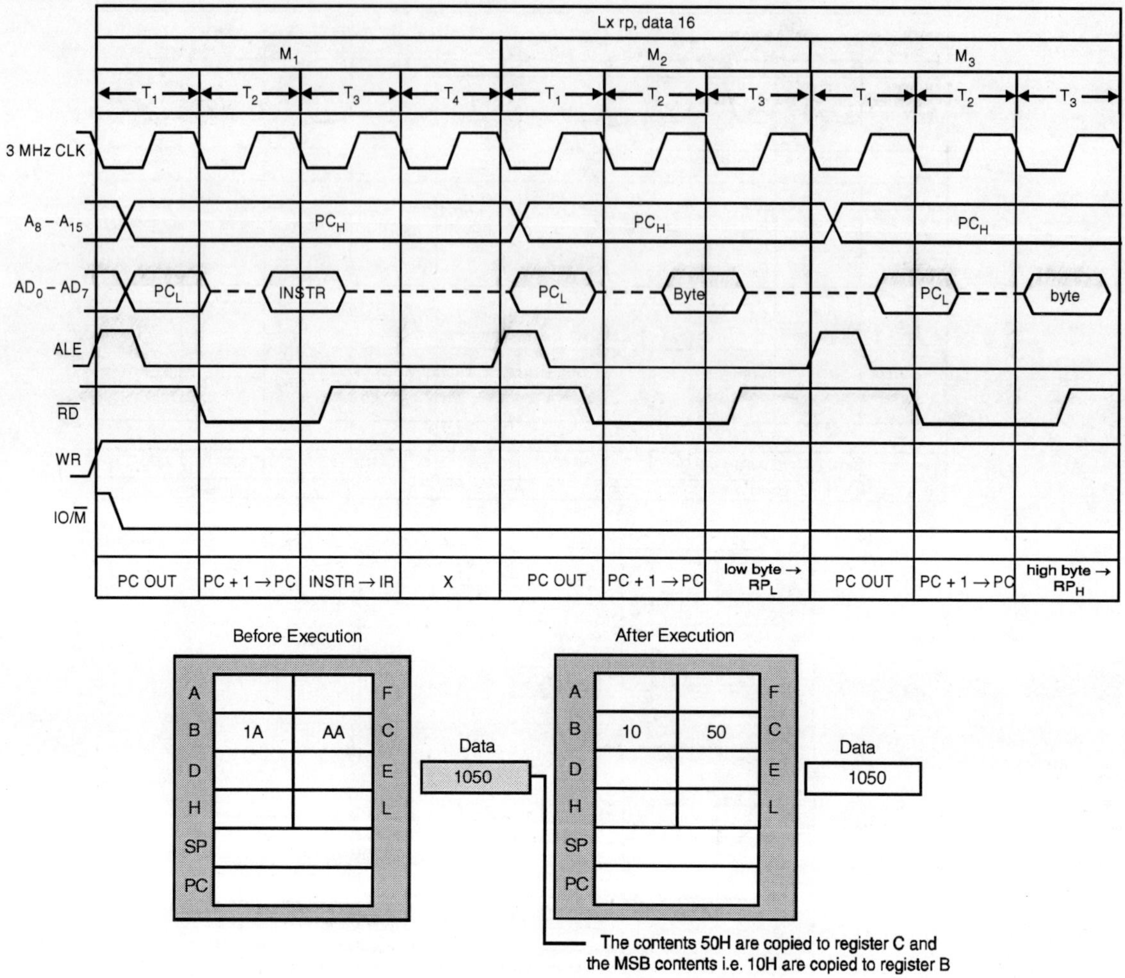

Fig. 2.58. LXI B, 1050H

Mnemonic : LXI rp, Data

Operation : This instruction loads 16-bit data specified with the instruction to the specified register pair or the stack pointer. The register pair can be any BC, DE, HL pair or stack pointer SP.

Number of Bytes Used : 3-byte

Addressing Mode : Immediate addressing mode

Flags Modified : None

Example : LXI B, 1050 H

The above instruction will load the BC register pair with data 1050H. 10H will be loaded in register B, and 50H in register C as shown in the Fig. 2.58.

0	0	R	P	0	0	0	1
Low-order data							
High order data							

(7) LHLD Addr (Load H and L Direct) : The content of the memory location whose address is specified in by byte 2 and byte 3 of the instruction is moved to register L. The content of the memory location at the succeeding address is moved to register H.

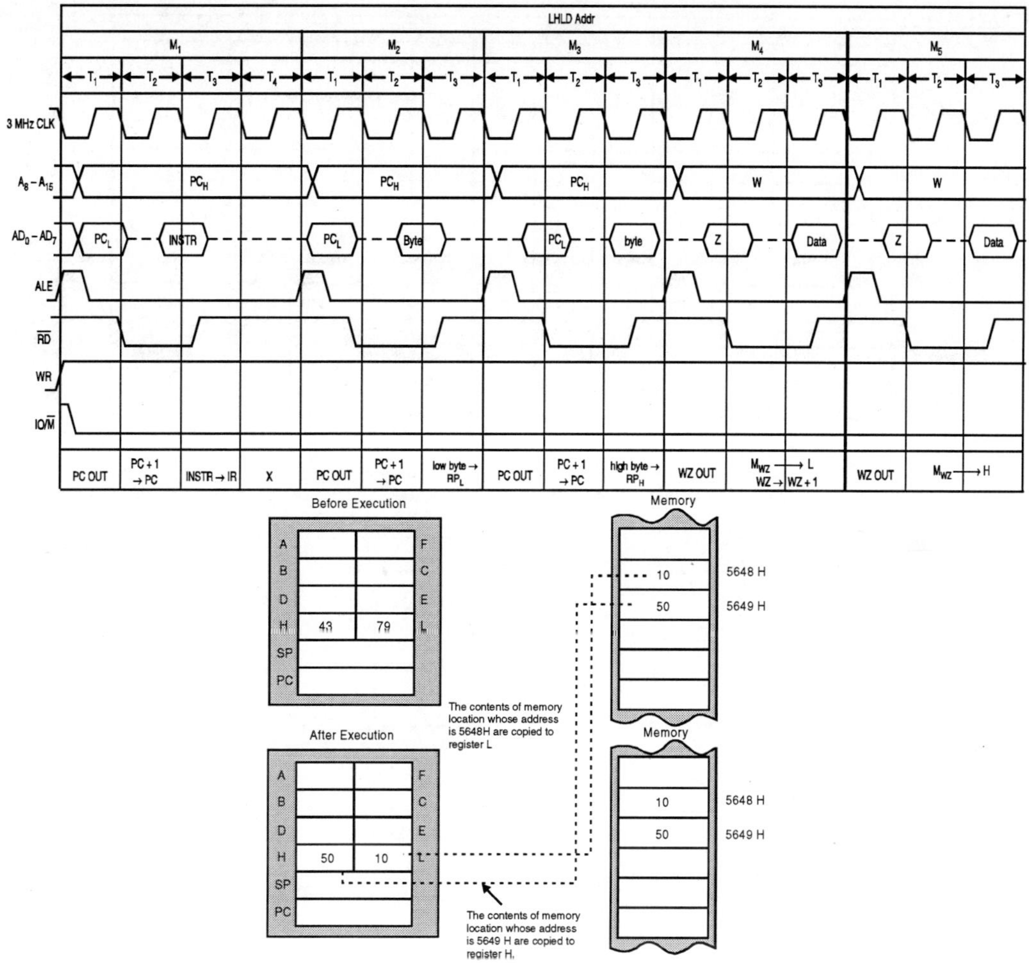

Fig. 2.59. LHLD 5648H

Mnemonic : LHLD address

Operation : This instruction copies the contents of memory location specified in instruction to the register H and L. The contents of memory location specified in the instruction are copied into the L register and the contents of the next memory location are copied to the H register.

Number of Bytes Used : 3-byte

Addressing Mode : Direct addressing mode

Flags Affected : None

Example : LHLD 5648H

Suppose H = 43H, L = 79H, and the contents of memory locations 5648 H and 5649h are 10H and 50H respectively.

The instruction LHLD 5648h will copy the contents of memory location 5648h (10H) to register L and contents of memory location 5649H (50H) to the register H as shown in Fig. 2.59.

<table>
<tr><td>0</td><td>0</td><td>1</td><td>0</td><td>1</td><td>0</td><td>1</td><td>0</td></tr>
<tr><td colspan="8">Low-order addr</td></tr>
<tr><td colspan="8">High-order addr</td></tr>
</table>

(8) STA Addr (Store Accumulator Direct) : The content of the accumulator is moved to the memory location whose address is specified in byte 2 and 3 of the instruction.

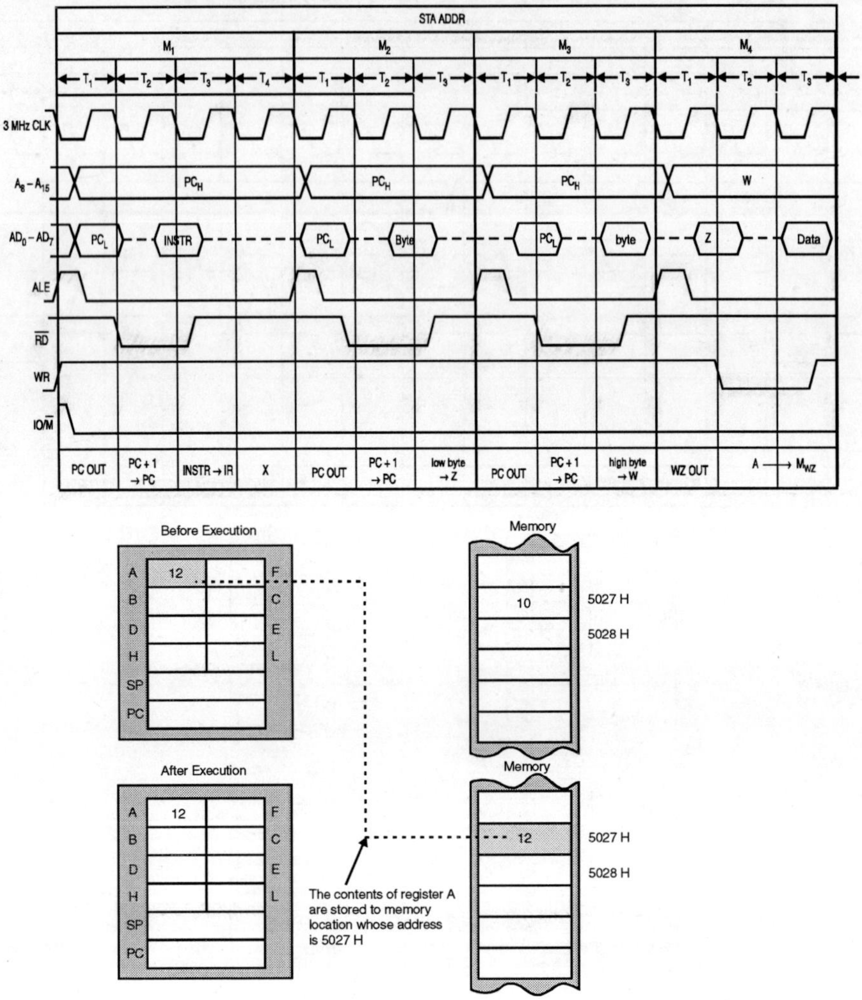

Fig. 2.60. STA 5027H

Mnemonic : STA address

Operation : This instruction copies the contents of the accumulator to the memory location whose address is specified in the instruction.

Addressing Mode : Direct addressing mode.

Flags Affected : None

Example : STA 5027H

The above instruction will load the contents of accumulator (12h) to the memory location specified in the instruction i.e. 5027H as shown in Fig. 2.60.

0	0	1	1	0	0	1	0
Low-order addr							
High-order addr							

(9) SHLD Addr (Store H and L Direct) : The content of register L is moved to the memory location whose address is specified in byte 2 and 3. The content of register H is moved to the succeeding memory location.

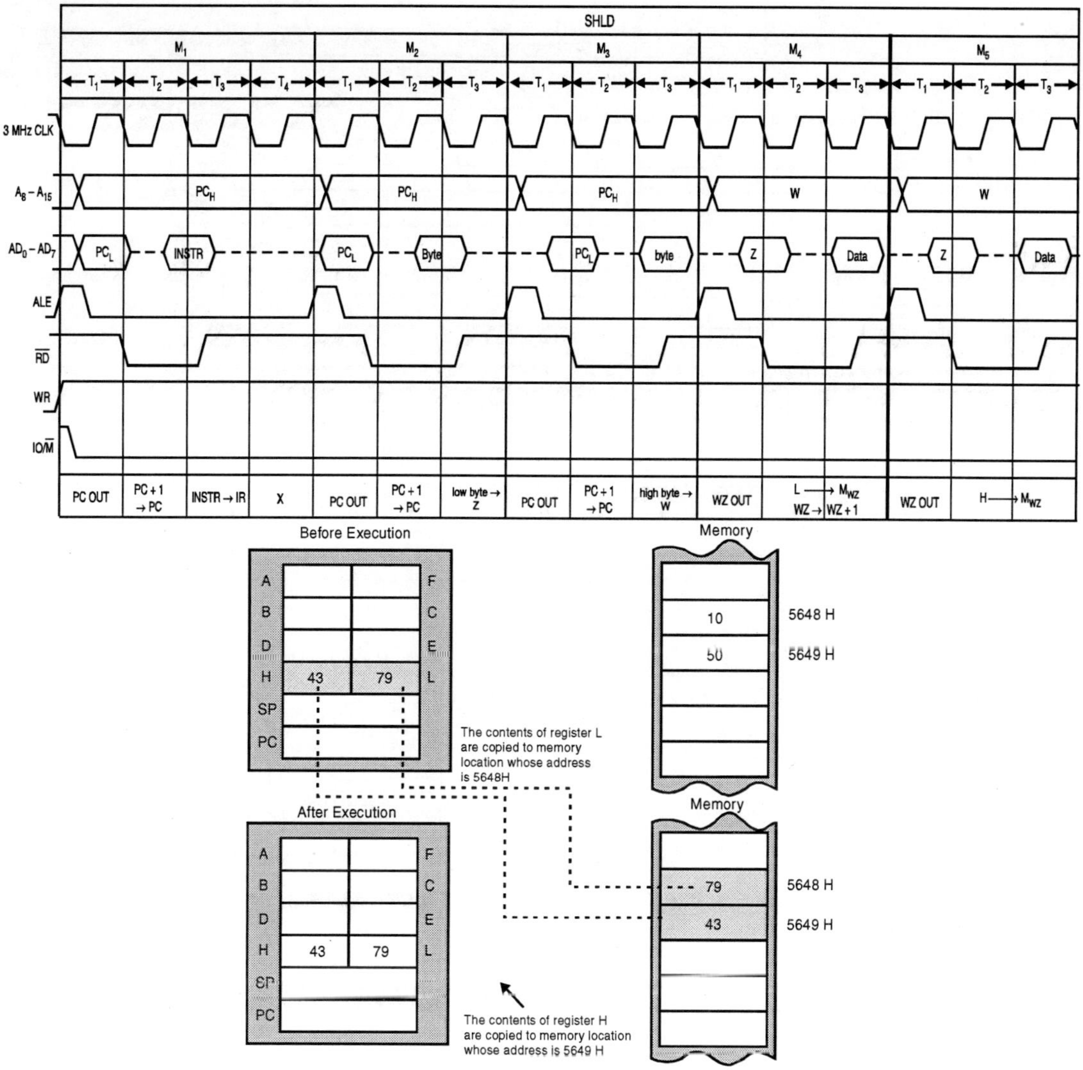

Fig. 2.61. SHLD 5648H

Mnemonic : SHLD address.

Operation : This instruction copies the contents of registers H and L to the memory location specified in instruction. The contents of L register are stored at the memory location whose address is specified and the contents of register H to the next memory location.

Addressing Mode : Direct addressing mode.

Flags Affected : None

Example : SHLD 5648H

Suppose H = 43H, L = 79H, the contents of memory locations 5648H and 5649H are 10H and 50H. The instruction SHLD 5648H, will copy the contents of register L (79H) to memory location 5648H and the contents of register H (43H) to memory location 5649H as shown in Fig. 2.61.

Format :

0	0	1	0	0	0	1	0
Low-order addr							
High-order addr							

(10) LDAX rp : (Load Accumulator Indirect) : The content of the memory location, whose address is in the register pair rp, is moved to register A.

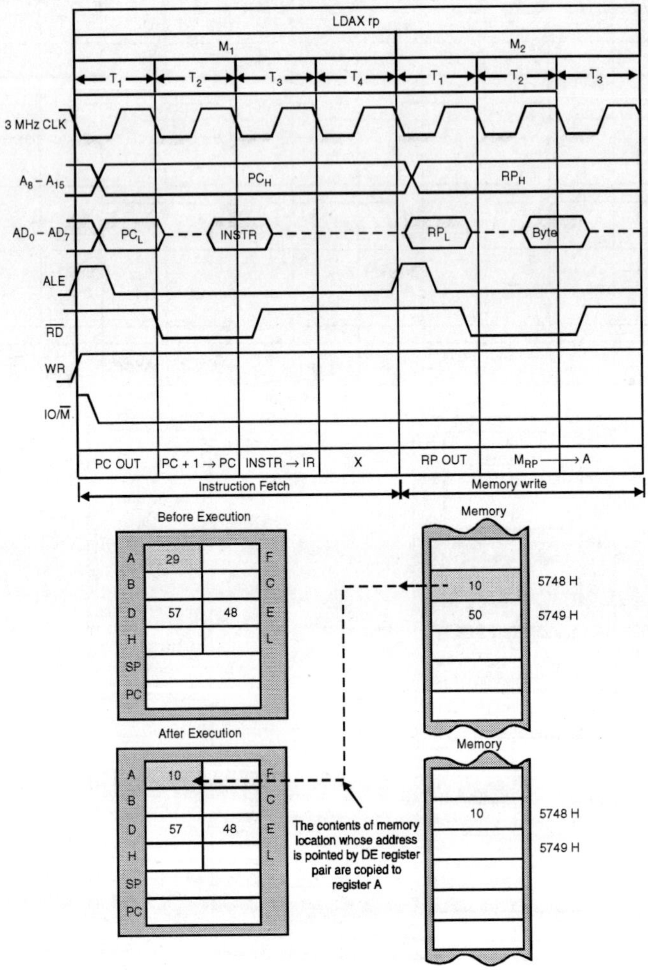

Fig. 2.62. LDAX D

Mnemonic : LDAX RP

Operation : This instruction copies contents of memory location whose address is pointed by the specified register pair to the accumulator. The register RP can be any valid register pair like BC or DE only. The contents of the memory location remain unchanged.

Addressing Mode : Register indirect addressing mode.

Flags Affected : None

Example : Suppose A = 29H, D = 57H, E = 48H.

The instruction LDAX D will copy the contents of the memory location, whose address is given by DE register pair i.e. 5748H. (10H) to the accumulator as shown in Fig. 2.62.

Format :

0	0	R	P	1	0	1	0

(11) STAX rp (Store Accumulator Indirect) : The content of register A is moved to the memory location, whose address is in the register pair RP.

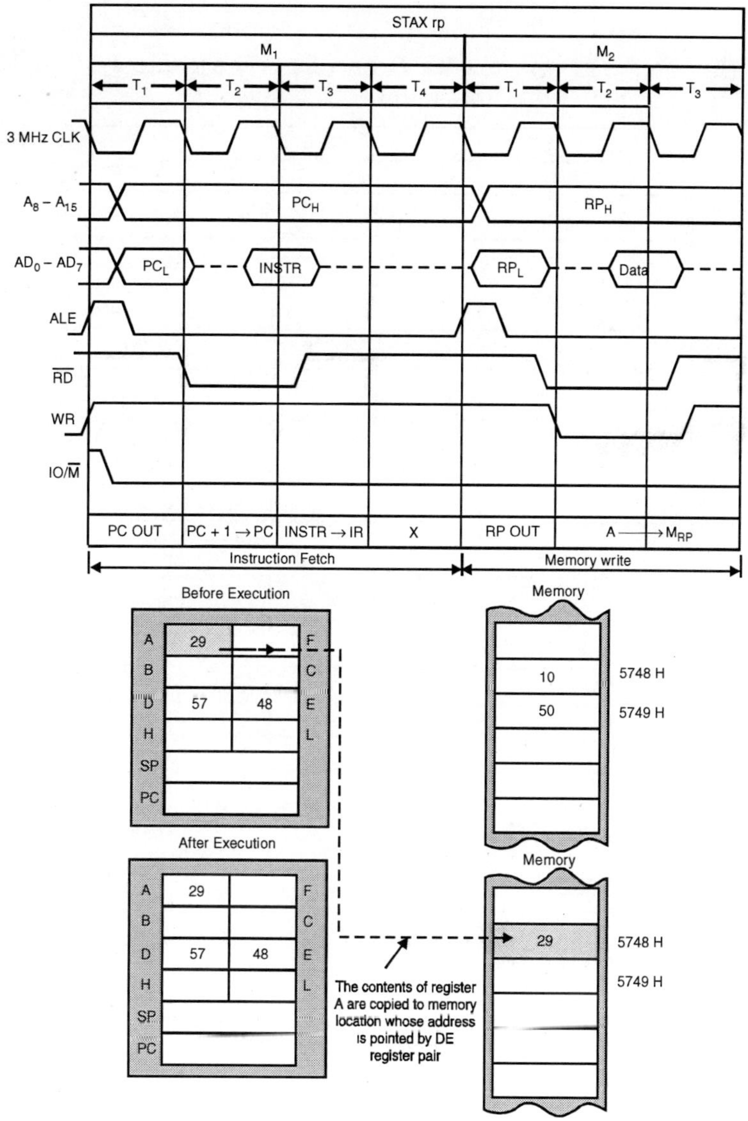

Fig. 2.63. STAX D

Mnemonic : STAX RP

Operation : The instruction copies the contents of the accumulator to memory location whose address is pointed by the specified register pair. The register RP can be a valid register pair like BC or DE only.

The contents of the accumulator remain unchanged.

Addressing Mode : Register indirect addressing mode.

Flags Affected : None

Example : Let A = 29H, D = 57H, E = 48H.

The instruction, STAX D will copy the contents of accumulator (29H) to the memory location, whose address is given by DE register pair i.e. 5748H as shown in Fig. 2.63.

Format :

0	0	R	P	0	0	1	0

(12) XCHG (Exchange H and L with D and E) : The content of register H and L are exchanged with the contents of register D and E.

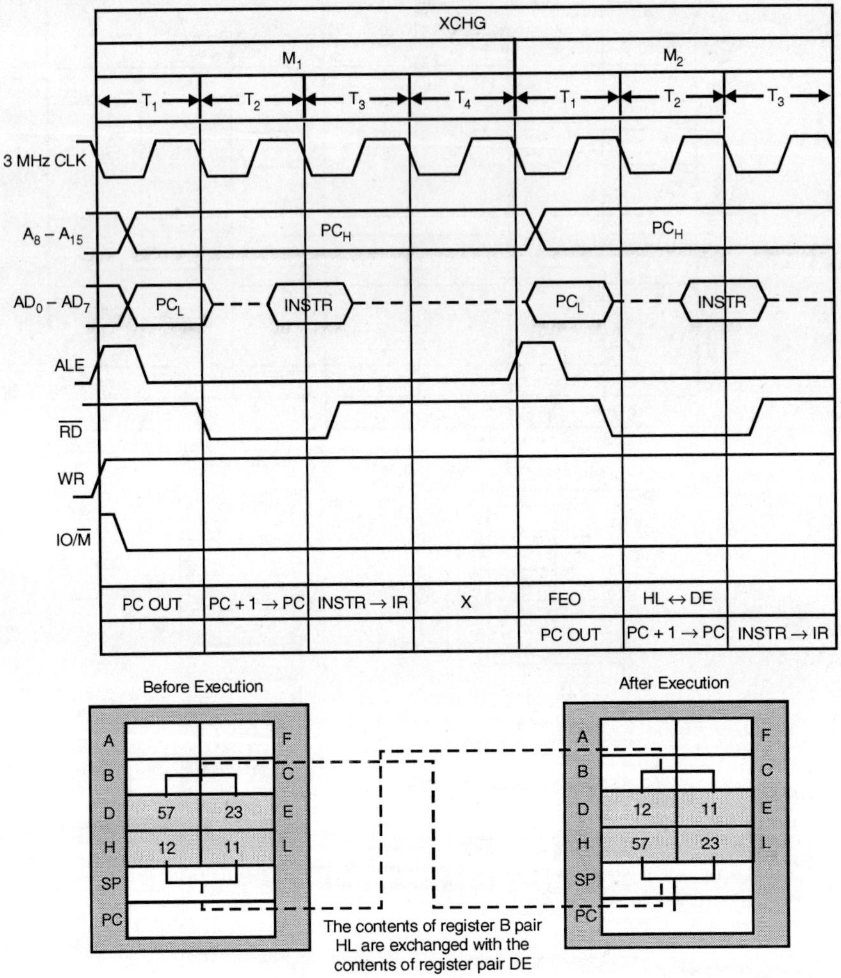

Fig. 2.64. XCHG

Mnemonic : XCHG.

Operation : The instruction exchanges the contents of register H with register D and register L with register E. The contents of register H are copied to the register D and the contents of register D are copied to register H. Similarly the contents of register L are copied to the register E and the contents of register E are copied to the register L.

Addressing Mode : Register mode.

Flags Affected : None

Example : Let H = 12H, L = 11H, D = 57H, E = 23H.

The instruction XCHG exchange the contents of register pair HL with the contents of register pair DE as shown in Fig. 2.64.

Format :

0	0	1	0	1	0	1	1

2.19.2. Arithmetic Group

This group of instructions perform arithmetic operations on data in register and memory. All instructions in this group affect the zero, sign, parity, carry and auxillary carry flags according to standard rules. All subtraction operations are performed via 2's complement arithmetic and set the carry flag to one to indicate a borrow and clear it to indicate no borrow.

(13) Add r (Add Register) : The content of register r is added to the content of the accumulator. The result is placed in the accumulator.

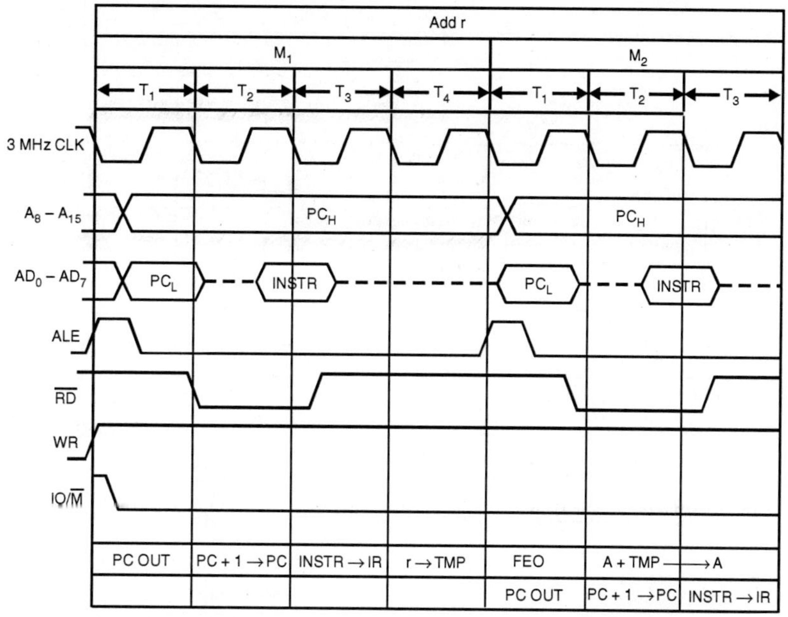

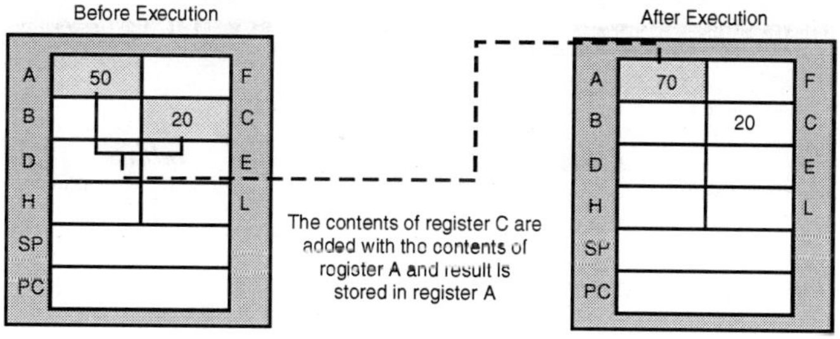

Fig. 2.65. ADD C

Mneomonic : Add R

Operation : This instruction adds the content of register R with the content of the accumulator. The result of addition is stored in the accumulator.

The register R can be any general purpose registers such as A, B, C, D, E, H and L.

Addressing Mode : Register addressing mode.

Flags Affected : None

Example : ADDC

Suppose A = 50H, C = 20H.

The instruction ADDC will add the contents of register C (20H) with the contents of the accumulator i.e. 50H. The result of addition (70H) will be stored in the accumulator as shown in Fig. 2.65.

Format :

1	0	0	0	0	S	S	S

(14) ADD M (Add Memory) : The content of the memory location whose address is contained in the H and L register added to the content of the accumulator. The result is placed in the accumulator.

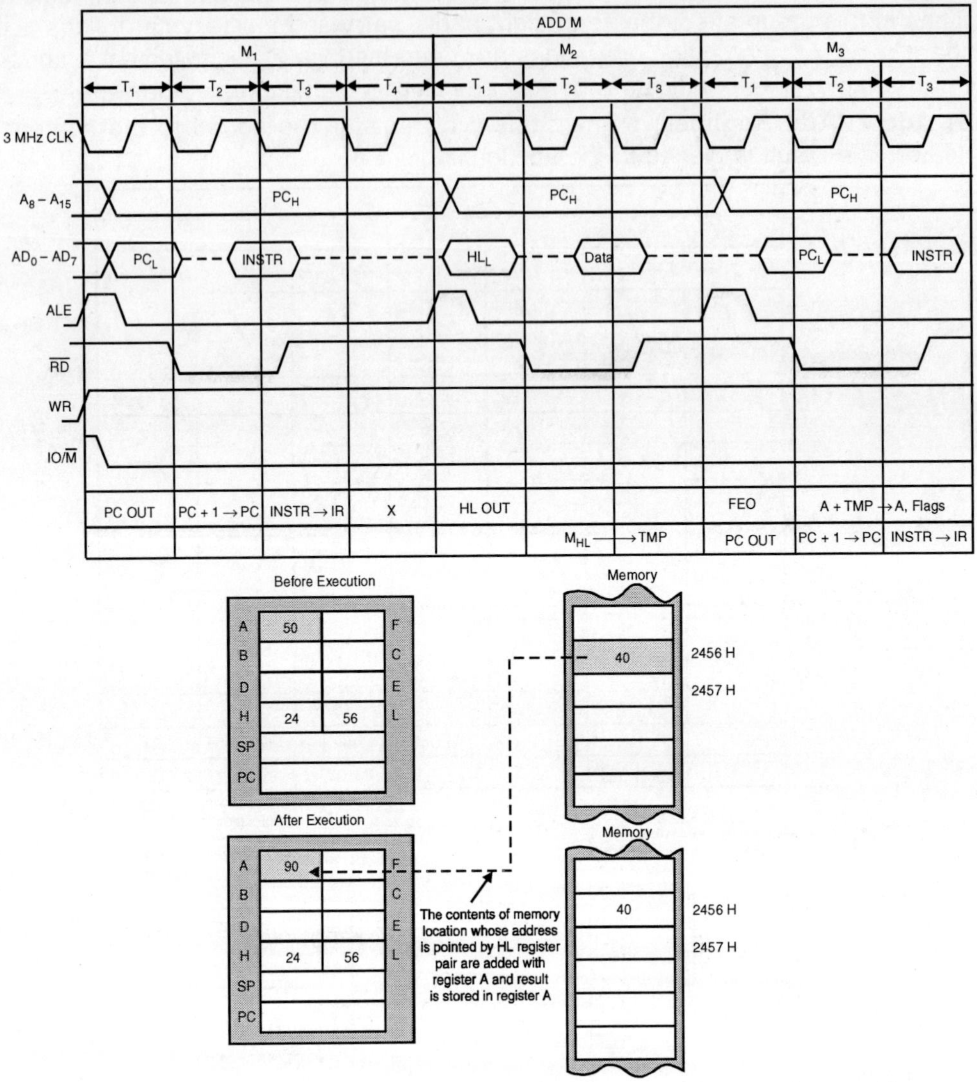

Fig. 2.66. ADD M

Mnemonic : ADD M

Operation : This instruction adds the contents of memory location pointed by the HL register pair with the contents to accumulator. The result is stored in accumulator.

Addressing Mode : Direct addressing mode

Flags Affected : All the flags

Example : Suppose A = 50H, H = 24H, L = 56H, contents ofthe memory location 2456H = 40H.

The instruction ADD M, will add the contents of memory location 2456 i.e. (40H) with the contents of the accumulator i.e. 50H. The result of addition (90H) will be stored in the accumulator as shown in Fig. 2.66.

Format :

1	0	0	0	0	1	1	0

(15) ADI Data (Add Immediate) : The content of the 2nd byte ofthe instruction is added to the content of the accumulator. The result is placed in the accumulator.

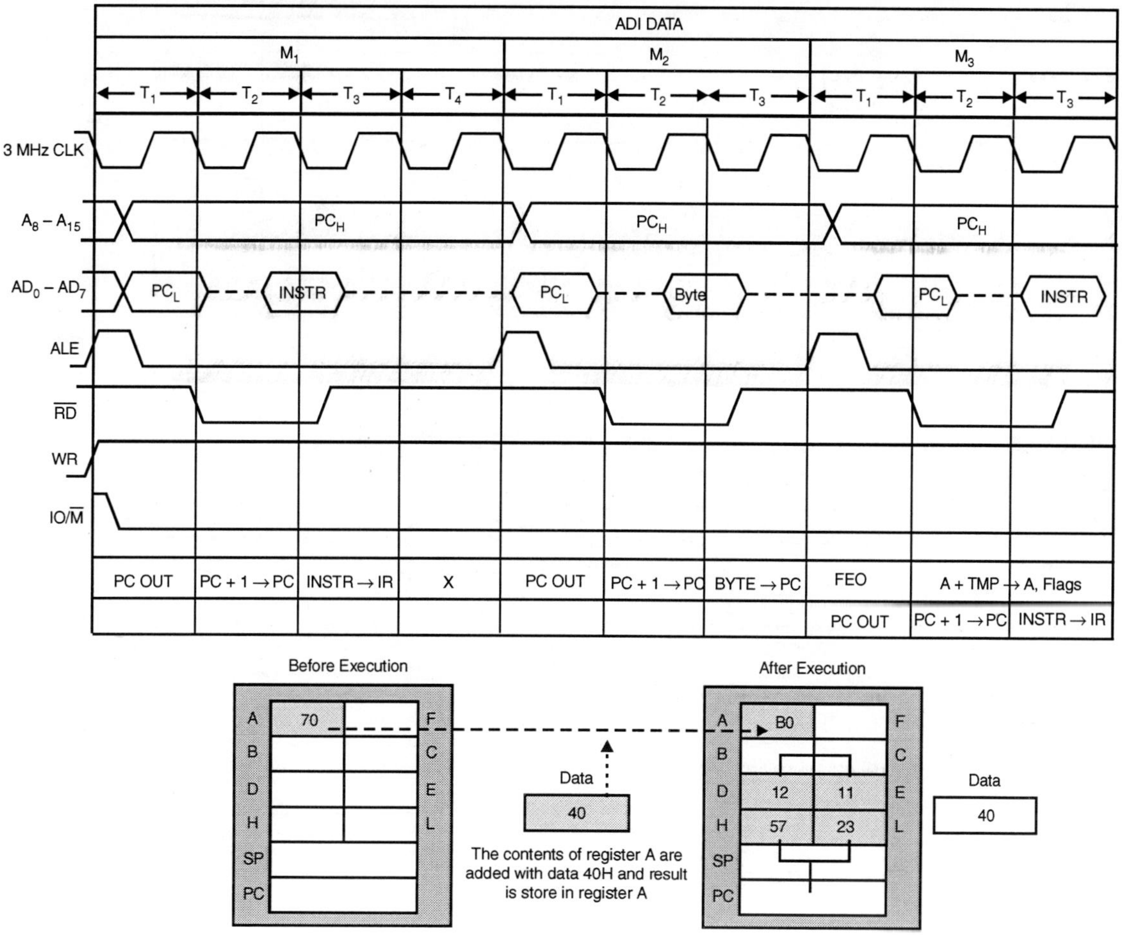

Fig. 2.67. ADI 40H

Mnemonic : ADI Data.

Operation : This instruction adds 8-bit data specified along with the contents of the accumulator. The result is in accumulator as shown in Fig. 2.67.

Addressing Mode : Immediate addressing mode

Flags Affected : All the flags affected

Example : ADI 40H

This instruction add the immediate data (40H) to contents of the accumulator. The result is in accumulator as shown in Fig. 2.67.

Format :

1	1	0	0	0	1	1	0
Data							

(16) ADC r (Add Register with Carry) : The content of register r and the content of the carry bit are added to the content of the accumulator. The result is placed in the accumulator.

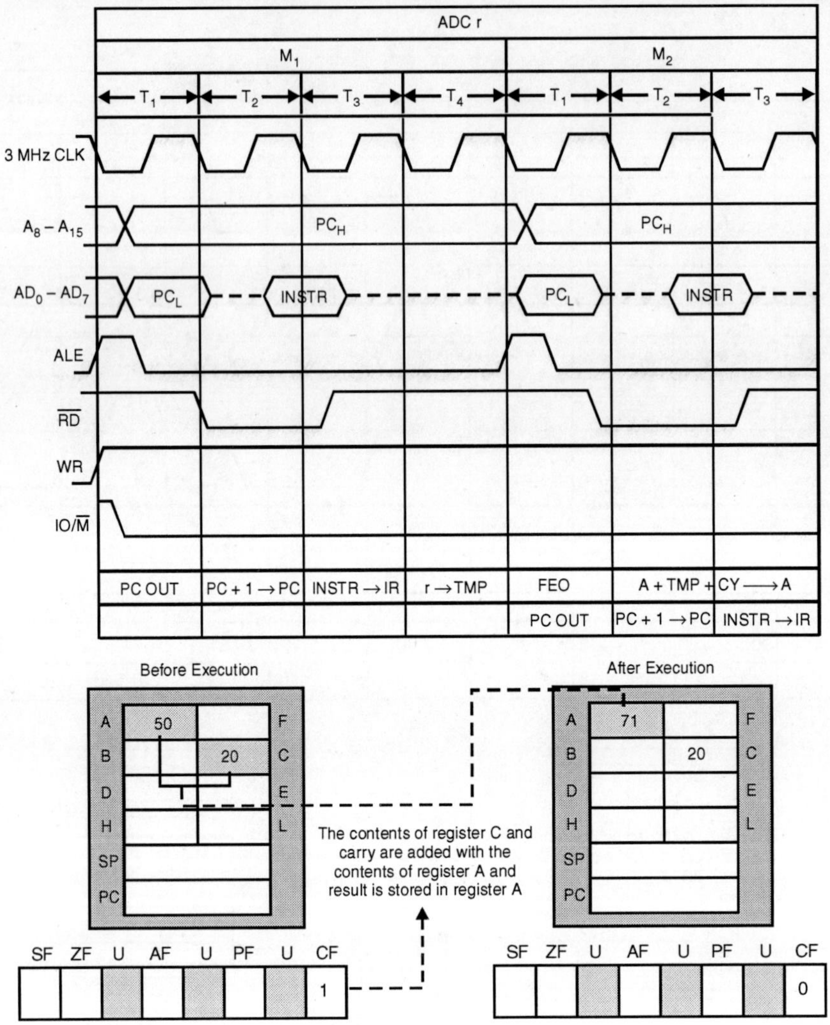

Fig. 2.68. ADC C

Menomic : ADC R

Operation : This instruction adds the contents of register R along with the carry to the content of the accumulator. The result of addition is stored in the accumulator. The register R can be any general purpose register such as A, B, C, D, E, H and L.

Addressing Mode : Immediate addressing mode.

Flags Affected : All flags are affected reflect the result

Example : ADC C

Example : Suppose a = 50H, C = 20H, CY = 1

Operation : The instruction ADC C will add the content of register C(20H) and contents of carry (01H) with the contents of the accumulator i.e. 50H. The result of addition (71H) will be stored in the accumulator as shown in Fig. 2.68.

Format :

1	0	0	0	1	S	S	S

(17) ADC M (Add Memory with Carry) : The content of the memory location whose address is contained in the H and L register and the content of the CX flag are added to the accumulator. The result is stored in accumulator.

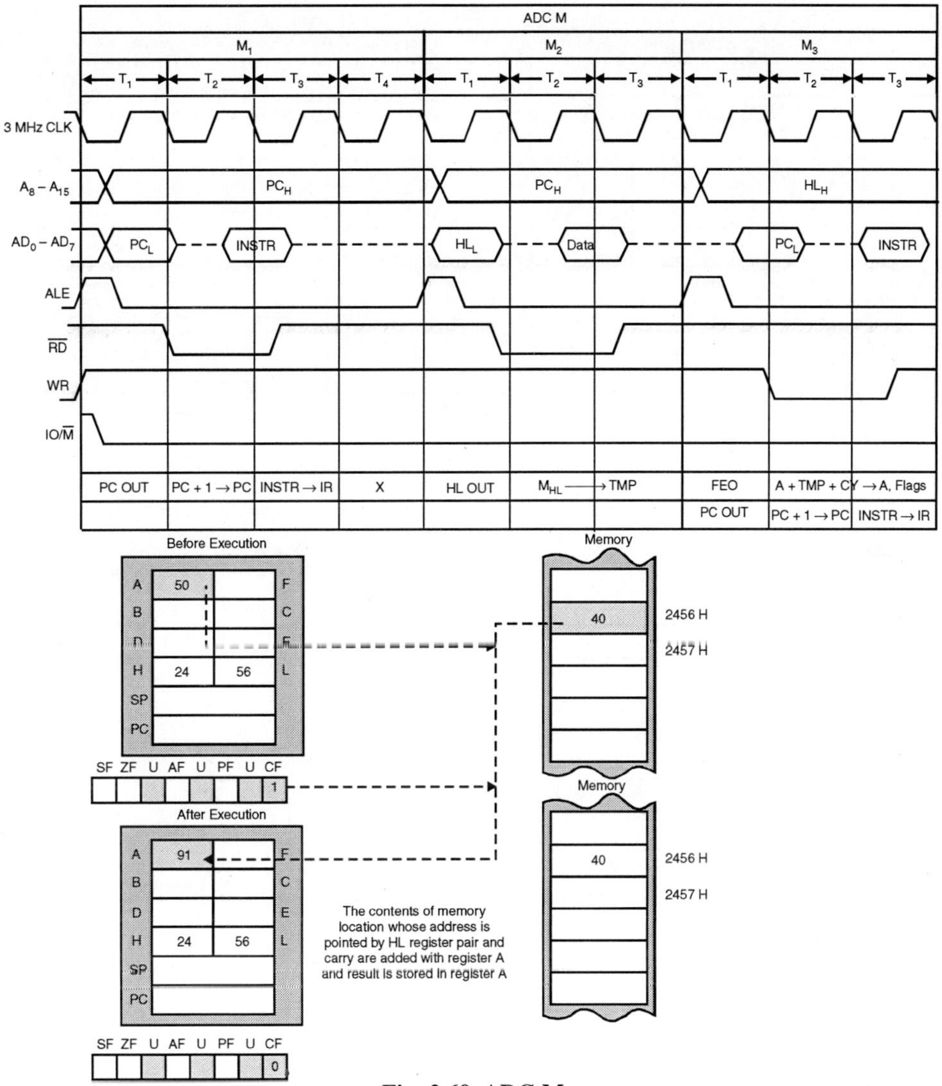

Fig. 2.69. ADC M

Menomonic : ADC M.

Operation : This instruction adds the contents of memory location pointed by the HL register pair and and contents to accumulator. The result is stored in accumulator.

Addressing Mode : Direct addressing mode

Flags Affected : All flags

Example : Suppose A = 50H, H = 24H, L = 56H, contents of memory location 2456H = 40H and the contents of carry = 1.

Operation : The instruction ADC M will add the contents of memory location 2456H i.e. (40H) and carry alongwith the contents of the accumulator i.e. 50H. The result of addition (91H) will be stored in the accumulator as shown in Fig. 2.69.

Format :

1	0	0	0	1	1	1	0

(18) ACI Data (Add Immediate with Carry) : The content of the 2nd byte of the instruction and the content of the CY flag are added to the contents of the accumulator. The result is placed in the accumulator.

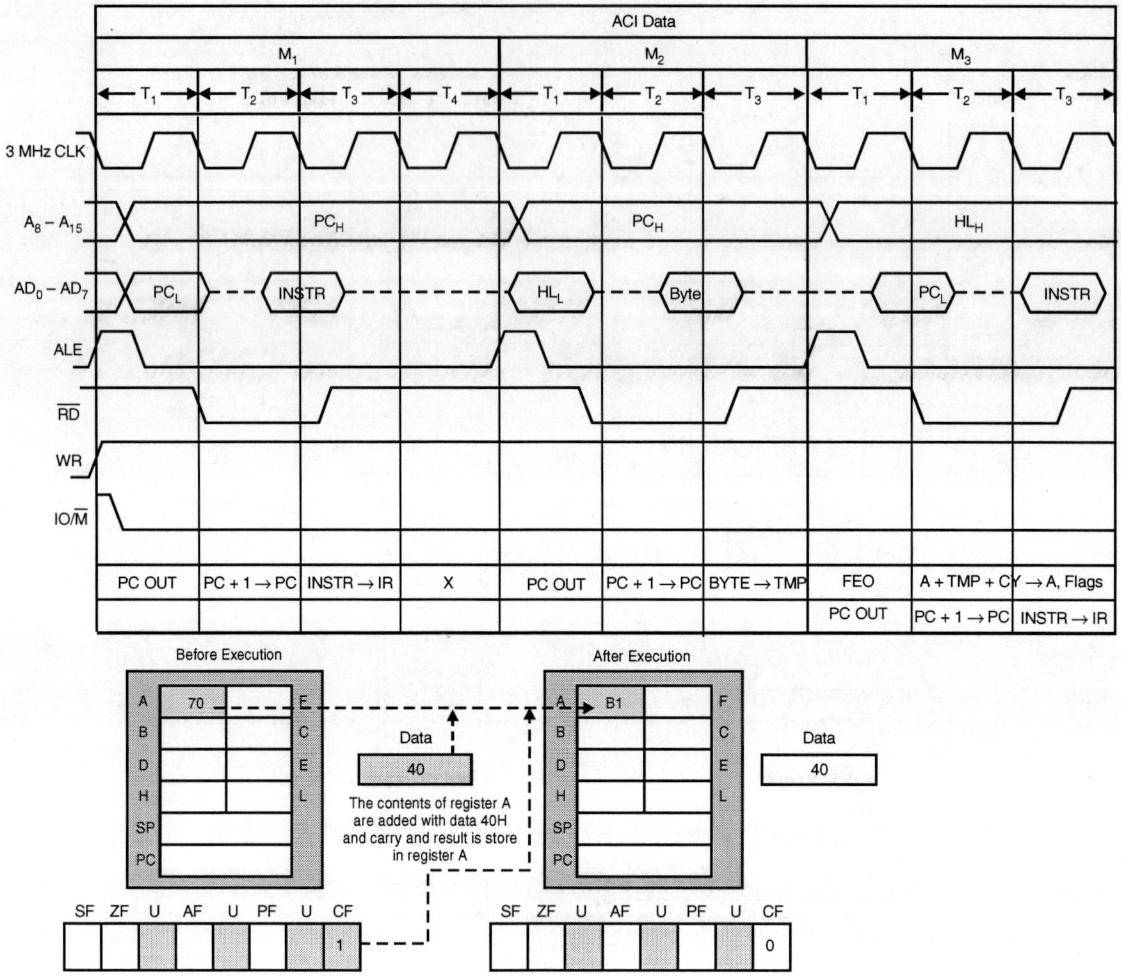

Fig. 2.70. ACI 40H

Mnemonic : ACI Data

Operation : This instruction adds the 8-bit data specified and content of carry along with the contents of accumulator. The result is stored in the accumulator.

Addressing Mode : Immediate addressing mode

Flags Modified : All the flags

Example : ACI 40H

This instruction adds the immediate data (40H) and carry (01H) to the contents of the accumulator. The result is stored in the accumulator as shown in Fig. 2.70.

Format :

1	1	0	0	1	1	1	0
Data							

(19) SUB r (Subtract Register) : The content of register r is subtracted from the content of the accumulator. The result is stored in the accumulator.

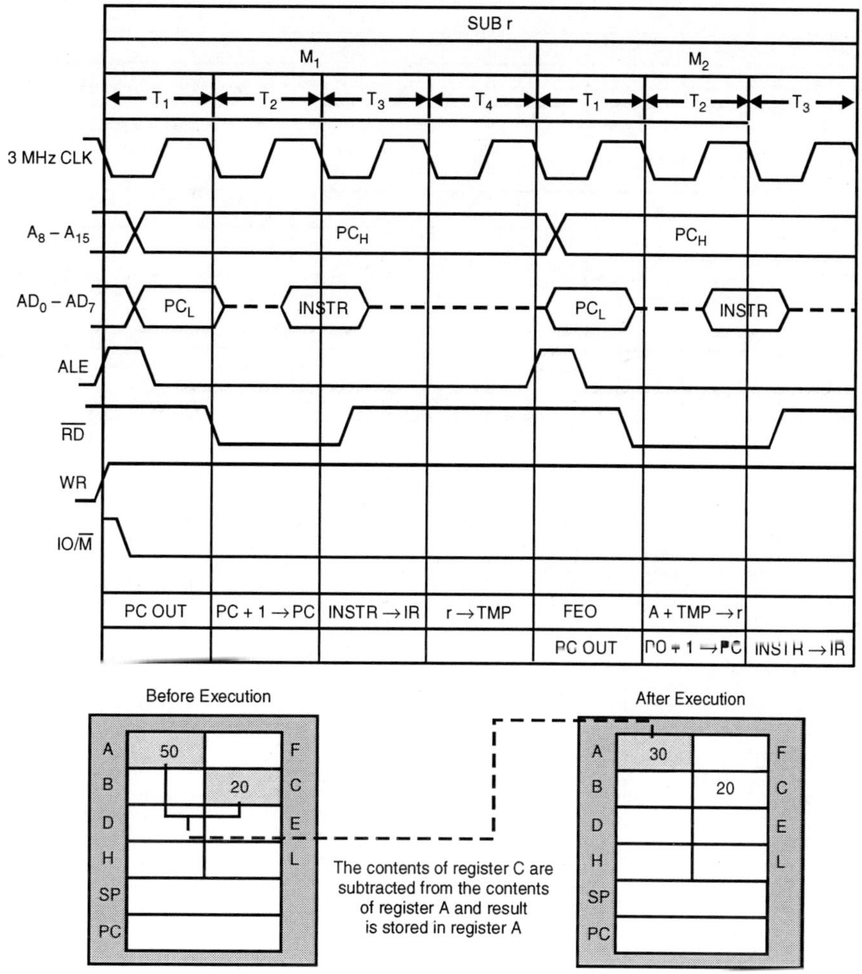

Fig. 2.71. SUB C

Mnemonic : SUB R.

Operation : This instruction subtracts the contents of register R from the contents of the accumulator. The result of subtraction is stored in the accumulator. The register R can be any general purpose register such as A, B, C, D, E, H and L.

Addressing Mode : Register addressing mode

Flags Affected : All flags modified

Example : SUB C

Suppose A = 50H, C = 20H. The instruction SUB C will subtract the contents of register C (20H) from the contents of the accumulator i.e. 50H. The result of subtraction (30H) will be stored in the accumulator as shown in Fig. 2.71.

Format :

1	0	0	1	0	S	S	S

(20) SUB M (Subtract Memory) : The content of the memory location whose address is contained in the H and L registers is subtracted from the content of the accumulator. The result is placed in the accumulator.

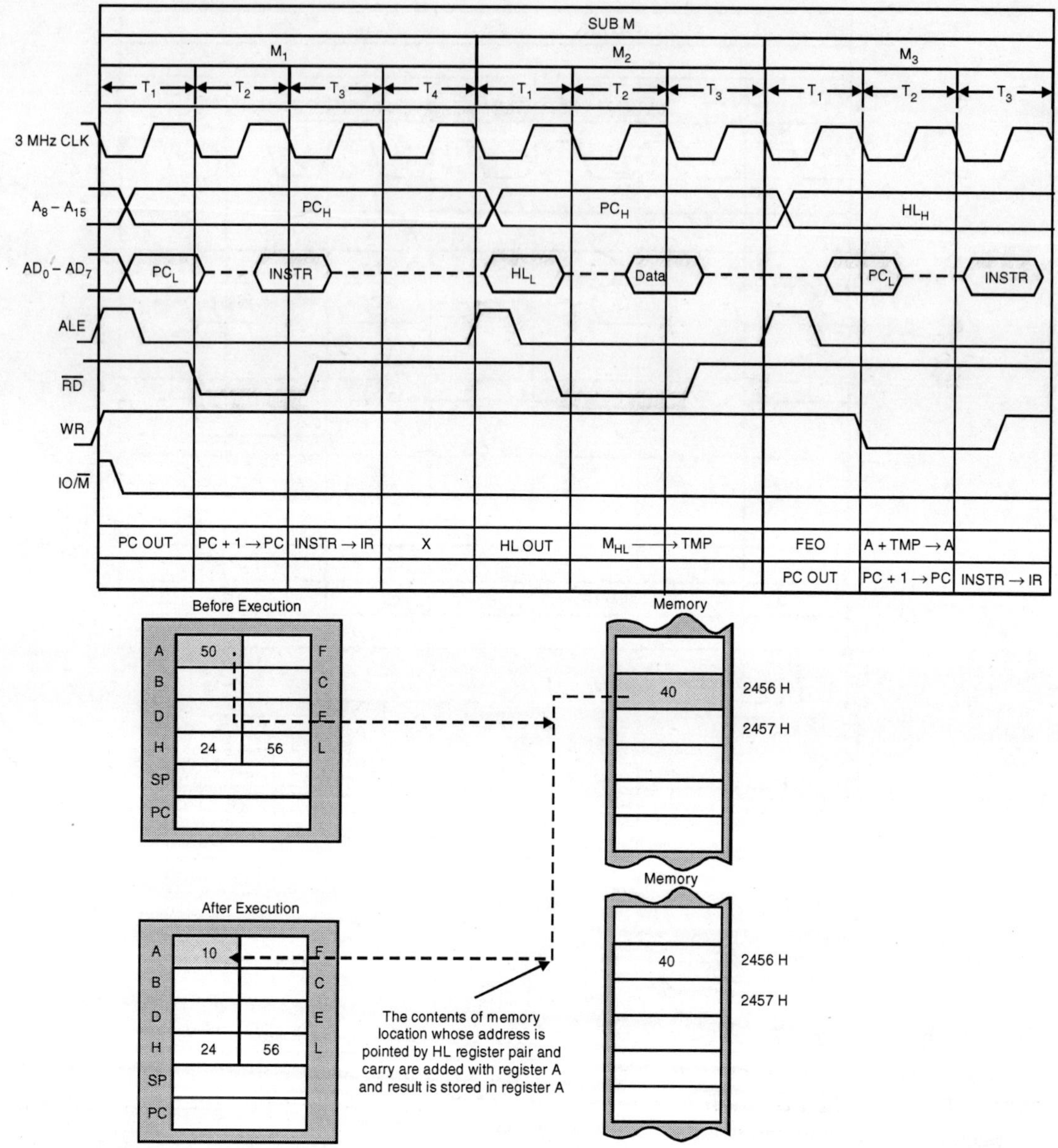

Fig. 2.72. SUB M

Mnemonic : SUB M.

Operation : This instruction subtracts the contents of memory location pointed by the HL register pair from the contents to accumulator. The result is stored in the accumulator.

Addressing Mode : Register indirect

Flags Modified : All the active flags

Example : SUB M.

Suppose A = 50H, H = 24H, L = 56H are contents of memory location 2456H = 40. The instruction SUB M will subtract the contents of memory location 2456H i.e. (40H) with the contents of the accumulator i.e. 50H. The result of subtraction (10H) will be stored in the accumulator as shown in Fig. 2.72.

Format :

1	0	0	1	0	1	1	0

(21) SUI Data (Subtract Immediate) : The content of 2nd byte of the instruction is subtracted from the content of the accumulator. The result is placed in the accumulator.

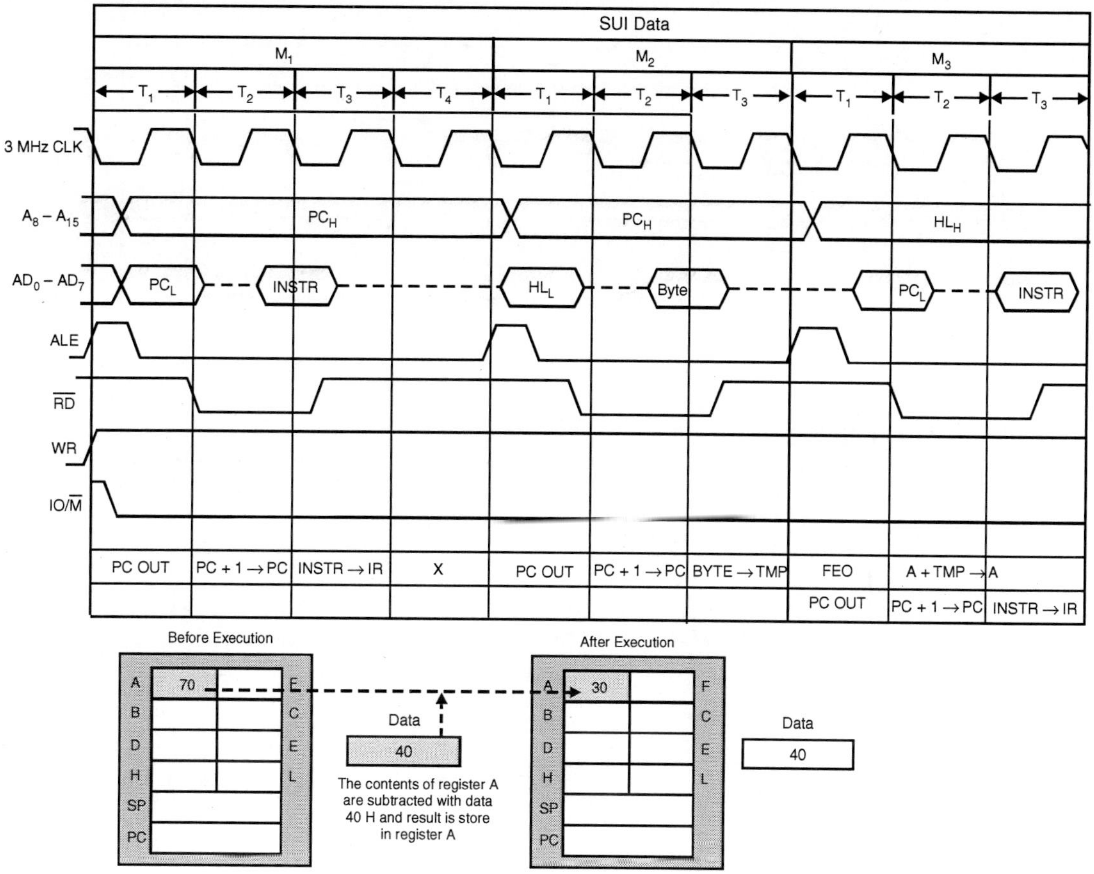

Fig. 2.73. SUI 40H

Mnemonic : SUI Data.

Operation : This instruction subtracts the 8-bit data specified from the contents of the accumulator.

Addressing Mode : Immediate addressing mode

Flags Affected : All the flags affected

Example : SUI 40H.

This instruction subtracts the immediate data (40H) from the contents of the accumulator. The result is in the accumulator as shown in Fig. 2.73.

Format :

1	1	0	1	0	1	1	0
Data							

(22) SBB r (Subtract Register with Borrow) : The content of register r and the content of the CY flag are both subtracted from the accumulator. The result is placed in the accumulator.

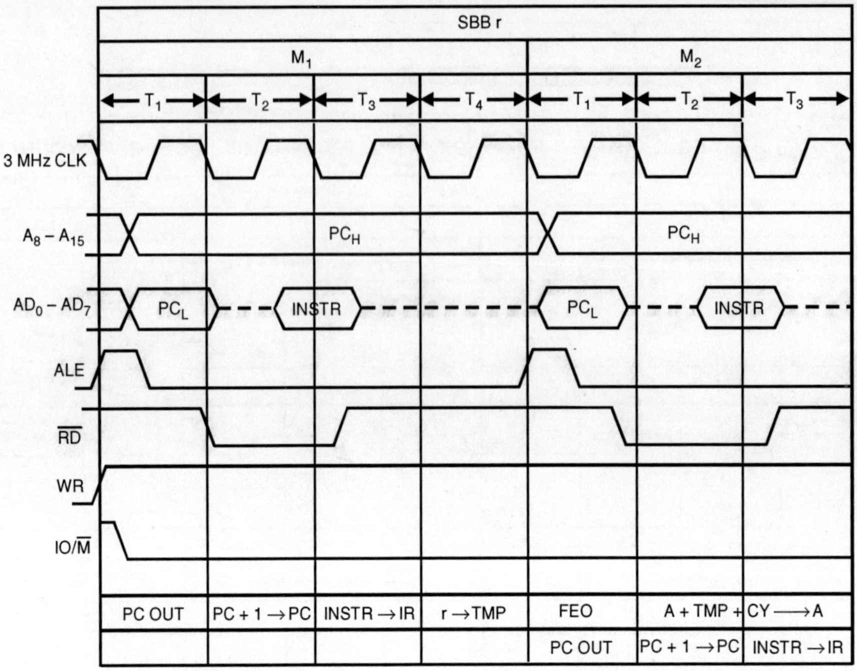

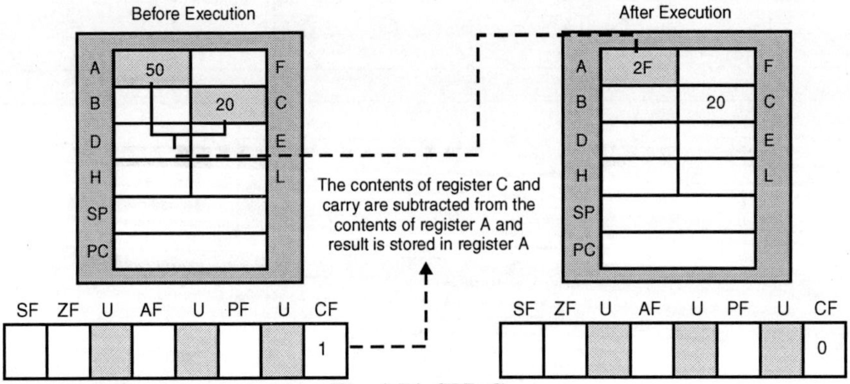

Fig. 2.74. SBB C

Mnemonic : SBB R.

Operation : The instruction subtracts the contents of register R along with the carry from the contents of the accumulator. The register R can be any general purpose register such as A, B, C, D, E, H and L.

Flags Modified : All flags affected.

Example : SBB C

Suppose A = 50H, C = 20H, CY = 1

The instruction SBB C will subtract the contents of register C (20H) and contents of carry (01H) from the contents of the accumulator i.e. 50H. The result of the subtraction (2FH) will be stored in the accumulator as shown in Fig. 2.74.

Format :

1	0	0	1	1	S	S	S

(23) SBB M (Subtract Memory with Borrow) : The content of the memory location whose address is contained in the H and L register and the content of the CY flag are subtracted from the accumulator.

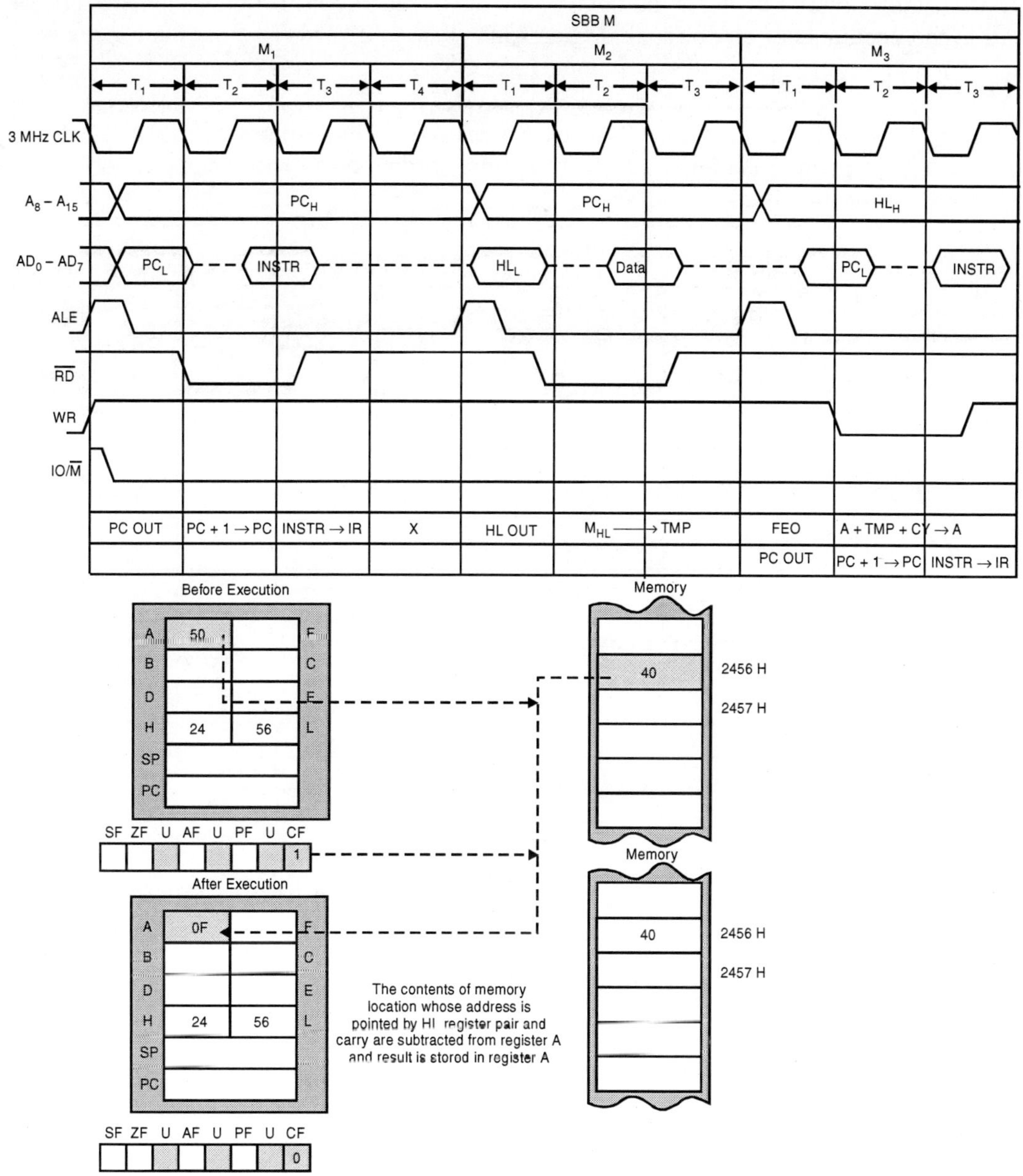

Fig. 2.75. SBB M

Mnemonic : SBB M.

Operation : The instruction subtracts the contents of memory location pointed by the HL register pair and the contents of carry from the contents to accumulator. The result is stored in the accumulator.

Addressing Mode : Direct addressing mode

Flags Affected : All active flags modified

Example : SBB M.

Suppose A = 50H, H = 24H, L = 56H, contents of the memory location 2456H = 40H and the contents of carry = 1.

The instruction SBB M will subtract the contents of memory location 2456H i.e. (40H) and carry from the contents of the accumulator i.e. 50H. The result of subtraction (0FH) will be stored in the accumulator as shown in Fig. 2.75.

Format :

1	0	0	1	1	1	1	0

(24) SBI Data (Subtract Immediate with Borrow) : The content of the 2nd byte of the instruction and the contents of the CY flag are both subtracted from the accumulator. The result is placed in the accumulator.

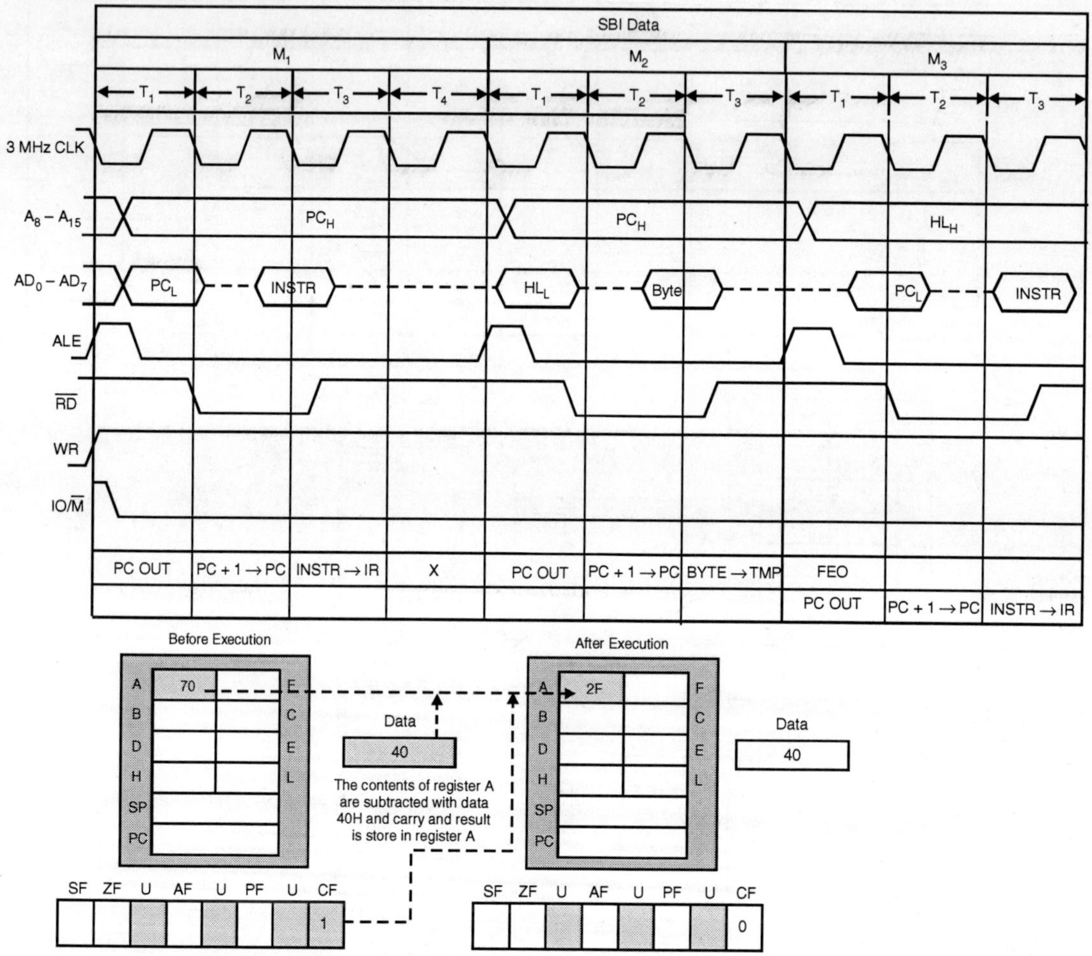

Fig. 2.76. SBI 40H

Mnemonic : SBI Data.

Operation : This instruction subtracts 8-bit data specified and contents of carry along with the contents of the accumulator. The result is stored in accumulator.

Addressing Mode : Immediate addressing mode

Flags Modified : All the flags modified

Example : SBI 40H.

The instruction subtracts the immediate data (40H) and carry (01H) from the contents of the accumulator.

The result is stored in the accumulator as shown in Fig. 2.76.

Format :

1	1	0	1	0	1	1	0
Data							

(25) INR r (Increment Register) : The content of register r is incremented by one.

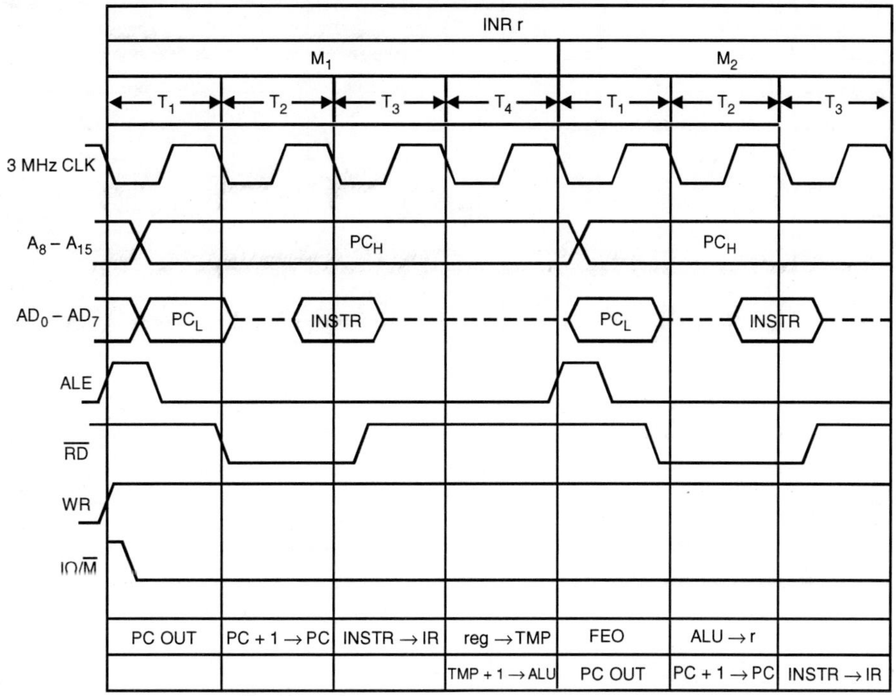

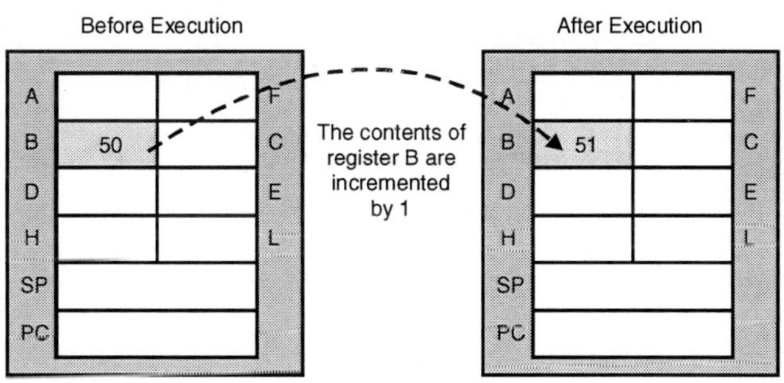

Fig. 2.77. INR B

Mnemonic : INR R.

Operation : This instruction increments the contents of the specified register R by 1. The result is stored in the same register.

Addressing Mode : Register addressing mode

Flags Modified : All flags except CY flag

Example : INR B.

Suppose B = 50H. Then the instruction INR B will increment the contents of register B by 1 as shown in Fig. 2.77.

Format :

0	0	D	D	0	1	0	0

(26) INR M (Increment Memory) : The content of the memory location whose address is contained the H and L registers is incremented by one.

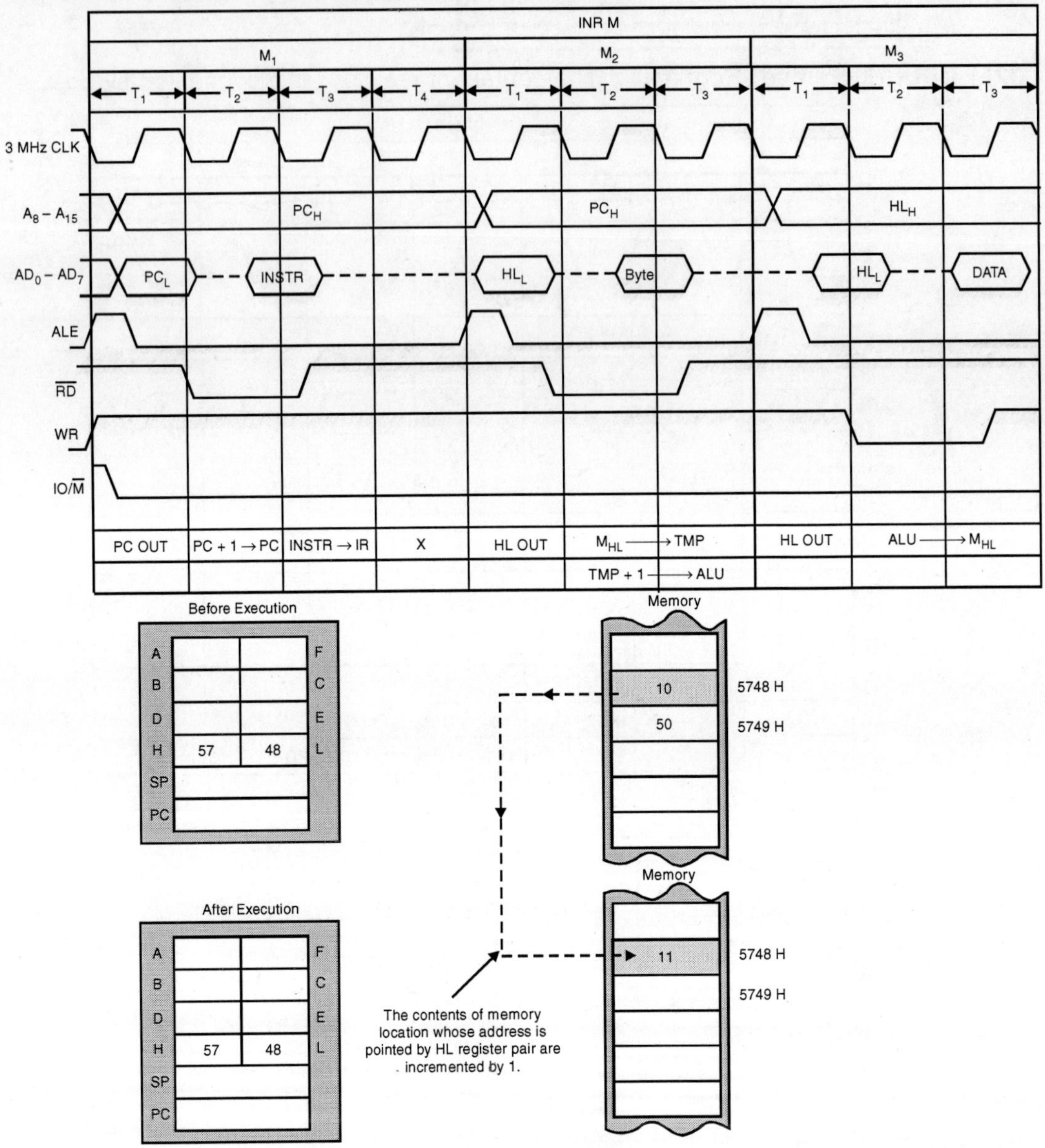

Fig. 2.78. INR M

Mnemonic : INR M

Operation : This instruction increments the contents of the memory location addressed by HL register pair by 1. The result is stored at same memory location.

Addressing Mode : Direct address mode.

Flags Modified : All flags excepts CY flag.

Example : Suppose H = 57H, L = 48H, the contents of memory location 5748H = 10H. The instruction INR M will increment the contents of memory location 5748 as shown in Fig. 2.78.

Format :

0	0	1	1	0	1	0	0

(27) DCR R (Decrement Register) : The content of register r is decremented by one.

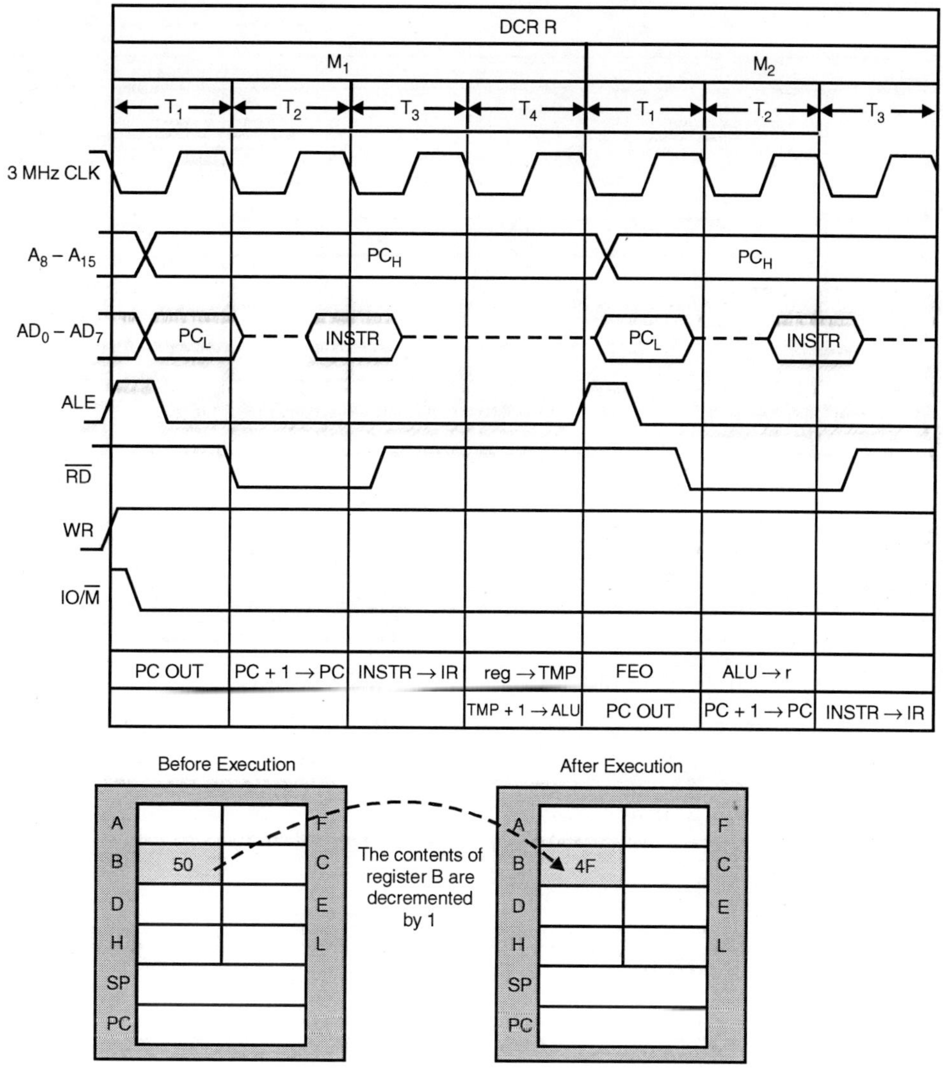

Fig. 2.79. DCR B

Mnemonic : DCR R

Operation : This instruction decrements the contents of the specified register R by 1. The result is stored in the same register.

Addressing Mode : Register addressing mode.

Flags Modified : All flags except CY flag.

Example : DCR B

Suppose B = 50H, the instruction DCR B will decrement the contents of register B by 1 as shown in Fig. 2.79.

Format :

0	0	D	D	0	1	0	1

(28) DCR M (Decrement Memory) : The content of the memory location whose address is contained in the H and L registers is decremented by one.

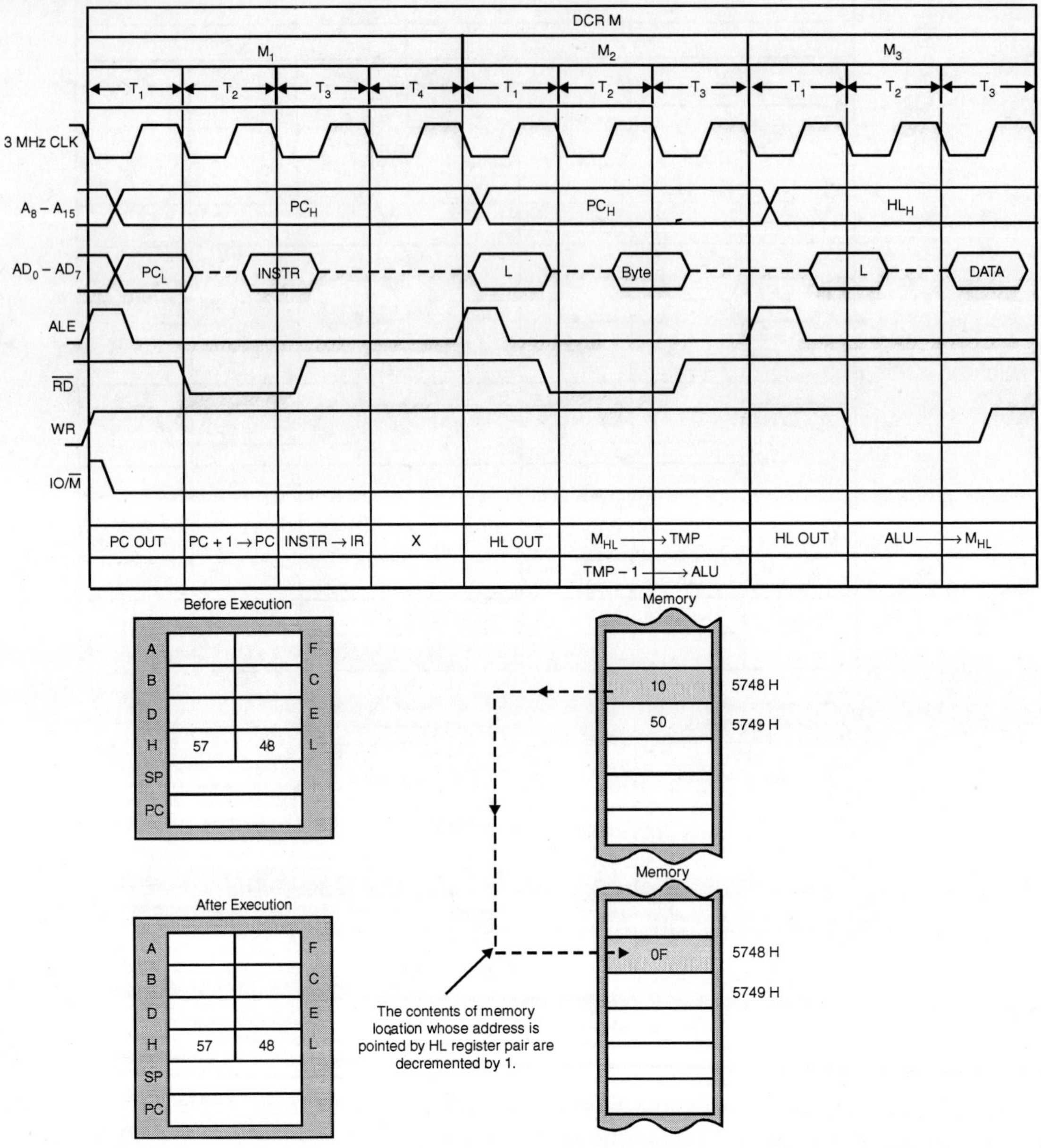

Fig. 2.80. DCR M

Mnemonic : DCR M.

Operation : The instruction decrements the contents of memory location addressed by HL register pair by 1. The result is stored back at the same memory location.

Addressing Mode : Direct addressing mode.

Flags Affected : All flags, except CY flag.

Example : DCR M.

Suppose H = 57H, L = 48H, the contents of memory locations 5748 = 10H. The instruction DCR M will decrement the contents of memory location 5748H as shown in Fig. 2.80.

Format :

0	0	1	1	0	1	0	1

(29) INX Rp (Increment Register Pair) : The content of the register pair rp is incremented by one.

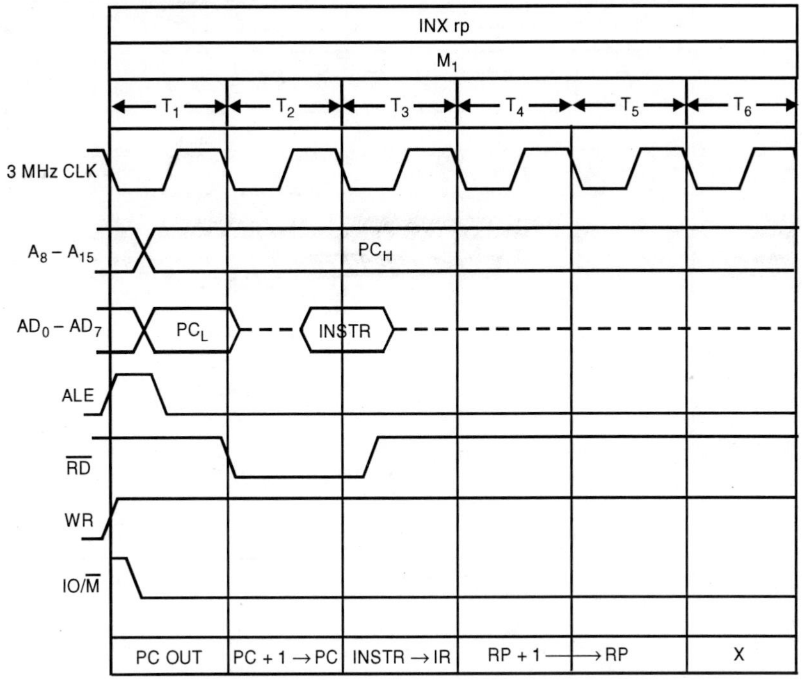

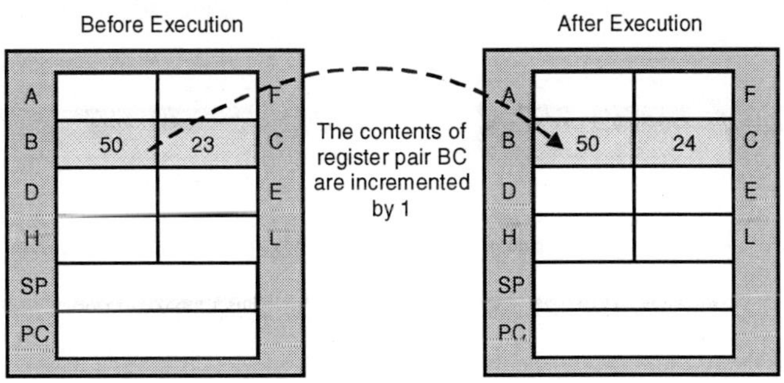

Fig. 2.81. INX B

Mnemonic : The instruction increments the contents of register pair specified in the instruction by 1. The register pair Rp can be any valid register pair like BC, DE, HL and SP.

Addressing Mode : Direct addressing mode.

Flags Modified : No flags are modified.

Example : INX B.

Suppose B = 50H, C = 23H. The instruction INX B will increment the contents of register pair BC by 1, as shown in Fig. 2.81.

Format :

0	0	R	P	1	1	1	1

(30) DCX Rp (Decrement Register Pair) : The content of the register pair rp is decremented by one.

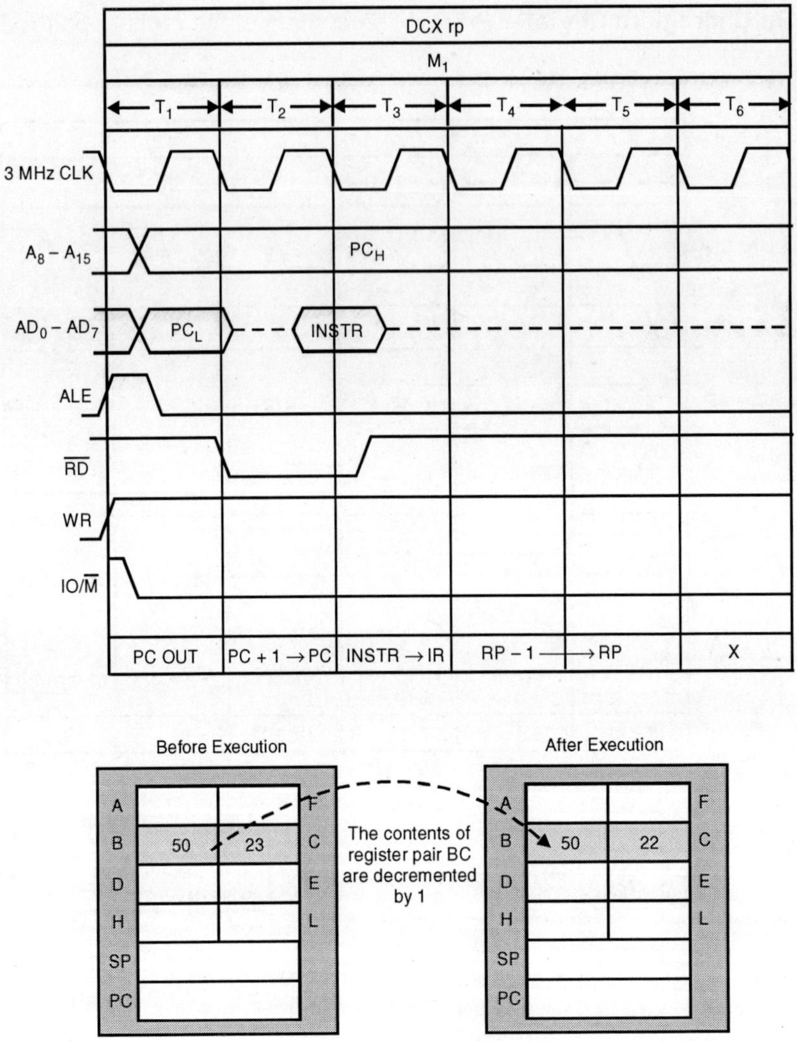

Fig. 2.82. DCX B

Mnemonic : DCX Rp.

Operation : This instruction decrements the contents of register pair specified in the instruction by 1. The register pair Rp can be any valid register pair like BC, DE, HL and SP.

Addressing Mode : Direct addressing mode.

Flags Modified : None.

Example : Suppose B = 50H, C = 23H. The instruction DCX B will decrement the contents of register pair BC by 1, as shown in Fig. 2.82.

Format :

0	0	R	P	1	0	1	1

(31) DAD rp (Add Regisger Pair to H and L) : The content of the register pair rp is added to the content of the register pair H and L. The result is placed in the register pair H and L.

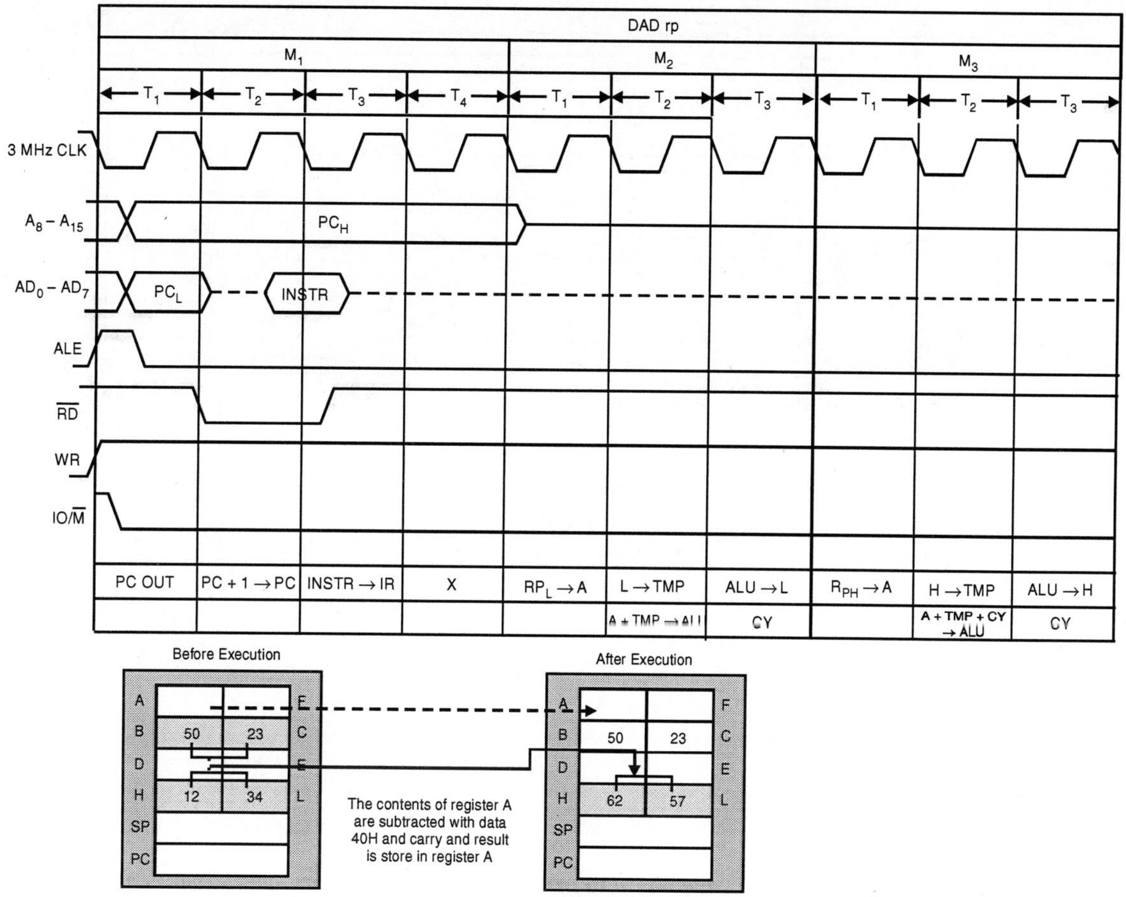

Fig. 2.83. DAD B

Mnemonic : DAD Rp.

Operation : The instruction adds the contents of specified register pair to HL pair and stores the result in HL pair. The register pair Rp can be SP, BC, DE or HL.

Addressing Mode : Register addressing mode.

Flags Modified : No flag is affected except CY flag.

Example : DAD B.

Suppose BC = 5023H, HL = 1234H. Then the instruction DAD B will add the contents of register pair BC (5023H) with the contents of register pair HL (1234H) and the result of addition 6257H is stored in the HL register pair as shown in Fig. 2.83.

Format :

0	0	R	P	1	0	0	1

(32) DAA (Decimal Adjust Accumulator).

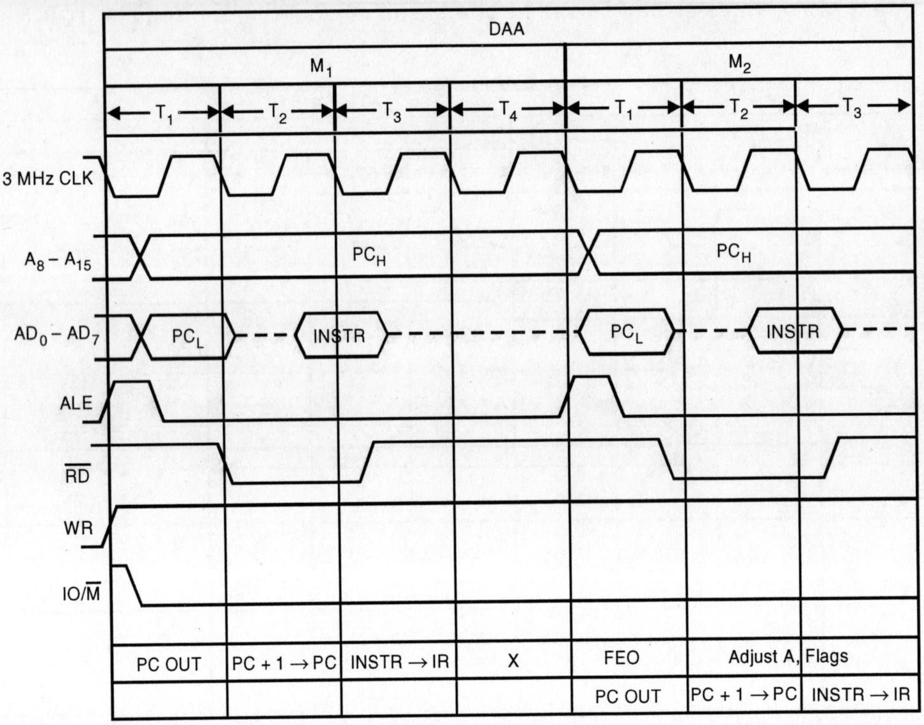

Fig. 2.84. DAA

Mnemonic : DAA.

Operation : The 8-bit number in the accumulator is adjusted to form to from two 4-bit BCD digits by the following process :

1. If the value of the least significant 4-bits of the accumulator is greater than 9 or if the AC flag is set, 6 is added to the accumulator.
2. If the value of the most significant 4-bits of the accumulator is now greater than 9 or if the CY flag is set, 6 is added to the most significant 4-bits of the accumulator.

Addressing Mode : Implied addressing mode.

Flags Modified : All flags.

Example :

Suppose A = 59H, valid BCD
B = 34H, valid BCD
A = 8DH invalid BCD addition of A and B.

DAA	1	0	0	0	1	1	0	1
+	0	0	0	0	0	1	1	0
	1	0	0	1	0	0	1	1
	9				3			H

A = 93H BCD

Format :

0	0	1	0	0	1	1	1

2.19.3. Logical Group

This group of instruction performs logical (boolean) operation on data in registers and memory and on condition flags. All instruction in this group affect the zero sign, parity, auxillary carry and carry flags according to the standard rules.

(33) ANA r :

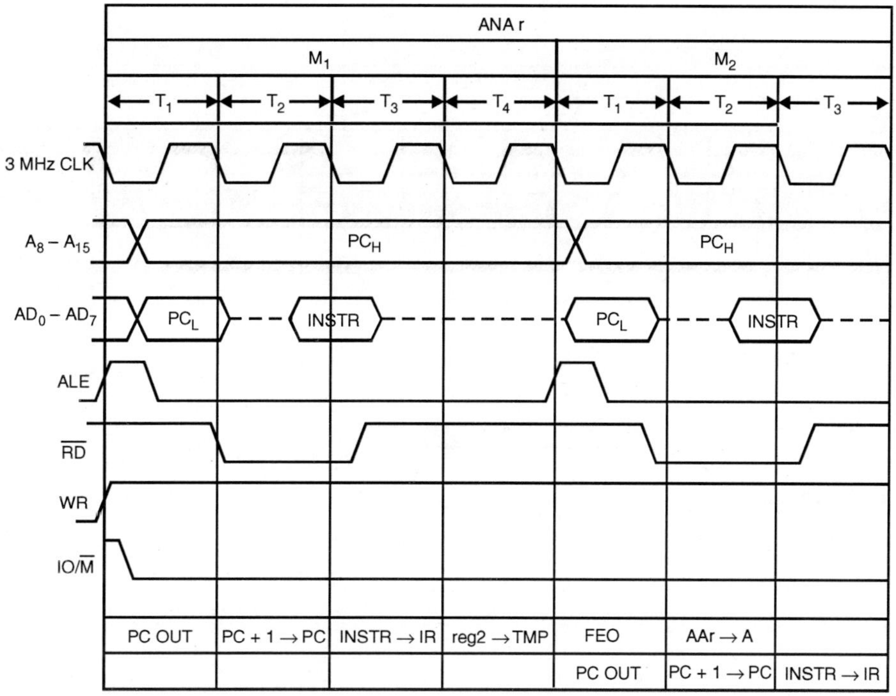

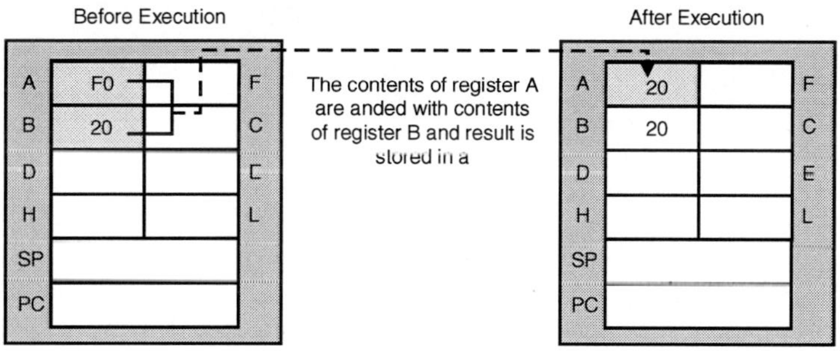

Fig. 2.85. ANA B

Mnemonic : ANA R.

Operation : The contents of the specified register are logically ANDed with the contents of the accumulator. The result is stored in the accumulator.

The operation of ANDing is performed bit by bit i.e. D_0 bit of accumulator is ANDed with bit D_0 of register and so on up to D7 bit of accumulator ANDed D_7 bit of register R can be any general purpose register such as A, B, C, D, E, H and L.

Addressing Mode : Register addressing mode.

Flags Modified : S, 7, P, flags are modified CY flag is reset and AC is set.

Example : Suppose the contents of register A = FOH and the contents of register B = 204. The contents of register B are ANDed with the contents of the accumulator. The result is stored in the accumulator as shown in Fig. 2.85.

Format :

1	0	1	0	0	S	S	S

(34) ANA M (And Memory) : The contents of the memory location whose address is contained in the H and L register is logically ANDed with the content of the accumulator.

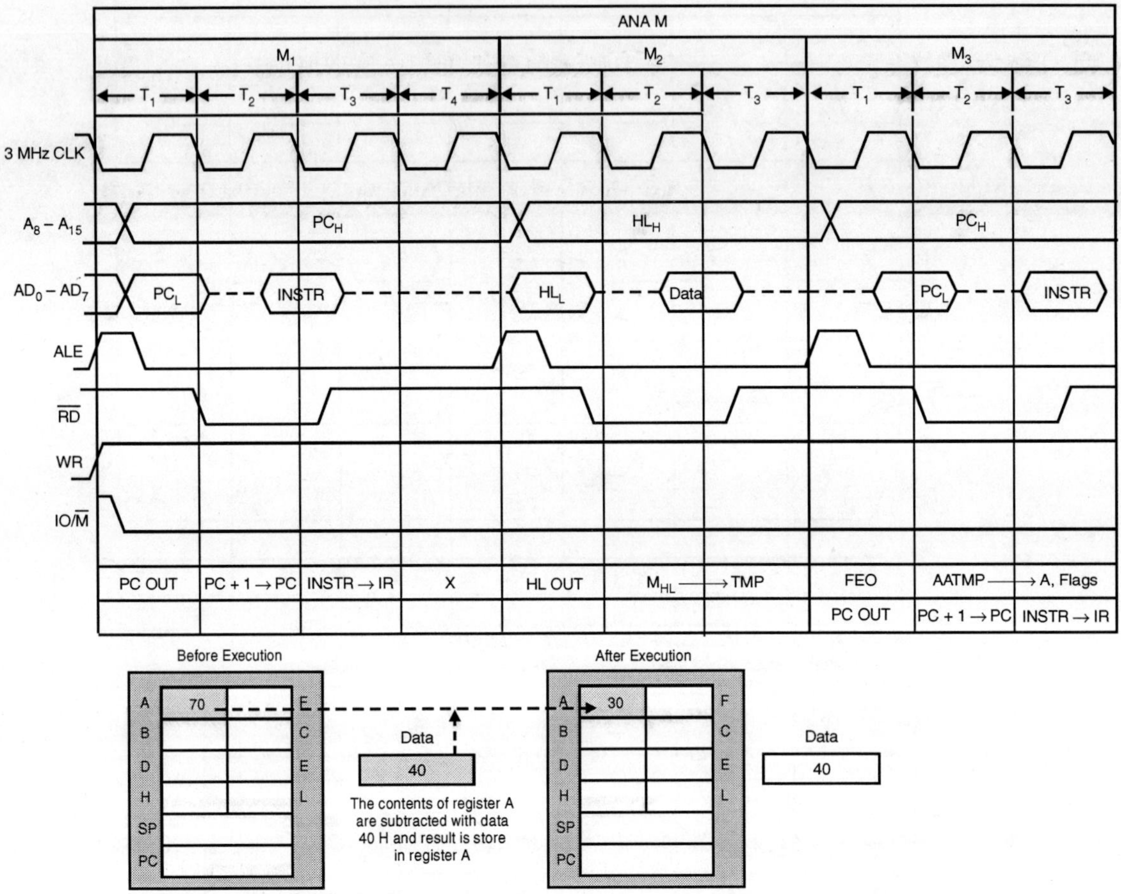

Fig. 2.86. ANA M

Mnemonic : ANA M.

Operation : The contents of the memory location pointed by the HL register pair are logically ANDed with the contents of the accumulator. The result is stored in the accumulator. The operation of ANDing is performed bit by bit i.e. D_0 bit of accumulator is ANDed with the bit D_0 of memory and so on up to D_7 bit of accumulator ANDed with D_7 bit of memory.

Addressing Mode : Direct addressing mode.

Flags modified : S, 7, P are modified. Carry flag is reset and AC is set.

Example : Suppose A = FOH, H = 57H, L = 48H and contents of memory location 5748H = 20H.

The contents of memory location 5748H are ANDed with the contents of the accumulator. The result is stored in the accumulator as shown in Fig. 2.86.

Format :

1	0	1	0	0	1	1	0

(35) ANI Data (And Immediate) : The content of the 2nd byte of the instruction is logically ANDed with the contents of the accumulator.

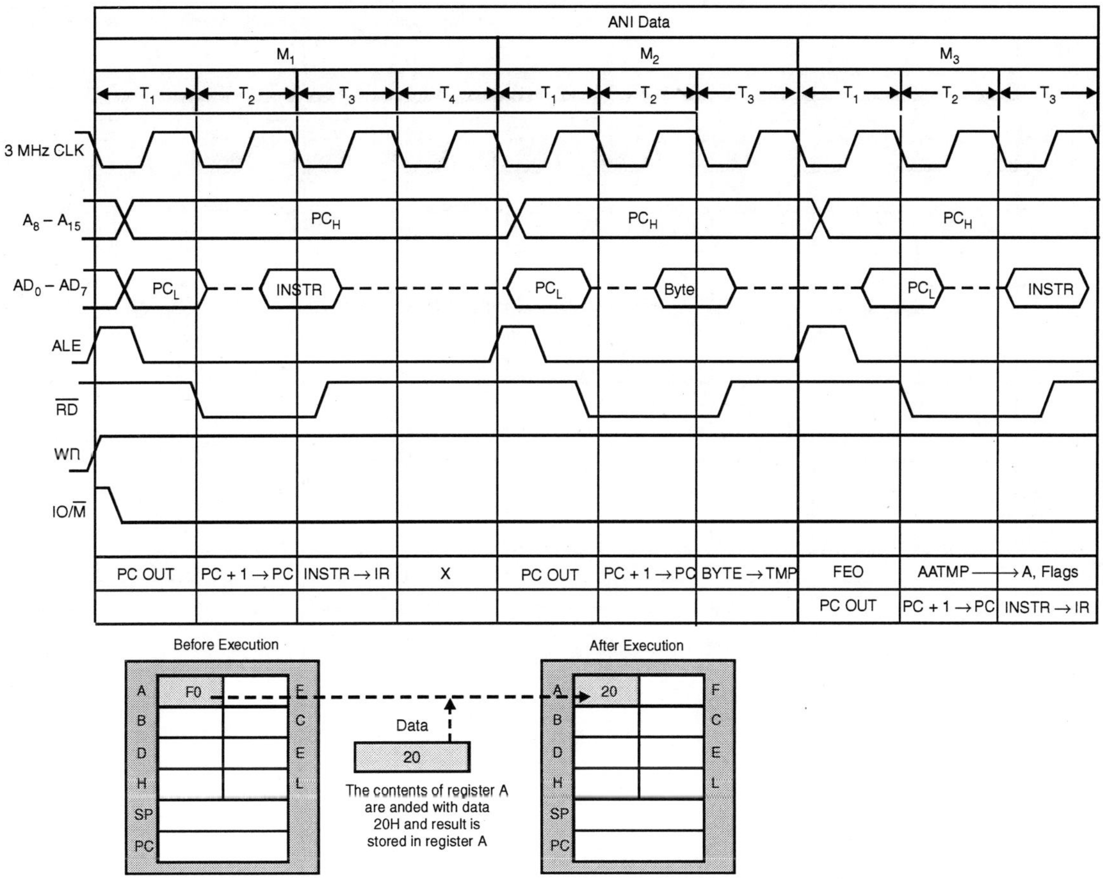

Fig. 2.87. ANI 20H

Mnemonic : ANI Data.

Operation : The contents of the 8-bit data specified in the instruction are logically ANDed with the contents of the accumulator. The result is stored in the accumulator. The operation of ANDing is performed bit by bit i.e. D_0 bit of accumulator is ANDed with bit D_0 of data and so on upto D_7 bit of accumulator ANDed with D_7 bit of data.

Addressing Mode : Immediate addressing mode.

Flags Modified : S, 7, P are modified. (X flags is reset and AC is set).

Example : ANI 20H.

Suppose the contents of register A = FOH. The immediate data contents (20H) are ANDed with the contents of the accumulator. The result is in accumulator as shown in Fig. 2.87.

Format :

1	1	1	0	0	1	1	0
Data							

(36) XRI Data (Exclusive OR immediate) : The content of 2nd byte of the instruction is X-OR'd with the content of the accumulator. The result is placed in the accumulator.

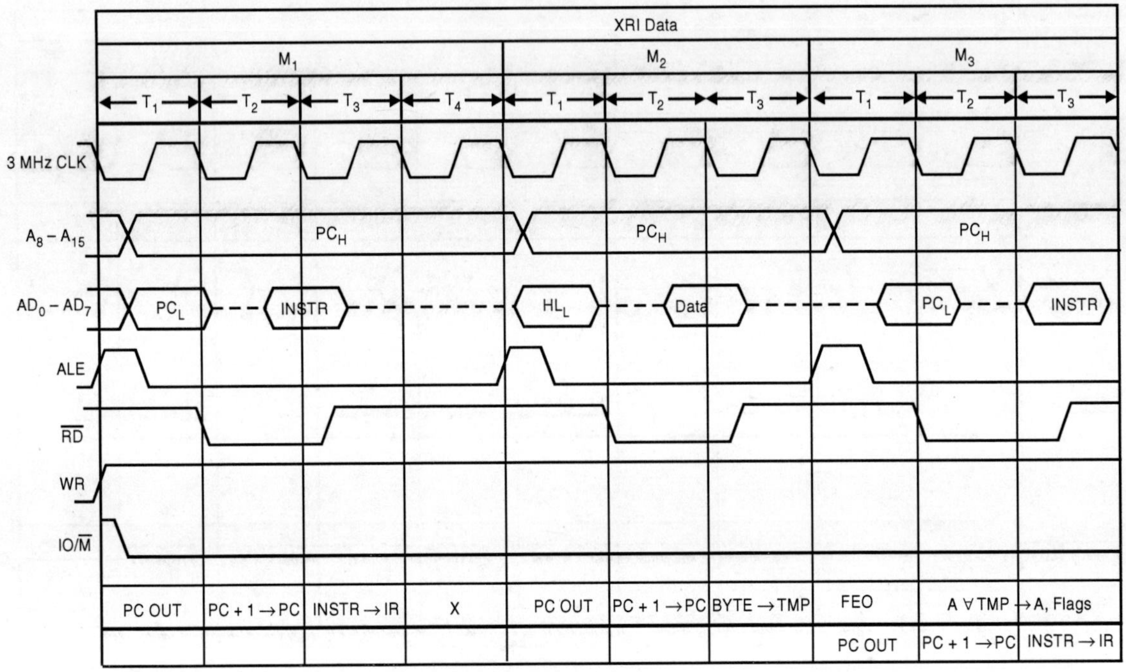

Fig. 2.88. XRI 20H

Mnemonic : XR1 Data

Algorithm : A ← A EX-OR Data

Operation : A = A EX-OR Data.

- ➢ The constants of the 8-bit data specified in the instruction are logically X-OR'd with the contents of Accumulator.
- ➢ The operation of X-ORing is performed bit by bit i.e. D_0 bit of accumulator is EX-OR with bit D_0 of data and so on upto D_7 bit of accumulator EX-OR'd with D_7 bit of data.

Addressing Mode : Immediate addressing mode.

Flags : S, Z, P are modified to reflect the result of operation. (array is reset and AC is set.)

Example : XRI 40H

Let the contents of register A = FOH.

The immediate data contents (40H) are EX-OR'd with the contents of the accumulator. The result is stored in the accumulator.

(37) ORA r : The content of register is inclusive-OR'd with content of the accumulator. The result is placed in the accumulator.

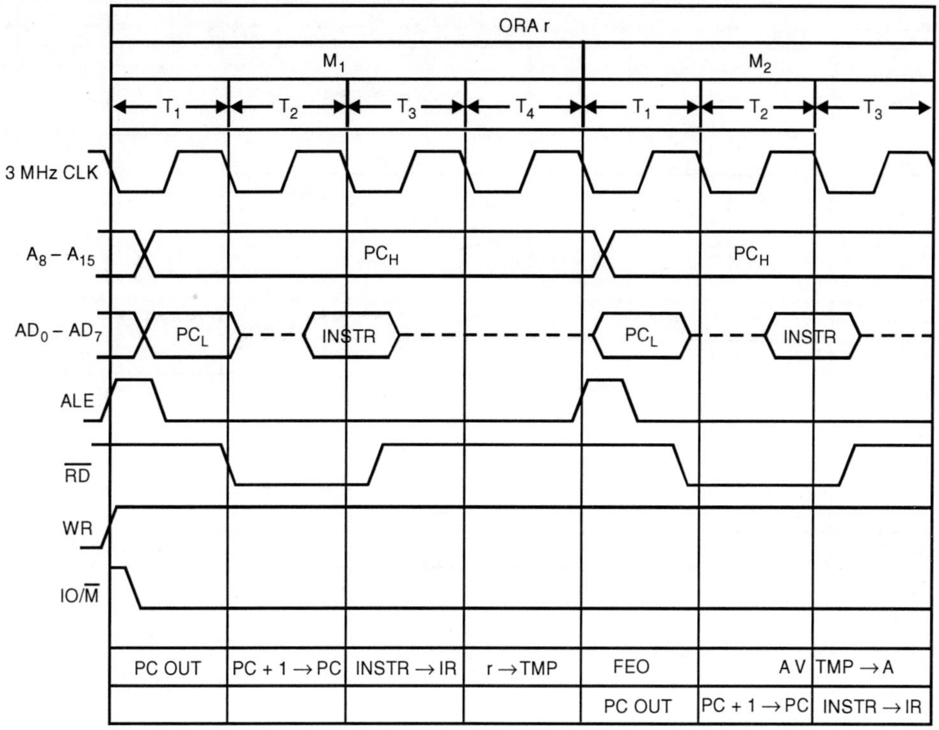

Fig. 2.89. ORA B

Mnemonic : ORA R

Operation : The contents of the specified register are logically ORed with the contents of accumulator.

- ➢ The operation ORing is performed bit by bit i.e. D_0 bit of accumulator is ORed with bit D_0 of register and so on upto D_7 bit of accumulator ORed with D_7 bit of register.
- ➢ R can be any general purpose registers such as A, B, C, D, E, H and L.

No. of Bytes : 1 byte.

Addressing mode : Register addressing mode.

(38) ORA M : (OR Memory) : The content of the memory location whose address is contained in the H and L registers is inclusive OR'd with the content of the accumulator.

Mnemonic : ORA M

Operation : A = AVM

- ➢ The contents of the memory lcoation pointed by the HL register pair are logically ORed with the contents OR accumulator. The result is stored in accumulator.
- ➢ The operation of ORing is performed bit by bit i.e. D_0 bit of accumulator is ORed with bit D_0 of register and so on upto D_7 bit of accumulator ORed with D_7 bit of register.

No. of Bytes : 1 byte

Addressing Mode : Direct addressing mode.

Flags : S, Z, P are modified. Carry is reset and AC is set.

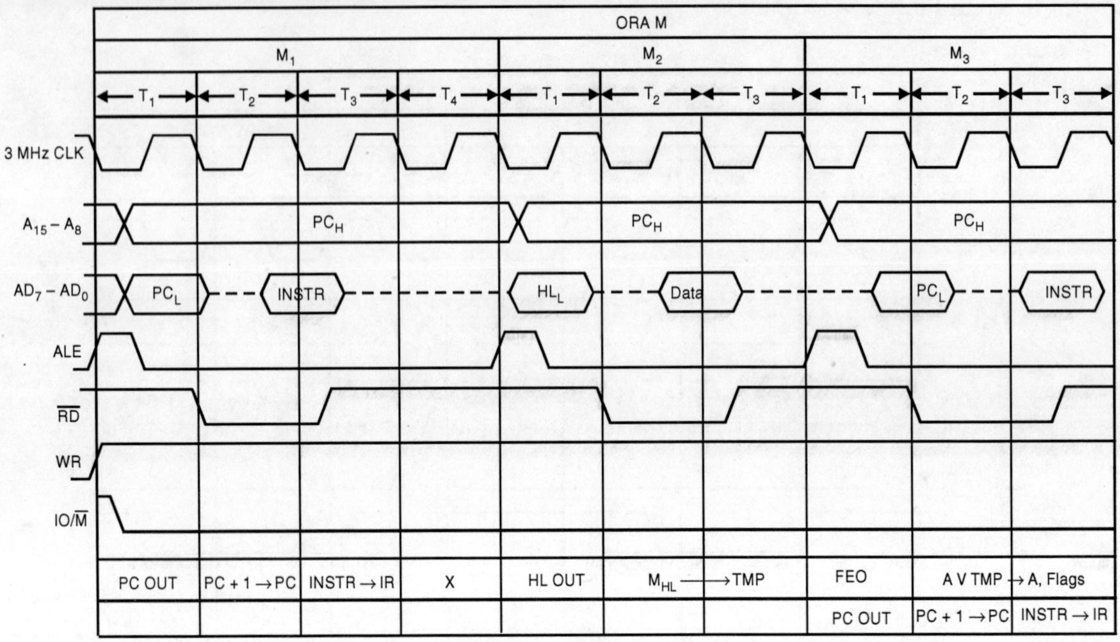

Fig. 2.90. ORA M

(39) ORI data : (OR immediate) : The content of the 2nd byte of the instruction is inclusive OR'd with the content of the accumulator. The result is placed in the accumulator.

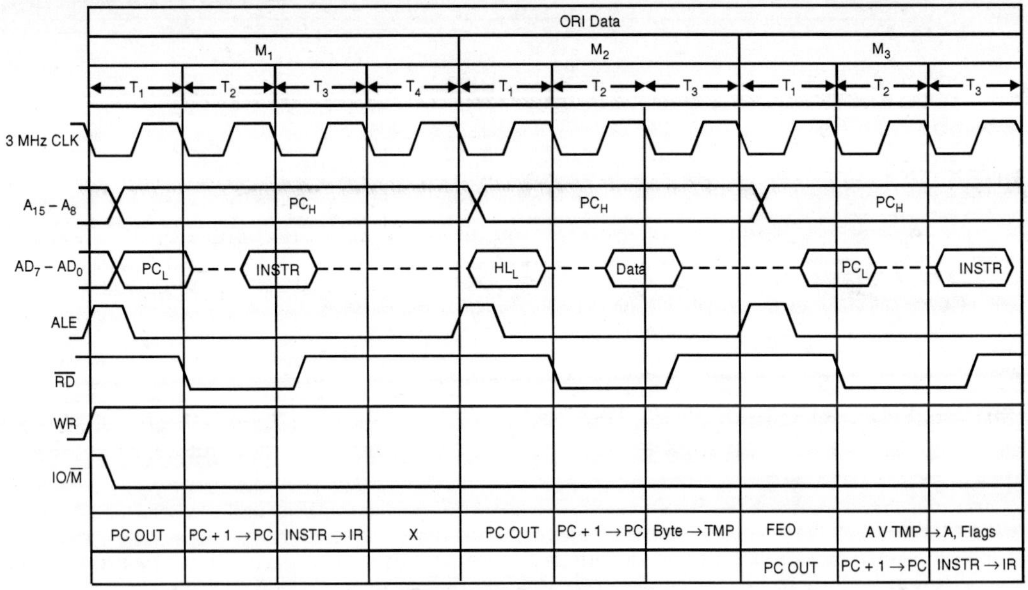

Fig. 2.91. ORI DATA

Mnemonic : ORI data.

Operation :

➢ The contents of the 8 bit data specified in the instruction are logically ORed with the contents of accumulator.

➢ The operation of ORing is performed bit by bit i.e. D0 bit of accumulator is ORed with bit D0 of data and so on upto D7 bit of accumulator. ORed with D7 bit of data.

Addressing Mode : Immediate addressing mode.

Flags : S, Z, P are modified to reflect the result of operation. Carry is reset and AC is set.

(40) CMP r (Compare Register) : The content of register r is subtracted from the accumulator. The accumulator remains unchanged.

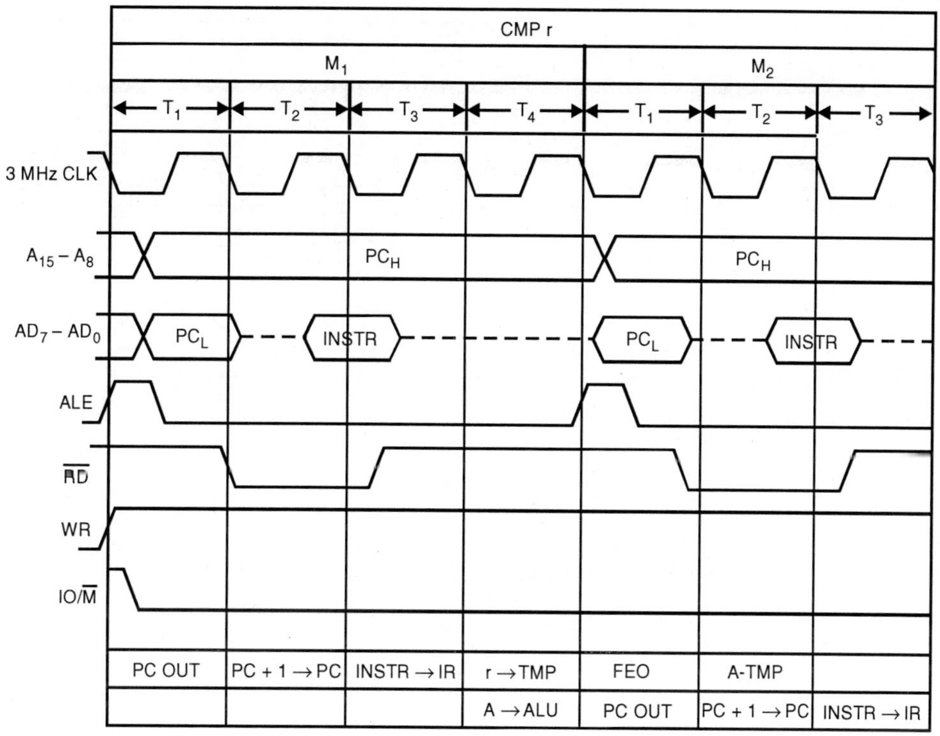

Fig. 2.92. CMP r

Mnemonic : CMP r.

Operation : This Instruction compares the contents of accumulator with the contents of register specified in the instruction. The comparison is done by subtracting the contents of the register from the contents of the accumulator. The result of subtraction is discarded.

The resut is not stored in either of the register or accumulator. The accumulator and register remains unchanged, only flags are updated.

If the contents of accumulator are equal to the contents of the register then the zero flag is set. If the contents of accumulator are less than the contents of the register then the carry flag is set. If the contents of accumulators more than the contents of the register then the carry flag and zero flag are reset.

Flags : All flags are updated.

Addressing mode : Register addressing mode.

(41) CMP M (Compare Memory) : The content of the memory location whose address is contained in the H and L registers is subtracted from the accumulator. The accumulator remains unchanged.

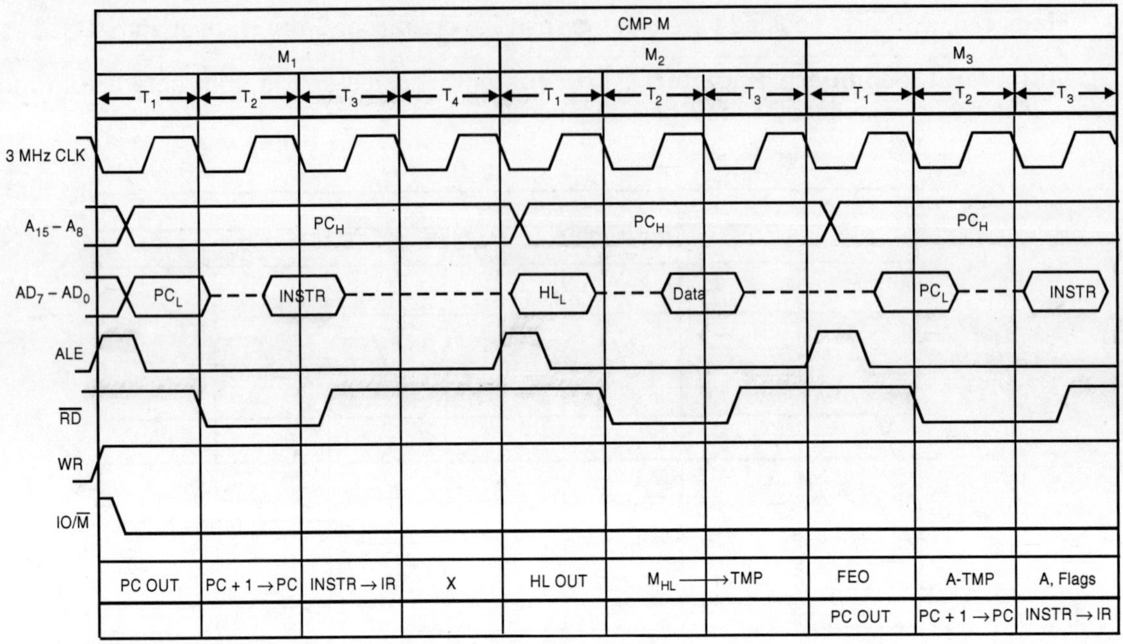

Fig. 2.93. CMP M

Mnemonic : CMP M.

Operation : This instruction compares the contents of accumulator with the contents of memory location specified by the HL register pair. The comparison is done by subtracting the contents of the memory location from the contents of the accumulator. The result of subtraction is discarded.

The result is not stored in either of the memory location or accumulator. The accumulator and register remain unchanged, only flags are updated. If the contents of accumulator are equal to the contents of the memory location then the zero flag is set.

If the contents of accumulator are less than the contents of the memory location then the carry flag is set. If the contents of accumulator are more than the contents of the memory location then the carry flag and zero flag are reset.

Flags : All flags are updated.

Addressing Mode : Direct addressing mode.

(42) RRC : The content of the accumulator is rotated right one position. The high order bit and the CY flag are both set to the value shifted out of the low order bit position.

Mnemonic : RRC

Algorithm : Shift all bits right, the bit that goes OFF is set to CF and the same bit is inserted into the left most position.

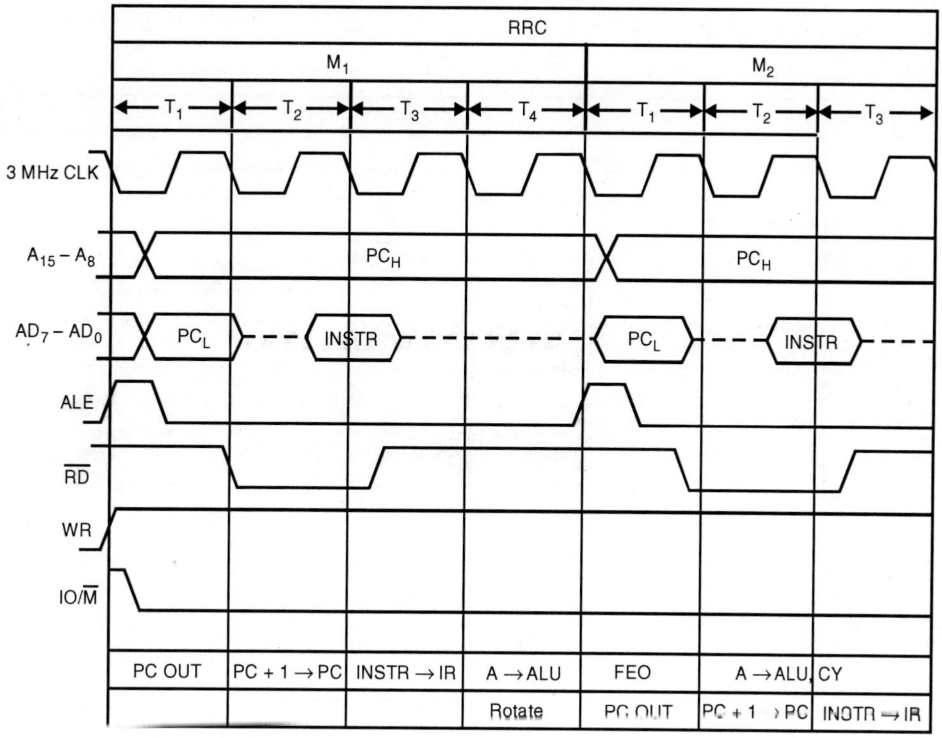

Fig. 2.94. RRC

Operation : 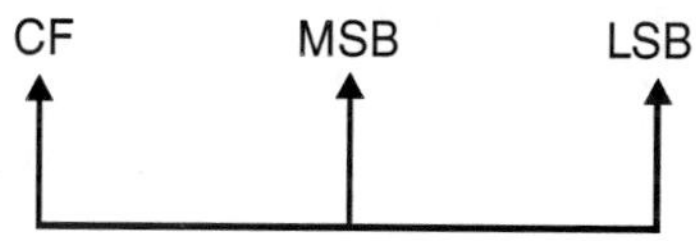

➢ This instruction rotates all the bits of the accumulator by 1 bit position to the right. The bit moved out OR the LSB is rotated around into the MSB.

➢ The data bit moved out of LSB is also copied into CF.

➢ Bit B0 is placed in bit B_7 as well as in carry.

No. of bytes : 1 byte opcode by RRC.

Addressing mode : Implied addressing mode.

Flags : only CY is affected.

Example : Let A = 11111110 and CY = 1.

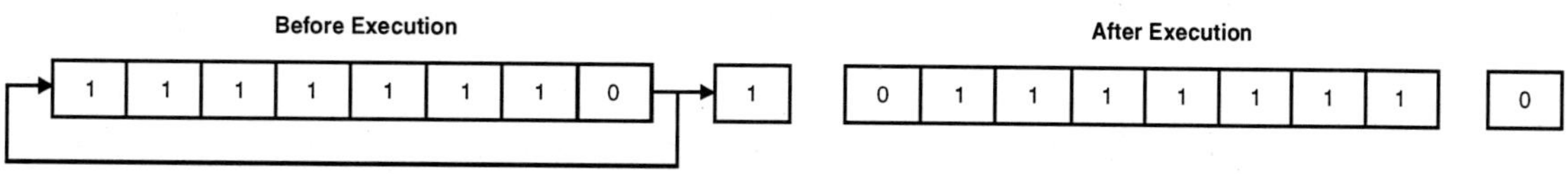

(43) RAL : The content of the accumulator is rotated left one position through the CY flag. The lower order bit is set equal to the CY flag and CY flag is set to the value shifted out of the high order bit.

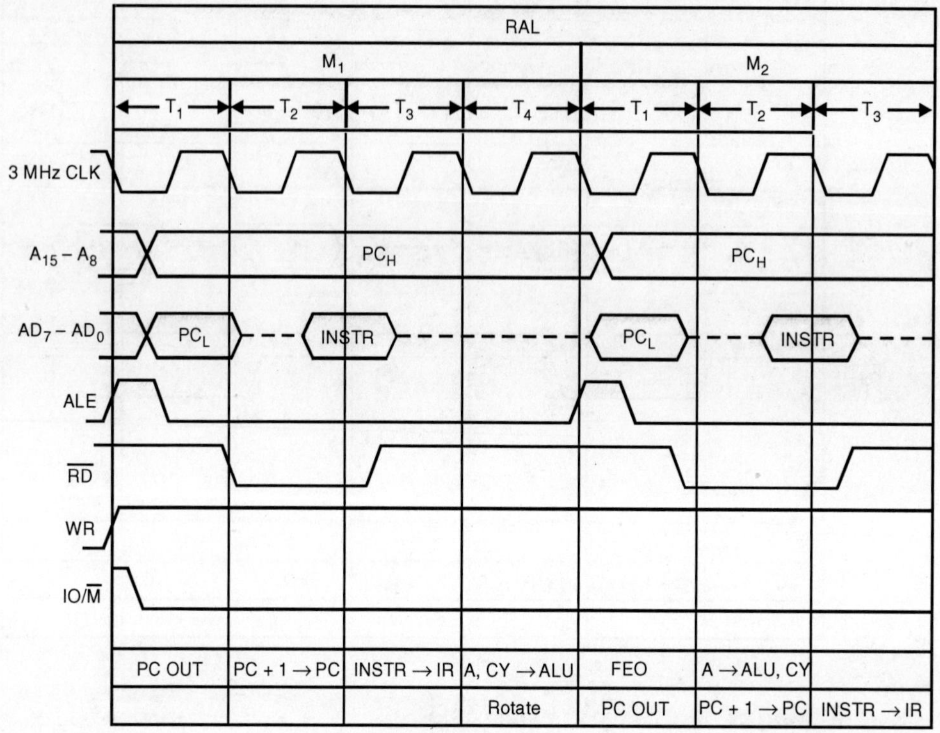

Fig. 2.95. RAL

Mnemonic : RAL

Algorithm : Shift all bits left, the bit that goes is set to CF and previous value of CY is inserted to the rightmost position.

Operation :

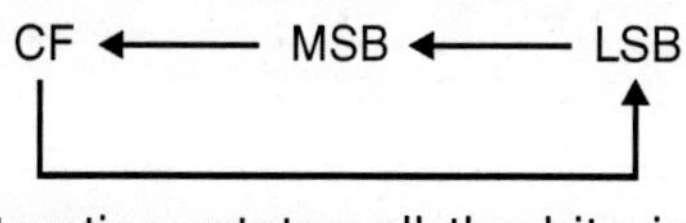

- ➢ This instruction rotates all the bits in the accumulator by one bit position to the left alongwith the carry.
- ➢ Bit D_7 is placed in carry and carry is placed in bit D_0.

No. of Bytes : 1 byte.

Addressing mode : Implied addressing mode.

Flags : Only CY is affected.

Example : RAL.

Let A = 1111 1110 and CY = 1.

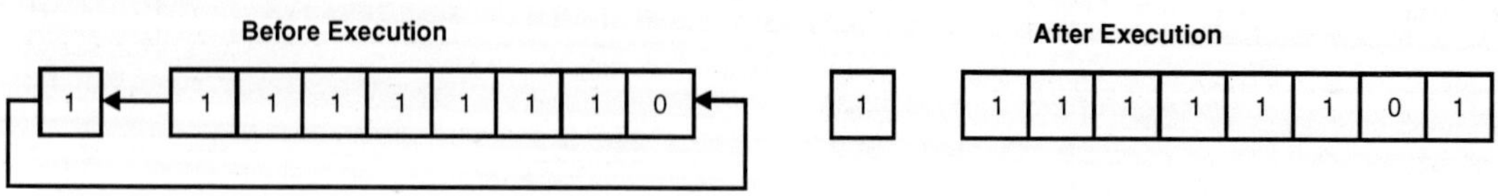

(44) RAR : The content of the accumulator is rotated right one position through the CY flag. The high order bit is set to the CY flag and the CY flag is set to the value shifted out of the lower order bit.

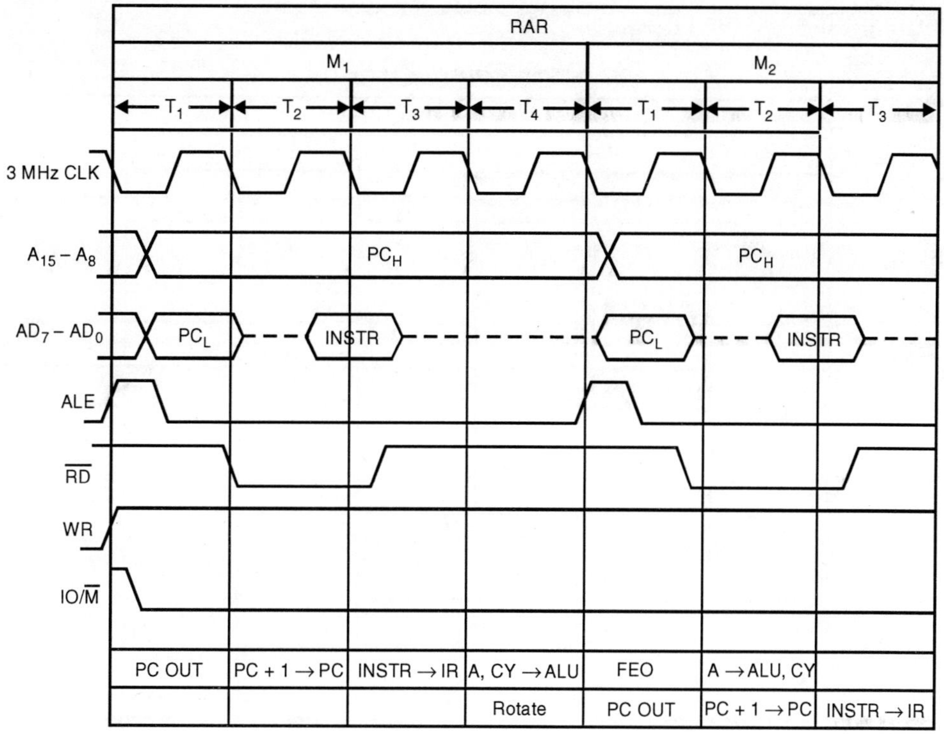

Fig. 2.96. RAR

Mnemonic : RAR

Algorithm : Shift all bits right, the bit that goes OFF is set to CF and previous value of CF is inserted to the left most position.

Operation : CF ⟶ MSB ⟶ LSB (LSB ⟶ CF)

➢ This instruction rotates all the bits of the accumulator by one bit position to right alongwith the carry.

➢ Bit D_0 is placed in carry and carry is placed in bit D_7.

No. of bytes : 1 byte

Addressing mode : Implied addressing mode.

Example : RAR

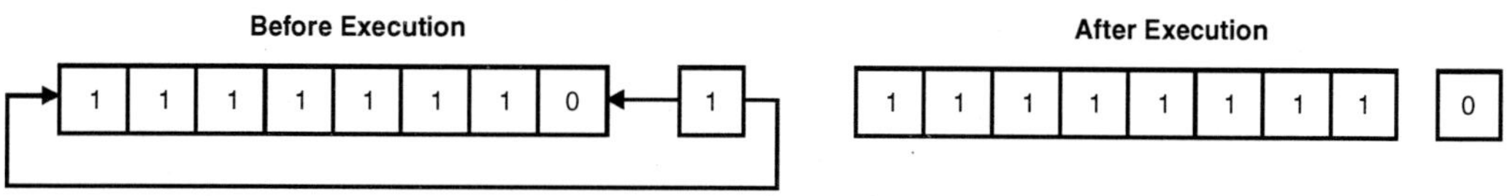

(45) CMA (Complement Accumulator) : The contents of the accumulator are complemented.

Mnemonic : CMA

Algorithm : If A = 0 then A = 1

If A = 1 then A = 0

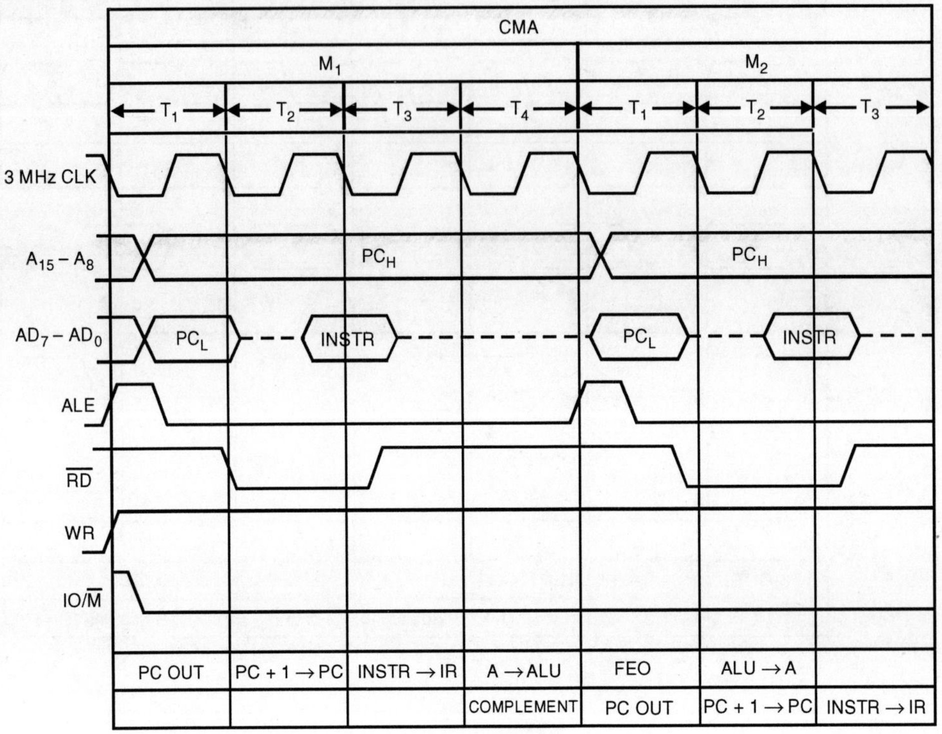

Fig. 2.97. CMA

Operation : $A = \overline{A}$

- This instruction inverts the value of contents of the accumulator.
- The complement is performing a NOT operation with each bit i.e. 0's will be replaced by 1's and 1's will be replaced by 0's.

No. of bytes : 1 byte.

Addressing mode : Implied addressing mode.

Flags affected : None

Example : Suppose A = 11110011

$\overline{A}$ = 00001100

(46) CMC (Complement Carry) : The CY flag is complemented. No other flags are affected.

Mnemonic : CMC

Algorithm : If CY = 1 then CY = 0

If CY = 0 then CY = 1

No. of bytes : 1 byte.

Addressing mode : Implied addressing mode.

Flags : Except carry no other flags affected.

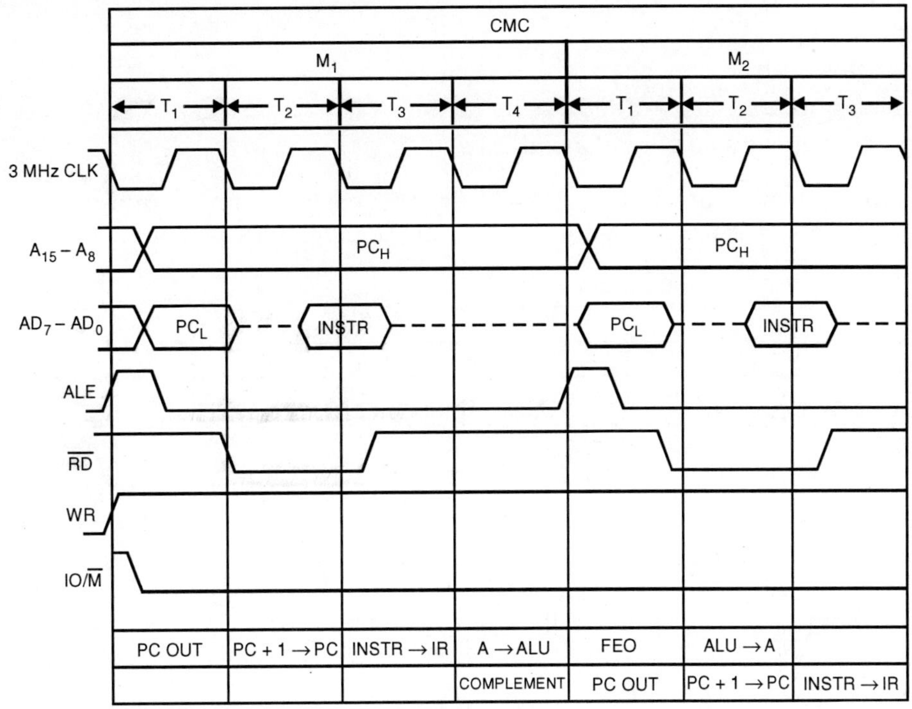

Fig. 2.98. CMC

(47) STC (Set Carry) : The CY flag is set to 1. No other flags are affected.

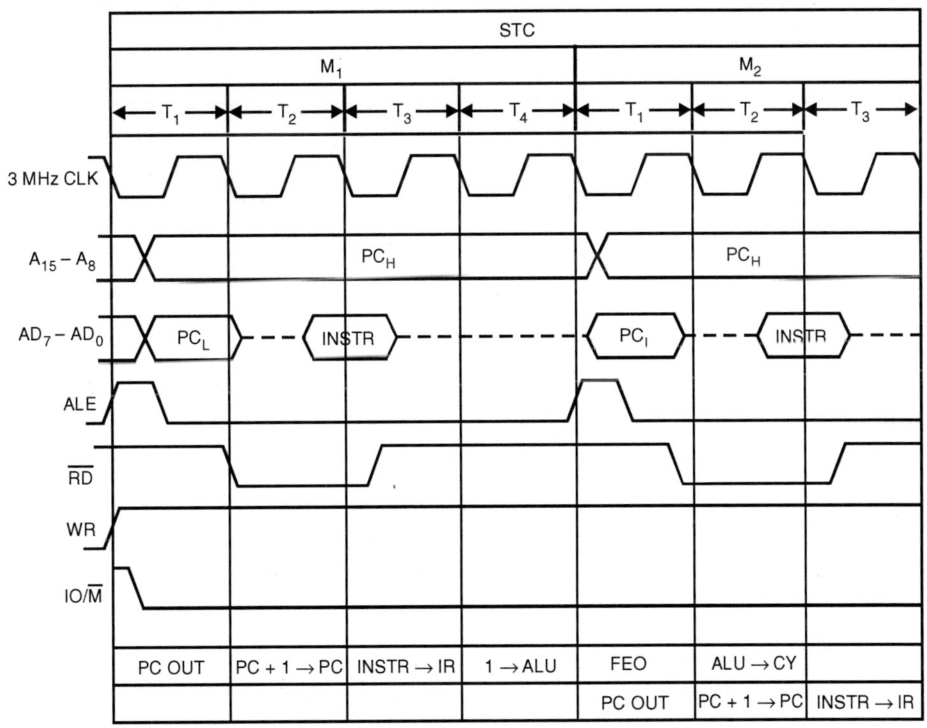

Fig. 2.99. STC

Mnemonic : STC

Algorithm : CY = 1

No. of bytes : 1 byte

Addressing mode : Implied addressing mode.

Flags : Except carry no other flags affected.

(48) JMP addr (Jump) : PC ← (byte 3) (byte 2) control is transferred to the instruction whose address is specified in byte 3 and byte 2 of the current instruction.

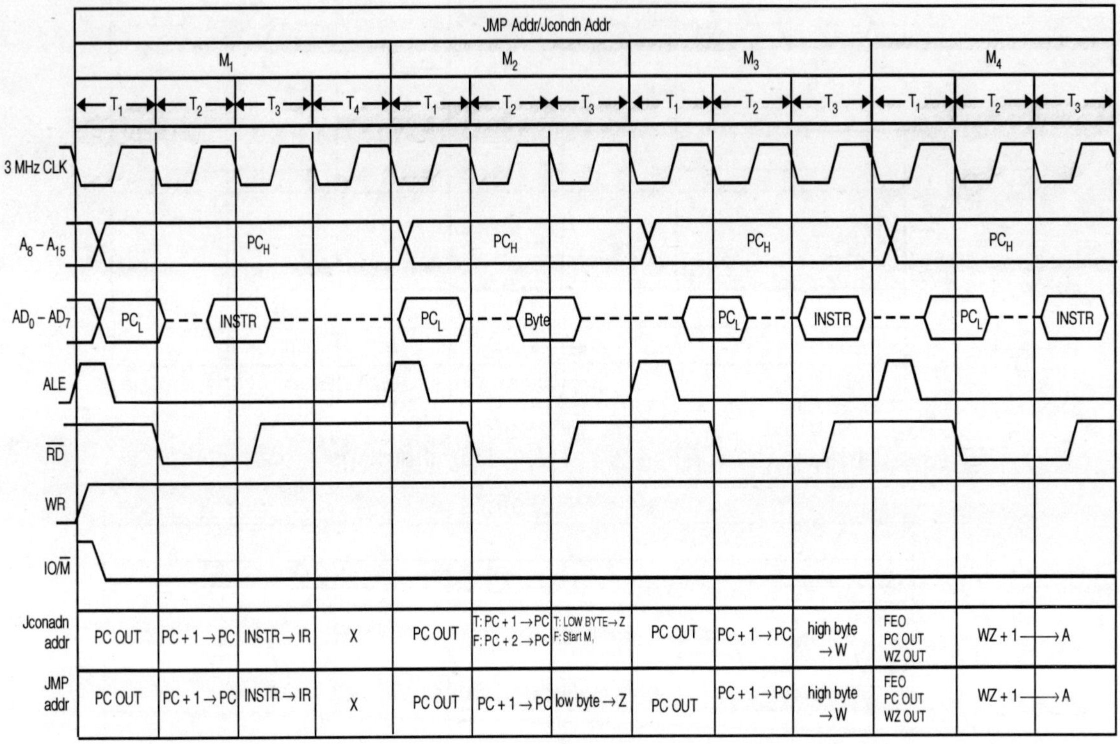

Fig. 2.100. JMP

Mnemonic : JMP address

Operation : This instruction loads the PC with the address given within the instruction and continues with the program execution from this location.

No. of bytes : 3 bytes

Addressing Mode : Immediate addressing

Flags : No flags are affected.

Example : JMP 200H

This instruction will load the PC with 200H and the processor will fetch the instruction from this address.

(49) Call addr (Call) :

Mnemonic : Call address.

Operation : This instruction is used to transfer the program control to a subprogram or subroutine, i.e. whenever this instruction is executed the program control is transferred to the address specified in the instruction. The current contents of the PC are pushed onto the stack. The stack pointer is decremented by 2. After the execution of the subroutine it is essential to transfer the program control back to the calling program. To do this the processor has to remember the address of the next instruction after the CALL instruction. Processor saves this address onto the stack.

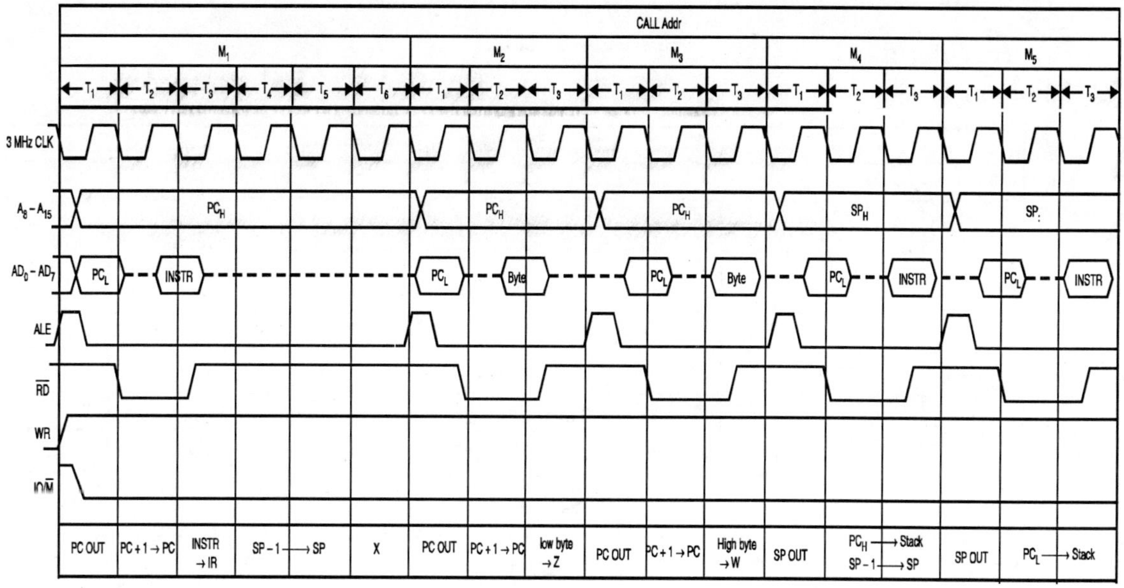

Fig. 2.101. CALL

No. of bytes : 3 bytes

Addressing Mode : Immediate address mode.

(50) Condition Addr (Condition call)

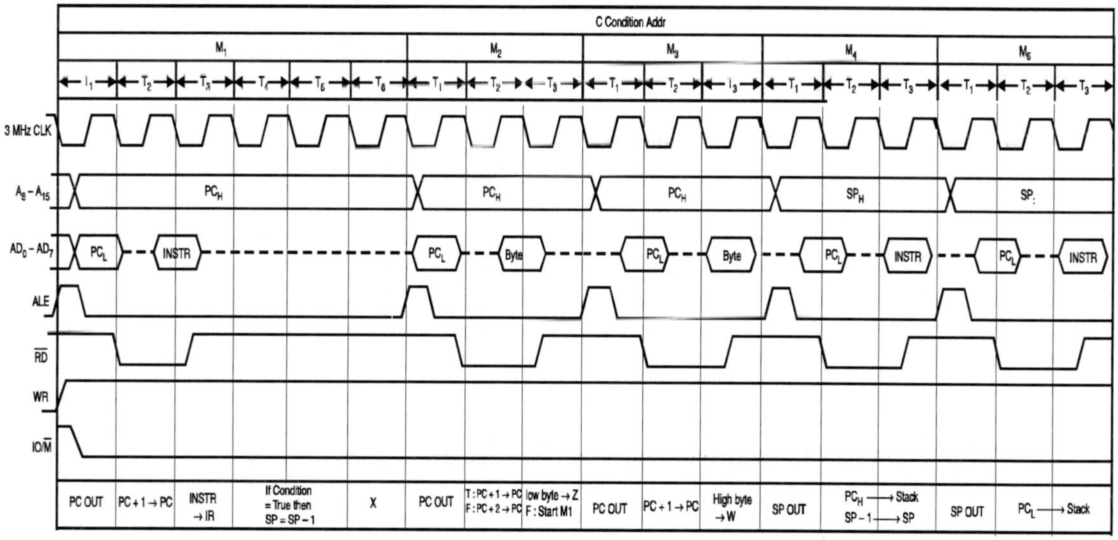

Fig. 2.102. Conditional Call

Mnemonic : C condition address

Operation : In conditional CALL instruction, when the condition is true, then a CALL at NEW address is made. If the condition is not satisfied then the instruction that after the CALL instruction is satisfied. Before CALL, the address of the instruction after the CALL instruction is stored on the stack and the stack pointer is decremented by 2.

No. of bytes : 3 bytes.

Flags : No flags are affected.

Addressing Mode : Immediate register indirect addressing.

(51) RET (Return)

Mnemonic : RET

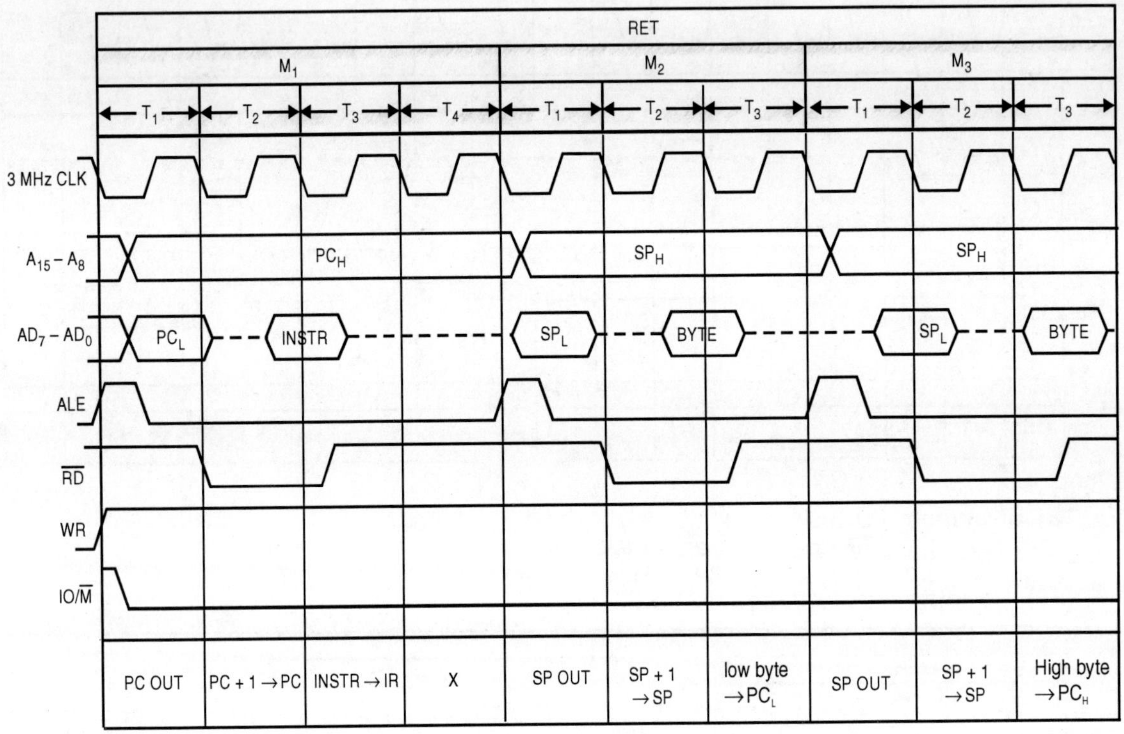

Fig. 2.103. RET

Operation : When this instruction is executed program control is transferred from the subroutine to the calling program. The return address is popped/taken from stack. This address is loaded in program counter and the program execution begins at address taken from the stack.

Thus the program control is transferred to the next instruction after CALL in the main program.

No. of bytes : 1 byte

Flags : No flags are affected.

Addressing mode : Register indirect addressing.

(52) R condition (Condition Return) :

Mnemonic : R condition

Algorithm : PCL ← (SP)

PCH ← (SP + 1)

SP ← SP + 2

Else

PC ← PC + 1

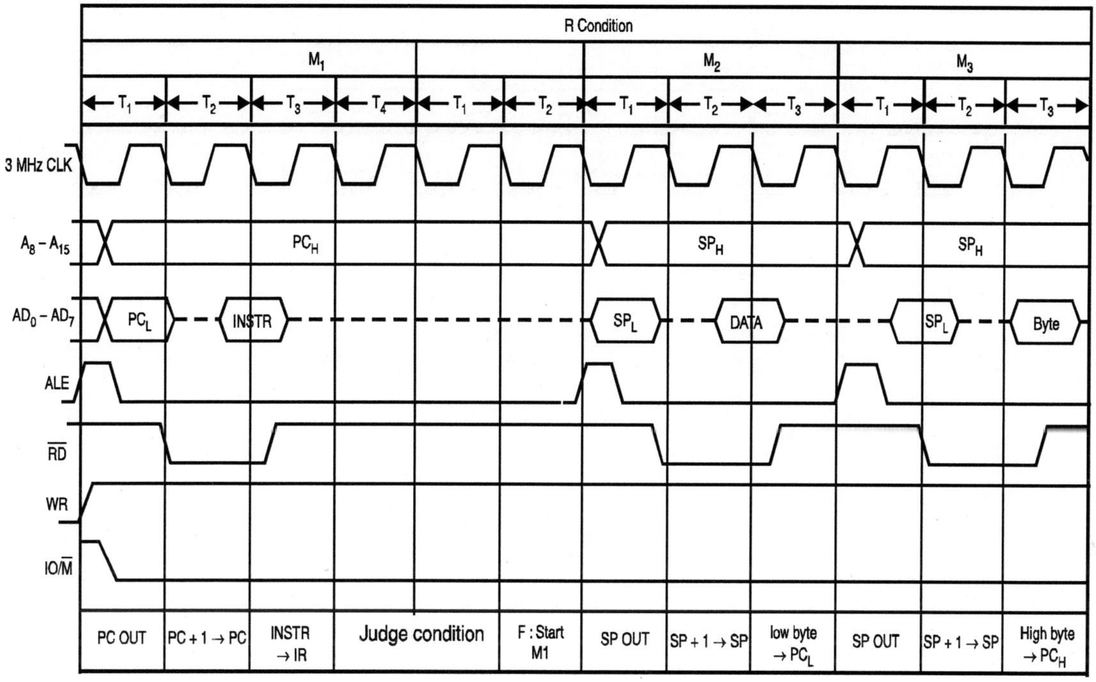

Fig. 2.104. Condition Return

Operation : In conditional RET instruction, when the condition is true, then the control is transferred to the main program.

No. of bytes : 1 byte

Flags : No flags are modified, flags are only checked.

Addressing mode : Register indirect addressing.

(53) PCHL : (Jump H and L indirect - move H and L to PC)

Mnemonic : PCHL

Algorithm : PC ¬ HL

Operation : PC = HL

The contents of H and L registers are transferred to PC. The H contents to high order 8 bits and L content to low order 8 bits of PC. This instruction is equivalent to a 1 byte unconditional JUMP instruction, provided the address of JUMP is specified by the HL register pair. The program sequence is transferred to address specified by the HL register pair.

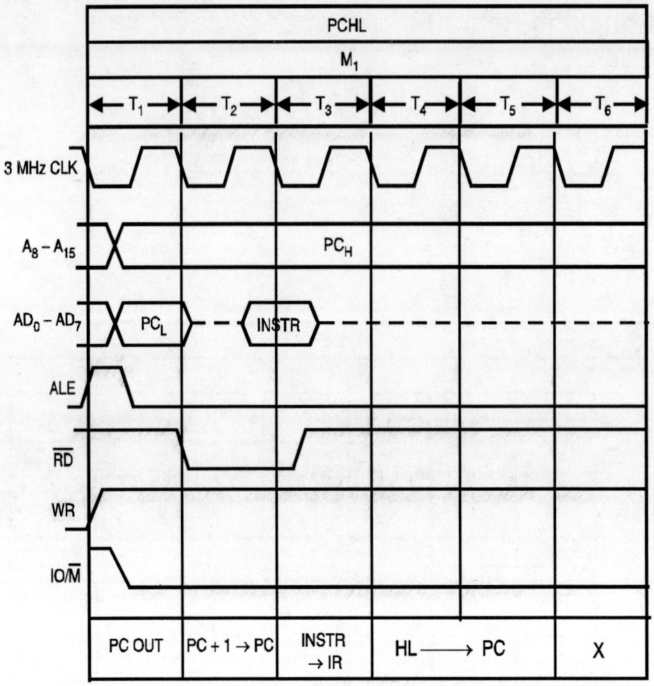

Fig. 2.105.

No. of bytes : 1 byte

Flags affected : None

Addressing Mode : Register addressing mode.

(54) Push RP (Push)

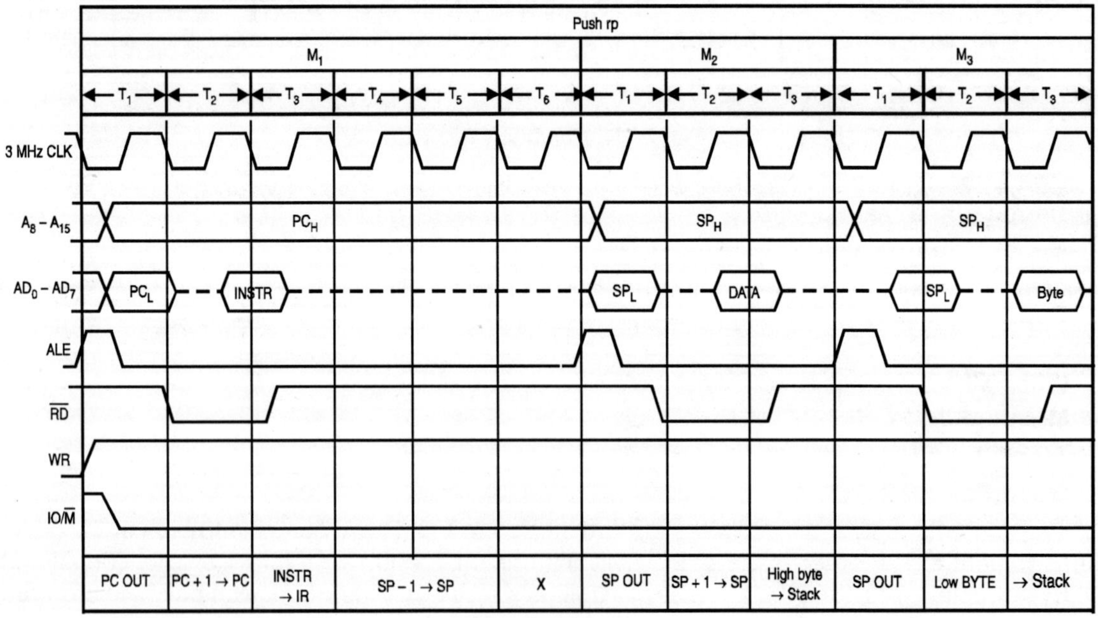

Fig. 2.106. Push RP

Mnemonic : PUSH RP

Operation : When this instruction is executed the contents of the specified register pair are copied onto the stack in the following sequence :

- The SP is decremented by one and the contents of higher order register of register pair are copied into that location.
- The SP is again decremented by 1 and the contents of lower order register of register pair are copied at that location.
- The register pair can be any valid register pair like BC, DE, HL.
- In instruction, only higher order register is specified for register pair.

No. of bytes : 1 byte.

Addressing mode : Register Indirect mode.

Flags modified : None

(55) PUSH PSW : (Push Processor Status Word)

Mnemonic : Push PSW

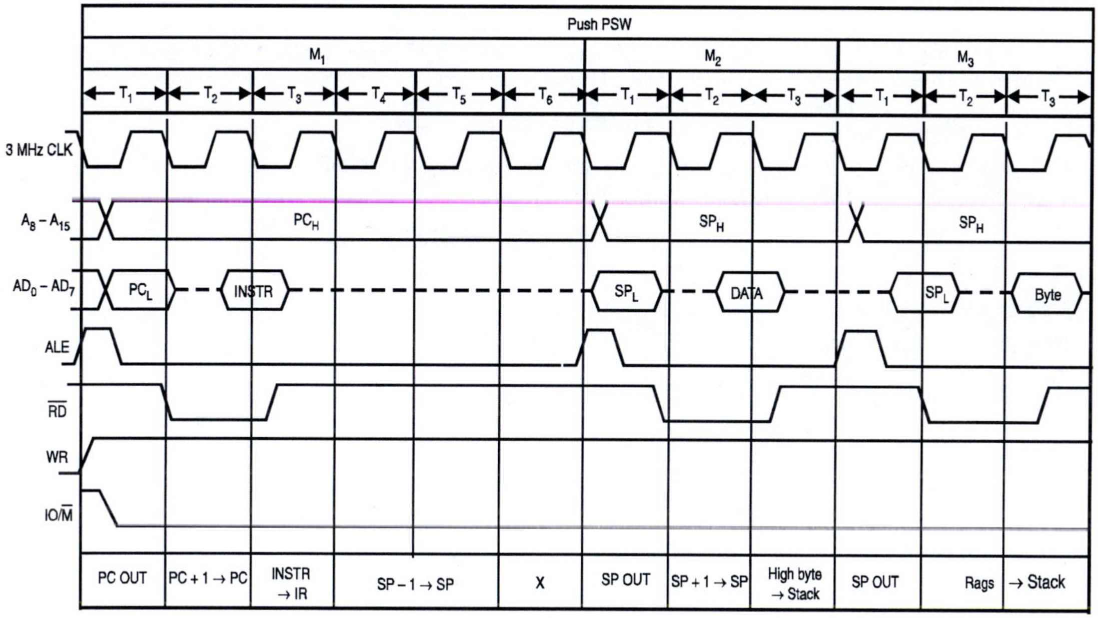

Fig. 2.107. PUSH PSW

Operation : The program status word or PSW of the processor. The A register acts as higher order register and flag register as low order register.

When this instruction is executed the contents of the specified register pair are copied onto the stack in the following sequence.

- The SP is decremented by one and the contents of the higher order register pair are copied into that location.
- The stack pointer is again decremented by 1 and the contents of lower order reg. of reg. pair are copied at that location.

No. of bytes : 1 byte

Addressing mode : Register Indirect mode

Flags modified : None

(56) XTHL : (Exchange Stack top with H and L)

Mnemonic : XTHL

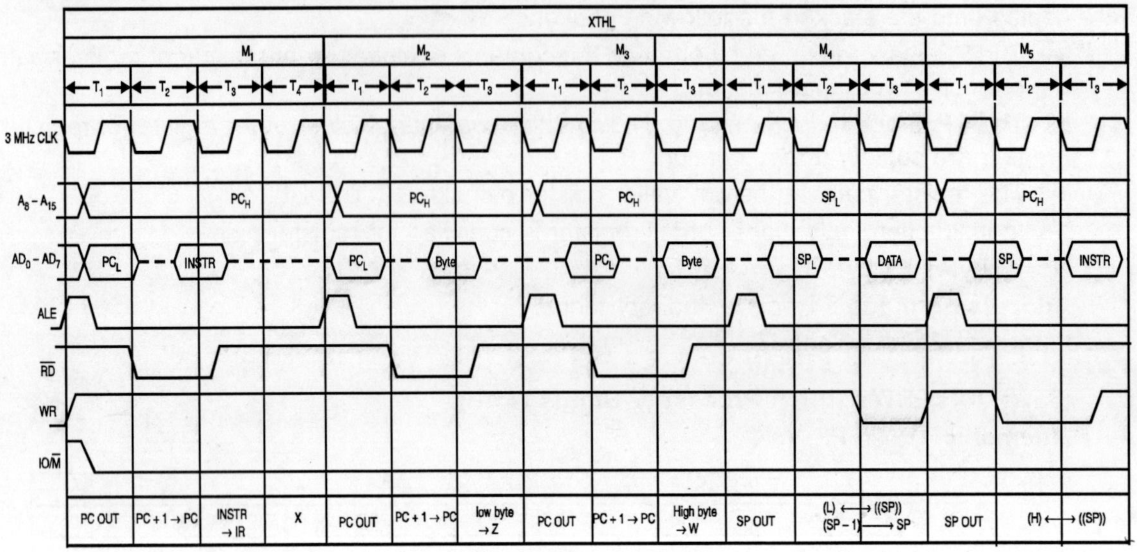

Fig. 2.108. XTHL

Operation : When this instruction is executed, the contents of L reg. are exchanged with the stack location pointed by the stack pointer.

The contents of H reg. are exchanged with the next stack location. The contents of the SP reg. are not altered.

No. of bytes : 1 byte

Flags modified : None

(57) SPHL (Move HL to SP)

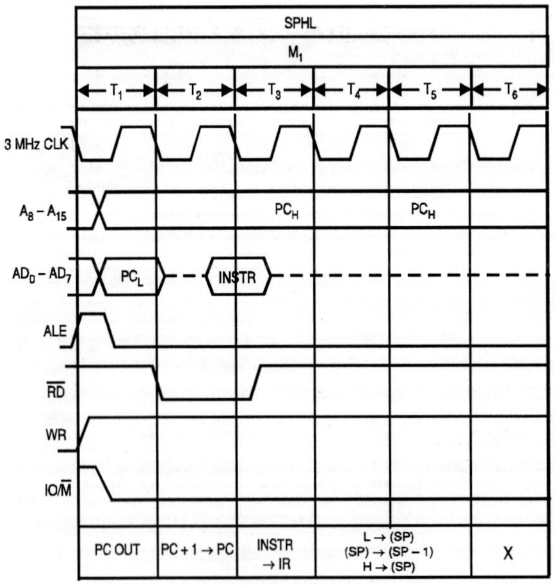

Fig. 2.109. SPHL

Mnemonic : SPHL

Operation : When this instruction is executed, the contents of HL pair transferred onto the stack pointer reg. H reg. contents are copied to high order byte of the SP and L reg. contents to low order byte of the SP. The contents of H and L are not altered.

No. of bytes : 1 byte

Addressing Mode : Register addressing Mode

Flags Modified : None

(58) In byte (Input)

Mnemonic : In 8-bit address

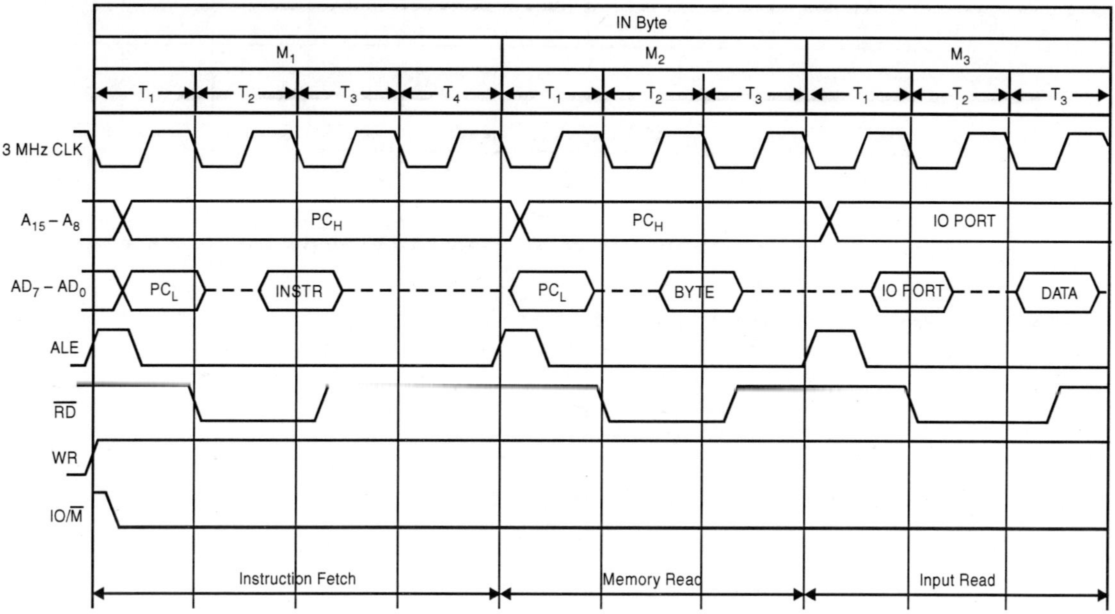

Fig. 2.110. IN Byte

Operation : This instruction is used to copy data from input port address specified into accumulator. When this instruction is executed the contents of port whose address is specified is copied into the accumulator.

No. of bytes : 2 bytes

Addressing Mode : Direct addressing.

Flags modified : None

(59) OUT Byte (Output)

Mnemonic : Out 8-bit address

Operation : This instruction is used to copy the contents of accumulator into the instruction to the iput port whose address is specified into the instruction.

No. of bytes : 2 bytes

Addressing Mode : Direct addressing

Flags affected : None

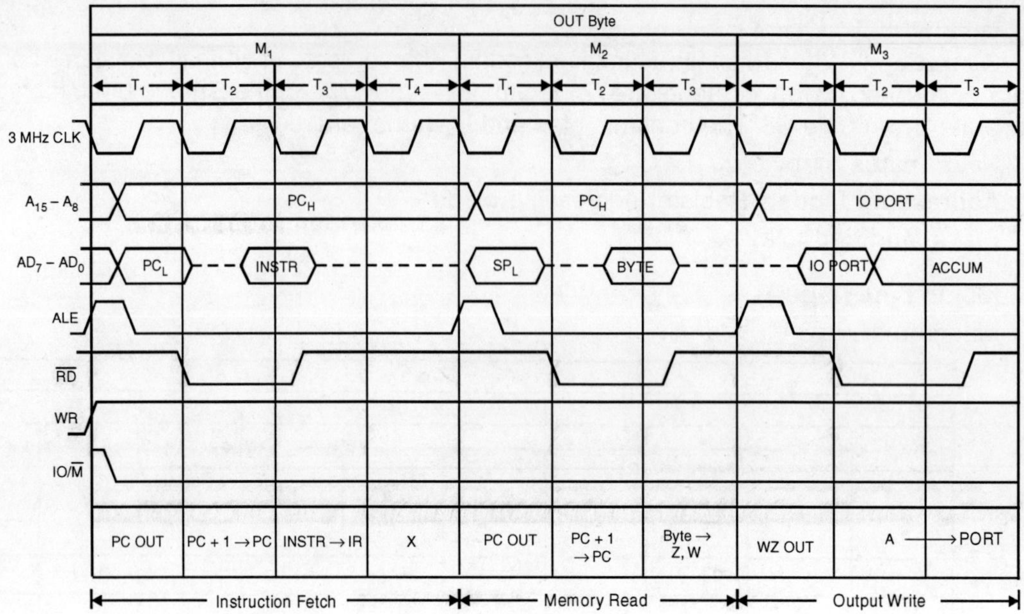

Fig. 2.111. OUT Byte

(60) DI : (Disable Interrupts)

Mnemonic : DI

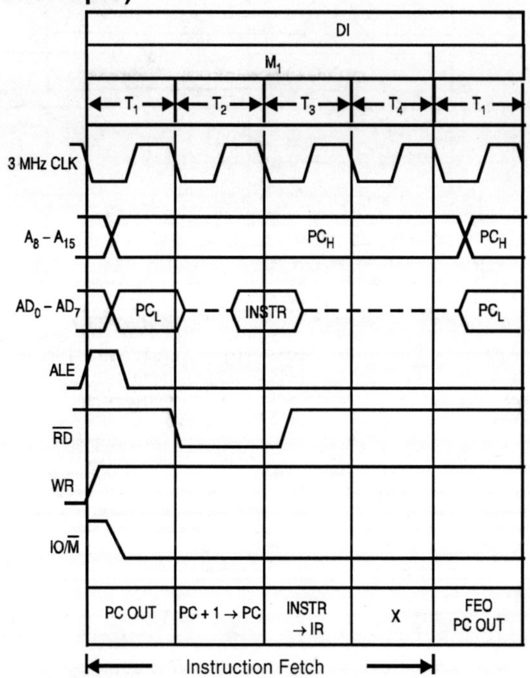

Fig. 2.112. DI

Operation : When this instruction is executed the interrupt enable flip-flop is reset so that all maskable interrupts are disabled.

Apart from TRAP all other maskable interrupts can be disabled.

No. of byte : 1 Byte

Flags modified : None

(61) RIM (Read Interrupt Masks)

Mnemonic : RIM

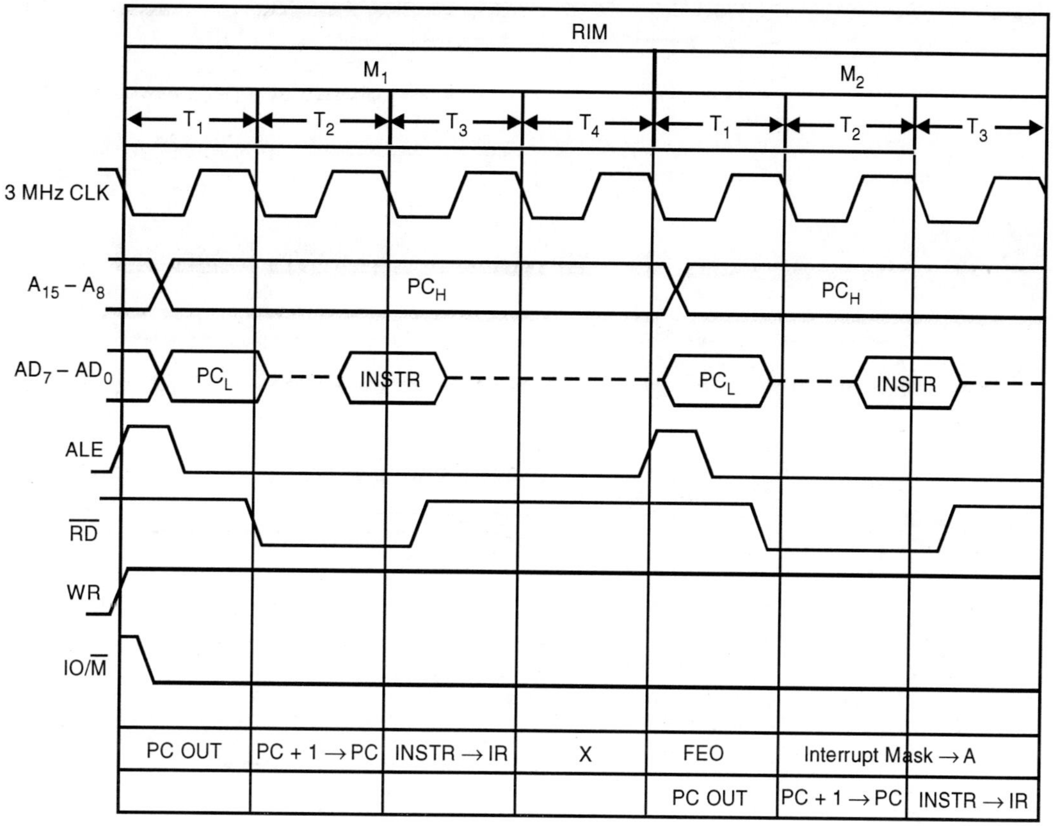

Fig. 2.113. RIM

Operation : The RIM instruction loads data into accumulator relating to interrupts and the serial input. This data contains the following information :

➢ Current interrupt mask status for the RSI 5.5, RST 6.5 and RST 7.5 hardware interrupts (1 = mask disabled)

➢ Current interrupt enable flag status (1 = interrupts enabled) except immediately following a TRAP interrupt.

➢ Hardware interrupts pending (i.e. signal received but not yet serviced), on the RST 5.5, 6.5 and 7.5 lines.

➢ Serial input data

Immediately following a TRAP interrupt the RIM instruction must be executed as a part of the service routine if you need to retrieve current interrupt status later. Bit 3 of accumulator is loaded with interrupt enable (IE Flag). Status that existed prior to the TRAP interrupt. Following an RST 5.5, 6.5, 7.5 or interrupt, the interrupt flag flip-flop reflects the current enable status. Bit 6 of the accumulator is loaded with the status of the RST 7.5 flip-flop, which is always set by an input on the RST 7.5 I/P line, even when interrupt has been previously masked.

Flags affected : None

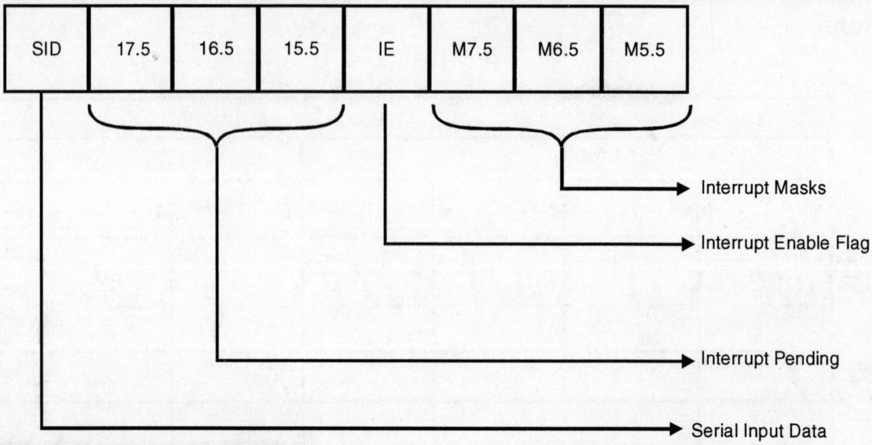

(62) SIM (Set Interrupt Masks)

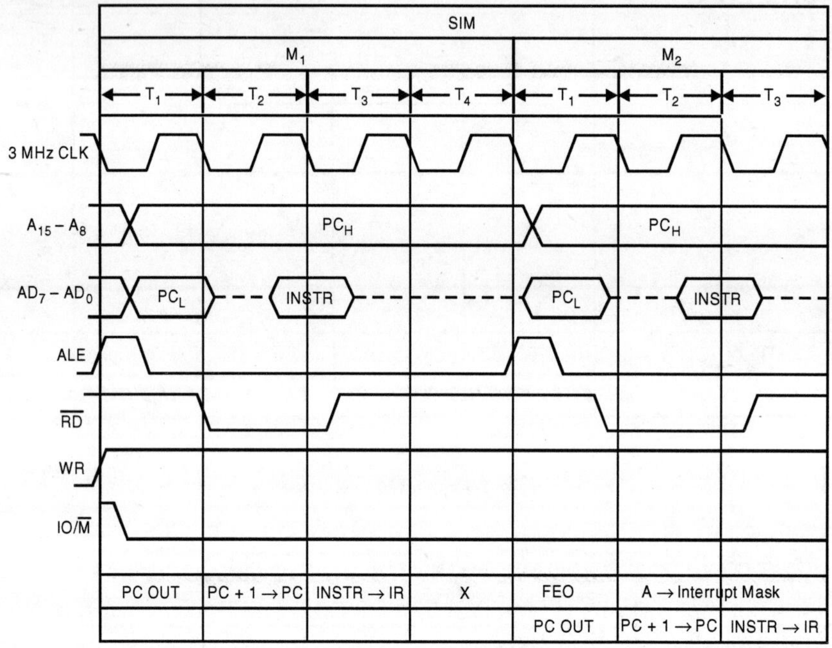

Fig. 2.114. SIM

Mnemonic : SIM

Operation : The execution of the SIM instruction uses the content of the accumulator to perform the following functions :

(a) Program the interrupt mask for the RST 5.5, 6.5 and 7.5 hardware interrupts.

(b) Reset the edge-triggered RST 7.5 input latch.

(c) Load the SOD output latch.

To program the interrupt masks, first set the accumulator bit 3 to 1 and set to 1 any bits 0, 1 and 2 which disable interrupts 5.5, 6.5, and 7.5 respectively. Then do a SIM instruction. If accumulator bit 3 is 0 when the SIM instruction is executed the interrupt mask register will not change. If accumulator bit 4 is 1 when the SIM instruction is executed, the RST 7.5 latch is then reset. The RST 7.5 is distinguished by the fact that its latch is always set by a rising edge on the RST 7.5 input pin, even if the jump to service routine is inhibited by masking. This latch remains high until cleared by a $\overline{\text{RESET IN}}$, by a SIM instruction with accumulator bit 4 high, or by an internal processor acknowledge to an RST 7.5 interrupt subsequent to the removal of the mask (by a SIM instruction). The $\overline{\text{RESET IN}}$ signal always sets all three RST mask bits.

If the accumulator bit 6 is at the level when the SIM instruction is executed, the state of accumulator bit 7 is loaded into the SOD latch and thus becomes available for interface to an external device. The SOD latch is unaffected by the SIM instruction if bit 6 is 0. SOD is always reset by the $\overline{\text{RESET IN}}$ signal.

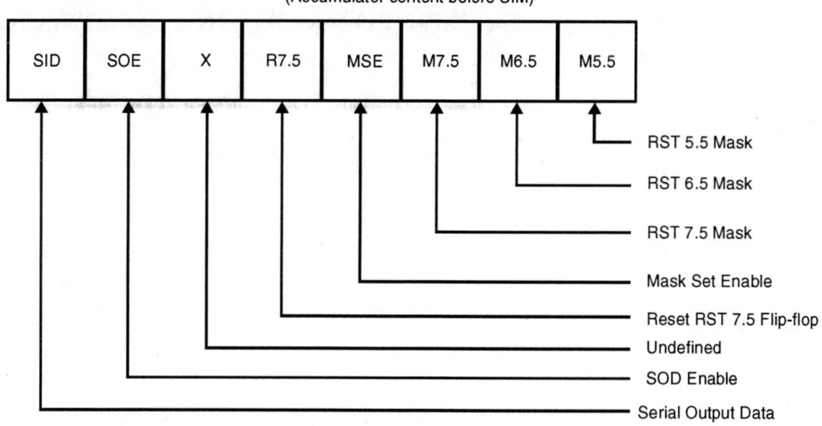

(63) NOP : No operation

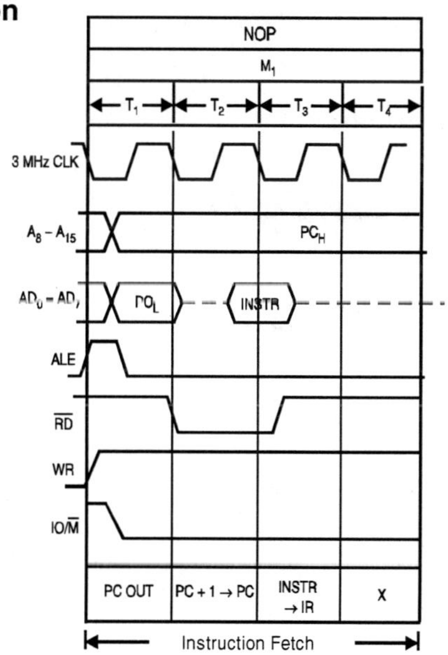

Fig. 2.115. NOP

Mnemonic : NOP

Operation : PC = PC + 1

When this instruction is executed no operation is performed. The register and flags are unaffected.

No. of bytes : 1 byte

Addressing mode : None

Flags modified : None

2.20. ASSEMBLY LANGUAGE PROGRAMMING OF 8085

The programming is a logical approach to instruct the microprocessor to perform operations in a given sequence. For repeating tasks the computer is the best, fast and accurate. In programming of a microprogramming-loop are set up by using the looping technique along with counting and indexing. Here some assembly language programs are developed-readers are advised to develop more programs for better skill. A program written by a programmer either in a note book or computer screens is referred as source code. When a source code of program is assembled, it is translated into machine language referred as object code.

Program 1

Write a program for adding two 8-bit numbers stored in memory, sum of both numbers is 8-bit is to be stored in a memory location ?

This program is for adding two 8-bit numbers stored in memory locations. Let the two numbers be 45H and 44H stored in memory locations 2051H and 2052H, respectively. The 8-bit sum of these two numbers is to be stored in memory location 2053H. Hence program involves the following steps:

1. Get first number 45H (0100 0101) from memory location 2051H into accumulator (A).
2. Add the second number 44H (01000100) form memory location 2052H to the accumulator.
3. Store the sum from accumulator to memory location 2053 H.
4. Terminate the program.

Address	Hex. Code	Label (s)	Mnemonic	Operands	Comments
2000	21,51,20		LXI	H, 2051H	Get address of first number in H-L pair
2003	7E		MOV	A, M	Get first number from memory location (2051H) into accumulator (A).
2004	23		INX	H	Increment content of H-L pair i.e., get the address of second number in H-L pair
2005	86		ADD	M	Add the content of memory location (2052H) specified by H-L pair to the content of accumulator.
2006	32, 53, 20		STA	2053H	Store the contents of accumulator i.e. sum into memory location 2053H.
2009	76		HLT		Terminate or stop and wait.

Data-to be added

First Number	: 45 H (0100 0101)
Second Number	: 44H (0100 0100)
Addition/Result	: 89 H (1000 1001)

Define Program and Procedure

Program is for adding two binary numbers each 8-bit or 1 Byte long—these Binary numbers are stored in memory locations. Since it is convenient and easy to use hexadecimal numbers-equivalent of binary number in hexadecimal is written by the programmer in source code of program.

Before we write instructions of a program it is better to write steps or algorithm of program-as written for this program (above point's i-iv). While writing/developing source code of a program each step of the algorithm must be covered, either in one or more instruction (s). Algorithm and source code become cross check of each other-lead to error free program.

As can be noticed from the format of the program, there are Address, Hex. code, Label, Mnemonic, Operands, and Comments fields. These fields are defined here with reference to the following instruction of program:

```
MOV A, M
```

It can be noticed from this instruction that MOV is an opcode or mnemonic, A and M are operands, here operand field consisting of destination (A) and source (M).

Address: The address field is used for the memory address of a code byte or data byte.

Hex. Code: The actual hexadecimal codes (Hex. code) of mnemonic and of data byte(s) are put in this field.

Labe(s) : A label is a name given to represent an address referred to in a jump or call instruction. A label is put in the label (s) field.

Mnemonic : This field contains mnemonics of the instructions-this field is also referred as opcode field. A mnemonic is short alphabetic code that assists the memory in remembering a microprocessor instruction. The mnemonic MOV is an example of this.

Operand : An operand may be a register, memory or an immediate data value. An instruction may contain zero, one or two operands. For example, above instruction MOV A, M, contains two operands, A (accumulator) and M (Memory location).

The choice of operand is usually determined by the addressing mode. The addressing mode tells the assembler where to find data in each operand i.e., memory, register, or as immediate value/data. The operands can be used in instructions as :

```
PUSH A          ; a register
MOV A, B        ; two registers
MVI A, 04H      ; one register, one data/value.
MOV A, M        ; one register, one memory location.
```

Comment : Comment for any instruction begins with a character semicolon (;) A line begins with a semicolon (;) the entire line will be treated as a comment. It is not taken as part of the program during execution of a program.

Execution of Program

An assembly language program for 8085 can be developed and executed either on microprocessor trainer kit/board or on personal computer. To develop program on personal computer-assembler software is to be installed on personal computer. However, developing program on microprocessor trainer kit is a preferable practice for basic training on assembly language programming of 8085. Here execution procedure of a program is mentioned with reference to 8085 trainer kit of VINYTICS.

1. Familiarize with 8085 trainer kit, and connect power supply to it. After putting on supply if display of the Kit shows-UP 85, it means Kit is fault free. If display does not show this message (-UP 85), press RESET key on keyboard of the kit; still if this message does not appear on display-Kit may have fault.
2. Edit the source code-edit Hex. Code into the memory of kit using keyboard of the kit in the following steps:

   Reset - Press the RESET key to initialize.

   Memory select - Press REL EXMEM to start, select and edit memory location-by editing desired memory address.

   Edit codes - Press NEXT to edit HEX. CODE, and continue pressing NEXT, till all the codes are edited.
3. Edit data value(s) in respective memory location address.
4. Execute the program-assemble, translate the source code into object code and an executable file is generated.

   Reset – press the GO (DEL GO/SEL) key.

   Starting Address – edit starting address the program.

   Execute – press FILL key to execute the program.
5. Result of the program – after execution, the result of the program based on data value(s) will be available in the specified memory location(s) or register(s). Result can be seen through the following steps :

   Reset – press RESET Key to initialize.

   Memory address – press REL EXMEM Key, and edit specified memory address of result.

   Result – After editing memory address of result, press NEXT Key to see the result.

Program 2

Write an assembly language program for addition of two 8-bit numbers; sum 16-bit ?

Let two numbers be 98H (1001 1000) and 9AH (1001 1010) and stored in memory locations 2051, and 2052, respectively. The memory locations 2053 and 2054 are reserved to store result.

Steps in program :

1. Get first number into accumulator.
2. Add second number with first in accumulator.
3. Put carries in a register.
4. Store result along with carry.

Address	Hex. Code	Label (s)	Mnemonic	Operands	Comments
2000	21,51,20		LXI	H, 2051 H	Get address of Ist number in H-L pair
2003	06,00		MV/I	B, 00	Carry (MSBs) in register B, initialize B
2005	7E		MOV	A, M	Get 1st number into accumulator.
2006	23		INX	H	Get address of 2nd number in H-L pair
2007	86		ADD	M	Add 1st and 2nd number
2008	D2, 0C, 20		JNC	NEXT	If no carry, go to label NEXT
200B	04		INR	B	If carry, increment B
200C	32, 53, 20	NEXT	STA	2053H	Store lower byte of sum in 2053H
200F	78		MOV	A, B	
2010	32, 54, 20		STA	2054H	Store MSBs of Sum in 2054H
2013	76		HLT		Stop

Data to be added

First Number : 98H (1001 1000) – 2051 H

Second Number : 9A H (1001 1010) – 2052 H

Addition/Result : 32 H (0011 0010) – 2053 H

01 H (0000 0001) – 2054 H

Program 3

Write an assembly language program for Subtraction of two 8-bit numbers ?

Let two numbers to be subtracted are F7H and 9AH, stored in 2051H and 2052H, respectively. The result of subtraction is to be stored at 2053H. Steps are :

1. Get the first number i.e. F7H from 2051H.
2. Subtract second number from first number in accumulator i.e. F7H – 9AH.
3. Store subtraction result in specified memory location i.e. 2053.

Address	Hex. Code	Label (s)	Mnemonic	Operands	Comments
2000	21, 51, 20		LXI	A, 2051H	Load H-L with 2051H
2003	7E		MOV	A, M	First number in accumulator
2004	23		INX	H	Increment H-L pair
2005	96		SUM	M	Subtract second number from first
2006	32, 53, 20		STA	2053H	Store result in 2053 H
2009	76		HLT		Stop

Data to be subtracted

First Number : F7H (1111 0111)-2051 H

Second Number : 9AH (1001 1010) – 2052 H

Subtraction/Result : 5DH (0101 1101) – 2053 H

Program 4

Write an assembly language program to subtract two 16-bit numbers ?

Let first number be 6A, 19H, LSBs, 19H in 2051 H and MSBs 6AH in 2052 H. Second number is 5C, 15H, LSBs, 15H in 2053 H and MSBs 5CH in 2054 H. Subtract second number from first number, store result LSBs in 2055H and MSBs in 2056H.

Address	Hex. Code	Label (s)	Mnemonic	Operands	Comments
2000	2A, 51, 20		LHLD	2051 H	First number $6A_19H$ in H-L pair
2003	EB		XCHG		Save first number $6A_19H$ in D-E pair
2004	2A, 53, 20		LHLD	2053 H	Second number 5C15H in H-L pair
2007	7B		MOV	A, E	Get 8-LSBs of first number in A.
2008	95		SUB	L	Subtract 8-LSBs of second number from A
2009	6F		MOV	L, A	Store 8-LSBs of result in L
200A	7A		MOV	A, D	Get 8-MSBs of first number in A
200B	9C		SBB	H	Subtract 8-MSBs of second number with borrow form A
200C	67		MOV	H, A	Store 8-MSBs of result in H
200D	22, 55, 20		SHLD	2055H	Store 16-bit result (8-LSBs and 8-MSBs) in memory locations 2055H and 2056 H
2010	76		HLT		Stop

Data to be subtracted

First number : 6A, 19H, 19H-2051 H, 6A-2052 H

Second Number : 5C15H, 15H 2053H, 5C-2054 H

Subtraction result : 0E, 04H, 04H-2055H, 0E-2056

Program 5

Write a ALP for addition of two 16-bit numbers ?

Let first number be A645H, LSBs 45H in 2051H and MSBs A6H in 2052H. Second number is 9B20H, LSBs 20H in 2053H and MSBs 9BH in 2054H. Addition result of both numbers is to be stored in memory – LSBs in 2055H and MSBs in 2056 H, and if carry in 2057 H.

Address	Hex. Code	Label (s)	Mnemonic	Operands	Comments
2000	2A, 51, 20		LHLD	2051 H	First number $6A_19H$ in H-L pair
2003	EB		XCHG		Save first number A_645H in D-E pair
2004	2A, 53, 20		LHLD	2053 H	Second number 9B20H in H-L pair
2007	06, 00		MVI	B, 00	Clear register b to save MSBs of the sum due to carry
2009	19		DAD	D	Addition of both 16-bit numbers
200A	D2, 0E, 20		JNC	NEXT	if no carry, go to the label NEXT
200D	04		INR	B	If carry, increment register B.
200E	22, 55, 20	NEXT	SHLD	2055H	Store 16-bit (lower) in 2055H and 2056 H
2011	78		MOV	A, B	Get MSBs of sum due to carry in accumulator A
2012	32, 57, 20		STA	2057 H	Store MSBs of sum due to carry in 2057 H
2015	76		HLT		Stop

Data to be added

First number : A645H, 45H-2051H, A6H-2052H

Second number : 9B20H, 20H-2053H, 9BH-2054H

Addition result : 014165H, 65-2055H, 41H-2056H, 01H- 2057H

MSBs of sum due to carry   Lower 16-bit of sum

Program 6

Write an assembly language program to find I's complement of an 8-bit numbers ?

Let the number to be I's complemented be 45H (0100 0101) in memory location 2051H.

Address	Hex. Code	Label (s)	Mnemonic	Operands	Comments
2000	3A,51,20		LDA	2051H	Data iri accumulator
2003	87		ADD	A	Add A to A-shift binary number to left by one bit
2004	32,52,20		STA	2052H	Store result
2007	76		HLT		Stop

Data-to be I's-.complemented

Data: 45H (0100 0101) - 2051H

Complemented result: BAH (1011 1010) - 2052H

Program 7

Write an assembly language program to find 2's complement of an 8-bit number ?

Let us find 2's complement of Program-6 above. The 2's complement of a binary number means adding 1 to its I's complement. Hence, digitally 2's complement of 45H (0100 0101).

Data 45H = 0100 0101

1's complement = 1011 1010 = BAH

Add 1 to I's complement = 0000 0001

2's complement of 45H = 1011 1011 = BBH

Address	Hex. Code	Label (s)	Mnemonic	Operands	Comments
2000	3A,51,20		LDA	2051H	Get number into accumulator
2003	2F		CMA		Complement number
2004	3C		INR	A	Add 1 (2's complement of number)
2005	32,52,20		STA	2052H	Store result
2008	76		HLT		Stop

Data-to be 2's complemented

Number: 45H (0100 0101) – 2051H

2's complemented result: BBH (1011 1011) – 2052 H

Program 8

Write an assembly language program for shifting an 8-bit binary number left by one bit ?

To shift a binary number left by one bit, the number is added to itself. Let the 45H (01000101) in 2051 H be the number to be shifted to left by one bit. The result is to be stored in 2052 H.

Address	Hex. Code	Label (s)	Mnemonic	Operands	Comments
2000	3A,51,20		LDA	2051H	Data in accumulator
2003	87		ADD	A	Add A to A-shift binary number to left by one bit
2004	32,52,20		STA	2052H	Store result
2007	76		HLT		Stop

Data to be shifted

Number : 45 H (0100 0101) - 2051 H

Shifted result : 8AH (1000 1010) - 2052 H

Program 9

Write an assembly language program to mask 4-LSBs of an 8-bit binary number ?

In a binary number bits to be masked need to be ANDed with equal number of 0 (zero) bits. To mask 4-LSBs of 45H (0100 0101), it is to be ANDed with FOH (1111 0000).

Address	Hex. Code	Label (s)	Mnemonic	Operands	Comments
2000	3A,51,20I		LDA	2051H	Number in accumulator.
2003	*E6,FO*		*ANI*	*FOH*	*Mask. off the 4-LSBs of nurnber*
2005	*32,52,20*		*STA*	*2052H*	*Store result*
2008	76		HLT		Stop

Program 10

Write an assembly language program to divide two"S-bit binary numbers ?

In the process of division, the divisor is subtracted from the MSBs of the dividend-if there is no borrow, the bit of the quotient is set to 1; otherwise O. A computer/microprocessor performs division by trial subtractions. In order to line up the dividend and quotient properly, the dividend is shifted left by one bit before each trial of subtraction. Let an 8-bit dividend (8AR - 2051H) be divided by another 8-bit divisor (12H - 2052H). The quotient is to be stored in 2053H and remainder in 2054H.

Address	Hex. Code	Label (s)	Mnemonic	Operands	Comments
2000H	3A, 52, 20		LDA	2052H	Get divisor into accumulator
2003H	47		MOV	B,A	Save divisor in B
2004H	3A, 51, 20		LDA	2051H	get dividend into accumulator
2007H	87		ADD	A	Add A to A-shift dividend and quotient left by one bit
2008H	90		SUB	B	Subtract divisor from dividend (A-B)
2009H	32,53,20		STA	2053H	Store quotient in 2053H and remainder in 2054H.
200AH	76		HLT		Stop

Data-to be divided

Dividend : 4"89BH, 9BH-2051H, 48H-2052H

Divisor : 1AH -2053H

Result Quotient : F2H - 2054H

Remainder : 07H - 2055H

Program 11

Write an assembly language program for dividing a 16-bit number by an 8-bit number ?

Let the 16-bit dividend be stored in memory locations 2051 H and 2052H. An 8-bit divisor is stored in memory at 2053H. The quotient and remainder are to be stored in memory locations 2054 H and 2056 H, respectively.

Address	Hex. Code	Label (s)	Mnemonic	Operands	Comments
2000	2A,51,20		LHLD	2051H	Dividend in H-L Pair
2003	3A,53,20		LDA	2053H	Divisor in Accumulator
2006	47		MOV	B,A	Save divisor in B
2007	.. OE,08		MVI	C,08	Count in register C, C = 08
2009	29	LOOP	DAD	H	Add H-L with itself-shift dividend and quotient by one bit to left
200A	7C		MOV	A,H	8-MSBs of dividend in accu-mulator
200B	90		SUB	B	Subtract divisor from MSBs of dividend.
200C	DA,11,20		JC	NEXT	Is most significant part of dividend >divisor? If no, go to NEXT.
200F	67		MOV	H,A	MSBs of dividend in register H
2010	2C		INR	L	Yes, increment quotient by 1
2011	OD	NEXT	DCR	C	Decrement count by 1
2012	C2,09,20		JNZ	LOOP	If count is not zero, jump back to LOOP
2015	22,54,20		SHLD	2054H	Store quotient in 2054H and remainder in 2055H
2018	76		HLT		Stop

Data-to be divided

Dividend : 489BH, 9BH-2051H, 48H-2052H

Divisor : 1AH -2053H

Result Quotient : F2H - 2054H

Remainder : 07H - 2055H

Program 12

Write an assembly language program for multiplying two 8-bit binary numbers, result 8-bit ?

Let the multiplier be 05H-2051H, and multiplicand 06H-2052H. Multiplication result is to be stored in 2053H. Multiplication process is same as adding the multiplicand to zero by number of times equal to the number of bits in multiplier. Steps in multiplication of two 8-bit binary numbers:

1. Get multiplicand and multiplier in H-L register pair.
2. Set accumulator (A) to zero.
3. Add multiplicand to accumulator.
4. Decrement the mulitplier.
5. Check zero flag-if not set, repeat from step 3 above.

6. Store result.
7. Stop.

Address	Hex. Code	Label (s)	Mnemonic	Operands	Comments
2000	2A,51,20		LHLD	2051H	Multiplier and multiplicand in H-L pair, L-multiplier, H-multiplicand
2003	3E,00		MVI	A, 00	Set accumulator to zero, A = 00
2005	84	LOOP	ADD	H	Add multiplicand to accumulator (A+ H) A
2006	2D		DCR	L	Decrement multiplier, (L - 1) = L
2007	C2		JNZ	LOOP	If zero flag not set, repeat
2008	32,53,20		STA	2053H	Store result
200B	76		HLT		Stop

Data-to be multiplied

Multiplier : 05H (0000 0101) - 2051H
Multiplicand : 06H (0000 0110) - 2052 H
Multiplication result : 1EH (0001 1110) - 2053 H

Program 13

Write an assembly language program for multiplying two 8-bit numbers, result 16-bit ?

In multiplication process each bit of multiplier is taken one-by-one 'and it is checked whether it is lor O. If the bit of the multiplier is 1, the multiplicand is added to the product and the product is shifted to left. When the bit of the multiplier is 0, the product is simply shifted to left by one bit.

Multiplicand : 8-LSBs - 2051H, 8-MSBs - 2052H Multiplier : 8-bit' - 2053 H
Result : 8-LSBs - 2054H, 8-MSBs - 2055H

Address	Hex. Code	Label (s)	Mnemonic	Operands	Comments
2000	2A,51,20		LHLD	2051H	Multiplicand in H-L pair
2003	EB		XCHG		Save multiplicand in D-E. pair
2004	3A,53,20		LDA	2053H	Get multiplier in accumulator
2007	21,00.00		LXI	H,OOOO	Initial value of product in H-L pair is zero.
200A	OE.08		MVI	C,08	Count in register C = 08
200C	29	LOOP:	DAD	H	Shift partial product left bv one bit.
200D	17		RAL		Rotate multiplier left by one bit. Is multiplier's bit = 1 ?
200E	D2,12,20		JNC	NEXT	No carry, go to NEXT.
2011	19		DAD	D	Product = Product + Multiplicand
2012	OD	NEXT:	DCR	C	Decrement count in C.
2013	C2,OC,20		JNZ	LOOP	If count is not zero, jump back to
2016	22,54,20	LOOP	SHLD	2054H	Store result.
2019	76		HLT		Stop.

Data-to be multiplied

Multiplicand: 0084H, 84H-205IH, 00H-2052H

Multiplier: 56H - 2053H

Multiplication result: 2C58H, 58H - 2054H, 2CH - 2055H

Program 14

Write an assembly language program for adding two8-bit BCD numbers; result 16-bit ?

In BCD addition, the two BCD numbers are taken by the microprocessor as binary numbers *only,* and result of addition of two nibbles of two BCD numbers may exceed 1001. If the addition of two nibbles of two BCD numbers becomes more than 1001, it is referred as out of range in *BCD* system. In such condition, the instruction DAA (Decimal Adjust Accumulator) is used just after the BCD addition-for correct BCD addition of numbers. In case addition of two BCD nibbles becomes more than 1001, or has resulted into carry from the addition of higher nibbles; DAA instruction adds 0110(6) to the resulted lower nibble or to the carry from higher nibbles addition, otherwise DAA does nothing.

Steps in BCD addition :

1. Get address of first 8-bit BCD number into H-L pair.
2. Initialize a register for storing final carry of the result.
3. Save first number into accumulater.
4. Get second 8-bit BCD number.
5. Add the two BCD numbers.
6. Decimal adjust the result.
7. Check for carry, if no carry skip, else increment reserved register for carry.
8. Store result.
9. Stop.

It can be observed from above BCD addition, that addition in binary of two BCD numbers is not correct. Whereas, addition of 0110(6) to addition-I, results in addition-2-which is a correct addition.

Address	Hex. Code	Label (s)	Mnemonic	Operands	Comments
2000	21,51,20		LXI	H,2051H	Address of first BCD number in H-L pair
2003	06,00		MVI	B,OO	Initialize register B for carry
2005	7E		MOV	A,M	Get first BCD number into accumulator.
2006	23		INX	H	Increment H-L pair, address of second BCD number in H-L pair
2007	86		ADD	M	Add first and second BCD numbers
2008	27, _		DAA		Decimal adjust accumulator
2009	D2,OD,20		JNC	NEXT	If no carry, go to NEXT
200C	04		INR	B	If carry, increment B
200D	23	NEXT	INX	H	Increment H-L pair, H-L pair points to 2053 H memory address
200E	77		MOV	M,A	Store lower byte of BCD addition in memory-2053H
200F	23		INX	H	Increment H-L pair
2010	70		MOV	M,B	Store carry i.e., uppder byte of addition in 5054 H
2011	76		HLT		Stop

Data for BCD addition

First Number : 25 (0010 0101) - 2051 H

Second Number : 53 (0101 0011) - 2052 H

Addition Result : 78 (0111 1000) - 2053 H

Flow chart of addition of two 8-bit BCD numbers is presented in Fig. 2.34.

Program 15

Write an assembly language program to find square of an 8-bit number; using look up table ?

A "look-up table" is a group of memory locations, where square of data are stored in a tabular form. Such as the squares of numbers from 00 to 09 (decimal) are stored in memory locations 2500H to 2509H - values of squares are stored in decimal. The data to be squared form the index-data is first transferred to accumulator and then to register L. This data in register L forms the LSBs of memory location where square of data. is stored. The MSBs of the address are moved to register H. This makes the address of memory location where square of the data resides in H-L pair. Now the square of the data is moved from memory to accumulator and stored in a desired location.

Address	Hex. Code	Label (s)	Mnemonic	Operands	Comments
2000	3A,51,20		LDA	2051H	Get data into accumulator.
2003	6F		MOV	L,A	Transfer data in register L.
2004	26,25		MVI	L,25H	Get higher byte of memory address of square of data in H register.
2006	7E		MOV	A,M	H-L contains the address of square of data, Put square of data in A.
2007	32,52,20		STA	2052H	Store square of data in memory.
200A	76		HLT		Stop.

Data to be squared

Number : Q8 - 2051 H

Square result : 64 - 2052 H

Address (in Hexadecimal)	Square (Decimal)
2500	00
2501	01
2502	04
2503	09
2504	16
2505	25
2506	36
2507	49
2508	64
2509	81

Program 16

Write an assembly language program to move a block of data from a section of memory to another section of memory.

Address	Hex. Code	Label (s)	Mnemonic	Operands	Comments
2000	21,50,20		LXI	H,2050H	Memory address of the count in H-L pair
2003	4E		MOV	C,M	Get count in register C
2004	21,51,20		LXI	H, 2051H	H-L pair points to source of data
2007	11,81,20		LXI	D,2081H	D-E pair points to destination of data to be moved.
200A	7E	LOOP	MOV	A,M	Get data byte into accumulator A from source
200B	EB		XCHG		Destination address in H-L pair
200C	77		MOV	M,A	Data byte in destination address of memory
200D	EB		XCHG		Source address in H-L pair
.200E	23		INX	H	H-L points to the source address of next data byte.
200F	13		INX	D	D-E points to the destination address of next data byte
2010	OD		DRC	C	Count decremented
2011	C2, OA, 20		JNZ	LOOP	Count not zero, jump back to LOOP.
2012	76		HLT		Stop

REVIEW QUESTIONS

1. What is the difference between an accumulator and any ordinary 8 BIT register?
2. What is the basic difference between SPR and GPR ? What is so special about special purpose registers.
3. Explain the block diagram for 8085 ,explaining each block in detail.
4. What is the main job of an instruction decoder .Design a circuit which is capable of decoding instructions.
5. Draw the block diagram of 8085 internal architecture and explain the functions of decoding and execution of an instruction?
6. List the internal registers in 8085 microprocessor and their abbreviations and lengths. Describe the primary function of each register.
7. Differentiate between NMI and MI interrupts
8. Explain how with external hardware TRAP can be masked
9. Interface a Speaker to SOD pin of 8085 Microprocessor.
10. Explain DMA function in 8085 microprocessor with timing diagrams
11. Explain the timing diagrams of 8085 when it is executing Memory mapped I/O and I/O mapped I/O instructions
12. Draw a logic schematic to generate four control signals using 8085 IO/ M, RD, WR signals. (1) MEMR (2) MEMW (3) IOR (4) IOW. Explain the functions of these control signals ?
13. Draw and explain the timing diagram for the execution of instruction

    LXI rp, 16 bit data

14. Explain the following instructions and show the status of PSW after each execution :

(1) LDAX B	(2) SHLD, fc90h
(3) DAD H	(4) XRA, A
(5) XTHL	(6) INR M
(7) DAA	(8) JM, 16 bit address
(9) PUSH D	(10) OUT, 8 Bit Port address

15. How many address lines are necessary to address 2MB of memory ? If the memory chip size is 2048 x 8 bits how many chips are required to make up 16 KB memory ? The memory address of last location of 1kb memory is given as FBFF h specify the starting address?
16. Draw & explain the pin configuration and the internal architecture of 8255 ? Write a BSR mode control word subroutine to set bits PC7 & PC3 and reset them after 10 milli seconds. The port address selected is 83h?
17. A Push Button key-board is connected to port A & 7 segment LED display is connected to port B of 8255. Write a program to monitor the key-board to sense a key pressed and display the no of key at 7 segment LED. Draw the interfacing circuit for the same ?
18. Write a subroutine to transmit ASCII character stored in reg B using SOD line as 1 bit O/P port?
19. WAP to enable all the interrupts in 8085?
20. What are different 8085 vectored interrupts and give the call locations for each interrupt ? Explain each of them?
21. Assume that the 8085 mp is completing an RST 7.5 interrupt request. Write an assembly language program to check if RST 6.5 is pending. If it is pending enable RST 6.5 without affecting any other interrupts otherwise return to the main program. Assume that the RST 6.5 is stored at location 7000h ?
22. What is RIM and SIM instructions. Discuss the bit pattern of the accumulator for both the instructions?
23. Write short notes on
    (1) Synchronous vs Asynchronous
    (2) Memory mapped I/O vs I/O mapped I/O.

❑❑❑

CHAPTER 3

8086 MICROPROCESSOR

INSIDE THIS CHAPTER

3.1. PIN CONFIGURATION OF 8086 IN DIFFERENT WORKING MODES

The Fig. 3.1 depicts the pin outs of 8086 microprocessor in its two operating modes the minimum and maximum modes. In this section, we will observe the pin functions of 8086 microprocessor.

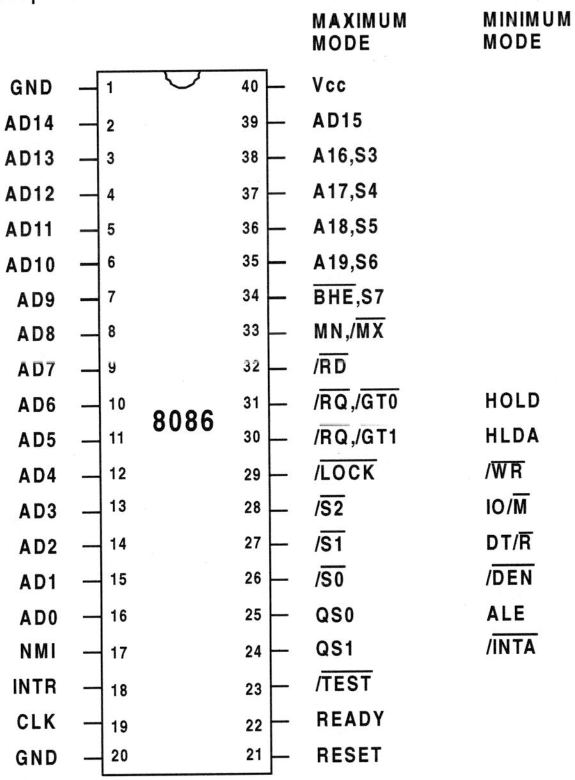

Fig. 3.1. Pin configuration for 8086 microprocessor.

Symbol	Pin No.	Type	Name and Function
AD_{15} –AD_0	2 – 16, 39	I/O	ADDRESS DATA BUS: These lines constitute the time multiplexed memory/IO address (T_1), and data (T_2,T_3,T_W,T_4) bus. A_0 is analogous to BHE for the lower byte of the data bus, pins D_7 – D_0.Itis LOW during T_1 when a byte is to be transferred on the lower portion of the bus in memory or I/O operations. Eight-bit oriented devices tied to the lower half would normally use A_0 to condition chip select functions. (See BHE.)
A_{19}/S_6, A_{18}/S_5, A_{17}/S_4, A_{16}/S_3	35–38	O	ADDRESS/STATUS: During T_1 these are the four most significant address lines for memory operations. During I/O operations these lines are LOW. During memory and I/O operations, status information is available on these lines during T_2,T_3,T_W,T_4. The status of the interrupt enable FLAG bit (S_5) is updated at the beginning of each CLK cycle. A_{17}/S_4 and A_{16}/S_3 are encoded as shown. This information indicates which relocation register is presently being used for data accessing. A_{17}/S_4 — A_{16}/S_3 — Characteristics 0 (LOW) — 0 — Alternate Data 0 — 1 — Stack 1 (HIGH) — 0 — Code or None 1 — 1 — Data S_6 is 0 (LOW)
$\overline{BHE}$ /S_7	34	O	BUSHIGH ENABLE/STATUS: During T_1 the bus high enable signal (BHE) should be used to enable data onto the most significant half of the data bus, pins D_{15} –D_8. Eight-bit oriented devices tied to the upper half of the bus would normally use BHE to condition chip select functions. BHE is LOW during T_1 for read, write, and interrupt acknowledge cycles when a byte is to be transferred on the high portion of the bus. The S_7 status information is available during T_2, T_3, and T_4. The signal is active LOW. It is LOW during T_1 for the first interrupt acknowledge cycle. BHE — A_0 — Characteristics 0 — 0 — Whole word 0 — 1 — Upper byte from/to odd address 1 — 0 — Lower byte from/to even address 1 — 1 — None
$\overline{RD}$	32	O	READ: Read strobe indicates that the processor is performing a memory or I/O read cycle, depending on the state of the S_2 pin. This signal is used to read devices which reside on the 8086 local bus. RD is active LOW during T_2,T_3 and T_W of any read cycle, and is guaranteed to remain HIGH in T_2 until the 8086 local bus has floated.
READY	22	I	READY: is the acknowledgement from the addressed memory or I/ O device that it will complete the data transfer. The READY signal from memory/IO is synchronized by the 8284A Clock Generator to form READY. This signal is active HIGH. The 8086 READY input is not synchronized. Correct operation is not guaranteed if the setup and hold times are not met.

INTR	18	I	INTERRUPT REQUEST: is a level triggered input which is sampled during the last clock cycle of each instruction to determine if the processor should enter into an interrupt acknowledge operation. A subroutine is vectored to an interrupt vector lookup table located in system memory. It can be internally masked by software resetting the interrupt enable bit. INTR is internally synchronized. This signal is active HIGH.
$\overline{TEST}$	23	I	TEST: input is examined by the "Wait" instruction. If the TEST input is LOW execution continues, otherwise the processor waits in an "Idle" state. This input is synchronized internally during each clock cycle on the leading edge of CLK.
NMI	17	I	NON-MASKABLE INTERRUPT: an edge triggered input which causes a type 2 interrupt. A subroutine is vectored to via an interrupt vector lookup table located in system memory. NMI is not maskable internally by software. A transition from LOW to HIGH initiates the interrupt at the end of the current instruction. This input is internally synchronized.
RESET	21	I	RESET: causes the processor to immediately terminate its present activity. The signal must be active HIGH for at least four clock cycles. It restarts execution, as described in the Instruction Set description, when RESET returns LOW. RESET is internally synchronized.
CLK	19	I	CLOCK: provides the basic timing for the processor and bus controller. It is asymmetric with a 33% duty cycle to provide optimized internal timing.
V_{CC}	40		V_{CC}: 5V power supply pin.
GND	1, 20		GROUND : The ground connection is the return for the power supply. [Note that 8086/8088 has two gnd pins, they must be connected together to the ground for proper operation].
MN/$\overline{MX}$	33	I	MINIMUM/MAXIMUM: indicates what mode the processor is to operate in. The two modes are discussed in the following sections.

The following pin function descriptions are for the 8086/8288 system in maximum mode (i.e., MN/MX ^{e} V_{SS}). Only the pin functions which are unique to maximum mode are described; all other pin functions are as described above.

$\overline{S}_2, \overline{S}_1, \overline{S}_0$	26–28	O	STATUS: active during T_4,T_1, and T_2 and is returned to the passive state (1, 1, 1) during T_3 or during T_W when READY is HIGH. This status is used by the 8288 Bus Controller to generate all memory and I/O access control signals. Any change by S_2,S_1,orS_0 during T_4 is used to indicate the beginning of a bus cycle, and the return to the passive state in T_3 or T_W is used to indicate the end of a bus cycle.
			These status lines are encoded as shown.

S_2	S_1	S_0	Characteristics
0 (LOW)	0	0	Interrupt Acknowledge
0	0	1	Read I/O Port
0	1	0	Write I/O Port
0	1	1	Halt
1 (HIGH)	0	0	Code Access
1	0	1	Read Memory
1	1	0	Write Memory
1	1	1	Passive

<table>
<tr>
<td>$\overline{RQ}/GT_0$,
$\overline{RQ}/GT_1$</td>
<td>30, 31</td>
<td>I/O</td>
<td>REQUEST/GRANT: pins are used by other local bus masters to force the processor to release the local bus at the end of the processor's current bus cycle. Each pin is bidirectional with RQ/GT_0 having higher priority than RQ/GT_1.RQ/GT pins have internal pull-up resistors and may be left unconnected. The request/grant sequence is as follows (see Page 2-24):
1. A pulse of 1 CLK wide from another local bus master indicates a local bus request ("hold") to the 8086 (pulse 1).
2. During a T_4 or T_1 clock cycle, a pulse 1 CLK wide from the 8086 to the requesting master (pulse 2), indicates that the 8086 has allowed the local bus to float and that it will enter the "hold acknowledge" state at the next CLK. The CPU's bus interface unit is disconnected logically from the local bus during "hold acknowledge".
3. A pulse 1 CLK wide from the requesting master indicates to the 8086 (pulse 3) that the "hold" request is about to end and that the 8086 can reclaim the local bus at the next CLK. Each master-master exchange of the local bus is a sequence of 3 pulses. There must be one dead CLK cycle after each bus exchange. Pulses are active LOW. If the request is made while the CPU is performing a memory cycle, it will release the local bus during T_4 of the cycle when all the following conditions are met:
1. Request occurs on or before T_2.
2. Current cycle is not the low byte of a word (on an odd address).
3. Current cycle is not the first acknowledge of an interrupt acknowledge sequence.
4. A locked instruction is not currently executing. If the local bus is idle when the request is made the two possible events will follow:
  1. Local bus will be released during the next clock.
  2. A memory cycle will start within 3 clocks. Now the four rules for a currently active memory cycle apply with condition number 1 already satisfied.</td>
</tr>
<tr>
<td>$\overline{LOCK}$</td>
<td>29</td>
<td>O</td>
<td>LOCK: output indicates that other system bus masters are not to gain control of the system bus while LOCK is active LOW. The LOCK signal is activated by the "LOCK" prefix instruction and remains active until the completion of the next instruction.</td>
</tr>
<tr>
<td>QS_1,QS_0</td>
<td>24, 25</td>
<td>O</td>
<td>QUEUE STATUS: The queue status is valid during the CLK cycle after which the queue operation is performed. QS_1 and QS_0 provide status to allow external tracking of the internal 8086 instruction queue.
QS_1 QS_0 Characteristics
0 (LOW)   0   No Operation
0   1   First Byte of Op Code from Queue
1 (HIGH)   0   Empty the Queue
1   1   Subsequent Byte from Queue</td>
</tr>
</table>

The following pin function descriptions are for the 8086 in minimum mode (i.e., MN/MX V_{CC}). Only the pin functions which are unique to minimum mode are described; all other pin functions are as described above.

$\overline{M}$/IO	28	O	STATUS LINE: logically equivalent to S_2 in the maximum mode. It is used to distinguish a memory access from an I/O access. M/IO becomes valid in the T_4 preceding a bus cycle and remains valid until the final T_4 of the cycle (M HIGH, IO LOW).
$\overline{WR}$	29	O	WRITE: indicates that the processor is performing a write memory or write I/O cycle, depending on the state of the M/IO signal. WR is active for T_2,T_3 and T_W of any write cycle. It is active LOW.
$\overline{INTA}$	24	O	INTA: is used as a read strobe for interrupt acknowledge cycles. It is active LOW during T_2,T_3 and T_W of each interrupt acknowledge cycle.
ALE	25	O	ADDRESS LATCH ENABLE : Provided by the processor to latch the address into the 8282/8283 address latch. It is a HIGH pulse active during T_1 of any bus cycle. Note that ALE is never floated.
DT/$\overline{R}$	27	O	DATA TRANSMIT/RECEIVE: Needed in minimum system that desires to use an 8286/8287 data bus transceiver. It is used to control the direction of data flow through the transceiver. Logically DT/R is equivalent to S_1 in the maximum mode, and its timing is the same as for M/IO.(T HIGH, R LOW.).
$\overline{DEN}$	26	O	DATA ENABLE: Provided as an output enable for the 8286/8287 in a minimum system which uses the transceiver. DEN is active LOW during each memory and I/O access and for INTA cycles. For a read or INTA cycle it is active from the middle of T_2 until the middle of T_4, while for a write cycle it is active from the beginning of T_2 until the middle of T_4.
HOLD, HLDA	31, 30	I/O	HOLD: Indicates that another master is requesting a local bus "hold." To be HLDA acknowledged, HOLD must be active HIGH. The processor receiving the "hold" request will issue HLDA (HIGH) as an acknowledgement in the middle of a T_4 or T_i clock cycle. Simultaneous with the issuance of HLDA the processor will float the local bus and control lines. After HOLD is detected as being LOW, the processor will LOWer the HLDA, and when the processor needs to run another cycle, it will again drive the local bus and control lines. Hold acknowledge (HLDA) and HOLD have internal pull-up resistors. The same rules as for RQ/GT apply regarding when the local bus will be released. HOLD is not an asynchronous input. External synchronization should be provided if the system cannot otherwise guarantee the setup time.

Question : Why has 8086 µp 2 ground pins ?

Answer : The 8086 has two ground pins in order to improve its ability to pull, multiple, potentially all of the bus pins to ground, in a period of time that meets the fall time and fan out requirements of the bus. Pull to ground (V_{OL}) requires more current than Pull to V_{OH} because of the design of LSTTL. Even though the internal design of the 87086 is MOS technology based the external data bus is designed (to interface with LSTTL. Out of the two gnd pins one is used as analog ground and another one is used for digital ground. Voltage breakdown won't occur if its grounded. The difference between analog and digital ground is that digital ground does not have transient currents. Another reason may be is to remove GROUND BOUNCE.

3.2. 8086 CPU ARCHITECTURE

The internal functions of the 8086 processor are partitioned logically into two processing units. The first is the Bus Interface Unit (BIU) and the second is the Execution Unit (EU) as shown in the block diagram in Fig. 3.2. These units can interact directly but for the most of the parts it performs as separate asynchronous operational processors. The bus interface unit provides the functions related to instruction fetching and queuing, operand fetch and store, and address relocation. This unit also provides the basic bus control. The overlap of instruction pre-fetching provided by this unit serves to increase processor performance through improved bus bandwidth utilization. Up to 6 bytes of the instruction stream can be queued while waiting for decoding and execution.

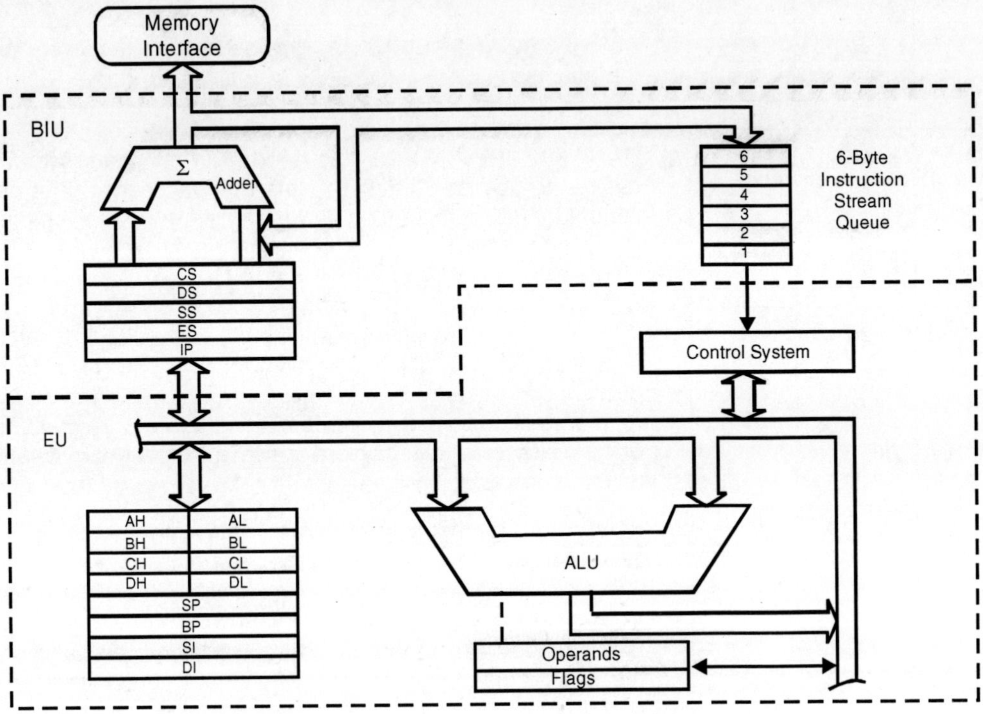

Fig. 3.2. Internal Architecture of 8086 Microprocessor.

The instruction stream queuing mechanism allows the BIU to keep the memory utilized very efficiently. Whenever there is space for at least 2 bytes in the queue, the BIU will attempt a word fetch memory cycle. This greatly reduces "dead time" on the memory bus. The queue acts as a First-In-First-Out (FIFO) buffer, from which the EU extracts instruction bytes as required. If the queue is empty (following a branch instruction, for example), the first byte into the queue immediately becomes available to the EU.

The execution unit receives pre-fetched instructions from the BIU queue and provides un-relocated operand addresses to the BIU. Memory operands are passed through the BIU for processing by the EU, which passes results to the BIU for storage. See the Instruction Set description for further register set and architectural descriptions.

8086 Microprocessor Features

- It is a 16-bit µp.
- 8086 has a 20 bit address bus can access upto 2^{20} memory locations (1 MB).

- It can support upto 64 K I/O ports.
- It provides 16-bit registers.
- It has multiplexed address and data bus AD_0–AD_{15} and A_{16}–A_{19}.
- It requires single phase clock with 33% duty cycle to provide internal timing.
- 8086 is designed to operate in two modes, minimum and maximum.
- It can prefetches upto 6 instruction bytes from memory and queues them in order to speed up instruction execution.
- It requires +5 V power supply.
- A 40 pin dual in line package.
- 8086 has two blocks BIU and EU.
- The BIU performs all bus operations such as instruction fetching, reading and writing operands for memory and calculating the addresses of the memory operands. The instruction bytes are transferred to the instruction queue.
- EU executes instructions from the instruction system byte queue.
- Both units operate asynchronously to give the 8086 an overlapping instruction fetch and execution mechanism which is called as Pipelining. This results in efficient use of the system bus and system performance.
- BIU contains instruction queue, segment registers, instruction pointer, address adder.
- EU contains Control circuitry, Instruction decoder, ALU, Pointer and Index register, Flag register.

Minimum and Maximum Modes

- The minimum mode is selected by applying logic 1 to the MN/$\overline{MX}$# input pin. This is a single microprocessor configuration.
- The maximum mode is selected by applying logic 0 to the MN/$\overline{MX}$# input pin. This is a multi micro processors configuration.

The 8086 CPU is organized as two separate processors, called the Bus Interface Unit (BIU) and the Execution Unit (EU). The BIU provides H/W functions, including generation of the memory and I/O addresses for the transfer of data between the outside world -outside the CPU, that is- and the EU. The EU receives program instruction codes and data from the BIU, executes these instructions, and store the results in the general registers. By passing the data back to the BIU, data can also be stored in a memory location or written to an output device. Note that the EU has no connection to the system buses. It receives and outputs all its data through the BIU.

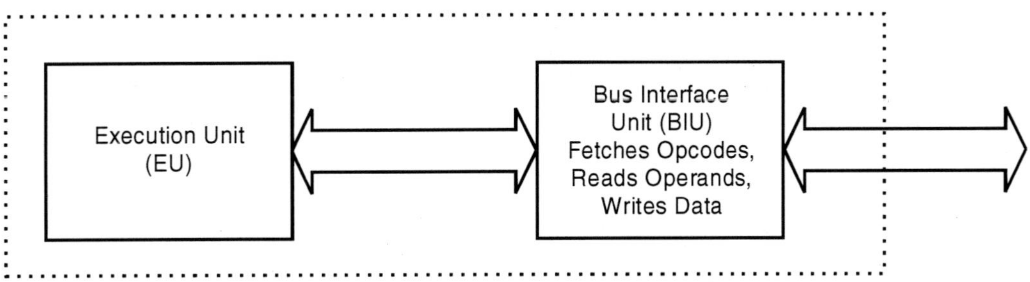

Fig. 3.3. 8086/8088 MPU.

The only difference between an 8088 microprocessor and an 8086 microprocessor is the BIU. In the 8088, the BIU data bus path is 8 bits wide versus the 8086's 16-bit data bus. Another difference is that the 8088 instruction queue is four bytes long instead of six.

The important point to note, however, is that because the EU is the same for each processor, the programming instructions are exactly the same for each. Programs written for the 8086 can be run on the 8088 without any changes.

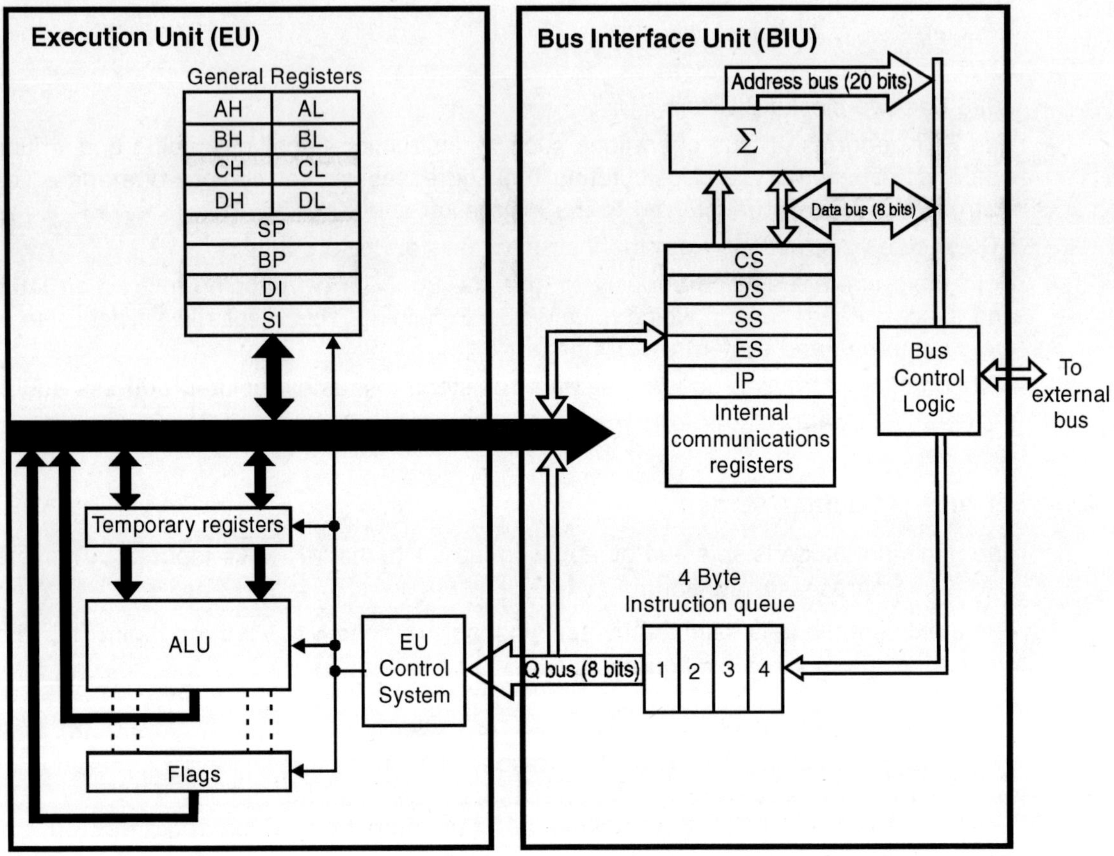

Fig. 3.4. 8088 Internal architecture.

Execution Unit

The Execution Unit executes all instructions, provides data and addresses to the Bus Interface Unit and manipulates the general registers and the Processor Status Word. The 16-bit ALU within the Execution Unit maintains the CPU status and control flags and manipulates the general registers and instruction operands. All registers and data paths in the Execution Unit are 16 bits wide for fast internal transfers.

The Execution Unit does not connect directly to the system bus. It obtains instructions from a queue maintained by the Bus Interface Unit. When an instruction requires access to memory or a peripheral device, the Execution Unit requests the Bus Interface Unit to read and write data. Addresses manipulated by the Execution Unit are 16 bits wide. The Bus Interface Unit, however, performs an address calculation that allows the Execution Unit to access the full megabyte of memory space.

To execute an instruction, the Execution Unit must first fetch the object code byte from the instruction queue and then execute the instruction. If the queue is empty when the Execution Unit is ready to fetch an instruction byte, the Execution Unit waits for the Bus Interface Unit to fetch the instruction byte.

The execution unit is responsible for decoding and executing all instructions.

- The EU extracts instructions from the top of the queue in the BIU, decodes them, generates operands if necessary, passes them to the BIU and requests it to perform the read or write byte cycles to memory or I/O and perform the operation specified by the instruction on the operands.
- During the execution of the instruction, the EU tests the status and control flags and updates them based on the results of executing the instruction.
- If the queue is empty, the EU waits for the next instruction byte to be fetched and shifted to top of the queue.
- When the EU executes a branch or jump instruction, it transfers control to a location corresponding to another set of sequential instructions.
- Whenever this happens, the BIU automatically resets the queue and then begins to fetch instructions from this new location to refill the queue.

Bus Interface Unit

The Bus Interface Unit executes all external bus cycles. This unit consists of the segment registers, the Instruction Pointer, the instruction code queue and several miscellaneous registers. The Bus Interface Unit transfers data to and from the Execution Unit on the ALU data bus.

The Bus Interface Unit generates a 20-bit physical address in a dedicated adder. The adder shifts a 16-bit segment value left 4 bits and then adds a 16-bit offset. This offset is derived from combinations of the pointer registers, the Instruction Pointer and immediate values. Any carry from this addition is ignored.

- It provides a full 16 bit bidirectional data bus and 20 bit address bus.
- The bus interface unit is responsible for performing all external bus operations.

Specifically it has the following functions :

- Instruction fetch, instruction queuing, Operand fetch and storage, Address relocation and bus control.
- The BIU uses a mechanism known as an instruction stream queue to implement a ***pipeline architecture***.
- This queue permits prefetch of upto six bytes of instruction code. Whenever the queue of the BIU is not full, it has room for at least two more bytes and at the same time the EU is not requesting it to read or write operands from memory, the BIU is free to look ahead in the program by prefetching the next sequential instruction.
- These prefetching instructions are held in its FIFO queue. With its 16 bit data bus, the BIU fetches two instruction bytes in a single memory cycle.
- After a byte is loaded at the input end of the queue, it automatically shifts up through the FIFO to the empty location nearest the output.
- The EU accesses the queue from the output end. It reads one instructon byte after the other from the output of the queue.

- These intervals of no bus activity, which may occur between bus cycles are known as ***Idle state.***
- If the BIU is already in the process of fetching an instruction when the EU request it to read or write operands from memory or I/O, the BIU first completes the instruction fetch bus cycle before initiating the operand read/write cycle.
- The BIU also contains a dedicated adder which is used to generate the 20bit physical address that is output on the address bus. This address is formed by adding an appended 16 bit segment address and a 16 bit offset address.
- For example : The physical address of the next instruction to be fetched is formed by combining the current contents of the code segment CS register and the current contents of the instruction pointer IP register.
- The BIU is also responsible for generating bus control signals such as those for memory read or write and I/O read or write.

3.3. FETCH AND EXECUTE

Although the 8086/88 still functions as a stored program computer, organization of the CPU into a separate BIU and EU allows the fetch and execute cycles to overlap. To see this, consider what happens when the 8086 or 8088 is first started.

1. The BIU outputs the contents of the instruction pointer register (IP) onto the address bus, causing the selected byte or word to be read into the BIU.
2. Register IP is incremented by 1 to prepare for the next instruction fetch.
3. Once inside the BIU, the instruction is passed to the queue. This is a first-in, first-out storage register sometimes likened to a "pipeline".
4. Assuming that the queue is initially empty, the EU immediately draws this instruction from the queue and begins execution.
5. While the EU is executing this instruction, the BIU proceeds to fetch a new instruction. Depending on the execution time of the first instruction, the BIU may fill the queue with several new instructions before the EU is ready to draw its next instruction.

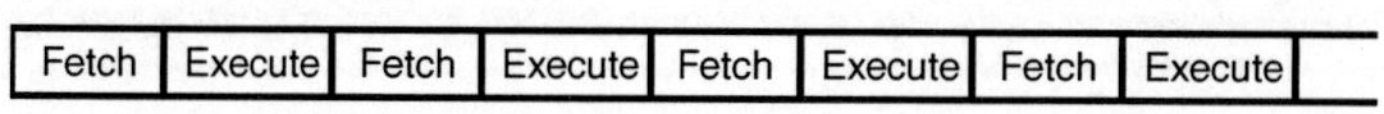

FETCH AND EXECUTE CYCLE

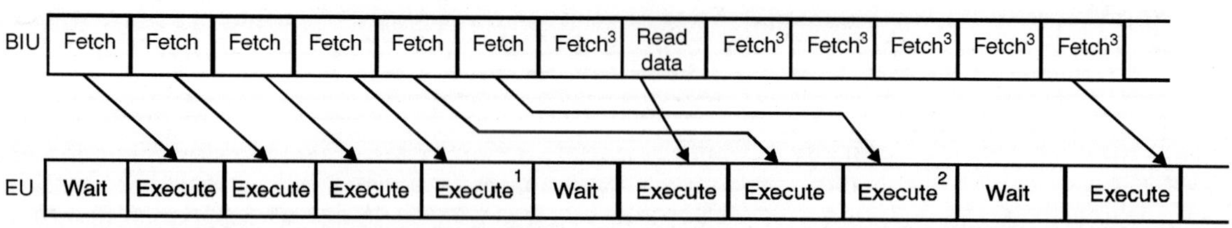

[1]This instruction requires a request for data not in the queue

[2]Jump instruction occurs

[3]These bytes are discarded

Fig. 3.5. Instruction Pipeline

The BIU is programmed to fetch a new instruction whenever the queue has room for one (with the 8088) or two (with the 8086) additional bytes. The advantage of this pipelined architecture is that the EU can execute instructions almost continually instead of having to wait for the BIU to fetch a new instruction.

There are three conditions that will cause the EU to enter a "wait" mode. The first occurs when an instruction requires access to a memory location not in the queue. The BIU must suspend fetching instructions and output the address of this memory location. After waiting for the memory access, the EU can resume executing instruction codes from the queue (and the BIU can resume filling the queue).

The second condition occurs when the instruction to be executed is a "jump" instruction. In this case control is to be transferred to a new (nonsequential) address. The queue, however, assumes that instructions will always be executed in sequence and thus will be holding the "wrong" instruction codes. The EU must wait while the instruction at the jump address is fetched. Note that any bytes presently in the queue must be discarded (they are overwritten).

One other condition can cause the BIU to suspend fetching instructions. This occurs during execution of instructions that are slow to execute. For example, the instruction AAM (ASCII Adjust for Multiplication) requires 83 clock cycles to complete. At four cycles per instruction fetch, the queue will be completely filled during the execution of this single instruction. The BIU will thus have to wait for the EU to pull over one or two bytes from the queue before resuming the fetch cycle.

A subtle advantage to the pipelined architecture should be mentioned. Because the next several instructions are usually in the queue, the BIU can access memory at a somewhat "leisurely" pace. This means that slow-mem parts can be used without affecting overall system performance.

3.4. INSIDE THE CPU

3.4.1. General Purpose Registers

8086 CPU has 8 general purpose registers, each register has its own name :

(i) **AX** - the accumulator register (divided into **AH / AL**).

(ii) **BX** - the base address register (divided into **BH / BL**).

(iii) **CX** - the count register (divided into **CH / CL**).

(iv) **DX** - the data register (divided into **DH / DL**).

(v) **SI** - source index register.

(vI) **DI** - destination index register.

(vii) **BP** - base pointer.

(viii) **SP** - stack pointer.

Despite the name of a register, it's the programmer who determines the usage for each general purpose register. The main purpose of a register is to keep a number (variable). The size of the above registers is 16 bit, it's something like: 0011000000111001b (in binary form), or 12345 in decimal (human) form.

4 general purpose registers (AX, BX, CX, DX) are made of two separate 8 bit registers, for example if AX= 0011000000111001b, then AH=00110000b and AL=00111001b. Therefore, when you modify any of the 8 bit registers 16 bit register is also updated, and vice-versa. The same is for other 3 registers, "H" is for high and "L" is for low part.

Because registers are located inside the CPU, they are much faster than memory. Accessing a memory location requires the use of a system bus, so it takes much longer. Accessing data in a register usually takes no time. Therefore, you should try to keep variables in the registers. Register sets are very small and most registers have special purposes which limit their use as variables, but they are still an excellent place to store temporary data of calculations.

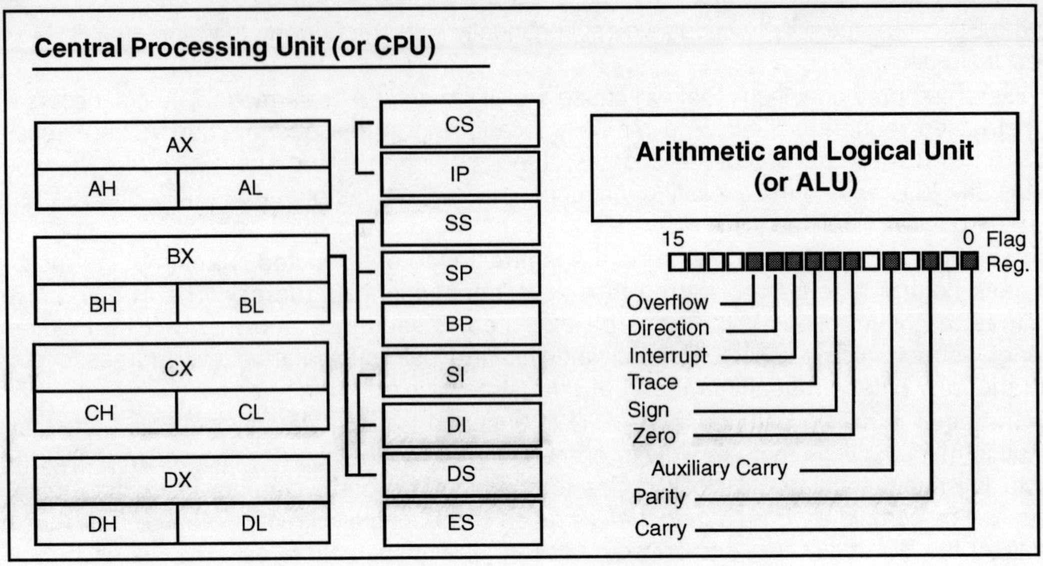

Fig. 3.6. Register organization for 8086 processor.

3.4.2. Segment Register

(a) **CS** - points at the segment containing the current program.

(b) **DS** - generally points at segment where variables are defined.

(c) **ES** - extra segment register, it's up to a coder to define its usage.

(d) **SS** - points at the segment containing the stack.

Although it is possible to store any data in the segment registers, this is never a good idea. The segment registers have a very special purpose - pointing at accessible blocks of memory.

Segment registers work together with general purpose register to access any memory value. For example if we would like to access memory at the physical address **12345h** (hexadecimal), we should set the **DS = 1230h** and **SI = 0045h**. This is good, since this way we can access much more memory than with a single register that is limited to 16 bit values. CPU makes a calculation of physical address by multiplying the segment register by 10h and adding general purpose register to it (1230h * 10h + 45h = 12345h):

The address formed with 2 registers is called an **effective address**.

By default **BX, SI** and **DI** registers work with **DS** segment register;

BP and **SP** work with **SS** segment register.

Other general purpose registers cannot form an effective address! Also, although **BX** can form an effective address, **BH** and **BL** cannot!

3.4.3. Special Purpose Registers

(a) **IP** - the instruction pointer.

(b) **Flags Register** - determines the current state of the processor.

IP register always works together with **CS** segment register and it points to currently executing instruction.

Flags Register is modified automatically by CPU after mathematical operations, this allows to determine the type of the result, and to determine conditions to transfer control to other parts of the program. Generally you cannot access these registers directly.

3.5. PROGRAMMING MODEL

The programming model for 8086/8088 microprocessor is shown in Fig. 3.7.

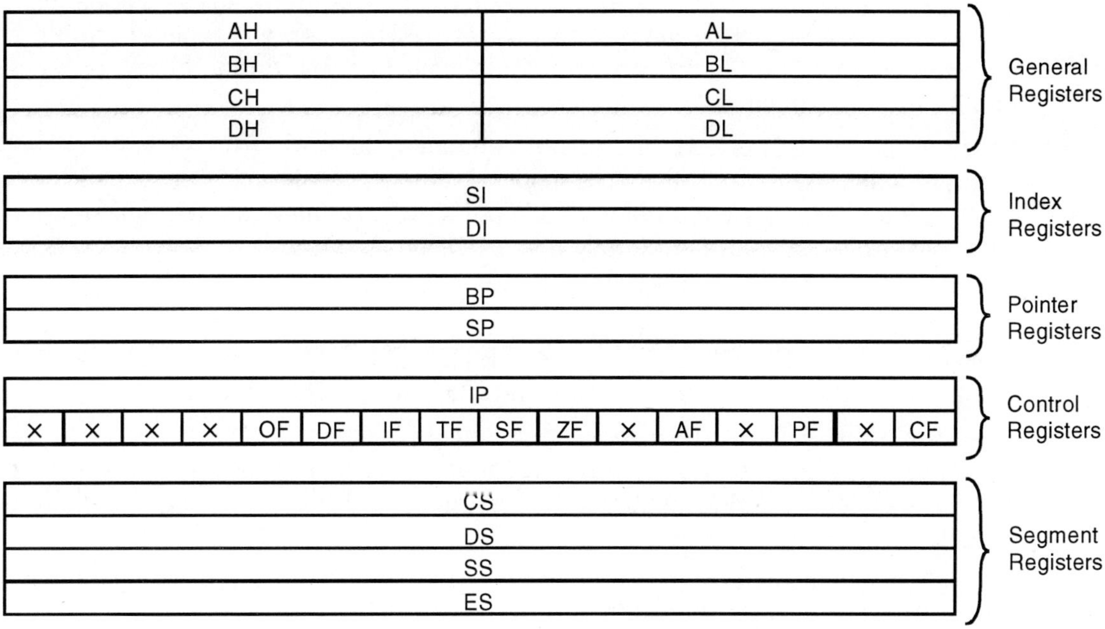

Fig. 3.7. Programming model for 8086 processor.

The data group consists of the accumulator and the BX, CX, and DX registers. Note that each can be accessed as a byte or a word. Thus BX refers to the 16-bit base register but BH refers only to the higher 8 bits of this register. The data registers are normally used for storing temporary results that will be acted on by subsequent instructions.

The pointer and index group are all 16-bit registers (you cannot access the low or high bytes alone). These registers are used as memory pointers. Sometimes a pointer reg will be interpreted as pointing to a memory byte and at other times a memory word. As you will see, the 8086/88 always stores words with the high-order byte in the high-order word address.

Register IP could be considered in the previous group, but this register has only one function -to point to the next instruction to be fetched to the BIU. Register IP is physically part of the BIU and not under direct control of the programmer as are the other pointer registers.

Six of the flags are status indicators, reflecting properties of the result of the last arithmetic or logical instructions. The 8086/88 has several instructions that can be used to transfer program control to a new memory location based on the state of the flags.

Three of the flags can be set or reset directly by the programmer and are used to control the operation of the processor. These are TF, IF, and DF.

The final group of registers is called the segment group. These registers are used by the BIU to determine the memory address output by the CPU when it is reading or writing from the memory unit. To fully understand these registers, we must first study the way the 8086/88 divides its memory into segments.

Most of the registers contain data/instruction offsets within 64 KB memory segment. There are four different 64 KB segments for instructions, stack, data and extra data. To specify where in 1 MB of processor memory these 4 segments are located the processor uses four segment registers:

(1) **Code segment (CS)** is a 16-bit register containing address of 64 KB segment with processor instructions. The processor uses CS segment for all accesses to instructions referenced by instruction pointer (IP) register. CS register cannot be changed directly. The CS register is automatically updated during far jump, far call and far return instructions.

(2) **Stack segment (SS)** is a 16-bit register containing address of 64 KB segment with program stack. By default, the processor assumes that all data referenced by the stack pointer (SP) and base pointer (BP) registers is located in the stack segment. SS register can be changed directly using POP instruction.

(3) **Data segment (DS)** is a 16-bit register containing address of 64 KB segment with program data. By default, the processor assumes that all data referenced by general registers (AX, BX, CX, DX) and index register (SI, DI) is located in the data segment. DS register can be changed directly using POP and LDS instructions.

(4) **Extra segment (ES)** is a 16-bit register containing address of 64KB segment, usually with program data. By default, the processor assumes that the DI register references the ES segment in string manipulation instructions. ES register can be changed directly using POP and LES instructions.

It is possible to change default segments used by general and index registers by prefixing instructions with a CS, SS, DS or ES prefix.

All general registers of the 8086 microprocessor can be used for arithmetic and logic operations. The general registers are:

(a) **Accumulator** register consists of 2 8-bit registers AL and AH, which can be combined together and used as a 16-bit register AX. AL in this case contains the low-order byte of the word, and AH contains the high-order byte. Accumulator can be used for I/O operations and string manipulation.

(b) **Base** register consists of 2 8-bit registers BL and BH, which can be combined together and used as a 16-bit register BX. BL in this case contains the low-order byte of the word, and BH contains the high-order byte. BX register usually contains a data pointer used for based, based indexed or register indirect addressing.

(c) **Count** register consists of 2 8-bit registers CL and CH, which can be combined together and used as a 16-bit register CX. When combined, CL register contains the low-order byte of the word, and CH contains the high-order byte. Count register can be used as a counter in string manipulation and shift/rotate instructions.

(d) **Data** register consists of 2 8-bit registers DL and DH, which can be combined together and used as a 16-bit register DX. When combined, DL register contains the low-order byte of the word, and DH contains the high-order byte. Data register can be used as a port number in I/O operations. In integer 32-bit multiply and divide instruction the DX register contains high-order word of the initial or resulting number.

The following registers are both general and index registers:

(a) **Stack Pointer (SP)** is a 16-bit register pointing to program stack.

(b) **Base Pointer (BP)** is a 16-bit register pointing to data in stack segment. BP register is usually used for based, based indexed or register indirect addressing.

(c) **Source Index (SI)** is a 16-bit register. SI is used for indexed, based indexed and register indirect addressing, as well as a source data address in string manipulation instructions.

(d) **Destination Index (DI)** is a 16-bit register. DI is used for indexed, based indexed and register indirect addressing, as well as a destination data address in string manipulation instructions.

3.6. OTHER REGISTERS

Instruction Pointer (IP) is a 16-bit register.

Program Status Word (PSW) : Flag register is a 16-bit register containing 9, 1-bit flags. Figure 3.8 gives the diagramatic representation of the flag register.

(i) **Overflow Flag (OF)** - set if the result is too large positive number, or is too small negative number to fit into destination operand.

(ii) **Direction Flag (DF)** - if set then string manipulation instructions will auto-decrement index registers. If cleared then the index registers will be auto-incremented.

(iii) **Interrupt-enable Flag (IF)** - setting this bit enables maskable interrupts.

(iv) **Single-step Flag (TF)** - if set then single-step interrupt will occur after the next instruction.

(v) **Sign Flag (SF)** - set if the most significant bit of the result is set.

(vi) **Zero Flag (ZF)** - set if the result is zero.

(vii) **Auxiliary carry Flag (AF)** - set if there was a carry from or borrow to bits 0-3 in the AL register.

(viii) **Parity Flag (PF)** - set if parity (the number of "1" bits) in the low-order byte of the result is even.

(ix) **Carry Flag (CF)** - set it there was a carry from or borrow to the most significant bil during last result calculation.

Register Name : **Processor Status Word**

Register Mnemonic : **PSW (FLAGS)**

Register Function : **Posts CPU status information.**

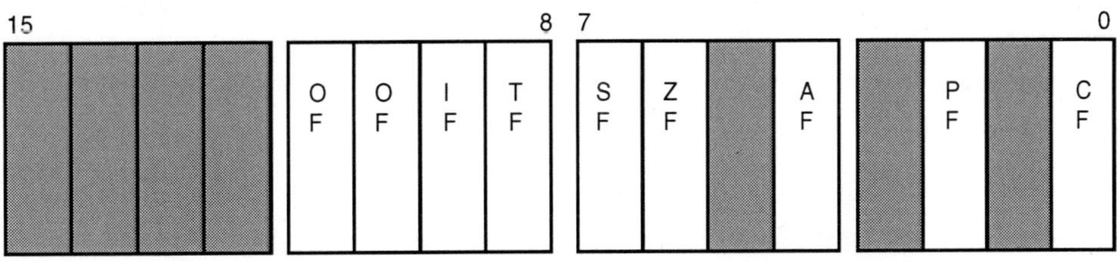

Fig. 3.8. PSW Format for 8086 Microprocessor.

Bit	Bit Name	Reset	Function
OF	Overflow Flag	0	If OF is set, an arithmetic overflow has occurred. If DF is set, string instructions are processed high address to low address.
DF	Direction Flag	0	If DF is clear, strings are processed low address to high address.
IF	Interrupt Enable Flag		If IF is set, the CPU recognizes maskable interrupt requests. If IF is clear, maskable interrupts are ignored.
TF	Trap Flag	0	If TF is set, the processor enters single-step mode.
SF	Sign Flag	0	If SF is set, the high-order bit of the result of an operation is 1, indicating it is negative.
ZF	Zero Flag	0	If ZF is set, the result of an operation is zero.
AF	Auxiliary Flag	0	If AF is set, there has been a carry from the lower byte
PF	Parity Flag	0	If PF is set, the result of an operation has even parity.
CF	Carry Flag	0	If CF is set there has been a carry out of, or a borrow into the high order bit of the result of an instruction.

3.7. MEMORY SEGMENTATION

Even though the 8086 is considered a 16-bit processor, (it has a 16-bit data bus width) its memory is still thought of in bytes. At first this might seem a disadvantage:

Why saddle a 16-bit microprocessor with an 8-bit memory?

Actually, there are a couple of good reasons. First, it allows the processor to work on bytes as well as words. This is especially important with I/O devices such as printers, terminals, and modems, all of which are designed to transfer ASCII-encoded (7- or 8-bit) data.

Second, many of the 8086's (and 8088's) operation codes are single bytes. Other instructions may require anywhere from two to seven bytes. By being able to access individual bytes, these odd-length instructions can be handled.

We have already seen that the 8086/88 has a 20-bit address bus, allowing it to output 2^{20}, or 1'048,576, different memory addresses. As you can see, 524,288 words can also be visualized.

As mentioned, the 8086 reads 16 bits from memory by simultaneously reading an odd-addressed byte and an even-addressed byte. For this reason the 8086 organizes its memory into an even-addressed bank and an odd-addressed bank.

With regard to this, you might wonder if all words must begin at an even address. Well, the answer is yes. However, there is a penalty to be paid. The CPU must perform two memory read cycles: one to fetch the low-order byte and a second to fetch the high-order byte. This slows down the processor but is transparent to the programmer.

The last few paragraphs apply only to the 8086. The 8088 with its 8-bit data bus interfaces to the 1 MB of memory as a single bank. When it is necessary to access a word (whether on an even- or an odd-addressed boundary) two memory read (or write) cycles are performed. In effect, the 8088 pays a performance penalty with every word access. Fortunately for the programmer, except for the slightly slower performance of the 8088, there is no difference between the two processors.

Memory Map

Still another view of the 8086/88 memory space could be as 16, 64 K-byte blocks beginning at hex address 000000h and ending at address 0FFFFFh. This division into 64 K-byte blocks is an arbitrary but convenient choice. This is because the most significant hex digit increments by

1 with each additional block. That is, address 20000h is 65.536 bytes higher in memory than address 10000h. Be sure to note that five hex digits are required to represent a memory address.

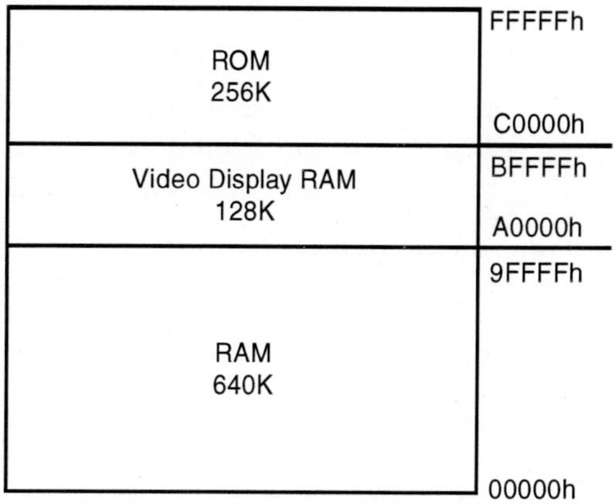

Fig. 3.9. Memory map for 8086 processor.

The Fig. 3.9 is called a memory map. This is because, like a road map, it is a guide showing how the system memory is allocated. This type of information is vital to the programmer, who must know exactly where his or her programs can be safely loaded.

Note that some memory locations are marked reserved and others dedicated. The dedicated locations are used for processing specific system interrupts and the reset function. Intel has also reserved several locations for future H/W and S/W products. If you make use of these memory locations, you risk incompatibility with these future products.

In the Intel family of microprocessors, memory is divided into segments. Since registers are 16-bit wide, each segment has its address space limited to 64 Kbytes. The 16-bit address used by the program is actually an offset from a segment base address, saved in the segment register. Pentium processors however, can operate in one of two modes; Real Mode and Protected Mode. In the real mode, the Pentium processor behaves exactly like a fast 8086 processor. In the protected mode, the Pentium processor supports both segmentation and pagination. Pagination is useful in implementing virtual memory, concept very used in operating system. However, pagination is transparent to the programmer and therefore is not covered at this level.

Segment registers are independent and segments can be contiguous, disjoint, partially overlapped or fully overlapped.

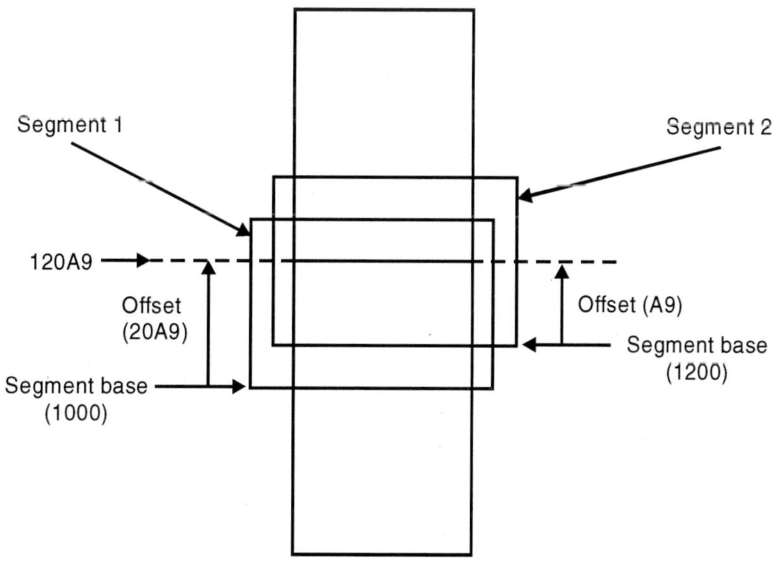

Fig. 3.10. Memory Segments.

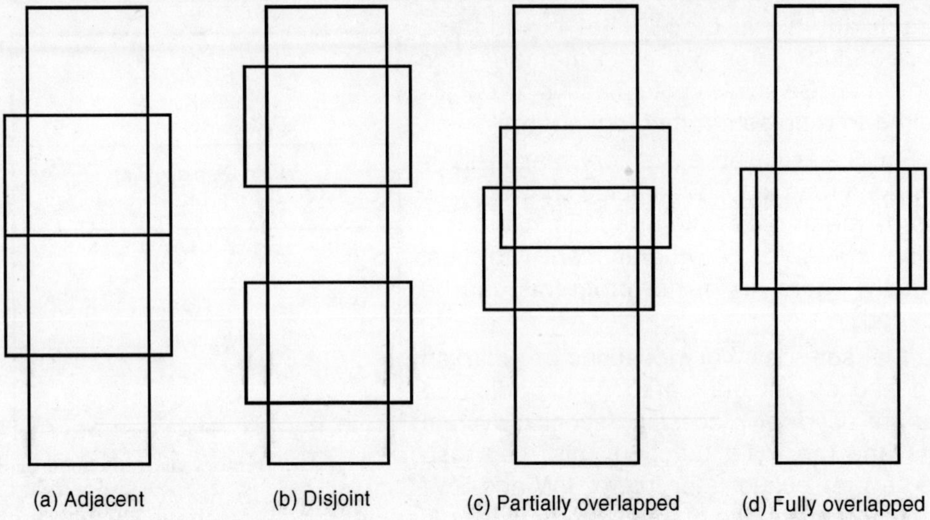

Fig. 3.11. Various ways of placing segments in the memory.

We generally can use Four types of segmentation schemes as shown in Fig. 3.11. These are :

(a) Adjacent memory segmentation.
(b) Disjoint memory segmentation.
(c) Partially overlapped memory segmentation
(d) Fully overlapped.

We commonly refer to adjacent or disjoint segmentation schemes where, when one segment ends than the other segment starts.

In adjacent the end of the memory segment is followed by the start of the next segment with immediate address that has to follow, whereas in the disjoint segment. There is some free space between the end and the start of the segment.

3.8. ADVANTAGES OF SEGMENTED MEMORY

Segmented memory can seem confusing at first. What you must remember is that the program op-codes will be fetched from the code segment, while program data variables will be stored in the data and extra segments. Stack operations use registers BP or SP and the stack segment. As we begin writing programs the consequences of these definitions will become clearer.

An immediate advantage of having separate data and code segments is that one program can work on several different sets of data. This is done by reloading register DS to point to the new data. Perhaps the greatest advantage of segmented memory is that programs that reference logical addresses only can be loaded and run anywhere in memory. This is because the logical addresses always range from 00000h to 0FFFFh, independent of the code segment base. Such programs are said to be relocatable, meaning that they will run at any location in memory. The requirements for writing relocatable programs are that no references be made to physical addresses, and no changes to the segment registers are allowed.

3.9. SEGMENT REGISTERS

Within the 1 MB of memory space the 8086/88 defines four 64 K-byte memory blocks called the code segment, stack segment, data segment, and extra segment. Each of these blocks of memory is used differently by the processor.

The code segment holds the program instruction codes. The data segment stores data for the program. The extra segment is an extra data segment (often used for shared data). The stack segment is used to store interrupt and subroutine return addresses.

You should realize that the concept of the segmented memory is a unique one. Older-generation microprocessors such as the 8-bit 8086 or Z-80 could access only one 64K-byte segment. This mean that the programs instruction, data and subroutine stack all had to share the same memory. This limited the amount of memory available for the program itself and led to disaster if the stack should happen to overwrite the data or program areas.

The four segment registers (CS, DS, ES, and SS) are used to "point" at location 0 (the base address) of each segment. This is a little "tricky" because the segment registers are only 16 bits wide, but the memory address is 20 bits wide. The BIU takes care of this problem by appending four 0's to the low-order bits of the segment register. In effect, this multiplies the segment register contents by 16.

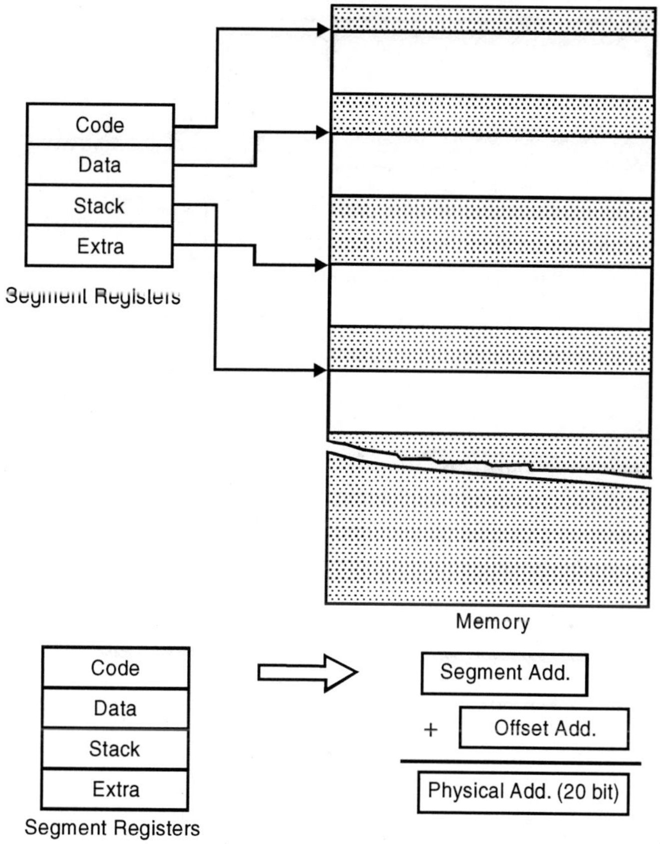

Fig. 3.12. Memory segmentation in 8086 processor.

The point to note is that the beginning segment address is not arbitrary -it must begin at an address divisible by 16. Another way if saying this is that the low-order hex digit must be 0.

Also note that the four segments need not be defined separately. Indeed, it is allowable for all four segments to completely overlap (CS = DS = ES = SS).

Memory locations not defined to be within one of the current segments cannot be accessed by the 8086/88 without first redefining one of the segment registers to include that location.

Thus at any given instant a maximum of 256 K (64 K * 4) bytes of memory can be utilized. As we will see, the contents of the segment registers can only be specified via S/W. As you might imagine, instructions to load these registers should be among the first given in any 8086/88 program.

3.10. LOGICAL AND PHYSICAL ADDRESS GENERATION

Addresses within a segment can range from address 00000h to address 0FFFFh. This corresponds to the 64K-byte length of the segment. An address within a segment is called an offset or logical address. A logical address gives the displacement from the address base of the segment to the desired location within it, as opposed to its "real" address, which maps directly anywhere into the 1 MB memory space. This "real" address is called the physical address.

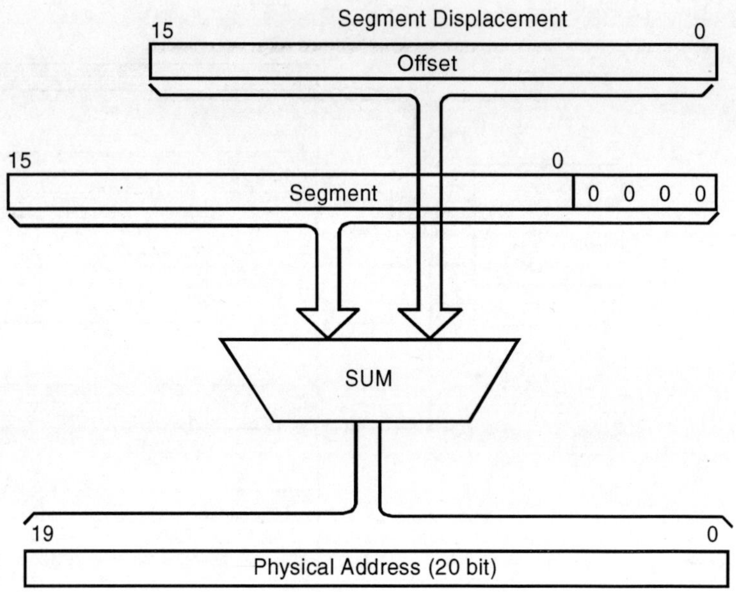

Fig. 3.13. Physical address computation.

What is the difference between the physical and the logical address?

The physical address is 20 bits long and corresponds to the actual binary code output by the BIU on the address bus lines. The logical address is an offset from location 0 of a given segment.

When two segments overlap it is certainly possible for two different logical addresses to map to the same physical address. This can have disastrous results when the data begins to overwrite the subroutine stack area, or vice versa. For this reason you must be very careful when segments are allowed to overlap.

You should also be careful when writing addresses on paper to do so clearly. To specify the logical address XXXX in the stack segment, use the convention SS:XXXX, which is equal to [SS] * 16 + XXXX.

3.11. MEMORY AND DATA ADDRESSING MODES

Addressing modes are a means, provided by assembly to the programmer, to access memory through different ways. This helps the programmer to write at his ease all of his programs. The different assembly addressing modes are introduced in this section.

3.11.1. Real Mode Memory Architecture

The segment register is offset by four bits and added to the offset to give a 20-bit address, known as the ***physical address***.

Physical Address = Segment Register*10H + Offset

Application programs use the ***logical address*** which consists of two components: a segment base and an offset or effective address, such as **BASE ADDRESS:OFFSET**. The base address is usually in a segment register, which is a pointer to the base of the segment. The offset is also known as the ***effective address***. Code, data and stack are each put in one segment. Each segment has its own base address saved in a segment register. The segment registers are:

- CS: code segment,
- SS: stack segment,
- DS : data segment
- ES : extra data segments.

3.11.2. 8086 Addressing Modes

Like the x86 processors described in the previous chapter, the 80x86 processors let you access memory in many different ways. The 80x86 memory addressing modes provide flexible access to memory, allowing you to easily access variables, arrays, records, pointers, and other complex data types. Mastery of the 80x86 addressing modes is the first step towards mastering 80x86 assembly language.

When Intel designed the original 8086 processor, they provided it with a flexible, though limited, set of memory addressing modes. Intel added several new addressing modes when it introduced the 8086 microprocessor. Note that the 80386 retained all the modes of the previous processors; the new modes are just an added bonus. If you need to write code that works on 80286 and earlier processors, you will not be able to take advantage of these new modes. However, if you intend to run your code on 80386sx or higher processors, you can use these new modes. Since many programmers still need to write programs that run on 80286 and earlier machines, it's important to separate the discussion of these two sets of addressing modes to avoid confusing them.

8086 Data Memory Addressing Modes

The 8086 provides 17 different ways to access memory. This may seem like quite difficult a bit at first, but fortunately most of the address modes are simple variants of one another so they're very easy to learn. And learn them you should! The key to good assembly language programming is the proper use of memory addressing modes.

The addressing modes provided by the 8086 family include displacement-only, base, displacement plus base, base plus indexed, and displacement plus base plus indexed. Variations on these five forms provide the 17 different addressing modes on the 8086. See, from 17 down to five. It's not so bad after all!

(1) 8086 Register Addressing Modes

Most 8086 instructions can operate on the 8086's general purpose register set. By specifying the name of the register as an operand to the instruction, you may access the contents of that register. Consider the 8086 mov (move) instruction:

```
mov  destination, source
```

This instruction copies the data from the source operand to the destination operand. The eight and 16 bit registers are certainly valid operands for this instruction. The only restriction is that both operands must be the same size. Now let's look at some actual 8086 mov instructions:

```
mov ax, bx ;Copies the value from BX into AX
mov dl, al ;Copies the value from AL into DL
mov si, dx ;Copies the value from DX into SI
mov sp, bp ;Copies the value from BP into SP
mov dh, cl ;Copies the value from CL into DH
mov ax, ax ;Yes, this is legal!
```

Remember, the registers are the best place to keep often used variables. As you'll see a little later, instructions using the registers are shorter and faster than those that access memory. Throughout this chapter you'll see the abbreviated operands reg and r/m (register/memory) used wherever you may use one of the 8086's general purpose registers.

In addition to the general purpose registers, many 8086 instructions (including the mov instruction) allow you to specify one of the segment registers as an operand. There are two restrictions on the use of the segment registers with the mov instruction. First of all, you may not specify cs as the destination operand, second, only one of the operands can be a segment register. You cannot move data from one segment register to another with a single mov instruction. To copy the value of cs to ds, you'd have to use some sequence like:

```
mov ax, cs
mov ds, ax
```

You should never use the segment registers as data registers to hold arbitrary values. They should only contain segment addresses. But more on that, later. Throughout this text you'll see the abbreviated operand sreg used wherever segment register operands are allowed (or required).

(2) The Displacement Only Addressing Mode

The most common addressing mode, and the one that's easiest to understand, is the displacement-only (or direct) addressing mode. The displacement-only addressing mode consists of a 16 bit constant that specifies the address of the target location. The instruction mov al,ds:[8088h] loads the al register with a copy of the byte at memory location 8088h. Likewise, the instruction mov ds:[1234h],dl stores the value in the dl register to memory location 1234h:

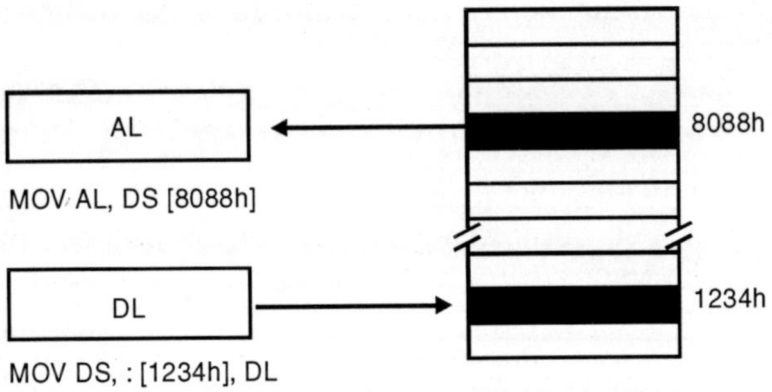

Fig. 3.14.

The displacement-only addressing mode is perfect for accessing simple variables. Of course, you'd probably prefer using names like "I" or "J" rather than "DS:[1234h]" or "DS:[8088h]". Well, fear not, you'll soon see it's possible to do just that.

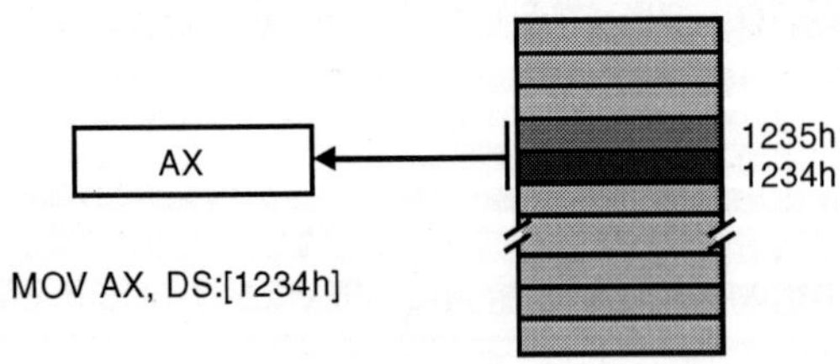

Fig. 3.15.

Intel named this the displacement-only addressing mode because a 16 bit constant (displacement) follows the mov opcode in memory. In that respect it is quite similar to the direct addressing mode on the x86 processors . There are some minor differences, however. First of all, a displacement is exactly that some distance from some other point. On the x86, a direct address can be thought of as a displacement from address zero. On the 80x86 processors, this displacement is an offset from the beginning of a segment (the data segment in this example). Don't worry if this doesn't make a lot of sense right now. For now, you can think of the displacement-only addressing mode as a direct addressing mode. The examples in this chapter will typically access bytes in memory. Don't forget, however, that you can also access words on the 8086 processors.

By default, all displacement-only values provide offsets into the data segment. If you want to provide an offset into a different segment, you must use a segment override prefix before your address. For example, to access location 1234h in the extra segment (es) you would use an instruction of the form mov ax,es:[1234h]. Likewise, to access this location in the code segment you would use the instruction mov ax, cs:[1234h]. The ds: prefix in the previous examples is not a segment override. The CPU uses the data segment register by default. These specific examples require ds: because of MASM's syntactical limitations.

(3) The Register Indirect Addressing Modes

The 80x86 CPUs let you access memory indirectly through a register using the register indirect addressing modes. There are four forms of this addressing mode on the 8086, best demonstrated by the following instructions:

```
mov al, [bx]
mov al, [bp]
mov al, [si]
mov al, [di]
```

As with the x86 [bx] addressing mode, these four addressing modes reference the byte at the offset found in the bx, bp, si, or di register, respectively. The [bx], [si], and [di] modes use the ds segment by default. The [bp] addressing mode uses the stack segment (ss) by default.

You can use the segment override prefix symbols if you wish to access data in different segments. The following instructions demonstrate the use of these overrides:

```
mov al, cs:[bx]
mov al, ds:[bp]
mov al, ss:[si]
mov al, es:[di]
```

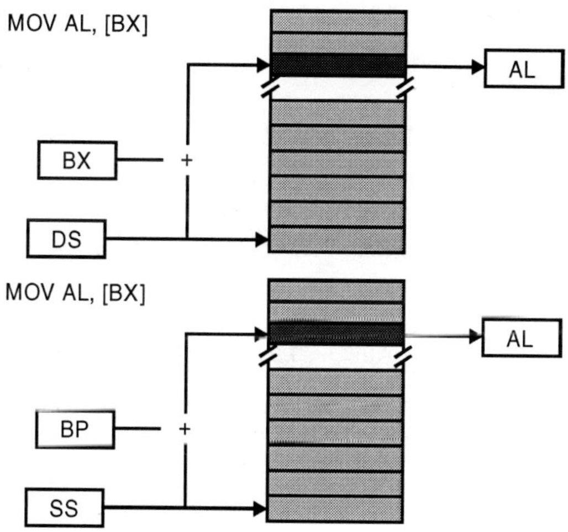

Fig. 3.16.

Intel refers to [bx] and [bp] as base addressing modes and bx and bp as base registers (in fact, bp stands for base pointer). Intel refers to the [si] and [di] addressing modes as indexed addressing modes (si stands for source index, di stands for destination index). However, these addressing modes are functionally equivalent. This text will call these forms register indirect modes to be consistent.

Note : The [si] and [di] addressing modes work exactly the same way, just substitute si and di for bx above.

(4) Immediate Addressing Mode

In this type of addressing, immediate data is a part of instruction, and appears in the form of successive bytes. For example :

```
mov BX, 2000H
```

In the above instruction, 2000H is the immediate data.

(5) Indexed Addressing Modes

The indexed addressing modes use the following syntax:

```
mov al, disp[bx]
mov al, disp[bp]
mov al, disp[si]
mov al, disp[di]
```

If bx contains 1000h, then the instruction mov cl,20h[bx] will load cl from memory location ds:1020h. Likewise, if bp contains 2020h, mov dh,1000h[bp] will load dh from location ss:3020.

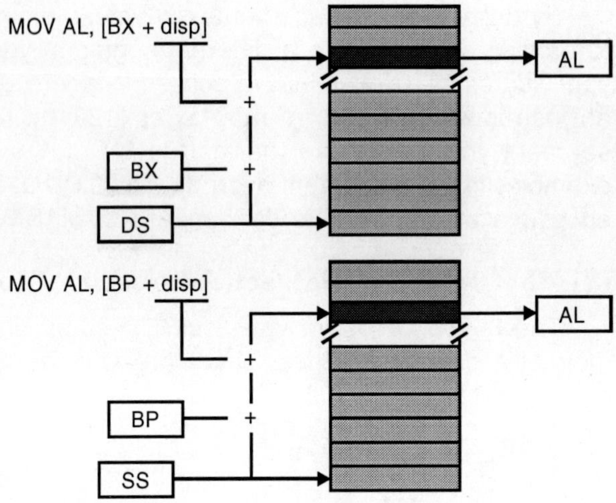

Fig. 3.17.

The offsets generated by these addressing modes are the sum of the constant and the specified register. The addressing modes involving bx, si, and di all use the data segment, the disp[bp] addressing mode uses the stack segment by default. As with the register indirect addressing modes, you can use the segment override prefixes to specify a different segment:

```
mov al, ss:disp[bx]
mov al, es:disp[bp]
mov al, cs:disp[si]
mov al, ss:disp[di]
```

You may substitute **si** or **di** in the figure above to obtain the [si+disp] and [di+disp] addressing modes.

Note that Intel still refers to these addressing modes as based addressing and indexed addressing. Intel's literature does not differentiate between these modes with or without the constant. If you look at how the hardware works, this is a reasonable definition. From the programmer's point of view, however, these addressing modes are useful for entirely different things. Which is why this text uses different terms to describe them. Unfortunately, there is very little consensus on the use of these terms in the 80x86 world.

(6) Based Indexed Addressing Modes

The based indexed addressing modes are simply combinations of the register indirect addressing modes. These addressing modes form the offset by adding together a base register (bx or bp) and an index register (si or di). The allowable forms for these addressing modes are

```
mov al, [bx][si]
mov al, [bx][di]
mov al, [bp][si]
mov al, [bp][di]
```

Suppose that bx contains 1000h and si contains 880h. Then the instruction

```
mov al, [bx] [si]
```

would load al from location DS:1880h. Likewise, if bp contains 1598h and di contains 1004, mov ax,[bp+di] will load the 16 bits in ax from locations SS:259C and SS:259D.

The addressing modes that do not involve bp use the data segment by default. Those that have bp as an operand use the stack segment by default.

You substitute di in the Fig. 3.18 to obtain the [bx+di] addressing mode.

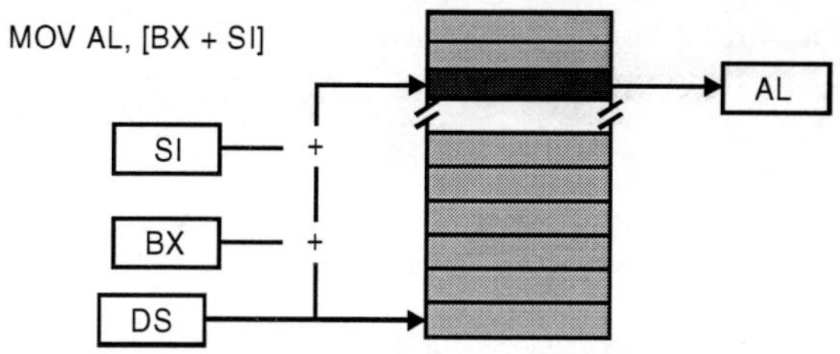

Fig. 3.18.

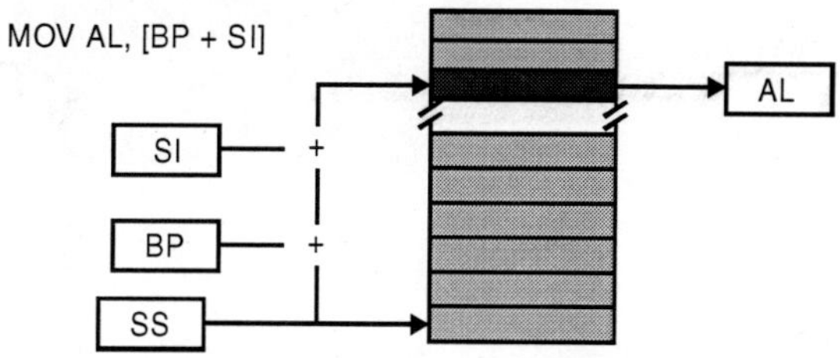

Fig. 3.19.

You substitute di in the figure above for the [bp+di] addressing mode.

(7) Based Indexed Plus Displacement Addressing Mode

These addressing modes are a slight modification of the base/indexed addressing modes with the addition of an eight bit or sixteen bit constant. The following are some examples of these addressing modes:

```
mov al, disp[bx][si]
mov al, disp[bx+di]
mov al, [bp+si+disp]
mov al, [bp][di][disp]
```

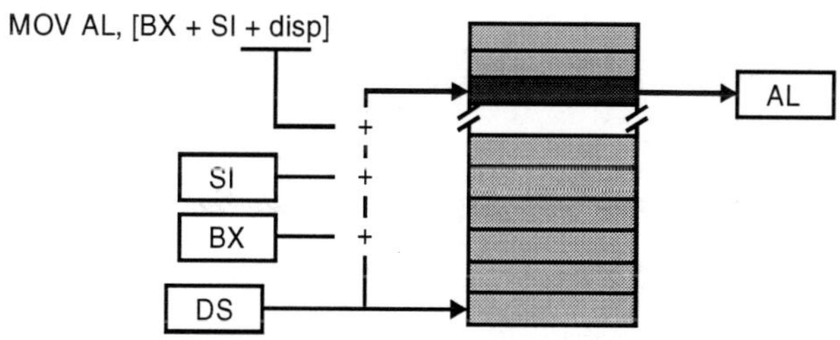

Fig. 3.20.

You may substitute di in the figure above to produce the [bx+di+disp] addressing mode.

You may substitute di in the figure above to produce the [bp+di+disp] addressing mode.

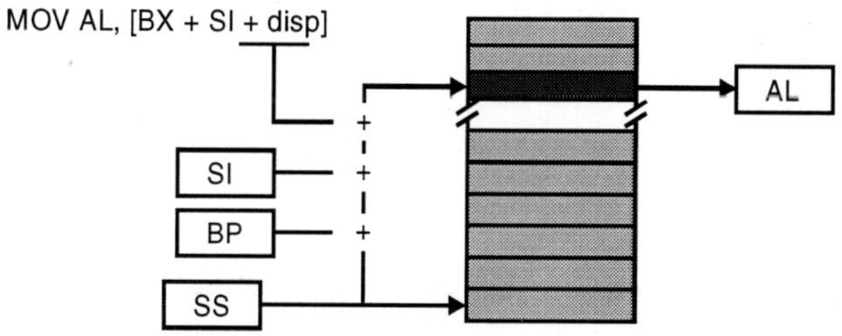

Fig. 3.21.

Suppose bp contains 1000h, bx contains 2000h, si contains 120h, and di contains 5. Then mov al,10h[bx+si] loads al from address DS:2130; mov ch,125h [bp+di] loads ch from location SS:112A; and mov bx,cs:2[bx][di] loads bx from location CS:2007.

An Easy Way to Remember the 8086 Memory Addressing Modes

There are a total of 17 different legal memory addressing modes on the 8086: disp, [bx], [bp], [si], [di], disp[bx], disp[bp], disp[si], disp[di], [bx][si], [bx][di], [bp][si], [bp][di], disp[bx][si], disp [bx][di], disp[bp][si], and disp[bp][di]. You could memorize all these forms so that you know which are valid (and, by omission, which forms are invalid). However, there is an easier way besides memorizing these 17 forms. Consider the chart:

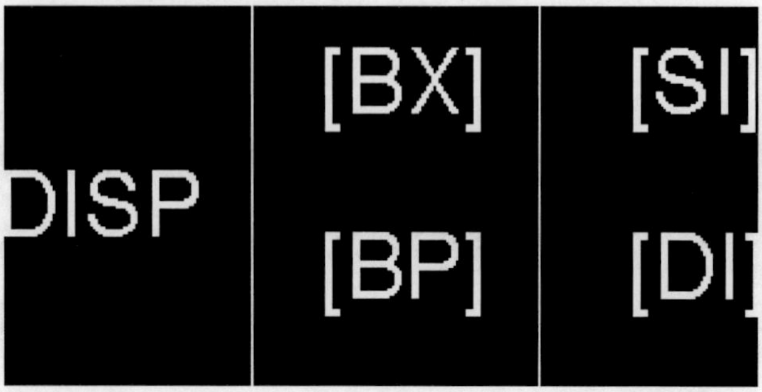

If you choose zero or one items from each of the columns and wind up with at least one item, you've got a valid 8086 memory addressing mode. Some examples:

- ➤ Choose disp from column one, nothing from column two, [di] from column 3, you get disp[di].
- ➤ Choose disp, [bx], and [di]. You get disp[bx][di].
- ➤ Skip column one & two, choose [si]. You get [si]
- ➤ Skip column one, choose [bx], then choose [di]. You get [bx][di]

Likewise, if you have an addressing mode that you cannot construct from this table, then it is not legal. For example, disp[dx][si] is illegal because you cannot obtain [dx] from any of the columns above.

Some Final Comments About 8086 Data Memory Addressing Modes

The effective address is the final offset produced by an addressing mode computation. For example, if bx contains 10h, the effective address for 10h[bx] is 20h. You will see the term effective address in almost any discussion of the 8086's addressing mode. There is even a special instruction load effective address (lea) that computes effective addresses.

Not all addressing modes are created equal! Different addressing modes may take differing amounts of time to compute the effective address. The exact difference varies from processor to processor. Generally, though, the more complex an addressing mode is, the longer it takes to compute the effective address. Complexity of an addressing mode is directly related to the number of terms in the addressing mode. For example, disp[bx][si] is more complex than [bx]. See the instruction set reference in the appendices for information regarding the cycle times of various addressing modes on the different 80x86 processors.

The displacement field in all addressing modes except displacement-only can be a signed eight bit constant or a signed 16 bit constant. If your offset is in the range –128...+127 the instruction will be shorter (and therefore faster) than an instruction with a displacement outside that range. The size of the value in the register does not affect the execution time or size. So if you can arrange to put a large number in the register(s) and use a small displacement, that is preferable over a large constant and small values in the register(s).

If the effective address calculation produces a value greater than 0FFFFh, the CPU ignores the overflow and the result wraps around back to zero. For example, if bx contains 10h, then the instruction mov al,0FFFFh[bx] will load the al register from location ds:0Fh, not from location ds:1000Fh.

In this discussion you've seen how these addressing modes operate. The preceding discussion didn't explain what you use them for. That will come a little later. As long as you know how each addressing mode performs its effective address calculation, you'll be fine.

Table 3.1 summarizes all addressing modes used by the 8086 processor and the real mode of the Pentium processor. The address is computed only when memory is accessed, no addressing is needed when the processor accesses an immediate operand making part of the instruction, or its own internal registers. Internal registers are accessed using a different mechanism.

TABLE 3.1. Addressing Modes.

Addressing Mode	Example	Source operand Assuming: DS = 1000H, SS = 2000H, BX = 0200H, SI = 0300H, BP = 0400H Address Generation	Phys. Address
Immediate	MOV AX, 0F7H	-	-
Register	MOV AX, BX	-	-
Direct	MOV AX, [1234H]	DS x 10H + 1234H	11234H
Register-Indirect	MOV AX, [BX]	DS x 10H + 0200H	10200H
Register-Indirect	MOV AX, [BP]	SS x 10H + 0400H	20400H
Based	MOVAX, [BX+ 06]	DS x 10H + 0200H + 0006H	10206H
Indexed	MOVAX, [SI + 06]	DS x 10H + 0300H + 0006H	10306H
Based-Indexed	MOV AX, [BX+SI+06]	DS x 10H + 0200H + 0300H + 0006H	10506H

3.12. SUMMARY OF THE DATA MEMORY ADDRESSING MODES

Addressing Mode	Operand	Default Segment
Register	Reg	None
Immediate	Data	None
Direct	[offset]	DS
Register Indirect	[BX]	DS
	[SI]	DS
	[DI]	DS
Based Relative	[BX]+disp	DS
	[BP]+disp	SS
Indexed Relative	[DI]+disp	DS
	[SI]+disp	DS
Based Indexed Relative	[BX][SI or DI]+disp	DS
	[BP][SI or DI]+disp	SS

3.12.1. 16 bit Segment Register Assignments

Type of Memory Reference	Default Segment	Alternate Segment	Offset
Instruction Fetch	CS	none	IP
Stack Operations	SS	none	SP, BP
General Data	DS	CS, ES, SS	BX, address
String Source	DS	CS, ES, SS	SI, DI, address
String Destination	ES	None	DI

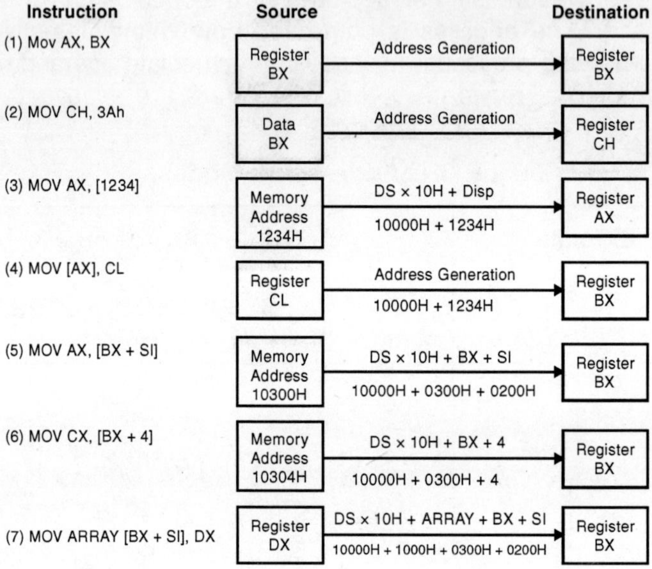

Fig. 3.22. Chart representing the generation of physical address for the possible addressing mode.

Program Memory-Addressing Modes

The program memory addressing modes are used with JMP and CALL instruction. These are also referred to as control transfer addressing modes or branched addressing modes.

For the control transfer instructions, the addressing modes depend upon whether the destination location is within the same segment or a different one. It also depends upon the method of passing the destination address to the processor. Basically there are two addressing modes for the control transfer instructions :

(a) Intersegment addressing mode

(b) Intrasegment address mode

The Fig. 3.23 below shows the modes for control transfer instructions :

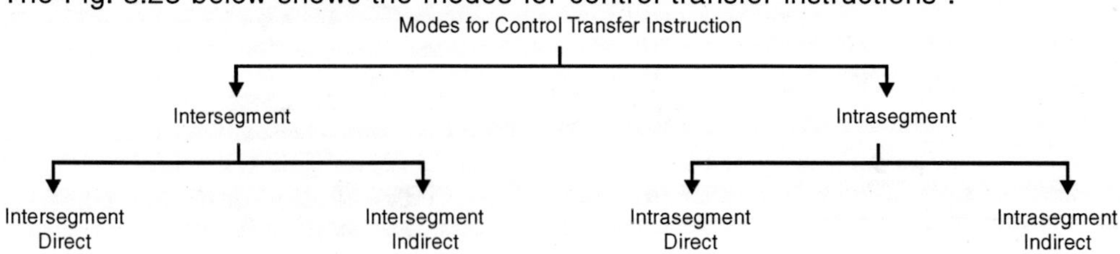

Fig. 3.23.

(i) Intersegment Mode : In this mode, the address to which the control is to be transferred lies in the same segment in which the control transfer instruction lies and appears directly in the instruction as an immediate displacement value. In this addressing mode, the displacement is computed relative to the content of the instruction pointer IP.

In case of jump instruction, if the signed displacement (D) is of 8 bits (i.e. –128 < D < +128) it is termed as a short jump and if it is 16 bits (i.e. –32768 < D < +32768) it is termed as a long jump.

(ii) Intrasegment Indirect Mode : In this mode, the displacement to which the control is to be transferred, is in the same segment in which the control transfer instruction lies, but it is passed to the instruction indirectly. In this mode, the branch address is found as the content of a register or a memory location. This addressing mode may be used in unconditional branch instruction.

(iii) Intersegment Direct Mode : In this mode, the address to which the control is to be transferred is in a different segment. This addressing mode provides a means of branching from one code segment to another code segment. Here the CS and IP of the destination address are specified directly in the instruction.

(iv) Intersegment Indirect Mode : In this mode, the address to which the control is to be transferred lies in a different segment and it is passed to the instruction indirectly, i.e. contents of a memory block containing 4-bytes, i.e. IP (LSB), IP (MSB), CS (LSB) and CS (MSB) sequentially. The starting address of the memory block may be referred using any of the addressing modes, except immediate mode.

Stack Memory Addressing Modes

The stack plays a vital role in all the microprocessors. It holds data temporarily and stores return addresses for procedures. The stack memory is a LIFO (last in, first out) memory, which describes the way that data are stored and removed from the stack.

We use PUSH instruction to place the data on to the stack, and we use POP instruction to remove the data from the stack.

The CALL instruction also uses the stack to hold the return address for procedures and we use an RET instruction to return the address from the stack.

The stack memory is maintained by two registers :

SP → Stack pointer

SS → Stack segment register.

The stack addressing mode is of two types :

(a) PUSH

(b) Pop

(a) PUSH : We can explain the working of this instruction with the help of the following example :

```
PUSH BX
```

This instruction places the contents of BX register onto the stack. We can show this with the following Fig. 3.24.

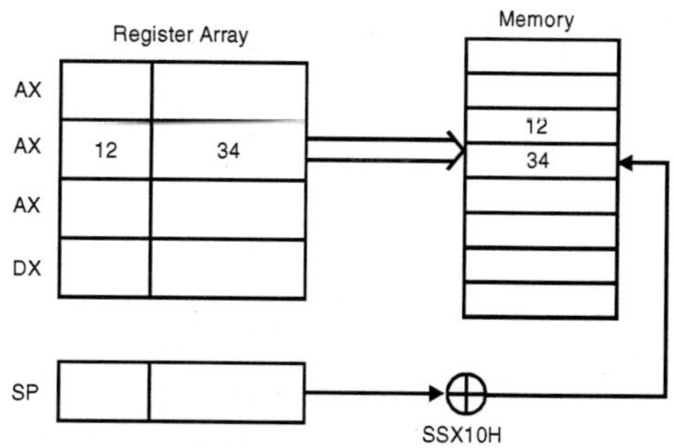

Fig. 3.24.

Whenever a word of data is pushed onto the stack as shown in Fig. 3.24, the high order 8 bits are placed in the location addressed by SP-1.

The low-order 8 bits are placed in the location addressed by SP-2. Hence the SP is decremented by 2 so that the next word of data is stored in the next available stack memory location. The SP register always points to an area of memory located within the stack segment. The SP register adds to SSX10H to form the stack memory address.

(b) POP : The working of POP instruction can be explained with the help of the following example "POPCX".

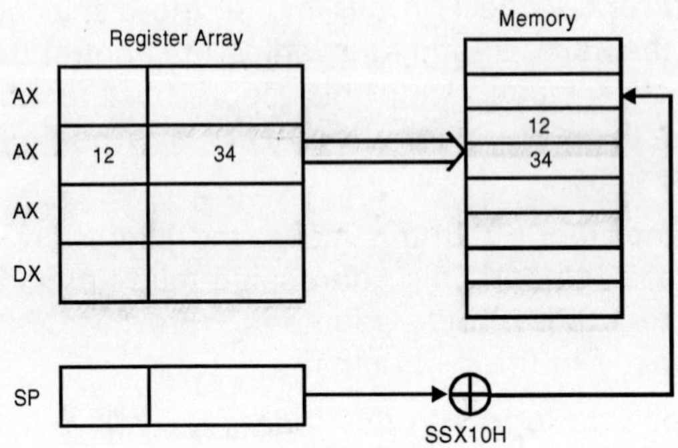

Fig. 3.25.

Whenever data is popped from the stack as shown in the Fig. 3.25, the low-order 8 bits are removed from the location addressed by SP. The high order 8 bits are removed from the location addressed by SP + 1. The SP register is then incremented by 2.

Example 3.1. What will be the physical address of the datum, given EA of a datum as 2359H and DS as 49030H ?

Solution.

Segment Reg.	Hex values
DS	490B0
EA	2359

Therefore the physical address will be (490B0 + 2359)

That is = **4B409H**

Example 3.2. The physical branch address is 5A23)H when (CS) = 5200H. What will be the physical address if (CS) is changed to 7800H ?

Solution. The physical address for CS = 5200H is

CS = 5200

Offset = xxxx

Physical adds 5A230H (given)

The offset = Physical adds – (segment adds in 4-bits) = 5A230-5200 = 8230H

Now if CS is changed to 7800H physical adds is

7800 + 8230 = FA30H

Memory Organization

The processor provides a 20-bit address to memory which locates the byte being referenced. The memory is organized as a linear array of up to 1 million bytes, addressed as 00000(H) to FFFFF(H). The memory is logically divided into code, data, extra data, and stack segments of up to 64K bytes each, with each segment falling on 16-byte boundaries.

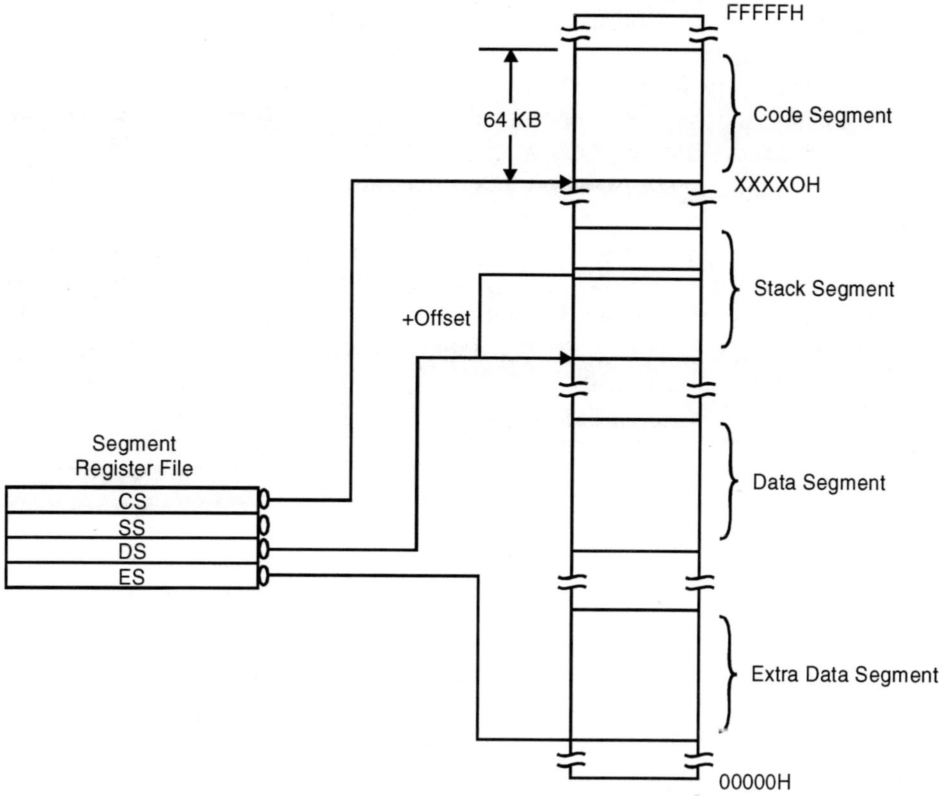

Fig. 3.26. Memory organization scheme used in 8086 microprocessor.

Memory Reference Need	Segment Register Used	Segment Selection Rule
Instructions	CODE (CS)	Automatic with all instruction prefetch.
Stack	STACK (SS)	All stack pushes and pops. Memory references relative to BP base register except data references.
Local Data	DATA (DS)	Data references when: relative to stack, destination of string operation, or explicitly overridden.
External (Global) Data	EXTRA (ES)	Destination of string operations: explicitly selected using a segment override.

All memory references are made relative to base addresses contained in high speed segment registers. The segment types were chosen based on the addressing needs of programs. The segment register to be selected is automatically chosen according to the rules of the following table. All information in one segment type share the same logical attributes (e.g. code or data). By structuring memory into relocatable areas of similar characteristics and by automatically selecting segment registers, programs are shorter, faster, and more structured.

Word (16-bit) operands can be located on even or odd address boundaries and are thus not constrained to even boundaries as is the case in many 16-bit computers. For address and data operands, the least significant byte of the word is stored in the lower valued address location and the most significant byte in the next higher address location. The BIU automatically performs the proper number of memory accesses, one if the word operand is on an even byte boundary and two if it is on an odd byte boundary. Except for the performance penalty, this double access is transparent to the software. This performance penalty does not occur for instruction fetches, only word operands.

Physically, the memory is organized as a high bank (D_{15} –D_8) and a low bank (D_7 –D_0) of 512K 8-bit bytes addressed in parallel by the processor's address lines A_{19} –A_1. Byte data with even addresses is transferred on the D_7 –D_0 bus lines while odd addressed byte data (A_0 HIGH) is transferred on the D_{15}–D_8 bus lines. The processor provides two enable signals, BHE and A_0, to selectively allow reading from or writing into either an odd byte location, even byte location, or both. The instruction stream is fetched from memory as words and is addressed internally by the processor to the byte level as necessary referencing word data the BIU requires one or two memory cycles depending on whether or not the starting byte of the word is on an even or odd address, respectively. Consequently, in referencing word operands performance can be optimized by locating data on even address boundaries. This is an especially useful technique for using the stack, since odd address references to the stack may adversely affect the context switching time for interrupt processing or task multiplexing.

Certain locations in memory are reserved for specific CPU operations (see Fig. 3.27). Locations from address FFFF0H through FFFFFH are reserved for operations including a jump to the initial program loading routine. Following RESET, the CPU will always begin execution at location FFFF0H where the jump must be. Locations 00000H through 003FFH are reserved for interrupt operations. Each of the 256 possible interrupt types has its service routine pointed to by a 4-byte pointer element consisting of a 16-bit segment address and a 16-bit offset address. The pointer elements are assumed to have been stored at the respective places in reserved memory prior to occurrence of interrupts.

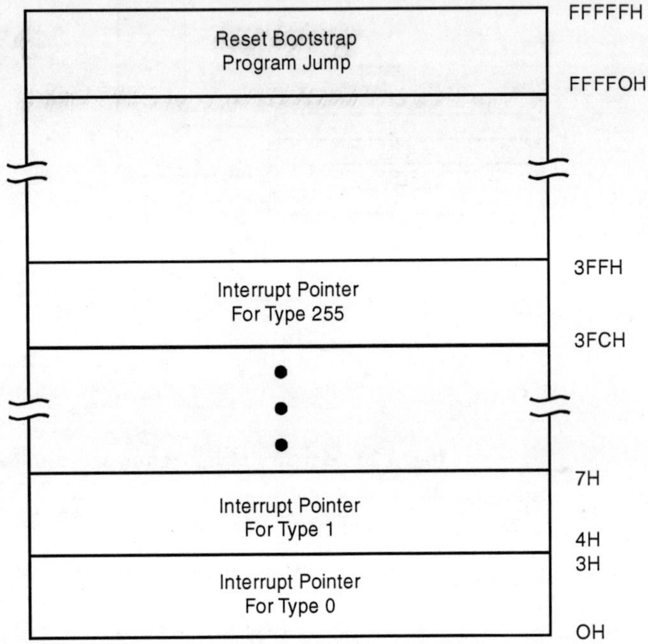

Fig. 3.27. Interrupt Structure for 8086 Processor.

3.13. TEMPLATE FOR DATA TRANSFER BETWEEN A REGISTER AND REGISTER/MEMORY

Shown below is the template used for generating the opcode for the instruction which moves data between a register and a memory location or between two registers. There is no instruction in 8086 to transfer between two memory locations directly. This template is not used for transferring immediate data to a register or memory location.

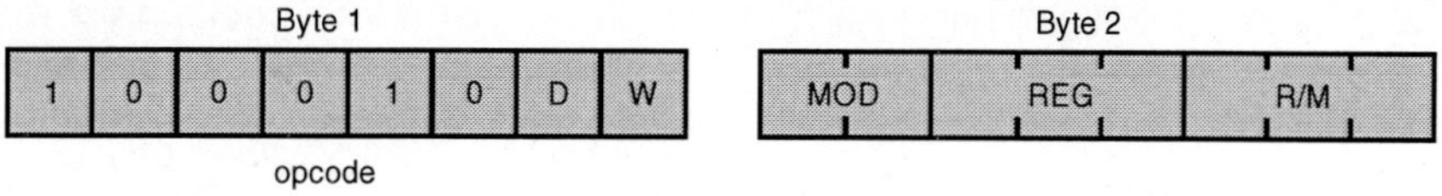

Fig. 3.28.

Many instructions in 8086 work on 16 bit as well as 8 bit data. In the template for such instructions, there will be a W bit, which indicates whether it is a word operation or a byte operation, W stands for word size operation.

W = 1 means it is a word operation.

W = 0 means it is a byte operation.

In an instruction we may specify a register a memory location, or two registers. If we are specifying only register, that will be indicated by the REG field in byte 2 of the template. Then it is necessary to indicate whether the register is a destination operand or a source operand. This is indicated by the D bit in the template. D stands for Destination register.

D = 1 means REG in byte 2 is destination register.

D = 0 means REG In byte 2 is source register.

The 3 bit REG field in byte 2 specifies an 8 bit or 16 bit register as shown below, depending on the W bit in byte 1. The REG field provides one of the operands, which to be in a register.

REG	Word operation	Byte operation
000	AX	AL
001	CX	CL
010	DX	DL
011	BX	BL
100	SP	AH
101	BP	CH
110	SI	DH
111	DI	BH

In 8086 there are 32 different ways of specifying the other operand which may be in a register (excluding segment registers) or a memory location. Accordingly, 5 bits are used in the opcode to indicate the operand, which may be in a register or a memory location. This is not applicable for immediate operand, which forms part of the instruction itself. The 5 bits are divided into a *2 bit* MOD field which indicates the mode in which the operand is specified, and a 3 bit R/M field which specifies a register or memory location.

Case of MOD = 11

If MOD = 11, then the operand is in a register. In such a case, the 3 bit field indicates the register, as shown below. The register selected depends on the W bit in byte 1 of the template.

REG	Word operation	Byte operation
000	AX	AL
001	CX	CL
010	DX	DL
011	BX	BL
100	SP	AH
101	BP	CH
110	SI	DH
111	DI	BH

Thus, of the 32 ways of specifying an operand, 8 are used for specifying a register, which could be an 8 bit register or a 16 bit register. So, there are 24 ways of specifying a memory location. When the MOD bits are 00, 01 or 10, we are specifying a memory location.

Case of MOD = 00

If MOD = 00, it indicates that the operand is the contents of a memory location. Thus the 3 bit R/M field specifies a memory location. The memory location is specified by one of the three addressing modes depending on the R/M bits, as shown in the table below.

R/M	Effective address	Addressing mode
000	[BX] + [SI]	Based Indexed
001	[BX] + [DI]	Based Indexed
010	[BP] + [SI]	Based Indexed
011	[BP] + [DI]	Based Indexed
100	[SI]	Register Indirect
101	[DI]	Register Indirect
110	Part of instruction	Direct Addressing
111	[BX]	Register Indirect

Case of MOD = 01

If MOD = 01, It indicates that the operand is the contents of a memory location. Thus the 3 bit R/M field specifies a memory location. The memory location is specified by one of the three addressing modes depending on the R/M bits, as shown in the table below. This mode uses 8 bit displacement, which is extended to 16 bits by treating the MS byte as 00H, before computing EA. In the table below, d8 stands for a 8 bit displacement.

R/M	Effective Address	Addressing mode
000	[BX] + [SI] + d8	Based Indexed with 8 bit displacement
001	[BX] + [DI] + d8	Based Indexed with 8 bit displacement
010	[BP] + [SI] + d8	Based Indexed with 8 bit displacement
011	[BP] + [DI] + d8	Based Indexed with 8 bit displacement
100	[SI] + d8	Indexed with 8 bit displacement
101	[DI] + d8	Indexed with 8 bit displacement
110	[BP] + d8	Based with 8 bit displacement
111	[BX] + d8	Based with 8 bit displacement

Case of MOD = 10

If MOD = 10, it indicates that the operand is the contents of a memory location. Thus the 3 bit R/M field specifies a memory location. The memory location is specified by one of the three addressing modes depending on the R/M bits, as shown in the table below. This mode uses 16 bit displacement, which is used in computing EA, In the table below, d16 stands for a 16 bit displacement.

R/M	Effective Address	Addressing mode
000	[BX] + [SI] + d16	Based Indexed with 16 bit displacement
001	[BX] + [DI] + d16	Based Indexed with 16 bit displacement
010	[BP] + [DI] + d16	Based Indexed with 16 bit displacement
011	[BP] + [DI] + d16	Based Indexed with 16 bit displacement
100	[SI] + d16	Indexed with 16 bit displacement
101	[DI] + d16	Indexed with 16 bit displacement
110	[BP] + d16	Based with 16 bit displacement
111	[BX] + d16	Based with 16 bit displacement

3.14. CODE GENERATION USING TEMPLATE

A number of examples are provided below, which illustrate the methods for generating the codes for the instructions to move data between a register and another register or memory location.

Example 1. Code for MOV AX, BX

In this example, we can specify either of AX or BX using the REG field. If we specify AX using REG field, then the D bit must be 1, indicating that AX is the destination register. The REG field must be 000 to indicate AX register. The W bit must be 1 to Indicate it is a word operation. In this case, BX is specified using MOD value of 11, R/M value of 011. Thus the 2 byte code for this instruction would be

Opcode	**D**	**W**	**MOD**	**REG**	**R/M**
1000 10	1	1	11	000	011 = 8B C3H

If we specify BX using REG field, then the D bit must be 0, indicating that BX is the source register. The REG field must be 011 to indicate BX register. The W bit must be 1 to indicate it is a word operation, in this case, AX is specified using MOD value of 11, and R/M value of 000. Thus the 2 byte code for this instruction would be

Opcode	**D**	**W**	**MOD**	**REG**	**R/M**
1000 10	0	1	11	011	000 = 89 D8H

Thus MOV AX,BX has two equally valid opcodes as shown above. An assembler generates one of these as the opcode. This is very much unlike in 8085 μP, where every instruction has a unique mnemonic.

Example 2. Code for MOV CL, [SI]

Here we have to specify CL using REG field. The D bit must be 1, indicating that CL is the destination register. The REG field must be 001 to indicate CL register. The W bit must be 0 to indicate it is a byte operation, [SI] is specified using MOD value of 00, and R/M value of 100. Thus the 2 byte code for this instruction would be

Opcode	**D**	**W**	**MOD**	**REG**	**R/M**
1000 10	1	0	00	001	100 = 8A 0CH

Example 3. Code for MOV 46H[BP],DX

Here we have to specify DX using REG field. The D bit must be 0, indicating that DX is the source register. The REG field must be 010 to indicate DX register. The W bit must be 1 to indicate it is a word operation, 46H[BP] is specified using MOD value of 01, and R/M value of 110, and a displacement of 46H, Thus the 3 byte code for this instruction would be

Opcode	**D**	**W**	**MOD**	**REG**	**R/M 8 bit disp**	
1000 10	0	1	01	010	110 46H	= 89 56 46H

Example 4. Code for MOV 0F246H [BP],DX

Here we have to specify DX using REG field. The D bit must be 0, indicating that DX is the source register. The REG field must be 010 to indicate DX register. The W bit must be 1 to indicate it is a word operation. F246H[BP] is specified using MOD value of 10, and R/M value of 1 10, and a displacement of F246H. Thus the 4 byte code for this instruction would be

Opcode	**D**	**W**	**MOD**	**REG**	**R/M 16 bit disp**	
1000 10	0	1	10	010 110	F2 46H	= 89 96 F2 46H

The .LST file, which is the optional list file generated by the assembler, displays the code as shown above. But in memory, the 8086 stores the LS byte at a lower address and the MS byte at a higher address. So the actual code in memory would be 89 96 46 F2H. In this text, the code for an. instruction will be given in the format provided by the .LST file.

Example 5. Code for MOV DS:0F246H[BP],DX

Here we have to specify DX using REG field. The D bit must be 0, indicating that DX is the source register. The REG field must be 010 to indicate DX register, The W bit must be 1 to indicate it is a word operation, F246H[BP] is specified with MOD = 10, and R/M = 110, and displacement = F246.H.

Whenever BP is used to generate the EA, the default segment would be SS. In this example, as we want the segment register to be DS, we have to provide the segment override prefix byte (SOP byte) to start with. The SOP byte is 001SR110, where SR value is provided as per table shown below.

SR	Segment Register
00	ES
01	CS
10	SS
11	DS

So to specify DS register, the SOP byte would be 001 11 110 = 3EH. Thus the 5 byte code for this instruction would be

SOP	**opcode**	**D**	**W**	**MOD**	**REG**	**R/M 16 bit disp**	
3E	1000 10	0	1	10	010	110 F2 46	= 3E 89 96 F2 46H

Suppose we want to code MOV SS:0F246H[BP],DX. This generate only a 4 byte code, without the SOP byte, as SS is already the default segment.

3.15. REVIEW OF FIELDS IN A TEMPLATE

A summary of REG bite, MOD bits, and R/M bits as well as SR bits, discussed in the previous chapter, is given below to facilitate generating the machine code for an instruction using templates.

R/M	Effective Address	Addressing mode
000	AX	AL
001	CX	CL
010	DX	DL
011	BX	BL
100	SP	AH
101	BP	CH
110	SI	DH
111	DI	BH

***Case of MOD* =11.** The operand is in a register, In such a case, the 3 bit R/M field indicates a register, as shown above.

***Case of MOD* = 00.** In this case, the operand is the contents of a memory location. Thus the 3 bit R/M field specifies a memory location as shown below.

R/M	Effective Address	Addressing mode
000	[BX] + [SI]	Based Indexed
001	[BX] + [DI]	Based Indexed
010	[BP] + [SI]	Based Indexed
011	[BP] + [DI]	Based Indexed
100	[SI]	Register Indirect
101	[DI]	Register Indirect
110	Part of Instruction	Direct addressing
111	[BX]	Register Indirect

***Case of MOD* = 01:** In this case, the operand is the contents of a memory location. Thus the 3 bit R/M field specifies a memory location as shown below where, d8 stands for a 8 bit displacement.

R/M	Effective Address	Addressing mode
000	[BX] + [SI] + d8	Based Index with 8 bit displacement
001	[BX] + [DI] + d8	Based Index with 8 bit displacement
010	[BP] + [SI] + d8	Based Index with 8 bit displacement
011	[BP] + [DI] + d8	Based Index with 8 bit displacement
100	[SI] + d8	Indexed with 8 bit displacement
101	[DI] + d8	Indexed with 8 bit displacement
110	[BP] + d8	Based with 8 bit displacement
111	[BX] + d8	Based with 8 bit displacement

***Case of MOD* = 10:** In this case, the operand Is the contents of a memory location. Thus the 3 bit R/M field specifies a memory location as shown below where, d16 stands for a 16 bit displacement.

R/M	Effective Address	Addressing mode
000	[BX] + [SI] + d16	Based Indexed with 16 bit displacement
001	[BX] + [DI] + d16	Based Indexed with 16 bit displacement
010	[BP] + [SI] + d16	Based Indexed with 16 bit displacement
011	[BP] + [DI] + d16	Based Indexed with 16 bit displacement
100	[SI] + d16	Indexed with 16 bit displacement
101	[DI] + d16	Indexed with 16 bit displacement
110	[BP] + d16	Based with 16 bit displacement
111	[BX] + d16	Based with 16 bit displacement

Segment Override Prefix: The Segment override prefix byte (SOP byte) is 001SR110, where SR indicates the segment override as indicated in the table shown below.

SR	Segment Register
00	ES
01	CS
10	SS
11	DS

Examples

- MOV BL,AL
- Opcode for MOV = 100010
- We'll encode AL so
  - D = 0 (AL source operand)
- W bit = 0 (8-bits)
- MOD = 11 (register mode)
- REG = 000 (code for AL)
- R/M = 011

OPCODE	D	W	MOD	REG	R/M
000000	0	1	10	000	001

```
MOV BL,AL => 10001000 11000011 = 88 C3h
ADD AX,[SI] => 00000011 00000100 = 03 04 h
ADD [BX][DI] + 1234h, AX => 00000001 10000001__ h => 01 81 34 12 h
```

Instructions

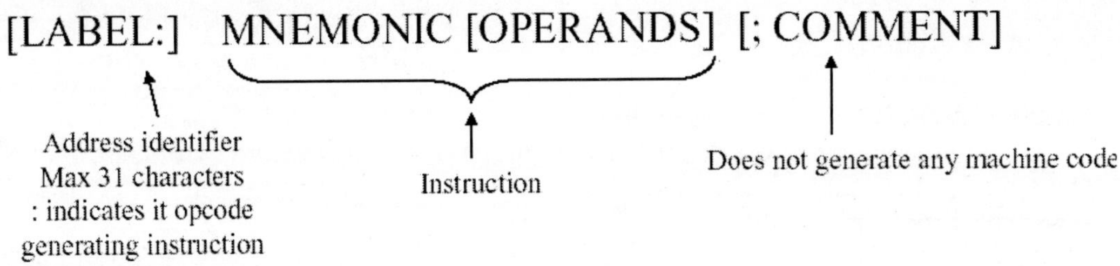

Example : MOV BL, CL

Process to decode the instruction :

(1) It is an 8-bit operation.

(2) It fits in the register/memory to/from register format.

(3) BL is the designation register and CL is the source reg.

	D_9	D_8	$D_7D_6D_5$	D_4D_3	D_2D_1 D_0
Opcode (6-bit)	D	W	Mo D	REG	R/M

- It is an 8-bit operation, hence 'w' bit is 0. If it would have been a 16-bit operation than 'w' bit would have been 1.

➤ If d = 1, then transfer of data is to the register shown by the REG field, i.e. the destination is a register. If d = 0, the source is a register shown by the REG field.

If we refer to the previous tables to search the REG to REG addressing in it, i.e. the last column with MOD 11.

According to the data sheet when MOD is 11, the R/M field is treated as REG field.

The REG field is used for source register and R/M field is used for the destination register if 'd' is 0.

If D is 1 the REG field is used for destination and the R/M field is used to indicate source. Hence the code can be written in decoded form as :

```
                    Opcode    D W MoD  REG   R/M
MOV BL, CL  ⇒    1 0 0 0 1 0  0 0  1 1  0 0 1  0 1 1
                                   ⇓
                                 88 CB
```

Example : MOV [SI], DL

The instruction belongs to the register to memory format. Hence we can write the code for the above instruction using the similar procedure as stated in the above example :

Solution. MOV [SI], DI

```
                    Opcode   D W   M o D  REG   R/M
MOV [SI], DI  ⇒   1 0 0 0 1 0  1 0  0 0  1 0 0 1  0 1 1
                                  ⇓
                       The machine code is = 8A4B
```

Example : MOV BP [SI], 0005H

```
Opcode      D W   Mo D   REG   R/M
1 1 0 0 0 1  1 1   0 0    0 0 0  0 1 0

0 0 0 0 0 1  0 0  1 0   0 0 0 0  0 0 0 0
                  ⇓
        Machine code is = C7020500
```

Example : ADD AX, BX

```
Opcode      D W   Mo D   REG   R/M
0 0 0 0 0 0  1 1   1 1    0 0 0  0 1 1
                ⇓
       Machine code = 03C3
```

Bus Operation

The 8086 has a combined address and data bus commonly referred to as a time multiplexed bus. This technique provides the most efficient use of pins on the processor while permitting the use of a standard 40-lead package. This "local bus" can be buffered directly and used throughout the system with address latching provided on memory and I/O modules. In addition, the bus can also be demultiplexed at the processor with a single set of address latches if a standard non-multiplexed bus is desired for the system.

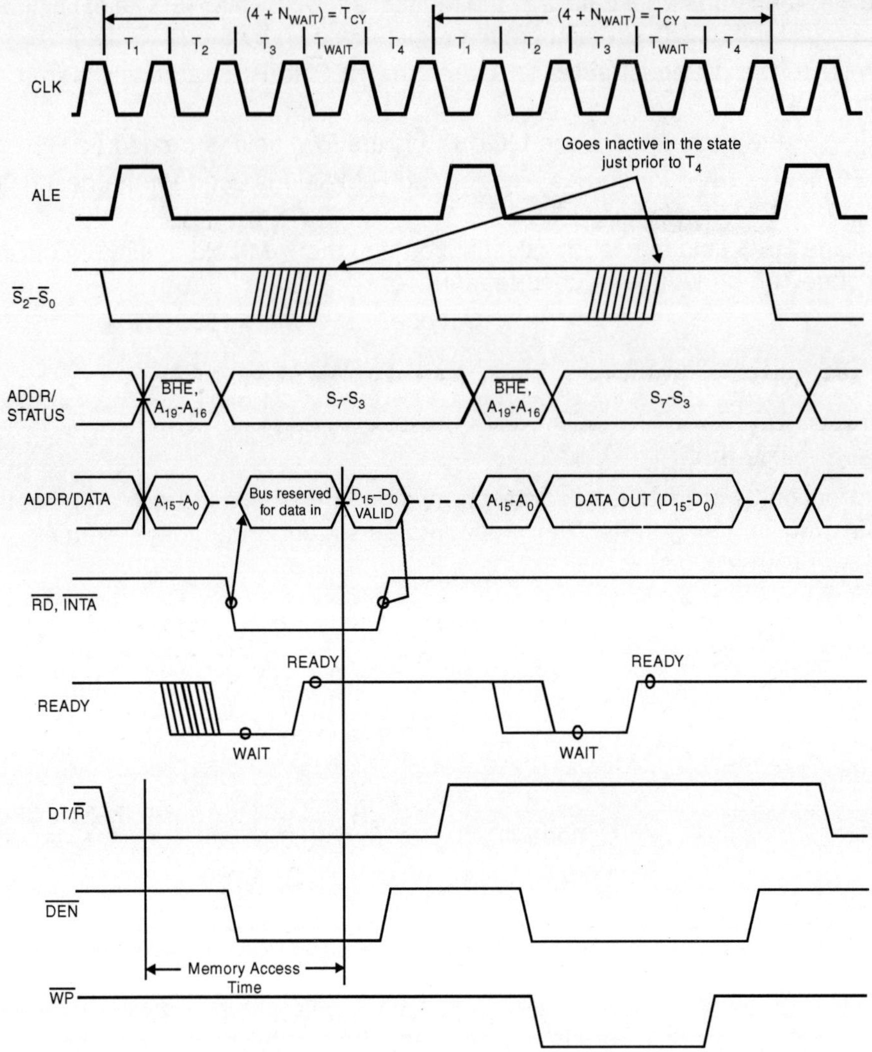

Fig. 3.29. Basic System Timing Diagram.

General Bus Operation

- The 8086 has a combined address and data bus commonly referred as a time multiplexed address and data bus.
- The main reason behind multiplexing address and data over the same pins is the maximum utilisation of processor pins and it facilitates the use of 40 pin standard DIP package.
- The bus can be demultiplexed using a few latches and transreceivers, whenever required.
- Basically, all the processor bus cycles consist of at least four clock cycles. These are referred to as T_1, T_2, T_3, T_4. The address is transmitted by the processor during T_1. It is present on the bus only for one cycle.
- The negative edge of this ALE pulse is used to separate the address and the data or status information. In maximum mode, the status lines S_0, S_1 and S_2 are used to indicate the type of operation.
- Status bits S_3 to S_7 are multiplexed with higher order address bits and the BHE signal. Address is valid during T_1 while status bits S_3 to S_7 are valid during T_2 through T_4.

Status bits S_3 through S_7 are **multiplexed** with high-order address bits and the BHE signal, and are therefore valid during T_2 through T_4. S_3 and S_4 indicate which segment register (see Instruction Set description) was used for this bus cycle in forming the address, according to the following table:

S_4	S_3	Characteristics
0 (LOW)	0	Alternate Data (extra segment)
0	1	Stack
1 (HIGH)	0	Code or None
1	1	Data

S_5 is a reflection of the PSW interrupt enable bit. S_6 = 0 and S_7 is a spare status bit.

I/O ADDRESSING In the 8086. I/O operations can address up to a maximum of 64K I/O byte registers or 32K I/O word registers. The I/O address appears in the same format as the memory address on bus lines $A_{15} - A_0$. The address lines $A_{19} - A_{16}$ are zero in I/O operations. The variable I/O instructions which use register DX as a pointer have full address capability while the direct I/O instructions directly address one or two of the 256 I/O byte locations in page 0 of the I/O address space.

I/O ports are addressed in the same manner as memory locations. Even addressed bytes are transferred on the $D_7 - D_0$ bus lines and odd addressed bytes on $D_{15} - D_8$. Care must be taken to assure that each register within an 8-bit peripheral located on the lower portion of the bus be addressed as even.

External Interface

PROCESSOR RESET AND INITIALIZATION Processor initialization or start up is accomplished with activation (HIGH) of the RESET pin. Tho 8086 RESET is required to be HIGH for greater than 4 CLK cycles. The 8086 will terminate operations on the high-going edge of RESET and will remain dormant as long as RESET is HIGH. The low-going transition of RESET triggers an internal reset sequence for approximately 10 CLK cycles. After this interval the 8086 operates normally beginning with the instruction in absolute location FFFF0H (see Fig. 3.28). The RESET input is internally synchronized to the processor clock. At initialization the HIGH-to-LOW transition of RESET must occur no sooner than 50 ms after power-up, to allow complete initialization of the 8086.

TABLE 3.2.

S.No.	Microprocessor 8086	Microprocessor 8088
1.	It has sixteen data lines. Therefore it has AD0-AD15 signals.	It has only eight data lines. Therefore, it has AD0-AD7 and A8-A15 signals.
2.	It has 6 BH signal to access higher byte.	As data bus is 8-bit wide, it does not have BHE signal.
3.	It has 6 byte instruction queue.	It has 4 byte instruction queue. Due to 8-bit data bus, instruction fetching is slow and 4 bytes are sufficient for queue.
4.	Its pin number 34 is BHE/S7. During T1 (first clock cycle) BHE should be used to enable data n to the most significant byte of the data bus. During T3-T4 and T4 status of this pin is logic 0. In maximum mode, 8087 monitors this pin to identify the CPU as an 8088 or an 8086 and accordingly sets its own queue length to 4 or 6 bytes.	Its pin number 34 is SSO. It acts as S0 in the minimum mode. In maximum mode, SSO, pin is always high.
5.	In minimum mode its pin 28 is assigned to signal M/O.	In minimum mode its pin 28 is assigned to signal IO/M

NMI asserted prior to the 2nd clock after the end of RESET will not be honored. If NMI is asserted after that point and during the internal reset sequence, the processor may execute one instruction before responding to the interrupt. A hold request active immediately after RESET will be honored before the first instruction fetch.

All 3-state outputs float to 3-state OFF during RESET. Status is active in the idle state for the first clock after RESET becomes active and then floats to 3-state OFF. ALE and HLDA are driven low.

NOTE: The differences between 8086 and 8088 are only in their BIU and not in EU, but the execution unit (EU) is the same for both. As EU is same, the programming instructions are exactly the same each. Programs written for the 8086 can be run on 8088 without any changes.

3.16. MEMORY ADDRESSING IN 8086

In 8086 the memory address space can be viewed as a sequence of one million bytes in which any byte may contain an 8-bit data element and any two consecutive bytes may contain a 16-bit data element, and there will not be constraint on byte or word address boundaries. The address space is physically connected to a 16-bit data bus by dividing the address space into two 8-bit banks of up to 512K bytes each, as depicted in Fig. 3.29(a).

As observed from Fig. 3.29(a). One bank is connected to the lower half of the 16-bit data bus (D_0-D_7) and contains even address bytes, i.e. when A_0 bit is low, the bank is selected. The other bank is connected to the upper half of the data bus (D_8-D_{15}) and contains odd address bytes. That means when A_0 is high and BHE (Bus High Enable) is low, the odd bank is selected. (*A specific byte within each bank is* selected by address lines A_1-A_{19}).

There are four possible ways to access the data from the memory. Such four ways are as listed below :

1. 8-bit data from lower (Even) address Bank
2. 8-bit data from higher (Odd) address Bank
3. 16-bit data starting from Even Address, and
4. 16-bit data starting from Odd Address.

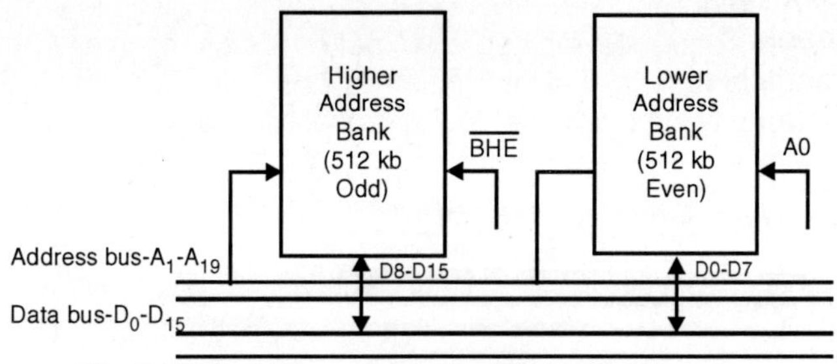

Fig. 3.29(*a*). Address Space in 8086 Microprocessor.

Net let us observe arrangement for the above four methods of assessing data from memory.

3.16.1. The 8-bit Data from Lower/Even Address Bank

In this case to access memory bytes from Even address, information is transferred over the lower half of the data bus (D_0-D_7), where A0 is output LOW and BHE is output HIGH enabling only the even address bank. The Fig. 3.29(b) depicts the arrangement for this method as below :

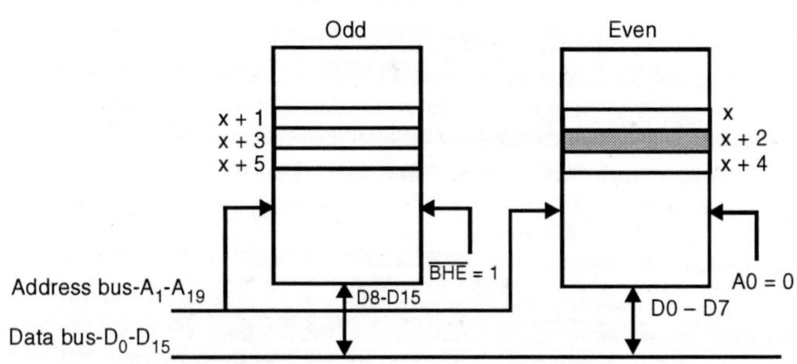

Fig. 3.29(*b*). 8-Bit Data Access From Lower/Even Addresss.

Now let us consider an example for loading a byte of data into CH register (i.e., higher order 8-bits of CX register) from the memory location with an even address. Here the data will be accessed from the even bank via the (D0-D7) DATA BUS. Even though this data is transferred into the 8086 over lower 8-bit lines, the 8086 processor will automatically redirects the data to the higher 8-bits of its internal 16-bit data path and hence to the CH-register. Such capability permits bytes input-output transfer via the AL register to access I/O device connected to either the upper half of the data bus or the lower half of the 16-bit data bus, as observed in Fig. 3.29(b).

3.16.2. 8-bit Data from Higher/Odd Address Bank

Here to access memory byte from an odd address information, is transferred over the higher half of the data bus (D_8-D_{15}). The BHE output low enables the upper memory bank, and whereas A0 is output high to disable lower memory bank. The arrangement is as depicted in Fig. 3.31.

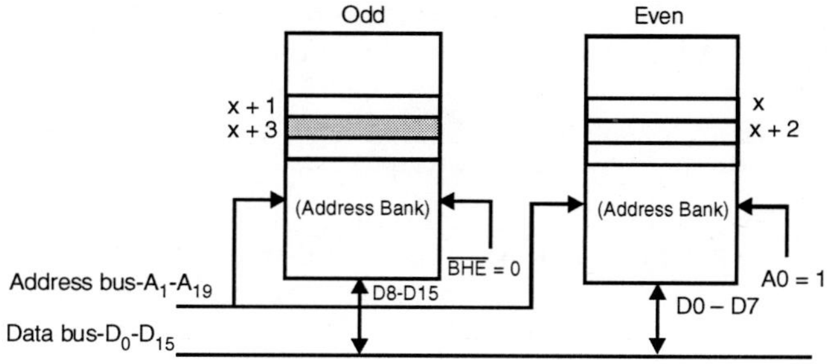

Fig. 3.31. 8-bit Data from Odd Address Bank.

3.16.3. The 16-bit Data Starting from Even-Address

Here, 16-bit data from an even address is accessed in a single bus cycle. The address lines A1-A19 select the appropriate byte within each bank. A0 low and BHE low enables both banks simultaneously. The arrangement is as shown in Fig. 3.32.

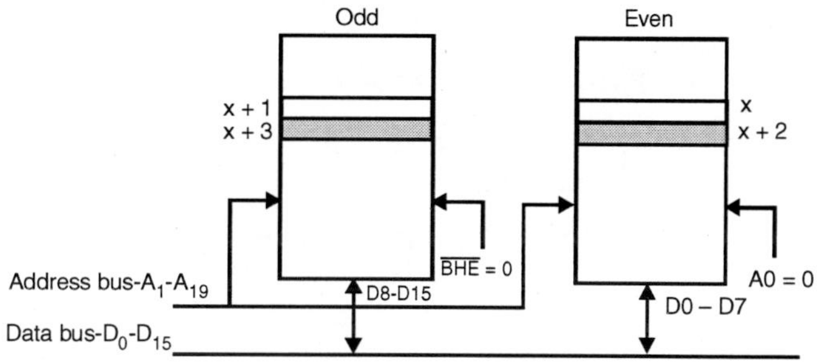

Fig. 3.32. 16-bit data access with even address.

3.16.4. The 16-bit Data Access starting from Odd Address

Here a 16-bits word located at an odd address (*two consecutive bytes with the least significant byte at an odd byte address)* is accessed using two bus cycles. For first bus cycle, lower byte will be accessed. The arrangement with odd address as 0005 is as shown in Fig. 3.33 (a).

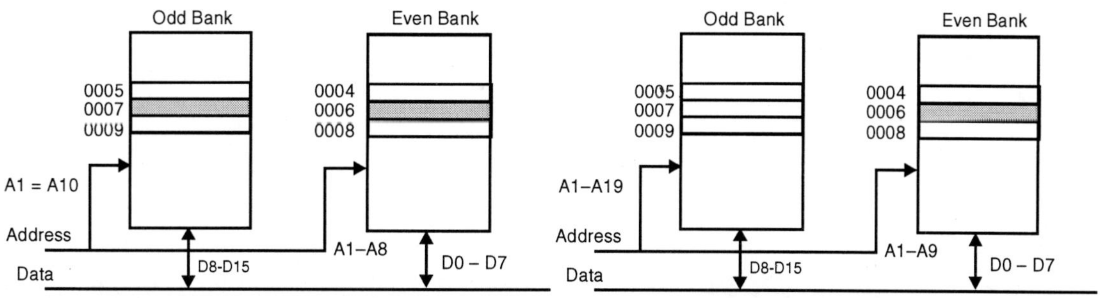

(a) First Access from odd address. ***(b) First Access from odd address***

Fig. 3.33. 16-bit Data Access with odd address.

Now during second bus cycle, the upper byte (*with the even address 0006H as observed in Fig. 3.33 (b)*) will be accessed. During first bus cycle, A_1-A_{19} address bus specifies the address and A_0 as 1 and BHE are low. That means the even memory bank will be disabled and odd memory bank is enabled and next, during second bus cycle, the address is incremented. That means, A_0 is zero and BHE is made high, and even memory bank is enabled and the odd memory bank is disabled, as observed in Fig. 3.33.

3.17. CLOCK GENERATOR 8284

In this section we will study the details of the clock generator 8284A. Apart we will observe the RESET signal and introduction of READY signal to 8086 microprocessor. The 8284A is an important ancillary component to the 8086/8088 microprocessor. Without clock generator, it required many additional circuits to generate the clock in an 8086/8088-based system. 8284A provides following basic functions or signals: clock generation, RESET synchronization, READY synchronization, and a TTL level peripheral clock signal which are very important essential signal to microprocessors. Now Fig. 3.34 depicts the details of pin out or functional pins of the clock generator 8284A.

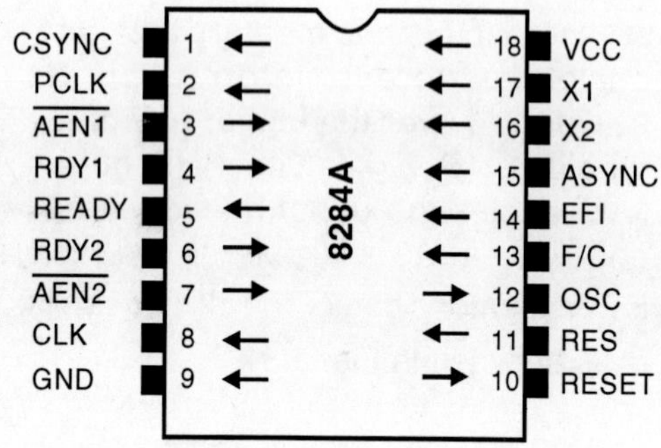

Fig. 3.34. Pin configuration of clock generator 8284

3.17.1. Pin Functions of 8284A

It is an 18-pin IC, which is specially designed to use with 8086/8088 microprocessors. Now let us observe the function of each pin of 8284A clock generator.

AENI and AEN2 : The **address enable** pins are provided to qualify the bus ready signals, RDY1 and RDY2, respectively, which are used to cause wait states, along with the RDY1 and RDY2 inputs. The wait states are generated by READY pin of the 8086/8088 microprocessors, which is controlled by these two inputs, RDY1 and RDY2 The bus ready inputs are provided in conjunction with the AEN1 and AEN2 pin to cause wait states in an 8086/8088 microprocessors, which is controlled by these two inputs.

ASYNC : The **ready synchronization** selection input selects either one or two stages of synchronization for the RDY1 and RDY2 inputs.

READY : The Ready is an output pin which connects to the 8086/8088 READY input. This signal is synchronized with the RDY1 and RDY2 inputs.

X1 and X2 : The crystal oscillator pin connect to an external crystal used as the timing source for the clock generator and all its functions.

EFI : The **external frequency** input is used when the F/C is pulled high. EFI supplies the timing whenever the F/C pin is high.

CLK : The **clock** output pin provides CLK input signal to the 8086/8088 microprocessor (*and other components in the system)* The CLK pin has an output signal that is one-third of the crystal or EFI input frequency and has a 33 percent duty cycle, which is required by the 8086/8088.

PCLK : The **peripheral clock** signal is one-sixth the crystal or EFI input frequency and has a 50 percent duty cycle. The PCLK output provides a clock signal to the peripheral equipment in the system.

OSC : The **oscillator** output is a TTL level signal that is at the same frequency as the crystal or EFI input. (*The OSC output* provides an EFI input to other 8284A clock generators in some multiple processor systems).

RES : The **reset** input is an active-low input to the 8284A. The RES pin is often connected to an RC network that provides power on resetting.

RESET : The reset output is connected to the 8086/8088 RESET input pin.

CSYNC : The **Clock synchronization** pin in used whenever the EFI input provides synchronization in systems with multiple processors. When the internal crystal oscillator is used, this pin must be grounded.

GND : The ground pin connects to ground.

Vcc : The Power supply pin connects to +5.0 V with a tolerance of ±10%.

F/C : The frequency/crystal select input choses the clocking source for the 8284A. If this pins is held high an external clock is provided to EFI input pin; and if it is held low, the internal crystal provides the timing signal.

3.17.2. The Operation of Clock Generator 8284A

The clock generator 8284A is a simple device, its internal logic diagram is as shown in Fig. 3.35.

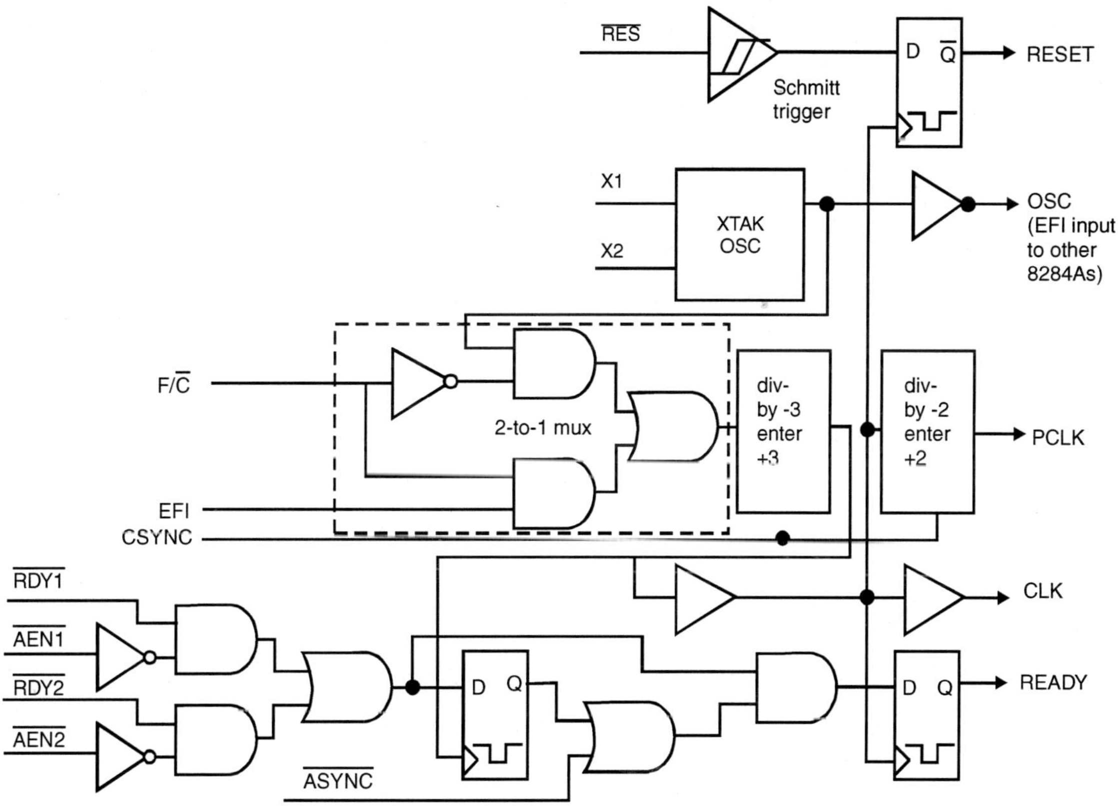

Fig. 3.35. Internal Architecture of 8284.

Now the two sections of operation those are of interested in 8284A clock generator are :

(a) Operation of the clock section, and

(b) Operation of the Reset section

(a) Operation of the Clock Section : The top half of the logic diagram in Fig. 3.35 represents the clock and reset synchronization section of the 8284A clock generator. As observed, the crystal oscillator has two inputs: XI and X2 when a crystal is attached to XI and X2, oscillator generates a square-wave signal at the same frequency as the crystal. The square-wave so generated signal is fed to an AND gate and also to an inverting buffer that provides the OSC output signal. (*The OSC signal is sometimes used as an EFI input to other 8284A circuits in a system).* The inspection of AND gate reveals that when F/C is a logic 0, the oscillator output is steered through to the divide-by-3 counter. If F/C is a logic 1, then EFI is steered through to the counter.

The output of the divide-by-3 counter generates timing for ready synchronization, a signal for another counter (divide-by-2), and the CLK signal to the 8086/8088 microprocessors. The CLK signal is also buffered before it leaves the clock generator. Observe that the output of the first counter feeds the second. These two cascaded counters provide the divide-by-6 output at PCLK, the peripheral clock output, the Fig. 3.36 below; shows the connection arrangement of 8284A to 8086/8088 microprocessor.

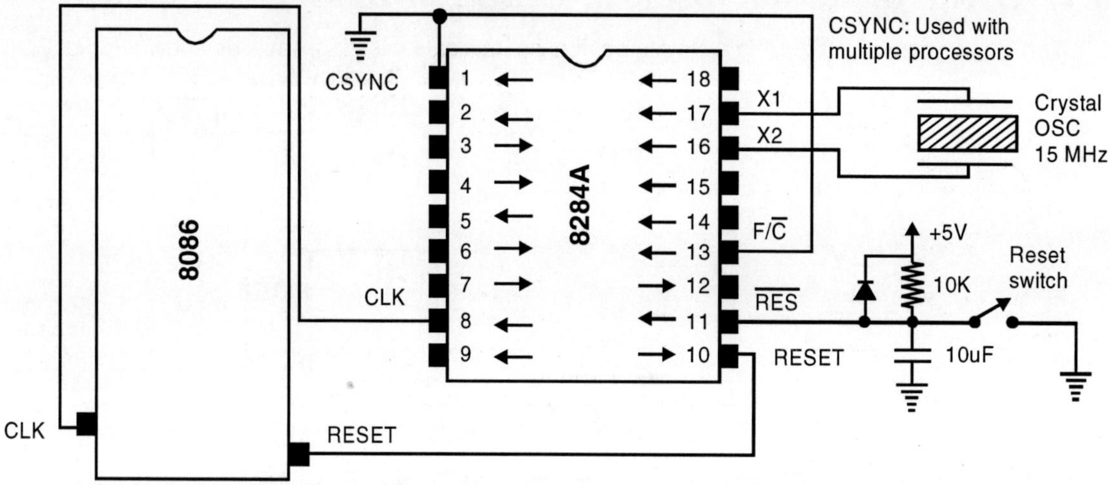

Fig. 3.36. Interfacing of clock generator and 8086/8088.

From Fig. 3.36 two points to be observed; i.e.,

1. That F/C and CSYNC are grounded to select the crystal oscillator, and
2. That a 15 MHz crystal provides the normal 5 MHz clock signal to the 8086/8088 as well as a 2.5 MHz peripheral clock signal.

(b) Operation of the Reset Section : The reset section of the 8284A is simple, and it consists of a Schmitt trigger buffer and a single D-type flip-flop circuit. The D-type flip-flop ensures that the timing requirement of the 8086/8088 RESET input are fulfilled. This circuit applies the RESET signal to the microprocessor on the negative edge (1-to-0 transition) of each clock. The 8086/8088 microprocessors sample RESET at the positive edge (0-to-1 transition) of the clocks; therefore (*this circuit meets the timing requirements of the 8086/8088).*

From Fig. 3.36 it can be observed that an RC circuit provides a logic 0 to the RES input pin when power is first applied to the system. After a short time, the RES input becomes a logic 1 because the capacitor charges toward +5.0 V through the resistor. A push-button switch allows the microprocessor to be reset by the operator.

Exact rest timing requires the RESET input to become a logic 1 no later than four clocks after system power is applied and to be held for at least 50 µs. (*The flip-flop* makes certain that RESET goes high in four clocks, and the RC time constant ensures that it stays high for at least 50 µs).

The Bus Buffering and Latching : To make use of 8086/8088 microprocessor with memory or I/O interfaces, the multiplexed buses of the µP has to be demultiplexed. Therefore we have to discuss about demultiplexing the buses, and mention how buses are buffered in the case of large system, depending on fan-out of the components. Now let us discuss about:

1. Demultiplexing the Buses of 8086/8088, and
2. Buffered System for 8086/8088.

3.18. THE READY AND WAIT STATES

The READY input causes wait states for slower memory and I/O components. Await state (T_w) is an extra clocking period, inserted between T_2 and T_3, that lengthens the bus cycle. When one wait state is inserted, then the memory access time, normally 460 ns with a 5 MHz clock, is lengthened by one clocking period (200 ns) to 660 ns. Now let us generator and show how to insert one or more wait states selectively into the bus cycle, and examine the READY input and the synchronization times it requires.

3.18.1. Ready Input

READY input is sampled at the end of T_2 and again, if applicable, in the middle of T_2 If READY is a logic 0 at the end of T_2, then T_3 is delayed and T_w is inserted between T_2 and T_3. READY is next sampled at the middle of Tw to determine if the next state is Tw or T_3. It is tested for a logic 0 on the 1-to-0 transition of the clock at the end of T_2 and for a 1 on the 0-to-1 transition of the clock in the middle of T_w. READY input 8086/8088 has some stringent timing requirement. The timing diagram in Fig. 3.37 depicts READY causing one wait state (T_w) along with the required setup and hold times from the system clock.

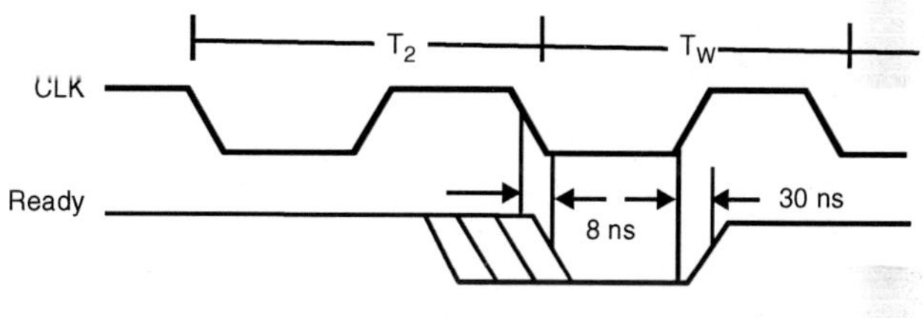

Fig. 3.37.

The timing requirement for such operations is fulfilled by the Internal READY synchronization circuitry of 8284A clock generator. When 82824A is used for **READY, the RDY** (*read input to the 8284A*) input occurs at the end of each T state.

3.18.2. RDY and 8284A

RDY is the synchronized ready input to the 8284A clock generator. Timing diagram for this input is as shown in Fig. 3.38. Even though it differs from the timing for the READY input to be 8086/8088, the internal 8284A circuitry guarantees the accuracy of the READY synchronization provided to the 8086/ 8088. While observing back in Fig. 3.35, the bottom half of the diagram is the READY synchronization circuitry. At the leftmost side, the RDY1 and AEN1 inputs are ANDed, as the RDY2 and

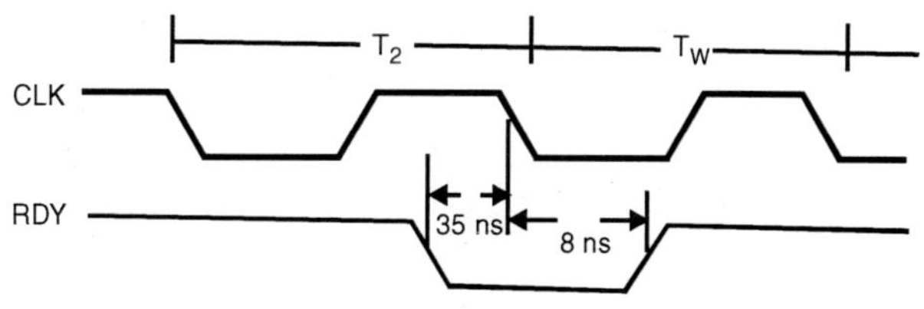

Fig. 3.38. Timing diagram for RDY input.

AEN2 inputs. The outputs of the AND gates are then ORed to generate the input to the one or two stages of synchronization. In order to obtain a logic 1 at the inputs to the flip-flops, RDY1 ANDed with AEN1 must be active or RDY2 ANDed with AEN2 must be active.

Next ASYNC input selects one stage of synchronization when it is the logic 1 and two stages when a logic 0. When one stage is selected, the RDY signal is kept from reaching the 8086/8088 READY pin until the next negative edge of the clock, and when two stages are selected, the first positive edge of the clock captures TDY in the first flip-flop. The output of this flip-flop is fed to the second flip-flop so that second flip-flop. The output of this flip-flop is fed to the second flip-flop so that the second flip-flop captures RDY on the next negative edge of the clock.

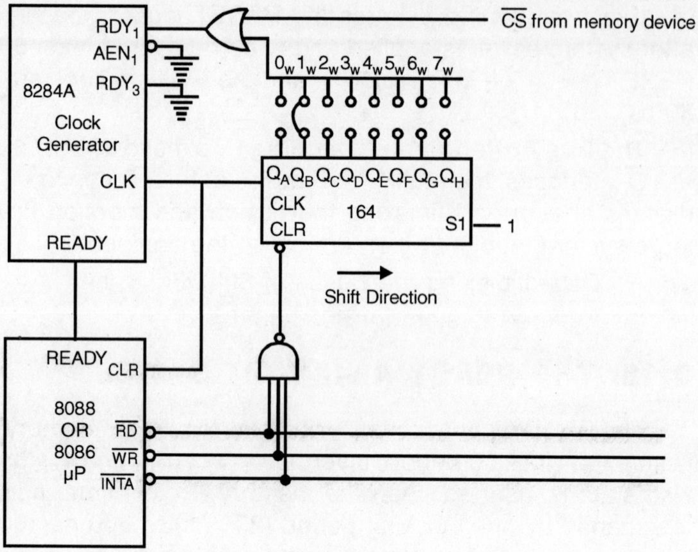

Fig. 3.39. Circuit diagram for generation of 0 to 7 wait states.

3.18.3. Wait State

The Fig. 3.39 depicts a circuit used to introduce any number of wait states for 8086/8088 micro-processors, where an 8-bit serial shift register (74LS164) shifts a logic 0 for one or more clock periods form one if its Q outputs through to the RDY_1 input of the 8284A. With suitable strapping, this circuit can provide various number of wait states.

It can also be observed that how, the shift register is cleared back point. The output of the register is forced high when the second flip-flop captures RD, WR and INTA pins are all at logic 1's. These three signals are high until state T_2, so the shift register shift four the first time when positive edge of the T_2 arrives. If one wait is desired, then output Q_B is connected to the OR gate. If two waits are desired, output Q_c is connected, and so forth.

In Fig. 3.39 it can observe that this circuit does not always generate wait states, but is enabled form the memory only for memory devices that require the insertion of waits. If the selection signals from a memory device is at logic 0, the device is selected; then this circuit will generate a wait state. Now Fig. 3.40 depicts the timing

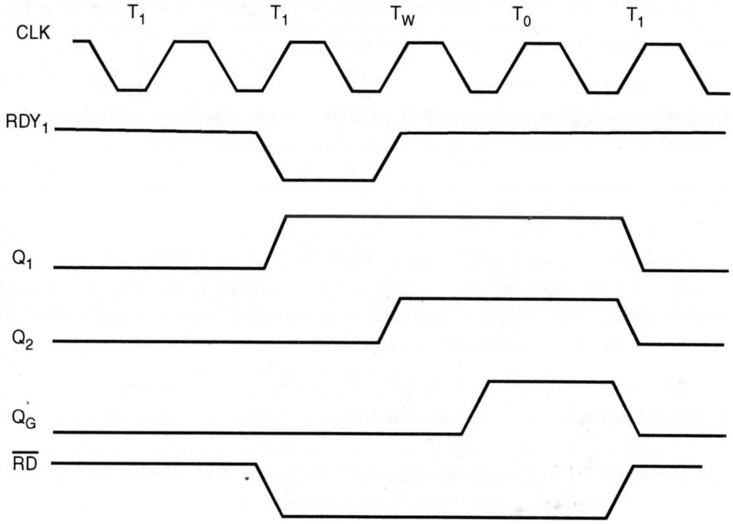

Fig. 3.40. Timing diagram for Wait state generator.

diagram for this shift register wait state generator when it is wired to insert for one wait state.

The timing diagrams of Fig. 3.40 also illustrates the internal contents of the shift registers flip-flops to present a more detailed view of its operation. (*here one wait state is generated.*)

3.19. DEMULTIPLEXING THE BUSES OF 8086/8088

The address/data bus on the 8086/8088 is multiplexed shared so as to reduce the number of pins required for the 8086/8088 integrated circuit. But this burdens the hardware designer with the task of extracting or demultiplexing information from these multiplexed pins. So can we not leave buses multiplexed ?

The memory and I/O require that address remains valid and stable throughout a read or write cycle. When the buses are multiplexed, the address changes at the memory and I/O, which causes them to read or write data in the wrong locations.

In all computer systems they have three buses: (1) an address bus that provides the memory and I/O with the memory address or the I/O port number, (2) a data bus that transfers data between the microprocessor and the memory and I/O in the system, and (3) a control bus that provides control signals to the memory and I/O. (These buses must be present in order to interface to memory and I/O).

Now let us observe the demultiplexing arrangement for the (a) 8086, (b) 8088 microprocessor.

(a) Demultiplexing the 8086 : The 8086 system requires separate address, data and control buses, which differs primarily in the number of multiplexed pins. (In the 8088, only AD_7-AD_0 and A_{19}/S_6-A_{16}/S_3 will be multiplexed). In 8086, the multiplexed pins include AD_{15}-AD_0-A_{19}/S_6-A_{16}/S_3, and BHE/S_7. All of these signals must be demultiplexed.

The Fig. 3.41 depicts a demultiplexed 8086 with all three buses: address (A_{19}-A_0 and BHE, data (D_{15}-D_0), and control (M/IO, RD and WR), as below.

In the circuit shown in Fig. 3.41 an additional 74LS373 latch has been added to demultiplex the address/data bus pins Ad15-AD8 and a BHE/S7 input has been added to the top 74LS273 to select the high-order memory bank in the 16-bit memory system of the 8086 if compared with that for 8088. The memory and I/O system observe the 8086 as a device with a 20-bit address bus (A_{19}-A_0), a 16-bit data (D_{15}-D_0) and a 3 line control bus (M/IO, RD and WR).

(b) Demultiplexing the 8088 : The Fig. 3.42 shows the 8088 microprocessor and the components required to demultiplexes its buses. Here, two 74LS373 transparent latches are used to demultiplex the address/data bus connections AD_7-AD_0 and the multiplexed address/status connections A_{19}/S_6-A_{16}/S_3.

The transparent latches, which are looking like wires whenever the address latch enable pin (ALE) becomes a logic 1 they pass the inputs to the outputs. After a small duration ALE returns to its logic 0 condition, which causes the latches to remember the inputs at the time of the change to a logic 0.

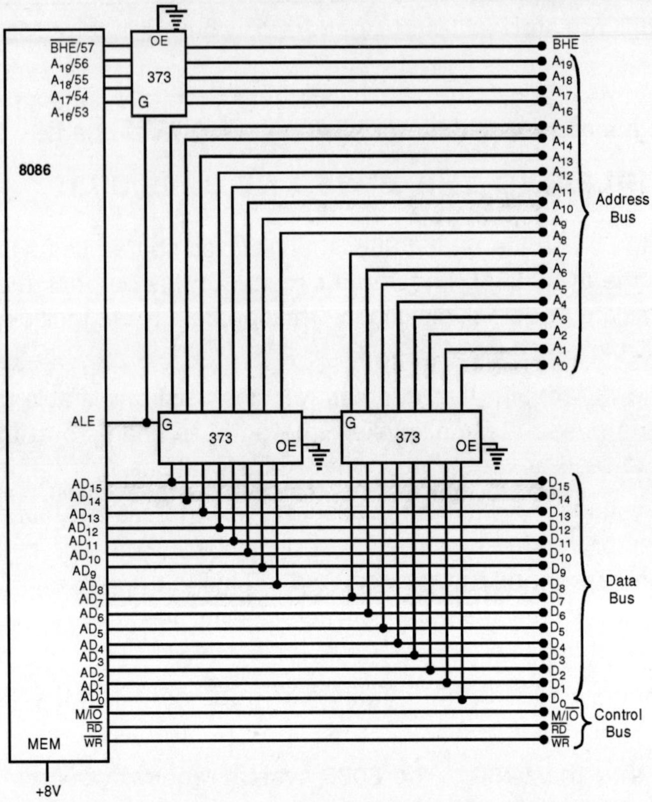

Fig. 3.41. Demultiplexing of address bus for 8086 microprocessor.

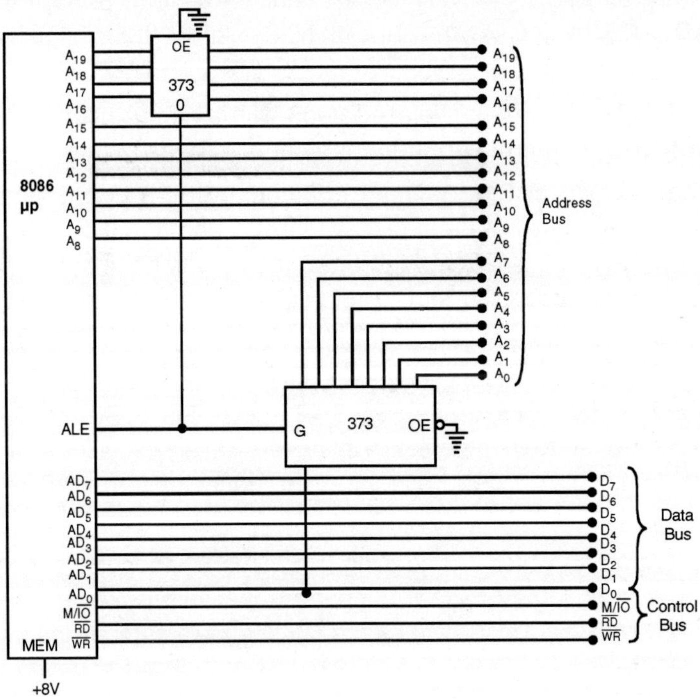

Fig. 3.42. Demultiplexing of address bus for 8088 microprocessor.

In this case, A_7-A_0 are stored in the bottom latch and A_8-A_{15}. These address connections allow the 8088 to address 1M byte of memory space. (*The fact that* the data bus is separate allows it to be connected to any 8-bit peripheral device or memory component).

Buffered System for 8086/8088

When more than 10 unit loads are attached to any bus pin, the entire 8086 or 8088 system must be buffered. The demultiplexed pins are already buffered by the 74LS373 latches, which have been designed to drive the high-capacitance buses encountered in microcomputer systems. The buffer's output current have been increased so that more TTL unit loads may be driven: a logic 0 output provides up to 32 mA of sink current, and a logic 1 output provides up to 5.2 mA of source current. That means a fully buffered signal will introduce a timing delay to the system. This causes no difficulty unless memory or I/O devices are used, which function at near the maximum speed of the bus.

Now let us observe the arrangement for fully buffered arrangement for

(a) 8086 and

(b) 8088 microprocessors.

(a) Fully Buffered 8086 : The Fig. 3.43 shows a fully buffered 8086 microprocessor. It address pins are already buffered by the 74LS373 address latches; its data bus employs two 74LS245 octal bi-directional bus buffers; and the control bus signals, and IO/M, M/IO, RD and WR use a 74LS244 buffer.

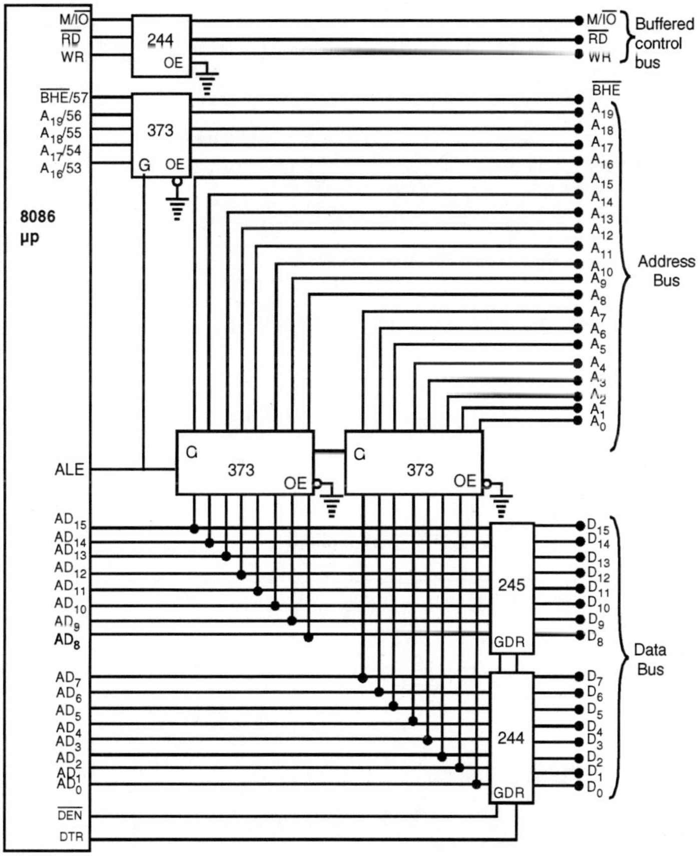

Fig. 3.43. Fully Buffered 8086 Microprocessor.

Fully buffered 8086 system requires one 74LS244, two 74LS245s, and three 74LS373s

Note: 8086 requires one more buffer than the 8088 because of the extra eight bit data bus connections, D_{15}-D_8. It also has a signal that is buffered for memory bar selection.

3.20. BUS TIMING DIAGRAMS

It Is required to know system bus timing before choosing a memory or I/O device for interfacing to the 8086 or 8088 microprocessors. Now, we discuss the bus signals and the basic read and write timing of the 8086/8088. (*We discuss only the times that affect memory and I/O interfacing*).

3.20.1. The Basic Bus Operation

In 8086/8088 microprocessor, the three buses, address, data and control, will function similar to those of other microprocessor outputs the memory address on the address bus,

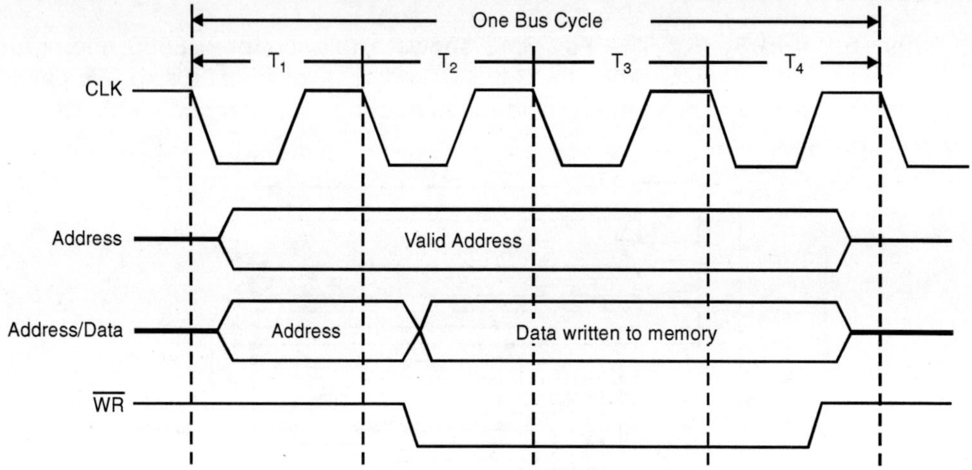

Fig. 3.44. Simplified 8086/8088 Write Bus Cycle.

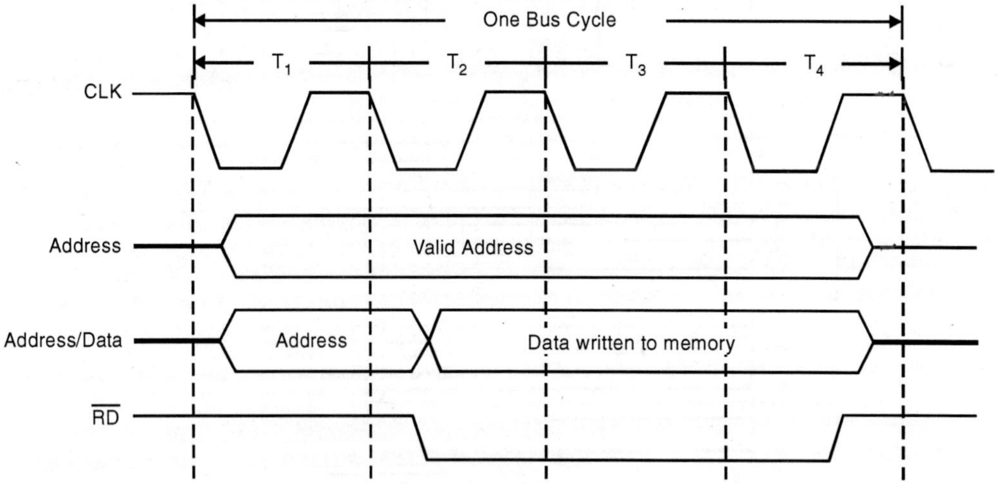

Fig. 3.45. Simplified 8086/8088 Read Bus Cycle.

outputs the data to be written into memory on the data bus, and issues a write (WR) to memory and IO/M = 0 for the 8088 and M/IO = 1 for the 8086. The simple write bus cycle for 8086/8088 is as shown in Fig. 3.44.

Now if data are read from the memory the microprocessor outputs the memory address on the address bus, issues a read memory signal, and accepts the data via the data bus. The simplified read bus cycle for 8086/8088 is as shown in Fig. 3.45.

3.20.2. Details of Timing in General

8086/8088 microprocessors use the memory and I/O in periods of time called bus cycles, each bus cycle equals four system-clocking periods (T-states). But some new microprocessors divide the bus cycle into as few as two clocking periods. When **the clock is operated at 5 MHz** (*The basic operating frequency for these two microprocessors),* then on 8086/8088 a bus cycle will complete in 800 ns. That is microprocessor reads or writes data between itself and memory or I/O at a maximum rate of 1.25 million times per second. *(Because of the internal queue, the 8086/8088 can execute 2.5 million instructions per second (MIPS) in bursts).*

During first clocking period in a bus cycle, TI, many things takes place. The address of the memory or I/O location is sent out via the address bus and the address/data bus connections. (*The address/data bus is multiplexed and sometimes contains memory addressing information, sometimes data).* Also output during TI are the control signals ALE, DT/R and IO/M (8088) or M/O (8086). The IO/M or M/O signal indicates whether the address bus contains a memory address or an I/O device (port) number.

Next during T_2, the 8086/8088 microprocessor issue the RD or WR signal, DEN, and, in the case of a write, the data to be written appears on the data bus. Such events cause the memory or I/O device to begin to perform a read or write. The DEN signal turns on the data bus buffers, if they are present in the system, so the memory or I/O can receive data to be written or so the microprocessor can accept the data read from the memory or I/O for a read operation. When this happens to be a write bus cycle, then the data are sent out to the memory or I/O through the data bus.

READY is sampled at the end of T_2, if it is low at this time, T_3 becomes a wait state (T_W). This clocking period helps to allow the memory time to access data. For the bus cycle happens to be a read bus cycle, the data bus is sampled at the end of T_3. Finally in T_4, all bus signals are deactivated in preparation for the next bus cycle, which is also the time when the 8086/8088 microprocessors sample the data bus connections for data that are read from memory of I/O. Apart at this point, the trailing edge of the WR signal transfers data to the memory or I/O, which activate and write the WR signal returns to a logic 1 level.

Now to let us observe the four T states, i.e., T_1, T_2, T_3 and T_4 in

(a) Read timing, and

(b) Write timing

For minimum mode timing operation

(a) Read Timing : The Fig. 3.46 depicts read timing for the 8088 microprocessors in detail when compared to Fig. 3.45. The 8086 read timing is identical except that the 8086 has 16 rather then 8 data bus bits. A close look at this timing diagram should allow to identify all the main events described for each T state. Most important item contained in the read timing diagram is the mount of time allowed the memory or I/O to read the data. The memory is chosen by its access time, which is the fixed amount of time the microprocessor allows it to access data for the read operation. Therefore, it is important to note that the memory we choose comply with the limitation of the systems.

NOTE: The microprocessor timing diagram does not, however, provide a straightforward memory access time. Instead, it is necessary to combine several times to arrive at the access time. To find memory access time in this diagram, we must first locate the point in T_3 when data are sampled. When we examine the timing diagram closely. We will observe a line that extends from the end of T_3 down to the data bus. The end of T_3 is where the microprocessor samples the data bus.

The Memory access time starts when the address appears on the memory address bus and continues until the microprocessor samples the memory data at T_3. Approximately three T states elapse between these times. The address does not appear until TCLAV time (110 ns if the clock is 5 MH5) after the start of TI. This means that TCLAV time must be subtracted from the three clocking states (600 ns) that separate the appearance of the address (TI) and the sampling of the data (T_3). One other time must also be subtracted : the data setup time (TDVCL), which occurs before T3. Memory access time is thus three clocking states minus the sum of TCLAV and TDVCL. Because TDVCL is 30 ns with a 5 MHz clock, the allowed memory access time is only 460 ms (asses time = 600 ns – 110 ns – 30 ns).

NOTE: In specific the memory devices chosen for connection to the 8086/8088 operated at 5 MHz must be able to access data in less than 460 ns because of the time delay introduced by the address decoders and buffers in the system. At least a 30 or 40 ns margin should exist for the operation of these circuits. That means, memory speed will not be slower than about 420 ns to operate correctly with the 8086/8088 microprocessors.

NOTE: Other timing factor that may affect memory operation is the width of the RD stroke. On the timing diagram, the read strobe width is given as TRLRH the time for which is 325 ns (5 MHz clock rate), which is wide enough for almost all memory devices manufactured with an access time of 400 ns or less.

Write Timing : The Fig. 3.47 depicts the write timing diagram for the 8088 microprocessor. Again, the 8086 is so nearly identical that it need not be presented here in a separate timing diagram. Main differences between read and write timing are minimal. The strobe is replaced by the $\overline{WR}$ strobe, the data bus contains information for the memory rather than information from the memory, and DT/$\overline{R}$ remains a logic 1 instead of a logic 0 throughout the bus cycle.

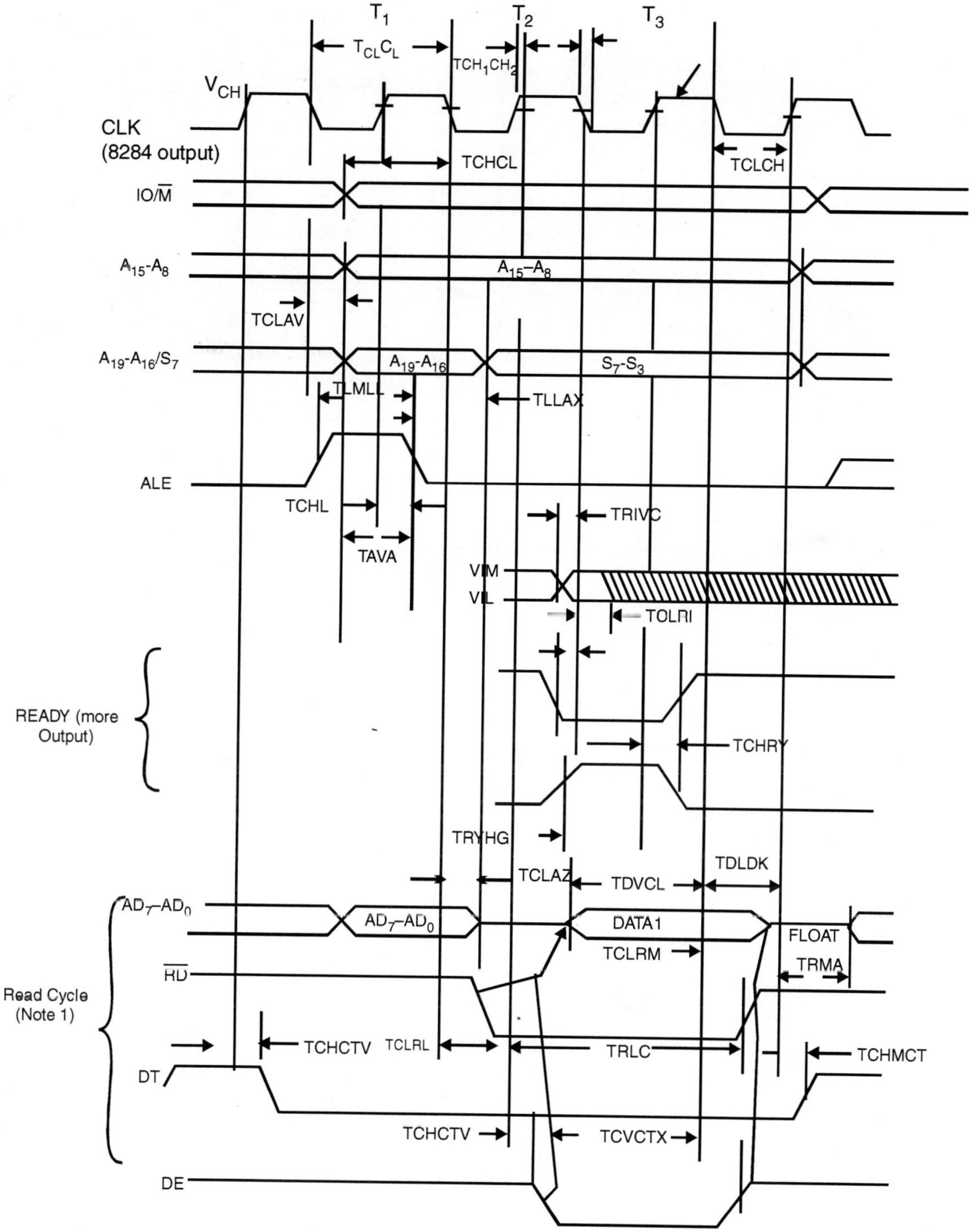

Fig. 3.46. Minimum Mode 8088 Bus Timing For Read Operation.

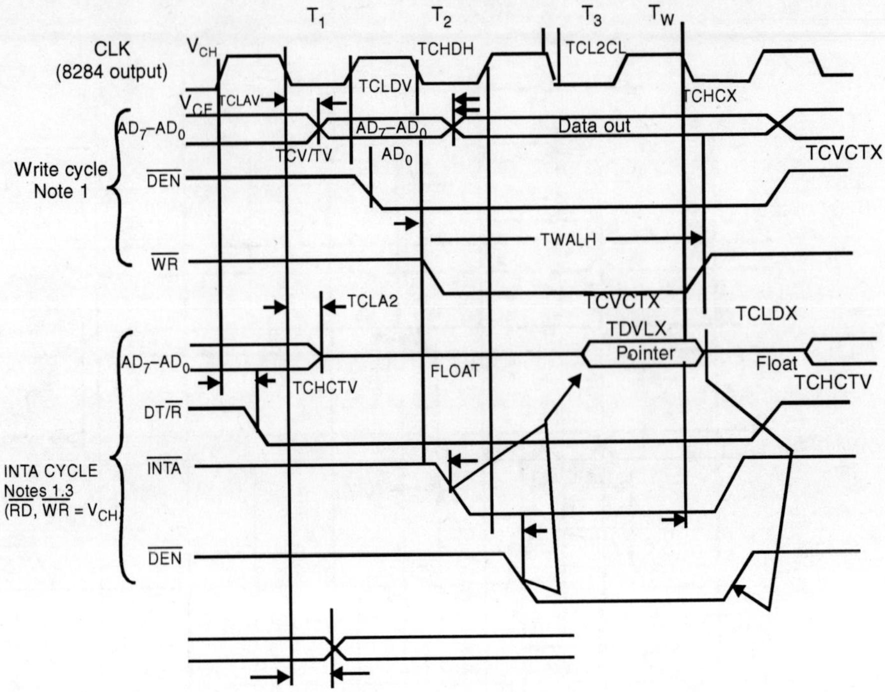

Fig. 3.47. Minimum Mode 8088 Bus timing For Write Operation.

Notes :

1. All signals switch between VOH and VOI unless otherwise specified.
2. RDY is sampled near the end of T_2, T_3, T_w to determine if T_w machines state are to be inserted.
3. Two INTA cycles run back-to-back. The 8088 local ADDR/Data bus is floating during both INTA cycles. Control signals are shown for the second INTA cycle.
4. Signals at 8284 are shown for reference only.
5. All timing measurement are made at 1.5 V unless otherwise noted.

While interfacing some memory devices, timing may be especially critical between the point at which WR becomes a logic 1 and the time when data are removed from the data bus. It is case because, as we can recall, memory data are written at the trailing edge of the WR stroke. As per the timing diagram, this critical period is TWHDX or 88 ns when the 8088 processor is operated with a 5 MHz clock. The hold time is often much less than this, and is in fact often 0 ns for memory devices. The width of the WR strobe is TWELWH or 3240 ns at a 5 MHz clock rate. This rate is also compatible with most memory devices that have an access time of 400 ns or less.

3.20.3. Instruction Execution Timing

The execution time for instructions varies and depend on the type of instruction and addressing modes. This basic instruction time can be determined by multiplying the number of clock cycles needed to execute the instruction by the clock's period. The total instruction execution time is the sum of the basic time plus the time required to calculate the effective address if a memory operand is accessed. The basic instruction time assumes that the instruction is not in the instruction queue then additional clock cycle must be added to find the total execution time.

Total instruction execution time will depend on

(a) **Number of memory references required by the instruction :** If operand is a single byte or word with an even address only one memory reference is required. However, if operand is a word with an odd address two memory references are required. Each memory reference requires four clock cycles, hence, each odd address word reference requires four extra clock cycles.

(b) **Addressing mode :** In case of two operand instructions, register addressing gives the faster operation than register-to-register and register-to-memory and memory-to-register operations. Between memory-to-register and register-to-memory, register-to-memory operation requires extra time to store the result into the memory.

(c) **Variable basic execution times :** Some instructions are data dependent and requires variable basic execution times. These are multiply, divide, shift, and rotate instructions.

(d) **Conditional branch instructions :** For conditional branch instructions, two different timings are specified. The smaller time should be considered when the condition is not met, and a branch is not taken. The larger time should be considered when the condition is met and the branch is taken. In the latter case, extra cycles are required to fetch the next instruction due to flushing of instruction queue.

3.21. MINIMUM AND MAXIMUM MODES

There are two available modes of operation for the 8086/8088 microprocessors :

(1) Minimum mode and (2) Maximum mode. Minimum mode operation is obtained by connecting the mode selection pin MN/MX to + 5.0 V, and maximum mode is selected by grounding this pin. Both modes enable different control structures for the 8086/8088 microprocessors.

3.21.1. Minimum Mode

In this mode the pin definition from 24 to 31 are to be observed, which is the main difference with respect to maximum mode, (refer Fig. 3.1). Now let us observe the pin definitions in minimum mode.

Pin 24	**(Interrupt Acknowledge) Output :** This indicates recognition or an interrupt request which consists of two negatives going pulses in two consecutive bus cycles. The first pulse informs the interface that its request has been recognized and upon receipt of the second pulse, the interface is to send the Interrupt type of the processor over the data bus.
Pin 25	**ALE (Address Latch Enable) Output :** This signal is provided by 8086 to demultiples AD_0-AD_{15} into AD_0–AD_{15} and D_0-D_{15} using external latches.
Pin 26	**(Data Enable) Output :** This signal indicates the transceivers that the CPU is ready to send or receive data.
Pin 27	**DT/R (Data transmit/Receive) Output :** This signal is helps to control data flow direction. High on this pin indicates that the 8086 is transmitting the data and low indicates that the 8086 is receiving the data.
Pin 28	**M/O Output :** It is used to distinguish memory data transfer (M/IO=HIGH) and I/O data transfer (M/IO=LOW).
Pin 29	**WR: Write Output :** WR is low whenever the 8086 is writing data into the memory or an I/O device.
Pin 30 and 31	**HOLD input, HLDA Output.** A HIGH on HOLD pin indicates that another master (DMA) is requesting to take over the system bus. On receiving HOLD signal processor outputs HLDA signal HLDA signal HIGH as an acknowledgement. At the same time, processor tristates the system bus. A low on HOLD gives the system bus control back to the processor. *(Processor then outputs low signal on HLDA).*

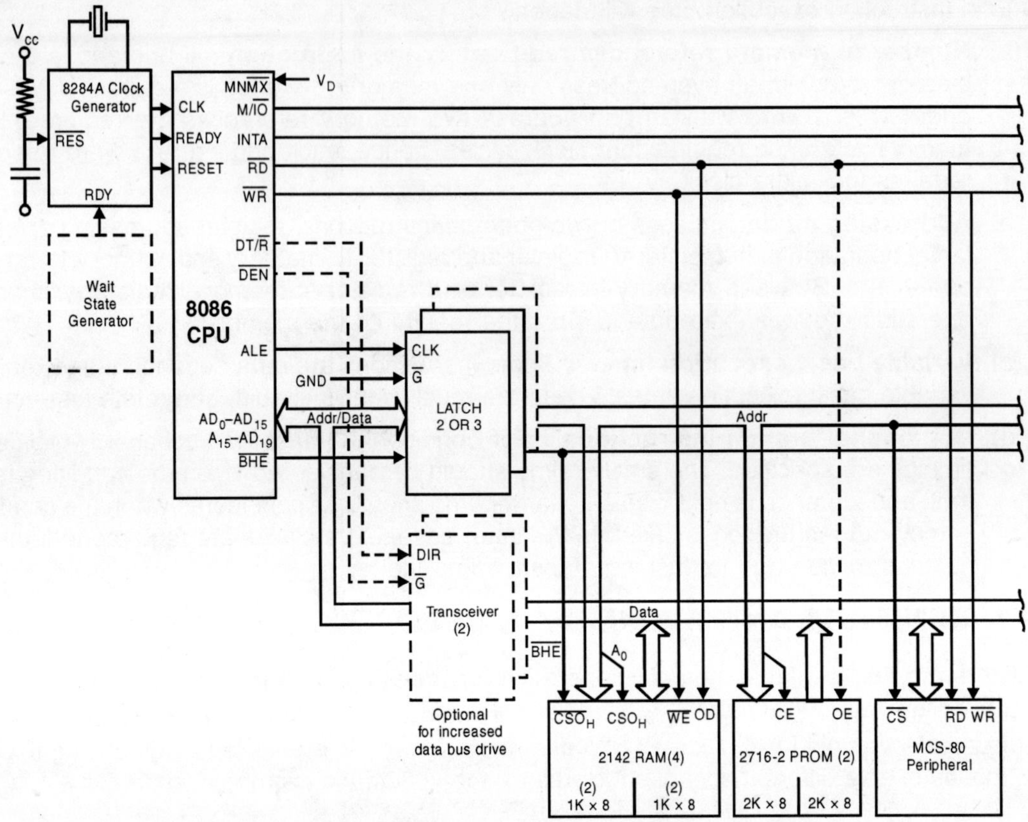

Fig. 3.48. Minimum Mode Configuration of 8086 Microprocessor

In minimum mode 8086 system, the microprocessor 8086 is operated in minimum mode by strapping its MN/$\overline{\text{MX}}$ pin to logic 1. In minimum, all the control signals are given out by the microprocessor chip itself.

There is a single microprocessor in minimum mode system. The remaining components in the system are latches, Tran receivers, clock generator, memory and I/O devices. Some type of chip selection logic may be required for selecting memory or I/O devices, depending upon the address map of the system.

The latches are generally buffered output D-Type flip-flops, like 74LS373 or 8282. They are used for separating the valid address from the multiplexed address/data signals and are controlled by the ALE signal generated by 8086.

Transceivers are the bidirectional buffers and some times they are called as data amplifiers. They are required to separate the valid data from the time multiplexed address/data signal. They are controlled by two signals, DEN and DT/ R. DEN signal indicates that the valid data is available on the data bus, while DT/R indicates the direction of data i.e. from or to the processor.

The system contains memory for the monitor and users program storage. Usually EPROMS are used for monitor storage and RAMS are used for users program storage. The clock generator generates the clock from the crystal oscillator and then shapes it and divides to make it more precise so that it can be used as an accurate timing reference for the system.

The clock generator also synchronizes some external signals with the system clock. Since the 8086 has 20 bit address bus and 16 data lines the CPU requires three octal address latches and two octal data buffers for the complete address and data operation. The system configuration is shown in Fig. 3.48.

The timing diagram can be categorized in two parts, the first is the timing diagram for read cycle and the second is the timing diagram for write cycle.

The read cycle begins in T_1 with the negative going edge of this signal, the valid Address Latch Enable (ALE) signal and also M/IO signal. During the negative going edge of this signal, the valid address is latched on the local bus.

The /BHE and A_0 signals address low, high or both bytes. From T_1 to T_4, the M/IO signal indicates a memory or I/O operation. At T_2, the address is removed from the local bus and is sent to the output.

The bus is then tristated. The read (RD) signal causes the addressed device to enable its data bus drivers.

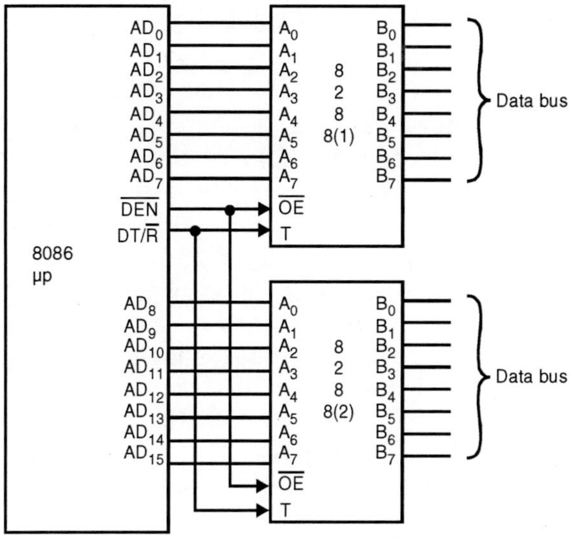

Fig. 3.49. Connection details of the 8286 with 8086.

After /RD goes low, the valid data is available on the data bus. The addressed device will derive the READY line high.

When the processor returns the read signal to high level, the addressed device will again tristate its bus drivers. A write cycle also begins with the assertion of ALE and the emission of the address. The M/IO signal is again asserted to indicate a memory I/O operation.

Minimum Mode 8086 System

- In a minimum mode 8086 system, the microprocessor 8086 is operated in minimum mode by strapping its MN/MX pin to logic 1.
- In this mode, all the control signals are given out by the microprocessor chip itself. There is a single microprocessor in the minimum mode system.
- The remaining components in the system are latches, transreceivers, clock generator, memory and I/O devices. Some type of chip selection logic may be required for selecting memory or I/O devices, depending upon the address map of the system.
- Latches are generally buffered output D-type flip-flops like 74LS373 or 8282. They are used for separating the valid address from the multiplexed address/dta signals and are controlled by the ALE signal generated by 8086.
- Transreceivers are the bidirectional buffers and some times they are called as data amplifiers.They are required to separate the valid data from the time multiplexed address/ data signals.
- They are controlled by two signals namely, DEN and DT/R.
- The DEN signal indicates the direction of data, i.e. from or to the processor. The system contains memory for the monitor and users program storage.
- Usually, EPROM are used for monitor storage, while RAM for users program storage. A system may contain I/O devices.
- The working of the minimum mode configuration system can be better described in terms of the timing diagrams rather than qualitatively describing the operations.
- The opcode fetch and read cycles are similar. Hence the timing diagram can be categorized in two parts, the first is the timing diagram for read cycle and the second is the timing diagram for write cycle.
- The read cycle begins in T_1 with the assertion of address latch enable (ALE) signal and

also M/IO signal. During the negative going edge of this signal, the valid address is latched on the local bus.

- The BHE and A_0 signals address low, high or both bytes. From T_1 to T_4, the M/IO signal indicates a memory or I/O operation.
- At T_2, the address is removed from the local bus and is sent to the output. The bus is then tristated. The read (RD) control signal is also activated in T_2.
- The read (RD) signal causes the address device to enable its data bus drivers. After RD goes low, the valid data is available on the data bus.
- The addressed device will drive the READY line high. when the processor returns the read signal to high level, the addressed device will again tristate its bus drivers.
- A write cycle also begins with the assertion of ALE and the emission of the address. The M/IO signal is again asserted to indicate a memory or I/O operation in T_2, after sending the address in T_1, the processor sends the data to be written to the addressed location.
- The data remains on the bus until middle of T_4 state. The WR becomes active at the beginning of T_2 (unlike RD is somewhat delayed in T_2 to provide time for flating).
- The BHE and A0 signals are used to select the proper byte or bytes of memory or I/O word to be read to write.
- The M/IO, RD and WR signals indicate the type of data transfers as specified in Table 3.3.
- **Hold Response Sequence :** The HOLD pin is checked at leading edge of each clock pulse. If it is received active by the processor before T_4 of the previous cycle or during T_1 state of the current cycle, the CPU activates HLDA in the next clock cycle and for succeeding bus cycles, the bus will be given to another requesting master.

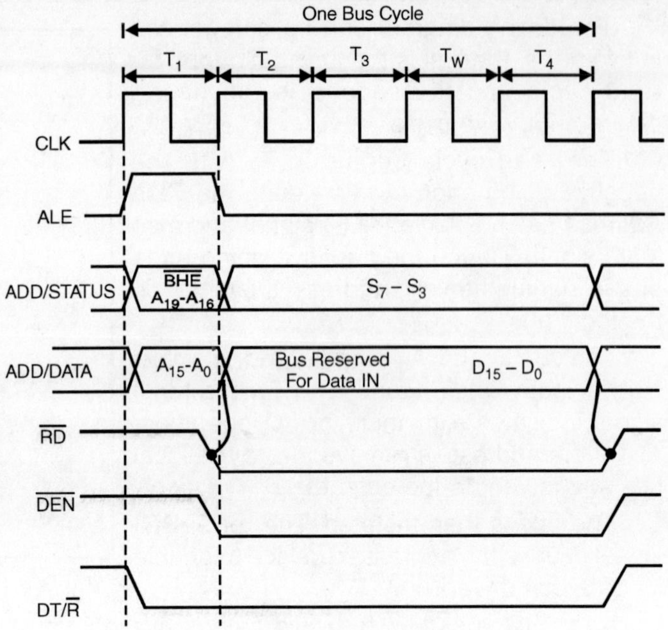

Fig. 3.50. (*a*) Timing Diagram for Read Cycle When 8086 is working in Minimum Mode.

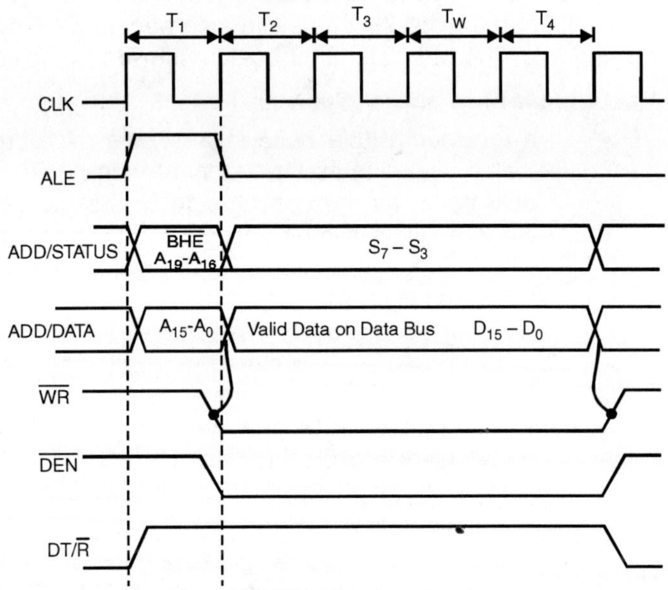

Fig. 3.50. (*b*) Timing Diagram for Write Cycle When 8086 is working in Minimum Mode.

➢ The control of the bus is not regained by the processor until the requesting master does not drop the HOLD pin low. When the request is dropped by the requesting master, the HLDA is dropped by the processor at the trailing edge of the next clock.

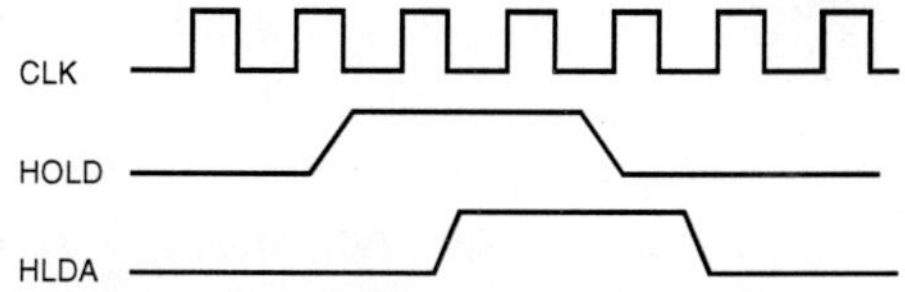

Fig. 3.51. Bus Request and Bus Grant Timings in Minimum MoSe system.

3.21.2. The Bus Timings for Minimum Mode

The bus timing diagrams for Input and Output transfers for 8086 microprocessor in minimum mode operation are as shown in Figs. 3.52 and 3.53.

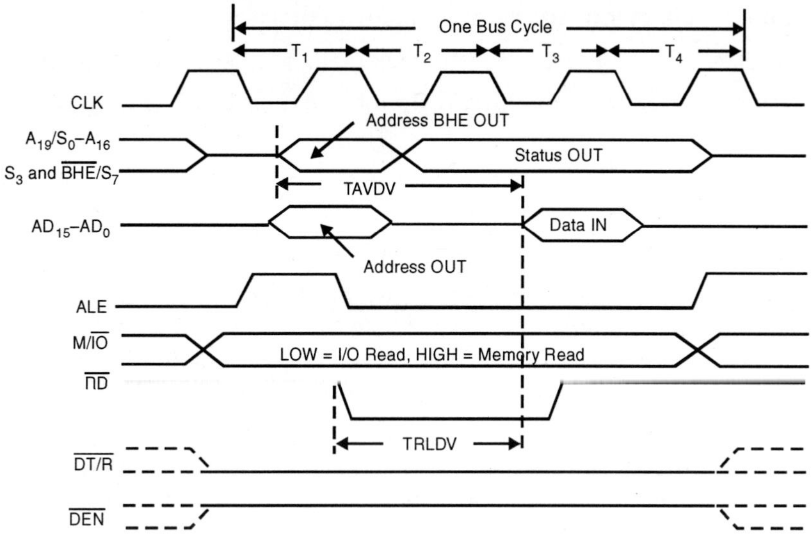

Fig. 3.52. Timing Diagram for Input to 8086 Microprocessor.

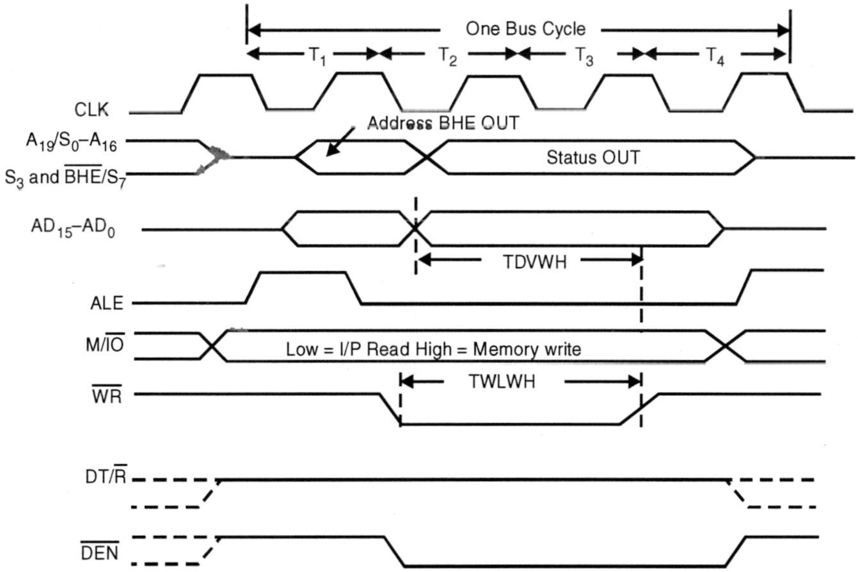

Fig. 3.53. Timming diagram for Output to 8086 Microprocessor.

"Now let us analyse the timing diagram of Figs. 3.52 and 3.53 in brief.

(i) If processor is ready to initiate the bus cycle, it applies a pulse to ALE during T_1. Before the falling edge of ALE, the address, BHE, M/IO, DEN and DT/ R must be stable i.e., DEN = high and DT/R = 0 for output or DT/R = 1 for output.

(ii) At trailing edge of ALE, ICs 72L8373 or 8282 latches the address.

(iii) During T_2 the address signals are disabled and $S_8 - S_7$ are available on A_{16}/S_3- A_{19}/S_6 and BHE/S_7. Also DEN is lowered to enable transceiver.

(iv) For input operation, RD is activated during T_2 and ADD to AD_{16} go in high impedance preparing for input.

(v) If memory or I/O interface" can perform the transfer immediately; there are no wait states and data is output on the bus during T_2 or T_3.

(vi) After the data is accepted by the processor, RD is raised high at the beginning of T_4.

(vii) Once detecting this transition during T_4, the memory or I/O device will disable its data signals.

(viii) For an output operation, processor applies WR = 0 and then the data on the data bus signals.

(ix) In T_4, WR is raised high and data signals are disabled.

(x) For either input or output operations, DEN is raised during T_4 to disable the transceiver and M/IO is set according to the next transfer at this or during next T_1 state. Thus length of bus cycle of 8086 is four clock cycle. When the bus is to be inactive after completion of bus cycle, then the gap between the successive cycles is filled by ideal stack clock cycles.

Now Fig. 3.49 depicts the internal block of 8282/8283 latches, where 8282 provides non inverting outputs while the 8283 version inverts the input data. Apart from their demultiplexing function, these chips also buffer the address lines, providing increased output driving capability. The output low level is specified as 0.45 V maximum with a sink current of 32 mA maximum. The high level is specified as 2.4 V minimum while supplying a 5 mA maximum high level load current.

When a system includes several interfaces then to increase current sourcing/sinking capacities it is' necessary to use drivers and receivers (transceiver) for data bus also. The Intel 8286 device is used to implement the transceiver block shown in Fig. 3.49 which contains 16 tristate elements, eight receivers, and eight drivers. That means two 8286s are required to service 16 data lines of 8086. The Fig. 3.49 shows the detailed connections of 8286.

The DT/R signal is connected to the T input, which controls the direction of the data flow. When this signal is low, receivers are enabled, so that 8086 can read data from memory or input device. To write data into memory or input device, the 8086's DT/I_i signal goes high. Since this drivers are enabled to transfer data from 8086 to the memory or the output device at the time of data transfer, to enable output of transceiver its OE should be low. This accomplished by connecting DEN signal of 8086 to the OE pin of 8286, since DEN-signal goes low when CPU is ready to send or receive data.

Then third component, other than the processor that appears in Fig. 3.48 is an 8284 clock generator. In addition to supplying a train of pulses at a constant frequency it *synchronizes ready (RDY) signal, which indicates* an *interface is ready to complete* a transfer, and reset (RES) signal, which initializes the system, with the clock pulses. As observed in Fig. 3.54, there are two input pins for frequency source which are :

(a) From a pulse generator which is connected to the external frequency input (EFI).

(b) From an oscillator which is connected across XI and X2.

As observed in Fig. 3.54, the F/C# signal selects one of the inputs. If F/C# input is 1, the EFI input determines the frequency; otherwise oscillator determines the **frequency.** *(In both the cases the output clock frequency is one third of the input frequency).* Next the status on the M/IO, RD and WR control bus lines will decide the type of data transfer. The Table 3.3 shows the details of data transfers as below.

TABLE 3.3.

M/IO	RD	WR	Operation
0	0	1	I/O Read
0	1	0	I/O write
1	0	1	Memory Read
1	1	0	Memory Write

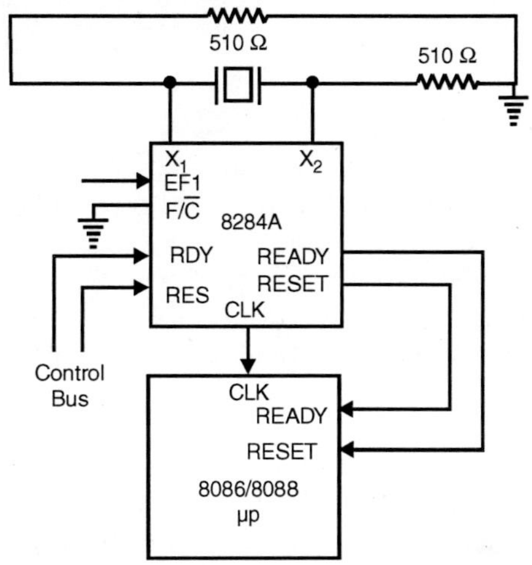

Fig. 3.54.

The HOLD and HLDA signals helps to interface other bus masters like DMA controller. Interrupt request (INTR) and interrupt acknowledge (INTA) are used to extend the intenupt handling capacity of the 8086 with the help of interrupt controller.

NOTE: If the memory or I/O device is not able to respond quickly during transfer, wait states (rw) are inserted between T_3 and T_4, by disabling the READY input of the 8086. The bus activity using wait state is same as during T_3.

3.21.3. Maximum Mode

In this mode, the pin MN / MX is grounded. Now to observe the condition of the processors in this mode the pin definition for 24 to 31 is explained briefly.

Pin 24 and 25 QS1 QSO (Output). These two output signals reflect status of the instruction queue. This status indicates the activity in the queue during the previous clock cycle, as in Table 3.4.

TABLE 3.4.

QSI	QSo	Status
0	0	No operation (que\1.e is idle)
0	1	First byte of an opcode
1	0	Queue is empty
1	1	Subsequent byte of an opcode

Pin 26, 27, 28 (Output) : These Three Status signals indicate The type of transfer to be take place during the current bus cycle as shown in Table 3.5.

TABLE 3.5.

$\overline{S_1}$	$\overline{S_2}$	$\overline{S_0}$	Machine Cycle
0	0	0	Interrupt Acknowledge
0	0	1	I/O Read
0	1	0	I/O Write
0	1	1	Halt
1	0	0	Instruction fetch
1	0	1	Memory read
1	1	0	Memory write
1	1	1	Inactive-Passive

Pin 29 LOCK. This signal indicates that an instruction with a LOCK prefix being executed and bus is not to be used by another potential most Pin 30 and 31 RQ / GT} and RQ / GTo. In maximum mode, HOLD and HLDA are replaced by RQ (Bus request)/ GTo (Bus Grant), and RQ/G_1 signals. By using bus request signal one more master can request for j system bus and processor communicate that the request is granted to 1 requesting master by using bus grant signal. However both signals: similar except the RQ /GTo has higher priority than RQ/GT_1}.

3.21.4. The Maximum Mode Configuration

The Fig. 3.55 illustrates the typical maximum mode configuration, where additional circuitry is required to translate the control signals. The additional circuitry converts the Status signals (82-80) into the I/O and memory transfer signals. It also generates the Control signals required to direct the data flow and for controlling 8282 latches and 8286 transceiver. The Intel 8288 bus controller is used to. Implement this control circuitry.

i8086 Circuit - Maximum Mode

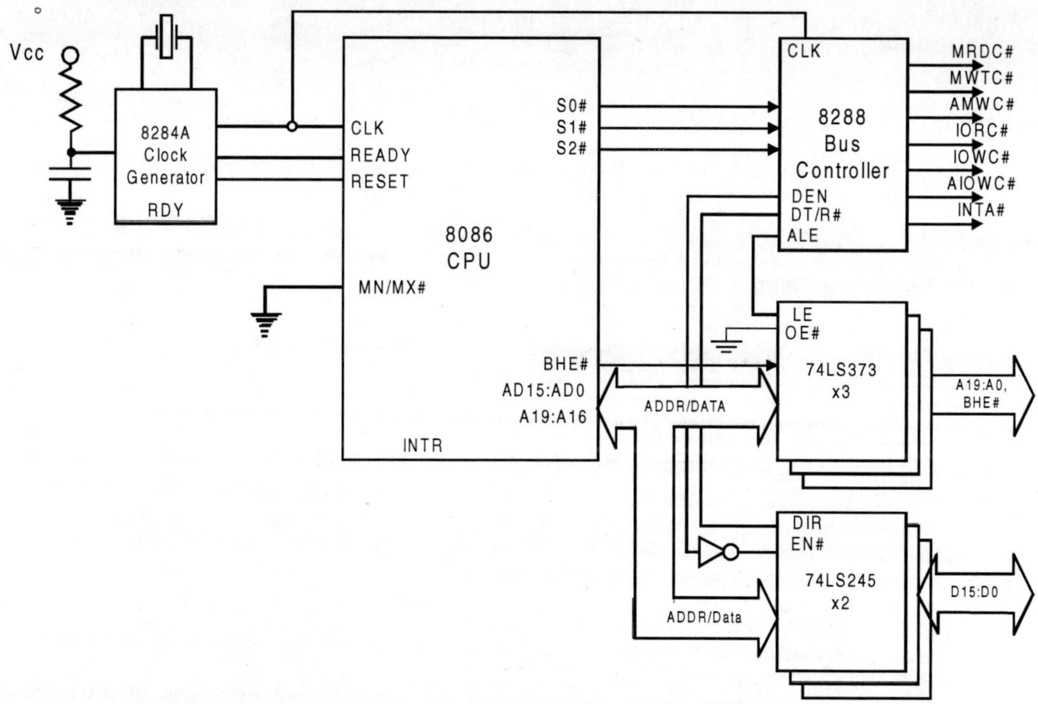

Fig. 3.55. Maximum Mode Configuration of 8086 Microprocessor

NOTE: # or / with any pin indicates that the pin is enable low.

Now Fig. 3.56 shows that 8288 is able to originate address latch enable signal to the 8282, the enable and direction signals to the 8286 transceivers, and the interrupt acknowledge signal to the interrupt controller. Apart it also decodes the S_2-S_0 signals to generate MRDC, MWrC, IORC, IOWC, MCE/PDEN, AEN, IOB, CEN, ALOWC and AMWC signals.

The commands are briefed as below :

(i) MRDC (Memory Read Command). It instructs memory to put the contents of the addressed location on the data bus.

(ii) MWTC (Memory Write Command). It instructs memory to accept the data on the data bus and load the data into the addressed memory location.

(iii) IORC (I/O Read Command). It instructs an I/O device to put the data contained in the addressed port on data bus.

(iv) IOWC (I/O Write Command). It instructs an I/O device to accept the data on the data Bus and load the data into the addressed port.

(v) MCE / PDEN (Master Cascade Enablel/Peripheral Data Enable). It controls the mode of operation of 8259.

(vi) AEN IOB and CEN. These pins helps in multiprocessor system, with a single processor in the' system, AEN and IOB are grounded and CEN is tied high.

(vii) AIOWC / AMWC (Advance I/O Write Command/Advance Memory Write Command). These signals are just similar to IOWC and MWTC except that they are activated one clock pulse earlier. This gives slow interfaces an extra clock cycle to prepare to input data.

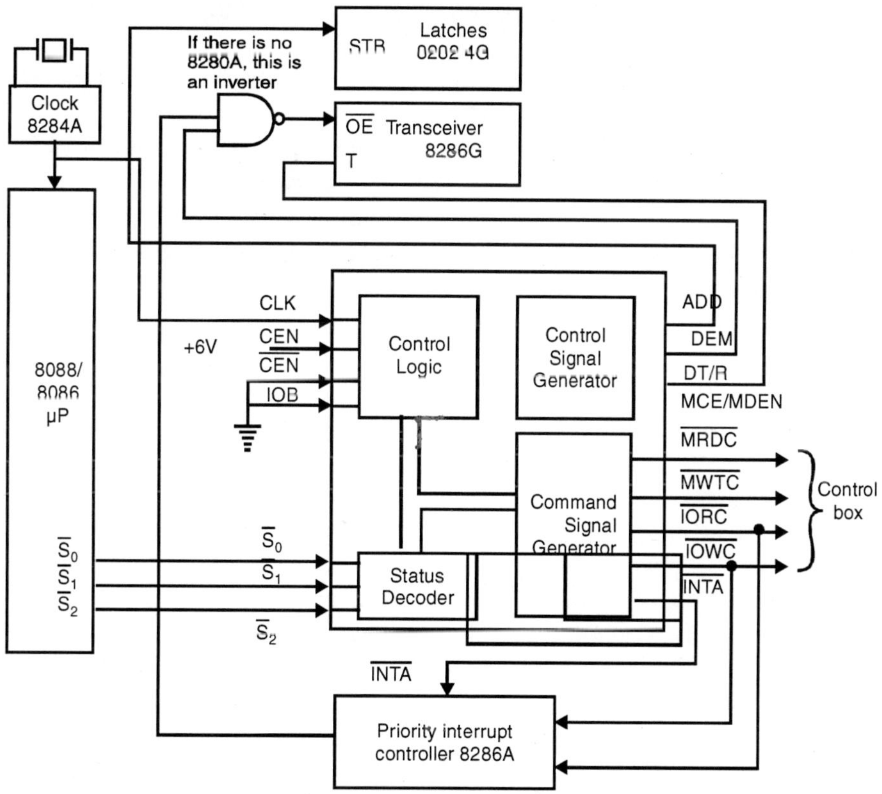

Fig. 3.56. The 8288 Bus Controller Interfaced with 8086 Microprocessor.

3.21.5. The Bus Timings or Maximum Mode

Maximum Mode 8086 System

- In the maximum mode, the 8086 is operated by strapping the MN/MX pin to ground.
- In this mode, the processor derives the status signal S_2, S_1, S_0. Another chip called bus controller drives the control signal using this status information.
- In the maximum mode, there may be more than one microprocessor in the system configuration.
- The components in the system are same as in the minimum mode system.
- The basic function of the bus controller chip IC8288, is to drive control signals like RD and WR (for memory and I/O devices), DEN, DT/R, ALE etc. using the information by the processor on the status lines.
- The bus controller chip has input lines S_2, S_1, S_0 and CLK. These inputs to 8288 are driven by CPU.
- It drives the outputs ALE, DEN, DT/R, MRDC, MWTC, AMWC, IORC, IOWC and AIOWC. The AEN, IOB and CEN pins are specially useful for multiprocessor systems.
- AEN and IOB are generally grounded. CEN pin is usually tied to +5 V. The significance of the MCE/PDEN output depends upon the status of the IOB pin.
- If IOB is grounded, it acts as master cascade enable to control cascade 8259A, else it acts as peripheral data enable used in the multiple bus configurations.
- INTA pin used to issue two interrupt acknowledge pulses to the interrupt controller or to an interrupting device.
- IORC, IOWC are I/O read command and I/O write command signals respectively. These signals enable an I/O interface to read or write the data from or to the address port.
- The MRDC, MWTC are memory read command and memory write command signals respectively and may be used as memory read or write signals.
- All these command signals instructs the memory to accept or send data from or to the bus.
- For both of these write command signals, the advanced signals namely AIOWC and AMWTC are available.
- Here the only difference between in timing diagram between minimum mode and maximum mode is the status signals used and the available control and advanced command signals.
- R_0, S_1, S_2 are set at the beginning of bus cycle 8288 bus controller will output a pulse as on the ALE and apply a required signal to its DT/R pin during T_1.
- In T_2, 8288 will set DEN = 1 thus enabling transceivers, and for an input it will activate MRDC or IORC. These signals are activated until T_4. For an output, the AMWC or AIOWC is activated from T_2 to T_4 and MWTC or IOWC is activated from T_3 to T_4.
- If reader input is not activated before T_3, wait state will be inserted between T_3 and T_4.
- **Timings for RQ/GT Signals :** The request/grant response sequence contains a series of three pulses. The request/grant pins are checked at each rising pulse of clock input.
- When a request is detected and if the condition for HOLD request are satisfied, the process issues a grant pulse over the RQ/GT pin immediately during T_4 (current) or T_1 (next) state.
- When the requesting master receives this pulse, it accepts the control of the bus, it sends a release pulse to the processor using RQ/GT pin.

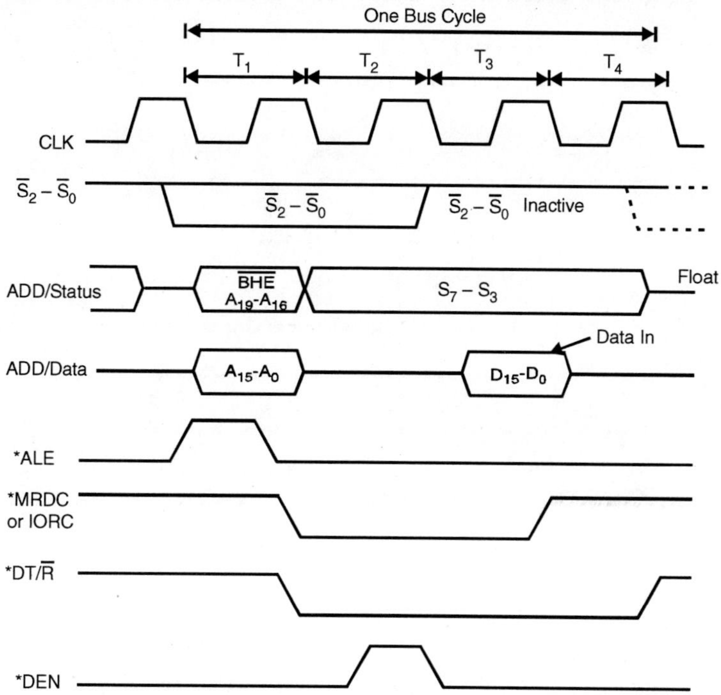

Fig. 3.57. Timing Diagram for Read Cycle (Input) when 8086 is in Maximum Mode

NOTE: * Here indicates that the signals are generated by 8288 bus controller outputs.

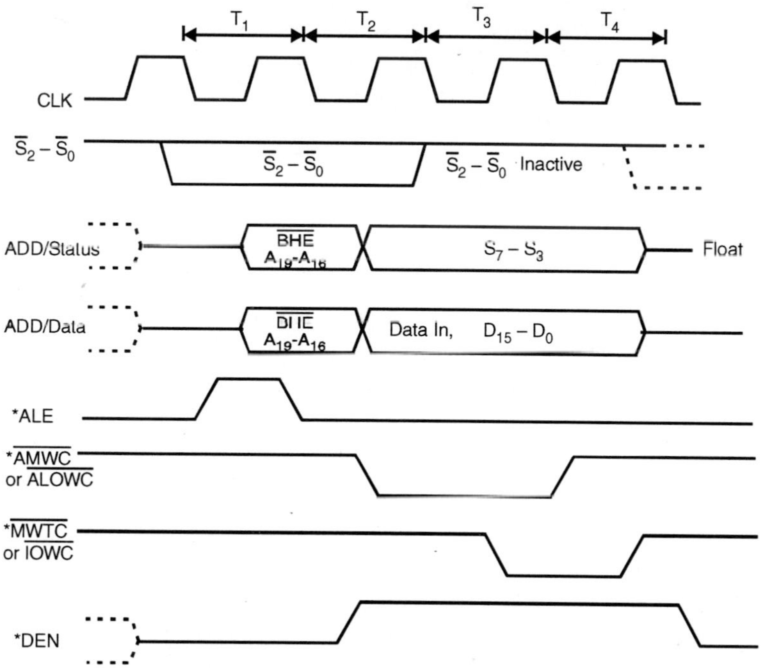

Fig. 3.58. Timing Diagram for Write Cycle (Output) when 8086 is in Maximum Mode

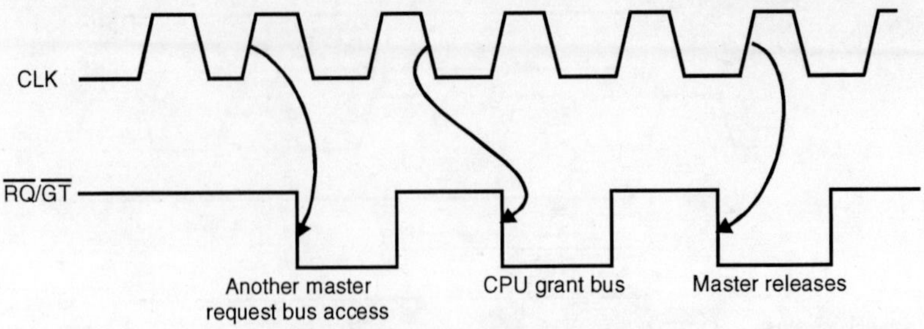

Fig. 3.59. $\overline{RQ}/\overline{GT}$ timing in maximum mode.

REVIEW QUESTIONS

1. List out the differences between 8086 and 8088 microprocessors.
2. Explain wether 8086/8088 are TTL compatible.
3. What is the purpose of status bits 53 and 54 ?
4. What happens when HOLD input to the 8086/8088 is placed at its logic 1 level?
5. What condition does a logic 0 an 8086/8088 RD pin indicate?
6. Briefly explain the operation of LOCK pin.
7. With the help of block diagrams explain the operation of 8284 clock generation.
8. Which bus connection on 8086 microprocessor are typically demultiplexed ?
9. Which TIL integrated circuit is often used to demultiplex the buses as 8086/8088 ?
10. Why buffers are often needed on the address data and control buses in a microprocessor system ?
11. Draw the block diagram showing 8086 in maximum mode configuration and explain.
12. Mention the signals which are different when 8086 in minimum mode and is maximum mode.
13. Explain the function of following pins in 8086 /IP :
    (a) NMI (b) TEST (c) DEN (d) Q50-Q51 (e) BHE
14. Draw and discuss the internal architecture of 8086 microprocessor?
15. Define a 16-bit Processor ?
16. Give the specifications of intel's 8086 microprocessor?
17. Give the historical development of the microprocessors with time?
18. Define Moore's Law?
19. What is the significance of BIU & EU in 8086 mp?
20. List all the registers associated with the four segment registers
21. List the internal registers in 8086 Microprocessor
22. What are the main blocks in BIU and EU
23. Explain the coordination between BIU an EU
24. Draw the functional block diagram of 8086 mp
25. Explain the bit pattern of PSW ?
26. What do we mean by pipelined architecture & explain how it is implemented in 8086 processor?
27. Explain the physical address formation in 8086?
28. Explain the concept of memory segmentation & what are its advantages?
29. Draw and explain the programming model of 8086 processors?
30. Explain all the features in 8284.
31. Why and when wait states are required. How do you insert wait states
32. How do you configure 8086 into minimum and maximum modes
33. Bring out the differences between 8086 and 8088 processors
34. What do you mean by pipelined architecture? How is it implemented in 8086 microprocessor?

35. Define the control transfer addressing modes present in 8086 processor?
36. Explain the physical address formation in 8086 microprocessor?
37. Explain the concept of memory segmentation? What are its advantages?
38. Describe the need for addressing modes in a microprocessor? Compare the addressing modes of 8085 with that of 8085?
39. Explain with examples the immediate addressing, register addressing and direct addressing modes?
40. Explain with examples the register indirect addressing mode?
41. Explain generation of 16-bit effective address in the case of based addressing with 8-bit displacement?
42. Explain with examples indexed addressing with displacement. Based indexed addressing and based indexed addressing with displacement?
43. Give all the valid addressing modes present in 8086 with an example?
44. Write down the addressing modes for the following instructions and calculate the physical address for the same making use of the given data Ax = fc39h, bx= 273f h, cx = 000a, dx = 4c2c, si = 30bc h, bp= 2000h, sp= 3000h, di= fc32h, cs= 4000h, ds =7000h, es= 5000h, ss = 1000h, displacement= 0010h.
    (1) Mov al, bh
    (2) Mov bl, [bp]
    (3) Mov cx, [1234]
    (4) Mov dx, 1234h
    (5) Mov cx, disp [si]
    (6) Mov al, [bx] [di],
    (7) mov cx, disp[bx] [si]
45. Describe how the various memory modes can be derived as special cases of based indexed addressing with displacement?
46. Explain with examples addressing of I/O ports?
47. Draw the register organization of 8086 and explain typical applications of each register?
48. Draw and discuss flag register of 8086 in brief?
49. Explain the function of op-code pre-fetch queue in 8086 microprocessor?
50. Draw and discuss the Read and Write cycle timing diagrams of 8086 microprocessor in minimum mode?
51. Draw and discuss the Read and Write cycle timing diagrams of 8086 microprocessor in maximum mode?
52. How will you synchronize an external phenomenon like energizing a relay with a program segment execution?
53. Bring out the architectural and signal differences between 8086 and 8088 microprocessor?
54. Explain the concept of memory Organization in 8086 processor?
55. Explain the function of each segment in detail with their size?
56. Define the concept dividing memory into banks?
57. What is the advantage of having memory banks?
58. Define the operating modes for 8086 processor?
59. Draw and explain the working of 8086 processor in minimum mode?
60. Draw and explain the timming diagram for RD & WR operations in MN- Mode of 8086 ?
61. Draw and explain the working of 8086 processor in maximum mode?
62. Draw and explain the timming diagram for RD & WR operations in MX- Mode of 8086 ?
63. Draw & explain the relevant pin diagram for 8086 in minimum mode?
64. Draw & explain the relevant pin diagram for 8086 in maximum mode?
65. Draw the basic system timming diagram for 8086 processor?
66. Expalin the function of each pin of 8086 in detail?
67. Draw a timing diagram for RD/WR cycle in MN mode by introducing a wait state for 5ms in the processor cycle?

68. Draw and explain the 8284 clock generator ?
69. Explain the concept of Bus buffering and latching ?
70. Draw and explain the bus controller 8288 ?
71. Name the GPRs of the 8086/8088.
72. What are different modes of addressing the 8086 ? Explain them.
73. What is the purpose of index registers ?
74. What is the purpose of the pointer registers ?
75. What is the purpose of the instruction pointers ?
76. How do the address/data buses in between the 8086 and 8088 ?
77. How long the address is valid on lines A_{19} to A_{16} ?
78. What is the purpose of the queue ?
79. How many words do the queue stores in the 8086 ?
80. How do we select the minimum versus maximum modes in the 8086 ?
81. During what time period(s) of the bus cycle is $\overline{DEN}$ active ?
82. What is the state of DT/$\overline{R}$ for a write bus cycle ?
83. Name the hardware interrupts of the 8086.
84. Find out the machine code for following instructions.
   (i) ADC AX, BX
   (ii) OR AX, [0500H]
   (iii) AND CX, [SI]
   (iv) TEST AX, 5555H
   (v) MUL [SI + 5]
   (vi) NEG 50 [BP]
   (vii) OUT DX, AX
   (viii) LES DI, [0700H]
   (ix) LEA SI, [BX + 500H]
   (x) SHL [BX + 2000], CL
   (xi) RET 0200H
   (xii) CALL 7000H
   (xiii) JMP 3000H:2000H
   (xiv) CALL [500H]
   (xv) DIV [5000H]
85. Which of the followings are legitimate instructions ? What do the legitimate instructions do ?
   (a) MOV ES, DS
   (b) MOV BL, BX
   (c) MOV CX, AX
   (d) MOV DI, SI
   (e) MOV AX, [BX]
   (f) MOV CX, MOON
   (g) MOV [BX + DI], CL
   (h) MOV AX, [BX + 4]
   (i) MOV AX, SUNY [BP]
   (j) MOV AX, SUNY [BX + DI]
   (k) MOV BL, 44H
   (l) MOV AX, 44H
   (m) MOV SI, 0H
   (n) MOV AL, 'A'
   (o) MOV AX, 'AB'
86. (a) Segment to segment transfer is not allowed.
   (b) Mixed sizes not allowed.
   (c) Code segment register cannot be used as destination register.
   (d) Copies (SI) into (DI)
   (e) Word size data memory with the offset (BX) to register (AX)
   (f) Word size contents from the memory labelled MOON is copied to CX.
   (g) Copies byte size data of (CL) into memory addressed by the (BX) + (DI).
   (h) Copies word size data into (AX) from memory addressed by (BX) + 4.
   (i) Copies word size data from memory addressed by SUNY + [BP] into AX.
   (j) Copies a word of data from memory addressed by SUNY + [BP + DI] into AX.
   (k) BL filled with 44H.
   (l) The (AH) = 00H and (AL) = 44H.
   (m) (SI) = 0000H
   (n) Moves ASCII 'A(=41H)' to AL
   (o) Moves ASCII 'BA(=4241H)' into AX.
87. What is wrong in MOV BL, CX and MOV DS, SS instructions ? Mixed byte size in MOV BL, CX is not allowed. Segment to getment moves not allowed.
88. What do the symbol [ ] indicate ? Indicates memory address.

89. Determine the memory address accessed by each of the following instructions in real mode operation, if DS = 0200H, BX = 0300H and DI = 400H.

(a) MOV AL, [1234] (b) MOV [DI], AL

(a) Effective address = 1234H, Physical address = 02000 + 1234 = 03234H

(b) Physical address = 02000 + 0400 = 02400H.

90. Determine the memory address accessed by each of the following instructions in real mode operation, if DS = 1000H, SS = 2000H, BP = 1000H and DI = 0100H.

(a) MOV AL, [BP + DI] (b) MOV CX, [DI]

(a) Physical address = SS × 10H + (BP + DI) = 20000 + 1000 + 100 = 21100H

(b) Physical address = DS × 10H + DI = 10000 + 100 = 10100H

91. Determine the memory address accessed by each of the following instructions in real mode operation. If DS = 1200H, BX = 0100H, and SI = 0500H.

(a) MOV [100H], DL (b) MOV DL, [BX + 100H]

(a) Physical address = DS × 10H + 100 = 12000 + 100 = 12100 H

(b) Physical address = DS × 10H + 0100 + 100 = 12000 + 0100 + 100 = 12200H

92. Determine the memory address accessed by each of the following instructions in real mode operation. If DS = 1100H, BX = 0200H, SUNY = 0250 H, and SI = 0500H.

(a) MOV CL, [BX + SI] (b) MOV CH, SUNY [BX + SI]

(a) Physical address = DS × 10H + BX + SI = 11000 + 0200 + 0500 = 11700H

(b) Physical address = DS × 10H + BX + SI + SUNY = 11000 + 0200 + 0500 + 0250 = 11950H

93. Determine the memory address accessed by each of the following instructions in real mode operation. If DS = 1300H, SS = 1400H, BP = 1500H and SI = 0100H

(a) MOV AL, [BP + SI – 200H] (b) MOV AL, [SI – 0100H]

(a) Physical address = SS × 10H + BP + SI – 200 = 14000 + 1500 + 0100 – 200 = 12400H

(b) Physical address = DS × 10H + SI – 100H = 13000 + 0100 – 100 = 13000H

94. Indicate the type of jump, if the JMP THREE instruction is stored at memory address 10000H and the address of the THREE is

(a) 10020H (b) 11000H (c) 0FFFFH (d) 300000H

(a) Short, (b) Near (c) Short and (d) Far

95. Which are illegal instructions for 8086 ?

(a) MOV ES, 4000H (b) MOV ES, AX (c) MOV DS, 0000H

Answer (a) and (c) are illegal instructions.

OBJECTIVE TYPE QUESTIONS

1. The PCLK output of 8284A is ______ MHz, if crystal oscillator is operating at 14 MHz.

2. A bus cycle is equal to clocking ______ periods.

3. The INTR input is ______ sensitive.

4. 8086/8088 use multiplexed address and data bus because ______ ?

(i) Multiplexing is supported by 8085 μP

(ii) 40 pins is a good size for IC

(iii) Multiplexing reduces number of liens between μP and auxiliary ICs.

(iv) All.

5. The SSO of 808~ pP indicate ______.

6. If READY pin is grounded, it will introduce ______ states into the bus cycle of 8086/8088 μP.

❑❑❑

CHAPTER 4

8086 INTERRUPT STRUCTURE

INSIDE THIS CHAPTER

4.1. TYPES OF INTERRUPTS

An 8086 interrupt can come from any of the three sources :

(a) External input applied to the nonmaskable interrupt pin (NMI).

(b) Interrupt caused due to some error conditions produced in 8086 by the execution of an instruction.

(c) Interrupt caused due to the execution of the interrupt instruction, INT.

An interrupt caused by a signal applied to one of these inputs is referred to as a hardware interrupt.

An interrupt caused by the execution of INT instruction is called a software interrupt and an interrupt caused by some internal failure is referred to as a error-condition interrupt.

At the end of each instruction cycle, the 8086 checks to see if any interrupts have been requested.

If an interrupt has been requested, the 8086 responds to the interrupt by stepping through the following series of actions :

(1) 8086 decrements the stack pointer by 2 and pushes the flag register on to the stack.

(2) 8086 disables the INTR interrupt input by clearing the interrupt flag (IF) in the flag register.

(3) 8086 resets the TRAP flag (TF) in the flag register.

(4) 8086 decrements the stack pointer by 2 and pushes the current code segment register contents on to the stack.

(5) 8086 decrements the stack pointer by 2 and pushes the current instruction pointer contents on to the stack.

(6) 8086 then does an indirect far jump to the start of the procedure one wrote to respond to the interrupt.

We can see the above explained steps very clearly with the help of the Fig. 4.1. The 8086 pushes the flag register on the stack, disables the INTR input and the single step function and then does an indirect far call to the interrupt service procedure. An IRET instruction at the end of the interrupt service procedure returns the execution to the main program.

When 8086 responds to an interrupt, it goes to four memory locations to get the CS and IP values for the start of the interrupt service procedure. In 8086 system, the first 1 Kbyte of memory, from 00000H to 003FFH, is set aside as a table for storing the starting addresses of interrupt service procedures. Since 4 bytes are required to store the CS and IP values for each interrupt service procedure, the table can hold starting addresses for up to 256 interrupt procedures. The starting address of an interrupt service procedure is often called interrupt vector or the interrupt pointer hence the table is called as interrupt vector table.

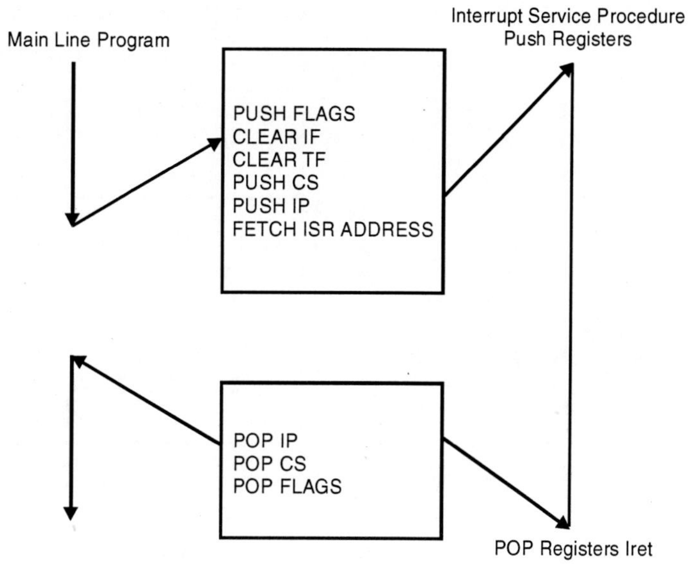

Fig. 4.1. 8086 Interrupt Response

Figure 4.2 shows how the 256 interrupt vectors are arranged in the table in memory.

The instruction pointer value is put in as the low word of the vector, and the code segment register is put in as the high word of the vector. Each doubleword interrupt is identified by a number 0 to 255. The lowest five types of interrupts are dedicated to specific interrupts such as :

(a) Divide by Zero interrupt
(b) Single Step interrupt
(c) Non Maskable Interrupt
(d) Breakpoint Interrupt
(e) Overflow Interrupt.

Interrupt types 5 to 31 are reserved by Intel for use in more complex microprocessors such as 80286, 80386, 80486 etc. the upper 224 interrupt types from 32 to 255 are available for the user to use for hardware or software interrupts.

As we can see in the Fig. 4.2, the vector for each interrupt type requires four memory locations. Therefore, when 8086 responds to a particular type interrupt, it automatically multiplies the type by 4 to produce the desired address in the vector table. 8086 then goes to that address in the table to get the starting address of the interrupt service procedure.

Interrupts result in a transfer of control to a new program location. A 256-element table containing address pointers to the interrupt service program locations resides in absolute locations 0 through 3FFH, which are reserved for this purpose. Each element in the table is 4 bytes in size and corresponds to an interrupt "type". An interrupting device supplies an 8-bit type number during the interrupt acknowledge sequence, which is used to "vector" through the appropriate element to the new interrupt service program location. All flags and both the Code Segment and Instruction Pointer register are saved as part of the INTA sequence. These are restored upon execution of an Interrupt Return (IRET) instruction.

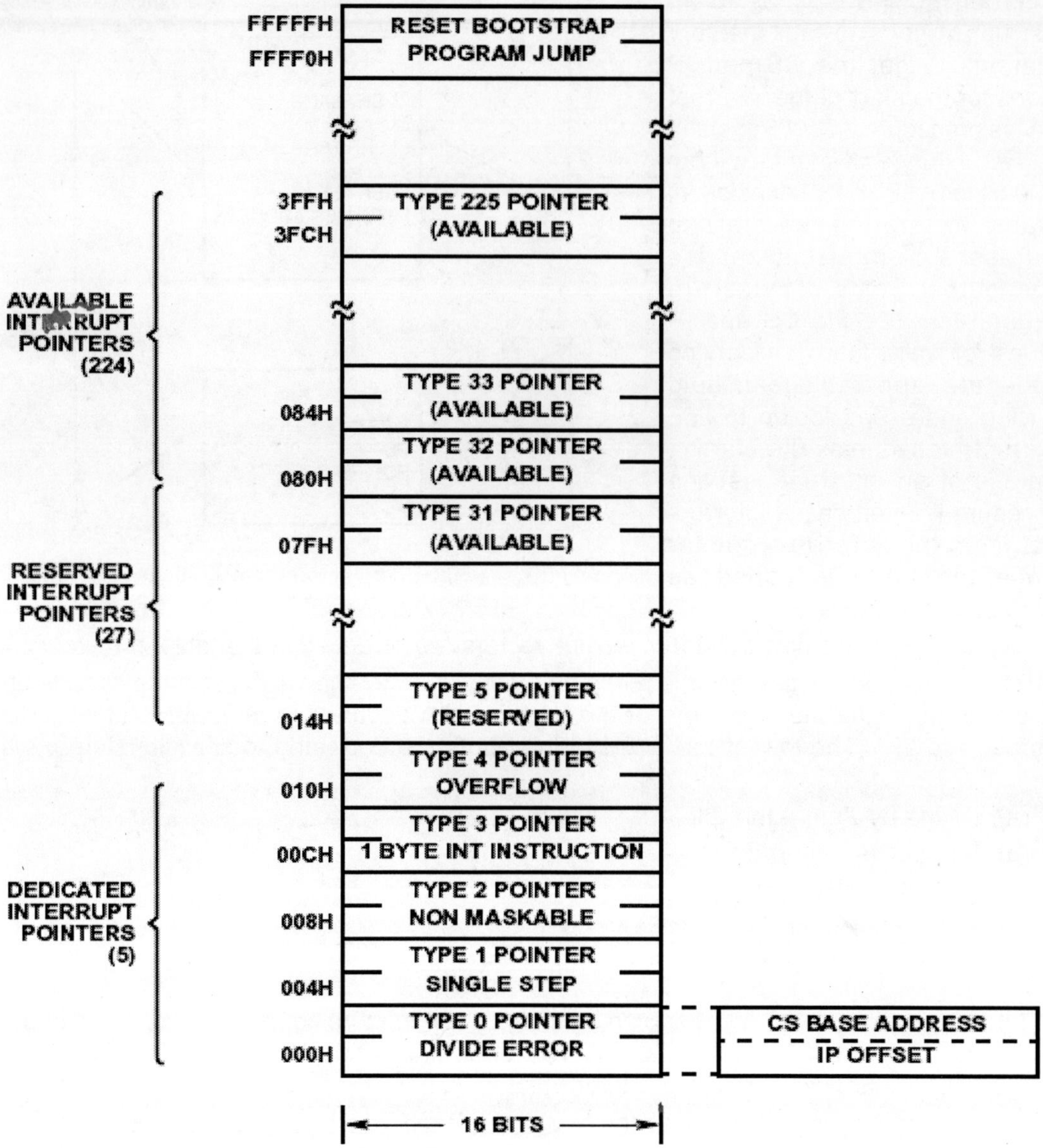

Fig. 4.2. 8086 Interrupt–Pointer Table.

4.2. 8086 INTERRUPT TYPES

(1) Divide by Zero Interrupt – Type 0

The 8086 will automatically do a type 0 interrupt if the result of a DIV operation or IDIV operation is too large to fit in the destination register. For a type 0 interrupt, the 8086 pushes the flag register onto the stack, resets the IF and TF flag and pushes the return address (CS and IP) on to the stack.

It then gets the CS value for the start of the interrupt service procedure from addresses 00002H in the interrupt pointer table and the IP value for the start of the procedure from addresses 00000H in the interrupt pointer table.

Since the 8086 type 0 response automatic and cannot be disabled in any way, one has to account for it in any program where one uses DIV or IDIV instruction.

One way is to make sure that the result will never be too large for the result register. Another way to account for type 0 interrupt is to simply write an interrupt service procedure which takes the desired action when an invalid division occurs.

(2) Single Step Interrupt – Type 1

In single step interrupt the system will execute one instruction and stop. We can then examine the contents of registers and memory locations. If they are correct then we can tell the system to go on and execute the next instruction. Hence we can say that when in single step mode, a system will stop after it executes each instruction and wait for the direction from the user.

The 8086 trap flag and type 1 interrupt response make it quite easy to implement a single step feature in 8086 based systems.

If the 8086 trap flag is set, the 8086 will automatically do a type 1 interrupt after each instruction executes.

For a type 1 interrupt, the 8086 pushes the flag register on to the stack, resets the IF and TF flag and pushes the return address (CS and IP) on to the stack. It then gets the CS value for the start of the interrupt service procedure from addresses 00006H in the interrupt pointer table and the IP value for the start of the procedure from addresses 00004H in the interrupt pointer table. The tasks involved in implementing single stepping are :

(a) Set the trap flag.

(b) Write an interrupt service procedure which saves all registers on to the stack, where they can be examined later on or displayed on the CRT.

(c) Load the starting address of the type 1 interrupt service procedure into the address 00004H and 00006H.

(3) Non-maskable Interrupt (NMI) – Type 2

The processor provides a single non-maskable interrupt pin (NMI) which has higher priority than the maskable interrupt request pin (INTR). A typical use would be to activate a power failure routine. The NMI is edge-triggered on a LOW- to-HIGH transition. The activation of this pin causes a type 2 interrupt. NMI is required to have a duration in the HIGH state of greater than two CLK cycles, but is not required to be synchronized to the clock. Any positive transition of NMI is latched on-chip and will be serviced at the end of the current instruction or between whole moves of a block-type instruction. Worst case response to NMI would be for multiply, divide, and variable shift instructions. There is no specification on the occurrence of the low-going edge; it may occur before, during or after the servicing of NMI. Another positive edge triggers another response if it occurs after the start of the NMI procedure. The signal must be free of logical spikes in general and be free of bounces on the low-going edge to avoid triggering extraneous responses.

For a type 2 interrupt, the 8086 pushes the flag register on to the stack, resets the IF and TF flag and pushes the return address (CS and IP) on to the stack. It then gets the CS value for the start of the interrupt service procedure from addresses 0000AH in the interrupt pointer table and the IP value for the start of the procedure from addresses 00008H in the interrupt pointer table.

(4) Break Point Interrupt - Type 3

The type 3 interrupt is produced by execution of the INT3 instruction. The main use of the type 3 interrupt is to implement a breakpoint function in a system. When one inserts a breakpoint, the system executes the instruction up to the breakpoint and then goes to the breakpoint procedure. Unlike the single-step feature, which stops executing after each instruction, the breakpoint features executes all the instructions up to the inser ted breakpoint and then stops execution.

For a type 3 interrupt, the 8086 pushes the flag register on to the stack, resets the IF and TF flag and pushes the return address (CS and IP) on to the stack. It then gets the CS value for the start of the interrupt service procedure from addresses 0000EH in the interrupt pointer table and the IP value for the start of the procedure from addresses 0000CH in the interrupt pointer table. A breakpoint interrupt service procedure usually saves all the register contents on the stack.

(5) Overflow Interrupt - Type 4

The 8086 overflow flag (OF) will be set if the signed result of the arithmetic operations on two signed numbers is too large to be represented in the destination registers or memory locations.

For example if we add the 8-bit signed numbers 01101100 (108) and the 8 bit signed number 0100001 (81), the result will be 10111101 (189). This would be the correct result if we were adding the unsigned binary numbers but it is not the correct signed result.

For signed operations the one in the most significant bit in the result indicates that the result is negative and in two's complement form the result, 10111101, then actually represents –67 decimal, which is not the correct result adding +108 and +89.

For a type 4 interrupt, the 8086 pushes the flag register on to the stack, resets the IF and TF flag and pushes the return address (CS and IP) on to the stack. It then gets the CS value for the start of the interrupt service procedure from addresses 00012H in the interrupt pointer table and the IP value for the start of the procedure from addresses 00010H in the interrupt pointer table. A breakpoint interrupt service procedure usually saves all the register contents on the stack.

4.3. MASKABLE INTERRUPT (INTR) INTERRUPTS—TYPES 0 THROUGH 255

The 8086 INT instruction can be used to cause the 8086 to do any one of the 256 possible interrupt types. The desired interrupt type is specified as a part of instruction. The instruction INT 32, for example, will cause the 8086 to do a type 32 interrupt response. The 8086 will pushes the flag register on to the stack, resets the IF and TF flag and pushes the return address (CS and IP) on to the stack. It will then get the CS and IP values for the start of the interrupt service procedure from the linterrupt pointer table in memory. The IP value for any interrupt type is always at an address of 4 times the interrupt type, and the CS value is at location 2 addresses higher. For a type 32 interrupt, then, the IP value will be put at 4x32 or 128 decimal (80H), and the CS value will be put at address 82h in the interrupt vector table.

The 8086 provides a single interrupt request input (INTR) which can be masked internally by software with the resetting of the interrupt enable flag (IF) status bit. The interrupt request signal is level triggered. It is internally synchronized during each clock cycle on the high-going edge of CLK. To be responded to, INTR must be present (HIGH) during the clock period preceding the end of the current instruction or the end of a whole move for a block type instruction. INTR may be removed anytime after the falling edge of the first INTA signal. During the interrupt response sequence further interrupts are disabled. The enable bit is reset as part of the response to any interrupt (INTR, NMI, software interrupt or single-step), although the FLAGS register which is automatically pushed onto the stack reflects the state of the processor prior to the interrupt. Until the old FLAGS register is restored, the enable bit will be zero unless specifically set by an instruction.

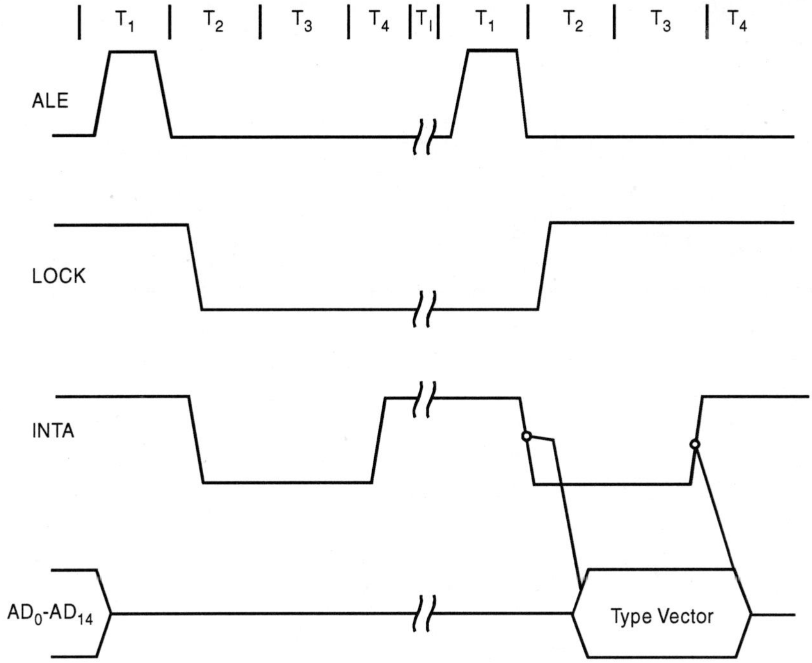

Fig. 4.3. 8086 interrupt acknowledge machine cycles.

During the response sequence the processor executes two successive (back-to-back) interrupt acknowledge cycles. The 8086 emits the LOCK signal (Max mode only) from T_2 of the first bus cycle until T_2 of the second. A local bus "hold" request will not be honored until the end of the second bus cycle. In the second bus cycle, a byte is supplied to the 80C86 by the 82C59A Interrupt Controller, which identifies the source (type) of the interrupt. This byte is multiplied by four and used as a pointer into the interrupt vector lookup table. An INTR signal left HIGH will be continually responded to within the limitations of the enable bit and sample period. The INTERRUPT RETURN instruction includes a FLAGS pop which returns the status of the original interrupt enable bit when it restores the FLAGS.

8086 uses INT instruction to cause the 8086 to do any of the 256 interrupts types possible. The interrupt type desired is specified as the part of instruction.

The IP (Instruction pointer) value for any interrupt type is always at an address of 4 times the interrupt type, and the code segment value is at a location 2 addresses higher.

Example 1. Say that 8086 has encountered an interrupt type 32, hence the value of

Instruction Pointer (IP) = 4 × 32

= 128 (decimal) or 80 H (Hexa)

Similarly the value of the

Code Segment (CS) = IP + 2

= 80 + 2 = 82 H

Hence the value of IP = 0080H and value of CS = 0082H for the interrupt type 32 in the interrupt vector table of 8086.

Example 2. Say that 8086 has encountered and interrupt type 52H, hence the value of

Instruction Pointer (IP) = 4 × 52 = 208 (decimal) or D0H

Similarly the value of

Code Segment (CS) = IP + 2

= D0 + 2

= D2H

Hence the value of IP = 00D0H and the value of CS = 00D2H for the interrupt vector type 52 in the interrupt vector table of 8086.

Applications of Interrupts

Timer Application : In this problem suppose we need to check the density of a solution say after 5 min. If we use a delay loop to count off the 5 min, we will have some difficult calculations to complete as to at what points in the program to check the density.

Hence to solve this problem we connect a simple 1Hz pulse source to an interrupt input, as shown in the Fig. 4.4 below.

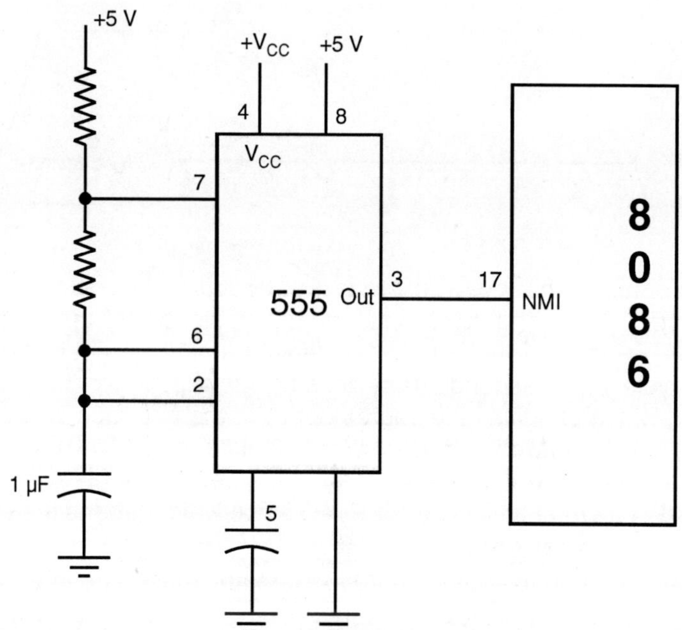

Fig. 4.4. 555 Times Used to Generate VHz Pulse.

The 555 timer is used to generate 1 Hz pulse and it will send an interrupt signal to the 8086 NMI input approximately once every second. An interrupt procedure is used to keep a count of how many NMI interrupts have occused. This count will be equal to the number of seconds that have passed.

Counting Application : One can use the interrupt input for counting application easily as shown in the Fig. 4.5 below

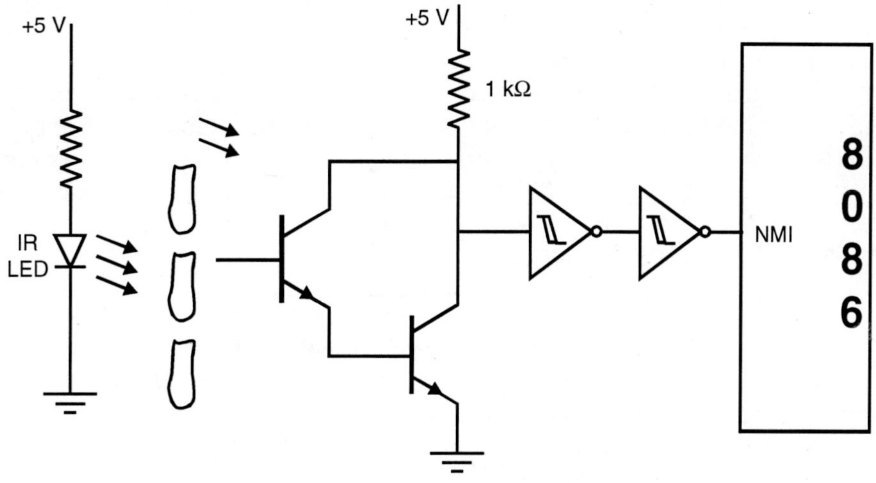

Fig. 4.5. Circuit for detecting precence of an object.

Suppose in our counting application we are using an 8086 to control bottling plant machine in a computerised factory. We want to count the number of bottles that come out ready of the plant, so that one can compare this count with the number of bottles fed in. This way one can determine if any bottles were lost in the machine or not.

We can do this count on interrupt basis very easily. All we have to do is to detect when a bottle passes out of the machine and then send an interrupt signal to an interrupt input on the 8086. The ISP (Interrupt Service Procedure) for that input can simply increment the board count stored in a named memory location.

To detect a bottle coming out of the automate machine, we can used an infrared LED, a phototransistor and two conditioning gates, as shown in the Fig. The LED is positioned over the track where the bottles come out, and the phototransistor is positioned below the track. When no bottle is between the LED and the phototransistor, the light from the LED will strike the phototransistor and turn it on. The collector of the phototransistor will than be low, as will the NMI input on 8086. When a bottle passes between the LED and the phototransistor, the light will not reach the phototransistor and it will turn off. It's collector will go high and so will the signal to the NMI input of 8086. The 742514 submitt trigger inverts are necessary to turn the slow-risetime signal from the phototransistor collector into a signal which meets the risetime requirements of the NMI input on the 8086.

When the 8086 recieves the low-to-high signal on its NMI input, it will automatically do type 2 interrupt response.

This type 2 interrupt service procedure has to do in this case is to increment the bottle count in a named memory location and return to the running machine.

The same technique may be used in number of areas such as to count people going into a stadium, or a railway station, or in a shrine temple etc.

Use of Interrupt to Produce a Real Time Clock

One more application of interrupt is to produce a real time clock. We may again use a 1 Hz interrupt input to generate a real time clock of seconds, minutes and hours. The time from this clock can be displayed and/or printed out on time cards. The generate this clock, a 1 Hz signal is applied to an interrupt input.

A seconds count, a minutes count and an hours count are kept in three successive memory locations. When an interrupt occurs, the seconds count is incremented by 1-IF the seconds count is not equal to 60, than execution is simply returned to the machine program. If the seconds count is equal to 60, then the seconds count is reset to 0 and the minutes count is incremented by 1. If the minutes count is not 60, then again the execution is simply returned to the mainline program. When the minutes count is 60, then the minutes count is reset to 0 and the hours count is incremented by 1. If the hours count is not 13, then execution is returned to the mainline program. When the hours coun is equal to 13 then it is reset to 1 and execution is returned again to the mainline problem.

REVIEW QUESTIONS

1. Define an interrupt and give the different types of interrupts in 8086 ?
2. Discuss the actions performed by 8086 when an interrupt is encountered by it ?
3. Write down the steps involved when an interrupt INT 44h is encountered in the main program and calculate address of ISR for this interrupt ?
4. Explain all the dedicated software interrupts present in 8086 µp ?
5. Discuss the various sources of interrupt in 8086 ?
6. Describe the interrupt vector table of 8086 microprocessor.
7. Describe the following interrupts of 8086 :
   (i) Type-0 (ii) Type-1 (iii) Type-2 (iv) Type-3 (v) Type-4
8. Draw the machine cycle for interrupt acknowledgement in 8086 microprocessor ?
9. Discuss the importance of NMI input pin in interrupts ?
10. Differentiate between maskable and non-maskable interrupts ?

❑❑❑

CHAPTER 5

MEMORY INTERFACING AND ADDRESS DECODING TECHNIQUES FOR 8086 MICROPROCESSOR

INSIDE THIS CHAPTER

5.1. INTRODUCTION

Every microprocessor-based system has a memory system, which may be simple or complex. Most of the systems contains mainly two types of memory: ROM, i.e., Read Only Memory and RAM, i.e., Random access memory or read/write memory. Interfacing of memory device with IP will be discussed in this chapter. The memory interfacing circuits are used to access memory often to read instruction code and data stored in memory, such read/write operations are monitored by control signals. The microprocessor activates such signals when required to read from and write into memory.

5.2. OPERATION AND TERMINOLOGY OF MEMORY

The memories are made up of registers, each register in the memory is one storage location and each location is identified by an address. The number of storage locations can vary from a few in some memories to hundreds of thousand in others. Each location can accommodate one or more bits. Normally, the total number of bits that a memory can store is its capacity. Most of the types the capacity is specified in terms of bytes. Each register consists of storage elements like flip-flop or capacitors in semiconductor memories and magnetic domain in magnetic storage, each of which stores one bit of data. A storage element is called cell. Figure 5.1 depicts simple Read/write operation is memory cells.

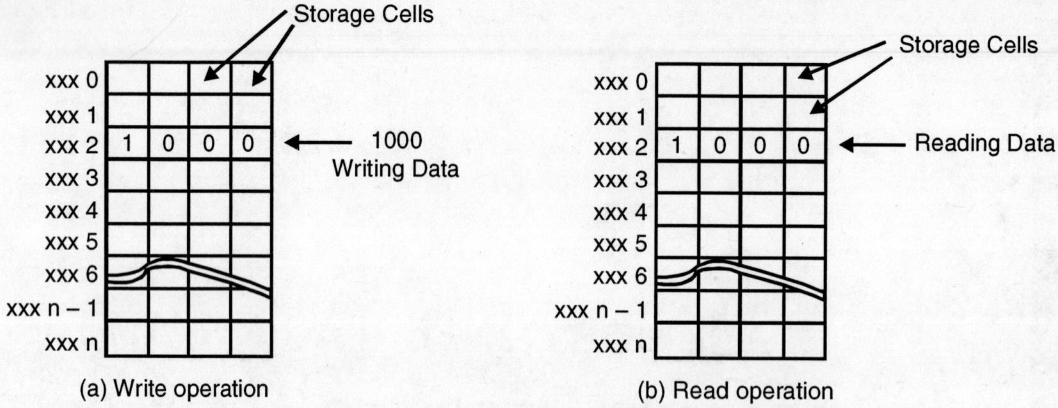

Fig. 5.1. Simple read/write operation in memory cells.

As observed from Fig. 5.1, the data stored in memory by a process called writing and are retrieved from the memory by a process called reading.

5.3. MEMORY DEVICES

Before trying to interface memory to the microprocessor, it is suitable to understand the operation of memory components. Now let us explain the function of the four common types of memory: read-only memory (ROM), flash memory (EEPROM), static random access memory (SRAM), and dynamic random access memory (DRAM). The classification of semiconductor memory is as shown in Fig. 5.2.

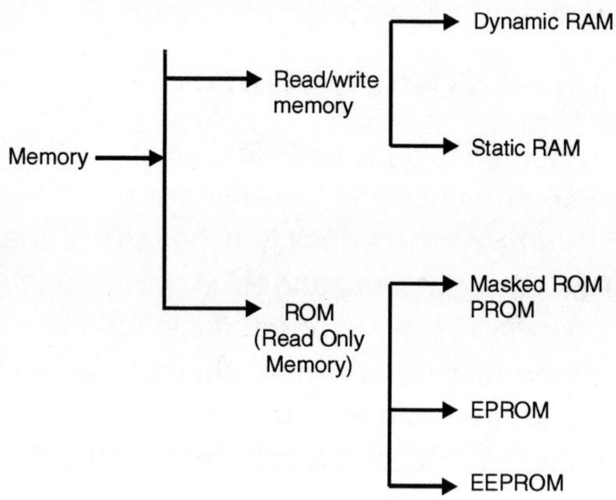

Fig. 5.2. Classification of semiconductor memory.

5.3.1. Memory Pin Connections

The Pin connections are common to all memory devices are the address inputs, data outputs or input/outputs, some type of selection input, and at least one control input use to select a read or write operation. Now Fig. 5.2 illustrates for ROM and RAM generio memory devices:

Before discussing different memory devices let us analyze the different pin connections for different purposes. The important connection such as :

(a) Address Connections

(b) Data Connections

(c) Selection Connections, and

(d) Control Connections

(a) Address Connections : All memory devices have address inputs that select a memory location within the memory device. The address inputs are almost always labelled from A_0 the least significant address input, to A_9, where subscript" can be any value but is always labelled as one less that the total number of address pins. For instant a memory device that has 10 address pins has its address pins labelled from A_0 to A_9. Number of address pins found on a memory device is determined by the number of memory locations found within it.

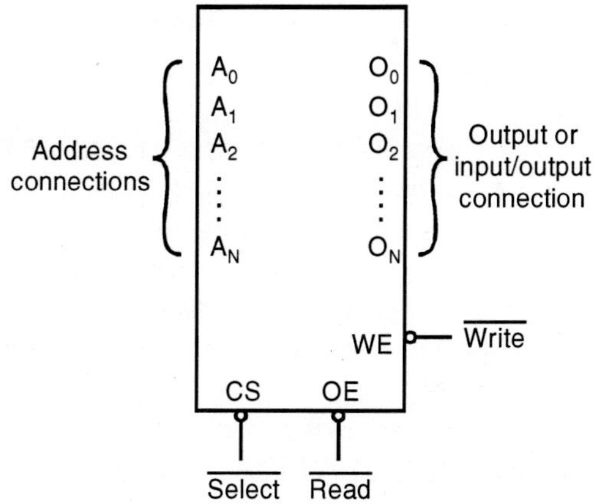

Fig. 5.3.

The more common memory devices have between 1 K (1,024) to 16 M (16,777,216) memory locations, with 256M memory location devices on the horizon. A 1 K memory device has 10 address pins ($A_0 - A_9$); therefore, 10 address inputs are required to select any of 1,024 memory locations. It takes a lO-bit binary number *(1,024 different combinations)* to select any single location on a 1,024-location device. When a memory device has 11 address connections (A_0-A_{10}) it has 2,048 (2K) internal memory locations. The number of memory locations can thus be extrapolated from the number of address pins. For instant, a 4K memory device has 12 address connections; an 8K device has 13, and so forth. A device that contains 1M locations requires a 20-bit address (A_0-A_{19}).

A 400H represents a 1K-byte section of the memory system, when a memory device is decoded to begin at memory address 10000H, and it is a 1K device, its. last location is at address 103FFH - one location less than 400H. One more important hexadecimal number to remember is a 1000H, because 1000H is 4K. A memory device containing a starting address of 14000H that is 4K bytes in size ends at location 14FFFH - one locations less than 1000H. A third number is 64K or 10000H. A memory that starts at location 3000H and ends at location 3FFFFH is a 64K byte memory. At the end since 1M of memory is not uncommon, a 1M memory contains 100000H memory locations.

(b) Data Connections : All memory devices have a set of data outputs or input/outputs. The device illustrated in Fig. 5.3 has a common set of input/output (I/O) connections. Nowadays memory devices have bidirectional common I/O pins.

The data connections are the points at which data are entered for storage or extracted for reading. Data pins on memory devices are almost always labelled Do through D7 for an 8-bit-wide memory device. In this sample memory device, there are eight I/O connections, which means that the memory device stores 8-bits of data in each of its memory locations. An 8-bit-wide memory device is often called a byte-wide memory. Although most devices are currently 8-bits wide, not all memory devices are 8-bits wide. Some devices are 16-bits, 4-bits, or just l-bit wide.

Catalogue listings of memory devices often refer to memory locations times bits per location. For instant, a memory device with 1K memory locations and 8-bits in each location is often listed as a 1K x 8 by the manufacturer. A 16K x 1 is a memory device containing 16K l-bit memory locations. The Memory devices are also often classified according to total bit capacity. For instant, a 1K x 8 bit memory device is sometimes listed as an 8K memory device, or a 64K x 4 memory device is listed as a 256K device. *(These variations occur from one manufacturer to an other*).

(c) Selection Connections : All memory device has an input - sometimes more than one - that selects or enables the memory device. Such kind of input is most often called a chip select (CS), chip enable (CE), or simply select (8) input. RAM memory generally has at least one CS or 8 input, and ROM at least one CE. If the CE, CS or S input active (*a logic 0 in this* *case, because of the over-bar*), the memory device performs a read or a write; if it is inactive (*a logic 1 in this case*), the memory device cannot do a read or a write because it is turned off or disabled. (*When more than one* cs *connection is present, all must be activated to read* or *write data).*

(d) Control Connections : Each memory devices have some form of control input or inputs. A ROM normally has only one control input, while a RAM often has one or two control inputs.

The control input, while a ROM is the output enable (OE) or gate G connection, which allows data to flow out of the output data pins of the ROM. If OE and the selection input are both active, then the output is enabled; if OE is inactive, the output is disabled at its high-impedance state. The OE connection enables and disables a set of three-state buffers located within the memory device and must be active to read data.

RAM memory device has either one or two control inputs, when there is one control input, it is often called R/W.This pin selects a read operation or a write operation only.

If the device is selected by the election input (or S). If the RAM has two control inputs, they are usually labelled WE (or W) and OE (or G). Here WE (write enable) must be active to perform a memory write operation, and OE must be active to perform a memory read operation. If these two controls (WE and OE) are present, they must never both be active at the same time. When both control inputs are inactive (logic 1's), then data are neither written nor read and the data connections are at their high-impedance state.

5.4. READ ONLY MEMORY (ROM)

The read-only will permanently store programs and data that are resident to the system and must not change when power is disconnected. Since ROM is permanently programmed so data are always present, even when power is disconnected. This type of memory is often called nonvolatile memory, available in many forms today. A device we call a ROM is purchased in mass quantities from a manufacturer and programmed during its fabrication at the factory itself. The EPROM *(erasable programmable read-only memory),* a type of ROM, is more commonly used when software must be changed often or when too limited a number are in demand to make the ROM economical. For a ROM to be practical, we usually must purchase at least 10,000 devices. An EPROM is programmed in the field on a device called an EPROM programmer. The EPROM is also erasable it exposed to high intensity ultraviolet light for about 20 minutes or less, depending on the type of EPROM.

The PROM memory devices are also available, but they are not as common today. The PROM (*programmable read-only memory*) is also programmed in the field by burning open tiny Nichrome or silicon oxide fuses, but once programmed it cannot be erased.

One more, newer type of read-mostly memory (RMM) is called the flash memory. The, flash memory, is also often called an EEPROM *(electrically erasable programmable ROM), EAROM (electrically alterable ROM),* or a NOVRAM (nonvolatile RAM). These memory devices are electrically erasable in the system, but require more time to erase than a normal RAM. The flash memory device is used to store setup information for systems such as the video card in the computer. It may also soon replace the EPROM In the computer for the BIOS memory. Some systems contains a password stored in the flash memory device.

5.4.1. Architecture of ROM

Figure 5.4 illustrates simple ROM for four byte diode ROM, which consists of only diodes and a decoder.

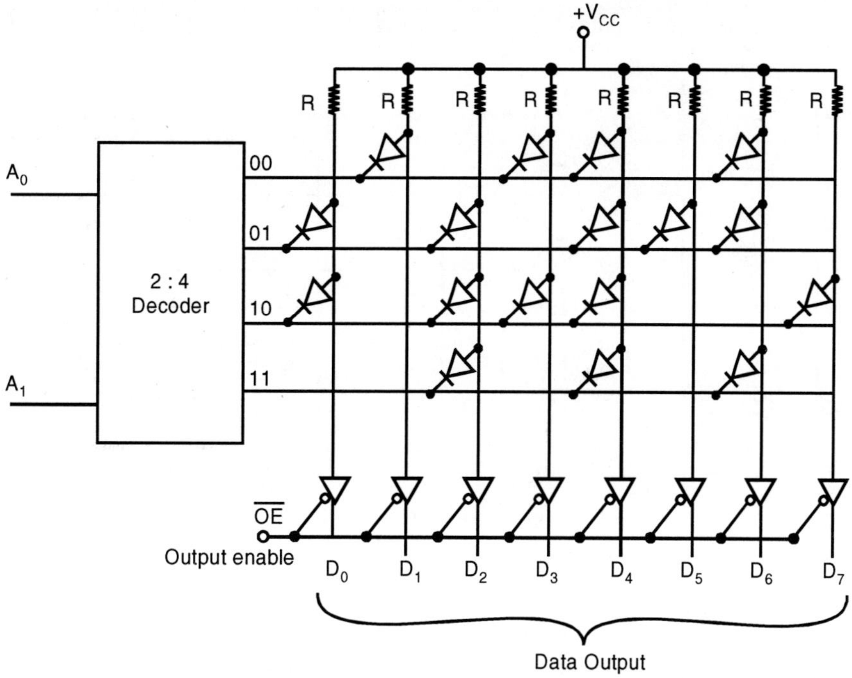

Fig. 5.4.

As observed in Fig. 5.4 address lines A_0 and A_1 are decoded by 2 : 4 decoder and used to select one of the four rows. As decoder output is active low, it places a logic 0 on the selected row. Each output data line goes to logic 0 if a diode connects the output data column to selected row. The data is available on the output data lines only when the output enable ($\overline{OE}$) signal is low. Now table 5.1 shows the contents of ROM at four locations as below :

TABLE 5.1. Contents of ROM

Address in Binary	The Binary Data								Data in Hex
	D_0	D_1	D_2	D_3	D_4	D_5	D_8	D_7	
00	1	0	1	0	0	1	0	1	A5
01	0	1	0	1	0	0	0	1	51
10	0	1	0	0	0	1	1	0	46
11	1	1	0	1	0	1	0	1	D5

ROMs use MOS technology instead of diode in present day. Figure 5.5 depicts four nibble ROM using MOS transistor, where diodes and pull up resistors are replaced by MOS transistors.

Address on the address lies (A_0 and A_1) is decoded by 2 : 4 decoder, which selects one of the four rows making it logic 0. The inverter connected at the output of decoder inverts the state of selected row (i.e., logic 1). Hence each output data line goes to logic 0 if a gate of MOS transistor is connected to row select lines. When gate of the MOS transistor is connected to the selected row, MOS transistor is turned on, which pulls the corresponding column data line to logic 0.

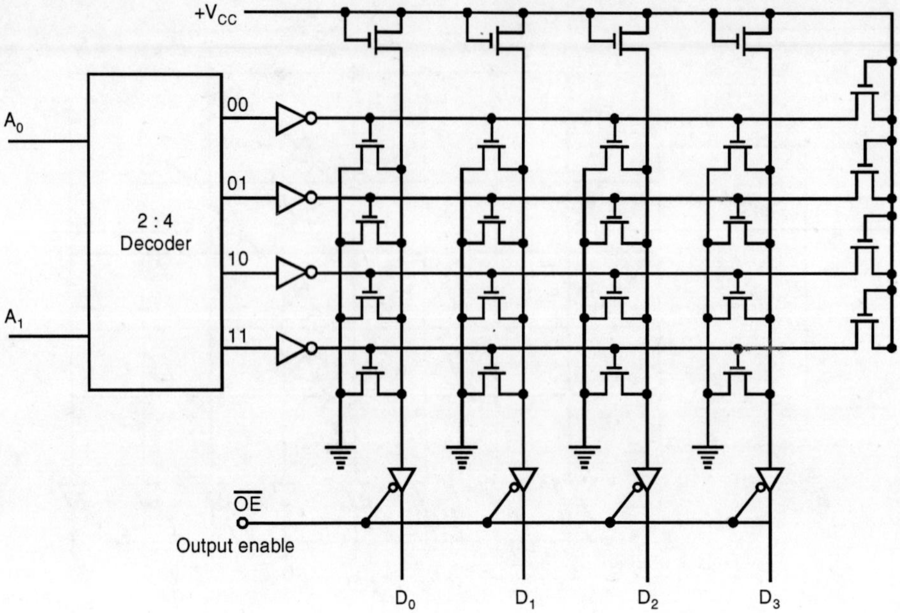

Fig. 5.5. 4-Nibble ROM using MOS Transistor.

Note. In integrated circuits, a thin metallized layer connects the gates of some transistors to the row select lines. The gate connections of MOS transistors depend on the data to be stored in the ROM. That means, according to the user truth table, manufacturer can deposit thin layer of metal to connect gates of the transistors. Once the pattern/mask is decided, it is possible to make thousands of such ROMs. Such ROMs are called Mask-programmed ROMs. The Masked ROMs are used in microprocessor based toys, TV games, home computers and other such high volume consumer products etc.

5.4.2. Programmable Read Only Memory (PROM)

The Fig. 5.6 depicts a four byte PROM, which has diodes in every bit position; hence the output 1s initially all 0s, and each diode, however has a fusible link. In series with diode. By addressing bit and applying proper current pulse at the corresponding output, we can blowout the use, storing logic 1 at that bit position. The fuse uses material like nichrome and polycrystalline. For blowing the fuse it is necessary to pass around 20 to 50 mA of current for period 5 to 20 ms. Blowing of fuses according to the truth table is called programming of ROM. The user can program PROMs with special PROM programmer. PROM programmer selectively burns the fuses according to pattern to be stored. This process

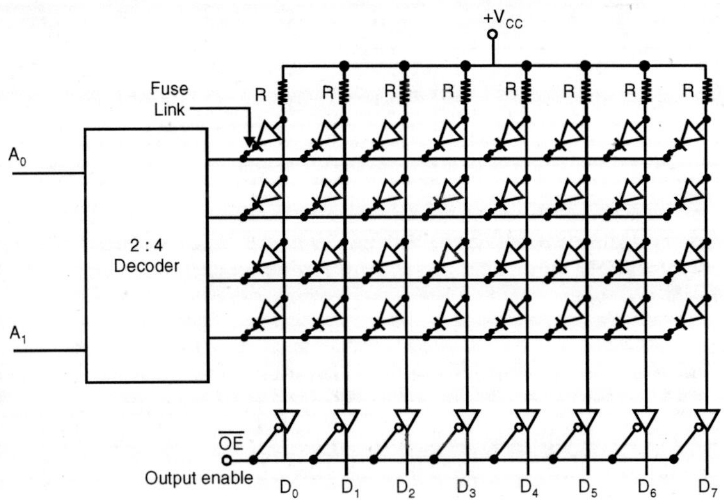

Fig. 5.6. 4-Byte P-ROM.

is also known as burning of PROM, which are one time programmable. Hence once programmed, the information stored is permanent.

5.4.3. Erasable Programmable Read Only Memory (EPROM)

The Erasable programmable ROMs use MOS circuitry, they store Is and Os as a packet of charge in a buried layer of the IC chip. There EPROMs can be programmed by the user with a special EPROM programmer. We can erase the stored data in the EPROMs by exposing the chip to ultraviolet light through its quartz window for 15 to 20 minutes. The Fig. 5.7 depicts the 2716 EPROM, is representative of most EPROMs.

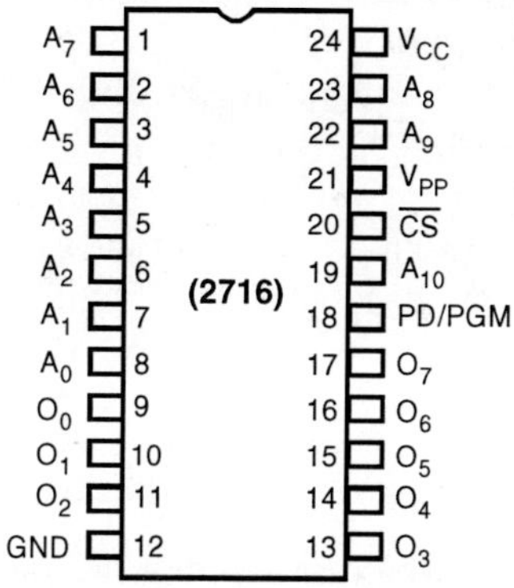

(a) Pin configuration

Pins MODE	PD/PGM (18)	$\overline{CS}$ (20)	V_{PP} (21)	V_{CC} (21)	Outputs (9-11, 13-17)
Read	V_{IL}	V_{IL}	+5	+5	D_{OUT}
Deselect	Don't care	V_{IH}	+5	+5	High Z
Power Down	VIL	VIL	+5	+5	High Z
Program	VIL	VIL	+5	+5	D_{IN}
Program Verify	VIL	VIL	+5	+5	D_{OUT}
Program Inhibit	VIL	VIL	+5	+5	High Z

(b) The Mode Selection

$A_0 - A_{10}$	Addresses
PD/PGM	Power Down/Program
DS	Chip Select
$O_0 - O_7$	Outputs

(c) Pin Names

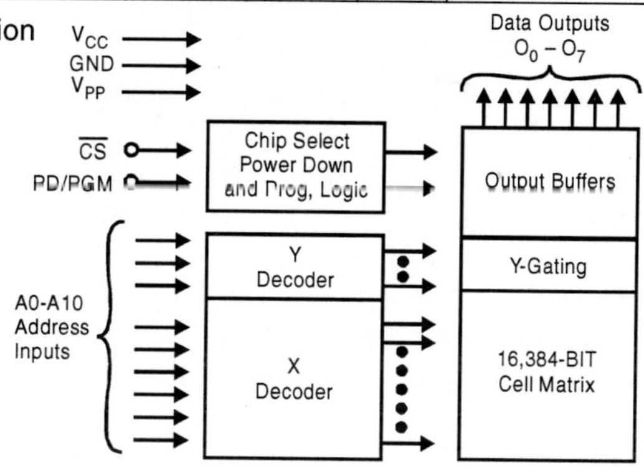

(d) Block Diagram

Fig. 5.7. EPROM 2716.

The device shown in Fig. 5.7 contains 11 address lines and 8 data outputs. The 2716 is a 2K × 8 memory device.

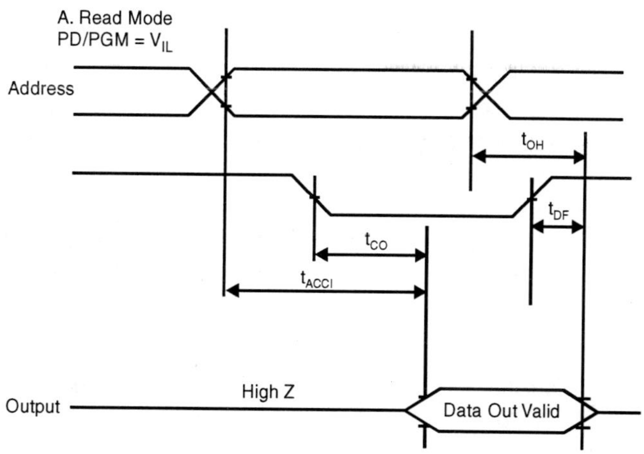

Fig. 5.8.

The Data only appear on output connections after a logic 0 is placed on both the CE and OE pin connections. If CE and OE are not both logic 0's the data output connections remain at their high-impedance or off states. Note that the Vpp pin must be placed at a logic 1 for data to be read from the EPROM. In some cases, the Vpp pin is in the same position as the WE pin on the SRAM. This can allow. a single socket to hold either an EPROM or an SRAM. Examples are the 2716 EPROM and the 6116 SRAM, both 2K × 8 devices that have the same pin-out, except for V pp on the EPROM and WE on the SRAM.

NOTE: Important piece of information provided by the timing diagram and data sheet is the memory access time-the time that it takes the memory to read information. As Fig. 5.8 illustrates. memory access time *(TAccl* is measured from the appearance of the address at the address inputs until the appearance of the data at the output connections. Which is based on the assumption that the CE input goes low at the same time that the address inputs become stable! Also, OE must be a logic 0 for the output connections to become active. Basic speed of the EPROM is 450 ns. Such type of memory component requires wait states to operate properly with the *8086/8088* microprocessors because of its rather long access time. If wait states are not desired, higher speed versions of the EPROM are available at an additional cost. *(Today, EPROM memory is available with access times of as little as 100 ns).*

5.4.4. Electrical Erasable Programmable Read Only Memory (EEPROM)

Electrically erasable programmable ROMs also use MOS circuitry which are similar to that of EPROM. Data is stored as charge or no charge on an insulated layer or on an insulated floating gate in the device. The insulating layer is made very thin ≈ 200 A. Hence a voltage as low as 20 to 25 V can be used to move charges across the thin barrier in either direction for programming or erasing. The EEPROM allows selective erasing at the register level rather than erasing all the information since the information can be changed by using electrical signals. EEPROM memory also has a special chip erase mode by which entire chip can be erased in 10 ms, which is quite small as compared to time required to erase EPROM, and it can be erased and reprogrammed with device right in the circuit. *(EEPROMs are most expensive and the least dense ROMs).*

5.4.5. NVRAM

It is non-volatile RAM created by the combination of a static RAM and an electrically erasable, PROM (EEPROM) in the same chip. The EEPROM and static RAM have exact mirror memories structure of each other, when powered it functions as a normal RAM, for power failure, it saves the stored data in the EEPROM. The entire contents of the RAM can be transferred in parallel to the EEPROM in less than 4 ms. When repowered it is possible to get entire contents of EEPROM back into the RAM. Hence EEPROM is used for temporary storage data during power failure. The STORE and RECALL signals are used to store data in EEPROM and transfer data back into RAM, respectively *(main advantage of NVRAM is that a battery backup is not required to save working data in case of power failure).*

The information such as time, data, monitor setting and other computer configurations are normally stored in the CMOS static RAM or NVRAM of a computer. Few standard NVRAMs and their capacities are mentioned here in Table 5.2.

TABLE 5.2

S.No.	NVRAM	Capacity
1.	M48Z2M1	16 MB x 8
2.	M48Z512A	4 MB × 8
3.	M48Z129	1MBx8
4.	M48Z02I12	16 KB x 8

5.4.6. Flash Memory

The Flash memory *(sometimes called "flash RAM")* is a type of constantly-powered non-volatile memory which can be erased and reprogrammed in units of memory called blocks, which is a variation of electrically erasable programmable read-only memory (EEPROM) which, unlike flash memory, is erased and rewritten at the byte level, which is slower than flash memory updating. Flash memory is frequently used to hold control code such as the Basic Input/Output System (BIOS) in a personal computer. If BIOS needs to be changed *(rewritten),* the flash memory can be written to in block *(rather than byte)* sizes, making it easy to update. *[flash memory is not useful as random access memory (RAM) because RAM needs to be addressable at the byte (not the block) level}.*

Flash memory is named because the microchip is organized so that a section of memory cells are erased in a single action or *"flash".* The erasure is caused by Fowler-Nordheim tunneling in which electrons pierce through a thin dielectric material to remove an electronic charge from a floating 'gate associated with each memory cell. *(Intel offers a form of flash memory that holds two bits (rather than one) in each memory cell, thus doubling the capacity of memory without a corresponding increase in price).*

The Flash memory is used in digital cellular phones, digital cameras, LAN switches, PC Cards for notebook computers, digital set-up boxes, embedded controllers, and other devices.

TABLE 5.3. Features of Hash Memory.

S. No.	Features	Explanation
1.	High Speed	Flash memories operate at higher speeds
2.	Low voltage	They operate at lower voltages thus save power and consume less energy.
3.	Durability	Flash memories are able to tolerate severe shock and vibration without losing data

5.5. RANDOM ACCESS MEMORY (RAM)

The Static RAM memory devices retain data for as long as DC power is applied, since no special action *(except power)* is required to retain stored data, these devices are called as static memories. They are also called as a volatile memories since they will not retain data without power. The main difference between a ROM and RAM is written under normal operation, while a ROM is programmed outside the computer and is only normally read. The SRAM stores temporary data and is used when the size of the read/write memory is relatively small. Today, a small memory is one that is less than 1M byte. The Fig. 5.9 illustrates 4016, which is a 2K x 8 read/write memory. This devices has 11 address inputs and 8 data input! Output connections, which is representative of all SRAM devices.

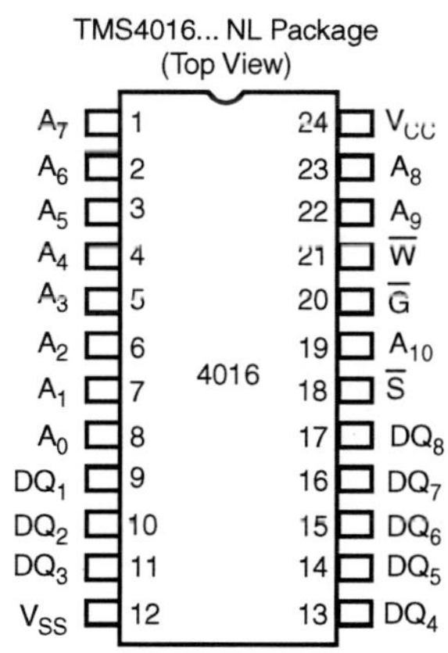

Pin Nomenclature	
$A_0 - A_{10}$	Address
$DQ_1 - DQ_8$	Data In/Data Out
$\overline{G}$	Address
$\overline{S}$	Address
V_{CC}	+5-V
V_{SS}	Ground
$\overline{W}$	Write Enable

Fig. 5.9. 4016, 2k × 8 RD/WR Memory.

NOTE : Control inputs of this RAM are slightly different from those presented earlier. The pin is labeled G, the CS pin S and the WE pin W. Despite the altered designations, the control pins function exactly the same as those outlined previously. *(Other manufacturers make this popular SRAM under the part numbers 2216 and 6116).*

5.5.1. TTL RAM Cell

The Fig. 5.10 illustrated a simplified schematic of a bipolar memory cell; which is implemented using TTL *(Transistor-Transistor-Logic)* multiple emitter technology. It stores 1 bit of information and is nothing but a flip-flop, which can store either 0 or 1 as long as power is applied, and it can set or reset to store either 1 or 0, respectively.

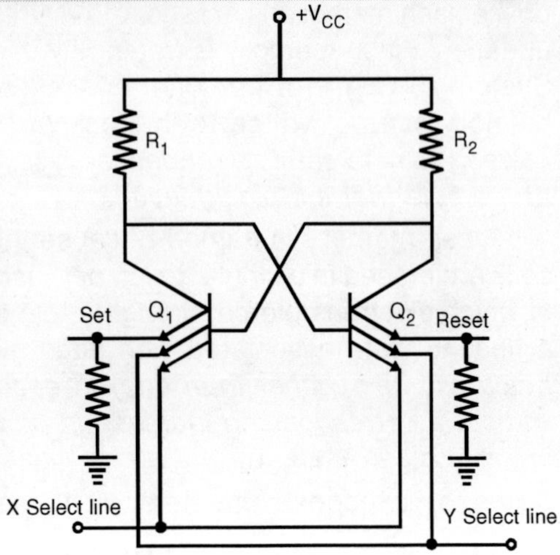

Fig. 5.10. TTL RAM cell.

In Fig. 5.10, X select and Y select input lines select a cell from matrix and the Q1 and Q2 are cross coupled inverters, hence one is always OFF while the other is ON. A "1" is stored in the cell if Q1 is conducting and Q2 is OFF. A "0" is stored in the cell if Q2 is conducting and Q1 is OFF. The state of the cell is changed to a "0" by pulsing a HIGH on the Q1 (SET) emitter. This turns OFF Q2. When Q1 is turned OFF, Q2 is turned ON. As long as Q2 is ON, its collector is LOW and Q1 is held OFF.

A "1" can be rewritten by pulsing the Q3 (reset) emitter high. Large number of these Memory Cells is organized on a row and column basis to form a memory chip. (Fig. 5.11)

This chip has 13 address lines, out of which the first seven address lines are connected to the column decoder to indicate one of the 128 columns, remaining 6 address lines are connected to the row decoder to indicate one of 64 rows. Where the decoded row and column cross, they select the desired individual memory cell *(Simple arithmetic shows that there are* 64 × 128 = 8192, *crossing, i.e., this memory has* 8192 *memory cells).*

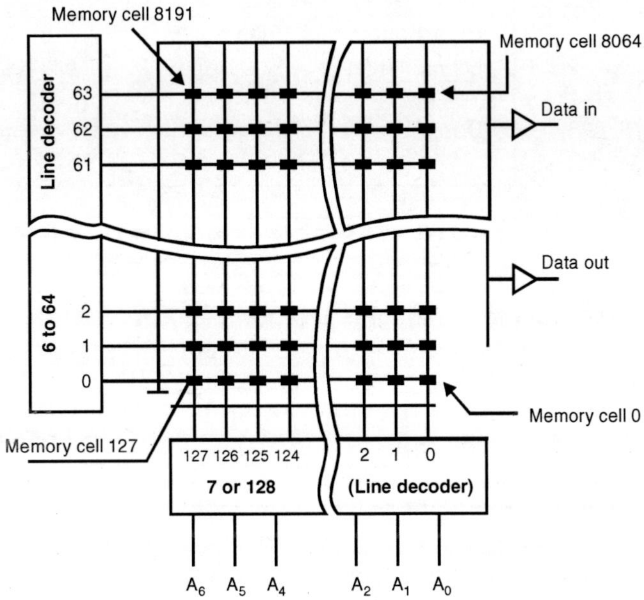

Fig. 5.11.

5.5.2. Dynamic RAM Memory (DRAM)

About the largest static RAM available today is a 128K × 8. The Dynamic RAMs, on the other hand, are available in much larger sizes : upto 16M × 1. In all other respects, DRAM is essentially the same. as SRAM, except that it retains data for only 2 or 4 ms on an integrated capacitor. After 2 or 4 ms, the contents of the DRAM must be completely rewritten (refreshed) because the capacitors, which store a logic 1 or logic 0, lose their charges. Figure 5.12 depicts the DRAM cell.

A dynamic RAM contains thousands of such memory cells shown in Fig. 5.12. When COLUMN *(Sense)* and ROW *(Control)* lines go high, the MOSFET conducts and charges the capacitor, thus the COLUMN and ROW lines go low, the MOSFET opens and the capacitor retains its charge, thus it stores 1 bit.

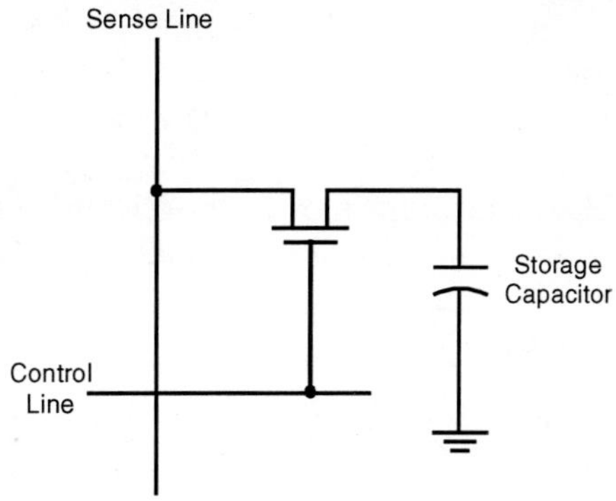

Fig. 5.12. DRAM Cell.

Since only a single MOSFET and capacitor are needed, the dynamic RAM contains more memory cells as compared to static RAM per unit area, the disadvantage of dynamic RAM is that it requires refreshing of charge on the capacitor after every few milliseconds. This complicates the system design, since it requires the extra hardware to control refreshing of dynamic RAMs.

The transistor acts as a switch in such type of cell. The DRAM consists of three tristate buffers i.e., Input buffer, Output buffer and refresh buffer.

Input and output buffers are enabled and disabled by controlling $R/\overline{W}$ line. When $R/\overline{W}$ line is LOW, input buffer is enabled and output buffer is disabled. When $R/\overline{W}$ line is HIGH, input buffer is disabled and output buffer is enabled. Now let us observe the three operations :

(a) Write Operation,

(b) Read Operation, and

(c) Refresh Operation.

(a) Write Operation

To enable write operation, $R/\overline{W}$ line is made LOW which enables input buffer and disables output buffers, as shown in Fig. 5.13 and Fig. 5.14.

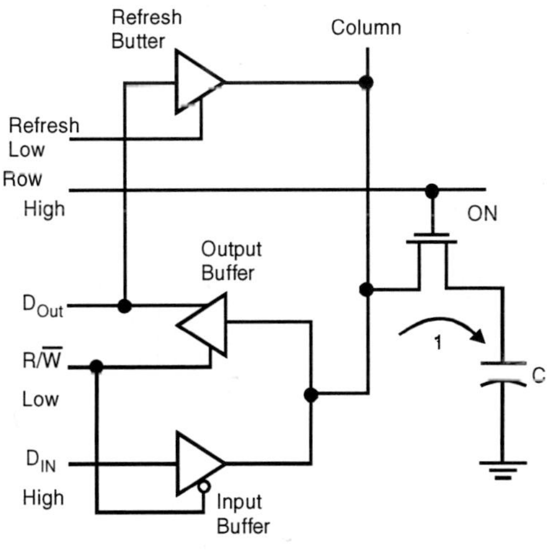

Fig. 5.13. Writing "1" into memory cell

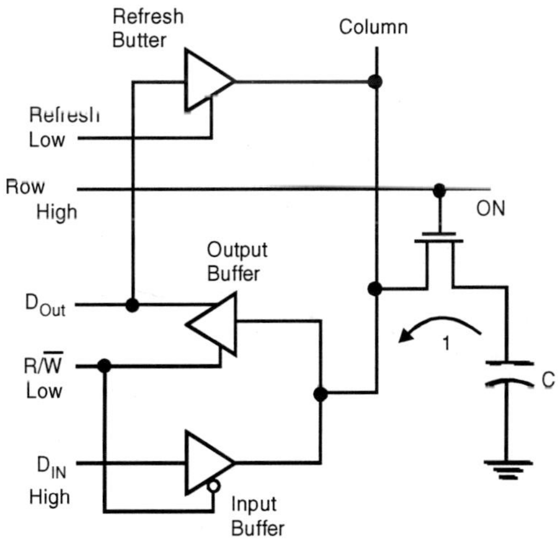

Fig. 5.14. Writing "0" into memory cell

As observed in Figs. 5.13 and 5.14 to write a 1 into the cell, the DIN line is HIGH, and the transistor is turned ON by a HIGH on the ROW line, which allows the capacitor to charge to a positive voltage. When 0 is to be stored, a LOW is applied to the DIN line. The capacitor remains uncharged, or if it is storing a 1, it discharges as indicated in Fig. 5.14. If the ROW line is made LOW, the transistor turns off and disconnects the capacitor from the data line, thus storing charge (either 1 or 0) on the capacitor.

(b) Read Operation

To read data from the cell, the $R/\overline{W}$ line is made HIGH, which enabies output buffer and disables input buffer. Then ROW line is made HIGH. It turns transistor ON and it connects the capacitor to the Dout line through output buffer. The Fig. 5.15 depicts the Read operation i.e., reading "1" from the memory cell.

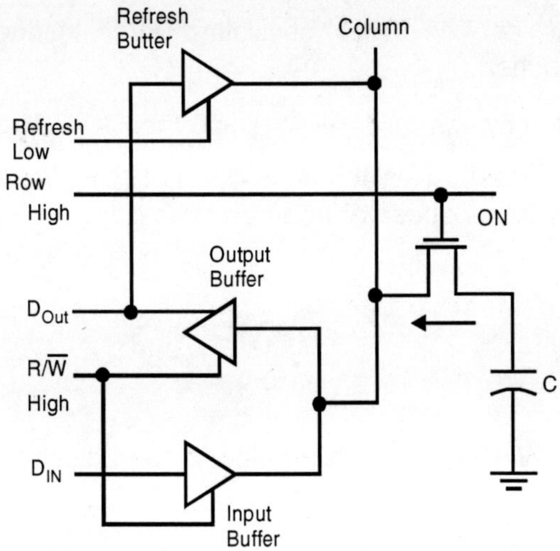

Fig. 5.15.

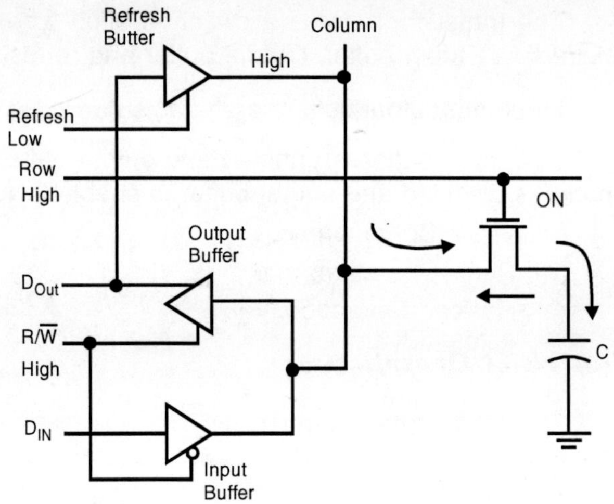

Fig. 5.16.

(c) Refresh Operation

Now to enable refresh operation $R/\overline{W}$ line, ROW line and REFRESH line are made HIGH, which turns ON transistor and connects capacitor to COLUMN line. All $R/\overline{W}$ is HIGH, output buffer is enabled and the stored data bit is applied to the input of refresh buffer. The enabled refresh buffer then produces a voltage on COLUMN line corresponding to the stored bit and thus replenishing the capacitor which is illustrated in Fig. 5.16 above.

5.5.3. Comparison between SRAM and DRAM

There are few points which provides comparisons between SRAM and DRAM, they are listed in Table 5.4.

TABLE 5.4. Comparison between SRAM and DRAM

S.No.	SRAM	DRAM
1.	Static RAM contains less dynamic.	RAM contains more memory cells as compared memory cells per unit area to static RAM per unit area.
2.	It has less access time hence faster memories.	Its access time is greater than static RAMs.
3.	Static RAM consists of number of Flip-flops.	Dynamic RAM stores the data as a charge on the capacitor. It consists of MOSFET and the capacitor for each cell.
4.	Refreshing circuitry is not needed.	Refreshing circuitry is needed to maintain the charge on the capacitors after every few milliseconds. Extra hardware is required to control refreshing, which
5.	Cost is more.	Cost is less.

5.5.4. Memory Chips and their Capacity

In market then are many memory chips are available. The user has to select the memory chips as required with respect to their memory capacity. Now the Table 5.5 provides the list of available memory chips with their capacities.

TABLE 5.5. Available Memory Chips.

S.No.	Memory Chip	Memory Capacity
1.	2114	IK x 4 Bit RAM
2.	6116	2K x 8 Bit RAM
3.	65116	2K x 8 Bit RAM
4.	6264	8K x 8 Bit RAM
5.	63256	32K x 8 Bit RAM
6.	84256	32K x 8 Bit RAM
7.	55257	32K x 8 Bit RAM
8.	4164	64K x 1 Bit Dynamic RAM
9.	41256	256K x 1 Bit Dynamic RAM
10.	6256	256K x 1 Bit Dynamic RAM
11.	511000	1 MB Dynamic RAM
12.	2316	2K x 8 Bit ROM
13.	2332	3K x 8 Bit ROM
14.	2364	8K x 8 Bit ROM
15.	65256	32K x 8 Bit ROM
16.	23512	64K x 8 Bit ROM
17.	521000	128K x 8 Bit ROM
18.	2708	1K x 8 Bit EPROM
19.	2716	2K x 8 Bit EPROM
20.	2732	4K x 8 Bit EPROM

S.No.	Memory Chip	Memory Capacity
21.	2764	8K x 8 Bit EPROM
22.	2816	2K x 8 Bit EPROM
23.	27128	16K x 8 Bit EPROM
24.	27256	32K x 8 Bit EPROM
25.	27512	64K x 8 Bit EPROM
26.	27F64	8K x 8 Bit Flash Memory
27.	28F256	32K x 8 Bit Flash Memory
28.	571000	128K x 8 Bit EPROM

5.6. READ AND WRITE MEMORY CYCLES

Now let us analyze the read and write memory cycles along with their timing parameters.

5.6.1. Read Cycle

The read cycle for static RAM is as shown in Fig. 5.17 which is drawn as the basis of different parameters of timing.

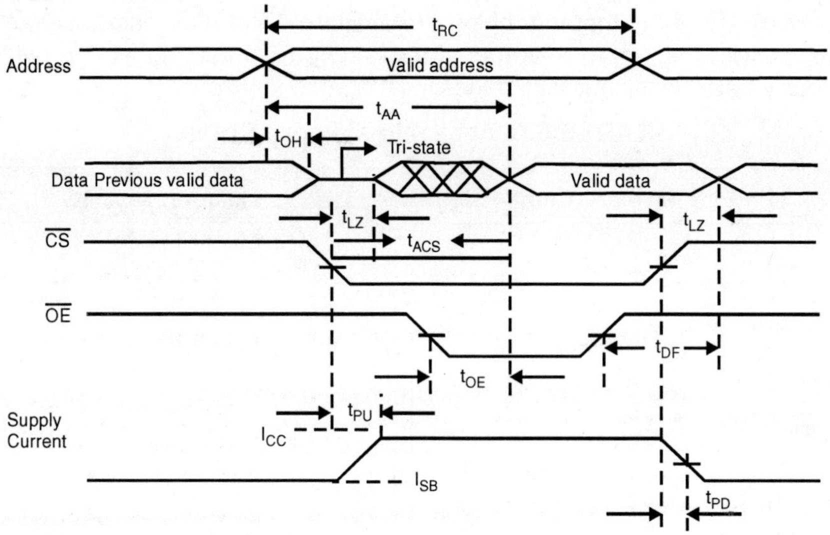

Fig. 5.17. Read cycle for S-RAM.

The different timing parameters are :

Read Cycle Time

Is the minimum time for which an address must be held stable on the address bus, in read cycle.

Address Access Time

Is the maximum specified time within which a valid new data is put on the data bus after an address is applied.

Output Hold from Address Change

The time for which a data from previously read address will be present on the output after a new address is applied.

Chip Selection to Output in Tri-state

The minimum time for which the data output will remain in tri-state after goes low.

Chip Select Access Time

Is the minimum time required for memory device to put valid data on the data bus after $\overline{CS}$ goes low.

Chip Deselection to Output in Tri-state

Is the minimum within which output returns to its high impedance state (tri-state) after $\overline{CS}$ goes high.

Output Enable to Output Delay

Is the minimum time required for the memory device to put valid data on the data bus after $\overline{CS}$ goes low.

Output Enable High to Output Float

Is the time for which data will remain valid on the data bus after $\overline{CS}$ goes high.

Chip Selection to Power Up Time

Is the time within which the memory device is powered upto its normal operating current after $\overline{OE}$ goes low.

Chip Deselection to Power Down Time

Is the time for which the memory device will remain powered up after $\overline{CS}$ goes high.

5.6.2. Write Cycle

The write cycle for static RAM is an shown in Fig. 5.18 where the timing diagram is drawn on the basis of different parameters of time.

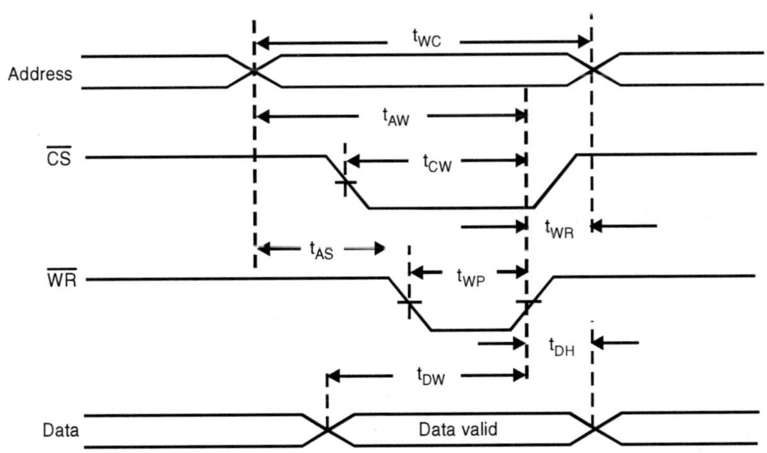

Fig. 5.18. Write cycle for S-RAM.

The different timing parameters are :

Write Cycle Time

Is the minimum time for which an address must be held stable on the address bus, in write cycle.

Address Valid to End of Write

Is the time at which address must be applied on the address bus before $\overline{WR}$ goes high.

Write Recovery Time

Is the time for which address will remain on address bus after $\overline{WR}$ goes high.

Address Setup Time

If address is applied, it is the time after which $\overline{WR}$ can be made low.

Chip Selection to the End of Write

Is the time at which the $\overline{CE}$ must be made low to select the device before $\overline{WR}$ goes high.

Write Pulse Width

Is the time for which WR goes low.

Data Valid to the End of Write

Is the minimum time for which data must be valid on the data bus before $\overline{WR}$ goes high.

Data Hold Time

Is the time for which data must be held valid after $\overline{WR}$ goes high.

Since 32K byte RAM and EPROM need 15 address lines, A_0 to A_{14} lines are used. A_0 and $\overline{BHE}$ are used to select even and odd memory banks respectively as shown in Fig. 5.18.

5.7. MEMORY AND ADDRESSING

There are several different types of memory in a micro. One is *Program* memory. This is where the program is located. Another is *Data* memory. This is where data, that might be used by the program, is located. The neat (or strange) thing is that they both reside in the same memory space and can be altered by the program. That's right, a program can actually alter itself if that was necessary. Two terms are used when talking about memory. *Reading (load)* is getting a value from memory and *Writing (store)* is putting a value into memory.

There are three buses (not the kind you ride in) associated with the memory subsystem. One is the **address bus**, the second is the **data bus**, and the third is the **control bus**. It's important for you to know exactly how all this works, because these busses transport data and addresses everywhere. All three are connected to the memory subsystem. Its also good to know the function of each to better understand what's happening. In the 8085 CPU, the address bus is 16 bits wide. It acts to select one of the unique 2^{16} (64K) memory locations. The control bus determines whether this will be a read or a write. In the case of an instruction fetch, the control bus is set up for a read operation. Data is read or written through the data bus, which is 8 bits wide. This is why all registers and memory are 8 bits wide, it's the width of the data bus on the 8085 CPU. A *bus* is just a group of connections that all share a common function. Instead of speaking of each bit or connection in the address separately, for example, all 16 are taken together and referred to simply as the address bus. The same is true for the control and data buses.

A byte is the most used number in a micro because each memory location or register is one byte wide. Memory has to be thought of as a sort of file cabinet with each location in it being a folder in the cabinet. In a file cabinet, you go through the tabs on the folders until you find the right one. To get to each memory location, a different method is used. Instead, a unique address is assigned to each location. In most micros this address is a word or 16 bits, or 4 digit hex. This allows for a maximum of 65536 (2^{16} or 64K) unique addresses or memory locations that can be accessed. These addresses are usually referred to by a 4 digit hex number. Memory usually starts at address 0000h and could go up to FFFFh (2^{16} or 64K or 65536 in total). To access these locations, a 16 bit address is presented to memory and the byte at that location is either read or written.

The Program Counter is what holds this address when the micro is executing instructions. The reason instructions are read sequentially, is because the program counter automatically increments after fetching the current instruction. It does this even before the current instruction is acted upon. The sequence is that the program counter's contents are placed on the memory address bus and the instruction is fetched from memory through the data bus, and immediately the program counter is incremented by 1. Then the micro looks at the instruction and starts processing it. If the instruction is not some kind of jump or call, the instruction is completed and the program counter is presented to the memory address bus again and the next instruction is fetched, the program counter is incremented and the process starts over. This is referred to, in computer jargon, as fetch, decode, and execute.

In the case of reading or writing data, the process is a little different. Data can be read from or written to memory in similar fashion to the fetch. But data does not need the decode and execute steps.

5.8. GENERAL MEMORY ORGANIZATION

In its most general sense the word *memory* refers to any device that stores information for later use. Under this definition, the memory of a computer can be divided into two categories. One category pertains to that part of the computer that holds the instructions and data that are presently being operated on, i.e., the part that the processor can access directly. The other category consists of the facilities that can store information, but the information must be transferred to the memory in the first category before it can be used by the processor. This book has referred to components in the former category as main *memory,* or simply *memory,* and components in the latter category as *mass storage.* This chapter will be concerned with main memory and how it is interfaced to the system bus.

As we have seen, the main memory is comprised of groups of bits called bytes and words that are addressed as units. Until now the processor-oriented meaning of "word" has been used which refers to the numbers of bits that can be simultaneously transferred over the data bus and manipulated as a unit by the processor. Most memory designers, however, use the term "word" to mean the smallest group of bits that is associated with an address. In an 8086/8088 system these groups are the 8-bit bytes. Therefore, when considering memory design in general, "word" will often be used in contexts where "byte" appeared in previous discussions. When a chance of confusion exists, the term "word" will be modified by the word "memory" or the word "processor." The number of bits in a memory word is called its *length* (e.g., the length of an 8086/8088 memory word is 8).

One of the principal attributes of a memory is whether or not it can retain its contents while its power is turned off. A memory that can do so is said to be *nonvolatile;* otherwise, it is said to be *volatile.* Whether or not a memory is volatile is particularly important if it contains instructions, because if it is volatile, these instructions will be lost during even momentary power failures and will have to be reloaded each time the computer is brought up. For most microcomputer applications, the system must contain at least some nonvolatile memory, but most of the memory can be volatile.

Another important attribute depends on the ways in which the memory can be accessed. If a memory unit can only be read from, it is called *read-only memory (ROM),* but if it can be both read from and written into, it is called *read/write memory. Random access* refers to storage for which all locations can be accessed in the same amount of time, as opposed to *sequential* (or *serial) access,* in which the locations must be referenced in order (e.g., as on a tape). The main memory of all modern computers is random access; however, when referring to main memory the phrase "random access" has historically been used as a synonym for read/ write and read/write memory has been referred to as *random access memory (RAM).* Thus, the two principal types of main memory are ROM and RAM.

The most prominent kind of nonvolatile read/write memory is *magnetic core* or simply *core, memory,* which is constructed of small, doughnut-shaped magnets. Volatile read/write memory is made from semiconductor ICs. Although core memory is nonvolatile, it requires a considerable amount of support electronics, which causes it to be relatively expensive. As a result, most microcomputer systems use semiconductor memory and rely on mass storage devices or memory with backup power to store information during shutdown. ROM memory, which is obviously nonvolatile, is also made from semiconductor ICs. Because it does not need to include write circuitry, ROM has the least complicated design, ROM is commonly used to store bootstrap loaders so that the system can be brought up without manually loading initialization instructions. It can, however, be used only to store data and programs that never change. Because core memory is seldom used in microcomputer systems it will not be considered in this book.

This chapter begins with a discussion of the overall design of memory. Sections 10-2 and 10-3 consider the two types of semiconductor read/write memory: static RAM and dynamic RAM. Section 10-4 considers the design of backup power for volatile memory so that it becomes, in effect, nonvolatile memory. The last section examines the various types of ROM.

The memory of a computer system normally consists of one or more PC boards that are connected to the system bus. On each board is a module that is addressed by the high-order hits on the address bus. As shown in Fig. 5.19, most systems include both ROM and RAM modules. It should be pointed out, however, that some small systems which are used as simple controllers may contain only ROM and/or may include the memory on the same board, or even the same IC chip, as the processor.

The general design of a memory module is shown in Fig. 5.20. It consists primarily of an interface and an array of memory IC devices, each of which contains an array of memory cells with each cell being the circuitry needed to store 1 bit. The cells in a device may be accessed separately or in groups, but either way the following relations must hold :

$$\text{No. of devices in a row} = \frac{\text{No. of bits in a word}}{\text{No. of cells in a group}}$$

$$\text{No. of devices in a column} = \frac{\text{No. of words in module}}{\text{No. of groups in a device}}$$

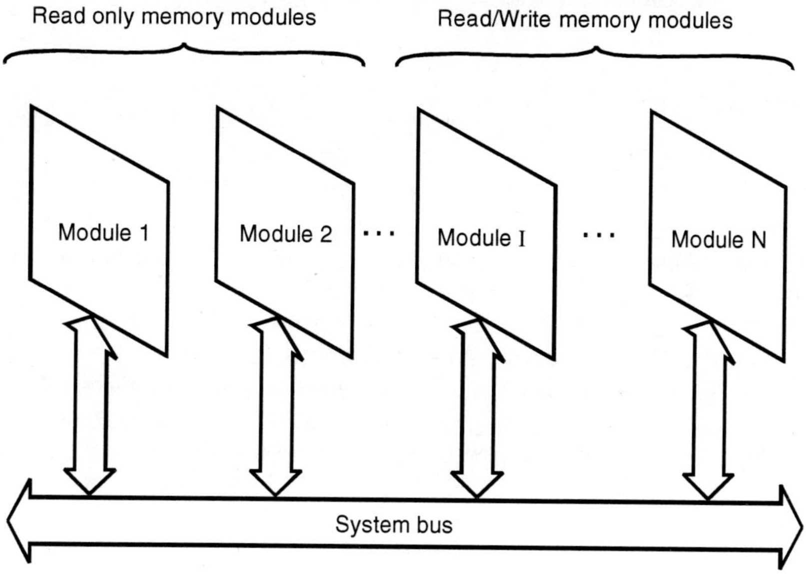

Fig. 5.19. Overall organization of memory.

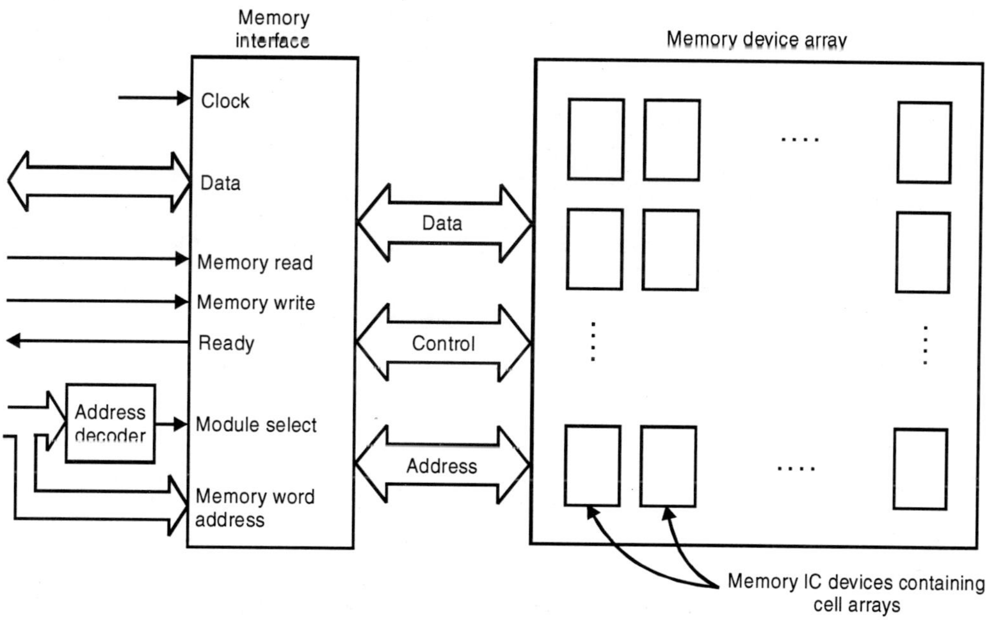

Fig. 5.20. Typical memory module design.

A memory device is said to be an M × *N device if it* contains *M* groups with each group consisting of *N* cells. Similarly, a *K* × *L module* is one that consists of *K* words, each having *L* bits. To demonstrate the application of the above relations, several memory module designs which utilize typical memory devices are summarized in Fig. 5.21.

The principal criteria involved in designing a memory are :

1. Cost.
2. Capacity.
3. Speed.
4. Power consumption.
5. Reliability.
6. Volatility and access capability.

The cost of a memory module normally consists of two components, one which is independent of the size of the module and is called *overhead,* and one which is proportional to the size and is called the *incremental cost.* The overhead cost is due primarily to the support and interface electronics in the module, and the incremental cost is most closely related to the cost of the memory devices. Both overhead and incremental costs are dependent on the number of pin connections and the complexity of the board design. Therefore, memory devices with large bit capacities and that require few supporting chips ordinarily have a cost advantage. Because the overhead tends to be the same regardless of the capacity of the module but must be included for each module, it is generally better to fulfill the overall memory capacity requirement using as few modules as possible. Another factor which enters into the overhead is the cost of the power supply. The fewer the number of supply voltages that are required, the less complicated the power supply and board design.

Memory Size	Type of RAM	No. of chips in each column	No. of chips in each row	No. of chips in the module
4 K × 8	1 K × 1	4	8	32
	4 K × 1	1	8	8
	256 × 4	16	2	32
	1 K × 4	4	2	8
4 K × 16	1 K × 1	4	16	64
	4 K × 1	16	16	16
	256 × 4	16	4	64
	1 K × 4	4	4	16
16 K × 8	1 K × 4	16	2	32
	4 K × 1	4	8	32
	8 K × 1	2	8	16
	16 K × 1	1	8	8
64 K × 8	16 K × 1	4	8	32
	64 K × 1	1	8	8

Fig. 5.21. Representative memory device arrays.

The speed of a memory is gauged by its *access time,* which is defined as the time delay between the receipt of a stable address input and the data output. This time depends on many factors and is even related to the bit capacity of the memory devices. High-speed transistors tend to require more chip space so that fewer cells can be included on each chip. Also, high-speed devices, which are normally made use of a bipolar technology, are more expensive and consume more power.

Power consumption can be very important in systems that must sometimes be powered up by batteries or solar cells (e.g., space vehicles). The technology used in constructing the IC determines the power required for each cell. The technology most often used in low-power applications is the complementary metal-oxide semiconductor (CMOS) technology. The main disadvantage associated with CMOS is that a fair amount of chip space is needed for each cell; thus the capacity of each device is reduced. Unfortunately, power consumption and speed tend to be proportional and it is difficult and expensive to achieve both low power and high speed. The technology that has been found to be a good compromise of speed, power consumption, capacity per device, and cost is the high-density MOS (HMOS) technology,

Because the reliability of an IC is high once it has been thoroughly tested, the reliability of a module is strongly related to the number of solder connections and board complexity. Therefore, reliability increases as the total pin count decreases, thus supplying an additional incentive to minimize the number of memory and supporting devices in a module.

Nonvolatility and access capability are pretty much dictated by the application. Unless an application requires nonvolatility there is no reason to provide it. If nonvolatile RAM is needed, then either core memory must be used or backup power must be available so that the volatile semiconductor memory can be maintained in a powered-up state, ROM is generally used whenever possible because it is less expensive, nonvolatile, more reliable, and impervious to noise, and its simple construction lends itself to high cell density,

Static RAM Devices

For static memory devices, a cell is commonly implemented using six MOS transistors, as shown in Fig. 5.22. Information is stored according to the states of transistors Q_1 and Q_2. This cross-coupled transistor pair is such that when one transistor is on, the other is off, and vice versa. A 1 is assigned to the state that exists when Q_2 is on and Q_1 is off, and a 0 is assigned to the opposite state. The transistors Q_3 and Q_4 serve as resistors and Q_5 and Q_6 act as enable gates. During a write operation, first the cell is selected by raising the voltage level on the select line. When this is done, transistors Q_5 and Q_6 act as short circuits, so that the Read/ Write 1 line is applied to the gate of Q_2 and the Read/Write 0 line is applied to the gate of Q_1. To write a 1 into the cell, a 1 is placed on Read/Write 1 and a 0 is placed on Read/Write 0; this causes Q_2 to be turned on and Q_1 to be turned off. On the other hand, if a 0 is to be written into the cell, a 1 is placed on Read/ Write 0 and a 0 is placed on Read/Write 1, In either case, once they are set the states of Q_1 and Q_2 will remain unchanged until the next write operation. The cell can be read simply by applying a voltage to the select line. When this is done, the state of Q_1 is applied to Read/Write 0 and the state of Q_2 is applied to Read/ Write 1.

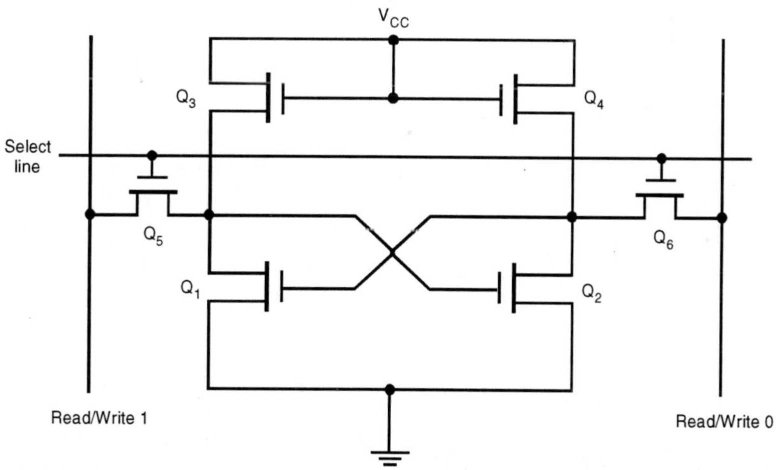

Fig. 5.22. Schematic diagram of a six-transistor static RAM cell.

The number of memory cells and their arrangement in a static memory device varies considerably. Common cells range from 256 × 4 to 16 K × 1. A 256 × 4 RAM contains 256 locations, each having 4 bits, while a 16 K × 1 configuration provides 16 K locations, each containing only 1 bit. The general organization of a static RAM is described by the block diagram of Fig. 5.23, which illustrates a 1 K × 1 RAM, The memory cells are organized in a matrix of 32 rows and 32 columns. Address bits A9 through A_0 are divided into row and column addresses and together specify one of the 1024 cells. Row address inputs (A_4-A_0) are decoded to select one of the 32 rows of storage cells. Column address inputs (A_9-A_5) not only select the column, but also enable the corresponding I/O circuits consisting of drivers and sense amplifiers. These circuits permit a stored bit to be output during a read operation and to be changed during a write operation. The read/write (R/W) control input specifies the type of operation, high for read and low for write. The chip enable input ($\overline{CE}$) is used to select the appropriate row of memory devices in the memory module.

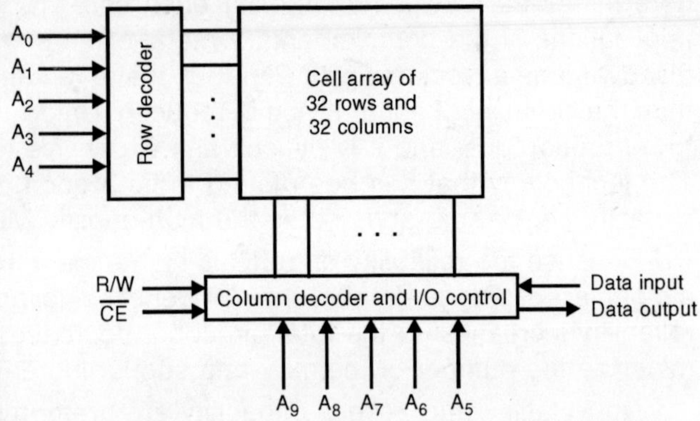

Fig. 5.23. Block diagram of a 1K × 1 memory device.

A 4 K × 8 memory device array which is constructed from 1 K × 1 devices is shown in Fig. 5.24, If a chip is enabled, a read or write operation will proceed as specified by the R/W control input. Otherwise, the read/write signal will not be recognized and the output is forced into a high-impedance state. This allows the data outputs of several memory chips to be directly tied together so that 2 K × 8, 4 K × 8, and so on, memories can be constructed from columns of 1 K × 1 chips with each column contributing one bit of the data byte. When this is done, the bit being output not only depends on the signals on the address lines, but also depends on which chips receive the chip enable signal. Each row in the array is connected to a row enable line and the row enable lines are activated by higher-order address bits, which for this example are bits A_0 and A_9. When a row is selected each device in the row will input or output a bit according to the signals on the A_9-A_0 lines. In summary, if the address contains 16 bits, A_{15}-A_{12} would select the module. A_{11} and A_{10} would select the row, and A_9-A_0 would select the bits in the devices which constitute the addressed byte.

Because each cell in a static RAM device must contain several transistors, each device generally contains fewer cells than a dynamic RAM chip of comparable design. Also, they tend to use more power since one of the two cross-coupled transistors is always on, thus consuming power continuously. The major advantage of static RAM is that it does not need to be refreshed.

As with most other digital devices, it is easy to connect semiconductor memory devices together because they have built-in supporting electronics. However, the timing constraints of the input signals are critical and the timing requirements vary from one device to another. To ensure proper operation, the control logic on a memory board must provide address inputs and control signals that satisfy the timing parameters as specified by the manufacturer of the memory devices being used. The input timing during a memory read operation is different from the input timing during a memory write operation.

The most important timing parameter to be considered in choosing a memory device is the *access time.* The maximum, time delay from an address input to a data output is longer than the delay between a chip enable and a data output, and consequently the former timing figure is normally considered to be the access time. The access time for commonly used MOS RAMs varies from 50 to 500 ns.

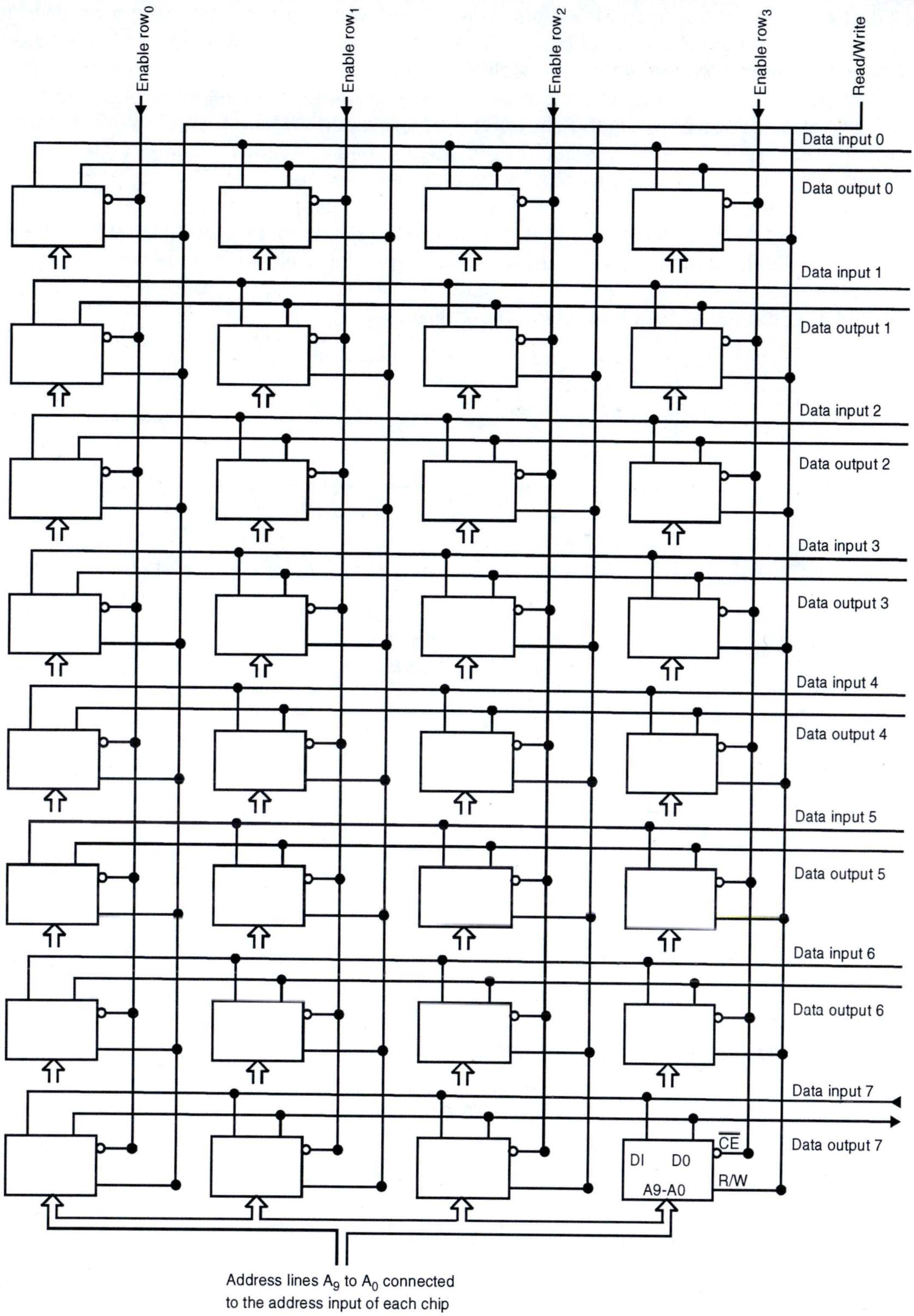

Fig. 5.24. 4k × 8 mem. device array constricted from 1k × 1 devices

For a read operation, once the output data is valid, the address input cannot be changed immediately to start another read operation. This is because the device needs a certain amount of time, called *read recovery time,* to complete its internal operations before the next memory operation. The sum of the access time and read recovery time is the *memory read cycle time.* This is the time needed between the start of a read operation and the start of the next memory cycle. The *memory write cycle time* can be similarly defined and may be different from the read cycle time. Figure 5.25(a) illustrates the timing of a memory read cycle. The address is applied at point A, which is the beginning of the read cycle, and must be held stable during the entire cycle. In order to reduce the access time, the chip enable input should be applied before point B. The data output becomes valid after point C and remains valid as long as the address and chip enable inputs hold. The R/W control input is not shown in the timing diagram for the read cycle, but should remain high throughout the entire cycle.

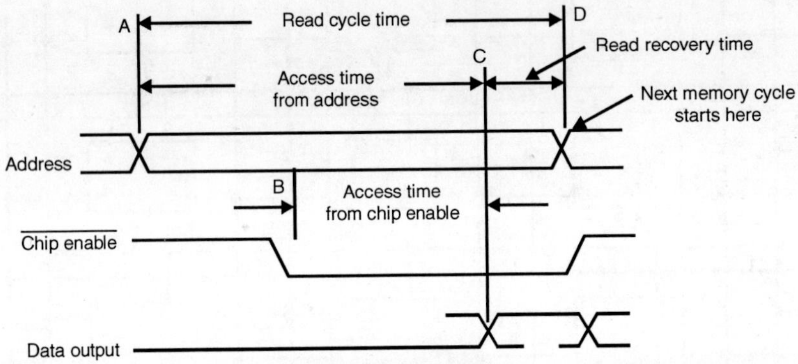

(a) Read Cycle.

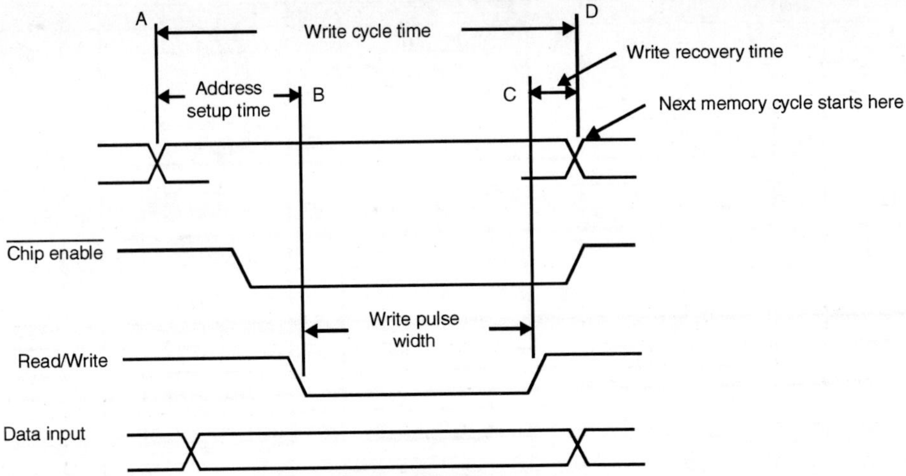

(b) Write Cycle.

Fig. 5.25.

A typical write cycle is shown in Fig. 5.25(*b*). In addition to the address and chip enable inputs, an active low write pulse on the R/W line and the data to be stored must he applied during the write cycle, The timing of data input is less restrictive and can be satisfied simply by holding the data input stable during the entire cycle. However, the application of the write pulse has two critical timing parameters: the address setup time and the write pulse width. The *address setup time* is the time required for the address to stabilize and is the time that must elapse before the write pulse can be applied. In Fig. 5.25(*b*) the address setup time is the time interval between points *A* and *B*. The *write pulse width* defines the amount of time that the write input must remain active low. The *write cycle time* is the time interval between points A and D and is the sum of address setup time, write pulse width, and write recovery time. Both the read and write recovery times may be zero for some memory devices.

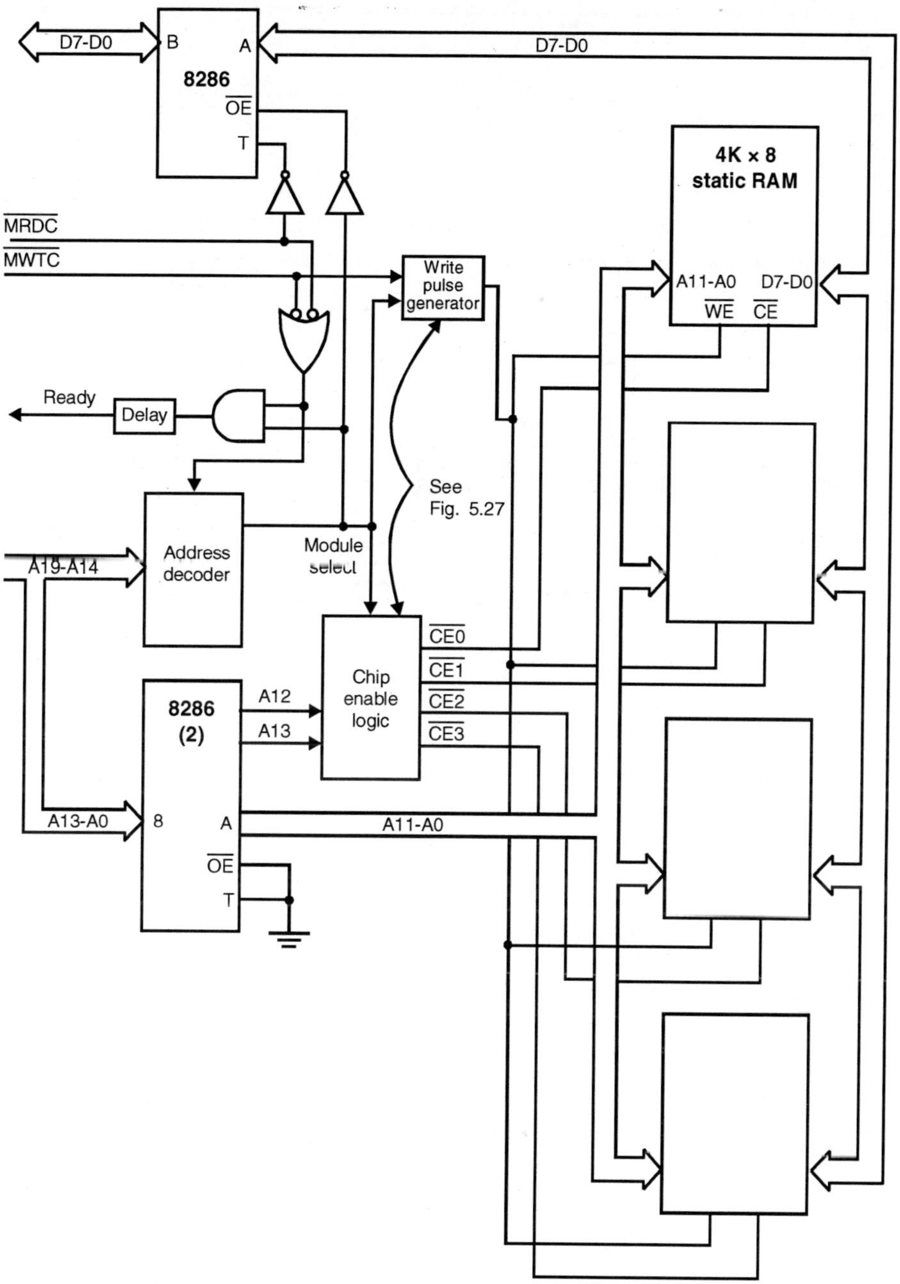

Fig. 5.26. 16k × 8 SRAM module for Mx mode.

It is important to note that the access time and cycle time discussed in this section are the minimum timing requirements for the memory devices themselves. The access time and cycle time for the memory system as a whole are considerably longer because of the delays resulting from the I/O control logic, system bus logic, and memory interface logic.

An example of the design of a 16 K × 8 static RAM memory module for a maximum mode 8088 is shown in Fig. 5.26. It is assumed that the CE and WE inputs, and the D7-D0 pins of the 4 K × 8 static RAM, have the following relationships functions :

$\overline{CE}$	$\overline{WE}$	Function	D7-D0
1	×	Not selected	High-impedance state
0	1	Read	Data put on lines
0	0	Write	High impedance and data are taken from lines

where 1 is high and 0 is low. When a memory device is not selected (i.e., $\overline{CE}$ is high) it will go into a standby state which permits it to operate at reduced power.

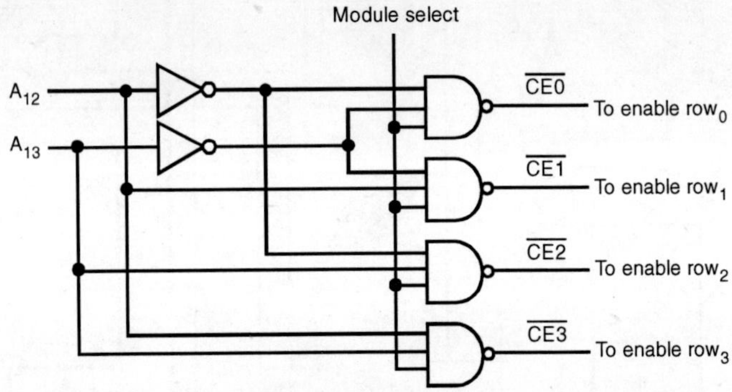

(a) Chip enable logic.

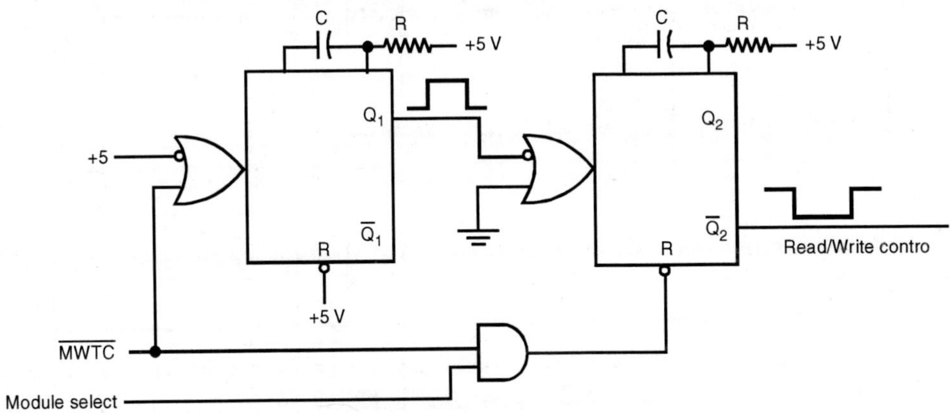

(b) Write pulse generator.

Fig. 5.27.

The incoming address bus splits into two parts, with lines A_{19}-A_{14} being used to select the module. A_{13} and A_{12} are input to the chip enable logic, which is detailed in Fig. 5.27(a). The chip enable logic has four outputs, $\overline{CE0}$ through $\overline{CE3}$, only one of which may be active at any one time, $\overline{CE0}$ is connected to the Chip Enable ($\overline{CE}$) input of the memory device associated with the lowest 4K addresses and is active when $A_{13} = A_{12} = 0$. Similarly, $\overline{CE1}$ is active when $A_{13} = 0$ and $A_{12} = 1$, $\overline{CE2}$ is active when $A_{13} = 1$ and $A_{12} = 0$, and $\overline{CE3}$ is active when $A_{13} = A_{12} = 1$. In all cases the module select line must be 1 before a $\overline{CE}$ output can be active. The module select line is active only when the module is selected and a memory read or write is recognized. The address lines A_{11}-A_0 are applied to the A_{11}-A_0 pins of all of the memory devices.

$\overline{MWTC}$ and the module select line are input to a write pulse generator which is constructed from two monostable multivibrators and is detailed in Fig. 5.27(*b*). The output of this circuit is connected to the Write Enable ($\overline{WE}$) pins of all the memory devices and causes the data on D_7-D_0 to be put into the addressed byte.

Note that in the device array shown in Fig. 5.26 there are separate data-in and data-out lines emerging from each device, but in Fig. 5.27, the data lines are bidirectional. Internally, there must be separate lines for reading from and writing to the cells, but it is possible to separate the bidirectional signals into read and write signals inside the device.

Dynamic RAM Devices

Just like static RAMs, the memory on a dynamic memory chip is organized in a matrix formed by rows and columns of memory cells. The simplest type of dynamic RAM cell contains only one transistor and one capacitor, as shown in Fig. 5.28. Whether a 1 or 0 is contained in a cell is determined by whether or not there is a charge on the capacitor. During a read operation, one of the row select lines is brought high by decoding the row address (low-order address bits). The activated row select line turns on the switch transistors Q for all cells in the selected row, This causes the refresh amplifier associated with each column to sense the voltage level on the corresponding capacitor and interpret it as a 0 or a 1. The column address (high-order address bits) enables one cell in the selected row for the output. During this process, the capacitors in an entire row are disturbed. In order to retain the stored information, the same row of cells is rewritten by the refresh amplifiers. A write operation is done similarly except that the data input is stored in the selected cell while the other cells in the same row are simply refreshed.

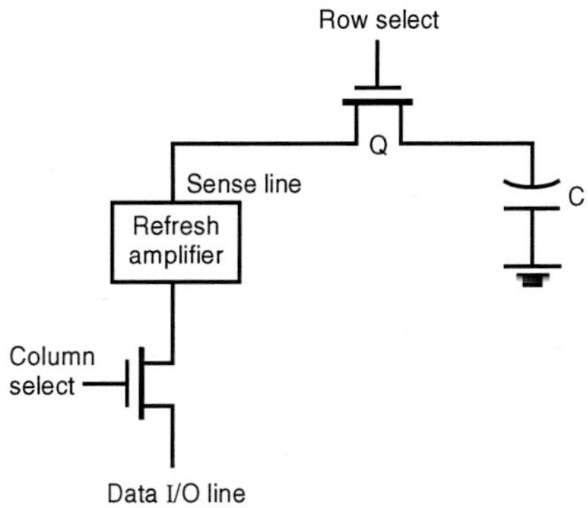

Fig. 5.28. Dynamic RAM Cell

As a result of the storage discharge through pn-junction leakage current, dynamic memory cells must be repeatedly read and restored; this process is called *memory refresh.* The storage discharge rate increases as the operating temperature rises, and the necessary time interval between refreshes ranges from 1 to 100 ms. When operating at 70°C, the typical refresh time interval is 2 ms. Although a row of cells is refreshed during a read or write, the randomness of memory references cannot guarantee that every word in a memory module is refreshed within the 2 ms time limit. A systematic way of accomplishing a memory refresh is through memory refresh cycles.

In a *memory refresh cycle,* a row address is sent to the memory chips, and a read operation is performed to refresh the selected row of cells. However, a refresh cycle differs from a regular memory read cycle in the following respects:

1. The address input to the memory chips does not come from the address bus, Instead, the row address is supplied by a binary counter called the *refresh address counter.*

This counter is incremented by one for each memory refresh cycle so that it sequences through all the row addresses. The column address is not involved because all elements in a row are refreshed simultaneously.

2. During a memory refresh cycle, all memory chips are enabled so that memory refresh is performed on every chip in the memory module simultaneously. This reduces the number of refresh cycles. In a regular read cycle, at most one row of memory chips is enabled.
3. In addition to the chip enable control input, normally a dynamic RAM has a data output enable control. These two control inputs are combined internally so that the data output is forced to its high-impedance mode unless both control inputs are activated. During a memory refresh cycle, the data output enable control is deactivated. This is necessary because all the chips in the same column are selected and their data outputs are tied together. On the other hand, during a regular memory read cycle, only one row of chips is selected; consequently, the data output enable signal to each row is activated.

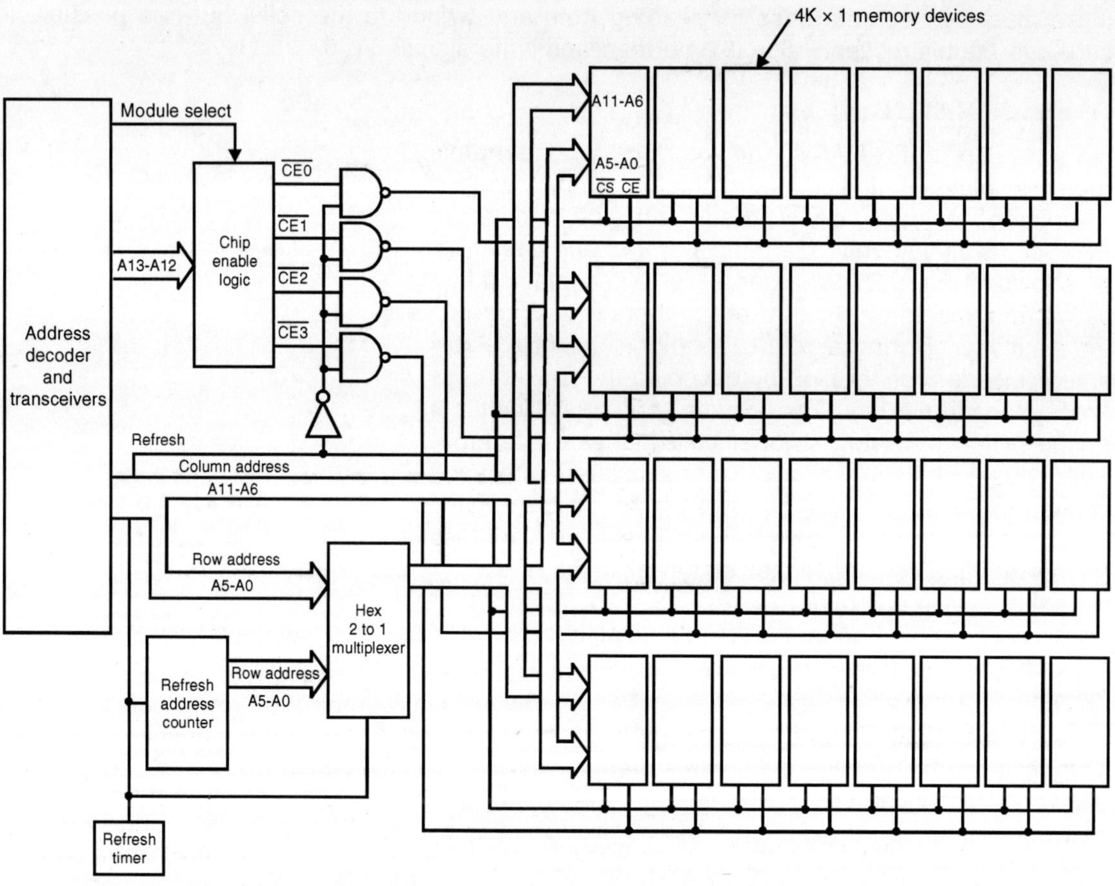

Fig. 5.29. Generation of Chip enable and refresh address signals

Consider a memory module of 16 K bytes implemented by 4 K × 1 dynamic RAMs. The memory device array consists of four rows and eight columns. Each chip has 64 rows and 64 columns of memory cells and has separate row address (6 bits) and column address (6 bits) pins. It is assumed that the chip enable and output enable pins are $\overline{CE}$ and $\overline{CS}$, respectively. The block diagram in Fig. 5.29 shows the logic needed to generate the chip enable and the refresh address signals during a memory refresh. The memory refresh cycle signal, which indicates that a refresh is in progress, can be generated by a refresh cycle timing generator in the memory module. If the current cycle is a refresh cycle, the 2-to-1 multiplexer selects the row address from the refresh address counter; otherwise, the row address comes from the address bus. Assuming that every cell in the chip array must be refreshed within a 2-ms time interval, a refresh cycle is required for every $2 \times 10^{-3}/64 = 31.25$ μs. At the end of each refresh cycle, the binary counter is incremented by 1 so that it points to the next row to be refreshed. During a refresh cycle, all chips on the module are enabled for performing a read operation by activating the chip enable signals. Their data outputs are forced into the high-impedance state by deactivating the output enable signals.

In addition to requiring refresh logic, a chief disadvantage in using dynamic RAMs is that during a refresh cycle the memory module cannot initiate a read or write cycle until the refresh cycle has been completed. As a result, a read or write request will take up to twice as long to complete if a refresh is in progress. Assuming a cycle time of 4(K) ns (for all memory cycles—refresh, read, write),

$$\frac{64 \times 400 \times 10^{-9}}{2 \times 10^{-3}} \times 100\% = 1.28\%$$

of the memory time is spent on refresh.

There are several reasons why dynamic RAMs are attractive to memory designers, especially when the memory is large. Three of the main reasons are :

1. High Density—For static RAM, a typical ceil requires six MOS transistors. The structure of a dynamic cell is much simpler and can be implemented with three, or even one MOS transistor. As a result, moro momory cells can be put into a single chip and the number of memory chips needed to implement a memory module is reduced. A common size for a dynamic RAM chip is 16K × 1, and 64K × 1 devices are also available.

2. Low Power Consumption—The power consumption per bit of dynamic RAM is considerably lower than that of static RAM. The power dissipation is less than 0.05 mW per bit for dynamic RAM and typically 0.2 mW per bit for static RAM. This feature reduces the system power requirements and lowers the cost. In addition, the power consumption of dynamic RAM is extremely low in standby mode; this makes it very attractive in the design of memory that is made nonvolatile through the use of a backup power source.

3. Economy—Dynamic RAM is less expensive per bit than static RAM. However, dynamic RAM requires more supporting circuitry, and, therefore, there is little or no economic advantage when building a small memory system.

For higher-density RAM, the row address and column address normally share the same set of pins, thus reducing the device's pin count. For these RAMs some manufacturers have produced single ICs that include the refresh support logic and other logic needed in controlling the row/column address pins. Towards this end, Intel has made available its 8203 dynamic RAM controller, which is specifically designed to support its 2117, 2118, and 2164 dynamic RAM memory devices. Here we will concentrate on the 8203's use with the 2164, which is a 64 K × 1 device. The block diagrams for the 2164 and 8203 are shown in Fig. 5.30. (The 8203 can be put in one of two modes according to its $16K/\overline{64K}$ pin; only the assignments for the 64K mode are shown.) The 2164 has four 128 × 128 cell arrays but only eight address pins, A_7-A_0. This means that the row address and the column address must share the same pins and be received one after the other. The row address is strobed by a negative-going pulse on the $\overline{RAS}$ pin and, with $\overline{RAS}$ still low, the column address is strobed by a negative-going pulse on the CAS pin. The most significant row and column address bits specify one of the four cell arrays.

During a memory refresh cycle, the address input A7 is not used and all four cell arrays perform their refreshes simultaneously. This allows the entire device to be refreshed in 128 cycles.

The timing diagrams for the read, write, and refresh-only cycles are shown in Fig. 5.31. For a read cycle, $\overline{\text{WE}}$ must be inactive before the $\overline{\text{CAS}}$ pulse is applied and remain inactive until the $\overline{\text{CAS}}$ pulse is over. After the column address is strobed. $\overline{\text{RAS}}$ is raised and with $\overline{\text{RAS}}$ high and $\overline{\text{CAS}}$ low the data bit is made available on DOUT, For a write cycle the DIN signal should be applied by the time $\overline{\text{CAS}}$ goes low, but after the $\overline{\text{WE}}$ pin goes low. The write is performed through the DIN pin while $\overline{\text{RAS}}$, $\overline{\text{CAS}}$ and $\overline{\text{WE}}$ are all low. The DOUT pin is held at its high-impedance state throughout the write cycle. For the refresh-only cycle only the row address is strobed and the $\overline{\text{CAS}}$ pin is held inactive. The DOUT pin is kept in its high-impedance state.

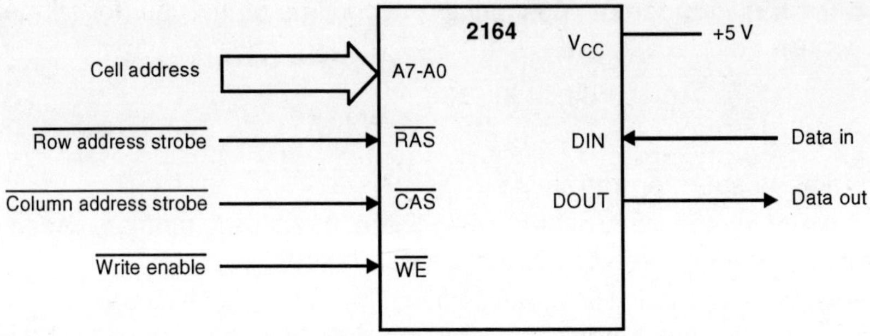

(a) 2164

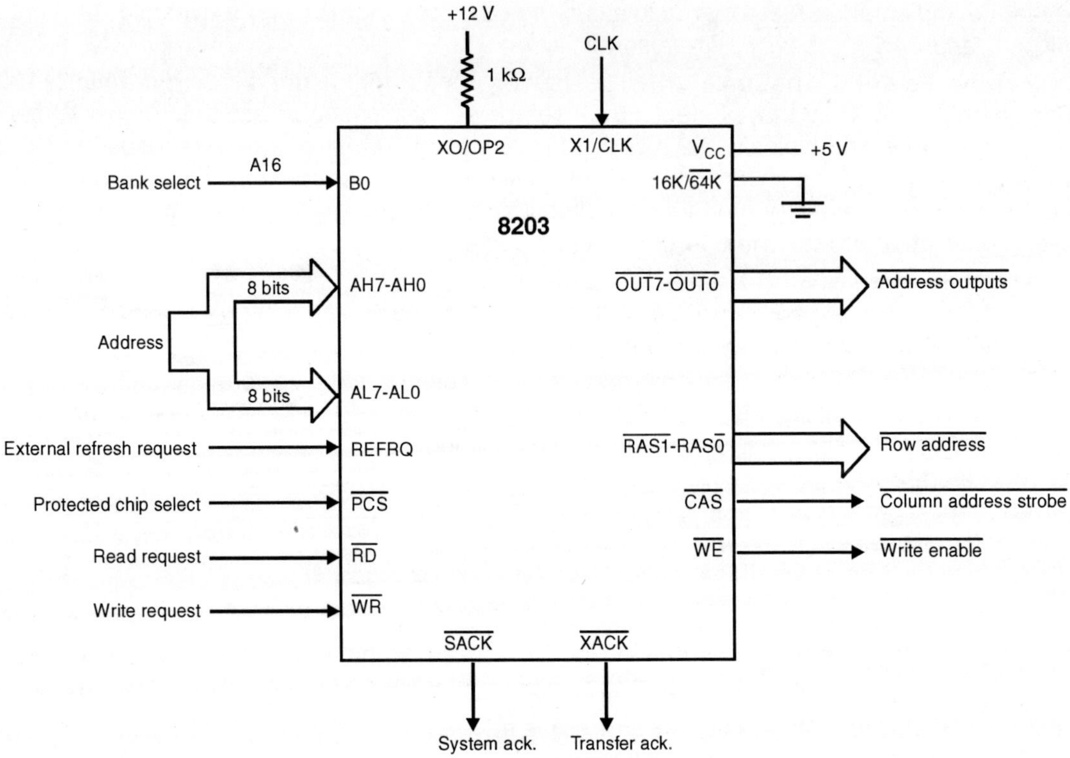

(b) 8203

Fig. 5.30.

The 8203 is designed to output signals whose timing meets the 2164 requirements. The $\overline{OUT7}$-$\overline{OUT0}$ lines provide the properly sequenced row and column addresses, $\overline{RAS1}$-$\overline{RAS0}$ provide the row addres strobes for up to two banks of 2164s, and $\overline{CAS}$ and $\overline{WE}$ supply the column address strobe and write enable signals for all of the 2164s in the module. (Note that the addresses output by the 8203 are inverted. This does not cause a problem, but it does mean that all 0s on the address lines will access the cells having row and column addresses that are all 1s.)

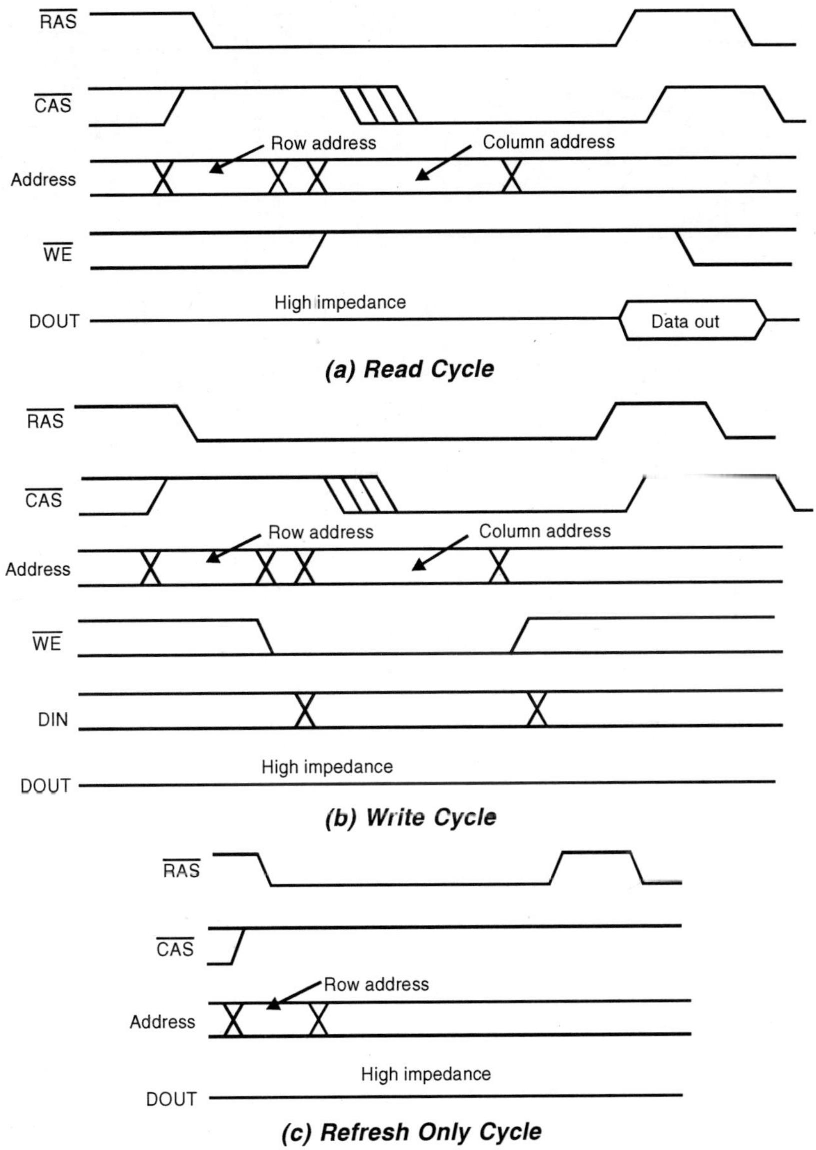

Fig. 5.31. Timing diagrams for the 2164.

The bank select input B0 determines the $\overline{RAS}$ pin to be activated. AL7-AL0 are used to generate the row address and AH7-AH0 the column address. Normally, the refresh cycle timing is generated inside the 8203, but the REFRQ pin allows the refresh cycles to be initiated by an external source. The module selection is through the $\overline{PCS}$ pin. It is called the protected chip select pin because once it becomes active the memory cycle cannot be aborted, even if it immediately returns to its inactive state. The $\overline{RD}$ and $\overline{WR}$ inputs specify whether a memory read or write is to be conducted.

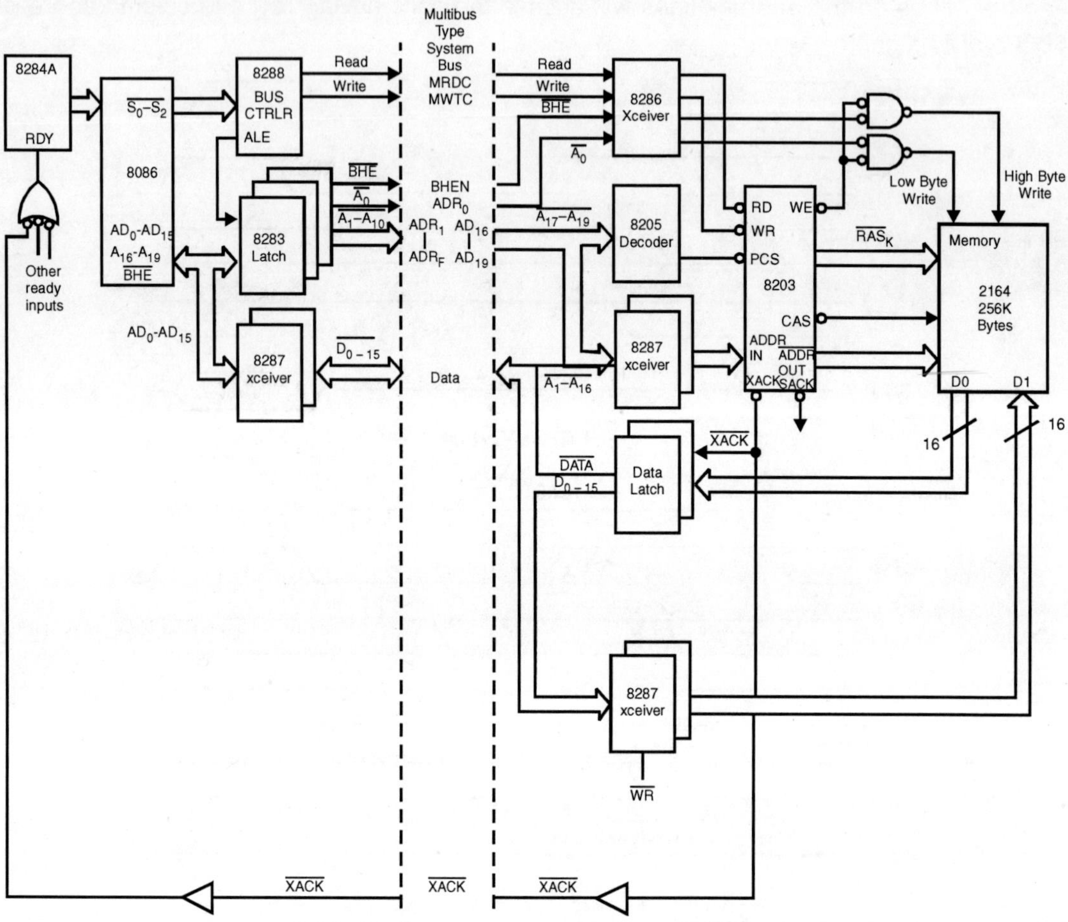

Fig. 5.32. Construction of a 256 K-byte mem. module for MX mode.

The $\overline{XACK}$ output is a strobe indicating that data are available during a read cycle or that the data have been written during a write cycle. It can be used to strobe data into data output latches and to send the ready signal to the processor. The $\overline{SACK}$ output signals the beginning of a memory access cycle and if a refresh cycle is taking place when a memory request is made, the $\overline{SACK}$ signal is delayed until the read or write cycle begins. If the memory device access time is known to be sufficiently low to guarantee that a read will be complete by the end of the T_3 clock cycle or a write will be complete by the end of the T_4 cycle, the $\overline{SACK}$ output may be used as the ready signal instead of the $\overline{XACK}$ output, thus saving wail states that might occur if $\overline{XACK}$ is used.

Either an oscillator must be connected across X0 and X1 or, if OP2 is connected to +12 V, an external clock signal must be applied to CLK. This signal may come from a bus clock line or a clock included in the memory module. Only + 5 V is needed for the main power supply, but if the OP2 input is used, a +12 V voltage is required. (REFRQ is actually a dual-purpose pin that can-be used for advanced reads, but this feature as well as the read-modify-write cycle will not be considered here.)

Figure 5.32 illustrates how an 8203 and thirty-two 2164s could be used to construct a 256 K-byte memory module for a maximum mode 8086 MULTIBUS system. This is an Intel design which assumes that the data and address buses are inverted and therefore uses 8283s and 8287s to interface to these buses instead of 8282s and 8286s. The memory device array has 16 columns so that words can be accessed. This requires that the $\overline{BHE}$ and A0 bus lines be used in conjunction with the $\overline{BHE}$ signal to determine whether only the low-order byte, only the high-order byte, or an entire word is being written.

Another way of reducing the number of chips needed in the support circuitry for dynamic RAM is to put a set of refresh logic on each memory device, thus permitting the device to refresh itself. Such a device is called an integrated RAM and, except for memory accesses sometimes being held up by refresh cycles, the device appears to the user to be a static RAM. An example of this approach is the Intel 2186/7, which is an 8 K × 8 integrated RAM. The 2186/7 has pin assignments that are essentially the same as for Intel's static RAM devices. In particular, it has $\overline{OE}$, $\overline{WE}$, and $\overline{CE}$ pins which serve the same purposes.

ROM Devices

ROM is such that once its contents are set they cannot be changed, at least not without the aid of special equipment. Its cell density can be higher than that of read/write memory because its construction is simpler (there is no need for write circuitry). It is also nonvolatile and reliable. However, it is restricted to use where writing is not necessary. As noted in the introduction to this chapter, most memory systems contain both ROM and RAM modules, with the ROM modules being used to store such things as bootstrap loaders and permanent data tables. ROM is also sometimes used to hold the resident monitor and other suitable parts of the operating system (thus obviating the need for a bootstrap loader). It may even contain permanently installed language interpreters.

There are basically four types of ROMs, the four types being distinguished by the way in which their contents are set, Setting the contents of a ROM is sometimes called programming, but should not be confused with the generation of instruction sequences discussed in the previous chapters. In one type of ROM the contents are determined by a masking operation that is performed while the chin is being manufactured. Such chips cannot be altered by the user and are referred to simply as ROMs. The contents of the second type can, if the proper equipment is available, be set by the user. They are called *programmable-read-only memories* (PROMs). As with the masked ROMs-, once a memory of this type is programmed, its contents can never be changed. The third and fourth types can not only be programmed by the user, but also by using special equipment can be erased and reprogrammed many times. They are called *erasable-programmable-read-only memories (EPROMs)* and *electrically-erasable-read-only memories (E^2ROMs),*

Because a masked ROM requires the manufacturer to first produce a mask according to a given set of contents, or *program,* it is expensive to produce the first ROM, but additional copies are inexpensive. Therefore, ROMs are used only after the development stage of a computer system is complete, when it is not likely that the contents of the ROMs will need to be changed. The only pins required on a ROM device are a set of address input pins, a set of data output pins, and a chip select pin. Figure 5.33 shows how two 2K × 8 ROMs could be used to build a 4K × 8 byte storage module. In addition to the logic shown, the usual module select address decoder, transceivers, and latches would be required.

Normally, PROMs are constructed of diode matrices. They are programmed by using external pins to select the diode links that are to be "burned" or "blown," thus causing the diode matrix to be permanently programmed. The Is are represented by the diode links left intact and the 0s by the diode links that are blown. Because the PROMs for all customers are identical when they are shipped, their price is not as dependent on quantity, but their circuitry is more complex and expensive due to their programmability. Therefore, for a few devices PROMs are cheaper than masked ROMs, but for large-quantity orders masked ROMs are cheaper. As a result, PROMs (or EPROMs) are used during development, but masked ROMs are used in the final mass-produced products.

PROMs are programmed bit by bit by placing the address of the location containing the bit on the address pins and forcing a current *into* the bit's data output pin while the voltage supply and appropriate control pins are pulsed. The magnitude of the current, the magnitudes and width of the pulses, and the manner in which the pulses are applied varies from one PROM to the next. Generally, the programming cycles are interspersed with verification cycles and the cycles are alternated for approximately twice the time needed to program the bit.

As opposed to a PROM, in which the contents are permanently programmed by "burning" out diode links, the contents of an EPROM are determined by charge distribution. EPROMs are programmed by charge injection, and once programmed the charge distribution is maintained until it is disturbed by some external energy source such as an ultraviolet light. Instead of completely ercasing an EPROM chip in a protective package as is done with other ICs, an EPROM is constructed by putting a quartz window over its circuitry, which will allow the external energy to pass. By exposing the memory to the external energy source for several minutes (10 to 50 minutes, depending on the specific device), the charge is redistributed to its natural state, thus destroying the old memory contents. The EPROM can then be reprogrammed.

As with PROMs, EPROMs are normally used during the development of a product and are replaced by masked ROMs once the design is complete. Because they can be erased, they have the advantage of being reusable. However, unlike PROMs, EPROMs may lose their contents over extended periods of time and should not he put in final products that are expected to last for years. How long one can depend on an EPROM retaining its contents depends on the environmental conditions it is subjected to, and may vary from a few months to several years.

An EPROM is programmed by applying an address to the address pins and high or low voltages to all of the data output pins, and then applying the proper voltages and pulses to the supply and control pins. As an example, to program an Intel 2764 PROM (8K × 8), 21 V is applied to the V_{pp} pin while the $\overline{CE}$ pin is held low. The address of the byte to be programmed is put on pins A12-A0 and the data byte is simultaneously applied to O7-O0. The data byte is then written into the addressed byte by pulsing $\overline{PGM}$ with +5 V. The contents of each byte should be verified after it has been written.

Even PROM and EPROM memory devices made by one manufacturer may have programming signal specifications that vary considerably. Particularly in the case of PROM programming, the programming specifications may be complex and must be followed precisely. Therefore, many manufacturers have made available units, called *PROM programmers,* especially designed to

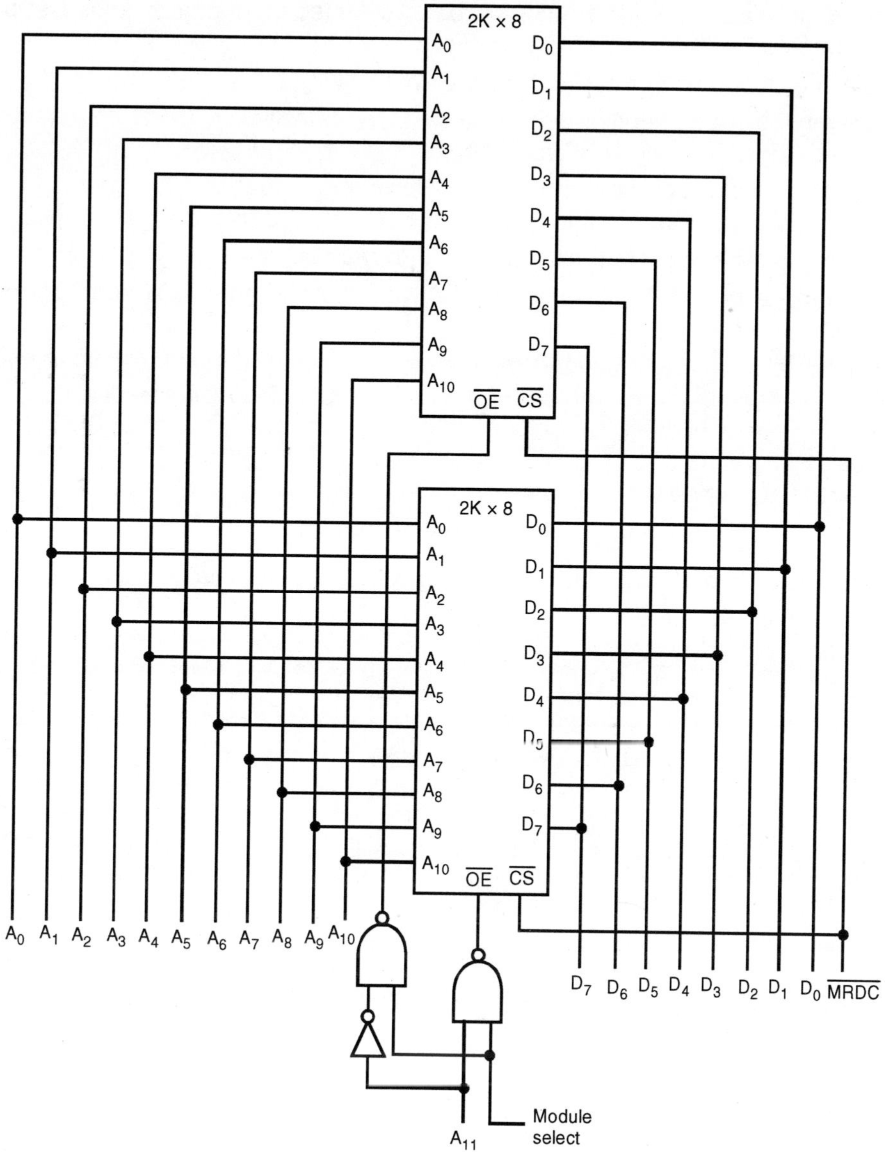

Fig. 5.33. Storage module of 4K × 8 build using 2K × 8 ROMs.

program PROMs and HPROMs. To accommodate the varying electrical requirements, some programmers are built to accept circuit modules which provide customized electrical outputs. Each module is called a *personality card* and provides all the electrical signals needed for programming a given type of memory chip. In addition to personality cards for the various PROMs and EPROMs, a PROM programmer may have a control module which includes a microprocessor. The control module permits a block of data from a microcomputer development system (MDS) to be written into a PROM (or EPROM) with a single command. A programmer may also include a display for verifying the contents of a PROM. In conjunction with the Intel Universal PROM programmer, commands are available under the ISIS-II operating system for performing the following operations:

1. Loading the data to be programmed from a selected input device (disk file, paper tape or system console) into the MDS memory.
2. Displaying or changing data in the MDS memory.
3. Programming a segment of a PROM with the data which are stored beginning at a specified address in the MDS memory.
4. Transferring a block of data in a PROM into memory so that the contents of the PROM may be examined through the system console or used to produce a duplicate PROM.
5. Transferring a block of data from a PROM into a disk file.
6. Comparing a block of data in a PROM with the contents of a segment of memory (program verification).

An E^2ROM is also reprogrammable, but offers the added advantage that each individual byte can be easily erased and reprogrammed. The principal disadvantage of an E^2ROM is that it tends to be more expensive.

I/O Interfacing Method

There are two types of I/O interfacing :

(1) I/O mapped I/O or accumulator I/O or peripheral I/O or isolated I/O.

(2) Memory mapped I/O

I/O mapped I/O	Memory mapped I/O
1. I/O mapped I/O is not treated as memory location.	Memory mapped I/O is treated as memory location.
2. It requires special instruction like IN/OUT to access I/O module.	It does not required any special instruction. Microprocessor can access I/O module by memory instruction. MOV AL, [BX]
3. It requires special instruction signals IO/W.	In does not require special instruction signals, $\overline{MEMR}, \overline{MEMW}$ signal can be used to access I/O module.
4. In I/O mapped I/O 8086 can access 64 K I/O parts.	In memory-mapped I/O, 8086 can access IMB memory locations or 1 MB I/O part.
5. In 8086, I/O mapped I/O requires 16 address lines.	In 8086, memory mapped I/O module requires 20 address lines.
6. Less hardware is required to decode 16 bit address.	More hardware is required to decode 20 bit address.
7. Arithmetical or logical operation cannot directly perform with the input data.	Arithmetic or logical operations can be directly performed with I/P data.

Advantages and Disadvantages of I/O Mapped I/O

I/O mapped I/O has the following advantages :

(1) Program becomes easier since different instructions are used for I/O.

(2) These I/O type instructions are usually less than memory type instructions used in the memory-mapped I/O.

(3) The I/O instructions isolate memory and I/O areas so that memory address is not affected by the I/O.

(4) The complexity of the device address decoder depends only on the length of the device address field in the I/O type instructions.

I/O mapped I/O has the following disadvantages :

(1) It is less powerful and less flexible than memory mapped I/O.

(2) Two additional control lines are required for issuing the $\overline{IO/R}$ and $\overline{IO/W}$ signals.

Advantages and Disadvantages of Memory Mapped I/O

Memory mapped I/O provides the following advantages :

(1) The port addresses are now 16-bits wide thereby it provides more input and output ports.

(2) The same powerful memory type instructions are used to access I/O module.

(3) Many CPU registers can exchange transfer with I/O devices.

(4) For a same application, memory mapped I/O simplifies program and increases speed transfer.

(5) There is also reduction in the CPU control line.

Memory mapped I/O provides the following disadvantages :

(1) The number of I/O ports connected is limited by number of available memory locations.

(2) Memory mapped I/O utilizes memory type instructions which is usually longer than I/O instructions.

(3) The length of program is increased.

(4) Because of wider port address, the interface hardware is also increased.

Address Decoding

Microprocessor are not used by the memory module to select particular memory locations. Unused address lines are decoded to generate chip select signals. There are two types of address decoding.

(1) Fully address decoding.

(2) Partial address decoding.

	Fully address decoding	Partial address decoding
1.	In this system, all unused lines are decoded to generate chip select.	All unused lines are not decoded to generate chip select.
2.	In this system, all unused address have a particular value either 1 or 0.	In this system, the value of undecoded address bit is don't care.
3.	In this system, each location has a fixed address.	The address of location is not fixed, each location may hve one or more undecoded lines.
4.	During read, write, output, other locations are not affected.	During read, write, output, other location or its imaginary location is affected. This type of memory is called fold—back memory.
5.	It does not reduce the size of a memory module.	It reduces the size of a memory module.
6.	It is expensive because hardware is complicated.	It is inexpensive as hardware is simple.

Important Points to be Considered for Memory Interfacing

(1) Decode logic for generation of I/O and memory control signals as shown in Fig. 5.34.

(2) 8086 memory access encoding.

$\overline{BHE}$	AO		Action
0	0	D15–D0	Access 16-bit word (even)
*0	1	D15–D0	Access byte (odd address)
*1	0	D7–D0	Access byte (even address)
1	1	–	NA

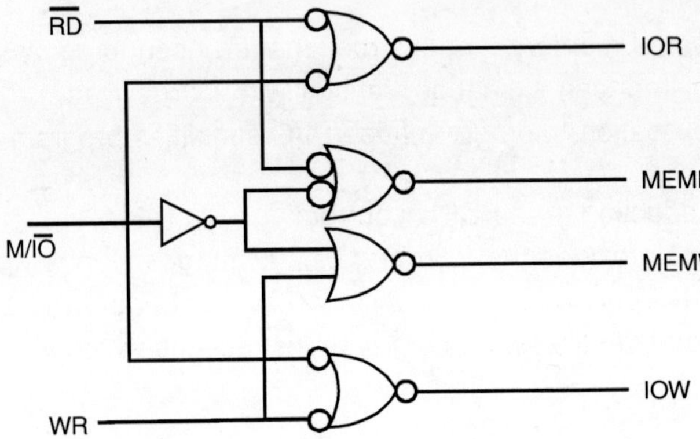

Fig. 5.34. Decoding logic for generation of I/O and memory control signals.

* In case of word at odd address.

$\overline{BHE}$	A0	
0	1	Access byte (odd address)
1	0	Access byte (even address)

(3) CPU state following reset

CPU	Contents
Flags	CLear
IP	0000
CS	FFFFh
DS	0000
SS	0000
ES	0000
Queue	Empty

After reset CS contains FFFFh and IP contains 0000H. Therefore, the physical address is FFFF0h.

Here instruction execution starts from FFFF0h which is normally a jump from some other location where a longer program resides. This program always resides in a ROM.

For example, we want to interface 4 chips of 2 k memory.

8k bytes of memory requires 13 address lines ... A0 – A12.

8 k bytes mean 01FFFh bytes.

Therefore, EPROM address starts from

FFFFFh-01FFFh = FE000H.

Thus our memory organization for the four chips of 2k will be as under :

1111 1110 0000 0000 0000	... FE000	Chip – 1
1111 1110 0111 1111 1111	... FE7FF	
1111 1110 1000 0000 0000	... FE800	Chip – 2
1111 1110 1111 1111 1111	... FEFFF	
1111 1111 0000 0000 0000	... FE000H	Chip – 3
1111 1110 0111 1111 1111	... FE7FFH	
1111 1111 1000 0000 0000	... FF800H	Chip – 4
1111 1111 1111 1111 1111	... FFFFFH	

Address range of EPROM is from FE000H to FFFFFFH.

(4) Decode the unused address lines to map the memory interface to a particular memory in the microprocessor's address space.

(5) For RAM interfacing A0 and $\overline{BHE}$ have to be used.

(6) Typoc of addrocc docoding

(i) Full (ii) Partial (iii) Block

(i) FUll decoding : All the unused lines are used in full decoding. For example, if a 16 kb EPROM is to start from 80000H, then the address range is,

1000 0000 0000 0000 0000

1000 0011 1111 1111 1111

Unused lines (1000 00) — Used Lines (remaining bits)

In the above quoted example, unused A19 to A14 lines are used to generate chip select ($\overline{CS}$) signal for EPROM as shown in Fig. 5.35(a).

(ii) **Partial decoding :** (i) All the unused lines are not used e.g. in the above case :

The EPROM chip will be enabled for addresses form 80000-83FFFH. For address 84000H, the location 80000H will be accessed and so on.

(iii) **Block decoding :** Same as full decoding except that in this case blocks of memory is enabled using the unused lines.

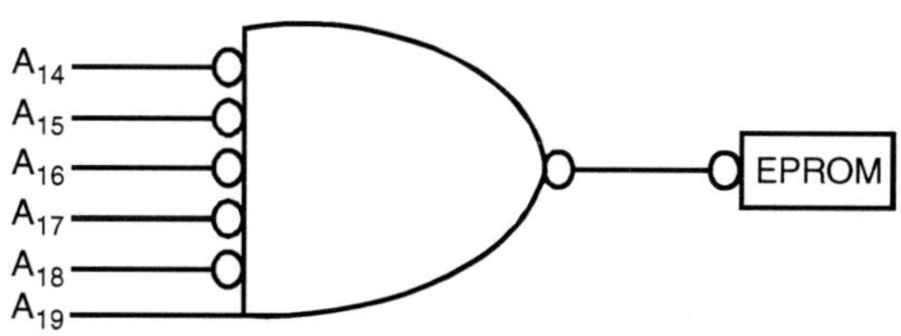

Fig. 5.35(*a*)

(7) A RAM always start after address 003FFH.

Fig. 5.35(*b*).

5.8.1. Memory Address Decoding

The processor can usually address a memory space that is much larger than the memory space covered by an individual memory chip. In order to splice a memory device into the address space of the processor, decoding is necessary. For example, the 8088 issues 20-bit addresses for a total of ***1MB*** of memory address space. However, the BIOS on a 2716 EPROM has only 2KB of memory and 11 address pins. A decoder can be used to decode the additional 9 address pins and allow the EPROM to be placed in ***any*** 2 KB section of the 1MB address space.

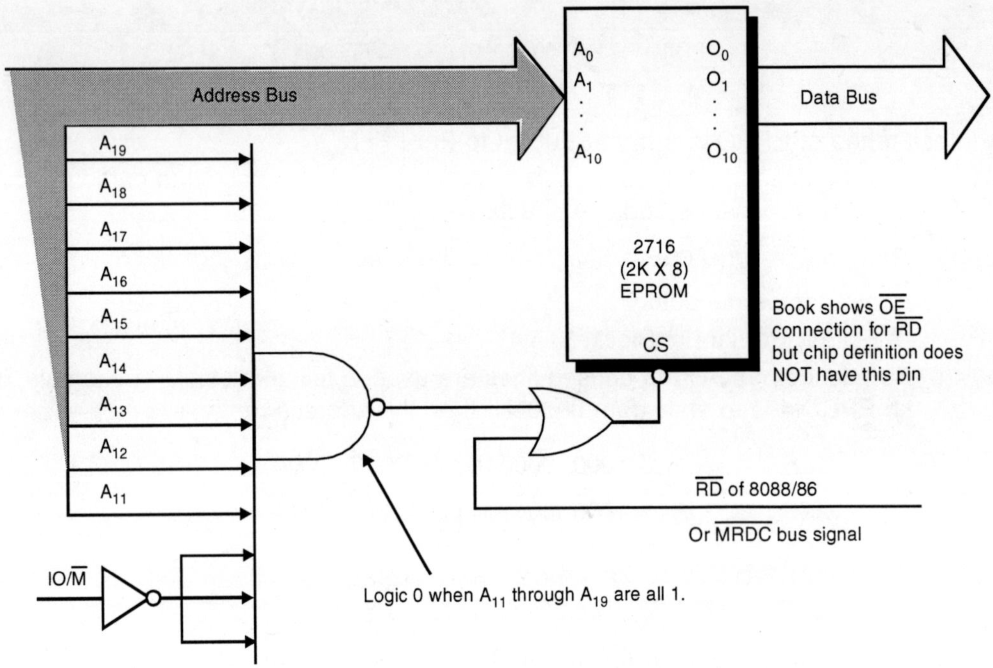

Fig. 5.36. Memory address decoding.

➤ Way to determine the address range that a device is mapped into:

$A_{19} - A_{11}$ $A_{10} - A_0$

1111 1111 1XXX XXXX XXXX

↓

1111 1111 1111 0000 0000 (FF800H)

↓ To

1111 1111 1111 1111 1111 (FFFFFH)

This 2KB memory segment maps into the ***reset location*** of the 8086/8088 (FFFF0H). NAND gate decoders are not often used. Rather the 3-to-8 Line Decoder (74LS138) is more common.

The 3-to-8 Line Decoder (74LS138)

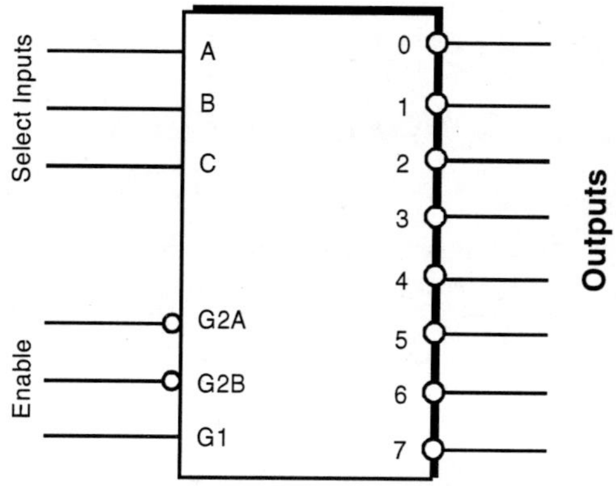

Inputs						Output							
Enable			Select										
G2A	G2B	G1	C	B	A	0	1	2	3	4	5	6	7
1	X	X	X	X	X	1	1	1	1	1	1	1	1
X	1	X	X	X	X	1	1	1	1	1	1	1	1
X	X	0	X	X	X	1	1	1	1	1	1	1	1
0	0	1	0	0	0	0	1	1	1	1	1	1	1
0	0	1	0	0	1	1	0	1	1	1	1	1	1
0	0	1	0	1	0	1	1	0	1	1	1	1	1
0	0	1	0	1	1	1	1	1	0	1	1	1	1
0	0	1	1	0	0	1	1	1	1	0	1	1	1
0	0	1	1	0	1	1	1	1	1	1	0	1	1
0	0	1	1	1	1	1	1	1	1	1	1	0	1
0	0	1	1	1	1	1	1	1	1	1	1	1	0

Fig. 5.37. Input/Output table.

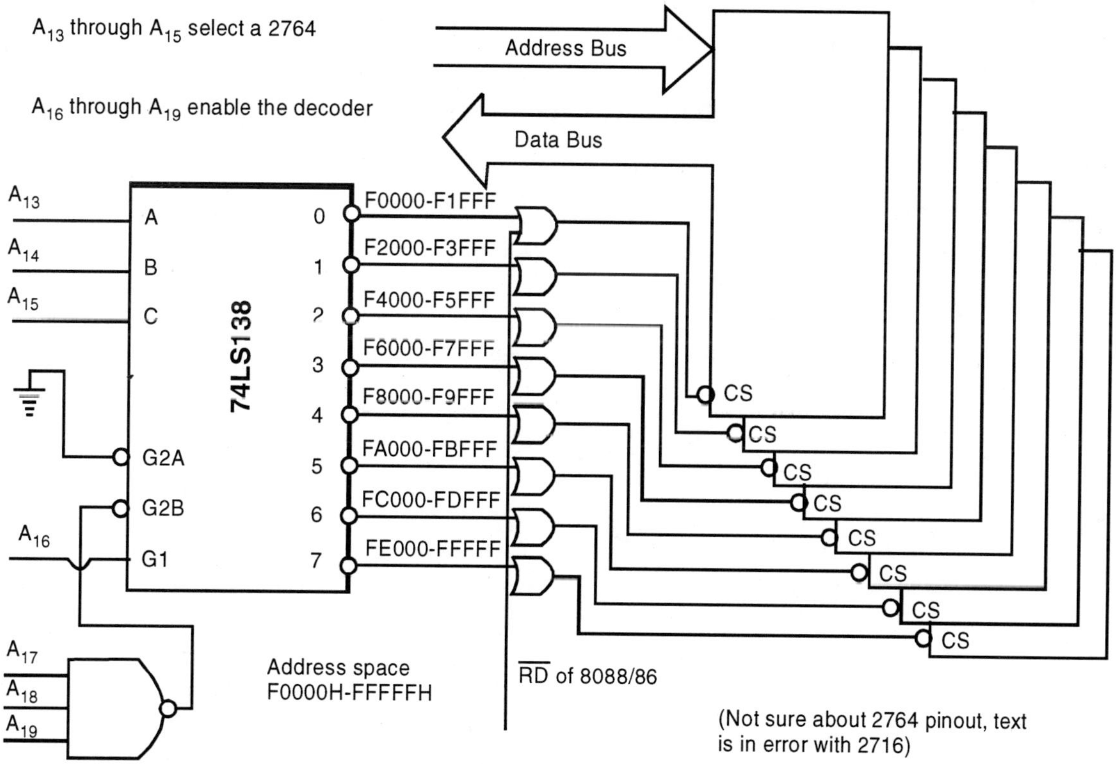

Fig. 5.38. EPROM interfacing.

➤ Note that all three Enables (G2A, G2B, and G1) must be active, e.g. low, low and high, respectively.

➤ Each output of the decoder can be attached to an 2764 EPROM (8K × 8).

The EPROMs cover a 64KB section of memory. Yet a third possibility is a ***PLD*** (Programmable Logic Device). PLDs come in three varieties:

(i) PLA (Programmable Logic Array)

(ii) PAL (Programmable Array Logic)

(iii) GAL (Gated Array Logic)

Programmed to decode address lines A19 – A13 onto 8 outputs

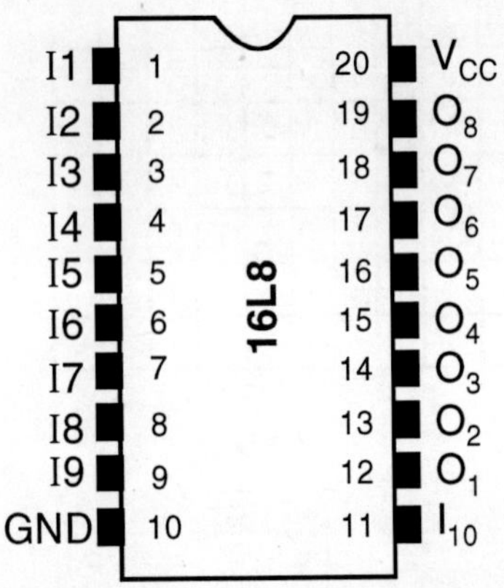


```
; pins 1   2   3   4   5   6   7   8  9  10
       A19 A18 A17 A16 A15 A14 A13 NC NC GND
; pins 11  12  13  14  15  16  17  18 19 20
       NC  O8  O7  O6  O5  O4  O3  O2 O1 VCC
```

Equations :

/O1 = A19 * A18 * A17 * A16 * /A15 * /A14 * /A13

/O2 = A19 * A18 * A17 * A16 * /A15 * /A14 * A13

/O3 = A19 * A18 * A17 * A16 * /A15 * A14 * A13

/O4 = A19 * A18 * A17 * A16 * A15 * /A14 * A13

/O5 = A19 * A18 * A17 * A16 * A15 * /A14 * /A13

/O6 = A19 * A18 * A17 * A16 * A15 * /A14 * A13

/O7 = A19 * A18 * A17 * A16 * A15 * A14 * /A13

Fig. 5.39. Decode mechanism.

PLDs have been around since the mid-1970s but have only recently appeared in memory systems (PALs have replaced PROM address decoders). PALs and PLAs are ***fuse-programmed*** (like the PROM). Some are erasable (like the EPROM). A PAL example (16L8) is shown in the Fig. 5.39 and is commonly used to decode the memory address, particularly for 32-bit addresses generated by the 80386DX and above, AMD 16L8 PAL decoder. It has 10 fixed inputs (Pins 1-9, 11), two fixed outputs (Pins 12 and 19) and 6 pins that can be either (Pins 13-18).

➤ AND/NOR device with logic expressions (outputs) with up to 16 ANDed inputs and 7 ORed product terms.

5.8.2. Semiconductor Memory Interfacing

S-RAM Interfacing. Semiconductor RAMs are basically classified into 2 categories

(a) Static RAM or (S-RAM)

(b) Dynamic RAM or (D-RAM)

Here we will consider the interfacing of static RAM and ROM with 8086 microprocessor.

The semiconductor memories are organised as twodimensional arrays of memory locations, for example 2 K × 8 or 2K byte memory or 4K X 8, 4K byte memory which contains 4096 locations, where each location contains 8-bit data. Only one of the 4096 locations can be selected at a time.

In general, to address a memory location out of 'N' memory locations, one would require at least 'n' bits of address i.e. 'n' address lines where n = $\log_2$ N.

Hence if the microprocessor has 'n' address lines, then it is able to address at most N locations of memory where $\boxed{2^n = N}$. If out of N locations only 'P' primary locations are to be interfaced, then the least significant 'P' address lines out of the available 'n' lines can be directly connected from the microprocessor 8086 to the memory chip while the remaining (n – p) higher order address lines may be used for address decoding. The output of the decoding circuit is connected with the $\overline{CS}$ pin of the memory chip.

5.8.2.1. The procedure of interfacing S-RAM with 8086 microprocessor is as given below

(1) Arrange the available memory chips so as to obtain 16-bit data bus width. The upper 8-bit bank is called as the "odd address memory bank" and the lower 8-bit bank is referred to as the "even address memory bank".

(2) Now one must connect the available memory address lines of memory chips with those of the 8086 microprocessor and connect the memory $\overline{RD}$ and $\overline{WR}$ inputs to the corresponding processor control signals.

(3) Connection of the 16-bit data bus of the memory bank with that of the microprocessor 8086 must be done.

(4) $\overline{BHE}$, A_0 and the rest of the address lines left are used for decoding the required chip select signals for the odd and even memory banks.

8086 Memory Interface

The 8086 memory interface has been discussed with the following example.

An 8086 microprocessor based Kit is to be designed with following features.

(a) 8086 CPU working at 4.77 MHz.

(b) 8087 Math co-processor is connected.

(c) 16 kb of EPROM for storing the system software.

(d) Using 8 KB of SRAM devices to have 32 KB of RAM.

(e) Two 8-bit parallel I/O ports to be available on the microprocessor kit, one for input and other for output. Design the system; give the memory map and draw a neat diagram of the whole system.

Solution. Since a math co-processor is to be connected, we have to use the 8086 in maximum mode.

To get the 8086 CPU working at 4.77 MHz, we use the 8284A clock generator with oscillator applied between X1 and X2 of the 8284A.

Memory design

For EPROM : Whenever the 8086 CPU is reset, its value is set of FFFFh and IP value is set to 0000H. This corresponds to physical address FFFF0h.

Therefore instruction execution reset starts from FFFF0h which is normamly a jump to other program. FFFF0h should contain random codes which may lead to hanging of the microprocessor. Hence FFFF0h is always a part of the ROM. The program to which the jump is made is also placed in the ROM. FFFFF0h is always a jump instruction since there are only 16 bytes from FFFF0h-FFFFFh and it is very difficult to place a single program in such a small place. From above discussion, we come to conclusion that FFFF0-FFFFFh should be always included in the ROM.

Now we want to address 16 k bytes of EPROM to 8086 microprocessor 16 k bytes means 3FFFh bytes.

Hence the EPROM memory should start from

FFFFFFh – 03FFFh = FC000h.

To address 16 k bytes, we require 14 address lines. Of the 16 k bytes, 8 k bytes will be at even addresses and 8 k bytes will be at odd addresses. Hence, we use 2 EPROM chips, each of 8 K bytes capacity. One for storing bytes at even address and other for storing odd address bytes.

A_{19}	A_{18}	A_{17}	A_{16}	A_{15}	A_{14}	A_{13}	A_{12}	A_{11}	A_{10}	A_9	A_8	A_7	A_6	A_5	A_4	A_3	A_2	A_1
1	1	1	1	1	1	0	0	0	0	0	0	0	0	0	0	0	0	0
1	1	1	1	1	1	1	1	1	1	1	1	1	1	1	1	1	1	1
used for chip select						Address within the 16K bytes												

The EPROM address range – FC000h – FFFFFh.

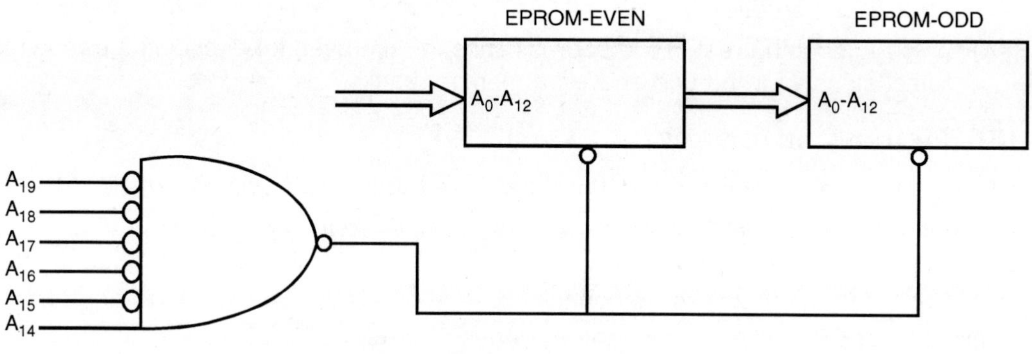

Fig. 5.40. EPROM Interfacing

As mentioned in point no. 4, A0 and $\overline{\text{BHE}}$ is not used for interfacing of EPROM. Both the 8 K bytes EPROM chips are selected whenever any address in the range FC000h–FFFFFh comes on the address bus. This can be achieved by using the fully decoding method of unused address lines. Note that A0 is not used for addressing the EPROM chips. This is because only 13 address lines are required to address an 8 K bytes memory chip. Now, a question arises in your mind is that why only A0 has to be omitted. The answer to this is that it is the A0 line which selects either even addressed chip or odd addressed chip. However, in case of EPROM chips, this not need to be considered. Bytes from both the chips are taken and only the appropriate byte mentioned in the instruction is used by the microprocessor. However this is the case only for ROM chips since only reading has to be done. The case for RAM chips is explained later. Thus we see that the microprocessor reads two bytes one each from either chip in every read cycle involving EPROM. Thus using only 13 address lines, we can able to read all the 16 K bytes.

For RAM : The interrupt vector table is stored in 00000-003FFh. Therefore RAM address can begin any where between 000400 to F0000h.

Now, we have to interface 32K bytes of RAM to the 8086 microprocessor. We will split 32 kilobytes into two blocks each of 16 k bytes. Each 16 k block contain 8 k bytes of even addressed bytes and 8 k bytes of odd addressed bytes. Therefore we will use 4 chips of 8 k bytes, 2 chips for even addressed bytes and 2 chips for odd addressed bytes. Now, as mentioned earlier, we have two memory blocks, each of 16 k bytes. This can be explain clearly, by the following figure.

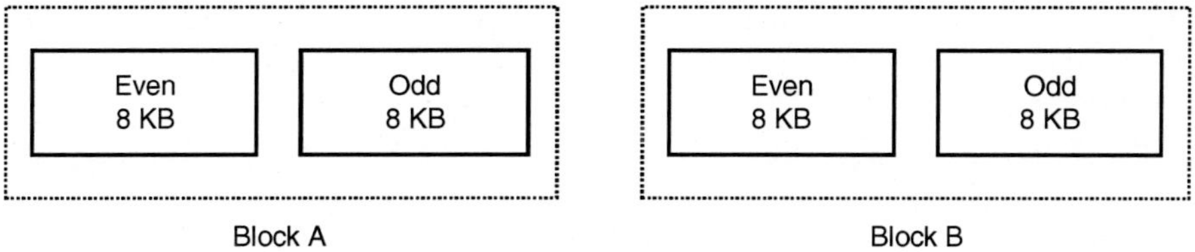

Fig. 5.41.

To acess a particular memory location in any bank, the entire block containing the bank would have to be enabled. The address range of both the blocks will have to be different. To address a particular block of 16 k bytes, 14 address lines are required. For example,

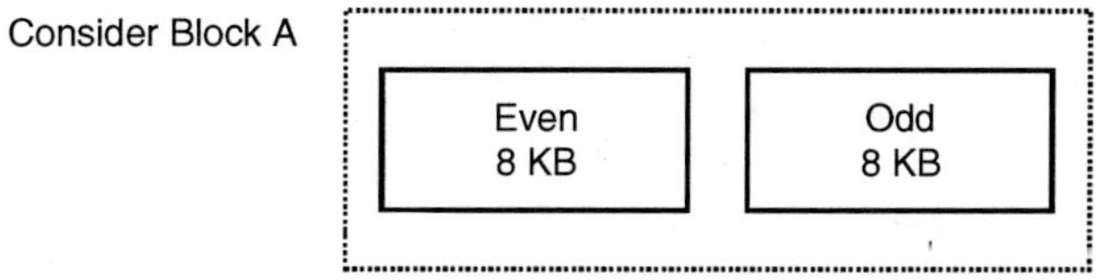

Fig. 5.42.

Now, depending upon the bit on AD0, either the even or odd bank will be selected. If AD0 = 0, the even bank is selected. Now, the $\overline{BHE}$ signal will be 0 whenever a byte or a word is being accessed at odd address. Also for an even addressed word, both the banks will have to be enabled at the same time.

From the above two points, we note that AD0 has to be given to $\overline{CS}$ of even bank and $\overline{BHE}$ has to be given to $\overline{CS}$ of odd bank

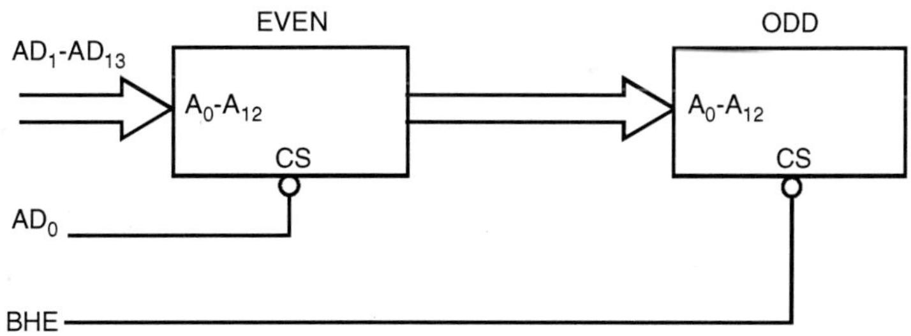

Fig. 5.43.

Thus we see that in this case also, only 12 lines are used to address each chip. However for each address coming on the address bus, number of locations accessed depends on the condition of A0 and $\overline{BHE}$.

➢ If only A0 is low, memory location from even bank is accessed.

➢ If only $\overline{BHE}$ is low, memory location from odd bank is accessed.

➢ If both are low, 2 memory locations are accessed from each bank.

Here, we see that addressing of the memory location is different as compared to that in EPROM. In EPROM, we did not have to bother about what had to selected. It was because only reading had to be performed.

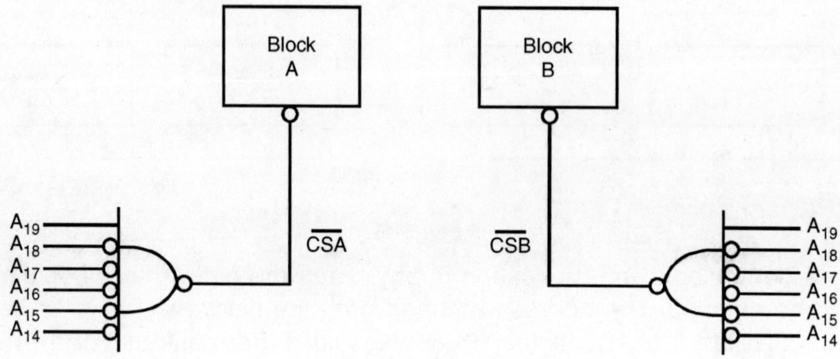

Fig. 5.44.

However in case of RAM, writing also has to be performed. Now let's find out the address for block A and block B. 16 K bytes require 14 address lines.

We will start the RAM address from 80000H

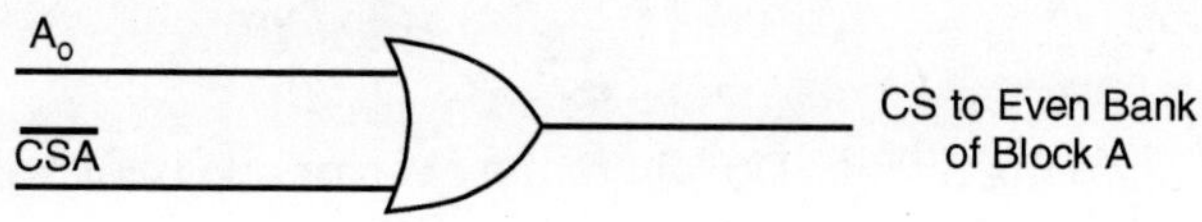

Fig. 5.45.

A_{19}	A_{18}	A_{17}	A_{16}	A_{15}	A_{14}	A_{13}	A_{12}	A_{11}	A_{10}	A_9	A_8	A_7	A_6	A_5	A_4	A_3	A_2	A_1
1	0	0	0	0	0	0	0	0	0	0	0	0	0	0	0	0	0	0
1	0	0	0	0	0	1	1	1	1	1	1	1	1	1	1	1	1	1

Block - A

A_{19}	A_{18}	A_{17}	A_{16}	A_{15}	A_{14}	A_{13}	A_{12}	A_{11}	A_{10}	A_9	A_8	A_7	A_6	A_5	A_4	A_3	A_2	A_1
1	0	0	0	0	1	0	0	0	0	0	0	0	0	0	0	0	0	0
1	0	0	0	0	1	1	1	1	1	1	1	1	1	1	1	1	1	1

Block - B

Block A 80000-83FFFh
Block B 84000-87FFFh

Now from the table we see that for block A, $A_{19} - A_{14}$ remains 100000 and for block B $A_{19} - A_{14}$ remains 100001.

This means that whenever $A_{19} - A_{14}$ is 10000, Block A is selected and whenever $A_{19} - A_4$ is 10001, block B is selected.

Note that $\overline{CS}$ signal for each block is generated using fully decoding method. Now a particular bank will be selected if and only if $\overline{CS}$ for the bank as well as $\overline{CS}$ for the block containing the bank low. For example, an even bank will be selected within block if A0 as well as $\overline{CS}$ for block A goes low.

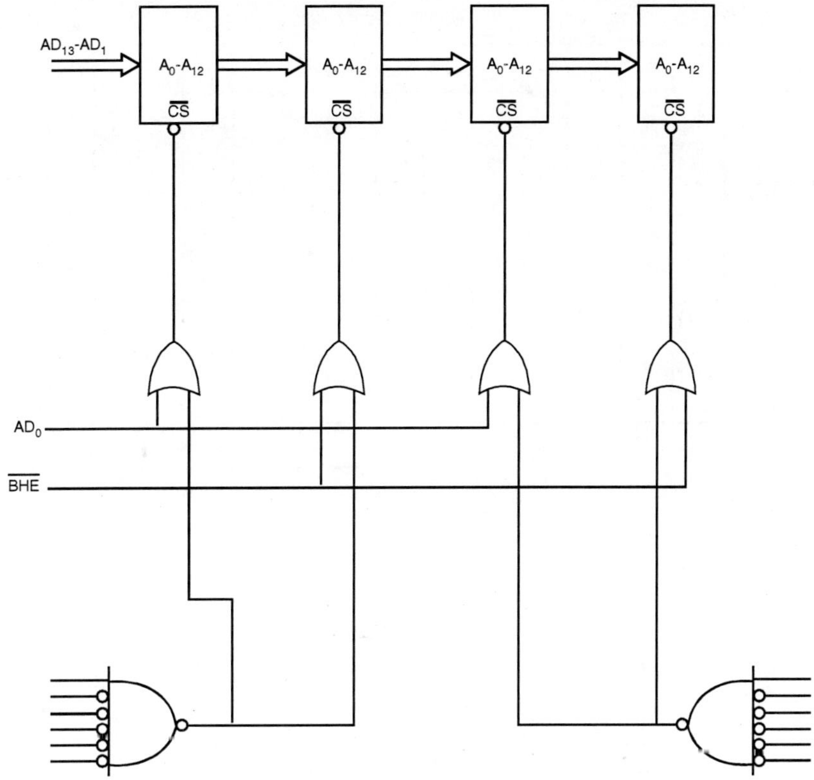

Fig. 5.46. RAM interfacing.

Thus RAM can be interfaced to 8086 as shown in Fig. 5.47.

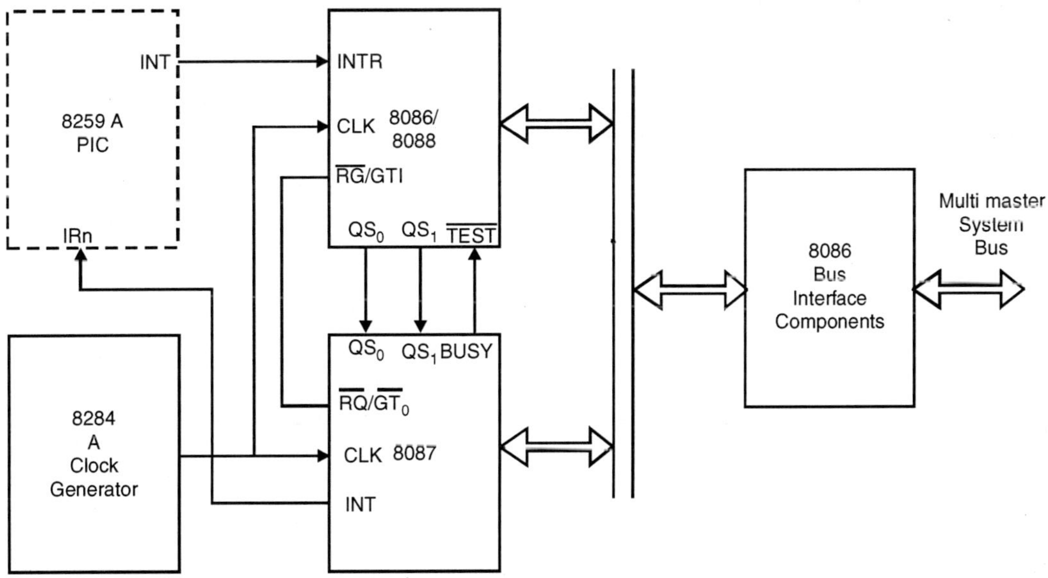

Fig. 5.47

Example 5.2. Interface two 4K X 8 EPROMS and two 4K X 8 RAM chips with 8086, microprocessor and draw the suitable circuit showing their interfacing ?

Memory Map

Address	A_{19}	A_{18}	A_{17}	A_{16}	A_{15}	A_{14}	A_{13}	A_{12}	A_{11}	A_{10}	A_9	A_8	A_7	A_6	A_5	A_4	A_3	A_2	A_1	A_0
FFFFFH	1	1	1	1	1	1	1	1	1	1	1	1	1	1	1	1	1	1	1	1
EPROM ADDRESS																				
FE000H	1	1	1	1	1	1	1	0	0	0	0	0	0	0	0	0	0	0	0	0
FDFFFH	1	1	1	1	1	1	0	1	1	1	1	1	1	1	1	1	1	1	1	1
RAM ADDRESS																				
FCOOOH	1	1	1	1	1	1	0	0	0	0	0	0	0	0	0	0	0	0	0	0

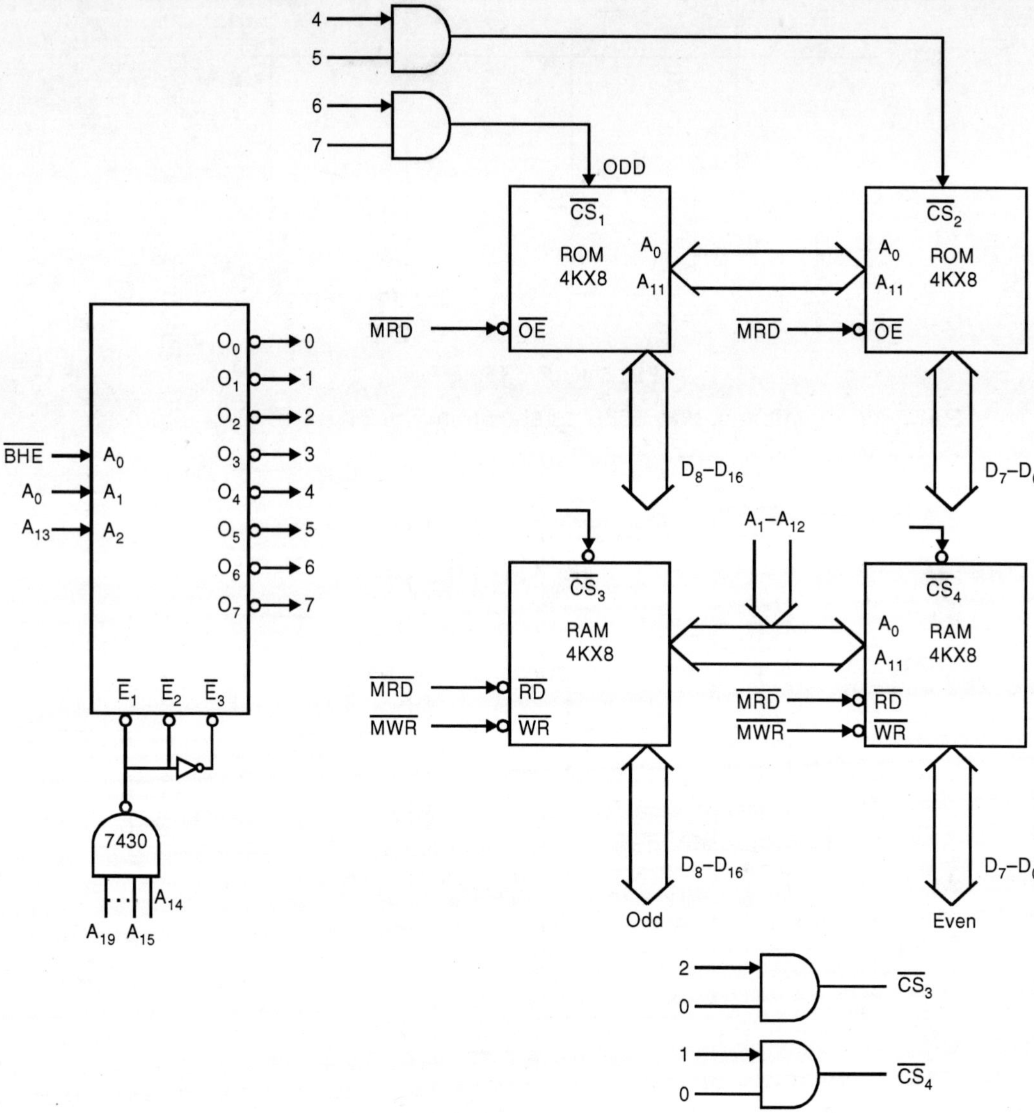

Fig. 5.48. Interfacing circuit.

Solution. The address of the RAM may be selected anywhere in the 1 MB address space of 8086, but to make the address space continuous we would follow the given procedure. After reset the IP and CS are initialised to for address FFFF0H. We must first calculate the total number of address lines required for 8K bytes of EPROM which is 13, as we have seen earlier that we have $\boxed{N = 2^n}$ hence we get

$$2^{13} = 8K$$

Address lines A_{13}-A_{19} are used for decoding to generate the chip select. The $\overline{BHE}$ signal goes low when a transfer is at odd address or higher byte of data is to be accessed.

The memory system here contains in total four 4K X 8 memory chps. The two 4K X 8 chips of RAM and ROM are arranged in parallel to obtain 16-bit data bus width. If A_0 is 0 i.e. the address is even and is in RAM, then the lower RAM chip is selected indicating 8-bit transfer at an even address.

If A_0 is 1, i.e. the address is odd and is in RAM the $\overline{BHE}$ goes low, the upper RAM chip is selected, further indicating that the 8-bit transfer is at an odd address. If the selected addresses are in ROM, the respective ROM chips are selected. If at a time A_0 and $\overline{BHE}$ both are 0, both the RAM or ROM chips are selected i.e. the data transfer is of 16-bits.

Example 5.3. Design an interface between CPU 8086 and two chips of 32K X 8 ROM and four chips of 32 K X 8 RAM according to the following memory map ?

ROM 1 and ROM 2 ⇒ F0000H–FFFFFH

RAM 1 and RAM 2 ⇒ D0000H–DFFFFH

RAM 3 and RAM 4 ⇒ E0000H–EFFFFH

Solution.

Address Map

Address	A_{19}	A_{18}	A_{17}	A_{16}	A_{15}	A_{14}	A_{13}	A_{12}	A_{11}	A_{10}	A_9	A_8	A_7	A_6	A_5	A_4	A_3	A_2	A_1	A_0
FOOOOH	1	1	1	1	0	0	0	0	0	0	0	0	0	0	0	0	0	0	0	0
ROM 1 and ROM 2																				
FFFFFH	1	1	1	1	1	1	1	1	1	1	1	1	1	1	1	1	1	1	1	1
DOOOOH	1	1	0	1	0	0	0	0	0	0	0	0	0	0	0	0	0	0	0	0
RAM 1 and RAM 2																				
DFFFFH	1	1	0	1	1	1	1	1	1	1	1	1	1	1	1	1	1	1	1	1
EOOOOH	1	1	1	0	0	0	0	0	0	0	0	0	0	0	0	0	0	0	0	0
RAM 3 and 4																				
EFFFFH	1	1	1	0	1	1	1	1	1	1	1	1	1	1	1	1	1	1	1	1

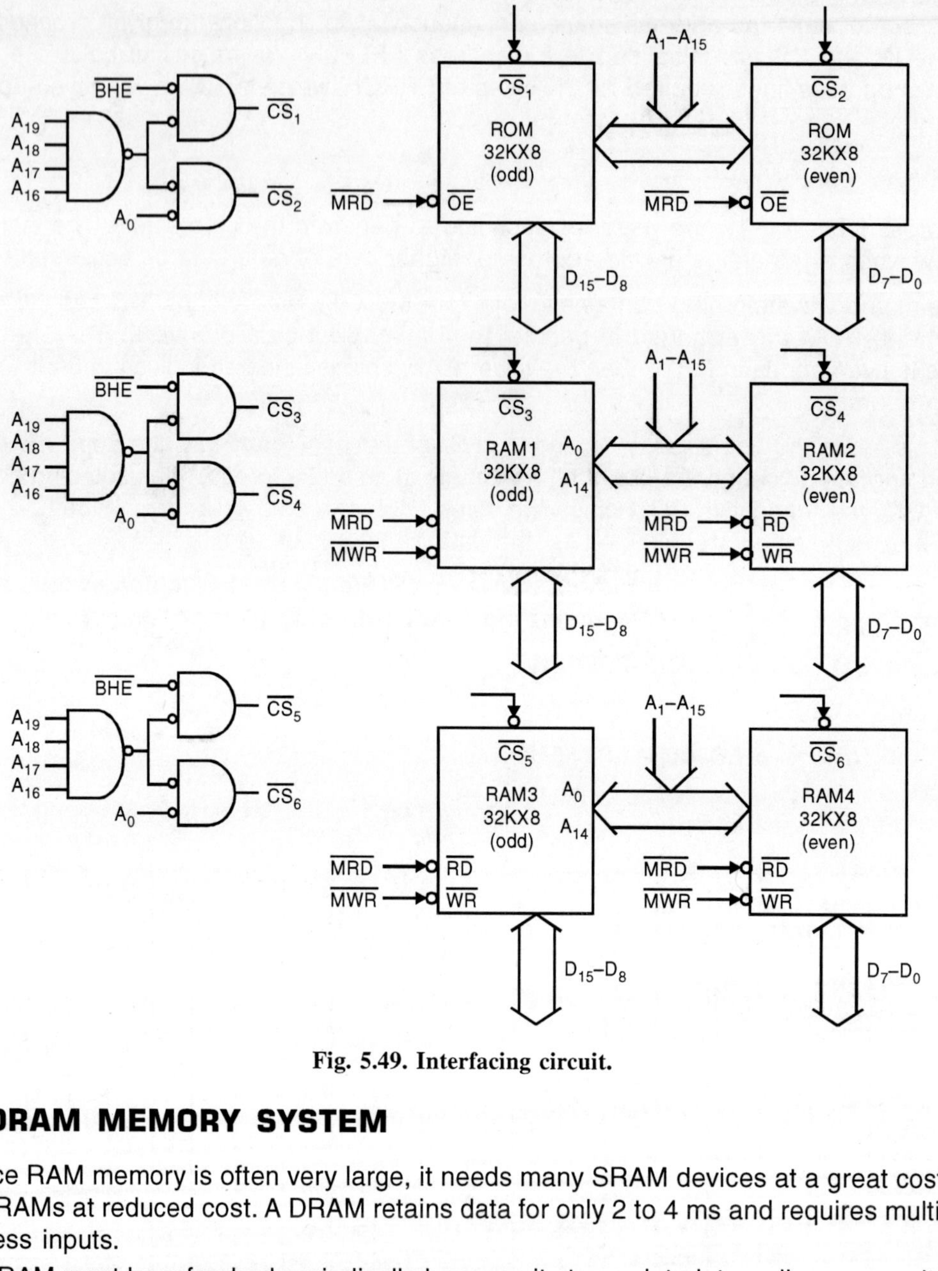

Fig. 5.49. Interfacing circuit.

5.9. DRAM MEMORY SYSTEM

Since RAM memory is often very large, it needs many SRAM devices at a great cost or just a few DRAMs at reduced cost. A DRAM retains data for only 2 to 4 ms and requires multiplexing of address inputs.

A DRAM must be refreshed periodically because it stores data internally on capacitors that lose their charge in a short period of time. In order to refresh a DRAM, the contents of a section of the memory must periodically be read or written. Any read or write automatically refreshes an entire section of the DRAM. *(The number of bits refreshed depends on the size of the memory components and its internal organization).*

Therefore, "DRAM contains more memory cells as compared to static RAM per unit area and they are available in much larger sizes: upto 16M × 1. Here, we see the interfacing of DRAM to the processor, *i.e.,* DRAM memory system. Fig. 5.50 depicts the pin diagram.

It is known that to address 64K memory registers we require 16 address lines. But in !11-\M to reduce total number of address lines, address is given in two phases, which allows : RAM to multiplex lower address and higher address. DRAM TMS4464 (64K × 4) has eight Address lines (A-A_1) as shown in Fig. 5.51. If column address strobe $\overline{CAS}$ is low these address lines give column address and when row address strobe $\overline{RAS}$ is low these addresses give row address. First, A_0-A_7 are placed on the address bus and strobes into an column row latch by asserting $\overline{RAS}$. Then, the address lines A_8-A_{15} are placed on the same eight address inputs and strobes into an internal column latch by asserting $\overline{CAS}$.

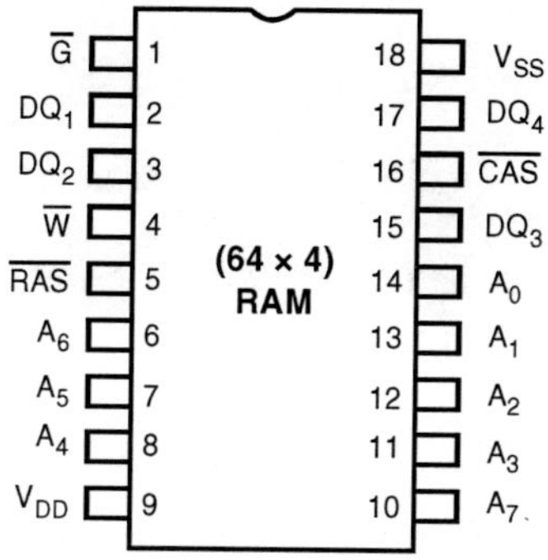

Fig. 5.50.

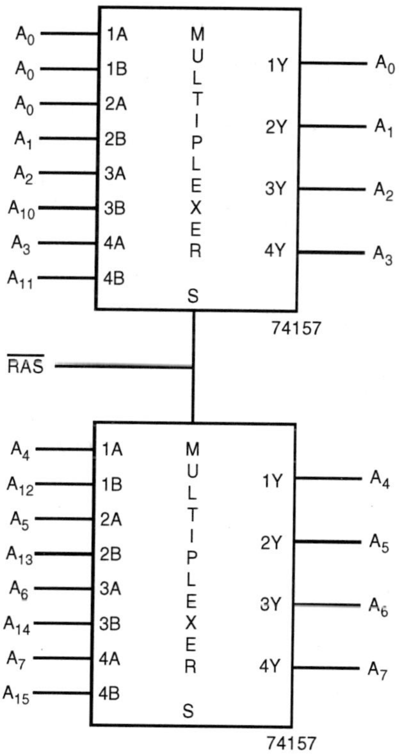

Fig. 5.51.

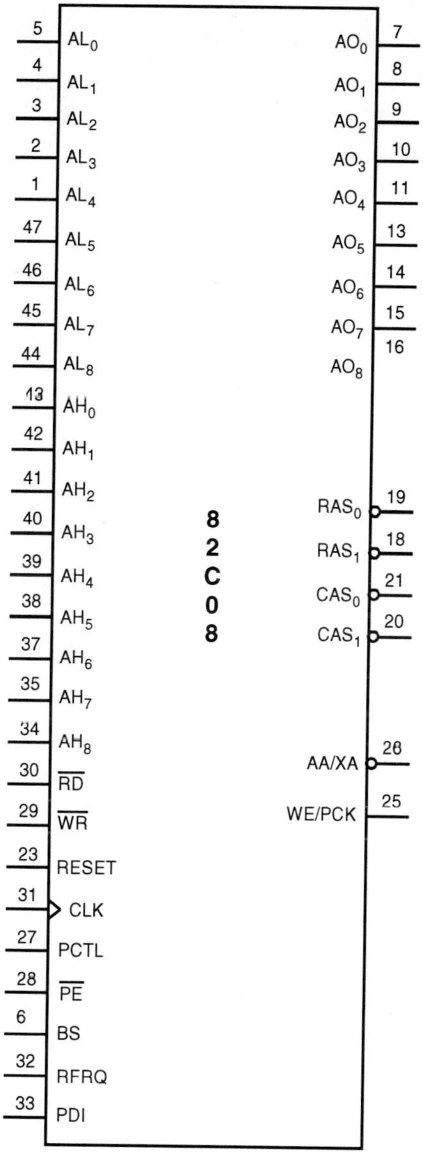

Fig. 5.52. Pin diagram of 82C08 DRAM controller (can control two bank of memory).

That means 16-bit address gets latched in two internal 8-bit latches. The internally Latched 16-bit access the contents of one of the 4-bit memory locations.

In many System DRAM. Controller integrated circuit is used to do address multiplexing and the generation of the RAM control signals, let us study the details of 82C08 The 82C08 can control upto 1M byte of memory for 8086, which contains an address multiplexer that multiplexes an 18-bit address onto 9 address connections. Figure 5.53 shows pin diagram of DRAM controller, where AL_0 - AL_{08} and AH_0 - AR_{08} are the address inputs for DRAM controller and A_0-A_8 are the address outputs from the DRAM controller.

As observed in Fig. 5.52, the *82C08,* DRAM *controller* has internal circuitry which generates the $\overline{CAS}$ and $\overline{RAS}$ signals for the DRAM. The 82C08 generates these signals internally using the CLK, S1 and $\overline{S0}$ signals. The processor provides S1 and S0 signals as $\overline{RD}$ and $\overline{WR}$ inputs to the 82C08 DRAM controller. The $\overline{AACK}/\overline{XACK}$ signal of 82C08 is an acknowledge output that is used to indicate the ready condition for the microprocessor.

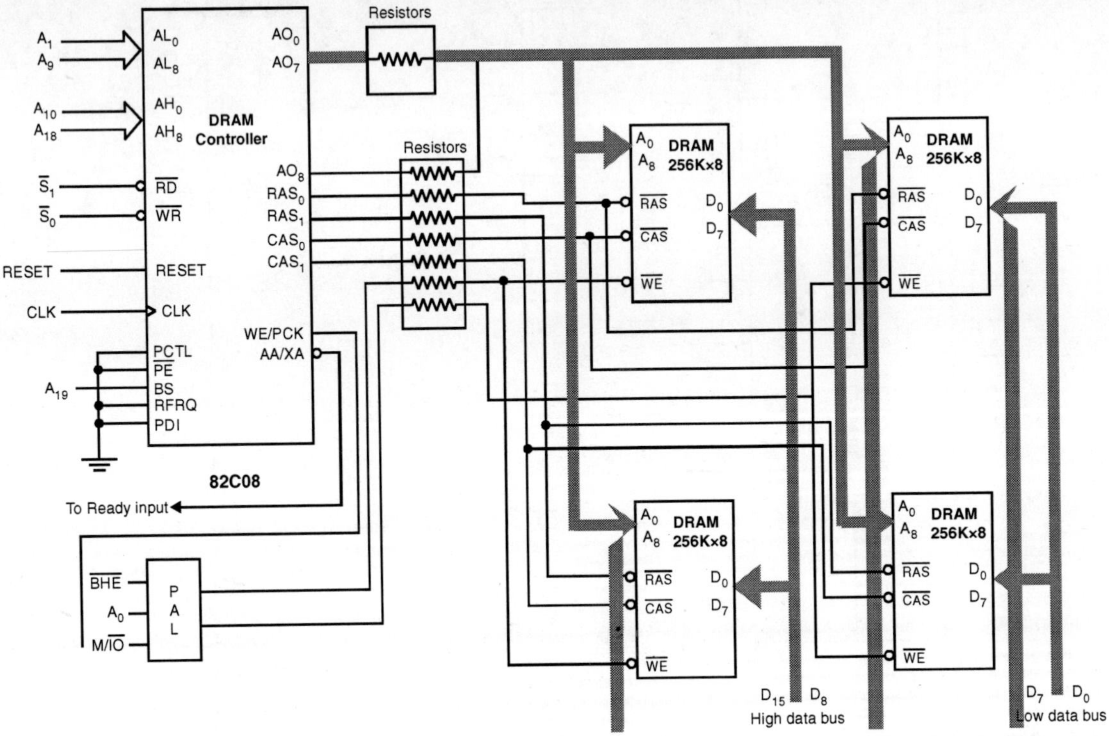

Fig. 5.53. DRAM controller.

This signal is normally connected to the READY input of the microprocessor or the clock generator READY input. Now Fig. 5.53 illustrates the interface of 1M byte DRAM memory for 8086. MP using DRAM controller.

Since DRAM controller accepts 18 address lines it can address up to 256 K memory locations. Therefore 1M byte DRAM memory is interfaced using a series of four 256 K × 8 bit DRAM modules. The PAL 16L8 is used to generate bank select signals and other control signals, as shown in the Fig. 5.53. The I/O signal selects the upper bank or the lower bank through the BS *(Bank select)* input to 82C08 DRAM controller.

5.10. DYNAMIC RAM INTERFACING

When we require a large capacity of memory in a system, the memory subsystem is generally designed using D-RAM or dynamic RAM. D-RAM has various advantages of D-RAM such as :

(1) Higher packaging density.

(2) Lower cost.

(3) Less power consumption

Some disadvantages of D-RAM cell are as stated below :

(1) The D-RAM cell uses a capacitor to store the charge as a representation of data but the reverse biased diode has leakage current that tends to discharge the capacitor giving rise to the possibility of data loss.

(2) To avoid the data loss we must refresh the D-RAM cell after a fixed time interval.

(3) During the refreshing of the D-RAM all the operations of the memory are suspended hence resulting in

  (a) Loss of time.

  (b) Reduced system performance.

(4) The refresh mechanism and additional hardware required makes the interfacing for D-RAM a complicated issue.

A dedicated hardware chip called as D-RAM controller is the most important part of the interfacing module. The (refresh cycle) is different from the memory read cycle in the following ways :

1. The memory address is not provided by the CPU address rather it is generated by a refresh mechanism counter called as refresh counter.
2. More than one memory chip may be enabled at a time so as to reduce the number of total memory refresh cycles.
3. Memory refresh is a independent regular activity initiated and carried out by the refresh mechanism whereas the memory read is either initiated by an external bus creator or a processor initiated operation.

D-RAM is arranged internally in a 2-dimensional matrix away so that it will have n-rows and n-columns. A typical 4 K bit D-RAM chip has an internally arranged bit array of dimensions 64 rows × 64 columns. Thus the row address and the column address would require require 6 bits each. These 6 bits will be generated by the refresh counter during the refresh cycles. A complete row of 64 cells is refreshed at the same time to minimise the same time. Hence the refresh counter needs to generate only row addresses. The row addresses are than multiplexed over lower order address lines. The address bus of the processor is connected to the address bus of D-RAM during normal operations. A refresh timer devices a pulse for refreshing action after each refresh interval.

Refresh interval is defined as the time for which a D-RAM cell can now data change level practically constant.

Procedure to calculate refresh interval : The refresh interval depends upon the manufacturing technology of a D-RAM cell and may range between 1 ms^{-3} ms. Now let us consider 3 ms as the typical refresh time interval.

$$\text{Refresh time (per row)} \Rightarrow t_r = \frac{3\times10^{-3}}{64}$$

Refresh frequency $f_r = \frac{64}{3 \times 10^{-3}} = 21.33 \times 10^3$ Hz

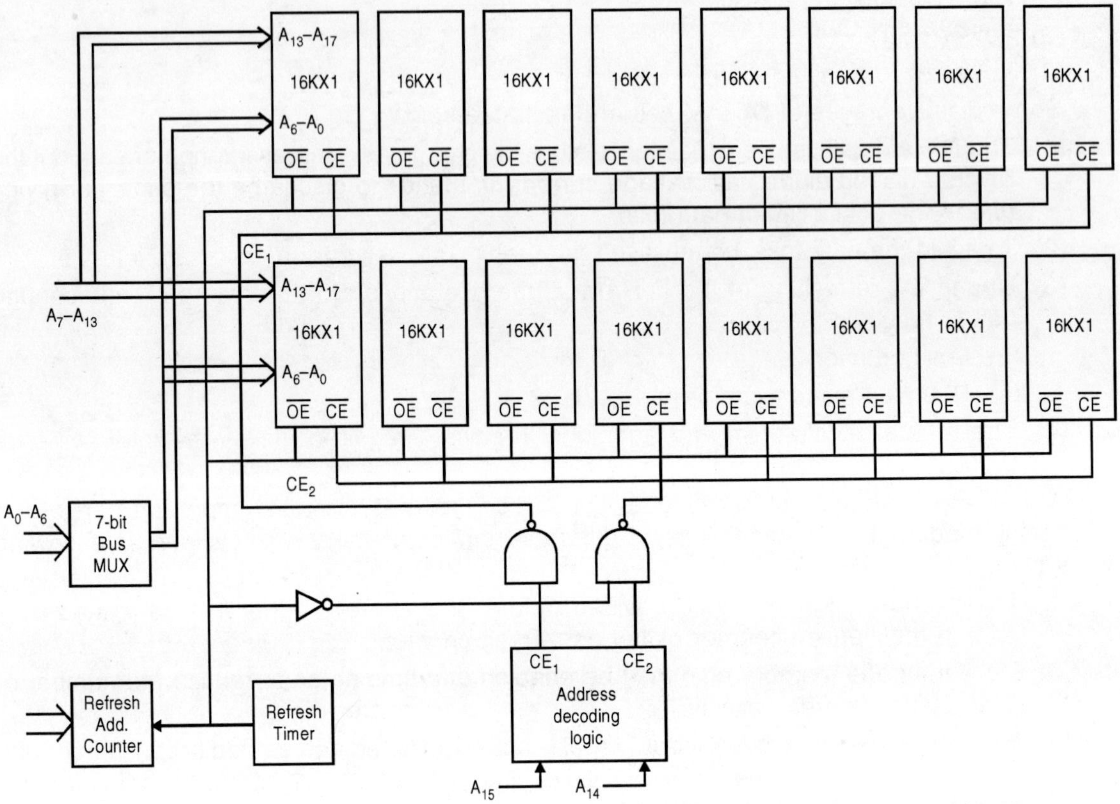

Fig. 5.54. D-RAM Refreshing logic.

The block diagram shows the refreshing logic and 8086 interfacing with D-RAM. Each of the chips used is a 16 K × 1 bit D-RAM cell array. The system shown contains two 16 K byte D-RAM units. The $\overline{OE}$ pin controls the output data buffers of the memory chips.

The CE pins are active high chip selects of memory chips.

Refresh cycle : If the refresh output of the refresh timer goes high, $\overline{OE}$ and CE also tend to go high. The high CE enables the memory chip for refreshing, while high $\overline{OE}$ prevents the data from appearing on the data bus. The 16 K × 1-bit D-RAM has an internal array of 128 × 128 cells requiring 7 bits for row addresses. The lower order seven lines A0-A6 are multiplexed with the refresh counter output A_{10}-A_{16}, as shown in the block diagram.

5.10.1. Memory Architecture

In order to build an *N-word* memory where each word is *M bits* wide (typically 1, 4 or 8 bits), a straightforward approach is to stack memory:

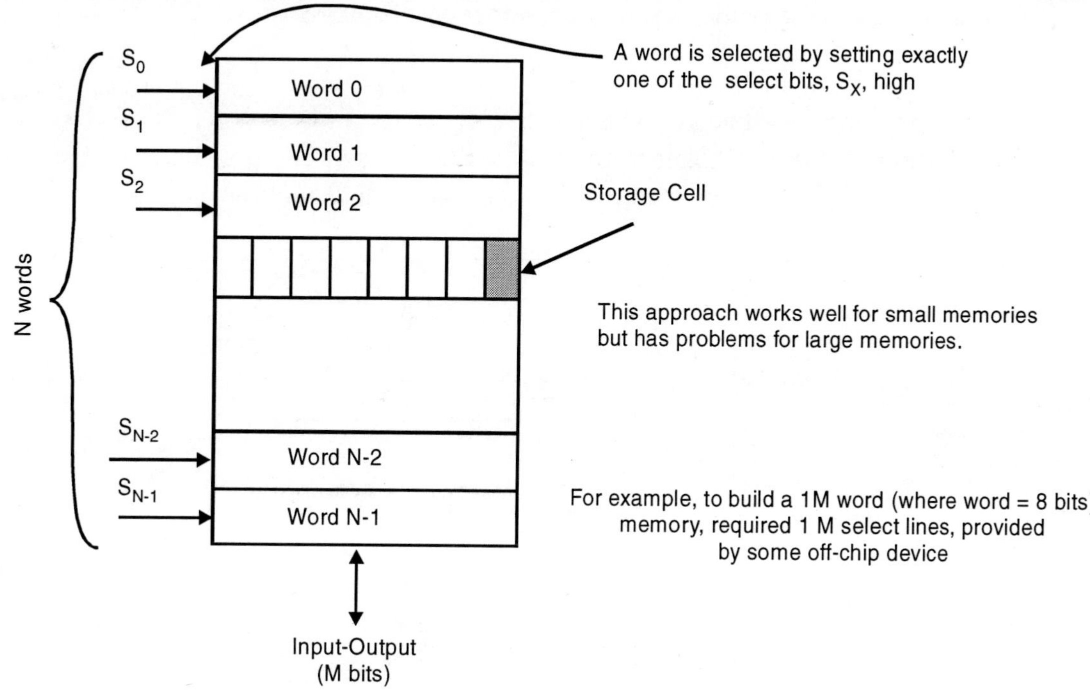

Fig. 5.55. Memory architecture for D-RAM.

This approach is not practical. What can we do ?

Add a decoder to solve the package problem as shown in the figure below :

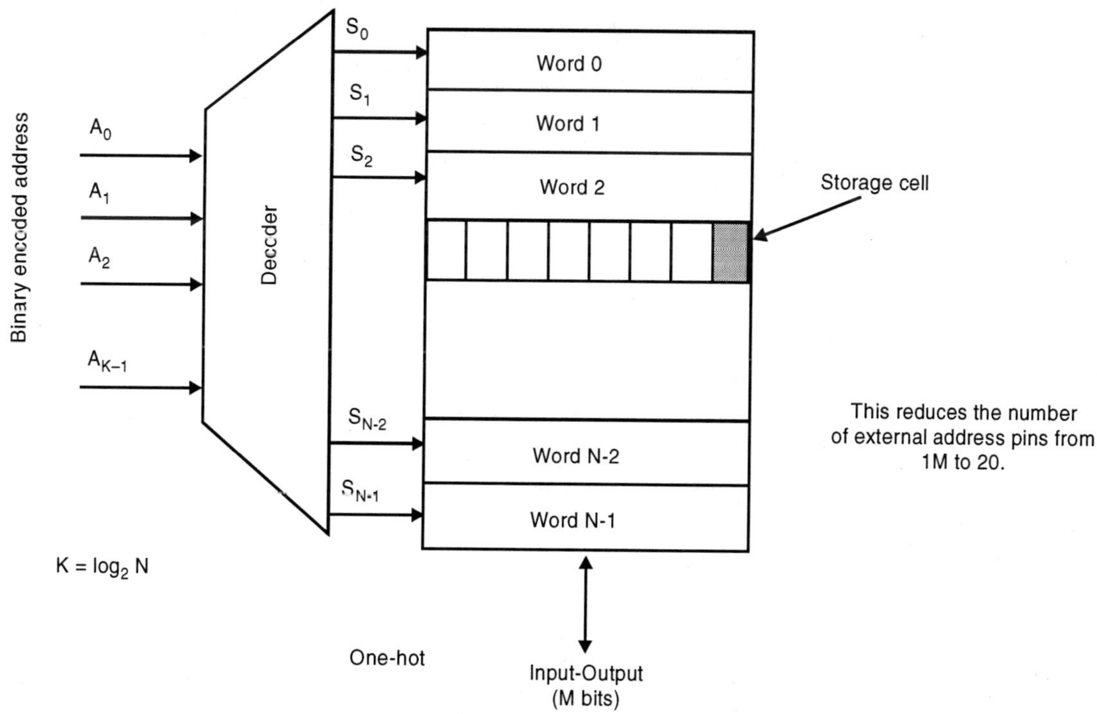

Fig. 5.56.

This does not address the *memory aspect ratio* problem:

The memory is 128,000 time higher than wide ($2^{20}/2^3$)!

Besides the bizarre shape factor, the design is *extremely slow* since the vertical wires are VERY long (delay is at least linear to length).

The vertical and horizontal dimensions are usually very similar, for an aspect ratio of *unity*.

Multiple words are stored in each row and selected simultaneously:

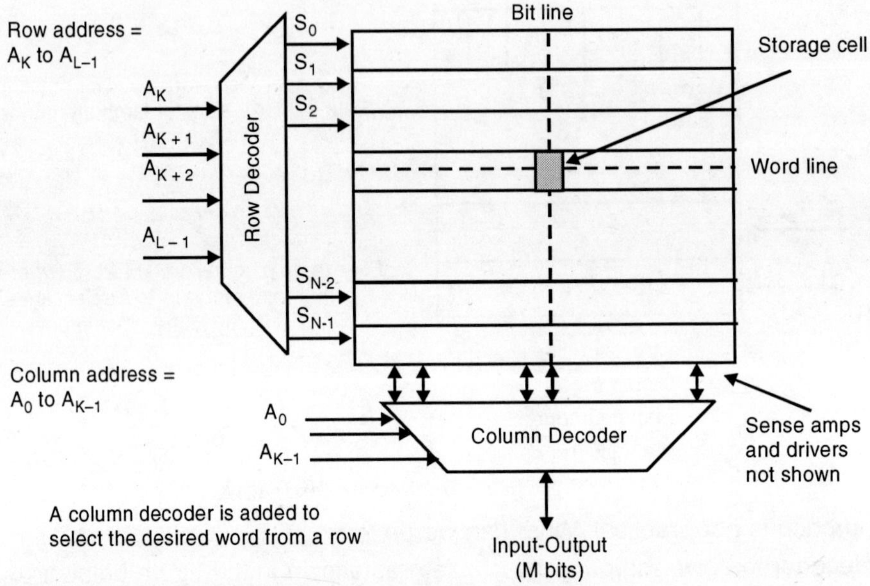

Fig. 5.57. Addressing mechanism.

This strategy works well for memories up to 64 K bits to 256 K bits.

Larger memories start to suffer excess delay along bit and word lines.

A *third dimension* is added to the address space to solve this problem:

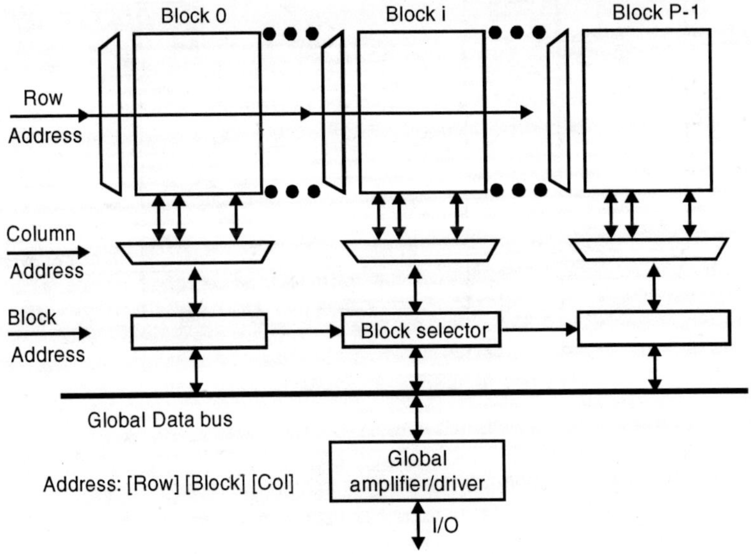

Fig. 5.58.

Dynamic RAM Controller (TMS 4500)

Features

1. Controls operation of 8K, 16K, 32K and 64K dynamic RAMs.
2. Creates static RAM apperance.
3. One package contains address multiplexers, refresh control and timing control.
4. TMS 4500 directly addresses and drives upto 256 K byte of memory without external devices.
5. TMS 4500 operates from microprocessor clock.
   — No crystals, delay lines or RC networks.
   — Eliminates arbitration delays.
6. Refresh may be internally or externally initiated.
7. Versatile :
   (a) Strap selected refresh rate.
   (b) Synchronous predictable refresh.
   (c) Selection of distributed, transparent and cycle steal refresh modes.
   (d) Interfaces easily to popular microprocessor.
8. Strap selected weight state generation for microprocessor/memory speed matching.
9. Ability to synchronise or interleave controller with microprocessor system. (including multiple controller).
10. Three state outputs allow multiport memory configuration.
11. Performance of TMS 4500 ranges of 150 ns, 200 ns or 250 ns.

Pin Configuration

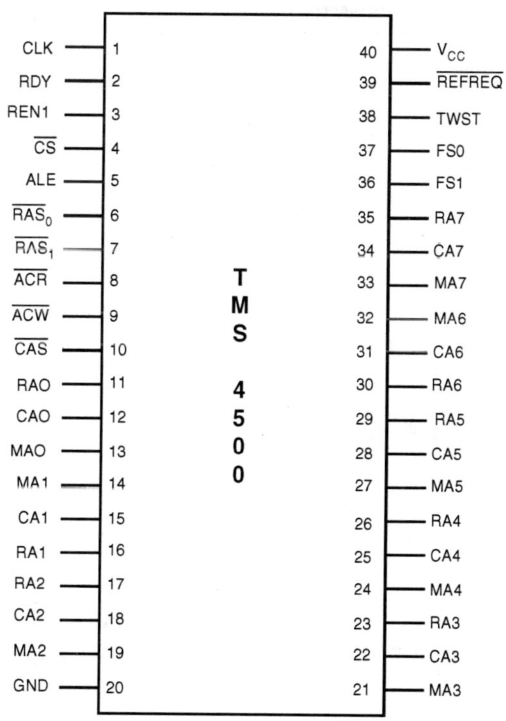

Fig. 5.59.

Functional Block Diagram

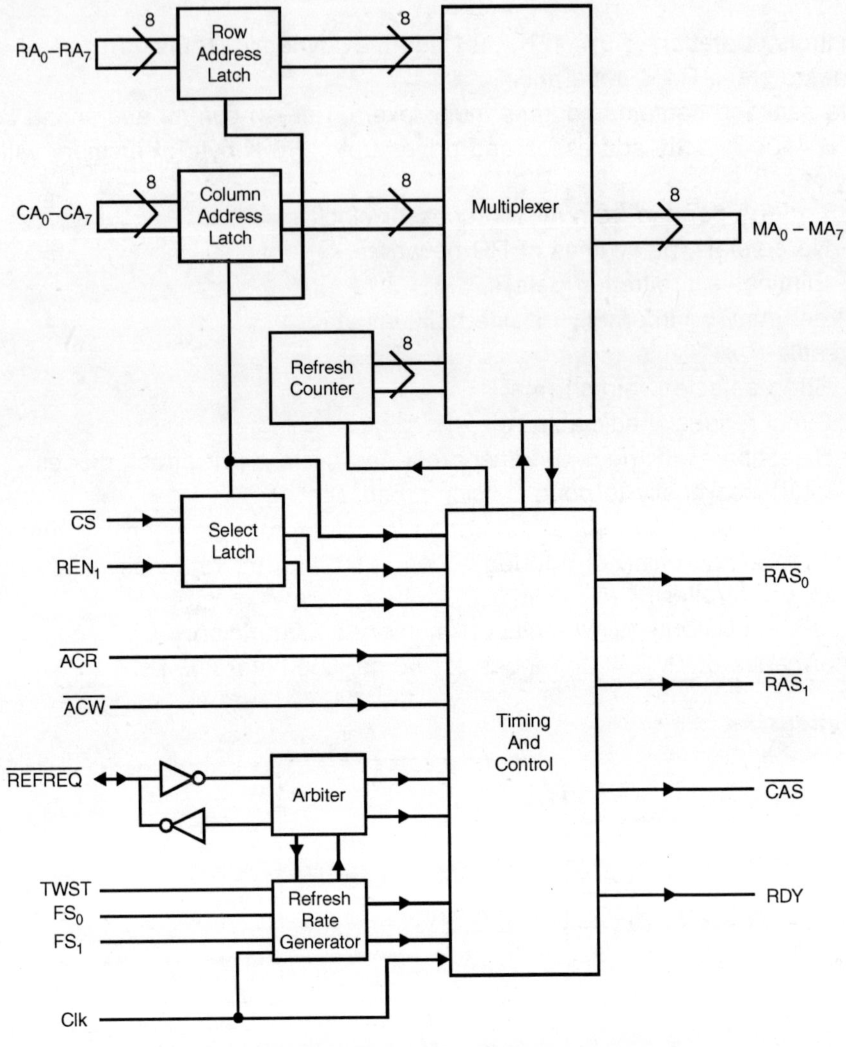

Fig. 5.60. Functional block diagram of TMS 4500.

Description

TMS 4500 is a monolithic DRAM system controller designed to provide address multiplexing, timing, control and refresh/access arbitration functions to simplify the interface of dynamic RAMs to microprocessor systems. The DRAM controller contains a 165 bit multiplexer that generates the address lines for the memory device from the 16 system address bits and provides the strobe signals required by memory to decode the address. A 8 bit refresh counter generates the 256 row addresses required for address. A refresh timer is provided that generates the necessary timings to refresh the dynamic memories and assure data retention. The TMS 4500 also contains refresh/access arbitration circuitry to resume conflicts between memory access requests and memory refresh cycles. The TMS 4500 is offered in a 40 pin dual in line plastic package and 44 pin square plastic carrier package. It is characterised for operation from 0°–70°C.

REVIEW QUESTIONS

1. State and explain the different instruction formats of 8086?
2. What do you mean by addressing modes? What are the different addressing modes supported by 8086? Explain each of them with suitable examples?
3. Explain the physical address formation in different addressing modes of 8086?
4. What are the different instruction types of 8086? Explain the addressing modes for control transfer instructions?
5. "A single instruction may use more than one addressing modes or some instructions may not require any addressing mode". Explain.
6. Bring out the developments in 8086 instruction set over 8085 instruction set in details?
7. What are the differences in interfacing RWMs while 8086 is in minimum and maximum modes
8. Sketch and explain the interface of 32K x 16 RWMs using a decoder in minimum mode. What is the maximum access time of ROMs such that it does not require wait states when 8086 operates at 8 MHz
9. Sketch and explain the timing diagrams in the above interface Question 8
10. Sketch and explain the 8086 bus activities during write machine cycle
11. What are the steps in interfacing peripherals with the micro processor
12. Sketch and explain the interface of PPI 8255 to the 8086 microprocessor in minimum mode. Interface 47 segment LEDs to display as a BCD counter
13. In the above question Q2 interface two keys UP and DOWN to the PPI. Write an 8086 assembly program segment such that when UP is pressed the counter counts up every second. Similarly when DOWN key is pressed the counter decrements every second
14. Sketch and explain the interface of 8279 to the 8086 microprocessor in minimum mode. Interface 8x8 key pad and 16x7 Seg LED display. Write an 8086 assembly program to read the key codes of keys and display –INDIA-?
15. Sketch and explain the interface of PIT 8254 to the 8086 microprocessor in minimum mode. Cascade two counters in the PIT. Write a program segment two get one minute delay,
16. Interface four 8K chips of static RAM and four 4K chips of EPROM with 8086. Interface two of the RAM chips in the zeroth segment so as to accommodate IVT inthem. Two EPROM chips should be interfaced so that the system is restartable as usual, and the remaining two EPROM chips should be interfaced at the starting of A000th segment. Use absolute decoding scheme.

    Interface eight 8K chips of RAM and four 8K chips of EPROM with 8086. Interface the RAM bank at a segment address 0B00H and the EPROM bank at a physical address F8000H. Do not alow any fold back space.
17. Interface eight 8K chips of RAM and four 8K × 4 chips of EPROM with 8086 at the further specified address map. You may use linear decoding scheme for minimising the required hardware.

Chips	*Segment Address*	*Starting Offset Address*
RAM1 and RAM2	0000H	0000H
RAM3 and RAM4	0500H	5000H
RAM5 and RAM6	7000H	2000H
RAM7 and RAM8	E000H	A000H
EPROM1 and EPROM2	F000H	0000H
EPROM3 and EPROM4	D000H	7000H

Will this system be practically useful ? Explain. If not then what minimum change do you suggest in this address map ?

18. Interface two 8K RAM chips and two 4K EPROM chips with 8088 so as to form a completely working system configuration.
19. Describe the procedure of interfacing static memories with a CPU ? Bring out the differences between interfacing the memories with 8086 and 8088.
20. Bring out the differences between static and dynamic RAM.
21. Design a 2-digit second counter using 74373 output ports.
22. Design a 3-digit pulse counter using 74373 output ports to count TTL compatible pulses using an input line of a 74245 input port.
23. Generate a square wave of period 1 sec using 74373 output ports.
24. Design a multiplexed display scheme to display seconds, minutes and hours counter using 8255 ports. Assume that a standard delay of 1 second is available.
25. Design a one unit 14-segment alphanumeric display and write a program to display an alphanumeric character of which the code is in AX.
26. Interface an 8255 with 8086 so as to have port A address 00, port B address 02, port C address 01 and CWR address 03.
27. Explain the different modes of operation of 8255.
28. Explain the control word format of 8255 in I/O and BSR mode.
29. Write a program to print message—'This is a printer test routine' to a printer using 8255 port initialised in model.
30. Interface an 8 × 8 keyboard using two 8255 ports and write a program to read the code of a pressed key.
31. Interface a typical 12-bit DAC with 8255 and write a program to generate a triangular waveform of period 10 ms. The CPU runs at 5 MHz clock frequency.
32. Using a typical 12-bit DAC generate a step waveform of duration 1 sec., maximum voltage 3 Volts and determine duration of each step suitably.
33. Draw and discuss analog and digital section of 7109 in brief.
34. Design a 7109 circuit to convert the analog voltage to digital equivalent at a rate of 30 samples per second.
35. Draw a typical stepper motor interface with 8255.
36. Write ALPs to rotate a 200 teeth, 4 phase stepper motor as specified below.
    (i) Five rotations clockwise and then five rotations anticlockwise.
    (ii) Rotate through angle 135° in 2 sec.
    (iii) Rotate the shaft at a speed of 10 rotations per minute.

❑❑❑

CHAPTER 6

8086 INSTRUCTION SET

INSIDE THIS CHAPTER

This chapter discusses the 80x86 real mode instruction set. Like any programming language, there are going to be several instructions you use all the time, some you use occasionally, and some you will rarely, if ever, use. This chapter organizes its presentation by instruction class rather than importance. Since beginning assembly language programmers do not have to learn the entire instruction set in order to write meaningful assembly language programs, you will probably not have to learn how every instruction operates. The following list describes the instructions this chapter discusses. There are many additional instructions, especially on the 80386 and later processors. These additional instructions make assembly language programming easier, but you do not need to know them to begin writing programs.

80x86 instructions can be (roughly) divided into eight different classes :

1. **Data movement instructions**

   `mov, lea, les, push, pop, pushf, popf`

2. **Conversions**

   `cbw, cwd, xlat`

3. **Arithmetic instructions**

   `add, inc sub, dec, cmp, neg, mul, imul, div, idiv`

4. **Logical, shift, rotate, and bit instructions**

   `and, or, xor, not, shl, shr, rcl, rcr`

5. **I/O instructions**

   `in, out`

6. **String instructions**

   `movs, stos, lods`

7. **Program flow control instructions**

   `jmp, call, ret, conditional jumps`

8. **Miscellaneous instructions**
   `clc, stc, cmc`

The following sections describe all the instructions in these groups and how they operate.

At one time a text such as this one would recommend against using the extended 80386 instruction set. After all, programs that use such instructions will not run properly on 80286 and earlier processors. Using these additional instructions could limit the number of machines your code would run on. However, the 80386 processor is on the verge of disappearing as this text is being written. You can safely assume that most systems will contain an 80386sx or later processor. This text often uses the 80386 instruction set in various example programs. Keep in mind, though, that this is only for convenience. There is no program that appears in this text that could not be recoded using only 8088 assembly language instructions.

A word of advice, particularly to those who learn only the instructions noted above: as you read about the 80x86 instruction set you will discover that the individual 80x86 instructions are not very complex and have simple semantics. However, as you approach the end of this chapter, you may discover that you haven't got a clue how to put these simple instructions together to form a complex program. Fear not, this is a common problem. Later chapters will describe how to form complex programs from these simple instructions.

6.1. THE PROCESSOR STATUS REGISTER (FLAGS)

The flags register maintains the current operating mode of the CPU and some instruction state information. The Fig. 6.1 shows the layout of the flags register:

The carry, parity, zero, sign, and overflow flags are special because you can test their status (zero or one) with the setcc and conditional jump instructions (see "The "Set on Condition" Instructions" and "The Conditional Jump Instructions"). The 80x86 uses these bits, the condition codes, to make decisions during program execution.

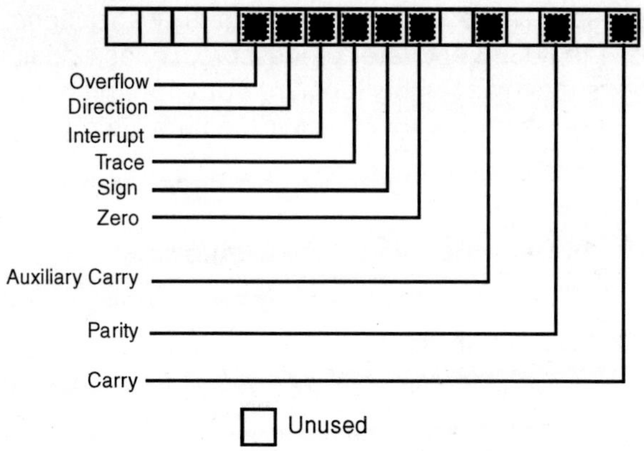

Fig. 6.1. Flag register for 8086.

Various arithmetic, logical, and miscellaneous instructions affect the overflow flag. After an arithmetic operation, this flag contains a one if the result does not fit in the signed destination operand. For example, if you attempt to add the 16 bit signed numbers 7FFFh and 0001h the result is too large so the CPU sets the overflow flag. If the result of the arithmetic operation does not produce a signed overflow, then the CPU clears this flag.

Since the logical operations generally apply to unsigned values, the 80x86 logical instructions simply clear the overflow flag. Other 80x86 instructions leave the overflow flag containing an arbitrary value.

The 80x86 string instructions use the direction flag. When the direction flag is clear, the 80x86 processes string elements from low addresses to high addresses; when set, the CPU processes strings in the opposite direction. See "String Instructions" for additional details.

The interrupt enable/disable flag controls the 80x86's ability to respond to external events known as interrupt requests. Some programs contain certain instruction sequences that the CPU must not interrupt. The interrupt enable/disable flag turns interrupts on or off to guarantee that the CPU does not interrupt those critical sections of code.

The trace flag enables or disables the 80x86 trace mode. Debuggers (such as CodeView) use this bit to enable or disable the single step/trace operation. When set, the CPU interrupts each instruction and passes control to the debugger software, allowing the debugger to single step through the application. If the trace bit is clear, then the 80x86 executes instructions without the interruption. The 80x86 CPUs do not provide any instructions that directly manipulate this flag. To set or clear the trace flag, you must:

- Push the flags onto the 80x86 stack,
- Pop the value into another register,
- Tweak the trace flag value,
- Push the result onto the stack, and
- Pop the flags off the stack.

If the result of some computation is negative, the 80x86 sets the sign flag. You can test this flag after an arithmetic operation to check for a negative result. Remember, a value is negative if its H.O. bit is one. Therefore, operations on unsigned values will set the sign flag if the result has a one in the H.O. position.

Various instructions set the zero flag when they generate a zero result. You'll often use this flag to see if two values are equal (e.g., after subtracting two numbers, they are equal if the result is zero). This flag is also useful after various logical operations to see if a specific bit in a register or memory location contains zero or one.

The auxiliary carry flag supports special binary coded decimal (BCD) operations. Since most programs don't deal with BCD numbers, you'll rarely use this flag and even then you'll not access it directly. The 80x86 CPUs do not provide any instructions that let you directly test, set, or clear this flag. Only the add, adc, sub, sbb, mul, imul, div, idiv, and BCD instructions manipulate this flag.

The parity flag is set according to the parity of the L.O. eight bits of any data operation. If an operation produces an even number of one bits, the CPU sets this flag. It clears this flag if the operation yields an odd number of one bits. This flag is useful in certain data communications programs, however, Intel provided it mainly to provide some compatibility with the older 8080 ***m*P.**

The carry flag has several purposes. First, it denotes an unsigned overflow (much like the overflow flag detects a signed overflow). You will also use it during multiprecision arithmetic and logical operations. Certain bit test, set, clear, and invert instructions on the 80386 directly affect this flag. Finally, since you can easily clear, set, invert, and test it, it is useful for various Boolean operations. The carry flag has many purposes and knowing when to use it, and for what purpose, can confuse beginning assembly language programmers. Fortunately, for any given instruction, the meaning of the carry flag is clear.

6.2. DATA TRANSFER INSTRUCTIONS

In this group we have a large number of instructions which perform data transfer of the following types.

(a) Move contents of a register, or contents of a memory location, or immediate data to a register or memory location.
(b) Data transfer between a segment register and a register or a memory location.
(c) Push and Pop instructions.
(d) Exchange Instructions.
(e) Data transfer with I/O ports.

6.3. DATA TO A REGISTER/MEMORY FROM A REGISTER/MEMORY IMMEDIATE DATA

The following instructions are described below :

1. Move data between two registers or between a register and a memory location.
2. Move immediate data to a register.
3. Move immediate data to a memory location.
4. Move data to accumulator using Direct addressing.
5. Move data from accumulator using Direct addressing.

6.3.1. Move Data Between Two Registers or Between a Register and a Memory Location

Template for this instruction is

Byte 1								Byte 2		
1	0	0	0	1	0	D	W	MOD	REG	R/M

As there are 10 bits in the template for identifying the operands, there are 2^{10}=1024 instructions in this category. This template was discussed extensively in the previous chapter also.

Example 1. Code for MOV AX,BX

This instruction moves the contents of BX register to AX register. It is a *2* byte instruction. In this example, we can specify either of AX or BX using the REG field. If we specify AX using REG field, then the D bit must be 1, indicating that AX is the destination register. The REG fieid must be (300 to indicate AX register. The W bit must be 1 to indicate it is a word operation. In this case, BX is specified using MOD value of 11, and R/M value of 011. Thus the *2* byte code for this instruction would be

opcode	**D**	**W**	**MOD**	**REG**	**R/M**	
100010	1	1	11	000	011	= 8B C3H

If we specify BX using REG field, then the D bit must be 0, indicating that BX is the source register. The REG field must be 011 to indicate BX register. The W bit must be 1 to indicate it is a word operation. In this case. AX is specified using MOD value of 11, and R/M value of 000. Thus the *2* byte code for this instruction would be

opcode	**D**	**W**	**MOD**	**REG**	**R/M**	
100010	0	1	11	011	000	=89D8H

Thus MOV AX, BX has two equally valid opcodes as shown above. An assembler generates one of these as the opcode.

Example 2. Code for MOV DS:0F546H[BP],DX

This instruction moves the contents of DX register to a memory location whose EA is given as the sum of the 16 bit offset value F546H and the contents of BP register. The PA is generated using DS as the segment register.

Here we have to specify DX using REG field. The D bit must be 0, indicating that DX is the source register. The REG field must be 010 to indicate DX register. The W bit must be 1 to indicate it is a word operation. F546H[BP] is specified using MOD value of 10, and R/M value of 110, and a displacement of F546H.

Whenever BP is used to generate the EA, the default segment would be SS. In this example, as we want the segment register to be DS, we have to provide the segment override prefix byte (SOP byte) to start with. The SOP byte is 001SR110, where SR indicates the segment override we desire. The SOP byte is not an instruction by itself. It only modifies the working of the instruction that follows the SOP byte.

To specify DS register, the SOP byte would be 001 11 110 = 3EH. Thus the 5 byte code for this instruction would be

SOP	Opcode	D	W	MOD	REG	R/M	16 bit disp	
3E	1000 10	0	1	10	010	110	F546	= 3E 89 96 F2 46 H

6.3.2. Move Immediate Data to a Register

The template is :

1	0	1	1	W	REG

As there are 4 bits in the template for identifying the operands, there are $2^4 = 16$ instructions in this category.

Example 1. Code for MOV AX,35H

This instruction moves the value 35H i.e. 0035H to AX register. It is a 3 byte instruction. In this example, we specify AX using the REG field. The REG field must be 000 to indicate AX register. W = 1, to indicate it is a word operation. Thus the 3 byte code for this instruction would be

Opcode	W	REG	16 bit data	
1011	1	000	00 35H	= B8 00 35H

But, in memory it will be stored as B8 25 00H.

Example 2. Code for MOV CL,0F2H

This instruction moves the 8 bit value F2H to CL. Here we specify CL using REG field. The W bit must be 0 to indicate it is a byte operation. Thus the 2 byte code for this instruction would be

Opcode	W	REG	8 bit data	
1011	0	001	F2H	= B1 F2H

6.3.3. Move Immediate Data to a Memory Location

Template for this instruction is

Byte 1								Byte 2				
1	1	0	0	0	1	1	W	MOD	0	0	0	R/M

As there are 6 bits in the template for identifying the memory operand, there are $2^6 = 64$ instructions in this category.

Example 1. Code for MOV DS:0F246H[BP],4567H

This instruction moves the value 4567H to a memory location whose EA is given as the sum of the 16 bit offset value F246H and the contents of BP register. The PA is generated using DS as the segment register.

The W hit must be 1 to indicate it is a word operation. F246H[BP] is specified using MOD value of 10, and R/M value of 110, and a displacement of F246H.

Whenever BP is used to generate the EA, the default segment would be SS. In this example, as we want the segment register to be DS, we have to provide, the segment override prefix byte (SOP byte) to start with. The SOP byte is 001SR110, where SR indicates the segment override we desire.

To specify DS register, the SOP byte would be 001 11 110 = 3EH. Thus the 7 byte code for this instruction would be

SOP	opcode	W	MOD	opcode	R/M	16 bit disp	16 bit data
3E	1100 011	1	10	000	110	F2 46	4567H

= 3E C7 86 F2 46 6745H

It will be stored in memory as 3E C7 86 46 F2 6745H.

Example 2. Code for MOV BYTE PTR 0F246H[BP], 34H.

This instruction moves the byte value 34 H to a memory location whose EA is given as the sum of the 16 bit offset value F246H and the contents of BP register. The PA is generated using SS as the segment register, which is the default segment register.

In this instruction, BYTE PTR is an assembler directive. An assembler directive is a direction provided to the assembler program to indicate the way the translation is to be done. It is not an instruction to the processor, i.e. it is not part of the instruction set of the processor.

In the above instruction, without such a directive, the assembler will not know whether to move the 16 bit value 0034H to the word location indicated by F246H[BP], or the 8 bit value 34H to the byte location indicated by F246H[BP]. So, the assembler has to be told clearly if it is a byte or a word operation. The other pointer directives are explained at the end of the chapter on BIOS and DOS services.

The W bit must be 0 to indicate it is a byte operation. F246H[BP] is specified using MOD value of 10, and R/M value of 110, and a displacement of F246H.

Thus the 5 byte code for this instruction would be

opcode	**W**	**MOD**	**opcode**	**R/M**	**16 bit disp**	**8 bit data**
1100011	0	10	000	110	F246	34H

= C6 86 F2 46 34H

It will be stored in memory as C6 86 46 F2 34H.

Example 3. Code for MOV WORD PTR 0F2H[BP], 24H.

This instruction moves 0024H to a memory location whose EA is given as the sum of the 16 bit offset value F2H and contents of BP register. The PA is generated using SS as the segment register, which is the default segment register,

The W bit must be 1 to indicate it is a word operation, F2H[BP] is specified using MOD value of 10, and R/M value of 110, and a displacement of 00 F2H. Thus the 6 byte code for this instruction would be

opcode	**W**	**MOD**	**opcode**	**R/M**	**16 bit disp**	**16 bit data**
1100011	1	10	000	110	00 F2	0024H

= C7 86 00F2 0024H

It will be stored in memory as C7 86 F2 00 24 00H.

6.3.4. Move Data to Accumulator using Direct Addressing

In 8086, AX is called as the Accumulator for 16 bit operations, and AL is called as the Accumulator for 8 bit operations. Accumulator is a very important register in 8086. This is because, for some operations like data transfer with an I/O port, or in multiplication and division operations, the Accumulator has to be used.

In view of the importance of the Accumulator, and the consequent frequent usage of the Accumulator in instructions, code for instructions involving the Accumulator is generally kept small. For example, data transfer instruction with Accumulator using direct addressing uses one byte less than instructions for data transfer with other registers using direct addressing. Template for moving data to Accumulator using direct addressing is

1	0	1	0	0	0	0	W

As there is only 1 bit in the template for identifying the size of operand, there are only 2 instructions in this category.

Example 1. Code for MOV AX,DS:24H

This instruction moves the word value at an offset of 0024H in the Data Segment to AX. In this example 0024H is the EA. The PA is generated using the DS register. Although in this instruction, the default segment register is DS, it has to be specified still. If we omit specifying the Segment register, the assembler treats it as immediate addressing mode. The assembler does not generate any SOP byte in this case, because DS is the default segment register. The W bit must be 1 to indicate it is a word operation. Thus the 3 byte code for this instruction would be

opcode	**W**	**16 bit EA**	
1010 000	1	00 24H	= A1 00 24H.

It will be stored in memory as A1 24 00H.

Example 2. Code for MOV AL,SS:44H.

This instruction moves the byte value at an offset of 0044H in the Stack Segment to AL. In this example 0044H is the EA. The PA is generated using the SS register. As SS is required to be used as the segment register, the first byte of the code will be the SOP byte with value 00110 110 = 36H. The W bit must be 0 to indicate it is a byte operation. Thus the 4 byte *code* for this instruction would be

SOP	**opcode**	**W**	**16 bit EA**	
36	1010000	0	00 44H	= 36A00044H.

It will be stored in memory as 36 A0 34 00H.

Example 3. MOV AX,LOC

Here, the source operand is specified in direct addressing mode. LOC is a symbolic memory location, specified using DW assembler directive in the data segment. This directive is discussed later. The assembler generates a 16 bit offset or EA in the instruction code, to specify this location. This is the more common method of specifying an operand using direct addressing.

Suppose the offset value corresponding to LOC is 1B57H. Then, the contents of the word at memory location DS: 1B57H is moved to AX. The 3 byte code for this instruction would be

opcode	**W**	**16 bit EA**	
1010 000	1	1B 57H	= A1 1B 57H.

It will be stored in memory as A3 57 1BH.

6.3.5. Move Data from Accumulator using Direct Addressing

Template for this instruction is

1	0	1	0	0	0	1	W

As there is only 1 bit in the template for identifying the size of operand, there are only 2 instructions in this category.

Example 1. Code for MOV SS:34H,AL.

This instruction moves the value in AL to a byte location at an offset of 0034H in the Stack Segment. In this example 0034H is the EA. The physical address is generated using the SS register. As SS is required to be used as the segment register, the first byte of the code will be the SOP byte with value 001 10 110 = 36H. The W bit must be 0 to indicate it is a byte operation. Thus the 4 byte code for this instruction would be

SOP	**opcode**	**W**	**16 bit EA**	
36	1010 001	0	00 34H	= 36 A2 00 34H

It will be stored in memory as 36 A2 34 00H.

Example 2. MOV LOC,AX.

Here, the destination operand is specified in Direct addressing mode, LOG is a symbolic memory location, specified using DW assembler directive in the data segment. This directive is discussed later. The assembler generates a 16 bit offset or EA in the instruction code, to specify this location. This is the more common method of specifying an operand using direct addressing.

Suppose the offset value corresponding to LOC is 1B57H. Then, AX is moved to the word at memory location DS:1B57H. The 3 byte code for this instruction would be

Opcode	W	16 bit EA	
1010 001	1	1B 57H	= A3 1B 57H.

It will be stored in memory as A3 57 1BH.

6.4. DATA TRANSFER BETWEEN A SEGMENT REGISTER AND A REGISTER OR A MEMORY LOCATION

In 8086, there is no instruction to load a segment register with an immediate data. A segment register can only be loaded from a memory location, or a register.

6.4.1. Move Data to a Segment Register From a Register or a Memory Location

Template for this instruction is

Byte 1

1	0	0	0	1	1	1	0

Byte 2

MOD	0	SR	R/M

As there are 7 bits in the template for identifying the operands, there are 128 instructions in this category.

Example 1. Code for MOV DS, BX.

It moves the contents of BX register to DS register. It is a 2 byte instruction. Here, BX is specified using MOD = 11, and R/M = 011. DS is specified using SR value of 11. Thus the 2 byte code for this instruction would be

Opcode	MOD	opcode	SR	R/M	
1000 1110	11	0	11	011	= 8E DBH

Example 2. Code for MOV SS, 4567H[BX+SI].

This instruction moves to SS the contents of a memory location, whose EA is the sum of the 16 bit displacement 1234H and the contents of BX and SI registers. The PA is generated using DS register. It is a 4 byte instruction. In this case, 4567H[BX+SI] is specified using MOD value of 10, and R/M value of 000. SS is specified using SR value of 10. Thus the 4 byte code for this instruction would be

Opcode	MOD	opcode	SR	R/M	16 bit disp	
1000 1110	10	0	10	000	12 34H	= 8E 90 4567H

This will be stored in memory as 8E 90 67 45H.

6.4.2. Move Data From a Segment Register to a Register or a Memory Location

Template for this instruction is Byte 1

Byte 1

1	0	0	0	1	1	0	0

Byte 2

MOD	0	SR	R/M

As there are 7 bits in the template for identifying the operands, there are 128 instructions in this category.

Example 1. Code for MOV BX, DS

This instruction moves the contents of DS segment register to BX register. It is a *2* byte instruction. In this case, BX is specified using MOD value of 11, and R/M value of 011. DS is specified using SR value of 11. Thus the 2 byte code for this instruction would be

Opcode	MOD	opcode	SR	R/M	
1000 1100	11	0	11	011	= 8C DBH

Example 2. Code for MOV 3456H[BX+SI],SS.

This instruction moves the contents of the segment register SS to a memory location, whose EA is the sum of the 16 bit displacement 3456H and the contents of BX and SI registers. The PA is generated using DS register. It is a 4 byte instruction. In this case, 3456H[BX+SI] is specified using MOD value of 10, and R/M value of 000. SS is specified using SR value of 10. Thus the 4 byte code for this instruction would be

opcode	MOD	opcode	SR	R/M	16 bit disp	
1000 1100	10	0	10	000	3456H	= 8C 90 3456H

This will be stored in memory as 8C 90 56 34H.

6.5. PUSH AND POP INSTRUCTIONS

The following instructions are described below.

1. Push contents of a register above the stack top.
2. Push contents of a memory location above the stack top.
3. Push contents of a segment register above the stack top.
4. Push contents of flags register above the stack top.
5. Pop to a register from the stack top.
6. Pop to a memory location from the stack top.
7. Pop to a segment register from the stack top.
8. Pop to flags register from the stack top.

6.5.1. Push Contents of a Register Above the Stack Top

Template for this instruction is

0	1	0	1	0	R/M

As there are 3 bits in the template for identifying the register, there are 8 instructions in this category. Notice that there is no W bit in the template. This is because the PUSH operation is always a word operation. Thus the REG bits specify a 16 bit register only.

Example 1. Code for PUSH SI

It is a 1 byte instruction, with the code 01010 110 = 56H. In this code SI is specified by the bits 110.

Stack is a LIFO (Last in, First out) data structure implemented in the RAM area. The SP is a 16 bit register which indicates the EA of the top most useful location in the stack. Suppose SP contents is 5678H. Then, it means that in Use Stack segment, the memory locations with offset addresses of 5678H, 5679H, ..., FFFFH are having useful information. Memory locations with offset addresses of 5677H, 5676H, ..., 0000H in the Stack segment are not having any useful information.

So, the SI contents are stored in 5677H and 5676H, with the MS byte of SI stored at 5677H. As the information we have stored in these locations are useful, the SP value will then change to 5676H.

	Before	*After*
(SI)	1234	1234
(SP)	5678	5676
(SS:5676)	C5	34
(SS:5677)	B2	1.2

6.5.2. Push Contents of a Memory Location Above the Top of the Stack

Template for this instruction is

Byte 1								Byte 2				
1	1	1	1	1	1	1	1	MOD	1	1	0	R/M

As there are 5 bits in the template for identifying the operands, there are 32 instructions in this category. Notice that there is no W bit in the template. This is because the PUSH operation is always a word operation.

Example 1. Code for PUSH 88H[SI].

This instruction pushes above the top of the stack, the contents of the word location whose EA is the sum of the contents of SI and the 16 bit offset 0088H. The MS byte of the word is stored at a higher address and the LS byte at a lower address. It is a 4 byte instruction, with the code as shown below.

opcode	**MOD**	**opcode**	**R/M**	**16 bit disp**	
11111111	10	110	100	00 88H	= FF B4 00 88H

It is stored in memory as FF B4 88 00H. Suppose in memory we have the following instruction code.

opcode	**MOD**	**opcode**	**R/M**	
1111 1111	11	110	110	= FF F6H

This is another possible code for the instruction PUSH SI. But, this is 2 bytes compared to the 1 byte code of 56H for the same instruction. So the assembler always translates PUSH SI as 56H instead of FF F6H.

6.5.3. Push Contents of a Segment Register Above the Top of the Stack

Template for this instruction is

0	0	0	SR	1	1	0

As there are 2 bits in the template for identifying the segment register, there are 4 instructions in this category.

Example 1. Code for PUSH DS.

It is a 1 byte instruction, with the code 000 11110 = 1EH. In this code DS is specified by the bits 11.

SI loaded with the contents of 5679H. As the information in these locations is no longer useful, the SP value will then change to 567AH.

	Before	*After*
(SI)	1234	B2C5
(SP)	5678	567A
(SS:5678)	C5	C5
(SS:5679)	B2	B2

6.5.4. Push Contents of Flags Register above the Stack Top

The mnemonic for this instruction is PUSHF and the template for this instruction is

	Before	*After*
(SI)	1234	1234
(SP)	5678	5676
(SS:5676)	C5	34
(SS:5677)	B2	12

6.5.5. Pop to a Register From the Stack Top

Template to a register from the stack top.

As there are 3 bits in the tempale for identifying the register, there are 8 instructions in this category. Notice that there is no W bit in the template. This is because the POP operation is always a word opewration. Thus the REG bits spcify a 16 bit register only.

Example 1. Code for POP SI

It is a 1 byte instruction, with the code 01011 110 = 5EH. IN this code SI is specified by the bits 110.

Stack is a LIFO (Last in, First out) data structure implemented in the RAM area. The SP is a 16 bit register which indicates the EA of the top most useful location in the stack segment, the memory locations with offset addresses of 5678H, 5679H, ..., FFFFH are having useful information. Memolry locations with offset addresses of 5677H, 5676H, ..., 0000H is the stack segment are not having any useful information.

So, the SI register is loaded from 5678H and 5679H, with the MS byte of Si loaded with the contents of 5679H. As the information in these locations is no longer useful, the SP value will then change to 567AH.

	Before	*After*
(SI)	1234	B2C5
(SP)	5678	567A
(SS:5678)	C5	C5
(SS:5679)	B2	12

6.5.6. Pop to a Memory Location From the Top of the Stack

Template for this instruction is

Byte 1

1	0	0	0	1	1	1	1

Byte 2

MOD	0	0	0	R/M

As there are 5 bits in the template for identifying the operands, there are 32 instructions in this category. Notice that there is no W bit in the template. This is because the POP operation is always a word operation.

Example 1. Code for POP 83H[SI].

This instruction pops the top of the stack information to the word location whose EA is the sum of the contents of SI and the 16 bit offset 0083H. The LS byte of the word is loaded from the top of the stack and the MS byte is loaded from the next higher address in the stack. It is a 4 byte instruction, with the code as shown below.

Opcode	**MOD**	**opcode**	**R/M**	**16 bit disp**	
1000 1111	10	000	100	0083H	= 8F 84 00 83H

It is stored in memory as 8F 84 83 00H. Suppose in memory we have the following instruction code.

opcode	MOD	opcode	R/M	
1000 1111	11	000	110	= 8F C6H

This is another possible code for the instruction POP SI. But, this is 2 bytes compared to the 1 byte code of 5EH for the same instruction. So the assembler always translates POP SI as 5EH instead of 8F C6H.

6.5.7. Pop to a Segment Register From the Stack Top

Template for this instruction is

0	0	0	SR	1	1	1

	Before	*After*
(FLAGS)	1234	B2C5
(SP)	5678	567A
(SS:5678)	C5	C5
(SS:5679)	B2	B2

As there are 2 bits in the template for identifying the segment register, there are 4 instructions in this category.

Example 1. Code for POP DS.

It is a 1 byte instruction, with the code 000 11 111 = 1FH. In this code DS is specified by the bits 11.

Before	*After*	
(DS)	1234	
(SP)	5678	567A
(SS:5678)	C5	C5
(SS:5679)	B2	B2

6.5.8. Pop to Flags Register from the Stack Top

The mnemonic for this instruction is POPF and the template for this instruction is

1	0	0	1	1	1	0	1	= 9DH

6.6. EXCHANGE INSTRUCTIONS

The following instructions are described in this section :

1. Exchange contents of a register with AX.
2. Exchange contents of a register with another register or memory location.

6.6.1. Exchange Contents of a Register with AX

The template for this instruction is

1	0	0	1	0	REG

As there are 3 bits in the template for identifying a register, there are eight instructions in this category. Notice that there is no W bit in the template.

This is because the exchange operation with AX is always a word operation.

Example 1. Code for XCHG AX, SI

It is a 1 byte instruction with the code 10010110 = 96H

It exchanges the contents of AX and SI.

Exchange contents of register with another register or memory location :

Template

Byte 1									Byte 2		
1	0	0	0	0	1	1	W		MOD	REG	R/M

Example 2. XCHG BL, 38H [SI]

It exchanges the contents of BL and a by location whose EA is the sum of the contents of SI and the 16-bit offset 0083H. The PA is generated using DS reg. The code is

Opcode	W	MOD	REG	R/M	16-bit Disp	
1000011	0	10	011	100	0038H	= 869C0038H

6.6.2. Data Transfer with I/O Ports

8086 will not directly communicate with input/output(I/O) devices, as the I/O devices are electromechanical and so are very slow compared to the speed of operation of the μP. So a μP will communicate with I/O ports which are semiconductor chips working at electronic speeds, which in turn will communicate with I/O devices. An I/O port is like a 8 bit register, which will be outside the μP, with 8 pins for communication with the fast μP and another 8 pins for communication with the slow I/O device.

The following instructions are described in this section.

- Move Data to Accumulator from an input port using
  - Fixed port addressing
  - Variable port addressing
- Move Data from Accumulator from an output port using
  - Fixed port addressing
  - Variable port addressing

6.6.3. Move Data to Accumulator from an Input using Fixed Port Addressing

The template for the instruction to move data to Accumulator from an input port using fixed addressing.

Notice the absence of REG bits in the template. This is because, only Accumulator can take part in data transfer with input ports. Also, notice the presence of W bit in the template. So, in this case we can move a word to AX or a byte to AL from an input port. In this instruction, a one byte port address will be provided, after the template byte shown above. In this type of addressing a port, the program can not change the port address as the execution of the program proceeds.

Example 1. Code for IN AL,83H

This instruction moves the contents of input port whose address is 83H to AL. It is a byte instruction with the following code.

	Before	*After*
(AL)	34	78
(i/p port 83)	78	78

Example 2. Code for IN AX, 83H

This instruction moves the contents of input port whose address is 83H to AL, and the contents of input port with address 84H to AH. It is a 2 byte instruction with the following code.

	Before	*After*
(AX)	1234	5678
(i/e port 83)	78	78
(i/e port 84)	56	56

6.6.4. Move Data to Accumulator from an Input using Variable Port Addressing

The template for the instruction to move data to Accumulator from an input port using variable port addressing is.

In this instruction, port address will not be provided explicitly in the instruction, after the template byte shown above. The address of the port is implied to be the contents of DX register always. This method of addressing an input port is called as variable port addressing a port, the program can change the port address as the execution of the program proceeds.

Example 1. Code for IN AL,DX

This instruction moves the contents of input port whose 16 bit port address is given in DX to AL. It is a 1 byte instruction with the following code.

	Before	*After*
(AL)	34	78
(DX)	CBDA	CBDA
(i/e port CBDA)	78	78

Example 2. Code for IN AX, DX

This instruction moves to AL the contents of input port whose 16 bit port address is given in DX, and to AH from the port with the next higher address.

It is a 1 byte instruction with the following code.

		Before	*After*
(AX)		1234	5678
(DX)		CBDA	CBDA
(i/e port CBDA)	78	78	
(i/e port CBDA)	56	56	

6.6.5. Move Data from Accumulator to an Output Port Using Fixed Port Addressing

The template for the instruction to move data from the Accumulator to an output port using fixed port addressing is.

Notice the absence of REG bits in the template. This is because; only Accumulator can take part in data transfer with output ports. Also, notice the presence of W bit in the template. So, in this case we can move a word from AX or a byte from AL to an output port. In this instruction, a one byte port address will be provided, after the template byte shown above.

Example 1. Code for OUT 83H, AL

This instruction moves the contents of AL to an output port whose address is 83H, and the contents of AH to an output port with address 84H. It is a 2 byte instruction with the following code.

	Before	*After*
(AX)	1234	5678
(i/e port 83)	78	34
(i/e port 84)	56	12

6.6.6. Move Data from Accumulator to an Output Port using Variable Port Addressing

The template for the instruction to move data from Accumulator an output port using variable port addressing is

In this instruction, port address will not be provided explicitly in the instruction, after the template byte shown above. The address of the port is implied to be the contents of DX register always.

Example 1. Code for OUT DX,AL

This instruction moves the contents of AL to an output port whose 16 bit port address is given in DX. It is a 1 byte instruction with the following code.

	Before	***After***
(AX)	**34**	**34**
(DX)	ACBC	ACBC
(o/p port ACBC)	78	34

Example 2. Code for OUT DX,AX

This instruction moves the contents of AL to an output port whose 16 bit port address is given in DX. It is a 1 byte instruction with the following code.

	Before	*After*
(AL)	34	34
(DX)	ACBC	ACBC
(o/p port ACBC)	78	34

Example 2. Code for OUT DX,AX

This instruction moves the contents of AL to an output port whose 16 bit port address is given in DX, and the content of AH to the port with the next higher address. It is a 1 byte instruction with the following code.

	Before	***After***
(AX)	1234	1234
(DX)	ACBC	ACBC
(o/p port ACBC)	78	34
(o/p port ACBD)	56	12

6.7. DATA CONVERSION INSTRUCTION

In this chapter the following instructions are explained.

1. Translate the value in AL.
2. Load effective address of a memory location into a register.
3. Load DS and another register with contents of 2 memory words.
4. Load ES and another register with contents of 2 memory words.
5. Load AH with LS byte of Flags register.
6. Store AH in LS byte of Flags register.

In these instructions, except for 'Store AH IN LS byte of Flags register', the flas are not affected.

6.7.1. Translate the Value in AL (XLAT)

The mnemonic for this instruction is XLAT. Another equivalent mnemonic is XLATB, which stands for 'Translate byte'. The template for this instruction is

1	1	0	1	0	1	1	1	= D7H

It is a 1 byte instruction with the code D7H. This instruction is equivalent to MOV AL,[AL][BX]. It is very useful for translating codes such as translating ASCII to EBCDIC code while transferring data from an IBM-PC to an IBM mainframe computer. It can be done as follows.

First of all, store in the data segment, say starting from offset location 4000H, the table of EBCDIC characters. For example, the character '6' has the ASCII code of 36H and EBCDIC code of F6H. Thus F6H is stored in location 4036H.

Now to convert the ASCII code of 36H to EBCDIC code of F6H, we just execute XLAT instruction with AL contents as 36H and BX contents as 4000H. This results in AL contents changing to F6H.

	Before	*After*
(AL)	36	F6
(BX)	4000	4000
(DS:36[BX])	F6	F6

6.7.2. Load Effective Address (EA) of a Memory Location into a Register

The template for this instruction is

Byte 1								Byte 2		
1	0	0	0	1	1	0	1	MOD	REG	R/M

As there are 8 bits in the template for identifying the operands, there are 256 instructions in this category. Notice the absence of W bit in the template. This is because, in this case only a word can be moved as EA to a register.

Example 1. Code for LEA BX,98H[SI].

It loads BX with the EA which is the sum of the contents of SI and the 16 bit offset 0088H. It is a 4 byte instruction with the following code.

opcode	**MOD**	**REG**	**R/M**	**16 bit disp**	
1000 1101	10	011	100	00 98	= 8D 9C 00 98H

Example 2. Code for LEA SI, 88H[BP].

It loads SI with the EA which is the sum of the contents of BP and the 8 bit offset 88H. It is a 3 byte instruction with the following code.

opcode	**MOD**	**REG**	**R/M**	**8 bit disp**	
1000 1101	01	110	110	48	= 8D 86 48H

	Before	*After*
(SI)	1234	5688 = 5600 + 88
(BP)	5600	5600

Example 3. Code for LEA BX, [SI].

It loads BX with the EA which is the contents of SI, It is a 2 byte instruction with the following code.

opcode	**MOD**	**REG**	**R/M**	
1000 1101	00	011	100	= 8D 1CH

	Before	***After***
(BX)	1234	5600
(SI)	5600	5600

LEA BX,[SI] instruction is same in its operation as MOV BX,SI. But the 2 byte code for MOV BX,SI will be different from the 2 byte code of LEA BX,[SI].

Example 4. Code for LEA BX,LOCN.

It loads BX with the offset of the variable LOCN from the start of the segment in which LOCN is defined. In other words, it loads BX with the EA of LOCN. The assembler generates the code using direct addressing for specifying LOCN. Thus it is a 4 byte instruction with the following code, where it is assumed that the EA of LOCN is 0002H.

opcode	**MOD**	**REG**	**R/M**	**16 bit EA**	
1000 1101	00	011	110	00 02H	= 8D 1E 00 02H

MOV BX, OFFSET LOCN instruction performs movement of EA of LOCN to BX and so is same in operation as LEA BX, LOCN. However, their opcodes will be different.

6.7.3. Load a Register and the Segment Register DS With Contents of Two Memory Words

The template for this instruction is

Byte 1								Byte 2		
1	1	0	0	0	1	0	1	MOD	REG	R/M

As there are 8 bits in the template for identifying the operands, there are 256 instructions in this category. Notice the absence of W bit in the template. This is because, in this case only words can be moved to a 16 bit register and DS.

Example 1. Code for LDS BX,82H[SI].

It loads BX with the contents of a word location whose EA is the sum of the contents of SI and the 16 bit offset 0082H. It then loads DS with the contents of the next word in memory. The PA is generated using DS. It is a 4 byte instruction with the following code.

Opcode	MOD	REG	R/M	16 bit disp	
1100 0101	10	011	100	00 82	= C5 9C 00 82H

	Before	***After***
(SI)	4600	4600
(DS:4682)	5683	5683
(DS:4684)	6789	6789
(BX)	1234	5683
(DS)	3456	6789

6.7.4. Load a Register and The Segment Register ES With Contents of Two Memory Words

The template for this instruction is

Byte 1								Byte 2		
1	1	0	0	0	1	0	0	MOD	REG	R/M

As there are 8 bits in the template for identifying the operands, there are 256 instructions in this category. Notice the absence of W bit in the template. This is because, in this case only words can be moved to a 16 bit register and ES.

Example 1. Code for LES BX,42H[SI].

It loads BX with the contents of a word location whose EA is the sum of the contents of SI and the 8 bit offset 42H. It then loads ES with the contents of the next word in memory. The PA is generated using DS. It is a 4 byte instruction with the following code.

opcode	MOD	REG	R/M	8 bit disp	
1100 0100	01	011	100	42	= C4 5C 42H

	Before	***After***
(SI)	4600	4600
(DS:4642)	5683	5683
(DS:4644)	6789	6789
(BX)	1234	5683
(ES)	3456	6789

6.7.5. Load a Register and the Segment Registers with Contents of Two Memory

As there are 8 bits in the template for identifying the operands, there are 256, instructions in this category. Notice the absences of W bit in the template. This is because, in this case only words can be moved to a 16 bit register and ES.

Example 1. Code for LES BX, 42H[SI]

It loads BX with the contents of a word location whose EA is the sum contents of SI and the 8 bit offset 42H. It then loads ES with the contents of the next word in memory. The PA is generated using DS. It is a 4 byte instruction with the following code.

	Before	***After***
(SI)	4600	4600
(DS:4642)	5683	5683
(DS:4644)	6789	6789
(BX)	1234	5683
(ES)	3456	6789

6.7.6. Load AH With the LS Byte of Flags Register

The mnemonic for this instruction is LAHF. The template for this instruction is

1	0	0	1	1	1	1	1	= 9FH

It is a 1 byte instruction with the code 9FH. It loads AH with the contents of the LS byte of Flags register.

	Before	***After***
(AH)	36	F6
(FLAGS)	12F6	12F6

6.7.7. Store AH Contents in the LS Byte of Flags Register

The mnemonic for this instruction is SAHF. The template for this instruction is

1	0	0	1	1	1	1	0	= 9EH

It is a 1 byte instruction with the code 9EH. It stores AH contents in the LS byte of flags register. This instruction is used for changing LS byte of Flags register to the desired value.

	Before	***After***
(AH)	36	36
(FLAGS)	12F6	1236

8086 has very powerful instructions in the Arithmetic group compared to 8085 microprocessor. It can perform add, subtract, multiply and divide lions on 8 or 16 bit quantities which can be unsigned or signed numbers.

6.8. ADD GROUP OF INSTRUCTIONS

In this group, we have instructions to

(a) Add contents of two registers, with or without carry.
(b) Add contents of a register and a memory location, with or without carry.
(c) Add immediate data to a register or a memory location, with or without carry.
(d) Increment contents of a register or a memory location.
(e) To perform, decimal adjustment after addition.
(f) To perform ASCII adjustment after addition

6.8.1. Add Contents of Two Registers or a Register and a Memory Location

Template for this instruction is

Byte 1

0	0	0	0	0	0	D	W

Byte 2

MOD	REG	R/M

As there are 10 bits in the template for identifying the operands, there are 1024 instructions in this category.

Example 1. Code for ADD AX, BX.

This instruction adds the contents of BX register to AX register, and stores the result in AX register. It is a 2 byte instruction.

In this example, we can specify either of AX or BX using the REG field. If we specify AX using REG field, then the D bit must be 1, indicating that AX is the destination register. The REG field must be 000 to indicate AX register. The W bit must be 1 to indicate it is a word operation. In this case, BX is specified using MOD value of 11, and R/M value of 011. Thus the 2 byte code for this instruction would be

opcode	**D**	**W**	**MOP**	**REG**	**R/M**	
0000 00	1	1	11	000	011	=03C3H

If we specify BX using REG field, then the D bit must be 0, indicating that BX is the source register. The REG field must be 011 to indicate BX register. The W bit must be 1 to indicate it is a word operation. In this case, AX is specified using MOD value of 11, and R/M value of 000. Thus the 2 byte code for this instruction would be

opcode	**D**	**W**	**MOD**	**REG**	**R/M**	
0000 00	0	1	11	011	000	= 01 D8H

Thus ADD AX.BX has two equally valid opcodes as shown above. An assembler generates one of these as the opcode.

After the addition operation, the status flags are affected depending on the result.

	Before	***After***
(AX)	122F	8326
(BX)	70F7	70F7
(FLAGS)	any value	Z = 0, S = 1, P = 0, C = 0, A = 1, O = 1

In the above example, Z flag becomes 0 as 8326H, the result of addition, is non zero. The S flag becomes 1 to indicate that the result is negative, as 8326H = 1000 0011 0010 0110 starts with a 1 in the MS bit position. The S flag will be meaningful for us only if we are working with signed numbers. The P flag becomes 0 to indicate that there are an odd number of 1's (there are 3 of them) in the LS byte of result. The C flag becomes 0, as no carry is generated in the addition. However, the A flag becomes 1 as a carry is generated during the addition of the

LS hex digits. Finally, the O flag is set to 1 as the addition of the two positive numbers in AX and BX has resulted in a negative answer.

Example 2. Code for ADD DS:0A256H[BP],DX.

This instruction adds the contents of DX register to a memory location whose EA is given as the sum of the 16 bit offset value A256H and the contents of BP register. The PA is generated using DS as the segment register. The result is stored in the same memory location.

Here we have to specify DX using REG field. The D bit must be 0, indicating that DX is the source register. The REG field must be 010 to indicate DX register. The W bit must be 1 to indicate it is a word operation, F256H[BP] is specified using MOD value of 10, and R/M value of 110, and a displacement of F256H.

Whenever BP is used to generate the EA, the default segment would be SS. In this example, as we want the segment register to he DS, we have to provide the segment override prefix byte (SOP byte) to start with. The SOP byte is 001 SRI 10, where SR indicates the segment override we desire.

To specify DS register, the SOP byte would be 001 11 110 = 3EH. Thus the 5 byte code for this instruction would be

SOP	**opcode**	**D**	**W**	**MOD**	**REG**	**R/M**	**16 bit disp**
3E	0000 00	0	1	10	010	110	A256

= 3E 01 96 A256H

	Before	***After***
(BP)	0100	0100
(DS:A256)	122F	8326
(DX)	70F7	70F7
(FLAGS)	any value	Z= 0, S= 1, P= 1, C= 0, A= 1, O= 1

6.8.2. Add Immediate Data to a Register or a Memory Location

Template for this instruction is

Byte 1

1	0	0	0	0	0	S	W

Byte 2

MOD	0	0	0	R/M

As there are 7 bits in the template for identifying the operands, there are 128 instructions in this category.

The S bit stands for Sign Extension. This bit will be 0 always for a byte operation (W = 0).

If it is a word operation (W = 1), the S bit must be a 1 only if the immediate data is <= 7FH. In such a case the assembler generates only 1 byte code for the immediate data. However, during execution, the one byte data will be sign extended to make it a 2 byte operand.

If the immediate data is > 7FH, the S bit must be a 0, and the assembler generates a 2 byte code for the immediate data.

Example 1. Code for ADD CX,98H.

As it is a word operation, with immediate data <= 7FH, the S bit will be 0, and the assembler generates 2 byte code for the immediate data. It is a 4 byte instruction.

opcode	**S**	**W**	**MOD**	**opcode**	**R/M**	**16 bit data**	
1000 00	0	1	11	000	001	0098H	= 81 C1 00 98H

After the addition operation, the status flags are affected depending on the result.

Example 2. Code for ADD CL, 88H

As it is a byte operation, the S bit will be 0. It is a 3 byte instruction.

opcode	S	W	MOD	opcode	R/M	8 bit data	
100000	0	0	11	000	001	88H	= 80 C1 88H

After the addition operation, the status flags are affected depending on the result.

Example 3. Code for ADD CX,38H.

As it is a word operation, with immediate data <= 7FH, the S bit will be 1, and the assembler generates 1 byte code for the immediate data. Thus it is a 3 byte instruction.

opcode	*S*	W	MOD	opcode	*WM*	8 bit data	
1000 00	I	1	11	000	001	38H	= 83 C1 38H

After the addition operation, the states flags are affected depending on the result.

6.8.3. Add Immediate Data to Accumulator

Template for this instruction is

0	0	0	0	0	1	0	W

As there is only one bit in the template for identifying the operands, there are 2 instructions in this category. The assembler always generates 1 byte or 2 bytes of code for the immediate data depending on whether it is a byte or a word operation.

Example 1. Code for ADD AX, 98H.

As it is a word operation, the assembler generates 2 byte code for the immediate data. So, it is a 3 byte instruction,

opcode	W	16 bit data	
0000 010	1	00 98H	= 05 00 98H

After the addition operation, the status flags are affected depending on the result.

Example 2. Code for ADD AL, 98H

As it is a byte operation, the assembler generates 1 byte code for the immediate data. So, it is a 2 byte instruction.

opcode	W	8 bit data	
0000010	0	98H	= 04 98H

After the addition operation, the status flags are affected depending on the result.

6.8.4. Add with Carry Contents of Two Registers or a Register and a Memory Location

It is required sometimes to perform addition of two numbers each of which is a multiword or multi byte number. Let us say, we want to add the numbers shown below.

```
       1
 1 2  3 4 5678H
+A B C 0 E012H
--------------
 B D F 5 368AH
```

Using the ADD instruction discussed earlier, we first of all add 5678H and E012H. We get 368A as the LS word of result, with a cany of 1 to the next stage. Now we are required to add 1234 and ABC0 along with the carry to get the correct MS word of the result. In view of this, instructions are provided in the instruction set to add two numbers along with the carry flag.

Template for adding two registers along with carry, or adding a register and a memory location along with carry is

Byte 1

0	0	0	1	0	0	D	W

Byte 2

MOD	REG	R/M

As there are 10 bits in the template for identifying the operands, there are 1024 instructions in this category. The mnemonic for this instruction is ADC.

Example 1. Code for ADC DS:0F246H[BP],DX.

This instruction adds the contents of DX register along with carry to a memory location whose EA is given as the sum of the 16 bit offset value F246H and the contents of BP register. The PA is generated using DS as the segment register. The result is stored in the same memory location.

Here we have to specify DX using REG field. The D hit must be 0, indicating that DX is the source register. The REG field must be 010 to indicate DX register. The W bit must be 1 to indicate it is a word operation, F246H[BP] is specified using MOD value of 110 and R/M value of 110, and a displacement. of F246H.

Whenever BP is used to generate the EA, the default segment would be 88. In this example, as we want the segment register to be DS, we have to provide the segment override prefix byte (SOP byte) to start with. The SOP byte is 001SR110, where SR indicates the segment override we desire.

To specify DS register, the SOP byte would be 001 11 110 = 3EH. Thus the 5 byte code for this instruction would be

SOP	Opcode	D	W	MOD	REG	R/M	16 bit disp
3E	0001 00	0	1	10	010	110	F2 46

= 3E 11 96 F2 46H

	Before	*After*
(BP)	0100	0100
(DS:F346)	122F	8327
(DX)	70F7	70F7
(FLAGS)	C = 1	Z = 0, S = 1, P = 1, C = 0, A = 1, O = 1

6.8.5. Add With Carry Immediate Data to a Register or a Memory Location

Template for this instruction is

Byte 1

1	0	0	0	0	0	S	W

Byte 2

MOD	0	1	0	R/M

As there are 7 bits in the template for identifying the operands, there are 128 instructions in this category.

The S bit stands for Sign Extension. This bit will be 0 always for a byte operation (W = 0).

If it is a word operation (W = 1), the S bit must be a 1 only if the Immediate dala is <= 7FH. In such a case the assembler generates only 1 byte code for the immediate data. However, during execution, the one byte data will be sign extended to make it a 2 byte operand.

If the immediate data is > 7FH, the S bit must be a 0, and the assembler generates a 2 byte code for the immediate data.

Example 1. Code for ADC CX,98H.

As it is a word operation, with immediate data > 7FH, the S bit will be 0, and the assembler generates 2 byte code for the immediate data. It is a 4 byte instruction,

opcode	S	W	MOD	opcode	R/M	16 bit data	
100000	0	1	11	010	001	00 98H	= 81 D1 00 98H

After the addition operation, the status flags are affected depending on the result.

Example 2. Code for ADC CL,98H.

As it is a byte operation, the S bit will be 0. It is a 3 byte instruction.

opcode	S	W	MOD	opcode	R/M	8 bit data	
1000 00	0	0	11	010	001	98H	= 80 D1 98H

After the addition operation, the status flags are affected depending on the result.

Example 3. Code for ADC CX, 58H.

As it is a word operation, with immediate data <= 7FH, the S bit will be 1, and the assembler generates 1 byte code for the immediate data. Thus it is a 3 byte instruction.

opcode	S	W	MOD	opcode	R/M	8 bit data	
1000 00	1	1	11	010	001	58H	= 83 D1 58H

After the addition operation, the status flags are affected depending on the result.

6.8.6. Add with Carry Immediate Data to Accumulator

Template for this instruction is

0	0	0	1	0	1	0	W

As there is only one bit in the template for identifying the operands, there are 2 instructions in this category. The assembler always generates 1 byte or 2 bytes of code for the immediate data depending on whether it is a byte or a word operation.

Example 1. Code for ADC AX,98H.

As it is a word operation, the assembler generates two byte code for the immediate data. So, it is a 3 byte instruction.

Opcode	W	16 bit data	
00010010	1	00 98H	= 15 00 98H

After the addition operation, the status flags are affected depending on the result.

Example 2. Code for ADC AL,98H.

As it is a byte operation, the assembler generates 1 byte code for the immediate data. So, it is a 2 byte instruction.

opcode	W	8 bit data	
0001 010	0	98H	= 14 98H

After the addition operation, the status flags are affected depending on the result.

6.8.7. Increment Contents of a Register or a Memory Location

This instruction increments the contents of a register or a memory location by 1. It can be a byte or a word operation. All the flags, except carry flag, are affected depending on the result. Template for this instruction is

Byte 1

1	1	1	1	1	1	1	W

Byte 2

MOD	0	0	0	R/M

As there are 6 bits in the template for identifying the operands, there are 64 instructions in this category. The mnemonic for this instruction is INC.

Example 1. Code for INC WORD PTR DS:0F346H[BP]

This instruction increments the contents of a word location whose EA is given as the sum of the 16 bit offset value F346H and the contents of BP register. The PA is generated using DS as the segment register.

The W bit must be 1 to indicate it is a word operation. F346H[BP] is specified using MOD value of 10, and R/M value of 110, and a displacement of F346H.

Whenever BP is used to generate the EA, the default segment would be SS. In this example, as we want the segment register to be DS, we have to provide the segment override prefix byte

(SOP byte) to start with. The SOP byte is 001SR110, where SR indicates the segment override we desire.

To specify DS register, the SOP byte would be 001 11 110 = 3EH. Thus the 5 byte code for this instruction would be

SOP	opcode	W	MOD	opcode	R/M	16 bit disp
3E	1111 111	1	10	000	110	F3 46H

= 3E FF 86 F3 46H

	Before	*After*
(BP)	0100	0100
(DS:F346)	122F	1230
(FLAGS)	any value	Z = 0, S = 0, P = 1, A = 1, O = 0

Example 2. Code for INC CL.

The W bit must be 0 to indicate it is a byte operation. CL is specified using MOD value of 11, and R/M value of 001.

It is a 2 byte instruction with the code

opcode	W	MOD	opcode	R/M	
1111 111	0	11	000	001	= FE C1H

	Before	*After*
(CL)	7F	80
(FLAGS)	any value	Z = 0, S = 1, P = 0, A = 1, O = 1

Similarly, INC CX can be coded as

opcode	W	MOD	opcode	R/M	
1111 111	1	11	000	001	= FF C1H

However, as incrementing of a word register is a frequently used instruction, for greater efficiency, the assembler will generate a single byte code as exolaiued next.

6.8.8. Increment the Contents of a Word Register

Template for this instruction is

0	1	0	0	0	REG

As there are 3 bits in the template for identifying the word register, there are 8 instructions in this category. The mnemonic for this instruction is also INC.

Example 1. Code for INC CX.

CX is specified using REG value of 001. Thus the 1 byte code for this instruction would be 41H as shown below.

opcode	REG	
0100 0	001	= 41H

	Before	*After*
(CX)	0100	0101
(FLAGS)	any value	Z = 0, S = 0, P = 1, A = 0, O = 0
(DS:F346)	122F	1230
(FLAGS)	any value	Z = 0, S = 0, P = 1, A = 1, O = 0

Example 2. Code for INC CL.

The W bit must be 0 to indicate it is a byte operation, CL is specified using MOD value of 11, and R/M value of 001.

It is a 2 byte instruction with the code

opcode	W	MOD	opcode	R/M	
1111 111	0	11	000	001	= FE C1H

	Before	After
(CL)	7F	80
(FLAGS)	any value	Z = 0, S = 1, P = 0, A = 1, O = 1

Similarly, INC CX can be coded as

opcode	W	MOD	opcode	R/M	
1111 111	1	11	000	001	= FF C1H

However, as incrementing of a word register is a frequently used instruction, for greater efficiency, the assembler will generate a single byte code as explained next.

6.8.9. Increment the Contents of a Word Register

Template for this instruction is

0	1	0	0	0	REG

As there are 3 bits in the template for identifying the word register, there are 8 instructions in this category. The mnemonic for this instruction is also INC.

Example 1. Code for INC CX.

CX is specified using REG value of 001. Thus the 1 byte code for this instruction would be 41H as shown below.

Opcode	REG	
0100 0	001	= 41H

	Before	After
(CX)	0100	0101
(FLAGS)	any value	Z = 0, S = 0, P = 1, A = 0, O = 0

6.8.10. Perform Decimal Adjustment After Addition

In a digital computer every thing is represented using 0's and 1 's oniy. For example, an instruction will have a code using only 0's and 1's. Data is also represented using 0's and 1's. And data can he of different types like unsigned numbers, signed numbers, floating point numbers, binary coded decimal (BCD) numbers etc. Thus a series of 0's and 1's will acquire a value based on the interpretation.

BCD numbers : Many a time, we are required to represent decimal numbers in a computer, and perform arithmetic on these numbers. For example, we may be required to total the marks a student has obtained in 5 different subjects. where obviously, the marks are awarded in decimal notation.

For this purpose BCD code is extensively used. There are 2 types of BCD numbers. They are Packed BCD numbers and Unpacked BCD numbers.

In packed BCD notation, 4 bits are used to code a digit, and so two digits of information is stored in a byte. For example, decimal 39 is represented in packed BCD as 0011 1001. Codes 1010, 1011, 1100, 1101, 1110, and 1111 are illegal in BCD notation. Similarly decimal 1024 is represented in packed BCD as 0001 0000 0010 0100. Same decimal 1024 is represented in binary as 0000 0100 0000 0000. As stated earlier, the value of a series of 0's and 1's depends on the interpretation.

In unpacked BCD notation, 8 bits are used to code a digit, and so only one digit of information is stored in a byte. In this notation, the MS 4 bits of a byte are 0000 always. For example, decimal 39 is represented in unpacked BCD as 0000 0011 0000 1001. Same decimal 39 is represented in binary as 0010 0111. As stated earlier, the value of a series of 0's and 1's depends on the interpretation.

DAA instruction : Suppose we want to add the two decimal numbers 38 and 45. They will be represented in packed BCD as 0011 1000 and 0100 0101. The addition results in 0111 1101, If we interpret this result as a packed BCD number, the answer is incorrect as well as illegal. This is because, 1101 is an illegal BCD number. This is where the DAA instruction proves its usefulness. All that is required to be done is to add the BCD numbers and store the result in AL and then execute DAA instruction. DAA stands for 'Decimal Adjust after Addition'.

Opcode for this instruction is 27H.

Working of DAA Instruction :

- If the LS hex digit in AL is <= 9 and auxiliary carry flag is 0, the LS hex digit value will not be altered.
- If the LS hex digit is >9, or if the auxiliary carry flag is set to 1, it adds 6 to the LS hex digit of AL and increments the MS hex digit position. The carry flag will be set to 1 if the MS hex digit was incremented from F to 0.
- If the MS hex digit is <= 9 and carry flag is 0, the MS hex digit will not be altered, and carry flag is reset to 0.
- If the MS hex digit is >9, or if CY flag is set to 1 it adds 6 to the MS hex digit of AL and sets carry flag to 1.

The working of DAA instruction depends on the contents of AL register. In effect, it adds 00H, 06H, 60H, or 66H to AL so that we get the correct packed BCD answer in AL.

If the LS hex digit in AL is <= 9 and auxiliary carry flag is 0, the LS hex digit value will not be altered.

If the LS hex digit is > 9, or if Auxiliary carry flag is set to 1, it adds 6 to the LS hex digit of AL and increments the MS hex, digit if this addition resulted in a carry to the MS digit position. In this process, the Carry flag will be set to 1 if the MS hex digit was incremented from F to 0.

Now, if the MS hex digit is <= 9 and Carry flag is 0, the MS hex digit will not be altered, and Carry flag is reset to 0,

If the MS hex digit is > 9, or if Carry flag is set to 1, it adds 6 to the MS hex digit of AL and sets Carry flag to 1.

Example 1. Let us say, we add 45 BCD and 38 BCD, and store the result 7DH in AL. In this case, C flag = 0 and A flag = 0. But as D is a invalid BCD code, the DAA instruction adds 06H to AL. Thus we get 83H in AL, which will now be interpreted as 83 BCD, the correct sum of 45 and 38.

Example 2. Let us say, we add 63 BCD and 88 BCD, and store the result EBH in AL. In this case, C flag = 0 and A flag = 0. But as both E and B are invalid BCD codes, the DAA instruction adds 66H to AL. Thus we get 51H in AL and C flag will be set to 1. This will now be interpreted as 51 BCD with C flag = 1, the correct sum of 63 and 88.

Example 3. Let us say, we add 53 BCD and 36 BCD, and store the result 89H in AL. In this case, C flag = 0 and A flag = 0. Since both 8 and 9 are valid BCD codes, the DAA instruction does not add anything to AL. In other words, it is same as adding 00 to AL. The result in AL will now be interpreted as 89 BCD, the correct sum of 53 and 36.

Example 4. Let us say, we add 99 BCD and 88 BCD, and store the result 21H in AL. In this case, C flag = 1 and A flag = 1. So, although both 2 and 1 are valid BCD codes, the DAA instruction adds 66H to AL. Thus we get 87H in AL and C flag will be set to 1. The result in AL will now be interpreted as 8? BCD with C flag = 1, the correct sum of 99 and 88.

Example 5. Let us say, we add 63 BCD and 42 BCD, and store the result ASH in AL. In this case, C flag = 0 and A flag = 0. But as A is a invalid BCD code, the DAA instruction adds 60H to AL. Thus we get 05H in AL, with C flag = 1, which will now be interpreted as 05 BCD with C flag = 1, the correct sum of 63 and 42.

Example 6. Let us say we add 63 BCD and 88 BCD and store the result EBH in A.

C Flag = 0, and A Flag = 0. But as both E and B are invalid BCD numbers, DAA instruction adds 66H to AL. Hence we get 51H in AL and C = 1.

Now the correct result is 51 BCD with CF = 1.

Example 2. Let us say we add 99 BCD and 88 BCD and store the result 21H in AL. Here CY = 1 and A Flag = 1. Although both 2 and 1 are valid BCD codes, the DAA instruction adds 66H to AL. Hence we get 87H in AL, with CY = 1. The result in AL will now be interpreted as 87 BCD with CY = 1.

6.8.11. Perform ASCII Adjustment After Addition

The data we enter form a keyboard will be in ASCII format. If we enter on the keyboard values like '7' and '8', its representation inside the computer will be 37H and 38H. Suppose we want to add the characters '7' and '8', and get the result as 01 05 in Unpacked BCD. This is easily made possible because of AAA instruction, which stands for 'ASCII Adjust after addition'.

The working of AAA instruction depends on the contents of AL register. It always make the MS hex digit of AL as 0. If the LS hex digit in AL is <=9 and Auxiliary carry flags is 0, the LS hex digit value will not be altered.

If the LS hex digit is >9, or if Auxiliary carry flags is set 1, it does the following.

1. Adds 6 to the LS hex digit position AL.
2. Carry and Auxiliary carry flags are set to 1 state.
3. Increments the contents of AH.

Example 1. Let us say, AL contains 37H, the ASCII code for '7', and BL contains 38H, the ASCII code for '8'. If we add them using ADD AL, BL we get in AL the value 6FH. Let us say, AH contains 00H. To get the correct answer, we execute AAA instruction. Since F is > 9, it adds 6 to the LS hex digit, makes MS hex digit as 0, and increments AH.

	Before	***After***
(AH)	00H	01H
(AL)	6FH	05H
Flags	Any Value	A=1, C=1

Thus we have got the correct unpacked BCD result of 01 05 in AH and AL.

Example 2. Let us say, AL contains 39H, the ASCII code for '9', and BL contains 38H, the ASCII code for '8'. If we add them using ADD AL, BL we get in AL the value 71H, with A flag =1. Let us say, AH contains 00H. To get the correct answer, we execute AAA instruction. Since A flag is 1, it adds 6 to the LS hex digit, makes MS hex digit as 0, and increments AH.

	Before	*After*
(AH)	00H	01H
(AL)	71H	07H
Flags	Any Value	A=1, C=1

Thus we have got the correct unpacked BCD result of 01 07 in AH and AL. from the discussion above it may be noticed that the AAA can also be used for getting the correct unpacked BCD answer after adding two unpacked BCD numbers. Of course, AH must be 00H to start with.

Example 3. Let us say, AL contains 07H, the unpacked BCD number 7, and BL contains 08H, the unpacked BCD number 8. If we add them using ADD AL, BL we get in AL the value 0FH. Let us say, AH contains 00H. To get the correct answer, we execute AA instruction. Since F is >9, it adds 6 to the LS hex digit, makes MS hex digit as 0, and increments AH.

	Before	*After*
(AH)	**00H**	**01H**
(AL)	0FH	05H
Flags	Any Value	A=1, C=1

Thus we have got the correct unpacked BCD result of 01 05 in AH and AL.

6.9. SUBTRACT GROUP OF INSTRUCTION

In this group, we have instructions to

(1) Subtract contents of two registers, with or without borrow
(2) Subtract contents of a register and a memory location, with or without borrow
(3) Subtract immediate data to a register or a memory location, with or without borrow
(4) Decrement contents of a register or a memory location
(5) To perform decimal adjustment after subtraction
(6) To perform ASCII adjustment after subtraction.

6.9.1. Subtract Contents of Two Registers or a Register and a Memory Location

Template for this instruction is

	Before	*After*
(AX)	**122F**	**A138**
(BX)	70F7	70F7
Flags	Any Value	Z=0, S=1, P=0, C=1, A=0, O=0

As there are 10 bits in the template for identifying the operands, there are 1024 instructions in this category.

Example 1. Code for Sub AX,BX

This instruction subtract the contents of BX register from AX register, and stores the result in AX register, It is a 2 byte instruction.

In this example, we can specify either of AX or BX using the REG field. If we specify AX using REG field, then the D bit must be 1, indicating that AX is the destination register. The REG field must be 000 to indicate AX register. The W bit must be 1 to indicate it is a word operation. In this case, BX is specified using MOD value of 11, and R/M value of 011. Thus the 2 byte code for this instruction would be

If we specify BX using REG field, then the D bit must be 0, indicating that BX is the source register. The REG field must be 011 to indicate BX register. The W bit must be 1 to indicate it is a word operation. In this case, AX is specified using MOD value of 11, and R/M value of 000. Thus the 2 byte code for this instruction would be

Thus SUB AX, BX has two equally valid opcodes as shown above. An assembler generates one of these as the opcode.

After the subtract operation, the status flags are affected depending on the result.

	Before	*After*
(AX)	**122F**	**A138**
(BX)	70F7	70F7
Flags	Any Value	Z=0, S=1, P=0, C=1, A=0, O=0

In the above example, Z flag becomes 0 as A138H, the result of subtraction, is non zero. The S flag becomes 1 to indicate that the result is negative, as A138H = 1010 0001 1000 starts with a 1 in the MS bit position. The S flag will be meaningful for us only if we are working with signed numbers. The P flag becomes 0 to indicate that there are an odd number of 1's (there are 3 of them) in the LS byte of result. The C flag becomes, 1 as we need to borrow to perform the subtraction. However, A flag becomes 0 as we do not need to borrow during the subtraction of the LS hex digits. Finally, the 0 flag is reset to 0 as subtraction of two positive numbers can never result in an overflow.

Example 2. Code for SUB DS:0F346H[BP], DX

This instruction subtract the contents of DX register from a memory location whose EA is given as the sum of the 16 bit offset value F346H and the contents of BP register. The PA is generated using Ds as the segment register. The result is stored in the same memory location.

Here we have to specify DX using REG field. The D bit must be 0, indicating that DX is the source register. The REG field must be 010 to indicate DX register. The W bit must be 1 to indicate it is a word operation. F346H [BP] is specified using MOD value of 10, and R/M value of 10, and a displacement of F346H.

Whenever BP is used to generate the EA, the default segment would be SS. In this example, as we want the segment register to be DS, we have to provide the segment override prefix byte (SOP byte) to start with. The SOP byte is 001SR110, where SR indicates the segment override we desire.

To specify DS register, the SOP byte would be 001 11 110 = 3EH. Thus the 5 byte code for this instruction would be

	Before	*After*
(BP)	0100	0100
(DS:F346)	122F	122F
(DX)	70F7	70F7
Flags	Any Value	Z=0, S=1, P=0, C=1, A=0, O=0

6.9.2. Subtract Immediate Data from a Register or a Memory Location

As there are 7 bits in the template for identifying the operands, there are 128 instructions in this category.

The S bit stands for Sign Extension. This bit will be 0 always or a byte operation (W=0).

If it is a word operation (W=1), the S bit must be a 1 only if the immediate data is <=7FH. In such a case the assembler generates only 1 byte code for the immediate data. However, during execution, the one byte data will be sign extended to make it a 2 byte operand.

If the immediate data is >7FH, the S bit must be a 0, and the assembler generates a 2 byte code for the immediate data.

Example 1. Code for SUB CX,83H

As it is a word operation, with immediate data >7FH, the S bit will be 0, and the assembler generates 2 byte code for the immediate data. It is a 4 byte instruction.

After the subtract operation, the status flags are affected depending on the result.

Example 2. Code for SUB CL,83H

AS it is a byte operation, the S bit will be 0. It is a 3 byte instruction.

After the subtract operation, the status flags are affected depending on the result.

Example 3. Code for SUB CX,43H

AS it is a word operation, with immediate data <=7FH, the S bit will be 1, and the assembler generates 1 byte code for the immediate data. Thus it is a 3 byte instruction.

After the subtract operation, the status flags are affected depending on the result.

6.9.3. Subtract Immediate Data From Accumulator

As there is only one bit in the template for identifying the operands, there are 2 instructions in this category. The assembler always generates 1 byte or 2 bytes of code for the immediate data depending on whether it is a byte or a word operation.

Example 1. Code For SUB AX,83H

AS it is a word operation, the assembler generates 2 byte code for the immediate data. So, it is a 3 bye instruction.

After the subtract operation, the status flags are affected depending on the result.

Example 2. Code for SUB AL,83H

AS it is a byte operation, the generates 1 byte code for the immediate data. So, it is a 2 byte instruction.

After the subtract operation, the status flags are affected depending on the result.

6.9.4. Subtract With Borrow Contents of Two Registers or a Register and a Memory Location

It is required sometimes to perform subtraction of two numbers each of which is a multi word byte number. Let us say, we want to subtract the numbers shown below.

```
         1
    1234 5678H
+ 0 BC0 E012H
--------------
  0 673 7666H
```

Using the SUB instruction discussed earlier, we first of all subtract 5678H and E012H. We get 7666H as the LS word of result, with a borrow of 1 from the next stage. Thus the carry flag is set to 1 in this subtraction. Now we are required to subtract 1234 and 0BC0 along with the borrow to get the correct MS word of the result. In view of this, instruction are provided in the instruction set to subtract two numbers along with the carry flag.

Template for subtracting two register along with borrow, or subtracting a register and a memory location along with borrow is.

As there are 10 bits in the template for identifying the operands, there are 1024 instruction in this category. The mnemonic for this instruction is SBB, which stands for subtract with borrow.

Example 1. Code for SBB DS:0F246H[BP],DX

This instruction subtract the contents of DX register along with borrow to a memory location whose EA is given as the sum of the 16 bit offset value F246H and the contents of BP register. The PA is generated using DS as the segment register. The result is stored in the same memory location.

Here we have to specify DX using REG field. The D bit must be 0, indicating that DX is the source register. The REG field must be 010 to indicate DX register. The W bit must be 1 to indicate it is a word operation. F246H[BP] is specified using MOD value of 10, and R/M value of 110, and a displacement of F246H.

Whenever BP is used to generate the EA, the default segment would be SS. In this example, as we want the segment register to be DS, we have to provide the segment override prefix byte (SOP Byte) to start with. The SOP byte is 001SR110, where SR indicates the segment override we desire.

To specify DS register, the SOP byte would be 001 11 110=3EH. Thus the 5 byte code for this instruction would be.

	Before	*After*
(BP)	0100	0100
(DS:F346)	1230	A138
(DX)	70F7	70F7
Flags	Any Value	Z=0, S=1, P=0, C=1, A=0, O=0

6.9.5. Subtract with Carry Immediate Data from a Register or a Memory Location

As there are 7 bits in the template for identifying the operands, there are 128 instructions in this category. The S bit stands for Sign Extension. This bit will be 0 always for a byte operation (W=0).

If it is a word operation (W=1), the S bit must be a 1 only if the immediate data is <=7FH. In such a case the assembler generates only 1 byte code for sign extended to make it a 2 byte operand.

If the immediate data is >7FH, The S bit must be a 0, and the assembler generates a 2 byte code for the immediate data.

Example 1. Code for SBB CX,83H

As it is a word operation, with immediate data >7FH, the S bit will be 0, and the assembler generates 2 byte code for the immediate data. It is a 4 byte instruction.

After the subtract operation, the status flags are affected depending on the result.

Example 2. Code for SBB CL, 83H

As it is a byte operation, the S bit will be 0. IT is a 3 byte instruction.

After the subtract operation, the status flags are affected depending on the result.

Example 3. Code for SBB CX,43H

As it is a word operation, with immediate data <=7FH, the S bit will be 1, and the assembler generates 1 byte code for the immediate data. Thus it is a 3 byte instruction.

After the subtract operation, the status flags are affected depending on the result.

6.9.6. Subtract Immediate Data from Accumulator Alongwith Borrow

As there is only one bit in the template for identifying the operands, there are 2 instructions in this category. The assembler always generates 1 byte or 2 bytes of code for the immediate data depending on whether it is a byte or a word operation.

Example 1. Code for SBB AX,83H

As it is a word operation, the assembler generates 2 byte code for the immediate data. So, it is a 3 byte instruction.

After the subtract operation, the status flags are affected depending on the result.

Example 2. Code for SBB AL,83H

As it is a byte operation, the assembler generates 1 byte code for the immediate data. So, it is a 2 byte instruction.

After the subtract operation, the status flags are affected depending on the result.

6.9.7. Decrement Contents of a Register or a Memory Location

This instruction decrements the contents of a register or a memory location by 1. It can be a byte or a word operation. All the flags, except carry flag, are affected depending on the result.

As there are 6 bits in the templates for identifying the operands, there are 64 instructions in this category. The mnemonic for this instruction is DEC.

Example 1. Code for DEC WORD PTR DS:0F246H[BP]

This instruction decrements the contents of a word location whose EA is given as the sum of the 16 bit offset value F246H and the contents of BP register. The PA is generated using DS as the segment register.

The W bit must be indicate it is a word operation. F246H[BP] is specified using MOD value of 10, and R/M value of 110, and a displacement of F246H.

Whenever BP is used to generate the EA, the default segment would be SS. In this example, as we want the segment register to be DS, we have to provide the segment override prefix byte (SOP byte) to start with. The SOP byte is 001SR110, where SR indicates the segment override we desire.

To specify DS register, the SOP byte would be 001 11 110=3EH. Thus the 5 byte code for this instruction would be

	Before	*After*
(BP)	**0100**	**0100**
(DS:F346)	120	122F
Flags	Any Value	Z=0, S=0, P=0, A=1, O=0

Example 2. Code for DEC CL

The W bit must be 0 to indicate it is a byte operation. CL is specified using MOD value of 11, and R/M value of 001.

It is a 2 byte instruction with the code

	Before	*After*
(CL)	**80**	**7F**
Flags	Any Value	Z=0, S=0, P=0, A=1, O=0

Similarly, DEC CX can be coded as

However, as decrementing of a word register is a frequently used instruction for greater efficiency, the assembler will generate a single byte code as explained next.

6.9.8. Decrement the Content of a Word Register

As there are 3 bits in the template for identifying the word register, there are 8 instructions in this category. The mnemonic for this instruction is also DEC.

Example 1. Code for DEC CX

CX is specified using REG value of 001. Thus the 1 byte code for this instruction would be 49H as shown below.

	Before	*After*
(CX)	**0102**	**0101**
Flags	Any Value	Z=0, S=0, P=0, A=0, O=0

6.9.9. Perform Decimal Adjustment After Subtraction

Suppose we want to subtract the two decimal numbers 45 and 38. They will be represented in packed BCD as 0100 0101 and 0011 1000. The subtraction result in 0000 1101. If we interpret this result as a packed BCD number, the answer is incorrect as well as illegal. This is because, 1101 is an illegal BCD number. This is where the DAS instruction proves its usefulness. All that is required to be done is to subtract the BCD numbers and store the result in AL, and then execute DAS instruction. DAS stands for decimal adjust after subtraction.

Opcode for this instruction is 2FH.

The working of DAS instruction depends on the contents of AL register. In effect, it subtract 00H, 06H, 60H, or 66H from AL so that we get the correct packed BCD answer in AL.

If the LS hex digit in AL is <=9 and Auxiliary carry flag is 0, the LS hex digit value will not be altered.

If the LS hex digit is > 9, or if Auxiliary carry flag is set to 1, it subtract 6 from the LS hex digit of AL.

Now, if the MS hex digit is <=9 and Carry flag is 0, the MS hex digit will not be altered, and carry flag is reset to 0.

If the MS hex digit is >9, or if carry flag is set to1, it subtracts 6 from the MS hex digit of AL and sets Carry flag to 1. In such a case, the result is interpreted to be negative and the value in AL represents the 10's complement of the absolute value of correct difference.

Example 1. Let us say, we subtract 45 BCD and 38 BCD, and store the result 0DH in AL. In this case, C flag=0 and A flag =1. But as D is invalid BCD code (or it could be because A flag =1), the DAS instruction subtracts 60H from AL. Thus we get 07H in AL, which will now be interpreted as 07 BCD, the correct difference of 45 and 38.

Example 2. Let us say, we subtract 63 BCD and 88 BCD, and store the result DBH in AL. In this case, C flag=1 and A flag =1. As both D and B are invalid BCD codes (or because C=1, A=1), the DAS instruction subtracts 66H from AL. Thus we get 75H in AL, and C flag will be set to 1. As C flag=1, it means result is negative. 75 BCD, is the 10's complement of 25 BCD, which is the magnitude of the correct difference.

Example 3. Let us say, we subtract 68 BCD and 85 BCD, and store the result E3H in AL. In this case, C flag=1 and A flag =0. Since E is an invalid BCD code the DAS instruction subtracts 60H from AL and stores the result 83H in AL. It sets C flag to 1. As C flag =1, it means is negative. 83 BCD, is the 10's complement of 17 BCD, which is the magnitude of the correct difference.

Example 4. Let us say, we subtract 99 BCD and 88 BCD, and store the result 11H in AL. In this case, C flag=0 and A flag =0. So nothing is done to the content of AL. IN other words, 00H is subtracted from AL. The result in AL will now be interpreted as 11 BCD, the correct difference of 99 and 88.

Opcode for this instruction is 27H.

6.9.10. Perform ASCII Adjustment After Subtraction

The data that we enter from a keyboard will be in ASCII format. If we enter the keyboard values like '9' and '8', its representation inside the computer will be 39H and 38H. Suppose we want to perform subtract operation on the characters '9' and '8', and get the result as 01 in Unpacked BCD. This is easily made possible because of AAS instruction, which stands for 'ASCII Adjust after Subtraction'.

Opcode for this instruction is 3FH.

The working of AAS instruction depends on the contents of AL register. It always makes the MS hex digit of AL as 0. If the LS hex digit in AL is <= 9 and Auxiliary carry flag is 0, the LS hex digit value will not be altered.

If the LS hex digit is > 9, or if Auxiliary carry flag is set to 1, it does the following.

1. Subtracts 6 from the LS hex digit position of AL.
2. Carry and Auxiliary cany flags are set to 1 state,
3. Decrements the contents of AH.

In such a case, the result is interpreted to be negative, and the value in LS hex position of AL represents the 10's complement of the absolute value of correct difference.

Example 1. Let us say, AL contains 39H, the ASCII code for '9', and BL contains 37H, the ASCII code for '7'. If we subtract them using SUB AL, BL we get in AL the value 02H. Let us say, AH contains 00H. If we execute AAS instruction, it does not alter the contents of AL in any way.

	Before	***After***
(AH)	00H	00H
(AL)	02H	02H
(FLAGS)	any value A= 0, C= 0	

Thus we have got the correct unpacked BCD result of 00 02 in AH and AL.

Example 2. Let us say, AL contains 35H, the ASCII code for '5', and BL contains 38H, the ASCII code for '8'. If we subtract them using SUB AL, BL we get in AL the value FDH, with C flag = 1 and A flag = 1. Let us say, AH contains 00H. To get the correct answer, we execute AAS instruction. Since A Flag is 1, it subtracts 6 from the LS hex digit, makes MS hex digit as 0, and decrements AH.

	Before	***After***
(AH)	00H	FFH
(AL)	FDH	07H
(FLAGS)	any value A = 1, C = 1	

As C flag = 1, it means result is negative. 7 BCD is the 10's complement of 3 BCD, which is the magnitude of the correct difference.

From the discussion above, it may be noticed that the AAS can also be used for getting the correct unpacked BCD answer after subtracting two unpacked BCD numbers. Of course, AH roust be 00H to start with.

Example 1. Let us say, AL contains 07H, the unpacked BCD number 7, and BL contains 04H, the unpacked BCD number 4. If we subtract them using SUB AL.BL we get in AL the value 03H. Let us say, AH contains 00H. Now, execution of AAS does not alter AL and AH values in any way.

	Before	***After***
(AH)	00H	00H
(AL)	03H	03H
(FLAGS)	any value A= 0, C= 0	

Thus we have got the correct unpacked BCD result of 00 03 in AH and AL.

Example 2. Let us say, AL contains 07H, the unpacked BCD number 7, and BL contains 09H, the unpacked BCD number 9. If we subtract them using SUB AL,BL we get in AL the value FEH. Let us say, AH contains 00H. To get the correct answer, we execute AAS instruction. Since H is > 9, it subtracts 6 from the IJS hex digit, makes MS hex digit as 0, and decrements AH.

	Before	***After***
(AH)	00H	FFH
(AL)	FEH	08H
(FLAGS)	any value A = 1, C = 1	

As C flag = 1, it means result is negative. 8 BCD Is the 10's complement of 2 BCD, which is the magnitude of the correct difference.

6.10. NEGATE INSTRUCTION

This instruction negates the contents of a register or a memory location. That is, a positive number will be converted to a negative number of same magnitude and vice versa. In effect, it replaces the number with its 2's complement.

6.10.1. Negate Contents of a Register or a Memory Location

It can work on word or byte data. All the flags are affected depending on the result. Overflow flag is set to 1 if 80H or 8000H is negated. Template for this instruction is

Byte 1								Byte 2				
1	1	1	1	0	1	1	W	MOD	0	1	1	R/M

As there are 6 bits in the template for identifying the operands, there are 64 instructions in this category. The mnemonic for this instruction is NEC.

Example 1. Code for NEG WORD PTR DS:0F246H[BP]

This instruction negates the contents of a word location whose EA is given as the sum of the 16 bit offset value F246H and the contents of BP register. The PA is generated using DS as the segment register.

The W bit must be 1 to indicate it is a word operation, F246H[BP] is specified using MOD value of 10, and R/M value of 110, and a displacement of F246H.

Whenever BP is used to generate the EA, the default segment would be SS. In this example, as we want the segment register to be DS, we have to provide the segment override prefix byte (SOP byte) to start with. The SOP byte is 001SR1 10, where SR indicates the segment override we desire.

To specify DS register, the SOP byte would be 001 11 110 = 3EH. Thus the 5 byte code for this instruction would be

SOP	Opcode W	MOD	opcode	R/M	16 bit disp	
3E	1111 011 1	10	011	110	F2 46	= 3E F7 9E F2 46H

	Before	*After*
(BP)	0100	0100
(DS:F346)	1230	EDD0

Example 2. Code for NEG CL.

The W bit must be 0 to indicate it is a byte operation. CL is specified using MOD value of 11, and R/M value of 001.

It is a 2 byte instruction with the code

Opcode W	MOD	Opcode	R/M	
1111 011 0	11	011	001	= F6 D9H

	Before	*After*
(CL)	60	A0

Example 3. Code for NEG CX.

The W bit must be 1 to indicate it is a word operation. CX is specified using MOD value of 11, and R/M value of 001. It is a 2 byte instruction with the code.

6.11. COMPARE INSTRUCTIONS

In this group, we have instructions to

(a) Compare contents of two registers

(b) Compare contents of a register and a memory location

(c) Compare immediate data with contents of a register or a memory location.

6.11.1. Compare Contents of Two Registers or a Register and a Memory Location

A compare instruction compares two operands, and affects the status flags values depending on the result of comparison. The compare instruction actually computes the value of operand_l minus operand_2. The original values of the operands are not changed. The result is stored in a register which is cot accessible to the programmer. Based on the result, the flags are affected.

It is similar to comparing the heights of two people. After the comparison, the heights remain unaltered. But, we would have come to one of the following conclusions.

(a) Both are of same height

(b) Personal is taller

(c) Personal is shorter.

Template for this instruction is

Byte 1								Byte 2		
0	0	1	1	1	0	D	W	MOD	REG	R/M

As there are 10 bits in the template for identifying the operands, there are 1024 instructions in this category. The mnemonic for this instruction is CMP.

Example 1. Code for CMP AX,BX.

This instruction compares the contents of AX register and BX register. It is a 2 byte instruction.

In this example, we can specify either of AX or BX using the REG field. If we specify AX using REG field, then the D bit must be 1, indicating that AX is ifae destination register. The REG field must be 000 to indicate AX register. The W bit must be 1 to indicate it is a word operation. In this case, BX is specified using MOD value of 11, and R/M value of Oil, Thus the 2 byte code for this instruction would be

Opcode	**D**	**W**	**MOD**	**REG**	**R/M**
0010 10	1	1	11	000	011

If we specify BX using REG field, then the D bit must be 0, indicating that BX is the source register. The REG field must be 011 to indicate BX register. The W bit must be 1 to indicate it is a word operation. In this case, AX Is specified using MOD value of 11, and R/M value of 000. Thus the 2 byte code for this instruction would be

opcode	**D**	**W**	**MOD**	**REG**	**R/M**	
0010 10	0	1	11	011	000	= 39 D8H

Thus CMP AX,BX has two equally valid opcodes as shown above. An assembler generates one of these as the opcode. After the compare operation, the status flags are affected depending on the result.

	Before	***After***
(AX)	122F	122F
(BX)	70F7	70F7
(FLAGS)	any value	Z = 0, S = 1, P = 0,
		C = 1, A = 0, O = 0

In the above example, the result of comparison is 122F.H - 70F7H = Al 38H. The Z flag is reset to 0 as the result is non zero. The S flag becomes 1 to indicate that the result is negative, as A138H = 1010 0001 0011 1000 starts with a 1 in the MS bit position. The S flag will be meaningful for us only if we are working with signed numbers. The P flag becomes 0 to indicate that there are an odd number of 1's (there are 3 of them) in the LS byte of result Hie C flag becomes 1, as we need to borrow to perform the subtraction. However, A flag becomes 0 as we do not need to borrow during the subtraction of the LS hex digits. Finally, the O flag is reset to 0 as subtraction of two positive numbers can never result in an overflow.

Example 2. Code for CMP DS:0F246H[BP],DX.

This instruction compares the contents of a memory location whose EA is given as the sum of the 16 bit offset value F246H and the contents of BP register with DX register contents.

Here we have to specify DX using REG field. The D bit must be 0, indicating that DX is the source register. The REG field must be 010 to indicate DX register. The W bit must be 1 to indicate it is a word operation. F246H[BP] is specified using MOD value of 10, and R/M value of 110, and a displacement of F246H.

Whenever BP is Used to generate the EA, the default segment would be SS. In this example, as we want the segment register to be DS, we have to provide the segment override prefix byte (SOP byte) to start with. The SOP byte is 001SR110, where SR indicates the segment override we desire. To specify DS register, the SOP byte would be 001 11 110 = 3EH. Thus the 5 byte code for this instruction would be

SOP	**Opcode**	**D**	**W**	**MOD**	**REG**	**R/M**	**16 bit disp**
3E	0011 10	0	1	10	010	110	F2 46

= 3E 39 96 F2 46H

	Before	***After***
(BP)	0100	0100
(DS:F346)	122F	122F
(DX)	70F7	70F7
(FLAGS)	any value	Z = 0, S = 1, P = 0, C = 1, A = 0, O = 0

6.11.2. Compare Contents of a Register or a Memory Location With Immediate Data

Template for this instruction is

Byte 1

1	0	0	0	0	0	S	W

Byte 2

MOD	1	1	1	R/M

As there are 7 bits in the template for identifying the operands, there are 128 instructions in this category.

The S bit stands for Sign Extension. This bit will be 0 always for a byte operation (W = 0).

If it is a word operation (W = 1), the S bit must be a 1 only if the immediate data is <= 7FH. In such a case the assembler generates only 1 byte code for the immediate data. However, during execution, the one byte data will be sign extended to make it a 2 byte operand.

If the immediate data is > 7FH, the S bit must be a 0, and the assembler generates a 2 byte code for the immediate data.

Example 1. Code for CMP CX,83H.

As it is a word operation, with immediate data > 7FH, the S bit will be 0, and the assembler generates 2 byte code for the immediate data. It is a 4 byte instruction.

opcode	**S**	**W**	**MOD**	**opcode**	**R/M**	**16 bit data**	
1000 00	0	1	11	111	001	00 83H	= 81 F9 00 83H

After the compare operation, the status flags are affected depending on the result.

Example 2. Code for CMP CL,88H.

As it is a byte operation, the S bit will be 0. It is a 3 byte instruction.

opcode	**S**	**W**	**MOD**	**opcode**	**R/M**	**16 bit data**	
1000 00	0	0	11	111	001	88H	= 80 F9 88H

After compare operation, status flags are affected depending on the result.

Example 3. Code for CMP CX, 48H.

As it is a word operation, with immediate data <= 7FH, the S bit will be 1, and the assembler generates 1 byte code for the immediate data. Thus it is a 3 byte instruction.

opcode	S	W	MOD	Opcode	R/M	8 bit data	
1000 00	1	1	11	111	001	48H	= 83 F9 48H

After the compare operation, the status flags are affected depending on the result.

6.11.3. Compare Accumulator Contents With Immediate Data

Template for this instruction is

0	0	1	1	1	1	0	W

As there is only one bit in the template for identifying the operands, there are 2 instructions in this category. The assembler always generates 1 byte or 2 bytes of code for the immediate data depending on whether it is a byte or a word operation.

Example 1. Code for CMP AX,88H.

As it is a word operation, the assembler generates 2 byte code for the immediate data. So, it is a 3 byte instruction.

opcode	W	16 bit data	
0011 110	1	00 88H	= 3D 00 88H

After the comparo operation, the status flags are affected depending on the result.

Example 2. Code for CMP AL,88H.

As it is a byte operation, the assembler generates 1 byte code for the immediate data. So, it is a 2 byte instruction.

opcode	W	8 bit data	
0011 110	0	88H	= 3C 88H

After the compare operation, the status flags are affected depending on the result.

6.12. DATA SIZE CONVERSION INSTRUCTIONS

In this category, we have instructions to convert the byte in AL to sign extended word in AX, and convert the word in AX to sign extended double word in DX and AX. These instructions do not affect the flags.

6.12.1. Convert Byte to Word

There are times when we want a byte to be converted into a word, of course without changing its value. This is done by sign extension which works as follows.

For a positive byte valued number (MS bit = 0), this is achieved by making the MS byte of the word as 00H and LS byte of the word as the byte to be converted. And for a negative byte valued number (MS bit = 1), this is achieved by making the MS byte of the word as FFH and LS byte of the word as the byte to be converted.

Only AL contents can be converted to its equivalent word using CBW instruction, CBW stands for 'convert byte to word'. AH will have the MS byte of the converted result. Opcode for this instruction is 98H.

Example 1.

	Before	*After*
(AH)	34H	00H
(AL)	02H	02H

Thus after CBW instruction we get 0002H in AX, which is the sign extended value of 92H in AL.

Example 2.

	Before	*After*
(AH)	34H	FFH
(AL)	92H	92H

Thus after CBW instruction we get FF92H in AX, which is the sign extended value of 92H in AL.

6.12.2. Convert Word to Double Word

There are times when we want a word to be converted into a double word, of course without changing its value. This is done by sign extension which works as follows.

For a positive word valued number (MS bit = 0), this is achieved by making the MS word of the double word as 0000H and LS word of the double word as the word to be converted. And for a negative word valued number (MS bit = 1), this is achieved by making the MS word of the double word as FFFFH and LS word of the double word as the word to be converted. Only AX contents can be converted to its equivalent double word using CWD instruction. CWD stands for 'convert word to double word'. DX will have the MS word of the converted result. Opcode for this instruction is 99H.

Example 2.

	Before	*After*
(DX)	1234H	0000H
(AL)	5678H	5678H

Thus after CWD instruction we get 0000 567SH in the register pair DX_AX, which is the sign extended value of 5678H in AX.

Example 2.

	Before	*After*
(DX)	1234H	FFFFH
(AL)	9678H	9678H

Thus after CWD instruction we get FFFF 9678H in the register pair DX_AX, which is the sign extended value of 9678H in AX.

6.13. MULTIPLY INSTRUCTIONS

8086 provides instructions to perform multiplication of two unsigned numbers or two signed numbers. The numbers we are multiplying can be of word size or byte size.

6.13.1. Perform Unsigned Multiplication

MUL is the mnemonic for the instruction to perform unsigned multiplication. Template for this instruction is

Byte 1								Byte 2				
1	1	1	1	0	1	1	W	MOD	1	0	0	R/M

As there are 6 bits in the template for identifying the operands, there are 64 instructions in this category.

For multiplication of byte valued numbers, one operand is implied to be in AL and the other is specified to be in a byte register or a byte location in memory. The 16 bit product will be stored in AX.

For multiplication of word valued numbers, one operand is implied to be in AX, and the other is specified to be in a word register or a word location in memory. The 32 bit product will be stored in the register pair DX_AX.

Carry and Overflow flags are set to 1 if AH < > 00H after 8 bit multiplication, or if DX < > 0000H after 16 bit multiplication.

Values of other flags are undefined after multiplication.

Example 1. Code for MUL WORD PTR DS:0F246H[BP].

This instruction multiplies the contents of a word location whose EA is given as the sum of the 16 bit offset value F246H and the contents of BP register with AX. The PA is generated using DS as the segment register. The result is stored in DX_AX.

The W bit must be 1 to indicate it is a word operation. F246H[BP] is specified using MOD value of 10, and R/M value of 110, and a displacement of F246H.

Whenever BP is used to generate the EA, the default segment would be SS. In this example, as we want the segment register to be DS, we have to provide the segment override prefix byte (SOP byte) to start with. The SOP byte is 001SR110, where SR indicates the segment override we desire. To specify DS register, the SOP byte would be 001 11 110 = 3EH. Thus the 5 byte code for this instruction would be

SOP	opcode	W	MOD	opcode	R/M	16 bit disp	
3E	1111 011	1	10	100	110	F2 46	= 3EF7 A6 F2 46H

	Before	*After*
(BP)	0100	0100
(DS:F346)	0030	0030
(AX)	9180	4800
(DX)	2345	001B

Thus the unsigned product of 0030H and 9180H is obtained in DX_AX as 001B 4800H.

Example 2. Code for MUL CL.

The W bit must be 0 to indicate it is a byte operation, CL is specified using MOD value of 11, and R/M value of 001. It has the 2 byte code

opcode	W	MOD	opcode	R/M	
1111 011	0	11	100	001	=F6E1H

	Before	*After*
(CL)	60	60
(AL)	A0	00
(AH)	45	3C

Thus the unsigned product of 60H and AOH is obtained in AX as 3C 00H.

Example 3. Code for MUL CX.

The W bit must be 1 to indicate it is a word operation. CX is specified using MOD value of 11, and R/M value of 001. It has the 2 byte code

opcode	W	MOD	opcode	R/M	
1111 011	1	11	100	001	= F7 E1H

	Before	After
(CX)	9180	9180
(AX)	0030	4800
(DX)	4567	001B

Thus the unsigned product of 0030H and 9180H is obtained in DX_AX as 001B 4800H.

6.13.2. Perform Signed Multiplication

IMUL is the mnemonic for the instruction to perform signed multiplication.

Template for this instruction is

Byte 1

1	1	1	1	0	1	1	W

Byte 2

MOD	1	0	1	R/M

As there are 6 bits in the template for identifying the operands, there are 64 instructions in this category,

For multiplication of byte valued numbers, one operand is implied to be in AL, and the other is specified to be in a byte register or a byte location in memory. The 16 bit product will be stored in AX.

For multiplication of word valued numbers, one operand is implied to be in AX, and the other is specified to be in a word register or a word location in memory. The 32 bit product will be stored in the register pair DX_AX.

Carry and Overflow flags are set to 1 if AH < > (00H or FFH) after 8 bit multiplication, or if DX < > (0000H or FFFFH) after 16 bit multiplication.

Values of other flags are undefined after multiplication.

Example 1. Code for IMUL WORD PTR DS:0F246H[BP].

This instruction multiplies the contents of a word location whose EA is given as the sum of the 16 bit offset value F246H and the contents of BP register with AX. The PA is generated using DS as the segment register. The result is stored in DX_AX.

The W bit must be 1 to indicate it is a word operation. F246H[BP] is specified using MOD value of 10, and R/M value of 110, and a displacement of F246H.

Whenever BP is used to generate the EA, the default segment would be SS. In this example, as we want the segment register to be DS, we have to provide the segment override prefix byte (SOP byte) to start with. The SOP byte is 001SR110, where SR indicates the segment override we desire.

To specify DS register, the SOP byte would be 001 11 110 = 3EH. Thus the 5 byte code for this instruction would be

SOP	**opcode**	**W**	**MOD**	**opcode**	**R/M**	**16 bit disp**
3E	1111 011	1	10	101	110	F2 46

= 3E F7 AE F2 46H

	Before	After
(BP)	0100	0100
(DS:F346)	0030	0030
(AX)	9180	4800
(DX)	2345	FFEB

Thus the signed product of 0030H (= +30H) and 9180H (= –6E80H) is obtained in DX_AX as FFEB 4800H (= – 0014 B800H).

Example 2. Code for IMUL CL.

The W bit must be 0 to indicate it is a byte operation. CL is specified using MOD value of 11, and R/M value of 001.

It is a 2 byte instruction with the code

opcode	W	MOD	opcode	R/M	
1111 011	0	11	101	001	=F6E9H

	Before	*After*
(CL)	60	60
(AL)	A0	00
(AH)	45	DC

Thus the signed product of 60H (= + 60H) and A0H (= -60H) is obtained In AX as DC 00H (= –24 00H).

Example 3. *Code* for IMUL CX.

The W bit must be 1 to indicate it is a word operation. CX is specified using MOD value of 11. and R/M value of 001. It is a 2 byte instruction with the code

opcode	W	MOD	opcode	R/M	
1111 011	1	11	101	001	=F7 E9H

	Before	*After*
(CX)	9180	9180
(AX)	0030	4800
(DX)	4567	FFEB

Thus the signed product of 0030H (= +30H) and 9180H (= –6E80H) is obtained in DX_AX as FFEB 4800H (= –0014 B800H).

6.13.3. Perform ASCII Adjustment After Multiplication

The data that we enter from a keyboard will be in ASCII format. If we enter on the keyboard values like '9' and '8', its representation inside the computer will be 39B and 38H. Suppose we want to perform multiplication of the characters '9' and '8', and get the result as 07 02 in Unpacked BCD. This is easily made possible because of AAM instruction, which stands for 'ASCII Adjust after multiplication', Opcode for this instruction is D4 0AH.

The working of AAM Instruction depends on the contents of AL register. Suppose AL content is 3FH. It is 63 BCD. It can be represented in unpacked BCD as 06 03 BCD. All that AAM instruction does is to convert 3FH in AL to 06 03 in AX. Note that the result in AX does not represent the ASCII characters '6' and '3'. It only represents unpacked BCD numbers.

Example : Let us say, AL contains 39H, the ASCII code for '9', and BL contains 37H, the ASCII code for '7'. If we subtract from AL and BL the immediate value 30H, we get 09H in AL and 07H in BL. If we now multiply these unpacked BCD numbers in AL and BL using MUL BL, we get 00 in AH and 3FH in AL. Now execution of AAM results in AH getting 06 and AL getting 03. This 06 03 is interpreted as the correct unpacked BCD product of the unpacked BCD numbers 09 and 07.

	Before	*After*
(AH)	00H	06H
(AL)	3FH	03H

If our interest is to get the result of multiplication as the ASCII characters '6' and *'3'* all that is needed is to add 3030H to AX.

It should be noted that after the multiplication is performed on unpacked BCD numbers, we have to convert the binary result to 2 digit unpacked BCD using the AAM instruction.

6.14. DIVIDE INSTRUCTIONS

8086 provides instructions to perform division of two unsigned numbers or two signed numbers. It is integer division only. That is, 9 divided by 4 gives a quotient of *2* and a remainder of 1. The divisor (the denominator) can be of word size or byte size. The dividend (numerator) is always double the size of the divisor.

6.14.1. Perform Unsigned Division

DIV is the mnemonic for the instruction to perform unsigned division. Template for this instruction is

Byte 1								Byte 2				
1	1	1	1	0	1	1	W	MOD				R/M

As there are 6 bits in the template for identifying the operands, there are 64 instructions in this category.

For byte valued divisor, the dividend is implied to be in AX, and the divisor is specified to be in a byte register or a byte location in memory. The 8 bit quotient will be stored in AL and the 8 bit remainder in AH. If the divisor is 0, or if the correct quotient is larger than FFH, the 8086 branches to an interrupt service subroutine, to take appropriate action to deal with this error. It could be an error message display on the terminal.

For word valued divisor, the dividend is implied to be in DX_AX register pair, and the divisor is specified to be in a word register or a word location in memory. The 16 bit quotient will be stored in AX and the 16 bit remainder in DX. If the divisor is 0, or if the correct quotient is larger than FFFFH, the 8086 branches to an interrupt service subroutine, to take appropriate action to deal with this error. It could be an error message display on the terminal.

Flags are undefined after execution of DIV instruction.

Example 1. Code for DIV WORD PTR DS:0F246H[BP].

This instruction divides the contents of the register pair DX_AX by ihe contents of a word location whose EA is given as the sum of the 16 bit offset value F246H and the contents of BP register. The PA is generated using DS as the segment register. The quotient is stored in AX and the remainder in DX.

The W bit must be 1 to indicate it is a word operation. F246H[BP] is specified using MOD value of 10, and R/M value of 110, and a displacement of F246H.

Whenever BP is used to generate the EA, the default segment would be SS. In this example, as we want the segment register to be DS, we have to provide the segment override prefix byte (SOP byte) to start with. The SOP byte is 001SR110, where SR indicates the segment override we desire.

To specify DS register, the SOP byte would be 001 11 110 = 3EH. Thus the 5 byte code for this instruction would be

SOP	opcode	W	MOD	opcode	R/M	16 bit disp
3E	1111 011	1	10	110	110	F2 46

= 3E F7 B6 F2 46H

	Before	*After*
(BP)	0100	0100
(DS:F346)	0030	0030
(AX)	9192	0308
(DX)	0000	0012

Thus after the unsigned division of 0000 9192H by 0030H, the quotient is obtained in AX as 0308H and remainder is obtained in DX as 0012H.

Example 2. Code for DIV CL.

The W bit must be 0 to indicate it is a byte operation. CL is specified using MOD value of 11, and R/M value of 001.

It is a 2 byte instruction with the code

Opcode	W	MOD	Opcode	R/M	
1111 011	0	11	110	001	= F6 F1H

	Before	*After*
(CL)	60	60
(AL)	A0	B9
(AH)	45	40

Thus after the unsigned division of 45A0H by 60H, the quotient is obtained in AL as B9H and remainder is obtained in AH as 40H.

Example 3. Code for DIV CX.

The W bit must be 1 to indicate it is a word operation. CX is specified using MOD value of 11, and R/M value of 001. It is a 2 byte instruction with the code.

opcode	W	MOD	Opcode	R/M	
1111 011	1	11	110	001	= F7 F1H

	Before	*After*
(CX)	0030	0030
(AX)	9192	0308
(DX)	0000	0012

Thus after the unsigned division of 0000 9192H by 0030H, the quotient is obtained in AX as 0308H and remainder is obtained in DX as 0012H.

6.14.2. Perform Signed Division

IDIV is the mnemonic for the instruction to perform signed division. Template for this instruction is

Byte 1								Byte 2				
1	1	1	1	0	1	1	W	MOD	1	1	1	R/M

As there are 6 bits in the template for identifying the operands, there are 64 instructions in this category.

For byte valued divisor, the dividend is implied to be in AX, and the divisor is specified to be in a byte register or a byte location in memory. The 8 bit quotient will be stored in AL and the 8 bit remainder in AH. If the divisor is 0, or if the correct quotient is more positive than 7FH, or more negative than 80H, the 8086 branches to an interrupt service subroutine, to take appropriate action to deal with this error. It could be an error message display on the terminal.

For word valued divisor, the dividend is implied to be in DX_AX register pair, and the divisor is specified to be in a word register or a word location in memory. The 16 bit quotient will be stored in AX and the 16 bit remainder in DX. If the divisor is 0, or if the correct quotient is more positive than 7FFFH, or more negative than 8000H, the 8086 branches to an interrupt service subroutine, to take appropriate action to deal with this error. It could be an error message display on the terminal.

Flags are undefined after execution of IDIV instruction.

Example 1. Code for IDIV WORD PTR DS:0F246H[BP].

This instruction divides the contents of the register pair DX_AX by the contents of a word location whose EA is given as the sum of the 16 bit offset value F246H and the contents of BP register. The PA is generated using DS as the segment register. The quotient is stored in AX and the remainder in DX.

The W bit must be 1 to indicate it Is a word operation, F246H[BP] is specified using MOD value of 10, and R/M value of 110, aad a displacement of F246H.

Whenever BP is used to generate the EA, the default segment would be SS. In this example, as we want the segment register to be DS, we have to provide the segment override prefix byte (SOP byte) to start with. The SOP byte is 001 SR 110, where SR indicates the segment override we desire.

To specify DS register, the SOP byte would be 001 11 110 = 3EH. Thus the 5 byte code for this instruction would be

SOP	opcode	W	MOD	opcode	R/M	16 bit disp
3E	1111 011	1	10	111	110	F2 46

= 3E F7 BE F2 46

	Before	*After*
(BP)	0100	0100
(DS:F346)	FFB0	FFBO
(AX)	OFB0	0300
(DX)	FFFF	FFE0

Thus after the signed division of FFFF OFEOH (= –0000 F020H) by FFBOH (= –0050H), the quotient is obtained in AX as 0300H, and remainder is obtained in DX as FFEOH (= –0020H).

Example 2. Code for IDIV CL.

The W bit must be 0 to indicate it is a byte operation. CL is specified using MOD value of 11, and R/M value of 001. It is a 2 byte instruction with the code.

Opcode	W	MOD	Opcode	R/M	
1111 011	0	11	111	001	= F6 F9H

	Before	*After*
(CL)	30	30
(AL)	EE	80
(AH)	E7	EE

Thus after the signed division of E7EEH (= –1812H) by 30H, the quotient is obtained in AL as 80H (= –80H) and remainder is obtained in AH as EEH (= –12).

Example 3. Code for IDIV CX.

The W bit must be 1 to indicate it is a word operation. CX is specified using MOD value of 11, and R/M value of 001. It is a 2 byte instruction with the code.

opcode	W	MOD	opcode	R/M	
1111 011	1	11	111	001	= F7 F9H

	Before	*After*
(CX)	FFB0	FFB0
(AX)	0FE0	0300
(DX)	FFFF	FFE0

Thus after the signed division of FFFF 0FE0H (= –0000 F020H) by FFB0H (= –0050H), the quotient is obtained in AX as 0300H, and remainder is obtained in DX as FFE0H (= –0020H).

6.14.3. Perform ASCII Adjustment Before Division

The data that we enter from a keyboard will be in ASCII format. If we enter on the keyboard values like '6' and '2', its representation inside the computer will be 36H arid 32H, Suppose we want to perform division of the characters '62' and '8', and get the quotient as 07 and remainder as 06 in Unpacked BCD. This is easily made possible because of AAD instruction, which stands for 'ASCII Adjust before division'. Opcode for this instruction is D5 0AH.

The working of AAM instruction depends on the contents of AX register. Suppose AX content is 06 02H. It can be interpreted as 62 in unpacked BCD. All that AAD instruction does is to convert 06 02H in AX to 00 3EH in AX. In effect, AAD converts the 2 digit unpacked BCD in AX to the equivalent binary in AX.

Example. Let us say, AH contains 36H, the ASCII code for '6', and AL contains 32H, the ASCII code for '2'. If we subtract from AX the immediate value 3030H, we get 06H in AH and 02H in AL. Now execution of AAD results in AX getting the value 003EH.

	Before	*After*
(AH)	06H	00H
(AL)	02H	3EH

Suppose CL contains 38H, the ASCII code for '8'. If we execute D1V CL, we get the remainder 06H in AH and the quotient 07H in AL. If our interest is to get the result of division as the ASCII characters '6' and '7', all that is needed is to add 3030H to AX.

It should be noted that before the division is performed on unpacked BCD numbers, we have to convert the 2 digit unpacked BCD to binary using the AAD instruction.

8086 has very powerful instructions in the Logical group compared to 8085 microprocessor. It can perform AND, OR, EX-OR, TEST, SHIFT, ROTATE etc. operations.

6.15. LOGICAL INSTRUCTIONS

6.15.1. And Instructions

The AND instruction performs bit-wise AND of the two operands. If X is a bit of operand_1 and Y is a bit of operand_2 in the same bit position, the AND operation is performed as per the following truth table.

Truth table for AND operation

X	Y	X AND Y
0	0	0
0	1	0
1	0	0
1	1	1

It can be noticed that X AND 1 = X, and X AND 0 = 0. Thus AND operation is used for selectively resetting to 0 some bits of operand_1. Bits of operand_1 which are ANDed with 0's are reset to 0, and bits of operand_1 which are ANDed with 1's are not changed. The AND instruction affects the flags as follows.

S any Z flags are updated based on the result. C and O flag are reset to 0. A flag is undefined. P flag updated based on the result. It is meaningful only for a byte operation.

In this group we have instructions to

(a) AND contents of two registers or a register and a memory location.

(b) AND immediate data with a register or a memory location.

(c) AND immediate data with accumulator.

6.15.2. AND Contents of Two Registers or a Register and a Memory Location

Template for this instruction is

Byte 1

0	0	1	0	0	0	0	W

Byte 2

MOD	REG	R/M

As there are 10 bits in the template for identifying the operands, there are 1024 instructions in this category.

Example 1. Code for AND AX,BX.

This instruction AMDs the contents of BX register with AX register, and stores the result in AX register. It is a 2 byte instruction.

In this example, we can specify either of AX or BX using the REG field. If we specify AX using REG field, then the D bit must be 1, indicating that AX is the destination register. The REG field must be 000 to indicate AX register. The W bit must be 1 to indicate it is a word operation. In this case, BX is specified using MOD value of 11, and R/M value of Oil. Thus the 2 byte code for this instruction would be

opcode	**D**	**W**	**MOD**	**REG**	**R/M**	
0010 00	1	1	11	000	011	= 23 C3H

If we specify BX using REG field, then the D bit must be 0, indicating that BX is the source register. The REG field must be 011 to indicate BX register, The W bit must be 1 to indicate it is a word operation. In this case, AX is specified using MOD value of 11, and R/M value of 000. Thus the *2* byte code for this instruction would be

opcode	**D**	**W**	**MOD**	**REG**	**R/M**	
0010 00	0	1	11	011	000	= 21 D8H

Thus AND AX, BX has two equally valid opcodes as shown above. An assembler generates one of these as the opcode.

	Before	***After***
(AX)	122F	1027
(BX)	70F7	70F7

Example 2. Code for AND DS:0F246H[BP],DX

This instruction ANDs the contents of DX register to a memory location whose EA is given as the sum of the 16 bit offset value F246H and the contents of BP register. The PA is generated using DS as the segment register. The result is stored in the same memory location.

Here we have to specify DX using REG field. The D bit must be 0, indicating that DX is the source register. The REG field must be 010 to indicate DX register. The W bit must be 1 to indicate it is a word operation, F246H[BP] is specified using MOD value of 10, and R/M value of 110, and a displacement of F246H.

Whenever BP is used to generate the EA, the default segment would be SS. In this example, as we want the segment register to be DS, we have to provide the segment override prefix byte (SOP byte) to start with. The SOP byte is 001SR110, where SR indicates the segment override we desire.

To specify DS register, the SOP byte would be 001 11 110 = 3EH. Thus the 5 byte code for this instruction would be

SOP	opcode	D	W	MOD	REG	R/M	16 bit disp
3E	001000	0	1	10	010	110	F2 46

= 3E 21 96 F2 46H

	Before	*After*
(BP)	0100	0100
(DS:F346)	122F	1027
(DX)	70F7	70F7

6.15.3. AND Immediate Data With a Register or a Memory Location

Template for this instruction is

Byte 1								Byte 2				
1	0	0	0	0	0	S	W	MOD	1	0	0	R/M

As there are 7 bits in the template for identifying the operands, there are 128 instructions in this category.

The S bit stands for Sign Extension. This bit will be 0 always for a byte operation (W = 0).

If it is a word operation (W = 1), the S bit must be a 1 only if the immediate data is <= 7FH. In such a case the assembler generates only 1 byte code for the immediate data. However, during execution, the one byte data will be sign extended to make it, a 2 byte operand.

If the immediate data is > 7FH, the S bit must be a 0, and the assembler generates a 2 byte code for the immediate data.

Example 1. Code for AND CX,88H.

As it is a word operation, with immediate data > 7FH, the S bit will be 0, and the assembler generates 2 byte code for the immediate data. It is a 4 instruction.

opcode	S	W	MOD	opcode	R/M	16 bit data	
100000	0	1	11	00	001	00 88H	= 81 E1 00 88H

Example 2. Code for AND CL,88H.

As it is a byte operation, the S bit will be 0. It is a 3 byte instruction.

opcode	S	W	MOD	opcode	R/M	8 bit data	
1000 00	0	0	11	00	001	88H	= 80 E1 88H

Example 3. Code for AND CX,48H.

As it is a word operation, with immediate data <= 7FH, the S bit will be 1, and the assembler generates 1 byte code for the immediate data. Thus it is a 3 byte instruction,

opcode	S	W	MOD	opcode	R/M	8 bit data
1000 00	1	1	11	001	48H	= 83 E1 48H

6.15.4. AND Immediate Data with Accumulator

Template for this instruction is

0	0	1	0	0	1	0	W

As there is only one bit in the template for identifying the operands, there are 2 instructions ia this category. The assembler always generates 1 byte or 2 bytes of code for the immediate data depending on whether it is a byte or a word operation.

Example 1. Code for AND AX,88H.

As it is a word operation, the assembler generates 2 byte code for the immediate data. So, it is a 3 byte instruction.

opcode	**W**	**16 bit data**	
0010 010	1	00 88H	= 25 00 88H

Example 2. Code for AND AL,83H.

As it is a byte operation, the assembler generates 1 byte code for the immediate data. So, it is a 2 byte instruction.

opcode	**W**	**8 bit data**	
0010 010	0	83H	= 24 83H

6.15.5. OR Instructions

The OR instruction performs bit-wise OR of the two operands. If X is a bit of operand_1, and Y is a bit of operand_2 in the same bit position, the OR operation is performed as per the following truth table.

Truth table for OR operation

X	Y	X OR Y
0	0	0
0	1	1
1	0	1
1	1	1

It can be noticed that X OR 1 = 1, and X OR 0 = X. Thus OR operation is used for selectively setting to 1 some bits of operand_1. Bits of operand_1 which are ORed with 1's are set to 1, and bits of operand_1 which are ORed with 0's are not changed. The OR instruction affects the flags as follows.

S and Z flags are updated based on the result. C and O flag are reset to 0. A flag is undefined, P flag updated based on the result. It is meaningful only for a byte operation.

In this group we have instructions to

(a) OR contents of two registers or a register and a memory location.

(b) OR immediate data with a register or a memory location.

(c) OR immediate data with accumulator.

6.15.6. OR Contents of Two Registers or a Register and a Memory Location

Template for this instruction is

Byte 1								Byte 2		
0	0	0	0	1	0	D	W	MOD	REG	R/M

As there are 10 bits in the template for identifying the operands, there are 1024 instructions in this category.

Example 1. Code for OR AX,BX.

This instruction ORs the contents of BX register with AX register, and stores the result in AX register. It is a 2 byte instruction.

In this example, we can specify either of AX or BX using the REG field. If we specify AX using REG field, then the D bit must be 1, indicating that AX is the destination register. The REG field must be 000 to indicate AX register.

The W bit must be 1 to indicate it is a word operation. In this case, BX is specified using MOD value of 11, and R/M value of Oil, Thus the 2 byte code for this instruction would be

opcode	**D**	**W**	**MOD**	**REG**	**R/M**	
0000 10	1	1	11	000	011	= 0B C3H

If we specify BX using REG field, then the D bit must be 0, indicating that BX is the source register. The REG field must be 011 to indicate BX register. The W bit must be 1 to indicate it is a word operation. In this case, AX is specified using MOD value of 11, and R/M value of 000. Thus the 2 byte code for this instruction would be

opcode	**D**	**W**	**MOD**	**REG**	**R/M**	
0000 10	0	1	11	011	000	=09 D8H

Thus OR AX,BX has two equally valid opcodes as shown above. An assembler generates one of these as the opcode.

	Before	***After***
(AX)	122F	72FF
(BX)	70F7	70F7

Example 2. Code for OR DS:0F246H[BP],DX.

This instruction ORs the contents of DX register to a memory location whose EA is given as the sum of the 16 bit offset value F246H and the contents of BP register. The PA is generated using DS as the segment register. The result is stored in the same memory location.

Here we have to specify DX using REG field. The D bit must be 0, indicating that DX is the source register. The REG field must be 010 to indicate DX register. The W bit must be 1 to indicate it is a word operation. F246H[BP] is specified using MOD value of 10, and R/M value of 110, and a displacement of F246H.

To specify DS register, the SOP byte would be 001 11 110 = 3EH. Thus the 5 byte code for this instruction would be

SOP	**opcode**	**D**	**W**	**MOD**	**REG**	**R/M**	**16 bit disp**
3E	0000 10	0	1	10	010	110	F2 46
							= 3E 09 96 F2 46H

	Before	***After***
(BP)	0100	0100
122F	72FF	
(DX)	70F7	70F7

6.15.7. OR Immediate Data With a Register or a Memory Location

Template for this instruction is

Byte 1

1	0	0	0	0	0	S	W

Byte 2

MOD	0	0	1	R/M

As there are 7 bits in the template for identifying the operands, there are 128 instructions In this category. The S bit stands for Sign Extension. This bit will be 0 always for a byte operation (W = 0).

If it is a word operation (W = 1), the S bit must be a 1 only if the immediate data is <= 7FH. In such a case the assembler generates only 1 byte code for the immediate data. However, during execution, the one byte data will be sign extended to make it a 2 byte operand.

If the immediate data is > 7FH, the S bit must be a 0, and the assembler generates a 2 byte code for the immediate data.

Example 1. Code for OR CX,88H.

As it is a word operation, with immediate data > 7FH, the S bit will be 0, and the assembler generates 2 byte code for the immediate data. It is a 4 byte instruction.

opcode	S	W	MOD	opcode	R/M	16 bit data	
1000 00	0	1	11	001	001	0088H	= 81 C9 00 88H

Example 2. Code for OR CL,88H.

As it is a byte operation, the S bit will be 0. It is a 3 byte instruction.

opcode	S	W	MOD	opcode	R/M	8 bit data	
1000 00	0	0	11	001	001	88H	= 80 C9 88H

Example 3. Code for OR CX,48H.

As it is a word operation, with immediate data <= 7FH, the S bit will be 1, and the assembler generates 1 byte code for the immediate data. Thus it is a 3 byte instruction.

opcode	S	W	MOD	opcode	R/M	8 bit data	
1000 00	1	1	11	001	001	48H	= 83 C9 48H

6.15.8. OR Immediate Data with Accumulator

Template for this instruction is

0	0	0	0	1	1	0	W

As there is only one bit In the template for identifying the operands, there are 2 instructions in this category. The assembler always generates 1 byte or 2 bytes of code for the immediate data depending on whether it is a byte or a word operation.

Example 1. Code for OR AX,88H.

As it is a word operation, the assembler generates 2 byte code for the immediate data. So, it is a 3 byte instruction.

opcode	W	16 bit data	
0000 110	1	00 88H	= 0D 00 88H

Example 2. Code for OR AL.88H.

As it is a byte operation, the assembler generates 1 byte code for immediate data. So, it is a 2 byte instruction.

opcode	W	8 bit data	
0000 110	0	88H	= 0C 88H

6.15.9. EX-OR Instructions

The Ex-OR instruction performs bit-wise Ex-OR of the two operands. This instruction has the mnemonic XOR. If X is a bit of operand_I and Y is a bit of operaad_2 in the same bit position, the Ex-OR operation is performed as per the following truth table.

Truth table for Ex-OR operation

X	Y	X Ex-OR Y
0	0	0
0	1	1
1	0	1
1	1	0

It can be noticed that X XOR 1 = X*, and X XOR 0 = X. Thus XOR operation is used for selectively inverting some bits of operand_1. Bits of operand_1 which are XORed with 1's are inverted, and bits of operand_1 which are XORed with 0's are not changed.

The XOR instruction affects the flags as follows :

S and Z flags are updated based on the result C and O flag are reset to 0. A flag is undefined. P flag updated based ON the result. It is meaningful only for a byte operation.

In this group we have instructions to

(a) Ex-OR contents of two registers or a register and a memory location.

(b) Bx-OR immediate data with a register or a memory location.

(c) Ex-OR immediate data with accumulator.

6.15.10. Ex-OR Contents of Two Registers or Register and a Memory Location

Template for this instruction is

Byte 1							
0	0	1	1	0	0	D	W

Byte 2		
MOD	REG	R/M

As there are 10 bits in the template for identifying the operands, there are 1024 instructions in this category.

Example 1. Code for XOR AX,BX.

This instruction XORs the contents of BX register with AX register, and stores the result in AX register. It is a 2 byte instruction.

In this example, we can specify either of AX or BX using the REG field. If we specify AX using REG field, then the D bit must be 1, indicating that AX is the destination register. The REG field must be 000 to indicate AX register. The W bit must be 1 to indicate it is a word operation. In this case, BX is specified using MOD value of 11, and R/M value of 011. Thus the 2 byte code for this instruction would be

opcode	P	W	MOD	REG	R/M	
0011 00	1	1	11	000	011	= 33 C3H

If we specify BX using REG field, then the D bit must be 0, indicating that BX is the source register. The REG field must be 011 to indicate BX register. The W bit must be 1 to indicate it is a word operation. In this case, AX is specified using MOD value of 11, and R/M value of 000. Thus the 2 byte code for this instruction would be

opcode	D	W	MOD	REG	R/M
0011 00	01	11	011	000	= 31 DSH

Thus XOR AX,BX has two equally valid opcodes as shown above. An assembler generates one of these as the opcode.

	Before	*After*
(AX)	122F	
(BX)	70F7	70F7

Example 2. Code for XOR DS:0F246H[BP],DX.

This Instruction XORs the contents of DX register to a memory location whose EA is given as the sura of the 16 bit offset value F246H and the contents of BP register. The PA is generated using DS as the segment register.

Here we have to specify DX using REG field. The D bit must be 0, indicating that DX is the source register. The REG field must be 010 to indicate DX register. The W bit must be 1 to indicate it is a word operation. F246H[BP] is specified using MOD value of 10, and R/M value of 1 10, and a displacement of F246H.

Whenever BP is used to generate the EA, the default segment would be SS. In this example, as we want the segment register to be DS, we have to provide the segment override prefix byte (SOP byte) to start with. The SOP byte is 001SR110, where SR indicates the segment override we desire.

To specify DS register, the SOP byte would be 001 11 110 = 3EH. Thus the 5 byte code for this instruction would be

SOP	opcode	D	W	MOD	REG	R/M	16 bit disp
3E	0011 00	0	1	10	010	110	F2 46

= 3E 31 96 F2 46H

	Before	After
(BP)	0100	0100
(DS:F346)	122F	62D8
(DX)	70F7	70F7

6.15.11. Ex-OR Immediate Data With a Register or a Memory Location

Template for this instruction is

Byte 1								Byte 2				
1	0	0	0	0	0	S	W	MOD	1	1	0	R/M

As there are 7 bits in the template for identifying the operands, there are 128 instructions in this category. The S bit stands for Sign Extension. This bit will be 0 always for a byte operation (W = 0).

If it is a word operation (W = 1), the S bit must be a 1 only if the immediate data is <= 7FH. In such a case the assembler generates only 1 byte code for the immediate data. However, during execution, the one byte data will be sign extended to make it a 2 byte operand.

If the immediate data is > 7FH, the S bit must be a 0, and the assembler generates a 2 byte code for the immediate data.

Example 1. Code for XOR CX,98H.

As it is a word operation, with immediate data > 7FH, the S bit will be 0, and the assembler generates *2* byte code for the immediate data. It is a 4 byte instruction.

opcode	S	W	MOD	opcode	R/M	16 bit data	
1000 00	0	1	11	110	001	00 98H	= 81 F1 00 98H

Example 2. Code for XOR CL,98H.

As it is a byte operation, the S bit will be 0. It is a 3 byte instruction.

opcode	S	W	MOD	opcode	R/M	8 bit data	
1000 00	0	0	11	110	001	98H	= 80 F1 98H

Example 3. Code for XOR CX,88H.

As it is a word operation, with immediate data <= 7FH, the S bit will be 1, and the assembler generates 1 byte code for the immediate data. Thus it is a 3 byte instruction,

opcode	S	W	MOD	opcode	R/M	8bit data	
1000 00	1	1	11	110	001	88H	= 83 F1 88H

6.15.12. Ex-OR Immediate Data With Accumulator

Template for this instruction is

0	0	1	1	0	1	0	W

As there is only one bit in the template for identifying the operands, there are 2 instructions in this category. The assembler always generates 1 byte or *2* bytes of code for the immediate data depending on whether it is a byte or a word operation.

Example 1. Code for XOR AX,98H.

As it is a word operation, the assembler generates 2 byte code for the immediate data. So, it is a 3 byte instruction.

opcode	W	16 bit data	
0011 010	1	00 98H	= 35 00 98H

Example 2. Code for XOR AL,98H.

As it is a byte operation, the assembler generates 1 byte code for the immediate data. So, it is a 2 byte instruction

Opcode	W	8 bit data	
0011 010	0	98 H	= 34 98H

6.15.13. Test Instructions

The TEST instruction performs bit-wise AND of the two operands. But the operand values remain unchanged after the TEST instruction, just like in a CMP instruction. The result of ANDing is stored in a register which is not accessible to the programmer. Only the flags are affected depending on the result of AND operation. In a CMP instruction the flags are affected depending on the result of Subtract operation.

If X is a bit of operand_1, and Y is a bit of operand_2 in the same bit position, the AND operation is performed as per the following truth table.

Truth table for AND operation

X	Y	X AND Y
0	0	0
0	1	0
1	0	0
1	1	1

It can be noticed that X AND 1 = X, and X AND 0 = 0. The test instruction affects the flags as follows.

S and Z flags are updated based on the result, C and O flag are reset to 0. A flag is undefined. P flag updated based on the result. It is meaningful only for a byte operation.

In this group we have instructions to

(a) AND contents of two registers or a register and a memory location, using TEST instruction.

(b) AND a register or a memory location with immediate data, using TEST instruction.

(c) AND accumulator contents with immediate data, using TEST instruction.

6.15.14. AND Contents of Two Registers or a Register and a Memory Location, using TEST Instruction

Template for this instruction is

Byte 1								Byte 2		
1	0	0	0	0	1	0	W	MOD	REG	R/M

As there are 9 bits in the template for identifying the operands, there are 512 instructions in this category.

Notice the absence of D bit in the template. This is because, in a TEST instruction operand_1 (destination operand) must be a register always. In other words, the REG field always specifies operand_1 (destination).

Example : Code for TEST AX9BX.

This instruction ANDs the contents of BX register with AX register, and stores the result in a register which is not accessible to the programmer. It is a 2 byte instruction.

la this example, we have to specify AX, as the destination, using the REG field. The REG field must be 000 to indicate AX register. The W bit must be 1 to indicate it is a word operation. In this case, BX is specified using MOD value of 11, and R/M value of 011. Thus the *2* byte code for this instruction would be

opcode	W	MOD	REG	R/M	
1000 010	1	11	000	011	= 85 C3H

	Before	*After*
(AX)	522F	522F
(BX)	F000	F000
(Flags)	any value	Z = 0, S = 0, P = 1, C = 0, O = 0, A = x

In the above example, the flags are affected based on the result of ANDing. Which is 5000H. The TEST instruction, in this cases is essentially testing the MS 4 bits of AX. Because, the MS 4 bits of AX is 0101 which is non zero, the *7,* flag is reset to 0.

If the AX value was 022F, the result of ANDing would be 0000H, and so Z flag would be set to 1, This explanation justifies the mnemonic, for this instruction.

6.15.15. AND a Register or a Memory Location with Data, using TEST Instruction

Template for this instruction is

Byte 1								Byte 2				
1	1	1	1	0	1	1	W	MOD	0	0	0	R/M

As there are 6 bits in the template for identifying the operands, there are 64 instructions in this category. The assembler codes the immediate data using 1 or 2 bytes depending on whether it is a byte or a word operation.

Example 1. Code for TEST CX,98H.

As it is a word operation, the assembler generates 2 byte code for the immediate data. It is a 4 byte instruction.

opcode	W	opcode	R/M	16 bit data		
1111 011	1	11	000	001	00B3H	= F7C1 00 98H

This instruction tests bits 0, 1 and 7 of CX register. If all these bits are 0's, the Z flag will be set to 1.

Example 2. Code for TEST CL,98H.

As it is a byte operation, the assembler uses one byte for the immediate data. It is a 3 byte instruction.

opcode	W	MOD	opcode	R/M	8 bit data	
1111 011	0	11	000	001	98H	=F6 C1 98H

This instruction tests bits 0,1, and 7 of CL register. If all these bits are 0's, the Z flag will be set to 1.

6.15.16. AND Accumulator Contents with Immediate Data, using TEST Instruction

Template for this instruction is

1	0	1	0	1	0	0	W

As there is only one bit in the template for identifying the operands, there are 2 instructions in this category. The assembler always generates 1 byte or 2 bytes of code for the immediate data depending on whether it is a byte or a word operation.

Example 1. Code for TEST AX,98H.

As it is a word operation, the assembler generates 2 byte code for the data. So, it is a 3 byte instruction.

opcode	W	16 bit data	
1010 100	1	00 98H	= A9 00 98H

Example 2. Code for TEST AL,98H.

As it is a byte operation, the assembler generates 1 byte code for the immediate data. So, it is a 2 byte instruction.

opcode	W	8 bit data	
1010 100	0	98H	= A8 98H

6.16. NOT INSTRUCTIONS

This instruction performs logical inversion of the contents of a register or a memory location. It changes 0's to 1's and 1's to 0's. In effect, it replaces the number with its 1's complement.

6.16.1. Perform Logical Inversion of the Contents of a Register or a Memory Location

If can work on word or byte data. It does not affect any flags. Template for this instruction is

Byte 1

1	1	1	1	0	1	1	W

Byto 2

MOD	0	1	0	R/M

As there are 6 bits in the template for identifying the operands, there 64 instructions in this category. The mnemonic for this instruction is NOT.

Example 1. Code for NOT WORD PTR DS:0F246H[BP].

This instruction inverts the contents of a word location whose EA is given as the sum of the 16 bit offset value F246H and the contents of BF register. The PA is generated using DS as the segment register.

The W hit must be 1 to indicate it is a word operation. F246H[BP] is specified using MOD value of 10, and R/M value of 110, and a displacement of F246H.

Whenever BP is used to generate the EA, the default segment would be SS. In this example, as we want the segment register to be DS, we have to provide the segment override prefix byte (SOP byte) to start with. The SOP byte is 001 SR110, where SR indicates the segment override we desire.

To specify DS register, the SOP byte would be 001 11 110 = 3EH. Thus the 5 byte code for this instruction would be

SOP	Opcode	W	MOD	opcode	R/M	16 bit disp
3E	1111 011	1	10	010	110	F2 46

= 3E F7 96 F2 46H

	Before	After
(BP)	0100	0100
(DS:F346)	1230	EDCF

Example 2. Code for NOT CL.

The W bit must be 0 to indicate it is a byte operation. CL is specified using MOD value of 11, and R/M value of 001.

It is a 2 byte instruction with the code

opcode	W	MOD	Opcode	R/M	
1111 011	0	11	010	001	= F6 D1H

	Before	After
(CL)	60	9F

Example 3. Code for NOT CX.

The W bit must be 1 to indicate it is a word operation. CX is specified using MOD value of 11, and R/M value of 001.

It is a 2 byte instruction with the code

opcode	W	MOD	opcode	R/M	
1111 011	1	11	010	001	= F7D1H

	Before	After
(CX)	2360	DC9F

6.17. ROTATE INSTRUCTIONS

In this group we have instructions to

(a) Rotate left contents of register or a memory location.

(b) Rotate right contents of register or a memory location.

(c) Rotate left including carry contents of register or a memory location.

(d) Rotate right including carry contents of register or a memory location.

6.17.1. Rotate Left Contents of a Register or a Memory Location

This instruction has the mnemonic ROL. It rotates left the contents of a register or a memory location. The operation can be on a byte or a word data. The rotation can be by one bit position or more.

If the rotation is by one bit position, it has to be clearly indicated in the instruction. If it is by more than one bit position, CL must be loaded with a count value to indicate the number of bit positions by which the rotation has to be made. This operation can be depicted as shown in the next page.

After performing rotate left operation by one bit position, the bit moved out from the MS bit position goes to the vacancy created in the LS bit position. Also, C flag gets a copy of the bit moved out from the MS bit position. In the case of rotation by several bit positions, C flag will get a copy of the most recently moved out bit from the MS bit position.

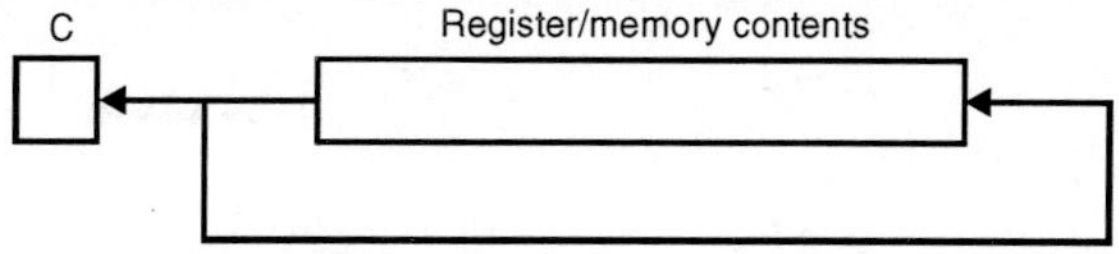

Only C and O flags are affected by the rotation. The O flag will be set to 1, if there is a change in the MS bit value.

The ROL instruction can be used for swapping the hex digits in a byte by using a count value of 4 in CL, or for swapping the bytes in a word by using a count value of 8 in CL. It can also be used to test the value of the MS bit.

Template for this instruction is

Byte 1							
1	1	0	1	0	0	V	W

Byte 2				
MOD	0	0	0	R/M

As there are 7 bits in the template for identifying the operands, there are 128 instructions in this category.

The V bit stands for count Value. This bit will be 0 always for a rotation by one bit position. If this bit is a 1, it means that CL contains the count value.

Example 1. Code for ROL CX,1.

It rotates left the contents of CX by one bit position. It is a 2 byte instruction.

opcode	**V**	**W**	**MOD**	**opcode**	**R/M**	
1101 00	0	1	11	000	001	= D1 C1H

Example 2. Code for ROL BYTE PTR DS:0F246H[BP],CL.

This instruction rotates left the contents of a memory location whose EA is given as the sum of the 16 bit offset value F246H and the contents of BP register. The PA is generated using DS as the segment register. The number of times the rotation is performed is given by the contents of CL register. The result is stored in the same memory location.

The W bit must be 0 to indicate it is a byte operation, F246H[BP] is specified using MOD value of 10, and R/M value of 110, and a displacement of F246H.

Whenever BP is used to generate the EA, the default segment would be SS. In this example, as we want the segment register to be DS, we have to provide the segment override prefix byte (SOP byte) to start with. The SOP byte is 001SR110, where SR indicates the segment override we desire.

To specify DS register, the SOP byte would be 001 11 110 = 3EH. Thus the 5 byte code for this instruction would be

SOP	**opcode**	**V**	**W**	**MOD**	**opcode**	**R/M**	**16 bit disp**
3E	1101 00	1	0	10	000	110	F246
							= 3E D2 86 F2 46H

	Before	***After***
(CL)	04	04
(BP)	0100	0100
(DS:F346)	2F	F2 with C = 0

Example 3. Code for ROL BX,CL.

If rotates left the contents of BX by a number of bit positions whose value is given in CL. It is a 2 byte instruction.

opcode	**V**	**W**	**MOD**	**opcode**	**R/M**	
1101 00	1	1	11	000	011	=D3 C3H

	Before	***After***
(CL)	08	08
(BX)	2360	6023 with C = 1

6.17.2. Rotate Right Contents of a Register or a Memory Location

This instruction has the mnemonic ROR. It rotates right the contents of a register or a memory location. The operation can be on a byte or a word data. The rotation can be by one bit position or more.

If the rotation is by one bit position, it has to be clearly indicated. In the instruction. If it is by more than one bit position, CL must be loaded with a count value to indicate the number of bit positions by which the rocation has to be made. This operation can be depicted as shown below.

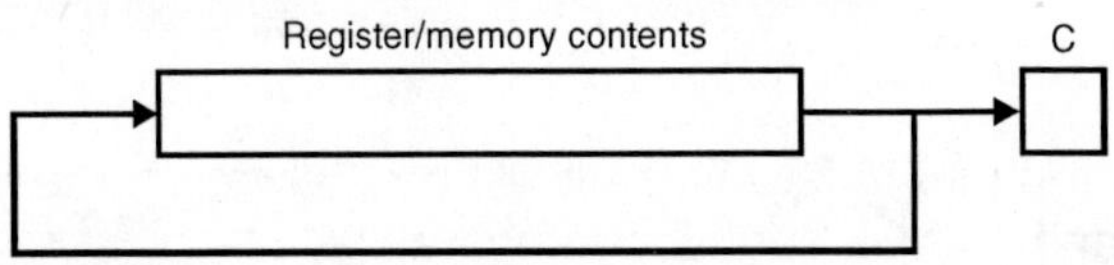

After performing rotate right operation by one bit position, the bit moved out from the LS bit position goes to the vacancy created in the MS bit position. Also, C flag gets a copy of the bit moved out from the LS bit position.

In the case of rotation by several bit positions, C flag will get a copy of the most recently moved out bit from the LS bit position.

Only C and O flags are affected by the rotation. The C flag will be set to 1, if there is a change in the MS bit value.

The ROR instruction can be used for swapping the hex digits in a byte by using a count value of 4 in CL, or for swapping the bytes in a word by using a count value of 8 in CL. It can also be used to test the value of the LS bit. Template for this instruction is

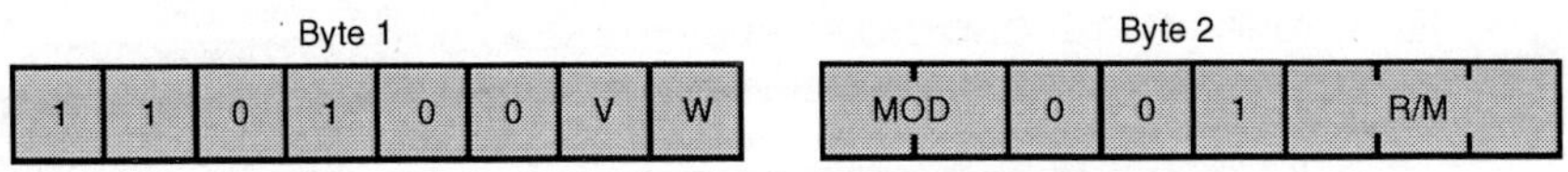

As there are 7 bits in the template for identifying the operands, there are 128 instructions in this category.

The V bit stands for count Value. This bit will be 0 always for a rotation by one bit position. If this bit is a 1, it means that CL contains the count value.

Example 1. Code for ROR CX,1.

It rotates right the contents of CX by one bit position. It is a 2 byte instruction.

opcode	**V**	**W**	**MOD**	**Opcode**	**R/M**	
1101 00	0	1	11	001	001	= D1 C9H

	Before	***After***
(CX)	2360	11B0 with C = 0

Example 2. Code for ROR BYTE PTR DS:0F246H[BP],CL.

This instruction rotates right the contents of a memory location whose EA is given as the sum of the 16 bit offset value F246H and the contents of BP register. The PA is generated using DS as the segment register. The number of times the rotation is performed is given by the contents of CL register. The result is stored in the same memory location.

The W bit must be 0 to indicate it is a byte operation. F246H[BP] is specified using MOD value of 10, and R/M value of 110, and a displacement of F246H.

Whenever BP is used to generate the EA, the default segment would be SS. In this example, as we want the segment register to be DS, we have to provide the segment override prefix byte (SOP byte) to start with. The SOP byte is 001SR110, where SR indicates the segment override we desire.

To specify DS register, the SOP byte would be 001 11 110 = 3EH. Thus the 5 byte code for this instruction would be

SOP	opcode	V	W	MOD	opcode	R/M	16 bit disp
3E	1101 00	1	0	10	001	110	F2 46

= 3E D2 8E F2 46H

	Before	After
(CL)	04	04
(BP)	0100	0100
(DS:F346)	2F	F2 with C = 1

Example 3. Code for ROR BX,CL.

It rotates right the contents of BX by a number of bit positions whose value is given in CL. It is a 2 byte instruction.

opcode	V	W	MOD	opcode	R/M	
1101 00	1	1	11	001	011	= D3 CBH

	Before	After
(CL)	08	08
(BX)	2360	6023 with C = 0

6.17.3. Rotate Left Including Carry on Contents of a Register or a Memory Location

This instruction has the mnemonic RCL. It rotates left the contents of a register or a memory location by involving the carry flag in the loop. The operation can be on a byte or a word data. The rotation can be by one bit position or more.

If the rotation is by one bit position, it has to be clearly indicated in the instruction. If it is by more than one bit position, CL must be loaded with a count value to indicate the number of bit positions by which the rotation has to be made. This operation can be depicted as shown below.

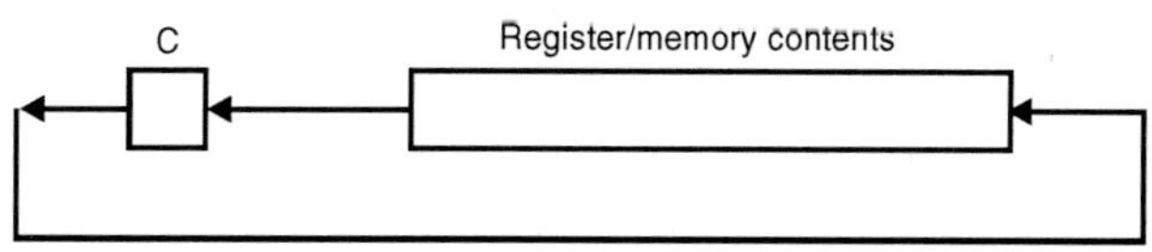

After performing rotate left operation by oae bit position, the bit moved out from the MS bit position goes to the cany flag, and in the process moves out the earlier carry bit to the vacancy created in the LS bit position. In the case of rotation by several bit positions, C flag will get a copy of the most recently moved out bit from the MS bit position.

Only C and O flags are affected by the rotation. The O flag will be set to 1, if there is a change in the MS bit value.

The RCL instruction is useful for testing the value of the MS bit of an operand. It is also used for moving the Carry flag value to the LS bit position of an operand. Template for this instruction is

Byte 1								Byte 2				
1	1	0	1	0	0	V	W	MOD	0	1	0	R/M

As there are 7 bits in the template for identifying the operands, there are 128 instructions in this category.

The V bit stands for count Value. This bit will be 0 always for a rotation by one bit position. If this bit is a 1, it means that CL contains the count value.

Example 1. Code for RCL CX,1.

It rotates left with carry the contents of CX by one bit position. It is a 2 byte instruction.

opcode	**V**	**W**	**MOD**	**opcode**	**R/M**	
1101 00	0	1	11	010	001	= D1 D1H

	Before	***After***
(CX)	2360	46C1
(C flag)	1	0

Example 2. Code for RCL BYTE PTR DS:0F246H[BP],CL.

This instruction rotates left with carry the contents of a memory location whose EA is given as the sum of the 16 bit offset value F246H and the contents of BP register. The PA is generated using DS as the segment register. The number of times the rotation is performed is given by the contents of CL register. The result is stored in the same memory location.

The W bit must be 0 to indicate it is a byte operation, F246H[BP] is specified using MOD value of 10, and R/M value of 110, and a displacement of F246H.

Whenever BP is used to generate the EA, the default segment would be SS. In this example, as we want the segment register to be DS, we have to provide the segment override prefix byte (SOP byte) to start with. The SOP byte is 001SR110, where SR indicates the segment override we desire.

To specify DS register, the SOP byte would be 001 11 110 = 3EH. Thus the 5 byte code for this instruction would be

SOP	**opcode**	**V**	**W**	**MOD**	**opcode**	**R/M**	**16 bit disp**
3E	1101 00	1	0	10	010	110	F2 46

= 3E D2 96 F2 46H

	Before	***After***
(CL)	04	04
(BP)	0100	0100
(DS:F346)	2F	F1
(C flag)	0	0

Example 3. Code for RCL BX,CL.

It rotates left with carry the contents of BX by a number of bit positions whose value is given in CL. It is a 2 byte instruction.

opcode	**V**	**W**	**MOD**	**opcode**	**R/M**	
1101 00	1	1	11	010	011	= D3 D3H

	Before	***After***
(CL)	03	03
(BX)	2360	1B04
(C flag)	1	1

6.17.4. Rotate Right Including on Carry Contents of a Register or a Memory Location

This instruction has the mnemonic RCR. It rotates right the contents of a register or a memory location by involving the carry flag in the loop. The operation can be on a byte or a word data. The rotation can be by one bit position or more.

If the rotation is by one bit position, it has to be clearly indicated in the instruction. If it is by more than one bit position, CL must be loaded with a count value to indicate the number of bit positions by which the rotation has to be made.

After performing rotate right operation by one bit position, the bit moved out from the LS bit position goes to the carry flag, and in the process moves out the earlier carry bit to the vacancy created in the MS bit position. In the case of rotation by several bit positions, C flag will get a copy of the most recently moved out bit from the LS bit position.

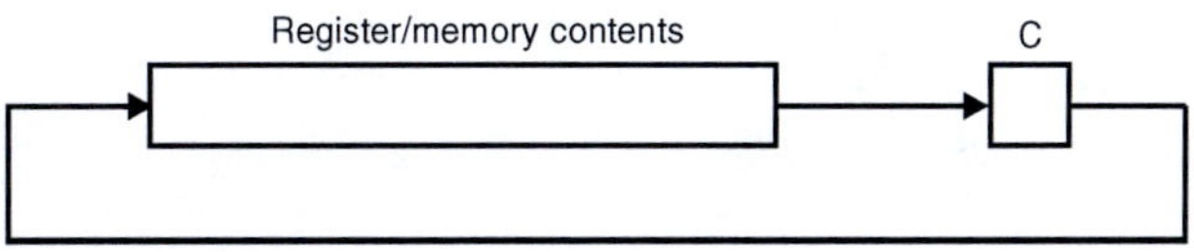

Only C and O flags are affected by the rotation. The O flag will be set to 1, if there is a change in the MS bit value.

The RCR instruction is useful for testing the value of the LS bit of an operand. It is also used for moving the Carry flag value to the MS bit position of an operand. Template for this instruction is

Byte 1								Byte 2				
1	1	0	1	0	0	V	W	MOD	0	1	1	R/M

As there are 7 bits in the template for identifying the operands, there are 128 instructions in this category.

The V bit stands for count Value. This bit will be 0 always for a rotation by one bit position. If this bit is a 1, it means that CL contains the count value.

Example 1. Code for RCR CX,1.

It rotates right with carry the contents of CX by one bit position. It is a 2 byte instruction.

opcode	**V**	**W**	**MOD**	**opcode**	**R/M**	
1101 00	0	1	11	011	001	= D1 D9H

	Before	***After***
(CX)	2360	91B0
(C flag)	1	0

Example 2. Code for RCR BYTE PTR DS:0F246H[BP],CL.

This instruction rotates right with carry the contents of a memory location whose EA is given as the sum of the 16 bit offset value F246H and the contents of BP register. The PA is generated using DS as the segment register. The number of times the rotation is performed is given by the contents of CL register. The result is stored in the same memory location.

The W bit must be 0 to indicate it is a byte operation. F246H[BP] is specified using MOD value of 10, and R/M value of 110, and a displacement of F246H.

Whenever BF is used to generate the EA, the default segment would be SS. In this example, as we want the segment register to be DS, we have to provide the segment override prefix byte (SOP byte) to start with. The SOP byte is 001 SR110, where SR indicates the segment override we desire.

To specify DS register, the SOP byte would be 001 11 110 = 3EH. Thus the 5 bte code for this instruction would be

SOP	opcode	V	W	MOD	opcode	R/M	16 bit disp
3E	1101 00	1	0	10	011	110	F2 46

= 3E D2 9E F2 46H

	Before	*After*
(CL)	04	04
(BP)	0100	0100
(DS:F346)	2F	E2
(C flag)	0	1

6.18. SHIFT INSTRUCTIONS

In this group, we have instructions to

(a) Shift Right the contents of a register or a memory location.

(b) Arithmetic Shift Right the contents of a register or a memory location.

(c) Shift Left the contents of a register or a memory location.

6.18.1. Shift Right Contents of a Register or a Memory Location

This instruction has the mnemonic SHR. It shifts right the contents of a register or a memory location. The operation can be on a byte or a word data. The shift can be by one bit position or more. This operation is used with unsigned numbers.

If the shift is by one bit position, it has to be clearly indicated in the instruction. If it is by more than one bit position, CL must be loaded with a count value to indicate the number of bit positions by which the shift has to be made. This operation can be depicted shown as below.

After performing shift right operation by one bit position, the bit moved out from the LS bit position goes to the carry flag. In the vacancy created at the MS bit position a 0 will be moved. In the case of shift by several bit positions, C flag will get a copy of the most recently moved out bit from the LS bit position.

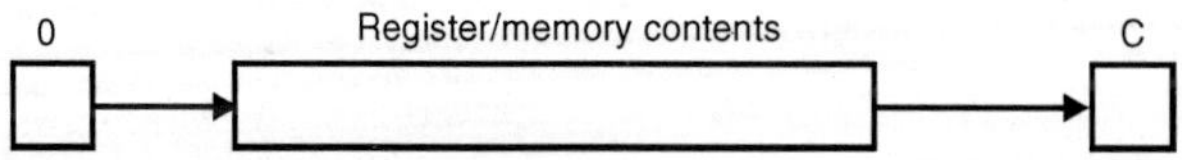

The SHR instruction affects the flags as follows.

S and Z flags are updated based on the result. C equals the most recently shifted bit from the LS position. O flag is set to 1 if there is a change in the MS bit value. A flag is undefined. P flag updated based on the result, it is meaningful only for a byte operation.

The SHR instruction is useful for dividing an unsigned binary number by a power of 2. Right shifting once will divide by 2, shifting twice will divide by 4, thrice will divide by 8, etc. When an odd number is shifted this way, the result will be truncated. For example, number *5* shifted right once will give 2 as the result.

Template for this instruction is

Byte 1								Byte 2				
1	1	0	1	0	0	V	W	MOD	1	0	1	R/M

As there are 7 bits in the template for identifying the operands, there are 128 instructions in this category.

The V bit stands for count Value. This bit will be 0 always for a shift by one bit position. If this bit is a 1, it means that CL contains the count value.

Example 1. Code for SHR CX,1.

It shifts right the contents of CX by one bit position. It is a 2 byte instruction.

opcode	V	W	MOD	opcode	R/M	
1101 00	0	1	11	01	001	= D1 E9H

	Before	*After*
(CX)	2460	1230

Example 2. Code for SHR BYTE PTR DS:0F246H[BP],CL.

This instruction shifts right the contents of a memory location whose EA is given as the sum of the 16 bit offset value F246H the contents of BP register. The PA is generated using DS as the segment register. The number of times the shift is performed is given by the contents of CL register. The result is stored in the same memory location.

The W bit must be 0 to indicate it is a byte operation. F246H[BP] is specified using MOD value of 10, and R/M value of 1 10, a displacement of F246H.

Whenever BP is used to generate the EA, the default segment would be SS. In this example, as we want the segment register to be DS, we have to provide the segment override prefix byte (SOP byte) to start with. The SOP byte is 001SR110, where SR indicates the segment override we desire.

To specify DS register, the SOP byte would be 001 11 110 = 3EH. Thus the 5 byte code for this instruction would be

SOP	opcode	V	W	MOD	opcode	R/M	16 bit disp
3E	1101 00	1	0	10	101	110	F2 46

= 3E D2 AE F2 46H

	Before	*After*
(CL)	04	04
(BP)	0100	0100
(DS:F346)	2F	02

6.18.2. Arithmetic Shift Right Contents of a Register or a Memory Location

This instruction has the mnemonic SAR. It shifts right the contents of a register or a memory location. The operation can be on a byte or a word data. The shift can be by one bit position or more. Arithmetic shifts are used with signed numbers. In arithmetic shifts, the sign of the number is not changed after the shift operation.

If the shift is by one bit position, it has to be clearly indicated in the instruction. If it is by more than one bit position, CL must be loaded with a count value to indicate the number of bit positions by which the shift has to be made. This operation can be depicted shown as below.

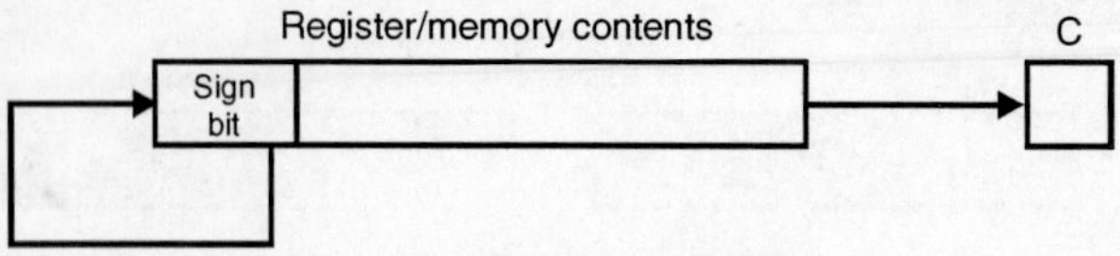

After performing arithmetic shift right operation by one bit position, the bit moved out from the LS bit position goes, to the carry flag. In the vacancy created at the MS bit position, once again the original MS bit value will be moved. Thus the sign of the number is not changed after the arithmetic shift. In the case of shift by several bit positions, C flag will get a copy of the most recently moved out bit from the LS bit position.

The SAR instruction affects the flags as follows.

S and Z flags are updated based on the result. C equals the most recently shifted bit from the LS position. O flag is reset to 0, as there cannot be any overflow, A flag is undefined. P flag updated based on the result. It is meaningful only for a byte operation.

The SAR instruction is useful for dividing a signed binary number by a power of 2. Arithmetic right shifting once will divide by 2, shifting twice will divide by 4, thrice will divide by 8, etc. When an odd number is shifted this way, the result will be truncated. For example, number 5 arithmetic right shifted once will give 2 as the result. Number-5 (byte FBH) arithmetic right shifted once will give 3 (byte EDH) as the result.

Template for this instruction is

Byte 1							
1	1	0	1	0	0	V	W

Byte 2				
MOD	1	1	1	R/M

As there are 7 bits in the template for identifying the operands, there are 128 instructions in this category.

The V bit stands for count Value. This bit will be 0 always for a shift by one bit position. If this bit is a 1, it means that CL contains the count value.

Example 1. Code for SAR CX,1.

It arithmetic shifts right the contents of CX by one bit position. It is a 2 byte instruction.

opcode	**V**	**W**	**MOD**	**opcode**	**R/M**	
1101 00	0	1	11	111	001	= D1 F9H

	Before	***After***
(CX)	FFFC	FFFE

Example 2. Code for SAR BYTE PTR DS:0F246H[BP],CL.

This instruction arithmetic shifts right the contents of a memory location whose EA is given as the sum of the 16 bit offset value F246H and the contents of BP register. The PA is generated using DS as the segment register. The number of times the shift is performed is given by the contents of CL register. The result is stored in the same memory location.

The W bit must be 0 to indicate it is a byte operation. F246H[BP] is specified using MOD value of 10, and R/M value of 110, and a displacement of F246H.

Whenever BP is used to generate the EA, the default segment would be SS. In this example, as we want the segment register to be DS, we have to provide the segment override prefix byte (SOP byte) to start with. The SOP byte is 001SR110, where SR indicates the segment override we desire.

To specify DS register, the SOP byte would be 001 111 10 = 3EH. Thus the 5 byte code for this instruction would be

SOP	opcode	V	W	MOD	opcode	R/M	16 bit disp
3E	1101 00	1	0	10	111	110	F2 46

= 3E D2 BE F2 46H

	Before	After
(CL)	04	04
(BP)	0100	0100
(DS:F346)	2F	02

6.18.3. Shift Left Contents of a Register or a Memory Location

This instruction has the mnemonic SHL. Another equivalent mnemonic for instruction is SAL, which stands for 'Shift Arithmetic Left'. This operation can be used with both unsigned as well as signed numbers. It shifts left the contents of a register or a memory location. The operation can be on a byte or a word data. The shift can be by one bit position or more.

If the shift is by one bit position, it has to be clearly indicated in the instruction. If it is by more than one bit position, CL must be loaded with a count value to indicate the number of bit positions by which the shift has to be made. This operation can be depicted shown as below.

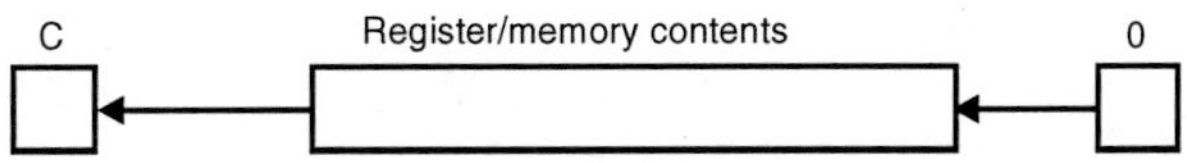

After performing shift left operation by one bit position, the bit moved out from the MS bit position goes to the carry flag. In the vacancy created at the LS bit position a 0 will be moved. In the case of shift by several bit positions, C flag will get a copy of the most recently moved out bit from the MS bit position.

The SHL/SAL instruction affects the flags as follows.

S and Z flags are updated based on the result. C equals the most recently shifted bit from the MS position. O flag is set to 1 if there is a change in the MS bit value. A flag is undefined. P flag updated based on the result. It is meaningful only for a byte operation.

The SHL/SAL instruction is useful for multiplying an unsigned binary number by a power of 2. Left shifting once will multiply by 2, shifting twice will multiply by 4, thrice will multiply by 8, etc. If the MS bit of the number is a 1, then after left shifting, the result will be incorrect. For example, byte 80H shifted left once will give 00H as the result, which is obviously incorrect.

Template for this instruction is

Byte 1								Byte 2				
1	1	0	1	0	0	V	W	MOD	1	0	0	R/M

As there are 7 bits in the template for identifying the operands, there are 1 28 instructions in this category.

The V bit stands for count Value. This bit will be 0 always for a shift by one bit position. If this bit is a 1, it means that CL contains the count value.

Example 1. Code for SHL CX,1 or SAL CX,1.

It shifts left the contents of CX by one bit position. It is a 2 byte instruction.

opcode	V	W	MOD	opcode	R/M	
1101 00	0	1	11	100	001	= D1 E1H

	Before	After
(CX)	1230	2460

Example 2. Code for SHL BYTE PTR DS:0F246H[BP],CL.

This instruction shifts left the contents of a memory location whose EA is given as the sum of the 16 bit offset value F246H and the contents of BP register. The PA is generated using DS as the segment register. The number of times the rotation is performed is given by the contents of CL register. The result is stored in the same memory location,

The W bit must be 0 to indicate it is a byte operation. F246H[BP] is specified using MOD value of 10, and R/M value of 110, and a displacement of F246H.

Whenever BP is used to generate the EA, the default segment would be SS. In this example, as we want the segment register to be DS, we have to provide the segment override prefix byte (SOP byte) to start with. The SOP byte is 001SR110, where SR indicates the segment override we desire.

To specify DS register, the SOP byte would be 001 11 110 = 3EH. Thus the 5 byte code for this instruction would be

SOP	opcode	V	W	MOD	opcode	R/M	16 bit disp
3E	1101 00	1	0	10	100	110	F2 46

= 3E D2 A6 F2 46H

	Before	After
(CL)	04	04
(BP)	0100	0100
(DS:F346)	2F	F0

In this chapter, instructions that operate on flags, instructions for external synchronization, etc. are discussed,

6.19. PROCESS CONTROL INSTRUCTION

Instructions to Set/Reset Flags

In this group, we have instruction to

(a) Set/Clear/cornplement carry flag.

(b) Set/Clear direction flag.

(c) Set/Clear interrupt enable flag.

6.19.1. Set the Carry Flag to 1 State

The mnemonic for the instruction is STC. Opcode for this instruction is F9H. This instruction execution results in setting of Carry Flag to the 1 state. Other flags are not affected.

6.19.2. Clear the Carry Flag to 0 State

The mnemonic for the instruction is CLC. Opcode for this instruction is F8H. This instruction execution results in clearing of Carry Flag to the 0 state. Other flags are not affected.

6.19.3. Complement the Carry Flag

The mnemonic for the instruction is CMC. Opcode for this instruction is F5H. This instruction execution results in inversion of Carry Flag. That is if C flag was 1 before this instruction, it changes to 0 after the instruction, and vice versa. Other flags are not affected.

6.19.4. Clear the Direction Flag to 0 State

The mnemonic for the instruction is CLD, Opcode for this instruction is FCH. This instruction execution results in clearing of Direction Flag to the 0 state. Other flags are not affected.

6.19.5. Set the Direction Flag to 1 State

The mnemonic for the instruction is STD. Opcode for this instruction is FDH. This instruction execution results in setting of Direction Flag to the 1 state. Other flags are not affected.

6.19.6. Clear the Interrupt Enable Flag to 0 State

The mnemonic for the instruction is CLL Opcode for this instruction is FAH. This instruction execution results in clearing of Interrupt enable Flag to the 0 state. Other flags are not affected. If the Interrupt enable flag is reset to 0, the 8086 will not get interrupted when INTR line is activated by an external device. However, NMI line can still cause an interrupt to 8086,

6.19.7. Set the Interrupt Enable Flag to 1 State

The mnemonic for the instruction is STL. Opcode for this instruction is FBH. This instruction execution results in setting of Interrupt enable Flag to the 1 state. Other flags are not affected.

Only if Interrupt enable flag is set to 1, the 8086 will get interrupted when the INTR line is activated. However, the 8086 will get interrupted when NMI line is activated, immaterial of the Interrupt enable flag status. Although Interrupt enable flag is set to 1, when STI instruction is executed, the 8086 can get interrupted because of intr pin only after the execution of the next instruction.

6.20. HALT INSTRUCTION

The mnemonic for the instruction is HLT. Opcode for this instruction is F4H. This instruction execution causes the 8086 to enter a Halt state where it stops fetching and executing instructions. The only way the 8086 can come out of this state is because of the activation of INTR, NMI, or RESET signal.

6.21. WAIT INSTRUCTION

The mnemonic for the instruction is WAIT. Opcode for this instruction is 9BH. The WAIT instruction Is used to synchronize the 8086 with external hardware such as the 8087 numeric coprocessor. TMs instruction is normally used just before an 8087. Instruction in our programs. In fact, a standard assembler like MASM from Micro Soft, automatically generates a WAIT instruction prefix for any 8087 instruction.

When this instruction is fetched, the IP is incremented by 1, and then 8086 senses the Test* input. If the Test* input is not active, the 8086 enters an idle state where it does not do any processing. However, it fetches the next instruction in the queue which is an 8087 instruction, and ignores that instruction. The 8086 stays in this Idle state until one of the following conditions occurs.

1. Test* is made low for outlast *5* clock cycles by the 8087
2. NMI interrupt line is activated
3. INTR interrupt line is activated, when I flag = 1.

If the Test* input Is activated for outlast 5 clock cycles, the 8086 comes out of the idle state, to execute the next 8086 instruction. This is useful in the following two cases.

1. When the result produced by 8087 is needed by the 8086 in its next instruction.

2. When there are a series of 8087 instructions in our program, obviously they are required to be fetched from the instruction queue as and when an 8087 instruction execution is completed. The WAIT instruction and Test input help in indicating when an 8087 instruction execution is completed.

If the 8086 is interrupted because of INTR or NMI activation, the 8086 returns to the idle state, after the Interrupt service routine is executed. This is because, the address of the WATT instruction is stored above the top of the stack, when 8086 branches to the interrupt service subroutine.

6.22. LOCK PREFIX

The mnemonic for the instruction is LOCK. Opcode for this instruction is FOH. LOCK is a prefix which can be written before an instruction. It is not an instruction by itself. It only modifies the execution of the instruction following the LOCK byte.

A microcomputer system can contain several microprocessors. Each microprocessor will have its own local buses for accessing local resources like local memory. The various microprocessors in the system are connected by a system bus. Thus each of the microprocessors can access global resources like disk drives or global memory.

The LOCK prefix for an instruction allows a microprocessor to make sure that another microprocessor not control of the system bus during execution of a critical instruction which uses the system bus. This is achieved by the 8086 activating Lock* signal, which is one of the signals in the maximum mode of operation. This signal is connected to a external bus controller, which prevents any other microprocessor from taking control of the system bus.

For example, XCHG AL,SEMA instruction requires two system bus cycles for the exchange operation, if SEMA is a global memory location. One memory access is for loading AL with SEMA value and the other for storing AL value in SEMA. If this instruction has to be fully executed before another microprocessor takes control of the system bus, we have to execute LOCK XCHG AL, SEMA which has the LOCK prefix byte to achieve the objective.

6.23. ESCAPE TO CO-PROCESSOR

The mnemonic for the instruction is ESC. Template for this instruction is

Byte 1							
1	1	0	1	1	x	x	x

Byte 2				
MOD	x	x	x	R/M

Instructions starting with 11011 will be treated by the 8086 as an instruction to a Coprocessor, such as the 8087 numeric Coprocessor which shares the address and data bus with the 8086. Instructions for the coprocessor is given by the 6 bit code, shown with x's in the above template. The 8087 decodes the 6 bit code in the ESC instruction to determine the action to be performed.

As the 8086 fetches instruction bytes, the 8087 also fetches them on the data bus, and puts them in its queue. The 8087 treats all normal 8086 instructions as 'No operation' instructions. In most cases, the 8086 treats ESC instructions as a NOP. In some cases, the 8086 accesses a data item in memory for the 8087.

6.24. STRING INSTRUCTIONS

8086 has very powerful instruction in the String manipulation group. It can perform Movement, comparison, load, Store and Scan operation on strings.

6.24.1. Move String Instruction

As can be noticed from the template, it can be a byte or a word operation. The mnemonic for byte operation is MOVSB, which stands for 'Move string byte'. It has the opcode A4H. For word operation the mnemonic is MOVSW, which stands for 'Move String Word'. It has the opcode A5H. This instruction copies the contents of a byte or word at location DS:SI to a byte or word at location ES:DI. It does not affect any flag.

Example 1. MOVSB instruction

Let us say, we have the string 'Nivedita' starting at an offset of 1213H in the data segment. We want to move this string, of size 8 characters to another memory area, starting at an offset 3451H in the Extra segment.

8086 provides MOVSB (stands for move string byte) instruction to perform this movement a byte at a time. This instruction should be executed 8 times in a loop to perform the desired string movement. MOVSB instruction expects that the source string is pointed by DS:SI. In other words, for the source string, the segment base address must be provided in DS, and the offset in SI. Similarly ES:DI must point to the destination string.

So, to facilitate the string movement we load SI with 123H and DI with 3451H. Then the MOVSB instruction is executed. This results in the movement of character 'N' to offset location 3451H in the Extra segment. Then automatically, SI and DI are incremented to 123H and 3452H respectively, so that next time through the loop, the MOVSB instruction moves the next character, and so on.

However, this automatically increment of SI and DI takes place only if the D (direction) flag is reset to 0 by executing the instruction CLD. If the D flag is set to 1 by executing the instruction STD, the SI and DI registers are automatically decremented after the execution of MOVSB instruction.

Thus the same string movement can be achieved by executing MOVSB a loop 8 times, by initially loading SI with 1238H and DI with 3458H, and setting D flag to 1.

It is possible to use "MOVS' instead of 'MOVSB' as the mnemonic. But then the assembler has to be told in some way that byte operation is desired. This is done by indicating the names of source and destination strings in the instruction itself, like in "MOVS SRC,DST". The assembler will code the instruction for a byte operation if SRC and DST are declared using DB(Define Byte) assembler directive. The assembler will code the instruction for a word operation if SRC and DST are declared using DW (Define Word) assembler directive.

6.24.2. Repeat prefix for MOVSB

It is not really necessary to execute the MOVSB instruction in a loop 8 times, because 8086 provides the REP (stands for repeat) prefix facility. The REP prefix is not an instruction by itself. It has the code F3H.

The result of execution of the instruction "REP MOVSB" is a follows.

1. If CX contents is 0, proceed with next instruction.
2. Execute MOVSB instruction.
3. Decrement CX by 1.
4. GO to step 1.

Thus the same string movement can be achieved by the execution of the following 2 instruction, instead of repeatedly executing MOVSB in a loop.

```
MOV     CX,08
REP     MOVSB
```

Example 2. MOVSW instruction

Let us say, we have a string of 4 words starting at an offset of 1232H in the Data segment. We want to move this string to another memory area, starting at an offset of 3451H in the Extra segment.

8086 provides MOVSW (stands for Move String Word) instruction to perform this movement a word at a time. This instruction should be executed 4 times in a loop to perform the desired string movement. MOVSW instruction expects that the source string is pointed by DS:SI. In other words, for the source string, the segment base address must be provided in DS, for the offset in SI. Similarly ES:DI must point to the destination string.

SO, to facilitate the string movement we load SI with 1231H and DI with movement of the first word to offset word location 3451H in the Extra segment. Then automatically, SI and DI are incremented by 2 to 1233H and 3453H respectively, so that next time through the loop, the MOVSW instruction moves the next word, and so on.

However, this automatic increment of SI and DI by 2 takes place only if the D (direction) flag is reset to 0 by executing the instruction CLD. If the D flag is set to 1 by executing the instruction STD, the SI and DI registers are automatically decremented by 2 after the execution of MOVSW instruction.

Thus the same string movement can be achieved by executing MOVSW in a loop 4 times, by initially loading SI with 1237H and DI with 3457H, and setting D flag to 1.

It is possible to use 'MOVS' instead of 'MOVSW' as the mnemonic. But then the assembler has to be told in some way that word operation is desired. This is done by indicating the names of the source and destination string in the instruction itself, like in 'MOVS SRC, DST'. The assembler will code the instruction for a byte operation if SRC and DST are declared using DB (define Byte) assembler directive. The assembler will code the instruction for a word operation if SRC and DST are declared using DW (define Word) assembler directive.

6.24.3. Repeat Prefix for MOVSW

It is not really necessary to execute the MOVSW instruction in a loop 4 times, because 8086 provides the REP (stands for Repeat) prefix facility. The REP prefix is not an ins t ruction by itself. It has the code F3H.

The result of execution of the instruction 'REP MOVSW' is as follows.

1. If CX contents is 0, proceed with next instruction.
2. Execute MOVSB instruction.
3. Decrement CX by 1.
4. **GO to step 1.**

Thus the same string movement can be achieved by the execution of the following 2 instruction, instead of repeatedly executing MOVSB in a loop.

```
MOV     CX,08
REP     MOVSB
```

6.24.4. Store String Instruction

Template for this instruction is

As can be noticed from the template, it can be a byte or a word operation. The mnemonic for byte operation is STOSB, which stands for 'Store string byte'. It has the opcode AAH. For word operation the mnemonic is STOSW, which stands for 'Store string word'. It has the opcode ABH.

This instruction copies the contents of the byte in AL or the word in AX to a byte or word at location ES:DI. It does not affect any flag.

Example 1. STOSB instruction

Let us say, we have in AL the ASCII code for the character 'N' and we want to copy this to the string location at an offset of 1231H in the Extra segment.

8086 provides STOSB (stands for Store String Byte) instruction to perform this byte movement. STOSB instruction expects that the destination string is pointed by ES:DI. In other words, for the destination string, the segment base address must be provided in ES, and the offset in DI.

So, to facilitate this movement we load DI with 1231H. Then the STOSB instruction is executed. This result in the movement of character 'N' in AL to offset location 1231H in the Extra segment. Then automatically, DI is incremented to 1232H.

However, this automatic increment of DI takes place only if the D (direction) flag is reset to 0 by executing the instruction CLD. If the D flag is set to 1 by executing the instruction STD, DI register is automatically decremented after the execution of STOSB instruction.

The REP prefix can be used with STOSB instruction, to copy the character in AL, into consecutive byte locations of a string. For example, the following instructions will store blanks in the first 20 byte locations of the string labeled as DST.

```
MOV     CX, 20
MOV     AL,'
MOV     DI, OFFSET DST
CLD
REP     STOSB
```

It is possible to use 'STOS' as mnemonic. But then the assembler has to be told in some way that byte operation is desired. This is done by indicating the name of the destination string in the instruction itself, like in 'STOS DST'. The assembler will code the instruction for a byte operation if DST is declared using DB (Define Byte) assembler directive. The assembler will code the instruction for a word operation if DST is declared using DW (Define Word) assembler directive.

Example 2. STOSW instruction

Let us say, we want to copy the word in AX to a string location at an offset of 1231H in the Extra segment.

8086 provides STOSW instruction expects that the destination string is pointed by ES:DI. In other words, for the destination string, the segment base address must be provided in ES, and the offset in DI.

So, to facilitate this movement we load DI with 1231H. Then the STOSW instruction is executed. This results in the movement of the word in AX to offset word location 1231H in the extra segment. Thus automatically, DI is incremented to 1233H.

However, this automatic increment of DI by 2 takes place only if the D (direction) flag is reset to 0 by executing the instruction CLD. If the D flag is set to 1 by executing the instruction STD, DI register is automatically decremented by 2 after the execution of STOSW instruction.

The REP prefix can be used with STOSW instruction, to copy the characters in AX into consecutive word locations of a string. For example, the following instructions will store blanks in the first 10 words locations, or first 20 byte locations, of the string labeled as DST.

```
MOV     CX,10
MOV     AX,'
MOV     DI, OFFSET DST
CLD
REP     STOSB
```

It is possible to use 'STOS' instead of 'STOSW' as the mnemonic. But then the assembler has to be told in some way that word operation is desired. This is done by indicating the name of the destination string in the instruction itself, like in 'STOS DST'. The assembler will code the instruction for a byte operation if DST is declared using DB (define byte) assembler directive. The assembler will code the instruction for a word operation if DST is declared using DW (define word) assembler directive.

6.24.5. Load String Instruction

Template for this instruction is

As can be noticed from the template, it can be a byte or a word operation. The mnemonic for byte operation is LODSB, which stands for 'Load string byte'. It has the opcode ACH. For word operation mnemonic is LODSW, which stands for "Load string word'. It has the opcode ADH.

This instruction copies the contents of the byte or word at location DS:SI into AL or AX. It does not affect any flag.

Example 1. LODSB instruction

Let us say we have at the byte location DS:1231H the ASCII code for the character 'N' and we want to copy this to AL.

8086 provides LODSB (stands for Load String Byte) instruction to perform this byte movement. LODSB instruction expects that the source string is pointed by DS:SI. In other words for the source string, the segment base address must be provided in DS, and the offset in SI.

So, to facilitate this movement we load SI with 1231H. Then the LODSB instruction is executed. This result in the movement of character 'N' to AL from Offset location 1231H in the Data segment. Then automatically, SI is incremented to 1232H.

However, this automatic increment of SI takes place only if the D(direction) flag is reset to 0 by executing the instruction CLD. If the D flag is set to 1 by executing the instruction STD, SI register is automatically decremented after the execution of LODSB instruction.

It is possible to use 'LODS' instead of 'LODSB' as the mnemonic. But then the assembler has to be told in some way that byte operation is desired. This is done by indicating the name of the destination string in the instruction itself, like in 'LODS SRC'. The assembler will code the instruction for a byte operation if SRC is declared using DB (Define Byte) assembler directive. The assembler will code the instruction for a word operation if SRC is declared using DW (Define Word) assembler directive.

Example 2. LODSW instruction

Let us say, we have at the word location DS:1231H the ASCII code for the characters 'Ni' and we want to copy this to AX. 8086 provides LODSW (stands for Load String Word) instruction to perform this word movement. LODSW instruction expects that the source string is pointed by DS:SI. In other words, for the source string the segment base address must be provided in DS, and the offset in SI.

So, to facilitate this movement we load SI with 1231H. Then the LODSW instruction is executed. This results in the movement of characters 'Ni' to AX from offset location 1231H in the Data segment. Then automatically, SI is incremented to 1233H.

However, This automatic increment of SI by 2 takes place only if the D (direction) flag is reset to 0 by executing the instruction CLD. If the D flag is set to 1 by executing the instruction STD, SI register is automatically decremented by 2 after the execution of LODSW instruction.

It is possible to use "LODS' instead of 'LODSW" as the mnemonic. But then the assembler has to be told in some way that word operation is desired. This is done by indicating the name of the destination string in the instruction itself, like in "LODS SRC'. The assembler will code the instruction for a byte operation if SRC is declared using DB (Define Byte) assembler directive. The assembler will code the instruction for a word operation if SRC is declared using DW (Define Word) assembler directive.

6.24.6. Compare Strings Instruction

Template for this instruction is

As can be noticed from the template, it can be a byte or a word operation. The mnemonic for byte operation is CMPSB, which stands for 'Compare string byte'. It has the opcode A6H. For word operation the mnemonic is CMPSW, which stands for 'Compare string word'. It has the opcode A7H.

This instruction compares the contents of the byte or word at location DS:SI with the byte or word at location ES:DI. It performs subtraction of the byte or word at ES:DI from the byte or word at DS:SI. The operand values are not changed. The status flags are affected depending on the result of subtraction.

If D flag = 0, the SI and DI registers are incremented by 1 or 2, depending on whether t is a byte or word operation. If D flag = 1, the SI and DI registers are decremented by 1 or 2 depending on whether it is a byte or word operation.

Example 1. CMPSB Instruction

Let us say, we have the string 'Nivedita' starting at an offset of 1231H in the Data segment. We have another string "Niranjan" starting at an offset of 3451H in the Extra segment. We want to compare these string, of size 8 characters.

8086 provides CMPSB (stands for Compare String Byte) instruction to perform this comparison a byte at a time. This instruction should be executed 8 times in a loop to perform the desired string comparison. CMPSB instruction expects that the source string is pointed by DS:SI. In other words, for the source string the segment base address must be provided in DS, and the offset in SI. Similarly ES:DI must point to the destination string.

So, to facilitate the string comparison we load SI with 1231H and DI with 3451H. Then the CMPSB instruction is executed. This result in the comparison of character 'N' at DS:1231H with the character 'N' at ES:3451H. Then automatically SI and DI are incremented to 1232H and 3452H respectively, so that next time through the loop, the CMPSB instruction compares the next character in the two strings and so on.

However, this automatic increment of SI and DI takes place only if the D (direction) flag is reset to 0 by executing the instruction CLD. If the D flag is set to 1 by executing the instruction STD, the SI and DI registers are automatically decremented after executlon of CMPSN instruction.

Thus the same string movement can be achieved by executing CMPSB in a loop 8 times, by initially loading SI With 1238H and DI with 3458H, and setting D flag to 1.

It is possible to use, 'CMPS" instead of "CMPSB" as the mnemonic. But then the assembler has to be told in some way that byte operation is desired. This is done by indicating the names of the source and destination strings in the instruction itself, like in 'CMPS SRC,DST". The assembler will code the instruction for a byte operation if SRC and DST are declared using DB (define Byte) assembler directive. The assembler will code the instruction for a word operation if SRC and DST are declared using DW (define Word) assembler directive.

6.24.7. REPE Prefix for CMPSB

Suppose we want to check whether two strings of size 8 characters are identical . Then it is not really necessary to execute the CMPSB instruction in a loop 8 times, because 8086 provides the REPE (stands for Repeat while equal) prefix facility. The REPE prefix is not an instruction by itself. It has the code F3H, which is the same as the code for REP prefix.

REPZ (stands for Repeat if result is Zero) is another equivalent mnemonic for the same operation. The user is free to use REPE or REPZ.

The REPE CMPSB instruction will cause CMPSB to be repeated as long as the compared bytes are equal (Z flag =1) and CX contents is not yet 0. Thus the repetition is terminated whenever the compared bytes are not equal or when CX contents becomes 0.

Thus the same comparison can be achieved by the execution of the following 2 instruction, instead or repeatedly executing CMPSB in a loop.

```
MOV      CX,08
REPE     CMPSB
```

6.24.8. REPNE prefix for CMPSB

The REPNE prefix is not an instruction by itself. It has the code F2H. The REPNE CMPSN instruction will cause CMPSB to be repeated as long as the compared bytes are not equal (Z flag =0) and CX contents is not yet 0. Thus the repetition is terminated whenever the compared bytes are equal or when CX contents becomes 0.

REPNZ (stands for Repeat if result is Not Zero) is another equivalent mnemonic for the same operation. The user is free to use REPNE or REPNZ.

Example 2. CMPSW instruction

Let us say, we have the word string "Ni", 've', 'di', 'ta' straining at an offset of 1231H in the Data Segment. We have another word string 'Ni', 'ra', 'nj', 'an' starting at an offset of 3451H in the Extra segment. We want to compare these string, of size 8 character, or 4 words.

8086 provides CMPSW (stands for Compare String Word) instruction to perform this comparison a word at a time. This Instruction should be executed 4 times in a loop to perform the desired string comparison. CMPSW instruction expects that the source string is pointed by DS:SI. In other words , for the source string the segment base address must be provided in DS, and the offset in SI. Similarly ES:DI must point to the destination string.

So, to facilitate the string comparison we load SI with 1231H and DI with 3451H. Then the CMPSW instruction is executed. This result in the comparison of characters 'Ni' at word location DS:1231H with the characters 'Ni' at word location ES:3451H. Then automatically, SI and DI are incremented at word location ES:3451H. Then automatically, SI and DI are incremented to 1233H and 3453 respectively, so that next time through the loop, the CMPSW instruction compares the next word in the two strings, and so on.

However, this automatic increment by 2 of SI and DI takes place only if the D(direction) flag is reset to 0 by executing the instruction CLD. If the D flag is set to 1 by executing the instruction STD, the SI and DI registers are automatically decremented by 2 after the execution of CMPSW instruction.

Thus the same string movement can be achieved by executing CMPSW in a loop 4 times, by initially loading SI with 1237H and DI with 3457H and DI with 3457H, and setting D flag to 1.

It is possible to use 'CMPS' instead of 'CMPSW' as the mnemonic. But then the assembler has to be told in some way that word operation is desired. This is done by indicating the names of the source and destination strings in the instruction itself, like in 'CMPS SRC,DST'. The assembler will code the instruction for a byte operation if SRC and DST are declared using DB (define byte) assembler directive. The assembler will code the instruction for a word operation if SRC and DST are declared using DW (Define Word) assembler directive.

6.24.9. REPE Prefix for CMPSW

Suppose we want to check whether two strings of size 4 characters are identical . Then it is not really necessary to execute the CMPSB instruction in a loop 4 times, because 8086 provides the REPE (stands for Repeat while equal) prefix facility. The REPE prefix is not an instruction by itself. It has the code F3H, which is the same as the code for REP prefix.

REPZ (stands for Repeat if result is Zero) is another equivalent mnemonic for the same operation. The user is free to use REPE or REPZ.

The REPE CMPSW instruction will cause CMPSW to be repeated as long as the compared bytes are equal (Z flag =1) and CX contents is not yet 0. Thus the repetition is terminated whenever the compared bytes are not equal or when CX contents becomes 0.

Thus the same comparison can be achieved by the execution of the following 2 instruction, instead or repeatedly executing CMPSB in a loop.

```
MOV     CX,04
REPE    CMPSW
```

6.24.10. REPNE Prefix for CMPSB

The REPNE prefix is not an instruction by itself. It has the code F2H. The REPNE CMPSW instruction will cause CMPSW to be repeated as long as the compared bytes are not equal (Z flag =0) and CX contents is not yet 0. Thus the repetition is terminated whenever the compared bytes are equal or when CX contents becomes 0. REPNZ (stands for Repeat if result is Not Zero) is another equivalent mnemonic for the same operation. The user is free to use REPNE or REPNZ.

6.24.11. Scan String Instruction

Template for this instruction is

As can be noticed from the template, it can be a byte or a word operation. The mnemonic for byte operation is SCASB, which stands for 'Scan string byte'. It has the opcode AEH. For word operation the mnemonic is SCASW, which stands for 'Scan string word'. It has the opcode AFH.

This instruction compares the contents of the byte or word in AL or AX with the byte or word at location ES:DI. It performs subtraction of the byte or word at ES:DI from the byte or word in AL or AX. The operand values are not changed. The status flags are affected depending on the result of subtraction.

If D flag = 0, the SI and DI registers are incremented by 1 or 2, depending on whether t is a byte or word operation. If D flag = 1, the SI and DI registers are decremented by 1 or 2 depending on whether it is a byte or word operation.

Example 1. SCASB Instruction

Let us say, we have the string 'Advitiya' starting at an offset of 3451H in the Extra segment. We want to scan this string of size 8 characters for the occurrence of the character 't'.

8086 provides SCASB (stands for Scan String Byte) instruction to perform this comparison a byte at a time. This instruction should be executed a maximum of 8 times in a loop to perform the desired string scan operation. SCASB instruction expects that the destination string is pointed by ES:DI.

So, to facilitate the string operation we load DI with 3451H. Then the SCASB instruction is executed. This result in the comparison of character 't' in AL with the character 'A' at ES:3451H. Then automatically DI is incremented to 3452H, so that next time through the loop, the SCASB instruction compares the next character in the string and so on.

However, this automatic increment of DI takes place only if the D (direction) flag is reset to 0 by executing the instruction CLD. If the D flag is set to 1 by executing the instruction STD, DI register is automatically decremented after execution of SCASB instruction.

Thus the same scan operation of the string can be achieved by executing SCASB in a loop 8 times, by initially loading DI With 3458H and setting D flag to 1.

It is possible to use, 'SCAS" instead of "SCASB" as the mnemonic. But then the assembler has to be told in some way that byte operation is desired. This is done by indicating the names of the source and destination strings in the instruction itself, like in 'SCAS DST". The assembler will code the instruction for a byte operation if DST is declared using DB (define Byte) assembler directive. The assembler will code the instruction for a word operation if DST is declared using DW (define Word) assembler directive.

6.24.12. REPE Prefix for SCASB

Suppose we want to check a particular byte is present in a string of size 8 characters. Then it is not really necessary to execute the SCASB instruction in a loop 8 times, because 8086 provides the REPNE (stands for Repeat while not equal) prefix facility. The REPNE prefix is not an instruction by itself. It has the code F2H.

REPNZ (stands for Repeat if result is Not Zero) is another equivalent mnemonic for the same operation. The user is free to use REPNE or REPNZ.

The REPNE SCASB instruction will cause SCASB to be repeated as long as the compared bytes are not equal (Z flag =0) and CX contents is not yet 0. Thus the repetition is terminated whenever the compared bytes are equal or when CX contents becomes 0.

Thus the same scan string operation can be achieved by the execution of the following 2 instructions, instead or repeatedly executing SCASB in a loop.

```
MOV      CX,08
REPNE    SCASB
```

6.24.13. REPE Prefix for SCASB

The REPE prefix is not an instruction by itself. It has the code F3H, same as that of REP prefix. The REPE SCASB instruction will cause SCASB to be repeated as long as the compared bytes are equal (Z flag =1) and CX contents is not yet 0. Thus the repetition is terminated whenever the compared bytes are equal or when CX contents becomes 0. REPZ (stands for Repeat if result is Zero) is another equivalent mnemonic for the same operation. The user is free to use REPE or REPZ.

Example 1. SCASW instruction

Let us say, we have the word string "Ad", 'vi', 'ti', 'ya' straining at an offset of 3451H in the Extra Segment. We wan to scan this string of size 8 character, or 4 words, for occurrence of the characters 'ti'.

8086 provides SCASW (stands for Scan String Word) instruction to perform this comparison a word at a time. This Instruction should be a maximum of 4 times in a loop to perform the desired string scan operation. SCASW instruction expects that the destination string is pointed by ES:DI.

So, to facilitate the scan operation of the string we load DI with 3451H. Then the SCASW instruction is executed. This result in the comparison of characters 'ti' in AX with the characters 'Ad' at ES:3451H. Then automatically, DI is incremented to 3453H, so that next time through the loop, the SCASW instruction compares the next word in the two strings, and so on.

However, this automatic increment of DI takes place only if the D (direction) flag is reset to 0 by executing the instruction CLD. If the D flag is set to 1 by executing the instruction STD, DI registers are automatically decremented by 2 after the execution of SCASW instruction.

Thus the same operating of the string can be achieved by executing SCASW in a loop 4 times, by initially loading DI with 3457H and setting D flag to 1.

It is possible to use 'SCAS' instead of 'SCASW' as the mnemonic. But then the assembler has to be told in some way that word operation is desired. This is done by indicating the names of the destination strings in the instruction itself, like in 'SCAS,DST'. The assembler will code the instruction for a byte operation if DST is declared using DB(define byte) assembler directive. The assembler will code the instruction for a word operation if DST is declared using DW (Define Word) assembler directive.

6.24.14. REPNE Prefix for SCASW

Suppose we want to check a particular byte is present in a string of size 8 characters or 4 words. Then it is not really necessary to execute the SCASW instruction in a loop 4 times, because 8086 provides the REPNE (stands for Repeat while not equal) prefix facility. The REPNE prefix is not an instruction by itself. It has the code F2H.

REPNZ (stands for Repeat if result is Not Zero) is another equivalent mnemonic for the same operation. The user is free to use REPNE or REPNZ.

The REPNE SCASW instruction will cause SCASW to be repeated as long as the compared bytes are not equal (Z flag =0) and CX contents is not yet 0. Thus the repetition is terminated whenever the compared bytes are equal or when CX contents becomes 0.

Thus the same scan string operation can be achieved by the execution of the following 2 instructions, instead or repeatedly executing SCASB in a loop.

```
MOV      CX,04
REPNE    SCASW
```

6.24.15. REPE prefix for SCASW

The REPE prefix is not an instruction by itself. It has the code F3H, same as that of REP prefix. The REPE SCASW instruction will cause SCASW to be repeated as long as the compared bytes are equal (Z flag =1) and CX contents is not yet 0. Thus the repetition is terminated whenever the compared words are equal or when CX contents becomes 0. REPZ (stands for Repeat if result is Zero) is another equivalent mnemonic for the same operation. The user is free to use REPE or REPZ.

6.25. BRANCH INSTRUCTIONS

Normally 8086 keeps on executing one instruction after another in a sequence. However, when a branch group of instruction is encountered by the 8086, a jump is effected, which can be a forward or a backward jump, if the condition for branching is satisfied.

8086 has very powerful instruction in the Branch group. It can perform a branch unconditionally, or based on a single flag value, or based on multiple flag values. There are instructions to branch to a subroutine and return from a subroutine etc.

6.25.1. Conditional Jumps Based on a Single Flag

Describe below is the instructions which perform a branch based on the value of a single status flag. The status flags used in these instructions are Zero, Sign, Overflow, Carry and Parity.

6.25.2. JE-Jump if Equal (Same as JZ-Jump if Zero)

The mnemonic for the instruction is JE. Another equivalent mnemonic is JZ. We generally use JE after a CMP instruction, and JZ after an arithmetic operation like addition. It is a 2 byte instruction, where the first byte provides the opcode and the second byte provides the displacement in 2's complement notation.

The 8 bit displacement indicates the number of bytes by which the jump tales place. The range for a 8 bit 2's complement number is form -128 (80H) to +127(7FH). If it is a positive number, the jump is to a forward location. If it is a negative number, the jump is to a backward location. Thus the memory location specified by the label in the JE instruction must be in the range of –128 to +127 bytes from the beginning of the next instruction after JE.

However, in Assembly language programming, the programmer just provides the label for the location where jump is desired. The assembler then generates the displacement in 2's complement notation. If the label is at a distance outside this range, an error message will be flashed by the assembler during assembly process. In such a case, a conditional branch instruction has to be combined with an unconditional branch instruction to circumvent the problem. This is explained later when the unconditional jump instruction , JMP, is discussed.

Opcode for JE instruction is 74H.

If the Z flag =1, this instruction execution results in a jump to the label specified in the instruction. If the Z flag-0, the execution continues with the next instruction after JE.

Example 1. Given below is an example for forward branch.

```
JE              SAME
ADD             CX,DX; this is executed next after JE SAME, IF Z=0
.
.
.
SAME:     SUB BX,AX; this is executed next after JE SAME, if Z=1
```

Example 2. Given below is an example for backward branch.

```
BACK:       SUB BX,AX; this is executed next after JE BACK, IF Z=1
.
.
.
JE        BACK
ADD       CX,DX; this is executed next after JE BACK, if Z=0
```

6.25.3. JNE-Jump if not Equal (Same as JNZ-Jump if not Zero)

The mnemonic for the instruction is JNE. Another equivalent mnemonic is JNZ. We generally use JNE after a CMP instruction, and JNZ after an arithmetic operation like addition. It is a 2 byte instruction, where the first byte provides the opcode and the second byte provides the displacement in 2's complement notation. The displacement has the same interpretation as was given in JE instruction earlier.

Opcode for this instruction is 75H.

If the Z flag =0, this instruction execution results in a jump to the label specified in the instruction. If the Z flag=1, the execution continues with the next instruction after JNE.

Example 1. Given below is an example for forward branch.

```
JNE      SAME
ADD      CX,DX; this is executed next after JNE DIFF, IF Z=1
.
.
.
DIFF:    SUB  BX,AX; this is executed next after JNE DIFF, if Z=0
```

Example 2. Given below is an example for backward branch.

```
BACK:      SUB BX,AX; this is executed next after JNE BACK, IF Z=0
.
.
.
JNE      BACK
ADD      CX,DX; this is executed next after JNE BACK, if Z=1
```

6.25.4. JS-Jump on Sign

The mnemonic for the instruction is JS. It is a 2 byte instruction, where the first byte provides the opcode and the second byte provides the displacement in 2's complement notation. The displacement has the same interpretation as was given in JE instruction earlier.

Opcode for this instruction is 78H.

If the S flag =1, this instruction execution results in a jump to the label specified in the instruction. If the S flag=0, the execution continues with the next instruction after JS.

Example 1. Given below is an example for forward branch.

```
JS       NEG
ADD      CX,DX; this is executed next after JS NEG, IF S=0
.
.
.
NEG:     SUB  BX,AX; this is executed next after JS NEG, if S=1
```

6.25.5. JNS-Jump on Not Sign

The mnemonic for the instruction is JNS. It is a 2 byte instruction, where the first byte provides the opcode and the second byte provides the displacement in 2's complement notation. The displacement has the same interpretation as was given in JE instruction earlier.

Opcode for this instruction is 79H.

If the S flag =0, this instruction execution results in a jump to the label specified in the instruction. If the S flag=1, the execution continues with the next instruction after JNS.

Example 1. Given below is an example for backward branch.

```
BACK:    SUB  BX,AX; this is executed next after JNS BACK, if S=0
.
.
.
JNE      BACK
ADD:     CX,DX; this is executed next after JNS BACK, if S=1
```

6.25.6. JO-Jump On Overflow

The mnemonic for the instruction is JO. It is a 2 byte instruction, where the first byte provides the opcode and the second byte provides the displacement in 2's complement notation. The displacement has the same interpretation as was given in JE instruction earlier.

Opcode for this instruction is 70H.

If the O flag =1, this instruction execution results in a jump to the label specified in the instruction. If the O flag=0, the execution continues with the next instruction after JO. The reader can visualize examples similar to those discussed for, say, JE instruction.

6.25.7. JNO-JUMP on Not Overflow

The mnemonic for the instruction is JNO. It is a 2 byte instruction, where the first byte provides the opcode and the second byte provides the displacement in 2's complement notation. The displacement has the same interpretation as was given in JE instruction earlier.

Opcode for this instruction is 71H.

If the O flag =0, this instruction execution results in a jump to the label specified in the instruction. If the O flag=1, the execution continues with the next instruction after JO. The reader can visualize examples similar to those discussed for, say, JE instruction.

6.25.8. JC-Jump if Carry (Same as JB-Jump if Below or JNAE-Jump if not Above or Equal)

The mnemonic for the instruction is JC. Other equivalent mnemonics are JB and JNAE. We generally use JB or JNAE after a CMP instruction, and JC after an arithmetic operation like addition. The terms Below and Above are used when we are comparing unsigned numbers. For example, the unsigned number 1100 is above the unsigned number 0101, or equivalently, 0101 is below the unsigned number 1100.

It is a 2 byte instruction, where the first byte provides the opcode and the second byte provides the displacement in 2's complement notation. The displacement has the same interpretation as was given in JE instruction earlier.

Opcode for this instruction is 72H.

If the C flag =1, this instruction execution results in a jump to the label specified in the instruction. If the C flag=0, the execution continues with the next instruction after JC. The reader can visualize examples similar to those discussed for, say, JE instruction.

6.25.9. JNC-Jump if Not carry (same as JAE-Jump if Above or Equal or JNB-Jump if not Below)

The mnemonic for the instruction is JNC. Other equivalent mnemonics are JAE and JNB. We generally use JAE or JNB after a CMP instruction, and JNC after an arithmetic operation like addition. The terms Below and Above are used when we are comparing unsigned numbers. For example, the unsigned number 1100 is above the unsigned number 0101, or equivalently, 0101 is below the unsigned number 1100.

It is a 2 byte instruction, where the first byte provides the opcode and the second byte provides the displacement in 2's complement notation. The displacement has the same interpretation as was given in JE instruction earlier.

Opcode for this instruction is 73H.

If the C flag =0, this instruction execution results in a jump to the label specified in the instruction. If the C flag=1, the execution continues with the next instruction after JNC. The reader can visualize examples similar to those discussed for, say, JE instruction.

6.25.10. JP-Jump if Parity (Same as JPE-Jump if Parity Even)

The mnemonic for the instruction is JP. Other equivalent mnemonics is JPE. It is a 2 byte instruction, where the first byte provides the opcode and the second byte provides the displacement in 2's complement notation. The displacement has the same interpretation as was given in JE instruction earlier.

Opcode for this instruction is 7AH.

If the P flag =1, this instruction execution results in a jump to the label specified in the instruction. If the P flag=0, the execution continues with the next instruction after JP. The reader can visualize examples similar to those discussed for, say, JE instruction.

6.25.11. JNP-Jump if Not Parity (JPO-Jump if Parity Odd)

The mnemonic for the instruction is JNP. Another equivalent mnemonics is JPO. It is a 2 byte instruction, where the first byte provides the opcode and the second byte provides the displacement in 2's complement notation. The displacement has the same interpretation as was given in JE instruction earlier.

Opcode for this instruction is 7BH.

If the P flag =0, this instruction execution results in a jump to the label specified in the instruction. If the P flag=1, the execution continues with the next instruction after JNP. The reader can visualize examples similar to those discussed for, say, JE instruction.

Conditional Jumps Based On More Than One Flag

Describe below is the instruction which branch based on the value of more than one flag, like JBE instruction which branches only when C flag =1 or Z flag =1. The conditional jump instructions JBE and JNBE work on unsigned numbers. The Instructions JLE, JNLE, JL, and JNL work on signed numbers.

6.25.12. JBE-Jump if below or Equal (JNA-Jump if not Above)

The mnemonic for the instruction is JBE. Another equivalent mnemonic is JNA. This instruction is generally used after a CMP instruction, to check if the first operand is below or equal to the second operand. The terms Below and above are used when we are comparing unsigned numbers. For example the unsigned number 1100 is above the unsigned number 0101, or equivalently 0101 is below the unsigned number 1100.

It is a 2 byte instruction, where the first byte provides the opcode and the second byte provides the displacement in 2's complement notation. The displacement has the same interpretation as was given in JE instruction earlier.

Opcode for this instruction is 76H.

This instruction results in a branch based on the values of C flag and Z flag. If C flag=1, it means the first operand is below the second operands in value, and hence a branch should take place. If Z flag =1, it means the numbers are equal, and hence a branch should take place.

Thus if C flag =1 or Z flag =1 this instruction execution results in a jump to the label specified in the instruction. If C flag =0 and Z flag=0 the execution continues with the next instruction after JBE.

Example 1. Given below is an example for backward branch.

```
Back: Sub BX,AX; this is executed after JBE BACK, If C=1 or Z=1
      .
      .
      .
      CMP  BL,89H
      JBE  BACK
      ADD  CX,DX; this is executed after JBE BACK, if c=0 and Z=0
```

6.25.13. JNBE-Jump if not below or equal (JA-Jump if Above)

The mnemonic for the instruction is JNBE. Another equivalent mnemonic is JA. This instruction is generally used after a CMP instruction, to check if the first operand is above the second operand in its value. The terms Below and Above are used when we are comparing unsigned numbers. For example the unsigned number 1100 is above the unsigned number 0101, or equivalently 0101 is below the unsigned number 1100.

It is a 2 byte instruction, where the first byte provides the opcode and the second byte provides the displacement in 2's complement notation. The displacement has the same interpretation as was given in JE instruction earlier.

Opcode for this instruction is 77H.

This instruction results in a branch based on the values of C flag and Z flag. If C flag=0, and Z flag=0, it means the first operand is above, the second operands in value, and hence a branch should take place. If Z flag =1, it means the numbers are equal, and hence a branch should take place. If C flag=1, the first number is below the second number, and so branch should not take place.

Thus if C flag =1 or Z flag =0, this instruction execution results in a jump to the label specified in the instruction. If C flag =1 and Z flag=1 the execution continues with the next instruction after JA. The reader can visualize examples similar to those discussed for, say JBE instruction.

6.25.14. JLE-Jump if Less or Equal (JNG-Jump if Not Greater)

The mnemonic for the instruction is JLE. Another equivalent mnemonic is JNG. This instruction is generally used after a CMP instruction, to check if the first operands is less than or equal to the second operand. The terms Less and Greater are used when we are comparing signed numbers. For example the signed number 1100 is less than the signed number 0101, or equivalently, 0101 is greater than the signed number 1100. This is because, 1100 represents -4 and 0101 represents +5, if the numbers are interpreted as signed numbers.

It is a 2 byte instruction, where the first byte provides the opcode and the second byte provides the displacement in 2's complement notation. The displacement has the same interpretation as was given in JE instruction earlier.

Opcode for this instruction is 7EH.

This instruction results in a branch based on the values of S flag, O flag and Z flag.

If S = 1 and O = 0, it means that operand_1 minus Operand_2 gave a negative result, and the result is correct as O flag is 0. Then as Operand_1 is less than Operand_2 the JLE instruction should result in a branch.

If S = 0 and O = 1, it means that operand_1 minus Operand_2 gave a positive result, and the result is wrong as O flag is 1. Thus in reality, Operand_1 is less than Operand_2 and so the JLE instruction should result in a branch.

If Z flag=1, it means the numbers are equal, and hence a branch should take place.

Thus if S flag is having a different value than O=flag, or if Z flag=1, this instruction execution results in a jump to the label specified in the instruction. If S flag has same value as O flag and Z flag=0, the execution continues with the next instruction after JLE.

Example. Given below is an example for backward branch. In this example S< > O stands for S flag unequal to O flag. If BL content is 93H, the branch takes place as S flag will be 0, O flag will be 1, and Z flag will be 0 after the execution of CMP BL, 89H. However if BL content is 7BH, the branch does

```
      BACK: SUB BX,AX; This is executed after JLE BACK, I f S< > O or Z=1
          .
          .
          .
          .
          CMP  BL,79H
      JBE BACK
      ADD CX,DX; This is executed after JLE BACK, If S=0 and Z =0 not
take place, because S flag will be 0, O flag will be 1, and Z flag=0
after the execution of the instruction CMP BL,89H.
```

6.25.15. JNLE-Jump if not less or Equal (JG-Jump if Greater)

The mnemonic for the instruction is JNLE. Another equivalent mnemonic is JG. This instruction is generally used after a CMP instruction, to check if the first operands is greater than the second operand. The terms Less and Greater are used when we are comparing signed numbers.

It is a 2 byte instruction, where the first byte provides the opcode and the second byte provides the displacement in 2's complement notation. The displacement has the same interpretation as was given in JE instruction earlier.

Opcode for this instruction is 7FH.

This instruction results in a branch based on the values of S flag, O flag and Z flag.

If S =1 , O=0 and Z=0, it means that operand_1 minus Operand_2 gave a positive non zero result, and the result is correct as O flag is 0. Then as Operand_1 is greater than Operand_2 the JG instruction should result in a branch.

If S =0, O=1 and Z=0, it means that operand_1 minus Operand_2 gave a negative non zero result, and the result is wrong as O flag is 1. Thus in reality, Operand_1 is greater than Operand_2 and so the JG instruction should result in a branch.

Thus if S flag is having a same value as O flag, or if Z flag=0, this instruction execution results in a jump to the label specified in the instruction. If S flag and O flag are having different values, or Z flag=1, the execution continues with the next instruction after JG. The reader can visualize examples similar to those discussed for, say JLE instruction.

6.25.16. JL-Jump if Less (JNGE-Jump if not Greater or Equal)

The mnemonic for the instruction is JL. Another equivalent mnemonic is JNGE. This instruction is generally used after a CMP instruction, to check if the first operands is less than the second operand. The terms Less and Greater are used when we are comparing signed numbers.

It is a 2 byte instruction, where the first byte provides the opcode and the second byte provides the displacement in 2's complement notation. The displacement has the same interpretation as was given in JE instruction earlier.

Opcode for this instruction is 7CH.

This instruction results in a branch based on the values of S flag, O flag.

If S =1 , O=0 and Z=0, it means that operand_1 minus Operand_2 gave a negative result, and the result is correct as O flag is 0. Then as Operand_1 is less than Operand_2 the JL instruction should result in a branch.

If S =0, O=1 it means that operand_1 minus Operand_2 gave a positive result, and the result is wrong as O flag is 1. Thus in reality, Operand_1 is less than Operand_2 and so the JL instruction should result in a branch.

Thus if S flag is having a different value than O flag, this instruction execution results in a jump to the label specified in the instruction. If S flag has same value as O flag, the execution continues with the next instruction after JL. The reader can visualize examples similar to those discussed for, say JLE instruction.

6.25.17. JNL-Jump if not less (JGE-Jump if Greater or Equal)

The mnemonic for the instruction is JNL. Another equivalent mnemonic is JGE. This instruction is generally used after a CMP instruction, to check if the first operands is greater than or equal to the second operand. The terms Less and Greater are used when we are comparing signed numbers.

It is a 2 byte instruction, where the first byte provides the opcode and the second byte provides the displacement in 2's complement notation. The displacement has the same interpretation as was given in JE instruction earlier.

Opcode for this instruction is 7DH.

This instruction results in a branch based on the values of S flag, O flag.

If S =0, O=0 it means that operand_1 minus Operand_2 gave a positive result, and the result is correct as O flag is 0. Then as Operand_1 is greater than or equal to Operand_2 and so the JGE instruction should result in a branch.

If S =1 and O=1, it means that Operand_1 minus Operand_2 gave a negative result, and the result is wrong as O flag is 1. Thus in reality, Operand_1 is greater than or Equal to Operand_2, and so the JGE instruction should result in a branch.

Thus if S flag is having a different value than O flag, this instruction execution results in a jump to the label specified in the instruction. If S flag and O flag are having different values, the execution continues with the next instruction after JGE. The reader can visualize examples similar to those discussed for, say JLE instruction.

6.25.18. Unconditional Jump Instruction

The mnemonic for the instruction is JMP. The execution of this instruction always in a branch. The branch could be within the same code segment, in which case it is called a NEAR jump or an INTRASEGMENT Jump.

NEAR jump instruction can provide the destination address using a label in a relative mode, with 1 or 2 byte of signed displacement, or the destination address is provide indirectly in a register or a memory location. A NEAR jump instruction using 1 byte displacement is generally termed as a SHORT jump. Thus the 3 types of NEAR jumps are

(a) NEAR jump with 1 byte signed displacement (or SHORT JUMP)

(b) NEAR jump with 2 byte signed displacement

(c) NEAR jump of INDIRECT type (Destination address in a register or a memory location)

If the branch is to a location in a code segment which has a different name from the segment in which JMP instruction is present, it is called a FAR jump or an INTERSEGMENT jump.

FAR jump instruction can provide the destination address using a label in a direct mode, with 2 words to indicate CS and IP value of the destination address, or the destination address is provided indirectly in 2 consecutive memory word locations. Thus the 2 types of FAR jumps are

(a) FAR DIRECT jump

(b) FAR INDIRECT Jump

Thus there are 5 templates for the jump instruction, which are discussed below. All these have the same mnemonic JMP.

6.25.19. JMP-Intrasegment Relative Jump with 2 Byte Displacement

The mnemonic for the instruction is JMP. It is a 2 byte instruction, where the first byte provides the opcode and the second byte provides the displacement in 2's complement notation. The displacement has the same interpretation as was given in JE instruction earlier. This is the most commonly used type of jump instruction.

Opcode for this instruction is EBH.

Example 1. Given below is an example for backward branch. It is assumed that (offset address of NEXT – Offset address of Back) is <=128. While translating JMP BACK, the assembler has earlier calculate the required displacement, as the assembler has earlier come across the label BACK. Only if we are required to go back by <=128 bytes, the JMP BACK is translated by the assembler as EBH, followed by the 1 byte displacement.

If we are required to go back by more than 128 bytes, the assembler automatically translate JMP BACK as E9H, followed by 2 byte displacement.

```
BACK :   SUB  BX,AX; this is executed next after JMP BACK.
     .
     .
     .
     JMP BACK
NEXT:    MOV  AX,CX
```

Example 2. Given below is an example for forward branch. It is assumed that (Offset address of SAME-Offset address of NEXT) is <=127. While translating JMP SHORT SAME, The assembler is informed by the programmer using the SHORT assembler directive that the required displacement will be within the range -128 to +127 bytes. Thus, the JMP SHORT SAME is translated by the assembler as EBH, followed by the 1 byte displacement.

If in this forward branch example, the jump instruction was written without the SHORT directive as JMP SAME, the assembler will have no idea of the required forward displacement. IN such a case, to be on the safer side, the assembler reserves 2 bytes for the displacement after opcode for JMP. So, a total of 3 bytes are reserved for the instruction by the assembler.

Finally, when the assembler comes across the label with the name SAME, it calculates the displacement for the forward jump in the instruction JMP SAME. If this forward displacement is <=127 bytes, the assembler realizes that only 1 byte was sufficient for the displacement. Then the assembler codes JMP SAME as EBH, followed by 1 byte for the displacement. The third byte reserved by the assembler is replaced with the one byte NOP instruction. This will never be executed as it is after the unconditional jump instruction.

If the forward displacement is >127 bytes, the assembler realizes that indeed 2 bytes are needed for the displacement. Then the assembler codes JMP SAME as E9H, followed by 2 bytes for the displacement. E9 is the code for NEAR jump with 2 byte displacement.

```
            JMP         SHORT SAME
NEXT              MOV         AX,CX
            .
            .
            .
SAME:       SUB   BX,AX; this is executed next after JMP SAME
```

6.25.20. JMP-Intrasegment Relative Short Jump Instruction

The mnemonic for the instruction is JMP. It is a 3 byte instruction, where the first byte provides the opcode and the next two bytes provides the 16 bit displacement in 2's complement notation. This is also a very commonly used type of jump instruction.

Opcode for this instruction is E9H.

The 16 bit displacement indicates the number of bytes by which the jump this place. The range for a 16 bit 2's complement number is from -32768 (88000H) to -32767(7F00H). If it is a positive number, the jump is usually to a forward location. If it is a negative number, the jump is usually to a backward location. Thus the memory location specified by the label in the JMP instruction should be generally in the range of -32768 to +32767 bytes from the beginning of the next instruction after JMP.

However, in Assembly language programming, the programmer just provides the label for the location where jump is desired. If it is a jump to a backward location, during translating of the jump instruction the assembler can easily calculate the required displacement. If we are required to go back <=128 bytes, the jump instruction is translated by the assembler as EBH, followed by the 1 byte displacement. EBH is the code for SHORT jump with 1 byte displacement. If we are required to go back by >128 bytes, the jump instruction is translated by the assembler as E9H, followed by 2 byte displacement. E9H is the code for NEAR jump with 2 byte displacement.

If the jump is to a forward location, the assembler will have no idea of the required displacement when it encounters the jump instruction. In such a case, to be on the safer side, the assembler reserves 2 bytes for the displacement after the opcode for JMP. SO, a total of 3 bytes are reserved for the instruction by the assembler.

Finally, when the assembler comes across the label of the destination for jump instruction, it calculates the displacement for the forward jump. If this forward displacement is <=127 bytes, the assembler realizes that only 1 byte was sufficient for the displacement. Then the assembler codes the jump instruction as EBH, followed by 1 byte for the displacement. EBH is the code for SHORT jump. The third byte reserved by the assembler is replaced with the one byte NOP instruction. This will never be executed as it is after the unconditional jump instruction.

If the forward displacement is >127 bytes, the assembler realizes that indeed 2 bytes are needed for the displacement. Then the assembler codes the jump instruction as E9H, followed by 2 bytes for the displacement. E9 is the code for NEAR jump with 2 byte displacement.

Suppose we have a forward jump instruction near the beginning of a 64K byte segment, say at offset location 000DH. The next instruction will be at 0010H, a distance of 3 bytes. Let us say, we want to branch to an address near the end of the segment, say to offset location FFF0H. Then the required forward displacement is greater than 32767 bytes. In this case, the assembler automatically translate the instruction as if it were a SHORT backward jump instruction with the

1 byte displacement value as -20H =E0H. Notice that 0010H+FFE0H = FFF0H, where FFE0H is the 16 bit sign extended value of E0H.

Suppose we have a backward jump instruction near the end of a 64 K byte segment, say, at offset location FFEEH. The next instruction will be at FFF0H, a distance of 2 bytes as it will be coded as a SHORT jump for the reason described below. Let us say, we want to branch to an address near the beginning of the segment, say to offset location 0010H. Then the required backward displacement is greater than 32768 bytes. In this case the assembler automatically translates the instruction as if it were a SHORT forward jump instruction with the 1 byte displacement value as 20H. Notice that FFF0H + 0020H = 0010H, where the carry generated is ignored.

The NEAR jump with 2 byte displacement is very useful when we want to perform conditional branch outside the permitted range of -128 to +127 bytes. Suppose our interest is to execute the instruction JNC REPEAT, where unfortunately the label REPETAT is outside the permitted range of -128 to +127 bytes. If the condition for branching is not satisfied the next instruction to be executed is let us say MOV AX,BX. This problem is circumvented using the unconditional jump as follows.

```
JC  NEXT        ; if C flag =1 , execute MOV AX,BX
JMP REPEAT      ; If C flag=0 , jump to REPEAT
NEXT:           MOV  AX,BX
```

6.25.21. JMP-Intrasegment Indirect Jump or Near Indirect Jump

The mnemonic for the instruction is JMP. Template for this instruction is

As there are 5 bits in the template for identifying the operands, there are 32 instructions in this category. Note that W bit is absent in the template since the jump address is given as the contents of a word register or a word location in memory. This method of jumping is not commonly used. This is because by looking at the program, it is not at all clear where we are branching. This form of jump instruction is used to select one of several jump destinations based on a computed value.

Example 1. Code for JMP DX

Suppose DX content is 1234H. Then JMP DX execution results in an unconditional branch to offset location 1234H in the same code segment. Note that 1234H is not a relative displacement. JMP DX is a 2 byte instruction with the code

Example 2. Code for JMP word ptr 2000H[BX]

Suppose BX content is 1234H. Let us say, the contents of word location 3234H is 5678H. 3234H is the sum of 1234H, which is BX contents and 2000H, which is the 16 bit displacement provided in the instruction. Then JMP WORD PTR 2000H[BX] execution results in an unconditional branch to offset location 5678H in the same segment. It is a 4 byte instruction with the code

6.25.22. JMP-Intersegment Direct Jump Or FAR direct Jump

The mnemonic for the instruction is JMP. As it is desired to branch to a different segment, it is necessary to provide the Segment base and offset value of the destination in the instruction. Thus it is a 5 byte instruction, where the first byte provides the opcode and the next 2 bytes provide the offset value, and the last 2 bytes provide the segment base value of the destination. FAR jumps are not very common in programs. If a FAR jump is desired, the direct type is commonly used.

Opcode for this instruction is EAH.

However, in assembly language programming the programmer just provides the label of the location where jumps is desired. Further, the assembler will be told that it is a FAR jump as in 'JMP FAR PTR LOC'. Then the assembler generates the 5 byte code for a FAR direct jump.

6.25.23. JMP-Intersegment Indirect Jump or FAR Indirect Jump

The mnemonic for the instruction is JMP. Template for this instruction is

As there are 5 bits in the template for identifying the operands, there are 32 instructions in this category. Note that W bit is absent in the template since the jump address is given as the contents of a double word in memory. This method of jumping is not commonly used. This is because by looking at the program, it is not at all clear where we are branching.

Example 1. Code for JMP dword ptr 2000H[BX]

Suppose BX content is 1234H.. Let us say, the contents of word location 3234H is 5678H. This 5678H provides the offset value. Let us say, the contents of the next word at location 3236H is ABCDH. Then ABCDH provides the segment base value. Thus JMP DWORD PTR 2000[BX] execution results in an unconditional branch to offset location 5678H in another code segment with segment base values as ABCDH. It is a 4 byte instruction with the code

6.26. INTERATION INSTRUCTION

8086 provides a very convenient way to implement loops in a program. These loop instructions can result in a branch both conditionally and unconditionally. Without such loop instructions, it requires more code and more execution time to achieve the same effect using other instruction. These iteration instructions are described below.

6.26.1. LOOP-Decrement CX and Branch if CX is not Equal to Zero

The mnemonic for instruction is LOOP. It is a 2 byte instruction, where the first byte provides the opcode and the second byte provides the displacement in 2's complement notation. The displacement has the same interpretation as was given in JE instruction earlier.

However, in Assembly language programming, the programmer just provides the label of the location where jump is desired. The assembler then generates the displacement in 2's complement notation. Thus the destination must be within a distance of -128 to +127 bytes from the beginning of the next instruction after LOOP instruction. If the label is at a distance outside this range, an error message will be flashed by the assembler during assembly process.

Although theoretically, it is possible to branch forward using a LOOP instruction, it is invariably used for a backward branch only. The execution of this instruction does not affect any of the flags.

Opcode for this instruction is E2H.

The execution of 'LOOP BACK' instruction will result in the following.

1. CX is decremented by 1.
2. If CX contents is not zero, a branch to BACK takes place.
3. If CX, contents is zero, execution of next instruction after LOOP takes place.

In a LOOP instruction, CX is always used as the implied counter. Thus LOOP BACK is equivalent to the following pair of instruction.

```
DEC     CX
JNE     BACK
```

Example in the example given below, the series of instructions starting from SUB AX,BX and ending in ADD BX,02 will be executed 4 times, as CX is loaded with 04 in the beginning.

```
MOV      CX,04
BACK:   SUB          BX,AX
         .
         .
         .
   ADD  BX,02
LOOP     BACK; Decrement CX by 1. If CX <>0, branch to BACK.
ADD CX,DX; This is executed next after LOOP Back, If CX=0
```

6.26.2. LOOPZ or LOOPE-Decrement CX and Branch if CX Is not Equal to zero and z flag=1

The mnemonic for the instruction is LOOPZ. It stands for 'LOOP while Z flag =1'. It is a 2 byte instruction, where the first byte provides the opcode and the second byte provides the displacement in 2's complement notation. The displacement has the same interpretation as was given in JE instruction earlier.

Another equivalent mnemonic for the instruction is LOOPE, which stands for 'Loop while equal'. We generally use LOOPE after a CMP instruction, and LOOPZ after an arithmetic like addition.

However, in Assembly language programming, the programmer just provides the label of the location where jump is desired. The assembler then generates the displacement in 2's complement notation. Thus the destination must be within a distance of -128 to +127 bytes from tho boginning of the next instruction after LOOPZ instruction. If the label is at a distance outside this range, an error message will be flashed by the assembler during assembly process.

Although theoretically, it is possible to branch forward using a LOOPZ instruction, it is invariably used for a backward branch only. The execution of this instruction does not affect any of the flags.

Opcode for this instruction is E1H.

The execution of 'LOOPZ BACK' instruction will result in the following.

1. CX is decremented by 1.
2. If CX contents is not zero and Z flag=1, a branch to BACK takes place.
3. If CX, contents is zero or A flag=0, execution of next instruction after LOOPZ takes place.

Example in the example given below, when LOOPE BACK is executed, jump to location BACK takes place if CX is < > 0 and Z flag =1. Otherwise, execution proceeds with the next instruction after LOOPE BACK.

```
MOV      CX,04
BACK:   SUB          BX,AX
         .
         .
         .
   CMP  BX,02
LOOP     BACK; Decrement CX If CX <>0 and Z=1, jump to back
ADD CX,DX; Executed after LOOP Back, If CX=0 or BX <>2
```

6.26.3. LOOPNZ or LOOPNE-Decrement CX and Branch If CX is not Equal to Zero and Z Flag =0

The mnemonic for the instruction is LOOPNZ. It stands for 'LOOP while Z flag =0'. It is a 2 byte instruction, where the first byte provides the opcode and the second byte provides the displacement in 2's complement notation. The displacement has the same interpretation as was given in JE instruction earlier.

Another equivalent mnemonic for the instruction is LOOPNE, which stands for 'Loop while not equal'. We generally use LOOPNE after a CMP instruction, and LOOPNZ after an arithmetic operation like addition.

However, in Assembly language programming, the programmer just provides the label of the location where jump is desired. The assembler then generates the displacement in 2's complement notation. Thus the destination must be within a distance of -128 to +127 bytes from the beginning of the next instruction after LOOPNZ instruction. If the label is at a distance outside this range, an error message will be flashed by the assembler during assembly process.

Although theoretically, it is possible to branch forward using a LOOPNZ instruction, it is invariably used for a backward branch only. The execution of this instruction does not affect any of the flags.

Opcode for this instruction is E0H.

The execution of 'LOOPNZ BACK' instruction will result in the following.

1. CX is decremented by 1.
2. If CX contents is not zero and Z flag=0, a branch to BACK takes place.
3. If CX contents is zero or Z flag=1, execution of next instruction after LOOPNZ takes place.

Example in the example given below, when LOOPNE BACK is executed, jump to location BACK takes place if CX is < > 0 and Z flag =0. Otherwise, execution proceeds with the next instruction after LOOPNE BACK.

```
MOV      CX,04
BACK:  SUB          BX,AX
         .
         .
         .
CMP      BX,02
LOOP     BACK; Decrement CX If CX <>0 and Z=0, jump to back
ADD CX,DX; Executed after LOOPNE Back, If CX=0 or BX <>2
```

6.26.4. JCXZ-Jump if CX is Equal to Zero

The mnemonic for the instruction is JCXZ. It stands for 'Jump if CX value is equal to zero'. It is a 2 byte instruction, where the first byte provides the opcode and the second byte provides the displacement in 2's complement notation. The displacement has the same interpretation as was given in JE instruction earlier.

However, in Assembly language programming, the programmer just provides the label of the location where jump is desired. The assembler then generates the displacement in 2's complement notation. Thus the destination must be within a distance of -128 to +127 bytes from the beginning of the next instruction after JCXZ instruction. If the label is at a distance outside this range, an error message will be flashed by the assembler during assembly process.

Opcode for this instruction is E3H.

JCXZ instruction execution does not affect any flag. The execution of the instruction is not affected by Z flag, and it is only affected by value of CX register. CX value is not altered after the execution. The execution of 'JCXZ BACK' instruction will result in the following.

1. If CX contents is zero, a branch to BACK takes place.
2. If CX, contents is not equal to zero, execution of next instruction after JCXZ takes place.

Example in the example given below, when JCXZ SKIP is executed, jump to location SKIP takes place if CX = 0. Otherwise, execution proceeds with the next instruction after Loop.

```
MOV      CX,04
JCXZ     SKIP
BACK:    SUB              BX,AX
         .
         .
         .
LOOP     BACK; Decrement CX If CX <>0, jump to back
SKIP     ADD CX,DX; Executed after JCXZ SKIP If CX=0, or is
         ; Executed after LOOP BACK, If CX=0
```

Thus the JCXZ is very useful to check if the iterative count value in CX is zero or not before the iteration starts. If the CX value is zero, the iteration is skipped.

6.26.5. Call Instruction

Very often we need to use a particular sequence of instructions at several points in our program. We can avoid writing the same sequence of instructions at these various points, by writing this series of instructions as a subprogram. In 8086 terminology, such subprograms are termed as Procedures.

Whenever the instructions in the procedure are required to be executed we branch to the procedure using the CALL instruction. CALL stands for 'Call a procedure'. After executing the instructions in procedure we return to the next instruction after the CALL instruction in the main program. This is achieved by the execution of RET (stands for Return from procedure) instruction in the procedure.

The difference between a JMP instruction and a CALL instruction is as follows. If a JMP instruction is executed, we jump to the destination location, and the execution carries on from there, without bothering to come back later to the instruction after the JMP. If a CALL instruction is executed, we jump to the destination location, and the execution carrics on from there, and we come back later to the instruction after CALL, by executing the RET instruction in the procedure.

The CALL instruction will first of all store above the top of stack, the address to which the branch should take place after executing all the instructions in the procedure. Only then the branch to the procedure takes place. The address saved above the top of stack is generally termed as the Return Address. Thus a CALL instruction can be visualized as Push Return Address on the stack, and then branch to the procedure.

The RET, instruction which is at the end of a procedure, pops out from the top of the stack the Return Address, so that a branch back to the calling program at the next instruction after CALL is achieved.

Using this approach, we can branch any number of times to the procedure, which is written only once. This saves a lot of memory space for the complete program.

Also usage of procedures helps in modular design for solving a programming problem. It makes the main program look very simple and compact.

The execution of the CALL instruction always results in a branch to a procedure. The branch could be within the same code segment, in which case it is called a NEAR call or an INTRASEGMENT call.

NEAR call instruction can provide the destination address directly using a label in a relative mode, with 2 byte of signed displacement, or the destination address is provided indirectly in a register or a memory location. Thus the two types of NEAR calls are

(a) NEAR call or Intrasegment call of direct type
(b) NEAR call or Intrasegment call of INDIRECT type (destination address in a register or a memory location)

If the procedure is in a code segment which has a different name from the segment in which CALL instruction is present, it is called a FAR call or an INTERSEGMENT call.

FAR call instruction can provide the procedure address using a label in a direct mode, with 2 words to indicate CS and IP value of the destination address, or the procedure address is provided indirectly in 2 consecutive memory locations. Thus the 2 type of FAR calls are

(a) FAR call or intersegment call of direct type
(b) FAR call or intersegment call of indirect type

Thus there are 4 templates for the call instruction, which are discussed below. All these have the same mnemonic CALL.

6.26.6. CALL-Intrasegmnet Call or NEAR Call of Direct Type

The mnemonic for the instruction is CALL. It is a 3 byte instruction, where the first byte provides the opcode and the next 2 bytes provide the 16 bit displacement in 2's complement notation. This is a very commonly used type of call instruction.

Opcode for this instruction is E8H.

The 16 bit displacements the number of bytes by which the jump takes place. The range for a 16 bit 2's complement number is from -32768 (8000H) to +32767 (7F00H). If it is a positive number, the jump if usually to a forward location. If it is a negative number, the jump is usually to a backward location. Thus the memory location specified by the label in the CALL instruction should be generally in the range of -32768 to +32767 bytes from the beginning of the next instruction after CALL. However, in Assembly language programming just provides the label for the location where jump is desired.

The execution of a NEAR DIRECT CALL instruction results in the following.

(a) The offset value of the next instruction after CALL, which is present in the instruction pointer (IP) register, is saved above the top of the stack.
(b) IP value is replaced by the sum of IP value and the 16 bit signed displacement provided in the instruction. Thus the new IP value is the offset address of the procedure, and hence branch to the procedure takes place.

In 8086, we do not have any conditional CALL instructions. Suppose our interest is to branch to the procedure REPEAT only if the C flag is zero. If the condition for branching is not satisfied, the next instruction to be executed is, let us say, MOV AX,BX. This problem is circumvented using the unconditional call instruction as follows.

```
               .
               .
               JC        NEXT       ; If C flag=1, execute MOV AX,BX
               CALL REPEAT     ; If C flag =0, jump to procedure REPEAT.
      NEXT:    MOV       AX,BX
               .
               .
```

6.26.7. CALL-Intrasegment Indirect Call or NEAR Indirect Call

The mnemonic for the instruction is CALL. Template for this instruction is

As there are 5 bits in the template for identifying the operands, there are 32 instructions in this category. Note that W bit is absent in the template since the call address is given as the contents of a word register or a word location in memory. This method of calling is not commonly used. This is because by looking at the program, it is not at all clear where we are branching. This form of CALL instruction is used to select one of several procedures based on a computed value.

The execution of a NEAR INDIRECT CALL instruction results in the following.

(a) The offset value of the next instruction after CALL, which is present in the instruction pointer (IP) register, is saved above the top of the stack.

(b) P value is replaced by contents of word register, or a word location in memory, which is provided in the instruction. Thus the new IP value is the offset address of the procedure, and hence branch to the procedure takes place.

Example 1. Code for CALL AX

Suppose AX content is 1234H. Then CALL AX execution result in a branch to the procedure at offset location 1234H in the same code segment. Note that 1234H is not a relative displacement. CALL AX is a 2 byte instruction with the code

Example 2. Code for CALL word ptr 200H[BX]

Suppose BX content is 1234H. Let us say, the contents of word location 3234H is 5678H. 3234H is the sum of 1234H, which is BX contents, and 2000H, which is the 16 bit displacement provided in the instruction. Then CALL WORD PTR 2000H[BX] execution result in a branch to the procedure at offset location 6578H in the same code segment. It is a 4 byte instruction with the code

6.26.8. Call-Intersegment Direct Call or FAR Direct Call

The mnemonic for the instruction is CALL. As it is desired to branch to a different segment, it is necessary to provide the Segment base and Offset values of the procedure in the instruction. Thus it is a 5 byte instruction, where the first byte provides the opcode and the next 2 byte provide the offset value, and the last 2 bytes provide the segment base value of the procedure. FAR calls are not very common in simple programs. If a FAR call is desired, the direct type is commonly used.

Opcode for this instruction is 9AH.

However, in assembly language programming, the programmer just provides the name of the procedure where jump is desired. Further the assembler will be told that it is a FAR call as in 'CALL FAR PTR LOC' , in case the procedure LOC is defined at a later point in the program. Then the assembler generates the 5 byte code for a FAR direct jump. The assembler directive 'FAR PTR' is not needed in the CALL instruction if the procedure LOC is defined in a different segment earlier to the CALL instruction.

The execution of a FAR DIRECT CALL instruction results in the following.

1. The segment value of the next instruction after CALL, which is present in the Code Segment(CS) register, is saved above the top of the stack.
2. The offset value of the next instruction after CALL, which is present in the Instruction Pointer (IP) register, is next saved above the top of the stack.

3. IP value is replaced by the 16 bit offset value provided in the second and third bytes of the instruction. Thus the new IP value is the offset address of the procedure.
4. CS value is replaced by the 16 bit segment base value provided in the fourth and fifth bytes of the instruction. Thus the new CS value is the segment base address of the procedure, and hence branch to the procedure takes place.

6.26.9. CALL-Intersegment Indirect Call or FAR Indirect Call

The mnemonic for the instruction is CALL. Template for this instruction is

As there are 5 bits in the template for identifying the operands, there are 32 instructions in this category. Note that W bit is absent in the template since the call address is given as the contents of a double word in memory. This method of calling is not commonly is not commonly used. This is because by looking at the program, it is not at all clear where we are branching. However, it can be useful when we wan to branch to one of several procedures based on a computed value.

The execution of a FAR INDIRECT CALL instruction results in the following.

1. The segment value of the next instruction after CALL, which is present in the Code Segment(CS) register, is saved above the top of the stack.
2. The offset value of the next instruction after CALL, which is present in the Instruction Pointer (IP) register, is next saved above the top of the stack.
3. IP value is replaced by the 16 bit offset value provided in the word location, whose address is given in the instruction. Thus the new IP value is the offset address of the procedure.
4. CS value is replaced by the 16 bit segment base value provided in the next word location. Thus the new CS value is the segment base address of the procedure, and hence branch to the procedure takes place.

Example 1. Code for CALL dword ptr 200H[BX]

Suppose BX content is 1234H. Let us say, the contents of word location 3234H is 5678H. This 5678H provides the offset value. Let us say, the contents of the next word at location 3236H is ABCDH. The ABCDH provides the segment value. Thus CALL DWORD PTR 2000H[BX] execution result in a branch to the procedure at offset location 6578H is another code segment. It is a 4 byte instruction with the code

6.26.10. Return Instruction

Very often we need to use a particular sequence of instructions at several points in our program. We can avoid writing the same sequence of instructions at these various points, by writing this series of instructions as a subprogram. In 8086 terminology, such subprograms are termed as Procedures.

Whenever the instructions in the procedure are required to be executed we branch to the procedure using the CALL instruction. CALL stands for 'Call a procedure'. The CALL instruction is already discussed.

The CALL instruction will first of all store above the top of stack, the address to which the branch should take place after executing all the instructions in the procedure. Only then the branch to the procedure takes place. The address saved above the top of stack is generally termed as the Return Address. Thus a CALL instruction can be visualized as Push Return Address on the stack, and then branch to the procedure.

The Return Address is the contents of Instruction Pointer (IP), in the case of a NEAR CALL, whether of Direct or Indirect type. The Return Address is the contents of instruction Pointer (IP) and Code Segment Register (CS), in the case of a FAR CALL, whether of direct or indirect type.

The RET (stands for Return from Procedure), instruction which is at the end of a procedure, pops out from the top of the stack the Return Address, to IP only or to both IP and CS, so that a branch back to the calling program at the next instruction after CALL is achieved.

The execution of the RET instruction always results in a branch in a branch back to the calling program, to the next instruction after the CALL. If the calling program is in the same segment as the called procedure, then a NEAR return instruction is used to achieve the object.

NEAR return instruction execution results in popping out the one word Return Address to IP. In addition to popping Return Address to IP, it can also optionally remove a specified number of bytes from the stack, This is achieved by adding a 16 bit immediate vale in the RET instruction to the SP register. Thus the 2 types of NEAR return instructions are

(a) NEAR Return or Intrasegment Return

(b) NEAR return or Intrasegment return with a 16 bit immediate value added to SP contents.

The execution of the RET instruction always results in a branch back to the calling program, to the next instruction after the CALL. If the calling program is in a different segment from the called procedure, then a FAR return or an INTERSEGMENT Return instruction is used to achieve the objective.

FAR return instruction execution results in popping out the 2 word return address to IP and CS. In addition to popping Return Address to IP and CS, it can also optionally remove a specified number of bytes from the stack. This is achieved by adding a 16 bit immediate value provided in the RET instruction to the SP register. Thus the 2 types of FAR return instructions are

(a) FAR Return or intersegment Return

(b) FAR Return or intersegment Return with a 16 bit immediate value added to SP contents.

Thus there are 4 templates for the Return instruction, which are discussed below. All these have the same mnemonic RET. The assembler codes a RET instruction as a NEAR or a FAR type, based on the type specification of the procedure. The type of a procedure with name REPEAT of NEAR type is declared as 'REPEAT PROC NEAR'. Similarly, a procedure with name BIN_DEC of FAR type is declared as 'BIN_DEC PROC FAR'. In case the type of procedure is not specified as in 'REPEAT PROC', the type will be assumed as NEAR type.

6.26.11. RET-Intrasegment return or NEAR Return

The mnemonic for the instruction is RET. It is a 1 byte instruction. This is a very commonly used type of return instruction.

Opcode for this instruction is C3H.

The execution of a NEAR RET instruction results in the following.

The offset value of the next instruction after CALL, which was earlier saved by the NEAR CALL instruction on the stack, is now popped from the top of stack to the IP register.

And so the branch back to the calling program takes place.

In 8086, we do not have any conditional Return instructions. Suppose our interest is to branch back to calling program in the middle of the procedure, if the C flag is zero. If the condition for Return is not satisfied, the next instruction to be executed in the procedure is, let us say, MOV AX,BX. This problem is circumvented using the unconditional Return instruction as follow.

```
    .
    .
    JC   NEXT ;If C flag=1, execute MOV AX,BX
    RET       ; If C flag=0, Return to calling program.
NEXT: MOV  AX,BX
    .
    .
    .
```

6.26.12. RET-Intrasegment Return or NEAR Return with Immediate Value Added to SP Register

The mnemonic for the instruction is RET. It is a 3 byte instruction, where the first byte provides the opcode, and the next two bytes provide the immediate value to be added to SP. An example for this instruction is 'RET 12H'.

Opcode for this instruction is C2H.

The execution of this instruction results in the following.

1. The offset value of the next instruction after CALL, which was earlier saved by the NEAR CALL instruction on the stack, is now popped from the top of stack to the IP register. And so the branch back to the calling program takes place.
2. The 16 bit immediate value provided in the instruction is added to the SP register contents. In other words, the immediate value provided in the RET instruction indicates the number of bytes that is removed from the top of the stack after popping out the Return Address.

6.26.13. RET-Intersegment Return or FAR Return

The mnemonic for the instruction is RET. It is a 1 byte instruction. This is a very commonly used type of FAR Return instruction.

Opcode for this instruction is CBH.

The execution of a FAR RET instruction results in the following.

The segment Base and offset value of the next instruction after CALL, which was earlier saved by the FAR CALL instruction on the stack, is now popped from the top of stack to the IP and CS registers. And so the branch back to the calling program takes place.

6.26.14. RET-Intersegment Return or FAR Return with Immediate Value Added to SP Register

The mnemonic for the instruction is RET. It is a 3 byte instruction, where the first byte provides the opcode, and the next two bytes provide the immediate value to be added to SP. An example for the instruction is 'RET 12H'.

Opcode for this instruction is CAH.

The execution of this instruction results in the following

The Segment Base and offset value of the next instruction after CALL, which was earlier saved by the FAR CALL instruction on the stack, is now popped from the top of stack to the IP and CS registers. And so the branch back to the calling program takes place.

The 16 bit immediate value provided in the instruction is added to the SP register contents. In other words, the Immediate value provided in the RET instruction indicates the number of bytes that is removed from the top of the stack after popping out the return address.

REVIEW QUESTIONS

1. Describe the need for templates in instruction coding of 8086.
2. Explain the various fields used in the template for the MOV instructions.
3. Generate the code for the following instructions using the template for MOV instruction. Generate alternate code also, if it exists.
   (a) MOV DX,BX
   (b) MOV DH, CL
   (c) MOV AL,[SI]
   (d) MOV [SI], CX
   (e) MOV DS:1234H[BP], DH
   (f) MOV DH, [SI+85H]
4. Draw your inference about the condition under which we will have two valid opcodes for the same instruction.
5. Explain with *Example* the instructions to move data between two registers or a register and a memory location.
6. Explain with *Example*s the instructions to move immediate data to a register/memory location.
7. Explain with *Example*s the instructions to perform data transfer between a segment register and a register/memory location.
8. Explain with *Example* the instructions to perform push and pop operations. How are they different from push and pop instruction of 8085?
9. Explain with *Example*s the exchange instructions.
10. Explain with *Example*s the input and out put instructions. How are they different from the input and output instructions of 8085?
11. Generate the code using appropriate templates for the following instructions.
    (a) MOV WORD PTR [SI+96H], 25H
    (b) XCHG AL,CH
    (c) XCHG ES:[SI+45H], AX
    (d) POP [SI]
    (e) PUSH CX
    (f) POPE
    (g) IN AX, 75H
    (h) OUT DX, AL
12. Discuss the utility of XLAT instruction. Write an 8086 assembly language program to convert excess 3 code to 8421 BCD code.
13. Explain with *Example*s the LEA instruction. Is LEA BX, [SI] same as MOV BX, SI? Discuss?
14. Explain with *Example* LDS and LES instructions.
15. Explain with *Example* LDS and Les instruction.
16. Generate the code using appropriate templates for the following instructions.
    (a) LEA BX,65H[SI] (b) LDS BX,C2H[SI] (c) LES BX,A2H[SI]
17. Write an 8086 assembly language program to complement the LS byte of flags register using LAHF and SAHF instructions.
18. Give an overview of AND instructions.
19. Give an overview of OR instructions.
20. Give an overview of Ex OR instructions.
21. Give an overview of TEST instructions. How is a TEST instruction different from an AND instruction?

22. Give an overview of SHIFT instructions
23. Give an overview of ROTATE instructions
24. Generate the code using appropriate templates for the following instructions, and explain their working using numerical examples.
    (a) AND AL,0ACH
    (b) OR WORD PTR 68H[SI], 87H
    (c) XOR DS:8678H[BP], CL
    (d) TEST AL,BH
    (e) TEST CX,25H
    (f) NOT WORD PTR DS:27H[BP]
    (g) ROL DX,CL
    (h) ROR BYTE PTR DS:89H[BP], CL
    (i) RCL BX,1
    (j) RCR word ptr DS:1243H[BP], CL
    (k) SHR BYTE PTR [SI], CL
    (l) SHL WORD PTR [SI], CL
25. Give an over view of the instructions that set/reset/complement flags.
26. How is the T flag set to 1 or reset to 0?
27. How can we set/reset A, S, Z and O flags?
28. Explain the use of WAIT instruction.
29. Explain the use of LOCK instruction.
30. Explain working of Halt and Escape instructions.
31. Explain MOVS instruction with examples.
32. Explain STOS instruction with examples. Describe use of REP prefix in this instruction.
33. Explain LODS instruction with examples.
34. Explain CMPS instruction with examples. Describe use of REPE and REPNE prefix in this instruction.
35. Explain SCAS instruction with examples. Describe use of REPE and REPNE prefix in this instruction.
36. Explain with examples conditional jump instruction which perform a jump based on the value of a single flag. What is the needed in the code to branch anywhere in the segment based on a condition?
37. Explain with examples conditional jump instruction which perform a jump based on the value of a More than one flag.
38. Explain the flag status under which the branch takes place for the following instructions.
    (a) JBE and JNBE
    (b) JLE and JNLE
    (c) JL and JNL
39. Explain with examples the intra segment unconditional jump instructions.
40. Explain with examples the inter segment unconditional jump instructions.
41. Explain with examples the LOOP, LOOPE and LOOPNE instructions.
42. Explain with examples the JCXZ Instruction.
43. Explain with examples the intra segment Call instructions
44. Explain with examples the inter segment Call instructions.
45. Explain with examples the intra segment Return instructions.
46. Explain with examples the inter segment Return instructions.

❑❑❑

CHAPTER 7

TECHNIQUES FOR ASSEMBLY LANGUAGE PROGRAMMING WITH DOS AND BIOS FUNCTIONS FOR 8086 MICROPROCESSOR

INSIDE THIS CHAPTER

7.1. OBJECTIVE

After completion of this chapter student will know program segments, application of assembly language and its development. Student will understand what is meant by program development and tools. Apart this student will be familiar with program formatting and Assembler directives.

7.2. INTRODUCTION TO ASSEMBLY LANGUAGE PROGRAMMING

The Assembly Language Programming continues to be critical in system programming the advancement in high level programming. The programs written in this language are compact and efficient, which has to be converted to machine code for execution and it is performed by the translator known as assembler. Assembly language programming needs good knowledge of machine architecture, operating system, and programming principles. Assembly language is case insensitive, therefore program can be coded either in uppercase, lowercase, or combination of lower/uppercase characters which will be discussed.

The program segments, procedures, and macro programming has to be discussed at great length. The program development tools' such as editor, assembler, linker, etc., are also to be discussed with the program development process.

Next it also covers software support available on the IBM-PCs with access methods and examples.

7.3. PROGRAM SEGMENTS

8086 microprocessor organizes main memory as segments of 64Kb size and also provides with four segment registers (CS, DS, SS and ES) to access these segments. The 8086 processor organizes the segments into four groups as mentioned below :

(a) Code Segment

(b) Data Segment

(c) Stack Segment

(d) Extra Segment

The segments code, data, stack, and extra are pointed to by the segment registers CS, DS, SS and ES respectively. Code segment is used to store the program code, the data, stack and extra segments are normally used to store data used in the program. The code segment can also be used to store the program instructions and data *i.e.,* the code and data segment register (CS, DS) points to the same segment.

7.3.1. Definition of Segment

The assembler directives SEGMENT and ENDS are made use to define a segment. The directive SEGMENT is made used to indicate the beginning of a logical segment and the directive ENDS informs the end of the segment to the assembler in its processes. The directives SEGMENT and ENDS must enclose the data or code of a segment and next both the SEGMENT and ENDS follows the name of the segment.

The General Form

```
Segment Name    SEGMENT [WORD/PUBLIC]
, Storage definition, allocation,
; Program code, and alignment directives
Segment Name ENDS
```

where

Segment Name is the name of the user defined segment. WORD/PUBLIC is the type specifiers.

7.3.2. Multiple Data and Code Segments

The data, stack and extra segments are normally used to store data used in the program and the code segment is used to store instructions of a program. The progra: having data and code which cannot fit into a segment of 64Kb must be define in multiple segments. The branch to a procedure located in a different segment is made by using FAR call. The outline of a segment mentioned below indicates the arrangement of the two code segments.

```
_MYCODEI SEGMENT
       ASSUME CS : _MYCODE1,          DS : DATA

       CALL FACTORIAL
MYCODEI ENDS.
MYCODE2 SEGMENT
       FACTORIAL PROC FAR
       ASSUME CS : _MYCODE2
```

```
        .RET
        "FACTORIAL ENDP
MYCODE2 ENDS
```

Note. In the case of multiple segments, the segment registers must be explicitly setup. The procedures defined in a different segment must be defined as the FAR procedure, so that the assembler recognizes a call to that procedure and codes the procedure call as FAR cell.

Number of code and data segments that can be referenced in a program is defined by the memory model. The data or procedure code located in the same segment is referenced using two byte address known as NEAR pointer. The data or a procedure code located in a different segment is referenced using four byte address *(explicit segment and offset value)* known as the FAR pointer. The memory model specifies the number of segments that can be used in a program. The available memory models which are shown in Table 7.1 as under.

TABLE 7.1.

Memory Model	Description
Tiny	Both code and data must fit in a single 64Kb segment Code and data reference are NEAR
Small	Code must fit within a single 64Kb segment Data must fit within a single 64Kb segment Code and data reference are NEAR
Medium	Program code can be larger than 64Kb segment Data must fit within a single 64Kb segment Code reference is PAR and data reference is NEAR
Compact	Code must fit within a single 64KB segment Program data can be larger than 64Kb segment Code reference is NEAK and data reference is FAR
Large	Both program code and data may be larger than 64Kb segment Code reference is PAR and data reference is FAR Single data array cannot be larger than 64Kb
Huge	Both program code and data may be larger than 64Kb segment Code reference is FAR and data reference is FAR Pointers to elements of an array are also PAR and data array may exceed 64Kb in size

7.3.3. End of Program

Physical end of the program is indicated by the directive END. The directive END is generally followed by the label, which indicates the point at which execution of the program begins. Such concept is illustrated as :

```
_CODE SEGMENT
        ASSUME CS : _CODE
START :
   _CODE ENDS
        END START
```

The directive END with label is used in the main program of a multiprogram module. *(Other than main program module, all other modules must end with the directive END without label).*

7.3.4. Program Termination

Directive END is not a command to terminate the execution of the program but it is just an assembler directive indicating the end of the source code in a program module. The program termination must be done explicitly using the program termination system call of the operating system which is explained with an example as

Example 1 :

```
^CODE SEGMENT
      ASSUME   CS   :   _CODE
START:
      ; Program code
      ; Program termination
      MOV AH, 4CH
      MOVAL, 004H
      [NT21H
   _CODE ENDS
      END START
; program terminate function ; return code ;
call DOS service
```

The instruction HLT may be used to stop execution of the program, however this does not transfer control back to the loader or the operating system shell for processing next command of the user. Instruction HLT is used as a program termination command in a microprocessor kit or subsystem designed using 8086 processor.

In case of the microcomputer (IMB-PC), the system does not hait after the execution of the program instead, it returns to the operating system shell which accepts the next command issued by the user.

Assembly language has very few restrictions rules; *(nearly everything is left to the discretion of the programmer),* while high level language details various semantically and syntactical restrictions.

7.4. APPLICATION OF AN ASSEMBLY LANGUAGE

Usually programmers will not write large application program in assembly language, but they write short, specific routines, subroutines in assembly language and call them from a high level language program. *(Assembly language subroutines can he written to handle operations that are not available in the higher level language).*

For instant consider writing a business application program in COBOL for IBM. PC :

There are two commands in the COBOL language definition to do. Check for the free space on the disk,

```
create a subdirectory,
write protect a file, and
create overlapping windows.
```

But all these can be done within the COBOL program by writing assembly language subroutines.

To develop an assembly language, the process or the steps to follow as listed below :

(a) Specify the source code as per the assembly language definition

(b) Assemble the program to create the object code

(c) Link the object code to create the executable code

(d) Test and debug the program.

The Fig. 7.1 depicts the process of assembly language programming in the form of flow chart, when each step will result in creating one or more files with extensions, as shown :

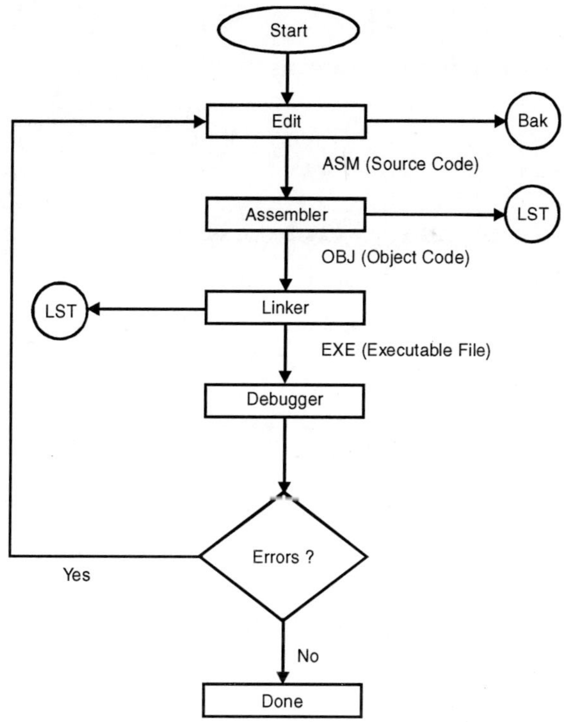

Fig. 7.1.

Now let us briefly analyze each steps to develop assembly language, program i.e.,

(i) Creating Source Code

(ii) Assembling the Program

(iii) Listing File

(iv) Test and Debug

(*i*) Creating Source Code

Source code consists of 8086/8088 program memories, suitable pseudo opcodes and assembler directives. A file can be created with a text editor and is given an extension ASM. The text editor may be any Word Processor (like EDLIN, NE) which can produce standard ASCII *(unformatted text containing no control characters)* code.

(*ii*) Assembling the Program

For assembling the program assemblers are available for the IBM-PC,

(a) Microsoft Macro Assembler (MASM) and

(b) Borland Turbo Assembler (TASM).

A part doing tedious task of producing the binary codes for instruction statements, an assembler also allows to refer to data items by name rather than by numerical addresses, which makes the program much more readable. In addition to program instruction, the source program also contain directives to the assembler. The Pseudo instructions are assembler directives entered into the source coding with the assembly language.

Once the program is written completely, it can be assembled to obtain the OBJ file by executing MASM or TASM. The assembly language program file name should mentioned along with the command, as shown below : MASM <file name.ASM>; or TASM <file name.ASM>

The <file name.ASM> file contains the assembly language program is assemble and assembler generates error messages if there are any errors such as Syntax errors. Such errors are listed along with the line number and no errors then OBJ file is created. *(To obtain the EXE file the user has to LINK the OBJ file).*

```
LINK <file name>; or TLINK <file name>;
```

When the file is smaller than 64K bytes, it can be converted from an execution file to a command file (*.COM*), which is slightly different from an execution file (.EXE). In a command file the program must be originated at location 100H before it can execute. Which this means that the program must be no longer than (*64K-100H*) on length. The command file requires less space in memory than the equivalent execution file. The system loads. COM file off the disk into the computer memory more quickly than the execution file.

Not to create a .COM file from a .EXE file, we require EXE2B. In converter EXE2BD: converts .EXE file to .COM or binary file.

Example 1. *EXE2BIN <fllename> <filename. com>*

Here, the <filename> with an EXE extension in converted to <filename> with .CO! Extension with above command.

(*iii*) Listing File

In general, Turbo assembler produces only one file as the result of assembly: an object *.OBJ* file with the same as the source (*.ASM*) file. If required ask TURBO assembler to produce a listing file with the extension. LST as well, simply required by typing two additional commas on the command line, as in following example :

Example 2. *TASM Hello,*

Assembles Hello. ASM and produces the Object tile Hello.OBJ

Example 3. *TASM Hello*; *Will Generate the listing file HELLO.LST along with the Hello.OBJ file. Similarly by using the command*

TASM Hello, Or 'TA$M/I

Hello generates listing file along with the .Obj file and so on.

(*iv*) Test and Debug

An executable program may be run under DOS or DEBUG with normal rule, run a program under DOS only when there is no error or it produces some not visible or audible result. When the program result is stored in registers or in memory, the result is visible. Therefore, it should be run using DEBUG or TD (Turbo Debugger) or code-view only. EXE files can be loaded into memory using DEBUG as :

Example 1: *DEBUG <filename.EXE>*

By using DEBUG it is possible to find the bugs in the program and next the memory it is possible to check and correct the errors using different commands in DEBUG. Now following Table 7.2 depicts few commands with format and action.

TABLE 7.2.

S.No.	Command		Format and Action
1.	G-GO	Format :	G [offset], [offset]
		Action :	Executes a program startinpt the current location. Offset values are temporary breakpoint instruction the processor stops and displays registers and flag contents.
2.	T-TRACE	Format :	T [Instruction Count]
		Action :	Executes one or more instructions and displays register and flag values for each of them.
		Example :	T; Executes only the next intstruction T5; Executes the next 5 instructions.
3.	P-PTRACE	Format :	P [Instruction Count]
		Action :	Same as Trace, but treats subroutine calls, interrupts, loop instructions, and repeat string 41 structions as a single instruction.
4.	Q-Quit	Format :	Q
		Action :	Exists to DOS
5.	D-DUMP	Format :	D seg : offset [offset] D seg : offset L byte count
		Action :	Displays the numeric contents of memory locations and their character equivalents (if any)

Now observe five more ideas/techniques to developed a program and debug

(a) Carefully define the program to solve the problem in hand and work out with best algorithm.

(b) If the program have several parts, write test and debug each part individually and then add parts one at a time.

(c) When a program or program section does not work then first recheck the algorithm. To make sure it really does what you want it to.

(d) If the algorithm seems correct, then check to make sure that you have used the correct instructions to implement the algorithm. Work out on paper the effect that a series of instructions will have on some sample data. (Such predictions on paper can later be compared with the actual results produced when the program section runs) macros resolved and header files included in the source program. Pre-processors output is then transferred to the next phase of the program translation. Normally, pre-processors are built into the assemblers or the compilers called as microprocessors. The model of a pre-processor is as shown in Fig. 7.2 below.

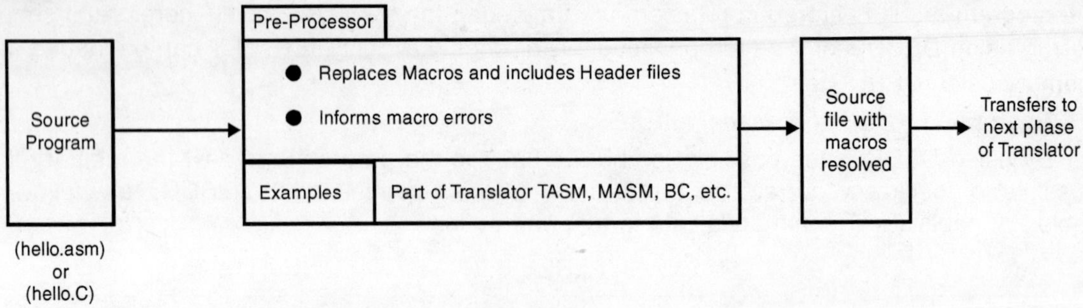

Fig. 7.2.

Assembler

An assembler is a software that translates assembly language program to machine language program and the input to the assembler is known as a source file. Assembler produces the output as relocatable object code and reports program errors during translation. The model of an assembler is as shown :

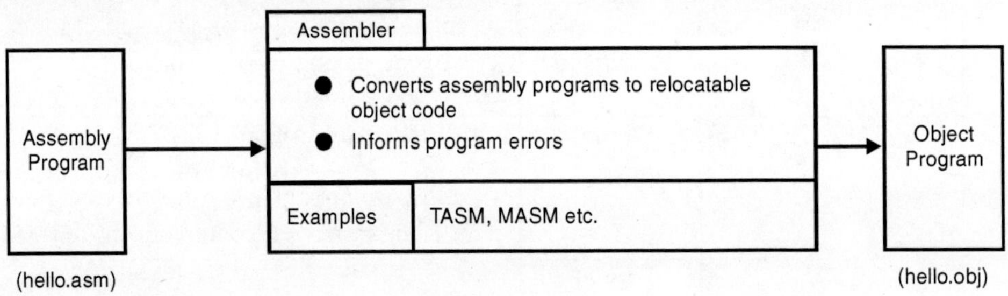

Fig. 7.3.

Compiler

Compiler is a software that translates high level language program to an equivalent object code in a relocatable form and the input to the compiler is a compiler is a source file. Compiler produces the output as a relocatable object code and reports program errors during translation. The model of the compiler is as shown :

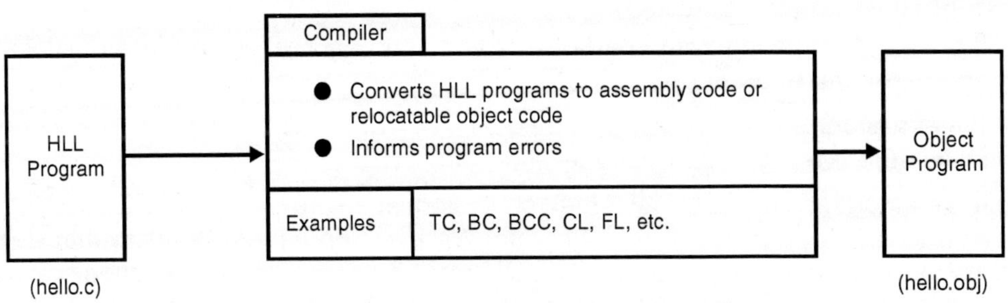

Fig. 7.4.

Linker

A linker is a software that combines relocatable object program modules and library functions into an executable file and combines library.

If you don't find a problem in the algorithm or the program instruction use debugger to help you localize the problem. Use single step or trace for short program sections. In the case of longer programs use breakpoints, which is often a faster technique to narrow the source of a problem down to a small region?

7.5. PROCESS FOR PROGRAM DEVELOPMENT

Borland program development tools are made used in explaining the program development process. The assembly program name HELLOA.ASM and C program name HELLOC.C are used in the explanation as source files.

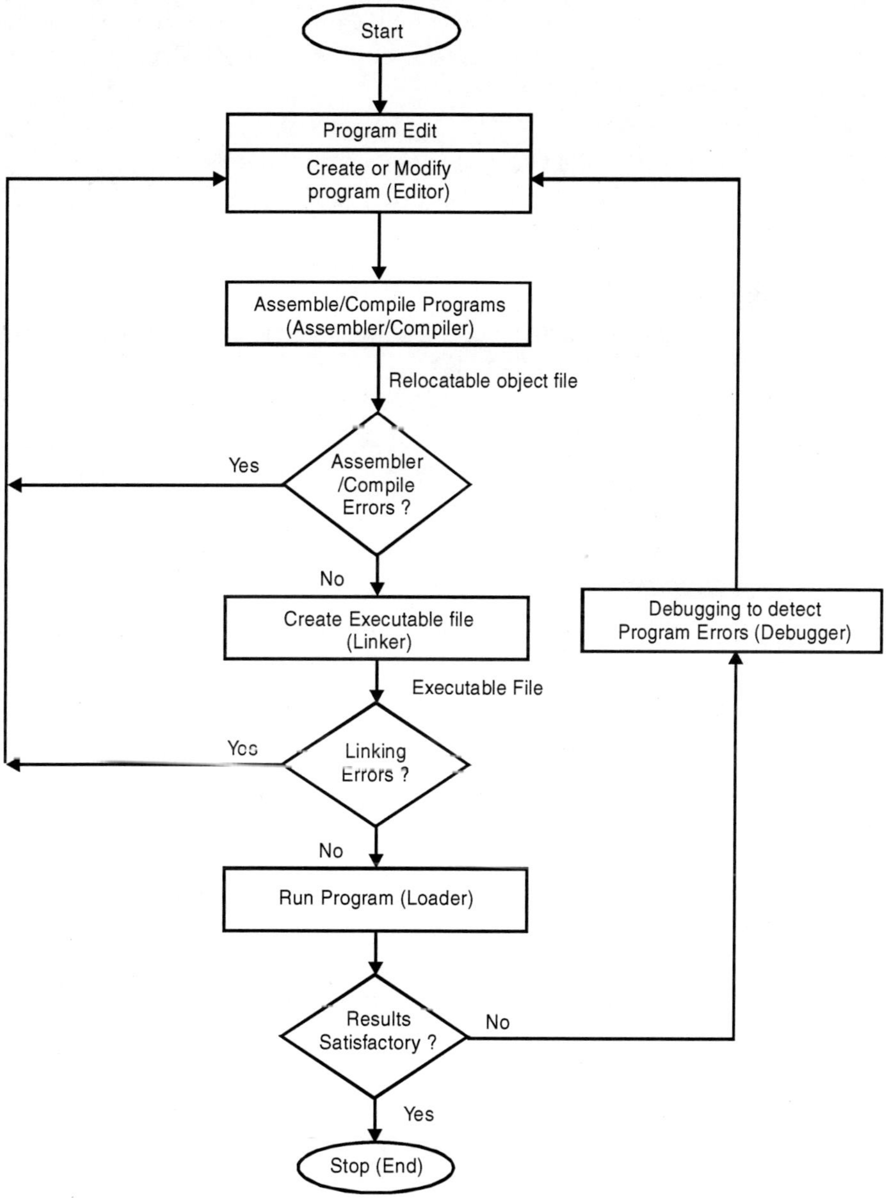

Fig. 7.5.

Now let us briefly analyse the 6 steps, which are involved in program development.

(a) Source File Creation

A file containing program statements in assembly or high level language is known as 'source file'. Which is created and edited using text editors. The assembly language source file has an extension ASM (*e.g. HELLO.ASM*) and C language source file has an extension (*e.g.* HELLO. C). (*The* C *source file can be created in an integrated program development editor*).

(b) Object Code Generation

Language translators are made use to translate source program into a relocatable object file. An assembler used in the translation of assembly language source code and the compiler is used in translation of high level language source code such as C. The commands for object code generation are as mentioned below.

Assembly source file: TASM HELLOA.ASM

Borland C source file : BCC HELLOC. C or compile option selection in IDE

An assembler generates a file HELLOA.OBJ and compiler generates HELLOC.OBJ if no errors are encountered during translation.

(c) Executable File Creation

A Linker is usco. To create executable files. The same linker is also used to translate assembly and C language object files into executable files.

Assembly Objcci file : TLINK HELLOA.OBJ

OR : TLINK HELLOA

OR : TLINK HELLOA

Borland C Object file : TLINK HELLOC

OR : TLINK HELLOC.OBJ

OR : TLINK HELLOC

Note. If the object file extension is not specified then OBJ is considered as dclaul extension.

Above commands generate an executable file HELLOA.EXE for assembly and generators an executable file HELLOC.EXE if no errors are encountered during linking.

(d) Program Running

The executable files can run by entering the name of the executable file mention below :

```
C>HELLOA
```

{The message displayed by the program on the screen is shown here}

```
C>HELLOC
```

{The message displayed by the program on the screen is shown here}

(e) Program Testing

The results generated by the program have to be tested for their validity and the result must confirm to the program specification. If there are any errors in the result, then the program must be debugged.

(f) Program Debugging

The errors in the program may be located using debuggers. The executable file of the program to be debugged must be created with program debug options. The Turbo assembler uses IZI option and linker uses N option to create an executable file which can be debugged at the source level.

Assembling : TASM/ZI HELLOA.ASM

Linkin~ : TLINKI VHELLOA.OBJ

7.6. DEVELOPMENT OF PROGRAM

While developing a program first it is required to know, what is the task of the program? It is just similar to thinking of a problem. Best method is to write what program has to do and the step/order in which it has to do. Now let us develop few steps in generally *(called as algorithm).* As seen in Fig. 6.9 a flow chart will provide a graphic shape to represent different types of program operation. The steps those required to convert an algorithm to assembly language are :

.. Set up and declare data structure for the algorithm required to work with it.

.. Write down the instructions required for initialization at starting of the code section.

.. Mention the instructions required to implement the major actions taken in the algorithm, and decide how the data must be positioned for these instructions.

.. Insert the instructions required to fetch the data in correct position.

7.6.1. Instruction Format for Assembler

In Fig. 6.4 model of assembler is shown the general format of assembler instruction will be :

```
Label :    Opcode and Operand, Mnemonic Operand, Operand; Comments
```

Spaces between label, opcode, operands, mnemonics and comments are arbitrary, except that at least one space one space must be inserted if no space would lead to an ambiguity *(e.g., between the mnemonic and first operand).* There may be no spaces within a mnemonic or identifier and spaces within string constants or comments will be included as space characters. Each statement in the program fields. Now let us define the terms will have its own

```
.. Label,
.. Opcode and Operand,
.. Mnemonic,
.. Operand, and
.. Comments
```

. **Label :** It is an identifier which is assigned the address of the first byte of the instruction in which it appears. Presence of a label in an instruction is optional, but, if present, the label provides a symbolic name that can be used in branch instructions to branch to the instruction. When there is no label, then the colon must not be entered. All labels begin with a letter or one of following special character: @, $, _ or ? A label may be any length from 1 to 35 characters, and appears in a program to identify the name of a memory location for storing data and for other purposes.

..**Opcode and Operands :** Opcode field is designed to hold the instruction opcode. To the right of the opcode field is the operand field, which contains information used by opcode. Mnemonic. All instructions has to contain a mnemonic, which specifies the operation to be executed.

. **Operand :** Presence of, operands depends on the instruction. But some instructions have no operands, some have one operand, and some have two. When there are two operands, they are separated by a comma.

. **Comments** : The comment file is for commenting the program and may contain any combination of characters, which is optional and if it is deleted the semicolon may also be deleted. *(A comment may appear on a line by itself provided that the first character on the line is a semicolon).*

7.6.2. Program Format and Assembler Directives

Following example shows a typical assembler program which is constructed for 8086/8088 microprocessor.

Example :

```
Line 1. MODEL SMALL  ; Select small model
Line 3. Data      Indicates data segment
...........................................)............data deelaroew,
      link k. Code.; Indicates start of code segment
      )...... Progrom bady
   Line n End ; End of file
```

In the above example. MODEL directive selects a standard memory model for the assembly language program. A memory model can be thought of a standard blue print O configuration, which determines the way segments are linked together, where each memory model has a different set of restrictions as to the maximum space available for code an.:. data. Important thing to know about models is that they affect the way that subroutine and data be reached by programs. In Table 6.1 summary of program memory models are listed.

Program running under DOS is divided into 3 primary segments, each of which hold.., different types of values.

Code Segment *(pointed to by CS)* contains program code,

the data segment *(pointed to by DS)* contains the program variables, and

the stack segment *(pointed to by SS)* contains the program stack.

."DATA" directive (*line* 2) indicates the start of the data segment, contains program variables. "CODE" directive *(line k)* indicates the start of the code segment and end directive *(line n)* indicates end of the program file.

7.7. THE ASSEMBLER DIRECTIVES

The Pseudo instructions are directives entered into the source code along with the assembly language program. The directives do not get translated into object code but are used as special instructions to the assembler to perform some special functions. Assemblers can recognize directives that assign memory space, assign address to tables, format the pages of the source code and so on. Pseudo instruction *(directives)* are normally placed in the opcode field. If any label or data are required by the directives, they are placed in the label or operand field as necessary. Now let us briefly observe the usages of different directives.

(1) .The SEGMENT and ENDS Directive

These are used to identify a group of data items or a group of instructions that we want to put together in a particular segment. A group of data statements or a group of instruction statements combined between *"SEGMENT"* and *"ENDS"* directives is known as a logical segment. When we set up a logical segment, we give a name of our choice.

Example 1 :

```
DATA- WHERE SEGMENT
DATA WHERE ENDS
```

Such statement will set up a logical segment named DATA WHERE. Here the name has been chosen as DATA WHERE helps to remember that this logical segment contains data statements. The statements *"CODE-WHERE SEGMENT"* and *"CODE-WHERE ENDS"* set up a logical segment name CODE-WHERE which contains instruction statements.

Normally logical segment is not given a physical starting address when it is declared. After the program is assembled and perhaps linked with other assembled program modules, it is assigned the physical address then it is loaded in memory to run.

(2) .ASSUME Directive

Assume directive helps to tell the assembler the name of the logical segment it should use for fl specified segment.

ASSUME CS : CODE-WHERE tells the assembler that the instructions for a program are in a logical segment named CODE-WHERE. ASSUME DS : DATA-WHERE tells the assembler that for any program instruction which refers to the data segment it should use the logical segment called DATA-WHERE.

Example 2 : *MO VAX; {BSJ*

An assembler will know that the memory location referenced to by [BX] is in the logl '"," segment DATA-WHERE. The assembler has to be told what to assume for any segment::.: used in the program, *(assembler can be told the name of the logical segment used for,~~ stack and extra segments also).*

(3) .END

END directive is put after the last statement of a program to tell the assembler that it is the end of the program module, then assembler will ignore any statement after the END directive; hence the END directive should be used at the very end of the program module. *(A carriage return is required after the END directive).*

(4) .EQU

An EQU *(EQUATE)* directive will help to give a name to some value or symbol. Every time the assembler finds the given name in the program it will replace the name with the value or symbol that is equated with the name.

Example 3 :

FACTOR. When the assembler coo.es. This introduction "LabelD2!::." it will code it as 'ADD AL, 03H'.

Advantage of using EQU in this way is that if FACTOR is used several times.:: program, and the value has to be changed, all that has to be done is to change the ~.~ statement and reassemble the program. Then the assemblers will automatically put the =-. value each time it finds the name FACTOR. When a constant is used in the program ins:~ of the EQU approach, then the value of the constant has to be found and changed a: the occurrences.

(5) .Data Definition Directives

The Data definition directives are used to allocate storage, based on the following predefined types are as shown in Table 7.3 below.

TABLE 7.3. Pre-defined Definition Directives

S. No.	Directive	Description	Byte	Attributes
1.	DB	Define byte	1	Byte
2.	DW	Define word	2	Word
3.	DD	Define double word	4	Double word
4.	DF, DP	Define Far Pointer	6	Far pointer
5.	DQ	Define Quard word	8	Quard word
6.	DT	Define Ten bytes	10	Ten byte

(6) .Define Byte [DB]

DB directive is used tQ define a byte-type variable, or to set aside one or more storat locations of type in memory. The syntax diagram shows that "name" is optional, and onl one initial value will be needed. (*If more values are supplied, they must be separated J commas*).

```
[name] DB initial Value [, initial value] ....
```

The initial value can be one or more 8-bit numeric values, a string constant, a constant expression, or a question mark (?). If the value is signed, it has a range of-128 to +127; if unsigned, the range is 0 to 255.

TABLE 7.4

1.	Char	DB	'A'	; ASCII character
2.	Signed	DB	-128	; Smallest Signed Value
3.	Signed2	DB	+127	; Largest Signed Value
4.	Unsigned 1	DB	0	; Smallest Unsigned Value
5.	Unsigned 2	DB	255	; Largest Unsigned Value

(7) .Multiple Values

A list of 8-bit numbers can be grouped under a single label, they are specified by listing their values separated by commas.

Example 4 : *List DB 10, 20, 30, 40*

Let list be stored at offset 0000 : That mean 10 is stored at offset 0000, 20 at offset 0001, 30 at offset 0002, and 40 at offset 0003.

The Characters and integers are one and the same, as far as the assembler is concerned. Each constant can use a different radix. If a list of items is defined, numeric, character and string constants may be freely mixed. *(If a hexadecimal number begins with a letter (A-F), a leading zero is added to prevent the assembler from interpreting it as a label).*

TABLE 7.5

1.	Charl	DB	'N	; A ASCII character
2.	Char	DB	41h	; Hexadecimal
3.	Dec	DB	65	; Decimal
4.	Bin	DB	0100001B	; Binary
5.	Oct	DB	107q	; Octal
6.	List!	DB	10, 20, 41h,	0101010B, 'C', 0FFH

Variable content may be left undefined by using a question mark (?). Or a numeric expression can initialize a variable with a value that is calculated at assemble time.

Example 5 :

```
Count  DB?
Ages   DB  ?,?,?,?
Row size  DB  10*10
```

The string can be assigned to a variable in which case the variable stands for the address of just the first byte of the string. Following example shows one string terminated by a null byte *(containing 0),* and another string with is length encoded in the first byte.

Example 6 :

```
C_String DB            "Good Afternoon", O
Pasca Cstring DB 14,   "Good Afternoon"
```

DB directive is ideal for storing one or more lines of text, where single variable can be continued on multiple lines without the necessity of supplying a label for each. Following string is terminated by an end_oCline sequence and a null byte

```
A_long_string DB "This is a long string that"
                 DB "clearly is going to take"
                 DB "several lines." Odh, Oah, 0
```

Assembler can automatically calculate the length of a string, where the "$" character is a special character which indicates the address of the current location. Following example shows how the '$' character can be used to find the length of a string

Example 7 :

```
a_string. DB "This is a string" a_string_length DB S - a-string
a_string_length is initialized to 16
```

(8) .Define Word [DW]

The values specified using DW can be 16-bit numeric numbers ranging from 0 to 65,535 (FFFFH). If signed, values are declared the acceptable range is –32,768 (8000H) to +32,767 (7FFFH). A character constant can be stored in the lower half of a word, as observed in example 8.

Example 8 :

```
NUM DW 1, 2, 3 ; Define 3 words
    DW 0, 65535    ; Lower and Higher Unsigned Values
DW -32768, + 32767    ; Lower and Higher Signed Values
DW 256*2             ; Constant Expression
DW 1000h, 4096, 'AB', O  ; Mixed Notation
DW?                  ; Single Un-initialized Word
```

(9) .Pointers

The offset of a variable or subroutine may be pointer, as shown in example 9.

Example 9 : Stored in another variable, called a

```
List   DW     256, 257, 258; Define 3 words;
P         DW list P points to list
PI        DW P ; A pointer to a pointer
```

Note. Assembler sets P to the offset of list PI contains the address of P.

(10) .Define Double Word [DD]

DD directive creates storage for one or more 32-bit double words. The syntax of which will be :

```
[Name] DD initial value [,initial value] ......
```

The initial value can be a binary number up to 0FFFFFFFFH. A segment-offset Address a 4-byte encoded real number or a decimal real number.

```
AP      DD      12562926H
RAT         DD        0ABCD111H
```

Define Quad Word [DQ]

```
[name] DQ initial value [, initial value]
```

This directive will help the assembler to declare a variable 4 words in length or to reserve 4 words of storage in memory.

```
Big-numDQ        123456789ABCDEF0H
```

(11) .Define Ten-Byte [DT]

```
[name] DT initial value [, initial value]
```

It helps to tell the assembler to define a variable which is 10 bytes in length, or to reserve 10 bytes of storage memory.

```
Pac_bcd DT    1234567890
```

(12) .DUP Operator

DUP operator only appears after a storage allocation directive *(DB, DW, DD, DQ, DT).* With DUP, you may repeat one or more values when allocating storage. *(It is mainly useful when allocating space for a table or array).*

```
NUM DB  20 DUP [0] ; 20 bytes, all equal to zero
    DB  20 DUP I?] ; 20 bytes, uninitialized
    DB  4 DUP ['ABC']  ; 12 bytes"ABCABCABCABC"
```

DUP operator can also be nested.

```
ATABLE      DB     4DUP (3 DUP [0], 2 DUP ['X'])
```

The assembler assigns values in memory as shown below :

```
00, 00, 00, X, X;
00, 00, 00, X, X;
00. 00, 00, X, X;
00, 00, 00, X, X;
```

(13) .Type Checking

If a variable is created using DB, DW, DD or any of the other data definition directives, assembler gives it a default attribute (byte, word, double word) based on its size. Such type is automatically checked when the variable is referred and an error results if the types do not match.

Example 10 :

```
MOV AL, COUNT; Error: operand sizes must match
COUNT      DW20H
```

(14) .Even Directive [Align on Even Address]

Since assembler assembles a section of data declarations or instruction statements, it used a location counter to keep track of how many bytes it is from the start of a segment at all time. EVEN directive tells the assembler to increment the location counter to the next address if it is not already at even address. 8086 J.1P can read a word from memory in one bus cycle if the word is at an even address position. When the word starts at an odd address, 8086 has to use two read cycles to obtain bytes of the word. Hence a series of words can be much more quickly read if they are at an even address.

If EVEN is used in a data segment, location counter will be incremented to the next even address (if not at an even location) and if EVEN is used in a code segment, the location counter will be incremented to the next even address (if note already at an even location). A NOP instruction is inserted in the location incremented completion.

Example 11 :

```
DATA- WHERE SEGMENT
AVG DB 9 DUP (?): Declare array of 9 bytes
```

Location Counter (LC) will point to 009: after assembler reads increment LC to next even. Address array of 100 words starting on even address

```
EVEN
INV DW 100 DUP (0): DATA-WHERE ENDS
```

(15) .Group

GROUP directive helps to tell the assembler, to group logical segments named after the directive into one logical group segment, which the segments to be accessed from the same group segment base. Next the assembler sends a message to the linker telling it to link the segments so that the segments are physically in the same 64K bytes segment. SMAL_SYS GROUP CODE_WHERE< DATA_WHERE STACK_WHERE An appropriate ASSUME statement to follow this can be as below : ASSUME CS: SMAL_SYS, DS : SMAL_SYS, SS : SMAL_SYS

(16) .LABEL

LABEL directive will help to give a symbolic name to the current value of the LC (Location Counter). LABEL directive has to be followed by a term which is specifies by the type associated with that symbolic name. When the label is going to be used as the destination for a jump or a CALL statement then the LABEL must be specified as 'type near' or as (type far).

Example 12 :

```
ENTRY-POINT  LABEL FAR ; Can be able to jump here from another
segment
```

Note. When the label is going to be used as reference of a data item, then the label must be specified as type byte, word, or doubled word.

Example 13 :

```
STACK WHERE SEGMENT
DW 100 DUP (0)
STACK_TOP LABEL WORD
; Sets aside 100 words for the stack
; Give a symbolic name to next location; after the last word in the
stack
```

```
STACK_WHERE ENDS
```

The ORG *(Originate)* directive allows to set the location counter to any desired value at any point in the program. Since the assembler assembles a section of data declarations or instruction statements, it uses the LC to keep track of the address of current location from the start of the segment at any time. Then the LC is automatically set to 0000H when the assembler starts reading a segment.

Example 14 :

ORG 2000H; makes assembler to set the LC to 2000H

A $ is often used as symbolical representation for current value of the LC as such $ actually represents the next available byte location where the assembler can put a data or code bye. (The $ is often used in ORG statements to tell the assembler to change the LC value relative to its current value).

***Example* 15 :**

ORG $ + 100

Tells the assembler and makes to increment the value of the LC by 100 from its current value. (A statement such as this might be used in a data segment to leave 100 bytes of space for future use).

(17) .OFFSET

The OFFSET is an operator which instructs assembler to determine the offset or displacement of a named data item *(variable)* from the start of the segment which contains it. *(This operator is usually used to load the offset of a variable into a register so that the variable can be accessed with one of the indexed addressing modes).*

Example 16:

MOV BX, OFFSET PRICES

Assembler will determine the offset of the variable PRICES from the start of the segment in which PRICES is defined and code this displacement as part of the instruction. When the instruction executes, this computed displacement will be loaded into BX. (Any instruction can follow, which is base addressing mode).

(18) .PTR

PTR *(Pointer)* operator helps to assign a specific type to a variable or to a label. It is necessary to do this in any instruction where the type of the operand is not clear.

Example 17:

(a) INC [BX]

Assembler will not be knowing whether to increment, the byte pointed to by BX or increment the word pointed to by BX. Therefore such ambiguity can be avoided by making use of PTR directive.

```
INC BYTE PTR [BX] ; This statement tells the assembler to
      increment the byte
               ; pointed to by [BX]
INC WORD PTR [BX] ; Increment the word pointed to by [BX]
```

(b) The PTR operator can be used to override the declared type of a variable. An array of words is declared with the statement WORDS

```
WORDS DW 2345H, OB987H, 7C41H
```

Usually an access will be in terms of words. However, if it is required to access a byte in an array, it could be done by writing an instruction with the PTR operation

```
MOV AL, BYTE PTR WORDS
```

(*c*) While indirect jump instruction is made PTR may be used. JMP [BX] instruction does not tell the assembler whether to code the instruction for a near jump or for a far jump.

For a near jump the instruction must be written as below

```
JMP WORD PTR [BX]
```

When far jump is required the above instruction must be

```
JMP DWORD PTR [BX]
```

Different type of PTR which are used as listed in Table 7.6 below

TABLE 7.6. Types of PTR

S. No.	Operator	Operand Byte
1.	Byte ptr	Byte (8 bits)
2.	Word ptr	Word (16 bits)
3.	Dword ptr	Double word (32 bits)
4.	Qword ptr	Quad word (64 bits)
5.	'J' byte ptr	Ten byte (80 bits)

(19) .PROC

PROC *{Procedure}* directive helps to identify the start of a procedure. The procedure directive follows a name given for the procedure. After the PROC directive the term NEAR or the term FAR is used to specify the type of the procedure.

Example 18 :

```
WRITE-DEC PROC FAR
```

The procedure name is WRITE-DEC and informs the assembler that the procedure is far (*in segment with a different name from where the procedure is called).* The PROC directive is used with the ENDP directive to bracket a procedure.

(20) .ENDP

The ENDP (END PROCEDURE) directive helps along with the name of the procedure to indicate the end of a procedure to the assembler. This directive, along with the procedure directive PROC, is made used to "bracket" a procedure.

Example 19: *Square-root PROC NEAR*

Body of the procedure Square-root ENDP

(21) .EXTRN

EXTRN directive is 1, 1 sed to inform the assembler that the names or labels following the directive are in some other assembly module.

Example 20 :

```
EXTRN DIVISOR : WORD
```

When implicitly not evident one must specify whether the label is near or far.

Example 21 :

```
EXTRN DIVIOR : FAR
```

This statement informs the assembler that DIVISOR is a label of type far in another assembly module. *(In the first example when DIVISOR is declared as type WORD, type NEAR is implicit).*

Now to call a procedure which is in a program module assembled at a different time from that which contains the CALL instruction, the assembler has to be inform that the procedure is external. Then the assembler will put information in the object code file so that the linker can connect the two modules together appropriately.

Note. The names or labels referred to as external in one module must be declared public with the PUBLIC directive in the module where they are defined. The EXTRN statements will SEGMENTENDS directive which identify the segment in which the external name or label will be found.

Example 22 :

```
PROCEDURE-    WHERE SEGMENT
EXTRN SQURT: FAR
EXTRM SUM: NEAR PROCEDURE-
WHERE ENDS
```

(22) .PUBLIC

LARGE programs are normally written as several separate modules where each modules is individually assembled, tested and debugged. If all modules are working correctly, their object code files are linked together to form a complete program. For the modules to link together correctly, any variable name or label referred to by any other module when the variable is not defined, must be declared PUBLIC in the module where it is defined. PUBLIC directive is used to inform the assembler that a specified name or label will be accessed from other modules.

Example 23:

```
PUBLIC MULTIPLICAND, MULTIPLIER.
PUBLIC SQRT, SUM
```

Such statement makes two variables MULTIPLICAND and MULTIPLIER available to other assembly modules. When an instruction in a module refers to a variable or label in another assembly module, the assembler must be told that it is external with the EXTRN directive.

(23) .ALIGN directive

General format is : ALIGN number, *(where number is 2, 4, 8 or 16)*

'ALIGN 16' directive forces assembler to align the next segment at an address that is divisible by 16. The assembler fills the unused bytes with 0 for data, and NOP instructions for code.

Note. DOS will load always the CODE segment first, and only then loads the DATA sigma::,: immaterial of the order in which they are defined in our program.

ALIGN 16 is used at the end of the CODE segment so as to force the beginning of DATA segment at an address divisible by 16. Thus the offset address of N1 will now be 0 in the DATA segment, which is convenient for viewing results using DEBUG utility.

ALIGN 2 directive may be used to start the DATA segment on a word boundary when the processor is 8086, which speeds up accessing of words.

ALIGN 4 directive can be used to start the DATA segment on a double word (32 bit boundary when the processor is 80386, or a later processor, which again speeds up assessing of double words. *(For 8088 processor, ALIGN is not of much advantage as it can only access* a *byte at a time).*

7.7.1. Summary of the Assembler Directives

Now to summarize the assembler directives it is formatted in the form of table as below in Table 7.7.

TABLE 7.7. Assembler Directives

S. No.	Directive	Action of the Directive
1.	ALIGN	aligns next variable or instruction to byte which is multiple of operand
2.	ASSUME	selects segment register(s) to be the default for all symbol in segment(s)
3.	COMMENT	indicates a comment
4.	DB	allocates and 'optionally initializes bytes of storage
5.	DW	allocates and optionally initializes words of storage
6.	DD	allocates and optionally initializes doublewords of storage
7.	DQ	allocates and optionally initializes quadwords of storage
8.	DT	allocates and optionally initializes 100 byte long storage units
9.	END	terminates assembly; optionally indicates program entry point
10.	ENDM	terminates a macro definition
11.	ENDP	marks end of procedure definition
12.	ENDS	marks end of segment or structure
13.	EQU	assigns expression to name
14.	EVEN	aligns next variable or instruction to even byte
15.	EXITM	terminates macro expansion
16.	EXTRN	indicates externally defined symbols
17.	LABEL	creates a new label with specified type and current location counter
18.	LOCAL	declares local variables in macro definition
19.	MACRO	starts macro definition
20.	MODEL	specifies mode for assembling the program
21.	ORG	sets location counter to argument
22.	PAGE	sets length and width of program listing; generates page break
23.	PROC	starts procedure definition
24.	PTR	assigns a specific type to a variable or to a level
25.	PUBLIC	identifies symbols to be visible outside module
26.	TITLE	defames the program listing title

7.8. VARIABLES SUFFIX AND OPERATORS

Now let us briefly define

(i) Variables

(ii) Suffix, and

(iii) Operator used is 8086 microprocessor.

(i) **Variable :** A variable is an identifier which is associated with the first byte of data item. In assembly language statement: COUNT DB 20H, COUNT is a variable.

Example 1 : *Array DB 10, 20, 30, 40, 50*

Here, array is the variable which is associated with the first byte of the data Item, i.e., 10.

(ii) Suffix : In assembly language programming base of the number of indicated by a suffix as mentioned below :

```
.B-Binary
.D-Decimal
.O-octal
.H-Hexadecimal
```

The default will be decimal. The first digit in a hexadecimal number must be 0 through 9; that means, if the most significant digit is a letter (A;.F), then it must be prefixed with a 0.

Example 2:

```
1010 B= 10102
967 D = 2967 = 296710
3F2AH=2F2A16
OB129H=B12916
```

(iii) **Operators :** The Arithmetic operators are : "+", "–", "*", and *"r"*.

The Logical operators : "AND", "OR", "NOT", and "XOR".

(The Logical operators are specially used for binary operands).

7.9. ASSEMBLING THE PROGRAM IN MASM

ADDSUB.ASM is created using an editor is just a text file which cannot be directly executed. We have to first assemble, and then link it, before executing. The assembly step involves translating the assembly language program *(called source code)* into machine code *(called object code),* and generating a. OBJ file, before execution.

Note. The examples are mentioned for Microsoft assembler .MASM 5.1 version.

The type 'MASM ADDSUB; <cr>' to perform translation of the source code in ADDSUB.ASM to the object code. The result of which will be placed in ADDSUB. OBJ file. '<cr>' stands for 'carriage return" which is done by typing the key on which 'Enter' is marked.

The same effect is obtained by typing 'MASM ADDSUB.ASM' '<cr>'.

In case of errors were encountered during the assembly process, it is suitable to generate a LIST file, which contains the source code; object code, Wong with details of errors encountered, if any, present.

To generate list file during assembly, we can type the following codes

```
'MASMIL ADDSUB; <cr>'
```

This results in generation of ADDSUB.OBJ as well as ADDSUB.LST files. Next using a text editor we can see the contents of ADDSUB.LST and locate the errors in the source file ADDSUB.ASM and rectify them.

Information regarding various options is possible during the assembly process can be. obtained by typing 'MASM/H <cr>'.

NOTE: In the assembly process, or later on, during the linking operation, we can come across warning messages and/or severe error messages. When we come across warning messages, we can ignore them if they are not serious in nature. However severe errors have to be always corrected.

7.9.1. Linking of the Program

Next step is to link the object module ADDSUB.OBJ, which contains the machine code. Now the linker converts the .OBJ file to .EXE file. The .EXE file is the one which can be executed. Linker is also used for combining more than one separately assembled modules into one executable program.

Type 'LINK ADSDSUB ; <cr>' to perform conversion of the object code in ADD SUB. OBJ to the executable code. The result will be placed in ADDSUB.EXE file.

The same effect can also be achieved by typing "LINK ADDSUB.OBJ; '<cr>'.

NOTE: In addition to .EXE file, a .COM file is also an executable file on the IBM.PC.

7.9.2. Testing of the Program

To test the program normally we use DOS DEBUG utility which enables the programmer to trace the execution of a program. The programmer may step through the program one instruction at a time, or step through several instructions at a time, or even run the entire program in one operation, also.

Type 'DEBUG ADDSUB.EXE <cr>'. The DEBUG prompt '.' appears on the CRT. The information regarding the various DEBUG commands can be obtB:ined by typing '? <cr>' at the '.' prompt of DEBUG.

7.9.3. Running the Entire Program in a Single Operation

At the '.' prompt, type 'G <cr>' then the entire program Upton, but not including the INT 3 break point instruction is executed.

To test the result type 'D DS"OL 20 <cr>'. This commands stands for 'Display memory contents starting from offset value 0 in the DS segment. The length or number of bytes to be displayed is 20H'. Whereas DEBUG always uses only hexadecimal values.

If we wish to display 20H i.e., 32 bytes of information then the first 8 bytes gives value of NUMBER 1, then the next 8 bytes giving value of NUMBER 2, the next 8 bytes giving value of their SUM, and finally last 8 bytes gives value of their difference.

Same can also be achieved by typing any of the following commands

D DS : 0, IF <cr>

D DS : 0, IF <cr>

D DS : 000, 00lf <cr>

D xxxx : 0, If <cr>, where xxxx is contents of DS as displayed by

DEBUG on the CRT.

Here'-, 0, 1£ stands for memory locations from offset value 0 to lf, a total of 20H or 32 bytes.

We exit DEBUG and return to DOS by typing 'Q <cr>' at the '.' prompt of DEBUG.

NOTE: In all these DEBUG commands we can use lower case characters also.

7.9.4. Running the Program in Single Step Operation

Normally we resort to single step operation when we find that the program has not yielded desired results after executing it in a single burst. In other words we use it for *'debugging'* the program, and hence the name DEBUG for this utility.

With the '.' prompt, type T <cr>'. Only first instruction of the program is executed, the value of various registers as well as the next instruction and its address will be displayed.

When the contents of any of the registers is not as per our calculations, we rectify our program logic, which is done by quitting DEBUG by typing 'Q <cr>' at the '.' prompt and then modifying program logic using a text editor.

For the register contents are as per our calculations, we go ahead and execute the next instruction by typing T <cr>' at the '.' prompt.

This process is repeated till the 'bugs' are identified and rectified in the program. We exit DEBUG and return to DOS by typing 'Q <cr>' at the '.' prompt of DEBUG.

7.9.5. Stepping Through the Program Several Instructions at a Time

This method is adopted when the program is quite large, and hence single stepping through the program will be tedious.

When we want to step through 5 instructions at a time, we type T5 <cr>' at the '.' prompt. Then the value of the various registers at the end of these 5 instructions will be displayed. The instruction to be executed next, as well as its address is also displayed.

When the contents of any of the registers is not as per the calculations, then we rectify our program logic. Which is done by quitting DEBUG by typing 'Q <cr>' at the '.' prompt and then modifying the program logic using a text editor.

Next if the register contents are as per our calculations; we can go ahead and execute several more instructions by typing Tn <cr>' at the '.' prompt, where n is any integer value of our choice.

This process is repeated till the 'bugs' are identified and rectified in the program. We exit DEBUG and return to DOS by typing 'Q <cr>' at the '.' prompt of DEBUG.

7.9.6. The Structure of an Assembly Language Program: HELLOSTU

When learning any new language, often the first program one writes is one that prints "Hello students" on the screen. We will now write, assemble, link, and execute an ASM "hellostudents" program. Fig. 7.6 shows the little program. The numbers in the left-hand column are to be used to describe the program.

Important note : The example below shows printed output being handled by a DOS function (#2).

```
1  TITLE Hello Students
2  ; FILE: h1lstu.asm
   ;
   ; This program prints "Hello students" and
   ; "See you around!"
   .MODEL SMALL

3  .DATA
   msg1     db      'Hello students',10,13,'$'
   msg2     db      'See you around!',10,13,'$'

4  .CODE
   main     PROC
            .STARTUP
5           ; DOS function # 9 prints string terminted by 'S'
            mov ah, 9
            mov dx, offset msg1 ; address of msg1 in dx
6           int 21h ; call DOS

            ; print msg2:
            mov ah, 9
7         mov dx, offset msg2
          int 21h

            mov ax, 4c00h    ; set up for exit
            int 21h          ; call DOS to exit
8  main     ENDP
   END
```

Fig. 7.6. Hello students Program.

Explanation :

Line 1 indicates that a title can be declared using the TITLE directive. In my programs I often capitalize assembler directives, but the assembler is entirely case-insensitive, so consistency about capitalization is not needed.

Line 2 : Comments are preceded by a semicolon and occur anywhere on a line.

Line 3 : The .MODEL SMALL directive indicates that data and code segments will each occupy no more than 64k of memory.

Line 4 : The .DATA segment indicates where a map of the data is stored. Symbolic names can be used to identify a sequence of data, but please remember that the names themselves are not stored in memory. Symbolic names are "variable names" in ASM. Later we will learn various means of "addressing" memory locations.

The actual data is just a series of, in this case, bytes (using the db directive). The data is in binary in the computer, but character data can be used in an assembler program. The 10 and 13 are codes for newline and carriage return, respectively. The '$' terminator is required by a particular DOS function we will be using.

Although two lines are identified (msg1, msg2), the actual data is just a continuous sequence of bytes in binary. This is the entire sequence in decimal:

72,101,108,108,111,32,119,111,114,100,10,13,36,83,101,101,32,121,11,117, 32,97,114,111,117,110,100,33,10,13,36

The above are just ASCII codes for the characters in the two strings. You can recognize some landmarks: 72 = 'H', 32 = space, 36='$', and 33='!'. We could have just use the sequence of numbers to get the same effect.

Line 5 : This is where the CPU instructions begin, in, of course, symbolic form.

Line 6 : Here are three lines of code that will cause the first string to be printed:

```
mov ah,9 ; the code for DOS function 9 is in ah
mov dx, offset msg1 ; address in dx
int 21h; run the function
```

INT 21h (h is for hexadecimal; 21h = 33 decimal, so INT 33 is also ok, but not when using the DEBUG program) calls the set of DOS "services." DOS looks in ah to find out what function to run ("DOS function number nine" – say, wasn't that a popular song from the late fifties? oops, it was "Love potion number nine"). We must pass the msg string to DOS, and we do it the same way we did in C, by passing the address of the string. Offset msg1 is the offset address of msg1.

Line 7 : In a similar way the second msg is printed.

Line 8 : The mov and int here are used to gracefully exit the program. If you are assembling with the new MASM or with TASM 5 or above you can replace these two lines with this directive:

```
.EXIT
```

Steps in Editing, Assembling, Linking, Executing

A. Edit

Open the DOS window (Command Prompt).

Use the EDIT editor to enter the program above:

>edit hllstu.asm

You may save the program in your folder, the name of this folder is your userid (You can also create subfolders). You can also save it on the drive.

B. Assemble, Link, and Execute Using Microsoft's MASM

You can download a copy of this. Details follow at the end. Our labs use Borland's Turbo Assemble (TASM). Here are steps for MASM (taken directly from my home machine). Input is shown in bold.

```
C:\fall2000asm>ml hllstu.asm
Microsoft (R) Macro Assembler Version 6.11d
Copyright (C) Microsoft Corp 1981-1995.  All rights reserved.

Assembling: hllstu.asm
Microsoft (R) Segmented Executable Linker Version 5.60.339 Dec 5 1994
Copyright (C) Microsoft Corp 1984-1993. All rights reserved.
LINK : warning L4017: /r : unrecognized option name; option ignored
Object Modules [.obj]: /r hllwld.obj
Run File [hllstu.exe]: "hllstu.exe"
List File [nul.map]: NUL
Libraries [.lib]:
Definitions File [nul.def]:
LINK : warning L4021: no stack segment

C:\fall2000asm>hllstu
Hello students
See you around!
```

The first LINK warning is a bug that can be ignored. The second refers to a true situation (no stack specified) which is of no concern (yet).

C. Assembling, Linking, Executing with Borland's Turbo Assembler (TASM 5.0)

To enter the editor window for tasm we must type the following command say the TASM lies in D directory hence we must first enter the directory and then the enter the TASM folder and type edit as shown in Fig. 7.7.

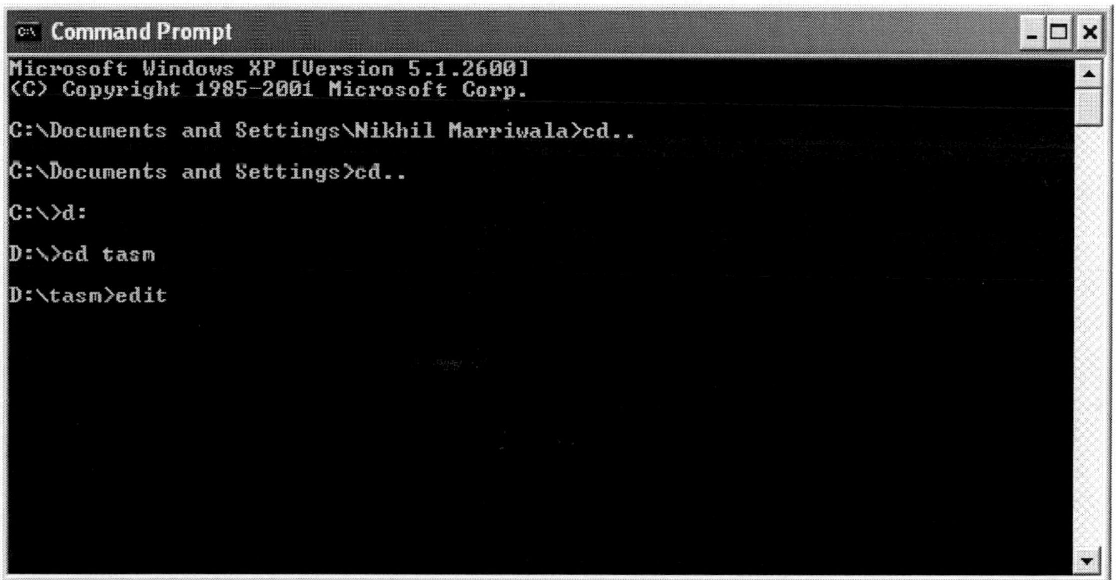

Fig. 7.7.

D:\tasm>edit

When we press enter we enter the edit window page where we will write the program that we want to execute as shown and we must then save the program by giving it some name with an extension .asm.

For example here in the example shown we have saved the program by the name 'hellstu.asm'.

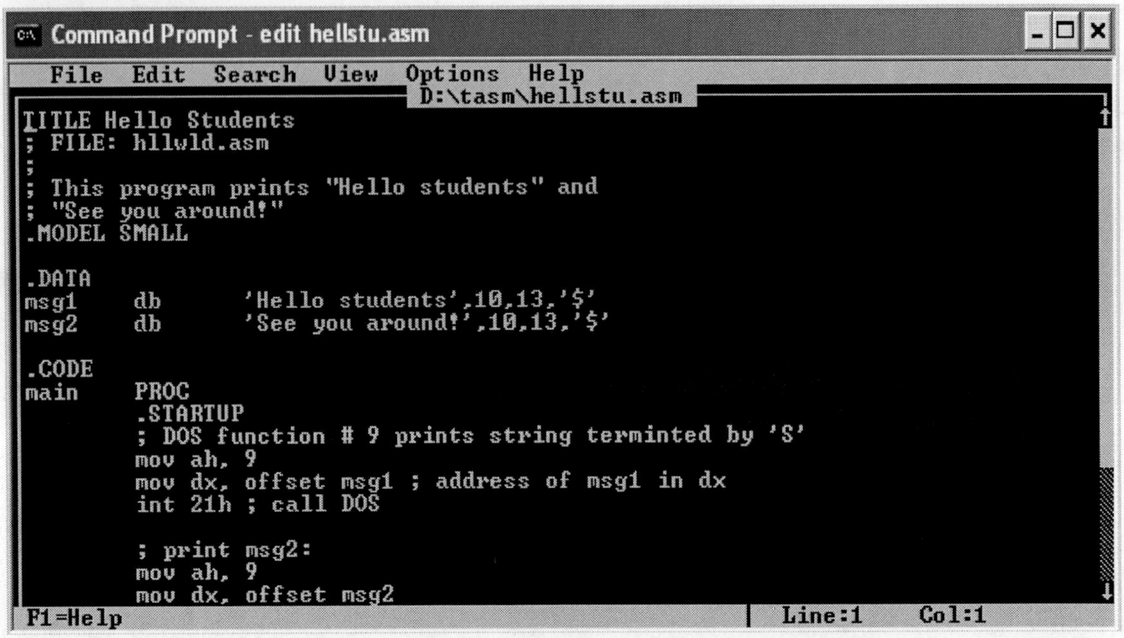

Fig. 7.8.

Next step is to then execute the program that we have written in the edit window by the following procedure as shown in Fig. 7.9

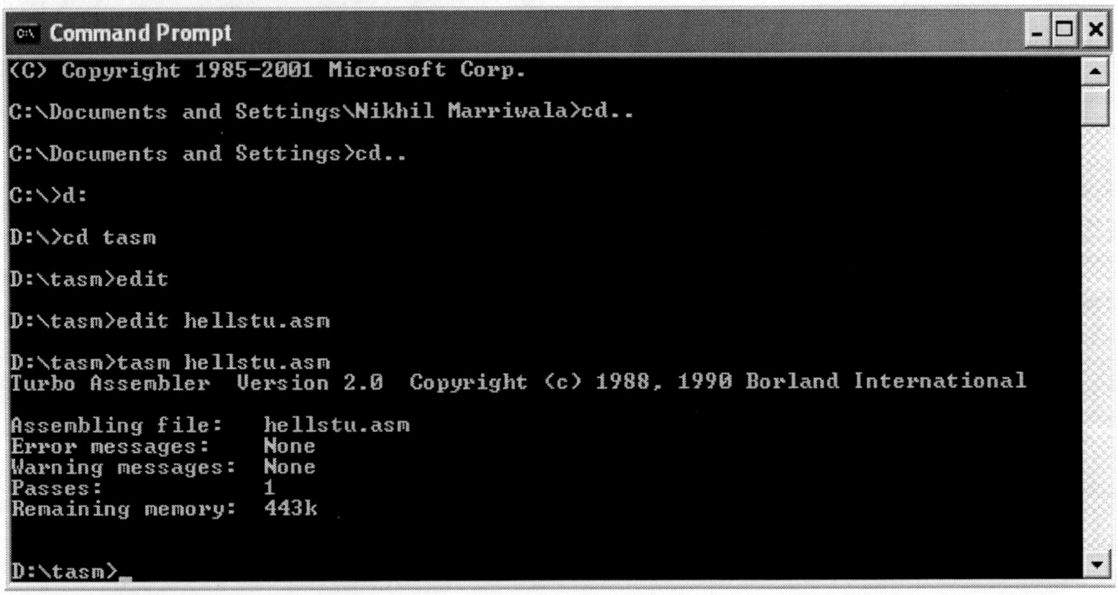

Fig. 7.9.

```
D:\tasm>tasm hellstu
Turbo Assembler Version 2.0 Copyright(c) 1988, 1990 Borland
International
```

The next step is to then perform the following action by writing the commands as written below :

```
D:\tasm>tlink hellstu
Turbo Link  Version 3.0 Copyright (c) 1987, 1990 Borland International
Warning: No stack
```

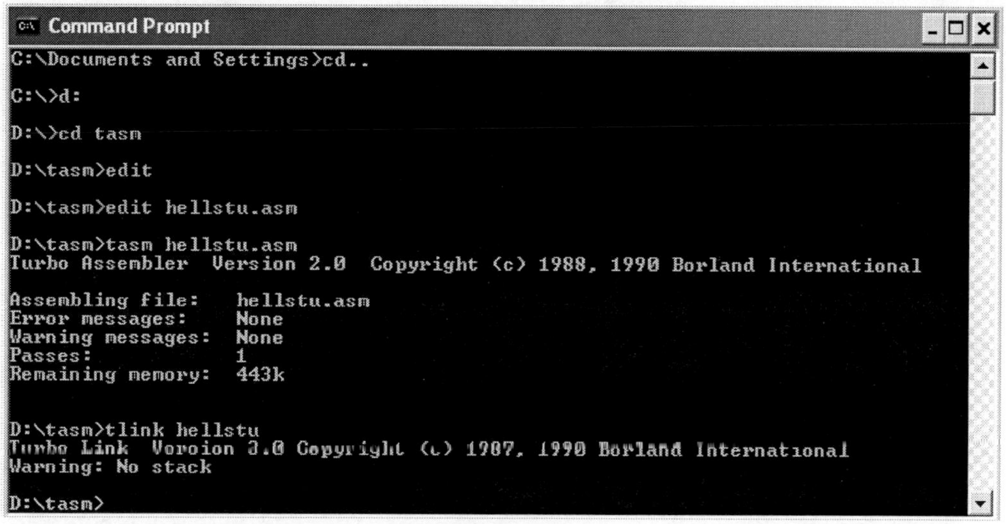

Fig. 7.10.

There are two methods to check for the result.

The first method is that we type the command

```
D:\tasm>td hellstu
```

As shown in the figure below and we will enter the debugger for the TASM.

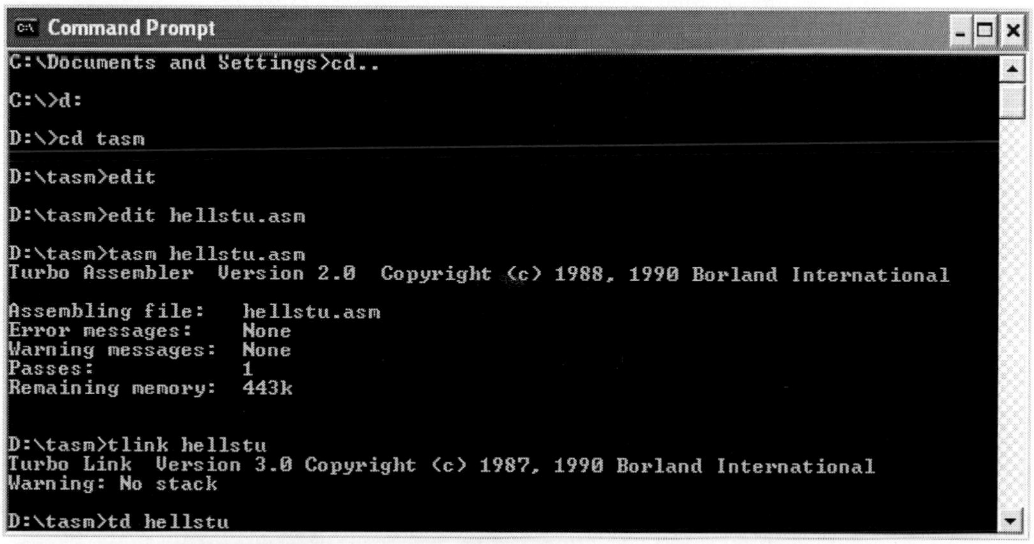

Fig. 7.11.

The debugger looks like the one that we have shown below.

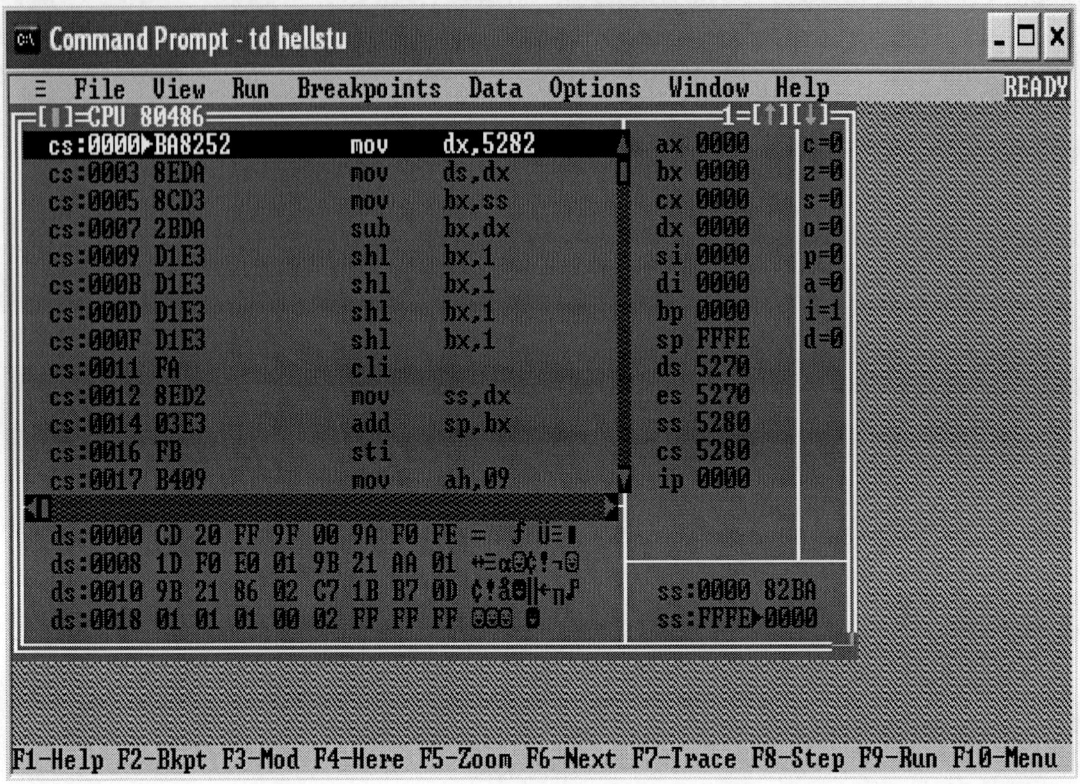

Fig. 7.12.

Here we can see the different registers of the 8086 and the Code segment contents representing the corresponding Opcodes for the various instructions accompanied with the program that we had written.

The debugger shell also shows us the status of the flag registers.

The process to execute the program is also done in two ways :

1. User can use F7 key on the keyboard and can execute the program in Single Step.
2. Or the user has another option that one may press F9 key and the complete program will be executed in one go.

The second method to see the result is that we simply type the following instruction hellostu ← will execute the program and show the result on DOS shell.

```
D:\tasm>hellstu
```

And we can can see the result displayed on the screen as shown in Fig. 7.13.

```
Hello students
See you around!
```

NOTE: To display the result on the screen i.e on the DOS shell we have to use some special function commands given with 8086 microprocessor with INT 21h as the interrupt command combined with some instructions such as mov ah, 02h etc. or mov ah, 09h.

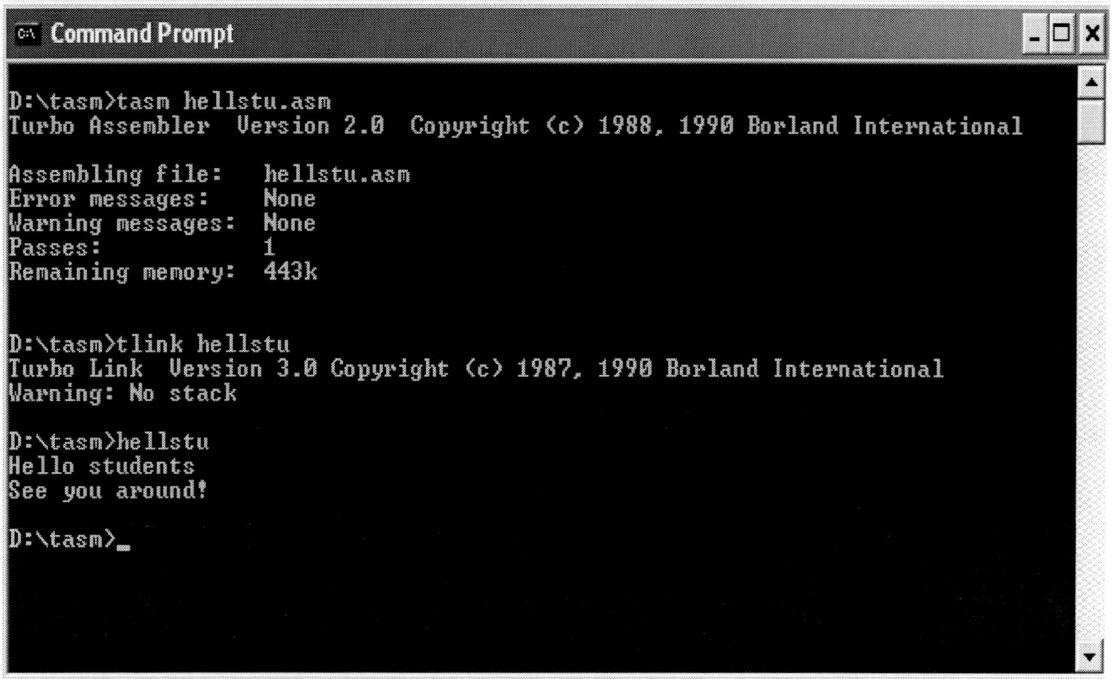

Fig. 7.13.

7.10. BASICS OF SIMPLE PROGRAMMING

The majority of people think that computers are some kind of complicated device that is impossible to learn and infinitely intelligent, able to *think* better than a person. The truth is much less glamorous.

A computer can only do what the programmer has *told* it to do, in the form of a *program*. A program is just a sequence of very simple commands that lead the computer to solve some problem. Once the program is written and debugged (you hardly ever get it right the first time), the computer can execute the instructions very fast, and always do it the same, every time, without a mistake.

And herein lies the power of a computer. Even though the program consists of very simple instructions, the overall result can be very impressive, due mostly to the speed at which the computer can process the instructions. Even though each step in the program is very simple, the sequence of instructions, executing at *millions* of steps per second, can appear to be very complicated, when taken as a whole. The *trick* is not to think of it as a whole, but as a *series of very simple steps, or commands.*

The microprocessor itself is usually a single integrated circuit (IC). Most microprocessors, or very small computers, (here after referred to simply as *micro's*) have much the same commands or instructions that they can perform. They vary mostly in the names used to describe each command. In a typical micro, there are commands to move data around, do simple math (add, subtract, multiply, and divide), bring data into the micro from the outside world, and send data out of the micro to the outside world. Sounds too simple....right?

A typical micro has three basic parts inside. They are the Program Counter (PC), Memory, and Input/Output (I/O). The Program Counter keeps track of which command is to be executed. The Memory contains the commands to be executed. The Input / Output handles the transfer of data to and from the outside world (outside the micro's physical package). The micro we'll be using is housed inside a 40 pin package, or *chip or IC*. There are many other actual parts inside our micro however, we will learn about each and every single one, one step at a time.

A Simple Program

As stated before, a program is a sequence or series of very simple commands or instructions. A real world example program might be the problem of crossing a busy street.

Step 1: Walk up to the traffic lights and stop.

Step 2: Look at the traffic light.

Step 3: Is your light green?

Step 4: If the light is red, goto step 2. (otherwise continue to step 5)

Step 5: Look to the left.

Step 6: Are there cars still passing by?

Step 7: If yes, goto step 5. (otherwise continue to step 8)

Step 8: Look to the right.

Step 9: Are there cars still passing by? (there shouldn't be any by now, but, you never know!)

Step 10: If yes, goto step 8. (otherwise continue to step 11)

Step 11: Proceed across the street, carefully!!

Now this may seem childish at first glance, but this is exactly what you do every time you cross a busy street, that has a traffic light (at least, I hope you do). This is also exactly how you would tell a micro to cross the street, if one could. This is what I mean by a sequence or series of very simple steps. Taken as a whole, the steps lead you across a busy intersection, which, if a computer did it, would seem very intelligent. It is intelligence, people are intelligent. A programmer that programmed these steps into a micro, would impart that intelligence to the micro.

The micro would not, however, in this case, know what to do when it got to the other side, since we didn't tell it. A person, on the other hand, could decide what to do next, at a moments notice, without any apparent programming. In the case of a person, though, there has been some programming, it's called past experiences.

Another program might be to fill a glass with water from a tap.

Step 1: Turn on the water.

Step 2: Put the glass under the tap.

Step 3: Look at the glass.

Step 4: Is it full?

Step 5: If no, goto step 3. (otherwise, continue to step 6)

Step 6: Remove the glass from under the tap

Step 7: Turn off the water.

This is a simpler program, with fewer steps, but it solves a problem, to fill a glass with water. In a micro, the problems are different but the logical steps to solve the problem are similar, that is, a series of very simple steps, leading to the solution of a larger problem.

Also notice that since the steps are numbered, 1 through 7, that is the order in which they're executed. The Program Counter, in this case, is you, reading each line, starting with 1 and ending with 7, doing what each one says. In a micro, the Program Counter automatically advances to the next step, after doing what the current step says, unless a *branch,* or *jump*, is encountered. A branch is an instruction that directs the Program Counter to go to a specific step, other than the next in the sequence. The branch in this example is step 5. Not only is this a branch, but it is a *conditional* branch. In other words, based on whether the glass is full or not, the branch is taken, or not. A micro has both branch and conditional branch instructions. Without this ability to reuse instructions, in a sort of looping action, a solution would take many more steps, if it would be possible at all.

The point of this lesson is to show how a simple set of instructions can solve a bigger problem. Taken as a whole, the solution could appear to be more complicated than any of the separate steps it took to solve it.

The most difficult problem to be solved in programming a micro is to define the problem you are trying to solve. Sounds silly but I assure you, it's not. This is the *Logical Thought Process* I mentioned earlier. It is having a good understanding of the problem you're trying to solve. You must understand the information I'm presenting in order pass the course. Trying to remember everything does not work at university.

Some Machine Language Programs

Problem 1. Write a program to add a data byte located at the offset 0400h in 3000h segment to another data byte available at 0700h in the same segment and then store the result at 0900h in the same segment?

Solution.

```
MOV AX, 3000h ;    initializing data segment with the value
MOV DS, AX        ;     3000h
MOV AX, [0400h]   ; Get the first data byte from the offset 0400h
ADD AX, [0700H]   ; Add the first data byte to the second from offset 0700h
MOV [0900H], AX   ; Store the result of addition in accumulator register
HLT             ; Halt the operation
```

In the above problem we have initialized the data segment with value in AX register because we cannot load the immediate data into a segment register so we have used an indirect method of loading the value 3000h in DS. We will have to repeat the same procedure each time we start the program or we move into different segments then we have to initialize the DS with the value given.

Problem 2. Write a program to move the contents of the memory location 0400h to register BX and also to CX. Add immediate byte 07h to the data residing in memory location, whose address is computed using DS= 3000h and offset= 0700h. Store the result of the addition in 0900h.

Solution.

```
MOV AX, 3000h   ; Initialize the Ds with 3000h
MOV DS, AX
MOV BX, [0400H] ; Get the contents of 0400h in BX
MOV CX, BX      ; Copy the same contents in CX
ADD [0700H], 07H   ; Add byte 07h  to the contents of 0700h
MOV DX, [0900H] ; Store the result in DX
MOV [0900H], DX ; Store the result in 0900h
HLT             ; Halt
```

Problem 3. Write a program to add the contents of the memory location 3000: 0400h to the contents of 4000: 0700h and store the result in 6000: 0900h?

Solution. In this problem we will have to refer to two different memory locations in different segments, hence we will have to initialize the data segment twice here with the given values.

```
MOV AX, 3000h     ; Initialize the Ds with 3000h
MOV DS, AX
MOV AX, [0400H]   ;Get the contents of 0400h in AX
MOV CX, 4000H
MOV DS, CX        ; Initialize the Ds with 4000h
MOV BX, [0700H]   ; Get the contents of 0700h in AX
ADD AX, BX        ; Perform Addition
MOV CX, 6000H     ; Initialize the Ds with 6000h
MOV DS, CX
MOV [0900H], AX   ; store the result in 0900h
HLT               ; Halt
```

Some Programs Written for TASM Software

Program 1

```
;TITLE:        ADD TWO NUMBERS
;OBJECTIVE:    THIS PROGRAM TAKES ANY TWO NUMBERS AS INPUT FROM THE
               USER THROUGH
;              THE INPUT DEVICE(KEYBOARD) & PRINTS THEIR SUM ON THE
               STANDARD
;              OUTPUT DEVICE (SCREEN)
;AUTHOR:       Nikhil Marriwala
.MODEL SMALL
.STACK
.DATA
.CODE
```

```
START: MOV AH,01        ;input first number
       INT 21H          ;
       MOV BL,AL        ;copying the contents of al i.e. number
                        ;inputted from keyboard to bl
       AND BL,0FH       ;converting ascii code of No 1 to hexa code
       MOV AH,01        ;input second number
       INT 21H          ;
       AND AL,0FH       ;converting ascii code of No 2 to hexa
       ADD AL,BL        ;adding the two numbers
       MOV DL,AL        ;copying the sum to register dl
       OR DL,30H        ;converting hexa sum to its ascii equivalent
       MOV AH,02        ;print the sum on the screen
                        ;
       INT 21H          ;
       MOV AH,4CH       ;exit to dos prompt
       INT 21H
       END              ;end of program
```

Program 2

```
;TITLE:       ADD TWO NUMBERS
;OBJECTIVE:   THIS PROGRAM TAKES ANY TWO NUMBERS AS INPUT FROM THE
              USER THROUGH
;             THE INPUT DEVICE(KEYBOARD) ON TWO DIFFERENT LINES AND
              PRINTS THEIR
;             SUM ON THE STANDARD OUTPUT DEVICE (SCREEN) ON A
              SEPARATE LINE
;AUTHOR:  Nikhil Marriwala
.MODEL SMALL
.STACK
.DATA
.CODE
START: MOV AH,01        ;input first number
       INT 21H          ;
       MOV BL,AL        ;copying the contents of al i.e. number
                        ;inputted from keyboard to bl
       AND BL,0FH       ;converting ascii code of No 1 to hexa code
       MOV DH,0DH       ;load dl register with carraige return code
       MOV DL,0AH       ;load dl register with line feed code
       MOV AH,02        ;to position the cursor on the next line
```

```
        INT 21H
        MOV AH,01           ;input second number
        INT 21H             ;
        AND AL,0FH          ;converting ascii code of No 2 to hexa
        ADD BL,AL           ;add the 2 no.s & store the sum in bl register
        MOV DL,0DH          ;load dl register with carraige return code
        MOV DL,0AH          ;load dl register with line feed code
        MOV AH,02           ;to position the cursor on the next line
        INT 21H             ;
        MOV DL,BL           ;copying the sum to register dl
        OR DL,30H           ;converting hexa sum to its ascii equivalent
        MOV AH,02           ;print the sum on the monitor
        INT 21H
        MOV AH,4CH          ;exit to dos prompt
        INT 21H
        END                 ;end of program
```

Program 3

```
;TITLE:       ADD TWO NUMBERS
;OBJECTIVE:   THIS PROGRAM TAKES ANY TWO NUMBERS AS INPUT FROM THE
;USER THROUGH THE INPUT DEVICE(KEYBOARD) AND PRINTS THEIR SUM ON THE
;STANDARD OUTPUT DEVICE (SCREEN) BY GIVING APPROPRIATE MESSAGES TO
;THE USER
;AUTHOR:      Nikhil Marriwala
.MODEL SMALL
.STACK
.DATA
MSG1 DB 'ENTER FIRST NO:$'
MSG2 DB 10,13,'ENTER SECOND NO:$'
MSG3 DB 10,13,'THE SUM IS:$'
.CODE
START:MOV AX,@DATA      ;initializing the data segment
      MOV DS,AX         ;"              "          "
      LEA DX,MSG1       ;prompts the user to enter the first number
      MOV AH,09         ;by displaying appropriate message
      INT 21H           ;
      MOV AH,01         ;input first number
      INT 21H           ;
```

```
        MOV BL,AL              ;copy contents of al (i.e no.1) to bl register
        AND BL,0FH             ;convert ascii code of No. 1 to hexa code
        LEA DX,MSG2            ;prompts the user to enter the second number
        MOV AH,09              ;by displaying appropriate message
        INT 21H                ;
        MOV AH,01              ;input second number
        INT 21H                ;
        AND AL,0FH             ;convert ascii code of No.2 to hexa
        ADD BL,AL              ;add 2 No.'s & store the sum in register bl
        LEA DX,MSG3            ;prints a message telling the user that the
                               ;sum is being outputed on the screen
        MOV AH,09              ;
        INT 21H                ;
        MOV DL,BL              ;copying the sum to register dl
        OR DL,30H              ;convert the hexa sum to its ascii equivalent
        MOV AH,02              ;prints the sum on the screen
        INT 21H                ;
        MOV AH,4CH             ;exit to dos prompt
        INT 21H
        END                    ;end of program
```

Program 4

```
   ;TITLE:       INITIALIZING AN ARRAY
   ;OBJECTIVE: THIS PROGRAM INITIALIZES 100 POSITIONS IN AN ARRAY AND
LOADS THEM WITH 0
   ;AUTHOR:      Nikhil Marriwala
   .MODEL SMALL
   .STACK
   .DATA
   .CODE
   START:MOV CX,64H            ;load the count of 100
         XOR AL,AL             ;load 00 in register al
   AGAIN:MOV [BX],AL           ;load 00 in the memory location whose offset
                               ;is provided in bx
         INC BX                ;increment bx to point to offset of next loc.
         LOOP AGAIN            ;repeat the loop 100 times
         MOV AH,4CH            ;exit to dos prompt
         INT 21H
         END                   ;end of program
```

Program 5

```
;TITLE:        ASCENDING ORDER
;OBJECTIVE:    THIS PROGRAM SORTS THE ELEMENTS OF A GIVEN ARRAY IN
               ASCENDING ORDER
;AUTHOR:       Nikhil Marriwala
.MODEL SMALL
.STACK
.DATA
ARRAY db 9,3,1,2,0
MSG1 DB 'BEFORE SORTING$'
MSG2 DB 10,13,'ARRAY:$'
MSG3 DB 10,13,'AFTER SORTING IN ASCENDING ORDER: $'
.CODE
START:MOV AX,@DATA   ;initializing the data segment
      MOV DS,AX      ;"            "      "
      LEA DX,MSG1    ;displays appropriate messages on the screen
      MOV AH,09      ;
      INT 21H        ;
      LEA DX,MSG2    ;tells the user that the contents of the array
                     ;before sorting are going to be displayed
      MOV AH,09      ;by printing appropriate message on screen
      INT 21H        ;
      CALL OP1       ;calls the procedure op1 which displays the
                     ;array elements before sorting
      MOV BX,OFFSET ARRAY;load reg. bx with the offset add. of array
      MOV DH,04H     ;load reg. dh with the count for outer loop
      MOV SI,04H     ;load reg. si with the count for inner loop
A1    :MOV CX,SI     ;load reg. cx as the count for the inner loop
      MOV DI,01H     ;di reg. is used to increment offset address
A2    :MOV AL,[BX]   ;move contents of memory location whose
                     ;offset is given by bx to reg. al
      CMP AL,[BX+DI] ;compare this with the content of the next
                     ;consecutive memory location
      JA SWAP        ;if content in the previous memory loc. is
                     ;greater than the content of the next memory
                     ;location then swap them
      INC DI         ;inc. di to point to second next memory loc.
      LOOP A2        ;repeat this inner loop till cx becomes zero
```

```
        INC BX          ;increment bx
        DEC SI          ;decrement si
        DEC DH          ;decrement dh
        JNZ A1          ;if its not zero repeat the entire loop again
        LEA DX,MSG3;tell user that sorted array will be displayed
        MOV AH,09        ;by displaying appropriate message
        INT 21H         ;
        CALL OP1        ;this procedure prints sorted array on screen
        JMP LAST        ;jump to the end of the program
                        ;swap larger and smaller values
SWAP :MOV DL,[BX+DI]     ;copy contents of memory loc. whose offset is
                         ;given by [BX+DI] to register dl
        MOV[BX+DI],AL   ;copy contents of reg. al into memory loc.
                        ;whose offset is given by [BX+DI]
       MOV [BX],DL       ;copy contents of register dl into the memory
                         ;location whose offset is given by [BX]
        JMP A2           ;jmp back to check for other greater values
OP1 PROC                 ;this procedure prints the contents of the
                        ;entire array
        MOV BX,OFFSET ARRAY     ;load bx with the offset address of array
        MOV CX,05H              ;load count in register cx
OP    :MOV DL,[BX]              ;move contents of memory loc. whose offset
                                 ;is given by [BX] into register dl
        OR DL,30H               ;convert hexa value to its ascii equivalent
        MOV AH,02                ;print the array element on screen
        INT 21H                 ;
        INC BX                  ;make bx point to offset of next memory
        LOOP OP                 ;loc.repeat the loop till cx becomes zero
RET
ENDP
LAST: MOV AH,4CH                ;exit to dos prompt
        INT 21H                 ;
        END                     ;end of program
```

Program 6

```
;TITLE:      BLINKING CHARACTER IN THE MIDDLE OF THE SCREEN.
;OBJECTIVE: THIS PROGRAM PRINTS A BLINKING CHARACTER IN THE MIDDLE OF
;            THE SCREEN.
;AUTHOR:     Nikhil Marriwala
```

```
.MODEL SMALL
.STACK
.DATA
.CODE
START:MOV AH,0          ;set video mode
      MOV AL,3          ;screen resolution 80*25
      INT 10H
      MOV AH,06         ;clearing window
      MOV CX,081AH      ;upper left corner
      MOV DX,1034H      ;lower right corner
      MOV BH,43H        ;cyan on red background
      MOV AL,00H        ;scroll all lines
      INT 10H
      MOV AH,02         ;set cursor position
      MOV DX,0C27H      ;center of screen
      XOR BH,BH         ;page 0
      INT 10H
      MOV AH,09         ;display character
      XOR BH,BH         ;page 0
      MOV BL,0C3H       ;blinking cyan character
      MOV CX,1          ;character count
      MOV AL,41H        ;char 'A'
      INT 10H
      MOV AH,4CH        ;exit to dos prompt
      INT 21H
      END               ;end of program
```

Program 7

```
;TITLE:      CLEAR THE SCREEN
;OBJECTIVE: THIS PROGRAM CLEARS/SCROLLS THE SCREEN AND POSITIONS
THE CURSOR
;            ON THE TOP LEFT CORNER
;AUTHOR:     Nikhil Marriwala
.MODEL SMALL
.STACK
.DATA
.CODE
START:XOR AL,AL         ;number of lines to scroll
```

```
        XOR CX,CX          ;ch=top row cl=top left column
        MOV DH,24          ;dh=bottom row
        MOV DL,79          ;dl=bottom left column
        MOV BH,07H         ;char attribute for the new line
        MOV AH,06          ;scroll page up
        INT 10H
        XOR DX,DX          ;set dh,dl i.e row and column
        XOR BH,BH          ;cursor position
        MOV AH,02          ;function 2 interrupt 10 sets the cursor
                           position
        INT 10H
        MOV AH,4CH         ;function to exit to dos prompt
        INT 21H
        END                ;end of program
```

Program 8

```
;TITLE:       DESCENDING ORDER
;OBJECTIVE:   THIS PROGRAM SORTS ELEMENTS OF A GIVEN ARRAY IN
              DESCENDING ORDER
;AUTHOR:      Nikhil Marriwala
.MODEL SMALL
.STACK
.DATA
ARRAY db 8,1,6,4,0
MSG1 DB 'BEFORE SORTING$'
MSG2 DB 10,13,'ARRAY:$'
MSG3 DB 10,13,'AFTER SORTING IN DESCENDING ORDER: $'
.CODE
START:MOV AX,@DATA   ;initializing the data segment
      MOV DS,AX      ;"                 "        "
      LEA DX,MSG1    ;displays appropriate messages on the screen
      MOV AH,09      ;
      INT 21H        ;
      LEA DX,MSG2    ;tells the user that the contents of the array
                     ;before sorting are going to be displayed
      MOV AH,09      ;by printing appropriate message on screen
      INT 21H        ;
      CALL OP1       ;calls the procedure op1 which displays the
                     ;array elements before sorting
```

```
        MOV BX,OFFSET ;load reg. bx with the offset add. of array
        ARRAY
        MOV DH,04H      ;load reg. dh with the count for outer loop
        MOV SI,04H      ;load reg. si with the count for inner loop
A1     :MOV CX,SI       ;load reg. cx as the count for the inner loop
        MOV DI,01H      ;di reg. is used to increment offset address
A2     :MOV AL,[BX]     ;move contents of memory location whose
                        ;offset is given by bx to reg. al
        CMP AL,[BX+DI]  ;compare this with the content of the next
                        ;consecutive memory location
        JB SWAP         ;if content in the previous memory loc. is
                        ;lesser than the content of the next memory
                        ;location then swap them
        INC DI          ;inc. di to point to second next memory loc.
        LOOP A2         ;repeat this inner loop till cx becomes zero
        INC BX          ;increment bx
        DEC SI          ;decrement si
        DEC DH          ;decrement dh
        JNZ A1          ;if its not zero repeat the entire loop again
        LEA DX,MSG3;tell user that sorted array will be displayed
        MOV AH,09       ;by displaying appropriate message
        INT 21H         ;
        CALL OP1        ;this procedure prints sorted array on screen
        JMP LAST        ;jump to the end of the program
                        ;swap larger and smaller values
SWAP   :MOV DL,[BX+DI]       ;copy contents of memory loc. whose offset is
                             ;given by [BX+DI] to register dl
        MOV[BX+DI],AL        ;copy contents of reg. al into memory loc.
                             ;whose offset is given by [BX+DI]
        MOV [BX],DL          ;copy contents of register dl into the memory
                             ;location whose offset is given by [BX]
        JMP A2               ;jmp back to check for other smaller values
OP1 PROC                     ;this procedure prints the contents of the
                             ;entire array
        MOV BX,OFFSET ARRAY  ;load bx with the offset address of array
        MOV CX,05H           ;load count in register cx
```

```
OP    :MOV DL,[BX]          ;move contents of memory loc. whose offset is
                            ;given by [BX] into register dl
       OR DL,30H            ;convert hexa value to its ascii equivalent
       MOV AH,02            ;print the array element on screen
       INT 21H              ;
      INC BX                ;make bx point to offset of next memory loc.
       LOOP OP              ;repeat the loop till cx becomes zero
RET
ENDP
LAST: MOV AH,4CH            ;exit to dos prompt
      INT 21H               ;
      END                    ;end of program
```

Program 9

```
;TITLE:            PRINT ALL THE ENGLISH ALPHABETS
;OBJECTIVE:        THIS PROGRAM PRINTS ALL THE ENGLISH ALPHABETS FROM
                   A-Z ON SCREEN
;AUTHOR    :       Nikhil Marriwala
.MODEL SMALL
.STACK
.DATA
.CODE
START:MOV CX,1AH            ;load cx with count of 26(26 alphabets)
      MOV DL,41H            ;load dl with ascii value of 'A'
AGAIN:CALL DISP             ;call procedure which displays the
                            characters
      INC DL                ;increment dl so that it hold ascii value
                            of
                            ;next alphabet
      LOOP AGAIN            ;repeat this loop 26 times
      JMP LAST              ;jump to the end of program
DISP PROC
      MOV AH,02             ;print the character/alphabet on the screen
      INT 21H
RET
ENDP
LAST:MOV AH,4CH             ;exit to dos prompt
     INT 21H
     END                    ;end of program
```

Program 10

```
;TITLE     : LOWER CASE TO UPPER CASE
;OBJECTIVE: THIS PROGRAM USES VARIOUS PROCEDURES TO CARRY OUT CASE
CONVERSION
;AUTHOR    : Nikhil Marriwala

.MODEL SMALL
.STACK
.DATA
MSG1 DB 'ERROR $'
.CODE
START:MOV AX,@DATA          ;initializing data segment
      MOV DS,AX
      CALL INPUT            ;take input from user by calling this
                            procedure
      CALL CHECK            ;check if entered data is a valid english
                        alphabet
      CALL CLEAR            ;clear the screen
      CALL POSN             ;position the cursor in the middle of the
                        screen
      CALL CONVERT          ;carry out code conversion
      JMP LAST              ;go to end of program
INPUT PROC                  ;takes input from user
      MOV AH,01             ;take input from user
      INT 21H
      MOV BL,AL             ;store the input in bl register
RET
ENDP
CLEAR PROC                  ;clears the screen
      XOR AL,AL             ;scroll all lines
      XOR CX,CX             ;ch=top row ,cl=top left column
      MOV DH,24             ;dh=bottom row
      MOV DL,79             ;dl=bottom left column
      MOV BH,07H            ;char attribute for new line
      MOV AH,06             ;scroll page up
      INT 10H
RET
ENDP
POSN PROC                   ;positions the cursor in the middle of the
                        screen
      MOV DX,0C27H        ;center of screen
      XOR BH,BH           ;page 0
```

```
        MOV AH,02           ;set cursor position
        INT 10H
    RET
    ENDP
    CHECK PROC              ;checks validity of input
          CMP BL,41H      ;checks if data's ascii value is less than 41h
          JB ERROR          ;if yes it reports error
          CMP BL,7AH        ;checks if data's ascii value is greater
                            than 7Ah
          JA ERROR          ;if yes it reports error
          CMP BL,5BH        ;check if >= 5Bh if yes then go to check1
          JAE CHECK1
          JMP LAST1         ;else no error
    CHECK1:CMP BL,5FH       ;now check if <=5Fh if yes
           JBE ERROR        ;report error
           JMP LAST1        ;else no error
    ERROR:LEA DX,MSG1       ;display error message on screen
          MOV AH,09
          INT 21H
          JMP LAST
    LAST1:RET
          ENDP
    CONVERT PROC            ;carry out code conversion after checking
                            validity
          CMP BL,61H        ;if the character is upper case display it
                            as it is
          JB NOCHANGE
          SUB BL,20H        ;if lower case then subtract 20h from its
                            ascii value
    NOCHANGE:MOV DX,BX      ;display the character on screen by using
                            function 2
             MOV AH,02      ;interrupt 21h
             INT 21H
    RET
    ENDP
    LAST:MOV AH,4CH         ;exit to dos prompt
         INT 21H
         END                ;end of program
```

Program 11

```
;TITLE:        FACTORIAL OF A NUMBER
;OBJECTIVE:    THIS PROGRAM ACCEPTS A NUMBER FROM THE USER THROUGH
               THE KEYBOARD
```

```
;               CALCULATES ITS FACTORIAL AND PRINTS IT ON THE SCREEN
;AUTHOR:        Nikhil Marriwala
.MODEL SMALL
.STACK
.DATA
MSG1 DB 'ENTER THE NUMBER:$'
MSG2 DB 10,13,'THE FACTORIAL IS:$'
.CODE
START:MOV AX,@DATA        ;initializing the data segment
      MOV DS,AX           ;"        "        "
      LEA DX,MSG1     ;prompts user to enter number by displaying
      MOV AH,09           ;appropriate message
      INT 21H             ;
      MOV AH,01           ;take input of the number
      INT 21H             ;
      XOR AH,AH           ;loading ah register with 00h
     AND AL,000Fh         ;convert ascii code of the No.to hexa code
      MOV CX,AX           ;loading the count with the number itself
      MOV AX,01h          ;loading ax register with 01h
AGAIN:MUL CX              ;multiplying the cx with ax
      LOOP AGAIN          ;decrementing cx and repeating the loop
      MOV CX,AX           ;copy factorial result into cx register
      LEA DX,MSG2     ;tell the user that the result will be displayed
      MOV AH,09           ;on the screen by printing app.message
      INT 21H             ;
      MOV DX,CX           ;copying the factorial into dx register
      OR DX,30H           ;converting the hexa code to ascii
equivalent
      MOV AH,02           ;print the factorial result onto screen
      INT 21H             ;
      MOV AH,4CH          ;exit to dos prompt
      INT 21H
      END                 ;end of program
```

Program 12

```
;TITLE:         SEARCH FOR A GIVEN NUMBER IN AN ARRAY
;OBJECTIVE:     THIS PROGRAM SEARCHES FOR A NUMBER GIVEN BY THE
                USER THROUGH A
;               KEYBOARD IN AN ARRAY IF THE NUMBER IS FOUND IT
            PRINTS APPROPRIATE
;               MESSAGE ON THE SCREEN IF ITS NOT FOUND IT PRINTS
                AN APP. MESSAGE
;               FOR THE SAME ON THE SCREEN
;AUTHOR:        Nikhil Marriwala
.MODEL SMALL
```

```
.STACK
.DATA
ARRAY DB  1,2,3,4,5,6,9
MSG1 DB 'ENTER THE NUMBER TO BE SEARCHED:$'
MSG2 DB 10,13,'THE NUMBER IS FOUND IN THE ARRAY$'
MSG3 DB 10,13,'THE NUMBER IS NOT FOUND IN THE ARRAY$'
.CODE
START:MOV AX,@DATA   ;initializing the data segment
      MOV DS,AX      ;"              "          "
      LEA DX,MSG1;prompts the user to enter the No. to be found
      MOV AH,09      ;by displaying appropriate message
      INT 21H        ;
      MOV AH,01      ;takes input from user
      INT 21H        ;
      MOV BX,OFFSET  ;load reg. bx with offset add. of memory loc.
       ARRAY
                     ; specified by ARRAY
      AND AL,0FH      ;convert ascii code of the No. to hexa
      MOV CX,07H      ;load cx register with the count i.e
                     ;the total No. of No.'s in the array for
                      ;which the loop has to be repeated
AGAIN:CMP AL,[BX]   ;compares the No. inputed with each element in
                    ; the array
      JE FOUND       ;if equal it jumps to the label found
      INC BX          ;if the first element is not equal it
                      ;increments bx thus pointing to next element in
                     ;the array
      LOOP AGAIN    ;repeats the loop till either
                    ;a)it finds the number or
                    ;b)the loop comes to end or
                    ;c)last element of the array has been checked
      LEA DX,MSG3   ;tells the user that the number isn't found
      MOV AH,09    ;by displaying appropriate message on the screen
      INT 21H      ;
      JMP LAST     ;the control ends the program
FOUND:LEA DX,MSG2  ;if the number is found this instruction tells
      MOV AH,09     ;the user the same by displaying an
      INT 21H       ;appropriate message

LAST :MOV AH,4CH   ;exit to dos prompt
      INT 21H
      END          ;end of program
```

Program 13

```
;TITLE:           MULTIPLY TWO NUMBERS
```

```
;OBJECTIVE:         THIS PROGRAM TAKES ANY TWO NUMBERS AS INPUT FROM
                    THE USER THROUGH
;                   THE INPUT DEVICE(KEYBOARD) & PRINTS THEIR PRODUCT
                ON THE STANDARD
;                   OUTPUT DEVICE (SCREEN)
;AUTHOR:            Nikhil Marriwala
.MODEL SMALL
.STACK
.DATA
.CODE
START:MOV AH,01   ;take input of first number from user
      INT 21H     ;
      MOV BL,AL   ;copy the contents of al i.e number
                  ;inputted from keyboard to bl
      AND BL,0FH  ;convert ascii code of Number 1 to hexa code
      MOV AH,01   ;take input of second number from user
      INT 21H     ;
      AND AL,0FH  ;convert the ascii code of Number 2 to hexa
      MUL BL      ;multiply the two numbers
      MOV DL,AL   ;copy the product to register dl
      OR DL,30H   ;convert hexa product to its ascii equivalent
      MOV AH,02   ;print the product on the screen
      INT 21H     ;
      MOV AH,4CH  ;exit to dos prompt
      INT 21H
      END         ;end of program
```

Program 14

```
;TITLE:             MULTIPLY TWO NUMBERS
;OBJECTIVE:         THIS PROGRAM TAKES ANY TWO NUMBERS AS INPUT FROM
                    THE USER THROUGH
;                   THE INPUT DEVICE(KEYBOARD) & PRINTS THEIR PRODUCT
                ON THE STANDARD
;                   OUTPUT DEVICE (SCREEN) GIVING APPROPRIATE MESSAGES
                TO THE USER.
;AUTHOR:            Nikhil Marriwala
.MODEL SMALL
.STACK
.DATA
MSG1 DB 'ENTER FIRST NO:$'
MSG2 DB 10,13,'ENTER SECOND NO:$'
MSG3 DB 10,13,'THE PRODUCT IS:$'
.CODE
START:MOV AX,@DATA   ;initializing the data segment
      MOV DS,AX      ;"              "          "
```

```
        LEA DX,MSG1;prints appropriate message to prompt the user
        MOV AH,09       ;to enter the first number
        INT 21H         ;
        MOV AH,01       ;take input from user first number
        INT 21H         ;
        MOV BL,AL       ;copy the contents of al i.e number inputted
                        ;from keyboard to bl
        AND BL,0FH      ;convert the ascii code of No.1 to hexa code
        LEA DX,MSG2;prints appropriate message to prompt the user
        MOV AH,09       ;to enter the second number
        INT 21H         ;
        MOV AH,01       ;take input from user second number
        INT 21H         ;
        AND AL,0FH      ;convert the ascii code of No.2 to hexa
        MUL BL          ;multiply the two numbers
        MOV BL,AL       ;copy the product to register bl
        LEA DX,MSG3;prints appropriate message telling the user
        MOV AH,09       ;that the product will be printed on the screen
        INT 21H         ;
        MOV DL,BL       ;copy the product to register dl
        OR DL,30H       ;convert hexa product to its ascii equivalent
        MOV AH,02       ;print the product on screen
        INT 21H         ;
        MOV AH,4CH
        INT 21H
        END
```

Program 15

```
;TITLE:            DISPLAY NAME USING INTERRUPT 10 FUNCTIONS
;OBJECTIVE:        THIS PROGRAM DISPLAYS A STRING FOR EG.NAME OF A
                   PERSON USING
;                  INTERRUPT 10 FUNCTIONS
;AUTHOR:           Nikhil Marriwala
.MODEL SMALL
.DATA
.STACK
.CODE
START:MOV AH,0             ;set video mode
      MOV AL,3             ;screen resolution 80*25
      INT 10H
      MOV AH,06            ;clearing window
      MOV CX,081AH         ;upper left corner
      MOV DX,1034H         ;lower right corner
      MOV BH,53H           ;cyan on majenta background
```

```
        MOV AL,00H          ;scroll all lines
        INT 10H
        MOV DX,0C27H        ;center of screen
        MOV AL,'S'          ;char 'S'
        CALL DISP           ;call procedure which displays this
                        character on the screen
        INC DX              ;increment cursor position
        MOV AL,'H'          ;load next character
        CALL DISP           ;call the procedure
        INC DX              ;increment cursor position
        MOV AL,'I'
        CALL DISP
        INC DX
        MOV AL,'N'
        CALL DISP
        INC DX
        MOV AL,'E'
        CALL DISP
        JMP LAST            ;jump to the end of the program
  DISP PROC
        MOV AH,02           ;set cursor position
        XOR BH,BH           ;page 0
        INT 10H
        MOV AH,09           ;display character
        XOR BH,BH           ;page 0
        MOV BL,0C3H         ;blinking cyan character
        MOV CX,1            ;character count
        INT 10H
  RET
  ENDP
  LAST: MOV AH,4CH          ;exit to dos prompt
        INT 21H
        END                 ;end of program
```

Program 16

```
  ;TITLE:           THE NUMBER OF LOGICAL ONES
  ;OBJECTIVE:       THIS PROGRAM COUNTS THE NUMBER OF LOGICAL ONES IN
                  A DATA BYTE
  ;                 STORED IN A REGISTER
  ;AUTHOR:          Nikhil Marriwala
  .MODEL SMALL
  .STACK
  .DATA
  MSG1 DB 'THE NUMBER OF LOGICAL ONES IN THE GIVEN DATA BYTE ARE:$'
```

```
.CODE
START:MOV AX,@DATA    ;initializing the data segment
      MOV DS,AX       ;"                "        "
      LEA DX,MSG1     ;tells the user that the number of ones in the
      MOV AH,09       ;given data byte shall be displayed by printing
      INT 21H         ;an appropriate message
      MOV AX,00FFH    ;load reg.ah->00h & al->ffh (i.e al has 8 1's)
      MOV CX,08H      ;load cx with count of 8(8bits in a byte)
      MOV DL,00H      ;make dl counter for the number of 1's
COUNT:RCL AL,1        ;rotating left through carry the contents of
                      ;the al register by 1 bit,so that the msb is
                      ;shifted to the carry
      JNC DO          ;if the carry is not set it means the bit
                      ;shifted into the carry was a zero hence do
not
                      ;increment dl and simply repeat the loop
      INC DL          ;if the carry is set it means the bit shifted
                      ;into the carry is a 1 and hence increment the
                      ;counter dl
DO   :LOOP COUNT      ;repeat the loop till cx becomes zero
      OR DL,30H       ;convert hexa sum to its ascii equivalent
      MOV AH,02       ;print the result on the screen
      INT 21H         ;
      MOV AH,4CH      ;exit to dos prompt
      INT 21H         ;
      END             ;end of program
```

Program 17

```
;TITLE      :     RAM CHECK
;OBJECTIVE  :     THIS PROGRAM CHECKS A 1K RANGE OF MEMORY FOR ITS
                  FUNCTIONING
;AUTHOR     :     Nikhil Marriwala

.MODEL SMALL
.DATA
MSG DB 'RAM CHECK OK $'
MSG1 DB 'RAM CHECK FAILED $'
.CODE
START:MOV AX,@DATA
      MOV DS,AX          ;initializing data segment
      XOR DL,DL
      XOR AL,AL
STAR :MOV BX,0100H       ;start checking from memory location 0100h
      MOV CX,1024     ;load count for 1k
```

```
AGAIN:MOV [BX],AL    ;load zeroes into all the mem.locs from 0100h
      CMP [BX],AL    ;check if zeroes have been written to
                     mem.locs
      JNZ ONE            ;if not jump to the given label
      INC BX
      LOOP AGAIN         ;continue the loop for 1k mem.locs
      INC DL
      MOV AL,0FFH        ;load all one's in al register
      CMP DL,2
      JNZ STAR           ;write ones and check if ones are written
                          into
                         ;the mem.locs
TWO:LEA DX,MSG           ;if all checks are successful
    MOV AH,09            ;print appropriate message of successful
    INT 21H              ;ram check
    JMP LAST             ;jump to end of program
ONE:LEA DX,MSG1          ;if some error is encountered
    MOV AH,09            ;print appropriate message
    INT 21H
LAST:MOV AH,4CH          ;exit to dos prompt
     INT 21H
     END                 ;end of program
```

Program 18

```
;TITLE:               REVERSE A BIT PATTERN
;OBJECTIVE:           THIS PROGRAM REVERSES THE BIT PATTERN OF A
DATA                     BYTE STORED IN A
;                     GIVEN REGISTER
;AUTHOR:              Nikhil Marriwala
.MODEL SMALL
.STACK
.DATA
.CODE
START:MOV AX,00ABH    ;load ax with the required bit pattern
      MOV CX,08H      ;load count of 8(8bits in a byte) in register cx
AGAIN:RCL AL,1        ;rotate left through carry the contents of
                    register
                      ;al so that the msb is shifted into the carry
      RCR DL,1        ;now shift the contents of carry flag into msb of
                      ;register dl through rotate right instruction
      LOOP AGAIN      ;repeat the loop 8 times for 8 bits
      MOV AH,4CH      ;exit to dos prompt
      INT 21H
      END             ;end of program
```

Program 19

```
;TITLE     : STRING INSTRUCTIONS
;OBJECTIVE: THIS PROGRAM COPIES A STRING FROM ONE LOCATION IN THE
MEMORY
;            TO ANOTHER
;AUTHOR    : Nikhil Marriwala

.MODEL SMALL
.STACK
.DATA
STRING1 DB 'GOOD GIRL $'
STRING2 DB '          $'
.CODE
START:MOV AX,@DATA        ;initializing data segment
      MOV DS,AX
      MOV ES,AX           ;initializing extra segment
      MOV CX,9            ;loading count for chars in the string
      LEA SI,STRING1      ;loading source index register with the
                          ;effective address of the source string
      LEA DI,STRING2      ;loading destination index register with the
                          ;effective address of destination string
      REP MOVSB           ;repeat the movement of the chars in the
                          ;string from source to destination till
                          ;count becomes zero
      MOV AH,4CH          ;exit to dos prompt
      INT 21H
      END                 ;end of program
```

Program 20

```
;TITLE:         SWAPPING OF TWO ARRAYS
;OBJECTIVE:     THIS PROGRAM INTERCHANGES THE CONTENTS OF TWO ARRAYS
;AUTHOR:        Nikhil Marriwala

.MODEL SMALL
.STACK
.DATA
ARRAY1 DB 1,2,3,4
ARRAY2 DB 5,6,7,8
MSG1 DB 'BEFORE SWAPPING:$'
MSG2 DB 10,13,'ARRAY1:$'
MSG3 DB 10,13,'ARRAY2:$'
MSG4 DB 10,13,'AFTER SWAPPING:$'
.CODE
START:MOV AX,@DATA        ;initializing the data segment
```

```
        MOV DS,AX             ;"              "          "
        LEA DX,MSG1           ;tells the user that the contents of the
        MOV AH,09             ;2 arrays before swapping will be displayed
        INT 21H               ;on screen,by printing appropriate message.
        LEA DX,MSG2           ;
        MOV AH,09             ;
        INT 21H               ;
        CALL ONE              ;calls the procedure ONE(prints the 1st array
                              ;before swapping)
        LEA DX,MSG3           ;
        MOV AH,09             ;
        INT 21H               ;
        CALL TWO              ;calls the procedure TWO(prints the 2nd array
                              ;before swapping)
        LEA DX,MSG4       ;tells the user that the contents of the
        MOV AH,09             ;2 arrays after swapping will be displayed
                          funtion 9 interrupt 21 prints the string
                          held by dx onto screen
        INT 21H               ;on screen,by printing appropriate message.
        CALL SWAP             ;calls the procedure SWAP(swaps the two
                           arrays)
        LEA DX,MSG2           ;
        MOV AH,09             ;
        INT 21H               ;
        CALL ONE              ;calls the procedure ONE(prints the 1st array
                              ;after swapping)
        LEA DX,MSG3           ;
        MOV AH,09             ;
        INT 21H               ;
        CALL TWO              ;calls the procedure TWO(prints the 2nd array
                              ;after swapping)
        JMP LAST              ;jump to the end of the program

SWAP PROC                     ;this procedure/subroutine swaps the contents
                              ;of the two arrays
    MOV BX,OFFSET ARRAY1      ;load bx with offset address of array1
    MOV SI,OFFSET ARRAY2      ;load si with offset address of array2
    MOV CX,04H                ;load count equal to the No. of No's in the
                              ;arrays
S1    :MOV AL,[BX]            ;copy successive elements of array1 into reg
                              ;al
        XCHG AL,[SI]          ;interchange contents of al with array2 ele.
        MOV [BX],AL           ;copy contents of al into array1
                              ;i.e interchange elements of array1,2
```

```
        INC BX                  ;make bx point to offset of next element of
                                ;array1
        INC SI                  ;make si point to offset of next element of
                                ;array2
        LOOP S1                 ;repeat this loop & interchange all the
                                ;elements until cx becomes zero
RET
ENDP
ONE   PROC                      ;this procedure prints the contents of array1
                                ;before and after the swapping has taken
                                place
MOV BX,OFFSET ARRAY1            ;load bx with the offset address of array1
        MOV CX,04H              ;load cx with the count
A1    :MOV DX,[BX]    ;load register dx with the elements of array1
        OR DX,30H               ;convert hexa value to its ascii equivalent
        MOV AH,02               ;print the element on screen
        INT 21H                 ;
        INC BX                  ;increment bx to point to offset of next element
                                ;in array1
        LOOP A1                 ;repeat this loop till cx becomes zero

RET
ENDP
TWO   PROC                      ;this procedure prints the contents of array1
                                ;before and after the swapping has taken place
 MOV SI,OFFSET ARRAY2           ;load si with the offset address of array2
        MOV CX,04H              ;load cx with the count
A2    :MOV DX,[SI]    ;load register dx with the elements of array2
        OR DX,30H               ;convert hexa sum to its ascii equivalent
        MOV AH,02               ;print the element on the screen
        INT 21H                 ;
        INC SI                  ;make si point to the offset of the next
                                element
                                ;in array2
        LOOP A2                 ;repeat the loop till cx becomes zero
RET
ENDP
LAST:MOV AH,4CH                 ;exit to dos prompt
      INT 21H
      END                       ;end of program
```

Program 20. ***Write an ALP in 8086 to find the average of two numbers***

```
    NAME    Average
    TITLE   8086/8088 ALP to find average of two numbers.
  .mode     small
  .stack    100
  .data
    Label   Mnemonics      Comments
    N0.1    DB 63H      ; First number storage
    No.2    DB 2EH      ; Second number storage
    Avg.    DB?         ; Average of two numbers.
    .code
    START   :MOV AX, @data; Initialises
                 MOV DX,AX      ;data segment
                 MOV AL, No 1   ; Get first number in AL
                 ADD AL, No.2   ; Add second to it
                 ADC AH, 00H    ; Put carry in AH
                 SAR AX, 1      ; Divide sum by 2
                 MOV AVG, AL    ; Copy result to memory
                 END START
```

Program 21. ***Write an ALP in 8086/8088 to find maximum number in the array of 10 numbers***

```
    NAME    Maximum number
    TITLE 8086 ALP to find maximum in the array.
  .model small
  .stack 100
  .data
    ARRAY DB 63H, 32H, 45H, 75H, 12H, 42H, 09H, 14H, 56H, 38H; Array
of ; ten numbers
    Labels      Mnemonics       Comments
       MAX     DB 0            ; Maximum number
    .code
    START : MOV AX, @data      ; [Initializes
                   MOV DS, AX      ; data segment]
                   XOR DI, DI      ; Initialize pointer
                MOV CL, 10         ; Initialize counter
                  LEA BX, ARRAY    ; Initialize base pointer for array
                  MOV AL, MAX      ; Get maximum number
    BACK: CMP AL, [BX+DI]      ; Compare number with maximum
                 JNC SKIP
                 MOV DL, [BX+DI]   ;    [If number > MAX
                 MOV AL, DL     ; MAX= number]
    SKIP:   INC DI                 ; Increment pointer
          DEC CI                   ; Decrement counter
          JNZ BACK                 ;IF count = 0 stop; otherwise go BACK
```

```
        MOV MAX, AL           ; Store maximum number
    END START
```

Note. Above two ALPs will not accept any input from keyboard and will not display result on screen. For display we need to use DOS.

Program 22. ***Write a recursive program in 8086/8088 ALP to find the sum of the first "n" integers.***

Recursive algorithm *(for summing the first tn' integers in PASCAL A)*

```
   BEGIN

IF N = 0 THEN SUM : = 0 ELSE
SUM = SUM + SUM (N:"'1);

END;

ALP : ; Program in assembly :. .MODEL SMALL
. DATA
N EQU X

. CODE
MAIN PROC
MOV BX, N
CALL SUM_OF_N
MOV AX, 4COOH
INT 21H
MAIN ENDP
SUM_OF_N_PROCEDURE
CPM BX, 00
JZ BX_ZERO
PUSH BX
DEC BX
CALL SUM_OF_N
POP BX
BX_ZERO: ADD AX, BX
   RET
SUM_OF_N END P
   END
```

Program 23. ***Write an ALP in 8086 to move a block of N bytes of data from source to destination and display the result***

Solution. To write this progam and make it clear, we divide the solution into three parts as

(i) ALP-explanation
(ii) ALP-Algorithm, and
(iii) The ALP

ALP Explanation

Let, Block 1 data is available from memory location block 1, and destination memory location be block 2. The number of bytes in block 1 be 10 i.e., 0AH. To start, initialize counter CX register.

Block 1 address be in SI register and Block 2 address in DI register. Clear the direction flag. By using string instruction move the data from the memory pointed by SI register to memory pointed to DI register, (it is assumed that the data is to be moved within the same segment). Therefore DS and ES are initialized to the same segment value. Finally execution of the problem, and the contents of destination memory location can be viewed by using debug DUMP command.

(i) ALP-Algorithm

Step No.	Action/Process
1.	Initialize data in the Source memory
2.	Initialize SI, and DI and CX registers with the source address, destination address and the count
3.	Initialize the DF (*Direction Flag*) to zero
4.	Use move string instruction to transfer the data
5.	Check whether all data is transferred.
6.	If not transferred go to step 4.
7.	Else terminate the program normally using interrupt 21 function 4CH.

(ii) The ALP

```
.MODEL SMALL
.DATA
BLOCK1 DB 01, 02, 03, 04, 05, 06, 07, 08, 09.0AH
BLOCK2 DB 10 DUP (?)
COUNT DW 0AH
.CODE
    MOV AX, @DATA            ;Initalize the DS and EX with the
                             segment address
    MOV DS, AX
    MOV ES, AX
    MOV SI.OFFSET BLOCK1     ;Initialize the S1 and D1 with the
                             address
    MOV DI, OFFSET BLOCK 2   ; of the source and destination
    MOV CX, COUNT            ; Initialize CX with the count
    CLD                      ; Initialize the DF=0
AGAIN: REP MOVSB             ;String move the data from source to
                              destination
                             ;till the CX register is zero
    MOV AH, 4CH              ;Terminate the program
    INT 21 H
    END
```

(Result : 01, 02, 03, p4, 05, 06, 07, 08, 0AH is moved to BLOCK 2 as required).

Program 24. ***Write an ALP n 8086/8088 to add 5 bytes of data in an array by making use of Procedure.***

Solution. Here let us write the program directly with normal format.

```
Label   Mnemonics and Operand     Comments
     .MODEL SMALL
     .DATA
             ARR DB 5,6,7,0A, OFFH ; Declaration of variable and data
             COUNT DW @-ARR      ; Count gives number of bytes in
                                   the ARR
             SUM DW ?
     .CODE
             MOV AX, @DATA
             MOV DS, AX
             XOR, AX, AX           ; Clearing AX register to 0
             MOV SI, OFFSET ARR
             MOV CX, COUNT         ; Initializing the CX register
                                     with count
     NEXT:   PUSH CX
             CALL NEAR PTR PROADD  ; Procedure which is with in the
                                   same
             INC SI                ; Segment is called
             POP CX                ; Procedure is repeated till the
             LOOP NEXT             ; Count in CX is zero
             MOV SUM, AX           ; Result is stored in memory
             MOV AH, 4CH           ; Normal termination to DOS
             INT 21H
     PROADD  PROCNEAR              ; Near procedure add the contents
             ADD AL, [SI]          ; In Al with contents pointed by
                                     the
             ADC AH, 00            ; Index register, and the carry
                                     is stored in AH
             RET
             PROADD ENDP
             END
```

Note. In this program, the procedure *"PROADD"* can refer to the variable AR, COUNT, and SUM directly, because the main calling program and the procedure are in the same file.

Program 25. *Write an ALP in 8086 to add two 16-digit packed BCD numbers.*

Solution. The Packed BCD numbers have two digits in each byte, therefore 16-digit packed BCD we need 8 locations, and count is initialized into CX register. The first set of BCD numbers are brought into AL register and added to the second set of BCD numbers. Observe that we get only two digit numbers at a time since we decimal adjust the result after addition only in AL register. The order in which the addition takes place is by bringing only lower order BCD first followed by higher order bytes. After decimal adjust is completed the result is stored back into memory. When all 8 bytes of BCD are added the result is stored, and the program is terminated, as usual.

Algorithm for ALP 25:

Step No.	Action/Process
1.	Initialize data memory with data
2.	Initialize data segment register with the data segment address.
3.	Initialize CX register with the count of 0008. (i.e., number of bytes that has to be added)
4.	Initialize SI.DI and BP with the address pointing to first operand, second operand and the result respectively.
5.	Clear the carry flag
6.	Move first operand byte into AL register.
7.	Add with carry the contents of AL to the second operand byte.
8.	Decimal adjust to obtain BCD number
9.	Increment to SI, DI and BP to point to the next byte of the first operand second operand and the result respectively
10.	Decrement CX register and go to step 6 is CX register is not zero.
11.	If the result is zero terminate the program.

The ALP for ALP 25

```
    . MODEL   SMALL
    .DATA

SOURCE1   BD  01H,    08H,    99H,    42H,    11H,    46H,    32H, 55H
SOURCE2   DB  55H,    44H,    66H,    41H,    83H,    21H,    41H, 84H
RESULT DB  8DUP(?)
    .CODE
Label       Mnemonic and Operand       Comments
        MOV AX, @DATA               ; Initialize the DS register
        MOV DS, AX                  ;
        MOV CX, 08                  ; Initialize the Counter with 08
        MOV SI, OFFSET SOURCE1; Initialize SI, DI and BP with address
        MOV DI, OFFSET SOURCE2      ;
        MOV BX, OFFSET RESULT;
        CLC                         ; Clear carry flag
BEG :       MOV AL, [SI]            ; Add two numbers after fetching
                                      the data first into
        ADC AL, [DI]                ; AL and then perform decimal adjust
```

```
DAA                     ;
MOV [BX], AL,           ; Save the result into memory
INC SI                  ; Point to the next data
INC DI                  ;
INC BX                  ; Point to the next location to save
LOOP BEG                ; Have all the bytes computed?
                        ; If 'NO' go to BEG
MOV AH, 4CH             ; if 'YES' terminate the program and
                          return to
INT 21H             ; DOS
END
```

[The Result : 56, 52, 65, 84, 94, 67, 73, 39, 56 in same sequence].

Program 26. ***Write an ALP sequence to convert a 16-bit packed BCD number beginning at PACKED to unpacked BCD number and store result beginning at UNPACKED location***

Solution. As mentioned in ALP 33, the Packed numbers have two digits in 8 bit byte, that means there are four digits in 16-bit packed BCD number. The unpacked numbers have one digit per byte, where the higher nibble is always zero. Therefore 16-bit packed number requires four location to store it as unpacked number. Each time the least significant number is separated and is ANDed with OFH, and such data is stored in memory, and other 120 bits are rotated right. The process continued four times to get the unpacked number at the end.

```
Label      Mnemonic and Operand        Comments
        .MODEL SMALL
        .DATA
           PACKED DW 1234 H          ; Declare packed number
           UNPACKED DB 4 DUP (?) : Allocate the space for
                                   unpacked Data
        .CODE                        ;
           MOV AX, @ DATA        ; Initialize ES with data area
           MOV DS, AX                ;
           MOV DI, OFFSET UNPACKED   ; Number is loaded
           MOV AX, PACKED            ; Packed number is loaded
                                       into AX
NEXT : MOV CX, 04                    ; The number of digits is
                                       initialized
           MOV BX, AX                ;
           PUSH CX                   ; Save CX on stack
           ADD AX, 0000              ; Add 00 to Ax register
           AAA                       ; Adjust to obtain unpacked
                                       number in AL
           MOV [DI], AL              ; Save the data
           INC DI                    ; Point to the next address
                                       to save unpacked
```

```
DATA
            MOV CL, 04          ;
            SHR BX, CL          ; Obtain the previous digit to
                                  first digit position
            MOV AX, BS          ;
            POP CX              ; Reload the count into CL
            LOOP NEXT           ; Loop till all 4 digits are
                      converted
            MOV AH, 4CH         ; Terminate the program
```

ALP-Explanation

Let, Block 1 data is available from memory location block I, and destination memory .DCation be block 2. The number of bytes in block 1 be 10 *i.e.,* OAH. To start, initialize

Counter CX register.

Block 1 address be in SI register and Block 2 address in DI register. Clear the direction flag. By using string instruction move the data from the memory pointed by SI register :0 memory pointed to DI register, *(it is assumed that the data is to be moved within the .;Dme segment).* Therefore DS and ES are initialized to the same segment value. Finally execution of the program and the contents of destination memory location can be viewed by using debug DUMP command.

(iii) ALP-Algorithm

Step No.	Action/Process
1.	Initialize data in the Source memory
2.	Initialize SI, and DI and CX registers with the source address. destination address and the count
3.	Initialize the DF *(Direction Flag)* to zero
4.	Use move string instruction to transfer the data
5.	Check whether all data is transferred
6.	If not transferred go to step 4
7.	Else terminate the program normally using interrupt 21 function 4CH

```
INT 21H
END
```

Result: 04, 03, 02, 01 is the output for the ALP8]

Problem 26A. ***Write a Program to print a-z in single line using TASM.***

```
.model small
.data
a db 41h
b db 61h
.code
mov ax, @data
mov ds, ax
mov si, offset a
```

```
mov cl, 26d
mov dl, [si]
p:mov ah, 02h
int 21h
inc dl
dec cl
jnz p
mov ah, 4ch
int 21h
end
```

Program 27. ***Write a Program to print a-z in different lines and display the result on the DOS shell.***

```
.model small
.data
var db 41h
.code
mov ax,@data
mov ds,ax
mov cl,0dh
disp:
mov dl,var
mov ah,02h
int 21h
inc dl
add dl,20h
mov ah,02h
int 21h
inc var
inc var
mov dl,0dh
int 21h
mov dl,0ah
int 21h
loop disp
mov ah,4ch
int 21h
end
```

Program 28. ***Write a Program to find out the greater number from an array of numbers.***

```
.model small
.data

  msg1 db 13 ,"arrar is:","$"
```

```
  msg2 db 10,13, "Greatest number is ","$"
  arr db 61d,72d,43d,24d,05d
  divsr db 10d

.code

  mov ax,@data
  mov ds,ax

   lea dx,msg1
   mov ah,09h
   int 21h

   mov ah,02h
   mov dl,0ah
   int 21h
   mov cl,05h
   mov si,offset arr
disp:
  mov al,[si]
  mov ah,00h
  div divsr
  or ax,3030h
  mov bx,ax
  mov dl,bl
  mov ah,02h
  int 21h
  mov dl,bh
  int 21h
  mov dl,0ah
  int 21h
  inc si
  loop disp
  mov cl,04h
  mov si, offset arr
  mov dl,[si]
find:
  inc si
  mov al,[si]
 cmp al,dl
 jbe next
 mov dl,al
next:
 dec cl
```

```
    jnz find
    mov bl,dl

    lea dx,msg2
    mov ah,09h
    int 21h
    mov ah,00h
    mov al,bl
    div divsr
    or ax,3030h
    mov bx,ax
    mov dl,bl
    mov ah,02h
    int 21h
    mov dl,bh
    int 21h
    mov dl,0dh
    int 21h
    mov ah,4ch
    int 21h
   end
```

Program 29. *Write a Program to display any message on the DOS shell.*

```
   .model small
   .stack
   .data
   message db "This program is presented by Nikhil Marriwala ","$"
   .code
    mov ax,@DATA
    mov ds,ax
    mov dX,OFFSET message
    mov ah,09h
    int 21h
    mov ah,4ch
    int 21h
    end
```

Program 30. *Write a Program to display digits of a number on the DOS shell.*

```
   dosseg
   .model small
   .stack 50h
   .code
    main proc
           mov ax,@data
```

```
        mov ds,ax
        mov ax,val1
        mov bx,0010h
        mov dx,0000h
     l1:
        div bx
        push ax
        mov dh,00h
        or dl,30h
        mov ah,02h
        int 21h
              mov dl,0h
              int 21h
              int 21h
        pop ax
        or ax,ax
        mov dx,0000h
        jnz l1
        mov ax,4c00h
        int 21h
      main endp
.data
     val1 dw 9876h
end main
```

Program 31. ***Write a Program to find out the factorial of a number.***

```
.model small
.data
msg1 db'Enter a number:$'
msg2 db'Factorial is:$'
.code
mov ax,@data
mov ds,ax
mov dx,offset msg1
mov ah,09h
int 21h
mov ah,01h
int 21h

mov bx,ax
and bx,0f0fh
mov ax,0001h
ab:mul bx
dec bx
jnz ab
mov cx,ax
```

```
mov dl,0dh
mov ah,02h
int 21h
mov dl,0ah
mov ah,02h
int 21h
mov dx,offset msg2
mov ah,09h
int 21h
mov dx,cx
or dx,3030h
mov ah,02h
int 21h
mov ah,4ch
int 21h
end
```

Program 32. *Write a Program to find out Fibonacci series.*

```
.model small
.data
.code
start:
mov si,0006h
mov bl,30h
mov cl,31h
mov dl,bl
mov ah,02h
int 21h
mov dl,0ah
mov ah,02h
int 21h
mov dl,0dh
mov ah,02h
int 21h
jump: mov al,bl
add al,cl
aaa
or al,30h
mov cl,bl
mov bl,al
mov dl,al
mov ah,02h
int 21h
mov dl,0Ah
mov ah,02h
int 21h
```

```
mov dl,0dh
mov ah,02h
int 21h
dec si
jnz jump
mov ah,4ch
int 21h
end
```

Program 33. ***Write a Program to find out no of even and odd numbers in a given series.***

```
.model small
.data
 ak db 05h, 03h, 04h, 01h, 02h
 b  db "even no. are", 13,10,'$'
 c  db 13,10," odd no. are",13,10,'$'
  .code
   start :
        mov ax,@data
        mov ds,ax
        mov bh,00h
        mov bl,00h
        mov si, offset ak
        mov cx,05h
 X:     mov dx,[si]
        ror dx,01
        jc       hj
        inc bh
        jmp      h
 hj:    inc bl
 h:     inc si
        dec cx
        jnz  x
        mov ah,09h
        mov dx,offset b
        int 21h
        or bh,30h
        mov dl,bh
        mov ah ,02h
        int 21h
        mov ah,09
        mov dx,offset c
        int 21h
        or bl,30h
        mov dl,bl
```

```
mov ah,02h
int 21h
mov ah,4ch
int 21h
end
```

Program 34. ***Write a Program to perform addition, subtraction, division and multiplication in single program for 32 bit microprocessor.***

```
.model small
.386
.stack
.data
.code
mov eax,@data
mov ds,eax
mov eax,00000145h
mov ebx,00000123h
add eax,ebx
mov edx,eax
clc
mov eax,00000145h
mov ebx,00000123h
sub eax,ebx
mov edx,eax
clc
mov eax,00000145h
mov ebx,00000123h
div ebx
mov edx,eax
clc
mov eax,00000145h
mov ebx,00000123h
mul ebx
mov ah,02h
mov edx,eax
mov ah, 4ch
int 21h
end
```

Program 35. ***Write a Program to display tables from 0 to 9 on the DOS shell.***

```
.model small
.data
a db "enter the no.:","$"
b db "the table is :","$"
.code
mov ax,@data
```

```
mov ds,ax
mov dx,offset a
mov ah,09h
int 21h
mov ah,01h
int 21h
mov bh,al
mov dx,offset b
mov ah,09h
int 21h
mov cl,0ah
mov ch,01h
d:
mov dl,0ah
mov ah,02h
int 21h
sub bh,30h
mov al,ch
mul bh
aam
or ax,3030h
mov bl,al
add bh,30h
mov dl,ah
mov ah,02h
int 21h
mov dl,bl
mov ah,02h
int 21h
inc ch
dec cl
jnz d
mov ah,4ch
int 21h
end
```

Program 36. ***Write a Program to display a pattern of '*' representing a right angled triangle on the screen.***

```
.model small
.data
    a db "enter","$"
.code
    mov ax, @data
    mov ds,ax
    mov dx, offset a
    mov ah, 09h
```

```
        int 21h
        mov ah, 01h
        int 21h
        mov bl, al
        mov dl, 0ah
        mov ah, 02h
        int 21h
        mov dl, 0dh
        mov ah, 02h
        int 21h
        and bl, 0fh
        mov dh, 01h
        mov bh, 01h
        mov cl, bl
   k:   mov ch, cl
   p:   dec ch
        jz pp
        mov dl, 20h
        mov ah, 02h
        int 21h
        jmp p
   pp:  mov dl, 0fh
        mov ah, 02h
        int 21h
        dec dh
        jnz pp
        mov dl, 0ah
        mov ah, 02h
        int 21h
        mov dl, 0dh
        mov ah, 02h
        int 21h
        mov dh, bh
        add dh, 02h
        mov bh, dh
        dec cl
        jnz k
   mov ah, 4ch
   int 21h
   end
```

Program 37. ***Write a Program to bubble sort a given series, for 32-bit µp.***

```
   .model small
```

```
.data
list dw 22h,18h,20h,21h,23h
.code
start:
    mov ax,@data
    mov ds,ax
    mov dx,05h
again0:
    mov cx,dx
    mov si,offset list

again1:
    mov ax,[si]
    cmp ax,[si+2]
    jg pr1
    xchg [si+2],ax
    xchg [si],ax
pr1:
    add si,02
    loop again1
    mov bx,ax
    dec dx
    jnz again0
    mov ah,4ch
    int 21h
    end
```

Program 38. *Write a Program to perform UNSIGNED COMPARISONS OF TWO NUMERICAL STRINGS.*

```
; 11/06/03
.model   small
.stack   100h
.data
stock1    DB      10, 20, 30, 255, 40, 50, 70, 250
stock2    DB      10, 20, 30, 127, 45, 30, 20, 250
strlen    EQU     $ - stock2
initial   DW      stock1
msg1     DB      'Stock 1 is bigger than stock 2.',0
msg2     DB      'Stock 2 is bigger than stock 1.',0
msg3     DB      'Divergence at position ',0
msg4     DB      'Stock 1 = ',0
msg5     DB      'Stock 2 = ',0
msg6     DB      'Come together at ',0
.code
```

```
include io.mac
main    PROC
        .startup
        mov cx, strlen
        mov ax, ds
        mov es, ax
        mov si, offset stock1
        PutInt initial
        nwln
        mov di, offset stock2
        cld
        repe cmpsb ; NOTICE: repE
        ; UNSIGNED COMPARISON
        jb print2
print1: PutStr msg1
        nwln
        jmp over
print2: PutStr msg2
        nwln
over:
        PutStr msg3
        mov ax, si
        sub ax, initial
        dec ax
        PutInt ax
        nwln
        mov bx, si
        dec bx ; puts it on top of stock1[3]
        mov al, byte ptr [bx]
        mov ah, 0 ; NOT CBW!!!
        PutStr msg4
        PutInt ax
        nwln
        mov bx, di
        dec bx ; puts it on top of stock2[3]
        mov al, byte ptr [bx]
        mov ah, 0
        PutStr msg5
        PutInt ax
        nwln

;       Now, let's find out where stocks come together
        repne cmpsb ; NOTICE: repNE
        mov ax, si
        sub ax, initial
```

```
        dec ax
        PutStr msg6
        PutInt ax
        nwln

        .exit
main    ENDP
END     main
```

Program 39. *Write a Program to perform SIGNED COMPARISONS OF TWO NUMERICAL STRINGS.*

```
; 11/06/03
.model small
.stack 100h
.data
stock1  DB      10, 20, 30, 255, 40, 50, 70, 250
stock2  DB      10, 20, 30, 127, 45, 30, 20, 250
strlen  EQU     $ - stock2
initial DW      stock1
msg1    DB      'Stock 1 is bigger than stock 2.',0
msg2    DB      'Stock 2 is bigger than stock 1.',0
msg3    DB      'Divergence at position ',0
msg4    DB      'Stock 1 = ',0
msg5    DB      'Stock 2 = ',0
.code
include io.mac
main    PROC
        .startup
        mov cx, strlen
        mov ax, ds
        mov es, ax
        mov si, offset stock1
        PutInt initial
        nwln
        mov di, offset stock2
        cld
        repe cmpsb
        ; SIGNED COMPARISON
        jl print2
print1: PutStr msg1
        nwln
        jmp over
print2: PutStr msg2
        nwln
```

```
over:
        PutStr msg3
        mov ax, si
        sub ax, initial
        dec ax
        PutInt ax
        nwln
        mov bx, si
        dec bx ; puts it on top of stock1[3]
        mov al, byte ptr [bx]
        cbw ; NOT mov ah, 0
        PutStr msg4
        PutInt ax
        nwln
        mov bx, di
        dec bx ; puts it on top of stock2[3]
        mov al, byte ptr [bx]
        cbw ; NOT mov ah, 0
        PutStr msg5
        PutInt ax
        nwln
         .exit
main    ENDP
END     main
```

Program 40. *Write a Program to perform signed multiplication of two strings.*

```
.model small
.stack 100h
.data
num1    DB      ?
num2    DB      ?
msg1    DB      "Number? ",0
msg2    DB      "Product = ",0
msg3    DB      "Answer will be more than 8 bits",10,13,0
msg4    DB      "Answer will be 8 bits or less",10,13,0
.code
include io.mac
main    PROC
        .startup
        PutStr msg1
        Getint AX
        nwln
        ; strip awy upper 8 bits:
        mov num1, al
        PutStr msg1
        Getint AX
```

```
        nwln
        mov num2, al

        ; since al contains already one operand:
        imul num1
        jnc eightbit
        PutStr msg3
        PutStr msg2
        PutInt ax
        nwln
        jmp finish
eightbit:
        PutStr msg4
        PutStr msg2
        PutInt AX
        nwln
finish:
        .exit
main    ENDP
END     main
```

Program 41. ***Write a Program to perform unsigned multiplication of the given strings.***

```
.model small
.stack 100h
.data
num1   DB   ?
num2   DB   ?
msg1   DB   "Number? ",0
msg2   DB   "Product = ",0
msg3   DB   "Answer will be more than 8 bits",10,13,0
msg4   DB   "Answer will be 8 bits or less",10,13,0
msg5  DB  "PutInt can't print answer as an unsigned number",10,13,0 .code
include io.mac
main    PROC
        .startup
        PutStr msg1
        Getint AX
        nwln
        ; strip awy upper 8 bits:
        mov num1, al
        PutStr msg1
        Getint AX
        nwln
        mov num2, al
```

```
        ; since al contains already one operand:
        mul num1
        jnc eightbit
        PutStr msg3
        ;if1 (if sign bit is on)
                add ax,0
                jns endif1
        then:
                PutStr msg5
         endif1:
        PutStr msg2
        PutInt ax
        nwln
        jmp finish
eightbit:
        PutStr msg4
        PutStr msg2
        PutInt AX
        nwln
finish:
        .exit
main    ENDP
END     main
```

7.11. SAMPLE ALP ON ASCII AND STRINGS

Here let us develop programs on ASCII codes.

ALP-ASCII (1) Algorithm Write an ALP in 8086/8088 to multiply an ASCII string of 16 bytes by a single ASCII digit and the result be a string of unpacked BCD digit.

Algorithm for ASCII (1) :

Step No	Explanation
1.	Initialize data in data memory area
2.	Initializo DS and ES with data area addresses
3.	Initialize SI & DI with thc offset addresses of multiplicand and the result Load multiplier into DL register
4.	Initialize CX register as counter
5.	Initialize CX register as counter
6.	Initialize the destination location with 00
7.	Convert multiplier from ASCII to a singled digit number decimal number
8.	Fetch string bytes from memory and convert it to decimal number

```
.MODE SMALL
   .DATA
```

```
STRI    DB  32 34 31 32 31 30 32 33 37 31 32 34 37 30 32
MULTIP DB  36
RESULT DB  16  DUP (?)

.CODE
Label       Mnemonics Operand          Comments
        MOV AX, @ DATA
        MOV DS, AX
        MOV ES, AX
        MOV SI, OFFSET STRI       ; Initialize with string Address
        MOV DI, OFFSET RESULT ; Initialize with result Address.
        MOV DL, MULTIP          ; Multiplier is loaded into Register
        MOV CX, 10H               ; Counter is initialized to 16 Dec.
        AND DL, OFH               ; Separate the lower-nibble of DL
LOOP1: MOV AL, [SI]               ; Access the string data and adjust the
        INC SI                    ; pointer to next string data
        AND AL, OFH               ;  Clear  the  higher  nibble  of
ASCII String
        MUL DL                    ; Multiply with the Multiplier and
                                    save the result
                                  ; after ASCII-adjust
        AAM
        ADD AL, [DI]
        AAA
        MOV [DI], AL              ; Store the result back in memory
        INC DI                    ; Adjust the pointer to the next
                                    memory address.
        MOV [DI], AH              ; and save the higher byte of the
                                    result in memory
        LOOP LOOP 1        ; Have all 16 ASCII string multiplied?
        MOV AH, 4CH        ; If no, go to loop
        INT 21H        ; Else terminate the program normally
        END
```

The Result for ASCII (1) : 02, 05, 00, 01, 03, 07, 00, 02, 09, 03, 00, 03, 05, 04, 04, 02, 01

```
Label          Mnemonics and Operand    Comments
               MOV AX, @ DATA     ; Initialize data segment
               MOV DS, AX              ;
               MOV SI, OFFSET VALUE ; Initialize SI with the address
                                      of the data
               MOV AL, [SI]            ; Move the first data in AL
               ADD AL, [SI + 1]        ; Add with the second data
```

```
            CALL WRITE-CHAR        ; Display data on screen
            MOV AH, 4CH            ; Terminate the program normally
            INT 21 H               ;
            WRITE-CHAR PROC
            MOV DH, AL             ; Save the result in DH register.
            AND AL, OFOH           ; Mark lower nibble and convert
                                     the higher
                                   ; nibble to ASCII value
            MOV AL, 04H
            ROR AL, CL
            CMP AL, 04H
            JB NEXT
            ADD AL, 07H
NEXT:       ADD AL, 30H
            PUSH DX
            MOV DL, AL             ; Display the digit on the screen
                               using DOS
                                   ; interrupt function 02
            MOV AH, 02H
            INT 21H
            POP DX
            AND DH, OFH
            JB NEXTI
            ADD AL, 07H
NEXT :      ADD DH, 30H
            MOV DL, DH             : Display it on the screen
            MOV AH, 02H
            INT 21H
            RET
            WRITE-CHAR ENDP        ; End the procedure and return
                                   to main program
            END
```

The Result for ASCII (2) : 96; is displayed on screen.

Program 42. *Write an ALP in 8086 to display the string information Science on the Screen.*

Explanation to write ALP for ALP string (1) : By using DOS interrupt, INT21; and function 09, it is possible to display a string of data on the screen. The function 09 needs the starting address of the function 09 needs the starting address of the string to be initialized in DI register, and using $ string in the data area has to be terminated.

```
   .MODEL SMALL
   .DATA

MSG DB "INFORMATION SCIENCE $"
CODE
   MOVE AX, @ DATA       ; Initialize the Data Segment
```

```
MOV DS, AX              ;
MOV DX, OFFSET MSG      ; Initialize DX with the starting of the message
MOV AH, 09              ; Call interrupt 21 function 09
INT 21H                 ; To display the string
MOV AH, 4CH             ; Terminate the program
INT 21H
END
```

The Result for ALP-string (1): INFORMATION SCIENCE is displayed.

Program 43. ***Write an ALP to generate a square wave of frequency 500 Hz and Amp. 5 V ?***

```
MOV AL, 80H      }
MOV DX, CSR      } initialise 8255 port A00 output port in mode 0
OUT DX, AL       }

MOV AL, 00H      }
MOV DX, Port A   } output 0 V to port A and cally delay
OUT DX, AL       }
CALL DELAY

MOV AL, FFH      }
MOV DX, Port A   } output 5 V to port A and call delay
OUT DX, AL       }
CALL Delay

Delay  : MOV BX, FFFF

Back   :  NOP       }
          NOP       }
          DEC BX    } Delay Routine
          JNZ Back  }
          Ret       }
```

Output :

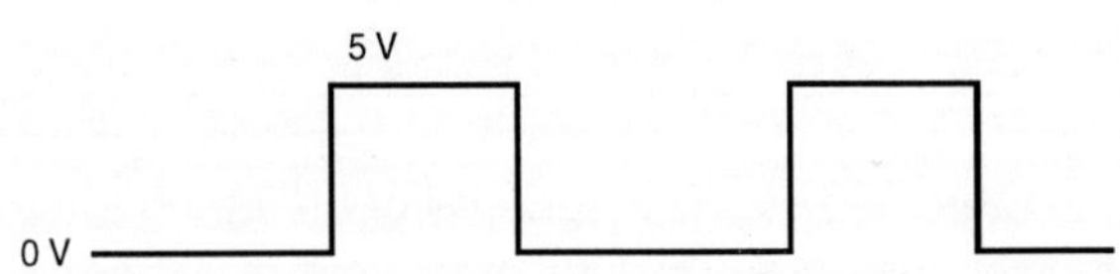

Program 44. ***Write an ALP to generate a rectangular wave of Freq. 500 Hz and amp 5 V ?***

```
MOV AL, 80H
MOV DX, CSR
OUT DX, AL
```

```
Again MOV AL, 00H
      MOV DX, Port A        } output 0 V to port A and call delay
      OUT DX, AL
      CALL AHEAD
      MOV AL, FFH
      MOV DX, Port A        } output 5 V to port A and call delay A2
      OUT DX, AL
      CALL Delay

Ahead : MOV BX, FFFF
          NOP
Back  :   MOV CX, FFFF
          NOP               } Delay for 0 V
          DEC CX
          JNZ Back
          DEC BX
          JNZ Again
          RET               } Delay for 5 V
A2 :      MOV BX, FFFF
          DEC BX
B1        JNZ B1
          RET
```

Output :

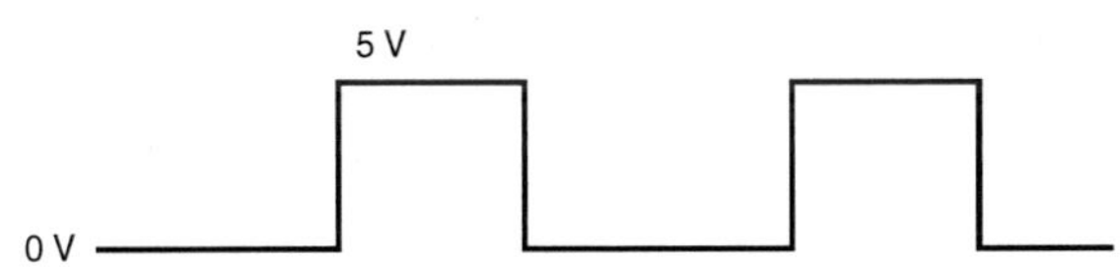

Program 45. ***Write an ALP to generate a triangular wave of frequency 500 Hz and Amp. 5 V ?***

```
MOV AL, 80H
MOV DX, CSR      } initialise 8255 port A as output port in mode 0
OUT DX, AL

      MOV AL, 00H
      MOV DX, Port A   } Start rising ramp from 0 V, increment Ramp till FF i.e. 5 V
Back: OUT DX, AL
```

```
        INC AL
        CMP AL, FFH
Again:  OUT DX, AL
        JNZ Back
        DEC AL              } Decrement ramp and repeat sequence till ramp reaches 0 V
        CMP AL, 00H
        JA Again
        JMP Back
```

Output :

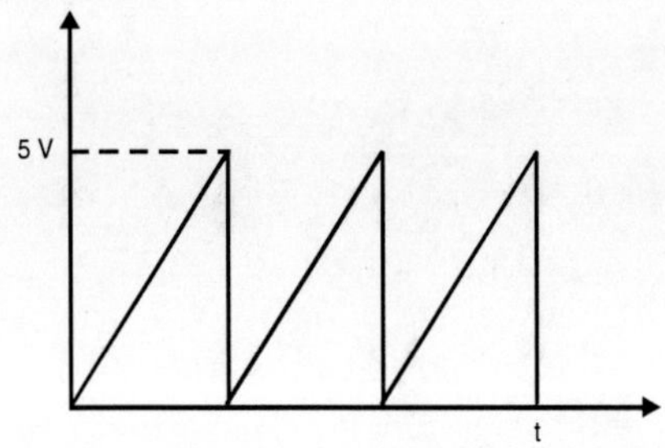

REVIEW QUESTIONS

1. What are the assembler directives and pseudo- ops?
2. Explain all the assembler directives, pseudo-ops and operators?
3. How does the cpu identify between 8-bit and 16-bit operations?
4. How is the addressing mode of an instruction communicated to the CPU?
5. Bring out the difference between the jump and loop instructions?
6. Describe the procedure for coding the intersegment and intrasegment jump and call instructions?
7. If BH = 0F3H what is the value of BH in hex after the instruction SAR BH, 1 IF AL = 78H and BL=73H explain how DAS instruction (after subtracting BL from AL) adjusts to the BCD result.
8. If CL=78H what is the value of CL after the instruction ROL CL, 3.
9. Why AAD is to be executed before DIV instruction while converting unpacked BCD to Binary number?
10. Under what conditions REPE MOVS executes.
11. Explain XLAT instruction to linearize transducer characteristics.
12. Explain intra segment and inter segment branch instructions with examples the instructions related to arithmetic and logical shift.
13. Explain EQU directive with example.
14. Explain SEGMENT directives with examples.
15. Explain coding template for 8086 instruction.
16. Explain all addressing modes with the assembler syntax and how effective address is calculated ?
17. Enlist the advantages of assembly language programming oven machine language programming?
18. What is an assembler?
19. What is a linker?
20. What are the DOS function calls?

21. Explain various DEBUG commands for troubleshooting executable programs?
22. Write an ALP to convert 4- digit hexadecimal number to decimal number?
23. Write an ALP to convert 4- digit octal number to decimal number?
24. Write an ALP to find out ASCII codes of alphanumeric characters from a look up table?
25. Write an ALP to change an already available ascending order byte string to descending order?
26. Write an ALP to perform a 16-bit increment operation using 8-bit instructions?
27. Write an ALP to find out average of a given string of data bytes neglecting fractions?
28. Write an ALP to find out decimal addition of 16 4-digit decimal numbers?
29. Write an ALP to a 4-digit decimal number to its binary equivalent?
30. Write an ALP to convert a given 16-bit Binary number Gray equivalent?
31. Write an ALP to find out transpose of a 3x3 matrix?
32. Write an ALP to display message "HAPPY BIRTHDAY" on the screen after a key A is pressed?
33. Write an ALP to open a file in drive C of your hard disk accept 256 byte entries each separated by comma and then store all these 256 byte into the opened file. Issue suitable messages wherever necessary?
34. Write an ALP to pint a message "PRINTER BUSY" onto a dot matrix printer?
35. Write an ALP that goes on accepting the keyboard entries and displays them on line on the CRT display. The control escapes to DOS prompt if enter key is pressed?
36. Write an ALP to set interrupt vector of 50 H to an address of routine ISR in segment code 1?
37. Write an ALP to set and get the system time. Assume arbitrary time for setting the system time?
38. Write an algorithm to compute Fibonacci numbers using a recursive procedure. Write 8086 assembly program for the above.
39. Write an algorithm and assembly program to convert an unpacked 4 digit number to Binary number.
40. Write an algorithm and assembly program to convert a 16 bit number to a maximum of 5 unpacked digits
41. Write an algorithm and assembly program to convert an unpacked 4 digit number to Binary number.
42. Write an algorithm and assembly program to find the square root of a 16 bit number using shift and subtract method.
43. Write an algorithm and assembly program to reverse the bits in a 16 bit number and check whether it is a palindrome.
44. Write an algorithm and assembly program for a cash bill of n materials. Rupees is a 4 digit and paisa is a 2 digit number which are stored in two different arrays. Find the total amount for the n materials. Subtract 10% discount on the total and give the actual amount to be paid. Hint Shift the total amount by one digit to get the 10% discount and get the actual amount.
45. Write a program to find out any power of a number such that $Z = X^N$. Where N is programmable and X is unsigned number.
46. Write a program to generate.
    (i) Sine Waveform (ii) Ramp Waveform (iii) Triangular Waveform
47. Write a program to measure frequency/Time period of the following functions.
    (i) Sine Waveform (ii) Square Waveform (iii) Triangular Waveform
48. Write a program to increase, decrease the speed of a stepper motor and reverse its direction of rotation ?
49. Write a programmable delay routine to cause a minimum delay = 2MS and a maximum delay = 20 minutes in the increments of 2 MS.
50. Write ALPs to trigger a triack with a +5V pulse of 200 msec as specified below.
    (i) At an angle 45° in each positive and negative half cycle.
    (ii) At an angle 30° in each positive half cycle.
    (iii) At an angle 20° in each negative half.

    Assume that after each zero crossing of a 50 Hz line signal an external zerocrossing circuit generates an interrupt to the CPU which runs at a frequency of 5 MHz.

❑❑❑

CHAPTER 8

INPUT/OUTPUT CHIPS AND THEIR INTERFACING WITH 8086 MICROPROCESSOR

INSIDE THIS CHAPTER

There are three basic ways to get data into or out of a memory. They are called programmed I/O, interrupt driven I/O, and direct memory access (DMA). Although programmed I/O is the slower of the three, it is used in simpler microprocessor systems where speed is unimportant. As the system becomes more complex, the interrupt approach becomes necessary. In the most advanced systems, DMA is needed because it is the only way to transfer large amounts of data in a short time.

8.1. PROGRAMMED I/O

Programmed I/O uses instructions to get data into or out of a CPU. To correctly time the data transfers, programmed I/O relies on clock timing or on handshaking (asynchronous). The latter type if used more often because it is simpler. This section given two examples of asynchronous programmed I/O.

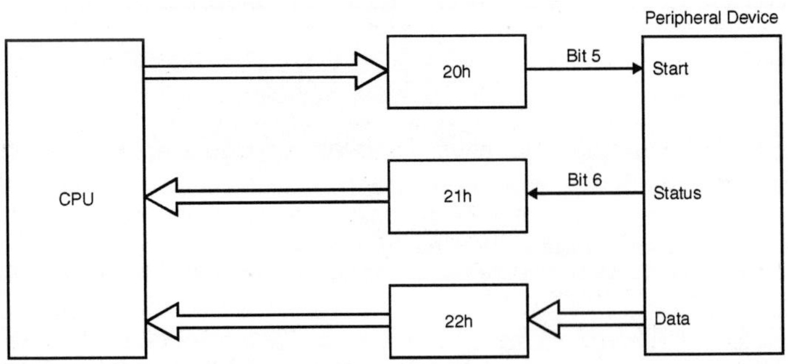

Fig. 8.1. Programmed Input

8.1.1. Programmed Input

Figure 8.1 shows a CPU connected to ports 20h, 21h and 22h. Bit 5 of port 20h is connecte to the START input of the peripheral device and bit 6 of 21 h to the status output. The CPU can send data to the CPU through input port 22h.

The basic idea is this. When the CPU is ready to input a data, it sends a high start bit to the peripheral device. when the device has the data ready for transfer, it second it high status bit to port 21h. Until the status bit is high, the CPU waits. As soon as the status bit goes high, the CPU inputs 1 byte of data.

Suppose we want to input 100 bytes and store them at address ARRAY onwards. Here is an example of programmed I/O written for an 8086 system.

```
title Input data from peripheral device
dosseg
.Model Small
.Stack 100h
.Code
Main Proc
   mov ax, @ data           ; Initialize ds register
   mov ds, ax
   mov cx, 64h
   mov si, offset array;
Back :    mov al, 20h.      ; Set start bit
          out 20h, al       ; Send high start bit
Wait :    in al, 21h        ; Get status bit
   and al, 40h              ; Isolate status bit
   jz wait                  ; Wait if device not ready
   int al, 22h              ; Input data
   mov [si], al             ; Store data
   inc si                   ; Update pointer
   mov al, 00               ; Reset start bit
   out 20h, al              ; Send low start bit
   loop back                ; Go back if notfinished
   mov ax, 4C00h            ; Return to dos
   int 21h
   main endp.
.data
array db 64h dup (0)
end main.
```

8.1.2. Output Example

Figure 8.2 shows a CPU connected to handshaking ports 20h and 21h. It is also connected to output port 22h. As before bit 5 of port 20h is the start bit, and bit 7 of port 21h is the status bit.

Here is the procedure for output operations. When the CPU is ready, it will latch the data into 22h. Then, the CPu sends a high start bit to indicate that valid data is waiting for transfer. After the peripheral device has loaded the data, it returns a high status bit.

Suppose we want to output 100 bytes from memory locations ARRAY onwards. Here is an example of a programmed output.

```
title output data to peripheral device
dosseg
.Model Small
.Stack 100h

.Code
Main proc
      mov ax, @ data; initialize ds register
      mov ds, ax
      mov cx, 64h; No. of bytes = 100
      mov si, offset array ; Initialize SI register
      mov al, [si]     ;
      out 22h, al      ; Latch data into 22h
      mov al, 20 h     ; Set start bit
   Wait : in al, 21h   ; Get status bit
      and al, 40h      ; Isolate status bit
      jz wait     ; Wait for high status
      inc si      ; Increment pointer
   mov al, 01h    ; Reset start bit
   out 20h, al    ; Send low start bit
   loop back      ; Go back if count ¹ 0
   mov ax, 4C00h ; Return to dos.
main endp
.data
array db 64h dup (0)
end main.
```

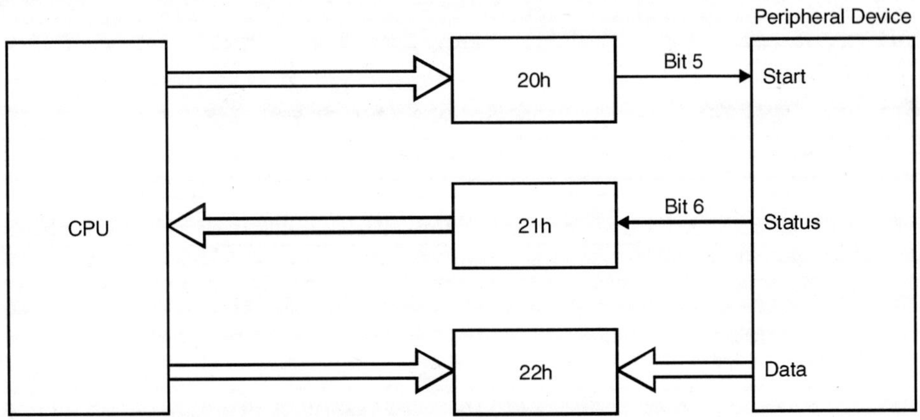

Fig. 8.2. Programmed output.

Incidentally, programmed I/O is sometimes referred to as polled I/O. In the examples given, we have used software to control the I/O transfers of a single peripheral devices and transfer data when each device is ready.

8.2. INTERRUPT I/O

A disadvantages of programmed I/O is that the computer needs to check the status bit by waiting in a loop. This type of I/O transfer is dependent on the speed of the external device. For a slow device, this waiting may slow down the computer's capability of processing other data. The interrupt I/O technique is efficient in this type of situation.

An interupt is an event that causes the CPU to initiate a fixed sequence, known as an interrupt sequence. Before an 8086 interrupt sequence can begin, the currently executing instruction must be completed unless the current instruction is a HLT or a wait instruction. For a prefixed instruction, the interrupt request is not recognized between the prefix and the instruction. In the case of the REP instruction, the interrupt request is not recognized after the primitive operation following the REP is completed, and the return address is the location of the REP prefix. For MOV and POP instructions in which the destination is a segment register, an interrupt request is not recognized until after the instruction following the MOV or POP instruction is executed. Provided that the segment register is filled first, this allows the contents of both a segment register and a pointer to be changed without interruption.

For the 8086, once the interrupt request has been recognized, the interrupt sequence consists of

(a) Establishing a type N

(b) Pushing the current contents of PSW, CS and IP onto stack.

(c) Clearing the IF and TF flags.

(d) Putting the contents of memory locations $4 \times N$ into IP and the contents of $4 \times N + 2$ into CS.

Thus, an interrupt request causes the normal program sequence to be suspended a branch to be made to the location indicated by the double word beginning at four times the type (i.e. the interrupt pointer). Control can be returned to the point at which the interrupt occurred by placing an IRET instruction at the end of the interrupt routine.

There are typically two types of interrupts : external interrupts and internal interrupts.

External interrupts are initiated through the computer's interrupt pins by external devices such as A/D convertor.

External interrupts can further be divided into two types : maskable and nonmaskable. A maskable interrupt is enabled or disabled by executing instructions such as EI or DI. If the computer's interrupt is disabled the computer ignores the maskable interrupt. Some processors, such as the the Intel 8086, have an interrupt flag bit in the processor status register. When the interrupt is doubled, the interrupt flag bit is 1, so no maskable interrupts are recognized by the processor. The interrupt flag bit is reset to zero when the interrupt is enabled.

The non-maskable interrupt has higher priority than the maskable interrupt. If both maskable and non-maskable interrupts are activated at the same time, the processor will service the nonmaskable interrupt. The nonmaskable interrupt is typically used as a power failure interrupt, processors normally use +5 V DC, which is transformed from 220 V AC. If the power falls below 16 V, AC, the Dc voltage of 5 volts cannot be maintained. However, it will take a few milliseconds before the AC power can drop this low. In these few milliseconds, the power failure-sensing circuitry can interrupt the processor. The interrupt service routine can be written to store critical data in nonvolatile memory such as battery-backed CMOS, RAM. The interrupted program can continue without any loss of data when the power returns.

8086 is also provided with a maskable handshake interrupt. This interrupt is usually implemented by using two pins—INTR and INTA. When the INTR pin is activated by an external device, the processor completes the current instruction, saves at least the current program counter onto stack and generates an interrupt acknowledge (INTA). In response to the INTA, the external device provides an instruction, such as call, using external hardware on the data bus of the computer. This instruction is then read and executed by the computer to branch to the desired subroutine.

Internal interrupts are activated internally by exceptional conditions such as overflow, division by zero, or execution of an illegal opcode. The user writes a service subroutine to take corrective measure and provide an indication to inform the user that an exceptional condition has occured.

8.2.1. Interrupt Address Vector

The technique used to find the starting address of the service routine (Commonly known as the interrupt address vector) varies from one processor to another. With some processors, the manufactured define the fixed starting address for each interrupt. For 8086 system, this address is 4 * N. The value of N can vary from 0 to 255.

8.2.2. Saving The CPU Registers

When a processor is interrupted, it saves at least the program counter on the stack so that processors save only one or two registers, such as the program counter and status register. Other processors save all CPU registers before going to the service routine. The user should know the specific registers the processor saves prior to executing the service routine. This will enable the user to use the appropriate return instruction at the end of the service routine to restore the original conditions upon return to the main program.

8.2.3. Interrupt priorities

A processor is typically provided with one or more interrupt pins on the chip. Therefore, a special mechanism is necssary to hand interrupts from several devices that share one of these interrupts lines. There are two ways of servicing multiple interrupts : polled and daisy chain techniques.

8.2.3.1. Polled Interrupts

Polled interrutps are handled by software, and therefore are slower, compared with daisy chaining. The processor responds to an interrupt by executing one-general purpose routine for all devices. The priorities of devices are determined by the order in which the routine polls each device. The processor checks the status of each device in the general service routine, starting with the highest priority device, to service an interrupt. Once the processor determines the source of interrupt, it branches to the service subroutine for the device. Figure 8.3 shows a typical configuration of the polled interrupt system.

In Figure 8.3 several external devices (device-1, device-2, ..., device N) are connected to a single interrupt line of the processor via an OR gate. When one or more devices activate the INT line high, the processor pushes the program counter and possibly some other register onto the stack p then branches to an address defind by the manufacturer of the processor. The user can write a program at this address to poll each device, starting with the highest priority device to find the source of the interrupt. Suppose the devices in figure 8.3 are A/D convertors. Each convertor, along with the associated logic for polling is shown in Fig. 8.4.

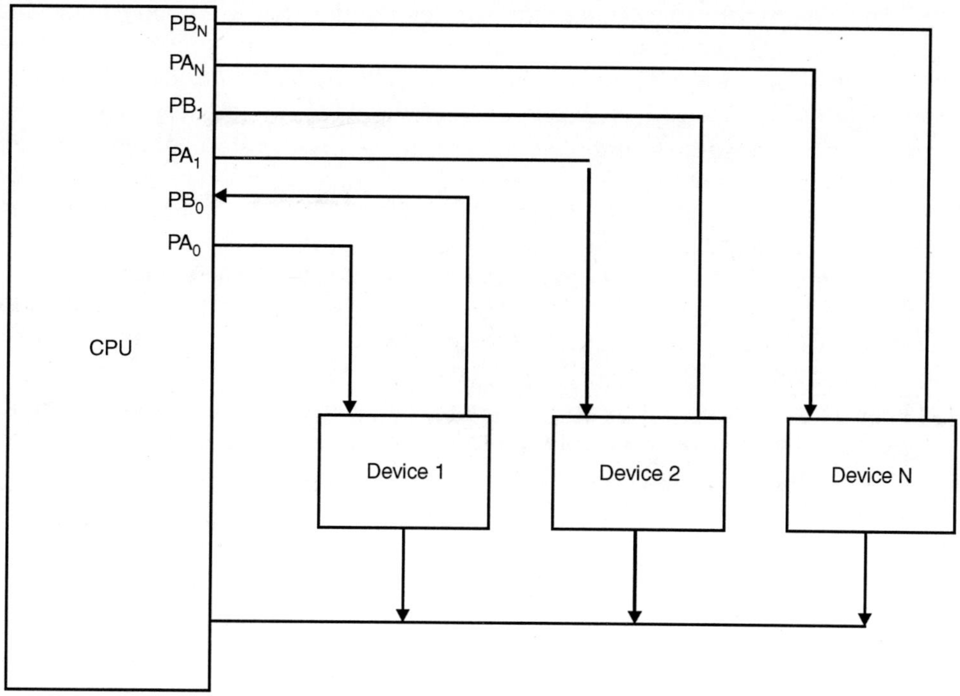

Fig. 8.3. Polled interrupt.

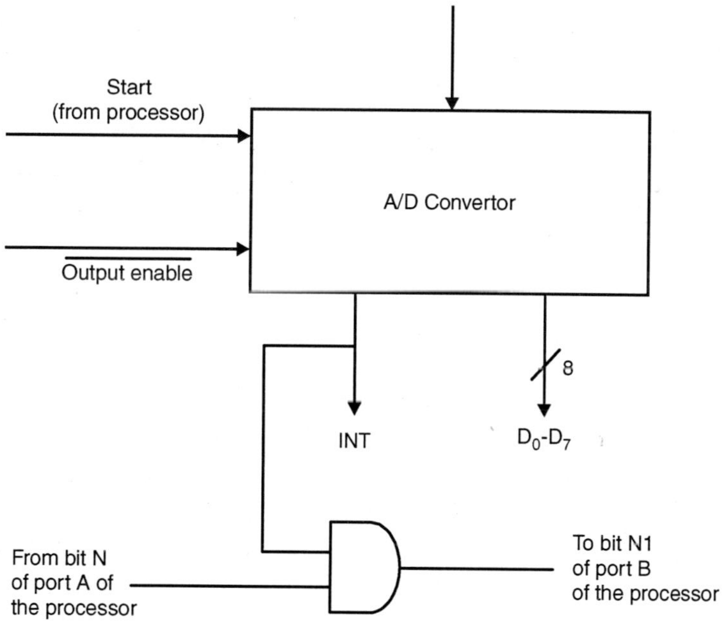

Fig. 8.4. Device N and associated logic for polled interrupt.

Assume that in Fig. 8.3 two A/D convertors (device 1 and device 2) are provided with the START pulse by the processor at nearly the same time. Suppose the user assigned device 2 with the higher priority. The user then sets up this priority mechanism in the general-service routine. For example when the busy signals from device 1 and/or device 2 become high, indicating the end of conversion the processor is interrupted. In response, the processor pushes at least the program counter on the stack, and loads the PC with the interrupt address vector defined by the manufacturer.

The general interrupt-service routine written at this address determines the source of the interrupt as follows : A1 is sent to PA1 for device 2, since the device has higher priority. If this device has generated the interrupt, the output (PB1) of the AND gate in Fig. 8.3 becomes high, indicating to the processor that device 2 generated the interrupt. If the output of the AND gate is 'O', the processor sends a high to PAO and checks the output (PBO) for high. Once the source of interrupt is determined, the processor can be programmed to jump to the service routine for that device. The service routine enable the A/D converter and input the convertor's outputs to the processor.

Polled interrupts are slow and for a large number of devices, the time required to poll each device may exceed the defined interrupt time to service. In such a case, a faster mechanism, such as the daisy chain approach, can be used.

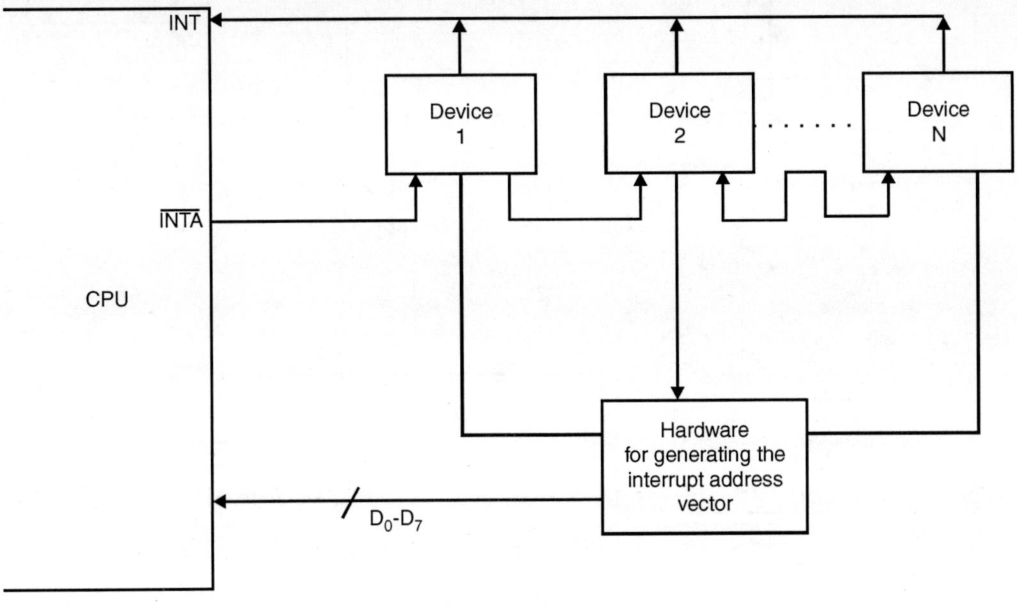

Fig. 8.5. Daisy Chain interrupt.

Suppose the processor in Fig. 8.5 sends a pulse to start the conversions of A/D convertor of devices 1 and 2 at nearly same time. When the Busy signal goes to high, the processor is interrupted through INT line. The processor pushes the PC and possibly some other registers. If then generates a low at the interrupt acknowledge ($\overline{INTA}$) for the highest priority (device 1 in Fig. 8.5). Device 1 has the highest priority it is the first device in the daisy chain configuration to receive $\overline{INTA}$. If A/D convertor 1 has generated the busy high, the output of the AND gate becomes high. This signal can be used to enable external hardware to provide the interrupt address vector on the processor's data lines. The processor then branches to the service routine. This program enables the convertor, and inputs the A/D output to the processor via port B. If A/D convertor #1 does not generate the busy high, however, the output of the AND gate in Fig. 8.6 becomes low (an input to device 2's logic) and the same sequence of operations takes place. In the daisy chain, each device has the same logic, with the exception of the last device which must accept the $\overline{INTA}$. Note that the outputs of all the devices are conencted to INT line via an OR gate.

8.3. DIRECT-MEMORY ACCESS

A floppy disk is an thin plastic disk about 5¼" inches in diameter, coated with magnetic code. A disk drive is a peripheral device that can either read or write data onto the disk, which can stored a million of bytes. The only practical way to transfer data to and from the disk is with direct memory access (DMA). As discussed earlier, the 8086 can turn over control of its buses to a DMA controller for a high-speed I/O transfers. In this way, large amounts of data can be transferred in a relatively short time.

Accumulator in the middle : The details of DMA transfer are too complicated to go into here, but we can discuss the basic idea. The IN instruction in the usual way to input data from peripheral devices. The accumulator is involved because it receives the input data.

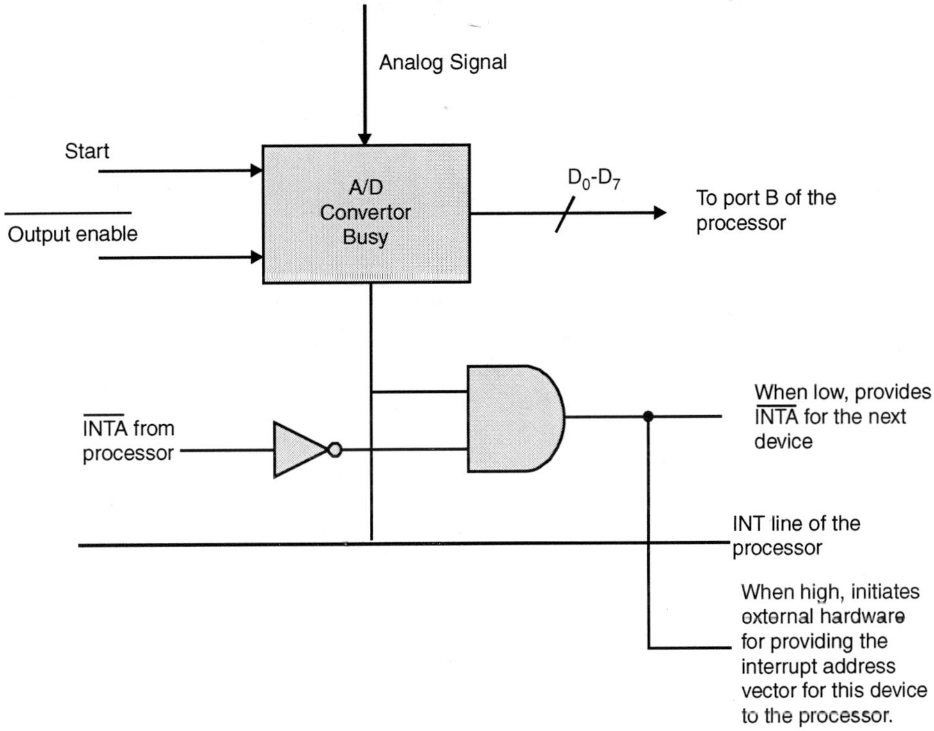

Fig. 8.6. Each device and the associated logic in daisy chain.

8.4. 8255 PROGRAMMABLE I/O CHIP

Description

The Intel 82C55A is a high performance CMOS version of the industry standard 8255A and is manufactured using a self-aligned silicon gate CMOS process. It is a general purpose programmable I/O device which may be used with many different microprocessors. There are 24 I/O pins which may be individually programmed in 2 groups of 12 and used in 3 major modes of operation. The high performance and industry standard configuration of the 82C55A make it compatible with the 80C86, 80C88 and other microprocessors.

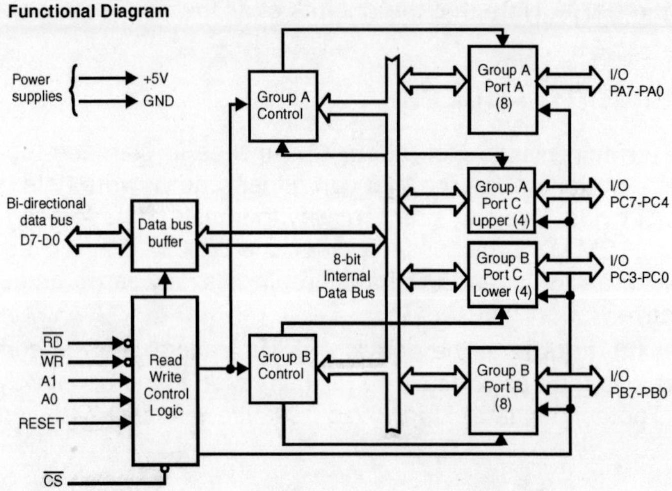

Fig. 8.7. Functional block diagram of 8255.

Static CMOS circuit design insures low operating power. TTL compatibility over the full military temperature range and bus hold circuitry eliminate the need for pull-up resistors.

8.4.1. Functional Description

Data Bus Buffer

This three-state bi-directional 8-bit buffer is used to interface the 82C55A to the system data bus. Data is transmitted or received by the buffer upon execution of input or output instructions by the CPU. Control words and status information are also transferred through the data bus buffer.

Read/Write and Control Logic

The function of this block is to manage all of the internal and external transfers of both Data and Control or Status words. It accepts inputs from the CPU Address and Control buses accepts inputs from the CPU Address and Control busses and in turn, issues commands to both of the Control Groups.

(CS) Chip Select : A "low" on this input pin enables the communication between the 82C55A and the CPU.

(RD) Read : A "low" on this input pin enables 82C55A to send the data or status information to the CPU on the data bus. In essence, it allows the CPU to "read from" the 82C55A.

(WR) Write : A "low" on this input pin enables the CPU to write data or control words into the 82C55A.

(A0 and A1) Port Select 0 and Port Select 1 : These input signals, in conjunction with the RD and WR inputs, control the selection of one of the three ports or the control word register. They are normally connected to the least significant bits of the address bus (A0 and A1).

(RESET) Reset : A "high" on this input initializes the control register to 9Bh and all ports (A, B, C) are set to the input mode. "Bus hold" devices internal to the 82C55A will hold the I/O port inputs to a logic "1" state with a maximum hold current of 400 mA.

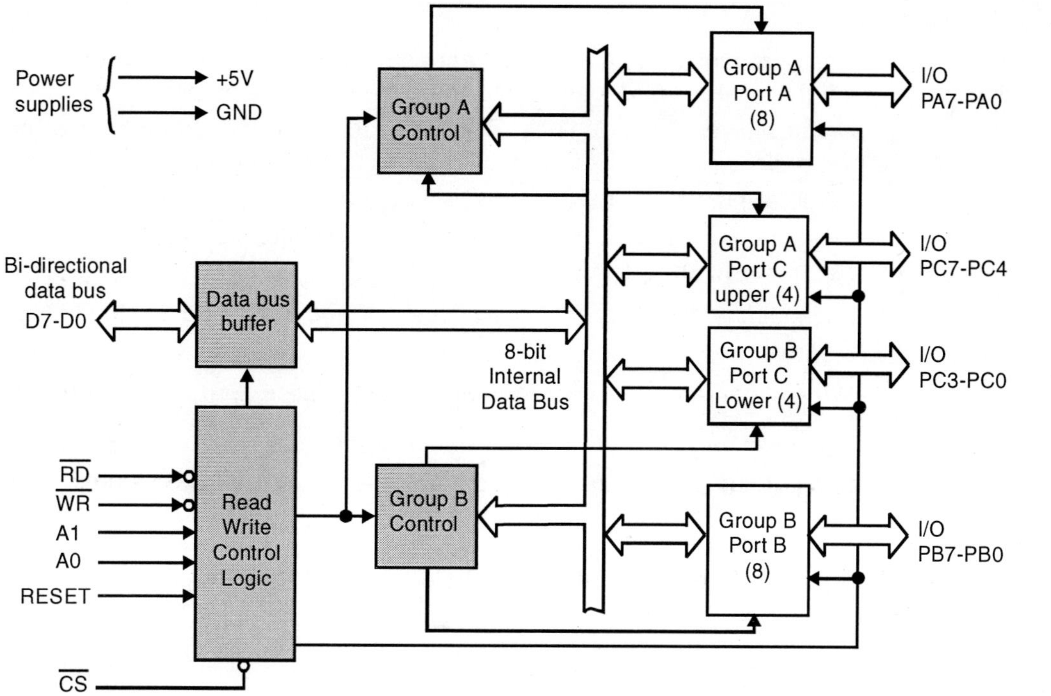

Fig. 8.8. 82C55A block diagram data bus buffer, read/write, groups A and B control logic functions.

Pin Diagram of 8255

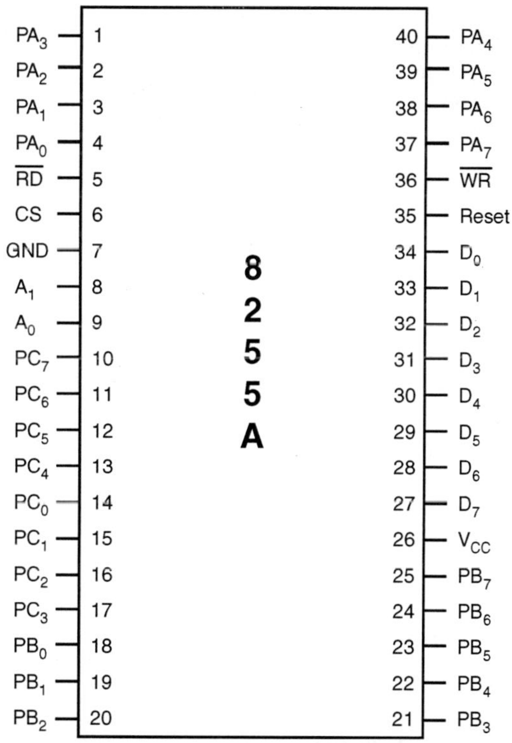

Fig. 8.9. Pin Digram of 8255

Pin Description

$PA_7 - PA_0$: These are eight part A lines that act as either latched output or buffered input line depending upon the control word loaded into the control word register.

$PC_7 - PC_4$: Upper nibble of part C lines. They may act as either output latches or input buffer lines. This part can also be used for generation of handshake line in mode 1 and mode 2.

$PC_3 - PC_0$: These are the lower part C lines.

$PB_7 - PB_0$: These are the 8 part B lines which are used as latched output lines or buffered input lines in the same way as part A.

$\overline{RD}$: This is the input line driven by the microprocessor and should be low to indicate read operation to 8255.

$\overline{WR}$: This is an input line driven by the microprocessor. A low on this line indicates write operation.

$\overline{CS}$: This is a chip select line. If this line goes low, it enables the 8255 to respond to $\overline{RD}$ and $\overline{WR}$ signal are neglected.

$A_1 - A_0$: These are the address input lines and are driven by the microprocessor. These lines ($A_1 - A_0$) with $\overline{RD}$, $\overline{WR}$, and $\overline{CS}$ form the following operations for 8255.

These address lines are used for addressing any one of the 4-registers, i.e. 3 parts and the control word register as shown in table below

TABLE 8.1

$\overline{RD}$	$\overline{WR}$	$\overline{CS}$	A_1	A_0	Input (Read Cycle)
0	1	0	0	0	→ Port A to data bus
0	1	0	0	1	→ Port B to data bus
0	1	0	1	0	→ Port C to data bus
0	1	0	1	1	→ CWR to data bus (Control Word Reg.)

TABLE 8.2

$\overline{RD}$	$\overline{WR}$	$\overline{CS}$	A_1	A_0	Output (Write Cycle)
1	0	0	0	0	→ Data bus to port A
1	0	0	0	1	→ Data bus to port B
1	0	0	1	0	→ Data bus to port C
1	0	0	1	1	→ Data bus to CWR

TABLE 8.3

$\overline{RD}$	$\overline{WR}$	$\overline{CS}$	A_1	A_0	Function
X	X	1	X	X	Data bus tristated
1	1	0	X	X	Data bus tristated

If 8255 is to be interfaced with lower order data bus, the A_0 and A_1 pins of 8255 are connected with A_1 and A_2 respectively in case of 8086 systems.

D_0 – D_7 : These are the data bus lines those carry data or control word to/from the microprocessor.

Reset : A logic high on this line clears the control word register of 8255. All parts are set as input ports by default after reset.

Group A and Group B Controls

The functional configuration of each port is programmed by the systems software. In essence, the CPU "outputs" a control word to the 82C55A. Information such as "mode", "bit set", "bit reset", etc., that initializes the functional configuration of the 82C55A. Each of the Control blocks (Group A and Group B) accepts "commands" from the Read/Write Control logic, receives "control words" from the internal data bus and issues the proper commands to its associated ports.

Control Group A - Port A and Port C upper (C7-C4) Control Group B - Port B and Port C lower (C3-C0)

The control word register can be both written and read as shown in the "Basic Operation" table. Figure 8.4 shows the control word format for both Read and Write operations. When the control word is read, bit D7 will always be a logic "1", as this implies control word mode information.

Ports A, B, and C

The 82C55A contains three 8-bit ports (A, B, and C). All can be configured to a wide variety of functional characteristics by the system software but each has its own special features or "personality" to further enhances the power and flexibility of the 82C55A.

Port A One 8-bit data output latch/buffer and one 8-bit data input latch. Both "pull-up" and "pull-down" bus-hold devices are present on Port A. (See Fig. 8.10.)

Port B One 8-bit data input/output latch/buffer and one 8-bit data input buffer. (See Fig. 8.10.)

Port C One 8-bit data output latch/buffer and one 8-bit data input buffer (no latch for input). This port can be divided into two 4-bit ports under the mode control. Each 4-bit port contains a 4-bit latch and it can be used for the control signal output and status signal inputs in conjunction with ports A and B. (Fig. 8.10)

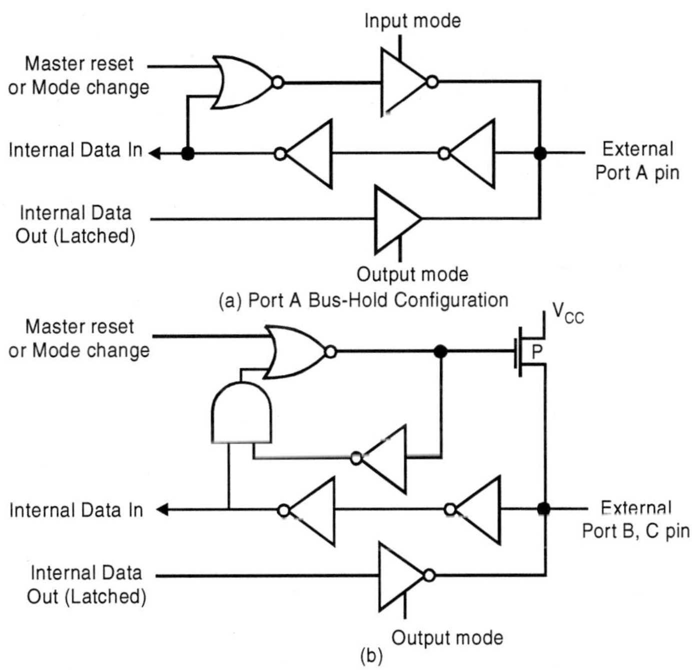

Fig. 8.10. Bus-hold configuration.

TABLE 8.4. 8255 a Basic Operation

A_1	A_0	$\overline{RD}$	$\overline{WR}$	$\overline{CS}$	Input Operation (Read)
0	0	0	1	0	Port A = Data bus
0	1	0	1	0	Port B = Data bus
1	0	0	1	0	Port C = Data bus
					Output operation (Write)
0	0	1	0	0	Data bus = Port A
0	1	1	0	0	Data bus = Port B
1	0	1	0	0	Data bus = Port C
1	1	1	0	0	Data bus = Control
					Disable Function
×	×	×	×	1	Data bus = 3 state
1	1	0	1	0	Illegal function
×	×	1	1	0	Data bus = 3 State

8.4.2. Group A and Group B Controls

The functional configuration of each port is programmed by the system software. In essence, the CPU outputs a control word to the 8255 A. The control word contains information such as "mode", "bit set", "bit reset", etc. that intializes the functional configuration of the 8255.

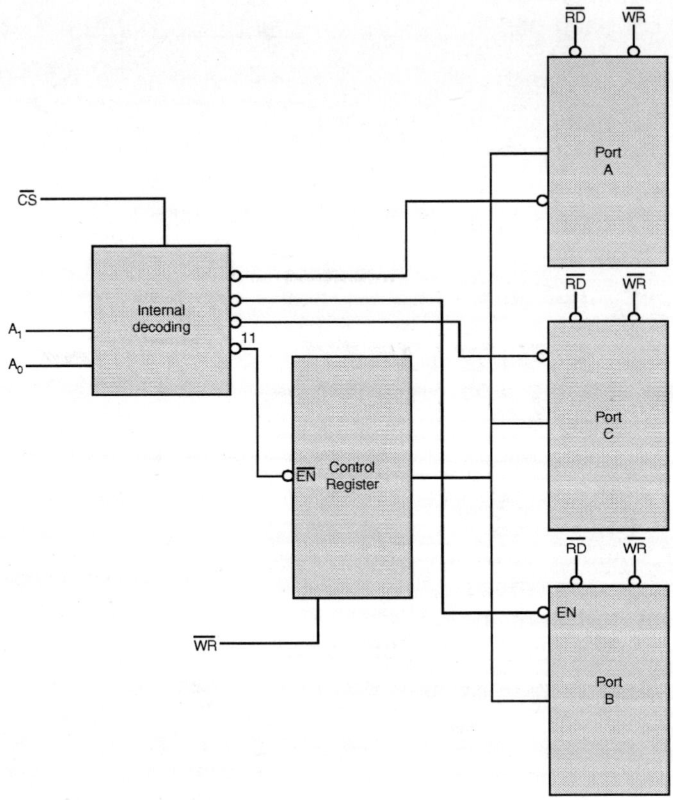

Fig. 8.11.

Each of the control blocks (Group A, and Group B) accepts "commands" from the Read/ Write control logic, receives control word from the internal data bus and issues the proper commands to its associated ports.

Control Group A - Port A and Port Cupper (C7 – C4)

Control Group B - Port B and Port Clower (C3 – C0).

Figure 8.11 shows a register called the control register. The contents of this register, called the control word, specify an I/O function for each port. This register can be accessed to write a control word when A0 and A1 are at logic 1, as mentioned previously. The register is not accessible for a read operation.

Three ports. No "handshaking" is required, data is simply written to or read from a specified port. Mode 0 basic functional definitions :

- Two 8-bit ports and two 4-bit ports.
- Any port can be input or output.
- Outputs are latched.
- Inputs are not latched.
- 16 different input/output configurations are possible in this mode.

8.4.3. Operational Description

Mode Selection

There are three basic modes of operation than can be selected by the system software:

Mode 0 - Basic Input/Output

Mode 1 - Strobed Input/Output

Mode 2 - Bi-directional Bus

When the reset input goes "high", all ports will be set to the input mode with all 24 port lines held at a logic "one" level by internal bus hold devices. After the reset is removed, the 82C55A can remain in the input mode with no additional initialization required. This eliminates the need to pullup or pull- down resistors in all-CMOS designs. The control word register will contain 9Bh. During the execution of the system program, any of the other modes may be selected using a single output instruction. This allows a single 82C55A to service a variety of peripheral devices with a simple software maintenance routine. Any port programmed as an output port is initialized to all zeros when the control word is written.

The modes for Port A and Port B can be separately defined, while Port C is divided into two portions as required by the Port A and Port B definitions. All of the

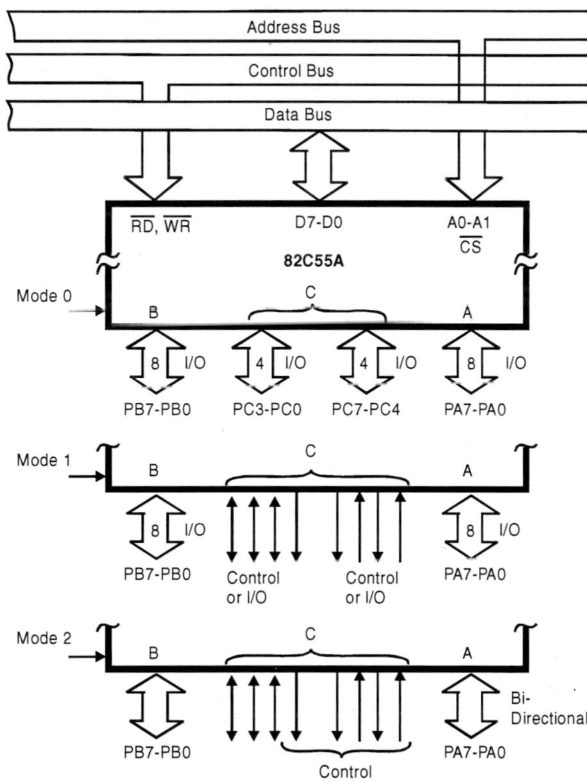

Fig. 8.12. Basic Mode definitions and Bus interface.

output registers, including the status flip-flops, will be reset whenever the mode is changed. Modes may be combined so that their functional definition can be "tailored" to almost any I/O structure. For instance: Group B can be programmed in Mode 0 to monitor simple switch closings or display computational results, Group A could be programmed in Mode 1 to monitor a keyboard or tape reader on an interrupt-driven basis.

The mode definitions and possible mode combinations may seem confusing at first, but after a cursory review of the complete device operation a simple, logical I/O approach will surface. The design of the 82C55A has taken into account things such as efficient PC board layout, control signal definition vs. PC layout and complete functional flexibility to support almost any peripheral device with no external logic. Such design represents the maximum use of the available pins.

Construction and Sending of Control Words of 8255

As shown in Fig. 8.13 we see the format of 8255 control word. The MSB of the control word tells the 8255 which control word we are sending it.

We use the mode definition control word format shown in Fig. 8.13 to tell the device what modes one wants to operate in.

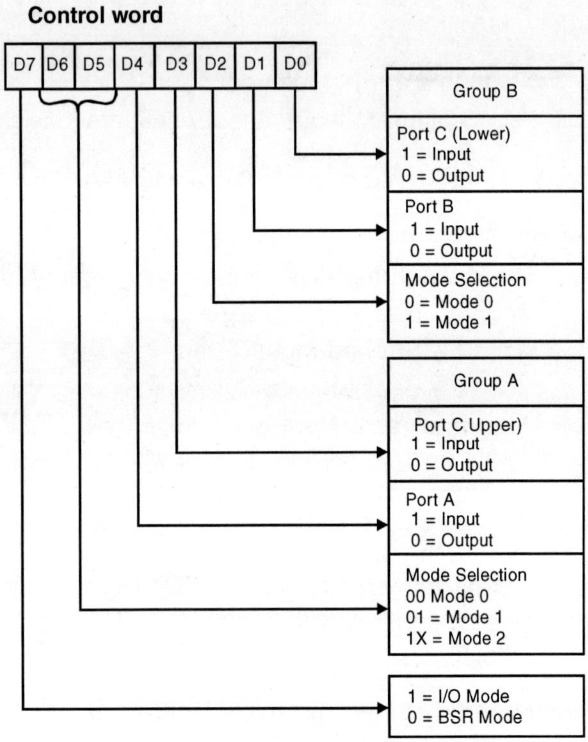

Fig. 8.13. Mode definition format.

Single Bit Set/Reset Feature (Fig. 8.14)

Any of the eight bits of Port C can be Set or Reset using a single Output instruction. This feature reduces software requirements in control-based applications. operation just as if they were output ports.

When Port C is being used as status/ control for Port A or B, these bits can be set or reset by using the Bit Set/Reset.

Word in the control register D7 = 0 is recognized as BSR control word and does not affect the controls transmitted previously. Hence the I/O operations of Ports A and B are not disturbed by BSR control word.

Example 1. Write a control word to set PC0 to PC5 of port C of 8255 ?

Solution. Following are the control words to set PC0-PC5 bits of Port C :

PC5	00001011 = 0B
PC4	00001001 = 09
PC3	00000111 = 07
PC2	00000101 = 05
PC1	00000011 = 03
PC0	00000001 = 01

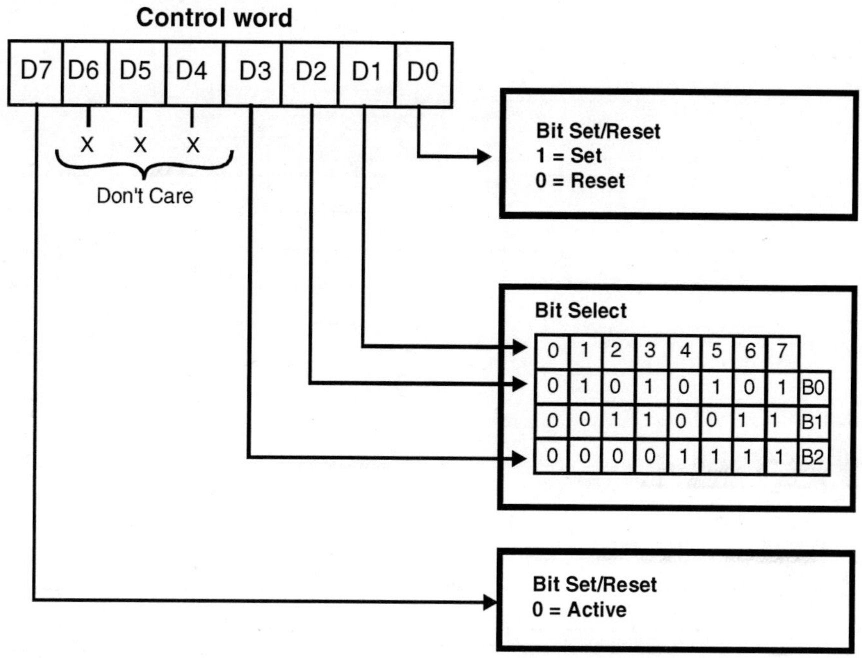

Fig. 8.14. Bit set/reset format.

Example 2. Write a control word to reset all bits of port C of 8255 chip ?

Solution. Following are the control words to reset all the bits of port C of 8255 PPI :

PC0	00000000 = 00H
PC1	00000010 = 02H
PC2	00000100 = 04H
PC3	00000110 = 06H
PC4	00001000 = 08H
PC5	00001010 = 0AH
PC6	00001100 = 0CH
PC7	00001110 = 0EH

8.4.4. Operating Modes

Mode 0 (Basic Input/Output) : This functional configuration provides simple input and output operations for each of the three ports. No handshaking is required, data is simply written to or read from a specific port.

Mode 0 Basic Functional Definitions:

➤ Two 8-bit ports and two 4-bit ports

➤ Any Port can be input or output

➤ Outputs are latched

➤ Input are not latched

➤ 16 different Input/Output configurations possible.

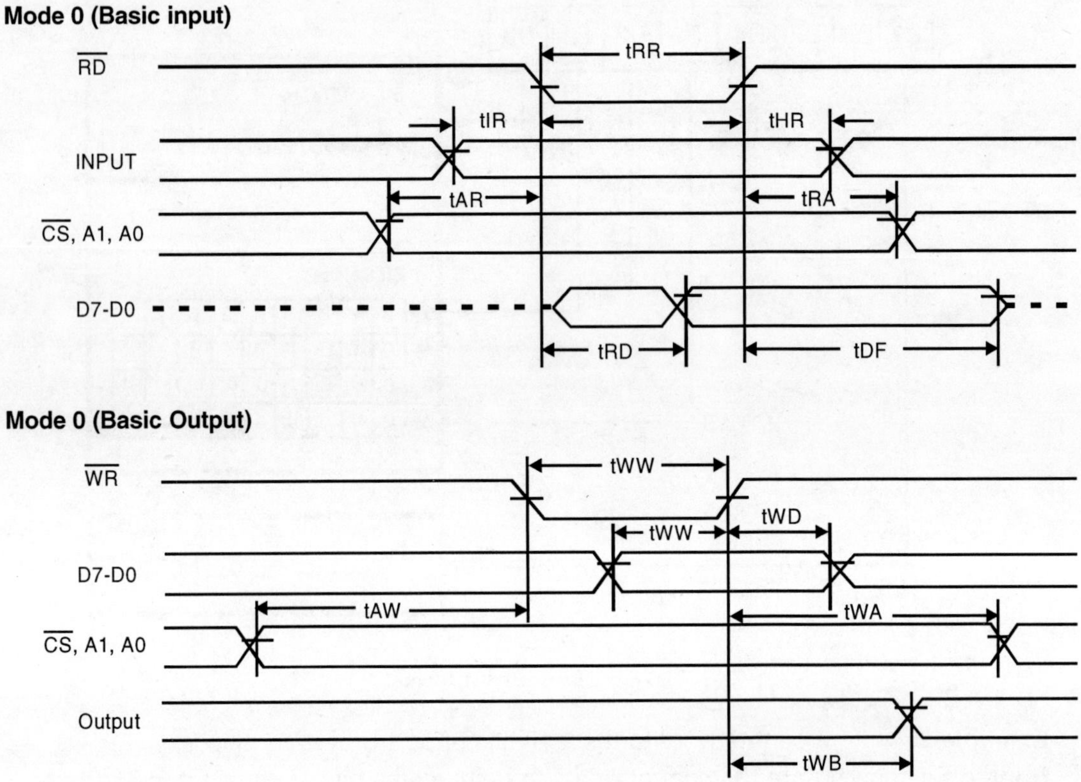

Fig. 8.15. Basic input/output cycle for Mode 0.

Mode 1 - (Strobed Input/Output) : This functional configuration provides a means for transferring I/O data to or from a specified port in conjunction with strobes or "hand shaking" signals. In mode 1, port A and port B use the lines on port C to generate or accept these "hand shaking" signals. Mode 1 Basic Function Definitions:

(i) Two Groups (Group A and Group B).
(ii) Each group contains one 8-bit port and one 4-bit control/data port
(iii) The 8-bit data port can be either input or output. Both inputs and outputs are latched.
(iv) The 4-bit port is used for control and status of the 8-bit port.

STB (Strobe Input)

A "low" on this input loads data into the input latch.

IBF (Input Buffer Full F/F)

A "high" on this output indicates that the data has been loaded into the input latch: in essence, and acknowledgment IBF is set by STB input being low and is reset by the rising edge of the RD input.

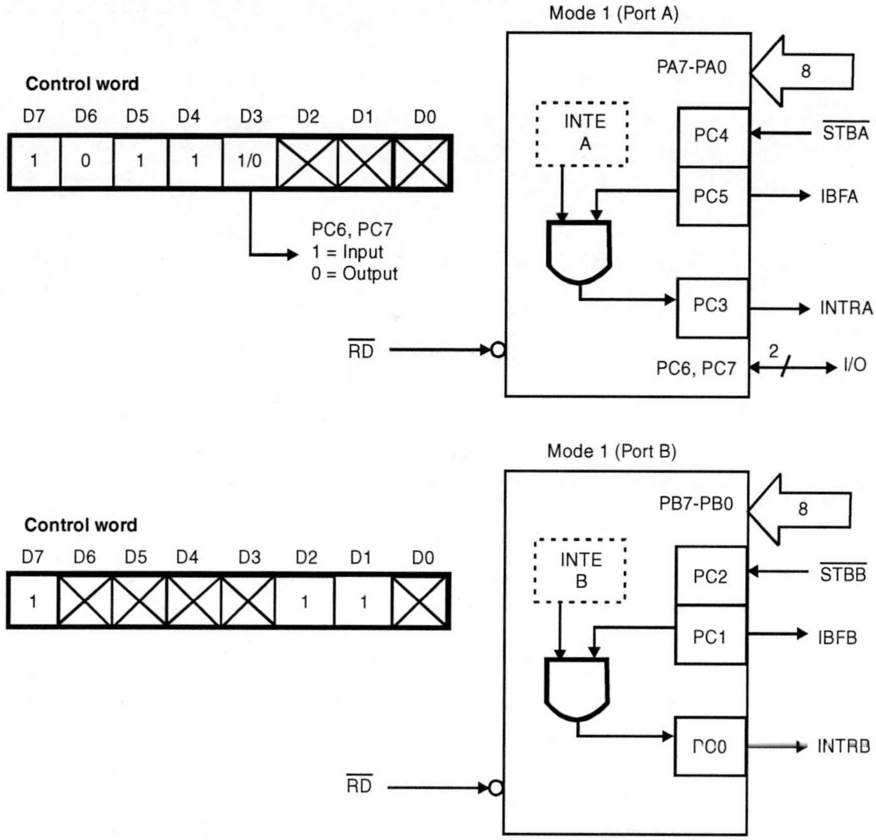

Fig. 8.16. Mode 1 input.

INTR (Interrupt Request)

A "high" on this output can be used to interrupt the CPU when an input device is requesting service. INTR is set by the condition: STB is a "one", IBF is a "one" and INTE is a "one". It is reset by the falling edge of RD. This procedure allows an input device to request service from the CPU by simply strobing its data into the port.

INTE A

Controlled by bit set/reset of PC4.

INTE B

Controlled by bit set/reset of PC2.

Output Control Signal Definition

OBF - Output Buffer Full F/F) : The OBF output will go "low" to indicate that the CPU has written data out to be specified port. This does not mean valid data is sent out of the part at this time since OBF can go true before data is available. Data is guaranteed valid at the rising edge of OBF. The OBF F/F will be set by the rising edge of the WR input and reset by ACK input being low.

ACK - (Acknowledge Input) : A "low" on this input informs the 82C55A that the data from Port A or Port B is ready to be accepted. In essence, a response from the peripheral device indicating that it is ready to accept data.

INTR - (Interrupt Request) : A "high" on this output can be used to interrupt the CPU when an output device has accepted data transmitted by the CPU. INTR is set when ACK is a "one", OBF is a "one" and INTE is a "one". It is reset by the falling edge of WR.

INTE A

Controlled by Bit Set/Reset of PC6.

INTE B

Controlled by Bit Set/Reset of PC2.

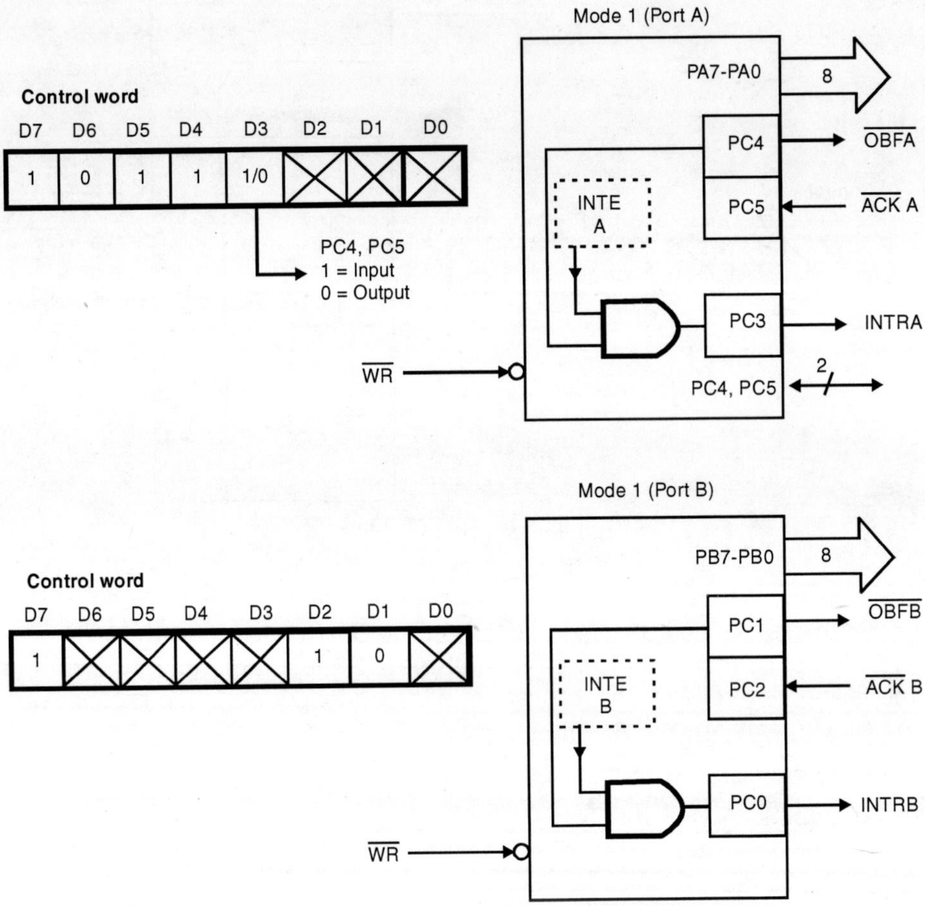

Fig. 8.17. Mode 1 output.

Programming the 8255 in Mode 1

The 8255 can be programmed to function using either status check I/O or interrupt I/O. The technique is similar to that of mode 0 combined with the BSR mode. The disadvantage of status check I/O with handshake is that the CPU is tied up in the loop.

The $INTE_A$ is set or reset in the BSR mode and the BSR control cword has no effect when ports A and B are set in mode 1.

Output Control Signal Definition

Figure 8.17(a) shows the control signals when ports A and B are configured as output ports. These signals are defined as follows :

OBF : Output Bufer Full : The OBF will go "low" to indicate that the CPU has written data out to the specified port. The OBF F/F will be set by the rising edge of the WR input and reset by ACK input being low.

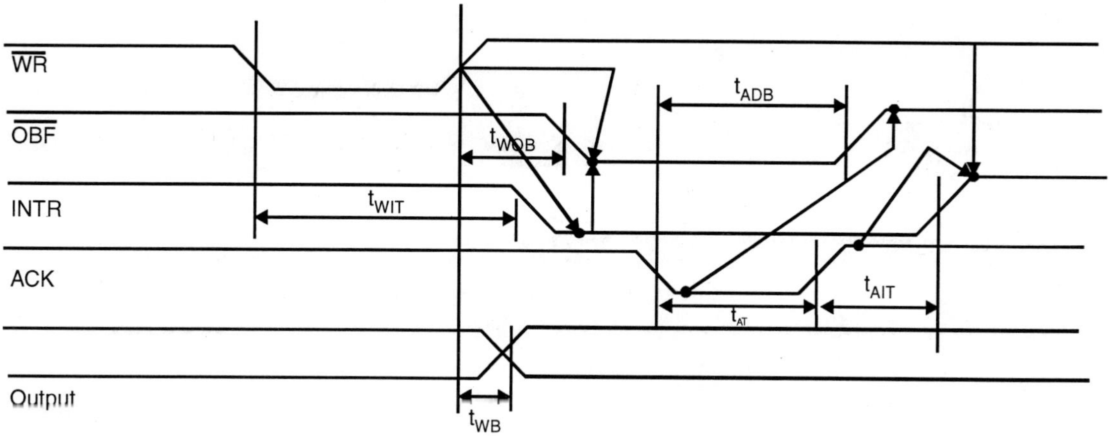

Fig. 8.18.

$\overline{ACK}$ (Acknowledgement input) : A 'low' on this input informs the 8255A that the data from port A or port B has been accepted. In essence, a response from the peripheral device indicating that it has received the data output by the CPU.

INTR (Interrupt Request) : A "high" on this output can be used to interrupt the CPU when an output device has accepted data transmitted by the CPU. INTR is set when $\overline{OBF}$, $\overline{ACK}$ and INTE are all one (Figure 8.18) and reset by falling edge of $\overline{WR}$.

INTE (Interrupt Enable) : This is an internal flip-flop to a port and needs to be set to generate the INTR signal. The two flip-flops $INTE_A$ and $INTE_B$ are controlled by bits PC_6 and PC_2 respectively, through the BSR mode.

PC_{4-5} : These two lines can be set either as input or output.

Interfacing of 8255

Figure 8.19 shows the interfacing of 8255 with DYNA 86 microprocessor kit.

Problem : Configure 8255 in mode 1 with port A and port B as output and input port respectively.

Solution :

1. Port Addresses

Default control wordPort A = 30h

Port B = 31h

Port C = 32h

Control register = 33h

2. Control Word

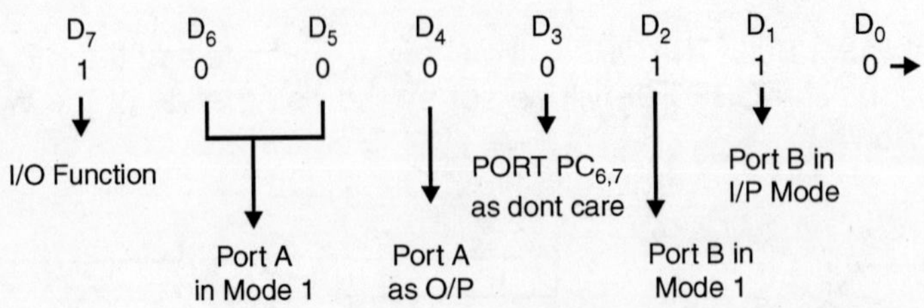

= A6h

3. Program

```
mov al, 07Fh  :    Unmask IRQ7
out ocwl, al  :    Send OCWI
sti           :    Enable interrupts
mov al, 0A6h  :    Port A O/P mode
              :    Port B I/P mode
out 33h, al
mov al, 05    :    Set PC2
out 33h, al
mov al, 0dh   :    Set PC6
out 33h, al
call save-reg :
jm monitor    :
```

4. Steps

(1) 7Fh is the command word given to unmask IRQ7.

(2) Observe the output operation of Port A in mode 1.

(a) Execute the program.

(b) See data 5Ah at port A.

(c) See PC7 i.e. $\overline{\text{OBF}}$ is low (*i.e.* buffer full)

(d) Now make PC6 ($\overline{\text{ACKA}}$) low by a wire. Falling edge of ACK makes PC_7 high and rising edge of $\overline{\text{ACKA}}$ makes PC_3 (i.e. INTRA) high.

(3) To observe input mode 1 operation of port B.

(a) Execute program

(b) See PC_2 (i.e. $\overline{\text{ACKA}}_B$) is high.

(c) Connect pins of port B to Vcc and GND according to data to be transmited.

(d) Make PC_2 ($\overline{ACKA_B}$) low. Falling edge of PC_2 makes PC_1 (*i.e.* IBFB) high and rising edge of PC_2 makes PC_0 high (i.e. $INTR_B$).

Problem 2. Figure 8.19 shows an interfacing circuit using the 8255 A in mode 1. Port A is designated as the input port for a Keyboard with interrupt I/O and port B is designated as the output port for a printer with status check I/O. Use personal computer I/O map *i.e.* Port A = 60h, Port B = 61h, Port C = 62h and control register = 63h.

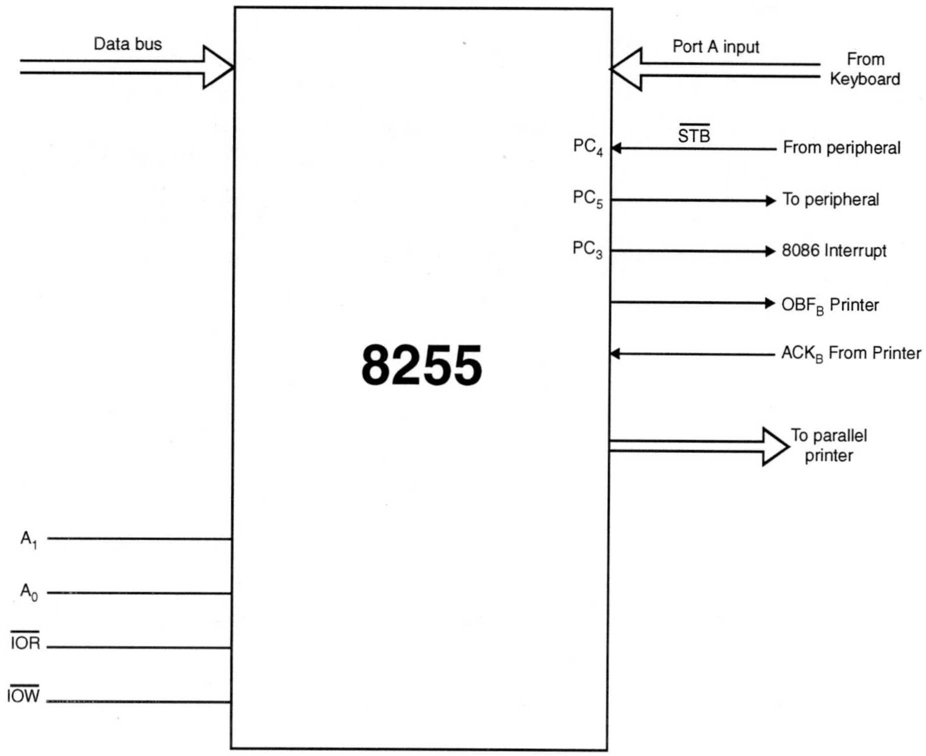

Fig. 8.19.

(1) Determine the control word to set up port A as input and port B as output in mode 1.

(2) Determine the BSR word to enable $INTE_A$.

(3) Determine the masking byte to verify the $\overline{OBF}$ line in status check I/O.

(4) Write program to

(i) Accept character from key board.

(ii) Send character to printer.

Solution :

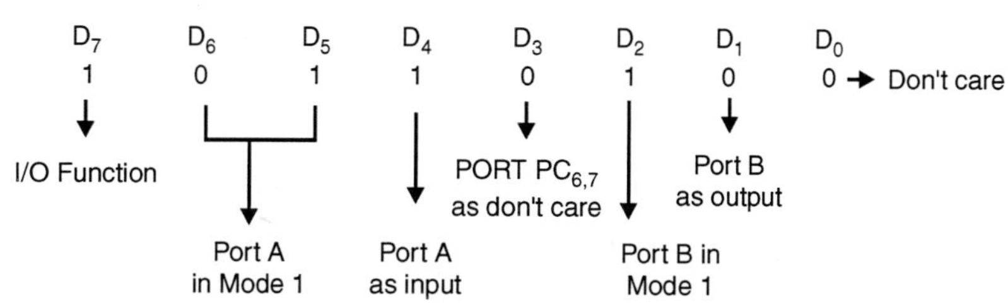

= B4h

(2) BSR Word = 0 000 100 1 = 09h

(0 = BSR Mode; 100 = PC4; 1 = Bit Set)

(3) Status word to check $\overline{OBF_B}$

× × × × × × $\overline{OBF_B}$ ×

masking byte = 02H

Program :

```
                          ; keyboard read procedure
                          ; procedure that reads the keyboard
                          ; encodes and returns with th8e
                          ; the ASCII character in Al.
5thbit equ 20h
read proc near
in al, 62h                ; Read port C
test al, 5thbit           ; Test IBF
jz read
in al, 60h                ; Read ASCII code
ret
read endp
                          ; Print procedure
                          ; procedure that trabsfers the ASCII coaded
                          ; Character from AH to the printer via port B.
stats equ 02h
print proc near
in al, 62h                ; get OBF
test al, status           ; test OBF
jz print                  ; If OBF = 0
mov al, ah                ; Get character
out 61h, ah               ; Send it
mov al, 8                 ; clear DS
out al, 9                 ; Set ds
out 63h, al               ; Set ds
ret
print endp.
```

Above "Keyboard read procedure" reads data from the keyboard each time a key is typed. This procedure read the key from port A and returns with the ASCII code in AL. To detect a key, port C is read the IBF bit (bit position PC5) is tested to see if the buffer is full. If the buffer is empty (IBF = 0) then the procedure keeps testing this bit waiting for a character to be typed on the keyboard.

Print procedure sends the ASCII coded character in AH to the printer.The procedure tests $\overline{OBF}$ to decide if the printer has removed the data from port B. If not, the procedure waits for the $\overline{ACK}$ signal to return from the printer. If $\overline{OBF}$ = 1, then the procedure sends the contents of AH to printer through port B and also sends the DS signal.

Problem : Configure port A in bidirectional mode (Mode 2) and port B in input mode (Mode 1).

Solution :

(i) Control Word 1 1 x x x 1 1 0 = C6h

(ii) (a) BSR word to set PC_6 = 0dh (as derived in earlier example)

(b) BSR word to set PC_4 = 09h (as derived in earlier example)

(iii) Program

```
mov al, 07Fh   ; Unmask IRQ7
out ocwl, al   ; Send OCW1
sti            ; Enable interrupts
mov al, 0C6h   ; Port A mode 2 port B in mode 1
out 33h, al
mov al, 0dh    ; Set PC6 i.e. INTE1 for o/p mode of operation of
port A.
out 33h, al
mov al, 7Ah    ; Data for port A
out 30h, al
mov al 09h     ; Set PC4 INTE2 for i/p mode of operation of Port A
out 33h, al=
mov al, 05h    ; Set PC2 i.e. INTEB.
out 33h, al
call save-reg
jmp monitor
```

Mode 2 (Strobed Bi-Directional Bus I/O)

The functional configuration provides a means for communicating with a peripheral device or structure on a single 8-bit

Output Operations

OBF (Output Buffer Full) : The OBF output will go "low" to indicate that the CPU has written data out to port A bus for both transmitting and receiving data (bi-directional bus I/O). "Handshaking" signals are provided to maintain proper bus flow discipline similar to Mode 1. Interrupt generation and enable/disable functions are also available.

Mode 2 Basic Functional Definitions :

(i) Used in Group A only

(ii) One 8-bit, bi-directional bus Port (Port A) and a 5-bit control Port (Port C)

(iii) Both inputs and outputs are latched

(iv) The 5-bit control port (Port C) is used for control and status for the 8-bit, bidirectional bus port (Port A)

INTR - (Interrupt Request) : A high on this output can be used to interrupt the CPU for both input or output operations.

ACK - (Acknowledge) : A "low" on this input enables the three-state output buffer of port A to send out the data. Otherwise, the output buffer will be in the high impedance state.

INTE 1 - (The INTE flip-flop associated with OBF) : Controlled by bit set/reset of PC4.

Input Operations

STB - (Strobe Input) : A "low" on this input loads data into the input latch.

IBF - (Input Buffer Full F/F) : A "high" on this output indicates that data has been loaded into the input latch.

INTE 2 - (The INTE flip-flop associated with IBF) : Controlled by bit set/reset of PC4.

Special Mode Combination Considerations

There are several combinations of modes possible. For any combination, some or all of Port C lines are used for control or status. The remaining bits are either inputs or outputs as defined by a "Set Mode" command.

During a read of Port C, the state of all the Port C lines, except the ACK and STB lines, will be placed on the data bus. In place of the ACK and STB line states, flag status will appear on the data bus in the PC2, PC4, and PC6 bit positions as illustrated by Fig. 8.20.

Through a "Write Port C" command, only the Port C pins programmed as outputs in a Mode 0 group can be written. No other pins can be affected by a "Write Port C" command, nor can the interrupt enable flags be accessed. To write to any Port C output programmed as an output in Mode 1 group or to change an interrupt enable flag, the "Set/Reset Port C Bit" command must be used.

With a "Set/Reset Port C Bit" command, any Port C line programmed as an output (including IBF and OBF) can be written, or an interrupt enable flag can be either set or reset. Port C lines programmed as inputs, including ACK and STB lines, associated with Port C fare not affected by a "Set/Reset Port C Bit" command. Writing to the corresponding Port C bit positions of the ACK and STB lines with the "Set Reset Port C Bit" command will affect the Group A and Group B interrupt enable flag.

INPUT CONFIGURATION

D7	D6	D5	D4	D3	D2	D1	D0
I/O	I/O	IBFA	INTEA	INTRA	INTEB	IBFB	INTRB
Group A					Group B		

OUTPUT CONFIGURATION

D7	D6	D5	D4	D3	D2	D1	D0
OBFA	INTEA	I/O	I/O	INTRA	INTEB	OBFB	INTRB
Group A					Group B		

MODE 1 STATUS WORD FORMAT

D7	D6	D5	D4	D3	D2	D1	D0
OBFA	INTE1	IBFA	INTE2	INTRA	X	X	X
Group A					Group B		

(Defined by Mode 0 or Mode 1 Selection)

Fig. 8.20.

8.4.5. Printer Interface

For most common printers , the data to be printed is sent to the printer as ASCII characters on eight parallel lines. The printer receives the characters to be printed and stores them in an internal RAM buffer. When the printer detects a carriage return character (0DH), it prints out the first row of characters from the print buffer. When the printer detects a second carriage return, it prints out the second row of characters and so on. This process continues until the desired characters have been printed.

Transfer of the ASCII characters from a microcomputer to the printer is done on a handshake basis because the microcomputer can send characters much faster than the printer can print them. The printer must in some way let the microcomputer know that its buffer is full and that it cannot accept anymore data till it prints out some of them. A common standard for interfacing with parallel printers is the Parallel interface used with the microcomputers as shown in the above Fig. 8.21.

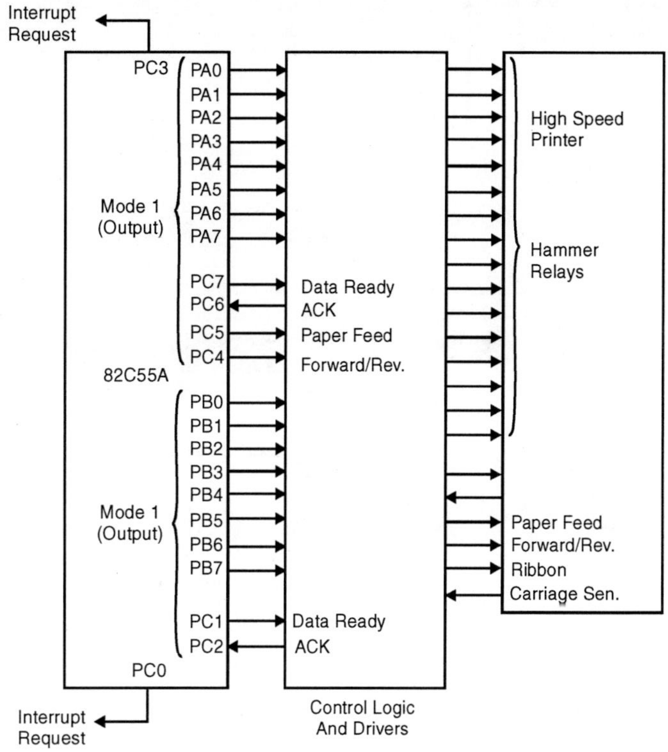

Fig. 8.21. Printer interface block diagram.

Example 3. Interface an 8255 chip with 8086 to work as an I/O port. Initialize port A as output port, Port B as I/P port and Port C as O/P port. Port A address should be 0740H. Write an ALP to sense switch positions SW_0–SW_7 connected at port B. The sensed pattern is to be displayed on port A, to which 8 LED's are connected, while port C lower displays number of on switches out of the total eight switches ?

Solution. The control word is decided as given as follows :

B7	B6	B5	B4	B3	B2	B1	B0	Control word
1	0	0	0	0	0	1	0	= 82 H
I/O mode	Port A in mode 0		Port A, O/P	Port C, O/P	Port B, mode 0	Port B, I/P	Port C, O/P	

82H is the control word. The control word format for BSR mode is as shown ahead :

8255 ports	I/O Address lines A_{15} A_{14} A_{13} A_{12} A_{11} A_{10} A_9 A_8 A_7 A_6 A_5 A_4 A_3 A_2 A_1 A_0	Hex port address
Port A	0 0 0 0 0 1 1 1 0 1 0 0 0 0 0 0	0740H
Port B	0 0 0 0 0 1 1 1 0 1 0 0 0 0 1 0	0742H
Port C	0 0 0 0 0 1 1 1 0 1 0 0 0 1 0 0	0744H
CWR	0 0 0 0 0 1 1 1 0 1 0 0 0 1 1 0	0746H

The ALP for the problem is shown below :

```
          MOV DX, 0746H;     Initialise CWR with
          MOV AL, 82H;       Control word 82H
          OUT DX, AL;
          SUB DX, 04;        Get address of port B in DX
          IN AL, DX;         Read port B for switch
          SUB DX, 02;        Positions into AL and get port A address
          MOV BL, 00H;       Initialise BL for switch count
          MOV CH, 08H;       Initialise CH for total switch number
Label 1 : ROL AL;            Rotate AL through carry to check
          JNC label 2
          INC BL
Label 2 : DEC CH
          JNZ label 1
          MOV AL, BL
          ADD DX, OH
          OUT DX, AL
          HLT
```

The interface circuit is as shown in Fig. 8.22.

The 8255 is to be interfared with lower order data bus; i.e. D_0-D_7. The A_0 and A_1 pins of 8255 are connected to A_1 and A_2 pins of the microprocessor respectively. We will use absolute decoding scheme that uses all the 16 address lines. For driving the device address pulse. Out of A_0–A_{15} lines, two address lines A_2 and A_1 are directly required by 8255 for three port and CWR address decoding. Hence only A_3 to A_{15} are used for decoding addresses. As shown in the interfacing circuit diagram, the 8086 is assumed to be in the maximum mode so that $\overline{IORD}$ and $\overline{IOWR}$ are readily available.

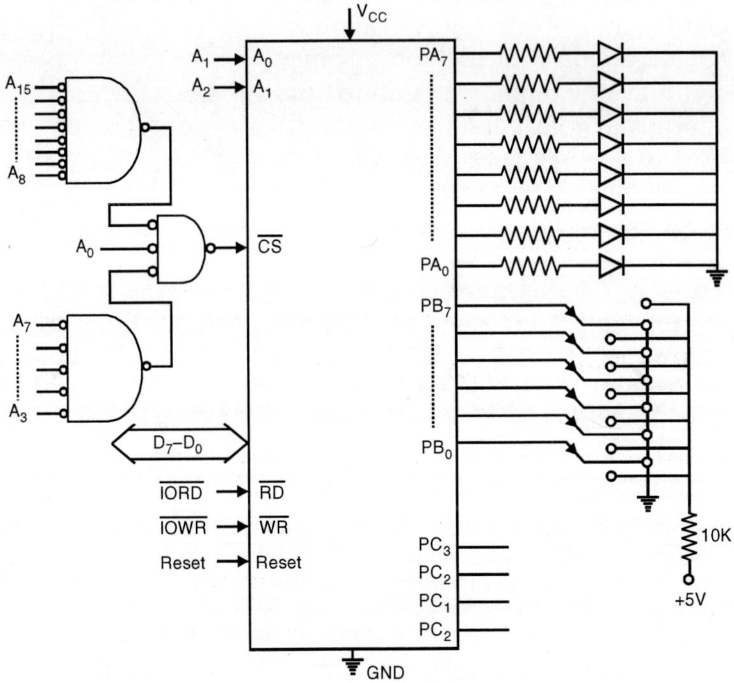

Fig. 8.22. Interfacing circuit showing 8-LED's interfaced.

Interfacing a Microprocessor to Keyboard

When you press a key on your computer, you are activating a switch. There are many different ways of making these switches. An overview of the construction and operation of some of the most common types.

1. **Mechanical key switches :** In mechanical-switch keys, two pieces of metal are pushed together when you press the key. The actual switch elements are often made of a phosphor-bronze alloy with gold plating on the contact areas. The key switch usually contains a spring to return the key to the nonpressed position and perhaps a small piece of foam to help damp out bouncing.
   (i) Some mechanical key switches now consist of a molded silicon dome with a small piece of conductive rubber foam short two trace on the printed-circuit board to produce the key pressed signal.
   (ii) Mechanical switches are relatively inexpensive but they have several disadvantages. First, they suffer from contact bounce. A pressed key may make and break contact several times before it makes solid contact.
   (iii) Second, the contacts may become oxidized or dirty with age so they are no longer make a dependable connection.
   ➢ Higher-quality mechanical switches typically have a rated life time of about 1 million keystrokes. The silicone dome type typically last 25 million keystrokes.

2. **Membrane key switches :** These switches are really a special type of mechanical switches. They consist of a three-layer plastic or rubber sandwich.
   ➢ The top layer has a conductive line of silver ink running under each key position. The bottom layer has a conductive line of silver ink running under each column of keys.
   ➢ When u press a key, you push the top ink line through the hole to contact the bottom ink line.
   ➢ The advantages of membrane keyboards is that they can be made as very thin, sealed units.
   ➢ They are often used on cash registers in fast food restaurants. The lifetime of membrane keyboards varies over a wide range.

3. **Capacitive key switches :** A capacitive keyswitch has two small metal plates on the printed circuit board and another metal plate on the bottom of a piece of foam.
   ➢ When u press the key, the movable plate is pushed closer to fixed plate. This changes the capacitance between the fixed plates. Sense amplifier circuitry detects this change in capacitance and produce a logic level signal that indicates a key has been pressed.
   ➢ The big advantages of a capacitive switch is that it has no mechanical contacts to become oxidized or dirty.
   ➢ A small disadvantage is the specified circuitry needed to detect the change in capacitance.
   ➢ Capacitive keyswitches typically have a rated lifetime of about 20 million keystrokes.

4. **Hall effect keyswitches :** This is another type of switch which has no mechanical contact. It takes advantage of the deflection of a moving charge by a magnetic field.

- A reference current is passed through a semiconductor crystal between two opposing faces. When a key is pressed, the crystal is moved through a magnetic field which has its flux lines perpendicular to the direction of current flow in the crystal.
- Moving the crystal through the magnetic field causes a small voltage to be developed between two of the other opposing faces of the crystal.
- This voltage is amplified and used to indicate that a key has been pressed. Hall effect sensors are also used to detect motion in many electrically controlled machines.
- Hall effect keyboards are more expensive because of the more complex switch mechanism, but they are very dependable and have typically rated lifetime of 100 million or more keystrokes.

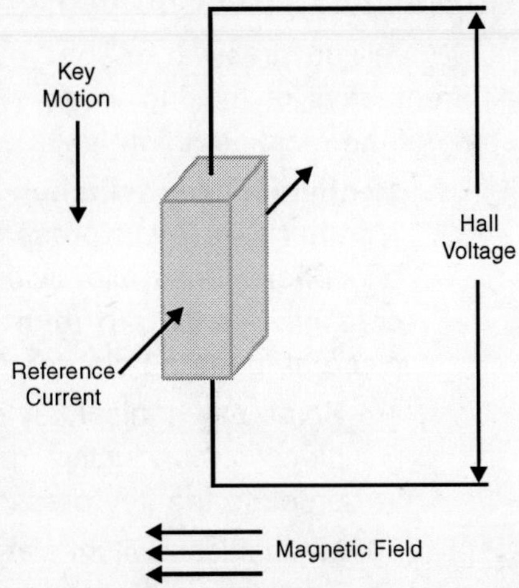

Fig. 8.23. Hall Effect.

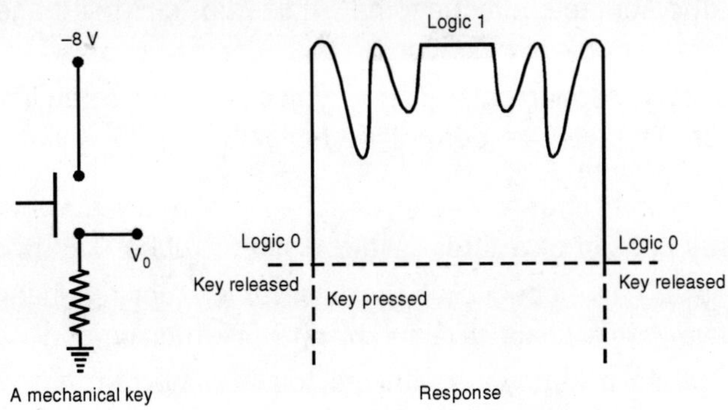

Fig. 8.24. Keyboard circuit connections and interfacing.

Example 4. Interface a 4 × 4 keyboard with 8086 using 8255, and write an ALP for detecting a key closure and return the key code in AL. The debouncing period for a key is 20 ms ?

Solution. The interface circuit for the problem is as shown in Fig. 8.25.

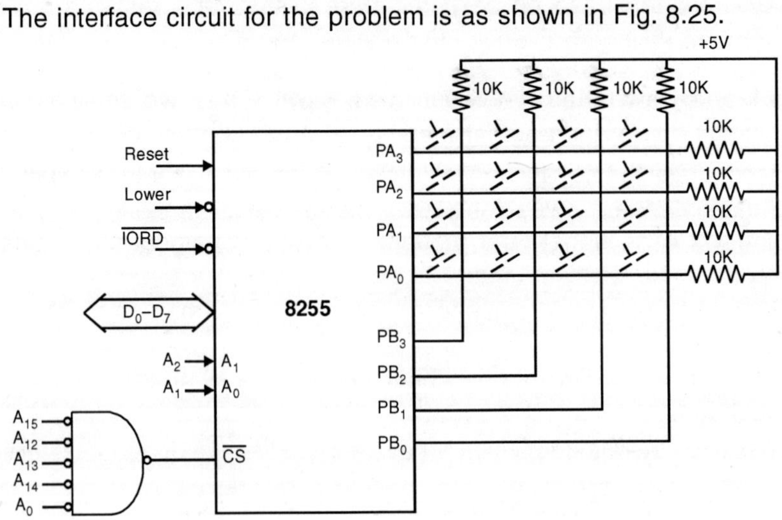

Fig. 8.25. Interfacing circuit for keyboard.

Here we use port A as output port for selecting a row of keys while port B is used as an input port. For sensing a closed key. Hence the keyboard lines are selected one by one through Port A and the Port B lines are polled continuously till a key closure is sensed. The higher order lines of Port A and Port B are left unused. The flow chart of the ALP is as shown in Fig. 8.26.

We suppose that we use simple mechanical switches. For keyboard, then to get the meaningful data from a keyboard requires three steps :

(1) Detect a keypress

(2) Debounce the keypress

(3) Encode the keypress

The three tasks can be done with a hardware, software or a combination of the two.

The rows of the matrix are connected to four output port lines. The column line of the matrix are connected to four input port lines.

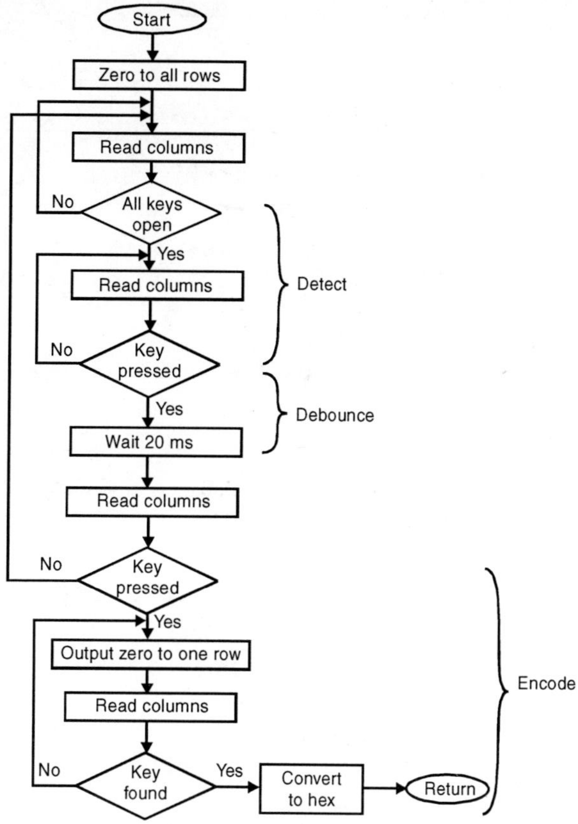

Fig. 8.26. Flow chart for key detection.

Explanation of the Flow Chart

When no keys are pressed, the column lines are held high i.e. they get +5 V. Pressing a key connects a row to a column. If a low is output on a row and key in that row is pressed, then the low will appear on the column of the pressed key, hence we come to know which key was pressed. An easy method to detect the key in the matrix arrangement is that we output 0's to all rows and check the columns to see if a pressed key has connected a low to a column. This is done to make sure a previous key has been released before looking for the next one. Once the columns are found to be all high, the program enters into the next loop. Here the program waits until a low appears on one of the columns, indicating that a key has been pressed. We use a 20 ms delay procedure to detect the task and then the debounce task is done.

After the debounce time, another check is made to see if the key is still pressed. If the columns are now all high, then no key is pressed and the initial detection was assumed to be caused by a noise pulse or a light brushing post a key. If any of the columns are still low, then the assumption is made that it was a valid keypress.

The last step is to determine the row and column of the pressed key and then convert this row and column information to the hex code for the pressed key. For this, a low is output to one row and the columns are read. If none of the columns is low, the pressed key is not in that row, so the low is rotated to the next row and the columns are checked again.

The process is repeated untill a low on a row produces a low on one of the columns. The pressed key then is in the row which is low at that time. Then the byte is read in from the I/P port which contains a 4-bit code which represents the column of the pressed key.

A look-up table can then be used to easily convert this row-column code to the desired hex value.

The ALP for this problem is as stated below

```
XX :      MOV AL, 82H        ; load the CWR
          MOV DX, 8006H
          OUT DX, AL
          MOV BL, 00H
          XOR AX, AX
          MOV DX, 8000H      ; Port address in AX
          OUT DX, AL         ; Ground all rows
          ADD DX, 02         ; Port B address in DX
Wait :    IN AL, DX
          AND AL, 0FH        ; Mask data lines D7-D4
          CMP AL, 0FH        ; Any key closed
          J7 wait            ; wait till key
          CALL debounce      ; closure else wait for 20 ms
          MOV AL, 0F7H;
          MOV BH, 04H
NEXT :    ROL AL, 01
          MOV CH, A1
          SUB DX, 02
          OUT DX, AL
          ADD DX, 02
          IN AL, DX
          ADD AL, 0FH
          MOV CL, 04H
NEXT 1 :  RCR
          JNC YY
          INC BL
          DCR CL
          JNZ NEXT 1
          MOV AL, CH
          DCR BH
          JNZ NEXT
          JMP wait
XX :      MOV AL, BL
          MOV AH, 4CH        ; Return to DOS prompt
          INT 21H
```

Debounce Procedure Near :

```
          MOV CL, 01C4H
Back :    NOP
```

```
DEC CL
JNZ back
RET
Debuonce END P
```

Example 5. Interface 16-bit 8255 ports with 8086. The address of Port A is FOH ?

Solution. For implementing a 16-bit port two 8255 chips are required. One will act for the lower 8-bit port i.e. D_0–D_7 while the other will act as the upper 8-bit port D_8–D_{15}. (Fig. 8.27)

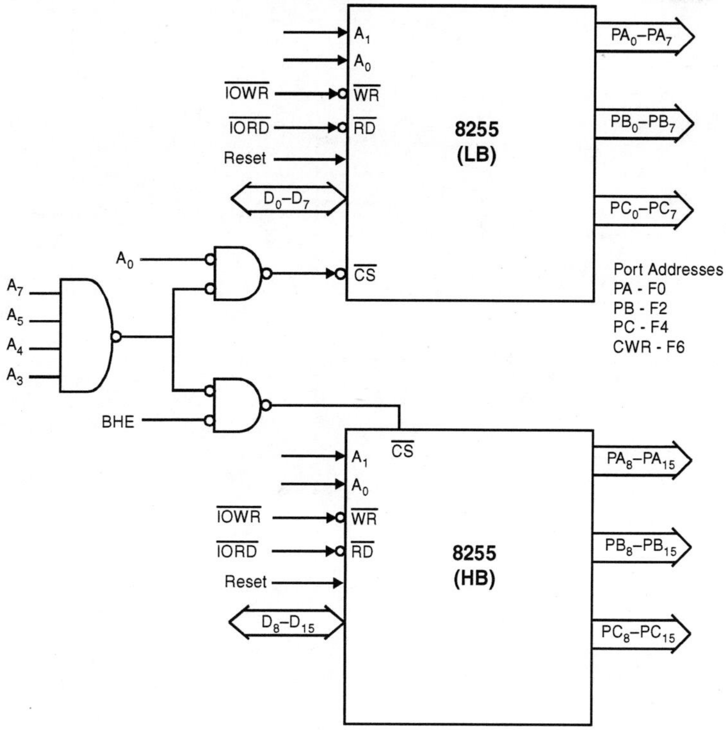

Fig. 8.27. Interface circuit.

Example 6. Interface an 8255 with 8086 at 80H as an I/O address of Port A. Interface five 7 segment displays with the 8255. Write an ALP to display 1, 2, 3, 4 and 5 over the 5 displays continuously as per their positions starting with 1 at the least significant position ?

Solution. We will first decide the codes and store them in a look up table as shown below :

Numbers to be displayed	PA7 dp	PA6 a	PA5 b	PA4 c	PA3 d	PA2 e	PA1 f	PA0 Code g
1	1	1	0	0	1	1	1	1 = CF
2	1	0	0	1	0	0	1	0 = 92
3	1	0	0	0	0	1	1	0 = 86
4	1	1	0	0	1	1	0	0 = CC
5	1	0	1	0	0	1	0	0 = A4

The interfacing circuit is as drawn in Fig. 8.28.

We have used the common cathode displays here. To glow a segment, logic "1" is applied on the corresponding line that drives a transistor to ground the common cathode pin of the display. For a common cathode display logic '1' is applied to a segment to glow it and a logic '0' is applied to blank the segment.

The ALP for the problem is as explained below :

```
START :  MOV CL, 05H
         MOV BX, 2000H
         MOV DS, BX
         MOV CH, 01H
         MOV AL, 80H
         OUT 86H, AL
         MOV DL, 01H
```

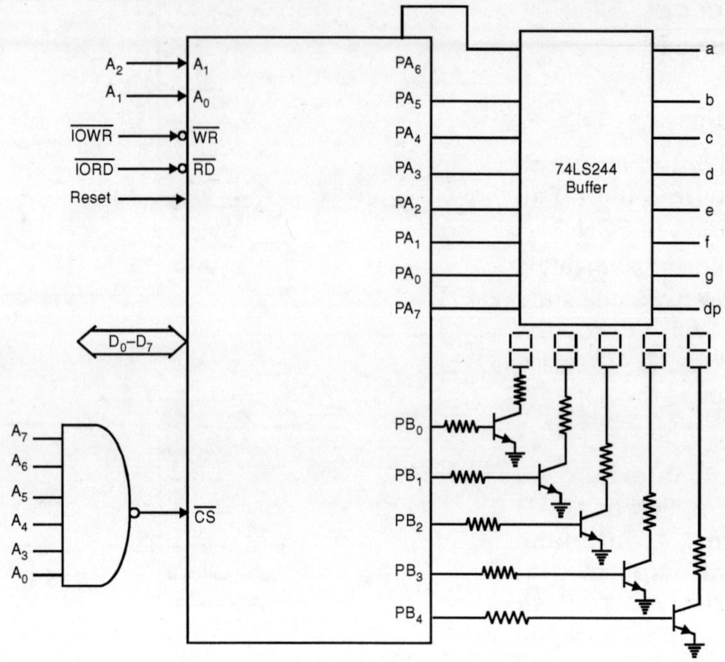

Fig. 8.28. Interfacing circuit.

```
NEXT :    MOV BX, 0000H
          MOV AL, CH
          XLAT
          OUT 80H, AL
          MOV AL, DL
          OUT 81H, AL
          ROL DL
          INC CH
          DEC CL
          JNZ NEXT
          JMP START
```

Interfacing to Alphanumeric Displays

- To give directions or data values to users, many microprocessor-controlled instruments and machines need to display letters of the alphabet and numbers. In systems where a large amount of data needs to be displayed a CRT is used to display the data. In system where only a small amount of data needs to be displayed, simple digit-type displays are often used.
- There are several technologies used to make these digit-oriented displays but we are discussing only the two major types.

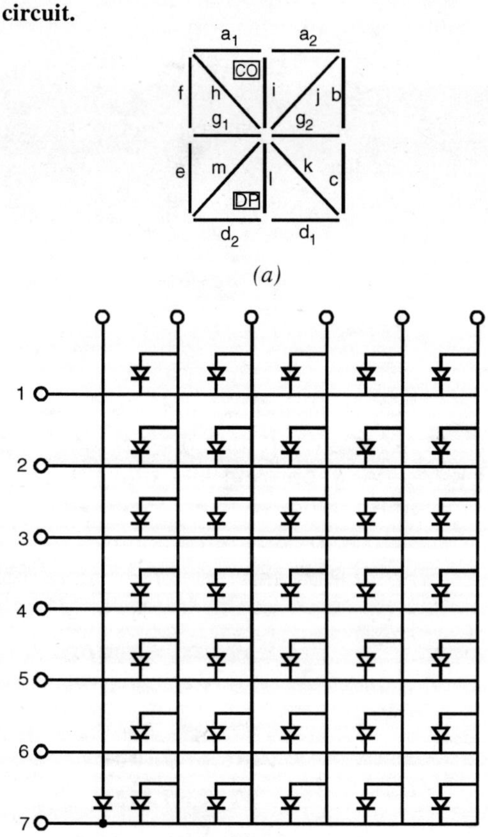

Fig. 8.29. 18 segment display.

- These are *light emitting diodes* (LED) and *liquid-crystal displays* (LCD).
- LCD displays use very low power, so they are often used in portable, battery-powered instruments. They do not emit their own light, they simply change the reflection of available light. Therefore, for an instrument that is to be used in low-light conditions, you have to include a light source for LCDs or use LEDs which emit their own light.
- Alphanumeric LED displays are meant/used for displaying only numbers and hexadecimal letters. To display numbers and the entire alphabet, 18-segment displays, or 5 by 7 dot-matrix displays can be used as shown in Fig. 8.29 (a), (b).

1. Directly Driving LED Displays : Figure 8.30 (a) shows a circuit that you might connect to a parallel port on a microcomputer to drive a single 7-segment, common-anode display. For a common-anode display, a segment is tuned on by applying a logic low to it.

- The 7447 converts a BCD code applied to its inputs to the pattern of lows required to display the number represented by the BCD code. This circuit connection is referred to as a *static display* because current is being passed through the display at all times.
- Each segment requires a current of between 5 and 30mA to light. Let's assume you want a current of 20mA. The voltage drop across the LED when it is lit is about 1.5V.

The output low voltage for the 7447 is a maximum of 0.4V at 40mA. So assume that it is about 0.2V at 20mA. Subtracting these two voltage drop from the supply voltage of 5V leaves 3.3V across the current limiting resistor. Dividing 3.3V by 20mA gives a value of 168D for the current-limiting resistor. The voltage drops across the LED and the output of 7447 are not exactly predictable and exact current through the LED is not critical as long as we don't exceed its maximum rating.

2. Software-Multiplexed LED Display :

- The circuit in fig works for driving just one or two LED digits with a parallel output port. However, this scheme has several problem if you want to drive, eight digits.
- The first problem is power consumption. For worst-case calculations, assume that all 8 digits are displaying the digit 8, so all 7 segments are all lit. Seven segment time 20 mA per segment gives a current of 140 mA per digit. Multiplying this by 8 digits gives a total current of 1120 mA or 1.12 A for 8 digits.
- A second problem of the static approach is that each display digit requires a separate 7447 decoder, each of which uses another 13 mA. The current required by the decoders and the LED displays might be several times the current required by the reset of the circuitry in the instrument.
- To solve the problem of the static display approach, we use a *multiplex method,* example for an explanation of the multiplexing.
- The Fig. 8.29 shows a circuit you can add to a couple of microcomputer ports to drive some common anode LED displays in a multiplexed manner. The circuit has only one 7447 and that the segment outputs of the 7447 are bused in parallel to the segment inputs of all the digits.
- The question that may occur to you on first seeing is: Aren't all the digits going to display the same number? The answer is that they would if all the digits were turned on at the same time. The tricky of multiplexing displays is that only one display digit is turned on at a time.
- The PNP transistor is series with the common anode of each digit acts as on/off switch for that digit. Here's how the multiplexing process works.
- The BCD code for digit 1 is first output from port B to the 7447. The 7447 outputs the corresponding 7-segment code on the segment bus lines. The transistor connected to

digit 1 is then turned on by outputting a low to the appropriate bit of port A. All the rest of the bits of port A are made high to make sure no other digits are turned on. After 1 or 2 ms, digit 1 is turned off by outputting all highs to port A.

- ➢ The BCD code for digit 2 is then output to the 7447 on port B, and a word to turn on digit 2 is output on port A.
- ➢ After 1 or 2 ms, digit 2 is turned off and the process is repeated for digit 3. the process is continued until all the digits have had a turn. Then digit 1 and the following digits are lit again in turn.
- ➢ A procedure which is called on an interrupt basis every 2 ms to keep these displays refreshed with some values stored in a table. With 8 digits and 2ms per digit, you get back to digit 1 every 16ms or about 60 times a second.
- ➢ This refresh rate is fast enough so that the digits will each appear to be lit all time. Refresh rates of 40 to 200 times a second are acceptable.

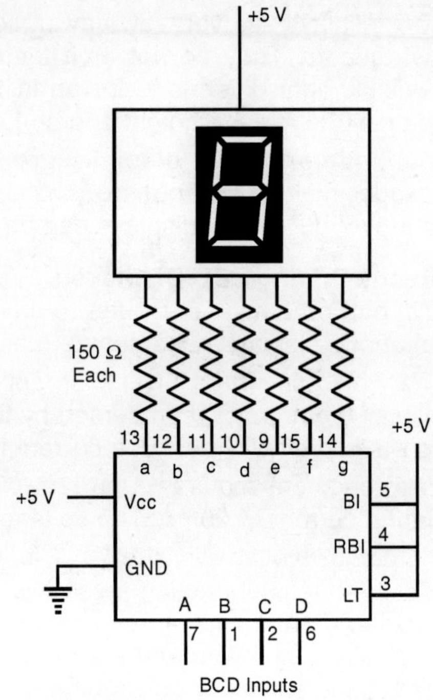

Fig. 8.30 (*a*). Circuit for driving single 7-segment LED display with 7447.

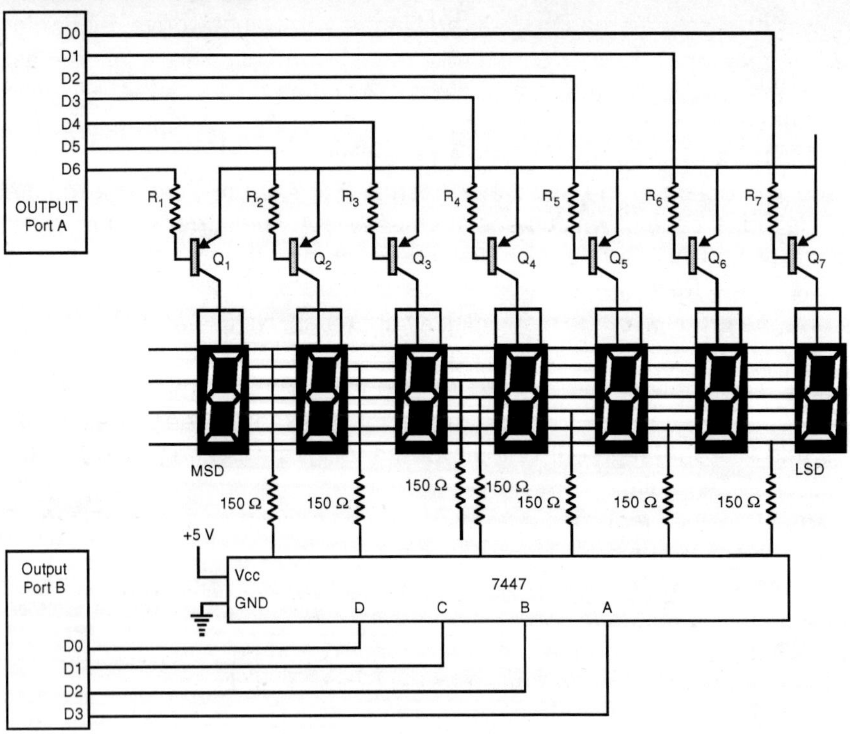

Fig. 8.30 (*b*). Liquid crystal display.

- The immediately obvious advantages of multiplexing the displays are that only one 7447 is required, and only one digit is lit at a time. We usually increase the current per segment to between 40 and 60 mA for multiplexed displays so that they will appear as bright as they would if they were not multiplexed. Even with this increased segment current, multiplexing gives a large saving in power and parts.
- The software-multiplexed approach we have just described can also be used to drive 18-segment LED devices and dot-matrix LED device. For these devices, however you replace the 7447 in fig with ROM which generates the required segment codes when the ASCII code for a character is applied to the address inputs of the ROM.

8.5. LCD INTERFACING WITH 8086

- Liquid Crystal displays are created by sandwiching a thin 10-12 μm layer of a liquid-crystal fluid between two glass plates. A transparent, electrically conductive film or backplane is put on the rear glass sheet. Transparent sections of conductive film in the shape of the desired characters are coated on the front glass plate.
- When a voltage is applied between a segment and the backplane, an electric field is created in the region under the segment. This electric field changes the transmission of light through the region under the segment film. There are two commonly available types of LCD : dynamic scattering and field-effect.

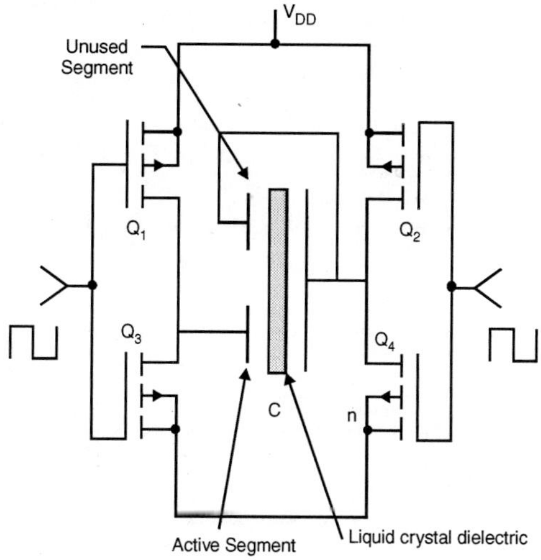

Fig. 8.31. CMOS Drive Circuit for LCD.

- The Dynamic scattering types of LCD: It scrambles the molecules where the field is present. This produces an etched-glass-looking light character on a dark background.
- Field-effect types use polarization to absorb light where the electric field is present. This produces dark characters on a silver-gray background.
- Most LCD's require a voltage of 2 or 3 V between the backplane and a segment to turn on the segment.
- We cannot just connect the backplane to ground and drive the segment with the outputs of a TTL decoder. The reason for this is a steady dc voltage of more than about 50mV is applied between a segment and the backplane.
- To prevent a dc buildup on the segments, the segment-drive signals for LCD must be square waves with a frequency of 30 to 150 Hz.
- Even if you pulse the TTL decoder, it still will not work because the output low voltage of TTL devices is greater than 50 mV.
- CMOS gates are often used to drive LCDs.
- The Following fig shows how two CMOS gate outputs can be connected to drive an LCD segment and backplane.
- The off segment receives the same drive signal as the backplane. No voltage and there between them, so no electric field is produced. The waveform for the on segment is 180 out of phase with the backplane signal, so the voltage between this segment and the backplane will always be +V.
- The logic for this signal, a square wave and its complement. To the driving gates, the segment-backplane sandwich appears as a somewhat leaky capacitor.

- The CMOS gates can be easily supply the current required to charge and discharge this small capacitance.
- Older inexpensive LCD displays turn on and off too slowly to be multiplexed the way we do LED display.
- LCD may require as much as 0.5s to turn on or off. To interface to those types we use a nonmultiplexed driver device.
- More expensive LCD can turn on and off faster, so they are often multiplexed using a variety of techniques.
- In the following section we show you how to interface a nonmultiplexed LCD to a microprocessor such as SDK-86.
- Intersil ICM7211M can be connected to drive a 4-digit, nonmultiplexed, 7-segment LCD display.
- The 7211M input can be connected to port pins or directly to microcomputer bus. We have connected the CS inputs to the Y2 output of the 74LS138 port decoder.
- According to the truth table the device will then be addressable as ports with a base address of FF10H. SDK-86 system address lines A2 is connected to the digit-select input (DS2) and system address lines Al is connected to the DS1 input. This gives digit 4 a system address of FF10H.

A_8-A_{15}	A_5-A_7	A_4	A_3	A_2	A_1	A_0	$M/\overline{IO}$	Y Output Selected	System Base Address	Device
1	0	0	0	X	X	0	0	00	F F 0 0	8259A #1
1	0	0	1	X	X	0	0	1	F F 0 8	8259A #2
1	0	1	0	X	X	0	0	2	F F 1 0	
1	0	1	1	X	X	0	0	3	F F 1 8	
1	0	0	0	X	X	1	0	4	F F 0 1	
1	0	0	1	X	X	1	0	5	F F 0 9	8254
1	0	1	0	X	X	1	0	6	F F 1 1	
1	0	1	1	X	X	1	0	7	F F 1 9	
All Other States								None		

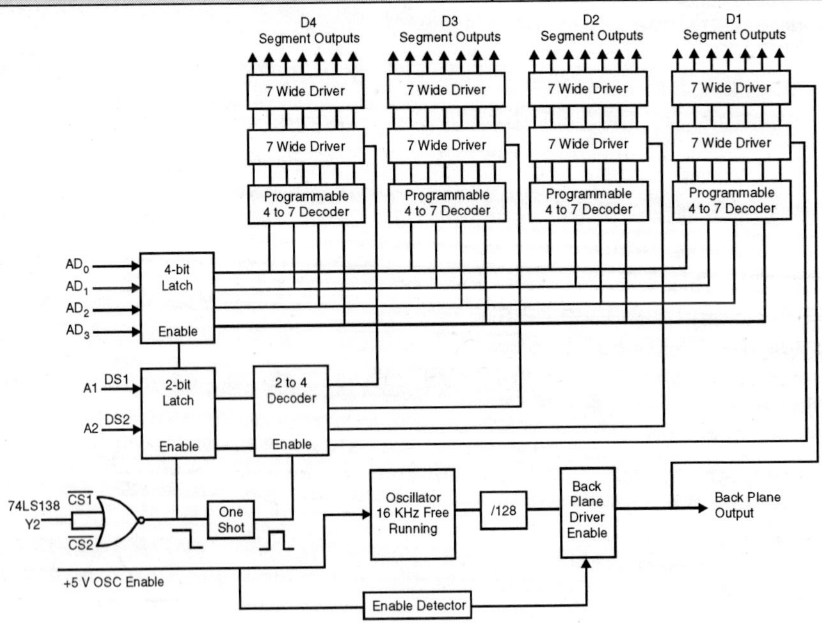

Fig. 8.32. Circuit for interfacing four LCD digits to an SDK-86 bus using ICM7211M.

8.6. 8254 PROGRAMMABLE INTERVAL TIMER/COUNTER

Functional Description

General the 8254 is a programmable interval timer/ counter designed for use with Intel microcomputer systems. It is a general purpose, multi-timing element that can be treated as an array of I/O ports in the system software.

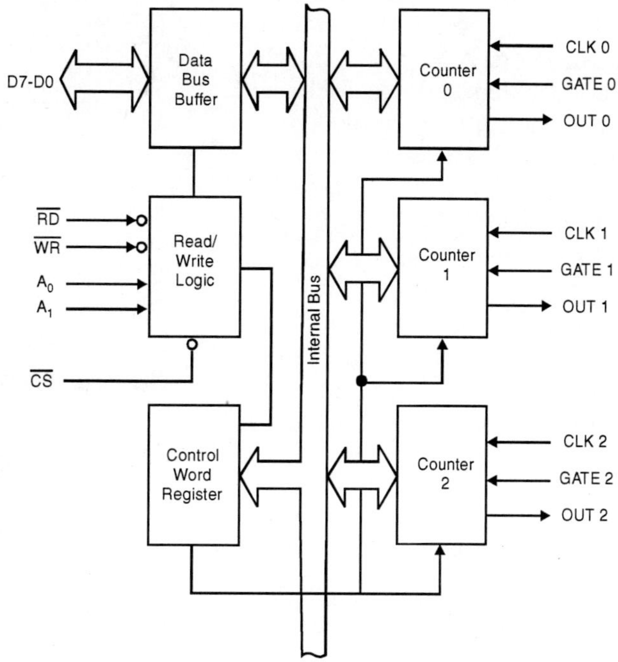

Fig. 8.33. Functional block diagram for 8254.

The 8254 solves one of the most common problems in any microcomputer system, the generation of accurate time delays under software control. Instead of setting up timing loops in software, the programmer configures the 8254 to match his requirements and programs one of the counters for the desired delay. After the desired delay, the 8254 will interrupt the CPU. Software overhead is minimal and variable length delays can easily be accommodated.

Some of the other counter/timer functions common to microcomputers which can be implemented with the 8254 are :

- ➤ Real time clock
- ➤ Event-counter
- ➤ Digital one-shot
- ➤ Programmable rate generator
- ➤ Square wave generator
- ➤ Binary rate multiplier
- ➤ Complex waveform generator
- ➤ Complex motor controller

8.6.1. Data Bus Buffer

This 3-state, bi-directional, 8-bit buffer is used to interface the 8254 to the system bus. (Fig. 8.34)

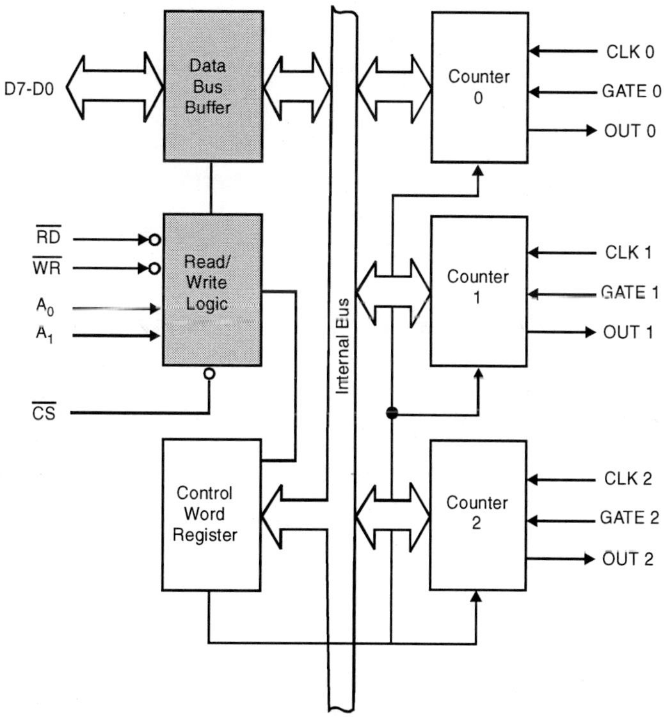

Fig. 8.34. Block diagram showing data bus buffer.

8.6.2. Read/Write Logic

The Read/Write Logic accepts inputs from the system bus and generates control signals for the other functional blocks of the 8254. A_1 and A_0 select one of the three counters of the

Control Word Register to be read from/written into. A "low" on the RD input tells the 8254 that the CPU is reading one of the counters. A "low" on the WR input tells the 8254 that the CPU is writing either a Control Word or an initial count. Both RD and WR are qualified by CS; RD and WR are ignored unless the 8254 has been selected by holding CS low.

Control Word Register : The Control Word Register (Fig. 8.35) is selected by the Read/Write Logic when A_1, A_0 = 11. If the CPU does a write operation to the 8254, the data is stored in the Control Word Register and is interpreted as a Control Word used to define the operation of the Counters.

The Control Word Register can only be written to; status, information is available with the Read-Back Command.

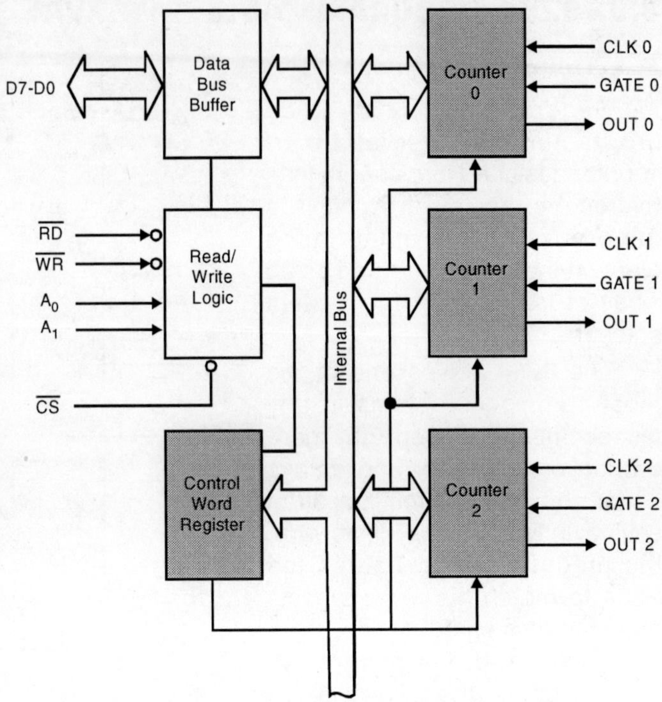

Fig. 8.35. Block diagram showing control word register and counter functions.

COUNTER 0, COUNTER 1, COUNTER 2

These three functional blocks are identical in operation, so only a single Counter will be described. The internal block diagram of a single counter is shown in Fig. 8.36.

The Counters are fully independent. Each Counter may operate in a different Mode.

The Control Word Register is shown in the figure; it is not a part of the Counter itself, but its contents determine how the Counter operates.

The status register, shown in Fig. 8.36, when latched, contains the current contents of the Control Word Register and status of the output and null count flag.

The actual counter is labeled CE (for "Counting Element"). It is a 16-bit presettable synchronous down counter.

OL_M and OL_L are two 8-bit latches. OL stands for "Output Latch"; the subscripts M and L stand for "Most significant byte" and "Least significant byte" respectively. Both are normally referred to as one unit and called just

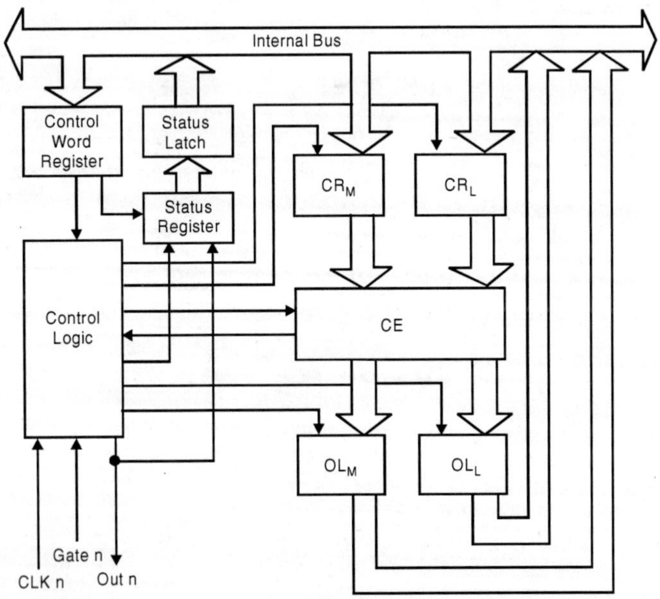

Fig. 8.36. Internal Block diagram of a single counter.

OL. These latches normally "follow" the CE, but if a suitable Counter Latch Command is sent to the 8254, the latches "latch" the present count until read by the CPU and then return to "following" the CE. One latch at a time is enabled by the counter's Control Logic to drive the internal bus. This is how the 16-bit Counter communicates over the 8-bit internal bus. Note that the CE itself cannot be read; whenever you read the count, it is the OL that is being read.

Similarly, there are two 8-bit registers called CR_M and CR_L (for "Count Register"). Both are normally referred to as one unit and called just CR. When a new count is written to the Counter, the count is stored in the CR and later transferred to the CE. The Control Logic allows one register at a time to be loaded from the internal bus. Both bytes are transferred to the CE simultaneously. CR_M and CR_L are cleared when the Counter is programmed. In this way, if the Counter has been programmed for one byte counts (either most significant byte only or least significant byte only) the other byte will be zero. Note that the CE cannot be written into; whenever a count is written, it is written into the CR.

The Control Logic is also shown in the diagram. CLK n, GATE n, and OUT n are all connected to the outside world through the Control Logic.

8.6.3. Interface 8254 System

The 8254 is a component of the Intel Microcomputer Systems and interfaces in the same manner as all other peripherals of the family. It is treated by the system's software as an array of peripheral I/O ports; three are counters and the fourth is a control register for MODE programming.

Basically, the select inputs A_0, A_1 connect to the A_0, A_1 address bus signals of the CPU. The CS can be drived directly from the address bus using a linear select method. Or it can be connected to the output of a decoder, such as an Intel 8205 for larger systems.

8.6.4. Operational Description

General

After power-up, the state of the 8254 is undefined. The Mode, count value, and output of all Counters are undefined.

How each Counter operates is determined when it is programmed. Each Counter must be programmed before it can be used. Unused counters need not be programmed.

Programming the 8254

Counters are programmed by writing a Control Word and then an initial count.

The Control Words are written into the Control Word Register, which is selected when $A_1A_0 = 11$. The Control Word itself specifies which Counter is being programmed.

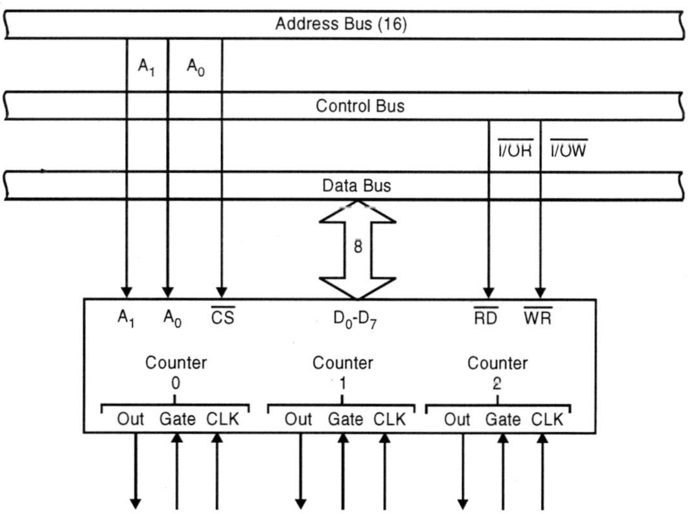

Fig. 8.37. System interface.

Control Word Format

D_7	D_6	D_5	D_4	D_3	D_2	D_1	D_0
SC1	SC0	RW1	RW0	M2	M1	M0	BCD

SCĐ Select Counter

SC1	SC0	
0	0	Select Counter 0
0	1	Select Counter 1
1	0	Select Counter 2
1	1	Read-Back Command (see Read Operations)

RWĐRead/Write

RW1	RW0	
0	0	Counter Latch Command (see Read Operations)
0	1	Read/Write least significant byte only
1	0	Read/Write most significant byte only
1	1	Read/Write least significant byte first, then most significant byte

MĐ Mode

M2	M1	M0	
0	0	0	Mode 0
0	0	1	Mode 1
X	1	0	Mode 2
X	1	1	Mode 3
1	0	0	Mode 4
1	0	1	Mode 5

BCD

0	Binary Counter 16-bits
1	Binary Coded Decimal (BCD) Counter (4 Decades)

By contrast, initial counts are written into the Counters, not the Control Word Register. The A_1,A_0 inputs are used to select the Counter to be written into. The format of the initial count is determined by the Control Word used.

8.6.5. Write Operations

The programming procedure for the 8254 is very flexible. Only two conventions need to be remembered:

1. For each Counter, the Control Word must be written before the initial count is written.
2. The initial count must follow the count format specified in the Control Word (least significant byte only, most significant byte only, or least significant byte and then most significant byte).

Since the Control Word Register and the three Counters have separate addresses (selected by the A_1, A_0 inputs), and each Control Word specifies the Counter it applies to (SC_0,SC_1 bits), no special instruction sequence is required. Any programming sequence that follows the conventions.

A new initial count may be written to a Counter at any time without affecting the Counter's programmed Mode in any way. Counting will be affected as described in the Mode definitions. The new count must follow the programmed count format.

If a Counter is programmed to read/write two-byte counts, the following precaution applies: A program must not transfer control between writing the first and second byte to another routine which also writes into that same Counter. Otherwise, the Counter will be loaded with an incorrect count.

8.6.6. Read Operations

It is often desirable to read the value of a Counter without disturbing the count in progress. This is easily done in the 8254.

There are three possible methods for reading the counters: a simple read operation, the Counter Latch Command, and the Read-Back Command. Each is explained below. The first method is to perform a simple read operation. To read the Counter, which is selected with the A1, A0 inputs, the CLK input of the selected Counter must be inhibited by using either the GATE input or external logic. Otherwise, the count may be in the process of changing when it is read, giving an undefined result.

Counter Latch Command : The second method uses the "Counter Latch Command". Like a Control Word, this command is written to the Control Word Register, which is selected when A_1,A_0 = 11. Also like a Control Word, the SC0, SC1 bits select one of the three Counters, but two other bits, D_5 and D_4, distinguish this command from a Control Word.

Mode Definitions. The following are defined for use in describing the operation of the 8254.

```
CLK Pulse:        a rising edge, then a falling edge, in
                  that order, of a Counter's CLK input.
Trigger:          a rising edge of a Counter's GATE input.
Counter loading: the transfer of a count from the CR to the CE
                  (refer to the ''Functional Description'')
```

Mode 0 : Interrupt on Terminal Count

Mode 0 is typically used for event counting. After the Control Word is written, OUT is initially low, and will remain low until the Counter reaches zero. OUT then goes high and remains high until a new count or a new Mode 0 Control Word is written into the Counter. GATE = 1 enables counting; GATE = 0 disables counting. GATE has no effect on OUT.

After the Control Word and initial count are written to a Counter, the initial count will be loaded on the next CLK pulse. This CLK pulse does not decrement the count, so for an initial count of N, OUT does not go high until N = 1 CLK pulses after the initial count is written.

If a new count is written to the Counter, it will be loaded on the next CLK pulse and counting will continue from the new count. If a two-byte count is written, the following happens:

1. Writing the first byte disables counting. OUT is set low immediately (no clock pulse required)
2. Writing the second byte allows the new count to be loaded on the next CLK pulse.

This allows the counting sequence to be synchronized by software. Again, OUT does not go high until N= 1 CLK pulses after the new count of N is written. If an initial count is written while GATE = 0, it will still be loaded on the next CLK pulse. When GATE goes high, OUT will go high N CLK pulses later; no CLK pulse is needed to load the Counter as this has already been done.

Note : The following conventions apply to all mode timing diagrams:

1. Counters are programmed for binary (not BCD) counting and for reading/writing least significant byte (LSB) only.
2. The counter is always selected (CS always low).
3. CW stands for "Control Word"; CW = 10 means a control word of 10 HEX is written to the counter.
4. LSB stands for "Least Significant Byte" of count.
5. Numbers below diagrams are count values. The lower number is the least significant byte. The upper number is the most significant byte. Since the counter is programmed to read/write LSB only, the most significant byte cannot be read.

   N stands for an undefined count. Vertical lines show transitions between count values

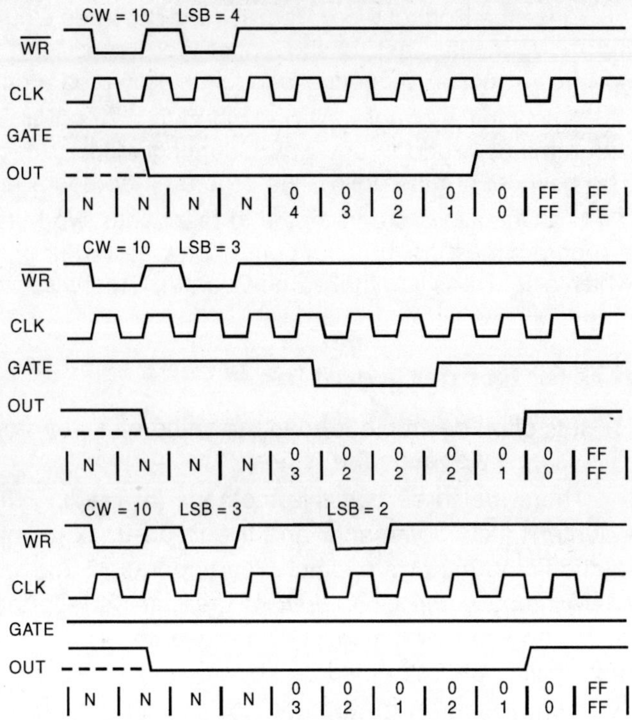

Fig. 8.38. Mode 0 Timing diagram.

Mode 1: Hardware Retriggerable One-shot

OUT will be initially high. OUT will go low on the CLK pulse following a trigger to begin the one-shot pulse, and will remain low until the Counter reaches zero. OUT will then go high and remain high until the CLK pulse after the next trigger. After writing the Control Word and initial count, the Counter is armed. A trigger results in loading the Counter and setting OUT low on the next CLK pulse, thus starting the one-shot pulse. An initial count of N will result in a one-shot pulse N CLK cycles in duration. The one-shot is retriggerable, hence OUT will remain low for N CLK pulses after any trigger. The one-shot pulse can be repeated without rewriting the same count into the counter. GATE has no effect on OUT. If a

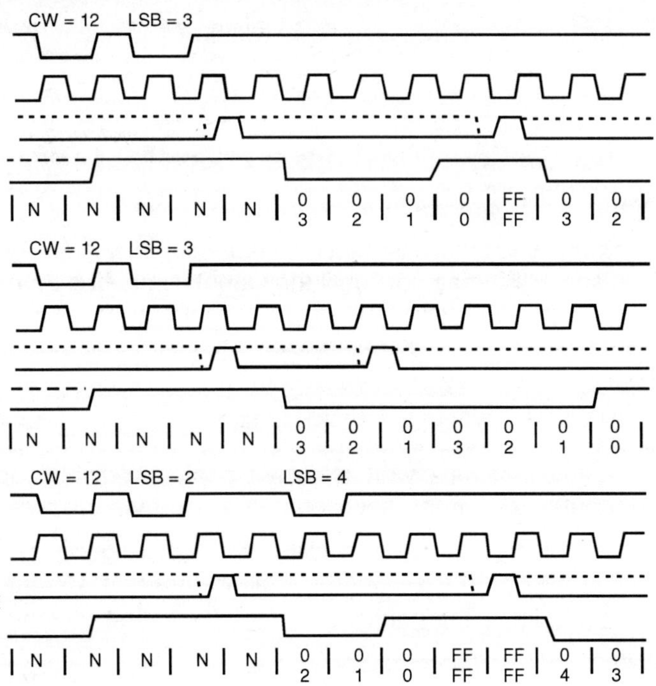

Fig. 8.39. Mode 1 Timing diagram.

new count is written to the Counter during a one-shot pulse, the current one-shot is not affected unless the counter is retriggered. In that case, the Counter is loaded with the new count and the one shot pulse continues until the new count expires.

Mode 2: Rate Generator

This Mode functions like a divide-by-N counter. It is typically used to generate a Real Time Clock interrupt. OUT will initially be high. When the initial count has decremented to 1, OUT goes low for one CLK pulse. OUT then goes high again, the Counter reloads the initial count and the process is repeated. Mode 2 is periodic; the same sequence is repeated indefinitely. For an initial count of N, the sequence repeats every N CLK cycles. GATE = **1 enables counting; GATE = 0 disables counting. If GATE goes low during an output pulse, OUT is set high immediately. A trigger reloads the Counter with the initial count on the next CLK pulse; OUT goes low N CLK pulses after the trigger. Thus the GATE input can be used to synchronize the Counter.**

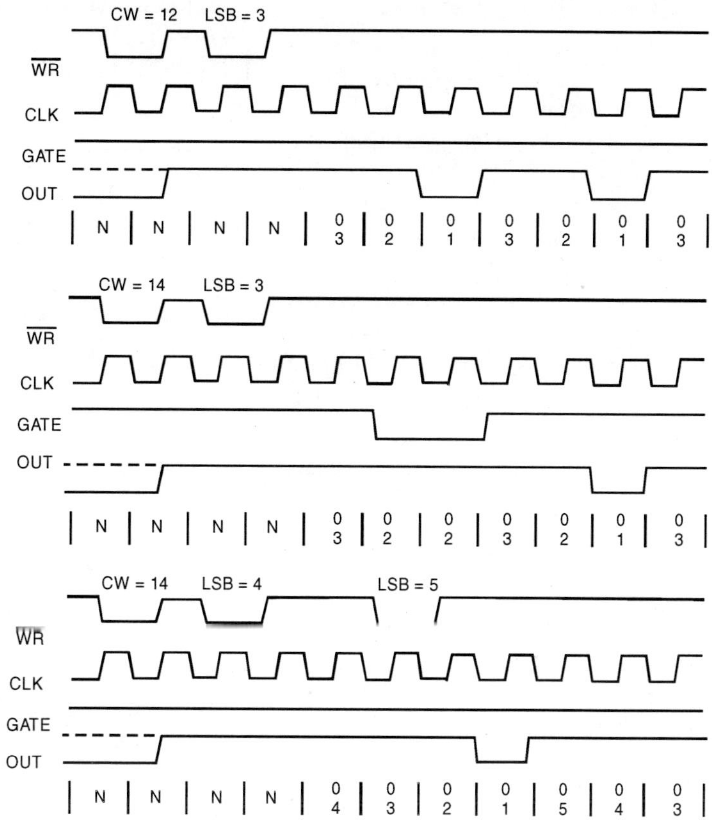

Fig. 8.40. Mode 2 Timing diagram.

After writing a Control Word and initial count, the Counter will be loaded on the next CLK pulse. OUT goes low N CLK Pulses after the initial count is written. This allows the Counter to be synchronized by software also.

Writing a new count while counting does not affect the current counting sequence. If a trigger is received after writing a new count but before the end of the current period, the Counter will be loaded with the new count on the next CLK pulse and counting will continue from the new count. Otherwise, the new count will be loaded at the end of the current counting cycle. In mode 2, a COUNT of 1 is illegal.

Mode 3: Square Wave Mode

Mode 3 is typically used for Baud rate generation. Mode 3 is similar to Mode 2 except for the duty cycle of OUT. OUT will initially be high. When half the initial count has expired, OUT goes low for the remainder of the count. Mode 3 is periodic; the sequence above is repeated indefinitely. An initial count of N results in a square wave with a period of N CLK cycles. GATE = 1 enables counting; GATE = 0 disables counting. If GATE goes low while OUT is low, OUT is set high immediately; no CLK pulse is required. A trigger reloads the Counter with the initial count on the next CLK pulse. Thus the GATE input can be used to synchronize the Counter.

After writing a Control Word and initial count, the Counter will be loaded on the next CLK pulse. This allows the Counter to be synchronized by software also.

Writing a new count while counting does not affect the current counting sequence. If a trigger is received after writing a new count but before the end of the current half-cycle of the square wave, the Counter will be loaded with the new count on the next CLK pulse and counting will continue from the new count. Otherwise, the new count will be loaded at the end of the current half-cycle.

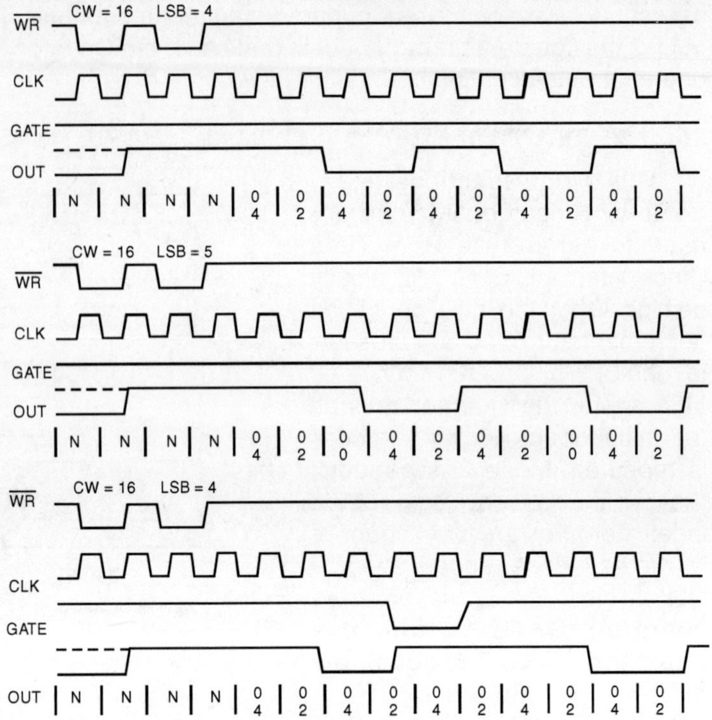

Fig. 8.41. Mode 3 Timing diagram.

Mode 3 is implemented as follows: **Even counts:** OUT is initially high. The initial count is loaded on one CLK pulse and then is decremented by two on succeeding CLK pulses. When the count expires OUT changes value and the Counter is reloaded with the initial count. The above process is repeated indefinitely.

Odd counts: OUT is initially high. The initial count minus one (an even number) is loaded on one CLK pulse and then is decremented by two on succeeding CLK pulses. One CLK pulse after the count expires, OUT goes low and the Counter is reloaded with the initial count minus one. Succeeding CLK pulses decrement the count by two. When the count expires, OUT goes high again and the Counter is reloaded with the initial count minus one. The above process is repeated indefinitely. So for odd counts, OUT will be high for (N = 1)/2 counts and low for (N = 1)/2 counts.

Mode 4: Software Triggered Strobe Out

It will be initially high. When the initial count expires, OUT will go low for one CLK pulse and then go high again.

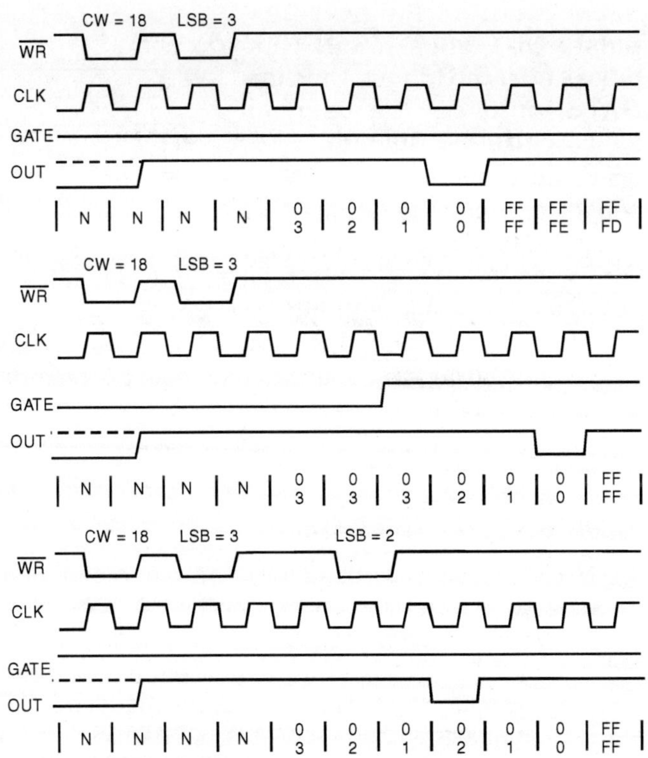

Fig. 8.42. Mode 4 Timing diagram.

The counting sequence is "triggered" by writing the initial count. GATE = **1 enables counting; GATE = 0 disables counting. GATE has no effect on OUT.**

After writing a Control Word and initial count, the Counter will be loaded on the next CLK pulse. This CLK pulse does not decrement the count, so for an initial count of N, OUT does not strobe low until N =1 CLK pulses after the initial count is written. If a new count is written during counting, it will be loaded on the next CLK pulse and counting will continue from the new count. If a two-byte count is written, the following happens:

1. Writing the first byte has no effect on counting.
2. Writing the second byte allows the new count to be loaded on the next CLK pulse.

This allows the sequence to be "retriggered" by software. OUT strobes low N = 1 CLK pulses after the new count of N is written.

Mode 5: Hardware Triggered Strobe

(RETRIGGERABLE) OUT will initially be high. Counting is triggered by a rising edge of GATE. When the initial count has expired, OUT will go low for one CLK pulse and then go high again.

After writing the Control Word and initial count, the counter will not be loaded until the CLK pulse after a trigger. This CLK pulse does not decrement the count, so for an initial count of N, OUT does not strobe low until N = **1 CLK pulses after a trigger.**

A trigger results in the Counter being loaded with the initial count on the next CLK pulse. The counting sequence is retriggerable. OUT will not strobe low for N = 1 CLK pulses after any trigger. GATE has no effect on OUT.

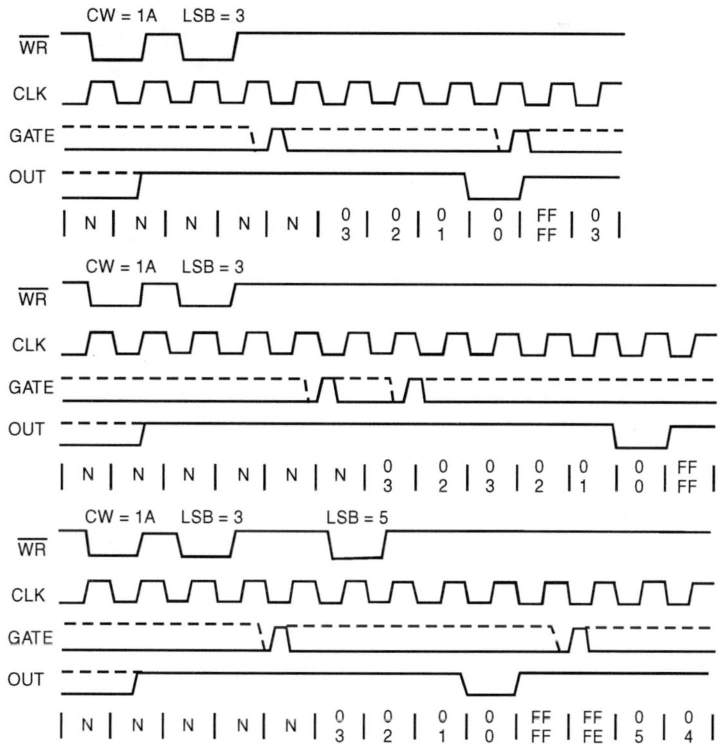

Fig. 8.43. Mode 5 Timing diagram.

If a new count is written during counting, the current counting sequence will not be affected. If a trigger occurs after the new count is written but before the current count expires, the Counter will be loaded with the new count on the next CLK pulse and counting will continue from there.

8.6.7. Operation Common to All Modes

PROGRAMMING When a Control Word is written to a Counter, all Control Logic is immediately reset and OUT goes to a known initial state; no CLK pulses are required for this.

Gate

The GATE input is always sampled on the rising edge of CLK. In Modes 0, 2, 3 and 4 the GATE input is level sensitive, and the logic level is sampled on the rising edge of CLK. In Modes

1, 2, 3, and 5 the GATE input is rising-edge sensitive. In these Modes, a rising edge of GATE (trigger) sets an edge-sensitive flip-flop in the Counter. This flip-flop is then sampled on the next rising edge of CLK; the flip-flop is reset immediately after it is sampled. In this way, a trigger will be detected no matter when it occurs and a high logic level does not have to be maintained until the next rising edge of CLK. Note that in Modes 2 and 3, the GATE input is both edge- and level-sensitive. In Modes 2 and 3, if a CLK source other than the system clock is used, GATE should be pulsed immediately following WR of a new count value.

COUNTER New counts are loaded and Counters are decremented on the falling edge of CLK.

The largest possible initial count is 0; this is equivalent to 2^{16} for binary counting and 10^4 for BCD counting.

The Counter does not stop when it reaches zero. In Modes 0, 1, 4 and 5 the Counter "wraps around" to the highest count, either FFFF hex for binary counting or 9999 for BCD counting, and continues counting. Modes 2 and 3 are periodic; the Counter reloads itself with the initial count and continues counting from there.

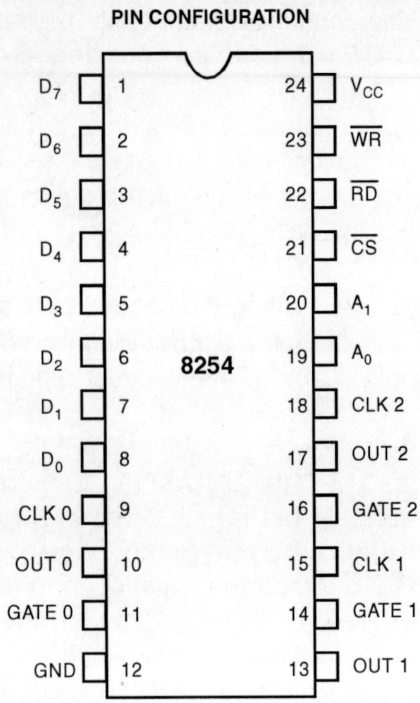

Fig. 8.44. Pin configuration of 8254.

Pin Description

Symbol	Pin No.	Type	Name and Function
D_7–D_0	1–8	I/O	DATA: Bi-directional three state data bus lines, connected to system data bus.
CLK 0	9	I	CLOCK 0: Clock input of Counter 0.
OUT 0	10	O	OUTPUT 0: Output of Counter 0.
GATE 0	11	I	GATE 0: Gate input of Counter 0.
GND	12		GROUND: Power supply connection.
V_{CC}	24		POWER: [a] 5V power supply connection.
WR	23	I	WRITE CONTROL: This input is low during CPU write operations.
RD	22	I	READ CONTROL: This input is low during CPU read operations.
CS	21	I	CHIP SELECT: A low on this input enables the 8254 to respond to RD and WR signals. RD and WR are ignored otherwise.
A_1,A_0	20–19	I	ADDRESS: Used to select one of the three Counters or the Control Word Register for read or write operations. Normally connected to the system address bus. A_1 A_0 Selects 0 0 Counter 0 0 1 Counter 1 1 0 Counter 2 1 1 Control Word Register
CLK 2	18	I	CLOCK 2: Clock input of Counter 2.
OUT 2	17	O	OUT 2: Output of Counter 2.
GATE 2	16	I	GATE 2: Gate input of Counter 2.
CLK 1	15	I	CLOCK 1: Clock input of Counter 1.
GATE 1	14	I	GATE 1: Gate input of Counter 1.
OUT 1	13	O	OUT 1: Output of Counter 1.

8.7. 8251 USART (UNIVERSAL SYNCHRONOUS ASYNCHRONOUS RECEIVER TRANSMITTER)

General Description

The 82C51A-2 is a USART (Universal Synchronous Asynchronous Receiver Transmitter) for serial data communication.

As a peripheral device of a microcomputer system, the MSM82C51A-2 receives parallel data from the CPU and transmits serial data after conversion. This device also receives serial data from the outside and transmits parallel data to the CPU after conversion.

Features :

- Wide power supply voltage range from 3 V to 6 V
- Wide temperature range from –40°C to 85°C
- Synchronous communication upto 64 Kbaud
- Asynchronous communication upto 38.4 Kbaud
- Transmitting/receiving operations under double buffered configuration.
- Error detection (parity, overrun and framing)

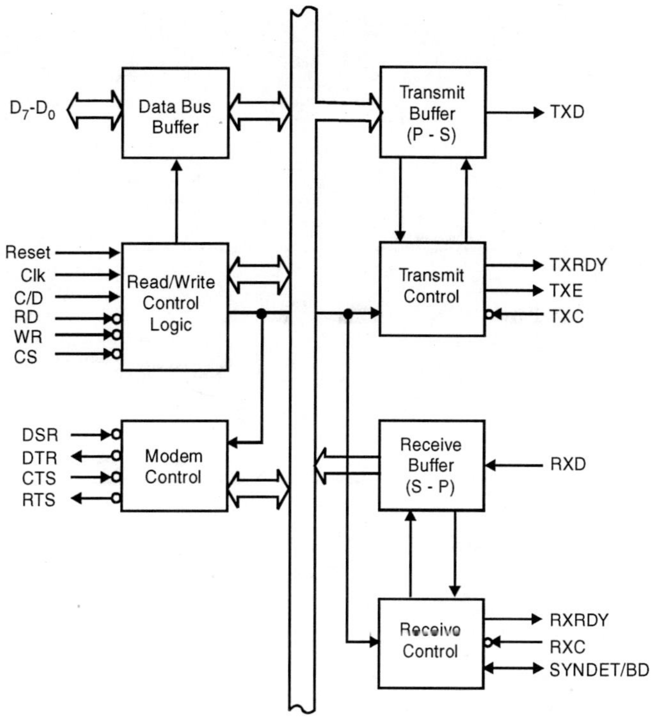

Fig. 8.45. 8251 Functional block diagram.

Functional Block Diagram

Function block diagram is shown in Fig. 8.45.

8.7.1. Function

Outline :

The MSM82C51A-2's functional configuration is programed by software.

Operation between the MSM82C51A-2 and a CPU is executed by program control. Table 8.6 shows the operation between a CPU and the device.

TABLE 8.6. Operation between MSM82C51A and CPU

CS	C/D	RD	WR	Operation
1	×	×	×	Data Bus 3-State
0	×	1	1	Data Bus 3-State
0	1	0	1	Status → CPU
0	1	1	0	Control Word ← CPU
0	0	0	1	Data → CPU
0	0	1	0	Data ← CPU

It is necessary to execute a function-setting sequence after resetting the MSM82C51A-2. Figure 8.46 shows the function-setting sequence.

If the function was set, the device is ready to receive a command, thus enabling the transfer of data by setting a necessary command, reading a status and reading/writing data.

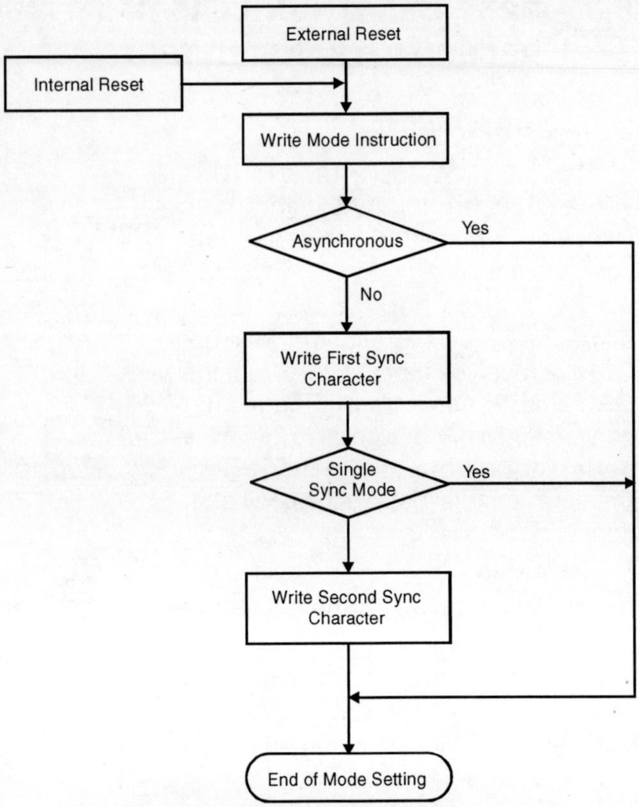

Fig. 8.46. Function-setting Sequence (Mode Instruction Sequence)

8.7.2. Control Words

There are two types of control word :

1. Mode instruction (setting of function)
2. Command (setting of operation)

(1) Mode Instruction

Mode instruction is used for setting the function of the MSM82C51A-2. Mode instruction will be in "wait for write" at either internal reset or external reset. That is, the writing of a control word after resetting will be recognized as a "mode instruction."

Items set by mode instruction are as follows :

- ➤ Synchronous/asynchronous mode
- ➤ Stop bit length (asynchronous mode)
- ➤ Character length
- ➤ Parity bit
- ➤ Baud rate factor (asynchronous mode)
- ➤ Internal/external synchronization (synchronous mode)
- ➤ Number of synchronous characters (Synchronous mode)

The bit configuration of mode instruction is shown in Figs. 8.47 and 8.48. In the case of synchronous mode, it is necessary to write one- or two- byte sync characters.

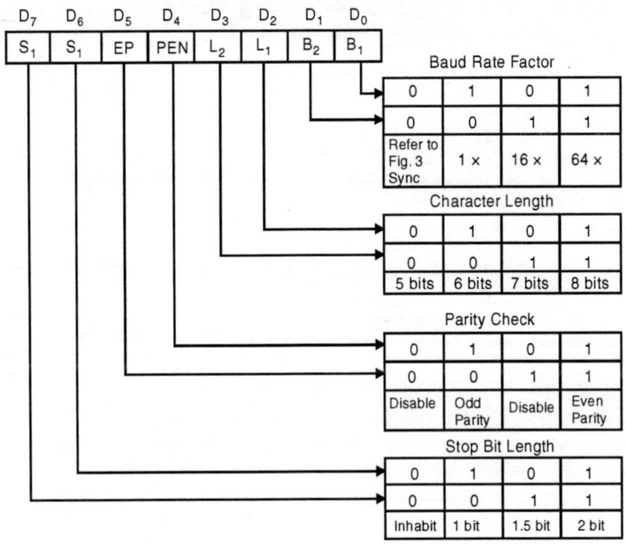

Fig. 8.47. Bit configuration of mode instruction (asynchronous).

If sync characters were written, a function will be set because the writing of sync characters constitutes part of mode instruction.

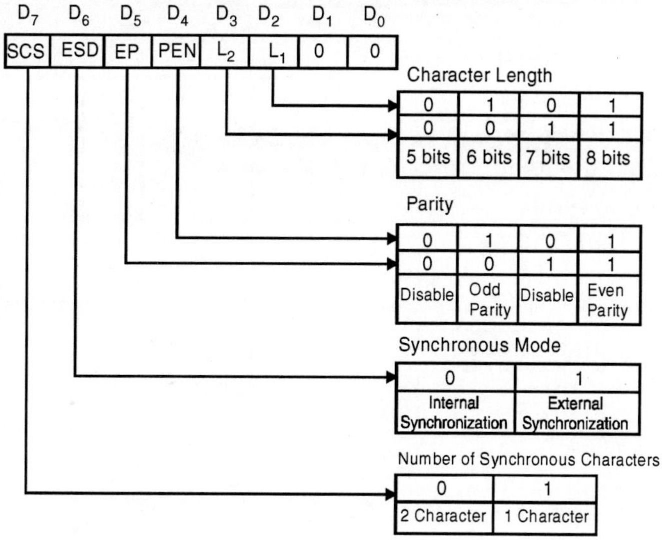

Fig. 8.48. Bit configuration of mode instruction (synchronous).

(2) Command

Command is used for setting the operation of the MSM82C51A-2.

It is possible to write a command whenever necessary after writing a mode instruction and sync characters.

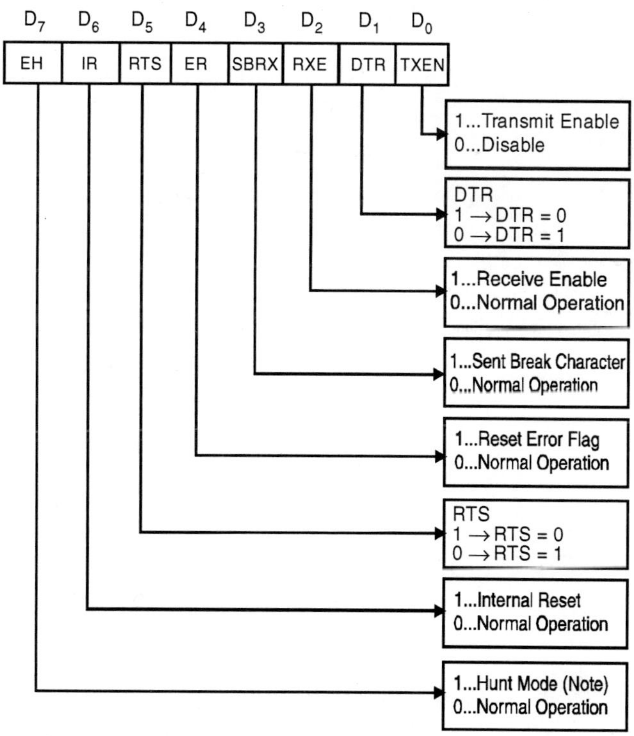

Note. Search mode for synchronous characters in synchronous mode

Fig. 8.49. Bit Configuration of Command.

Items to be set by command are as follows:

- ➤ Transmit Enable/Disable
- ➤ Receive Enable/Disable
- ➤ DTR, RTS Output of data
- ➤ Resetting of error flag.
- ➤ Sending to break characters.
- ➤ Internal resetting
- ➤ Hunt mode (synchronous mode)

The bit configuration of a command is shown in Fig. 8.49.

Status Word

It is possible to see the internal status of MSM82C51A-2 by reading a status word. The bit configuration of status word is shown in Fig. 8.50.

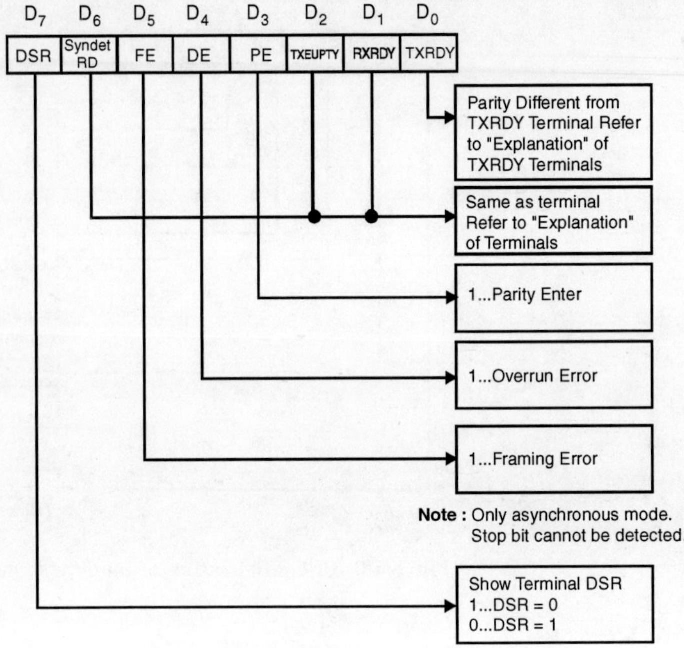

Fig. 8.50. Bit configuration of Status word.

Standby Status

It is possible to put the MSM82C51A-2 in "standby status" When the following conditions have been satisfied the MSM82C51A-2 is in "standby status."

(1) CS terminal is fixed at Vcc level.

(2) Input pins other CS , D_0 to D_7, RD, WR and C/D are fixed at Vcc or GND level (including SYNDET in external synchronous mode).

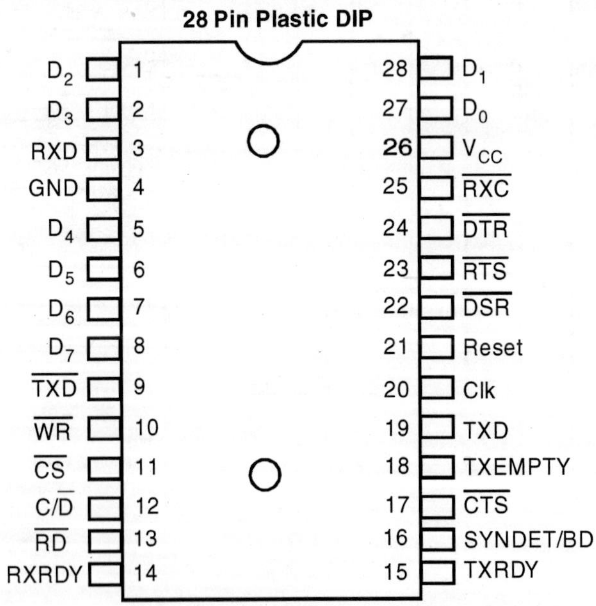

Fig. 8.51. Pin configuration 8251.

8.7.3. Pin Description

D_0 to D_7 (I/O terminal)

This is bidirectional data bus which receive control words and transmits data from the CPU and sends status words and received data to CPU.

RESET (Input terminal)

A "High" on this input forces the MSM82C51A-2 into "reset status." The device waits for the writing of "mode instruction."

The min. reset width is six clock inputs during the operating status of CLK.

CLK (Input terminal)

CLK signal is used to generate internal device timing. CLK signal is independent of RXC or TXC. However, the frequency of CLK must be greater than 30 times the RXC and TXC at Synchronous mode and Asynchronous "x1" mode, and must be greater than 5 times at Asynchronous "x16" and "x64" mode.

WR (Input terminal)

This is the "active low" input terminal which receives a signal for writing transmit data and control words from the CPU into the MSM82C51A-2.

RD (Input terminal)

This is the "active low" input terminal which receives a signal for reading receive data and status words from the MSM82C51A-2.

C/D (Input terminal)

This is an input terminal which receives a signal for selecting data or command words and status words when the MSM82C51A-2 is accessed by the CPU.

If C/D = low, data will be accessed.

If C/D = high, command word or status word will be accessed.

CS (Input terminal)

This is the "active low" input terminal which selects the MSM82C51A-2 at low level when the CPU accesses.

Note : The device won't be in "standby status"; only setting CS = High. Refer to "Explanation of Standby Status."

TXD (Output terminal)

This is an output terminal for transmitting data from which serial-converted data is sent out. The device is in "mark status" (high level) after resetting or during a status when transmit is disabled. It is also possible to set the device in "break status" (low level) by a command.

TXRDY (Output terminal)

This is an output terminal which indicates that the MSM82C51A-2 is ready to accept a transmitted data character. But the terminal is always at low level if CTS = high or the device was set in "TX disable status" by a command.

Note. TXRDY status word indicates that transmit data character is receivable, regardless of CTS or command.

If the CPU writes a data character, TXRDY will be reset by the leading edge or WR signal.

TXEMPTY (Output terminal)

This is an output terminal which indicates that the MSM82C51A-2 has transmitted all the characters and had no data character.

In "synchronous mode," the terminal is at high level, if transmit data characters are no longer remaining sync characters are automatically transmitted. If the CPU writes a data character, TXEMPTY will be reset by the leading edge of WR signal.

Note. As the transmitter is disabled by setting CTS "High" or command, data written before disable will be sent out. Then TXD and TXEMPTY will be "High".

Even if a data is written after disable, that data is not sent out and TXE will be "High". After the transmitter is enabled, it sent out.

TXC (Input terminal)

This is a clock input signal which determines the transfer speed of transmitted data. In "synchronous mode," the baud rate will be the same as the frequency of TXC.

In "asynchronous mode", it is possible to select the baud rate factor by mode instruction. It can be 1, 1/16 or 1/64 the TXC.

The falling edge of TXC shifts the serial data out of the MSM82C51A-2.

RXD (Input terminal)

This is a terminal which receives serial data.

RXRDY (Output terminal)

This is a terminal which indicates that the MSM82C51A-2 contains a character that is ready to READ.

If the CPU reads a data character, RXRDY will be reset by the leading edge of RD signal. Unless the CPU reads a data character before the next one is received completely, the preceding data will be lost. In such a case, an overrun error flag status word will be set.

RXC (Input terminal)

This is a clock input signal which determines the transfer speed of received data. In "synchronous mode," the baud rate is the same as the frequency of RXC.

In "asynchronous mode," it is possible to select the baud rate factor by mode instruction. It can be 1, 1/16, 1/64 the RXC.

SYNDET/BD (Input or Output terminal)

This is a terminal whose function changes according to mode.

In "internal synchronous mode" this terminal is at high level, if sync characters are received and synchronized. If a status word is read, the terminal will be reset.

In "external synchronous mode," this is an input terminal. A "High" on this input forces the MSM82C51A-2 to start receiving data characters.

In "asynchronous mode," this is an output terminal which generates "high level"output upon the detection of a "break" character if receiver data contains a "low-level" space between the stop bits of two continuous characters. The terminal will be reset, if RXD is at high level. After Reset is active, the terminal will be output at low level.

DSR (Input terminal)

This is an input port for MODEM interface. The input status of the terminal can be recognized by the CPU reading status words.

DTR (Output terminal)

This is an output port for MODEM interface. It is possible to set the status of DTR by a command.

CTS (Input terminal)

This is an input terminal for MODEM interface which is used for controlling a transmit circuit. The terminal controls data transmission if the device is set in "TX Enable" status by a command. Data is transmitable if the terminal is at low level.

RTS (Output terminal)

This is an output port for MODEM interface. It is possible to set the status RTS by a command.

TIMING CHART

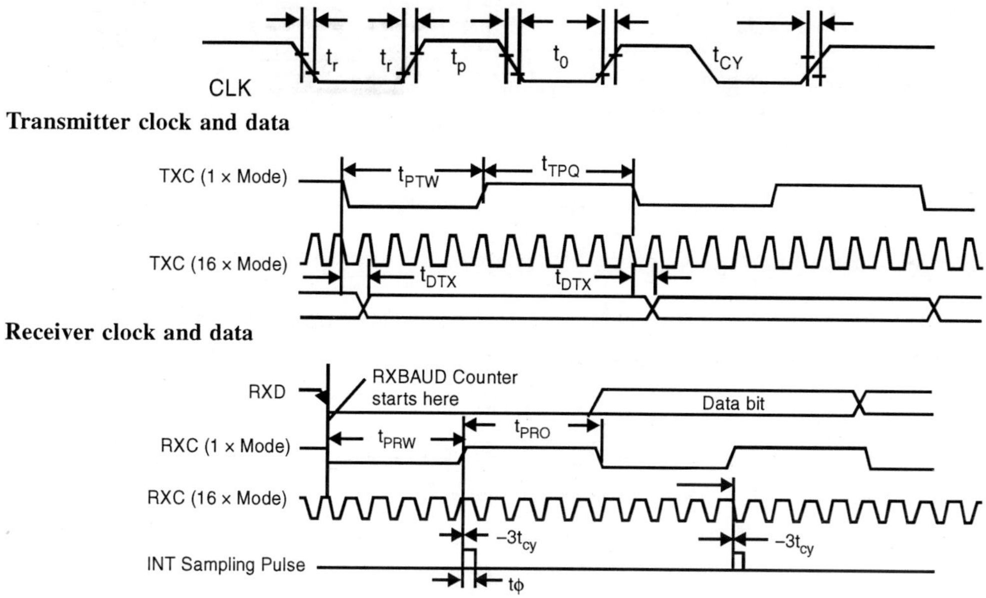

Fig. 8.52. Timing chart.

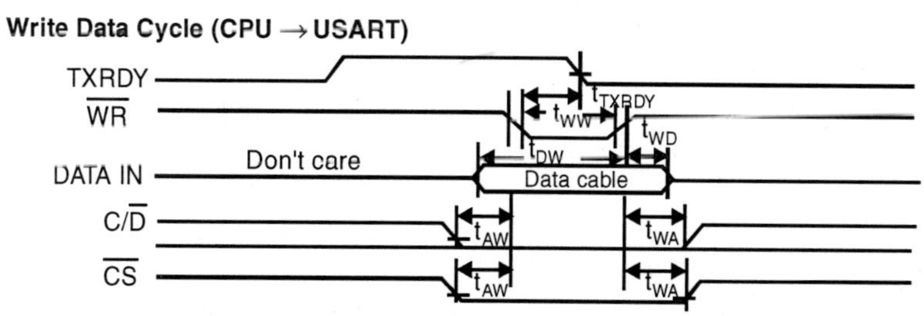

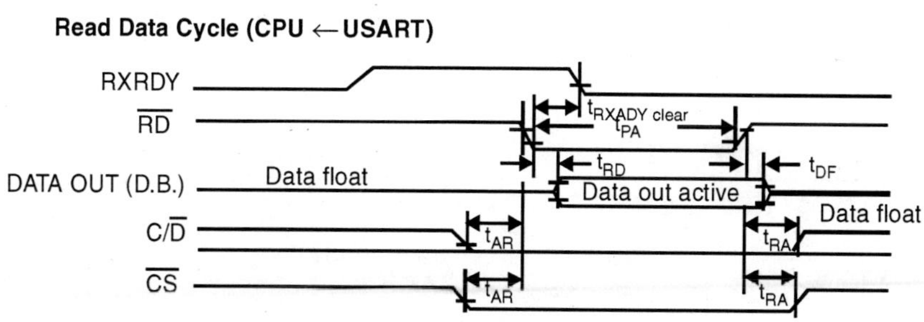

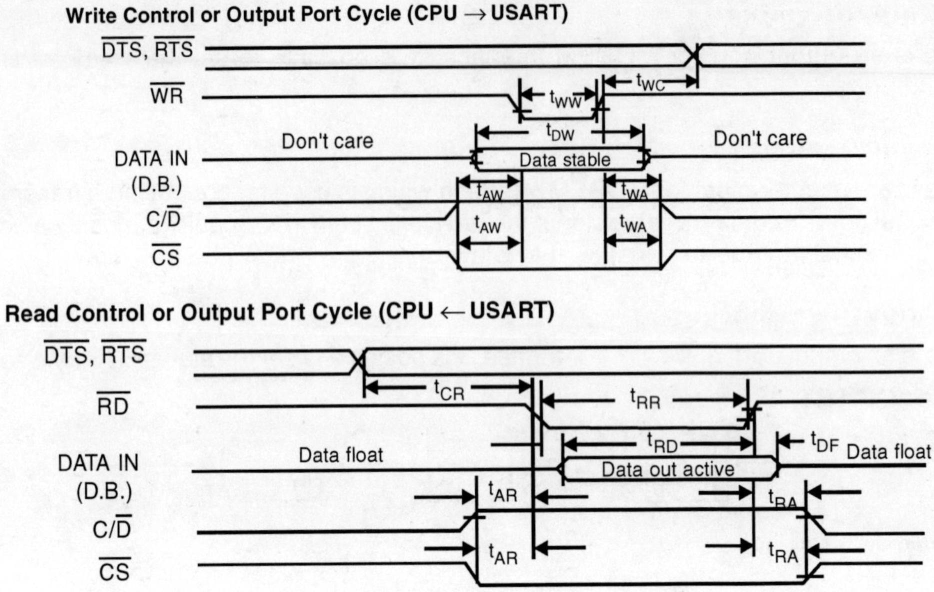

Fig. 8.53. Write control or output port cycle (CPU → USART).

8.8. PROGRAMMABLE DMA CONTROLLER (8237A-5)

The 8237A Multimode Direct Memory Access (DMA) Controller is a peripheral interface circuit for microprocessor systems. It is designed to improve system performance by allowing external

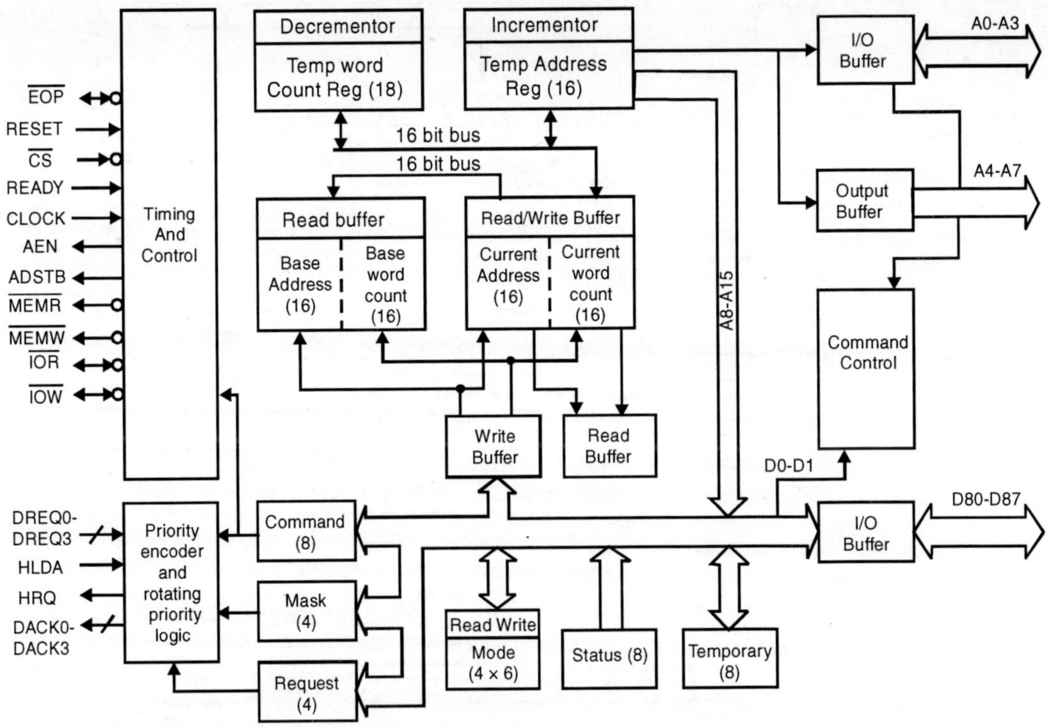

Fig. 8.54. Block Diagram (8237).

devices to directly transfer information from the system memory. Memory-to-memory transfer capability is also provided. The 8237A offers a wide variety of programmable control features to enhance data throughput and system optimization and to allow dynamic reconfiguration under program control.

The 8237A is designed to be used in conjunction with an external 8-bit address latch. It contains four independent channels and may be expanded to any number of channels by cascading additional controller chips. The three basic transfer modes allow programmability of the types of DMA service by the user. Each channel can be individually programmed to Autoinitialize to its original condition following an End of Process (EOP). Each channel has a full 64K address and word count capability

8.8.1. Functional Description

The 8237A block diagram includes the major logic blocks and all of the internal registers. The data interconnection paths are also shown. The various control signals are not shown between the blocks. The 8237A contains 344 bits of internal memory in the form of registers. A detailed description of the registers and their functions can be found under Register Description.

Name	Size	Number
Base Address Registers	16 bits	4
Base Word Count Registers	16 bits	4
Current Address Registers	16 bits	4
Current Word Count Registers	16 bits	4
Temporary Address Register	16 bits	1
Temporary Word Count Register	16 bits	1
Status Register	8 bits	1
Command Register	8 bits	1
Temporary Register	8 bits	1
Mode Registers	6 bits	4
Mask Register	4 bits	1
Request Register	4 bits	1

The 8237A contains three basic blocks of control logic. The Timing Control block generates internal timing and external control signals for the 8237A. The Program Command Control block decodes the various commands given to the 8237A by the microprocessor prior to servicing a DMA Request. It also decodes the Mode Control word used to select the type of DMA during the servicing. The Priority Encoder block resolves priority contention between DMA channels requesting service simultaneously. The Timing Control block derives internal timing from the clock input. In 8237A systems, this input will usually be the W2 TTL clock from an 8224 or CLK from an 8085AH or 8284A. 33% duty cycle clock generators, however, may not meet the clock high time requirement of the 8237A of the same frequency. For example, 82C84A-5 CLK output violates the clock high time requirement of 8237A-5. In this case 82C84A CLK can simply be inverted to meet 8237A-5 clock high and low time requirements. For 8085AH-2 systems above 3.9 MHz, the 8085 CLK(OUT) does not satisfy 8237A-5 clock LOW and HIGH time requirements. In this case, an external clock should be used to drive the 8237A-5.

DMA Operation : The 8237A is designed to operate in two major cycles. These are called Idle and Active cycles. Each device cycle is made up of a number of states. The 8237A can assume seven separate states, each composed of one full clock period. State I (SI) is the inactive state. It is entered when the 8237A has no valid DMA requests pending. While in SI, the DMA controller is inactive but may be in the Program Condition, being programmed by the processor. State S0 (S0) is the first state of a DMA service. The 8237A has requested a hold but the processor has not yet returned an acknowledge. The 8237A may still be programmed until it receives HLDA from the CPU. An acknowledge from the CPU will signal that DMA transfers may begin. S1, S2, S3 and S4 are the working states of the DMA service. If more time is needed to complete a transfer than is available with normal timing, wait states (SW) can be inserted between S2 or S3 and S4 by the use of the Ready line on the 8237A. Note that the data is transferred directly from the I/O device to memory (or vice versa) with IOR and MEMW (or MEMR and IOW) being active at the same time. The data is not read into or driven out of the 8237A in I/O-to-memory or memory-to-I/O DMA transfers.

Memory-to-memory transfers require a read-from and a write-to-memory to complete each transfer. The states, which resemble the normal working states, use two digit numbers for identification. Eight states are required for a single transfer. The first four states (S11, S12, S13, S14) are used for the read-from-memory half and the last four states (S21, S22, S23, S24) for the write-to-memory half of the transfer.

IDLE CYCLE When no channel is requesting service, the 8237A will enter the Idle cycle and perform "SI" states. In this cycle the 8237A will sample the DREQ lines every clock cycle to determine if any channel is requesting a DMA service. The device will also sample CS, looking for an attempt by the microprocessor to write or read the internal registers of the 8237A. When CS is low and HLDA is low, the 8237A enters the Program Condition. The CPU can now establish, change or inspect the internal definition of the part by reading from or writing to the internal registers. Address lines A0 – A3 are inputs to the device and select which registers will be read or written. The IOR and IOW lines are used to select and time reads or writes. Due to the number and size of the internal registers, an internal flip-flop is used to generate an additional bit of address. This bit is used to determine the upper or lower byte of the 16-bit Address and Word Count registers. The flip-flop is reset by Master Clear or Reset. A separate software command can also reset this flip-flop.

Special software commands can be executed by the 8237A in the Program Condition. These commands are decoded as sets of addresses with the CS and IOW. The commands do not make use of the data bus. Instructions include Clear First/ Last Flip-Flop and Master Clear.

Active Cycle : When the 8237A is in the Idle cycle and a non-masked channel requests a DMA service, the device will output an HRQ to the microprocessor and enter the Active cycle. It is in this cycle that the DMA service will take place, in one of four modes:

Single Transfer Mode : In Single Transfer mode the device is programmed to make one transfer only. The word count will be decremented and the address decremented or incremented following

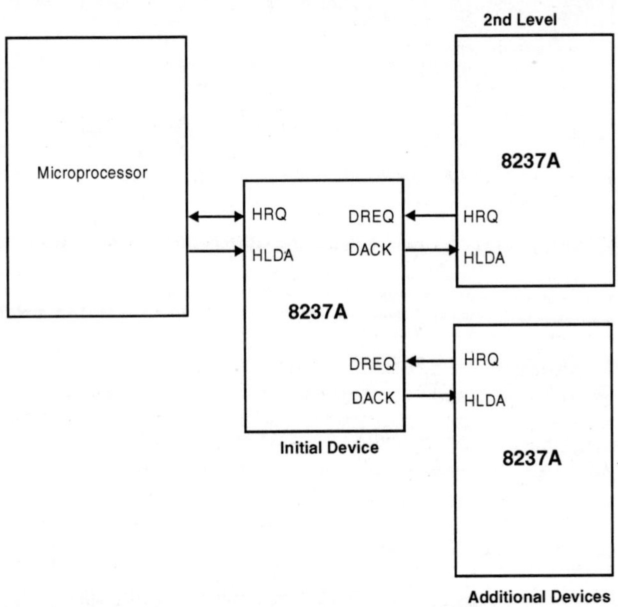

Fig. 8.55. Cascaded 8237A's.

each transfer. When the word count "rolls over" from zero to FFFFH, a Terminal Count (TC) will cause an Auto- initialize if the channel has been programmed to do so.

DREQ must be held active until DACK becomes active in order to be recognized. If DREQ is held active throughout the single transfer, HRQ will go inactive and release the bus to the system. It will again go active and, upon receipt of a new HLDA, another single transfer will be performed. In 8080A, 8085AH, 8088, or 8086 system, this will ensure one full machine cycle execution between DMA transfers. Details of timing between the 8237A and other bus control protocols will depend upon the characteristics of the microprocessor involved.

Block Transfer Mode : In Block Transfer mode the device is activated by DREQ to continue making transfers during the service until a TC, caused by word count going to FFFFH, or an external End of 8237A signals of its own. These could conflict with the outputs of the active channel in the added device. The 8237A will respond to DREQ and DACK but all other outputs except HRQ will be disabled. The ready input is ignored.

Figure 8.55 shows two additional devices cascaded into an initial device using two of the previous channels. This forms a two level DMA system. More 8237As could be added at the second level by using the remaining channels of the first level. Additional devices can also be added by cascading into the channels of the second level device, forming a third level.

Transfer Types : Each of the three active transfer modes can perform three different types of transfers. These are Read, Write and Verify. Write transfers move data from an I/O device to the memory by activating MEMW and IOR. Read transfers move data from memory to an I/O device by activating MEMR and IOW. Verify transfers are pseudo transfers. The 8237A operates as in Read or Write transfers generating addresses, and responding to EOP, etc. However, the memory and I/O control lines all remain inactive. The ready input is ignored in verify mode.

Memory-to-Memory : To perform block moves of data from one memory address space to another with a minimum of program effort and time, the 8237A includes a memory-to-memory transfer feature. Programming a bit in the Command register selects channels 0 and 1 to operate as memory-to- memory transfer channels. The transfer is initiated by setting the software DREQ for channel 0. The 8237A requests a DMA service in the normal manner. After HLDA is true, the device, using four state transfers in Block Transfer mode, reads data from the memory. The channel 0 Current Address register is the source for the address used and is decremented or incremented in the normal manner. The data byte read from the memory is stored in the 8237A internal Temporary register. Channel 1 then performs a four-state transfer of the data from the Temporary register to memory using the address in its Current Address register and incrementing or decrementing it in the normal manner. The channel 1 current Word Count is decremented. When the word count of channel 1 goes to FFFFH, a TC is generated causing an EOP output terminating the service.

Channel 0 may be programmed to retain the same address for all transfers. This allows a single word to be written to a block of memory.

The 8237A will respond to external EOP signals during memory-to-memory transfers. Data comparators in block search schemes may use this input to terminate the service when a match is found. Memory-to-memory operations can be detected as an active AEN with no DACK outputs.

Autoinitialize : By programming a bit in the Mode register, a channel may be set up as an Autoinitialize channel. During Autoinitialize initialization, the original values of the Current Address and Current Word Count registers are automatically restored from the Base Address and Base Word count registers of that channel following EOP. The base registers are loaded simultaneously with the current registers by the microprocessor and remain unchanged throughout the DMA service. The mask bit is not altered when the channel is in Autoinitialize.

Following Autoinitialize the channel is ready to perform another DMA service, without CPU intervention, as soon as a valid DREQ is detected. In order to Autoinitialize both channels in a memory-to-memory transfer, both word counts should be programmed identically. If interrupted externally, EOP pulses should be applied in both bus cycles.

Priority : The 8237A has two types of priority encoding available as software selectable options. The first is Fixed Priority which fixes the channels in priority order based upon the descending value of their number. The channel with the lowest priority is 3 followed by 2, 1 and the highest priority channel, 0. After the recognition of any one channel for service, the other channels are prevented from interfering with that service until it is completed.

After completion of a service, HRQ will go inactive and the 8237A will wait for HLDA to go low before activating HRQ to service another channel.

The second scheme is Rotating Priority. The last channel to get service becomes the lowest priority channel with the others rotating accordingly.

With Rotating Priority in a single chip DMA system, any device requesting service is guaranteed to be recognized after no more than three higher priority services have occurred. This prevents any one channel from monopolizing the system.

Compressed Timing : In order to achieve even greater throughput where system characteristics permit, the 8237A can compress the transfer time to two clock cycles.

Address Generation : In order to reduce pin count, the 8237A multiplexes the eight higher order address bits on the data lines. State S1 is used to output the higher order address bits to an external latch from which they may be placed on the address bus. The falling edge of Address Strobe (ADSTB) is used to load these bits from the data lines to the latch. Address Enable (AEN) is used to enable the bits onto the address bus through a three-state enable. The lower order address bits are output by the 8237A directly. Lines A_0 – A_7 should be connected to the address bus.

During Block and Demand Transfer mode services, which include multiple transfers, the addresses generated will be sequential. For many transfers the data held in the external address latch will remain the same. This data need only change when a carry or borrow from A_7 to A_8 takes place in the normal sequence of addresses. To save time and speed transfers, the 8237A executes S1 states only when updating of A_8–A_{15} in the latch is necessary. This means for long services, S1 states and Address Strobes may occur only once every 256 transfers, a savings of 255 clock cycles for each 256 transfers.

Register Description

Current Address Register : Each channel has a 16-bit Current Address register. This register holds the value of the address used during DMA transfers. The address is automatically incremented or decremented after each transfer and the intermediate values of the address are stored in the Current Address register during the transfer. This register is written or read by the microprocessor in successive 8-bit bytes. It may also be reinitialized by an Autoinitialize back to its original value. Autoinitialize takes place only after an EOP.

Current Word Register : Each channel has a 16-bit Current Word Count register. This register determines the number of transfers to be performed. The actual number of transfers will be one more than the number programmed in the Current Word Count register (i.e., programming a count of 100 will result in 101 transfers). The word count is decremented after each transfer. The intermediate value of the word count is stored in the register during the transfer. When the value in the register goes from zero to FFFFH, a TC will be generated. This register is loaded or read in successive 8-bit bytes by the microprocessor in the Program

Condition. Following the end of a DMA service it may also be reinitialized by an Autoinitialization back to its original value. Auto-initialize can occur only when an EOP occurs. If it is not Autoinitialized, this register will have a count of FFFFH after TC.

Base Address and Base Word Count Registers : Each channel has a pair of Base Address and Base Word Count registers. These 16-bit registers store the original value of their associated current registers. During Autoinitialize these values are used to restore the current registers to their original values. The base registers are written simultaneously with their corresponding current register in 8-bit bytes in the Program Condition by the microprocessor. These registers cannot be read by the microprocessor.

Command Register : This 8-bit register controls the operation of the 8237A. It is programmed by the microprocessor in the Program Condition and is cleared by Reset or a Master Clear instruction. The following table lists the function of the command bits.

Mode Register : Each channel has a 6-bit Mode register associated with it. When the register is being written to by the microprocessor in the Program Condition, bits 0 and 1 determine which channel Mode register is to be written.

Request Register : The 8237A can respond to requests for DMA service which are initiated by software as well as by a DREQ. Each channel has a request bit associated with it in the 4-bit Request register. These are non-maskable and subject to prioritization by the Priority Encoder network. Each register bit is set or reset separately under software control or is cleared upon generation of a TC or external EOP. The entire register is cleared by a Reset. To set or reset a bit, the software loads the proper form of the data word. See Fig. 8.58 for register address coding. In order to make a software request, the channel must be in Block Mode.

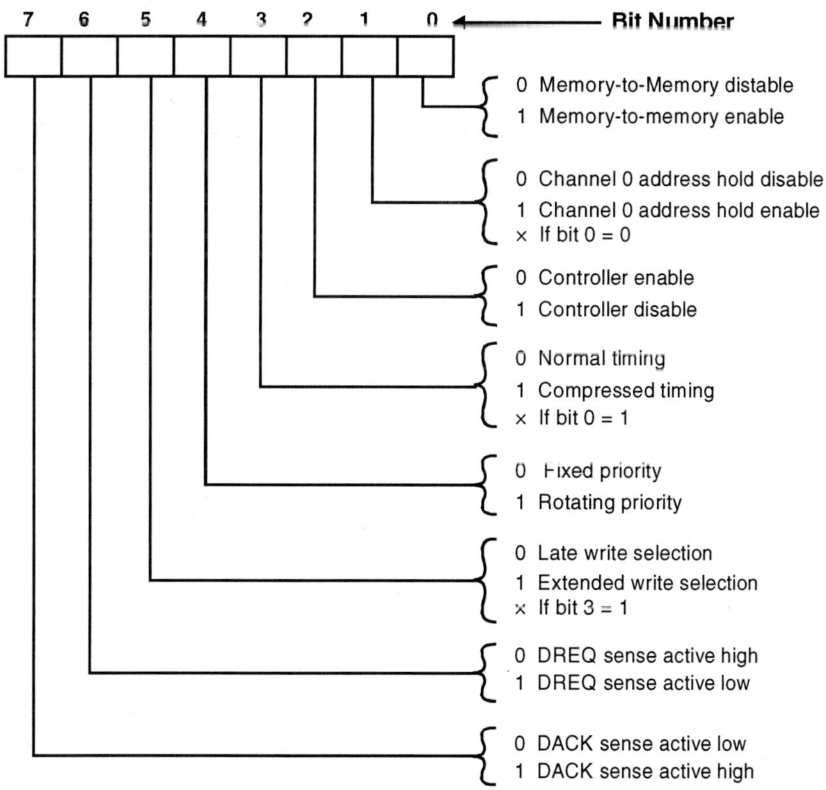

Fig. 8.56. Command register.

Mode Register

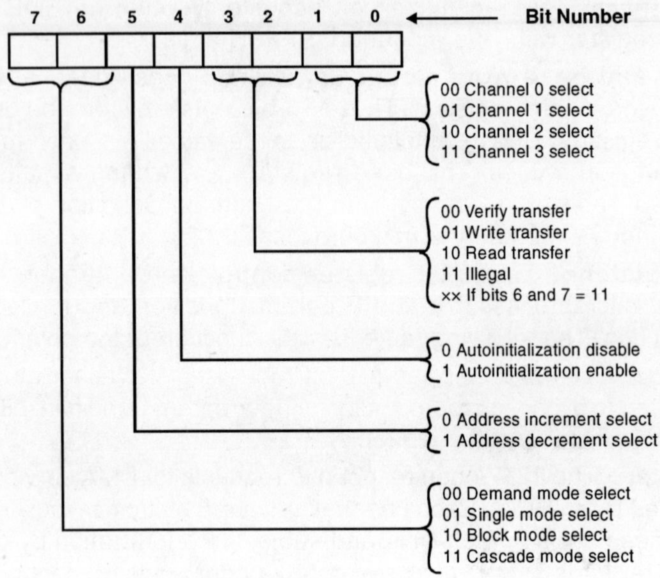

Fig. 8.57. Mode register.

Request Registers

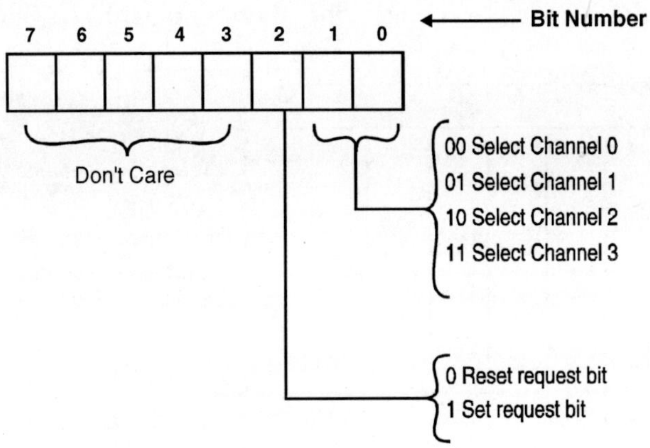

Fig. 8.58. Request register.

Mask Register : Each channel has associated with it a mask bit which can be set to disable the incoming DREQ. Each mask bit is set when its associated channel produces an EOP if the channel is not programmed for Autoinitialize. Each bit of the 4-bit Mask register may also be set or cleared separately under software control.

The entire register is also set by a Reset. This disables all DMA requests until a clear Mask register instruction

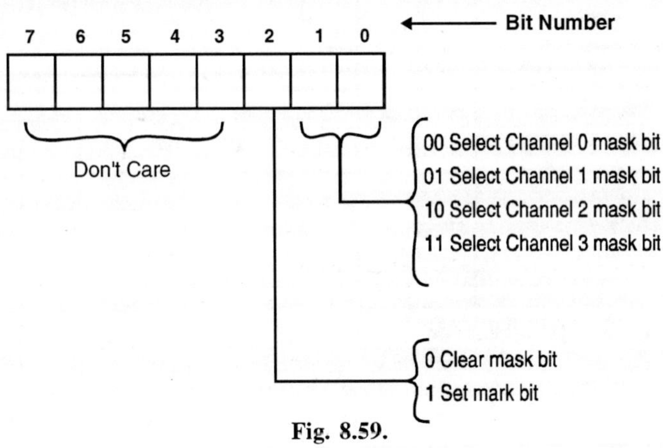

Fig. 8.59.

allows them to occur. The instruction to separately set or clear the mask bits is similar in form to that used with the Request register.

All four bits of the Mask register may also be written with a single command.

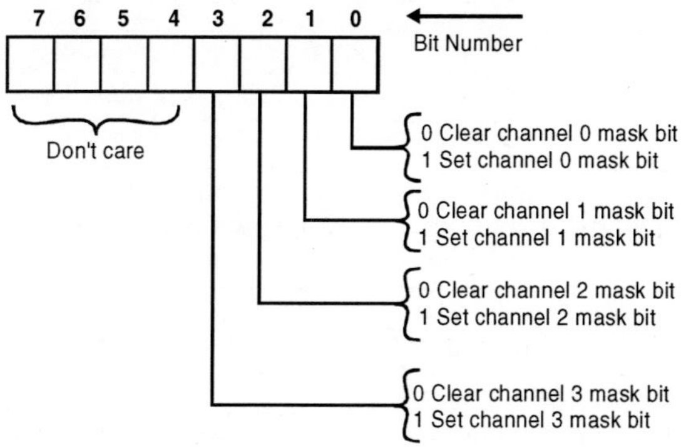

Register	Operation	Signals						
		$\overline{CS}$	$\overline{IOR}$	$\overline{IOW}$	A3	A2	A1	A0
Command	Write	0	1	0	1	0	0	0
Mode	Write	0	1	0	1	0	1	1
Request	Write	0	1	0	1	0	0	1
Mask	Set/Reset	0	1	0	1	0	1	0
Mask	Write	0	1	0	1	1	1	1
Temporary	Read	0	1	0	1	0	0	0
Status	Read	0	0	1	1	0	0	0

Fig. 8.60. Definition of register codes.

Status Register : The Status register is available to be read out of the 8237A by the microprocessor. It contains information about the status of the devices at this point. This information includes which channels have reached a terminal count and which channels have pending DMA requests. Bits 0 – 3 are set every time a TC is reached by that channel or an external EOP is applied. These bits are cleared upon Reset and on each Status Read. Bits 4 – 7 are set whenever their corresponding channel is requesting service.

Temporary Register : The Temporary register is used to hold data during memory-to-memory transfers. Following the completion of the transfers, the last word moved can be read by the microprocessor in the Program Condition. The Temporary register always contains the last byte transferred in the previous memory-to-memory operation, unless cleared by a Reset.

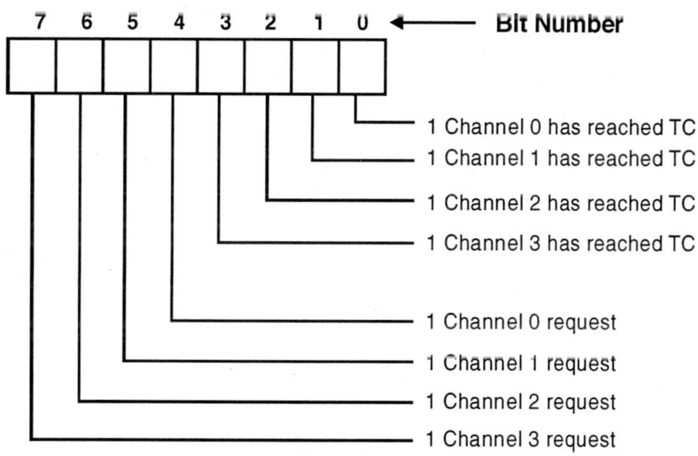

Fig. 8.61.

Software Commands : These are additional special software commands which can be executed in the Program Condition. They do not depend on any specific bit pattern on the data bus. The three software commands are:

Clear First/Last Flip-Flop : This command must be executed prior to writing or reading new address or word count information to the 8237A. This initializes the flip-flop to a known state so that subsequent accesses to register contents by the microprocessor will address upper and lower bytes in the correct sequence.

Master Clear : This software instruction has the same effect as the hardware Reset. The Command, Status, Request, Temporary, and Internal First/Last Flip-Flop registers are cleared and the Mask register is set. The 8237A will enter the Idle cycle.

Clear Mask Register : This command clears the mask bits of all four channels, enabling them to accept DMA requests.

TABLE 8.7. Pin Description.

Symbol	Type	Name and Function
V_{CC}		POWER: a 5V supply.
V_{SS}		GROUND: Ground.
CLK	I	CLOCK INPUT: Clock Input controls the internal operations of the 8237A and its rate of data transfers. The input may be driven at up to 5 MHz for the 8237A-5.
CS	I	CHIP SELECT: Chip Select is an active low input used to select the 8237A as an I/O device during the Idle cycle. This allows CPU communication on the data bus.
RESET	I	RESET: Reset is an active high input which clears the Command, Status, Request and Temporary registers. It also clears the first/ last flip/flop and sets the Mask register. Following a Reset the device is in the Idle cycle.
READY	I	READY: Ready is an input used to extend the memory read and write pulses from the 8237A to accommodate slow memories or I/O peripheral devices. Ready must not make transitions during its specified setup/ hold time.
HLDA	I	HOLD ACKNOWLEDGE: The active high Hold Acknowledge from the CPU indicates that it has relinquished control of the system busses.
DREQ0 – DREQ3	I	DMA REQUEST: The DMA Request lines are individual asynchronous channel request inputs used by peripheral circuits to obtain DMA service. In fixed Priority, DREQ0 has the highest priority and DREQ3 has the lowest priority. A request is generated by activating the DREQ line of a channel. DACK will acknowledge the recognition of DREQ signal. Polarity of DREQ is programmable. Reset initializes these lines to active high. DREQ must be maintained until the corresponding DACK goes active.
DB0 – DB7	I/O	DATA BUS: The Data Bus lines are bidirectional three-state signals connected to the system data bus. The outputs are enabled in the Program condition during the I/O Read to output the contents of an Address register, a Status register, the Temporary register or a Word Count register to the CPU. The outputs are disabled and the inputs are read during an I/O Write cycle when the CPU is programming the 8237A control registers. During DMA cycles the most significant 8 bits of the address are output onto the data bus to be strobed into an external latch by ADSTB. In memory-to-memory operations, data from the memory comes into the 8237A on the data bus during the read-from- memory transfer. In the write-to-memory transfer, the data bus outputs place the data into the new memory location.
IOR	I/O	I/O READ: I/O Read is a bidirectional active low three-state line. In the Idle cycle, it is an input control signal used by the CPU to read the control registers. In the Active cycle, it is an output control signal used by the 8237A to access data from a peripheral during a DMA Write transfer.

Symbol	Type	Name and Function
IOW	I/O	I/O WRITE: I/O Write is a bidirectional active low three-state line. In the Idle cycle, it is an input control signal used by the CPU to load information into the 8237A. In the Active cycle, it is an output control signal used by the 8237A to load data to the peripheral during a DMA Read transfer.
EOP	I/O	END OF PROCESS: End of Process is an active low bidirectional signal. Information concerning the completion of DMA services is available at the bidirectional EOP pin. The 8237A allows an external signal to terminate an active DMA service. This is accomplished by pulling the EOP input low with an external EOP signal. The 8237A also generates a pulse when the terminal count (TC) for any channel is reached. This generates an EOP signal which is output through the EOP line. The reception of EOP, either internal or external, will cause the 8237A to terminate the service, reset the request, and, if Autoinitialize is enabled, to write the base registers to the current registers of that channel. The mask bit and TC bit in the status word will be set for the currently active channel by EOP unless the channel is programmed for Autoinitialize. In that case, the mask bit remains unchanged. During memory-to-memory transfers, EOP will be output when the TC for channel 1 occurs. EOP should be tied high with a pull-up resistor if it is not used to prevent erroneous end of process inputs.
A0 – A3	I/O	ADDRESS: The four least significant address lines are bidirectional three-state signals. In the Idle cycle they are inputs and are used by the CPU to address the register to be loaded or read. In the Active cycle they are outputs and provide the lower 4 bits of the output address.
A4–A7	O	ADDRESS: The four most significant address lines are three-state outputs and provide 4 bits of address. These lines are enabled only during the DMA service.
HRQ	O	HOLD REQUEST: This is the Hold Request to the CPU and is used to request control of the system bus. If the corresponding mask bit is clear, the presence of any valid DREQ causes 8237A to issue the HRQ.
DACK0 – DACK3	O	DMA ACKNOWLEDGE: DMA Acknowledge is used to notify the individual peripherals when one has been granted a DMA cycle. The sense of these lines is programmable. Reset initializes them to active low.
AEN	O	ADDRESS ENABLE: Address Enable enables the 8-bit latch containing the upper 8 address bits onto the system address bus. AEN can also be used to disable other system bus drivers during DMA transfers. AEN is active HIGH.
ADSTB	O	ADDRESS STROBE: The active high, Address Strobe is used to strobe the upper address byte into an external latch.
MEMR	O	MEMORY READ: The Memory Read signal is an active low three-state output used to access data from the selected memory location during a DMA Read or a memory-to-memory transfer.
MEMW	O	MEMORY WRITE: The Memory Write is an active low three-state output used to write data to the selected memory location during a DMA Write or a memory-to-memory transfer.
PIN5	I	PIN5: This pin should always be at a logic HIGH level. An internal pull-up resistor will establish a logic high when the pin is left floating. It is recommended however, that PIN5 be connected to V_{CC}.

8.9. PRIORITY INTERRUPT CONTROLLER 82C59A

Description

The Intel 82C59A is a high performance CMOS Priority Interrupt Controller manufactured using an advanced 2 mm CMOS process. The 82C59A is designed to relieve the system CPU from the task of polling in a multilevel priority system. The high speed and industry standard configuration of the 82C59A make it compatible with microprocessors such as 80C286, 80286, 80C86/88, 8086/88, 8080/85 and NSC800.

The 82C59A can handle up to eight vectored priority interrupting sources and is cascadable to 64 without additional circuitry. Individual interrupting sources can be masked or prioritized to allow custom system configuration. Two modes of operation make the 82C59A compatible with both 8080/85 and 80C86/88/286 formats.

Functional Description

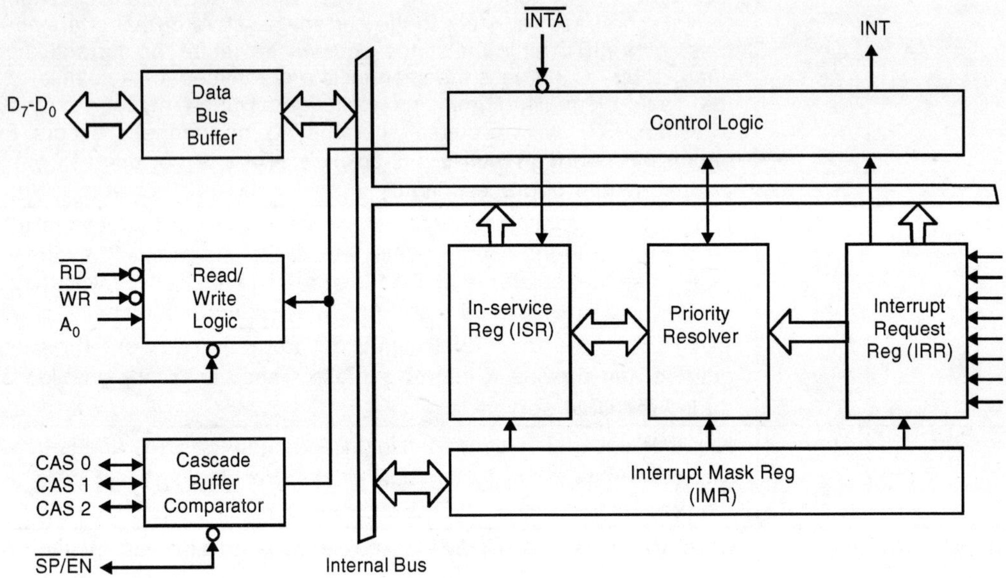

Fig. 8.62. Block diagram 8259.

Interrupts in Microcomputer Systems

Microcomputer system design requires that I/O devices such as keyboards, displays, sensors and other components receive servicing in an efficient manner so that large amounts of the total system tasks can be assumed by the microcomputer with little or no effect on throughput.

The most common method of servicing such devices is the Polled approach. This is where the processor must test each device in sequence and in effect "ask" each one if it needs servicing. It is easy to see that a large portion of the main program is looping through this continuous polling cycle and that such a method would have a serious, detrimental effect on system throughput, thus,

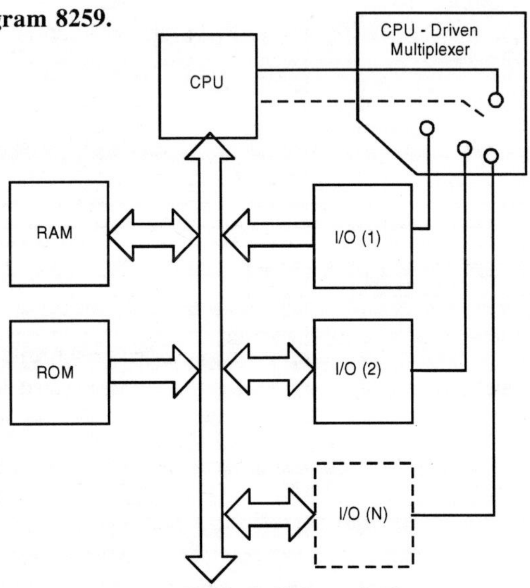

Fig. 8.63. Polled method.

limiting the tasks that could be assumed by the microcomputer and reducing the cost effectiveness of using such devices.

A more desirable method would be one that would allow the microprocessor to be executing its main program and only stop to service peripheral devices when it is told to do so by the device itself. In effect, the method would provide an external asynchronous input that would inform the processor that it should complete whatever instruction that is currently being executed and fetch a new routine that will service the requesting device. Once this servicing is complete, however, the processor would resume exactly where it left off.

This is the Interrupt-driven method. It is easy to see that system throughput would drastically increase, and thus, more tasks could be assumed by the microcomputer to further enhance its cost effectiveness.

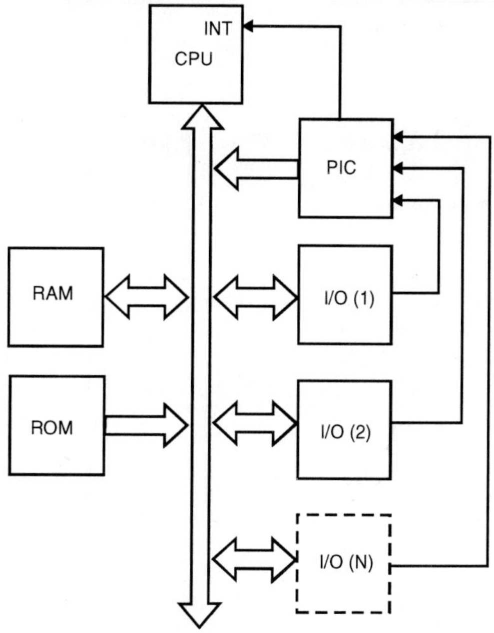

Fig. 8.64. Interrupt method.

The Programmable Interrupt Controller (PIC) functions as an overall manager in an Interrupt-Driven system. It accepts requests from the peripheral equipment, determines which of the incoming requests is of the highest importance (priority), ascertains whether the incoming request has a higher priority value than the level currently being serviced, and issues an interrupt to the CPU based on this determination. Each peripheral device or structure usually has a special program or "routine" that is associated with its specific functional or operational requirements; this is referred to as a "service routine". The PIC, after issuing an interrupt to the CPU, must somehow input information into the CPU that can "point" the Program Counter to the service routine associated with the requesting device. This "pointer" is an address in a vectoring table and will often be referred to, in this document, as vectoring data.

82C59A Functional Description

The 82C59A is a device specifically designed for use in real time, interrupt driven microcomputer systems. It manages eight levels of requests and has built-in features for expandability to other 82C59As (up to 64 levels). It is programmed by system software as an I/O peripheral. A selection of priority modes is available to the programmer so that the manner in which the requests are processed by the 82C59A can be configured to match system requirements.

The priority modes can be changed or reconfigured dynamically at any time during main program operation. This means that the complete interrupt structure can be defined as required, based on the total system environment.

The interrupts at the IR input lines are handled by two registers in cascade, the Interrupt Request Register (IRR) and the In-Service Register (ISR). The IRR is used to indicate all the interrupt levels which are requesting service, and the ISR is used to store all the interrupt levels which are currently being serviced.

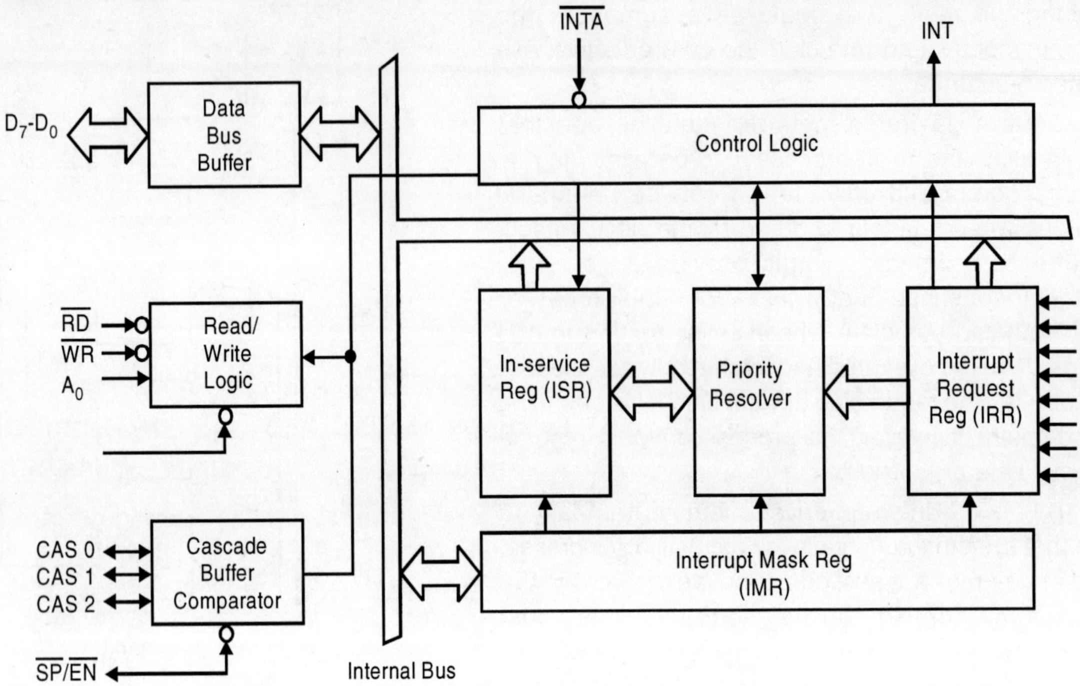

Fig. 8.65. 82C59A functional diagram.

Priority Resolver

This logic block determines the priorities of the bits set in the IRR. The highest priority is selected and strobed into the corresponding bit of the ISR during the INTA sequence.

Interrupt Mask Register (IMR)

The IMR stores the bits which disable the interrupt lines to be masked. The IMR operates on the output of the IRR. Masking of a higher priority input will not affect the interrupt request lines of lower priority.

Interrupt (INT)

This output goes directly to the CPU interrupt input. The VOH level on this line is designed to be fully compatible with the 8080, 8085, 8086/88, 80C86/88, 80286 and 80C286 input levels.

Interrupt Acknowledge (INTA)

INTA pulses will cause the 82C59A to release vectoring information onto the data bus. The format of this data depends on the system mode (mPM) of the 82C59A.

Data Bus Buffer

This 3-state, bidirectional 8-bit buffer is used to interface the 82C59A to the System Data Bus. Control words and status information are transferred through the Data Bus Buffer.

Read/Write Control Logic

The function of this block is to accept output commands from the CPU. It contains the Initialization Command Word (ICW) registers and Operation Command Word (OCW) registers which store the various control formats for device operation. This function block also allows the status of the 82C59A to be transferred onto the Data Bus.

Chip Select (CS)

A LOW on this input enables the 82C59A. No reading or writing of the device will occur unless the device is selected.

Write (WR)

A LOW on this input enables the CPU to write control words (ICWs and OCWs) to the 82C59A.

Read (RD)

A LOW on this input enables the 82C59A to send the status of the Interrupt Request Register (IRR), In-Service Register (ISR), the Interrupt Mask Register (IMR), or the interrupt level (in the poll mode) onto the Data Bus.

A0

This input signal is used in conjunction with WR and RD signals to write commands into the various command registers, as well as to read the various status registers of the chip. This line can be tied directly to one of the system address lines.

The Cascade Buffer/Comparator

This function block stores and compares the IDs of all 82C59As used in the system. The associated three I/O pins (CAS0-2) are outputs when the 82C59A is used as a master and are inputs when the 82C59A is used as a slave. As a master, the 82C59A sends the ID of the interrupting slave device onto the CAS0-2 lines. The slave, thus selected will send its preprogrammed subroutine address onto the Data Bus during the next one or two consecutive INTA pulses. (See section "Cascading the 82C59A".)

Interrupt Sequence

The powerful features of the 82C59A in a microcomputer system are its programmability and the interrupt routine addressing capability. The latter allows direct or indirect jumping to the specified interrupt routine requested without any polling of the interrupting devices. The normal sequence of events during an interrupt depends on the type of CPU being used.

These events occur in an 8080/8085 system :

1. One or more of the INTERRUPT REQUEST lines (IR0 - IR7) are raised high, setting the corresponding IRR bit(s).
2. The 82C59A evaluates those requests in the priority resolver and sends an interrupt (INT) to the CPU, if appropriate.
3. The CPU acknowledges the INT and responds with an INTA pulse.
4. Upon receiving an INTA from the CPU group, the highest priority ISR bit is set, and the corresponding IRR bit is reset. The 82C59A will also release a CALL instruction code (11001101) onto the 8-bit data bus through D0 - D7.
5. This CALL instruction will initiate two additional INTA pulses to be sent to 82C59A from the CPU group.
6. These two INTA pulses allow the 82C59A to release its preprogrammed subroutine address onto the data bus. The lower 8-bit address is released at the first INTA pulse and the higher 8-bit address is released at the second INTA pulse.
7. This completes the 3-byte CALL instruction released by the 82C59A. In the AEOI mode, the ISR bit is reset at the end of the third INTA pulse. Otherwise, the ISR bit remains set until an appropriate EOI command is issued at the end of the interrupt sequence.

   The events occurring in an 80C86/88/286 system are the same until step 4.

8. The 82C59A does not drive the data bus during the first INTA pulse.
9. The 80C86/88/286 CPU will initiate a second INTA pulse. During this INTA pulse, the appropriate ISR bit is set and the corresponding bit in the IRR is reset. The 82C59A outputs the 8-bit pointer onto the data bus to be read by the CPU.

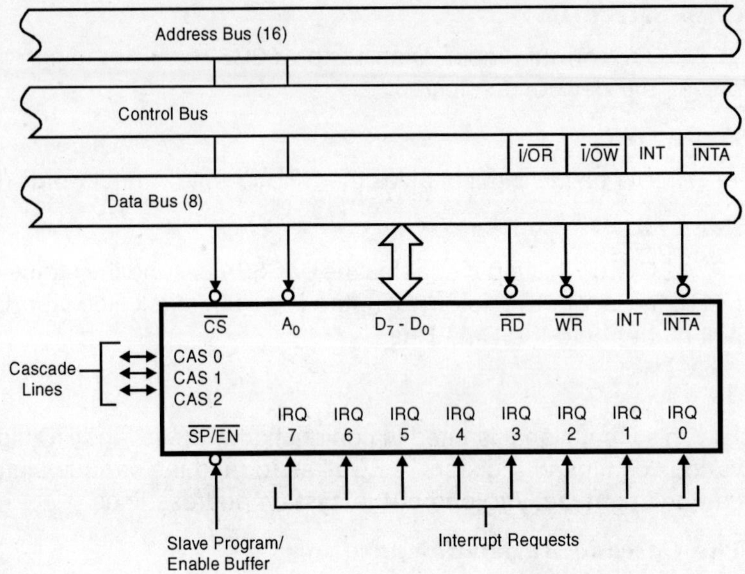

Fig. 8.66. 82C59A standard system bus interface.

This completes the interrupt cycle. In the AEOI mode, the ISR bit is reset at the end of the second INTA pulse. Otherwise, the ISR bit remains set until an appropriate EOI command is issued at the end of the interrupt subroutine. If no interrupt request is present at step 4 of either sequence (i.e., the request was too short in duration), the 82C59A will issue an interrupt level 7. If a slave is programmed on IR bit 7, the CAS lines remain inactive and vector addresses are output from the master 82C59A.

Programming the 82C59A

The 82C59A accepts two types of command words generated by the CPU:

1. Initialization Command Words (ICWs) : Before normal operation can begin, each 82C59A in the system must be brought to a starting point by a sequence of 2 to 4 bytes timed by WR pulses.

2. Operation Command Words (OCWs) : These are the command words which command the 82C59A to operate in various interrupt modes. Among these modes are :

(a) Fully nested mode.
(b) Rotating priority mode.
(c) Special mask mode.
(d) Polled mode.

The OCWs can be written into the 82C59A anytime after initialization.

Initialization Command Words (ICWs)

General

Whenever a command is issued with A0 = 0 and D4 = 1, this is interpreted as Initialization Command Word 1 (ICW1). ICW1 starts the initialization sequence during which the following automatically occur :

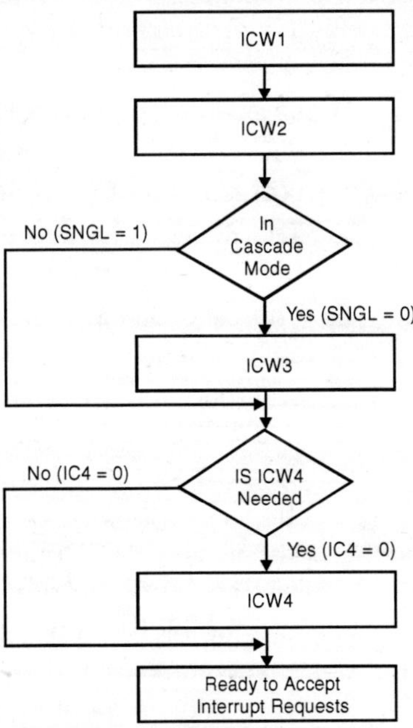

Fig. 8.67. 82C59A initialization sequence.

(a) The edge sense circuit is reset, which means that following initialization, an interrupt request (IR) input must make a low-to-high transition to generate an interrupt.

(b) The Interrupt Mask Register is cleared.

(c) IR7 input is assigned priority 7.

(d) Special Mask Mode is cleared and Status Read is set to IRR.

(e) If IC4 = 0, then all functions selected in ICW4 are set to zero. (Non-Buffered mode, no Auto-EOI,8080/85 system).

Note : Master/Slave in ICW4 is only used in the buffered mode.

Initialization Command Words 1 and 2 (ICW1, lCW2)

A5 - A15 : Page starting address of service routines. In an 8080/85 system the 8 request levels will generate CALLS to 8 locations equally spaced in memory. These can be programmed to be spaced at intervals of 4 or 8 memory locations, thus, the 8 routines will occupy a page of 32 or 64 bytes, respectively.

The address format is 2 bytes long (A0 - A15). When the routine interval is 4, A0 - A4 are automatically inserted by the 82C59A, while A5 - A15 are programmed externally. When the routine interval is 8, A0 - A5 are automatically inserted by the 82C59A while A6 - A15 are programmed externally.

The 8-byte interval will maintain compatibility with current software, while the 4-byte interval is best for a compact jump table.

In an 80C86/88/286 system, A15 - A11 are inserted in the five most significant bits of the vectoring byte and the 82C59A sets the three least significant bits according to the interrupt level. A10 - A5 are ignored and ADI (Address interval) has no effect.

LTIM : If LTIM = 1, then the 82C59A will operate in the level interrupt mode. Edge detect logic on the interrupt inputs will be disabled.

ADI : ALL address interval. ADI = 1 then interval = 4; ADI = 0 then interval = 8.

SNGL : Single. Means that this is the only 82C59A in the system. If SNGL = 1, no ICW3 will be issued.

IC4 : If this bit is set - ICW4 has to be issued. If ICW4 is not needed, set IC4 = 0.

This word is read only when there is more than one 82C59A in the system and cascading is used, in which case SNGL = 0. It will load the 8-bit slave register. The function of this register are :

(a) In the master mode (either when SP = 1, or in buffered mode when M/S = 1 in ICW4) a "1" is set for each slave in the bit corresponding to the appropriate IR line for the slave. The master then will release byte 1 of the call sequence (for 8080/85 system) and will enable the corresponding slave to release bytes 2 and 3 (for 80C86/88/286, only byte 2) through the cascade lines.

(b) In the slave mode (either when SP = 0, or if BUF = 1 and M/S = 0 in ICW4), bits 2 - 0 identify the slave. The slave compares its cascade input with these bits and if they are equal, bytes 2 and 3 of the call sequence (or just byte 2 for 80C86/88/286) are released by it on the Data Bus.

Note : The slave address must correspond to the IR line it is connected to in the master ID.

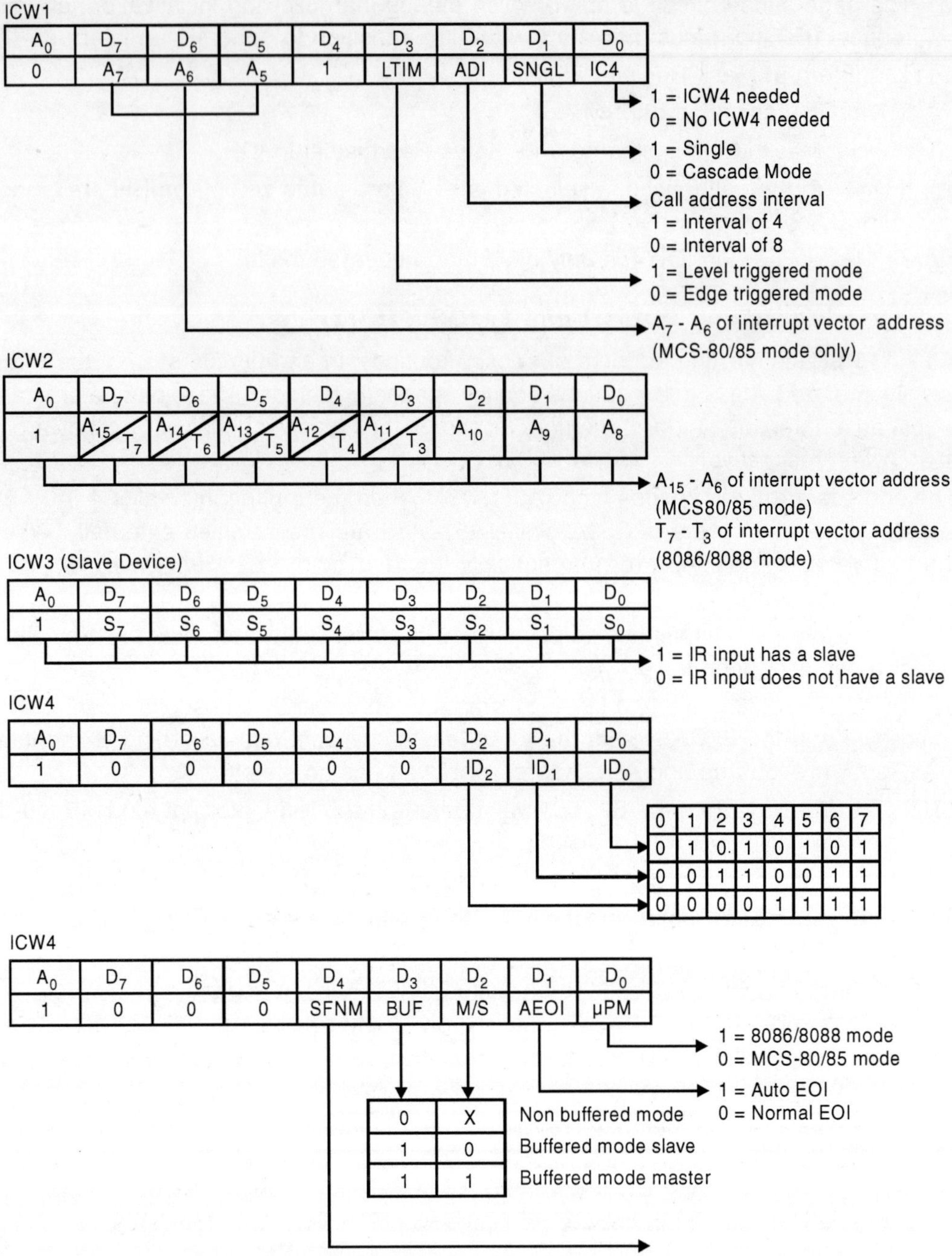

E : Slave ID is equal to the corresponding IR input

Fig. 8.68. 82C59A initialization command word format.

Initialization Command Word 4 (ICW4)

SFNM : If SFNM = 1, the special fully nested mode is programmed.

BUF : If BUF = 1, the buffered mode is programmed. In buffered mode, SP/EN becomes an enable output and the master/slave is determined by M/S.

M/S : If buffered mode is selected: M/S = 1 means the 82C59A is programmed to be a master, M/S = 0 means the 82C59A is programmed to be a slave. If BUF = 0, M/S has no function.

AEOI : If AEOI = 1, the automatic end of interrupt mode is programmed.

mPM : Microprocessor mode: mPM = 0 sets the 82C59A for 8080/85 system operation, mPM = 1 sets the 82C59A for 80C86/88/286 system operation.

Operation Command Words (OCWs)

After the Initialization Command Words (ICWs) are programmed into the 82C59A, the device is ready to accept interrupt requests at its input lines. However, during the 82C59A operation, a selection of algorithms can command the 82C59A to operate in various modes through the Operation Command Words (OCWs).

Operation Command Words (OCWs)

A0	D7	D6	D5	D4	D3	D2	D1	D0
OCW1								
1	M7	M6	M5	M4	M3	M2	M1	M0
OCW2								
0	R	SL	EOI	0	0	L2	L1	L0
OCW3								
0	0	ESMM	SMM	0	1	P	RR	RIS

Fig. 8.69. Operation command words 82C59A.

Operation Command Word 1 (OCW1)

OCW1 sets and clears the mask bits in the Interrupt Mask Register (IMR) M7 - M0 represent the eight mask bits. M = 1 indicates the channel is masked (inhibited), M = 0 indicates the channel is enabled.

Operation Command Word 2 (OCW2)

R, SL, EOI : These three bits control the Rotate and End of Interrupt modes and combinations of the two. A chart of these combinations can be found on the Operation Command Word Format.

L2, L1, L0 : These bits determine the interrupt level acted upon when the SL bit is active.

Operation Command Word 3 (OCW3)

ESMM : Enable Special Mask Mode. When this bit is set to 1 it enables the SMM bit to set or reset the Special Mask Mode. When ESMM = 0, the SMM bit becomes a "don't care".

SMM - Special Mask Mode : If ESMM = 1 and SMM = 1, the 82C59A will enter Special Mask Mode. If ESMM = 1 and SMM = 0, the 82C59A will revert to normal mask mode. When ESMM = 0, SMM has no effect.

Fully Nested Mode

This mode is entered after initialization unless another mode is programmed. The interrupt requests are ordered in priority from 0 through 7 (0 highest). When an interrupt is acknowledged the highest priority request is determined and its vector is placed on the bus. Additionally, a bit of the Interrupt Service register (IS0 - 7) is set. This bit remains set until the microprocessor issues an End of Interrupt (EOI) command immediately before returning from the service routine, or if the AEOI (Automatic End of Interrupt) bit is set, until the trailing edge of the last INTA. While the IS bit is set, all further interrupts of the same or lower priority are inhibited, while higher levels will generate an interrupt (which will be acknowledged only if the microprocessor internal interrupt enable flip-flop has been re-enabled through software).

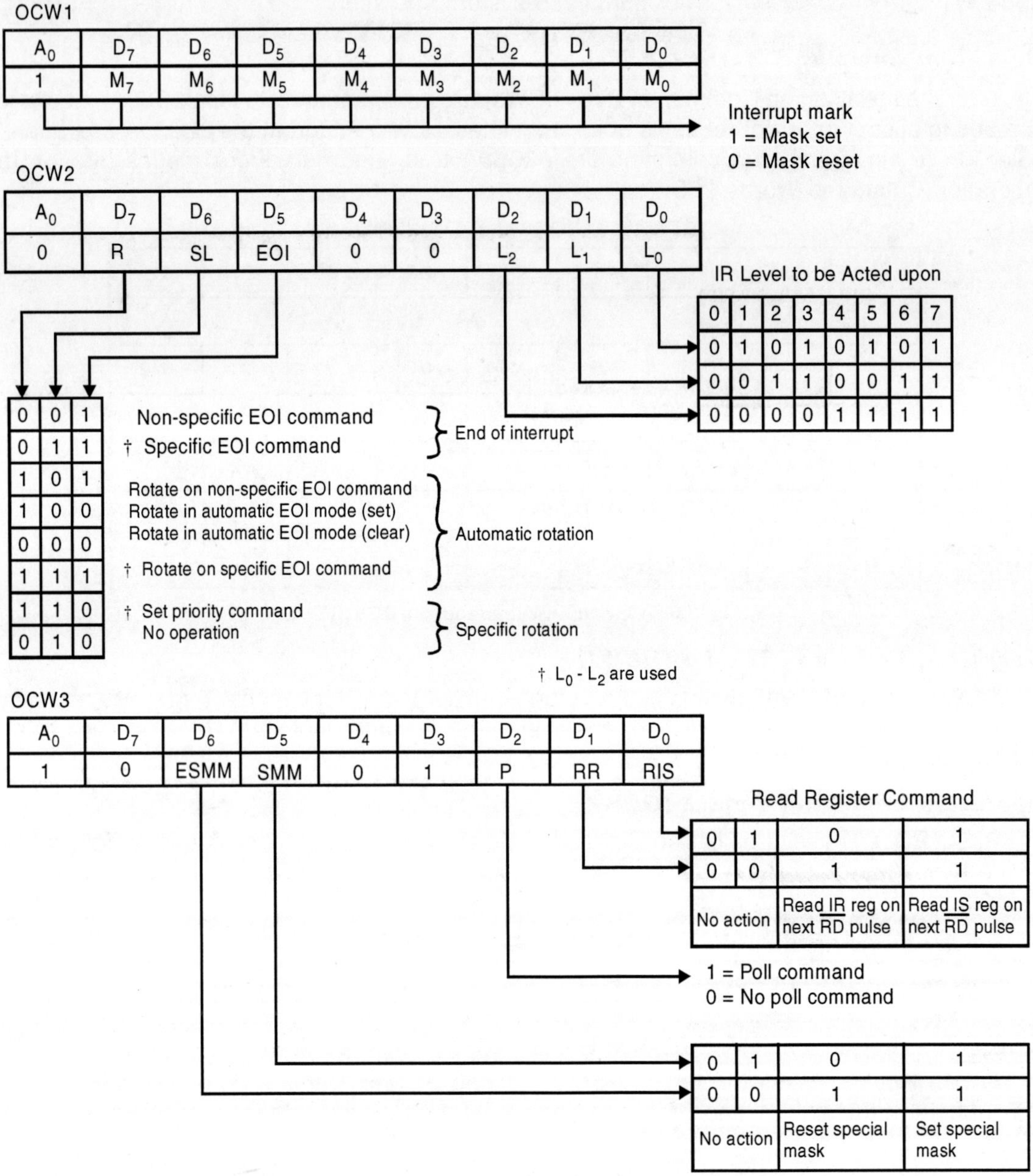

Fig. 8.70. 82C59A operation command word format.

After the initialization sequence, IR0 has the highest priority and IR7 the lowest. Priorities can be changed, as will be explained in the rotating priority mode or via the set priority command.

End of Interrupt (EOI)

The In-Service (IS) bit can be reset either automatically following the trailing edge of the last in sequence INTA pulse (when AEOI bit in ICW1 is set) or by a command word that must be issued to the 82C59A before returning from a service routine (EOI Command). An EOI command must be issued twice if servicing a slave in the Cascade mode, once for the master and once for the corresponding slave.

There are two forms of EOI command: Specific and Non-Specific. When the 82C59A is operated in modes which preserve the fully nested structure, it can determine which IS bit to reset on EOI. When a Non-Specific command is issued the 82C59A will automatically reset the highest IS bit of those that are set, since in the fully nested mode the highest IS level was necessarily the last level acknowledged and serviced. A non-specific EOI can be issued with OCW2 (EOI = 1, SL = 0, R = 0).

When a mode is used which may disturb the fully nested structure, the 82C59A may no longer be able to determine the last level acknowledged. In this case a Specific End of Interrupt must be issued which includes as part of the command the IS level to be reset. A specific EOI can be issued with OCW2 (EOI = 1, SL = 1, R = 0, and L0 - L2 is the binary level of the IS bit to be reset).

An IRR bit that is masked by an IMR bit will not be cleared by a non-specific EOI if the 82C59A is in the Special Mask Mode.

Automatic End of Interrupt (AEOI) Mode

If AEOI = 1 in ICW4, then the 82C59A will operate in AEOI mode continuously until reprogrammed by ICW4. In this mode the 82C59A will automatically perform a non-specific EOI operation at the trailing edge of the last interrupt acknowledge pulse (third pulse in 8080/85, second in 80C86/88/286). Note that from a system standpoint, this mode should be used only when a nested multilevel interrupt structure is not required within a single 82C59A.

Automatic Rotation (Equal Priority Devices)

In some applications there are a number of interrupting devices of equal priority. In this mode a device, after being serviced, receives the lowest priority, so a device requesting an interrupt will have to wait, in the worst case until each of 7 other devices are serviced at most once. For example, if the priority and "in service" status is:

Before Rotate (IR4 the highest priority requiring service)

	IS7	IS6	IS5	IS4	IS3	IS2	IS1	IS0
"1S" Status	0	0	0	0	0	0	0	0
Priority Status	7	6	5	4	3	2	1	0
	↑ Lowest							↑ Highest

Fig. 8.71.

After Rotate (IR4 was serviced, all other priorities rotated correspondingly).

	IS7	IS6	IS5	IS4	IS3	IS2	IS1	IS0
"1S" Status	0	1	0	0	0	0	0	0
Priority Status	2	1	0	7	6	5	4	3
	↑ Highest							↑ Lowest

Fig. 8.72.

There are two ways to accomplish Automatic Rotation using OCW2, the Rotation on Non-Specific EOI Command (R = 1, SL = 0, EOI = 1) and the Rotate in Automatic EOI Mode which is set by (R = 1, SL = 0, EOI = 0) and cleared by (R = 0, SL = 0, EOI = 0).

Specific Rotation (Specific Priority)

The programmer can change priorities by programming the lowest priority and thus, fixing all other priorities; i.e., if IR5 is programmed as the lowest priority device, then IR6 will have the highest one.

The Set Priority command is issued in OCW2 where R = 1, SL = 1, L0 - L2 is the binary priority level code of the lowest priority device.

Observe that in this mode internal status is updated by software control during OCW2. However, it is independent of the End of Interrupt (EOI) command (also executed by OCW2). Priority changes can be executed during an EOI command by using the Rotate on Specific EOI command in OCW2 (R = 1, SL = 1, EOI = 1, and L0 - L2 = IR level to receive lowest priority).

Interrupt Masks

Special Mask Mode

Some applications may require an interrupt service routine to dynamically alter the system priority structure during its execution under software control. For example, the routine may wish to inhibit lower priority requests for a portion of its execution but enable some of them for another portion.

That is where the Special Mask Mode comes in. In the Special Mask Mode, when a mask bit is set in OCW1, it inhibits further interrupts at that level and enables interrupts from all other levels (lower as well as higher) that are not masked. Thus, any interrupts may be selectively enabled by loading the mask register.

The Special Mask Mode is set by OCW3 where ESMM = 1, SMM = 1, and cleared where ESMM = 1, SMM = 0.

Poll Command

In this mode, the INT output is not used or the microprocessor internal Interrupt Enable flip flop is reset, disabling its interrupt input. Service to devices is achieved by software using a Poll command.

The Poll command is issued by setting P = 1 in OCW3. The 82C59A treats the next RD pulse to the 82C59A (i.e., RD = 0, CS = 0) as an interrupt acknowledge, sets the appropriate IS bit if there is a request, and reads the priority level. Interrupt is frozen from WR to RD.

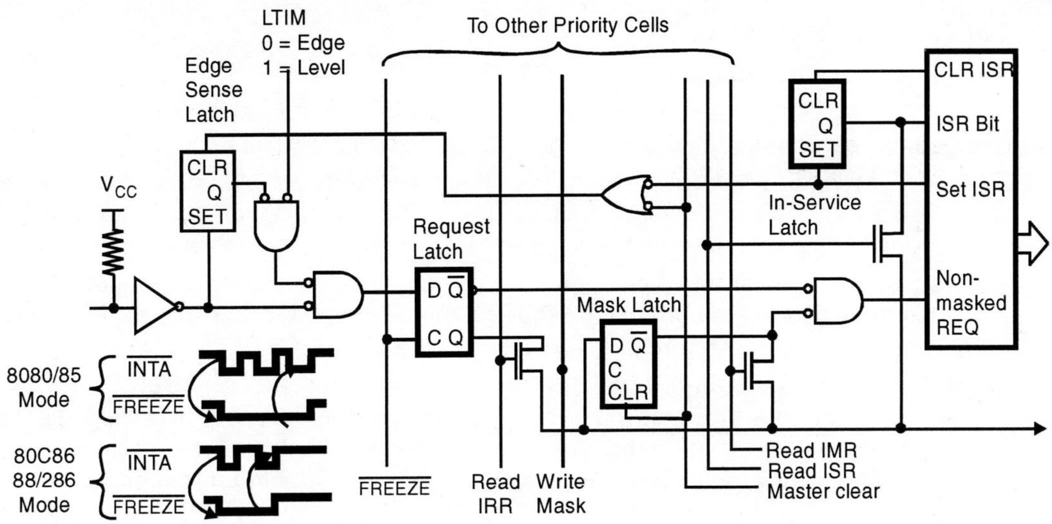

Fig. 8.73. Priority cell simplified logic diagram.

Reading the 82C59A Status

The input status of several internal registers can be read to update the user information on the system. The following registers can be read via OCW3 (IRR and ISR) or OCW1 (IMR).

Interrupt Request Register (IRR) : 8-bit register which contains the levels requesting an interrupt to be acknowledged. The highest request level is reset from the IRR when an interrupt is acknowledged. IRR is not affected by IMR.

In-Service Register (ISR) : 8-bit register which contains the priority levels that are being serviced. The ISR is updated when an End of Interrupt Command is issued.

The word enabled onto the data bus during RD is :

D7	D6	D5	D4	D3	D2	D1	D0
1	–	–	–	–	W2	W1	W0

W0 - W2 : Binary code of the highest priority level requesting service.

I : Equal to a "1" if there is an interrupt.

This mode is useful if there is a routine command common to several levels so that the INTA sequence is not needed (saves ROM space). Another application is to use the poll mode to expand the number of priority levels to more than 64.

Interrupt Mask Register : 8-bit register which contains the interrupt request lines which are masked.

The IRR can be read when, prior to the RD pulse, a Read Register Command is issued with OCW3 (RR = 1, RIS = 0). The ISR can be read when, prior to the RD pulse, a Read Register Command is issued with OCW3 (RR = 1, RIS = 1). There is no need to write an OCW3 before every status read operation, as long as the status read corresponds with the previous one, i.e., the 82C59A "remembers" whether the IRR or ISR has been previously selected by the OCW3. This is not true when poll is used. In the poll mode, the 82C59A treats the RD following a "poll write" operation as an INTA. After initialization, the 82C59A is set to IRR.

For reading the IMR, no OCW3 is needed. The output data bus will contain the IMR whenever RD is active and A0 = 1 (OCW1). Polling overrides status read when P = 1, RR = 1 in OCW3.

Edge and Level Triggered Modes

This mode is programmed using bit 3 in ICW1.

If LTIM = "0", an interrupt request will be recognized by a low to high transition on an IR input. The IR input can remain high without generating another interrupt.

If LTIM = "1", an interrupt request will be recognized by a "high" level on an IR input, and there is no need for an edge detection. The interrupt request must be removed before the EOI command is issued or the CPU interrupt is enabled to prevent a second interrupt from occurring.

The priority cell diagram shows a conceptual circuit of the level sensitive and edge sensitive input circuitry of the 82C59A. Be sure to note that the request latch is a transparent D type latch

In both the edge and level triggered modes the IR inputs must remain high until after the falling edge of the first INTA. If the IR input goes low before this time a DEFAULT IR7 will occur when the CPU acknowledges the interrupt. This can be a useful safeguard for detecting interrupts caused by spurious noise glitches on the IR inputs. To implement this feature the IR7 routine is used for "clean up" simply executing a return instruction, thus, ignoring the interrupt. If IR7 is needed for other purposes a default IR7 can still be detected by reading the ISR. A normal IR7 interrupt will set the corresponding ISR bit, a default IR7 won't. If a default IR7 routine occurs during a normal IR7 routine, the ISR will remain set. In this case it is necessary to keep track of whether or not the IR7 routine was previously entered. If another IR7 occurs it is a default.

In power sensitive applications, it is advisable to place the 82C59A in the edge-triggered mode with the IR lines normally high. This will minimize the current through the internal pull-up resistors on the IR pins.

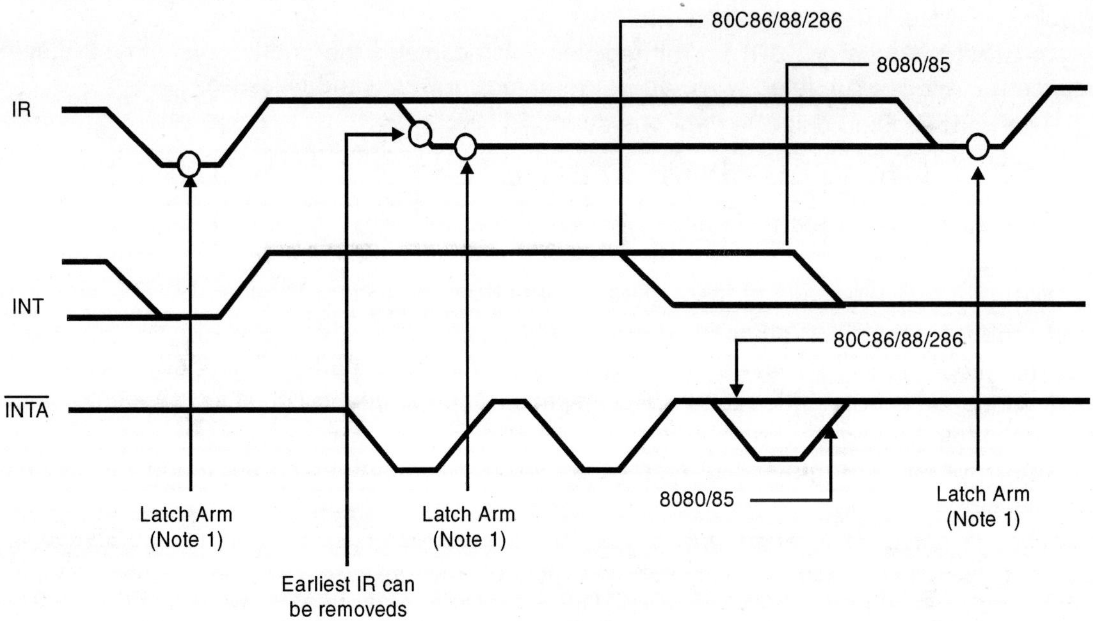

Fig. 8.74. IR triggering timing requirements.

The Special Fully Nested Mode

This mode will be used in the case of a big system where cascading is used, and the priority has to be conserved within each slave. In this case the special fully nested mode will be programmed to the master (using ICW4). This mode is similar to the normal nested mode with the following exceptions:

(a) When an interrupt request from a certain slave is in service, this slave is not locked out from the master's priority logic and further interrupt requests from higher priority IRs within the slave will be recognized by the master and will initiate interrupts to the processor. (In the normal nested mode a slave is masked out when its request is in service and no higher requests from the same slave can be serviced.

(b) When exiting the Interrupt Service routine the software has to check whether the interrupt serviced was the only one from that slave. This is done by sending a non-specific End of Interrupt (EOI) command to the slave and then reading its In-Service register and checking for zero. If it is empty, a non-specified EOI can be sent to the master, too. If not, no EOI should be sent.

Buffered Mode

When the 82C59A is used in a large system where bus driving buffers are required on the data bus and the cascading mode is used, there exists the problem of enabling buffers. The buffered mode will structure the 82C59A to send an enable signal on SP/EN to enable the buffers. In this mode, whenever the 82C59A's data bus outputs are enabled, the SP/EN output becomes active.

This modification forces the use of software programming to determine whether the 82C59A is a master or a slave. Bit 3 in ICW4 programs the buffered mode, and bit 2 in ICW4 determines whether it is a master or a slave.

Cascade Mode

The 82C59A can be easily interconnected in a system of one master with up to eight slaves to handle up to 64 priority levels.

The master controls the slaves through the 3 line cascade bus (CAS2 - 0). The cascade bus acts like chip selects to the slaves during the INTA sequence.

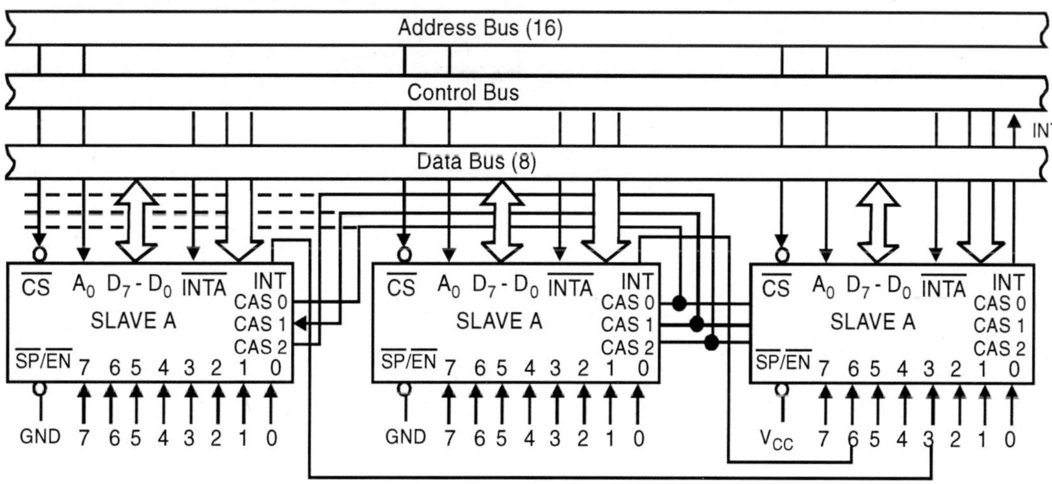

Fig. 8.75. Cascading the 82C59A.

In a cascade configuration, the slave interrupt outputs (INT) are connected to the master interrupt request inputs. When a slave request line is activated and afterwards acknowledged, the master will enable the corresponding slave to release the device routine address during bytes 2 and 3 of INTA (Byte 2 only for 80C86/88/286).

The cascade bus lines are normally low and will contain the slave address code from the leading edge of the first INTA pulse to the trailing edge of the last INTA pulse. Each 82C59A in the system must follow a separate initialization sequence and can be programmed to work in a different mode. An EOI command must be issued twice: once for the master and once for the corresponding slave. Chip select decoding is required to activate each 82C59A.

Auto EOI is supported in the slave mode for the 82C59A. The cascade lines of the Master 82C59A are activated only for slave inputs, non-slave inputs leave the cascade line inactive (low). Therefore, it is necessary to use a slave address of 0 (zero) only after all other addresses are used.

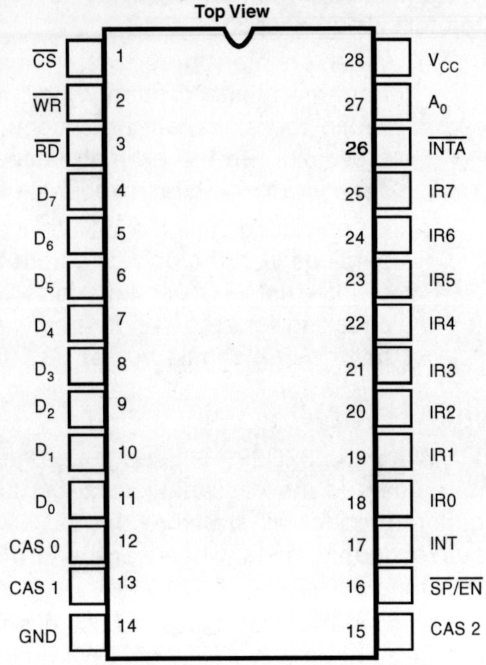

Fig. 8.76. Pin configuration.

8.10. ANALOGUE-TO-DIGITAL CONVERSION

The situation with Analogue-to-Digital Converters (ADCs) is the same as that for DACs, with some more tricks occasioned by the fact that the analogue signal might change at a faster rate than the ADC can handle, i.e. that the speed of the converter may not be infinite compared to the speed of the signal. There is therefore the problem of how to digitise a static signal (compared to the converter speed) and then the problem of converting a dynamic signal.

"DAC + Logic" Converters

The most common breed of ADC runs by using a DAC and an analogue comparator. The simplest converter is therefore as shown in this diagram where the block marked "LOGIC" can be one of a variety of designs which I shall discuss below. Essentially the strategy of all these converters is to put a value on the DAC, compare the value with the unknown and then update the DAC for another "guess". This is repeated until the optimum value is obtained. The differences between the converters essentially lie in the "guess-strategy" or conversion algorithm.

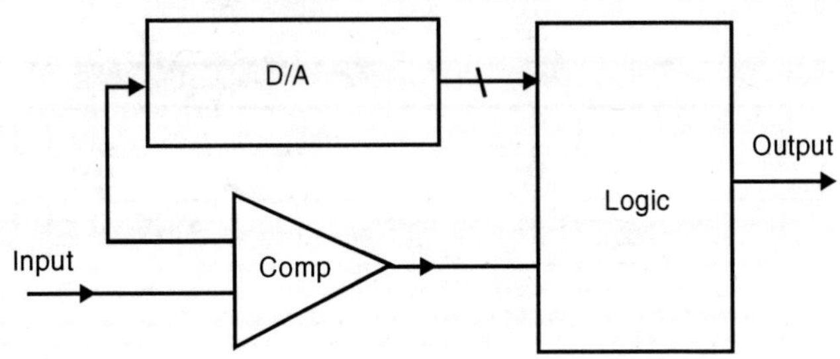

Fig. 8.77. A Generalised "DAC + Logic" ADC.

8.10.1. Counting Strategy

The simplest conversion algorithm is one which uses a counter which is zeroed and then counts up until the comparator output changes. Notice that the maximum allowed count rate is determined by the sum of the DAC settling time *t*, the comparator settling time t_s, and the logic reaction time *t*. The latter is usually negligible at all ultimate speeds. The converter is obviously slow, taking a maximum of 2 cycles to convert to n bits. The conversion time is also variable - which can be a disadvantage. The maximum conversion rate is given by :

$$f = \frac{1}{2^n (t_{da} + t_s + t_l)}$$

Putting some figures in for *t* = 200 nS, *t* = 100 nS and *t* = 10 nS - reasonable figures gives a maximum time of 256*310 nS = 79 μS for an 8-bit conversion or a rate of 126 KHz. The general problem is that for more bits (ie more resolution) the settling times of the DAC and comparator go up, not down, and therefore for larger numbers of bits the conversion times go up faster than 2 (which is already going up very fast).

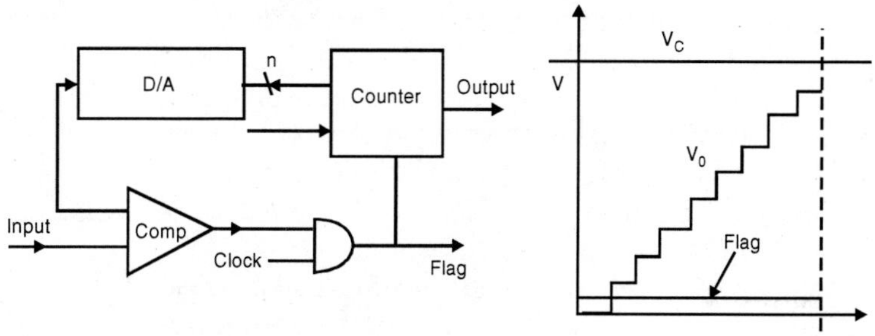

Fig. 8.78. A "Counting" converter.

The primary disadvantage of this counter is that it has to count from zero every time a conversion is made. However if we allow the counter to be bidirectional and count "up" for comparator "true" and "down" otherwise, we have a tracking converter which will always be within 1 count of the correct answer assuming that the rate of change of the voltage input does not exceed the rate at which the counter can adjust. If we are accurately monitoring (slow converter rate) an output which changes very slowly, this turns into an advantageous arrangement since the value is always available without intervention, whereas if a full conversion is necessary there will always be some necessary delay before the results are available. However the popularity of the "continuous conversion" converter is not high at the moment because there are in general better solutions to the problem.

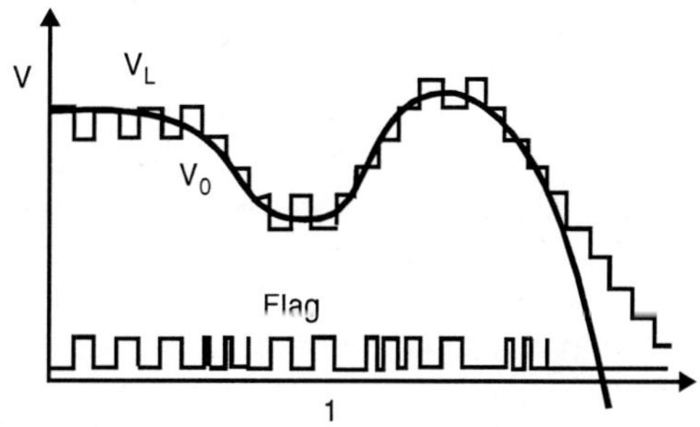

Fig. 8.79. Continuous conversion using a Bi-directional counter.

8.10.2. Successive Approximation

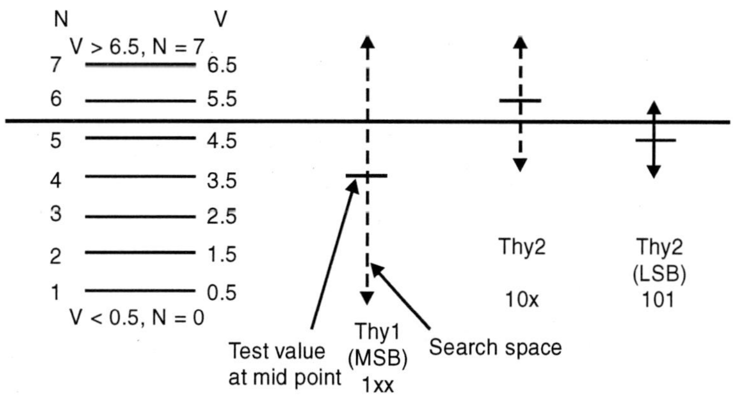

Fig. 8.80. Binary search strategy.

The second major type of algorithm is the so-called successive approximation, which is really a binary search of the space for the result. The value is first checked for presence in the upper half-space or the lower (>127 or <128 for an 8-bit converter). Having located the appropriate half-space, the process is then repeated on that space and so on. Thus to digitise a voltage of 5 volts in a space from 0-7 (actually levels of 0.5V - 6.5V remember) requires the steps as shown above. A flow chart of the operation is also given below where *V* is the mid-point of the *sp* current space being tested, i.e. 0.5x the space size, and V_{est} the current estimate of the voltage with zeros in the bit positions which have not yet been tested. First the total possible range of values is divided into two and the unknown is compared to this value. If the comparator output indicates that the result is over the half-way value, then the half space is subdivided into two equal spaces (at the 3/4 mark) and the value is compared to that. Depending on the result the upper or lower Binary Search Strategy—Problems quarter is divided and so on. The process terminates when the remaining space cannot be subdivided because it is only one bit wide The advantage of this process is that it produces a result in a defined length of time, *n* cycles for an n-bit converter, and that this time is much smaller than the counting algorithm since 2 >> *n*.

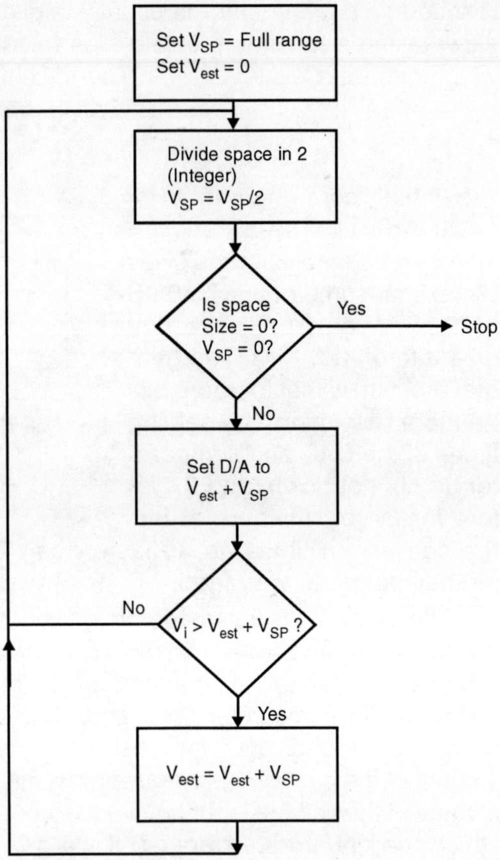

Fig. 8.81. Binary search strategy.

The disadvantage is that if the value to be measured changes during the *n* measurement period, then the result can be wrong. this can be illustrated by the accompanying example where the voltage changes from 5 volts to 2 volts suddenly during the conversion.

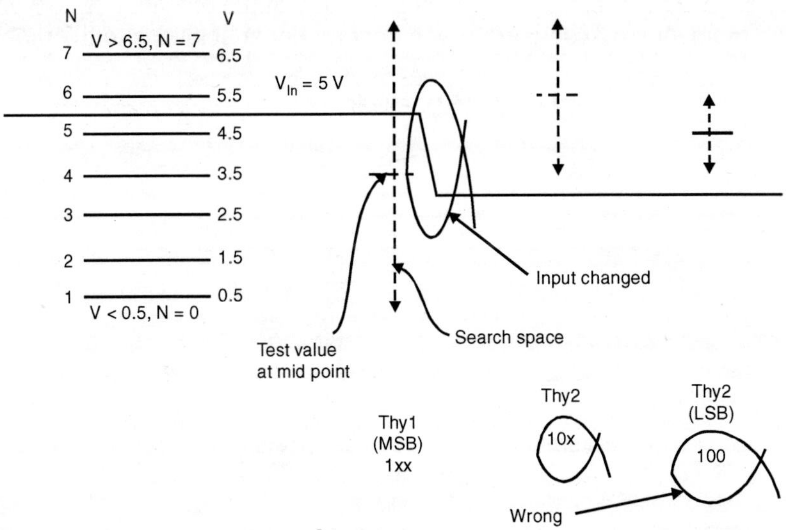

Fig. 8.82.

"Successive approximation" converters are by far the most popular converters for microprocessors in circumstances where a small number of bits of accuracy are required (small #14). For larger numbers of bits a number of factors combine to make other solutions more attractive.

8.10.3. Calibration

One should never trust an ADC to be accurate because it might drift with time. ADC should therefore (like any other lab. instrument) be calibrated periodically. This can be accomplished in the traditional manner by manually connecting sources of known potential to it, or by incorporating a calibration cycle into the normal operation of the converter and installing calibrated sources in the system (which then need periodic checking against higher standard sources until one gets a traceback chain to the NRC standards).

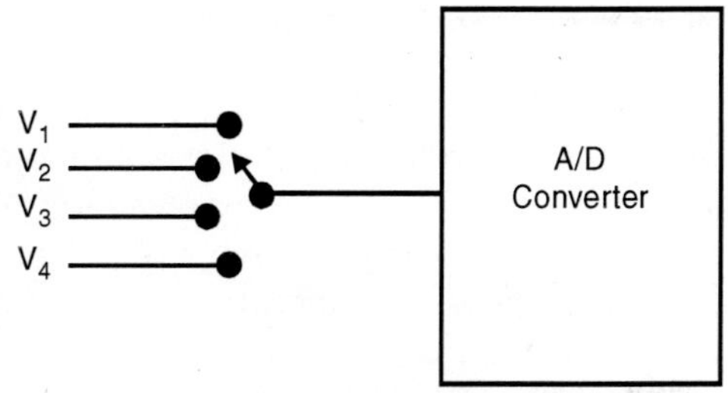

Fig. 8.83. Calibration of an A/D Converter.

In circumstances where you cannot take the calibration on trust (i.e. always!), an input multiplexer and a calibration sequence is mandatory. The simplest input multiplexer consists of a set of relays which the computer can operate to select one of a number of voltages for monitoring. More complex solutions involve solid-state multiplexers.

By making a couple of the voltages applied to the multiplexer known (i.e. standards) and assuming that the convertor is linear, i.e. $N - mV + c$, then the use of two known voltages will allow a "2-point" calibration.

However note that there is a hidden problem here. If a number N is recorded then the number is really $N \pm 1/2$ since N has to be integer, so that the equation for calibration becomes:

$$V = \frac{(V_1 + V_2)N + V_2N_1 + V_1N_2}{N_1 + N_2} \pm \frac{|V_1 + V_2| + V_1|N + N_1| + V_2|N + N_2|}{|N_1 + N_2|}$$

Note. I'm assuming that the calibration voltages are on either side of the unknown as shown in the diagram, i.e. that we are *interpolating* not *extrapolating*.

In order to minimise the errors and maximise the allowed range we should choose V and V as far apart as 12 possible - preferably near the ends of the range i.e. just less than 2 counts apart.

The calibration errors decrease the accuracy of the measurement and it may well be necessary to work to more bits than you think in order to get the answer that you want. Remember that I haven't included the inaccuracies of the converter yet!.

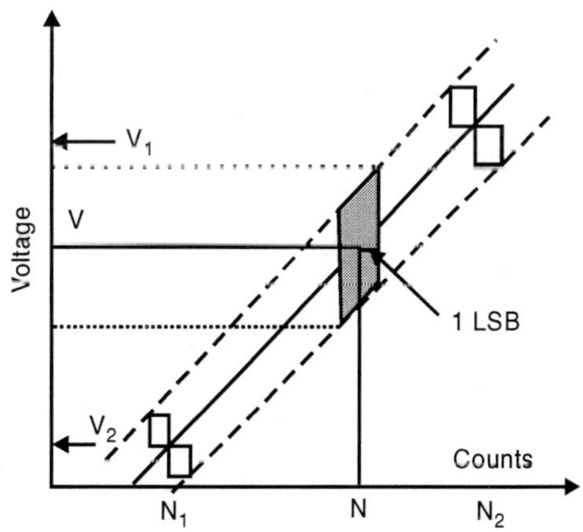

Fig. 8.84. Calibration of A A/D Converter - 2 point plot.

A practical consideration is not to let any calibration point be at a limiting value of the converter. Supposing we allowed zero volts to equal zero counts in a unipolar (only converts voltage of one sign, conceptual end of range at 0 V and 0 counts) and then consider a zero drift. If a 0 V input is indicated by a count of +1 instead of 0, then a drift in either direction can be monitored by putting 0 V in as a calibration point and taking the result into the calibration equations. However if the drift is such that 0 V should be represented by –1 counts we have a problem as we cannot represent numbers of less than zero. Thus the calibration points should not be extrema of the converter values under any conceivable set of drifts.

There are also other forms of error which can afflict an ADC but I want to delay discussing those until I have described another major form of converter - the dual-slope. This converter relies on the transformation of a measurement of voltage to a measurement of time.

8.10.4. Dual-Slope ADC

Consider this circuit.

In operation the integrator is first zeroed (close SW2), then attached to the input (SW1 up) for a fixed time *M* counts of the clock (frequency 1/*t*). At the end of that time it is attached to the reference voltage (SW1 down) and the number of counts *N* which accumulate before the integrator reaches zero volts output and the comparator output changes are determined. The equations of operation are therefore:

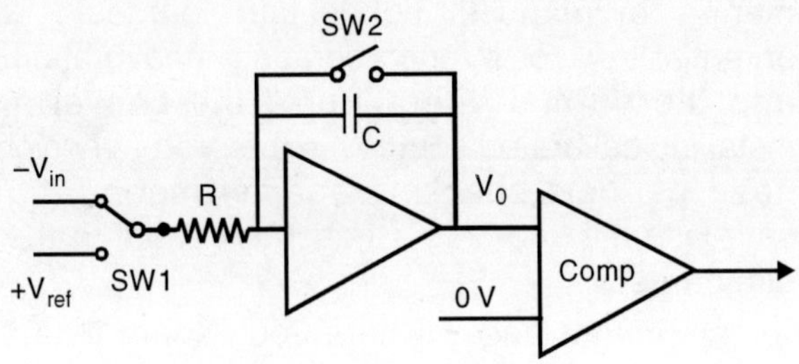

Fig. 8.85. Simple dual slope A/D converter.

$$V_x = \frac{V_{in}(Mt)}{RC} = \frac{V_{ref}(Nt)}{RC}$$

$$V_{in} = V_{ref}\frac{N}{M}$$

The unknown voltage is then just *V *N/M* and is *ref* reasonably independent of everything else. The main problem with a simple dual slope ADC is in returning the converter to an exact zero before the start of each conversion as interpreted by an imperfect comparator.

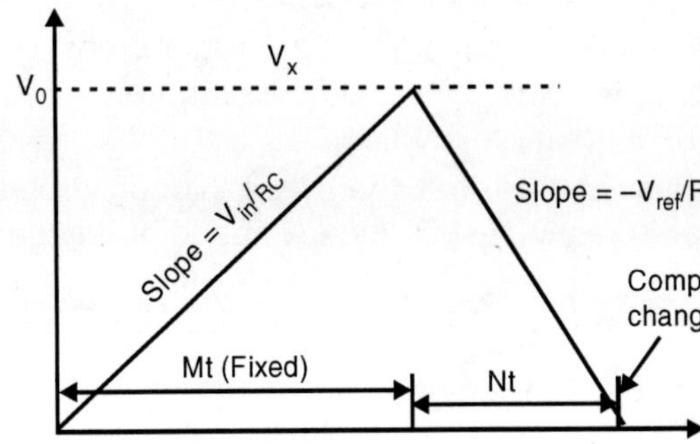

Fig. 8.86. Dual slope A/D converter output and timing.

This may be circumvented by changing the comparison level from zero volts to a small, stable, but not necessarily known, voltage (*V*), which in our example must be more positive than any combination of imperfections in the comparison.

The first timing interval then starts when the voltage crosses the level and the second interval finishes when the level is recrossed at the end.

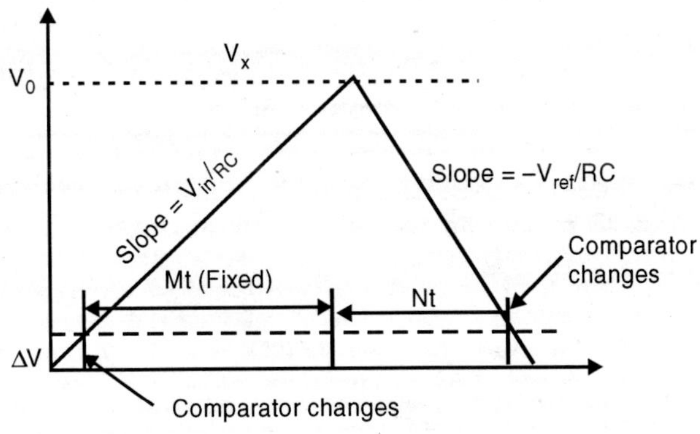

Fig. 8.87. Dual slope A/D converter -zero offset.

This brings with it a further problem that a very small input voltage will necessarily have a VERY long conversion time (takes a long time for the integrator output to reach (=V), and a zero voltage an infinite time because the integrator output never reaches the comparator threshold. In order to circumvent that a small bias current is added to the integrator (using −*V* and *RN*) to force the output to go positive even with an input of zero. Of course this *ref* means that zero volts is not zero output counts any more and a calibration step is required where a zero is established. Notice however that despite the increase in complexity, all that 24 is required out of all components except the reference voltage is stability over one measurement sequence, not accuracy. Those of you with an inventive mind might like to consider the problems of polarity of input.

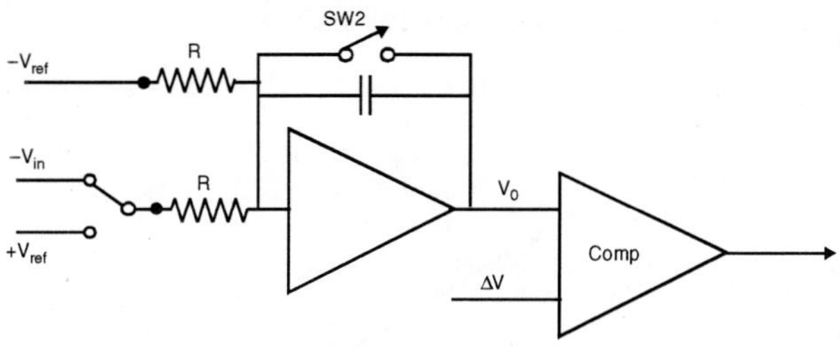

Fig. 8.88. Dual slope A/D converter - Full circuit.

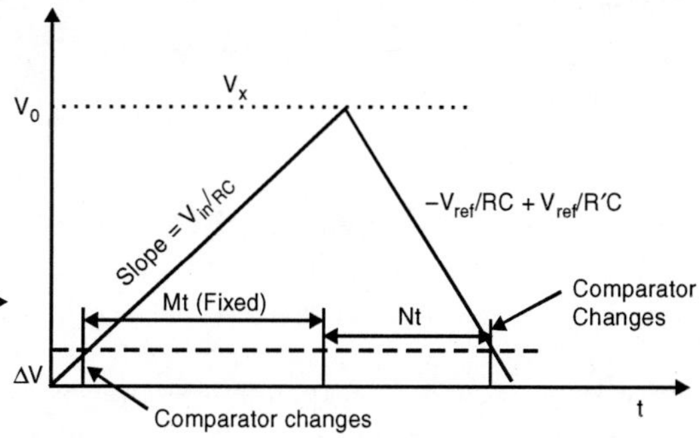

Fig. 8.89. Dual slope A/D converter - Full circuit timing.

Further Considerations

There are two properties of the dual-slope converter (and a large number of related converters with slightly different but broadly similar operating philosophies) which make it peculiarly suitable for some purposes. The first is the *integrating* property and the second the *equal-interval* property.

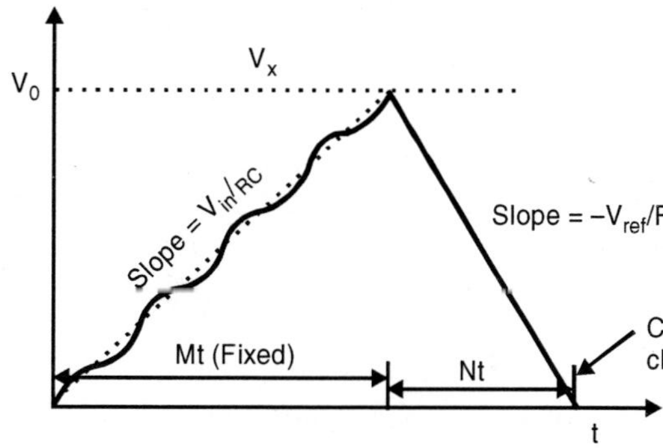

Fig. 8.90. Dual-slope converter and a varying voltage.

The *integrating* property distinguishes the dual-slope from successive approximation converters which digitise a "spot" value. The input is integrated over a fixed time period (*N* clock cycles). This helps considerably with the problem of voltages changing whilst being measured because the converter effectively integrates for a fixed time period and then measures the result. This property can be turned to great advantage when measuring voltages which are fluctuating at a known precise frequency by making the first integration period an integer number of cycles long, e.g. if the input is d.c. + 60 Hz ripple then a first integration time of 1/60 S or any multiple thereof, will cause a true average to be taken because the component at 60 Hz (actually any component at 60*n Hz) will average exactly to zero in 1/60 S. The example is not trivial because most voltmeters use precisely this principle to eliminate line "hum" (60 Hz) from their measurements. In physical experimentation the property may be turned to advantage as a "filter" to eliminate unwanted problems. To illustrate. Consider an optical system employing a chopper to interrupt the radiation.

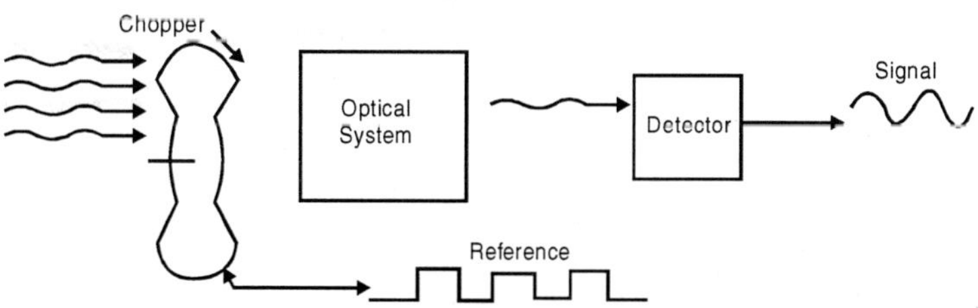

Fig. 8.91. A chopped optical system.

The signal from an optical detector is detected by a phasesensitive detector and the processing system is shown in the diagram. Using a dual-slope system whose first period is some integer number of cycles long (assuming that you know the frequency) will result in an exact d.c. average being output without further intervention.

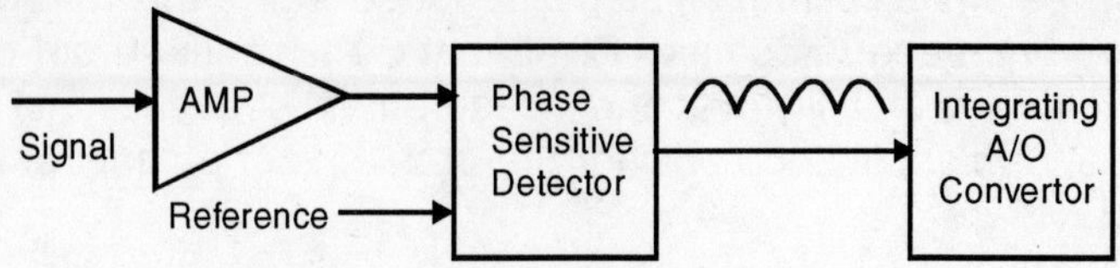

Fig. 8.92. A electronic filter system.

The second advantage is *equal interval*. This will be explained in more detail later but the important property to realise is that no count is preferred over any other count because the integrator does not know anything about the count period. Consider the output of the integrator at the end of the cycle as the comparator is about to change state. The slope of the integrator ramp-down is *–V/RC ref* which is independent of the input and the analogue circuit has no way of knowing the current count value. Therefore every count looks like every other count and they are indistinguishable from each other. Thus the input voltage change required to go from count 1006101 should be exactly the same as that for 100061001. This is in contrast to successive approximation ADCs where some states may be wider than others due to resistor tolerancing—the equivalent of DNL in a DAC. The equal interval property is equivalent to a DNL of zero.

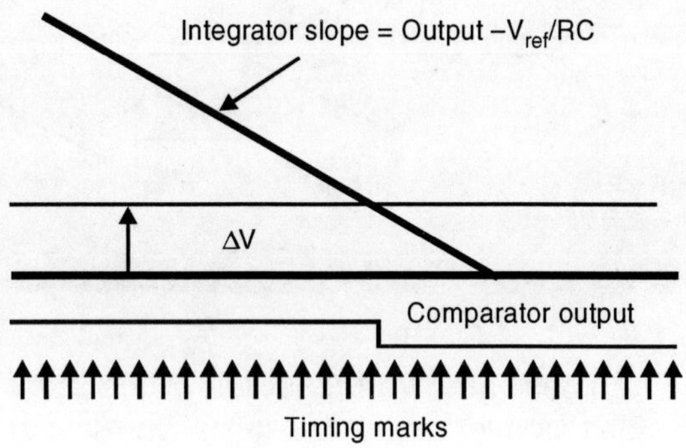

Fig. 8.93. Dual-slope converter at the end of the cycle.

The Voltage-to-Frequency (V-f) Converter

A further example of an averaging converter is the voltage-to-frequency converter which uses a linear voltage controlled oscillator to produce a frequency output. Consider a simple integrating system with a switched current source as shown here. If, every time the comparator "flips" we switch on the current source for a time *t* and drain a charge *It* from the capacitor, the total charge collected in a time *T* is given by:

$$Q = \int_0^T \frac{V_t}{R} dt + \frac{\bar{V}_t T}{R} + NIt$$

where *N* is the number of "discharges" in the integration time *T*. The average frequency is therefore *f = N/T*.

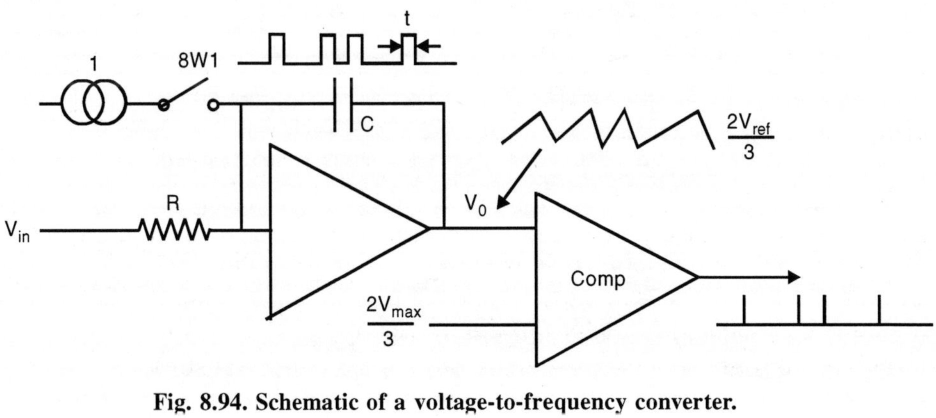

Fig. 8.94. Schematic of a voltage-to-frequency converter.

We can express the relationship between the average voltage and the average frequency as :

$$f = \left(\frac{1}{RIt}\right)\bar{V}_t + mV + c$$

Then, using calibration techniques we can easily recover the voltage by counting the output of the system for a fixed time. Note that I have included an offset term, *c*, in the equation, Counting a Simple V-f Output even though the ideal circuit does not have such a term—however ideal circuits do not exist in reality. If we record *N* counts then, because of the interaction of the "edges" of our count interval with the actual frequency, we count ±1 of the average relationship. This imposes a resolution limit on our measurement but one which we can make arbitrarily small by making *N* as large as possible.

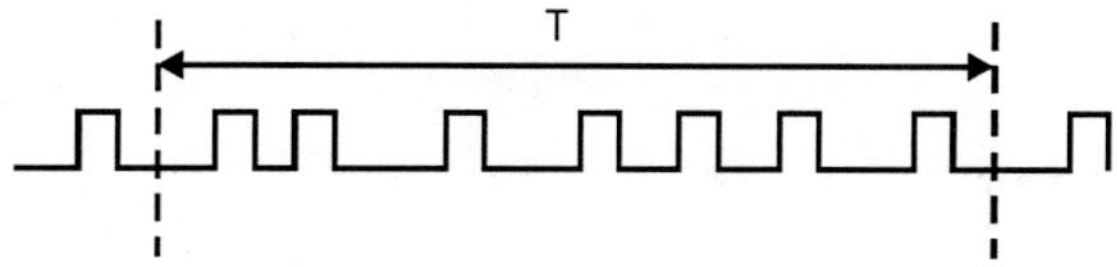

Fig. 8.95. Counting a simple V-f output.

So what limits the values of *N*? boredom and the fact that we require some time relationship for our input, i.e. we want to measure the input this second and not this hour. V-f converters are normally limited in output frequency to about 100 kHz which imposes a limit of 1:10 in resolution (0.001%) for a one second measurement which is still inconveniently 5 long. What is even more galling is that the converters can be linear to better than 1:10 and 6 we are losing precision due to the ±1.

A solution to this problem is to measure the period of the waveform and then take a reciprocal to get the frequency. However we may run into the other problem that the period is too short for a good measurement. If we knew that the period was too short then we could measure say 10 periods and get the resolution that way. However there is a slightly more complex solution which solves all of these problems—if there is a handy computer.

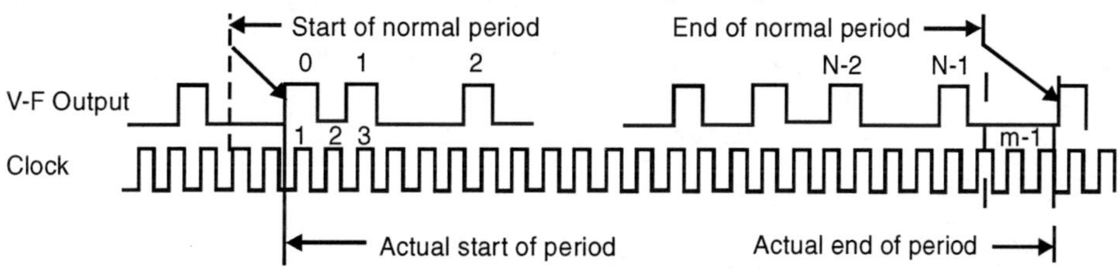

Fig. 8.96. Multiple period counting a V-f system.

The trick is to make the total measurement period approximately constant but actually synchronous with the V-f output as shown in the diagram. The start of the nominal timing period opens a "gate". The actual timing starts from the first +ve transition of the V-f output after the gate opens. Similarly the end of the nominal timing period opens another "gate" and the actual timing ends at the next +ve edge of the V-f output. Within the actual timing period the number of V-f cycles *N* and the number of clock cycles *m* is counted. Then we know that there are $m\pm1$ clock cycles of frequency *F* in exactly *N* cycles of the voltage-to-frequency converter and the output frequency is $NF/(m\pm1)$. Since *F* may now be MHz, *m* can be extremely large and the resolution in a fixed time period is improved at the expense of some computing—hence the computer. As always there is a problem and the problem in this case is that *N* is actually variable and should not drop below 100 for a 1% uncertainty in the time location of the integration time. It is therefore necessary to offset the converter (make *c* in the linearity relation non-zero) in order to circumvent the problem. However here we come upon an interesting property of the converter—it is more linear in a restricted central range than it is over a very large dynamic range - and therefore by restricting the frequency range to say a 2:1 ratio (e.g. 1000-2000 Hz) the potential resolution actually increases.

By the further use of multiple counters - and a computer—a continuous averager can be constructed which conserves all quantities (*N* and *m*) whilst allowing all the above properties. This allows a precise average over a longer time period to be determined in the post-processing phase of the data analysis.

8.10.5. The Delta-Sigma Converter

An alternative way of interpreting the output from a voltage-to-frequency converter is to regard the switch waveform (the set of pulses going to the current source switch) as continuous and sample it on a regular basis. Evidently if the frequency of switching is zero then the result will always be zero. If the output has a frequency of *N* pulses/sec each of length *t*, then the probability of getting a one as the result of a regular sampling at *M* samples/sec is *Nt/M* which is proportional to *N* which is in turn, the frequency of the output. If we regard the sampling output as the regular output of a one-bit averagable converter, then we can average the ones and zeros (as real numbers, of course) to get the average result.

Two things combine to make this a very interesting possibility :

(a) The use of digital filters to "average" the output in a manner which gives the converter some very good properties, such as transient response and frequency response. We shall touch briefly on digital filtering later in the course.

(b) The fact that the input circuit can be pretty well anything so long as the "charge balance" condition is maintained. Briefly stated, the charge-balance condition is that the charge being put onto a capacitor from a source must be balanced by the charge being pulsed off the capacitor by the converter. The use of a more elaborate circuit than a simple integrator and converter endows the converter with more flexibility in its response.

All this may seem somewhat esoteric, but these techniques are very powerful and an example is that 18-bit conversion is possible at 48 kHz update rates for a very modest outlay using this method called "Delta-Sigma Conversion".

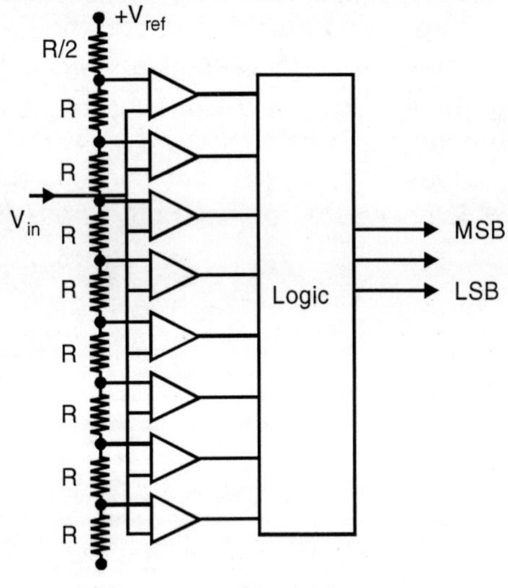

Fig. 8.97. A "Flash" Converter.

8.10.6. The "Flash Converter"

Every converter has its advantages—and the flash converter, converts in a flash, i.e. very fast. It is a brother to the successive approximation converter but whereas the successive approximation converter compares to a "guess" each time, the flash converter compares to all guesses simultaneously and then decides the "best" value from the results.

Note that a flash converter is very extravagant in hardware—but if you have to convert at 10 M samples/Sec you have to give up something!

8.11. INTERFACING ANALOG TO DIGITAL DATA CONVERTERS

This topic is aimed at the study of 8-bit and 12-bit analog to digital converters and their interfacing with 8086. In most of the cases, the PIO 8255 is used for interfacing the analog to digital converters with a microprocessor. We have already studied 8255 interfacing with 8086 as an I/O port, in the previous section. This section will only emphasize the interfacing techniques of analog to digital converters with 8255.

The analog to digital converter is treated as an input device by the microprocessor, that sends an initialising signal to the ADC to start the analog to digital data conversation process. The start of conversion signal is a pulse of a specific duration. The process of analog to digital conversion is a slow process, and the microprocessor has to wait for the digital data (EOC) signal to inform the microprocessor that the conversion is over and the result is ready at the output buffer of the ADC. These tasks of issuing an SOC pulse to ADC, reading EOC signal from the ADC and reading the digital output of the ADC are carried out by the CPU using 8255 I/O ports.

The time taken by the ADC from the active edge of SOC pulse (the edge at which the conversion process actually starts) till the active edge of EOC signal is called as the conversion delay of the ADC. Or broadly speaking the time taken by the converter to calculate the equivalent digital data output from the instant of the start of conversion is called conversion delay. It may range any where from a few microseconds in case of fast ADCs to even a few hundred milliseconds in case of slow ADCs. A number of ADCs are available in the market. The selection of ADC for a particular application is done, keeping in mind the required speed, resolution and the cost factor. The available ADCs in the market use different conversion techniques for the conversion of analog signals to digital signals. Successive approximation and dual slope integration techniques are the most popular techniques used in the integrated ADC chips. Whatever may be the technique used for conversion, a general algorithm for ADC interfacing contains the following steps.

1. Ensure the stability of analog input, applied to the ADC.
2. Issue start of conversion (SOC) pulse to ADC.
3. Read end of conversion (EOC) signal to mark the end of conversion process.
4. Read digital data output of the ADC as equivalent digital output.

It may be noted that analog input voltage must be constant at the input of the ADC right from the start of conversion till the end of conversion to get correct results. This may be ensured by a sample and hold circuit which samples the analog signal and holds it constant for a specified time duration. The microprocessor may issue a hold signal to the sample and hold circuit. If the applied input changes before the complete conversion process is over, the digital equivalent of the analog input calculated by the ADC may not be correct.

In this section, we are going to study a few ADC chips and their interfacing techniques with 8086. The first is the ADC0808 an 8-bit ADC and the second is ICL7109, Intersil's 12 bit dual slope ADC. Before proceeding with the interfacing part each of the chip is discussed in significant details so that the interfacing circuits and the algorithms can clearly be understood.

ADC 0808/0809

The analog to digital converter chips 0808 and 0809 are 8-bit CMOS, *successive approximation converters.* Successive approximation technique is one of the fast techniques for analog to digital conversion. The conversion delay is 100 μs at a clock frequency of 640 kHz, which is quite low as compared to other converters. These converters do not need any external zero or full scale adjustments as they are already taken care of by internal circuits. These converters internally have a 3 : 8 analog multiplexer so that at a time eight different analog inputs can be connected to the chips. Out of these eight inputs only one can be selected for conversion by using address lines ADD A, ADD B and ADD C, as shown. Using these address inputs, multichannel data acquisition systems can be designed using a single ADC. The CPU may drive these lines using output port lines in case of multichannel applications. In case of single input applications, these may be hardwired to select the proper input.

Analog I/P selected	Address lines		
	C	B	A
I/P 0	0	0	0
I/P 1	0	0	1
I/P 2	0	1	0
I/P 3	0	1	1
I/P 4	1	0	0
I/P 5	1	0	1
I/P 6	1	1	0
I/P 7	1	1	1

These are unipolar analog to digital converters, i.e. they are able to convert only positive analog input voltages to their digital equivalents. These chips do not contain any internal sample and hold circuit. If one needs a sample and hold circuit for the conversion of fast signals into equivalent digital quantities, it has to be externally connected at each of the analog inputs. Figure 8.98 (a) and (b) show the block diagrams and pin diagrams for ADC 0808/0809.

I/P0-I/P7	**Analog inputs**
ADD A, B, C	Address lines for selecting analog inputs
O7-O0	Digital 8-bit output with O7 MSB and O0 LSB
SOC	Start of conversion signal pin
EOC	End of conversion signal pin
OE	Output latch enable pin, if high enables output
CLK	Clock input for ADC
VCC, GND	Supply pins +5 V and GND
Vref+ and Vref–	Reference voltage positive (+5 Volts max.) and Reference voltage negative (0 V minimum)

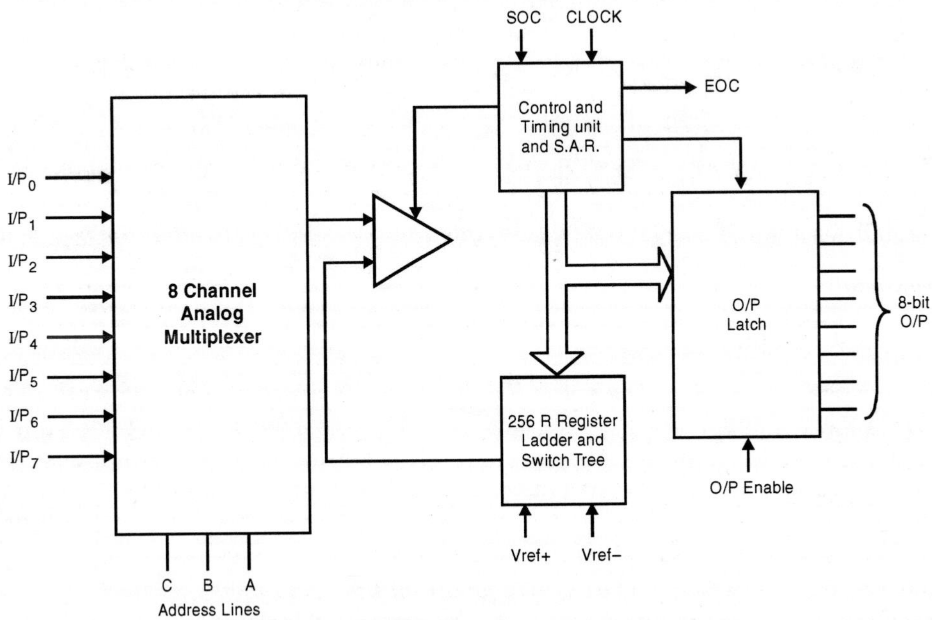

Fig. 8.98. ***(a) Block Diagram of ADC 0808/0809.***

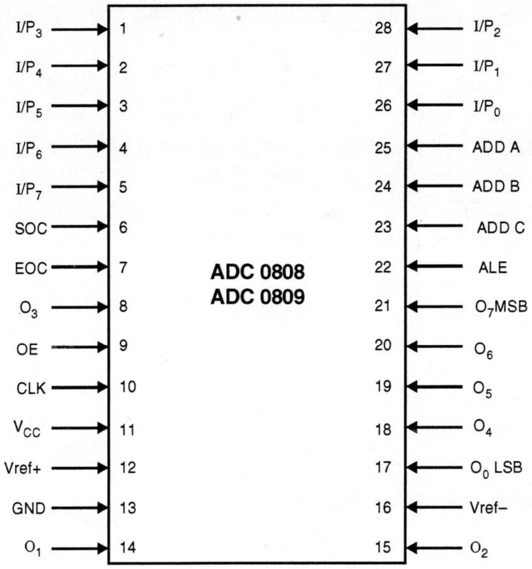

$I/P_0 - I/P_7$	–	Analog inputs
ADD A, B, C	–	Address lines for selecting analog inputs
$O_7 - O_0$	–	Digital 8-bit output with O_7 MSB and O_0 LSB
SOC	–	Start of conversion signal pin
EOC	–	End of conversion signal pin
OE	–	Output latch enable pin if 1 enable output
CLK	–	Clock input for ADD
V_{CC} GND	–	Supply pins +5 V and GND
$Vref^+$ and $Vref^-$	–	Reference voltage positive +5 Volts maximum and Reference voltage negative 0V minimum

Fig. 8.98. ***(b) Pin Diagram of ADC 0808/0809***

Some electrical specifications of the ADC 0808/0809 are given below.

Minimum SOC pulse width	100 ns
Minimum ALE pulse width	100 ns
Clock frequency	10 to 1280 kHz
Conversion time	100 μs at 640 kHz
Resolution	8-bit
Error	+/– 1 LSB
Vref+	Not more than +5 V
Vref–	Not less than GND
+Vcc supply	+5 V DC
Logical 1 i/p voltage	minimum Vcc–1.5 V
Logical 0 i/p voltage	maximum 1.5 V
Logical 1 o/p voltage	minimum Vcc–0.4 V
Logical 0 o/p voltage	maximum 0.45 V

The timing diagram of different signals of ADC0808 is shown in Fig. 8.99.

Till now we have studied the necessary details of the analog to digital converter chips 0808/0809. Now we consider some interfacing examples of these chips with 8086 so that the working of these ADCs will be absolutely clear along with the required algorithms for interfacing.

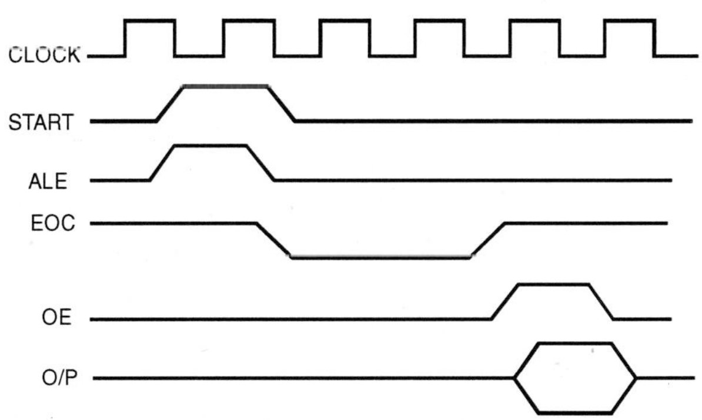

Fig. 8.99. Timing diagram of ADC 0808

Example 1. Interface ADC 0808 with 8086 using 8255 ports. Use port A of 8255 for transferring digital data output of ADC to the CPU and Port C for control signals. Assume that an analog input is present at

I/P_2 of the ADC and a clock input of suitable frequency is available for ADC. Draw the schematic diagram and write required ALP.

Solution. Figure 8.100 shows the interfacing connections of ADC 0808 with 8086 using 8255. The analog input I/P_2 is used and therefore address pins A, B, C should be 0, 1, 0 respectively to select IP_2. The OE and ALE pins are already kept at +5 V to select the ADC and enable the outputs. Port C upper acts as the input port to receive the EOC signal while port C lower acts as the output port to send SOC to the ADC. Port A acts as a 8-bit input data port to receive the digital data output from the ADC. The 8255 control word is written as follows :

D7	D6	D5	D4	D3	D2	D1	D0	Control word
1	0	0	1	1	0	0	0	= 98 H

The required ALp is given as follows :

```
MOV AL, 98 H          ; Initialise 8255 as
OUT CWR, AL           ; discussed above
MOV AL, 02H           ; Select I/P2 as analog
OUT Port B, AL        ; input
MOV AL, 00H           ; Give start of conversion
```

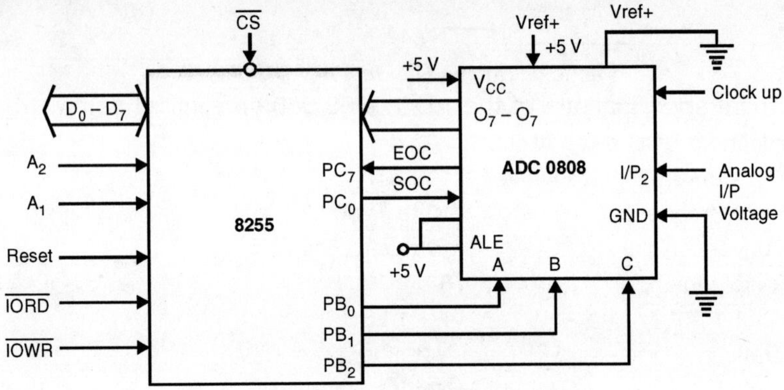

Fig. 8.100. Interfacing 0808 with 8086.

Program : ALP for Problem

```
OUT Port C, AL          ; pulse to the ADC
MOV AL, 01 H            ;
OUT port C, AL          ;
MOV AL, 00H             ;
OUT Port C, AL          ;
WAIT : IN AL, Port C    ; Check for EOC by
RCR                     ; reading port C upper and
JNC WAIT                ; rotating through carry.
IN AL, Port A           ; If EOC, read digital equivalent in AL.
HLT    ; Stop
```

ADC 7109—A Dual Slope 12 bit ADC

Intersil's ICL 7109 is a dual slope integrating analog to digital converter. This 12 bit ADC is designed to work with 8-bit and 16 bit or higher order microprocessors with great ease. As compared to other 12 bit ADCs, it is a very low cost option, useful for slow practical applications, as the method used for analog to digital conversion is dual slope integration.

The 12 bit data output, polarity and over-range signals can be directly accessed under the software control of two byte enable inputs $\overline{LBEN}$ and $\overline{HBEN}$. The $RUN/\overline{HOLD}$ and STATUS outputs allow monitoring and control of the conversion process. The salient features of the ADC 7109 are as given as follows :

1. 12-bit data output equivalent to analog input alongwith polarity, over range and under range outputs.
2. It can be operated in parallel or serial output mode.
3. Differential input and differential reference.
4. Low noise (Typical 15 µV p-p)
5. Low input current (Typical 1 pA).
6. Can operate upto 30 conversions per second, with an external crystal or RC circuit used to decide the operating clock frequency.

The pin diagram of the ADC is given here followed by brief signal descriptions, in Fig. 8.101.

The internal circuit of ICL 7109 is slightly complicated. Hence it is divided in two parts namely—analog section and digital section for better understanding. The following text explains the internal operation in terms of the separate sections.

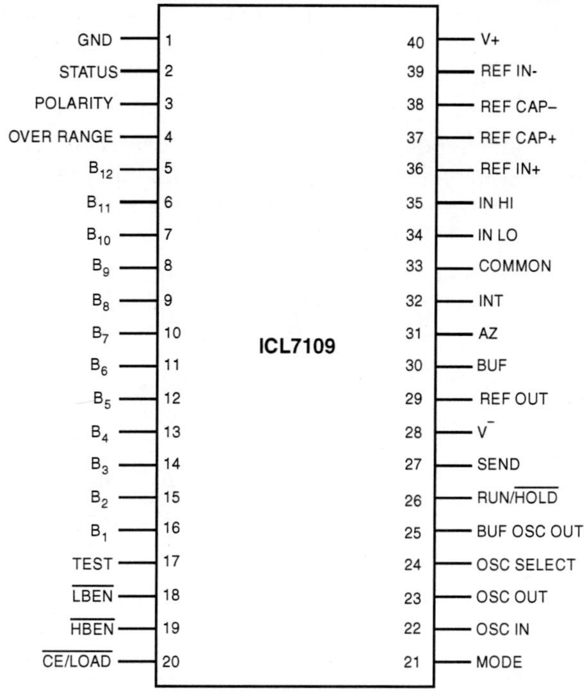

Fig. 8.101. Pin Configuration of ICL 7109ADC.

Analog Section. The block diagram of the internal analog section of ICL 7109 is shown in Fig. 8.102. If RUN/$\overline{\text{HOLD}}$ is either left open or connected to +VCC, the ADC starts conversion at the rate determined by the clock frequency, to be decided either by a crystal or by an RC circuit. The total conversion cycle is divided into three phases namely autozero phase, signal integrate phase and deintegrate phase.

Autozero phase. During autozero phase, input high (IN HI) and input low (IN LO) are disconnected from the external input signal and shorted to analog common. Then the reference capacitor is charged to reference voltage. A feedback loop is closed around the system to charge the autozero capacitor CAZ to compensate for the offset voltages in the buffer amplifier, integrator and comparator.

Signal Integrate Phase. In this phase the autozero capacitor CAZ is opened out. The external input signal is connected with the internal circuit. The differential input voltage between IN LO and IN HI pins is then integrated by the internal integrator for a fixed period of 2048 clock cycles.

De-integrate Phase. This is the final phase of the analog to digital conversion. The input low is internally connected to analog common, and input high is connected across the previously charged reference capacitor. The capacitor then discharges through the internal circuit of the chip. Hence integrator output returns to zero crossing with a fixed slope. This time taken by the integrator output to return to zero is proportional to the input signal.

Digital Section. The digital section includes the clock oscillator, 12-bit binary counter, output latches, TTL compatible output drivers, polarity, over-range and their control logics and UART handshake logic as organised in Fig. 8.103.

The digital section uses a positive logic system wherein logical 'low' corresponds to a low voltage and logical 'High' corresponds to a high voltage. The actual ranges for logical 'low' and 'High' can be obtained from the data sheets.

Intersil's Component Data Catalogue may be referred for the detailed information about the chip and its functioning. This text is only aimed at giving some brief introduction of the chip before we proceed for its interfacing with 8086.

Typical Component Value Selection. For the proper working of ICL 7109, it is necessary that the component values of the practical circuit are properly chosen. All the component values are recommended in the data manual but, the three components, viz. Rint, Cint and CAz should be selected suitably as all of them directly determine the required input voltage range and operating clock frequency. The full scale input voltage range is double the voltage between REF IN- and REF IN+. The clock frequency should be integral multiple of the power supply frequency (50 Hz) to achieve the optimum power supply frequency rejection. The formulae for component values selection are given as follows :

$$R_{int} = \frac{\text{Full scale i/p voltage}}{20\ \mu\text{A}}$$

$$C_{int} = \frac{2048 \times (\text{Clock period}) \times 20\ \mu\text{A}}{\text{Integrator o/p voltage swing}}$$

The ADC 7109 can either be driven with a clock frequency, derived from a crystal or from an RC circuit. If the clock frequency is derived from the crystal then the actual operating frequency is given by f.

$$f = \frac{\text{Crystal frequency}}{58}$$

Otherwise, if the ADC is driven by a clock frequency derived from an RC circuit it is given by the formula.

$$f = \frac{0.45}{RC}$$

Interfacing with 8086

Figure 8.104 shows the interfacing of ICL 7109 with 8086 using 8255. The assembly language program reads the digital equivalent output from ADC 7109 and stores it in register CX. Note that the ADC gives only 12-bit output, hence the most significant nibble of CX must be masked.

The parallel I/O port chip 8255 is used to interface ICL 7109 with 8086. Being a 12-bit ADC, it will require 12 I/O lines for data outputs. Port A and Port C lower are used as input ports to read digital data output from the ADC. The pins $\overline{LBEN}$ and $\overline{HBEN}$ are permanently grounded to enable all the data lines from B1 to B12 at a time. The lines $\overline{CE/LOAD}$ and $\overline{RUN/HOLD}$ are controlled using port C upper of 8255, that is used as an output port and finally port B is used to read the STATUS, POL and OR signals and hence is used as an input port. The polarity (POL) and overrange data (OR) will be available in DL register with D0 corresponding to OR and D1 corresponding to POL. The digital data is read from the ports if STATUS goes low, i.e. conversion is over. Figure 8.101 shows the algorithm for the analog to digital conversion with ADC 7109. After the execution of the program the digital equivalent of the analog input is in register CX, while the overrange and polarity information is in DL. From the above description 8255 control word comes out to be 93H.

The above circuit components are designed for 4096 mV full scale analog input range. As specified in the manual the R_{int} for this range is 200 K. The reference input voltage between REF IN- and REF IN+ will be half of full scale analog input voltage.

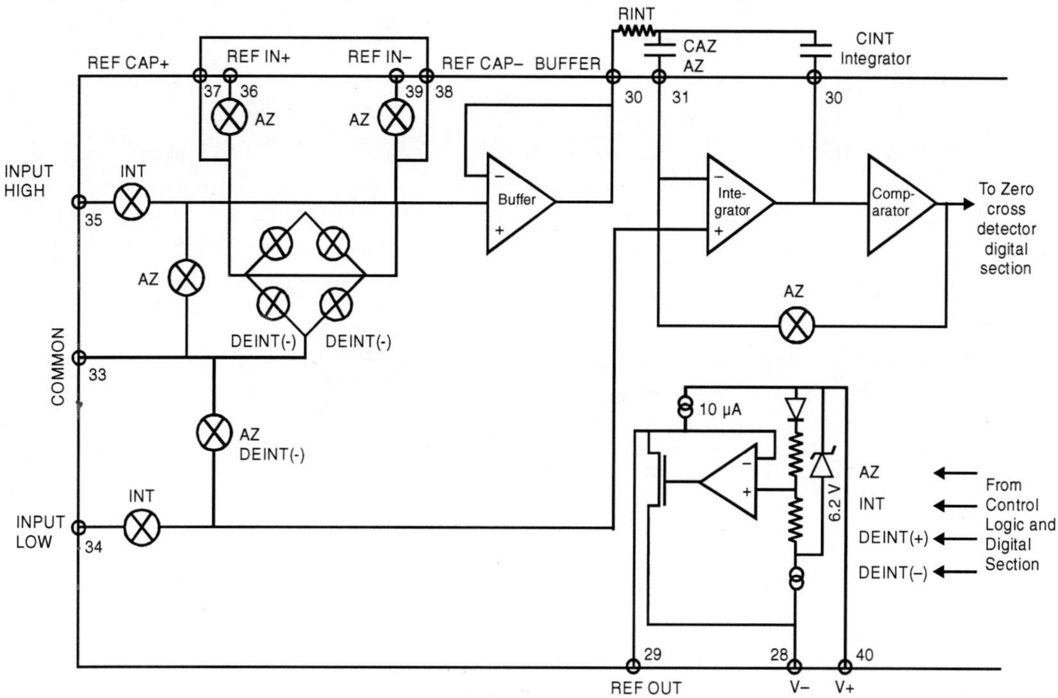

Fig. 8.102. Analog section of 7109.

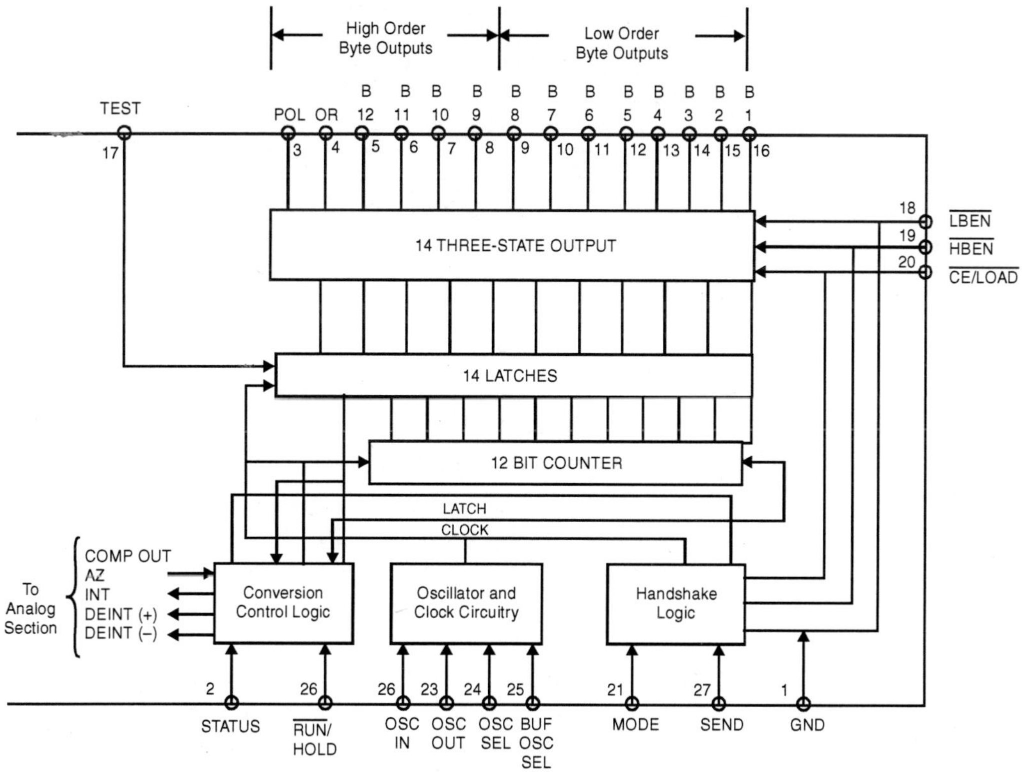

Fig. 8.103. Digital section of ICL 7109.

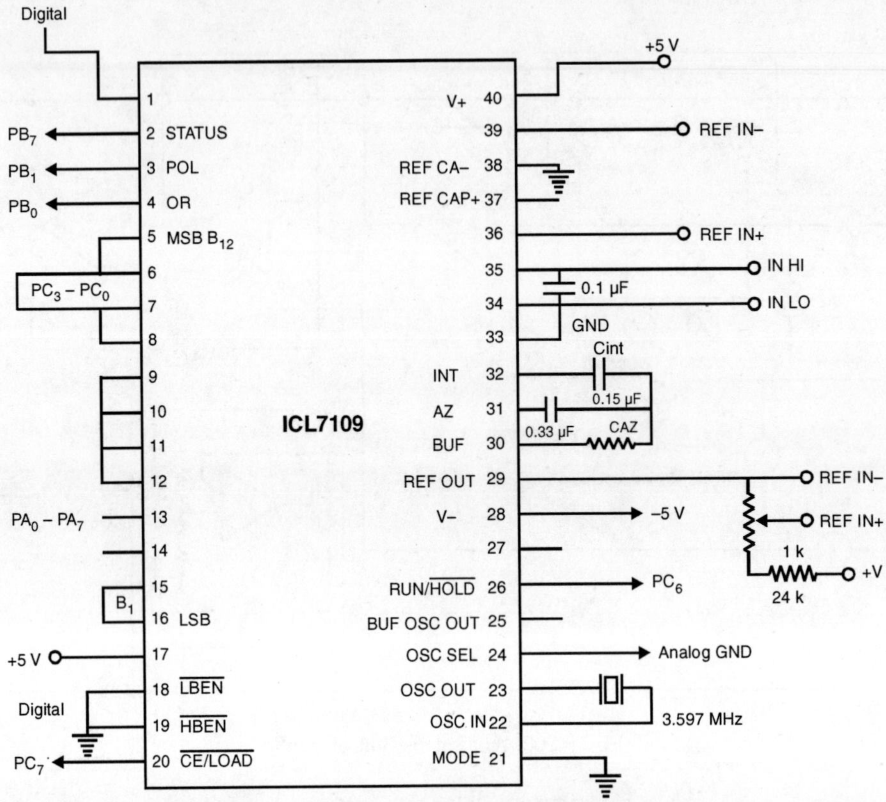

Fig. 8.104. Interfacing 7109 with 8086 through 8255.

Program : ALP for Interfacing ICL 7109 with 8086

```
    ; This program issues a RUN signal to ICL7109, checks for
    ; STATUS (EOC) and reads digital output with POL and OR outputs.
ASSUME    CS: CODE
CODE      SEGMENT
START :   MOV AL, 93H ; Initialization of
          OUT CWR, AL ; 8255
          MOV AL, 40 H      ; RUN (PC6) to go high and
          OUT Port C, AL    ; CE(PC7) to go low, for start of conversion
          WAIT : IN AL, Port B  ; Read STATUS signal
          RCL AL, 01        ; Check STATUS using carry flag.
          JC WAIT           ; Wait till carry (STATUS) goes
          IN AL, Port A     ; low i.e. conversion is over
          MOV CL, AL
          IN AL, Port C     ; Read digital data output and store the
          AND AL, 0FH   ; lower byte in CL and higher byte in CH. Mask
          MOV CH, AL    ; higher bits of CH as of only 12 bits are of
                                 interest.
          IN AL, Port ; Store OR and POL in D0 and D1 in
```

```
        MOV DL, AL        ; DL register
        MOV AH, 4CH ; Return to DOS
        INT 21H
CODE    ENDS
        END START
```

8.12. DIGITAL-TO-ANALOGUE CONVERSION

The problem of digital-to-analogue conversion and vice versa is the problem of how to simulate a system which may take an infinite sequence of values of voltage at an infinite sequence of times, with a system which has a finite set of voltage values and may be updated only at a finite rate. This mismatch is at the heart of, and the cause of, most of the problems with conversion from the electronic world of volts, amps and time to the computer world of clocks and finite word lengths.

In general it may be said that the more money you pay, the closer to the real world you can get in terms of resolution in voltage and time. The question to be addressed is whether this is an altogether desirable state or if a little thought can produce a result which is as good, or in some cases better than, the expensive solution using only a few cheap parts.

To give a simple example: Suppose we wish to slowly move a mirror using a piezo-electric transducer and it is important that the motion be linear, i.e. not stepwise. We could spend a tremendous amount of money and get a very accurate Digital-to-Analog Converter (DAC) and feed a count sequence to it as shown here.

As an alternative we could use a less accurate DAC and follow it by an integrator which will necessarily produce a smooth output as shown in the next diagram.

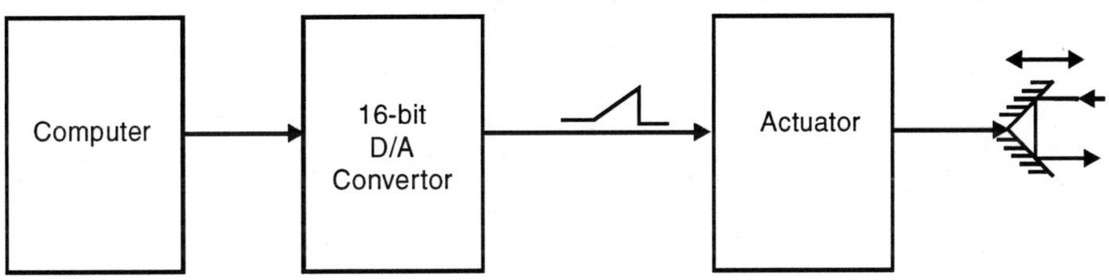

Fig. 8.105. Moving a Mirror - D/A converter.

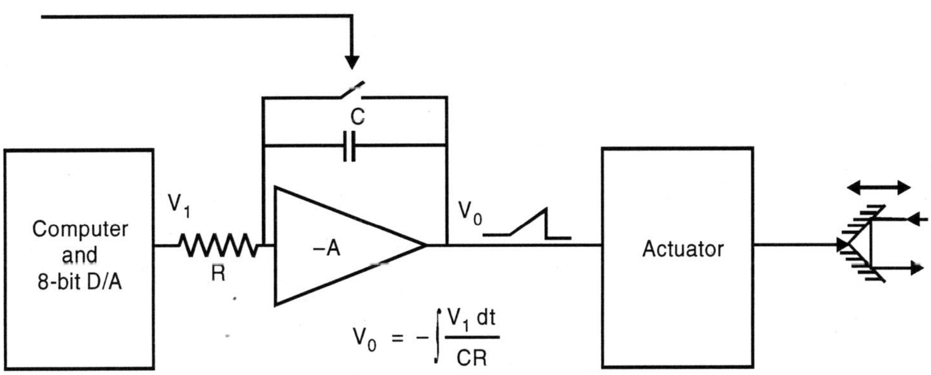

Fig. 8.106. Moving a Mirror Integrator.

Now in order to reset the integrator we may need a couple of extra parts but the result is much nearer perfection than can be obtained reasonably. To give some figures: A very good

DAC can produce 2 states or a resolution of 0.2 mV in a 10 V swing. A simple 16 integrator can easily better that. However we have lost some flexibility in that we cannot now arbitrarily choose the rate of the ramp, being limited to 256 discrete rates if we use a simple 8-bit DAC in our design. However often when moving a mirror in an optical system 256 rates are plenty. In a practical system we also should give some thought to measuring where the mirror is and how to return the mirror to the rest position - but this is only an example.

A further advantage of the integrator system is that it relieves the computer of the ramp timing and therefore makes the time constraints of the program easier. Again with figures: If we ramp through 10v in 1 sec at 2 resolution, we require updates after every 15 μS, which is a substantial job for most micros (an 8088 can just about make it with optimised code) whereas our integrator solution is a "set and forget" thing where we only need to check at the limits - if at all.

Thus a little, or much, thought given to the strategy of any converters will probably repay itself. However in order to have constructive thoughts it is necessary to know a bit about how things work. I therefore plan to run quickly over the main philosophies of DACs and then to discuss their relative strengths and weaknesses in a fair degree of detail—this is necessary because many a good experiment has been ruined, or been less than the best, because of inadequate thought at the planning stage.

8.12.1. Practical Devices

DACs come in a number of styles with various properties. The key parameters for the unit are :

Resolution - What is the smallest average change in output voltage which can be effected?

Accuracy - What is the relationship between the output and some standard of voltage? This parameter needs several other parameters such as linearity to explain properly.

Speed - How fast can the output change? Important but less easy to quantify is: what does it do while changing?

Integrating Converters

The simplest illustration of the above is a converter relying on an integrator and time :

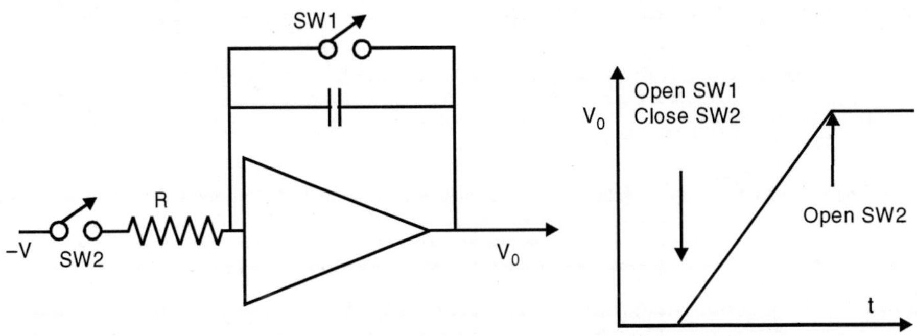

Fig. 8.107. An integrating D/A converter.

The computer closes SW1 for a long enough time to discharge the capacitor and then opens SW1 and closes SW2 for a finite time, t_1 after which the voltage at the output is:

$$V_0 = \frac{Vt_1}{CR}$$

Notice that the resolution is determined essentially by the timing resolution which can be made arbitrarily fine by the use of a short basic time interval counted for many counts. The accuracy is determined by the constancy, linearity and perfection of V,C and R. The speed is slow to very slow. Notice that the converter does very odd things while changing its value - thus to change from 1v to 1.1v the voltage-time graph as shown in Fig. 8.108.

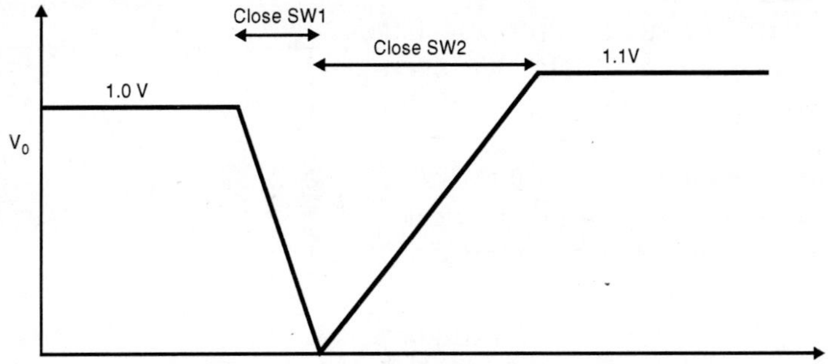

Fig. 8.108. Integrating D/A - Changing the value.

Just to finally confuse the issue, notice that since no op-amp is perfect and no capacitor is perfect there is a tendency for the voltage to drift from the set value after some period of inactivity.

Some of the problems can be alleviated by using two voltage sources of opposite polarity and then running the converter from the appropriate reference. However the value will still steadily drift from the correct one after some time. This may not be a problem if the system is in a servo-loop where the exact value is under some other control. Using bi-directional references the movement from 1 to 1.1 volt is a smooth ramp of defined slope.

Although it is the simplest converter and has some intrinsic merit—the counting converter is very little used.

8.12.2. Switched Converters

The commonest form of DAC converter is the switched sort which is illustrated by the following diagram:

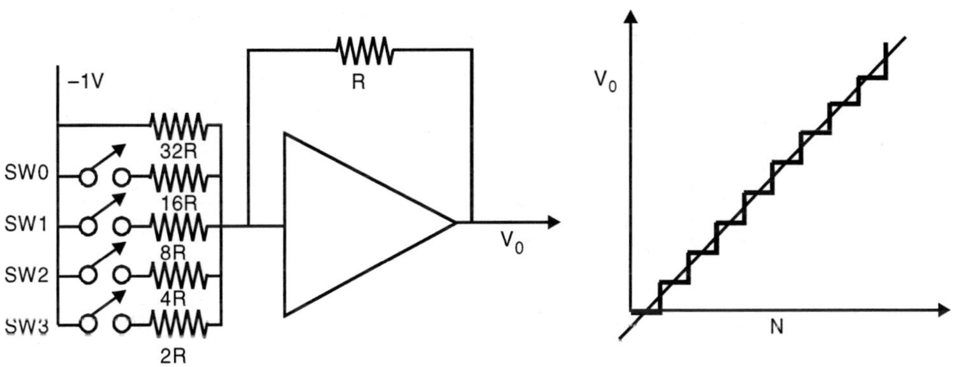

Fig. 8.109. Summed source D/A converter.

The setting of the switches SW3-SW0 sets up a voltage of 0 - 0.9375 V in increments of 0.0625 volts. The first thing that we can see which is a problem here is that if we plot a graph of V vs N where N is the number on the converter and then run a "best straight line" through the results, we find that the line does not pass through the origin. An alternative statement is that the fit to a line through the origin is within +0 or –1 LSB of the line. We can make a better fit by offsetting the whole line by +0.0313 V with an extra resistor which gives a fit of ±1/2 LSB to a straight line through the origin. We can now see clearly the problem of mapping an infinite set of voltages (the straight line) onto a finite set of values (the staircase) and the clincher is that an input of 0 comes out as +0.0313 V and is within the accuracy statement of the device!

Even worse things happen if we use real resistors - say 5% accuracy. If the four top resistors are –5% then the output for state 7 is 0.4688*1.05=0.4922v. If the bottom resistor is +5% then the output for 8 is 0.5313*0.95=0.5047V. The difference between 7 and 8 on the output is then

0.0125 V instead of the expected 0.0625 V - a substantial, even worrying change. Some of the problems inherent in this design can be overcome by the use of more sophisticated networks and resistor chains instead of the simple scheme used above.

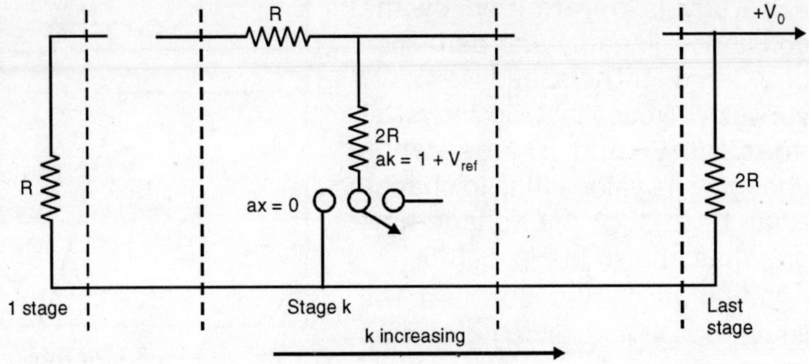

Fig. 8.110. An R-2R ladder.

One common alternative network is the "R-2R ladder" which uses only resistors of value R and 2R (or just one value if 2R = R + R). The problems of matching a large number of resistors of the same value are much lower than the problems of matching resistors of different values such as in the scheme above. A sequence of stages is then set up so that each stage produces an appropriate weighting for the system (see accounts in Millman: Microelectronics for details).

$$V_0 = \frac{V_{ref}}{3} \int_{k+1}^{k-n-1} \frac{a_k}{2^k}, \quad a_k = \{0,1\}$$

Another scheme which is gaining in popularity in integrated circuits is to use a resistor chain as shown here. This scheme simply involves the selection of the appropriate "tap" off the resistor chain with a set of switches. Its strength is that the output voltage must at least increase with increasing N which the circuits above are not constrained to do and it uses large numbers of identical parts which is terrible in discrete circuits but heaven for makers of integrated circuits. Notice the use of the end resistors to offset the converter output.

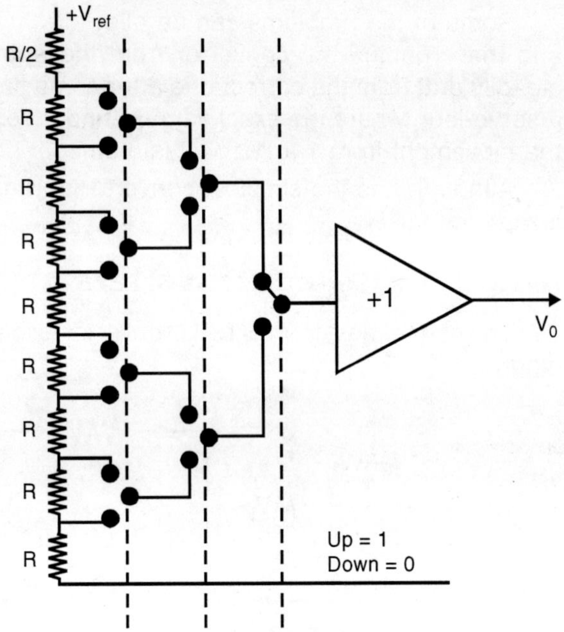

Fig. 8.111.

8.13. INTERFACING DIGITAL TO ANALOG CONVERTERS

The digital to analog converters convert binary numbers into their analog equivalent voltages. The DAC find applications in areas like digitally controlled gains, motor speed controls, programmable gain amplifiers, etc. This text explains the generally available DAC integrated circuit and their interfacing techniques with the microprocessor.

AD 7523 8-bit Multiplying DAC

Intersil's AD 7523 is a 16 pin DIP, multiplying digital to analog converter, containing R-2R ladder (R = 10 K) for digital to analog conversion along with single pole double throw nMOS switches to connect the digital inputs to the ladder.

Pin diagram of AD7523 is shown in Fig. 8.112. The supply range extends from +5 V to +15 V, while Vref may be any where between –10 V to +10 V. The maximum analog output voltage will be +10 V, when all the digital inputs are at logic high state. Usually a Zener is connected between OUT1 and OUT2 to save the DAC from negative transients. An operational amplifier is used as a current to voltage converter at the output of AD 7523 to convert the current output of AD 7523 to a proportional output voltage. It also offers additional drive capability to the DAC output. An external feedback resistor acts to control the gain. One may not connect any external feedback resistor, if no gain control is required. The following example explain the interfacing of AD 7523 with 8086.

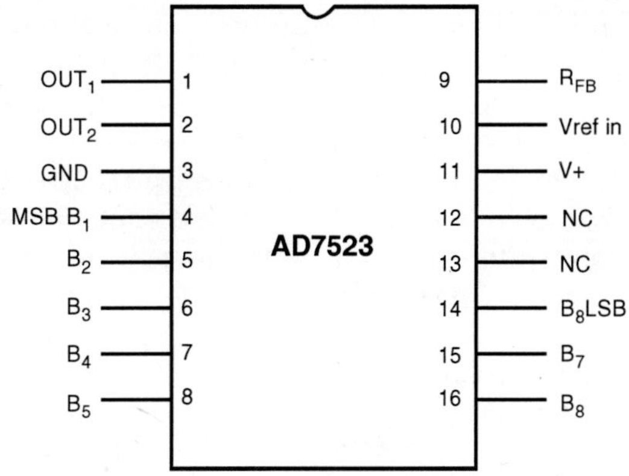

Fig. 8.112. Pin diagram of AD7523.

Example 2. Interface DAC AD 7523 with an 8086 CPU running at 8 MHz and write an assembly language program to generate a sawtooth waveform of period 1 ms with V_{max} 5 V.

Solution. Figure 8.113 shows the interfacing circuit of AD 7523 with 8086 using 8255. Program gives an ALP to generate a sawtooth waveform using this circuit.

Program : ALP for Generating sawtooth waveform using AD 7523

```
ASSUME CS : CODE
CODE   SEGMENT
START: MOV AL, 80 H       ; Initialise port A as output
       OUT CWR, AL        ; port.
AGAIN: MOV AL, 00H        ; Start the ramp from 0 V
BACK : OUT Port A, AL     ; Input 00H to DAC
       INC AL             ; Increment AL to increase ramp output
       CMP AL, 0F2H       ; Is upper limit reached ?
       JB BACK            ; If not, then increment the ramp.
       JMP AGAIN          ; Else start again from 00H
       ENDS
       END START
```

In the above program port A is initialized as the output port for sending the digital data as input to DAC. The ramp starts from the 0 V (analog), hence AL starts with 00H. To increment the ramp, the content of AL is incremented during each execution of the loop till it reaches F2H. After that the sawtooth wave again starts from 00H, i.e. 0 V (analog), and the procedure is repeated. Note that the ramp

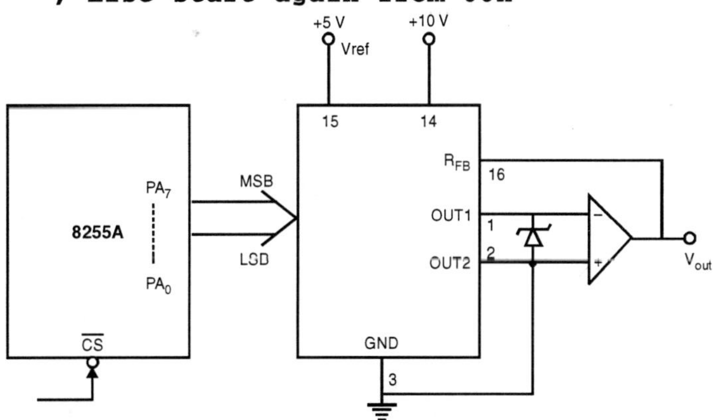

Fig. 8.113. Interfacing of AD7523

period given by this program is precisely 1.000625 ms. Here the count F2H has been calculated by dividing the required delay of 1 ms by the time required for the execution of the loop once. The ramp slope can be controlled (reduced) by calling a controllable delay after the OUT instruction. It may be noted that the meeting the frequency i.e. time and amplitude requirement exactly is slightly difficult in such applications.

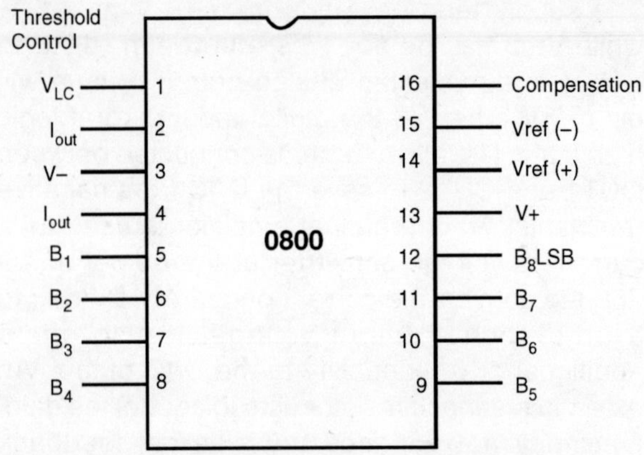

Fig. 8.114. Pin Diagram of DAC 0800

DAC0800 8-bit Digital to Analog Converter

The DAC 0800 is a monolithic 8-bit DAC manufactured by National Semiconductor. It has settling time around 100 ms and can operate on a range of power supply voltages, i.e. from 4.5 V to +18 V. Usually the supply V+ is 5 V or +12 V. The V– pin can be kept at a minimum of –12 V. The pin diagram of DAC 0800 is shown in Fig. 8.114. The pin definitions are self explanatory. Figure 8.115 shows interfacing of 0800 DAC with 8086.

Example 3. Write an assembly language program to generate a triangular wave of frequency 500 hz using the interfacing circuit given in Fig. 8.115. The 8086 system operates at 8 MHz. The amplitude of the triangular wave should be +5 V.

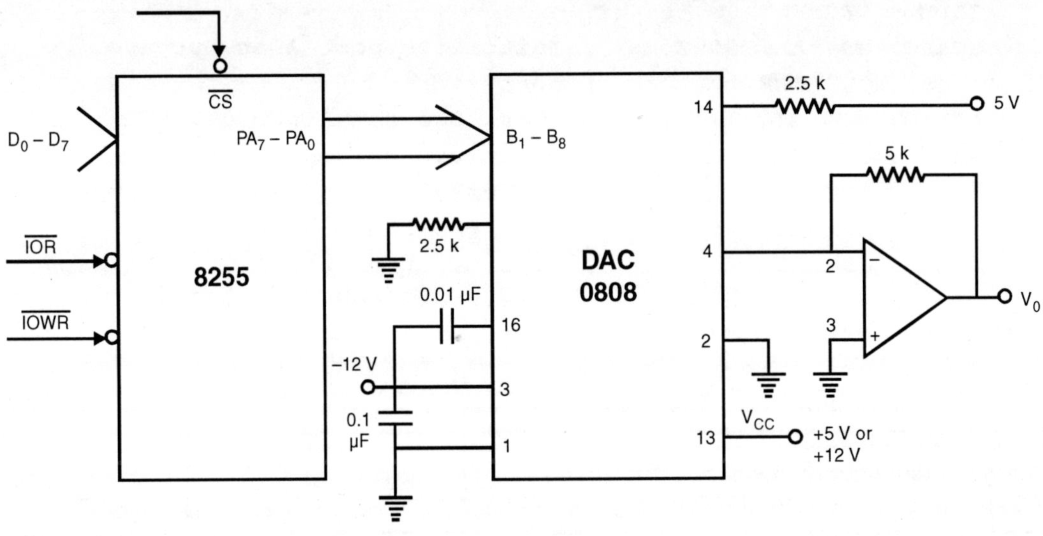

Fig. 8.115. Interfacing DAC0800 with 8086.

Solution. The V_{ref+} should be tied to +5 V to generate a wave of +5 V amplitude. The required frequency of the output is 500 Hz, i.e. the period is 2 ms. Assuming the wave to be generated is symmetric, the waveform will rise for 1 ms and fall for 1 ms. This will be repeated continuously. In the previous program, we have already written an instruction sequence for period 1 ms. Using the same instruction sequence one can derive this triangular waveform. The ALP is given as follows.

Program : ALP for Generating a Triangular Wave Using DAC 0800

```
ASSUME          CS : CODE
CODE      SEGMENT
START :   MOV AL, 80H        ; Initialise 8255 ports
          OUT CWR, AL        ; suitably
          MOV AL, 00H        ; Start rising ramp from
BACK :    OUT Port A, AL     ; 0 V by sending 00H to DAC
          INC AL             ; Increment ramp till 5 V
          CMP AL, FFH        ; i.e. FFH
          JB BACK            ; If it is FFH then,
BACK1 :   OUT Port A, AL     ; Output it and start the falling
          DEC AL             ; ramp by decrementing the
          CMP AL, 00         ; counter till it reaches
          JA BACK1           ; zero. Then start again
          JMP BACK           ; for the next cycle.
CODE      ENDS
          END START
```

The technique of interfacing 12-bit DACs with 8086 is also similar. If 8-bit ports are used for interfacing a 12-bit DAC, two successive 8-bit OUT instructions are required to apply input to the DAC. If a 16-bit port is used for interfacing a 12-bit DAC, a single OUT instruction is sufficient to apply the 12-bit input to the DAC. 12-bit DACs generate more precisely.

8.13.1. Stochastic Converters

Most of the practical DACs available use one of the above schemes with some refinements to salvage manufacturers pride in producing a different product. However several other schemes are possible and I will just illustrate them with one - the stochastic converter. This converter consists of a single switch followed by a low-pass filter.

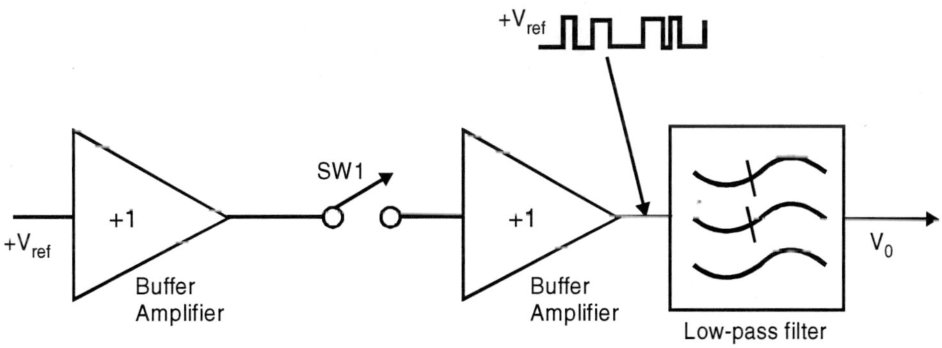

Fig. 8.116. Stochastic converter.

The method of operation is to operate the switch SW1 so that the fraction of time *f* which it spends in the "on" position is given by $f = V/V$ when *V* is the required output. *o ref* is the simplest strategy for this scheme is to take a time period of *n* cycles of a clock *N* and turn the switch on for *m* of them such that $f = m/n$—note that again the result is a set of discrete values. However if the update time is *t* secs then the output will have a harmonic content at harmonics of $1/nt$. A better strategy therefore is to update at the same rate but the update have a "probability"

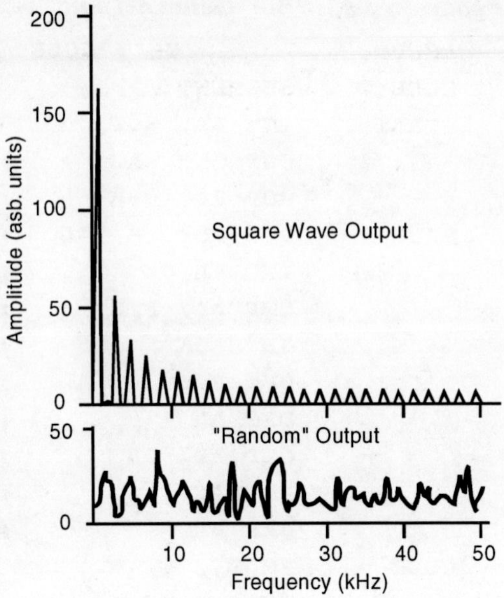

Fig. 8.117.

of being "on" equal to *f*. There are several well-known ways of doing this but the simplest is to pick a random number between 0 and 1 and turn the switch "on" if the number <*f* and "off" otherwise. The output will now be a random sequence which will not have any particular frequencies embedded in it (hopefully).

A particular point to note is that if the random number takes a finite but variable time to generate then it will be necessary to use an external clock (real-time clock) to fix the update rate.

In order to demonstrate the difference between the two approaches more thoroughly, here is an example: I used a sampling rate of 128 kHz and then made a 7-bit converter by taking 128 time slices. I am going to try to set the output up for 1/2 scale - 1/2 the time on and half the time off. The simplest strategy would be to put the switch on for half the time and then off for the other half. This produces a square wave of frequency 1 kHz whose frequency analysis is in the upper graph of the diagram. An alternate strategy is to use a random number generator generating values between 0 and 1. For every time slot we generate a random number and if it is greater than 0.5 we put the switch on and if it is less we turn it off. The frequency analysis of the output is shown in the lower graph on the same vertical and horizontal scale. It is obviously easier to reduce the frequency content of the random sequence than the regular sequence because of the absence of peaks with high energy. Colloquially we can say that the energy in the first case is concentrated in the peaks whereas in the second case it is spread throughout the spectrum.

8.13.2. Non-Linear Systems

So far we have discussed the DAC as though it is always designed as a linear device. By linear we mean that the relationship between the input and the output is of the form

$$V_0 = mN + c$$

where *m* and *c* are constants. However this is not always desirable and several converters have been designed to emulate curves rather than a straight line. One such system is a "companding" converter. These are used in telecommunications and often follow the "Bell μ-255" law which can be stated as:

$$V_0 = \frac{V_{fs}}{255}\left(\exp\left[\frac{N/(2^m+1)}{0.18}\right]+1\right)$$

where *V* is the full-scale output. This can also be expressed in the alternative form relating *fs* the number N to the output V_0.

$$N = (2^m+1)0.18 \ln\left(1+255\frac{V_0}{V_{fs}}\right)$$

The advantage of these schemes is that they can accurately convey the size of a large signal without compromising the resolution of a small signal. This implies that they can handle signals with a large "dynamic range" The term refers to the ratio of the largest to the smallest expected useful values for the quantity. A voltage range from 0.1-1 V has a dynamic range of 10x. A voltage range of 0-1 V has infinite dynamic range. However if you would regard the smallest "significant" signal as 1 mV then the dynamic range is the ratio of the maximum to the minimum signal - 10 in this case. It is often useful to restrict dynamic 3 range in order to make things more easily measurable and sometimes necessary to cope with a very large dynamic range in a finite size of result, e.g. by using range switching or a log amplifier.

In digital-to-analogue terms the dynamic range of a linear converter is usually something like 2:1 where n is the number n of bits in the converter. If we use a companding converter as we are discussing now the dynamic range can be larger. The Bell μ-255 law allows a range of about 4000:1 in 8 bits whereas a linear converter could only manage 256:1. (In fact our realisation of the μ-255 law doesn't do quite as well as the theory because of the straight line segment approach)

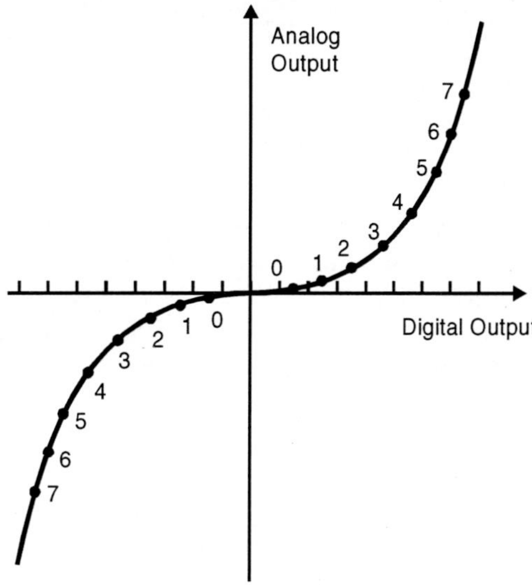

Fig. 8.118. Bell system μ-225 companding approximation.

Dynamic range is often expressed in "dB" which is 20log (ratio). Thus 256:1 is 48 dB.

A practical realisation of a companding DAC is shown in this figure where the most significant four bits of the input number define which segment (chord) of the converter is used, including the sign, and the last four bits determine where on the chord the output is. The curved characteristic is therefore approximated by 16 straight line segments.

Terms and Definitions

At this point we introduce some of the terms that one reads in specification sheets. Manufacturers specify a large number of parameters for DACs. Some will be relevant in a given situation—some will not. This list cannot be exhaustive, but at least it tries to set down some of the main terms and discuss some of the issues associated with them.

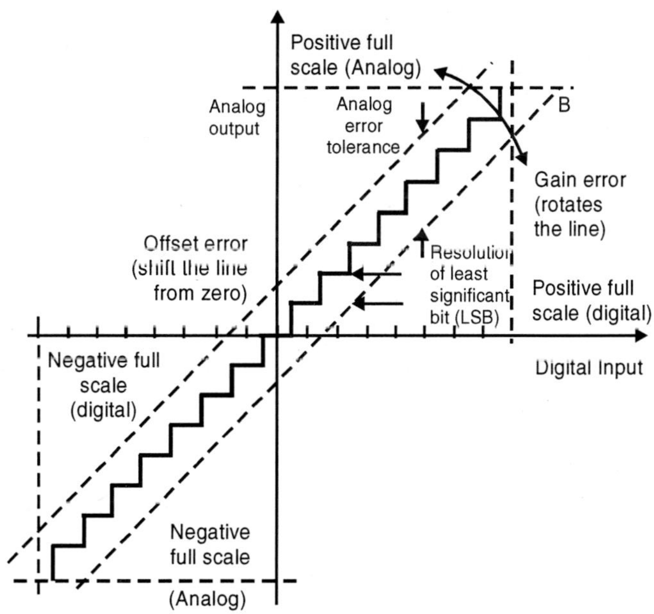

Fig. 8.119. Specification of a digital-to-analog converter.

Full Scale

Essentially a DAC can produce an output which is a scaled version of an input (although the input may simply be a reference voltage). If this is a unipolar output we could expect zero input code to correspond to zero output and 2-1 to correspond to the maximum output which would be full-scale. A bipolar output would be approximately symmetrical although note that since the number of codes is even and zero output is one of the codes, there is going to be a slight asymmetry in the output range. The mapping of the input codes to the bipolar output may be as an offset, where 0 input corresponds to maximum negative output and 2-1 to maximum positive output, 2s-complement, or by magnitude and a separate sign bit.

Multiplying DACs

As mentioned above, the action of a DAC can be considered to be the multiplication of an input by a fraction. Some DACs actually achieve this in practice and therefore become digital multipliers. Such DACs may be 1-quadrant (+ve input and fractions only), 2-quadrant (usually bipolar input and +ve fraction but could be the other way round) or 4-quadrant (bipolar input and multiplier)

Monotonicity

In an ideal DAC, the change in output from one input code to the next is the same. We have already noted the difficulty with maintaining this relationship in practical cases. It is very desirable that the output stay the same or rise with each increment in the input code, even if the rise is not the theoretical step size. This can be accomplished if the converter steps are always within ±1/2LSB of the correct size. Such a converter is said to be "monotonic" and is achieved for converters with DNL (see below) of <1/2LSB.

Gain Error

This is the difference between the actual analog output range and the ideal value from the data sheet. It speaks to the average slope of the characteristic. A gain error is only serious if it drifts or if you are relying on the theoretical value, ie no trimming. The gain most commonly changes with temperature and time. In addition the actual output is also dependent upon the stability of the reference input where one is used.

Zero Error

This is the difference from true zero when the appropriate code for zero output is given to the converter. Zero error is more likely to be a problem than gain error and can be particularly troublesome if it drifts.

Settling Time

The time taken from the change in the input code for the converter to come within some specified error band of the resulting output. Should be specified for a Full-scale change.

Glitch

A switching transient appearing on the output during a code transition. Its values can be expressed in volts or amps (as appropriate to voltage and current converters) and time or in terms of the charge transferred.

Power Supply Sensitivity

The change in output of the converter due to a change in the power supply voltage(s). Usually expressed as %full scale/% supply change.

Compliance and Output Resistance

These parameters (for current and voltage DACs respectively) determine how near they come to being ideal sources under differing load conditions. In practical design it is better to isolate any load changes from the DAC using a buffer amplifier and then, since the load is fixed, these parameters are folded into the gain error.

Accuracy

This is really a system specification since it actually specifies how the output relates to the absolute standard of voltage (or current). This involves the converter, the reference and the traceability of their various calibrations to an international standard. However even assuming that the full-scale is accurate, the accuracy of intermediate values depends upon the Integral Non-Linearity of the converter.

Resolution

The resolution of the DAC is the smallest output change which can be made in the output. This can be a hard thing to describe in general, but for a monotonic DAC with reasonable Integral Non-Linearity it is bound up with the Differential Non-Linearity.

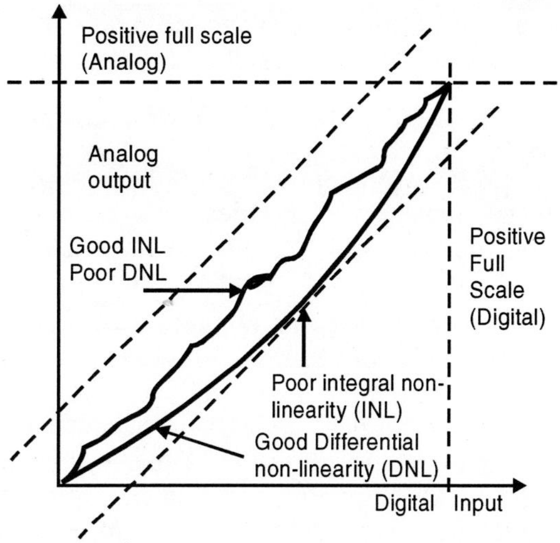

Fig. 8.120. Illustration of integral non-linearity and differential non-linearity in a digital-to-analog converter

Non-Linearity

Two types of non-linearity in the mapping of input code to output are recognised **Integral Non-Linearity (INL)** is the degree to which the line deviates from the ideal straight line through all the states (or more usefully the line joining the zero and full-scale points). Good INL means that the converter has good accuracy and a low distortion of a digital waveform being converted (ie if asked to convert a sine wave a converter with good INL will produce less harmonic content than one with a poor INL)

Differential Non-Linearity (DNL) is the degree to which the steps between the codes are uniform. Good DNL implies good resolution, because all the steps are almost exactly the same size, and a good noise performance.

Notice that good DNL does not mean good INL because there could be a systematic variation in the steps which produces a curve. Even though the deviation for each step is small, the effect is cumulative. However good INL tends to imply good, but not necessarily excellent, DNL.

Crudely speaking INL controls the "straightness" of the transfer characteristic whereas DNL controls the "smoothness".

INL and DNL are becoming more important as converters become more sophisticated. However at the time of writing not all manufacturers are good about distinguishing between them and providing the necessary data to assist you in your choices.

Transition Issues

The various converters have very different reactions if you move from one value to another and this is a point to watch if your equipment has a response time anything like the "settling time", i.e. the time that it takes a converter to settle to a new value. The stochastic and integrating (bidirectional) converters however are smooth in their transitions if a bit slow.

One serious word of caution about reading specifications of converters—always remember that all manufacturers have one eye on the competition and are keen to show you how their product is better than anybody else's. Always be sure that you know what the manufacturer means by the terms in his specification, the words may not mean the same thing to both of you, and always beware of specifications which are important but not quoted. If in doubt - test it before committing yourself and even if not in doubt, test it.

8.14. SENSORS OF PHYSICAL PARAMETERS

So far we have confined our attention in terms of sensors to the problem of analogue-to-digital conversion of voltages but there are a large number of other parameters which we need to measure which are not, per se, voltages. Examples such as shaft rotation, temperature, pressure and humidity come to mind. We therefore need to look seriously at the question of transducers.

Before looking at any methods of conversion in any detail, consider one or two significant points: In order to be used in a digital system any physical parameter has to be converted into a digital form and, unless the conversion can be done directly, an intermediate form has to be used. Direct digital conversion is always preferable where possible because it reduces the potential for error in the analogue systems. By far the most prevalent intermediate forms are voltage and time (current is assumed the same as voltage here as they are readily interchangeable). Of these two forms voltage can be measured to a very finite limit of resolution and accuracy whereas time can be measured to any conceivable resolution and accuracy, provided you have enough patience. The order of preference for conversion is therefore:

- direct conversion to digital form
- convert to time (or frequency)
- convert to voltage, current and then digitise.

8.14.1. Optical Motor Shaft Encoder

To control machines in industry, the microprocessor in these machines need information about the position, direction of rotation, speed of rotation of various shafts. The circuit which provides this information is called as **shaft encoders.** Two types of encoders are

(1) Absolute shaft encoders

(2) Incremental.

(1) Absolute Encoders : These encoders have a binary coded disk, on the rotating shaft light sections of the disk are transparent, and dark opaque. An LED is mounted on one side of each track and a photo transistor on the other side of each track, opposite the LED. Output from 4 photo transistors will produce one of the binary codes. The phototransistor output can be conditioned with schimit trigger buffers and connected to I/P port lines.

Each code represents an absolute angular position of the shaft in its rotation.

With a 4-bit disk, 360° is divided up into 16 ports so that the position of the shaft can be determined to 22.5°.

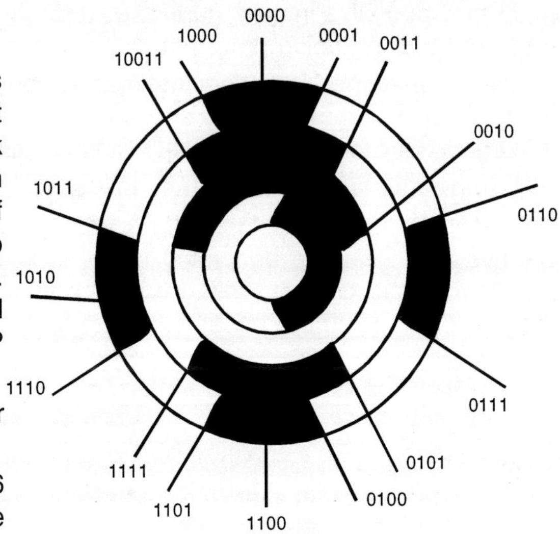

Fig. 8.121.

Stepper Motor

Device used to obtain an accurate position control of rotating shafts. A stepper motor employs rotation of its shaft in terms of steps rather than continuous rotation.

A sequence of pulses is needed to be applied to the windings of stepper motor to rotate the shaft. With a pulse applied to the winding I/P the rotor rotates by one teeth position or angle x.

$$x = 360°/\text{No. of rotor teeth}$$

Note. The number of pulse required for one complete rotation of the shaft of the stepper motor is equal to its number of internal teeth on its rotor.

A typical stepper motor has many parameters like torque 3 kg-cm, operating voltage 12 V, current rotating 1.2 A, step angle 1.8°. Hence 200 steps/revolution.

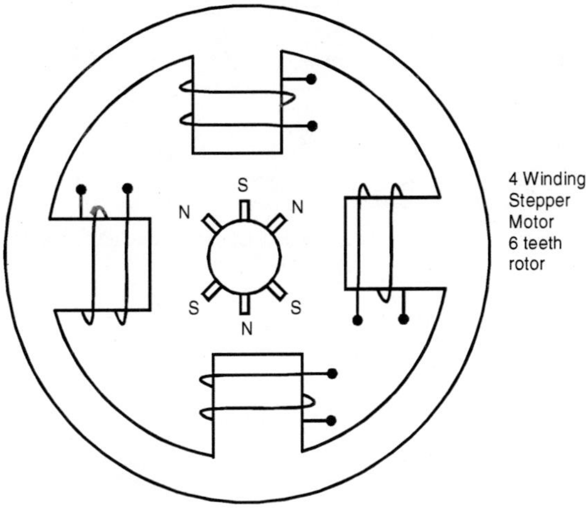

Fig. 8.122. 4 winding stepper motor 6 teeth rotor.

Winding Arrangement of Stepper Motor

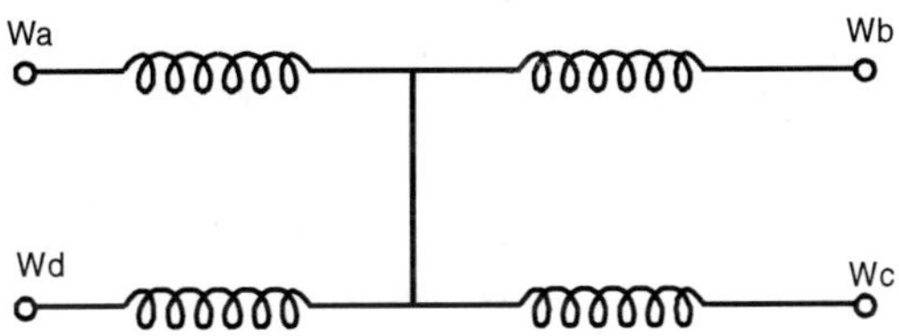

Fig. 8.123. 4 winding stepper motor 6 teeth rotor.

Interfacing Stepper Motor Winding

```
ALP    Mov AL, 80 H
       OUT CWR, AL
       Mov AL, 88H
       Mov Cx, 100
Again  out port A, AL
       CALL DELAY
       ROL AL, 01
       DEC CX
       JN2 Again
       MOV AH, 88H
       MOV CX, 100
Repeat OUT port A, AL
       Call delay
```

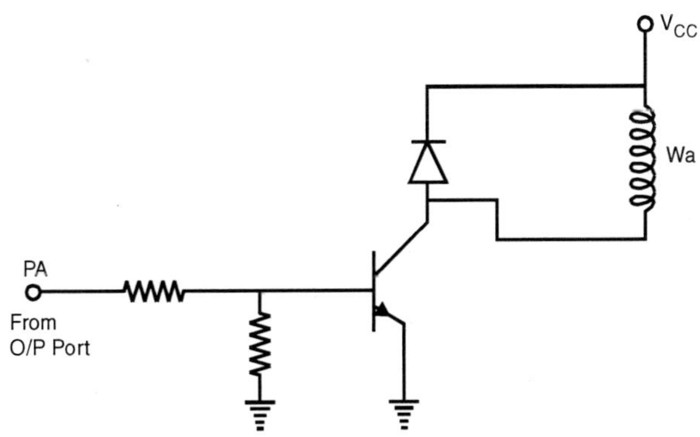

Fig. 8.124.

```
    ROR AL, 01
    DEC CX
    JNZ Repeat
    Mov AH, 4CH
    JNT 21H
    ENDS
    END
Delay Loop
    Mov CX, count
    DEC CX
    NOP
    JNZ Again
```

$$\text{Count} = \frac{\text{Number of rotor teeth}}{360}$$

$$\boxed{C = \frac{\text{No. of teeth rotor}}{360} \times 0}$$

A stepper motor is stepped from one position to the next by changing the currents through the fields in the motor.

4-Phase Stepper Motor Interface Circuit

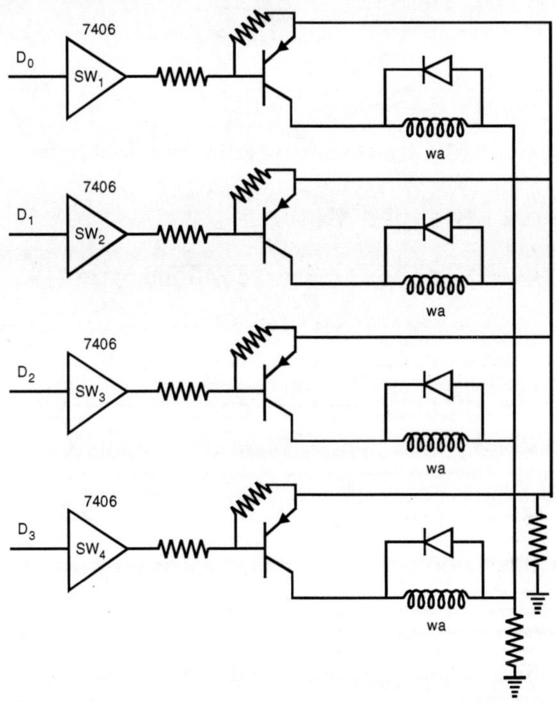

Fig. 8.125. 4-phase stepper motor interface circuit.

7406 buffers are inverting, hence a high on output port pin produces a low on buffer output. This low turns on PNP driver transistor and supplies current to the winding.

SW1 and S≅1 are on. Turning OFF SW2 and turning on SW4 will cause motor to rotate one step.

Changing to SW4 and SW3 on will cause the motor to rotate another 0° clockwise.

Changing to SW3 and SW2 on will cause another step.

Changing to SW2 and SW1 one will cause another step clockwise.

Step	Switch			
	Sw_4	Sw_3	Sw_2	Sw_1
1	0	0	1	1
2	1	0	0	1
3	1	1	9	9
4	0	1	1	9
1	0	0	1	1

Clockwise ↓ Counter Clockwise ↑

0011 → Initial
ROR 1001 → Next Step
0011 0011 → Initial → ROR → 1001 1001

8.14.2. Microcomputer based Process Control

Process control involves :

(1) Measuring system variables such as motor speed, temp, flow of reactants, level in tank, thickness of material etc.

The output of the controller then adjusts the value of each variable until it is equal to a predetermined value or set point.

The system controller must maintain each variable close to this **set point,** and must compensate quickly and accurately for any change in the variable caused.

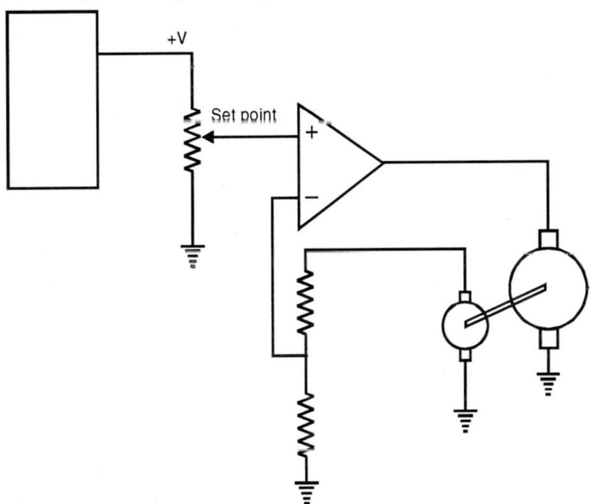

Fig. 8.126. Microcomputer based process control.

To the shaft of the motor a DC generator is attached which puts out a voltage directly α to the speed of the motor. The output voltage is typically a few volts per 1000 rpm.

A fraction of the output voltage from the tachometer is fed back to the inverting input of the power amplifier driving the motor.

A +ve voltage is applied to the non inverting input of the amplifier as a set point. When the power is turned on, the motor accelerates until the voltage fed back from the tachometer to the inverting input of the amplifier is nearly equal to the set-point voltage.

If the load on the motor is increased, the motor will initially slow down, and the voltage output from the tachometer will decrease. This will increase the differentiate in voltage between the input of the amplifier and cause it to drive more current to the motor. The increased current will increase the speed of the motor to nearly the speed it not before the increased load was added.

Block diagram of the Based Process Control

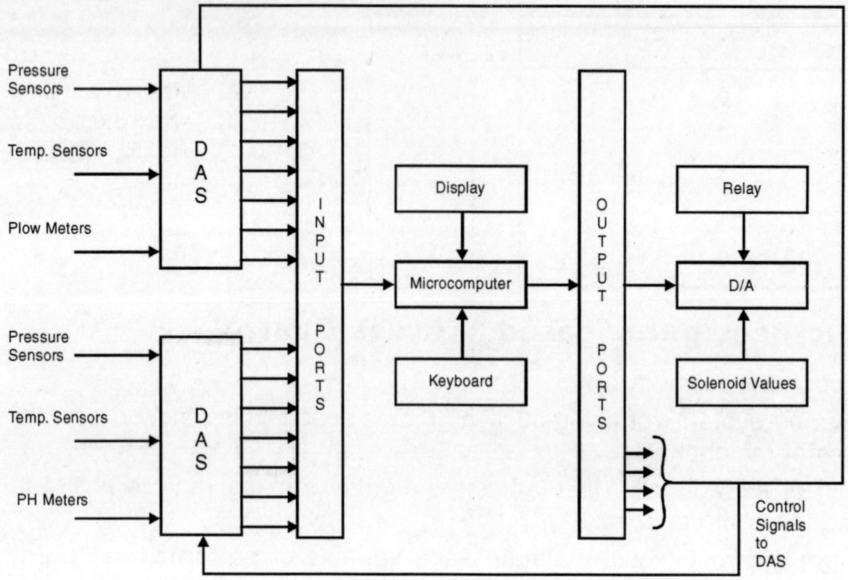

Fig. 8.127. Block diagram of process control.

Data Acquisition System (DAS) converts the analog signals from various sensors to digital values that can be read in and processed by the microprocessor.

A keyboard and display system allow the user to enter set point values, to read the current values of process variables, and to issue commands.

Relays, D/A and solenoid values are used to control process variables under program direction A programmable timer in the system determines the rate at which control loops are serviced.

8.14.3. Sensing Temperature

Let us now consider another major area of sensing - temperature. Virtually all measures of temperature use a form of a/d conversion of a voltage which is related to the temperature in some way and there are three major forms of sensor:

Thermocouples are the primary calibration source for most temperature measurements. If one junction of the thermocouple is kept at 273 K (well strictly speaking at the triple point of water —but the melting point of ice is more common) and the materials of the thermocouple are known then the voltage out as a function of temperature is known to a very high accuracy. However there are several problems with thermocouple systems which make them awkward to use :

1. They require a reference temperature of 273 K provided by an ice-bath. There are ways of getting around the problem but these render the primary calibration property a bit suspicious.
2. The output voltage is small (and how!). The maximum voltage that can be got out of a thermocouple is of order mV and in order to measure fractional degrees resolution to μV is required. Thus considerable analogue processing is required before the output can be digitised.
3. They require great care with materials in complex apparatus. This means keeping similar metal contacts at all joints, ensuring an absence of thermal gradients in connectors, etc.,
4. They are also slightly non-linear but that is just a matter of using an appropriate correction curve and is a slight inconvenience. Overall thermocouples are good primary standards

and not too good for general use. *Resistance thermometers* share some of the disadvantages of thermocouples but they do lose the requirement for a reference point. They have the advantage of being reproducible standards but have two disadvantages:

(i) They require the measurement of a small change of a large resistance. This is in contrast to a thermocouple where the requirement is to measure something which is small, period. They require both amplification and offsetting or an enormous dynamic range. Therefore they require considerable analogue processing.

(ii) The materials problem with an additional twist. Because we are measuring resistance we need to be wary of extra resistance in the circuit and that necessitates the use of a 4-wire system as shown here.

At this point it might be worth pointing out that a specialised form of a/d converter can work as a bridge circuit for measuring resistance directly. Notice that a reference voltage is not required for this circuit but a somewhat specialised form of comparator and ladder may be. This technique is sometimes used for resistance thermometry because of the very small changes in resistance being detected (temperature coefficient of resistance 0.0038/K).

The last category of temperature sensors is the *thermistor* type which have a large temperature-sensitive part of their resistance. Their advantage is that the signal is large, no reference is required but their disadvantage is their inaccuracy. They are also, in general, not interchangeable. The resistance vs temperature relation for a particular type of thermistor is

$$\frac{1}{T} = A + B\ln(R) + C(\ln(R))^3$$

which is non-linear—and how! If we put the thermistor into a circuit thus :

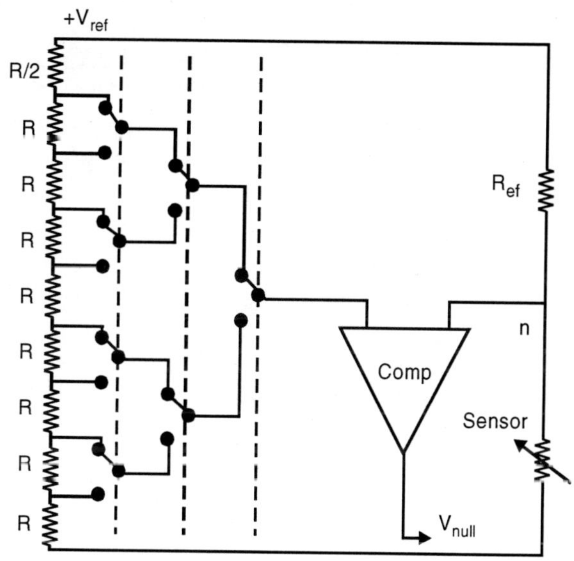

Fig. 8.128. Bridge-Nulling A/D Converter.

We shall have good resolution. Therefore we should always set up the circuit so that the operating range is in the centre. Calibration is tedious, using either a lot of measurements or a complex relationship, but once done then a table look-up or other "unkinker" will give a temperature readout. Note that in general thermistors are not interchangeable and therefore this must be done for each one. However some firms (for a price) do make interchangeable versions which make life easier.

8.14.4. Linearisation

Many sensors as we have seen are non-linear and it is very useful to be able to linearise the output so that subsequent processing can be simplified. There are many techniques for linearisation and I'll just put forward a few of the most useful.

Analytic equation : If the non-linear equation is known, then it can always be applied to the input to get the linear quantity back again. A simple example is the generation of the power in a heater by squaring the voltage drop across it.

Look-up Table : Good and fast if you can afford the storage. It has the advantage that it can cope with any non-linearity including non-analytic ones. However it can get tedious initially entering the values and the tables can get large if good resolution is required.

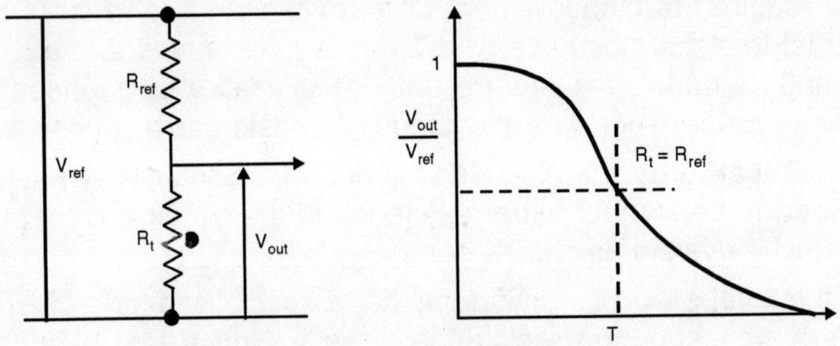

Fig. 8.129. Thermistor circuit and output.

Piece-wise Linearisation : A very acceptable technique for small non-linearities. Essentially the range is broken up into small enough sub-ranges so that a straight line in the sub-range is an adequate approximation. We've already seen a piece-wise approximation in the companding a/d converter. The advantages are that the precision can be good at a modest storage requirement and the speed is intermediate between the analytic solution and the lookup table.

REVIEW QUESTIONS

1. Define programmed I/O. What are the drawbacks of programmed I/O.
2. Explain interrupt driven I/O. What are the merits of it over programmed I/O.
3. Define polling of interrupts. How is this technique used to control the priority of interrupts ?
4. Explain, how the DMA transfer takes place in 8086 ?
5. With the help of diagram explain operation of 8237 DMA controller.
6. Define bus cycle, bus grant, and cycle stealing with reference to DMA.
7. Distinguish between serial and parallel communication.
8. Draw the block diagram of 8255A and briefly explain the function of each unit.
9. Explain BSR mode operation of 8255A.
10. Explain the interfacing of A/D and D/A converters to an 8255A.
11. Why the data are sent to a printer on a handshake basis ?
12. The system base address of an 8255A is FFF9H. What are the system addresses for the three ports and the control register of this 8255A ?
13. Write the mode set control word needed to intialize an 8255A as : Port A – handshake input, Port B – handshake output, and port C – bits PC6 and PC7 as outputs.
14. Explain handshaking between a printer and centronics parallel port interface.
15. What do you mean by asynchronous and synchronous communication ?
16. Draw the block diagram of an 8251A USART, and explain each unit.
17. Distinguish between UART and USART.
18. Explain, how synchronous serial data communication is more efficient than asychronous communication.
19. "Two ground pins on an RS-232C connector are not jumpered together." Give your logic.
20. Write the mode word and command word to initialize an 8251A for :

    (a) Baud rate factor of 64 (b) 7 bits/character

(c) Even parity

(d) One stop bit

(e) Transmit and receive interrupt enabled

(f) $\overline{DTR}$ and $\overline{RTS}$ asserted

(g) Error flags reset

(h) No hunt mode and no break character.

21. Why do we need to use 8254 in a microprocessor based system ?

22. Draw the block diagram of 8254, and explain each unit in brief.

23. How 8254 can be used to generate square wave ?

24. Name the operating modes of 8254. How can 8254 be used to generate a negative pulse of time equal to one clock period ?

25. What do you mean by read-back the count from a counter of 8254 ?

26. Why are A/D and D/A conversions required in a digital system ?

27. Define A/D converter. Name few A/D converters.

28. Explain successive approximation A/D converter.

29. Define digital-to-analog conversion. Explain any D/A converter.

30. How 8259A can be used for multiple interrupts priority management ?

31. "An IRET instruction rather than the RET instruction is used at the end of an interrupt service procedure" from an interfacing device ?

32. How an 8086 will respond if it receives an NMI interrupt signal, if it is responding an interrupt on INTR input from an interfacing device ?

33. Draw and briefly explain the internal architecture of 8259A programmable interupt controller.

34. How 8259A devices are cascaded, briefly explain with diagram ?

❑❑❑

CHAPTER 9

THE 80286/287 MICROPROCESSORS

INSIDE THIS CHAPTER

9.1. INTRODUCTION

The 80286 is a new VLSI microprocessor system with exceptional capabilities for supporting large-system applications. Based on a new-generation CPU (the Intel 80286), this powerful micro system is designed to support multi-user reprogrammable and real-time multitasking applications. Its dedicated system support circuits simplify system hardware; sophisticated hardware and software tools reduce both the time and the cost of product development.

The 80286 is a virtual-memory microprocessor with on-chip memory management and protection. The 80286 micro system offers a total-solution approach, enabling you to develop high-speed, interactive, multi-user, multitasking and multiprocessor systems more rapidly and at higher performance than ever before.

(1) Reliability and system up-time are becoming increasingly important in all applications. Information must be protected from misuse or accidental loss. The 80286 includes a sophisticated and flexible four-level protection mechanism that isolates layers of operating system programs from application programs to maintain a high degree of system integrity.

(2) The 80286 provides 16 megabytes of physical address space to support today's application requirements. This large physical memory enables the 80286 to keep many large programs and data structures simultaneously in memory for high-speed access.

(3) For applications with dynamically changing memory requirements, such as multi-user business systems, the 80286 CPU provides on-chip memory management and virtual memory support. On an 80286-based system, each user can have up to a gigabyte (2^(30) bytes) of virtual-address space. This large address space virtually eliminates restrictions on the number or size of programs that may be part of the system.

(4) Large multi-user or real-time multitasking systems are easily supported by the 80286. High-performance features, such as a very high-speed task switch, fast interrupt-response time, inter-task protection, and a quick and direct operating system interface, make the 80286 highly suited to multi-user/multitasking applications.

(5) The 80286 has two operating modes : Real-Address mode and Protected-Address mode. In Real-Address mode, the 80286 is fully compatible with the 8086, 8088, 80186, and 80188 microprocessors; all of the extensive libraries of 8086 and 8088 software execute four to six times faster on the 80286, without any modification.

(6) In Protected-Address mode, the advanced memory management and protection features of the 80286 become available, without any reduction in performance. Upgrading 8086 and 8088 application programs to use these new memory management and protection features usually requires only reassembly or recompilation (some programs may require minor modification). This compatibility between 80286 and 8086 processor families reduces both the time and the cost of software development.

The 80286 is the most powerful 16-bit processor in the 8086 series of microprocessors, which includes the 8086, the 8088, the 80186, the 80188, and the 80286. It is designed for applications that require very high performance. It is also an excellent choice for sophisticated "high end" applications that will benefit from its advanced architectural features :

(1) Memory management

(2) protection mechanisms

(3) task management

(4) Virtual memory support.

The 80286 provides, on a single VLSI chip, computational and architectural characteristics normally associated with much larger minicomputers. Because the 80286 represents an extension of the 8086 architecture, some of this overview material may be new and unfamiliar to previous users of the 8086 and similar microprocessors. But the 80286 is also an evolutionary development, with the new architecture superimposed upon the industry standard 8086 in such a way as to affect only the design and programming of operating systems and other such system software.

9.1.1. General Attributes

The 80286 base architecture has many features in common with the architecture of other members of the 8086 family, such as byte addressable memory, I/O interfacing hardware, interrupt vectoring, and support for both multiprocessing and processor extensions. The entire family has a common set of addressing modes and basic instructions. The 80286 base architecture also includes a number of extensions which add to the versatility of the computer.

The 80286 processor can function in two modes of operation . In one of these modes only the base architecture is available to programmers, whereas in the other mode a number of very powerful advanced features have been added, including support for virtual memory, multitasking, and a sophisticated protection mechanism.

The 80286 base architecture was designed to support programming in high-level languages, such as Pascal, C or PL/M. The register set and instructions are well suited to compiler-generated code. The addressing modes allow efficient addressing of complex data structures, such as static and dynamic arrays, records, and arrays within records, which are commonly supported by high-level languages. The data types supported by the architecture include, along with bytes and words, high level language constructs such as strings, BCD, and floating point.

The memory architecture of the 80286 was designed to support modular programming techniques. Memory is divided into segments, which may be of arbitrary size, that can be used to contain procedures and data structures.

Segmentation has several advantages over more conventional linear memory architectures. It supports structured software, since segments can contain meaningful program units and data, and more compact code, since references within a segment can be shorter (and locality of reference usually insures that the next few references will be within the same segment). Segmentation also lends itself to efficient implementation of sophisticated memory management, virtual memory, and memory protection.

In addition, new instructions have been added to the base architecture to give hardware support for procedure invocations, parameter passing, and array bounds checking.

9.1.2. Modes of Operation

The 80286 can be operated in either of two different modes : Real Address Mode or Protected Virtual Address Mode (also referred to as Protected Mode). In either mode of operation, the 80286 represents an upwardly compatible addition to the 8086 family of processors.

In Real Address Mode, the 80286 operates essentially as a very high-performance 8086. Programs written for the 8086 or the 80186 can be executed in this mode without any modification. Such upward compatibility extends even to the object code level; for example, an 8086 program stored in read-only memory will execute successfully in 80286 Real Address Mode. An 80286 operating in Real Address Mode provides a number of instructions not found on the 8086. These additional instructions, also present with the 80186, allow for efficient subroutine linkage, parameter validation, index calculations, and block I/O transfers.

The advanced architectural features and full capabilities of the 80286 are realized in its native Protected Mode. Among these features are sophisticated mechanisms to support data protection, system integrity, task concurrency, and memory management, including virtual storage. Nevertheless, even in Protected Mode, the 80286 remains upwardly compatible with most 8086 and 80186 application programs. Most 8086 applications programs can be re-compiled or re-assembled and executed on the 80286 in Protected Mode.

9.1.3. Advanced Features

In addition to these common features, Protected Mode provides a number of advanced features, including a greatly extended physical and logical address space, new instructions, and support for additional hardware-recognized data structures. The Protected Mode 80286 includes a sophisticated memory management and multilevel protection mechanism. Full hardware support is included for multitasking and task switching operations.

9.1.4. Memory Management

The memory architecture of the Protected Mode 80286 represents a significant advance over that of the 8086. The physical address space has been increased from 1 megabyte to 16 megabytes (2^(24) bytes), while the virtual address space (i.e., the address space visible to a program) has been increased from 1 megabyte to 1 gigabyte (2^(30) bytes). Moreover, separate virtual address spaces are provided for each task in a multi-tasking system.

The 80286 supports on-chip memory management instead of relying on an external memory management unit. The one-chip solution is preferable because no software is required to manage an external memory management unit, performance is much better, and hardware designs are significantly simpler. Mechanisms have been included in the 80286 architecture to allow the efficient implementation of virtual memory systems. (In virtual memory systems, the user regards the combination of main and external storage as a single large memory. The user can write large programs without worrying about the physical memory limitations of the system.

To accomplish this, the operating system places some of the user programs and data in external storage and brings them into main memory only as they are needed.) All instructions that can cause a segment-not-present fault are fully restartable. Thus, a not-present segment can be loaded from external storage, and the task can be restarted at the point where the fault occurred.

The 80286, like all members of the 8086 series, supports a segmented memory architecture. The 80286 also fully integrates memory segmentation into a comprehensive protection scheme. This protection scheme includes hardware-enforced length and type checking to protect segments from inadvertent misuse.

9.1.5. Task Management

The 80286 is designed to support multi-tasking systems. The architecture provides direct support for the concept of a task. For example, task state segments are hardware recognized and hardware-manipulated structures that contain information on the current state of all tasks in the system. Very efficient context-switching (task-switching) can be invoked with a single instruction. Separate logical address spaces are provided for each task in the system. Finally, mechanisms exist to support inter task communication, synchronization, memory sharing, and task scheduling.

9.1.6. Protection Mechanisms

The 80286 allows the system designer to define a comprehensive protection policy to be applied, uniformly and continuously, to all ongoing operations of the system. Such a policy may be desirable to ensure system reliability, privacy of data, rapid error recovery, and separation of multiple users.

The 80286 protection mechanisms are based on the notion of a “hierarchy of trust.” Four privilege levels are distinguished, ranging from Level 0 (most trusted) to Level 3 (least trusted). Level 0 is usually reserved for the operating system kernel. The four levels may be visualized as concentric rings, with the most privileged level in the center.

This four-level scheme offers system reliability, flexibility, and design options not possible with the typical two-level (supervisor/user) separation provided by other processors. A four-level division is capable of separating kernel, executive, system services, and application software, each with different privileges.

At any one time, a task executes at one of the four levels. Moreover, all data segments and code segments are also assigned to privilege levels. A task executing at one level cannot access data at a more privileged level, nor can it call a procedure at a less privileged level (i.e., trust a less privileged procedure to do work for it). Thus, both access to data and transfer of control are restricted in appropriate ways.

A complete separation can exist between the logical address spaces local to different tasks, providing users with automatic protection against accidental or malicious interference by other users. The hardware also provides immediate detection of a number of fault and error conditions, a feature that can be useful in the development and maintenance of software.

Finally, these protection mechanisms require relatively little system overhead because they are integrated into the memory management and protection hardware of the processor itself.

9.1.7. Support for Operating Systems

Most operating systems involve some degree of concurrency, with multiple tasks vying for system resources. The task management mechanisms described above provide the 80286 with inherent support for such multi-tasking systems. Moreover, the advanced memory management features of the 80286 allow the implementation of sophisticated virtual memory systems.

Operating system implementors have found that a multi-level approach to system services provides better security and more reliable systems. For example, a very secure kernel might implement critical functions such as task scheduling and resource allocation, while less fundamental functions (such asI/O) are built around the kernel. This layered approach also makes program development and enhancement simpler and facilitates error detection and debugging. The 80286 supports the layered approach through its four-level privilege scheme.

9.2. MEMORY ORGANIZATION AND SEGMENTATION

The main memory of an 80286 system makes up its physical address space.

This address space is organized as a sequence of 8-bit quantities, called bytes. Each byte is assigned a unique address ranging from 0 up to a maximum of 2^(20) (1 megabyte) in Real Address Mode, and up to 2^(24) (16 megabytes) in Protected Mode.

A virtual address space is the organization of memory as viewed by a program. Virtual address space is also organized in units of bytes. In Real Address Mode, as with the 8086 itself, programs view physical memory directly, inasmuch as they manipulate pure physical addresses. Thus, the virtual address space is identical to the physical address space (1 megabyte).

In Protected Mode, however, programs have no direct access to physical addresses. Instead, memory is viewed as a much larger virtual address space of 2^(30) bytes (1 gigabyte). This 1 gigabyte virtual address is mapped onto the Protected Mode's 16-megabyte physical address space by the address translation mechanisms.

The programmer views the virtual address space on the 80286 as a collection of up to sixteen thousand linear subspaces, each with a specified size or length. Each of these linear address spaces is called a segment. A segment is a logical unit of contiguous memory. Segment sizes may range from one byte up to 64K (65,536) bytes.

80286 memory segmentation supports the logical structure of programs and data in memory. Programs are not written as single linear sequences of instructions and data, but rather as modules of code and data. For example, program code may include a main routine and several separate procedures. Data may also be organized into various data structures, some private and some shared with other programs in the system. Run-time stacks constitute yet another data requirement. Each of these several modules of code and data, moreover, may be very different in size or vary dynamically with program execution. Segmentation supports this logical structure. Each meaningful module of a program may be separately contained in individual segments. The degree of modularization, of course, depends on the requirements of a particular application. Use of segmentation benefits almost all applications. Programs execute faster and require less space. Segmentation also simplifies the design of structured software.

9.2.1. Registers

The 80286 contains a total of fourteen registers that are of interest to the application programmer. As shown in Fig. 9.1, these registers may be grouped into four basic categories :

(a) **General registers.** These eight 16-bit general-purpose registers are used primarily to contain operands for arithmetic and logical operations.

(b) **Segment registers.** These four special-purpose registers determine, at any given time, which segments of memory are currently addressable.

(c) **Status and Control registers.** These three special-purpose registers are used to record and alter certain aspects of the 80286 processor state.

9.2.2. General Registers

The general registers of the 80286 are the 16-bit registers AX, BX, CX, DX, SP, BP, SI, and DI. These registers are used interchangeably to contain the operands of logical and arithmetic operations. Some instructions and addressing modes, however, dedicate certain general registers to specific uses. BX and BP are often used to contain the base address of data structures in memory (for example, the starting address of an array); for this reason, they are often referred to as the base registers. Similarly, SI and DI are often used to contain an index value that will be incremented to step through a data structure; these two registers are called the index registers. Finally, SP and BP are used for stack manipulation. Both SP and BP normally contain offsets into the current stack. SP generally contains the offset of the top of the stack and BP contains the offset or base address of the current stack frame.

As shown in Fig. 9.1, eight byte registers overlap four of the 16-bit general registers. These registers are named AH, BH, CH, and DH (high bytes); and AL, BL, CL, and DL (low bytes); they overlap AX, BX, CX, and DX. These registers can be used either in their entirety or as individual 8-bit registers. This dual interpretation simplifies the handling of both 8- and 16-bit data elements.

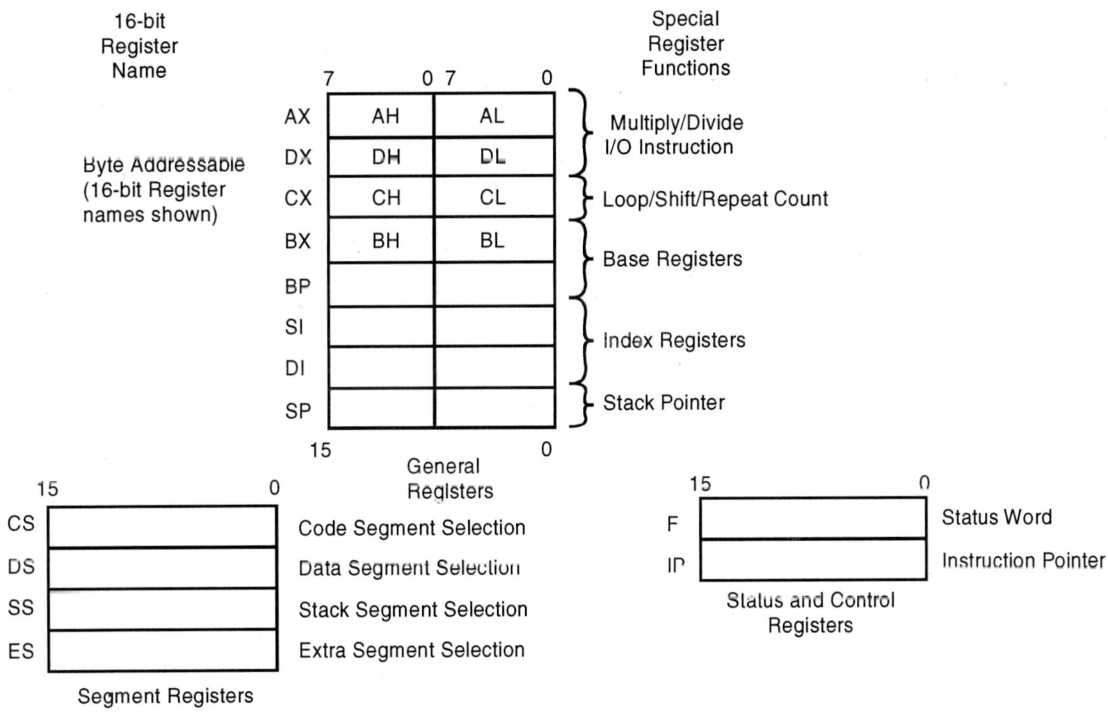

Fig. 9.1. Register Set of 80286.

9.2.3. Memory Segmentation and Segment Registers

Complete programs generally consist of many different code modules (or segments), and different types of data segments. However, at any given time during program execution, only a small subset of a program's segments are actually in use. Generally, this subset will include code, data, and possibly a stack. The 80286 architecture takes advantage of this by providing mechanisms to support direct access to the working set of a program's execution environment and access to additional segments on demand.

At any given instant, four segments of memory are immediately accessible to an executing 80286 program. The segment registers DS, ES, SS, and CS are used to identify these four current segments. Each of these registers specifies a particular kind of segment, as characterized by the associated mnemonics ("code," "stack," "data," or "extra") shown in Fig. 9.1.

An executing program is provided with concurrent access to the four individual segments of memory "a code segment, a stack segment, and two data segments" by means of the four segment registers. Each may be said to select a segment, since it uniquely determines the one particular segment from among the numerous segments in memory, which is to be immediately accessible at highest speed. Thus, the 16-bit contents of a segment register is called a segment selector.

Once a segment is selected, a base address is associated with it. To address an element within a segment, a 16-bit offset from the segment's base address must be supplied. The 16-bit segment selector and the 16-bit offset taken together form the high and low order halves, respectively, of a 32-bit virtual address pointer. Once a segment is selected, only the lower 16-bits of the pointer, called the offset, generally need to be specified by an instruction. Simple rules define which segment register is used to form an address when only a 16-bit offset is specified.

An executing program requires, first of all, that its instructions reside somewhere in memory. The segment of memory containing the currently executing sequence of instructions is known as the current code segment; it is specified by means of the CS register. All instructions are fetched from this code segment, using as an offset the contents of the instruction pointer (IP). The CS :IP register combination therefore forms the full 32-bit pointer for the next sequential program instruction. The CS register is manipulated indirectly. Transitions from one code segment to another (e.g., a procedure call) are effected implicitly as the result of control-transfer instructions, interrupts, and trap operations.

Stacks play a fundamental role in the 80286 architecture; subroutine calls, for example, involve a number of implicit stack operations. Thus, an executing program will generally require a region of memory for its stack. The segment containing this region is known as the current stack segment, and it is specified by means of the SS register. All stack operations are performed within this segment, usually in terms of address offsets contained in the stack pointer (SP) and stack frame base (BP) registers. Unlike CS, the SS register can be loaded explicitly for dynamic stack definition.

Beyond their code and stack requirements, most programs must also fetch and store data in memory. The DS and ES registers allow the specification of two data segments, each addressable by the currently executing program. Accessibility to two separate data areas supports differentiation and access requirements like local procedure data and global process data. An operand within a data segment is addressed by specifying its offset either directly in an instruction or indirectly via index and/or base registers.

Depending on the data structure (e.g., the way data is parceled into one or more segments), a program may require access to multiple data segments. To access additional segments, the DS and ES registers can be loaded under program control during the course of a program's execution. This simply requires loading the appropriate data pointer prior to accessing the data.

The interpretation of segment selector values depends on the operating mode of the processor. In Real Address Mode, a segment selector is a physical address. In Protected Mode, a segment selector selects a segment of the user's virtual address space. An intervening level of logical-to-physical address translation converts the logical address to a physical memory address. In general, considerations of selector formats and the details of memory mapping need not concern the application programmer.

9.2.4. Index, Pointer, and Base Registers

Five of the general-purpose registers are available for offset address calculations. These five registers, shown in figure 9.1, are SP, BP, BX, SI, and DI. SP is called a pointer register; BP and BX are called base registers; SI and DI are called index registers.

As described in the previous section, segment registers define the set of four segments currently addressable by a program. A pointer, base, or index register may contain an offset value relative to the start of one of these segments; it thereby points to a particular operand's location within that segment. To allow for efficient computations of effective address offsets, all base and index registers may participate interchangeably as operands in most arithmetical operations.

Stack operations are facilitated by the stack pointer (SP) and stack frame base (BP) registers. By specifying offsets into the current stack segment, each of these registers provides access to data on the stack. The SP register is the customary top-of-stack pointer, addressing the uppermost datum on a push-down stack. It is referenced implicitly by PUSH and POP operations, subroutine calls, and interrupt operations. The BP register provides yet another offset into the stack segment. The existence of this stack relative base register, in conjunction with certain addressing modes, is particularly useful for accessing data structures, variables and dynamically allocated work space within the stack.

Stacks in the 80286 are implemented in memory and are located by the stack segment register (SS) and the stack pointer register (SP). A system may have an unlimited number of stacks, and a stack may be up to 64K bytes long, the maximum length of a segment.

One stack is directly addressable at a time; this is the current stack, often referred to simply as "the" stack. SP contains the current top of the stack (TOS). In other words, SP contains the offset to the top of the push down stack from the stack segment's base address. Note, however, that the stack's base address (contained in SS) is not the "bottom" of the stack.

80286 stack entries are 16 bits wide. Instructions operate on the stack by adding and removing stack items one word at a time. An item is pushed onto the stack by decrementing SP by 2 and writing the item at the new TOS. An item is popped off the stack by copying it from TOS and then incrementing SP by 2. In other words, the stack grows down in memory toward its base address. Stack operations never move items on the stack; nor do they erase them. The top of the stack changes only as a result of updating the stack pointer.

The stack frame base pointer (BP) is often used to access elements on the stack relative to a fixed point on the stack rather than relative to the current TOS. It typically identifies the base address of the current stack frame established for the current procedure. If an index register is used relative to BP (e.g., base + index addressing mode using BP as the base), the offset will be calculated automatically in the current stack segment.

Accessing data structures in data segments is facilitated by the BX register, which has the same function in addressing operands within data segments that BP does for stack segments. They are called base registers because they may contain an offset to the base of a data structure.

Operations on data are also facilitated by the SI and DI registers. By specifying an offset relative to the start of the currently addressable data segment, an index register can be used to address an operand in the segment. If an index register is used in conjunction with the BX base register (i.e., base + index addressing) to form an offset address, the data is also assumed to reside in the current data segment. As a rule, data referenced through an index register or BX is presumed to reside in the current data segment. That is, if an instruction invokes addressing for one of its operands using either BX, DI, SI, or BX with SI or DI, the contents of the register(s) (BX, DI, or SI) implicitly specify an offset in the current data segment. As previously mentioned, data referenced via SP, BP or BP with SI or DI implicitly specify an operand in the current stack segment.

There are two exceptions to the rules listed above. The first concerns the operation of certain 80286 string instructions. For the most flexibility, these instructions assume that the DI register addresses destination strings not in the data segment, but rather in the extra segment (ES register). This allows movement of strings between different segments. This has led to the descriptive names "source index" and "destination index." In all cases other than string instructions, however, the SI and DI registers may be used interchangeably to reference either source or destination operands.

A second more general override capability allows the programmer complete control of which segment is used for a specific operation. Segment-override prefixes, allow the index and base registers to address data in any of the four currently addressable segments.

Implied Segment Usage by Index, Pointer, and Base Registers.

Register	Implied Segment
SP	SS
BP	SS
BX	DS
DI	DS, ES for String Operations
BP + SI, DI	SS
BX + SI, DI	DS

Note. All implied Segment usage, except SP to SS and DI to ES for String Operations, may be explicitly specified with a segment override prefix for any of the four segments. The prefix precedes the instruction for which explicit reference is desired.

9.2.6. Status and Control Registers

Two status and control registers are of immediate concern to applications programmers : the instruction pointer and the FLAGS registers. The instruction pointer register (IP) contains the offset address, relative to the start of the current code segment, of the next sequential instruction to be executed. Together, the CS :IP registers thus define a 32-bit program-counter. The instruction pointer is not directly visible to the programmer; it is controlled implicitly, by interrupts, traps, and control-transfer operations.

The FLAGS register encompasses eleven flag fields, mostly one-bit wide as shown in Fig. 9.2, Six of the flags are status flags that record processor status information. The status flags are affected by the execution of arithmetic and logical instructions. The carry flag is also modifiable with instructions that will clear, set or complement this flag bit. The carry flag (CF) generally indicates a carry or borrow out of the most significant bit of an 8- or 16-bit operand after performing an arithmetic operation; this flag is also useful for bit manipulation operations involving the shift and rotate instructions.

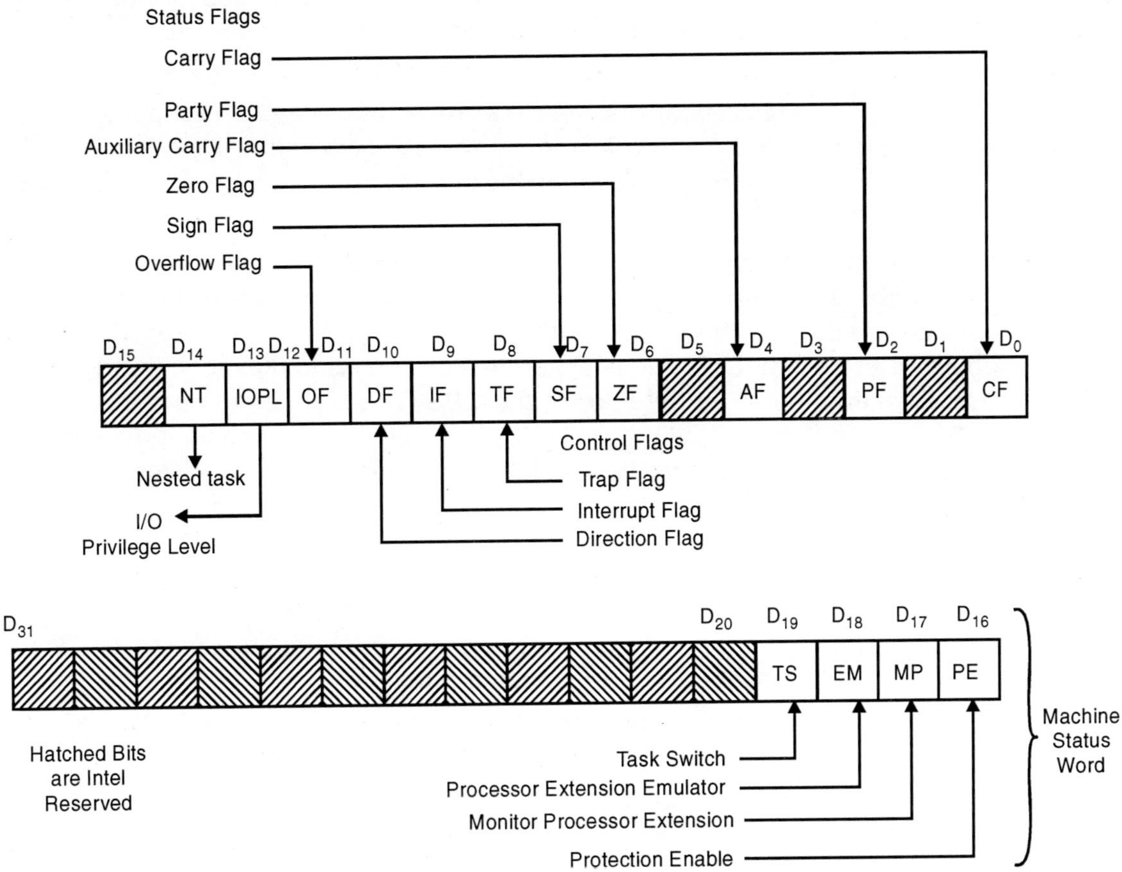

Fig. 9.2. 80286 Flag Register.

The effect on the remaining status flags, when defined for a particular instruction, is generally as follows :

(a) The zero flag (ZF) indicates a zero result when set.

(b) The sign flag (SF) indicates whether the result was negative (SF=1) or positive (SF=0); when set,

(c) The overflow flag (OF) indicates whether an operation results in a carry into the high order bit of the result but not a carry out of the high-order bit, or vice versa

(d) The parity flag (PF) indicates whether the modulo 2 sum of the low-order eight bits of the operation is even (PF=0) or odd (PF=1) parity. The auxiliary carry flag (AF) represents a carry out of or borrow into the least significant 4-bit digit when performing binary coded decimal (BCD) arithmetic. The FLAGS register also contains three control flags that are used, under program control, to direct certain processor operations.

(e) The interrupt-enable flag (IF), if set, enables external interrupts; otherwise, interrupts are disabled.

(f) The trap flag (TF), if set, puts the processor into a single-step mode for debugging purposes where the target program is automatically interrupted to a user supplied debug routine after the execution of each target program instruction.

(g) The direction flag (DF) controls the forward or backward direction of string operations : 0 = forward or auto increment the address register(s) (SI, DI or SI and DI), 1 = backward or auto-decrement the address register(s) (SI, DI or SI and DI).

In general, the interrupt enable flag may be set or reset with special instructions (STI = set, CLI = clear) or by placing the flags on the stack, modifying the stack, and returning the flag image from the stack to the flag register. If operating in Protected Mode, the ability to alter the IF bit is subject to protection checks to prevent non-privileged programs from effecting the interrupt state of the CPU. This applies to both instruction and stack options for modifying the IF bit. The TF flag may only be modified by copying the flag register to the stack, setting the TF bit in the stack image, and returning the modified stack image to the flag register. The trap interrupt occurs on completion of the next instruction. Entry to the single step routine saves the flag register on the stack with the TF bit set, and resets the TF bit in the register. After completion of the single step routine, the TF bit is automatically set on return to the program being single stepped to interrupt the program again after completion of the next instruction. Use of TF is not inhibited by the protection mechanism in Protected Mode.

The DF flag, like the IF flag, is controlled by instructions (CLD = clear, STD = set) or flag register modification through the stack. Typically, routines that use string instructions will save the flags on the stack, modify DF as necessary via the instructions provided, and restore DF to its original state by restoring the Flag register from the stack before returning. Access or control of the DF flag is not inhibited by the protection mechanism in Protected Mode.

The Special Fields bits are only relevant in Protected Mode. Real Address Mode programs should treat these bits as don't-care's, making no assumption about their status. Attempts to modify the IOPL and NT fields are subject to protection checking in Protected Mode. In general, the application's programmer will not be able to and should not attempt to modify these bits.

9.3. INTERNAL ARCHITECTURE OF 80286

The internal architecture of 80286 can be divided into four functional parts as stated below :

(1) Address Unit (AU)
(2) Bus Unit (BU)
(3) Instruction Unit (IU)
(4) Execution Unit (EU)

The block diagram shown in the Fig. 9.3 illustrates the overall architecture of 80286.

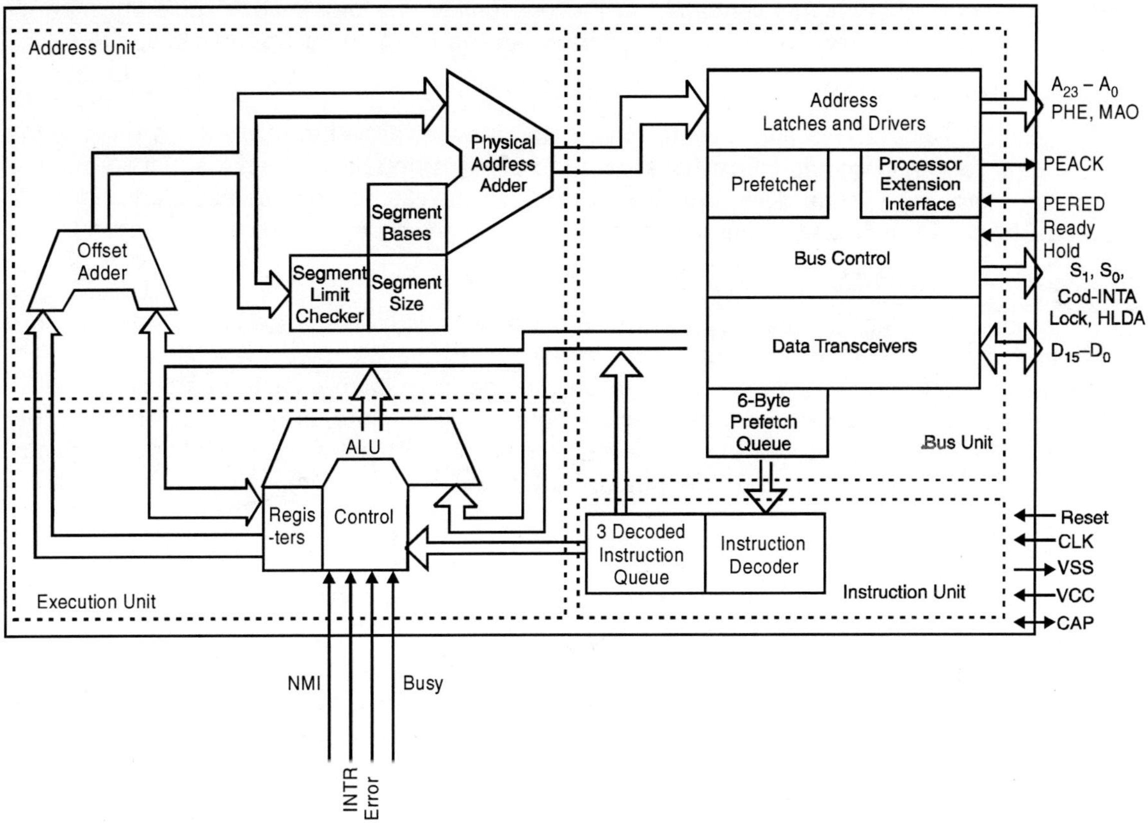

Fig. 9.3. Internal architecture of 80286.

The functions of different blocks of the 80286 are as stated below :

Address Unit (AU)

The address unit is responsible for calculating the physical address of instructions and data that the CPU wants to access. Also the address lines derived by this unit may be used to address different peripherals.

Bus Unit (BU)

The physical address computed by the address unit is handed over to the bus unit (BU) of the 80286. The address latches and drivers in the bus unit transmit the physical address thus formed over the address bus A0- A23. One major function of the bus unit is to fetch the instruction bytes from the memory. In fact, the instructions are fetched in advance and stored in a queue to enable faster execution of the instructions. This concept is known as instruction pipelining. Thus for fetching the next instruction, the CPU need not wait till the completion of execution of the previous instruction. Rather, when one instruction is getting executed, the subsequent instruction is being pre-fetched, decoded and kept ready for execution. The bus unit also contains a bus control module that controls the pre-fetcher module. These fetched instructions are arranged in a 6-Byte prefetch Queue. Thus usually the CPU prefetches the instructions, to enhance the speed of execution.

But in case of a conditional branch, depending upon the success of the condition, the pre-fetched instructions will be flushed out of the queue and further pre-fetching may be carried out, if required. Another major module in the bus unit is the processor extension interface module which takes care of communication between the CPU and a Coprocessor.

Instruction Unit (IU)

The 6-Byte prefetch queue forwards the instructions arranged in it to the instruction unit. The instruction unit accepts instructions from the pre-fetch queue and an instruction decoder decodes them one by one. The decoded instructions are latched onto a decoded instruction queue. The data transreceivers interface and control the internal data bus with the system bus.

Execution Unit (EU)

The output of the decoding circuit drives a control circuit in the execution unit, which is responsible for executing the instructions received from the decoded instruction queue. The decoded instruction queue sends the data part of the instruction over the data bus. The EU contains the register bank, used for storing the data as scratch pad, or used as special purpose registers. The ALU, the heart of the EU, carries out all the arithmetic and logical operations and sends the results either over the data bus or back to the register bank.

9.3.1. Signal Descriptions of 80286

80286 is available in 68-pin LCC (Lead Less Chip carrier), 68-pin PLCC (Plastic Leaded Chip Carrier) and 68-pin PGA (Pin Grid Array) packages. There is no differences in pin allotments of PLCC and LCC packages only except in PLCC.

The Fig. 9.4 (a) and (b) for PLCC/LCC and PGA packages describe the pin configuration for 80286 processor.

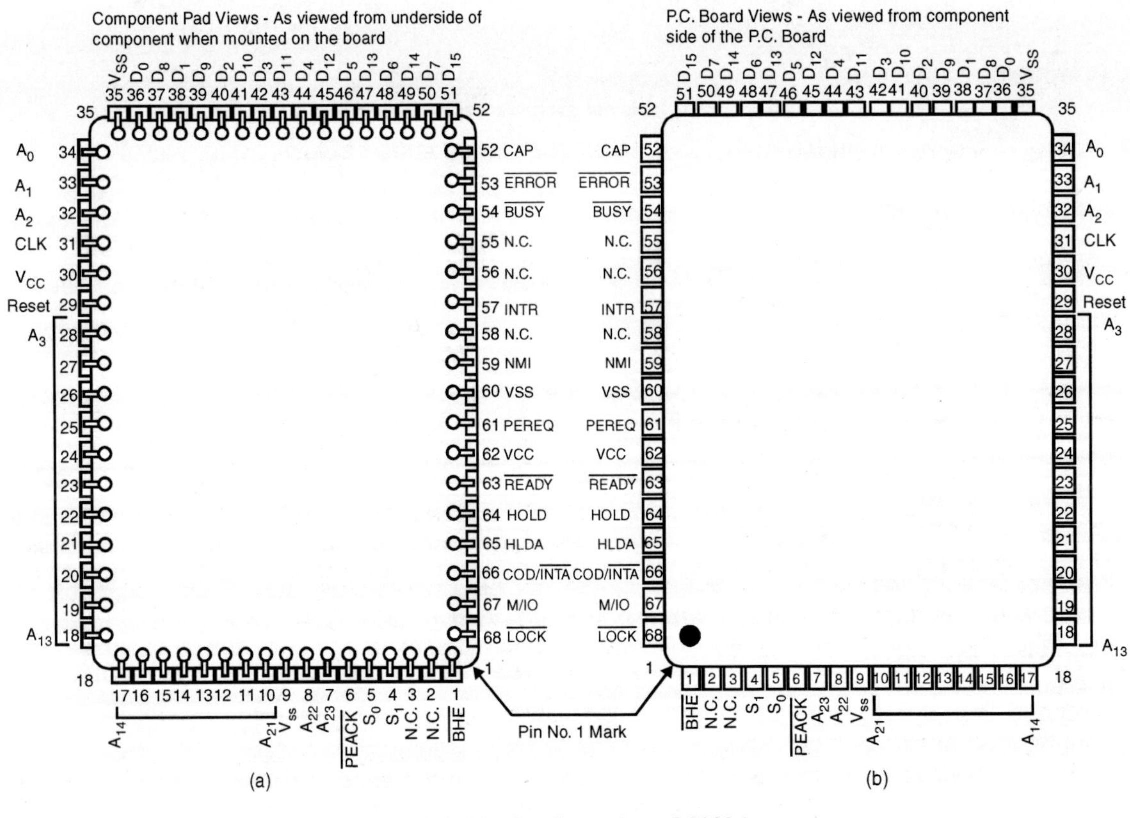

Fig. 9.4. Pin Configuration of 80286.

The signal description for 80286 processor is briefly explained below :

(1) **CLK :** This is the system clock input pin. The clock frequency applied ate this pin is divided by two, internally and is used for deriving fundamental timings for basic operations of the circuit. The clock is generated using 8284 clock generator.

(2) **D_{15}-D_0 :** These are the bidirectional sixteen bit data bus lines for 80286.

(3) **A_{23}-A_0 :** These are the physical address output lines used to address memory or I/O devices.

(4) **$\overline{BHE}$:** This output signal, as in 8086 processor, indicates that there is a transfer of data on the higher order data bus **D_{15}-D_8.**

(5) **$\overline{S_1}$, $\overline{S_0}$:** These are the active-low status output signals which indicate initiation of a bus cycle and with M/IO and COD/INTA, they define the type of the bus cycle as shown in Table 9.1.

TABLE 9.1. 80C286 Cycle status Definition (Intel Corp.)

80C286 Cycle status Definition				
COD/INTA	M/IO	S_1	S_0	**Bus Cycle**
0(LOW)	0	0	0	Interrupt acknowledge
0	0	0	1	Will not occur
0	0	1	0	Will not occur
0	0	1	1	None; not a status cycle
0	1	0	0	IF A_1 = 1 Then halt; else shutdown
0	1	0	1	Memory data read
0	1	1	0	Memory data write
0	1	1	1	None; not a status cycle
1(HIGH)	0	0	0	Will not occur
1	0	0	1	I/O read
1	0	1	0	I/O write
1	0	1	1	None; not a status cycle
1	1	0	0	Will not occur
1	1	0	1	Memory instruction read
1	1	1	0	Will not occur
1	1	1	0	None; not a status cycle

(6) **M/$\overline{IO}$:** This output line differentiates memory operations from I/O operations. IF there is a 0 at this pin it indicates that an I/O cycle or INTA cycle is in process and if there is 1 at this pin it indicates that a memory or a HALT cycle is in progress.

(7) **COD/$\overline{INTA}$:** This output signal, in combination with M/IO signal and S1-S0 distinguishes different memory, I/O and INTA cycles.

(8) **$\overline{LOCK}$:** This active low output pin is used to prevent the other masters from gaining the control of the bus for the current and following bus cycles.

(9) **$\overline{READY}$:** This active low input pin is used to insert wait states in a bus cycle, for interfacing low speed peripherals.

(10) **HOLD/HLDA :** This pair of pins is used by the external bus masters to request for the control of the system bus (HOLD) and to check whether the main processor has granted the control (HLDA) or not.

(11) **INTR :** Through thiis active high input pin, 80286 to suspend the current instruction execution and serve the interrupt request.

(12) **NMI :** The non–maskable interrupt request is an active-high, edge triggered input that is equivalent to an INTR signal of type 2.

(13) **VSS :** This pin is a system ground pin of 80286 processor.

(14) **VCC :** This pin is used to apply +5v power supply voltage to the internal circuit of 80286.

(15) **RESET :** The active high RESET input clears the internal logic of 80286, and reinitializes it.

(16) **CAP :** A 0.047 μf, 12v capacitor must be connected between this input pin and the ground to filter the output of the internal substrate bias generator.

(17) **PEREQ and** [illegible] **:** The processor Extension Request refers to Coprocessor. This pair of pins extend the memory management and protection capabilities of 80286 to processor extension 80287. The PEACK active low output indicates to the processor extension that the requested operand is being transffred.

9.4. REAL ADDRESSING MODE

As we mentioned earlier, the 80286 CPU operates in two modes :

(a) Real address mode and

(b) Protected virtual mode.

In real addressing mode of operation of 80286,it just acts as a fast 8086. The instruction set is upward compatible with that of 80166. The 80286 addresses only 1 Mbytes of physical memory using A_0-A_{19}. The lines A_{20} – A_{23} are not used by the internal circuit of 80286 in this mode. The registers and addressing modes will be discussed later in this chapter.

In real address mode, while addressing the physical memory, the 80286 uses BHE along with A_0-$A_{19.}$ The 20-bit physical address is again formed in the same way as that in 8086. The contents of segment registers are uses as segment base addresses. The other registers, depending upon the addressing mode, contain the offset addresses. The address formation in real address mode is shown in Fig. 9.5. An interesting question may be raised at this point : In real address mode, are the speeds of 8086 and 80286 are identical. Because of extra pipelining and other circuit level improvements, in real address mode also, the 80286 operates at a much faster rate than 8086, although functionally they work in identical fashion.

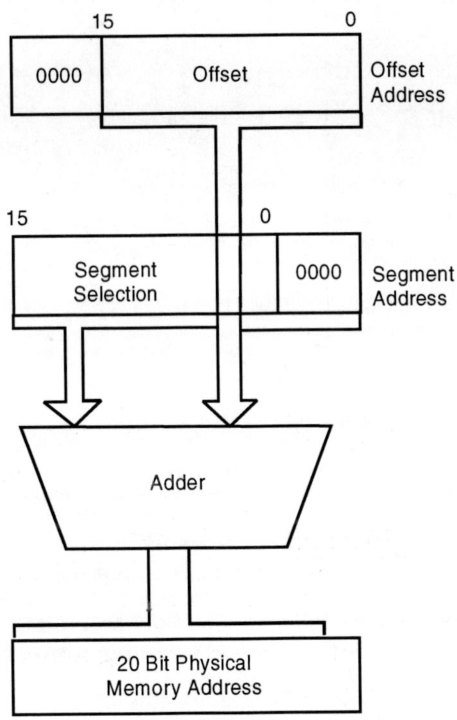

Fig. 9.5. Real Address mode Address Calculation.

9.5. PHYSICAL ADDRESS CALCULATION IN PVAM

In PVAM, the 80286 uses the 16-bit content of a segment register as a selector to address a description stored in physical memory. The descriptor is a block of contiguous memory locations containing information of a segment like segment base address, segment limit, segment type privilege level, segment arability in physical memory, descriptor type and segment use by anther task. The base address i.e. the starting location of a segment is important descriptor information. The segment limit indicates the maximum size of a segment. Thus using the base address of a segment and the segment limit, one can determine the last location in the segment. Similarly, each segment has a level and its privilege level, which indicate the importance of the segment. The privilege level indicates the privilege measure of a segment. A segment with lower privilege level will not be allowed to access anther segment having higher privilege, thus offering protection to the segment from the unauthorized accesses. Moreover, a certain segment may or may not be present in the physical memory at a given time instant. This information is also stored in a descriptor. Finally, important information i.e. whether the segment has been accessed by another task in the past, is also stored in the segment descriptor. This information helps in the deciding whether the segment should be unswapped from the main memory. The segment base address is a 24-bit pointer that addresses the first location in that segment. This 24-bit segment base address is added with 16-bit offset to calculate a 24-bit physical address. The maximum segment size will be of 64 Kbyte, since the offset is only of 16 bits. Fig. 9.6 shows the calculation of physical address in PVAM. The descriptors are automatically referred to by the CPU when a segment register is loaded with a selector.

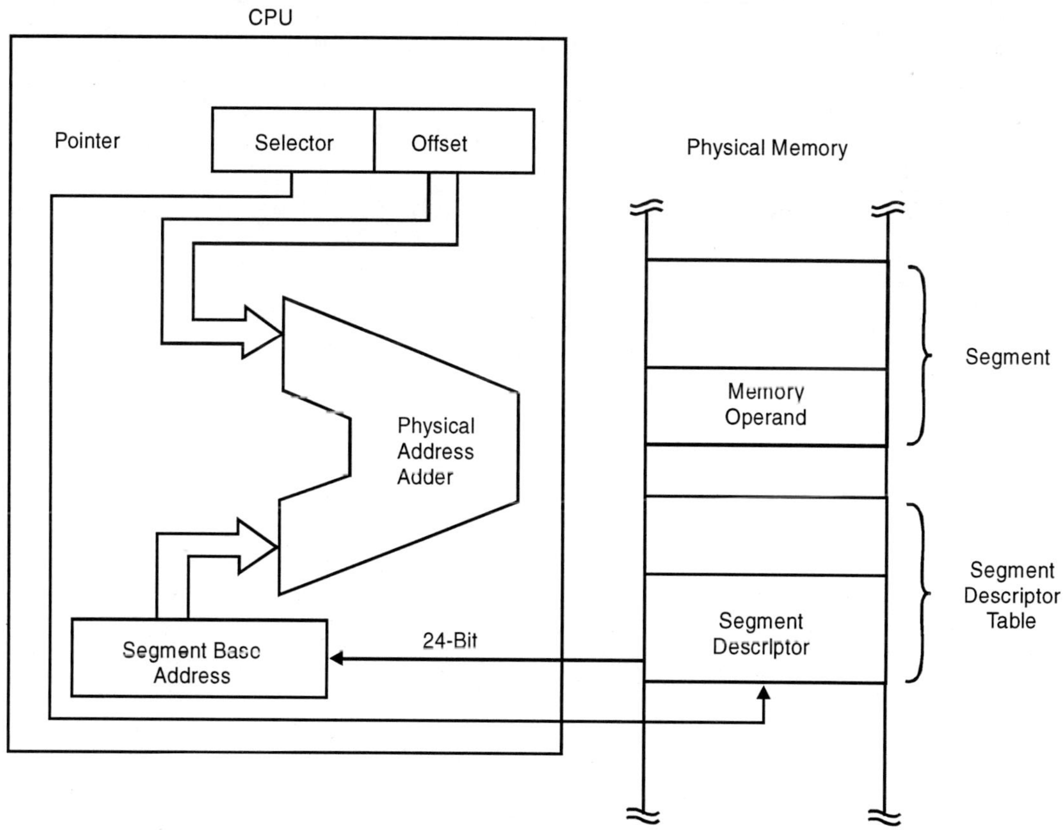

Fig. 9.6. Physical Address Calculation in PVAM.

9.5.1. Description and their Types

in general description carry all the relevant information regarding a segment and its access rights. Besides this information, special types of descriptions which are used to carry out additional functions like transfer of control and task switching, may have additional information. The 80286 has segment descriptors for code, stack and data segment as basic descriptors. In additional to this, it has system control descriptors for special system data segment and control transfer operations. Fig. 9.7 shows the structure of a code or data segment descriptor. Each descriptor is 8-bytes long. The information stored in the 8-bytes of a descriptor can be used by the operating system to support the implementation of memory management and protection schemes. The definition of access rights byte of code or data segment descriptor is given in Table 9.2.

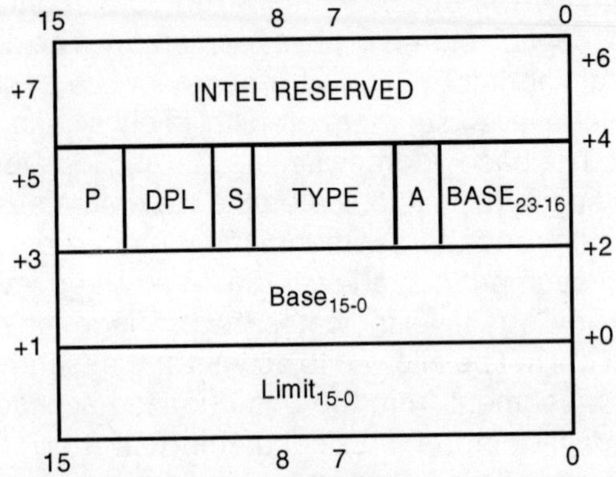

Fig. 9.7. Data or Code Segment Descriptor.

TABLE 9.2. Access Rights Byte Definition (Intel Corp.)

Bit Position		Name	Function	
7	Present (P)	P=1	Segment is mapped into physical memory	
		P=0	No mapping to physical memory exits, base and limit are not used.	
6-5	Descriptor Privilege Level (DPL)		Segment Privilege attribute used in privilege tests.	
4	Segment Descriptor (S)	S=1	Code or Data (includes Stacks) Segment Descriptor	
		S=0	System Segment Descriptor or Gate Descriptor	
3	Executable (E)	E=0	Data Segment descriptor type is	If Data Segment (S=1, E=0)
2	Expansion Direction (ED)	ED=0	Expand up segment, offsets must be< limit.	
		ED=1	Expand down segment, offsets must be>limit	
1	Writable	W=0	Data segment may not be written into.	
		W=1	Data segment may be written into.	
3	Executable (E)	E=1	Code segment descriptor type is	If Code Segment S=1, E=1)
2	Confirming	C=1	Code segment may only be executed when CPL>DPL and CPL remains unchanged.	
1	Readable	R=0	Code segment may not be read.	
		R=1	Code segment may be read.	
0	Accessed (A)	A=0	Segment has not been accessed.	
		A=1	Segment selector has been loaded into segment Register or used by selector test instruction.	

A code or data segment descriptor contains 16-bit segment limit,24-bit segment base address, 8-bit access rights type and the remaining 16-bits are reserved by Intel for upward comp ability. Any segment access, violating the specified access rights in the descriptor, prevent memory cycle and generates an exception interrupt. Code segment descriptors are used to refer to code segment and data segment descriptor are used to refer data segments. The descriptor is either a code segment descriptor or a data segments descriptor if the S bit in the access rights byte is '1'. The system segment descriptor are selected by clearing the S bit, i.e. S=0. The code segments are distinguished from data segment by the E bit. Thus E=1 indicates code segment while E=0 indicates data segment. The present (P) bit indicates whether the segment is available in physical memory or not be accessed (A) bit indicates whether it is accessed previously or not, for both code and data segments. The descriptor privilege level fields indicate the accessibility of the descriptor. The CPU will set A bit while accessing the descriptor. If P is 0 and access to the segment is tried, a 'not present' exception is generated.

A data segment may by either read only or read –write depending upon the W-bit of the corresponding descriptor. Foe read only segments, W is 0 and for read-write segments W is 1. The data segments (S=1 and E=0) can have data area on both the sides as determined by the ED bit. For normal data segment, the access is upward, i.e. ED=0 and for stack data segment, the access is downward, i.e. The limit field is interrupted depending upon the ED bit. A code segment identified by S=1 and E=1 may be executable or execute-read, as determined by the readable (R) bit. If R is 0, the code segment is executable only; otherwise, it is readable as well as executable. The code segment is never written. A code segment has a confirming attribute that allows the use of the code segment by different program which execute at different privilege levels. The DPL of such a code segment defines the range of a privilege levels at which the segment may be executed. The following text will be elaborate the use of DPL field in such descriptor. The limit field determines the size of code /data segment.

System Segment Descriptors : In addition to code and data segment descriptors, the other types of descriptors with S=0 are used by 80286 to store system data and execution state of a task (for multitasking systems). These are called as system segment descriptors. The system segment descriptors are of seven types. The types 1 to 3 are called system descriptors and the types 4 to 7 are called gate descriptors as they are used to control the access of entry points within the code to be executed.

Type 1-13 System Segment Descriptors : Fig. 9.8(a) and (b) show the type 1-3 system segment descriptors format and the corresponding bit definition respectively.

This descriptors contains 16-bit segment limit,24-bit segment base address,and an access rights byte that contains P-bit,a 2-bit DPL field,S-bit (0) and a 4-bit types field. The fourth world of the descriptor is reserved by Intel for comp ability with future processors.

Type 4-7 Gate Descriptors : The gate descriptors control the access to entry points of the code to be executed. There are four types of gates descriptors, viz. call gate, task gate, interrupt gate and trap gate. The gate descriptors contain information regarding the destination of control transfer, required stack manipulation, whether it is present in memory or not, privilege level and its. Gate descriptors provide a mechanism to keep track of source and Destination of control Transfer. Hence, the CPU can perform protection checks and control the entry points of destinations codes.

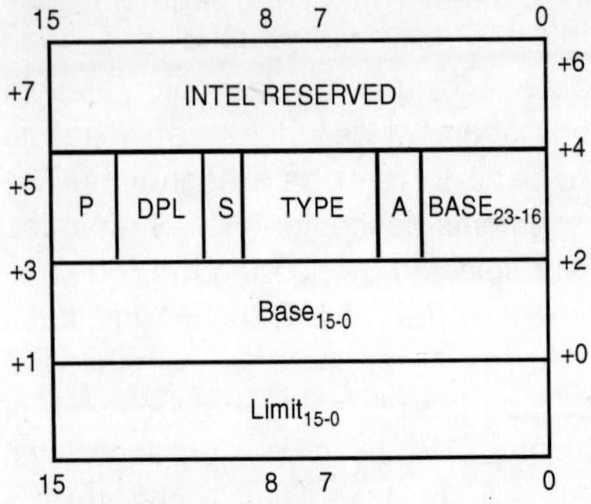

*Must be set to 0 for compatibility with 80386

(*a*) **System Segment Descriptor**

Name	Value	Description
Type	1 2 3	Available Task Segment (TSS) Local Descriptor Table Busy Task State Segment (TSS)
P	0 1	Descriptor Contents are not Valid Descriptor Contents are Valid
DPL	0-3	Descriptor Privilege Level
BASE	24-bit number	Base address of special system data segment in real memory
LIMIT	24-bit number	Offset of last byte in segment

(*b*) **System Segment descriptor bit Definitions (Intel Corp.)**

Fig. 9.8.

Call gates are used to alert the privilege levels. Task gates are used to switch from one task to another. Interrupt and trap gates are used to specify corresponding service routines. Fig. 9.9 (a) and (b) show the gate descriptor format and the corresponding bit definitions respectively. If a destination selector does not refer to a correct descriptor type, exception 13 is generated. The task gate does not use the definition offset field. The task gate may only.

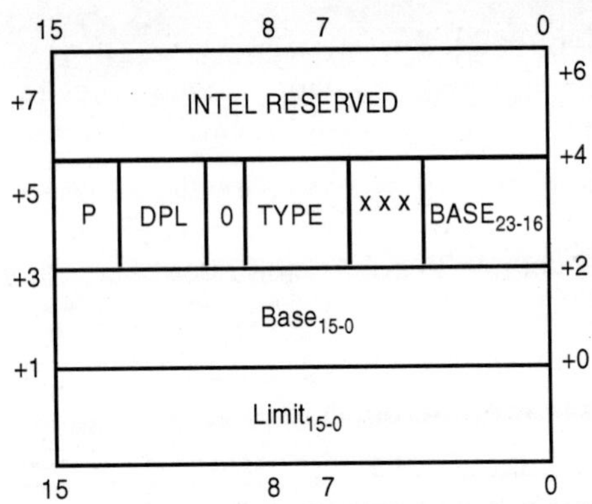

*Must be set to 0 for compatibility with 80386
(x is don't care)

(*a*) **Gate Descriptor Format**

Name	Value	Description
Type	4 5 6 7	-Call Gate -Task Gate -Interrupt Gate -Trap Gate
P	0 1	Descriptor Contents are not Valid Descriptor Contents are Valid
DPL	0-3	Descriptor Privilege Level
World Count	0-31	Number of words to copy from callers stack to called procedures stack only used with call gate
DESTI-NATION SELECTOR	16-bit selector	Selector to the target code segment (Call, Interrupt of Trap Gate) Selector to the Target task state segment (Task) Gate)
DESTI-NATION OFFSET	16-bit offset	Entry point within the target code segment

(*b*) **BIT Definition of Gates Descriptor Format (Intel Corp.)**

Fig. 9.9.

Refer to a task state segment. The world count field is only used by the a call gate descriptor to indicate the number of bytes to be transferred from the stack of the calling routine to the stack of the called routine, when a control transfer changes the privilege levels. The access right byte format and its definition are same as the code or data segment descriptor.

9.5.2. Segment Descriptor Cache Registers

A concept of caching was introduced in 80286 to minimize the time required for fetching the frequently required descriptor information from the main memory. The caching is nothing but maintaining the most frequently required data for execution in a high speed memory called cache memory. A 6–byte segment descriptor cache register is assigned to each of the four segment, i.e. CS, DS, SS, and ES. A segment descriptor is automatically loaded in segment descriptor cache registers, whenever the associated segment register is loaded with a selector. Once a cache register, instead of referring to the main memory for the descriptor again and again. These cache registers are not available for programming. They automatically change when a segment registers is reload. Fig. 9.10 shows the 6-bytes format of the cache registers and the corresponding segment registers.

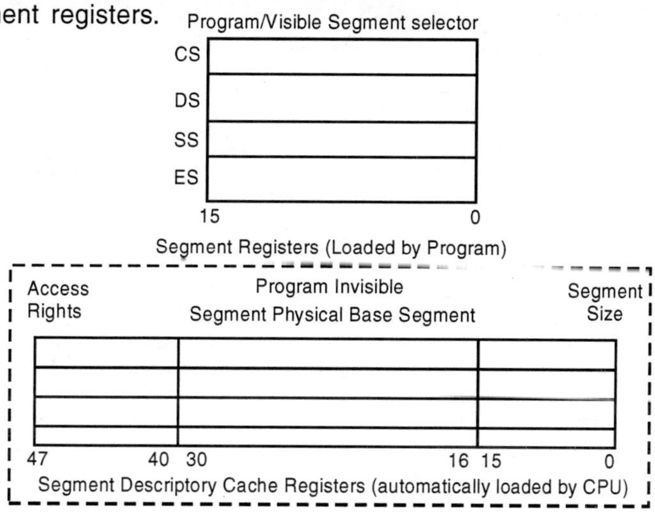

Fig. 9.10. Descriptor Cache Register and their Formats (Intel Corp.).

Selector Fields : In protected mode, the contents of the segment registers are called as selectors. The selectors contain three fields in its 16-bit format. The 2-bit field D_0-D_1 is called as RPL field, i.e. requested privilege level, that describes the desired privilege of the segment. The D_2 bit indicates the descriptor table type, i.e. local descriptor table, if it is 1 and global descriptor table, if it is 0. The index field D_3-D_{15} points to the required descriptor base in the descriptor table. Figure 9.11 shows the selector field format.

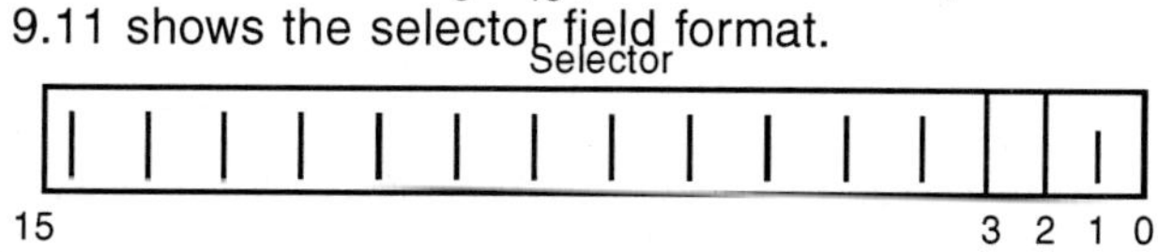

BITS	NAME	FUNCTION
1-0	REQUISTED PRIVILIGE	INDICATES SELECTOR PRIVILEGE LEVEL DESIRED
2	TABLE INDICATER (T1)	T1=0 USE GLOBAL DESCRIPTOR TABLE (GDT)
		T1=1 USE LOCAL DESCRIPTOR TABLE (LDT)
15-3	INDEX	SELECT DESCRIPET ENTRY IN TABLE

Fig. 9.11. Selector Field and Definitions (Intel Corp.).

Local and Global Descriptor Table : Every descriptor required by a task is either in a local descriptor table (LDT) or a global descriptor table (GDT). A descriptor table is an array of 8K descriptors. The upper 13 bits of a selector field (i.e. index) point to a particular entry in a descriptor table. This means that there may be 8K descriptors in a descriptor table. Thus a Descriptor table, either global or local requires 8K* 8=64 Kbytes of memory. Obviously, to point to 8K descriptors, a 13-bit address (2^{13}=8Kbyte) is required. Each selector can address a segment of size 64kbytes. There can be at most,8K local and 8K global descriptors outside any of the descriptor tables. A global descriptor table (GDT) contains global descriptors common to all the task. A local descriptor table (LDT) contains descriptors specify to a particular task. All the tasks may have their private LDTs. The GDT may contain all the descriptor types except interrupt and trap descriptors. The LDT contains segment, task gate and call gate descriptors. A segment cannot be accessed, if its descriptor does not exist in either LDT or GDT at that instant.

The LGDT (load global descriptor table) and LLDT (load local descriptor table) instructions load the base and limit fields of GDT and LDT respectively. The LGDT and LLDT instructions are privilege and may be executed only at privilege level 0. The LLDT instruction load a selector which refers to a local descriptor table containing the base address and limit for an LDT as shown in Fig. 9.8 and (b) Figure 9.12 elaborates global and local descriptor table definition. Figure 9.13 shows a global descriptor data type.

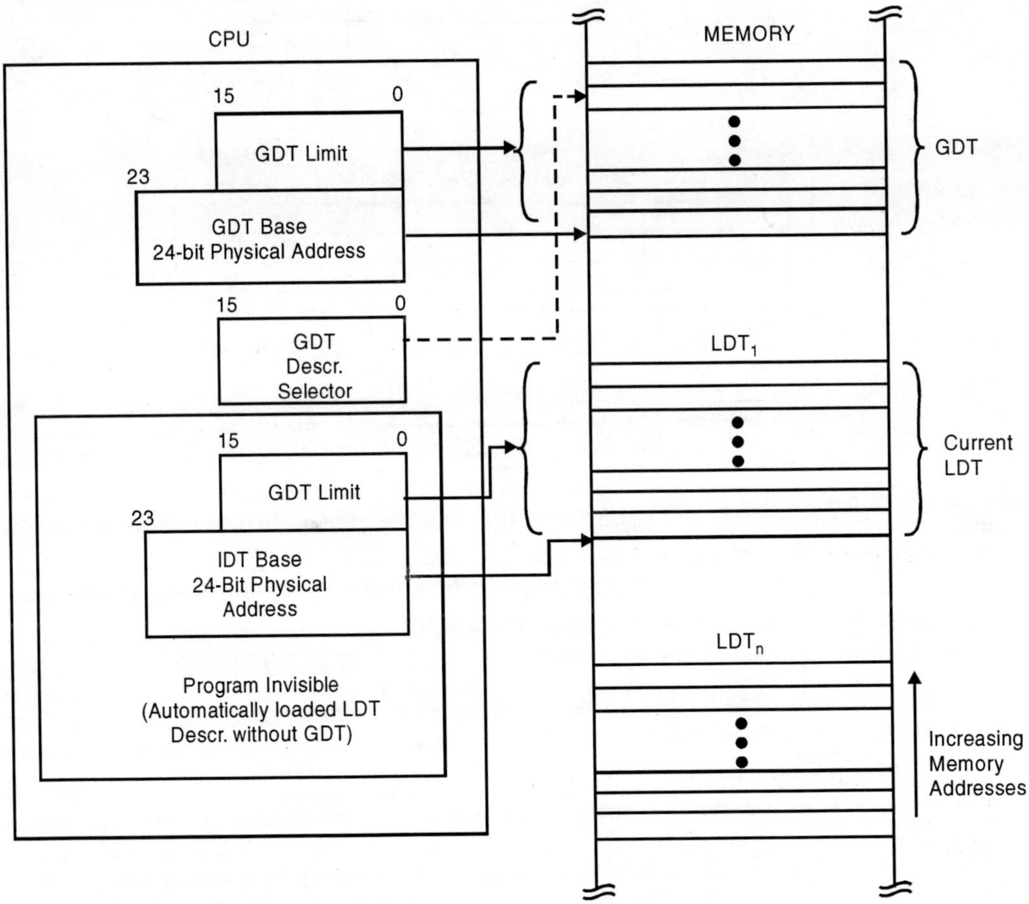

Fig. 9.12. Local and Global descriptor Table Definition.

Interrupt Descriptor Table : Besides the local and global descriptor tables, the 80286 has a third type of descriptor table called as interrupt descriptor table (IDT). These are used to store task gates, interrupt gates and trap gates. The IDT has a 24-bit address and a 16-bit limit registers in the CPU. The LIDT (load interrupt descriptor table register) instruction loads these internal registers with a 6-byte value in the same way as LGDT instruction. The IDT data format is shown in Fig. 9.13. The IDT 80286 is able to handle upto 256 interrupt descriptor. Fig. 9.14 shows the arrangement of interrupt gates for different interrupts in physical memory. The IDT entries can be referred to by using INT instructions or external interrupt or exceptions. Six bytes are required for each interrupt in an interrupt descriptor table.

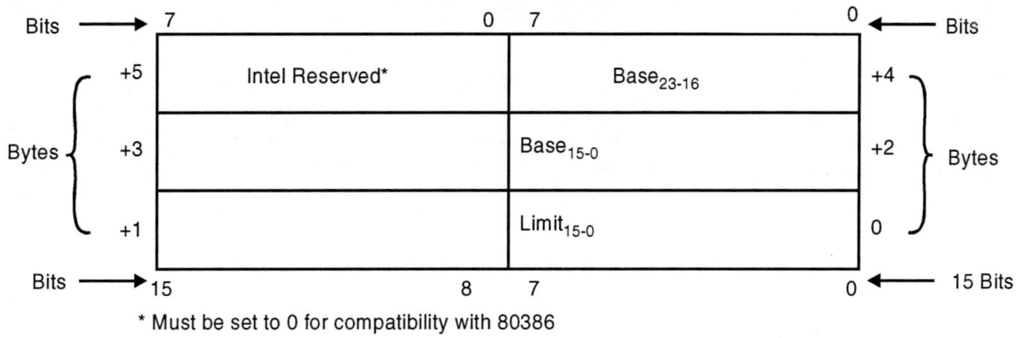

Fig. 9.13. Global and Interrupt Descriptor Data Type.

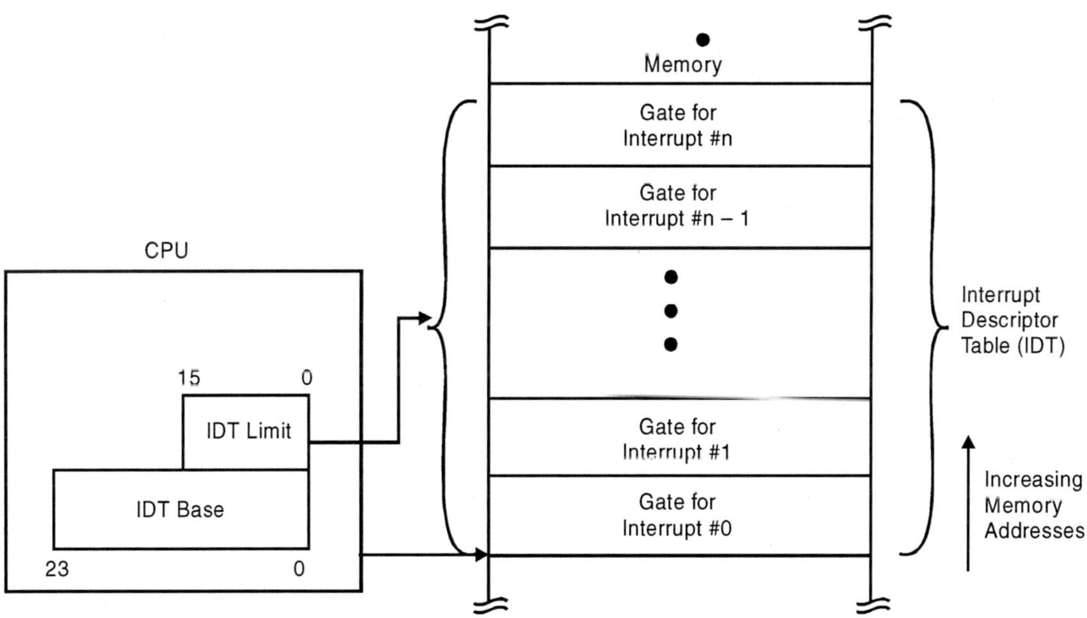

Fig. 9.14. Interrupt Descriptor Table Organisation.

9.6. PRIVILIGE

The 80286 supports a four level hierarchical privilege mechanism to control the access to descriptors and hence to the corresponding segment of the task. The control of the access to descriptors results in the prevention of unwanted or undue access to any of the code or data segment or unintentional interference in the higher privilege level tasks. Level 0 is the highest privilege level while level 4 being the least privileged level. The privilege levels provide3 protection within a task. The operating system, interrupt handlers and other system software can be protected from unauthorized accesses in virtual address space of each task using the privilege mechanism. Each task in system has a separate stack for each of privilege levels. The privilege mechanism offers or denies access to a segment at the behest of the privilege bits of the corresponding descriptor. The task privileging controls the use of instructions and descriptors. Fig. 9.15 shows the four level privilege mechanism. The capabilities of the privilege mechanism are explored using the privileged instructions.

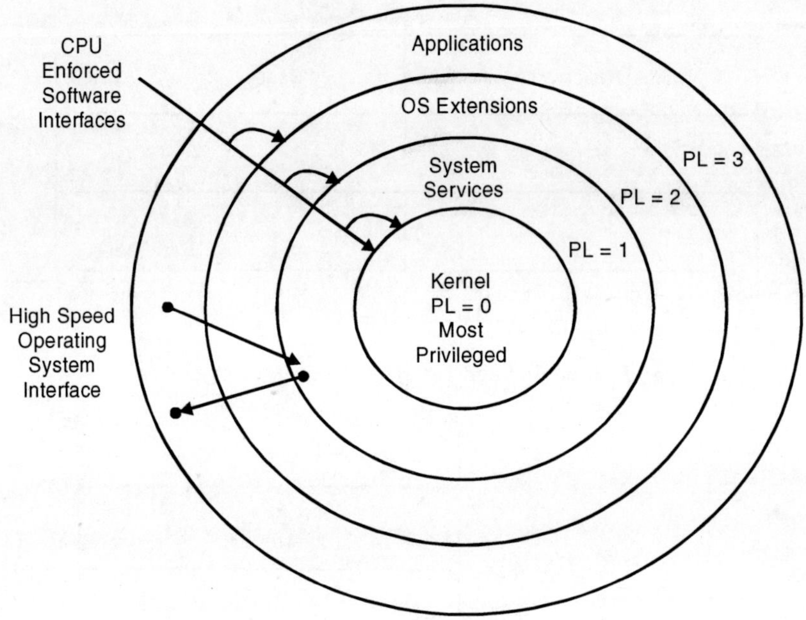

Fig. 9.15. Four Level Privilege Mechanism.

9.6.1. Task Privilege

Each task is assigned privilege level, which indicates the priority or privilege of that task. Any one of the four privilege levels may be used to execute a task. The task privilege level at that instant is called the current privilege level (CPL). The CPL is defined by the lower order two bits of the CS registers for an executables segment. Once the CPL is selected, it cannot be changed during the execution normally in a single code segment. It can only be changed by at the selected CPL value specified by the CS within TSS, if it is initiated via a task switch operation. A task executing at level 0, the most privileged level, can access all the data segment defined in GDT and the LDT of the task. Obviously, a task executing at level 3, the least privileged level, will have the most limited assesses to data and other descriptors. Figure 9.16 shows task state segment (TSS) and TSS registers are used by a task privilege.

9.6.2. Descriptor Privilege

The descriptor privilege is specified by the DPL field of the access rights byte. The DPL specifies the least task privilege level (CPL) that may be used to refer the descriptor. Hence the task with privilege level 0 can refer to all the lower level 3 can refer to only level 3 descriptors. This rule applies to all the descriptors except LDT descriptors.

9.6.3. Selector Privilege

This privilege is specified by the RPL field of a segment register (selector). A selector RPL may use a less trusted privilege than the current privilege level for further use. This is called as effective privilege level (EPL) of the task. The effective privilege level is thus the maximum of RPL and CPL (i.e. numeric maximum and privilege maximum). The RPL is used to ensure that the pointer parameters passed to a more privileged procedure are not given the access of data at the privilege higher than the caller routine. The pointer testing instructions are used for this purpose.

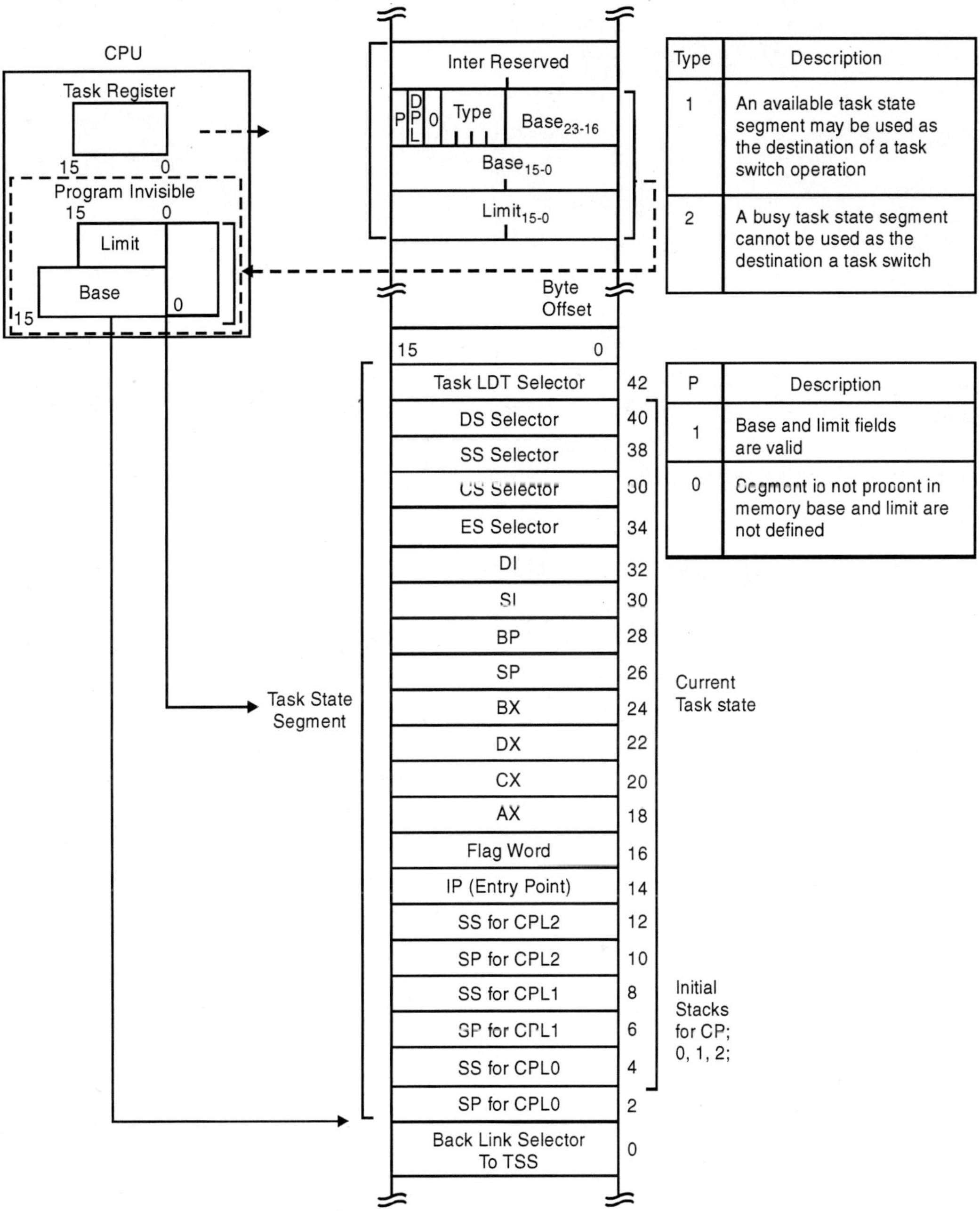

Fig. 9.16. Task State Segment and TSS Registers.

TABLE 9.3. Control Transfer Descriptor Types (Intel Corp.).

Control Transfer System	Operating system	Descriptor Referenced	Descriptor Table
Intersegment within the same Privilege level	JMP, CALL, RET, IRET*	Code Segment	GDDLDT
Intersegment to the same or higher	CALL	Call gate	GDDLDT
Privilege level interrupt within Task may change CPL.	Interrupt Instruction, Exception, External Interrupt	Trap or Interrupt Gate	IDT
Intersegment to a lower	RET, IRET*	Code segment	GDDLDT
Privilege level (change Task CPL)	CALL, JMP	Task State Segment	GDT
Task Switch	CALL, JMP IRET*	Task Gate	GDDLDT
	Interrupt Instruction Exception, external Interrupt	Task Gate	IDT

*NT (Nested Task bit of flag word)=0 **NT (Nested Task bit of flag word=1

However, a Readable confirming code segment can be read from any privilege level. If privilege test is negative or an improper segment is referenced, an exception 13 is generated. If the referenced segment is not present in physical memory, an exception 11 is generated.

Loading of SS register always refers to data segment descriptors (stack data) that are writable. The DPL and RPL must be equal to CPL to prevent unwanted cross referencing of stack data. The negative privilege tests and improper descriptor references generate exception 13. If the stack data segment to be referred to is not present in physical memory, an exception 12 is generated.

9.6.4. Privileging Level Alteration

If a task needs to changes its privilege level during the control transfers within it, the stack must be manipulated correspondingly. The current SS :SP fore task privilege levels 0,1 and 2 are stored in the task state segment. If the control is to be transferred using JMP and CALL the new SS : SP contents are loaded and the previous stack pointer is pushed to the new stack. While returning to the original privilege level the stack is restored as a part of control transfer operation after execution of RET or IRET instructions. For subroutine calls which use stack for passing parameters to subroutines and then cross the privileged levels, a fixed number of words are copied from the privilege stack to the current stack (Refer to word count field of the gate descriptors). The corresponding RET instruction with a stack adjustment value will correctly retain the previous stack pointer.

9.7. PROTECTION

As it is obvious from the forgoing discussion, the 80286 can utilize its privilege mechanism for protecting its data or code from the unwanted accesses. The 80286 has instruction designed to exploit its protection capabilities. The 80286 supports the following three basic mechanisms to provide protection.

(1) Restricted use of segment (segment load check) this is accomplished with the help of read/ write privileges. The segment usages are restricted by classifying the corresponding descriptors under LDT (local descriptor table) and GDT (Global descriptor table).

(2) Restricted Accesses to Segment (operation reference check) This is accomplished using descriptor usages limitations and the rules of privilege check, i.e. DPL, TPL and CPL.

(3) Privileged instructions or operations (Privileged instruction check) These are to be executed or carried out at certain privilege levels determined by CPL and I/O privilege level (IOPL) as defined by the flag register.

The three types of checks discussed above, if the result is negative, generate exceptions as shown in Table 9.4, Table 9.5 and Table 9.6 respectively. The IRET and POPF instructions do not perform any of their functions, if CPL is not of the required privilege level. For example If remains unaffected if CPL is greater than 0.No exception is generated for this condition.

TABLE 9.4. Segment Register Load Check Exception.

Error Description	Exception Number
Descriptor table limit exceeded	13
Segment descriptor not-present	11 or 12
Invalid descriptor/ segment type segment register load	
-Read only data segment load to SS	
-Special Control descriptor load to DS, ES, SS	13
-Data segment load to GS	
-Read/Execute code segment	
Load to SS	

TABLE 9.5. Operand Reference Check Exceptions.

Error Description	Exception Number
Write into code segment	13
Read from execute-only Segment	13
Write to read-only data segment	13
Segment limit exceeded	

Exceptions in Protected Mode : In protected mode, 80286 specially generate some exceptions and interrupt as a result of violation of the corresponding condition. These are listed in Table 9.7. Some these exception are rest arable, i.e. execution can further be continued, if the cause of generation is eliminated. The interrupt service routines for them can read an error code and return address from the stack. The return address identifies the selector and points to the falling instruction. For a processor extension segment overrun exception, the return address does not points at the ESC instruction at which the exception was generated but the processor extension registers contain the address of the failing instruction. Though some of the exceptions are restorable, their generation indicates fatal error somewhere and hence the restart is not advisable. These checks are performed for all the instruction and operations. A 'Not Present' exception generation interrupt 11 or 12 and is

TABLE 9.6. Privilege Instruction Check.

Error Description	Exception Number
CPL # when executing the following Instructions.	
LIDT, LLDT, LGDT, LTR, LMSW	13
CTS, HLT	
CPL> IOPL when executing the following Instructions.	
INS, IN, OUTS, OUT, STI, CLI	13
LOCK	

TABLE 9.7. Privilege Mode exceptions.

Interrupt Vector	Function	Return Address of Falling Instruction?	Always Restart Able?	Error Code On Stack
8	Double exception detected	Yes	No	Yes
9	Processor extension segment overrun	No	No	No
10	Invalid Task State Segment	Yes	Yes	Yes
11	Segment not present	Yes	Yes	Yes
12	Stack Segment overrun or stack segment not present	Yes	Yes	Yes
13	Great Protection	Yes	No	Yes

9.8. SPECIAL OPERATION

The 80286 carries out five operations, which should be studied in details before we start with the bus cycle and instruction set. These are—

(1) Processor reset and initialization.

(2) Task Switch operation.

(3) Pointer testing instruction.

(4) Protected mode initialization.

(5) How to enter protected code.

(6) Halt.

9.8.1. Processor Reset and initialization

The processor is reset by applying a high or RESET input that terminates all the execution and internal bus activities till RESET remains high. At the trailling edge of RESET, the 80286 starts internal initialization. (that requires 34 clock states) and then starts executing instruction from the phycial address FFFFF0(H). The other registers are initialized after RESET as shown in Table 9.8. The HOLD must be not activate during the time from the leading edge RESET signal to at least 24 clock cycles after trailling edge of the RESET signal. After reset,80286 is always in real address mode.

TABLE 9.8. Register Initialization after RESET.

FLAG	0002 H
MSW	FFF0 H
IP	FFF0 H
CS	F000 H
DS, ES, SS	0000 H

9.8.2. Task Switch Operation

The 80286 supports multitasking, i.e. more than one task may be ready for execution at a time. A job may be divided in to a number of tasks. These tasks are to be executed one by one using 80286, for completion of the job. A number of the task allocation strategies like first come first serve, shortest task first, time sharing etc. have been experimented by operating system designers. In case of time sharing technique the time of the CPU is divided in to equal duration slices. Each of the tasks in the queue in then allocated a fixed time slice, then it is removed from the queue of the task to be executed. Otherwise whatever is its state at the end of the allotted time slice, it is saved beck with all the requires details and is made to wait for its next turn. The CPU is allotted to the next task in the queue for an identical time slice. Thus each task will receive the attention of the CPU sequentially, after a fixed duration time slice. After the first cycle is over, the first task will again be scheduled for the cycle and the processor continues. The previous task that was incomplete may be completed during its coming turns of the allotted CPU time slice. This switch-over operation from one task to another is called as task switch operation.

The 80286 Internal architecture provides a task switch operation to save the execution state of a task (that includes registers, address space, and link to the previous task) and to load new task to be executed. The execution of the new task commences after its executions state is loaded. The task switch operation is carried out using a JMP or CALL to a new segment of the new task that refers to the corresponding task state segment (TSS) or task gate descriptor in the GDT or LDT. A software interrupt instruction, exception or external interrupt (hardwired setting of the time slice externally) can also be used to carry out task switch operation. However a corresponding task gate descriptor must be selected in the associated IDT to use any type of interrupt or exception for this purpose. A task gate descriptor contains a TSS selector. The TSS descriptor specifies a segment containing the new task execution state.

Each task has a TSS for it. The TSS currently under execution is pointed to by a special function registers called as task Register TR. The selector in TR refers to a TSS descriptor that Defines current TSS. An internal hidden base and limit registers are loaded automatically when the TR is loaded with new selector. The IRET instruction is internally executed to return the control to the main task that called the current task or was itself interrupted. The NT (nested task flag) bit of the flag register controls the function of IRET instruction. If NT is 1, the IRET instruction gets back the execution state of the previous task. Otherwise the IRET instruction lets the current task to continue after popping the required values from stack.

If a CALL, JMP or INT instruction is used to start the task switching operation, the new TSS is marked busy and the back link field of the new TSS is set to the selector of old TSS. The NT flag is set by CALL or INT initiated task switch operations. An interrupt does not clear NT. The NT flag is either set or cleared by POPF or IRET instructions. The task management can be marked busy by changing the descriptor type 1 to type 3. Any attempt to refer a busy task state segment generates an exception 13. Figure 9.16 shows the TSS and the corresponding registers.

9.8.3. Pointer Testing Instruction

The pointer testing instructions of 80286 use the memory management hardware to verify whether the loaded selector value refers to a valid segment without generating any exception. The ZF indicates that the selector loaded or the referred segment may or may not generate an exception. Table 9.9 shows the pointer testing instructions and their details.

TABLE 9.9. Pointer testing instructions of 80286.

Instruction	Operands	Functions
APRL	Selector, Register	Adjust requested privilege level : adjust the RPL of the selector to the numeric maximum of current selector RPL value and the RPL value in the registr. Set zero flag if selector RPL was changed by ARPL.
VERR	Selector	VERify for read: sets the zero flag if the segment referred to by the selector can be read.
VERW	Selector	VERify for write sets the zero flag if the segment referred to by the selector can be written.
LSL	Register, Selector	Load segment limit : reads the segment limit into the register if privilege rules and descriptor type allow. Set zero flag if successful.
LAR	Register, Selector	Load access rights : reads the descriptor access rights byte into the register if privilege rules allow. Set zero flag if successful.

9.8.3.1. Protected Mode Initialization

After 80286 is reset, it starts the execution in real address mode. The address lines A_{20} - A_{23} are pulled down to zero level for addressing peripherals and memory. Initially the CS :IP value is set at F000H : FFF0H to allot a segment of 64K bytes for initializations. The Initialization of protected mode is carried out in real mode by setting the internal registers of 80286 suitably. The GDT and IDT base registers must point to valid GDT and IDT, before 80286 enters the protected mode.

To enter into protected mode, 80286 executes LMSW instruction that sets PE flag. The 80286 then clears the instruction queue that may contain opcodes fetched in real mode.

The operating system assumes the initial state in protected mode and accordingly initializes the internal registers of 80286. after entering the protected mode, the 80286 executes a jump instruction that directs the control to a selector that refers to the initial TSS. The jump instruction initializes the task register, LDT register and the segment registers.

9.8.3.2. How to Enter Protected Virtual Address Mode (PVAM)

80286 enters in real address mode, after the reset. The execution of instruction LIDT (Load interrupt descriptor table base) prepares the 80286 for protected virtual address mode. This instruction loads the 24-bit interrupt table and 16-bit limit from memory into the interrupt descriptor table register. This instruction also can set the base and limit of interrupt vector table in real mode.

Then the PE flag of MSW is set to enter the PVAM using the LMSW (load machine status word).

9.8.3.3. How to Enter Protected Virtual Address Mode (PVAM)

This instruction stops program execution and prevents the CPU from restarting till it is interrupted or reset is asserted. If the CPU is interrupted in the halt state the execution starts from the next instruction after HLT. On the other hand, if the CPU is reset, the execution starts from the physical address FFFFF0H.

9.9. 80286 BUS INTERFACE

The 80286 provides a set of signals to interface the memory and I/O devices with it. This set contains 24 address lines to address 16Mbytes of physical memory, 16 data lines to enable 16-bit data transfer at a time and 8 control signals to coordinate the transfers. The supporting chip 82284 provides synchronized RESET and READY signals, along with the system clock. The 82288 derives various control signals from the signals provided by 80286 to encode the bus cycles.

The 80286 is able to address 16 Mbytes of memory, that is addressed in the same way as 8086, i.e. in the terms of even address bank and odd address bank, using the signals A0 and BHE. The 80286 is able to address 64K 8-bit output and 64K 8-bit input devices or 32K 16-bit input and 32K 16-bit output devices. The I/O devices are also addressed using even address and odd address banks technique, using A0 and BHE. The upper byte of the 16-bit data that is to be transferred to/from an I/O device is transferred on D8-D15 to an odd address, while the lower byte of the 16-bit data is transferred on D0-D7 to an even address.

The 80286 divides the input clock frequency applied at input CLK pin by two to derive the actual operating clock frequency PCLK, that determines the bus timings. According to the type of bus operation, the 80286 bus cycles are of six types, viz. memory read, memory write, I/O read, I/O write, interrupt acknowledge and halt.

The 80286 bus at a particular instant may be in either of these four states :

(i) Idle state (Ti)

(ii) Perform command state (Tc)

(iii) Send status state (Ts)

(iv) Hold state (Th)

In the idle state, the local bus remains idle, while in perform command state and sent status state, it perform memory/IO read /write operations. When a device desires to have a DMA data transfer the hold state indicates that the local bus is being relinquished for another bus master. These bus states are shown in Fig. 9.17.

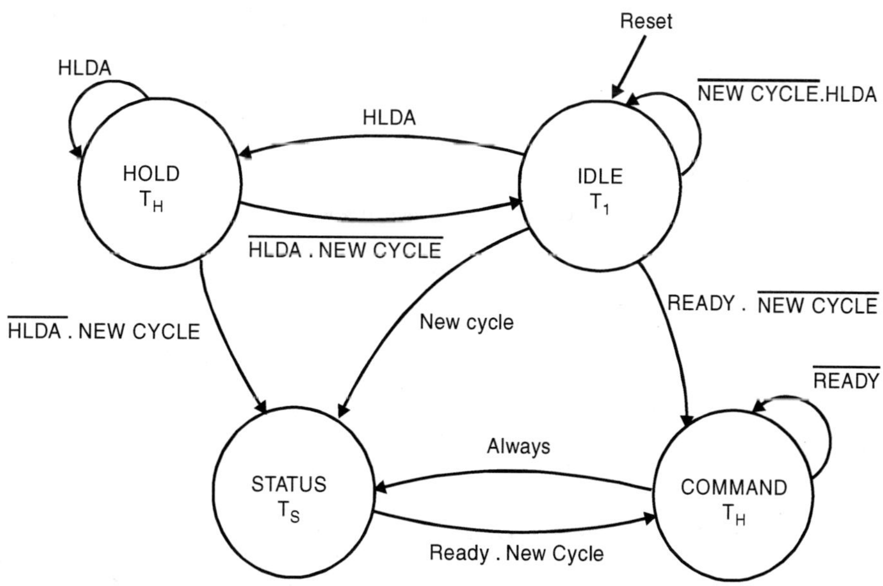

Fig. 9.17. 80286 Bus states and their Co-relation.

The 80286 uses pipelining techniques to speed up the data access procedure. Each individual bus operations is three cycle long. Suppose the CPU, an instant the initiated a memory read bus operation. This operation will continue for this cycle. Suppose also that the next operation is a memory write operation. Then the address of the next memory location to be written, as a part of the next bus cycle will be available during the current bus cycle. Thus the next bus cycle will be initiated after two clock cycles of the current bus cycle. This is the pipeline operation, i.e. the first clock of the second bus operation, which is memory write operation overlaps with last clock of the previous operation remains valid for the first processor clock cycle.

9.10. BASIC BUS OPERATIONS

The bus controller 82288 derives ALE, RD and WR commands, DT/R and DEN signals to control the data transfer to/from the 80286. The bus controller uses the CPU output signals S_1,S_0 and m/IO as inputs. The basic bus operation here is studied only in terms of the signals derived by 80286. The basic bus cycle is shown in Fig 9.18, that the elaborates two successive read cycle by 80286. These basic timing of 80286 can be controlled, using two command control options, namely, command execution (input READY) and command delay (CMDL input) of 82C288.

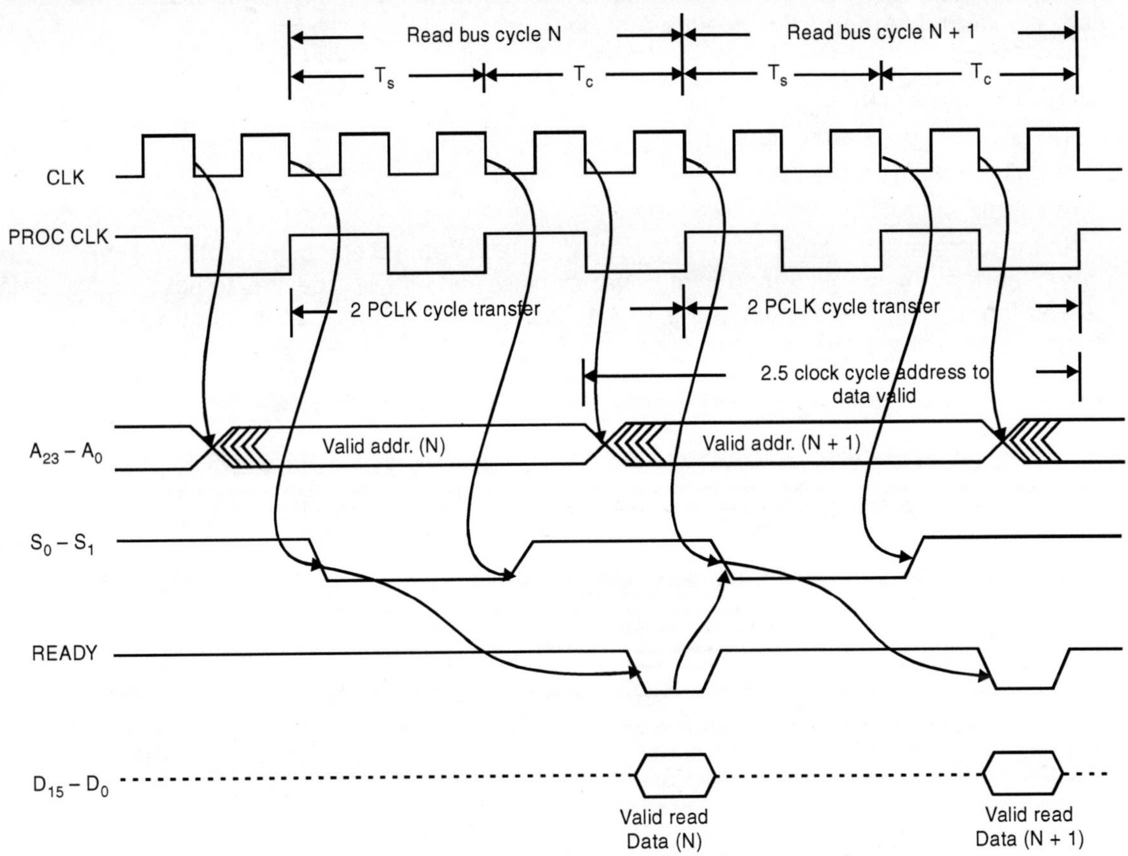

Fig. 9.18. Basic Bus Cycle of 80286.

The command exception option inserts the wait states in the basic bus cycle using external hardware to interface 80286 with low speed peripherals. The command delay option stretches the interval between read/write data setup to system bus command active interval for any of the bus operation. The 82C288 checks the CMDLY input at each trailing edge of CLK. If is high the 82C288 will not activate the command signal. Once the basic 82286 bus cycle. The CMDLY signal does not affect ALE, DLE or DT/R . Figure 9.20, Fig. 9.21 and Fig. 9.22 show the different sequences of the basic operations and the behavior of data control signal for each operation.

9.11. FETCH CYCLES OF 80286

The 80286 architecture implements pipelined fetching of instructions. In other words, the 80286 per fetches the initial bytes of the next instruction, While executing the current instruction. These prefetched instruction bytes are arranged in a 6-byte preftech queue, from where they are accepted further for decoding and execution, as has been discussed in section 9.2.2. This fetch operation is carried only if at least two bytes of the queue are empty. The fetch operations usually fetch two bytes of the program code at a time, sequentially, independent of byte-wise instruction alignment, in the physical memory. The prefetcher fetches only one byte at a time if the word fetch operation begins at an add address. In case of a branch type instruction, where the sequential instruction execution may be hampered, the prefetching operating stops and the already prefetched queue is flushed. The 80286 continues the perfetching till HLT instruction is executed. Thus the 80286 may blindly prefetch six bytes even after the HLT instruction has been fetched. In protected mode the 80286 prefetcher can not cross the segment overrun exception cannot be generated. The prefetcher stops at the last word in the segment, otherwise exception 13 is generated.

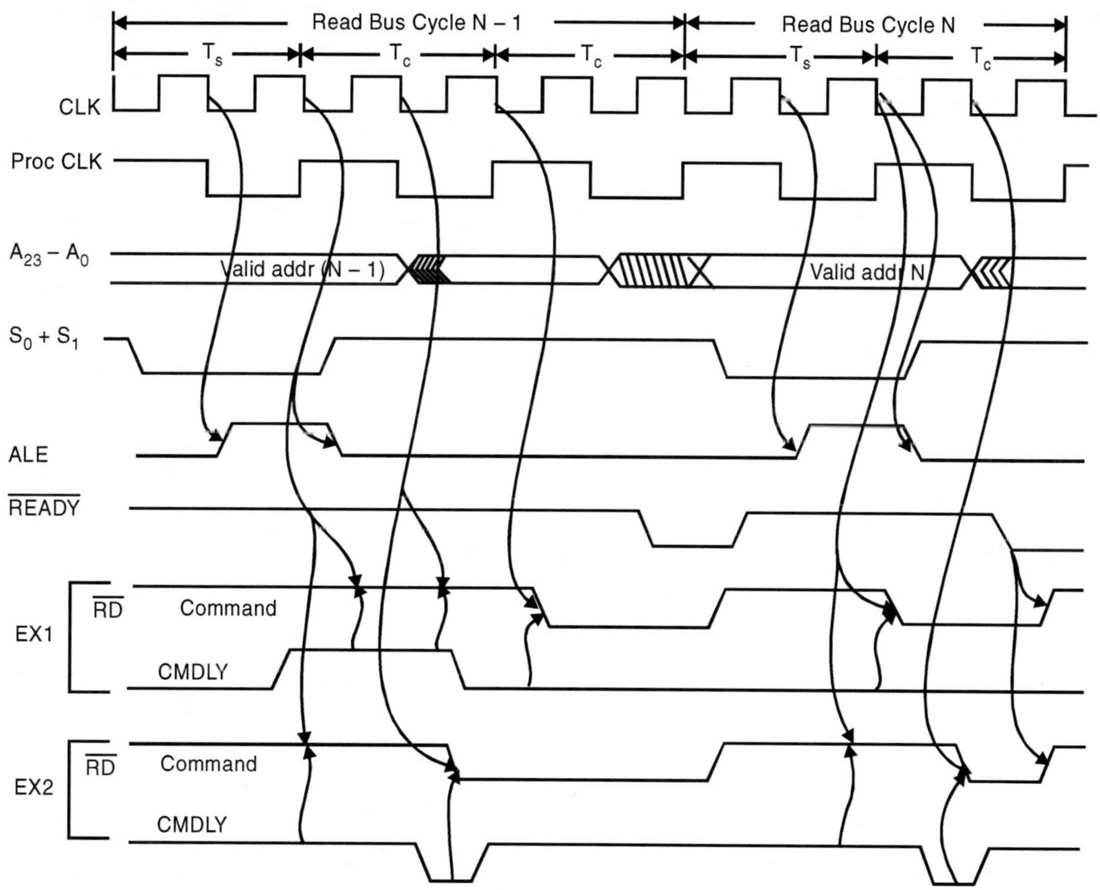

Fig. 9.19. Use of CMDLY to control Command Signal (Intel Corp.).

9.12. 80286 MINIMUM SYSTEM CONFIGURATION

Unlike 80286 system the numeric data processor (processor extension) 80287 is an integral (but optional) part of the system i.e. 80286 is always in the maximum mode. Also the interrupt controller 82516A, clock generate 82C284 and bus controller 82C288 are the unavoidable members of the family of supporting chips of 80286. All these components along with a processor extension 80287 from a integrated processing system. All the data transfers to/from memory or I/O are carried out by 80286. The 80286 also controls the data transfer

And instruction execution of 80287. The 80287 adds its instruction set to the instruction set of 80286. as if 80286-80287 couplet is an integrated processor. The execution by 80287 is transparent to the users, but has all the protection features of 80286 at its service.

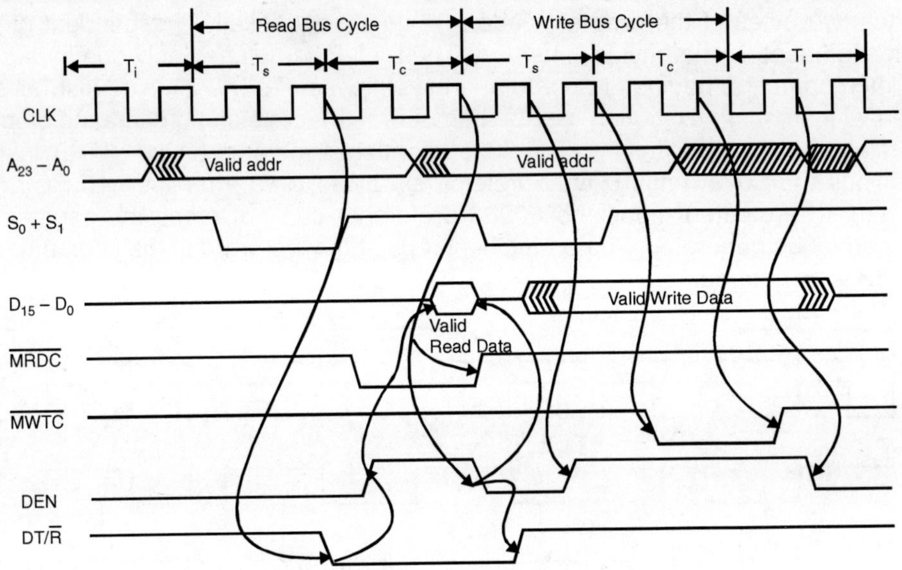

Fig. 9.20. Successive Read and Write Cycles (Intel Corp.).

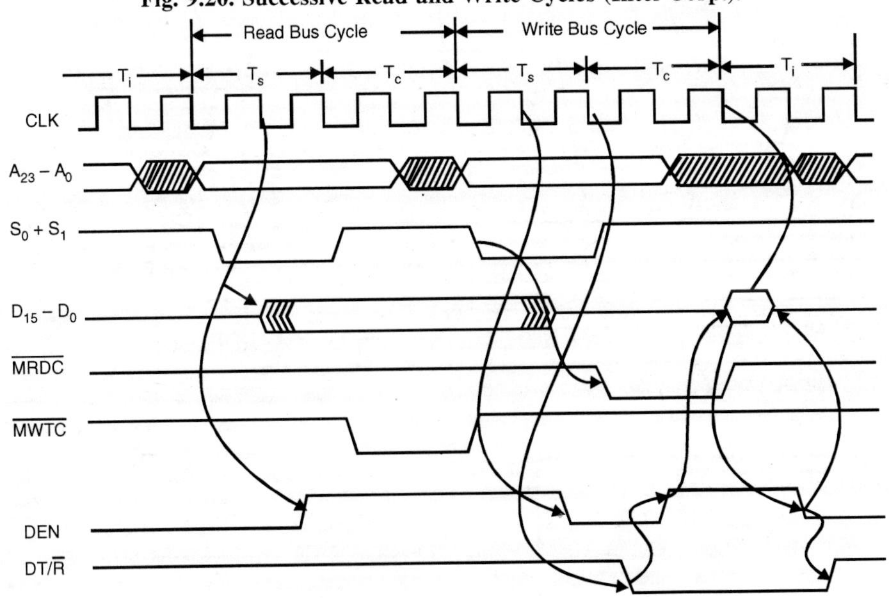

Fig. 9.21. Successive Write and Read Cycles (Intel Corp.).

As already discussed, the addresses available on the local bus of the 80286 are for the next, i.e. (N+1)th fetch operations. The chip select decoding and addresses transmission for data transfer of Nth cycle may overlap with fetch operation for the (N+1)th cycle. The decode logic is an 80286 system is usually designed using PROM or PLA. The decode logic uses the overlap between address and data of the 80286 bus cycles to generate to the advance memory I/O select signals. The COD/INTA and M/IO signals are applied to the decoding logic to differentiate between interrupt, I/O, code fetch and data bus cycle. Figure 9.23 shows a minimum 80286 system configuration.

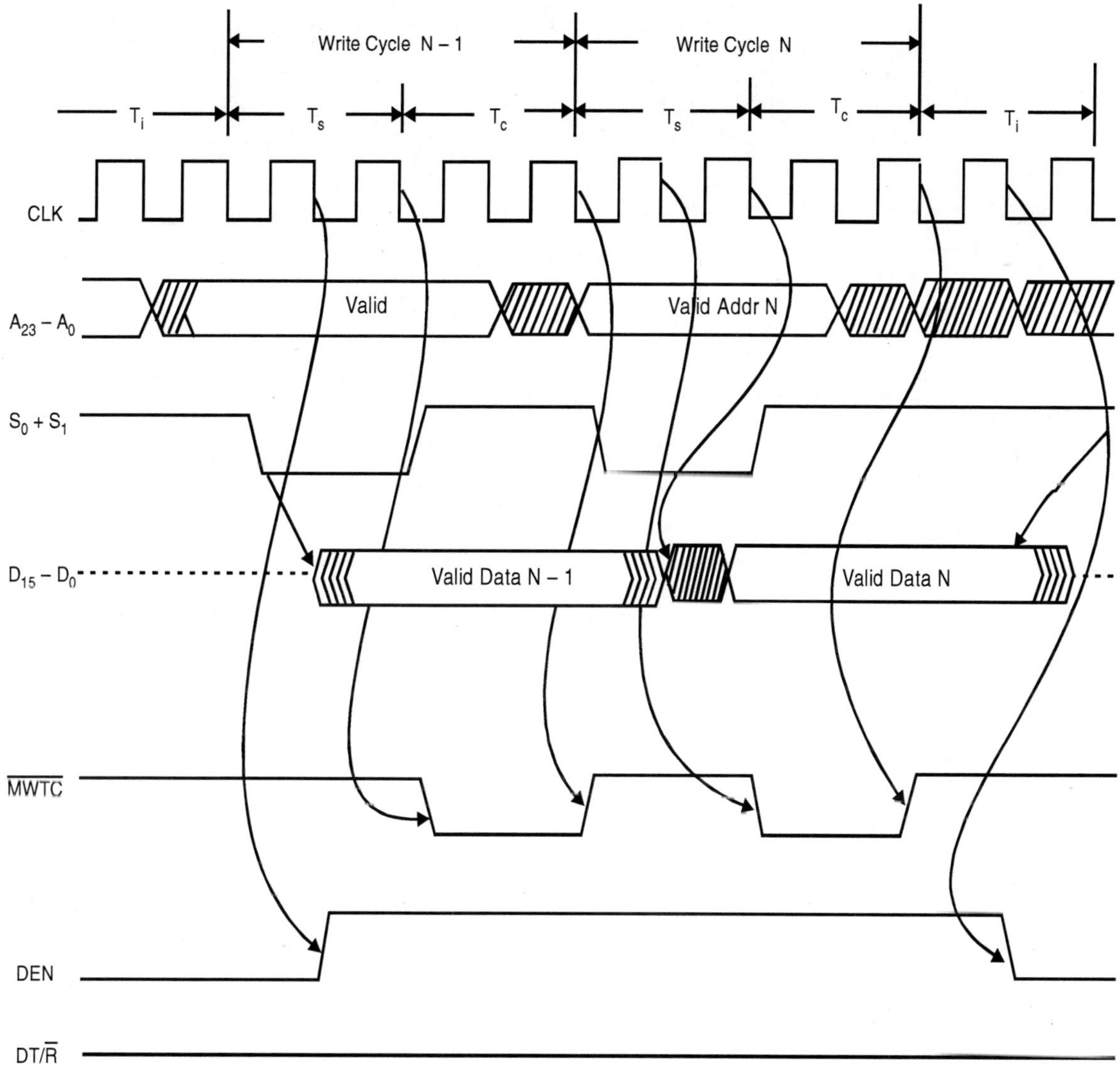

Fig. 9.22. Successive Write-Write Cycles (Intel Corp.).

The addition of just a simple chip 82C286 called as bus arbiter, to the configuration of Fig. 9.23 gives, the multibus structure of Fig. 9.24. The ALE signal of 82C288 in Fig. 9.26 is multibus system.

The addition of one or more bus controllers, additional latches and transceivers to the circuit in Fig. 9.24 generates a bus for local memory and peripheral interfacing besides the multibus.

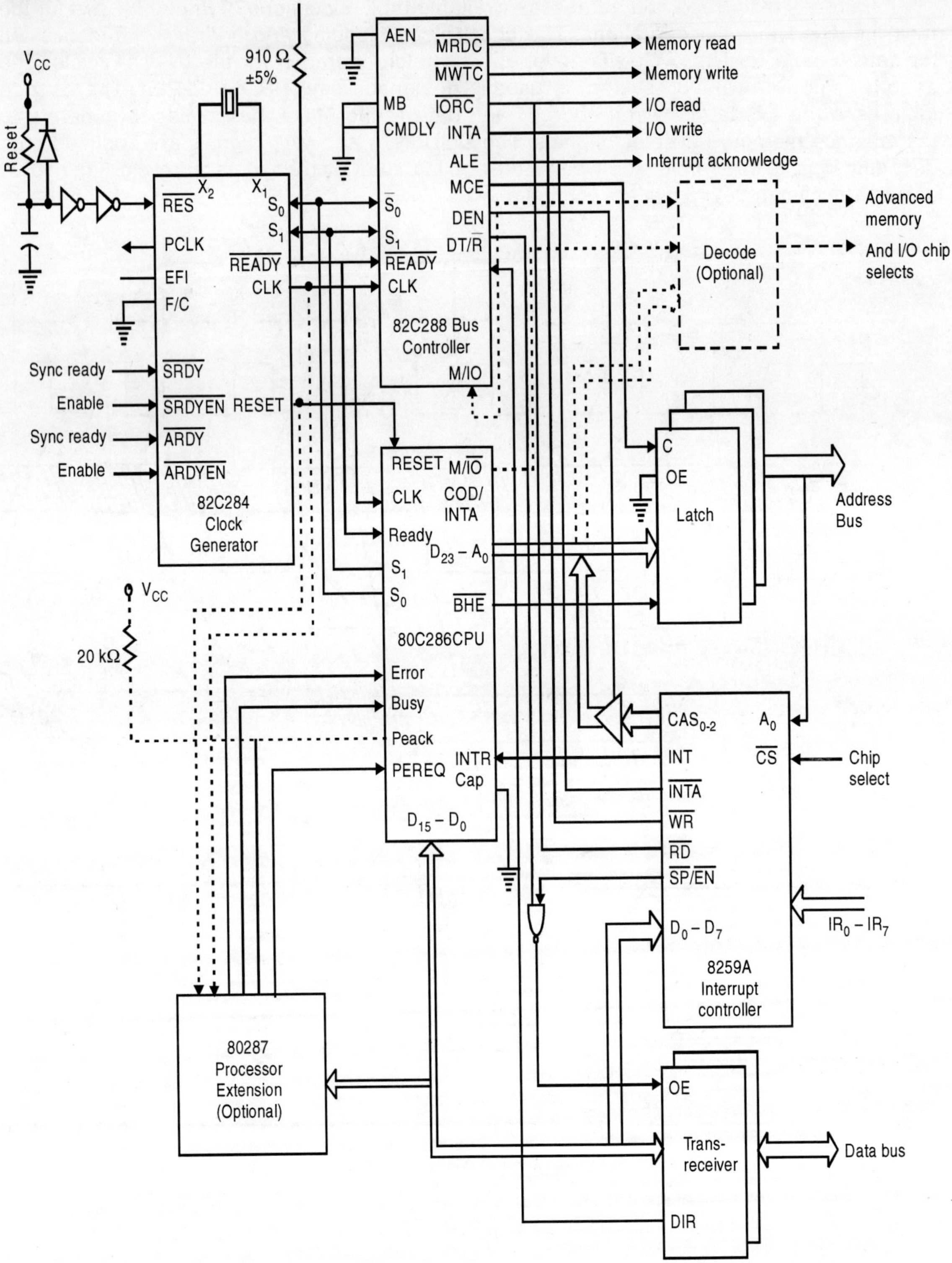

Fig. 9.23. 80286 Minimum System Configuration (Intel Corp.).

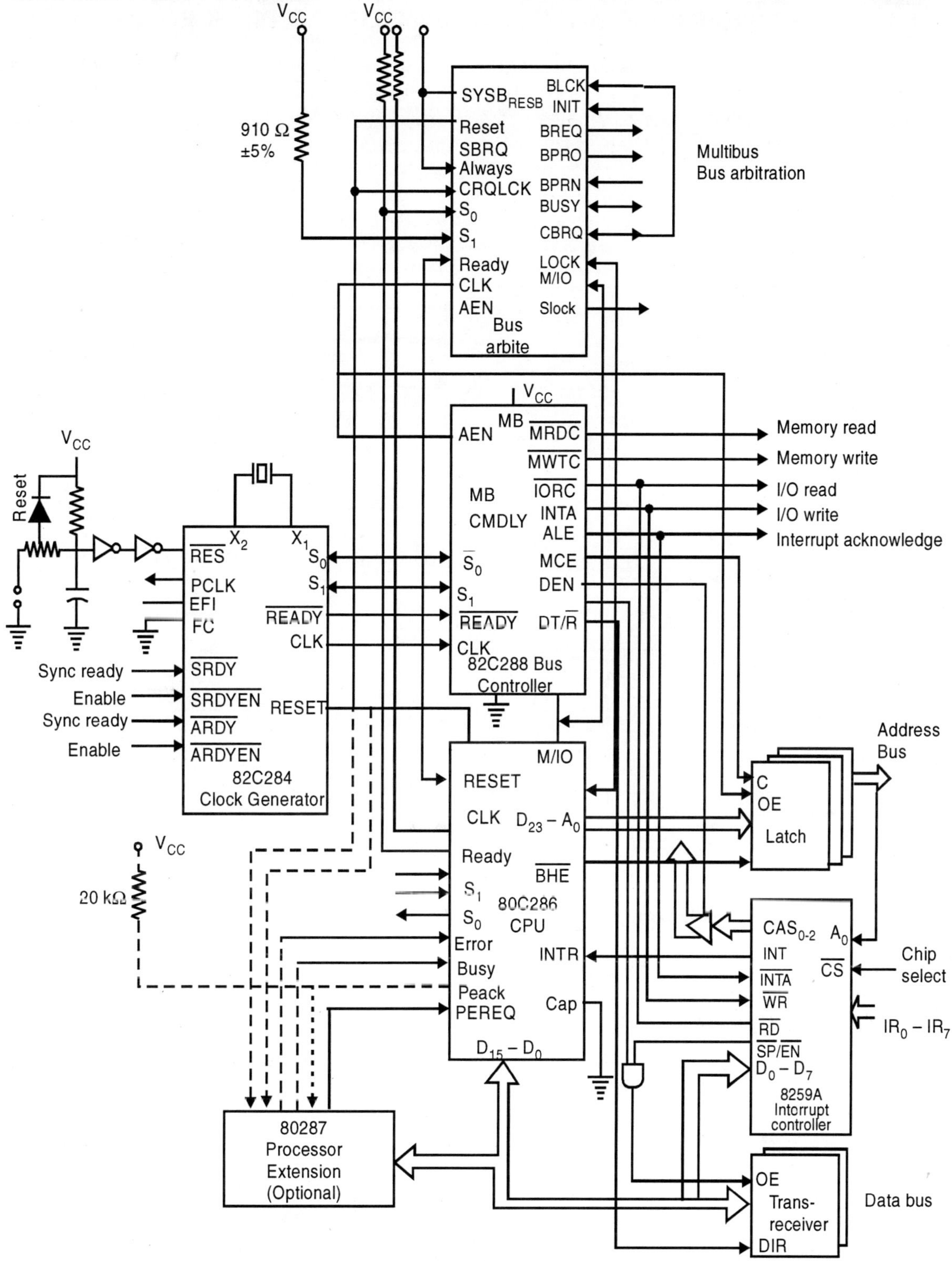

Fig. 9.24. 80286 Multibus System Interface (Intel Corp.).

9.13. INTERFACING MEMORY AND I/O DEVICES WITH 80286

The memory interfacing technique with 80286 is similar to that with 8068 in maximum mode. The maximum mode 8086 system has 8288 and 82816 as bus controller and bus arbiter respectively, which derive and control the system bus. In 80286 System compatible bus controller 80288 and bus arbiter 82889 are used for the same purpose.

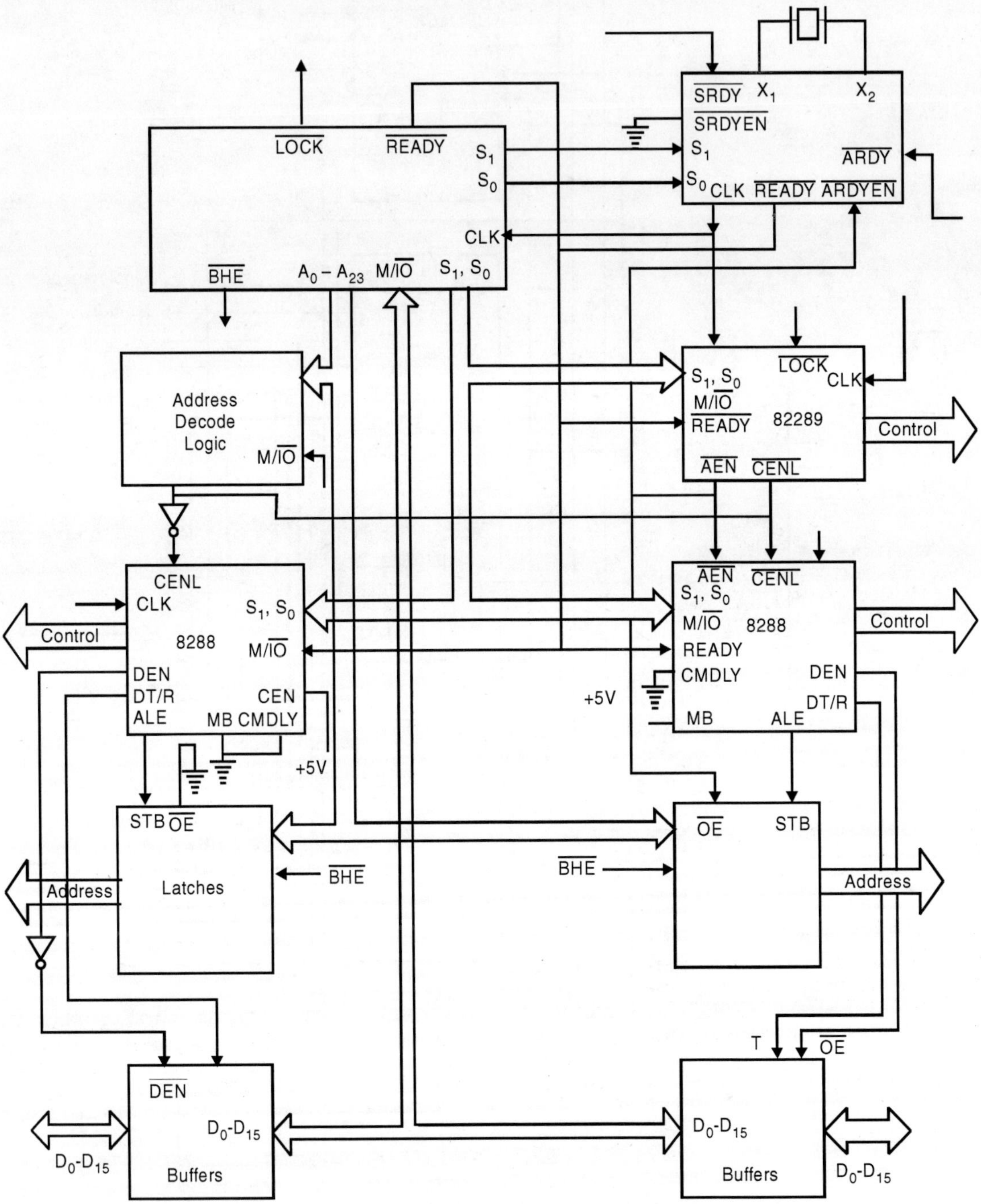

Fig. 9.25. Deriving System Bus and Local Bus (Intel Corp.).

The 80286 may use its local bus address, its local memory,I/O devices and a system bus to share with the other masters as shown in Fig. 9.25. This configuration use two 82288 bus controller one each for local and system bus. The bus arbiter is required for arbitration of system bus only. The ARDYEN and SRDYEN inputs are the enable pins for ARDY and SRDY which generate ready input to 80286. The ARDYEN pin is to be activated, if the 80286 is to use system bus. The SRYDEN pin is to be grounded, if the local bus is to be kept dedicated to 80286. If the pin MBYTES (multibus mode select) of 82288 is tied high, it places the bus controller in multibus mode, else the bus controller is in single bus mode. As the commands are not to be delayed, the CMDLY pin is grounded. The MBYTES inputs select the function of CEN/AEN pin. If MYBTES is high the pin serves as AEN else it serves as CEN. The CENL pin is used for selecting one of the available 82288s. The SYS/RESB input of 82289 enables the system bus is to be used.

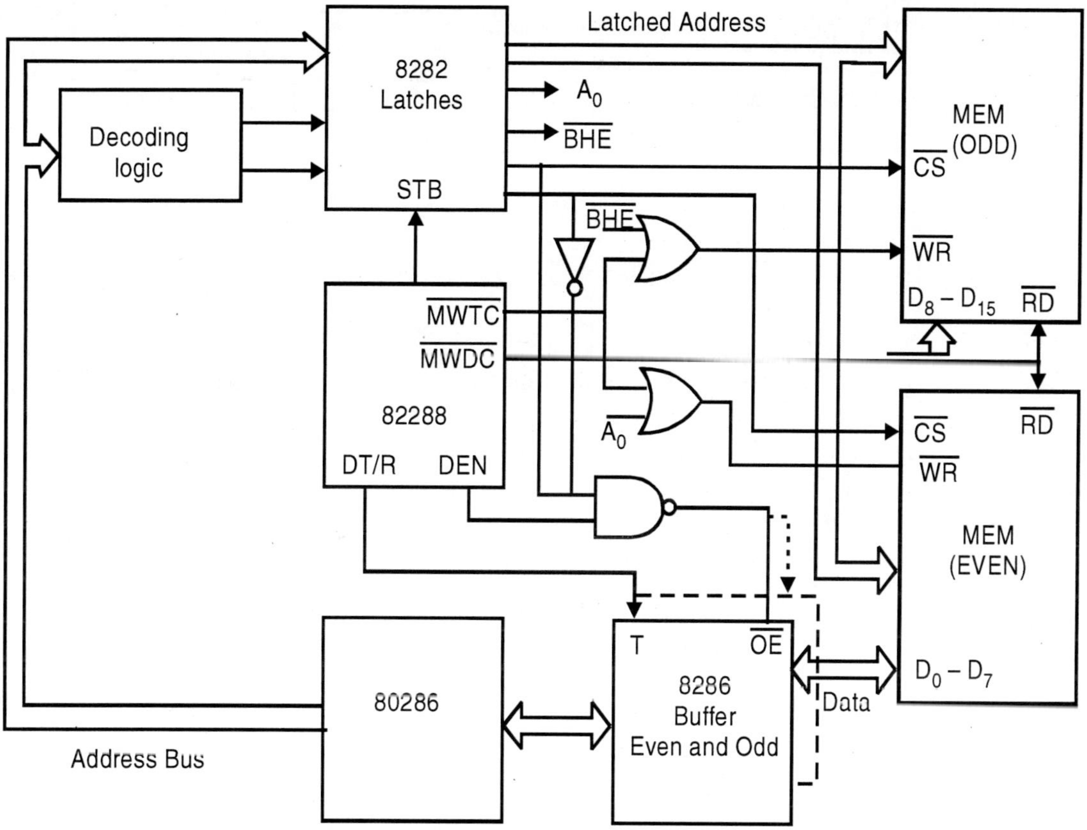

Fig. 9.26. Memory Interfacing using ALE signal (Intel Corp.).

With this much information regarding the buses, one may go for interfacing the memory with 80286. As already said, the 80286 also addresses the memory in terms of even and odd banks. Figure 9.26 shows a bank of memory interfaced with 80286 are not multiplexed hence no latches are required in an 80286 system. Rather the addresses of the next bus cycle are displayed in advance, hence latch are required for latching the addresses and decode the select signals. For this the ALE derived by 82288 is used for latching the addresses. The problem with this method is that it does not allow insertion of wait states for interfacing low speed devices. Figure 9.27 shows an interfacing scheme to interface slower memory devices without compromising for the data transfer rate.

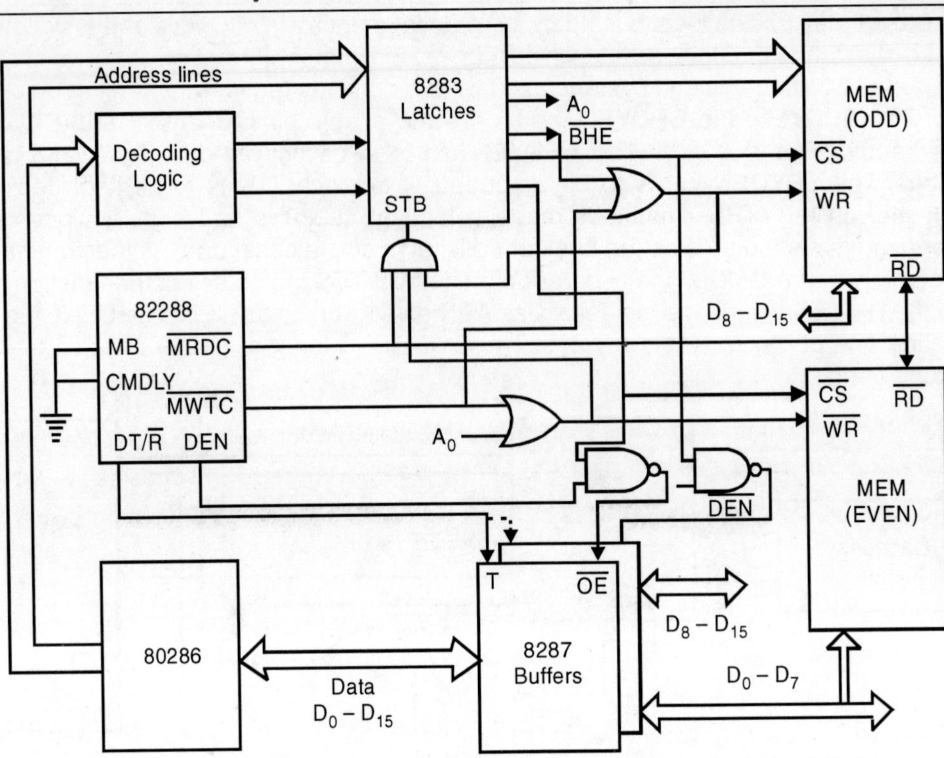

Fig. 9.27. Memory Interfacing using Strobe Logic (Intel Corp.).

For interfacing I/O devices with 80286, a separate set of latches may be used that may be stored using IORD and IOWR signals. Generally, the ALE is not using for latching the I/O addresses. The storable data transfer allows various sloe devices to be interfaced with 80286. The 80259 can address at the most 64K, 8-bit output ports or 32K 16-bit input and 32K 16bit output ports. The ports addresses are either 8-bit (usually specified in the instruction) or 16-bit (stored in DX register).The 8-bit port addresses are zero extended or in other words unused higher addressing lines during addressing an I/O device are pulled low by 80286. The I/O port addresses 00F8H to 00FFH are reserved by intel, hence these should not be used while designing practical system around 80286.

9.14. PRIORITY OF BUS USE BY 80286

The 80286 uses its local bus different purposes during its operation. Rather the internal units of 80286 ask for the access of the local bus whenever they require it. If at any instant of time, only one of the units of 80286 ask for the access of the bus, there is no problem. However, if there is more than one request for the access of the bus, contention may arise. To avoid this problem, the internal architecture of 80286 has allotted priorities to different usages of the bus. The higher priority usages may supercede the lower priority usages and the units may according gain the control of the bus. The relative priorities of the usage are as listed below, starting from the highest one to the lowest one.

Highest Priority

1. Transfer with LOCK activated.
2. The second byte transfer of the 2-byte transfer at an odd address.
3. The second or third transfer cycle of a processor extension data transfer.

4. HOLD request.
5. Processor extension data transfer using PEREQ.
6. Data transfer performed by EU for instruction execution.

Lowest Priority

7. A perfetch operation to fetch and arranges the next instruction bytes in queue.

Once the Processor halts or enters a shut down state, the local bur goes into tristate condition. Only RESET or NMI may pull 80286 out of these states, unconditionally. If is set, the INTR or coprocessor segment overrun exception will pull the 80286 out of halt or shut down. Processor enters the halt state as a result of execution of a HLT instruction. It enters shut down due to multiple violations of the protection norms.

9.15. BUS HOLD AND HLDA SEQUENCE

The 80286 bus is relinquished for another bus master if a valid bus hold request is received at the HOLD input pin. As a response to a valid bus request, the bus is pushed into the T_H State. The status lines are relinquished by 80286 during $T_{H,}$ but are pulled up internally by 82C284. The addresses M,/IO and COD/INTA are also relinquished by the bus arbiter. The sequence of control transfer from 80286 to another master is shown in Fig. 9.28. if during any operation, another local bus master requests the local bus from the 80286 using a HOLD signal, the 80286 complete the current operation, remains idle for one PCLK and then enters T_H sending HLDA high.

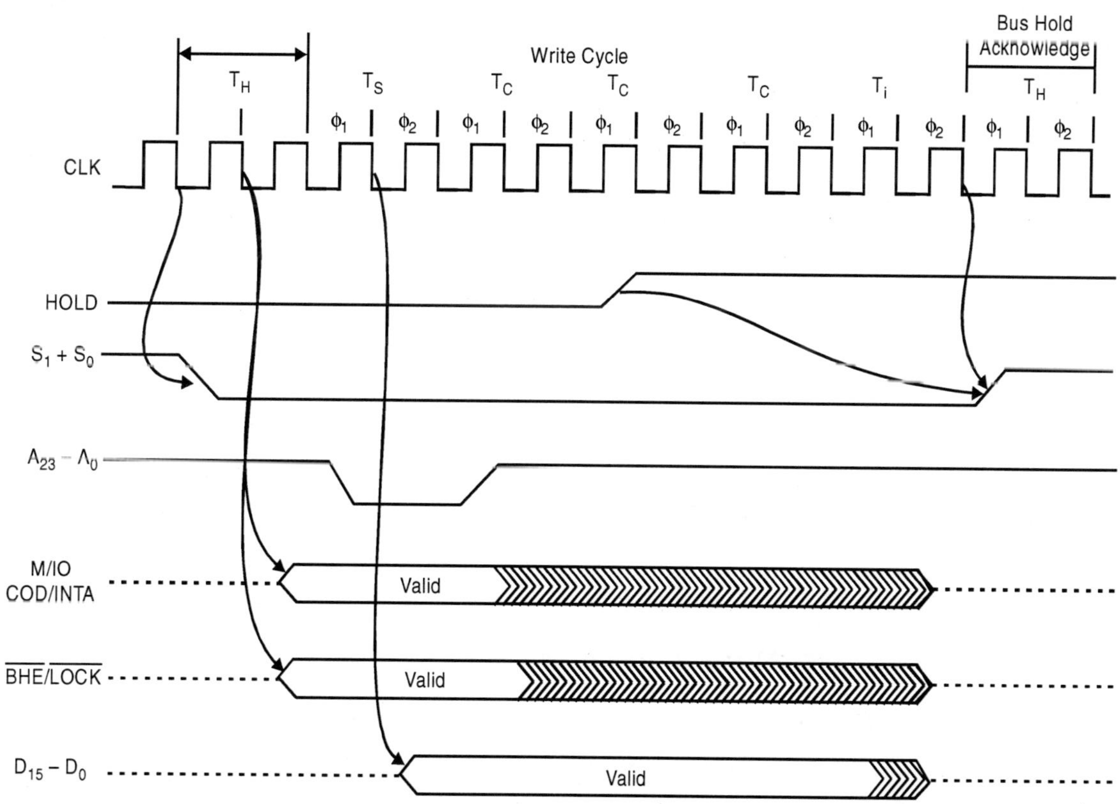

Fig. 9.28. HOLD and HLDA Sequence of 80286 (Intel Corp.).

9.16. 80286 SUPPORTED DATA TYPES

The 80286 supports the following seven data types :

(1) **Integer : 8-bit or 16-bit signed binary operands using 2's complement representation.**

(2) **Ordinal (Unsigned) : 8-bit or 16-bit Unsigned numeric value in binary.**

(3) **Pointer : 32-bit pointers consisting of two 16-bit parts for segment selector and offset.**

(4) **String : A data string of maximum 64kbytes or 32K words.**

(5) **BCD : BCD representations and operations on decimal digit 0-9.**

(6) **Packed BCD : Two digit number represented by using BCD symbols.**

The 80286 supported data types are shown in Fig. 9. 29.

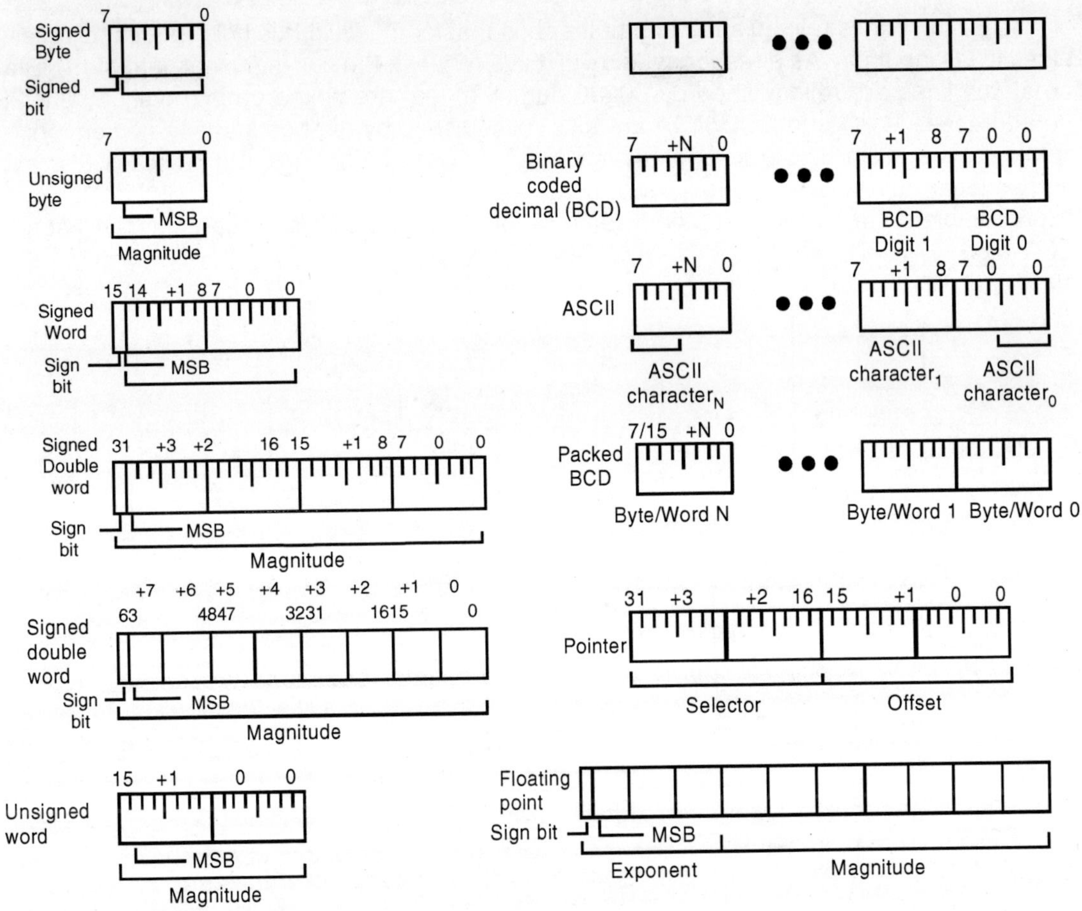

Fig. 9.29. Data Types supported by 80286 (Intel Corp.).

9.17. INTRODUCTION TO THE 80287 NUMERIC PROCESSOR EXTENSION

This Section of the chapter describes the 80287 Numeric Processor Extension (NPX) for the 80286 microprocessor. Below is a brief overview of 80286 concepts.

The 80287 NPX is a high-performance numerics processing element that extends the 80286 architecture by adding significant numeric capabilities and direct support for floating-point, extended-integer, and BCD data types. The 80286 CPU with 80287 NPX easily supports powerful and accurate numeric applications through its implementation of the proposed IEEE 754 Standard for Binary Floating-Point Arithmetic.

The 80287 Numeric Processor Extension (NPX) is highly compatible with its predecessor, the earlier Intel 8087 NPX. The 8087 NPX was designed for use in 8086-family systems. The 8086 was the first microprocessor family to partition the processing unit to permit high-performance numeric capabilities. The 8087 NPX for this processor family implemented a complete numeric processing environment in compliance with the proposed IEEE 754 Floating-Point Standard.

With the 80287 Numeric Processor Extension, high-speed numeric computations have been extended to 80286 high-performance multi-tasking and multi-user systems. Multiple tasks using the numeric processor extension are afforded the full protection of the 80286 memory management and protection features.

Ease of Use

The 80287 NPX offers more than raw execution speed for computation-intensive tasks. The 80287 brings the functionality and power of accurate numeric computation into the hands of the general user.

Like the 8087 NPX that preceded it, the 80287 is explicitly designed to deliver stable, accurate results when programmed using straightforward "pencil and paper" algorithms. The IEEE 754 standard specifically addresses this issue, recognizing the fundamental importance of making numeric computations both easy and safe to use.

For example, most computers can overflow when two single-precision floating-point numbers are multiplied together and then divided by a third, even if the final result is a perfectly valid 32-bit number. The 80287 delivers the correctly rounded result. Other typical examples of undesirable machine behavior in straightforward calculations occur when solving for the roots of a quadratic equation :

$$\frac{-b \pm {}^{1}(b^2 - 4ac)}{2a}$$

for computing financial rate of return, which involves the expression : (1+i)^(n). On most machines, straightforward algorithms will not deliver consistently correct results (and will not indicate when they are incorrect). To obtain correct results on traditional machines under all conditions usually requires sophisticated numerical techniques that are foreign to most programmers. General application programmers using straightforward algorithms will produce much more reliable programs using the 80287. This simple fact greatly reduces the software investment required to develop safe, accurate computation-based products.

Beyond traditional numerics support for scientific applications, the 80287 has built-in facilities for commercial computing. It can process decimal numbers of up to 18 digits without round-off errors, performing exact arithmetic on integers as large as 2^(64) or 10^(18). Exact arithmetic is vital in accounting applications where rounding errors may introduce monetary losses that cannot be reconciled.

The NPX contains a number of optional facilities that can be invoked by sophisticated users. These advanced features include two models of infinity, directed rounding, gradual underflow, and either automatic or programmed exception-handling facilities.

These automatic exception-handling facilities permit a high degree of flexibility in numeric processing software, without burdening the programmer. While performing numeric calculations, the NPX automatically detects exception conditions that can potentially damage a calculation. By default, on-chip exception handlers may be invoked to field these exceptions so that a reasonable result is produced, and execution may proceed without program interruption. Alternatively, the NPX can signal the CPU, invoking a software exception handler whenever various types of exceptions are detected.

Applications

The NPX's versatility and performance make it appropriate to a broad array of numeric applications. In general, applications that exhibit any of the following characteristics can benefit by implementing numeric processing on the 80287 :

(1) Numeric data vary over a wide range of values, or include nonintegral values.

(2) Algorithms produce very large or very small intermediate results.

(3) Computations must be very precise; i.e., a large number of significant digits must be maintained.

(4) Performance requirements exceed the capacity of traditional microprocessors.

(5) Consistently safe, reliable results must be delivered using a programming staff that is not expert in numerical techniques.

Note : Also that the 80287 can reduce software development costs and improve the performance of systems that use not only real numbers, but operate on multiprecision binary or decimal integer values as well.

A few examples, which show how the 80287 might be used in specific numerics applications, are described below. In many cases, these types of systems have been implemented in the past with minicomputers. The advent of the 80287 brings the size and cost savings of microprocessor technology to these applications for the first time.

(1) **Business data processing :** The NPX's ability to accept decimal operands and produce exact decimal results of up to 18 digits greatly simplifies accounting programming. Financial calculations that use power functions can take advantage of the 80287's exponentiation and logarithmic instructions.

(2) **Process control :** The 80287 solves dynamic range problems automatically, and its extended precision allows control functions to be fine-tuned for more accurate and efficient performance. Control algorithms implemented with the NPX also contribute to improved reliability and safety, while the 80287's speed can be exploited in real-time operations.

(3) **Computer numerical control (CNC) :** The 80287 can move and position machine tool heads with accuracy in real-time. Axis positioning also benefits from the hardware trigonometric support provided by the 80287.

(4) **Robotics :** 'Coupling small size and modest power requirements with powerful computational abilities, the NPX is ideal for on-board six-axis positioning.

(5) Navigation"Very small, lightweight, and accurate inertial guidance systems can be implemented with the 80287. Its built-in trigonometric functions can speed and simplify the calculation of position from.

(7) Graphics terminals"The 80287 can be used in graphics terminals to locally perform many functions that normally demand the attention of a main computer; these include rotation, scaling, and interpolation. By also using an 82720 Graphics Display Controller to perform high speed data transfers, very powerful and highly self-sufficient terminals can be built from a relatively small number of 80286 family parts.

(8) **Data acquisition :** The 80287 can be used to scan, scale, and reduce large quantities of data as it is collected, thereby lowering storage requirements and time required to process the data for analysis.

The preceding examples are oriented toward traditional numerics applications. There are, in addition, many other types of systems that do not appear to the end user as computational, but can employ the 80287 to advantage. Indeed, the 80287 presents the imaginative system designer with an opportunity similar to that created by the introduction of the microprocessor itself. Many applications can be viewed as numerically-based if sufficient computational power is available to support this view. This is analogous to the thousands of successful products that have been built around "buried" microprocessors, even though the products themselves bear little resemblance to computers.

Upgradability

The architecture of the 80286 CPU is specifically adapted to allow easy upgradability to use an 80287, simply by plugging in the 80287 NPX. For this reason, designers of 80286 systems may wish to incorporate the 80287 NPX into their designs in order to offer two levels of price and performance at little additional cost.

Two features of the 80286 CPU make the design and support of upgradable 80286 systems particularly simple :

(1) The 80286 can be programmed to recognize the presence of an 80287 NPX; that is, software can recognize whether it is running on an 80286 or an 80287 system.

(2) After determining whether the 80287 NPX is available, the 80286 CPU can be instructed to let the NPX execute all numeric instructions. If an 80287 NPX is not available, the 80286 CPU can emulate all 80287 numeric instructions in software. This emulation is completely transparent to the application software"the same object code may be used by both 80286 and 80287 systems. No relinking or recompiling of application software is necessary; the same code will simply execute faster on the 80287 than on the 80286 system.

To facilitate this design of upgradable 80286 systems, Intel provides a software emulator for the 80287 that provides the functional equivalent of the 80287 hardware, implemented in software on the 80286. Except for timing, the operation of this 80287 emulator (E80287) is the same as for the 80287 NPX hardware. When the emulator is combined as part of the systems software, the 80286 system with 80287 emulation and the 80286 with 80287 hardware are virtually indistinguishable to an application program.

This capability makes it easy for software developers to maintain a single set of programs for both systems. System manufacturers can offer the NPX as a simple plug-in performance option without necessitating any changes in the user's software.

Programming Interface

The 80286/80287 pair is programmed as a single processor; all of the 80287 registers appear to a programmer as extensions of the basic 80286 register set. The 80286 has a class of instructions known as ESCAPE instructions, all having a common format. These ESC instructions are numeric instructions for the 80287 NPX. These numeric instructions for the 80287 are simply encoded into the instruction stream along with 80286 instructions.

All of the CPU memory-addressing modes may be used in programming the NPX, allowing convenient access to record structures, numeric arrays, and other memory-based data structures. All of the memory management and protection features of the CPU are extended to the NPX as well.

Numeric processing in the 80287 centers around the NPX register stack. Programmers can treat these eight 80-bit registers as either a fixed register set, with instructions operating on explicitly-designated registers, or a classical stack, with instructions operating on the top one or two stack elements.

Internally, the 80287 holds all numbers in a uniform 80-bit temporary-real format. Operands that may be represented in memory as 16-, 32-, or 64-bit integers, 32-, 64-, or 80-bit floating-point numbers, or 18-digit packed BCD numbers, are automatically converted into temporary-real format as they are loaded into the NPX registers. Computation results are subsequently converted back into one of these destination data formats when they are stored into memory from the NPX registers. Software routines for the 80287 may be written in ASM286 Assembler or any of the following higher-level languages :

PL/M-286
PASCAL-286
FORTRAN-286
C-286

In addition, all of the development tools supporting the 8086 and 8087 can also be used to develop software for the 80286 and 80287 operating in Real-Address mode.

All of these high-level languages provide programmers with access to the computational power and speed of the 80287 without requiring an understanding of the architecture of the 80286 and 80287 chips. Such architectural considerations as concurrency and data synchronization are handled automatically by these high-level languages. From the programmer's perspective, the 80287 can be considered just an extension of the 80286 processor. All interaction between the 80286 and the 80287 processors on the hardware level is handled automatically by the 80286 and is transparent to the software.

9.17.1. Internal Architecture of 80287

The internal architecture of 80287 is shown in figure 9.30. The architecture of 80287 is divided into three sections :

(1) Bus Control Logic
(2) Data Interface and Control unit
(3) Floating Point Unit

The control logic provides and controls the interface between the internal 80287 us and the 80286 bus via a data buffer. The data interface and control unit contains :

(a) Status words
(b) Control words
(c) TAG word and
(d) Error point interface

The status word reflects the current status of 80287. The control word selects one of the processing options provided by it and is to be programmed by the CPU. The TAG word optimizes the NDP performance by maintaining a record of empty and non-empty register locations. The error pointer points to the source of exception generated. The instruction decoder and sequencer decodes and forwards the instructions for further execution by the floating point unit.

The floating point unit is an actual processing section of the NDP. The data bus interface and data alignment and operand checking section checks the alignment and validity of the data. If any error is used for storing operand, data and are arranged by the 80287.

The eight 80-bit registers are used for storing operand data and are arranged as a stack. The data bus in floating point unit is of 84 bits, out of which the lower 68 bits are significant data bits, the next 16 bits are used for exponent. The exponential operand registers are used to store the operands in exponential form during the operation. The 80-bit registers maintain 80-bit operands required for 80287 operations. The barrel shifter arranges and presents the data to be shifted successively whenever required for the execution.

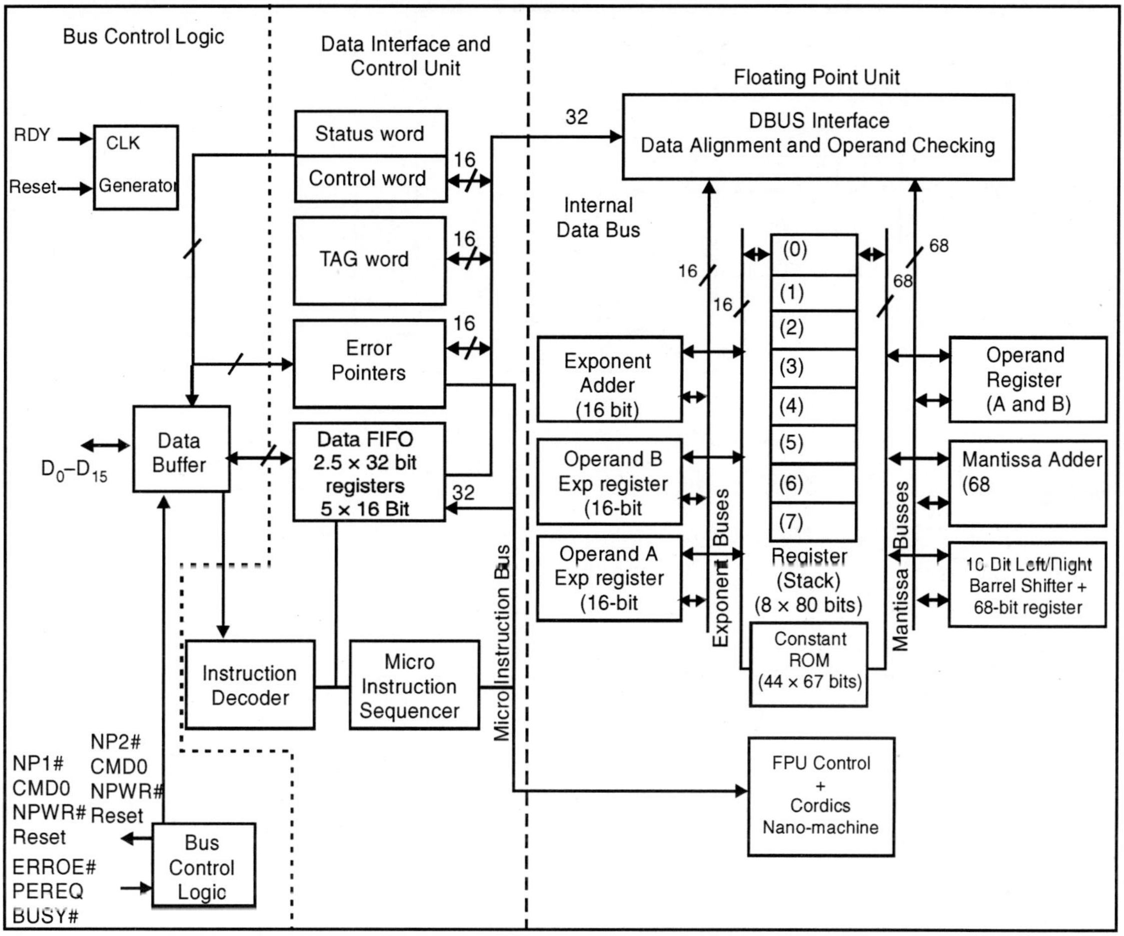

Fig. 9.30. Internal Architecture of 80287.

9.17.2. Register Set of 80287

The 80287 has a set of 80-bit registers, that can be used either as a stack or a set of general registers. When operating as a stack, it operates on the top one or two registers. While operating as a register set, they may be used only with the instructions designed for them. These registers are divided into three fields :

(a) Sign (1-bit)

(b) Exponent (15-bits)

(c) Significand (64-bits)

Corresponding to each of these eight registers, there is a two bit TAG field to indicate the status of the contents. The TAG word register presents all the TAG fields to the CPU. The instructions may address data registers either implicitly or explicitly. An internal status register field, 'TOP' is used to address any one of the eight registers imlicitly. While explicitly addressing the registers, they may be addressed relative to 'TOP'.

The status word of 80287 is shown in the Fig. 9.31.

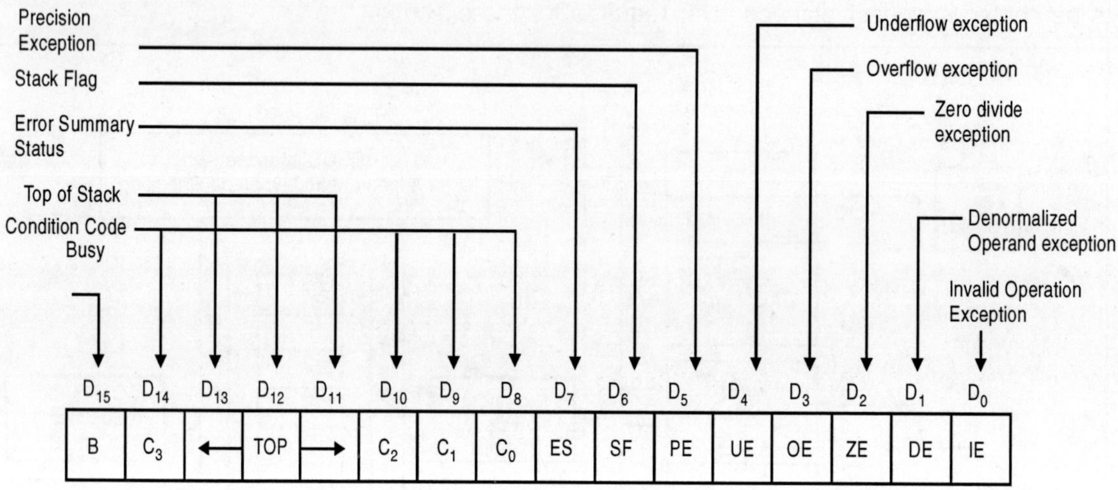

Notes :

1. ES is set if any unmasked exception bit is set, cleared otherwise.
2. See Table 8.1 for condition code interpretation.
3. TOP values

   000 = Register 0 is Top of Stack

   001 = Register 1 is Top of Stack

   •

   •

   •

   111 = Register 7 is Top of Stack

Fig. 9.31. Status Word of 80287.

9.17.2.1. Status and Control Words of 80287

Status word is a set of 16 flags which are modified depending upon the current status of 80287. The bit definitions for the various flag bits are as explained below :

(i) **B Flag (D_{15}) :** The BUSY flag has the same status as ES flag. This is just for maintaining the compatibility with 8087. This is not at all related with the BUSY output of 80287.

(ii) **TOP (D_{13} - D_{11}) :** These bits point to one of the eight stack registers as stack top.

(iii) **C_3, C_2, C_1 and C_0 (D_{14} and D_{10} - D_8) :** These condition code bits are similar to the flags of a CPU. These are modified depending upon the result of execution of arithmetic instructions.

(iv) **ES (D_7) :** Error summary bit is set, if an unmasked exception is generated. If this is set, the ERROR signal is activated.

(v) **SF (D_6) :** The stack flag is set, if the operation goes invalid due to stack overflow or underflow. If this is set, and $C_1 = 1$, the stack has overflown and if this is set and $C_1 = 0$, the stack has underflown.

(vi) **Exception Flags (D_5 - D_0) :** These exception flags are used to show the generation of an exception while 80287 is executing. This can also be written by using FLDENV or FRSTOR instructions.

9.17.2.2. Control Word

The control word is used to select one of the processing options amongst the ones provided by 80287. The various bits of the control word are discussed as follows :

(i) **Masking Bits (D_5 - D_0) :** These are six masking bits used to mask the six exceptions shown in the status register. If this is '1' the respective exception is masked.

(ii) **Precision Control Bits (D_8 - D_9) :** These are used to set the internal precision of 80287. These bits affect the ADD, SUB, DIV, MUL and SQRT results. For other instructions the precision is decided by the opcode or the extended precision format.

(iii) **Rounding Control Bits (D_{10} - D_{11}) :** These bits are used to set rounding or chopping, as specified in IEEE standard. Rounding control bits affect, only the instructions which perform rounding operation, for example, in arithmetic and transcendental instructions.

(iv) **Infinity Control Bit (D_{12}) :** Infinity control bit is meaningless in case of 80287 TMXL but can be programmed for compatibility with 80287. this is initialized to zero after reset

The control word format for 80287 is shown in figure 9.32 with definitions in short.

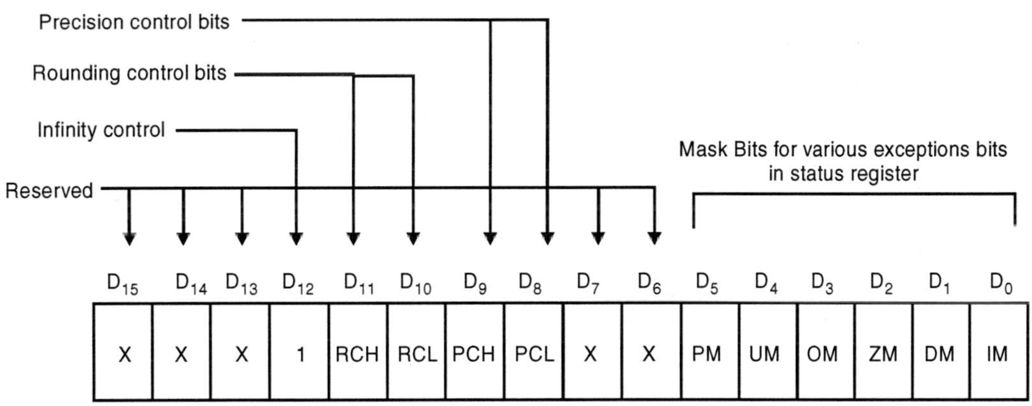

RCH	RCL		PCH	PCL	
0	0	-Round to nearest even	0	0	24-bit (Single)
0	1	-Round Down (to $-\infty$)	0	1	Reserved
1	0	-Round Up (to $+\infty$)	1	0	53-bit (Double)
1	1	-Truncate towards zero (Chop)	1	1	64-bit (Extended)

Fig. 9.32. Control Word of 80287.

9.17.3. Signal Description of 80287

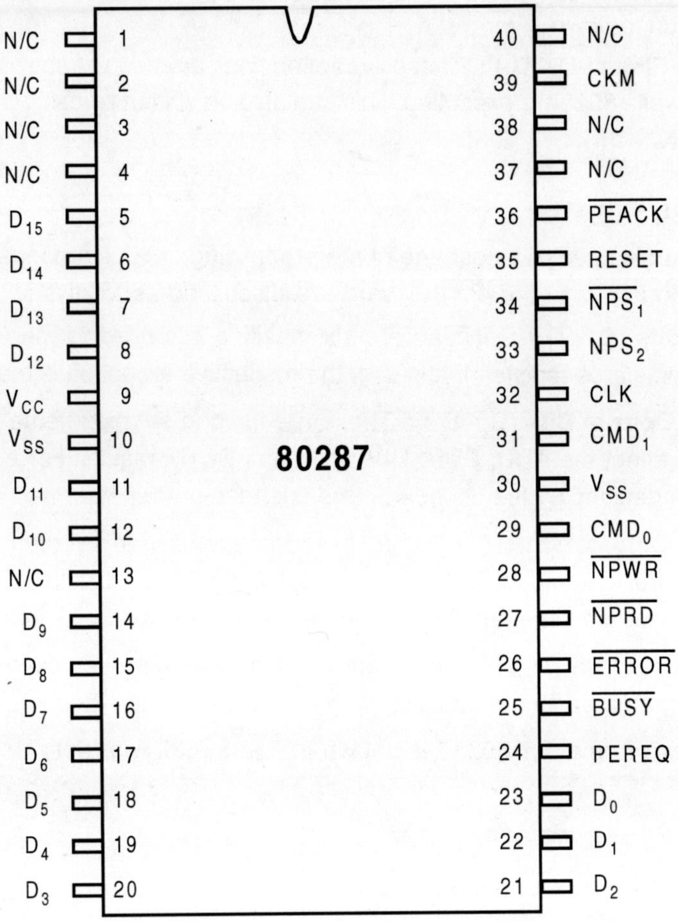

Fig. 9.33. Pin Configuration of 80287.

Fig. 9.33 shows the pin configuration of 80287. The pin description for each pin is as explained :

(1) D_0 - D_{15} : This is a 16-bit data bus to be connected with 80286 data bus.

(2) CLK : This is an input pin that accepts the clock required for deriving the basic system timings.

(3) RESET : This is a reset pin. A high on this pin resets the processor 80287.

(4) NPWR : Numeric processor write active-low pin, if activated, enables a data transfer from 80286 to 80287.

(5) NPRD : Numeric processor Read active-low input pin, if activated, enables a data transfer from 80287 to 80286.

(6) CMD_0 - CMD_1 : The CPU uses these active-high input pins along with select pins to control the operations of 80287.

(7) ERROR : The error status is active high output pin and it represents the ES bit of the internal status register. If this is active, it indicates that an exception has occurred. This is to be connected with the ERROR pin of 80286.

(8) **PEREQ** : Processor Extension Request pin which is active high output pin an dit indicates to the 80286 that the NDP is ready for data transfer.

(9) **BUSY** : This is active low output pin and it indicates to the CPU that it is busy with the execution of an instruction. This is connected to the TEST pin of 80286.

(10) **PEACK :** Processor Extension Acknowledge is an active-low input pin and is used by the CPU to acknowledge a reciept of a valid PEREQ signal.

(11) **CKM** : Clock Mode is an input pin and is kept high, the CLK input is directly used for deriving the internal timings. Else, it is divided by two. This must be stable at reset to decide the mode of operation properly.

9.17.4. Interfacing 80287 with 80286

Figure 9.34 shows the interface circuit between 80286 and 80287. The 80287 establishes its interface with an 80286 system using a set of 10 pins, namely PEREQ, PEACK#, BUSY#, ERROR#, NPS1#, NPS0#, CMD0, CMD1, NPRD# and NPWR#.

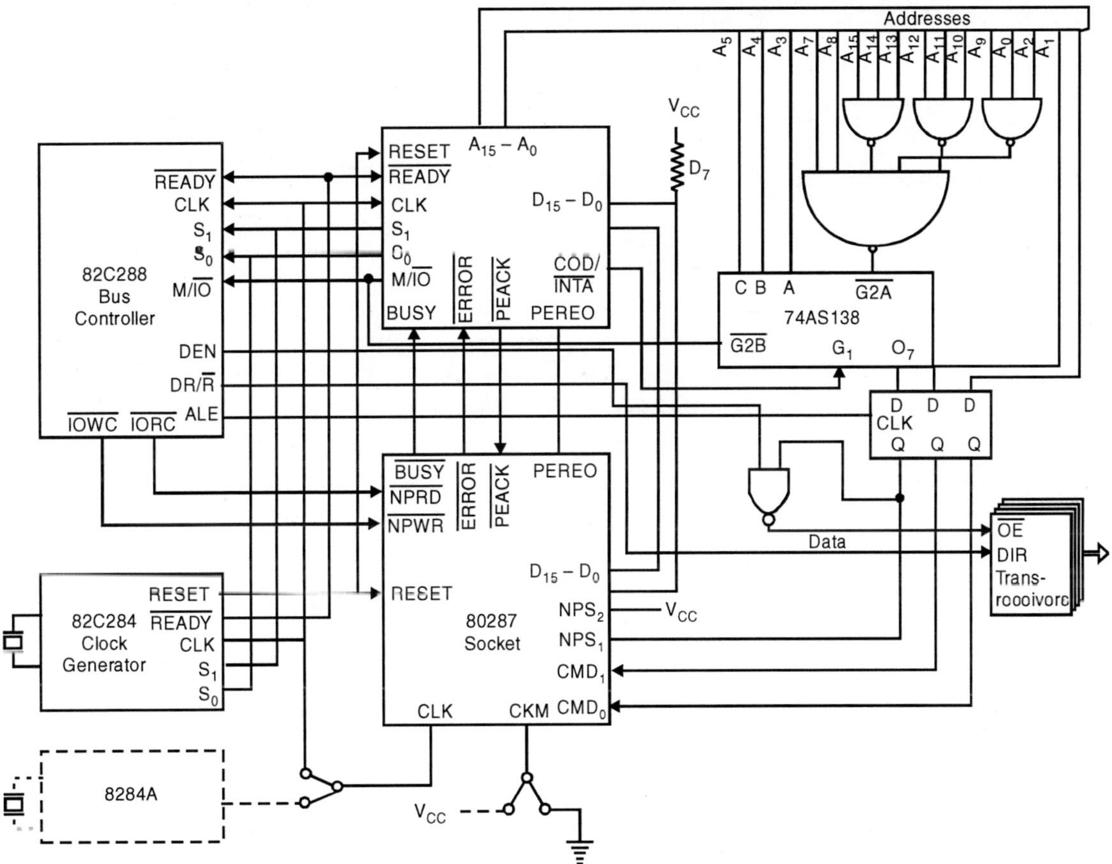

Fig. 9.34. Interface between 80286 and 80287.

The functions of all these pins have already been explained. The 80287 is asynchronous in operation with 80286. The 80286 activates NSP1# and NSP2# signals to start the bus cycle of 80287. In the same clock period in which the NSP1# and NSP2# are activated, the NDP checks

NPRD# and NPWR# signals to check whether it is a read or write cycle. Also it checks CND0 and CMD1 to decide whether it is an opcode, operand or control/status register transfer. The NDP activates its BUSY output after it receives a valid read or write command. The PREQ signal is used by 80287 to inform the CPU that it is ready for a data transfer. Here the processor or extension refers to the coprocessor 80287. When the data transfer is over, the CPU activates PEACK# pin which results in deactivating the PEREQ# pin by 80287.

9.17.5. Data Types Supported by 80287

The 80287 supports, in all, seven data types. The data operands are stored in memory with the least significant byte at the lowest memory address and so on. The operands are referred to by this lowest address. Instructions which read data from memory automatically convert them in the standard formats acceptable to 80287. The standard data formats of 80287 are shown in the below.

S. No.	Format	Range	Size		Definition 1st byte — Last byte
1.	Word Integer	$\pm 10^4$	16 bits		D_{15} ——— D_0 2's complement form
2.	Short Integer	$\pm 10^9$	32 bits		D_{31} ——— D_0 2's complement form
3.	Long integer	$\pm 10^{18}$	64 bits		D_{63} ——— D_0 2's complement form
4.	Single precision	$\pm 10^{\pm 38}$	24 bits	D_{31} Sign	D_{30} — D_{24} Biased exponent; D_{23} — D_0 significand
5.	Double Precision	$\pm 10^{\pm 308}$	53 bits	D_{63} sign	D_{62} — D_{53} Biased exponent; D_{52} — D_0 Significand
6.	Extended precision	$\pm 10^{\pm 4932}$	64 bits	D_{79} Sign	D_{78} — D_{64} Biased exponent; D_{71} — D_0 significand
7.	Packed BCD	$\pm 10^{18}$	18 digits	D_{79} Sign	D_{78} — D_{72} ⟷ x; D_{71} — D_0 ⟷ 18 digits; four bits per digit

9.17.6. Instruction Set Summary

The instructions of 80287 fit in one of the five formats as follows. All these instructions are 2-Byte instructions. The CPU, after its fetch operation, identifies these instructions from the five most significant bits of the first byte. The same addressing modes as that of 80286 can be used to specify the memory operands.

The MOD and R/M fields have their definitions and significances in similar way as in the case of 8086 processor. The instruction formats available in 80287 are listed as follows. Here the OP field represents the opcode bit field. MF field represents the memory format of the operands.

(1)

D_{15} — D_{11}	D_{10} D_9	D_8	D_7 D_6	D_5	D_4 D_3	D_2 D_1 D_0	Optional
11011	OPA	1	MOD	1	OPB	R/M	Displacement

(2)

D_{15} — D_{11}	D_{10} D_9	D_8	D_7 D_6	D_5 D_4 D_3	D_2 D_1 D_0	Optional
11011	MF	OPA	MOD	OPB	R/M	Displacement

MF (Memory Format) OPA and OPB are two parts of opcode.

00-32 bit real

01-32 bit integer

10-64 bit real

11-65 bit integer

(3)

D_{15} — D_{11}	D_{10}	D_9	D_8	D_7	D_6	D_5 D_4 D_3	D_2 D_1 D_0
11011	d	P	OPA	1	1	OPB	ST(i)

d - destination

if d = 0, destination is ST (0) (0th stack register).

if d = 1, destination is ST (i) (ith stack register).

Here D3 is the reserved bit R.

if R XOR d = 0, the format is-Destination opcode source.

if R XOR d = 1, the format is-Source Opcode Destination

P-Pop stack

if P = 0, Don't pop stack

if P = 1, Pop stack

ST(i) - 000 to 111 for stack top ST(0) to ST(7)

(4)

D_{15} — D_{11}	D_{10}	D_9	D_8	D_7	D_6	D_5	D_4——D_0
11011	0	0	1	1	1	1	Opcode

(5)

D_{15} — D_{11}	D_{10}	D_9	D_8	D_7	D_6	D_5	D_4——D_0
11011	0	0	1	1	1	1	Opcode

In all the above formats, the code 11011 corresponding to bits D15-D11 informs the CPU in execute an ESCAPE sequence and initiate a NDP bus cycle.

Complete Instruction set summary for 80287 processor is as follows.

Table Intel 287™ XL MCP Extension to the CPU's Instruction Set

Instruction	Encoding Byte 0	Encoding Byte 1	Optional Bytes 2-3	Clock Count Range 32-Bit Real	32-Bit Integer	64-Bit Real	16-Bit Integer
Data Transfer							
FLD-Load							
Integer/real memory to ST(0)	ESC MP 1	MOD 000 R/M	SIB/DISP	36	61-68	45	61-65
Long Integer momery to ST(0)	ESC 111	MOD 101 R/M	SIB/DISP			76-87	
Extended real memory to ST(0)	ESC 011	MOD 101 R/M	SIB/DISP			48	

Instruction	Encoding Byte 0	Encoding Byte 1	Optional Bytes 2-3	Clock Count Range 32-Bit Real	32-Bit Integer	64-Bit Real	16-Bit Integer
BCD memory to ST(0)	ESC 111	MOD 100 R/M	SIB/DISP				270-279
ST(i) memory to ST(0)	ESC 001	11000 ST(i)				21	
FST-Store and Pop							
ST(0) to Integer/real memory	ESC MF 1	MOD 011 R/M	SIB/DISP	51	86-100	56	88-101
ST(0) to ST(i)	ESC MF 1	MOD 010 R/M			18		
FSTP-Sotre and Pop							
ST(0) to Integer/real memory	ESC MF 1	MOD 011 R/M	SIB/DISP	51	86-100	56	88-101
ST(0) to Long Integer memory	ESC 111	MOD 111 R/M	SIB/DISP			91-108	
ST(0) to extend real	ESC 011	MOD 111 R/M	SIB/DISP			61	
ST(0) to BCD memory	ESC 111	MOD 110 R/M	SIB/DISP			520-542	
ST(0) to ST(i)	ESC 101	11001 (ST(i)			19		
FXCH-Exchange							
ST(i) to ST(0)	ESC 001	11001 ST(i)					25
FCOM-Compare							
Integer/real memory to ST(0)	ESC MF 0	MOD 010 R/M	SIB/DISP	42	72-79	51	71-75
ST(i) to ST(0)	ESC 000	11010 ST(i)				31	
FCOMP-Compare and Pop							
Integer/real memory to ST	ESC MF 0	MOD 011 R/M	SIB/DISP	42	72-79	51	71-77
ST(i) to ST(0)	ESC 000	11011 ST(i)					33
FCOMPP-Compare and pop twice							
ST(i) to ST(0)	ESC 110	1101 1001				33	
FTST-Test ST(0)	ESC 001	1110 0100					35
FUCOM-Unordered Compare	ESC 101	11100 ST(i)			31		
FUCOMP-Unordered compare and pop	ESC 101	11101 ST(i)				33	
FUCOMPP-Unordered Compare and pep twice	ESC 010	0110 1001				33	
FXAM-Examine ST(0)	ESC 001	11100101				37-45	
CONTANTS							
FLDZ-Load + 0.0 into ST(0)	ESC 001	1110 1110				27	
FLD1-Load + 1.0 into ST(0)	ESC 001	1110 1000				31	
FLDP1-Load pi into ST(0)		ESC 001	1110 1001				47
FLDL2T-Load $Log_2(10)$ into ST(0)	ESC 001	11101001				47	
CONSTANTS (Compared)							
FLDL2E-Load $Log_2(e)$ into ST(0)	ESC 001	1110 1010				47	
FLDLG2-Load $Log_{10}(2)$ into ST(0)	ESC 001	1110 1100				48	
FLDLG2-Load $Log_e(2)$ into ST0)	ESC 001	1110 1101				48	
ARITHMETIC							
Integer/real memory with ST(0)	ESC MF 0	MOD 000 R/M	SIB/DISP	40-48	73-789	49-79	71-85
ST(i) and ST(0)	ESC dP 0	11000 ST(i)					30-38

Instruction	Encoding		Optional Bytes 2-3	Clock Count Range			
	Byte 0	Byte 1		32-Bit Real	32-Bit Integer	64-Bit Real	16-Bit Integer
FSUB-Subtract							
Integral/real memory to ST(0)	ESC MF 0	MOD 10R R/M	SIB/DISP	40-48	73-96	49-77	71-83
ST(i) and ST(0)	ESC dP 0	1110R R/M				33-41	
FMUL-Multiply							
ST(0) to Integer/real memory	ESC MF 0	MOD 001 R/M	SIB/DISP	43-51	77-88	52-77	76-87
ST(i) and ST(0)	ESC dP 0	1100 1 R/M			25-53		
FDIV-Divided							
ST(0) to integer/real memory	ESC MF 0	MOD 11R R/M	SIB/DISP	105	135-143	114	136-140
ST(i) and ST(0)	ESC dP 0	1111 R R/M				95	
FSQRTI-Square root	ESC 001	1111 1010				129-136	
FSCALE-Scale ST(0) by ST(1)	ESC 001	1111 1101			74-95		
FPREM-Partial remainder of ST(0) + ST(1)	ESC 001	1111 1000				81-162	
FPREM1-Partial remainder of (IEEE)	ESC 001	1111 010				102-198	
FRNDINT-Round ST(0) to integer	ESC 001	1111 1100				73-87	
FXTRACT-Extract components of ST(0)	ESC 001	1111 0100				75-83	
FABS-Absolute value of ST(0)	ESC 001	1110 0001				29	
FCHS-Change sign of ST(0)	ESC 001	1110 0000			31-37		
TRANSCENDENTAL							
FCCS-cosine of ST(0)	ESC 001	1111 1111				130-779	
FPTAN-Partial tangent of ST(0)	ESC 001	1111 0010			198-504		
FPATAN-Partial arctangent	ESC 001	1111 0011			321-494		
FSIN-sine of ST(0)	ESC 001	1111 1110				124-678	
FSINCOS-sine and cosine of ST(0)	ESC 001	1111 1011			201-815		
F2XM1-2SST(0) - 1	ESC 001	1111 0000				215-483	
FYL2X-ST(1) * $\log_2$ (ST(0))	ESC 001	1111 0001			127-545		
FYL2XP1-ST(1)*$\log_2$(ST(0)+ 1.0)	ESC 001	1111 1001			264-554		
PROCESSOR CONTROL							
FINIT-Initialize MCP	ESC 001	1110 0011				25	
FSETPM-Set protected mode	ESC 001	1110 0100			12		
FRSETPM-Reset protected mode	ESC 001	1111 0100			12		
FSTSW AX-Stone status word	ESC 111	1110 0000			18		
FLDCW-Load control word	ESC 001	MOD 101 R/M	SIB/DISP			33	
FSTCW-Store control word	ESC 101	MOD 111 R/M	SIB/DISP			18	
FSTSW-Store status word	ESC 101	MOD 111 R/M	SIB/DISP			18	
FCLEX-Clear exception	ESC-101	1110 0010					8
FSTENV-Store environment	ESC 001	MOD 110 R/M	SIB/DISP			192-193	
FLDENV-Load environment	ESC 001	MOD 100 R/M	SIB/DISP			85	
FSAVE-Save state	ESC 101	MOD 110 R/M	SIB/DISP			521-522	
FRSTOR-Restore state	ESC 101	MOD 100 R/M	SIB/DISP			396	
FINSCPT-Increment stack pointer	ESC 001	1111 0111				28	
FDECSTP-Decrement stack	ESC 001	1111 0110				29	
FFREE-Free ST(i)	ESC 101	1100 0ST(i)				25	
FNOP-No operations	ESC 001	1101 0000				19	

REVIEW QUESTIONS

1. Draw and discuss the internal block diagram of 80286 microprocessor?
2. Explain in detail the various feature of 80286 processor?
3. Explain the concept of virtual memory?
4. Draw and discuss the register organization of 80286?
5. Draw and discuss the flag register of 80286?
6. What are the different interrupts available in 80286?
7. Draw and discuss the machine control word of 80286?
8. What are the salient features of 80286 in protected virtual address mode?
9. What are the salient features of 80286 in real address mode?
10. Explain the physical address formation in protected virtual address mode?
11. What do u mean by a descriptor?
12. Draw and discuss the structure of a general 80286 descriptor?
13. Explain the following terms :
    (a) Task Privilege
    (b) Descriptor Privilege
    (c) Selector Privilege
14. Discuss different types of descriptors supported by 80286 and give their typical functions?
15. What do you mean by boundary testing instructions?
16. How many descriptors can be accessed at most by 80286?
17. How much memory can be accessed by a descriptor?
18. Draw and discuss different data types supported by 80286?
19. Draw and discuss the minimum configuration of 80286?
20. Draw and discuss basic read, write and fetch cycles of 80286?
21. Enlist the priority of bus usage in 80286?
22. Draw and discuss the internal architecture of 80287?
23. Enlist different instruction formats of 80287?
24. Enlist the different data types supported by 80287?
25. Draw and discuss the interface between 80286 and 80287?
26. Explain with the help of a diagram the pin configuration of 80286 processor?
27. Explain with the help of a diagram the pin configuration of 80287 processor?

❑❑❑

CHAPTER 10

THE 80386/387/486 MICROPROCESSORS

INSIDE THIS CHAPTER

10.1. THE 80386 MICROPROCESSOR

The block diagram of an 80386 is shown in Fig. 10.1.

The 80386 includes a Bus Interface Unit for reading and providing data and instructions, with a Prefetch Queue, an IU for controlling the EU with its registers, as well as an AU for generating memory and I/O addresses

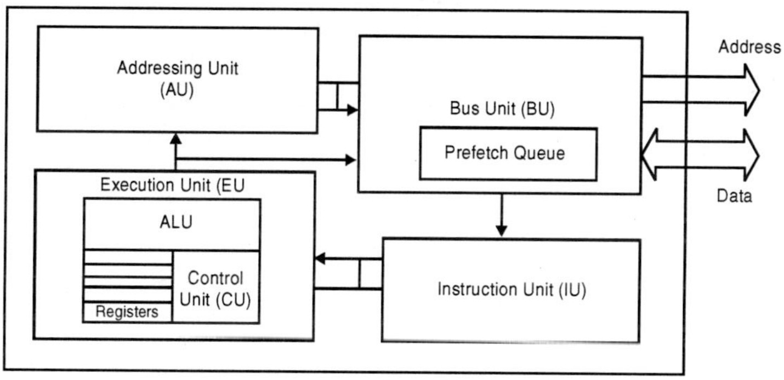

Fig. 10.1. Block diagram of 80386.

The features of the 80386 are:

(i) 32-bit general and offset registers
(ii) 16-byte prefetch queue
(iii) Memory Management Unit with a Segmentation Unit and a Paging Unit
(iv) 32-bit Address and Data Bus
(v) 4-Gbyte Physical address space
(vi) 64-Tbyte virtual address space
(vii) i387 numerical coprocessor with IEEE standard-754-1985 for floating point arithmetic
(viii) 64K 8-, 16-, or 32-bit ports
(ix) Implementation of real, protected and virtual 8086 modes

Some of the elements that help to give the 80386 a performance improvement over earlier generation processors are its expanded bus width, its prefetch queue, its numeric coprocessor and its generally improved instruction set.

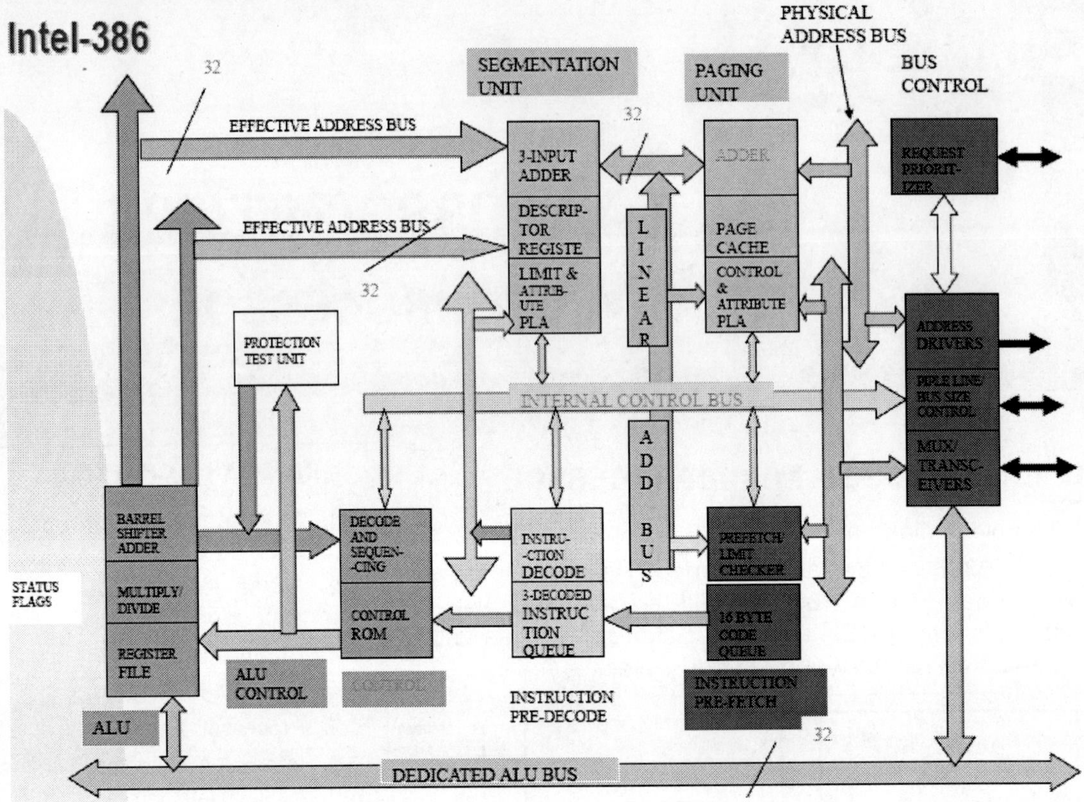

Fig. 10.2. 80386 internal architecture

The 386 DX and 386 SX can both run Windows 3.X in enhanced mode, which is formally called 386 Enhanced Mode.

Enhanced mode is where the windows software takes advantage of the 386's ability to continually switch between Real and Protected mode.

The features of the 386 SX include:

- Restricted to 16 bit external bus, but internally identical to the 386 DX with 32 bit internal registers.
- 24 bit memory address bus which addresses only 16 MB of RAM.
- Available at 16 mhz to 33 mhz
- Can run 386 specific software same as the DX
- Not pin compatible with 286 processor sockets
- form factor is a 100 pin plastic QFP.

80386 SL

Due to the need for light portable laptop and notebook computers, the 80386 SL was created. This processor is a mixture of the 386 DX and 386 SX; it retains the 32 bit address bus, an external 16 bit data bus, a 32 bit internal data path, and maintains the virtual memory capabilities of the SX and DX. The features of the 386 SL include:

- Low power consumption
- A cache controller designed to control a 16 Kb to 64 KB external processor cache.
- A system management interrupt (SMI) with power management features for battery conservation including several sleep modes.
- Support for Lotus/Intel/Microsoft (LIM) expanded memory functions
- 25 MHz clock speed

The Internal architecture of 80386 is divided into three sections :-

(1) Central Processing Unit
   (a) Execution Unit
   (b) Instruction Unit
(2) Memory Management Unit
   (a) Segmentation Unit
   (b) Paging Unit
(3) Bus Interface Unit
   (a) Pre-fetch Unit
   (b) Decode Unit

The Execution unit : It has 8 general purpose registers and 8 special purpose registers which are used for handling data or calculating offset addresses. The instruction unit decodes the opcode bytes received from the 16-byte insruction code queue and arranges them in in a 3-instruction decoded-instruction queue. Then it passes it to control section for deriving the necessary control signals. The barrel shifter increases the speed of all shift and rotato oporationc. 32-bit multiplications can be executed within one microsecond by the multiply/divide logic.

Segmentation Unit : This Unit allows the use of two address components, viz. segment and offset for relocability and sharing of code and data segment. The unit allows a segments of size 4Gbytes at maximum. It provides a four level protection mechanism for protecting and isolating system the systems code and data from those of the application program.

Paging Unit : The paging unit works under the control of segmentation unit, i.e. each segment is further divided into pages. The virtual memory is also organized in terms of segments and pages by the memory management unit. This unit converts linear addresses into physical addresses.

- The Instruction unit decodes the opcode bytes received from the 16-byte instruction code queue and arranges them in a 3- instruction decoded instruction queue.
- After decoding them pass it to the control section for deriving the necessary control signals. The barrel shifter increases the speed of all shift and rotate operations.
- The multiply / divide logic implements the bit-shift-rotate algorithms to complete the operations in minimum time.
- Even 32- bit multiplications can be executed within one microsecond by the multiply / divide logic.
- The Memory management unit consists of a Segmentation unit and a Paging unit.
- Segmentation unit allows the use of two address components, viz. segment and offset for relocability and sharing of code and data.
- Segmentation unit allows segments of size 4Gbytes at max.
- The Paging unit organizes the physical memory in terms of pages of 4kbytes size each.
- Paging unit works under the control of the segmentation unit, i.e. each segment is

further divided into pages. The virtual memory is also organizes in terms of segments and pages by the memory management unit.

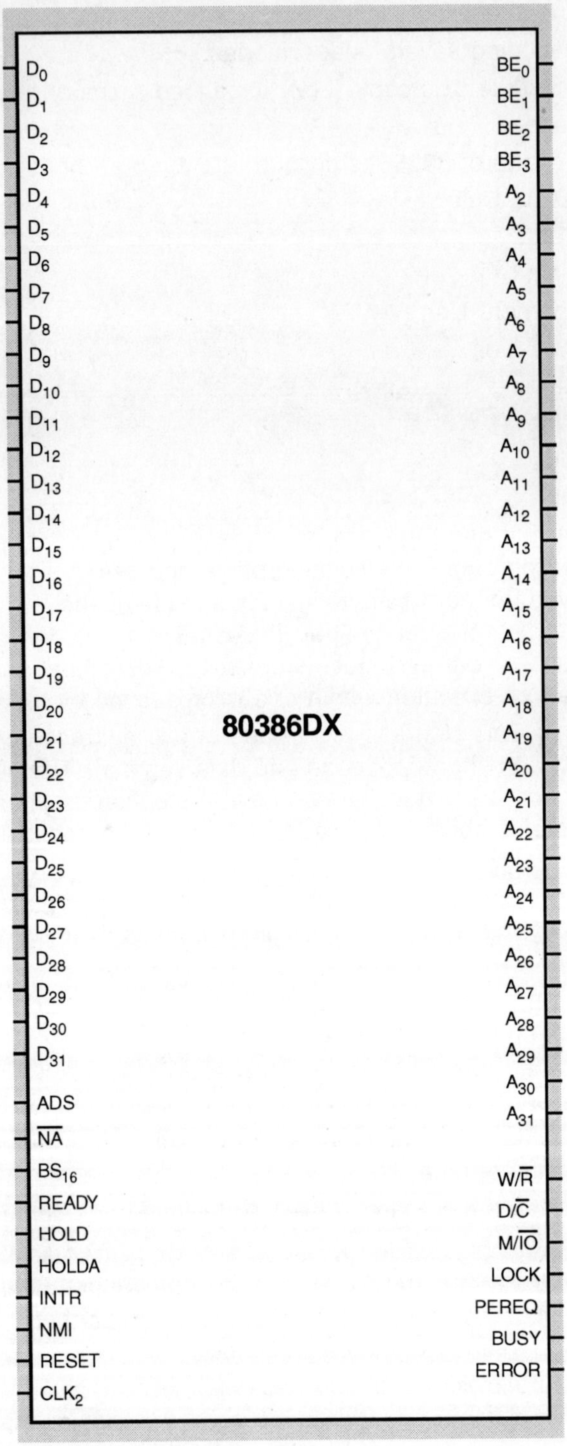

Fig. 10.3. Pin diagram of 80386 microprocessor

- The Segmentation unit provides a 4 level protection mechanism for protecting and isolating the system code and data from those of the application program.
- Paging unit converts linear addresses into physical addresses.
- The control and attribute PLA checks the privileges at the page level. Each of the pages maintains the paging information of the task. The limit and attribute PLA checks segment limits and attributes at segment level to avoid invalid accesses to code and data in the memory segments.
- The Bus control unit has a prioritize!" to resolve the priority of the various bus requests. This controls the access of the bus. The address driver drives the bus enable and address signal $A_0 - A_{31}$. The pipeline and dynamic bus sizing unit handle the related control signals.
- The data buffers interface the internal data bus with the system bus.

Signal Descriptions of 80386

- **CLK_2 :** The input pin provides the basic system clock timing for the operation of 80386.
- **$D_0 - D_{31}$:** These 32 lines act as bidirectional data bus during different access cycles.
- **$A_{31} - A_2$:** These are upper 30 bit of the 32-bit address bus.
- **BE_0 to BE_3 :** The 32- bit data bus supported by 80386 and the memory system of 80386 can be viewed as a 4-byte wide memory access mechanism. The 4 byte enable lines BE_0 to BE_3 may be used for enabling these 4 blanks. Using these 4 enable signal lines, the CPU may transfer 1 byte / 2 / 3 / 4 byte of data simultaneously.
- **ADS# :** Tho address status output pin indicates that the address bus and bus cycle definition pins(W/R#, D/C#, M/IO#, BE0# to BE3#) are carrying the respective valid signals. The 80383 does not have any ALE signals and so this signals may be used for latching the address to external latches.
- **READY# :** The ready signals indicates to the CPU that the previous bus cycle has been terminated and the bus is ready for the next cycle. The signal is used to insert WAIT states in a bus cycle and is useful for interfacing of slow devices with CPU.
- **VCC :** These are system power supply lines.
- **VSS :** These return lines for the power supply.
- **BS_{16}# :** The bus size - 16 input pin allows the interfacing of 16 bit devices with the 32 bit widc 80386 data bus. Successive 16 bit bus cycles may be executed to rcad a 32 bit data from a peripheral.
- **HOLD :** The bus hold input pin cnables the other bus masters to gain control of the system bus if it is asserted.
- **HLDA :** The bus hold acknowledge output indicates that a valid bus hold request has been received and the bus has been relinquished by the CPU.
- **BUSY# :** The busy input signal indicates to the CPU that the coprocessor is busy with the allocated task.
- **ERROR# :** The error input pin indicates to the CPU that the coprocessor has encountered an error while executing its instruction.
- **PEREQ :** The processor extension request output signal indicates to the CPU to fetch a data word for the coprocessor.
- **INTR :** This interrupt pin is a maskable interrupt, that can be masked using the IF of the flag register.

- **NMI :** A valid request signal at the non-maskable interrupt request input pin internally generates a non- maskable interrupt of type2.
- **RESET :** A high at this input pin suspends the current operation and restart the execution from the starting location.
- **N/C :** No connection pins are expected to be left open while connecting the 80386 in the circuit.

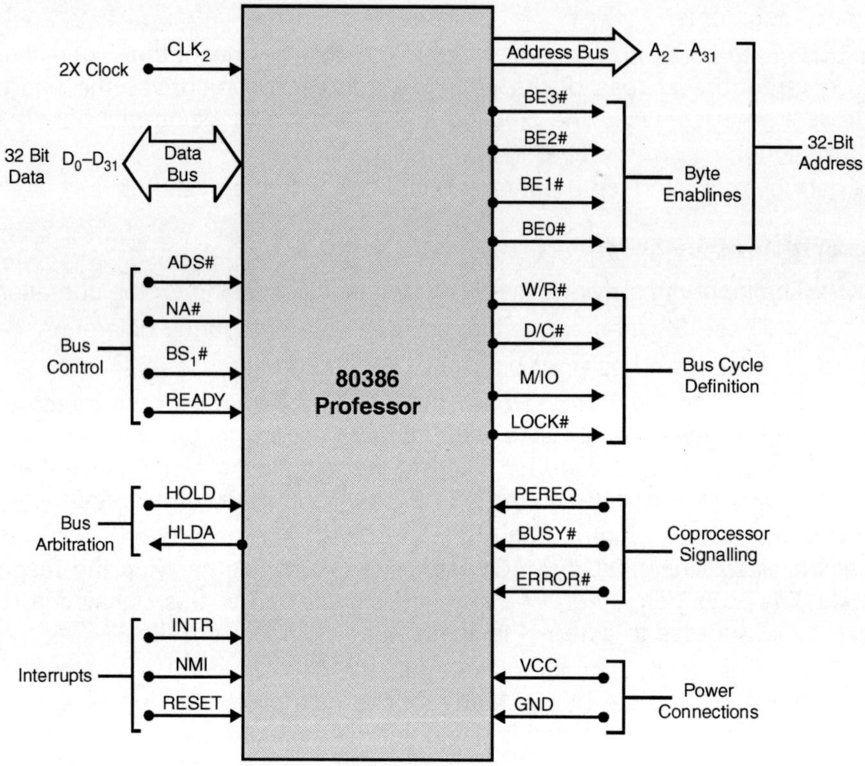

Fig. 10.4. Pin configuration of 80386

Register Organisation

- The 80386 has eight 32-bit general purpose registers which may be used as either 8 bit or 16 bit registers.
- A 32-bit register known as an extended register, is represented by the register name with prefix E.

  Example : A 32 bit register corresponding to AX is EAX, similarly BX is EBX etc.
- The 16 bit registers BP, SP, SI and DI in 8086 are now available with their extended size of 32 bit and are names as EBP, ESP, ESI and EDI.
- AX represents the lower 16 bit of the 32 bit register EAX.
- BP, SP, SI, DI represents the lower 16 bit of their 32 bit counterparts, and can be used as independent 16 bit registers.
- The six segment registers available in 80386 are CS, SS, DS, ES, FS and GS.
- The CS and SS are the code and the stack segment registers respectively, while DS, ES, FS, GS are 4 data segment registers.
- A 16 bit instruction pointer IP is available along with 32 bit counterpart EIP.

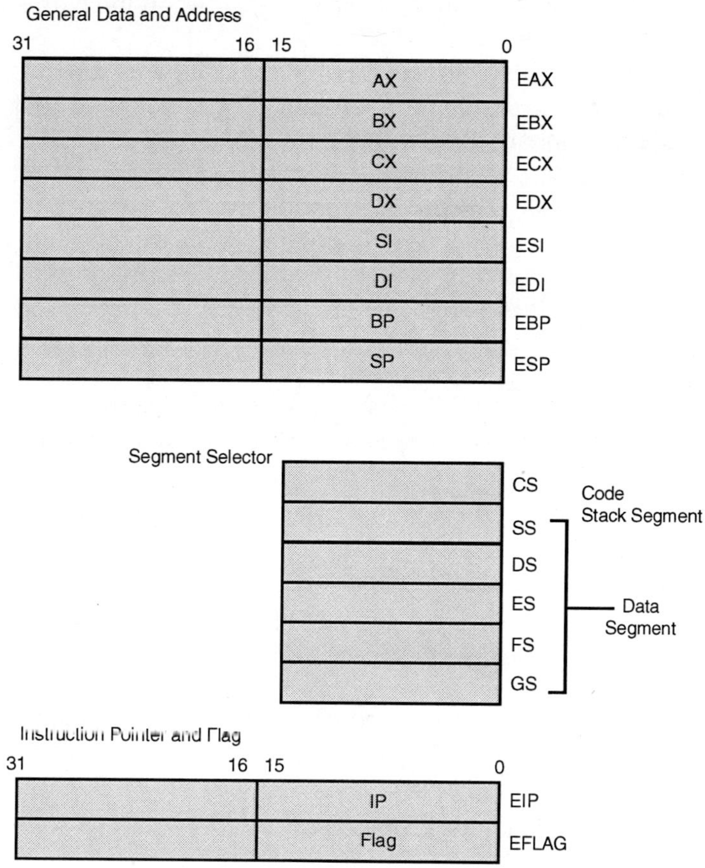

Fig. 10.5. Register organisation of 80386

- **Flag Register of 80386 :** The Flag register of 80386 is a 32 bit register. Out of the 32 bits, Intel has reserved bits D18 to D31, D5 and D3, while DI is always set at I. Two extra new flags are added to the 80286 flag to derive the flag register of 80386. They are VM and RF flags.
- **VM - Virtual Mode Flag :** If this flag is set, the 80386 enters the virtual 8086 mode within the protection mode. This is to be set only when the 80386 is in protected mode. In this mode, if any privileged instruction is executed an exception 13 is generated. This bit can be set using IRET instruction or any task switch operation only in the protected mode.
- **RF- Resume Flag :** This flag is used with the debug register breakpoints. It is checked at the starting of every instruction cycle and if it is set, any debug fault is ignored during the instruction cycle. The RF is automatically reset after successful execution of every instruction, except for IRET and POPF instructions. Also, it is not automatically cleared after the successful execution of JMP, CALL and INT instruction causing a task switch. These instruction are used to set the RF to the value specified by the memory data available at the stack.
- **Segment Descriptor Registers :** This registers are not available for programmers, rather they are internally used to store the descriptor information, like attributes, limit and base addresses of segments. The six segment registers have corresponding six 73 bit descriptor registers. Each of them contains 32 bit base address, 32 bit base limit

and 9 bit attributes. These are automatically loaded when the corresponding segments are loaded with selectors.

- **Control Registers :** The 80386 has three 32 bit control registers CRo, CR2 and CRs to hold global machine status independent of the executed task. Load and store instructions are available to access these registers.
- **System Address Registers :** Four special registers are defined to refer to the descriptor tables supported by 80386. The 80386 supports four types of descriptor table, viz. global descriptor table (GOT), interrupt descriptor table (IDT), local descriptor table (LDT) and task state segment descriptor (TSS).
- **Debug and Test Registers :** Intel has provide a set of 8 debug registers for hardware debugging. Out of these eight registers DR0 to DR7, two registers DR_0 and DR_5 are Intel reserved.

  The initial four registers DRo to DR3 store four program controllable breakpoint addresses. while DReand DR? respectively hold breakpoint status and breakpoint control information.

  Two more test register are provided by 80386 for page cacheing namely test control and test status register.
- **Addressing Modes :** The 80386 Supports overall eleven addressing modes to facilitate efficient execution of higher level language programs.

  In case of all those modes, the 80386 can now have 32-bit immediate or 32- bit register operands or displacements.

  The 80386 has a family of scaled modes. In case of scaled modes, any of the index register values can be multiplied by a valid scale factor to obtain the displacement.

  The valid scale factor are 1, 2, 4 and 8.

  The different scaled modes are as follows.
- **Scaled Indexed Mode :** Contents of the an index register are multiplied by a scale factor that may be added further to get the operand offset.
- **Based Scaled Indexed Mode:** Contents of the an index register are multiplied by a scale factor and then added to base register to obtain the offset.
- **Based Scaled Indexed Mode with Displacement :** The Contents of the an index register are multiplied by a scaling factor and the result is added to a base register and a displacement to get the offset of an operand.

Real Address Mode of 80386

- After reset, the 80386 starts from memory location FFFFFFF0H under the real address mode. In the real mode, 80386 works as a fast 8086 with 32-bit registers and data types.
- In real mode, the default operand size is 16 bit but 32-bit operands and addressing modes may be used with the help of override prefixes.
- The segment size in real mode is 64k, hence the 32-bit effective addressing must be less than OOOOFFFFFH. The real mode initializes the 80386 and prepares it for protected mode.
- **Memory Addressing in Real Mode :** In the real mode, the 80386 can address at the most 1 Mbytes of physical memory using address lines A_0-A_{19}.
- Paging unit is disabled in real addressing mode, and hence the real addresses are the same as the physical addresses.

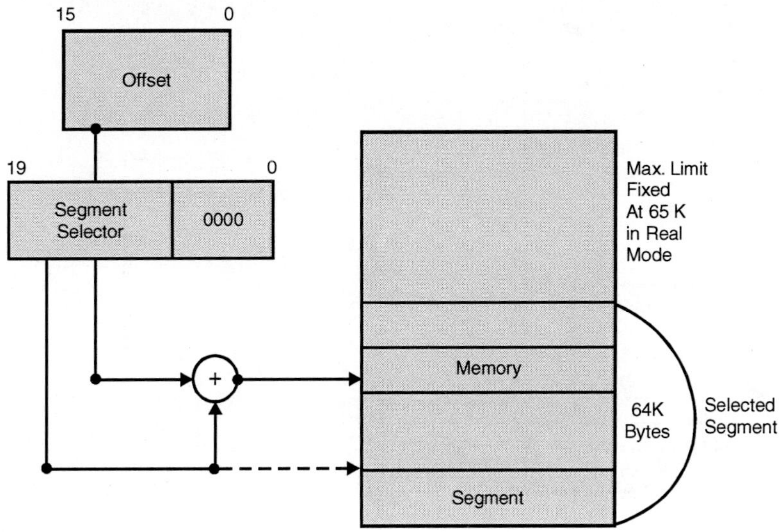

Fig. 10.6. Physical Address Formation in Real Mode of 80386

- To form a physical memory address, appropriate segment registers contents (16-bits) are shifted left by four positions and then added to the 16-bit offset address formed using one of the addressing modes, in the same way as in the 80386 real address mode.
- The segment in 80386 real mode can 'be read, write or executed, i.e. no protection Is available.
- Any fetch or access past the end of the segment limit generate exception 13 in real address mode.
- The segments in 80386 real mode may be overlapped or non-overlapped.
- The interrupt vector table of 80386 has been allocated I Kbyte space starting from OOOOOH to 003FFH.

Protected Mode of 80386

- All the capabilities of 80386 arc available for utilizatlon in its protected modc of operation.
- The 80386 in protected mode support all the software written for 80286 and 8086 to be executed under the control of memory managemcnt and protection abilities of 80386.
- The protected mode allows the use of additional instruction, addressing modes and capabilities of 80386.

Addressing in Protected Mode : In this mode, the contents of segment registers are used as selectors to address descriptors which contain the segment limit, base address and access rights byte of the segment.

- The effective address (offset) is added with segment base address to calculate linear address. This linear address is further used as physical address, if the paging unit is disabled. otherwise the paging unit converts the linear address into physical address.
- The paging unit is a memory management unit enabled only in protected mode. The paging mechanism allows handling of large segments of memory in terms of pages of 4 Kbyte size.

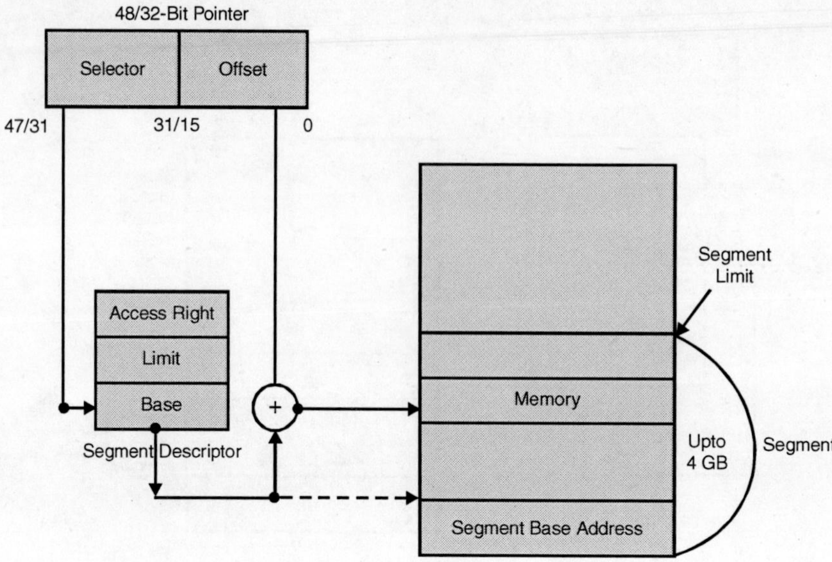

Fig. 10.6. Physical address formation in protected mode of 80386

Memory-Management Registers

Four registers of the 80386 locate the data structures that control segmented memory management:

GDTR Global Descriptor Table Register

LDTR Local Descriptor Table Register

These registers point to the segment descriptor tables GDT and LDT.

IDTR Interrupt Descriptor Table Register

This register points to a table of entry points for interrupt handlers (the IDT).

TR Task Register

This register points to the information needed by the processor to define the current task.

Control Registers

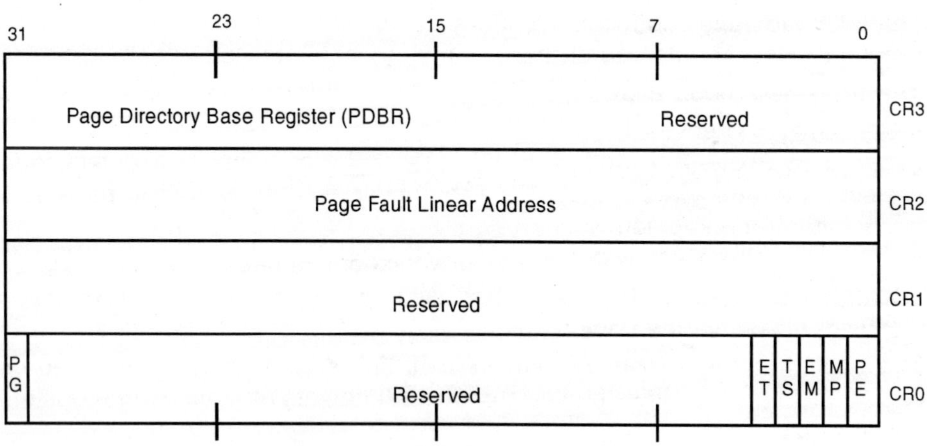

Fig. 10.7. Control register for 80386

Figure 10.7 shows the format of the 80386 control registers CR0, CR2, and CR3. These registers are accessible to systems programmers only via variants of the MOV instruction, which allow them to be loaded from or stored in general registers; for example:

```
MOV EAX, CR0
MOV CR3, EBX
```

CR0 contains system control flags, which control or indicate conditions that apply to the system as a whole, not to an individual task.

- ➢ EM (Emulation, bit 2):- EM indicates whether coprocessor functions are to be emulated.
- ➢ ET (Extension Type, bit 4):- ET indicates the type of coprocessor present in the system (80287 or 80387).
- ➢ MP controls the function of the WAIT instruction, which is used to coordinate a coprocessor.
- ➢ PE (Protection Enable, bit 0):- Setting PE causes the processor to begin executing in protected mode. Resetting PE returns to real-address mode. PG (Paging, bit 31)
- ➢ PG indicates whether the processor uses page tables to translate linear addresses into physical addresses. Refer to Chapter 5 for a description of page translation; refer to Chapter 10 for a discussion of how to set PG.
- ➢ TS (Task Switched, bit 3):- The processor sets TS with every task switch and tests TS when interpreting coprocessor instructions.
- ➢ CR2 is used for handling page faults when PG is set. The processor stores in CR2 the linear address that triggers the fault.
- ➢ CR3 is used when PG is set. CR3 enables the processor to locate the page table directory for the current task.

Debug Register

The debug registers bring advanced debugging abilities to the 80386, including data breakpoints and the ability to set instruction breakpoints without modifying code segments.

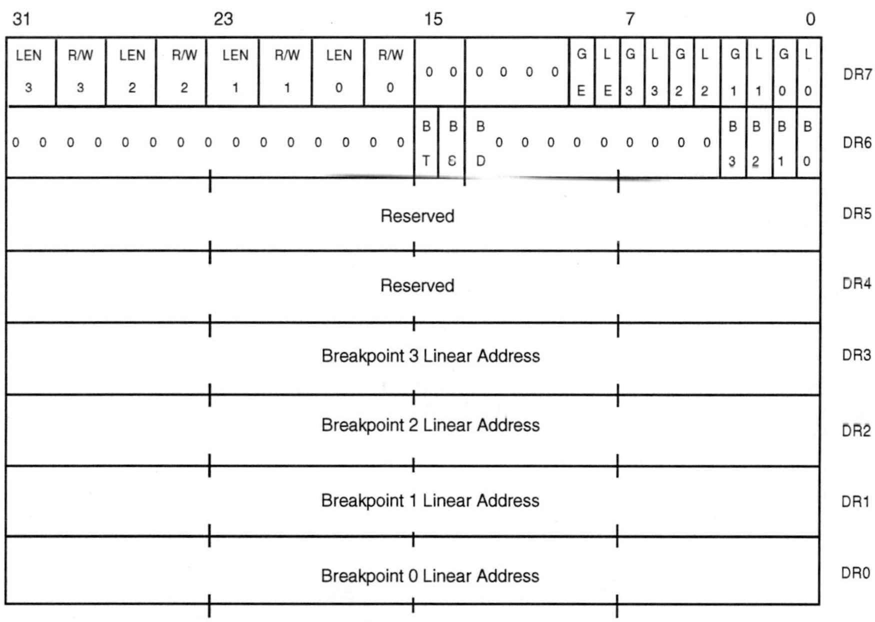

Fig. 10.8. Debug registers for 80386

Test Registers

The test registers are not a standard part of the 80386 architecture. They are provided solely to enable confidence testing of the translation lookaside buffer (TLB), the cache used for storing information from page tables.

80386 Operating Modes

The 80386 and the Pentium support three operating modes: protected mode, real-address mode, and system management mode. We'll discuss these when we talk about the Pentium.

10.2. 80386 REGISTER SET AND INSTRUCTION SET

The i386 Application Programming Registers are shown below :

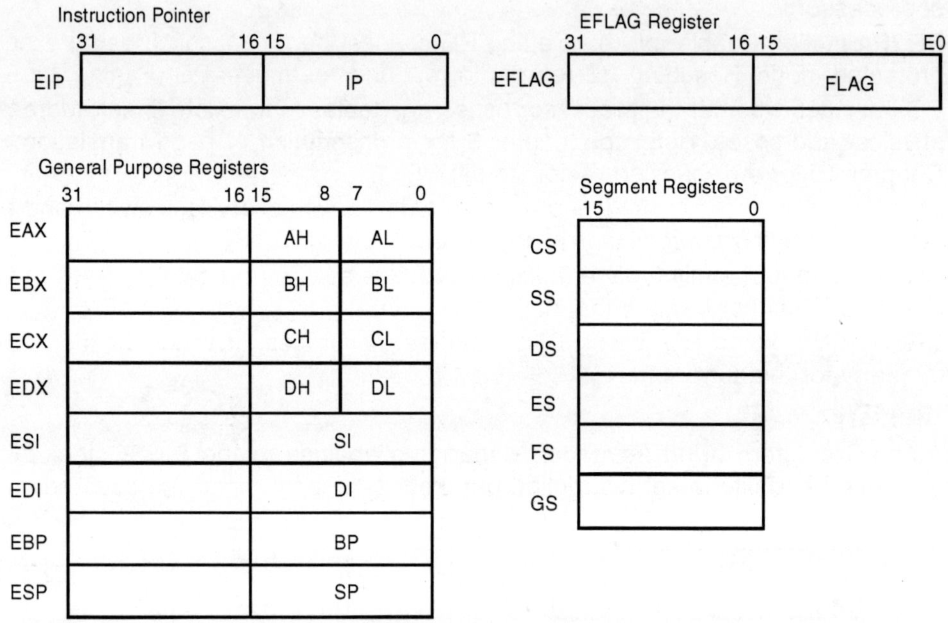

Fig. 10.9. Register set of 80386.

The 80386 maintains compatibility with the 8086, and 80286, so even though its general registers are 32-bit, they can also be used as 16- or 8-bit registers. The table above illustrates this. Note that this programming model of the 80386 also applies to the Pentium.

The 32-bit EIP register can support programs up to 4Gbytes, whereas the 8086 and 80286, with their 16-bit IP register, could only support program segments of 64kbytes. The CS register enables larger programs. Note that the CS can be changed under program control, but that the instruction pointer cannot be written to directly by a program. It can only be changed by jumps, calls, returns or interrupts. Note that far calls and jumps change the value of the CS register as well as the value of the instruction pointer.

Stack Segment and Stack Pointer

Usually every program has its own stack segment. As on the 8086 the stack grows downwards, that is, the value of the stack pointer decreases with a PUSH instruction and increases with a POP instruction. In the i386, if data is stored on the stack the value of ESP is reduced by 4, because the i386 always writes a complete double word (2*16 bits = 4*8 bits).

When the i386 operates in 16-bit mode, only 2 bytes are written to the stack, and the value of SP is only reduced by 2, with each push.

Data Segments

The i386 adds two more data segment registers, called FS and GS.

The i386 has four control registers and four memory mangement registers for protected mode, as well as 8 debug registers. These registers are particularly useful in a multitasking environment. The debug registers can be useful in locating errors in a given task.

32-bit register	16-bit register	8-bit register	Name	Main Use
EAX	AX	AH, AL	Accumulator	Multiplication/division/I/O, fast shifts
EBX	BX	BH, BL	Base Register	Pointer to Base Address in data segment
ECX	CX	CH, CL	Count Register	Count value for repetitions, shifts, rotates
EDX	DX	DH, DL	Data Register	Multiplication, Division, I/O
EBP	BP		Base Pointer	Pointer to Base Address in Stack Segment
ESI	SI		Source Index	Source String and index pointer
EDI	DI		Destination Index	Destination String and index pointer
ESP	SP		Stack Pointer	
	CS		Code Segment	
	DS		Data Segment	
	SS		Stack Segment	
	ES		Extra Segment	
	FS		Extra Segment	
	GS		Extra Segment	
EIP	IP		Instruction Pointer	Instructions offset
EFLAG	Flag		Flags	Processor status

Segmentation

Descriptor Tables

These descriptor tables and registers are manipulated by the operating system to ensure the correct operation of the processor, and hence the correct execution of the program. Three types of the 80386 descriptor tables are listed as follows :

- Global Descriptor Table (GDT)
- Local Descriptor Table (LOT)
- Interrupt Descriptor Table (IDT)

Descriptors : The 80386 descriptors have a 20-bit segment limit and 32-bit segment address. The descriptor of 80386 are 8-byte quantities access right or attribute bits along with the base and limit of the segments.

Descriptor Attribute Bits : The A (accessed) attributed bit indicates whether the segment has been accessed by the CPU or not.

(i) The TYPE field decides the descriptor type and hence the segment type.

(ii) The S bit decides whether it is a system descriptor (S = 0) or code/data segment descriptor (S = 1).

(iii) The DPL field specifies the descriptor privilege level.

(iv) The D bit specifies the code segment operation size. If D = 1, the segment is a 32-bit operand segment, else, it is a 16-bit operand segment.

(v) The P bit (present) signifies whether the segment is present in the physical memory or not. If P = 1, the segment is present in the physical memory.

(vi) The G (granularity) bit indicates whether the segment is page addressable. The zero bit must remain zero for compatibility with future process.

(vii) The AVL (available) field specifies whether the descriptor is for user or for operating system.

The 80386 has five types of descriptors listed as follows:

1. Code or Data Segment Descriptors.
2. System Descriptors.
3. Local descriptors.
4. TSS (Task State Segment) Descriptors.
5. GATE Descriptors.

The 80386 provides a four level protection mechanism exactly in the same way as the 80286 does.

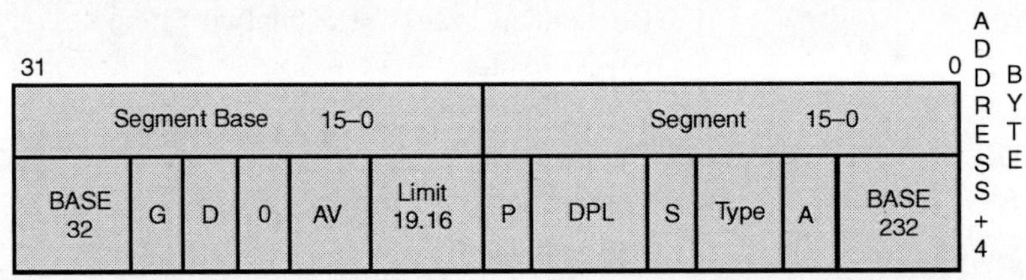

Base = Base address of the segment
Limit = The length of the segment
P = Present Bit, –1 = Present, 0 = not present
S = Segment Descriptor, –0 = System Descriptor, 1 = Code or data segment descriptor
Type = Type of segment
G = Granularity Bit –1 = Segment length is page granular, 0 = Segment length is byte granular
D = Default operation size
0 = Bit must be zero
AVL = Available field for user or OS

Fig. 10.10. Structure of a Descriptor

Paging

PAGING OPERATION : Paging is one of the memory management techniques used for virtual memory multitasking operating system. The segmentation scheme may divide the physical memory into a variable size segments but the paging divides the memory into a fixed size pages. The segments are supposed to be the logical segments of the program, but the pages do not have any logical relation with the program. The pages are just fixed size portions of the program module or data. The advantage of paging scheme is that the complete segment of a task need not be in the physical memory at any time. Only a few pages of the segments, which are required currently for the execution need to be available in the physical memory. Thus the memory requirement of the task is substantially reduced, relinquishing the available memory for other tasks. Whenever the other pages of task are required for execution, they may be fetched from the secondary storage. The previous page which are executed, need not be available in the memory, and hence the space occupied by them may be relinquished for other tasks. Thus paging mechanism provides an effective technique to manage the physical memory for multitasking systems.

Paging Unit : The paging unit of 80386 uses a two level table mechanism to convert a linear address provided by segmentation unit into physical addresses. The paging unit converts the complete map of a task into pages, each of size 4K. The task is further handled in terms of its page, rather than segments. The paging unit handles every task in terms of three components namely page directory, page tables and page itself.

Paging Descriptor Base Register : The control register CR2 is used to store the 32-bit linear address at which the previous page fault was detected. The CR3 is used as page directory physical base address register, to store the physical starting address of the page directory. The lower 12 bit of the CR3 are always zero to ensure the page size aligned directory. A move operation to CRa automatically loads the page table entry caches and a task switch operation, to load CRo suitably.

Page Directory : This is at the most 4Kbytes in size. Each directory entry is of 4 bytes, thus a total of 1024 entries are allowed in a directory. The upper 10 bits of the linear address are used as an index to the corresponding page directory entry. The page directory entries point to page tables.

Page Tables : Each page table is of 4Kbytes in size and many contain a maximum of 1024 entries. The page table entries contain the starting address of the page and the statistical information about the page. The upper 20 bit page frame address is combined with the lower 12 bit of the linear address. The address bits A_{12}- A_{21} are used to select the 1024 page table entries. The page table can be shared between the tasks. The P bit of the above entries indicate, if the entry can be used in address translation. If P = 1, the entry can be used in address translation, otherwise it cannot be used. The P bit of the currently executed page is always high. The accessed bit A is set by 80386 before any access to the page. If A= 1, the page is accessed, else unaccessed. The D bit (Dirty bit) is set before a write operation to the page is carried out. The D-bit is undefined for page director entries. The OS reserved bits are defined by the operating system software. The User/Supervisor (U/S) bit and read/write bit are used to provide protection. These bits are decoded to provide protection under the 4 level protection model. The level 0 is supposed to have the highest privilege, while the level 3 is supposed to have the least privilege. This protection provide by the paging unit is transparent to the segmentation unit.

Page Table 31...1	0 Reserve	0	0	D	A	0	0	U S	R W	P

Fig. 10.11. Page Directory Entry

Page Frame Address 31...1	OS Reserve	0	0	D	A	0	0	U S	R W	P

Fig. 10.12. DBA Physical directory base address

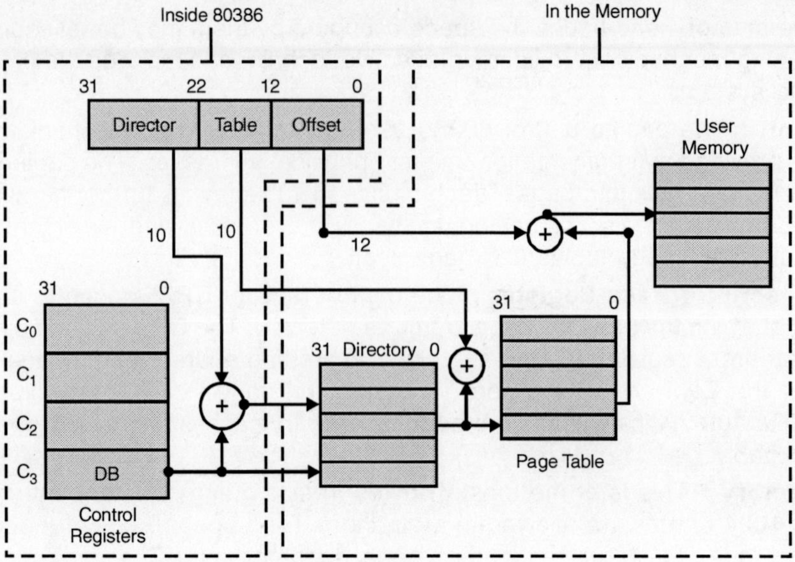

Fig. 10.13. DBA Physical Directory Base Address.

Virtual 8086 Mode

In its protected mode of operation, 80386DX provides a virtual 8086 operating environment to execute the 8086 programs. The real mode can also used to execute the 8086 programs along with the capabilities of 80386, like protection and a few additional instructions. Once the

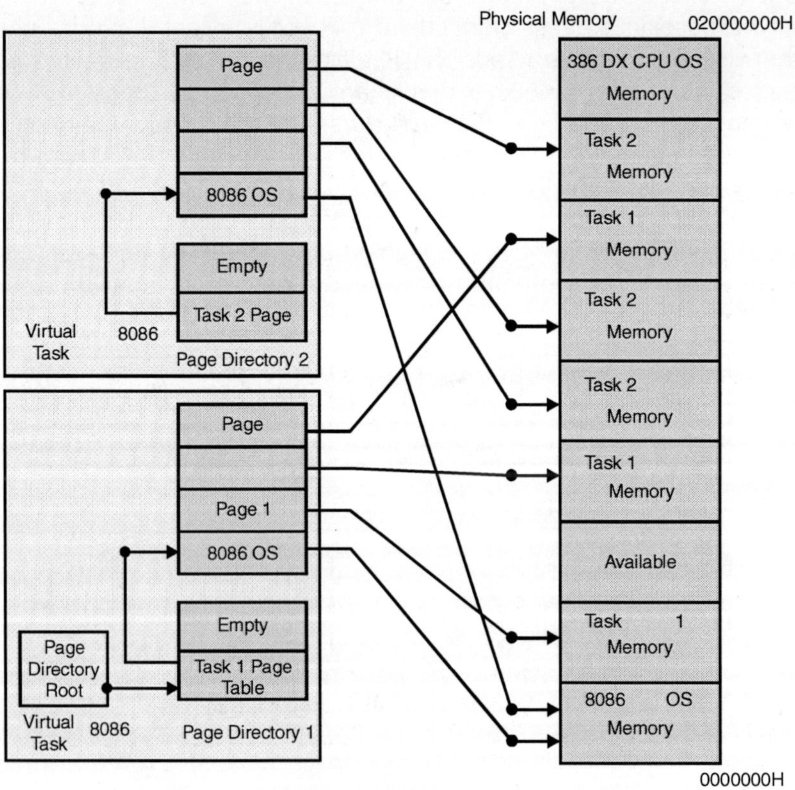

Fig. 10.14. Memory Management in Virtual 8086

80386 enters the protected mode from the real mode, it cannot return back to the real mode without a reset operation. Thus, the virtual 8086 mode of operation of 80386, offers an advantage of executing 8086 programs while in protected mode. The address forming mechanism in virtual 8086 mode is exactly identical with that of 8086 real mode. In virtual mode, 8086 can address IMbytes of physical memory that may be anywhere in the 4Gbytes address space of the protected mode of 80386. Like 80386 real mode, the addresses in virtual 8086 mode lie within IMbytes of memory. In virtual mode, the paging mechanism and protection capabilities are available at the service of the programmers. The 80386 supports multiprogramming, hence more than one programmer may be use the CPU at a time.

Paging unit may not be necessarily enable in virtual mode, but may be needed to run the 8086 programs which require more than 1 Mbyts of memory for memory management function. In virtual mode, the paging unit allows only 256 pages, each of 4K bytes size. Each of the pages may be located anywhere in the maximum 4Gbytes physical memory. The virtual mode allows the multiprogramming of 8086 applications. The virtual 8086 mode executes all the programs at privilege level 3. Any of the other programmes may deny access to the virtual mode programs or data. However, the real mode programs are executed at the highest privilege level, i.e. level 0. The virtual mode may be entered using an IRET instruction at CPL = 0 or a task switch at any CPL, executing any task whose TSS is having a flag image with VM flag set to 1. The IRET instruction may be used to set the VM flag and consequently enter the virtual mode. The PUSHF and POPF instructions are unable to read or set the VM bit, as they do not access it. Even in the virtual mode, all the interrupts and exceptions are handled by the protected mode interrupt handler. To return to the protected mode from the virtual mode, any interrupt or execution may be used. As a part of interrupt service routine, the VM bit may be reset to zero to pull back the 80386 into protected mode.

Summary of 80386

This 80386 is a 32bit processor that supports, 8bit/32bit data operands. The 80386 instruction set is upward compatible with all its predecessors. The 80386 can run 8086 applications under protected mode in its virtual 8086 mode of operation.

80386 32-Bit Data Bus

The 80386 was Intel's first 32-bit microprocessor. The expansion of the external and internal data buses to 32-bit represented a big leap forward in performance because the processor could access and process much more data in each clock cycle.

10.3. COPROCESSORS

Coprocessors produce an increase in performance for certain applications. In general a coprocessor is mathematical coprocessor that supports the CPU in calculating complicated mathematical expressions in hardware. The i387 is the coprocessor for the i386. The i387 provides hardware support for floating point arithmetic. The i386 can execute all mathematical expressions on its own using software emulation of the i387. However the hardware implementation of floating point processing in the i387 means that the i387 floating point operations run at much higher speed.

The 80386 represents a good performance improvement over our earlier simple processor model. The 80386 is a classic CISC (Complex Instruction Set Computer) CPU. Looking at the 80386 as a representative CISC we can distinguish the following features :

- Extensive (Complex) instructions

- ➤ Complex and efficient machine instructions
- ➤ Micro-encoding of the machine instructions
- ➤ Extensive addressing capabilities for memory operations
- ➤ Relatively few, but very useful CPU registers

The 80386 (and the Pentium processors) maintain code compatibility all the way back to the original 8088 and 8086 processors that were used in the first IBM PC. The instruction set of the 80386 includes some very powerful instructions, but these come at a price: because the instructions are very powerful they require more decoding effort. In classic microcoded CISC machines the instructions are broken down into smaller steps.

In a microprogrammed CISC the processor fetches the instructions via the bus interface into a prefetch queue, which transfers them to a decoding unit. The decoding unit breaks the machine instruction into many elementary micro-instructions and apples them to a microcode queue. The micro-instructions are transferred from the microcode queue to the control and execution unit which drives the ALU and the registers.

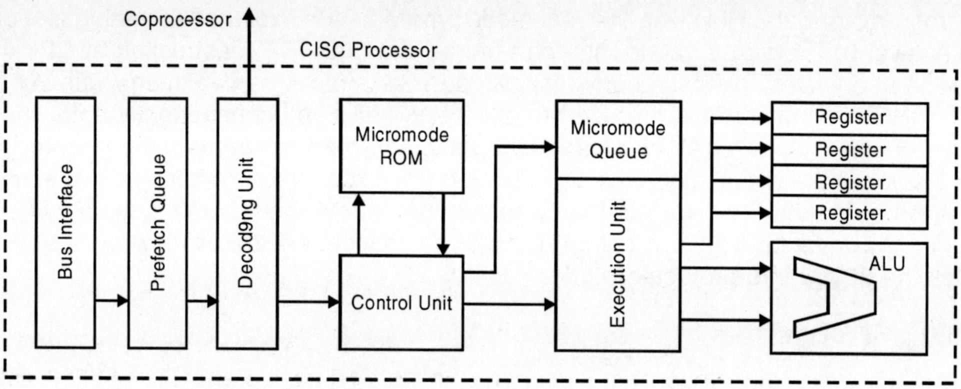

Fig. 10.7. ISC processor block diagram.

The processor decoding unit must first decode the instruction that has been read: i.e. split the instruction into the actual instruction, the type of address, extent and type of relevant register and so on. This information is contained in the op code, prefix operand and displacement data fields of the machine instruction. This is illustrated below:

Prefix(es)	Opcode	Operand(s)	Displacement/Data

In cases where the instructions are quite complex the decode time can be quite significant – sometimes as long as the execution time itself.

Most CISC processors use the microcoding technique. Microcoding makes backward compatibility a bit easier – for new instructions just add a bigger microcode ROM, with the old instructions preserved as a subset of the new one. The big disadvantage of microcoding is that the sequence of microinstructions takes more time to execute.

80387 : Math Co-processor for 80386

80387 Registers : The additional registers consist of

(1) Eight individually-addressable 80-bit numeric registers, organized as register stack.

(2) Three sixteen-bit registers containing:

(a) the NPX status word (b) the NPX control word

(c) the tag word

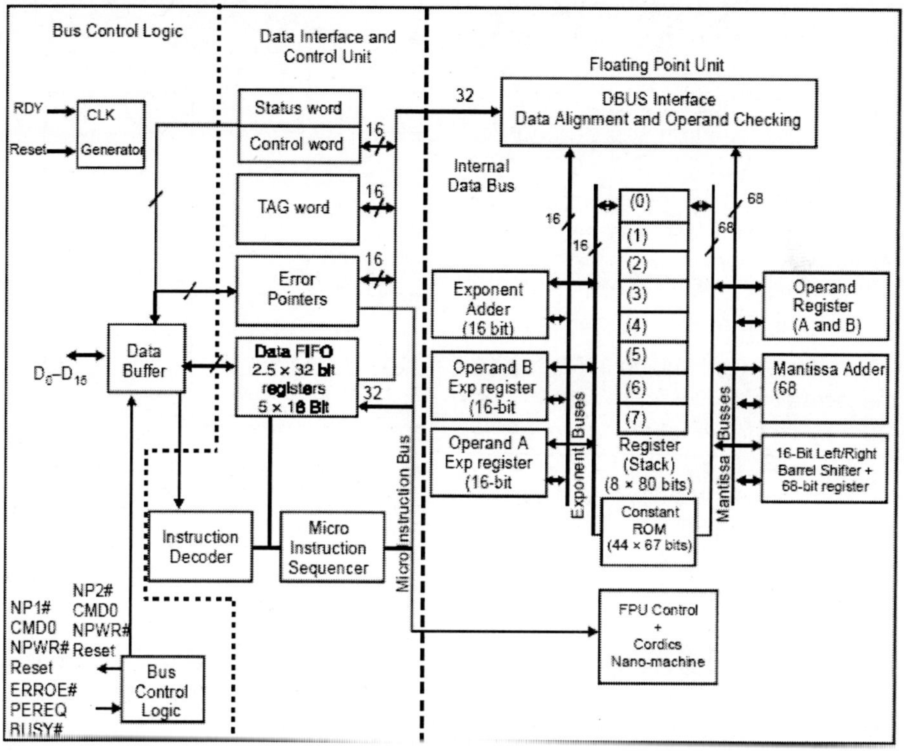

Fig. 10.15. Internal architecture of 80387 coprocessor.

(3) Two 48-bit registers containing pointers to the current instruction and operand (these registers are actually located in the 80386)

All of the NPX numeric instructions focus on the contents of these NPX registers.

The NPX Register Stack

The 80387 register stack is shown in Fig. 10.16. Each of the eight numeric registers in the 80387's register stack is 80 bits wide and is divided into fields corresponding to the NPX's extended real data type. Numeric instructions address the data registers relative to the register on the top of the stack. At any point in time, this top-of-stack register is indicated by the TOP (stack TOP) field in the NPX status word. Load or push operations decrement TOP by one and load a value into the new top register. A store-and-pop operation stores the value from the current TOP register and then increments TOP by one. Like 80386 stacks in memory, the 80387 register stack grows down toward lower-addressed registers.

Many numeric instructions have several addressing modes that permit the programmer to implicitly operate on the top of the stack, or to explicitly operate on specific registers relative to the TOP. The ASM386 Assembler supports these register addressing modes, using the expression ST(0), or simply ST, to represent the current Stack Top and ST(i) to specify the ith register from TOP in the stack. For example, if TOP contains 01 IB (register 3 is the top of the stack), the following statement would add the contents of two registers in the stack (registers 3 and 5):

The stack organization and top-relative addressing of the numeric registerssimplify subroutine programming by allowing routines to pass parameters on the register stack. By using the stack to pass parameters rather than using "dedicated" registers, calling routines gain more flexibility in how they use the stack. As long as the stack is not full, each routine simply loads the parameters onto the stack before calling a particular subroutine to perform a numeric calculation. The subroutine then addresses its parameters as ST, ST(I), etc., even though TOP may, for example, refer to physical register 3 in one invocation and physical register 5 in another.

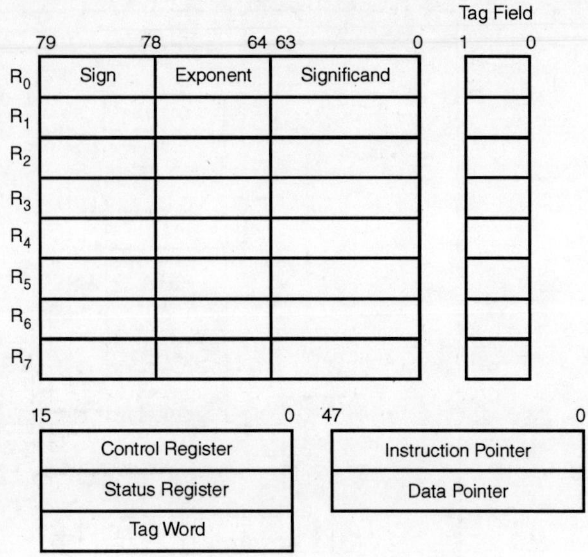

Fig. 10.16. 80387 Register Set

The NPX Status Word

The 16-bit status word shown in Fig. 10.16 reflects the overall state of the 80387. This status word may be stored into memory using the FSTSW/FNSTSW, FSTENV/FNSTENV. and FSAVE/FNSAVE instructions, and can be transferred into the 80386 AX register with the FSTSW AX/FNSTSW AX instructions, allowing the NPX status to be inspected by the CPU.

The B-bit (bit 15) is included for 8087 compatibility only. It reflects the contents of the ES bit (bit 7 of the status word), not the status of the BUSY# output of the 80387. The four NPX condition code bits (C{3}-C{0}) are similar to the flags in a CPU: the 80387 updates these bits to reflect the outcome of arithmetic operations. These condition code bits are used principally for conditional branching. The FSTSW AX instruction stores the NPX status word directly into the CPU AX register, allowing these condition codes lo be inspected efficiently by 80386 code. The 80386 SAHF instruction can copy C{3}-C{0} directly to 80386 flag bits to simplify conditional branching.

Bits 12-14 of the status word point to the 80387 register that is the current Top of Stack (TOP). The significance of the stack top has been described in the prior section on the register stack.

Figure 10.17 shows the six exception flags in bits 0-5 of the status word. Bit 7 is the exception summary status (ES) bit. ES is set if any unmasked exception bits are set, and is cleared otherwise. If this bit is set, the ERROR# signal is asserted. Bits 0-5 indicate whether the NPX has detected one of six possible exception conditions since these status bits were last cleared or reset. They are "sticky" bits, and can only be cleared by the instructions FINIT, FCLEX, FLDENV, FSAVE and FRSTOR.

Bit 6 is the stack fault (SF) bit. This bit distinguishes invalid operations due to stack overflow or underflow from other kinds of invalid operations. When SF is set, bit 9 (C{1}) distinguishes between stack overflow (C{1} = 1) and underflow (C{1} = 0).

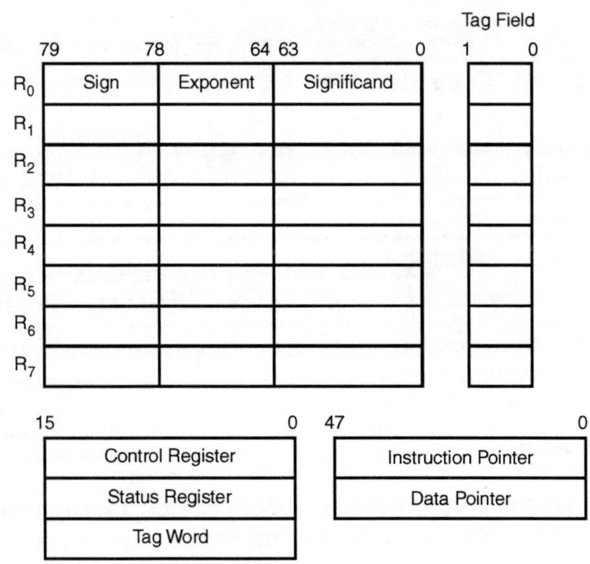

Fig. 10.17. 80387 Status word

Control Word

The NPX provides the programmer with several processing options, which are selected by loading a word from memory into the control word. Figure 10.18 shows the format and encoding of the fields in the control word.

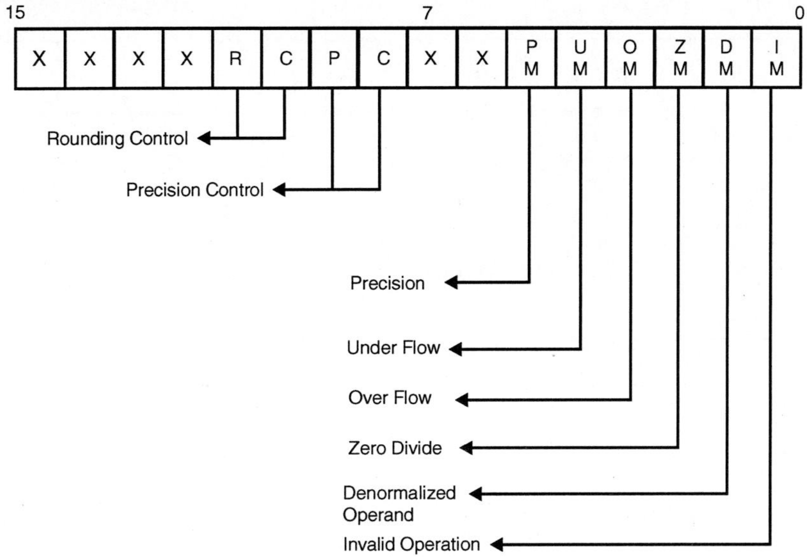

Note :

Precision Control	**Rounding Control**
00 – 24 Bits (Single Precision)	00 – Round to Near nest or even
01 – (Reserved)	01 – Round Down (Toward –1)
10 – 53 Bits (Double Precision)	10 – Round up (Toward +1)
11 – 64 Bits (Extended Precision)	11 – Chop (Truncate toward zero)

Fig. 10.18. 80387 control word format

The low-order byte of this control word configures the 80387 exception masking. Bits 0-5 of the control word contain individual masks for each of the six exception conditions recognized by the 80387. The high-order byte of the control word configures the 80387 processing options, including

(a) Precision control

(b) Rounding control

The precision-control bits (bits 8-9) can be used to set the 80387 internal operating precision at less than the default precision (64-bit significand). These control bits can be used to provide compatibility with the earlier-generation arithmetic processors having less precision than the 80387. The precision-control bits affect the results of only the following five arithmetic instructions: ADD, SUB(R), MUL, DIV(R), and SQRT. No other operations are affected by PC.

The rounding-control bits (bits 10-11) provide for the common ground-to-nearest mode, as well as directed rounding and true chop. Rounding control affects only the arithmetic instructions.

The NPX Instruction and Data Pointers

The instruction and data pointers provide support for programmed exception-handlers. These registers are actually located in the 80386, but appear to be located in the 80387 because they are accessed by the ESC instructions FEDENV, FSTENV, FSAVE, and FRSTOR. Whenever the

80386 decodes an ESC instruction, it saves the instruction address, the operand address (if present), and the instruction opcode. When stored in memory, the instruction and data pointers appear in one of four formats, depending on the operating mode of the 80386 (protected mode or real-address mode) and depending on the operand-size attribute in effect (32-bit operand or 16-bit operand). When the 80386 is in virtual-8086 mode, the real-address mode formats are used.

15							0
TAG (7)	TAG (6)	TAG (5)	TAG (4)	TAG (3)	TAG (2)	TAG (1)	TAG (0)

Tag Values :
00 = Valid
01 = Zero
10 = Invalid or infinity
11 = Empty

Fig. 10.19. Tag word format

31 — 16	15 — 0	
Reserved	Control Word	0H
Reserved	Status Word	4H
Reserved	Tag Word	8H
IP Offset		CH
Opcode	CS Selector	10H
Data Operand Offset		14H
Reserved	Operand Selector	18H

Fig. 10.20. Protected mode 80387 instruction and data pointer image in memory

31 — 16	15 — 0	
Reserved	Control Word	0H
Reserved	Status Word	4H
Reserved	Tag Word	8H
Reserved	Instruction Pointer	CH
Instruction Pointer (31, 16-0) Opcode		10H
Opcode	CS Selector	14H
Operand Pointer (31, 16-0)		18H

Fig. 10.21. Read Mode 80387, Instruction and data pointer image in memory.

15 — 0	
Control Word	0H
Status Word	2H
Tag Word	4H
IP Offset	6H
CB Selector	8H
Operand Offset	AH
Operand Selector	CH

Fig. 10.22. Protected mode 80387 Instruction and data pointer image in memory.

15 — 0	
Control Word	0H
Status Word	2H
Tag Word	4H
Instruction Pointer 15 – 0	6H
IP Opcode (10 – 0)	8H
Operand Pointer (15 – 0)	AH
Opcode '0' 00000000000	CH

Fig. 10.23. Real Mode 80387 instructional data pointer image in memory.

Figure 10.17-10.22 through 10.17-10.23 show these pointers as they are stored following an FSTENV instruction.

The FSTENV and FSAVE instructions store this data into memory, allowing exception handlers to determine the precise nature of any numeric exceptions that may be encountered.

The instruction address saved in the 80386 (as in the 80287) points to any prefixes that preceded the instruction. This is different from the 8087, forwhich the instruction address points only to the ESC instruction opcode.

Note that the processor control instructions F1N1T, FLDCW, FSTCW, FSTSW. FCLIiX, FSTENV, FEDENV, FSAVE, FRSTOR, and FWA1T do not affect the data pointer. Note also that, except for the instructions just mentioned, the value of the data pointer is undefined if the prior ESC instruction did not have a memory operand.

The NPX Tag Word

The tag word indicates the contents of each register in the register stack, as shown in Figure 10.19. The tag word is used by the NPX itself to distinguish between empty and nonempty register locations. Programmers of exception handlers may use this tag information to check the contents of a numeric register without performing complex decoding of the actual data in the register. The tag values from the tag word correspond to physical registers 0-7. Programmers must use the current top-of-stack (TOP) pointer stored in the NPX status word to associate these tag values with the relative stack registers ST(0) through ST(7). The exact values of the tags are generated during execution of the FSTENV and FSAVE instructions according to the actual contents of the nonempty stack locations. During execution of other instructions, the 80387 updates the TW only to indicate whether a stack location is empty or nonempty.

Instruction Pipelining

The main reason for pipelining instructions is that it allows some operations to be carried out in parallel. Let's look at our Fetch-Decode-Execute model in a little more detail. These are the steps that need to be carried out when a microprocessor executes an instruction:

- Read the instruction from memory or the prefetch queue (instruction fetch phase)
- Decode the instruction (decode phase)
- Where necessary, fetch the operands (operand fetch phase)
- Execute the instruction (execute phase)
- Write back the result (write-back phase)

Note that some implementations combine the decode and operand fetch phases, ending up with four stages, while others expand out some of the phases here to make a longer sequence. When the pipeline gets longer, each stage is even simpler so it can be pushed even faster but the overall logic gets more complex.

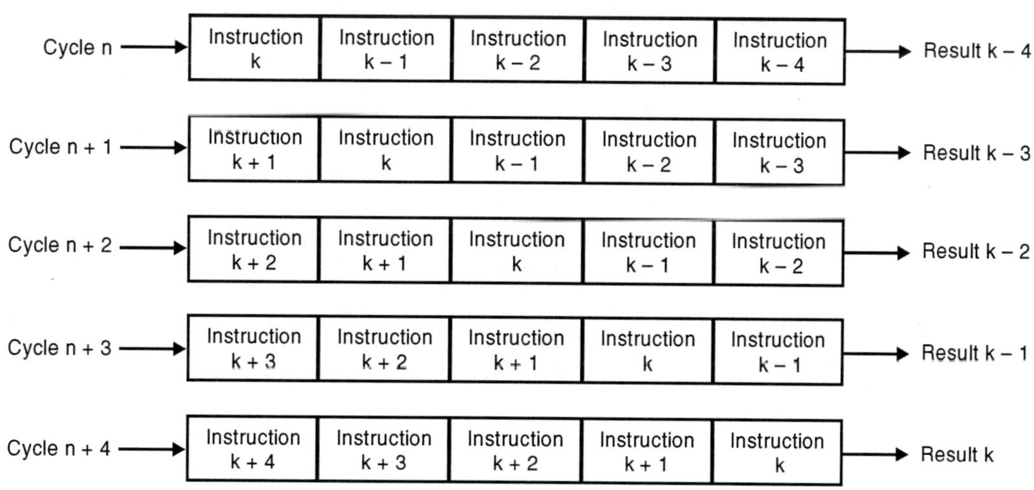

Fig. 10.24. Instruction pipelining

The aim of pipelining is to achieve single clock cycle instruction execution. The important thing to note is that each instruction is not actually executed completely in one cycle, but rather that each instruction can be completed in one cycle. We haven't increased the execution speed at which one instruction is executed, what we have done is increased the throughput by allowing a number of tasks to execute simultaneously. Since computers execute billions of instructions it's throughput that matters. The illustration below shows the throughput of a number of instructions through a pipeline.

This is a 5-stage pipeline running under best-case conditions. You can see that we have realised our aim of executing one instruction per clock cycle.

Integer Pipelines, U and V

(1) Superscalar processors use more than one pipeline
(2) Under best case conditions the Pentium can complete two instructions in every clock
(3) IA-32 instructions have to be 'paired' according to Intel rules
   (a) Pipeline u can execute any IA-32 instruction
   (b) Pipeline v can execute 'simple' instructions
   (c) Pipeline u gets filled first
   (d) If the second instruction is NOT part of a pair - it waits for the next slot
(4) All pairing & decoding decisions are done in hardware-software support not required but helps performance.

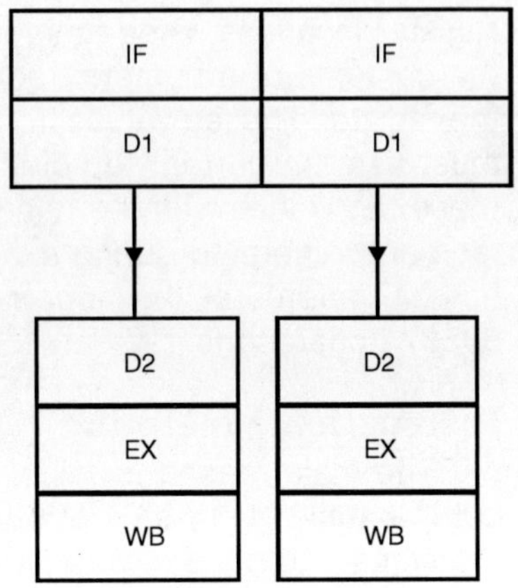

Fig. 10.25. u, v Pipeline

10.4. THE 80486 PROCESSOR

The i486 was the first x86 to incorporate RISC elements into its design to improve performance. In addition to maintaining compatibility with the 80386 and earlier x86 processors it added the following features :

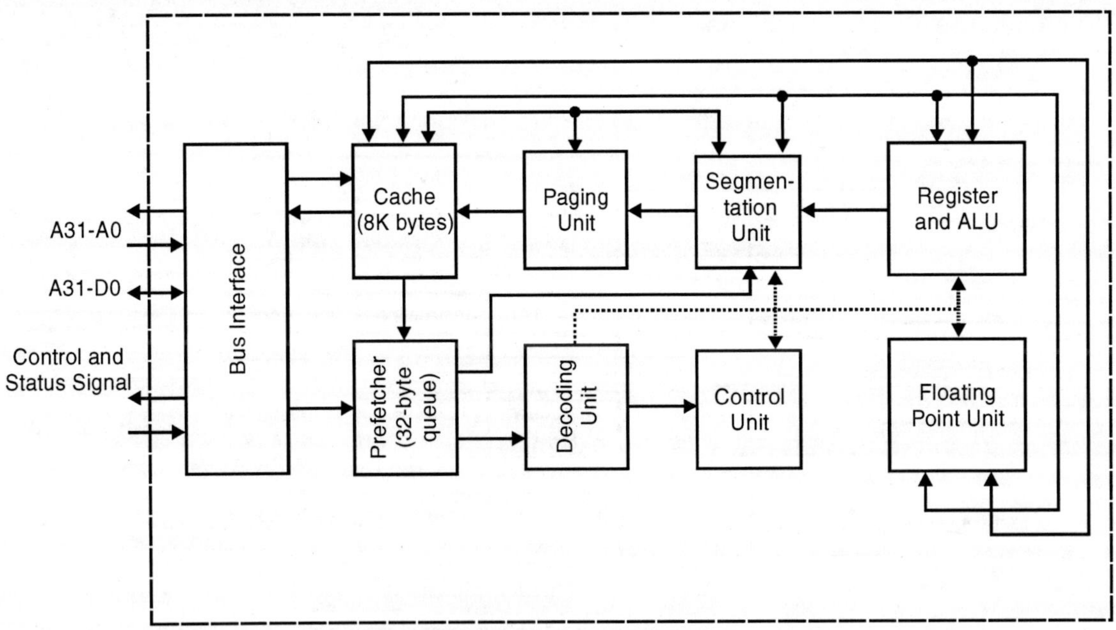

Fig. 10.26. 80486 Block diagram.

(i) Improved 80386 CPU (6 extra instructions)
(ii) Hard-wired implementation of frequently used instructions (as in RISCs)
(iii) A 5 stage instruction pipeline
(iv) An 8K Cache Memory + cache controller (previously a separate device)
(v) An on-chip Floating Point coprocessor
(vi) Longer Prefetch Queue (32-bytes as opposed to 16 on the 80386)
(vii) Higher frequency operation
(viii) About a million transistors

Like the 80386 it uses real, protected and virtual 8086 modes and its Memory Management Unit includes a Segmentation Unit and a Paging Unit. A block diagram of the 80486 processor is given below:

10.5. PROCESSOR FEATURES

All Intel486 processors consist of a 32-bit integer processing unit, an on-chip cache, and a memory management unit. These ensure full binary compatibility with the 8086, 8088, 80186, 80286, Intel386™ SX, and Intel386 DX processors, and with all versions of Intel486 processors. All Intel486 processors offer the following features:

(i) *32-bit RISC integer core.* The Intel486 processor performs a complete set of arithmetic and logical operations on 8-, 16-, and 32-bit data types using a full-width ALU and eight general purpose registers.

(ii) *Single Cycle Execution.* Many instructions execute in a single clock cycle.

(iii) *Instruction Pipelining.* The fetching, decoding, address translation, and execution of instructions are overlapped within the Intel486 processor.

(iv) *On-Chip Floating-Point Unit.* The IntelDX2 and Intel DX4 processors support the 32-, 64-, and 80-bit formats specified in IEEE standard 754.

(v) *On-Chip Cache with Cache Consistency Support.* An 8-Kbyte (16-Kbyte on the IntelDX4 processor) internal cache is used for both data and instructions. Cache hits provide zero wait state access times for data within the cache. Bus activity is tracked to detect alterations in the memory represented by the internal cache. The internal cache can be invalidated or flushed so that an external cache controller can maintain cache consistency.

(vi) *External Cache Control.* Write-back and flush controls for an external cache are provided so the processor can maintain cache consistency.

(vii) *On-Chip Memory Management Unit.* Address management and memory space protection mechanisms maintain the integrity of memory in a multi-tasking and virtual memory environment. The memory management unit supports both segmentation and paging.

(viii) *Burst Cycles.* Burst transfers allow a new doubleword to be read from memory on each bus clock cycle. This capability is especially useful for instruction prefetch and for filling the internal cache.

(ix) *Write Buffers.* The processor contains four write buffers to enhance the performance of consecutive writes to memory. The processor can continue internal operations after a write to these buffers, without waiting for the write to be completed on the external bus.

(x) *Bus Backoff.* If another bus master needs control of the bus during a processor-initiated bus cycle, the Intel486 processor floats its bus signals, then restarts the cycle when the bus becomes available again.

(xi) *Instruction Restart.* Programs can continue execution following an exception that is generated by an unsuccessful attempt to access memory. This feature is important for supporting demand-paged virtual memory applications.

(xii) *Dynamic Bus Sizing.* External controllers can dynamically alter the effective width of the data bus. Bus widths of 8, 16, or 32 bits can be used (the 8-bit and 32-bit bus widths are not available on the Ultra-Low Power Intel486 GX processor).

(xiii) *Boundary Scan (JTAG).* Boundary Scan provides in-circuit testing of components on printed circuit boards. The Intel Boundary Scan implementation conforms with the IEEE Standard Test Access Port and Boundary Scan Architecture.

Modes of Operation

Protected mode : native mode (Windows, Linux)

Real-address mode : native MS-DOS

System management mode : power management, system security, diagnostics

Virtual-8086 mode : Hybrid of Protected, each program has its own 8086 computer

The Intel Architecture supports three operating modes: protected mode, real-address mode, and system management mode. The operating mode determines which instructions and architectural features are accessible :

Protected mode : The native state of the processor. In this mode all instructions and architectural features are available, providing the highest performance and capability. This is the recommended mode for all new applications and operating systems.

Virtual-8086 mode : Among the capa-bilities of protected mode is the ability to directly execute "real-address mode" 8086 software in a protected, multi-tasking environment. This feature is called virtual-8086 mode, although it is not actually a processor mode. Virtual-8086 mode is actually a protected mode attribute that can be enabled for any task.

Real-address mode : Provides the programming environment of the Intel 8086 processor with a few extensions (such as the ability to switch to protected or system management mode). The processor is placed in real-address mode following power-up or a reset.

System management mode : A standard architectural feature unique to all Intel processors, beginning with the Intel386 SL processor. This mode provides an operating system or executive with a transparent mechanism for implementing platform-specific functions such as power management and system security.

The processor enters SMM when the external SMM interrupt pin (SM1#) is activated or an SMI is received from the advanced programmable interrupt controller (APIC). In SMM, the processor switches to a separate address space while saving the entire context of the currently running program or task.

SMM-specific code may then be executed transparently. Upon returning from SMM, the processor is placed back into its state prior to the system management interrupt.

With the 32 bit address bus, the 80386 can address upto 4G bytes of physical memory. The physical memory is organised in terms of segments of 4G bytes at maximum.

The 80386 CPU supports 16K number of segments and thus the total virtual space of 4Gbytes * 16K = 64 terrabytes. The memory management section of 80386 supports the virtual memory, paging and four levels of protection, maintaining full compatibility with 80286.

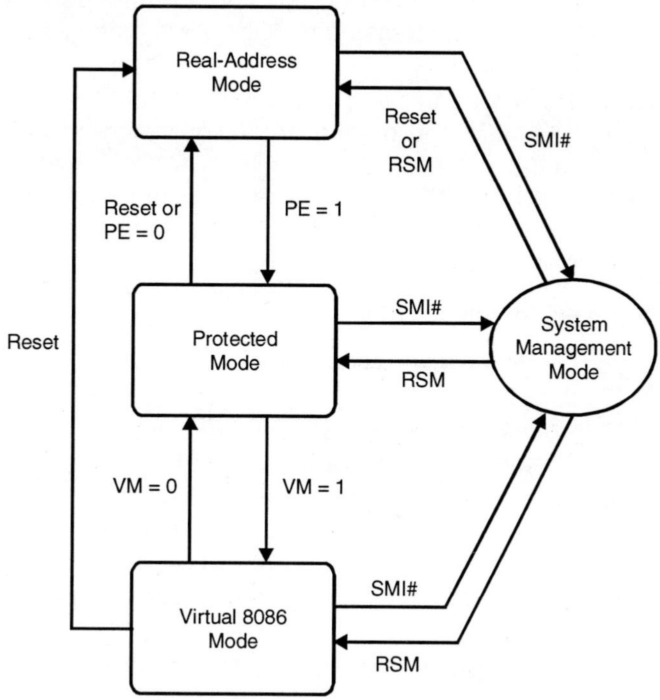

Fig. 10.27. Modes of operation.

The 80386 offers a set of 8 debug registers DR0-DR7 for hardware debugging and control. The 80386 has on chip address translation cache. The concept of paging is introduced in 80386 that enables it to organise the available physical memory in terms of pages of size 4K bytes each, under the segmented memory.

The 80386 can be supported by 80387 for mathematical data processing.

10.5.1. Operating Modes and Compatibility

The Intel486 processor can run in modes that give it object-code compatibility with software written for the 8086, 80286, and Intel386 processor families. The operating mode is set in software as one of the following:

(i) **Real Mode.** When the processor is powered up or reset, it is initialized in Real Mode. This mode has the same base architecture as the 8086 processor but allows access to the 32-bit register set of the Intel486 processor. The address mechanism, maximum memory size (1 Mbyte), and interrupt handling are identical to the Real Mode of the 80286 processor. Nearly all Intel486 processor instructions are available, but the default operand size is 16 bits; in order to use the 32-bit registers and addressing modes, override instruction prefixes must be used. The primary purpose of Real Mode is to set up the processor for Protected Mode operation.

(ii) **Protected Mode (also called Protected Virtual Address Mode).** The complete capabilities of the Intel486 processor become available when programs are run in Protected Mode. In addition to segmentation protection, paging can be used in Protected Mode. The linear address space is four gigabytes and virtual memory programs of up to 64 terabytes can be run. All existing 8086, 80286, and Intel386 processor software can be run under the Intel486 processor's hardware-assisted protection mechanism. The addressing mechanism is more sophisticated in Protected Mode than in Real Mode.

(iii) **Virtual 8086 Mode.** A sub-mode of Protected Mode, allows 8086 programs to be run with the segmentation and paging protection mechanisms of Protected Mode. This mode offers more flexibility than the Real Mode for running 8086 programs. Using this mode, the Intel486 processor can execute 8086 operating systems and applications simultaneously with an Intel486 operating system and both 80286 and Intel486 processor applications.

10.6. INTERNAL ARCHITECTURE

10.6.1. Overview

The Intel486™ SX processor and the Ultra-Low Power Intel486 SX have a 32-bit architecture with on-chip memory management and level-1 cache.

The IntelDX2™ and IntelDX4™ processors also have a 32-bit architecture with on-chip memory management and cache, but add clock multiplier and floating point units. The Intel486 SX, Ultra- Low Power Intel486 SX, and Intel486 DX processors support dynamic bus sizing for the external data bus; that is, the bus size can be specified as 8-, 16-, or 32-bits wide.

Internally, the ultra-low power processors are similar to the Intel486 SX, but add a specialized clock control unit. Although the Ultra-Low Power Intel486 SX supports dynamic bus sizing, the Ultra-Low Power Intel486 GX supports only a 16-bit external data bus. The Ultra-Low Power Intel486 GX also has advanced power management features.

Table 10.1 lists the functional units of the embedded Intel486 processors.

TABLE 10.1. Intel486™ Processor Family Functional Units.

Functional Unit	IntelDX2™ and IntelDX4™ Processors	Intel486 SX™ Processor	Ultra-Low Power Intel486 SX and Ultra-Low Power Intel486 GX Processors
Bus Interface	3	3	3
Cache (L1)	3	3	3
Instruction Prefetch	3	3	3
Instruction Decode	3	3	3
Control	3	3	3
Integer and Datapath	3	3	3
Segmentation	3	3	3
Paging	3	3	3
Floating-Point	3		
Clock Multiplier	3		
Clock Control			3

Figure 10.28 is a block diagram of the embedded IntelDX2 and IntelDX4 processors. Note that the cache unit is 8 Kbytes for the IntelDX2 processor and 16 Kbytes for the IntelDX4.

Figure 10.29 is a block diagram of the embedded Intel486 SX processor.

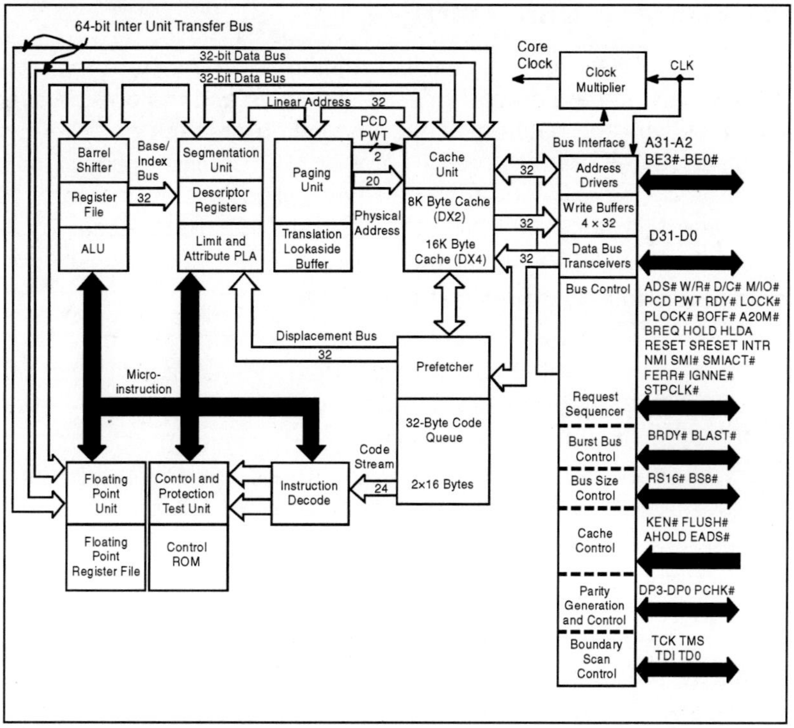

Fig. 10.28. IntelDX2™ and IntelDX4™ Processors Block Diagram.

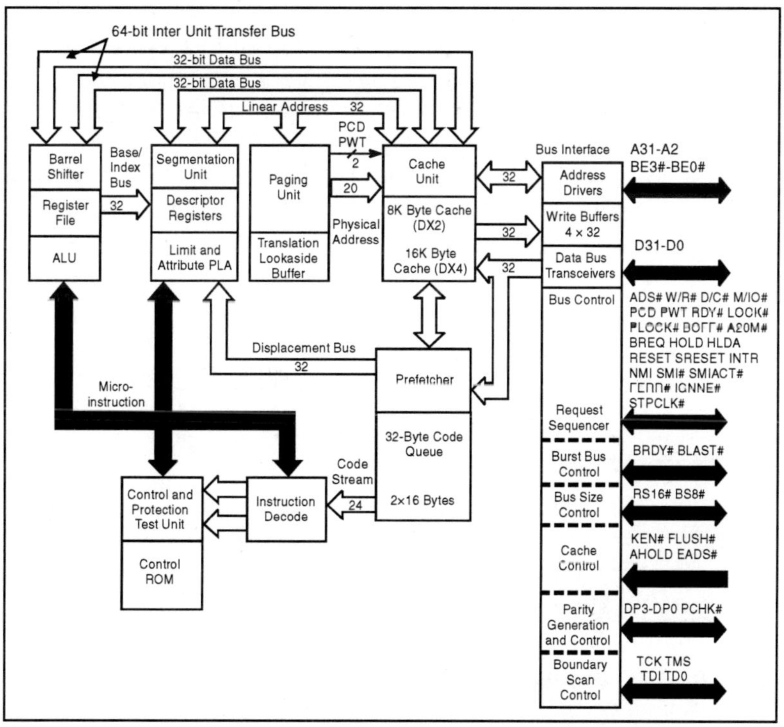

Fig. 10.29. Intel486™ SX Processor Block Diagram.

The functional units of the Intel486 processor are described in the following sections.

10.6.2. Bus Interface Unit (BIU)

The bus interface unit (BIU) prioritizes and coordinates data transfers, instruction prefetches, and control functions between the processor's internal units and the outside system. Internally, the BIU communicates with the cache and the instruction prefetch units through three 32-bit buses, as shown in Fig. 10.1. Externally, the BIU provides the processor bus signals, as described in Chapter 3. Except for cycle definition signals, all external bus cycles, memory reads, instruction prefetches, cache line fills, etc., look like conventional microprocessor cycles to external hardware, with all cycles having the same bus timing.

The BIU contains the following architectural features:

(1) **Address Transceivers and Drivers.** The A31–A2 address signals are driven on the processor bus, together with their corresponding byte-enable signals, BE3#–BE0#. The high-order 28 address signals are bidirectional, allowing external logic to drive cache invalidation addresses into the processor.

(2) **Data Bus Transceivers.** The D31–D0 data signals are driven onto and received from the processor bus (for the Ultra-Low Power Intel486 GX processor, signals D15–D0 comprise the data bus transceivers).

(3) **Bus Size Control.** Three sizes of external data bus can be used: 32, 16, and 8 bits wide. Two inputs (BS16# and BS8#) from external logic specify the width to be used. Bus size can be changed on a cycle-by-cycle basis. The Ultra-Low Power Intel486 GX does not support dynamic bus sizing; its external data bus is 16 bits wide.

(4) **Write Buffering.** Up to four write requests can be buffered, allowing many internal operations to continue without waiting for write cycles to be completed on the processor bus.

(5) **Bus Cycles and Bus Control.** A large selection of bus cycles and control functions are supported, including burst transfers, non-burst transfers (single- and multiple-cycle), bus arbitration (bus request, bus hold, bus hold acknowledge, bus locking, bus pseudo-locking, and bus backoff), floating-point error signalling, interrupts, and reset. Two software- controlled outputs enable page caching on a cycle-by-cycle basis. One input and one output are provided for controlling burst read transfers.

(6) **Parity Generation and Control.** Even parity is generated on writes to the processor and checked on reads. An error signal indicates a read parity error.

(7) **Cache Control.** Cache control and consistency operations are supported. Four inputs (KEN#, Flush#, PCD, and PWT) allow the external system to control the consistency of data stored in the internal cache unit. Two special bus cycles allow the processor to control the consistency of external cache.

10.6.3. Data Transfers

To support the cache, the BIU reads 16-byte cacheable transfers of operands, instructions, and other data on the processor bus and passes them to the cache unit. When cache contents are updated from an internal source, such as a register, the BIU writes the updated cache information to the external system. Non-cacheable read transfers are passed through the cache to the integer or floating-point units.

During instruction prefetch, the BIU reads instructions on the processor bus and passes them to both the instruction prefetch unit and the cache. The instruction prefetch unit may then obtain its inputs directly from the cache.

10.6.4. Write Buffers

The BIU has temporary storage for buffering up to four 32-bit write transfers to memory. Addresses, data, or control information can be buffered. Single I/O-mapped writes are not buffered, although multiple I/O writes may be buffered. The buffers can accept memory writes as quickly as one per clock. Once a write request is buffered, the internal unit that generated the request is free to continue processing. If no higher-priority request is pending and the bus is free, the transfer is propagated as an immediate write cycle to the processor bus. When all four write buffers are full, any subsequent write transfer stalls inside the processor until a write buffer becomes available.

The BIU can re-order pending reads in front of buffered writes. This is done because pending reads can prevent an internal unit from continuing, whereas buffered writes need not have a detrimental effect on processing speed.

Writes are propagated to the processor bus in the first-in-first-out order in which they are received from the internal unit. However, a subsequently generated read request (data or instruction) may be re-ordered in front of buffered writes. As a protection against reading invalid data, this re-ordering of reads in front of buffered writes occurs only if all buffered writes are cache hits. Because an external read is generated only for a cache miss, and is re-ordered in front of buffered writes only if all such buffered writes are cache hits, any read generated on the external bus with this protection never reads a location that is about to be written by a buffered write. This re-ordering can only happen once for a given set of buffered writes, because the data returned by the read cycle could otherwise replace data about to be written from the write buffers.

To ensure that no more than one such re-ordering is done for a given set of buffered writes, all buffered writes are re-flagged as cache misses when a read request is re-ordered ahead of them. Buffered writes thus marked are propagated to the processor bus before the next read request is acted upon. Invalidation of data in the internal cache also causes all pending writes to be flagged as cache misses. Disabling the cache unit disables the write buffers, which eliminates any possibility of re-ordering bus cycles.

10.6.5. Locked Cycles

The processor can generate signals to lock a contiguous series of bus cycles. These cycles can then be performed without interference from other bus masters, if external logic observes these lock signals. One example of a locked operation is a semaphor read-modify-write update, where a resource control register is updated. No other operations should be allowed on the bus until the entire locked semaphor update is completed.

When a locked read cycle is generated, the internal cache is not read. All pending writes in the buffer are completed first. Only then is the read part of the locked operation performed, the data modified, the result placed in a write buffer, and a write cycle performed on the processor bus. This sequence of operations ensures that all writes are performed in the order in which they were generated.

10.6.6. I/O Transfers

Transfers to and from I/O locations have some restrictions to ensure data integrity:

- **Caching.** I/O reads are never cached.
- **Read Re-ordering.** I/O reads are never re-ordered ahead of buffered writes to memory. This ensures that the processor has completed updating all memory locations before reading status from a device.

➢ **Writes.** Single I/O writes are never buffered. When processing an OUT instruction, internal execution stops until all buffered writes and the I/O write are completed on the processor bus. This allows time for external logic to drive a cache invalidate cycle or mask interrupts before the processor executes the next instruction. The processor completes updating all memory locations before writing to the I/O location. Repeated OUT instructions may be buffered.

The write buffers and the cache unit determine I/O device recovery time. In the Intel386 processor, back-to-back write recovery time could be guaranteed to exceed a certain value by inserting a jump to the next instruction that writes to the I/O device. This forced an instruction prefetch cycle that could only be performed after the preceding write was completed. This technique is not used in the Intel486 processor because a prefetch can be satisfied internally by the cache and recovery time may be too short. The same effect is achieved in the Intel486 processor by explicitly generating a read to an area of memory that is not cacheable. Because the Intel486 processor does not buffer single I/O writes, such a read is not done until the I/O write is completed.

10.7. CACHE UNIT

The cache unit stores copies of recently read instructions, operands, and other data. When the processor requests information already in the cache, called a cache hit, no processor-bus cycle is required. When the processor requests information not in the cache, called a cache miss, the information is read into the cache in one or more 16-byte cacheable data transfers, called cache line fills. An internal write request to an area currently in the cache causes two distinct actions if the cache is using a write-through policy: the cache is updated, and the write is also passed through the cache to memory. If the cache is using a write-back policy, then the internal write request only causes the cache to be updated and the write is stored for future main memory updating.

The cache transfers data to other units on two 32-bit buses, as shown in Fig. 10.1. The cache receives linear addresses on a 32-bit bus and the corresponding physical addresses on a 20-bit bus. The cache and instruction prefetch units are closely coupled. 16-Byte blocks of instructions in the cache can be passed quickly to the instruction prefetch unit. Both units read information in 16-byte blocks.

The cache can be accessed as often as once each clock. The cache acts on physical addresses, which minimizes the number of times the cache must be flushed. When both the cache and the cache write-through functions are disabled, the cache may be used as a high-speed RAM.

10.7.1. Cache Structure

The cache has a four-way set associative organization. There are four possible cache locations to store data from a given area of memory. Four-way association is a compromise between the speed of a direct-mapped cache during cache hits and the high cache-hit ratio of a fully associative cache. As shown in Fig. 10.30, the 8-Kbyte data block is divided into four data ways, each containing 128 16-byte sets, or cache lines (the DX4 processor has 256 16-byte sets). Each cache line holds data from 16 successive byte addresses in memory, beginning with an address divisible by 16.

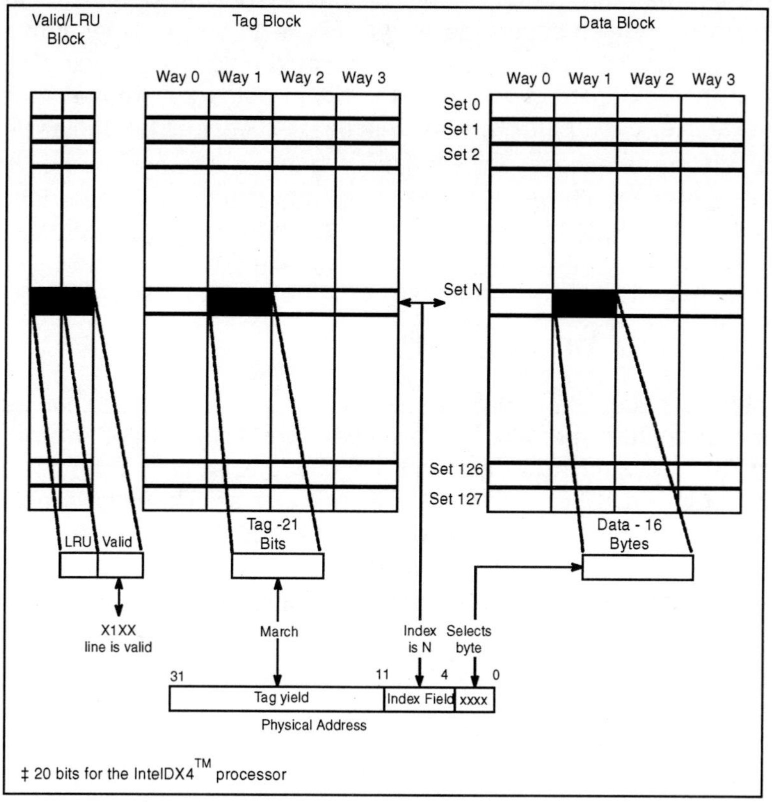

Fig. 10.30. Cache Organization.

Cache addressing is performed by dividing the high-order 28 bits of the physical address into three parts, as shown in Fig. 10.30. The 7 bits of the index field specify the set number, one of 128, within the cache. The high-order 21 bits (20 on the IntelDX4 processor) are the tag field; these bits are compared with tags for each cache line in the indexed set, and they indicate whether a 16-byte cache line is stored for that physical address. The low-order 4 bits of the physical address select the byte within the cache line. Finally, a 4-bit valid field, one for each way within a given set, indicates whether the cached data at that physical address is currently valid.

10.8. INSTRUCTION PREFETCH UNIT

When the BIU is not performing bus cycles to execute an instruction, the instruction prefetch unit uses the BIU to prefetch instructions. By reading instructions before they are needed, the processor rarely needs to wait for an instruction prefetch cycle on the processor bus.

Instruction prefetch cycles read 16-byte blocks of instructions, starting at addresses numerically greater than the last-fetched instruction. The prefetch unit, which has a direct connection (not shown in Fig. 10.1) to the paging unit, generates the starting address. The 16-byte prefetched blocks are read into both the prefetch and cache units simultaneously. The prefetch queue in the prefetch unit stores 32 bytes of instructions. As each instruction is fetched from the queue, the code part is sent to the instruction decode unit and (depending on the instruction) the displacement part is sent to the segmentation unit, where it is used for address calculation. If loops are encountered in the program being executed, the prefetch unit gets copies of previously executed instructions from the cache.

The prefetch unit has the lowest priority for processor bus access. Assuming zero wait-state memory access, prefetch activity never delays execution. However, if there is no pending data transfer, prefetching may use bus cycles that would otherwise be idle. The prefetch unit is flushed whenever the next instruction needed is not in numerical sequence with the previous instruction; for example, during jumps, task switches, exceptions, and interrupts.

The prefetch unit never accesses beyond the end of a code segment and it never accesses a page that is not present. However, prefetching may cause problems for some hardware mechanisms. For example, prefetching may cause an interrupt when program execution nears the end of memory. To keep prefetching from reading past a given address, instructions should come no closer to that address than one byte plus one aligned 16-byte block.

10.9. INSTRUCTION DECODE UNIT

The instruction decode unit receives instructions from the instruction prefetch unit and translates them in a two-stage process into low-level control signals and microcode entry points, as shown in Fig. 10.1. Most instructions can be decoded at a rate of one per clock. Stage 1 of the decode, shown in Fig. 10.8, initiates a memory access. This allows execution of a two-instruction sequence that loads and operates on data in just two clocks.

The decode unit simultaneously processes instruction prefix bytes, opcodes, modR/M bytes, and displacements. The outputs include hardwired microinstructions to the segmentation, integer, and floating-point units. The instruction decode unit is flushed whenever the instruction prefetch unit is flushed.

10.10. CONTROL UNIT

The control unit interprets the instruction word and microcode entry points received from the instruction decode unit. The control unit has outputs with which it controls the integer and floatingpoint processing units. It also controls segmentation because segment selection may be specified by instructions.

The control unit contains the processor's microcode. Many instructions have only one line of microcode, so they can execute in an average of one clock cycle. Fig. 3.8 shows how execution fits into the internal pipelining mechanism.

10.11. INTEGER (DATAPATH) UNIT

The integer and datapath unit identifies where data is stored and performs all of the arithmetic and logical operations available in the Intel386 processor's instruction set, plus a few new instructions. It has eight 32-bit general-purpose registers, several specialized registers, an ALU, and a barrel shifter. Single load, store, addition, subtraction, logic, and shift instructions execute in one clock.

Two 32-bit bidirectional buses connect the integer and floating-point units. These buses are used together for transferring 64-bit operands. The same buses also connect the processing units with the cache unit. The contents of the general purpose registers are sent to the segmentation unit on a separate 32-bit bus for generation of effective addresses.

10.12. FLOATING-POINT UNIT

The floating-point unit executes the same instruction set as the 387 math coprocessor. The unit contains a push-down register stack and dedicated hardware for interpreting the 32-, 64-, and 80- bit formats as specified in IEEE Standard 754. An output signal passed through to the processor bus indicates floating-point errors to the external system, which in turn can assert an input to the processor indicating that the processor should ignore these errors and continue normal operations.

10.12.1. IntelDX2™ and IntelDX4™ Processor On-Chip Floating-Point Unit

The IntelDX2 and IntelDX4 processors incorporate the basic Intel486 processor 32-bit architecture, with on-chip memory management and cache memory units. They also have an on-chip floating-point unit (FPU) that operates in parallel with the arithmetic and logic unit. The FPU provides arithmetic instructions for a variety of numeric data types and executes numerous built-in transcendental functions (e.g., tangent, sine, cosine, and log functions). The floating-point unit fully conforms to the ANSI/IEEE standard 754-1985 for floating-point arithmetic.

All software written for the Intel386 processor, Intel387 math coprocessor and previous members of the 86/87 architectural family runs on these processors without modifications.

10.13. SEGMENTATION UNIT

A segment is a protected, independent address space. Segmentation is used to enforce isolation among application programs, to invoke recovery procedures, and to isolate the effects of programming errors.

The segmentation unit translates a segmented address issued by a program, called a logical address, into an unsegmented address, called a linear address. The locations of segments in the linear address space are stored in data structures called segment descriptors. The segmentation unit performs its address calculations using segment descriptors and displacements (offsets) extracted from instructions. Linear addresses are sent to the paging and cache units. When a segment is accessed for the first time, its segment descriptor is copied into a processor register. A program can have as many as 16,383 segments. Up to six segment descriptors can be held in processor registers at a time. Figure 10.31 shows the relationships between logical, linear, and physical addresses.

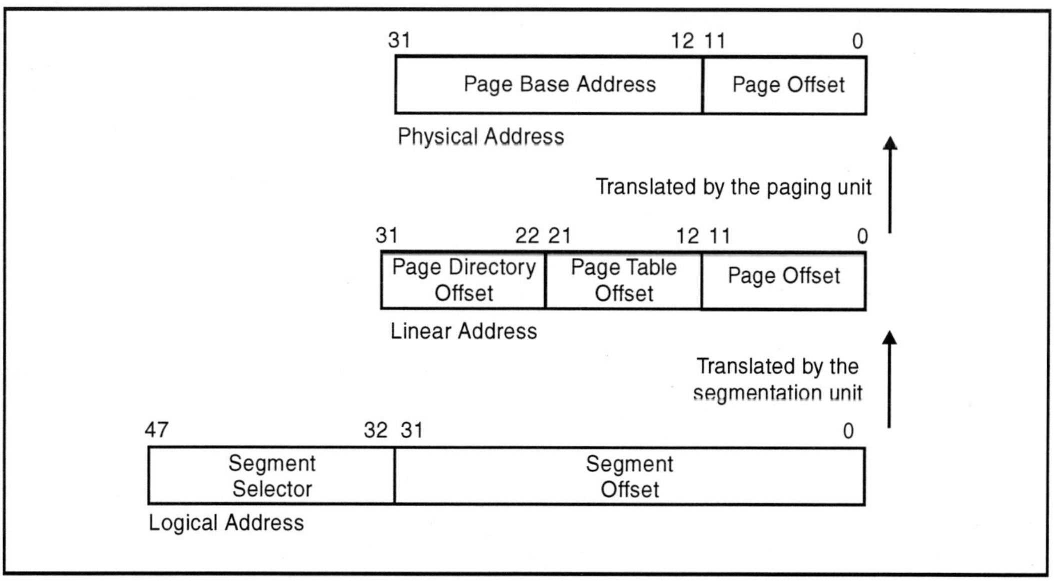

Fig. 10.31. Segmentation and Paging Address Formats.

10.14. PAGING UNIT

The paging unit allows access to data structures larger than the available memory space by keeping them partly in memory and partly on disk. Paging divides the linear address space into 4-Kbyte blocks called pages. Paging uses data structures in memory called page tables for mapping a linear address to a physical address. The cache uses physical addresses and puts them on the processor bus. The paging unit also identifies problems, such as accesses to a page that is not resident in memory, and raises exceptions called page faults. When a page fault occurs, the operating system has a chance to bring the required page into memory from disk. If necessary, it can free space in memory by sending another page out to disk. If paging is not enabled, the physical address is identical to the linear address.

The paging unit includes a translation lookaside buffer (TLB) that stores the 32 most recently used page table entries. Figure 10.32 shows the TLB data structures. The paging unit looks up linear addresses in the TLB. If the paging unit does not find a linear address in the TLB, the unit generates requests to fill the TLB with the correct physical address contained in a page table in memory. Only when the correct page table entry is in the TLB does the bus cycle take place. When the paging unit maps a page in the linear address space to a page in physical memory, it maps only the upper 20 bits of the linear address. The lowest 12 bits of the physical address come unchanged from the linear address.

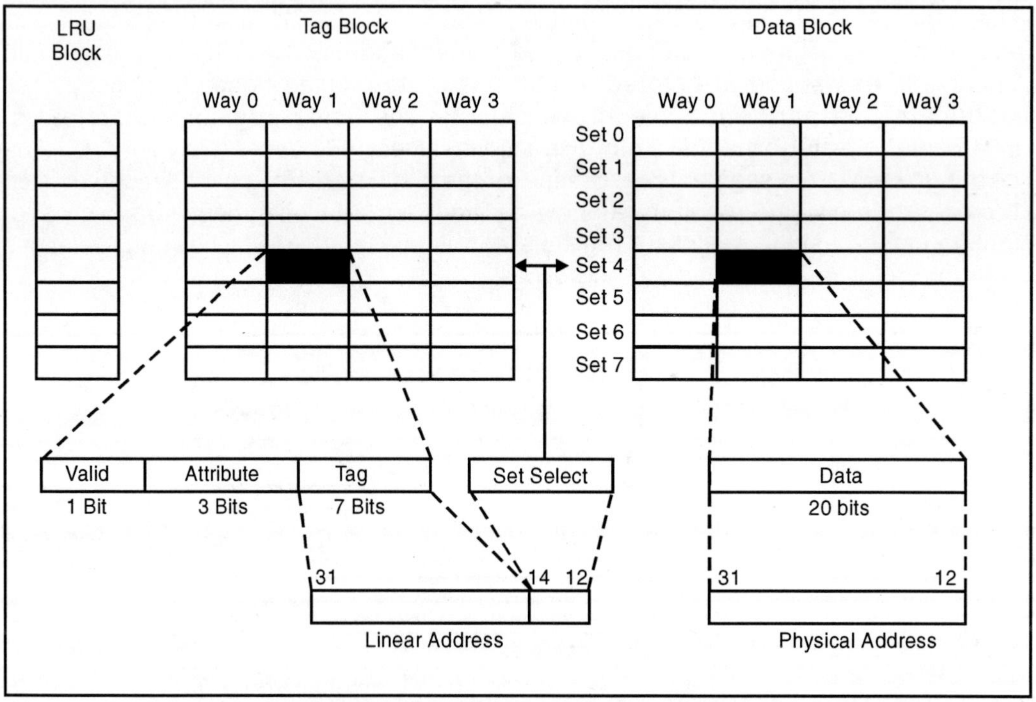

Fig. 10.32. Translation Lookaside Buffer.

Most programs access only a small number of pages during any short span of time. When this is true, the pages stay in memory and the address translation information stays in the TLB. In typical systems, the TLB satisfies 99% of the requests to access the page tables. The TLB uses a pseudoLRU algorithm, similar to the cache, as a content-replacement strategy.

The TLB is flushed whenever the page directory base register (CR3) is loaded. Page faults can occur during either a page directory read or a page table read. The cache can be used to supply data for the TLB, although this may not be desirable when external logic monitors TLB updates.

Unlike segmentation, paging is invisible to application programs and does not provide the same kind of protection against programs altering data outside a restricted part of memory. Paging is visible to the operating system, which uses it to satisfy application program memory requirements.

10.14.1. Instruction Pipelining

Not every instruction involves all internal units. When an instruction needs the participation of several units, each unit operates in parallel with others on instructions at different stages of execution. Although each instruction is processed sequentially, several instructions are at varying stages of execution in the processor at any given time. This is called *instruction pipelining*. Instruction prefetch, instruction decode, microcode execution, integer operations, floating-point operations, segmentation, paging, cache management, and bus interface operations are all performed simultaneously. Figure 10.33 shows some of this parallelism for a single instruction: the instruction fetch, two-stage decode, execution, and register write-back of the execution result. Each stage in this pipeline can occur in one clock cycle.

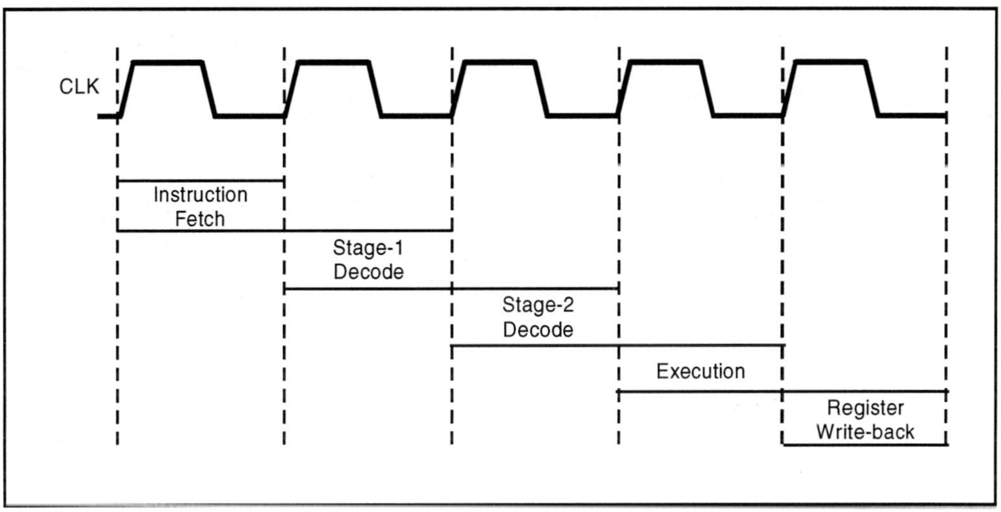

Fig. 10.33. Internal Pipelining.

The internal pipelining on the Intel486 processor offers an important performance advantage over many single-clock RISC processors: in the Intel486 processor, data can be loaded from the cache with one instruction and used by the next instruction in the next clock. This performance advantage results from the stage-1 decode step, which initiates memory accesses before the execution cycle. Because most compilers and application programs follow load instructions with instructions that operate on the loaded data, this method optimizes the execution of existing binary code.

The method has a performance trade-off: an instruction sequence that changes register contents and then uses that register in the next instruction to access memory takes three clocks rather than two. This trade-off is only a minor disadvantage, however, since most instructions that access memory use the stable contents of the stack pointer or frame pointer, and the additional clock is not used very often. Compilers often place an unrelated instruction between one that changes an addressing register and one that uses the register. Such code is compatible with the Intel386 processor, and the Intel486 processor provides special stack increment/decrement hardware and an extra register port to execute back-to-back stack push/pop instructions in a single clock.

80486 Microprocessor

The 32-bit 80486 is the next evolutionary step up from the 80386. One of the most obvious feature included in a 80486 is a built in math coprocessor. This coprocessor is essentially the same as the 80387 processor used with a 80386, but being integrated on the chip allows it to execute math instructions about three times as fast as a 80386/387 combination. 80486 is an SKbyte code and data cache. To make room for the additional signals, the 80486 is packaged in a 168 pin, pin grid array package instead of the 132 pin PGA used for the 80386.

Pin Definitions

A_{31}–A_2 : Address outputs A_{31}-A_2 provide the memory and I/O with the address during normal operation. During a cache line invalidation A_{31}–A_4 are used to drive the microprocessor.

$A_{20}M_3$: The address bit 20 mask causes the 80486 to wrap its address around from location 000FFFFFH to 00000000H as in 8086. This provides a memory system that functions like the 1M byte real memory system in the 8086 processors.

ADS : The address data strobe become logic zero to indicate that the address bus contains a valid memory address.

AHOLD : The address hold input causes the microprocessor to place its address bus connections at their high-impedance state, with the remainder of the buses staying active. It is often used by another bus master to gain access for a cache invalidation cycle.

BREQ : This bus request output indicates that the 486 has generated an internal bus request.

BE_3-BE_0 : Byte enable outputs select a bank of the memory system when information is transferred between the microprocessor and its memory and I/O.

The BEs signal enables D_{31} - D_{24}, BE_2 enables D_{23}–D_{16}, BE_1 enables D_{15} – D_8 and BE0 enables D_7–D_0.

Blast : The burst last output shows that the burst bus cycle is complete on the next activation of BRDY# signal.

BOFF : The Back-off input causes the microprocessor to place its buses at their high impedance state during the next cycle. The microprocessor remains in the bus hold state until the BOFF# pin is placed at a logic 1 level.

NMI : The non-maskable interrupt input requests a type 2 interrupt.

BRDY : The burst ready input is used to signal the microprocessor that a burst cycle is complete.

KEN : The cache enable input causes the current bus to be stored in the internal.

LOCK : The lock output becomes a logic 0 for any instruction that is prefixed with the lock prefix.

W/R : Current bus cycle is either a read or a write.

IGNNE : The ignore numeric error input causes the coprocessor to ignore floating point error and to continue processing data. The signal does not affect the state of the FERR pin.

FLUSH : The cache flush input forces the microprocessor to erase the contents of its 8K byte internal cache.

EADS : The external address strobe input is used with AHOLD to signal that an external address is used to perform a cache invalidation cycle.

FERR : The floating point error output indicates that the floating point coprocessor has detected an error condition. It is used to maintain compatibility with DOS software.

BS_8 : The bus size 8, input causes the 80486 to structure itself with an 8-bit data bus to access byte-wide memory and I/O components.

BS_{16} : The bus size 16, input causes the 80486 to structure itself with an 16-bit data bus to access word-wide memory and I/O components.

PCHK : The parity check output indicates that a parity error was detected during a read operation on the $DP_3 - DP_0$ pin.

PLOCK : The pseudo-lock output indicates that current operation requires more than one bus cycle to perform. This signal becomes a logic 0 for arithmetic coprocessor operations that access 64 or 80 bit memory data.

PWT : The page write through output indicates the state of the PWT attribute bit in the page table entry or the page directory entry.

RDY : The ready input indicates that a non-burst bus cycle is complete. The RDY signal must be returned or the microprocessor places wait states into its timing until RDY is asserted.

M/IO : Memory/IO# defines whether the address bus contains a memory address or an I/O port number. It is also combined with the W/ R signal to generate memory and I/O read and write control signals.

80486 Signal Group

The 80486 data bus, address bus, byte enable, ADS#, RDY#, INTR, RESET, NMI, M/IO#, D/C#, W/R#, LOCO, HOLD, HLDA and BSi6# signals function as we described for 80386. The 80486 requires 1 clock instead of 2 clock required by 80386. A new signal group on the 486 is the PARITY group DPo-DPs and PCHK#. These signals allow the 80486 to implement parity detection/generation for memory reads and memory writes. During a memory write operation, the 80486 generates an even parity bit for each byte and outputs these bits on the DP_0-DP_3 lines. These bits will store in a separate parity memory bank. During a read operation the stored parity bits will be read from the parity memory and applied to the DP_0-DP_3 pins. The 80486 checks the parities of the data bytes read and compares them with the DPo-DPs signals. If a parity error is found, the 80486 asserts the PCHK# signal. Another new signals group consists of the BURST ready signal BRDY# and BURST last signal BLAST#. These signals are used to control burst-mode memory reads and writes. A normal 80486 memory read operation to read a line into the cache requires 2 clock cycles. However, if a series of reads is being done from successive memory locations, the reads can be done in burst mode with only 1 clock cycle per read. To start the process the 80486 sends out the first address and asserts the BLAST# signal high. When the external DRAM controller has the first data bus, it asserts the BRDY# signal. The 80486 reads the data word and outputs the next address. Since the data words are at successive addresses, only the lower address bits need to be changed. If the DRAM controller is operating in. the page or the static column modes then it will only have to output a new column address to the DRAM. In this mode the DRAM will be able to output the new data word within 1 clock cycle.

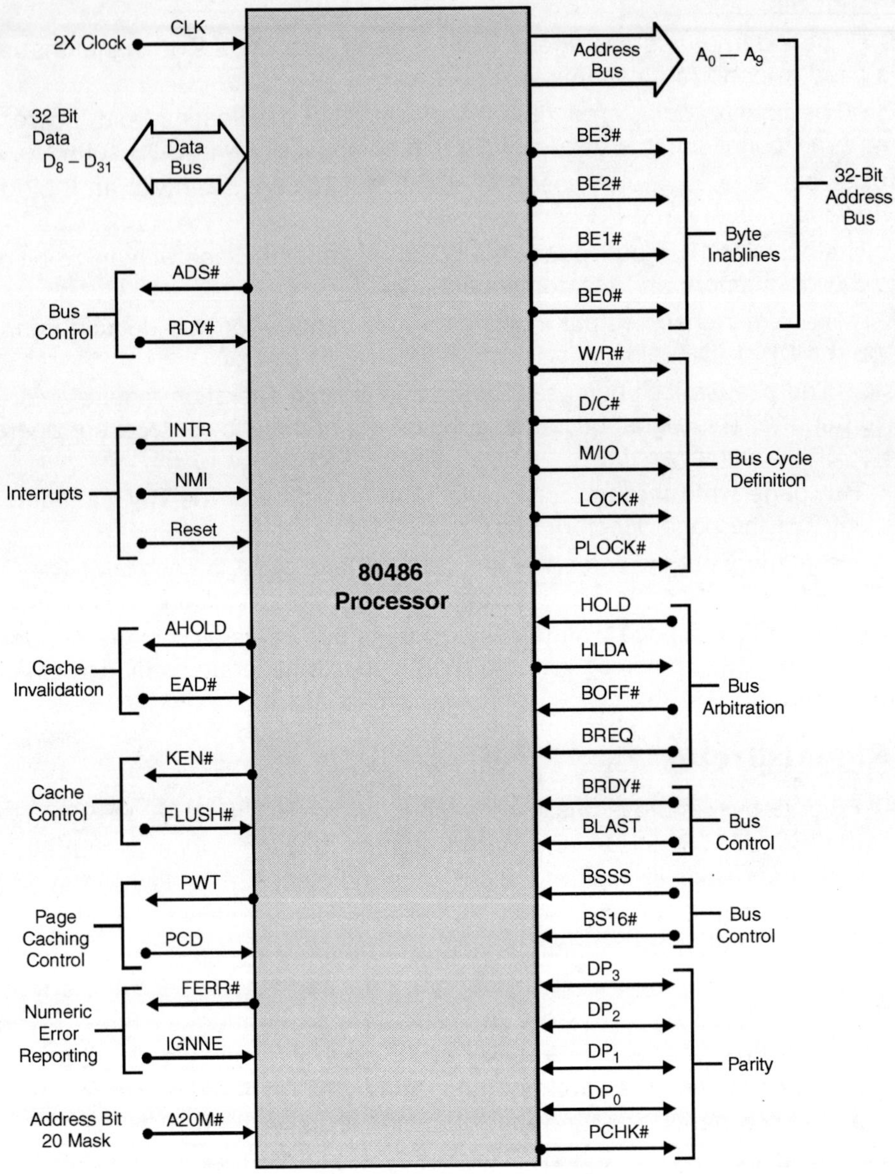

Fig. 10.34. Pin diagram of 80486.

When the processor has read the required number of data words, it asserts the BLAST# signal low to terminate the burst mode. The final signal we want to discuss here are the bus request output signal BREQ, the backoff input signal BOFF#, the HOLD signal and the hold-acknowledge signal HLDA. These signals are used to control sharing the local 486 bus by multiple processors (bus master). When a master on the bus need to use the bus, it asserts its BERQ signal. An external parity circuit will evaluate requests to use the bus and grant bus use to the highest - priority master. To ask the 486 to release the bus, the bus controller asserts the 486 HOLD input or BOFF# input. If the HOLD input is asserted, the 486 will finish the current bus cycle, float its buses and assert the HLDA signal. To prevent another master from taking over the bus during a critical operation, the 486 can assert its LOCK# or PLOCK# signal.

EFLAG Register of The 80486

The extended flag register EFLAG is illustrated in the figure. The only new flag bit is the AC alignment check, used to indicate that the microprocessor has accessed a word at an odd address or a double word boundary. Efficient software and execution require that data be stored at word or doubleword boundaries.

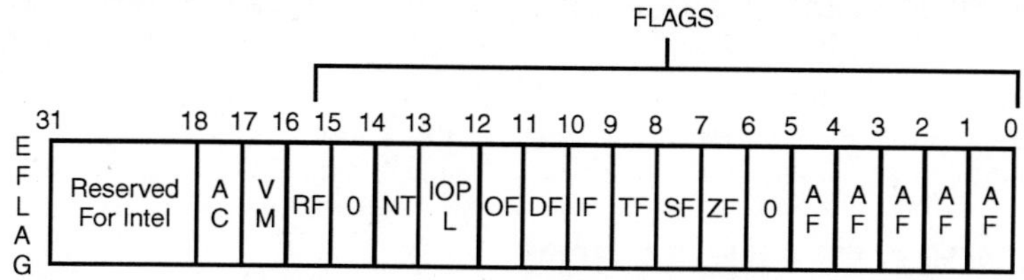

CF : Carry Flag; AF = Auxiliary carry; ZF : Zero Flag; SF : Sign Flag; TF : Trap Flag; IE : Interrupt Enable; AC : Alignment Check; DF : Direct Flag; OF : Over Flow; IOPL : I/O Privilege Level; NT : Nested Task Flag; RF : Resume Flag; VM : Virtual Mode

Fig. 10.35. Flag Register of 80486.

80486 Memory System

The memory system for the 486 is identical to 386 microprocessor. The 486 contains 4G bytes of memory beginning at location 00000000H and ending at FFFFFFFFH. The major change to the memory system is internal to 486 in the form of 8K byte cache memory, which speeds the execution of instructions and the acquisition of data. Another addition is the parity checker/ generator built into the 80486 microprocessor. Parity Checker / Generator : Parity is often used to determine if data are correctly read from a memory location. INTEL has incorporated an internal parity generator / decoder. Parity is generated by the 80486 during each write cycle. Parity is generated as even parity and a parity bit is provided for each byte of memory. The parity check bits appear on pins DPO-DP3, which are also parity inputs as well as parity outputs. These are typically stored in memory during each write cycle and read from memory during each read cycle. On a read, the microprocessor checks parity and generates a parity check error, if it occurs on the PCHK# pin. A parity error causes no change in processing unless the user applies the PCHK signal to an interrupt input. Interrupts are often used to signal a parity error in DS-based computer systems. This is same as 80386, except the parity bit storage. If parity is not used, Intel recommends that the DPO - DP3 pins be pulled up to +5v.

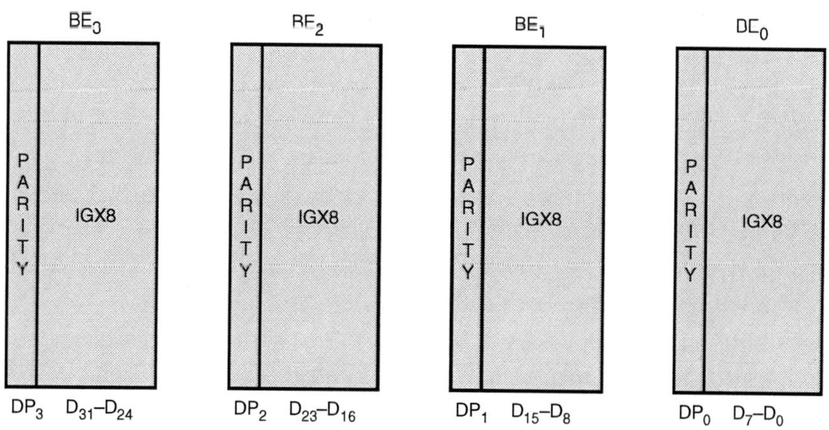

Fig. 10.36. 80486 memory system.

Cache Memory : The cache memory system stores data used by a program and also the instructions of the program. The cache is organised as a 4 way set associative cache with each location containing 16 bytes or 4 doublewords of data. Control register CRO is used to control the cache with two new control bits not present in the 80386 microprocessor. The CD (cache disable), NW (non-cache write through) bits are new to the 80486 and are used to control the 8K byte cache. If the CD bit is a logic 1, all cache operations are inhibited. This setting is only used for debugging software and normally remains cleared. The NW bit is used to inhibit cache write-through operation. As with CD, cache write through is inhibited only for testing. For normal operations CD = 0 and NW = 0. Because the cache is new to 80486 microprocessor and the cache is filled using burst cycle.

10.15. MEMORY ORGANIZATION

Memory on the Intel486 processor is divided up into 8-bit quantities (bytes), 16-bit quantities (words), and 32-bit quantities (dwords). Words are stored in two consecutive bytes in memory with the low-order byte at the lowest address, the high order byte at the high address. Dwords are stored in four consecutive bytes in memory with the low-order byte at the lowest address, the high-order byte at the highest address. The address of a word or dword is the byte address of the low-order byte.

In addition to these basic data types, the Intel486 processor supports two larger units of memory: pages and segments. Memory can be divided up into one or more variable-length segments, which can be swapped to disk or shared between programs. Memory can also be organized into one or more 4-Kbyte pages. Both segmentation and paging can be combined, gaining the advantages of both systems. The Intel486 processor supports both pages and segments in order to provide maximum flexibility to the system designer. Segmentation and paging are complementary. Segmentation is useful for organizing memory in logical modules, and as such is a tool for the application programmer, while pages are useful for the system programmer for managing the physical memory of a system.

10.15.1. Address Spaces

The Intel486 processor has three distinct address spaces: logical, linear, and physical. A logical address (also known as a virtual address) consists of a selector and an offset. A selector is the contents of a segment register. An offset is formed by summing all of the addressing components (BASE, INDEX, DISPLACEMENT), "32-BIT MEMORY ADDRESSING MODES," into an effective address. Because each task on the Intel486 processor has a maximum of 16 K (2^{14} – 1) selectors, and offsets can be 4 Gbytes (2^{32} bits), this gives a total of 246 bits or 64 terabytes of logical address space per task. The programmer sees this virtual address space.

The segmentation unit translates the logical address space into a 32-bit linear address space. If the paging unit is not enabled then the 32-bit linear address corresponds to the physical address. The paging unit translates the linear address space into the physical address space. The physical address is what appears on the address pins.

The primary difference between Real Mode and Protected Mode is how the segmentation unit performs the translation of the logical address into the linear address. In Real Mode, the segmentation unit shifts the selector left four bits and adds the result to the offset to form the linear address. While in Protected Mode every selector has a linear base address associated with it. The linear base address is stored in one of two operating system tables (i.e., the Local

Descriptor Table or Global Descriptor Table). The selector's linear base address is added to the offset to form the final linear address.

Figure 10.37 shows the relationship between the various address spaces.

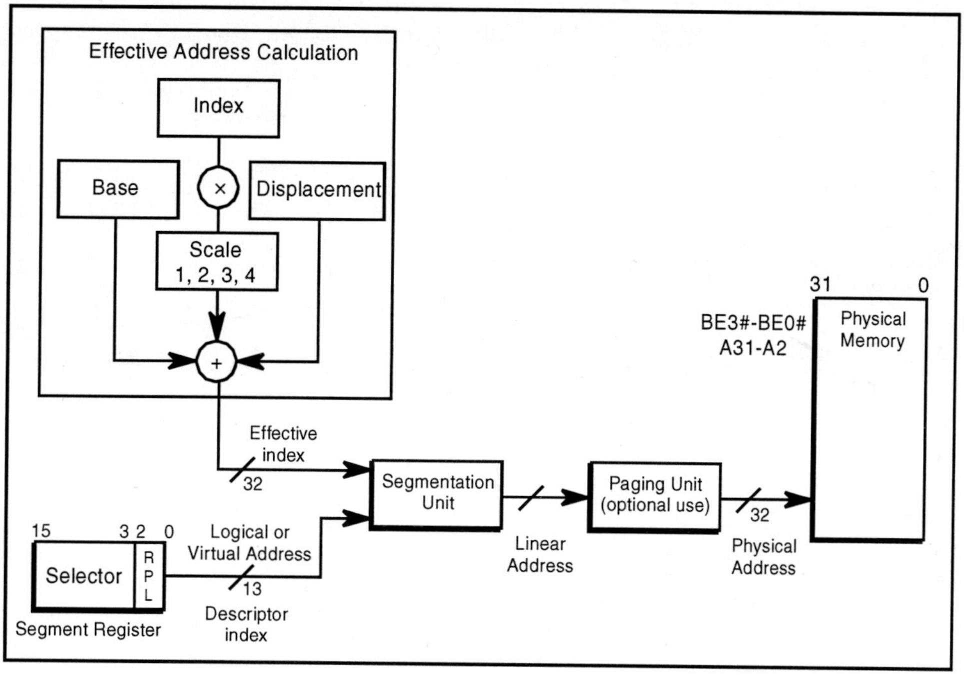

Fig. 10.37. Address Translation.

10.15.2. Segment Register Usage

The main data structure used to organize memory is the segment. On the Intel486 processor, segments are variable sized blocks of linear addresses which have certain attributes associated with them. There are two main types of segments: code and data. The segments are of variable size and can be as small as 1 byte or as large as 4 Gbytes (2^{32} bytes).

In order to provide compact instruction encoding, and increase Intel486 processor performance, instructions do not need to explicitly specify which segment register is used. A default segment register is automatically chosen according to the rules of Table 10.3. In general, data references use the selector contained in the DS register; stack references use the SS register and Instruction fetches use the CS register. The contents of the Instruction Pointer provide the offset. Special segment override prefixes allow the explicit use of a given segment register, and override the implicit rules listed in Table 10.3. The override prefixes also allow the use of the ES, FS and GS segment registers.

There are no restrictions regarding the overlapping of the base addresses of any segments. Thus, all 6 segments could have the base address set to zero and create a system with a 4-Gbyte linear address space. This creates a system where the virtual address space is the same as the linear address space.

10.15.3. I/O Space

The Intel486 processor has two distinct physical address spaces: Memory and I/O. Generally, peripherals are placed in I/O space, although the Intel486 processor also supports memory-mapped peripherals. The I/O space consists of 64 Kbytes. It can be divided into 64 K 8-bit ports, 32 K 16 bit ports, or 16 K 32-bit ports, or any combination of ports that add up to less than 64 Kbytes. The 64 K I/O address space refers to physical memory rather than linear address, because I/O instructions do not go through the segmentation or paging hardware. The M/IO# pin acts as an additional address line, thus allowing the system designer to easily determine which address space the processor is accessing.

TABLE 10.2. Segment Register Selection Rules.

Type of Memory Reference	Implied (Default) Segment Use	Segment Override Prefixes Possible
Code Fetch	CS	None
Destination of PUSH, PUSHF, INT, CALL, PUSHA Instructions	SS	None
Source of POP, POPA, POPF, IRET, RET instructions	SS	None
Destination of STOS, MOVS, REP STOS, REP MOVS Instructions (DI is Base Register)	ES	None
Other Data References, with Effective Address using Base Register of:		
[EAX]	DS	
[EBX]	DS	
[ECX]	DS	
[EDX]	DS	All
[ESI]	DS	
[EDI]	DS	
[EBP]	SS	
[ESP]	SS	

The I/O ports are accessed via the IN and OUT I/O instructions, with the port address supplied as an immediate 8-bit constant in the instruction or in the DX register. All 8and 16-bit port addresses are zero extended on the upper address lines. The I/O instructions cause the M/IO# pin to be driven low.

I/O port addresses 00F8H through 00FFH are reserved for use by Intel.

I/O instruction code is cacheable.

I/O data is not cacheable.

I/O transfers (data or code) can be bursted.

80486 Memory Management

The 80486 contains the same memory-management system as the 80386. This includes a paging unit to allow any 4K byte block of physical memory to be assigned to any 4K byte block of linear memory. The only difference between 80386 and 80486 memorymanagement system is paging. The 80486 paging system can disabled caching for section of translation memory pages, while the 80386 could not.

If these are compared with 80386 entries, the addition of two new control bits is observed (PWT and PCD). The page write through and page cache disable bits control caching. The PWT controls how the cache functions for a write operation of the external cache memory. It does not control writing to the internal cache. The logic level of this bit is found on the PWT pin of the 80486 microprocessor. Externally, it can be used to dictate the write through policy of the external caching. The PCD bit controls the on-chip cache. If the PCD = 0, the on-chip cache is enabled for the current page of memory. Note that 80386 page table entries place a logic 0 in the PCD bit position, enabling caching. If PCD = 1, the on-chip cache is disable. Caching is disable regard less of condition of KEN#, CD, and NW.

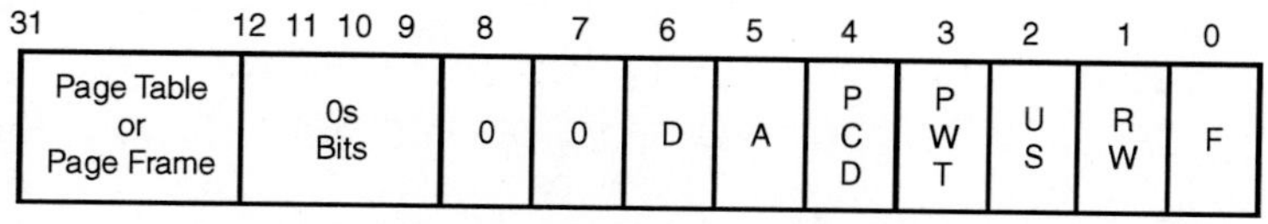

Fig. 10.38. Page Directory or Page Table Entry for the 80486 Microprocessor

Cache Test Registers

The 80486 cache test registers are TR3, TR4, TR5. Cache data register (TR3) is used to access either the cache fill buffer for a write test operation or the cache read buffer for a cache read test operation. In order to fill or read a cache line (128 bits wide), TR3 must be written or read four times. The contents of the set select field in TR5 determine which internal cache line is written or read through TR3. The 7 bit test field selects one of the 128 different 16 byte wide cache lines. The entry select bits of TR5 select an entry in the set or the 32 bit location in the read buffer. The control bits in TR5 enable the fill buffer or read buffer operation (00). Perform a cache write (01), Perform a cache read (10). Flush the cache (11). The cache status register (TR4) hold the cache tag, LRU bits and a valid bit. This register is loaded with the tag and valid bit before a cache a cache write operation and contains the tag, valid bit, LRU bits, and 4 valid bits on a cache test read. Cache is tested each time that the microprocessor is reset if the AHOLD pin is high for 2 clocks prior to the RESET pin going low. This causes the 486 to completely test itself with a built in self test or BIST. The BIST uses TR3, TR4, TR5 to completely test the internal cache. Its outcome is reported in register EAX. If EAX is a zero, the microprocessor, the coprocessor and cache have passed the self test. The value of EAX can be tested after reset to determine if an error is detected. In most of the cases we do not directly access the test register unless we wish to perform our own tests on the cache orTLB.

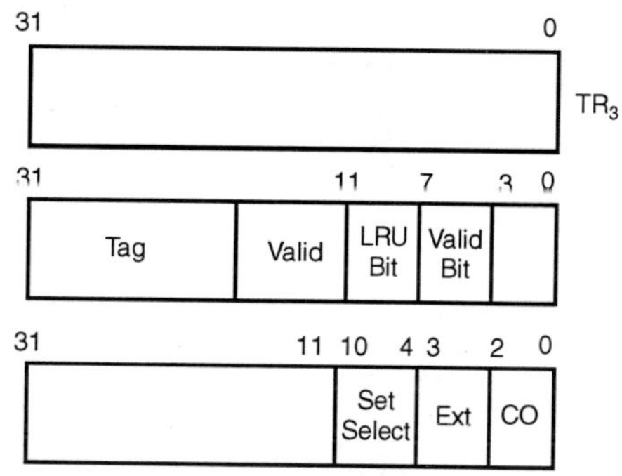

Fig. 10.39. Cache test register of the 80486 microprocessor

80487

x87 is a floating point-related subset (originally an extension) of the x86 architecture instruction set. The term derives from the originally separate floating point coprocessors that had names ending in "87". Like other extensions to the basic instruction set, these instructions are not strictly needed to construct working programs, but provide hardware and microcode implementations of common numerical tasks, allowing these tasks to be performed much faster than corresponding machine code routines can. The x87 instruction set includes instructions for basic floating point operations such as addition, subtraction and comparison, but also for more complex numerical operations, such as the computation of the tangent function and its inverse, for example.

Most x86 processors since the Intel 80486DX have had these x87 instructions implemented in the main CPU but the term is sometimes still used to refer to that part of the instruction set. Before x87 instructions were standard in PCs, compilers or programmers had to use rather slow library calls to perform floating-point operations, a method that is still common in (low-cost) embedded systems.

Intel 80487 was introduced in 1991 as a math co-processor for Intel 80486SX family of processors. The i487SX was marketed as a floating point unit coprocessor for Intel i486SX machines. It was actually a full-blown i486DX chip. In fact, the 80487 was more than just a co-processor. It was a full-fledged 80486 processor with integrated Floating Point Unit. When installed into upgrade socket on 486 motherboards, the 80486SX processor was disabled, and the 80487 was used for both integer and floating-point calculations. After upgrade the original 80486SX CPU could be removed from its socket if desired. Intel 80487 has the same performance as 80486DX CPU. The main difference between them is the pinout.

Fig. 10.40. 80487 chip.

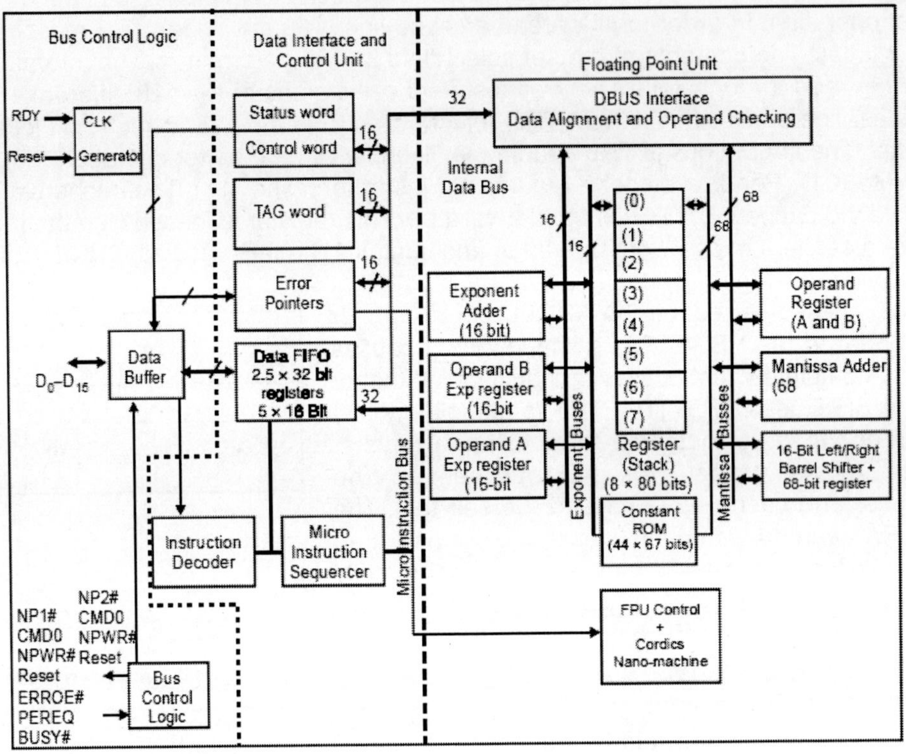

Fig. 10.41. Internal architecture of 80387 coprocessor.

169-pin ceramic PGA

Intel 80487 co-p'rocessor didn't have speed marked on the chip. The co-processor could upgrade 80486SX CPUs running at speeds from 16 MHz to 25 MHz, possibly 33 MHz too.

Description : The x87 registers form an 8-level deep non-strict stack structure ranging from ST(0) to ST(7) with registers that can be directly accessed by either operand, using an offset relative to the top, as well as pushed and popped. (This scheme may be compared to how a stack frame may be both pushed, popped and indexed.)

There are instructions to push, calculate, and pop values on top of this stack; monadic operations (FSQRT, FPTAN etc.) then implicitly address the topmost ST(0) while dyadic operations (FADD, FMUL, FCOM, etc.) implicitly address ST(0) and ST(1). The non-strict stack-model also allows dyadic operations to use ST(0) together with a direct *memory operand or* with an *explicitly* specified stack-register, ST(x), in a role similar to a traditional accumulator (a combined destination and left operand).

This can also be reversed on an instruction-by-instruction basis with ST(0) as the unmodified operand and ST(x) as the *destination.* Furthermore, the contents in ST(0) can be exchanged with another stack register using an instruction called FXCH ST(x).

These properties makes the x87 stack usable as seven freely addressable registers plus a dedicated accumulator (or as seven independent accumulators). This is especially applicable on superscalar x86 processors (such as the Pentium of 1993 and later) where these exchange instructions (codes D9C8..D9CFh) are optimized down to a zero clock penalty by using one of the integer paths for FXCH ST(x) in parallel with the FPU instruction. Despite being natural and convenient for human assembly language programmers, some compiler writers have found it complicated to construct automatic code generators that schedule x87 code effectively.

The x87 provides single precision, double precision and 80-bit double-extended precision binary floating-point arithmetic as per the IEEE 754-1985 standard. By default, the x87 processors all use 80-bit double-extended precision internally (to allow for sustained precision over many calculations). A given sequence of arithmetic operations may thus behave slightly differently compared to a strict single-precision or double-precision IEEE 754 FPU. This may sometimes be problematic for some semi-numerical calculations relying on knowledge of exact FPU precision for correct operation. To avoid such problems, the x87 can be configured via a special configuration/status register to automatically round to single or double precision after each operation. Since the introduction of SSE2, the x87 instructions are not as essential as they once were, except for high-precision calculations demanding the 64-bit mantissa precision available in the 80-bit format.

10.16. INTERRUPTS

10.16.1. Interrupts and Exceptions

Interrupts and exceptions alter the normal program flow, in order to handle external events, to report errors or exceptional conditions. The difference between interrupts and exceptions is that interrupts are used to handle asynchronous external events while exceptions handle instruction faults. Although a program can generate a software interrupt via an INT N instruction, the Intel486 processors treat software interrupts as exceptions.

Hardware interrupts occur as the result of an external event and are classified into two types: maskable or non-maskable. Interrupts are serviced after the execution of the current instruction. After the interrupt handler is finished servicing the interrupt, execution proceeds with the instruction immediately after the interrupted instruction. Section 3.15.3, "Maskable Interrupt" and Section 3.15.4, "Non-Maskable Interrupt" discuss the differences between Maskable and NonMaskable interrupts.

Exceptions are classified as faults, traps, or aborts, depending on the way they are reported, and whether or not restart of the instruction causing the exception is supported. Faults are exceptions that are detected and serviced before the execution of the faulting instruction. A fault would occur in a virtual memory system when the processor referenced a page or a segment that was not present. The operating system would fetch the page or segment from disk, and then the Intel486 processor would restart the instruction. Traps are exceptions that are reported immediately after the execution of the instruction that caused the problem. User defined interrupts are examples of traps. Aborts are exceptions that do not permit the precise location of the instruction causing the exception to be determined. Aborts are used to report severe errors, such as a hardware error or illegal values in system tables.

Thus, when an interrupt service routine has been completed, execution proceeds from the instruction immediately following the interrupted instruction. On the other hand, the return address from an exception fault routine will always point at the instruction causing the exception and include any leading instruction prefixes. Tables 3-5 and 3-6 summarize the possible interrupts for Intel486 processors and shows where the return address points.

Intel486 processors can handle up to 256 different interrupts and/or exceptions. In order to service the interrupts, a table with up to 256 interrupt vectors must be defined. The interrupt vectors are simply pointers to the appropriate interrupt service routine. In Real Mode, the vectors are 4-byte quantities, a Code Segment plus a 16-bit offset; in Protected Mode, the interrupt vectors are 8-byte quantities, which are put in an Interrupt Descriptor Table. Of the 256 possible interrupts, 32 are reserved for use by Intel, the remaining 224 are free to be used by the system designer.

10.16.2. Interrupt Processing

When an interrupt occurs, the following actions happen. First, the current program address and the Flags are saved on the stack to allow resumption of the interrupted program. Next, an 8-bit vector is supplied to the Intel486 processor which identifies the appropriate entry in the interrupt table. The table contains the starting address of the interrupt service routine. Then, the user supplied interrupt service routine is executed. Finally, when an IRET instruction is executed the old Intel486 processor state is restored and program execution resumes at the appropriate instruction.

The 8-bit interrupt vector is supplied to the Intel486 processor in several different ways: exceptions supply the interrupt vector internally; software INT instructions contain or imply the vector; maskable hardware interrupts supply the 8-bit vector via the interrupt acknowledge bus sequence. Non-maskable hardware interrupts are assigned to interrupt vector 2.

TABLE 10.3. Interrupt Vector Assignments.

Function	Interrupt Number	Instruction that Can Cause Exception	Return Address Points to Faulting Instruction	Type
Divide Error	0	DIV, IDIV	YES	FAULT
Debug Exception	1	Any instruction	YES	TRAP†
NMI Interrupt	2	INT 2 or NMI	NO	NMI
One Byte Interrupt	3	INT	NO	TRAP
Interrupt on Overflow	4	INTO	NO	TRAP
Array Bounds Check	5	BOUND	YES	FAULT
Invalid OP-Code	6	Any illegal instruction	YES	FAULT
Device Not Available	7	ESC, WAIT	YES	FAULT
Double Fault	8	Any instruction that can generate an exception		ABORT
Intel Reserved	9			
Invalid TSS	10	JMP, CALL, IRET, INT	YES	FAULT
Segment Not Present	1	Segment Register Instructions	YES	FAULT
Stack Fault	12	Stack References	YES	FAULT
General Protection Fault	13	Any Memory Reference	YES	FAULT
Page Fault	14	Any Memory Access or Code Fetch	YES	FAULT
Intel Reserved	15			
Alignment Check Interrupt	17	Unaligned Memory Access	YES	FAULT
Intel Reserved	18–31			
Two Byte Interrupt	0–255	INT n	NO	TRAP

†Some debug exceptions may report both traps on the previous instruction, and faults on the next instruction.

10.16.3. Maskable Interrupt

Maskable interrupts are the most common way used by the Intel486 processor to respond to asynchronous external hardware events. A hardware interrupt occurs when the INTR is pulled high and the Interrupt Flag bit (IF) is enabled. The Intel486 processor only responds to interrupts between instructions, (REPeat String instructions, have an "interrupt window," between memory moves, which allows interrupts during long string moves). When an interrupt occurs, the Intel486 processor reads an 8-bit vector supplied by the hardware which identifies the source of the interrupt, (one of 224 user defined interrupts).

The IF bit in the EFLAG registers is reset when an interrupt is being serviced. This effectively disables servicing additional interrupts during an interrupt service routine. However, the IF may be set explicitly by the interrupt handler, to allow the nesting of interrupts. When an IRET instruction is executed, the original state of the IF is restored.

TABLE 10.4. FPU Interrupt Vector Assignments.

Function	Interrupt Number	Instruction that Can Cause Exception	Return Address Points to Faulting Instruction	Type
Floating-Point Error	16	Floating-point, WAIT	YES	FAULT

10.16.4. Non-Maskable Interrupt

Non-maskable interrupts provide a method of servicing very high priority interrupts. A common example of the use of a non-maskable interrupt (NMI) would be to activate a power failure routine or SMI# to activate a power saving mode. When the NMI input is pulled high, it causes an interrupt with an internally supplied vector value of 2. Unlike a normal hardware interrupt, no interrupt acknowledgment sequence is performed for an NMI.

While executing the NMI servicing procedure, the Intel486 processor will not service further NMI requests until an interrupt return (IRET) instruction is executed or the processor is reset (RSM in the case of SMI#). If NMI occurs while currently servicing an NMI, its presence will be saved for servicing after executing the first IRET instruction. The IF bit is cleared at the beginning of an NMI interrupt to inhibit further INTR interrupts.

10.16.5. Software Interrupts

A third type of interrupt/exception for the Intel486 processor is the software interrupt. An INT n instruction causes the processor to execute the interrupt service routine pointed to by the *n*th vector in the interrupt table.

A special case of the two byte software interrupt INT n is the one byte INT 3, or breakpoint interrupt. By inserting this one byte instruction in a program, the user can set breakpoints in his program as a debugging tool.

A final type of software interrupt is the single step interrupt. It is, "Single-Step Trap."

REVIEW QUESTIONS

1. Explain the features of 80386 processor and draw a block diagram of it?
2. Draw and expalin the register set of 80386 processor in detail ?
3. Expalin with the help of a diagram the function of control and debug registers present in 80386 processor?
4. Explain the function of pre-fetch queue in 80386 processor?
5. Give some of the methods to improve the performance of 80386 processor?
6. Explain the concept of extensive pipelining used in 80386 processor?

7. Explain the concept of speculative execution?
8. Draw the block diagram of 80486 processor and explain it in detail?
9. Give some of the features of 80486 processor?
10. Explain the cache access in 80486 processor?
11. Comment on the cache organization structure of 80486 processor?
12. Explain in detail the different blocks of 80486 processor?
13. Give the instruction pipelining concept used in 80486?
14. Define all the interrupts present in 80486 processor?
15. Write down the steps that a processor does whenever it encounters an interrupt?
16. List all the additional features that the 80386 microprocessor has over 8086
17. What is the main difference between the 80386 DX and 80386 SX microprocessor
18. How much the physical memory can 80386 address in real mode and in protected mode?
19. How are the tasks in 80386 system protected from each other.
20. How is an 80386 switched into virtual 8086 mode during task switch.
21. Describe three major additions or improvements that the 80486 processor has over 80386 processor.
22. Draw and discuss the register set of 80386 and explain a typical function of each of the registers in brief.
23. Draw and discuss the flag register of 80386 in detail.
24. Explain the use of each of the following registers of 80386.
    (i) Segment Descriptor registers (ii) Control register
    (iii) Debug and Test Registers (iv) System Address registers
25. Explain the different additional addressing modes supported by 80385 over 80286.
26. Enlist the different data types supported by 80386.
27. Explain the physical address formation in real address mode of 80386.
28. Explain the physical address formation in PVAM of 80386.
29. Draw and discuss the structures of the different descriptors supported by 80386.
30. What do you mean by a descriptor table ?
31. How many maximum descriptros can be accessed by 80386 for a single task ? What is the memory addressing capability of a single descriptor ? Justify the virtual memory addressing capability of 80386 per task.
32. Discuss the different descriptor types supported by 80386.
33. What are the different exceptions generated by 80386 ?
34. What do you mean by paging ? What are its advantages and disadvantages ?
35. Draw and discuss the paging mechanism of 80386 in details.
36. What are the differences between the logical addresses, linear addresses and physical addresses ?
37. Explain the procedure or converting a linear address into a physical address.
38. What is translation look aside buffer ? How does it speed up the execution of the programs ?
39. Write a short note on virtual 8086 mode of 80386.
40. Draw and discuss the architecture of 80387.
41. Enlist the different data types supported by 80387.
42. Discuss the register set of 80387.
43. Draw and discuss the interface of 80387 with 80386.
44. Enlist the salient features of 80486.
45. Draw and discuss the flag register of 80486.
46. Enlist four major architectural advancement in 80486 over 80386.
47. What is the use of TEST and DEBUG facility in 80486 ?
48. What do youmean by cache memory ? How does it speed up the program execution ?
49. Explain the cache management unit of 80486.
50. Write short notes on the following :
    (i) Cache maintenance operations
    (ii) Paging unit
51. Enlist the data types supported by 80486.
52. Enlist the different functional groups of signals provided by 80486.

❑❑❑

CHAPTER 11

THE PENTIUM PROCESSOR

11.1. P6 FAMILY OF PROCESSORS

INSIDE THIS CHAPTER

(1) The P6 family of processors use a dynamic execution micro-architecture. This three-way superscalar, pipelined micro-architecture features a decoupled, multi-stage super-pipeline, which trades less work per pipe stage for more stages.

(2) A P6 family processor, for example, has twelve stages with a pipe stage time 33 percent less than the Pentium processor, which helps achieve a higher clock rate on any given manufacturing process.

(3) The approach used in the P6 family micro-architecture removes the constraint of linear instruction sequencing between the traditional "fetch" and "execute" phases, and opens up a wide instruction window using an instruction pool.

(4) This approach allows the "execute" phase of the processor to have much more visibility into the program instruction stream so that better scheduling may take place.

(5) It requires the instruction "fetch/decode" phase of the processor to be much more efficient in terms of predicting program flow. Optimized scheduling requires the fundamental "execute" phase to be replaced by decoupled "dispatch/execute" and "retire" phases.

(6) This allows instructions to be started in any order but always be completed in the original program order. Processors in the P6 family may be thought of as three independent engines coupled with an instruction pool as shown in Fig. 11.1.

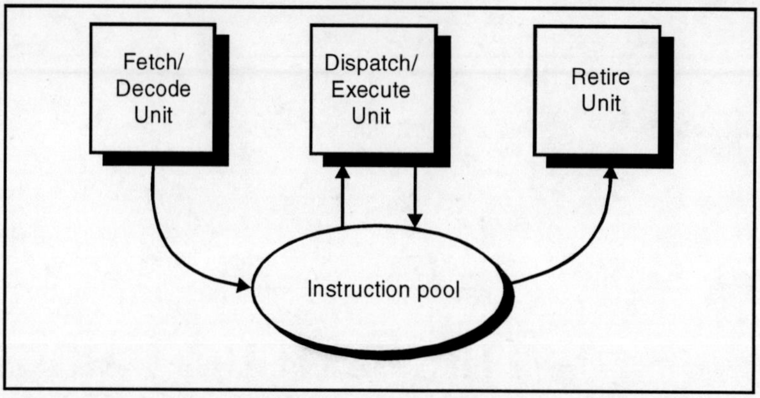

Fig. 11.1. Three engines of communication using an instruction pool.

11.2. P6 FAMILY OF PROCESSORS : PENTIUM PRO, PENTIUM II AND PENTIUM III

Major features :

(1) Dynamic program execution
(2) Three 12-stage pipelines
(3) On-chip FPU
(4) Separate L1 code and data caches with write-back strategy
(5) Out-of-order completion of instructions
(6) Data forwarding
(7) Dynamic Branch Prediction
(8) Speculative Program Execution
(9) Multiprocessing with up to four processors – with no additional logic.
(10) Address bus widened to 36 bits – physical address space of 64Gbytes – enabled by new bit in CR4
(11) New functions CMOVcc for conditional MOVs
(12) Integrated L2-cache, within the same package, connected over a dedicated bus running at the full (or half) CPU clock
(13) Pentium II & Pentium III: New MMX instructions for enhanced floating point operation when doing graphics
(14) Pentium III: SIMD extensions for enhanced high speed graphics
(15) Dramatic increases in clock frequencies supported
(16) A block diagram of the Pentium Pro is given ahead.
(17) Three-way superscalar, pipelined micro-architecture.
(18) Decoupled, multi-stage super pipeline,
   - Pentium II has twelve stages (with a pipe stage time 33 percent less than the Pentium processor)

     ==> a higher clock rate on any given manufacturing process

     ==> less work per pipe stage for more stages.

(19) The Pentium Pro, Pentium II and III processors use basically the same "dynamic execution" (i.e. out-of-order superscalar) microarchitecture principles.

(20) A wide instruction window using an "instruction pool".
- ➤ "Execute" phase is replaced by decoupled "issue", "execute", and "retire" phases. instruction execution is started in any order but always be retired in the original program order.
- ➤ Processors in the P6 family may be thought of as three independent engines coupled with an instruction pool.

11.3. DETAILED DESCRIPTION OF THE PENTIUM® PRO PROCESSOR MICROARCHITECTURE

Figure 11.2 shows a functional block diagram of the Pentium Pro processor microarchitecture. In this diagram, the following blocks make up the four processing units and the memory subsystem shown in Fig. 11.2:

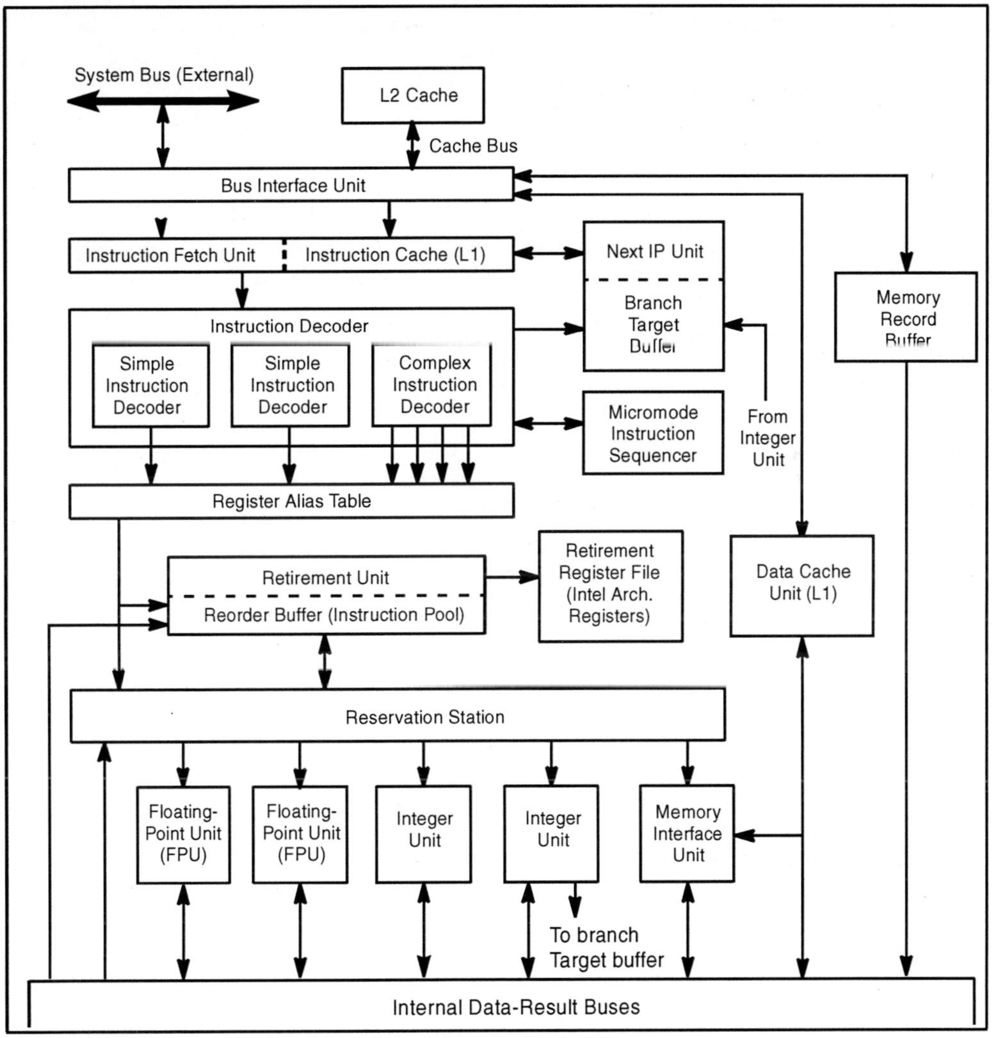

Fig. 11.2. Functional Block Diagram of the Pentium® Pro Processor Microarchitecture.

- ➤ Memory subsystem—System bus, L2 cache, bus interface unit, instruction cache (L1), data cache unit (L1), memory interface unit, and memory reorder buffer.

- Fetch/decode unit—Instruction fetch unit, branch target buffer, instruction decoder, microcode sequencer, and register alias table.
- Instruction pool—Reorder buffer
- Dispatch/execute unit—Reservation station, two integer units, two floating-point units, and two address generation units.
- Retire unit—Retire unit and retirement register file.

11.3.1. Memory Subsystem

The memory subsystem for the Pentium Pro processor consists of main system memory, the primary cache (L1), and the secondary cache (L2). The bus interface unit accesses system memory through the external system bus.

This 64-bit bus is a transaction-oriented bus, meaning that each bus access is handled as separate request and response operations. While the bus inter- face unit is waiting for a response to one bus request, it can issue numerous additional requests.

The bus interface unit accesses the close-coupled L2 cache through a 64-bit cache bus. This bus is also transactional oriented, supporting up to four concurrent cache accesses, and operates at the full clock speed of the processor.

Access to the L1 caches is through internal buses, also at full clock speed. The 8-KByte L1 instruction cache is four-way set associative; the 8-KByte L1 data cache is dual-ported and two- way set associative, supporting one load and one store operation per cycle.

Coherency between the caches and system memory are maintained using the MESI (modified, exclusive, shared, invalid) cache protocol. This protocol fosters cache coherency in single- and multiple-processor systems. It is also able to detect coherency problems created by self-modifying code.

Memory requests from the processor's execution units go through the memory interface unit and the memory order buffer. These units have been designed to support a smooth flow of memory access requests through the cache and system memory hierarchy to prevent memory access blocking. The L1 data cache automatically forwards a cache miss on to the L2 cache, and then, if necessary, the bus interface unit forwards an L2 cache miss to system memory.

Memory requests to the L2 cache or system memory go through the memory reorder buffer, which functions as a scheduling and dispatch station. This unit keeps track of all memory requests and is able to reorder some requests to prevent blocks and improve throughput. For example, the memory reorder buffer allows loads to pass stores. It also issues speculative loads. (Stores are always dispatched in order, and speculative stores are never issued.)

11.3.2. The Fetch/Decode Unit

The fetch/decode unit reads a stream of Intel Architecture instructions from the L1 instruction cache and decodes them into a series of micro-operations called "micro-ops." This micro-op stream (still in the order of the original instruction stream) is then sent to the instruction pool.

The instruction fetch unit fetches one 32-byte cache line per clock from the instruction cache. It marks the beginning and end of the Intel Architecture instructions in the cache lines and trans- mits 16 aligned bytes to the decoder.

The instruction fetch unit computes the instruction pointer, based on inputs from the branch target buffer, the exception/interrupt status, and branch-misprediction indications from the integer execution units. The most important part of this process is the branch prediction performed by the branch target buffer. Using an extension of Yeh's algorithm, the 512 entry branch target buffer looks many instructions ahead of the retirement program counter. Within this instruction window there may be numerous branches, procedure calls, and returns that must be correctly predicted if the dispatch/execute unit is to do useful work.

The instruction decoder contains three parallel decoders: two simple-instruction decoders and one complex instruction decoder. Each decoder converts an Intel Architecture instruction into one or more triadic micro-ops (two logical sources and one logical destination per micro-op). Micro-ops are primitive instructions that are executed by the processor's six parallel execution units.

Many Intel Architecture instructions are converted directly into single micro-ops by the simple instruction decoders, and some instructions are decoded into from one to four micro-ops. The more complex Intel Architecture instructions are decoded into sequences of preprogrammed micro-ops obtained from the microcode instruction sequencer. The instruction decoders also handle the decoding of instruction prefixes and looping operations. The instruction decoder can generate up to six micro-ops per clock cycle (one each for the simple instruction decoders and four for the complex instruction decoder).

The Intel Architecture's register set can cause resource stalls due to register dependencies. To solve this problem, the processor provides 40 internal, general-purpose registers, which are used for the actual computations. These registers can handle both integer and floating-point values. To allocate the internal registers, the enqueued micro-ops from the instruction decoder are sent to the register alias table unit, where references to the logical Intel Architecture registers are converted into internal physical register references.

In the final step of the decoding process, the allocator in the register alias table unit adds status bits and flags to the micro-ops to prepare them for out-of-order execution and sends the resulting micro-ops to the instruction pool.

11.3.3. Instruction Pool (Reorder Buffer)

Prior to entering the instruction pool (known formally as the reorder buffer), the micro-op instruction stream is in the same order as the Intel Architecture instruction stream that was sent to the instruction decoder. No reordering of instructions has taken place.

The reorder buffer is an array of content-addressable memory, arranged into 40 micro-op registers. It contains micro-ops that are waiting to be executed, as well as those that have already been executed but not yet committed to machine state. The dispatch/execute unit can execute instructions from the reorder buffer in any order.

11.3.4. Dispatch/Execute Unit

The dispatch/execute unit is an out-of-order unit that schedules and executes the micro-ops stored in the reorder buffer according to data dependencies and resource availability and temporarily stores the results of these speculative executions.

The scheduling and dispatching of micro-ops from the reorder buffer is handled by the reserva- tion station. It continuously scans the reorder buffer for micro-ops that are ready to be executed (that is, all the source operands are available) and dispatches them to the available execution units. The results of a micro-op execution are returned to the reorder buffer and stored along with the micro-op until it is retired. This scheduling and dispatching process supports classic out-of-order execution, where micro-ops are dispatched to the execution units strictly according to data-flow constraints and execution resource availability, without regard to the original ordering of the instructions.

When two or more micro-ops of the same type (for example, integer opera- tions) are available at the same time, they are executed in a pseudo FIFO order in the reorder buffer.

Execution of micro-ops is handled by two integer units, two floating-point units, and one memory-interface unit, allowing up to five micro-ops can be scheduled per clock.

The two integer units can handle two integer micro-ops in parallel. One of the integer units is designed to handle branch micro-ops. This unit has the ability to detect branch mispredictions and signal the branch target buffer to restart the pipeline. This operation is handled as follows. The instruction decoder tags each branch micro-op with both branch destination addresses (the predicted destination and the fall-through destination). When the integer unit executes the branch micro-op, it is able to determine whether the predicted or the fall-through destination was taken. If the predicted branch is taken, then speculatively executed micro-ops are marked usable and execution continues along the predicted instruction path. If the predicted branch was not taken, a jump execution unit in the integer unit changes the status of all of the micro-ops following the branch to remove them from the instruction pool. It then provides the proper branch destination to the branch target buffer, which in turn restarts the pipeline from the new target address.

The memory interface unit handles load and store micro-ops. A load access only needs to specify the memory address, so it can be encoded in one micro-op. A store access needs to specify both an address and the data to be written, so it is encoded in two micro-ops. The part of the memory interface unit that handles stores has two ports allowing it to process the address and the data micro-op in parallel. The memory interface unit can thus execute both a load and a store in parallel in one clock cycle.

The floating-point execution units are similar to those found in the Pentium processor. Several new floating-point instructions have been added to the Pentium Pro processor to streamline conditional branches and moves.

11.3.5. Retirement Unit

The retirement unit commits the results of speculatively executed micro-ops to permanent machine state and removes the micro-ops from the reorder buffer. Like the reservation station, the retirement unit continuously checks the status of micro-ops in the reorder buffer, looking for ones that have been executed and no longer have any dependencies with other micro-ops in the instruction pool. It then retires completed micro-ops in their original program order, taking into accounts interrupts, exceptions, breakpoints and branch mispredictions.

The retirement unit can retire three micro-ops per clock. In retiring a micro-op, it writes the results to the retirement register file and/or memory. The retirement register file contains the Intel Architecture registers (eight general-purpose registers and eight floating-point data registers). After the results have been committed to machine state, the micro-op is removed from the reorder buffer.

11.3.6. P-II Processor Pipeline

(1) **The FETCH/DECODE unit**: An in-order unit that takes as input the user program instruction stream from the instruction cache, and decodes them into a series of μoperations (μops) that represent the dataflow of that instruction stream.

(2) **The pre-fetch** is speculative.

(3) **The DISPATCH/EXECUTE unit**: An out-of-order unit that accepts the dataflow stream, schedules execution of the μops subject to data dependencies and resource availability and temporarily stores the results of these speculative executions.

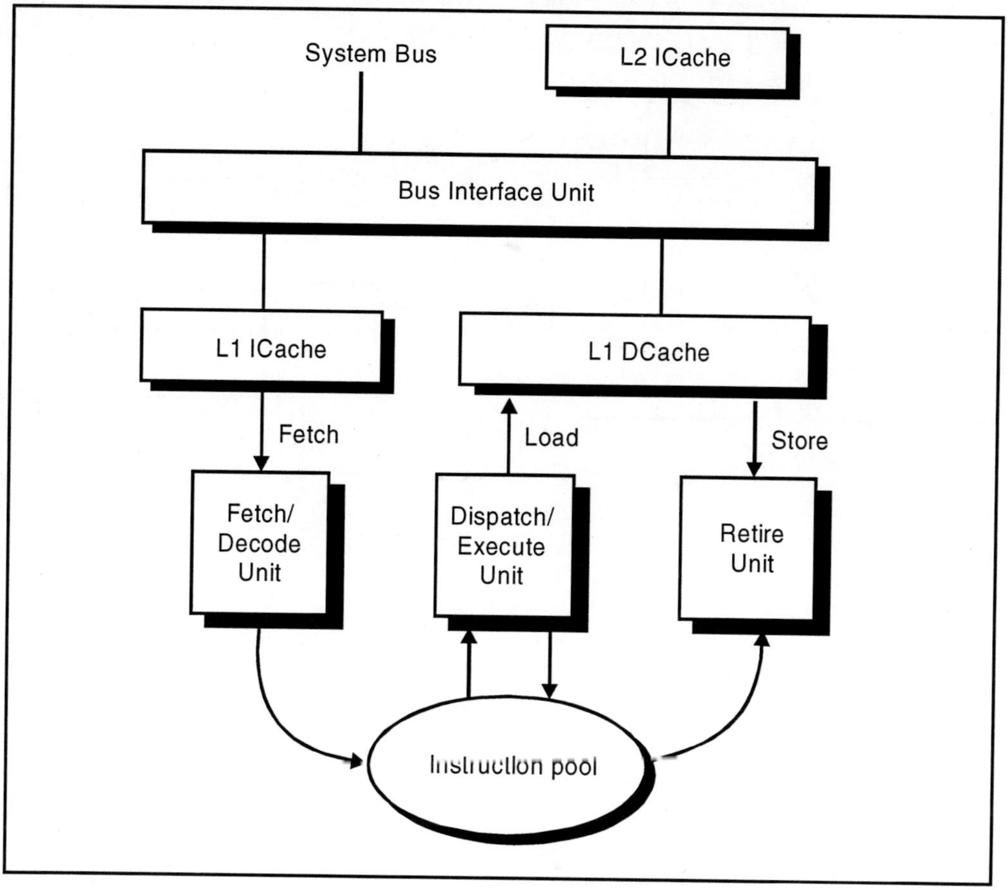

Fig. 11.3. P-II Processor Pipeline.

(4) **The RETIRE unit :** An in-order unit that knows how and when to commit ("retire") the temporary, speculative results to permanent architectural state.

(5) **The BUS INTERFACE unit :** A partially ordered unit responsible for connecting the three internal units to the real world. The bus interface unit communicates directly with the L2 (second level) cache supporting up to four concurrent cache accesses. The bus interface unit also controls a transaction bus, with MESI snooping protocol, to system memory.

11.4. INTERNAL ARCHITECTURE P-II ARCHITECTURE

Pentium II: The In-Order Section

(i) The instruction fetch unit (IFU) accesses a non-blocking I-cache and contains Next IP unit.

(ii) The Next IP unit provides the I-cache index (based on inputs from the BTB), trap/interrupt status, and branch-misprediction indications from the integer FUs.

(iii) Branch prediction:

(iv) BTB contains 512 entries, maintains branch history information and the predicted branch target address.

- ➢ Branch misprediction penalty: at least 11 cycles, on average 15 cycles
- ➢ The instruction decoder unit (IDU) is composed of three, separate decoders

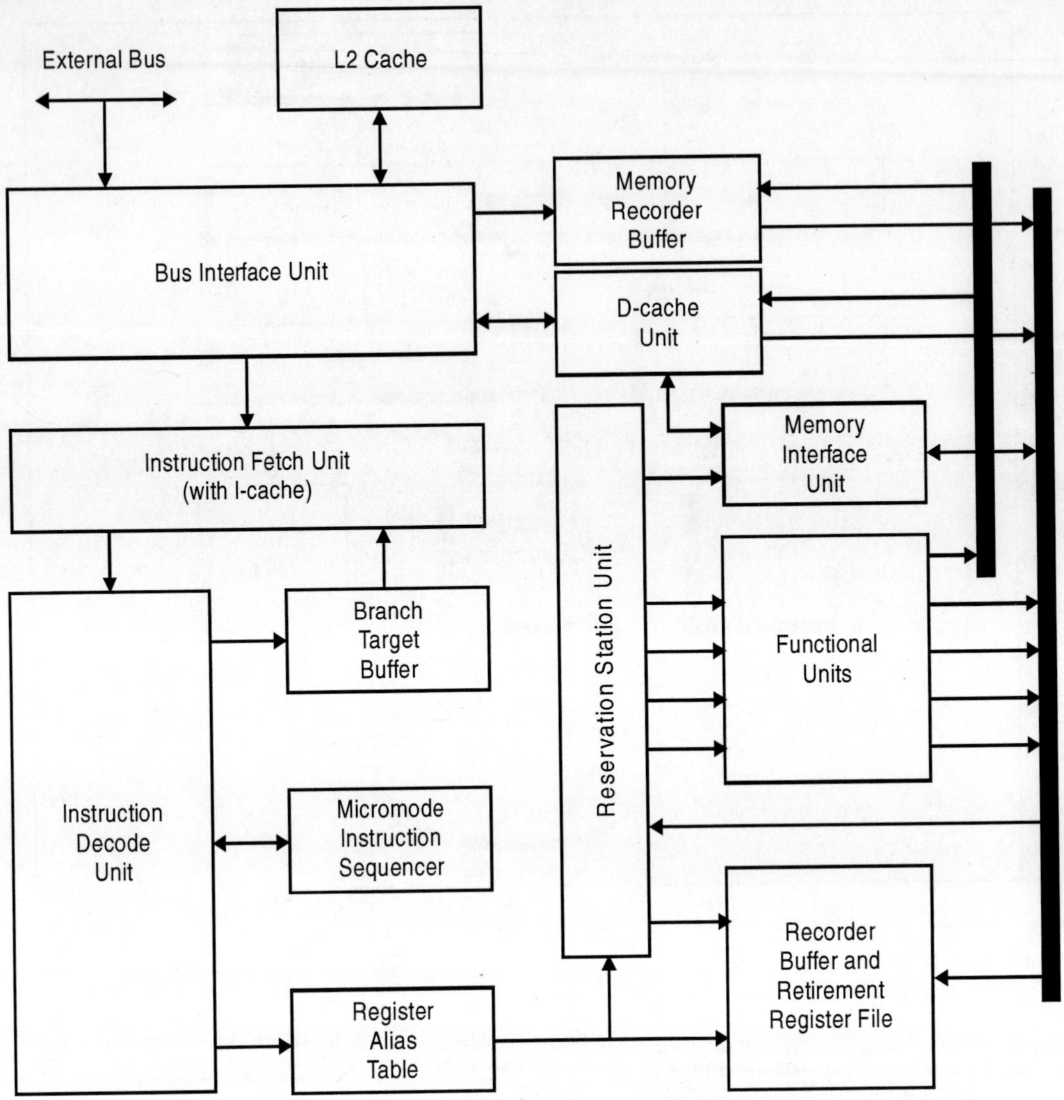

Fig. 11.4. Internal Architecture P-II Architecture.

(v) A decoder breaks the IA-32 instruction down to mops, each comprised of an opcode, two source and one destination operand. These mops are of fixed length.

- ➤ Most IA-32 instructions are converted directly into single micro ops (by any of the three decoders),
- ➢ some instructions are decoded into one-to-four mops (by the general decoder),
- ➢ more complex instructions are used as indices into the microcode instruction sequencer (MIS) which will generate the appropriate stream of mops.

(vi) The mops are send to the register alias table (RAT) where register renaming is performed, i.e., the logical IA-32 based register references are converted into references to physical registers.

(vii) Then, with added status information, mops continue to the reorder buffer (ROB) and to the reservation station unit (RSU).

11.4.1. PENTIUM –III Internal Architecture

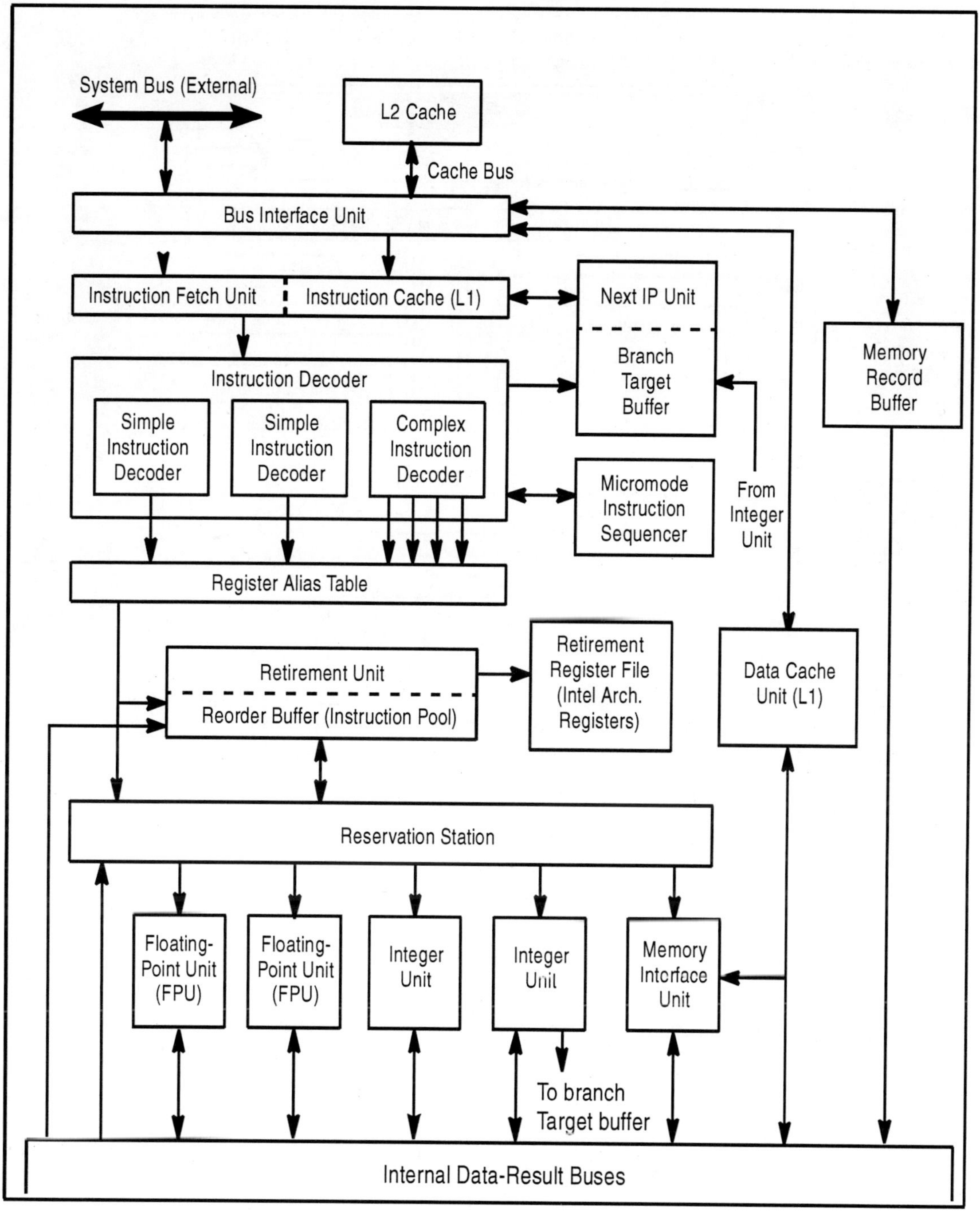

Fig. 11.5. Internal Architecture P-III Architecture.

11.4.2. PENTIUM-III Internal Architecture

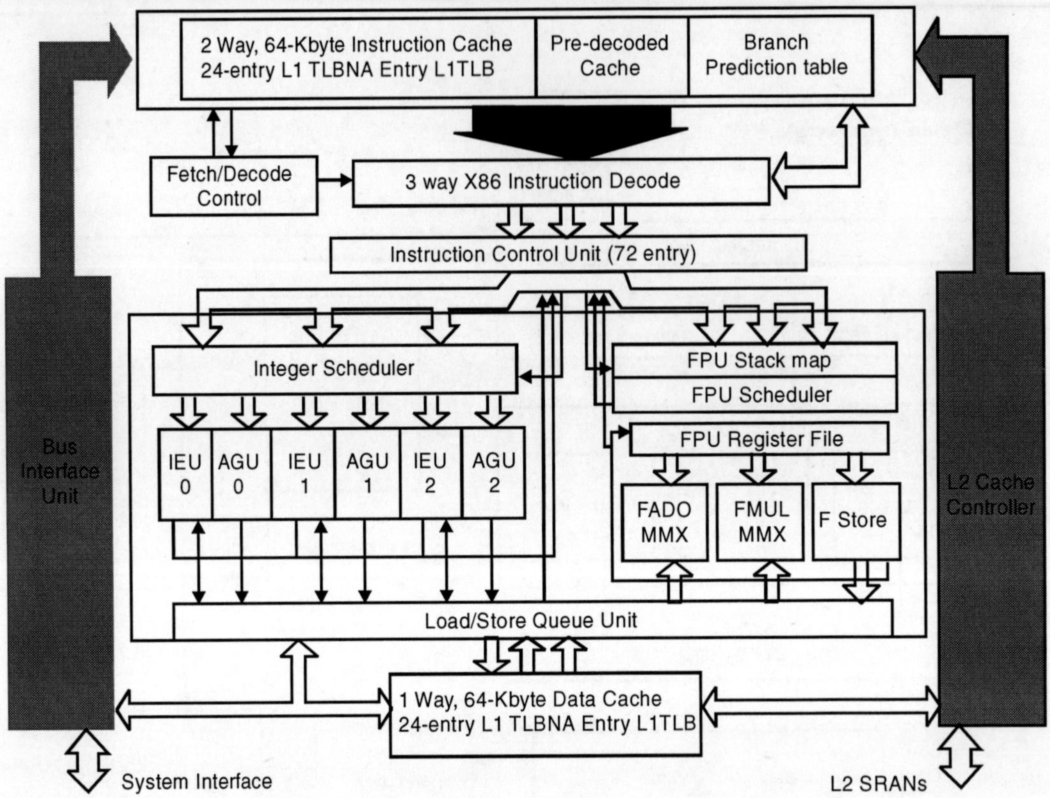

Fig. 11.6. Internal Architecture P-III Architecture.

11.4.3. The Fetch/Decode Unit

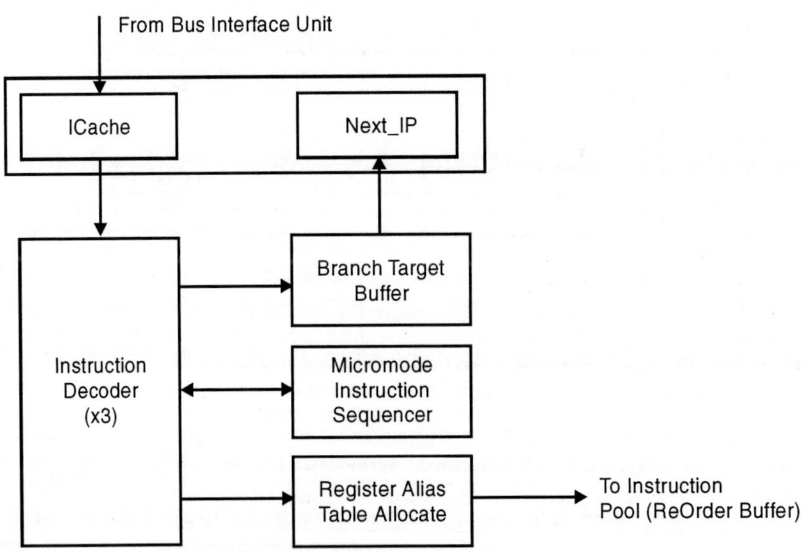

Fig. 11.7. The Fetch/Decode Unit.

The Fetch/Decode Unit

(i) The L1 Instruction Cache is a local instruction cache. The Next_IP unit provides the L1 Instruction Cache index, based on inputs from the Branch Target Buffer (BTB), trap/ interrupt status, and branch-misprediction indications from the integer execution section.

(ii) The L1 Instruction Cache fetches the cache line corresponding to the index from the Next IP, and the next line, and presents 16 aligned bytes to the decoder.

(iii) The perfected bytes are rotated so that they are justified for the instruction decoders (ID). The beginning and end of the Intel Architecture instructions are marked.

(iv) Three parallel decoders accept this stream of marked bytes, and proceed to find and decode the Intel Architecture instructions contained therein.

(v) The decoder converts the Intel Architecture instructions into triadic μops (two logical sources, one logical destination per μop).

(vi) Most Intel Architecture instructions are converted directly into single μops, some instructions are decoded into one-to-four μops and the complex instructions require microcode

(vii) The μops are queued, and sent to the Register Alias Table (RAT) unit, where the logical Intel Architecture-based register references are converted into references to physical registers in P6 family processors physical register references, and to the Allocator stage, which adds status information to the μops and enters them into the instruction pool.

(viii) The instruction pool is implemented as an array of Content Addressable Memory called the Re Order Buffer (ROB).

11.4.4. Dispatch/ Execute Unit

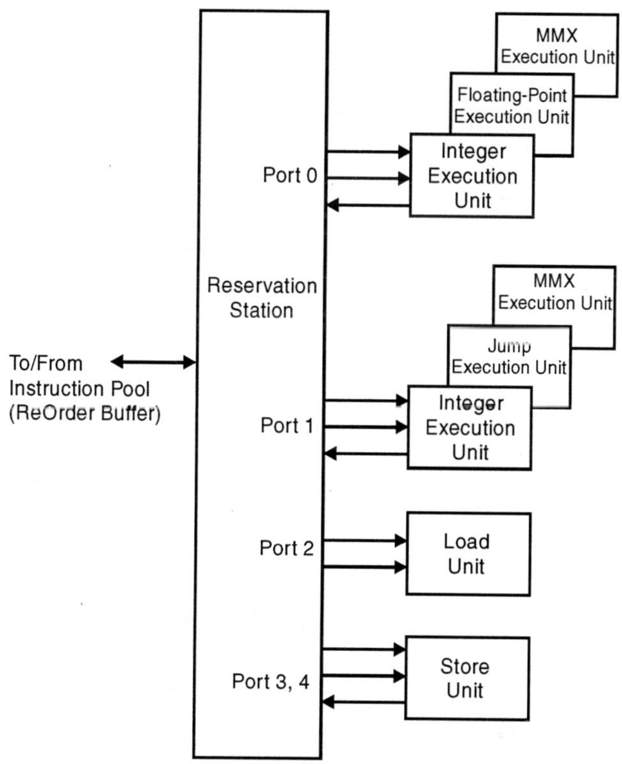

Fig. 11.8. Dispatch & Execute unit.

Dispatch and Execute Unit

(i) The Dispatch unit selects μops from the instruction pool depending upon their status. If the status indicates that a μop has all of its operands then the dispatch unit checks to see if the execution resource needed by that μop is also available. If both are true, the Reservation Station removes that μop and sends it to the resource where it is executed.

(ii) The results of the μop are later returned to the pool. There are five ports on the Reservation Station, and the multiple resources are accessed

(iii) The Pentium II processor can schedule at a peak rate of 5 μops per clock, one to each resource port, but a sustained rate of 3 μops per clock is more typical.

(iv) The activity of this scheduling process is the out-of-order process; μops are dispatched to the execution resources strictly according to dataflow constraints and resource availability, without regard to the original ordering of the program.

(v) Note that many of the μops are branches. The Branch Target Buffer will correctly predict most of these branches but it can't correctly predict them all.

11.4.5. Inside Retire Unit

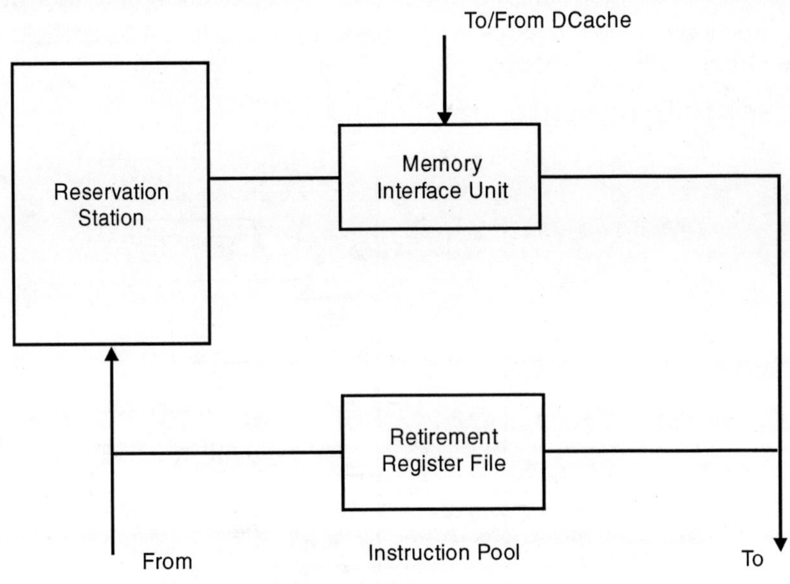

Fig. 11.9. Retire Unit.

Retire Unit

(i) The Retire Unit is also checking the status of μops in the instruction pool. It is looking for μops that have executed and can be removed from the pool. Once removed, the original architectural target of the μops is written as per the original Intel Architecture instruction.

(ii) The Retire Unit must not only notice which μops are complete, it must also re-impose the original program order on them. It must also do this in the face of interrupts, traps, faults, breakpoints and mis-predictions.

(iii) The Retire Unit must first read the instruction pool to find the potential candidates for retirement and determine which of these candidates are next in the original program order.

(iv) Then it writes the results of this cycle's retirements to the Retirement Register File (RRF). The Retire Unit is capable of retiring 3 μops per clock.

11.4.6. Inside The BIU

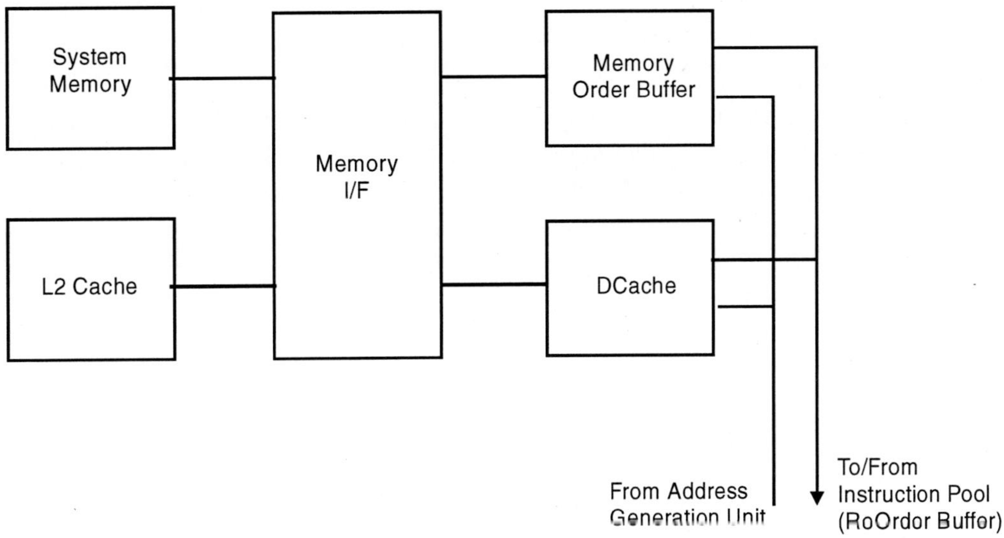

Fig. 11.10. Bus Interface Unit.

Bus Interface Unit

(i) There are two types of memory access: loads and stores. Loads only need to specify the memory address to be accessed, the width of the data being retrieved, and the destination register.

(ii) Loads are encoded into a single μop. Stores need to provide a memory address, a data width, and the data to be written. Stores therefore require two μops, one to generate the address and one to generate the data.

(iii) These μops must later re-combine for the store to complete.

(iv) Stores are never performed speculatively since there is no transparent way to undo

(v) Stores are also never re-ordered among themselves. A store is dispatched only when both the address and the data are available and there are no older stores awaiting dispatch.

Pipeline Operation

Microcode unit can use BOTH u and v pipelines: Pentium microcode much quicker than i486 Instruction fetch uses both prefetch buffers along with BTB (Branch Target Buffer) logic DI, D2 - decoding units: also implement pairing rules EX - execution stage WB - instructions can modify processor state and results can be written.

Pipeline Stage 1

(a) Determines next value of instruction pointer

(b) Dynamic branch prediction in the Branch Target Buffer is realized by using a two level adaptive training algorithm proposed by Yeh and Patt.

(c) First level is a 4 bit history of branches

(d) Second Level is behavior for a specific pattern of branch history updated using Lee and Smithfs Saturating Up-Down counter

Pipeline Stages 2-4

(a) Fetch cache lines (2 cache lines of 32 bytes each)

(b) Marks the boundaries of each instruction

(c) (x86 instructions vary in length from 8 to 120 bits)

(d) Passes 16 aligned bytes to the decoder

Pipeline Stages 5-6

(a) x86 instructions are decoded into a series of micro-ops

(b) Appends flags and status bits to prepare for out of order issue

(c) Three decoders working in parallel

(d) Two simple decoders (perform mostly integer operations)

(e) One complex decoder (for complex instructions)

(f) End result is micro-ops that always contain three operands

(g) Two sources plus a destination

(h) They are longer than true RISC instructions to accommodate floating point operations

Pipeline Stages 5-6

(a) Most instructions translate into one to 4 micro-ops

(b) If more than 4 micro-ops are required then instruction is passed to the Microcode Instruction Sequencer (MS)

(c) Static branch prediction is made by the Branch Address Calculator (BAG). It makes predictions under two conditions:

(d) It sees an absolute branch the BTB did not make a prediction on

(e) It sees a conditional branch with a target address whose direction is gbackwardsh

Pipeline Stages 7-8

(a) References to x86 logical registers are reassigned to physical registers by the Register Alias Table (RAT)

(b) X86 has only 8 integer-type and 8 floating point General Purpose Registers Visible to programmers

(c) Register renaming adds 40 extra General Purpose Registers wide enough to handle integer and floating point values as well as additional status information appended in stage 8

(d) Physical designators become part of the micro-op overall state from this point on

(e) Micro-ops are passed to Reorder Buffer

(f) 40-entry array of content-addressable memory that»fs arranged as a circular FIFO buffer

(g) Micro-ops are held here during various states of completion with with status bits (such as a dependency flag and a time stamp that records the micro-ops original position in the instruction stream) and additional information such as which execution unit can handle that type of micro-op

(h) Micro-ops are passed to reservation station

Pipeline Stages 9-10

(a) All entries of the Reservation Station (RS) are identical and can hold any type of micro-ops

(b) The Reservation Station has 20 entries

(c) Reservation Station dispatches micro-op to one of five execution units
(d) Two Integer units
(e) Two Address generation Units for load/store operations
(f) One Floating Point Unit
(g) Although reservation station can dispatch up to five of this micro-ops per clock cycle, three per clock cycle is more common since units may be busy
(h) Dispatching is determined by checking status bits in ROB for micro-op dependencies and checking whether an execution unit for this type of micro-op is available
(i) If more than one micro-op is ready for dispatching the reservation station picks those that were generated in sequence (this does not yield in order execution since instruction stream has been rearranged by dependencies and branches)

Pipeline Stage 11 Execution

(a) May consist of many stages depending on micro-op complexity (ex: floating point operations).
(b) Most integer instructions (which constitute a majority in most programs) execute in one cycle.
(c) A load operation needs to know only the memory location and width of the data so it is only one micro-op.
(d) A store must generate the memory address and the data so the decoder breaks the operation into two micro-ops that can execute in parallel (the Address Generation Unit handling stores has two ports).
(e) After a micro-op has executed its status flag is changed to indicate completion.
(f) When retirement unit finds a completed micro-op, it verifies the micro-op can indeed be retired. Consider an interrupt invalidating any micro-ops that were completed speciilatively after the micro-op that was interrupted. In this case the micro engine recovers by discarding speculative results stored in the ROB.
(g) Retire unit must put micro-ops back in original order and must watch for interrupts, traps, faults, breakpoints and mispredicted branches.

Pipeline Stage 12

(a) If the micro-op is ready then the retire unit stores the results in the Retirement Register File. At this point the physical state is committed to the logical state. The Retirement Register File is the architectural set of eight integer and eight floating point General Purpose registers
(b) The last execution stage may overlap with the first retirement stage.

Instruction Fetch and Branch Prediction

(a) Two 32B prefetch buffers operate along with the Branch Target Buffer, BTB.
(b) When a branch instruction is fetched, the BTB predicts whether the branch will be taken or not.
   - If Prediction = NOT TAKEN - continue linearly
   - If Prediction = TAKEN - enable other prefetch buffer
(c) If prediction wrong: flush pipelines and start again
(d) BTB uses past history to make predictions
(e) Correct predictions reduce pipeline stalls

11.5. SOFTWARE/PROGRAMMING MODEL

Addressable Memory

- Protected mode
  - 64 GB
  - 36-bit address
- Real-address and Virtual-8086 modes
  - 1 MB space
  - 20-bit address

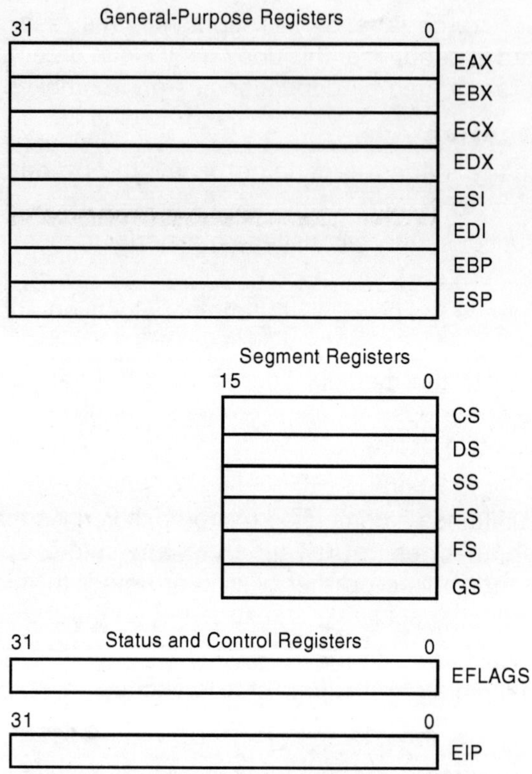

Fig. 11.11. Software/Programming Model.

11.5.1. General-Purpose Registers

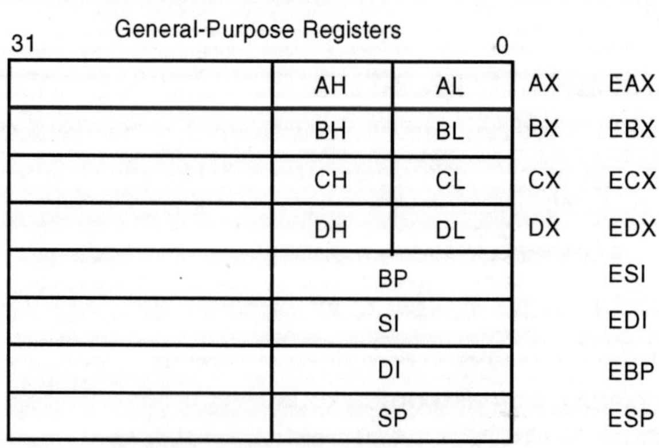

Fig. 11.12. General-Purpose Registers.

Accessing Parts of Registers

➢ Use 8-bit name, 16-bit name, or 32-bit name
Applies to EAX, EBX, ECX, and EDX

32-bit	16-bit	8-bit (high)	8-bit (low)
EAX	AX	AH	AL
EBX	BX	BH	BL
ECX	CX	CH	CL
EDX	DX	DH	DL

Index and Base Registers

➢ Some registers have only a 16-bit name for their lower half:

11.5.2. Some Specialized Register Uses

➢ General-Purpose
- EAX – accumulator
- ECX – loop counter
- ESP – stack pointer
- ESI, EDI – index registers
- EBP – extended frame pointer (stack)

➢ Segment
- CS – code segment
- DS – data segment
- SS – stack segment
- ES, FS, GS - additional segments

Some Specialized Register Uses (2 of 2)

➢ EIP – instruction pointer
➢ EFLAGS
- status and control flags
- each flag is a single binary bit

➢ Instruction Pointer

➢ The instruction pointer (EIP) register contains the offset in the current code segment for the next instruction to be executed. It is advanced from one instruction boundary to the next in straight-line code or it is moved ahead or backwards by a number of instructions when executing JMP, J*cc*, CALL, RET, and IRET instructions. The EIP register cannot be accessed directly by software; it is controlled implicitly by control transfer instructions (such as JMP, J*cc*, CALL, and RET), interrupts, and exceptions. The only way to read the EIP register is to execute a CALL instruction and then read the value of the return.

11.6. FLAG REGISTER

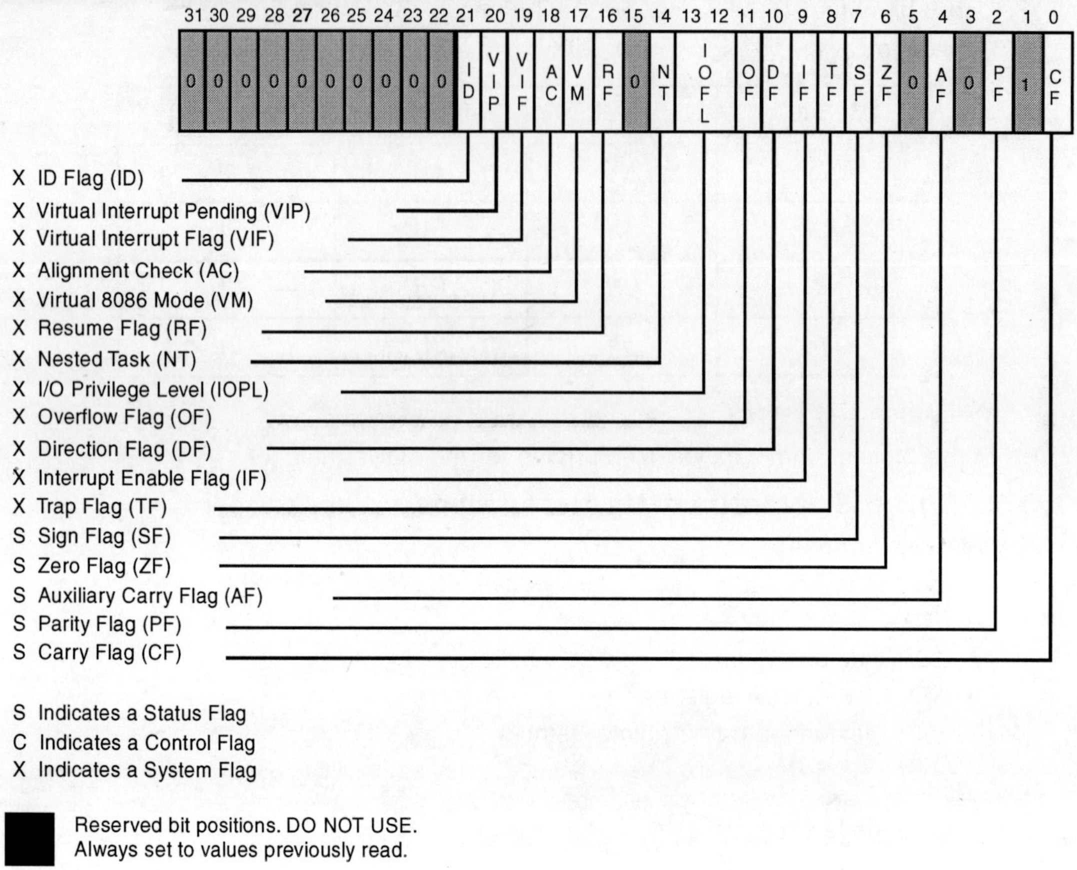

Fig. 11.13. Flag register of pentium processor.

➢ Status Flags

The status flags (bits 0, 2, 4, 6, 7, and 11) of the EFLAGS register indicate the results of arithmetic instructions, such as the ADD, SUB, MUL, and DIV instructions. The functions of the status flags are as follows:

(1) **CF (bit 0) Carry flag.** Set if an arithmetic operation generates a carry or a borrow out of the most-significant bit of the result; cleared otherwise. This flag indicates an overflow condition for unsigned-integer arithmetic. It is also used in multiple-precision arithmetic.

(2) **PF (bit 2) Parity flag.** Set if the least-significant byte of the result contains an even number of 1 bits; cleared otherwise.

(3) **AF (bit 4) Adjust flag.** Set if an arithmetic operation generates a carry or a borrowout of bit 3 of the result; cleared otherwise. This flag is used in binary coded decimal (BCD) arithmetic.

(4) **ZF (bit 6) Zero flag.** Set if the result is zero; cleared otherwise.

(5) **SF (bit 7) Sign flag.** Set equal to the most-significant bit of the result, which is the sign bit of a signed integer. (0 indicates a positive value and 1 indicates a negative value.)

(6) **OF (bit 11) Overflow flag.** This flag indicates Set if the integer result is too large a positive number or too small a negative number (excluding the sign-bit) to fit in the destination operand; cleared otherwise an overflow condition for signed-integer (two's complement) arithmetic.

(7) **DF FLAG.** The direction flag (DF, located in bit 10 of the EFLAGS register) controls the string instructions (MOVS, CMPS, SCAS, LODS, and STOS). Setting the DF flag causes the string instructions to auto-decrement (that is, to process strings from high addresses to low addresses). Clearing the DF flag causes the string instructions to auto-increment (process strings from low addresses to high addresses.

System Flags and IOPL Field

(8) **IF (bit 9) Interrupt enable flag.** Controls the response of the processor to maskable interrupt requests. Set to respond to maskable interrupts; cleared to inhibit maskable interrupts.

(9) **TF (bit 8) Trap flag.** Set to enable single-step mode for debugging; clear to disable single-step mode.

(10) **IOPL (bits 12 and 13) I/O privilege level field.** Indicates the I/O privilege level of the currently running program or task. The current privilege level (CPL) of the currently running program or task must be less than or equal to the I/O privilege level to access the I/O address space. This field can only be modified by the POPF and IRET instructions when operating at a CPL of 0.

(11) **NT (bit 14) Nested task flag.** Controls the chaining of interrupted and called tasks. Set when the current task is linked to the previously executed task; cleared when the current task is not linked to another task.

(12) **RF (bit 16) Resume flag.** Controls the processor's response to debug exceptions.

(13) **VM (bit 17) Virtual-8086 mode flag.** Set to enable virtual-8086 mode; clear to return to protected mode.

(14) **AC (bit 18) Alignment check flag.** Set this flag and the AM bit in the CR0 register to enable alignment checking of memory references; clear the AC flag and/or the AM bit to disable alignment checking.

(15) **VIF (bit 19) Virtual Interrupt flag.** Virtual image of the IF flag. Used in conjunction with the VIP flag. (To use this flag and the VIP flag the virtual mode extensions are enabled by setting the VME flag in control register CR4.)

(16) **VIP (bit 20) Virtual interrupt pending flag.** Set to indicate to that an interrupt is pending; clear when no interrupts are pending. (Software sets and clears this flag. The processor only reads it.) Used in conjunction with the VIF flag.

(17) **ID (bit 21) Identification flag.** The ability of a program to set or clear this flag indicates support for the CPUID instruction

11.7. THE PENTIUM MICROPROCESSOR IN CIRCUIT

The Pentium's memory and I/O interfaces are more advanced than the simple models we looked at earlier. They're also very different from the 8086 bus interfaces. The figure below shows the logic symbol for a Pentium processor.

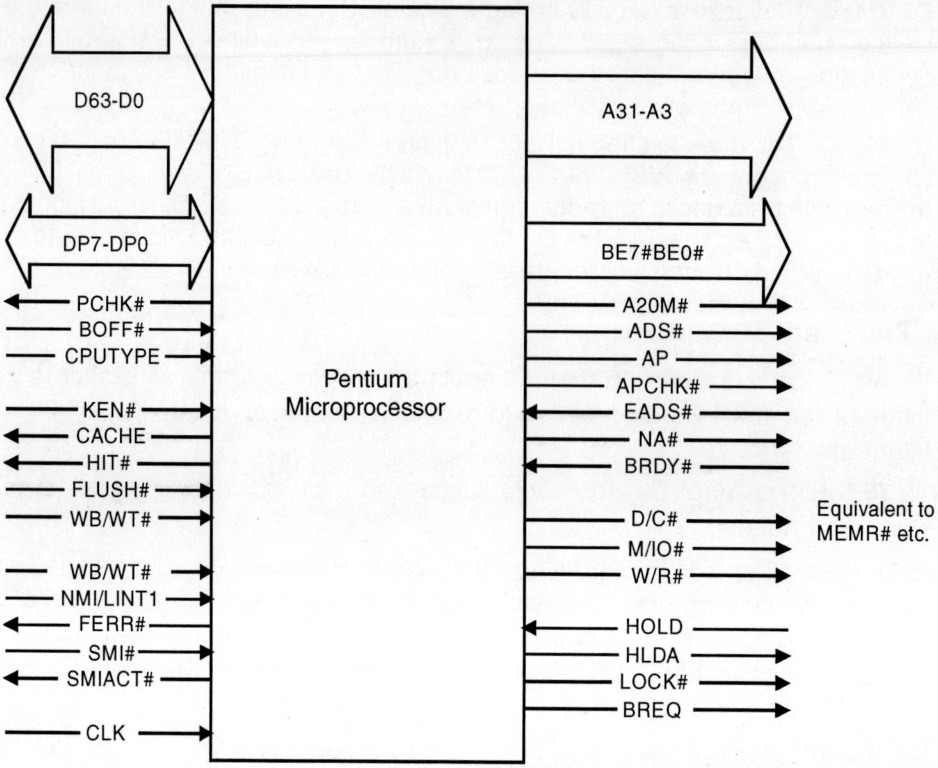

Fig. 11.14. Pin Configuration of the Pentium Pro.

Note that not all pins of the Pentium have been shown in the figure. You can find a more complete diagram in "The Indispensable PC Hardware Book" by Hans-Peter Messmer. Note that a number of revisions of the Pentium processor were released. Messmer has the latest version.

Pin Functions

The table below describes some of the more important features of the external interface signals used in the Pentium

Symbol	Type	Name and Function
A20/M#	I	When the **address bit 20 mask** pin is asserted the Pentium emulates the address wraparound of the 8086. This signal is only used in real mode.
A31:A3	O, I/O	**Address Bus** connections drive out to external memory or I/O devices and decoders. Used along with the byte enables (BE7#-BE0#). The external system can drive inquire addresses into the processor on A31-A5.
ADS#	O	The **Address Status** indicates that a new valid bus cycle is being driven by the Pentium
AP	I/O	The **Address Parity** signal gives even parity information on all Pentium address cycles. Also used along with APCHK# and EADS# for inquire cycles.
APCHK#	O	The **Address Parity Check** status signal is asserted 2 clock cycles after EADS# if the Pentium has detected address parity errors on an inquire cycle.
BE7#	O	The Byte Enable signals are used with A31-A3 to indicate which bytes are to

- BE0#	(Reset: BE3#-BE0#:I/O)	be written to external memory, or which bytes were requested in the current cycle. Same timing as A31-A3. During Reset, BE3#-BE0# are given an APIC-ID, which is interpretd by the on-chip APIC for dual processor support.
BF1, BF0	I	During reset, these pins are sampled and fix the ratio of the external and internal clocks
BOFF#	I	The backoff input aborts all uncompleted bus cycles. The Pentium floats the pins normally affected by a Bus hold a clock cycle after BOFF#. When BOFF# is removed the Pentium restarts the aborted cycle(s) in their entirety.
BRDY#	I	The burst ready input indicates that the external device has completed the read or write (BRDY#=0) or needs more time (BRDY#=1). The signal is sampled in the T2, T12 and T2P cycles.
BREQ	O	The bus request output tells the external system that the Pentium has internally generated a bus request. This happens even if the Pentium is not driving its bus at the moment.
BUSCHK#	I	The **Bus Check** input allows the external system to signal that a bus cycle was unsuccessful. Can cause a machine check exception, depending on the state of CR4.
CACHE#	O	For Pentium initiated cycles, the **Cache** signal indicates if the current read cycle is internally cacheable or indicates a burst writeback cycle for a write cycle.
CLK	I	The **clock** input provides fundamental timing for Pentium cycles. All timingo except the JTAG test signals are referred to the Clock input.
CPUTYPE	I	In a one-processor system, CPUTYPE is connected to GND. In a dual processor system, the primary processor has CPUTYPE also connected to GND and the secondary processor has CPUTYPE tied to VCC (high).
D63-D0	I/O	The 64 **data** lines of the processor. Lines D7-D0 are the least significant byte, D63-D56 are the most significant byte.
DP7-DP0	I/O	These are **data parity** bits. There's one bit for each byte of the data bus (DP7->D63-D56..DP0->D7-D0). During writes the Pentium drives out even parity in the same cycle; during reads the external system has to drive even parity back into the Pentium.
PCHK#	O	**Parity Check** indicates the result of a parity check on a data read.
PEN#	I	**Parity Enable** (along with CR4.MCE) determines if a machine cycle exception is taken as a result of parity error on a read cycle.
D/C#,	O	**Data/Code** (1=Data, 0=Code/special cycle), **Memory/IO** (1=memory, 0=IO
M/IO#, W/R#	(D/C#=0,	cycle), and **write/read** (1=Write, 0=Read) define the current bus cycle. They're
(Dual: I/O)	M/IO#=0,	equivalent to MRDC# on the 8086. Special cycles W/R#=1) are interpreted along with the BE7#-BE0# signals, which identify the exact type of special cycle.
EADS#	I	This indicates that a **valid external address** has been driven into the Pentium for an inquire cycle.
FLUSH#	I	If this signal goes low, the Pentium writes back to memory all changed cache lines and invalidates its code and data internal caches.

HIT#	O	This signal gives the result of an inquiry cycle. A low indicates a hit and a high indicates a miss.
HOLD, HLDA	I, O	Bus Hold Request and Bus Hold Acknowledge are used for bus arbitration. An external bus master asserts HOLD; The Pentium tristates most of its output and input/output lines and asserts HLDA after completing all outstanding bus cycles; The external master releases HOLD when its finished; The Pentium leaves bus hold and release HLDA.
LOCK#	O	During **lock**, the Pentium does not recognize HOLD requests.
KEN#	I	The **cache enable** input is used to determine if the current cycle is cacheable.
N/A#	I	The next address input indicates that the external system is ready to accept a new address even if all the current bus cycles have not completed. Used to speed up memory access cycles.
INTR/LINT0	I	A High input indicates a maskable interrupt request from an external device. Only if the IE Flag is set in the EFLAG register will the processor accept the interrupt. LINT0 means local interrupt 0 when the on-chip APIC is used.
NMI/LINT1	I	The Non-Maskable Interrupt. LINT1 means local interrupt 1 when the on-chip APIC is used.
RESET	I	Reset forces the Pentium to begin execution at a known state. All internal caches are invalidated. The Pentium samples some other signals when Reset changes from high to low, to determine what mode it should enter after a reset (e.g. tristates test mode).
SMI#	I	**System Management Interrupt** – causes the Pentium to enter System Management Mode
SMIACT#	O	An active **system management interrupt** active output indicates that the processor is in System Management Mode (SMM).
WB/WT#	I	The **writeback/writethrough#** input allows a data cache line to be defined as writeback (1) or writethrough (0) on a line-by-line basis. WB/WT# determines whether a cache line is initially in the S or E state in the data cache. Used with the PWT bit.

Pentium-IV Processor

Pentium 4 and new Celeron processors use Intel's seventh generation architecture, also called Netburst. Its overall look you can see on Fig. 11.14.

Here are the basic differences between Pentium 4 architecture and the architecture from other CPUs.

Externally, Pentium 4 transfers four data per clock cycle. This technique is called QDR (Quad Data Rate) and makes the local bus to have a performance four times its actual clock rate, see table below. On Figure 11.14 this is shown on "3.2 GB/s System Interface";

Since this slide was produced when the very first Pentium 4 was released, it mentions the "400 MHz" system bus.

Real Clock	Performance	Transfer Rate
100 MHz	400 MHz	3.2 GB/s
133 MHz	533 MHz	4.2 GB/s
200 MHz	800 MHz	6.4 GB/s
266 MHz	1,066 MHz	8.5 GB/s

The datapath between the L2 memory cache ("L2 cache and control" on Figure 11.14) and L1 data cache ("L1 D-Cache and D-TLB" on Figure 11.14) is 256-bit wide. On previous processors from Intel this datapath was of only 64 bits. So this communication can be four times faster than processors from previous generations when running the same clock. The datapath between L2 memory cache ("l2 cache and control" on Figure 11.14) and the pre-fetch unit ("BTB & I-TLB" on Figure 11.14), however, continues to be 64-bit wide.

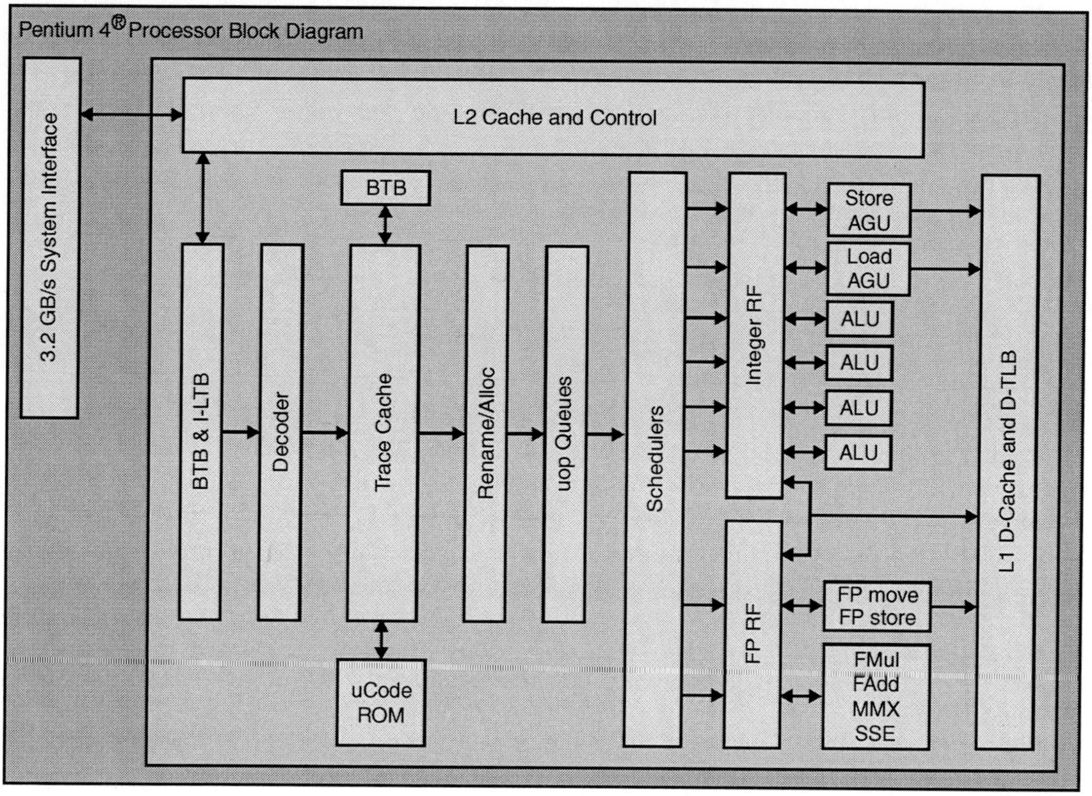

Fig. 11.14. Pentium 4 block diagram

The L1 instruction cache was relocated. Instead of being before the fetch unit, the L1 instruction cache is now after the decode unit, with a new name, "Trace Cache". This trace cache can hold up to 12 K micro instructions. Since each micro instruction is 100-bit wide, the trace cache is of 150 KB (12 K × 100/8). On of most common mistakes people make when commenting Pentium 4 architecture is saying that Pentium 4 doesn't have any instruction cache at all. That's absolutely not true. It is there, but with a different name and a different location.

On Pentium 4 there are 128 internal registers, on Intel's 6th generation processors (like Pentium 2 and Pentium 3) there were only 40 internal registers. These registers are in the Register Renaming Unit (a.k.a RAT, Register Alias Table, shown as "Rename/Alloc" on Fig. 11.14).

Of course this is just a summary for those who already has some knowledge on the architecture from other processors. If all this look like Greek to you, don't worry. We will explain everything you need to know about Pentium 4 architecture in an easy to follow language on the next pages.

Pentium 4 Pipeline

Pipeline is a list of all stages a given instruction must go thru in order to be fully executed. On 6th generation Intel processors, like Pentium 3, their pipeline had 11 stages. Pentium 4 has 20 stage! So, on a Pentium 4 processor a given instruction takes much longer to be executed then on a Pentium 3, for instance! If you take the new 90 nm Pentium 4 generation processors, codenamed "Prescott", the case is even worse because they use a 31-stage pipeline! Holy cow!

This was done in order to increase the processor clock rate. By having more stages each individual stage can be constructed using fewer transistors.

With fewer transistors is easier to achieve clock rates. In fact, Pentium 4 is only faster than Pentium 3 because it works at a higher clock rate. Under the same clock rate, a Pentium 3 CPU would be faster than a Pentium 4 because of the size of the pipeline.

Because of that, Intel has already announced that their 8th generation processors will use Pentium M architecture, which is based on Intel's 6th generation architecture (Pentium 3 architecture) and not on Netburst (Pentium 4) architecture. This architecture, called Core, can be studied on our Inside Core Micro architecture tutorial.

On Figure 11.15 you can see Pentium 4 20-stage pipeline. So far Inlet didn't disclosure Prescott's 31-stage pipeline, so we can't talk about it.

1	2	3	4	5	6	7	8	9	10	11	12	13	14	15	16	17	18	19	20
TC Nxt IP		TC Fetch		Drive	Alloc	Rename		Que	Sch	Sch	Sch	Disp	Disp	RF	RF	Ex	Figs	Br Ck	Drive

Fig. 11.15. Pentium 4 pipeline.

Here is a basic explanation of each stage, which explains how a given instruction is processed by Pentium 4 processors. If you think is too complex for you, don't worry. This is just a summary of what we will be explaining on the next pages.

TC Nxt IP : Trace cache next instruction pointer. This stage looks at branch target buffer (BTB) for the next microinstruction to be executed. This step takes two stages.

TC Fetch : Trace cache fetch. Loads, from the trace cache, this microinstruction. This step takes two stages.

Drive : Sends the microinstruction to be processed to the resource allocator and register renaming circuit.

Alloc : Allocate. Checks which CPU resources will be needed by the microinstruction – for example, the memory load and store buffers.

Rename : If the program uses one of the eight standard ×86 registers it will be renamed into one of the 128 internal registers present on Pentium 4. This step takes two stages.

Que : Queue. The microinstructions are put in queues accordingly to their types (for example, integer or floating point). They are held in the queue until there is an open slot of the same type in the scheduler.

Sch : Schedule : Microinstructions are scheduled to be executed accordingly to its type (integer, floating point, etc). Before arriving to this stage, all instructions are in order, i.e., on the same order they appear on the program. At this stage, the scheduler, reorders the instructions in order to keep all execution units full. For example, if there is one floating point unit going to be available, the scheduler will look for a floating point instruction to send it ti this unit, even if the next instruction on the program is an integer one. The scheduler is the heart of the out-of-order engine of Inlet 7th generation processors. This step takes three stages.

Disp : Dispatch : Sends the microinstructions to their corresponding execution engines. This step takes two stages.

RF : Register file ; The internal registers, stored in the instructions pool, are read. This step takes two stages.

Ex : Executes : Microinstructions are executed.

Flgs : Flags : The microprocessor flags are updated.

Br Ck : Branch check : Checks if the branck taken by the program is the same predicted by the branch prediction circuit.

Drive : Sends the results of this check to the branch target buffer (BTB) present on the processor's entrance.

Memory Cache and Fetch Unit

Pentium 4's L2 memory cache can be of 256 KB, 512 KB, 1 MB or 2 MB, depending on the model. L1 data cache is of 8 KB or 16 KB (on 90 nm models.)

As we explained before, the L1 instruction cache was moved from before the fetch unit to after the decode unit using a new name. "trace cache". So, instead of storing program instructions to be loaded by the fetch unit, the trace cache stores microinstructions already decoded by the decode unit. The trace cache can store up to 12 K microinstructions and since Pentium 4 microinstructions are 100-bit wide, the trace cache is of 150 KB (12,288 × 100 / 8).

The idea behind this architecture is really interesting. In the case of a loop on the program (a loop is a part of a program that needs to be repeated several times), the instructions to be executed will be already decoded, because they are stored already decoded on the trace cache. On other processors, the instructions need to be loaded from L1 instruction cache and decoded again, even if they were decoded a few moments before.

The trace cache has also its own BTB (Branch Target Buffer) of 512 entries. BTB is a small memory that lists all identified branches on the program.

As for the fetch unit, its BTB was increased to 4,096 entries. On Inlet 6th generation processors, like Pentium 3, this buffer was of 512 entries and on Inlet 5th generation processors, like the First Pentium processor, this buffer was of 256 entries only.

On Fig. 11.16 you see the block diagram for what we were discussing. TLB means Translation Lookaside Buffer.

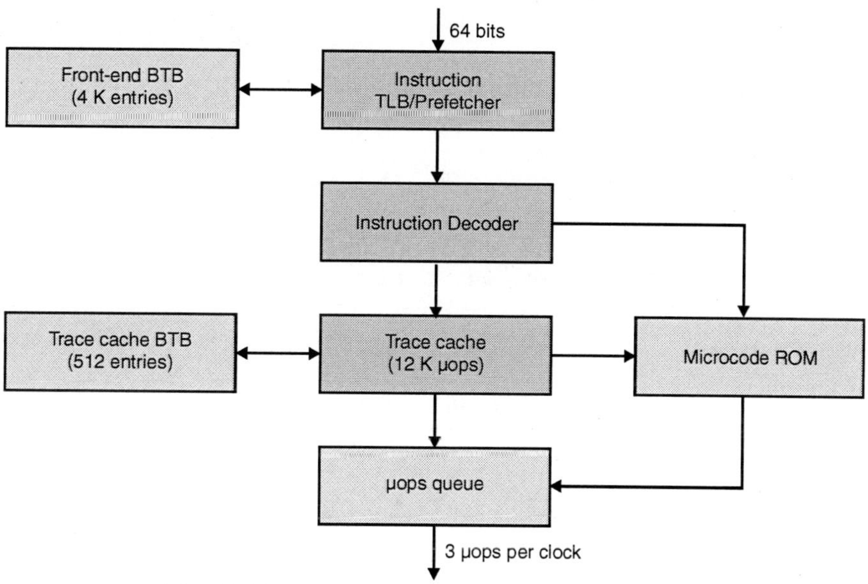

Fig. 11.16. Fetch and decode units and trace cache.

Decoder

Since previous generation (6th generation), Inlet processors are use a hybrid CISC/RISC architecture. The processor must accept CISC instructions, also known as ×86 instructions, since all software available today is written using this kind of instructions. A RISC-only CPU couldn't be create for the PC because it wouldn't run software we have available today' like Windows and Office.

So, the solution used by all processors available on the market today from both Intel and AMD is to use a CISC/RISC decoder. Internally the CPU processes RISC-like instructions, but its front-end accepts only CISC ×86 instructions.

CISC ×86 instructions are referred as "instructions" as the internal RISC instructions are referred as "microinstructions" or μops".

These RISC microinstructions, however, cannot be accessed directly, so we couldn't create software based on these instructions to bypass the decoder. Also, each CPU uses its own RISC instructions, which are not public documented and are incompatible with microinstructions from other CPUs. I.e., Pentium 3 microinstructions are different from Pentium 4 microinstructions, which are different from Athlon 64 microinstructions.

Depending on the complexity of the ×86 instruction, it has to be converted into several RISC microinstructions.

Pentium 4 decoder can decode one ×86 instruction per clock cycle, as long as the instruction decodes in up to four microinstructions. If the ×86 instruction to be decoded is complex and will be translated in more than four microinstructions, it is routed to a ROM memory ("microcode ROM" on Figure 3) that has a list of all complex instructions and how they should be translated. This ROM memory is also called MIS (Microcode Instruction Sequencer).

As we said earlier, after being decoded microinstructions are sent to the trace cache, and from there they go to a microinstructions queue. The trace can put up to three microinstructions on the queue per clock cycle, however Intel doesn't tell the depth (size) of this queue.

From there, the instructions go to the Allocator and register Renamer. The queue can also deliver up to three microinstructions per clock cycle to the allocator.

Allocator and Register Renamer

What the allocator does :

- Reserves one of the 126 reorder buffers (ROB) to the current microinstructions, in order to keep track of the microinstruction completion status. This allows the microinstructions to be executed out-of-order, since the CPU will be able to put them in order again by using this table.
- Reserves on of the 128 register files (RF) in order to store there the data resulted from the microinstruction processing.
- If the microinstruction is a load or a store, i.e. it will read (load) or write (store) data from/to RAM memory, it will reserve one of the 48 load buffers or one of the 24 store buffers accordingly.
- Reserves an entry on the memory or general queue, depending on the kind of microinstruction it is.

After that the microinstruction goes to the register renaming stage. CISC ×86 architecture has only eight 32-bit registers (EAX, EBX, ESI, EDI and ESP). This number is simply too low, especially because modern CPUs can execute code out-of-order, what would "kill" the contents of a given register, crashing the program.

So, at this stage, the processor changes the name and contents of the registers used by the program into one of the 128 internal registers available, allowing the instruction to run at the same time of another instruction that uses the exact same standard register, or even out-of-order, i.e., this allows the second instruction to run before the first instruction even if they mess with the same register.

It is interesting to note that Pentium 4 has actually 256 internal registers, 128 registers for integer instructions and 128 registers for floating point and SSE instructions.

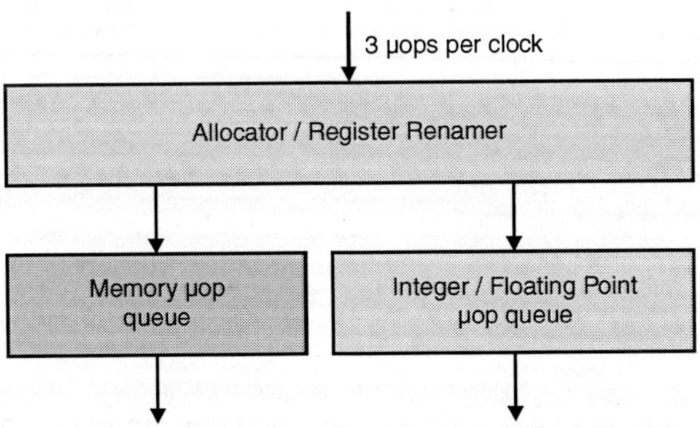

Fig. 11.17. Allocator and register Renamer

Pentium 4 renamer is capable of processing three microinstruction per clock cycle.

From the renamer the microinstructions go to a queue, accordingly to its type : memory queue, for memory-related microinstructions, or Integer/Floating Point Queue, for all other instruction types.

Scheduler

The scheduler has four units. It analyses each microinstruction and puts it on a scheduler unit accordingly to its type :

(i) **Memory scheduler unit :** For memory-related microinstructions. These are the microinstructions that came from the memory microinstruction queue.

(ii) **Fast scheduler unit :** This unit is for simple microinstructions.

(a) **Slow / General FP scheduler unit :** This unit is for other microinstructions and for complex floating point microinstructions.

(b) **Simple FP scheduler unit :** This unit is for simple floating point microinstructions.

So, the scheduler sorts the microinstructions accordingly to their type. Then the scheduler can dispatch each microinstruction directly to the correct execution unit to be processed.

The scheduler is the heart of the out-of-order engine on Pentium 4. Until now all microinstructions were delivered on the same order they were decoded. On the scheduler the microinstructions can be dispatched on a totally different order to the execution engines. The goal of the scheduler it to keep all CPU execution engines busy all the time.

The execution units are connected to the scheduler thru four dispatch ports, numbered 0 thru 3, as you can see on Fig. 11.18.

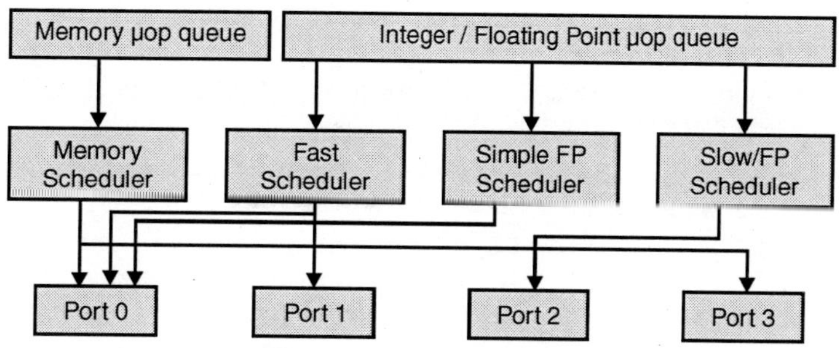

Fig. 11.18. Scheduler units.

Dispatch and Execution Units

As we've seen Pentium 4 has four dispatch ports numbered 0 thru 3. Each port is connected to one, two or three execution units, as you can see on Fig. 11.19.

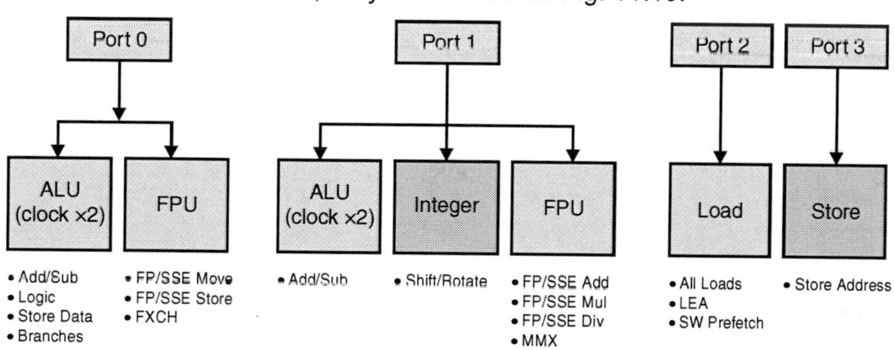

Fig. 11.19. Dispatch and execution units.

The units marked as "clock ×2" can execute two microinstructions per clock cycle. Ports 0 and 1 can send two microinstructions per clock cycle to these units. So the maximum number of microinstructions that can be dispatched per clock cycle is six :

(i) Two microinstructions on port 0;

(ii) Two microinstructions on port 1;
(iii) One microinstruction on port 2;
(iv) One microinstruction on port 3.

Keep in mind that complex instructions may take several clock cycles to be processed. Let's take an example of port 1, where the complete floating point unit is located. While this unit is processing a very complex instruction that takes several clock ticks to be executed, port 1 dispatch unit won't stall : it will keep sending simple instructions to the ALU (Arithmetic and Logic Unit) while the FPU is busy.

So, even thought the maximum dispatch rate is six microinstructions, actually the CPU can have up to seven microinstructions being processed at the same time.

Actually that's why ports 0 and 1 have then one execution unit attached. If you pay attention, Intel put on the same port one fast unit together with at least one complex (and slow) unit. So, while the complex unit is busy processing data, the other unit can keep receiving microinstructions from its corresponding dispatch port. As we mentioned before, the idea is to keep all execution units busy all the time.

The two double-speed ALUs can process two microinstructions per clock cycle. The other units need at least one clock cycle to process the microinstructions they receive. So, Pentium 4 architecture is optimized for simple instructions.

As you can see on Figure 11.19, dispatch ports 2 and 3 are dedicated to memory operations : load (read data from memory) and store (write data to memory), respectively. As for memory operation, it is interesting to note that port 0 is also used during store operations (see Figure 5 and the list of operations on Figure 11.19). On such operations, port 3 is used to send the memory address, while port 0 is used to send the data to be stored at this address. This data can be generated by either the ALU or the FPU, depending on the kind of data to be stored (integer or floating point/SSE).

On Figure 11.19 you have a complete list of the kinds of instructions each execution unit deals with. FXCH and LEA (Load Effective Address) are two ×86 instructions. Actually Intel's implementation for FXCH instruction on Pentium 4 caused a great deal of surprise to all experts, because on processors from previous generation (Pentium 3) and processors from AMD this instruction can be executed at zero clock cycle, while on Pentium 4 it takes some clock cycles to be executed.

REVIEW QUESTIONS

1. Describe the P-6 Family of processors?
2. Explain various features possessed by the P-6 family processors?
3. With the help of a diagram explain in detail the Pentium Pro Architecture ?
4. Define Branch prediction and explain the concept used in the P-6 family processors?
5. Differentiate between Pentium pro and Pentium-II processor?
6. Explain in detail about the Pentium pro processor architecture with each of its internal units?
7. Draw the block diagram of P-II processor pipeline and explain it in detail?
8. Draw the internal architecture of P-II processor and explain its each unit in detail?
9. Draw the internal architecture of P-III processor and explain its each unit in detail?
10. Give the software model for the Pentium processor?
11. Describe the register set used in the Pentium processor?
12. Draw & explain the PSW format of the Pentium processor?
13. Explain the functions of different pins possessed by the Pentium processor?
14. Explain the concept of memory banks in the P-6 family processor?
15. Draw & explain with the help of a timing diagram the Single Transfer Cycles?
16. Draw & explain with the help of a timing diagram the Burst cycles in the Pentium processor?
17. Draw & explain with the help of a timing diagram Inquiry Cycles and Internal Snooping?
18. Explain the concept of dual processing in the Pentium processor?

❑❑❑

CHAPTER 12

APPLICATION OF MICROPROCESSORS

INSIDE THIS CHAPTER

So far in this book, we have shown you how a basic microcomputer functions and how to interface a wide variety of devices to the basic microcomputer. Now it's time to show you how some of these pieces are put together to make a microcomputer-based instrument. The first instrument we have chosen is a "smart" scale such as you might see at the checkout stand in your local grocery store.

12.1. OVERVIEW OF SMART-SCALE OPERATION

Figure 12.1 shows a block diagram of our smart scale. A load cell converts the applied weight of, for example, a bunch of carrots to a proportional electrical signal. This small signal is amplified and converted to a digital value which can be read in by the microprocessor and sent to the attached display. The user then enters the price per pound with the keyboard, and this price per pound is shown on the display. When the user presses the compute key on the keyboard, the microprocessor multiplies the weight times the price per pound and displays the computed price. After holding the price display long enough for the user to read it, the scale goes back to reading in the weight and displaying it. To save the user from having to type the computed price into the cash register, an output from the scale could be connected directly into the cash register circuitry. Also, a speech synthesizer could be added to verbally tell the customer the weight, price per pound, and total price.

Smart scales such as this have many applications other than weighing carrots. A modified version of this scale is used in company mail rooms to weigh packages and calculate the postage required to send them to different postal zones. The output of the scale is usually connected to a postage meter, which then automatically prints out the required postage sticker. Another application of smart scales is to count coins in a bank or gambling casino. For this application, the user simply enters the type of coin being weighed. A conversion factor in the program computes the total number of coins and the total dollar amount. Still another application of a scale such as this is in packaging items for sale. Suppose, for example, that we are manufacturing wood-screws and that we want to package 100 of them per box. We can pass the boxes over the load cell on a conveyor belt and fill them from a chute until the weight and therefore the court, reaches some entered value. The point here is that the combination of intelligence and some simple interface circuitry gives you an instrument with as many uses as your imagination can come up with.

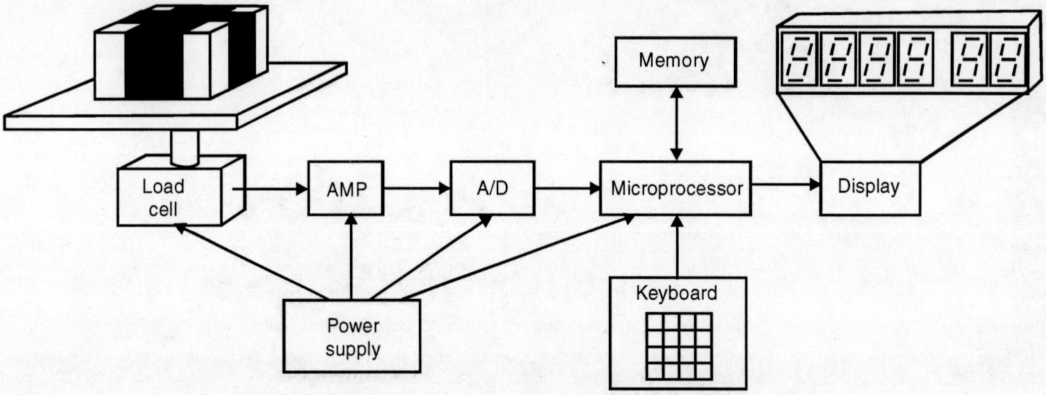

Fig. 12.1. Block Diagram of microprocessor based smart scale.

12.2. SMART-SCALE INPUT CIRCUITRY

We added a piece of plywood to the top of the load cell to keep the carrots from falling off. This load cell has an accuracy of about 1 part in 1000 or 0.01 lb over the 0 to 10 lb range for which it was designed.

The load cell consists of four 350 W resistors connected in a bridge configulation. A stable 10.00 V excitation voltage is applied to the top of the bridge. With no load on the cell, the outputs from the bridge are at about the same voltage, 5 V. When a load is applied to the bridge, the resistance of one of the lower resistors will be changed. This produces a small differential output voltage from the bridge. The maximum differential output voltage for this 10 lb load cell mum is 2 mV per volt of excitation, so with 10.00 V excitation as shown, the maximum differential output voltage is 20 mV.

To amplify this small differential signal, we use a National LM363 instrumentation amplifier. This device contains all the circuitry shown for the instrumentation amplifier . The closed loop gain of the amplifier is programmable with jumpers on pins 2, 3 and 4 for fixed values of 5, 100 and 500. We have jumpered it for a gain of 100 so that the 20 mV maximum signal from the load cell will give a maximum voltage of 2.00 V to the A/D converter input. A precision voltage divider on the output of the amplifier divides this signal in half so that a weight of 10.00 lb produces an output voltage of 1,000 V. This scaling simplifies the display of the weight after it is read into the microprocessor. The 0.1 mF capacitor between pins 15 and 16 of the amplifier reduces the bandwidth of the amplifier to about 7.5 Hz. This removes 60 Hz and any high frequency noise that might have been induced in the signal lines.

The MC 14433 A/D converter used here is an inexpensive dual slope device intended for use in 3½ digit digital voltmeters etc. Because the load cell output changes slowly, a fast converter isn't needed here. The voltage across an LM329 6.9 V precision reference diode is amplified by IC4 to produce the 10.00 V excitation voltage for the load cell and a 2.000 V reference for the A/D. With a 2.000 V reference voltage, the full scale input voltage for the A/D is 2.000 V. Conversion rate and multiplexing frequency for the converter are determined by an internal oscillator and R11. An R11 of 300 kW gives a clock frequency for the converter are determined by an internal oscillator and R11. An R11 of 300 kW gives a clock frequency of 66 kHz, a multiplex frequency of 0.8 kHz, and about four conversions per second. Accuracy of the converter is ±0.05 percent and ±1 count, which is comparable to the accuracy of the load cell. In other words, the last digit of the displayed weight may be off by 1 or 2 counts. As we described in a previous section, the output from this converter is in multiplex BCD form.

12.2.1. An Algorithm for the Smart Scale

Figure 12.3 shows the flow chart for our smart scale. Note that, we indicated by the double-ended boxes in the flowchart, most major parts of the program are written as procedures. This is an example of the structured, modular programming approach we have stressed throughout the book. Here's how it all works.

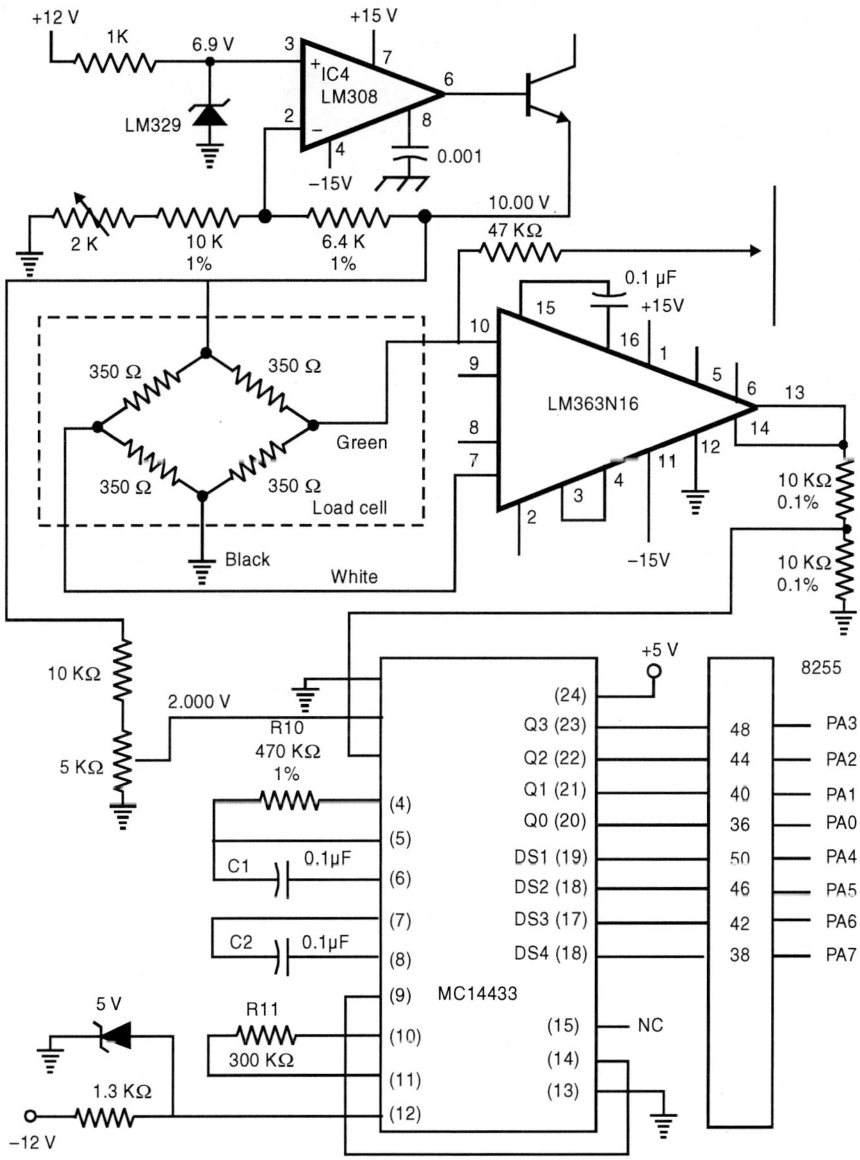

Fig. 12.2. Circuit diagram for Load cell interface circuitry and A/D converter for smart scale.

The output of the A/D is in multiplexed BCD form. The converter outputs the BCD code for a digit on its Q0-Q3 lines and outputs the strobe for that digit on the corresponding digit strobe line, DS1-DS4. To read the data for a digit, that digit strobe is polled until it goes high: then the BCD code for that digit is read in. After the four BCD values are read in from the converter, a display procedure is called to display these values on the address field displays of the SDK-86. The letters "Lb" are displayed in the data field displays.

Next, a check is made to see if any keys have been pressed by the user. If a key has been pressed, the letters "SP", which represent selling price, are displayed in the address field. Key codes are read from the 8279 as entered and displayed on the data field display. Keys can be pressed until the desired price per pound shows on the display. When a nonnumeric key is pressed, it is assumed that the entered price per pound is correct, and the program goes on to compute the total price.

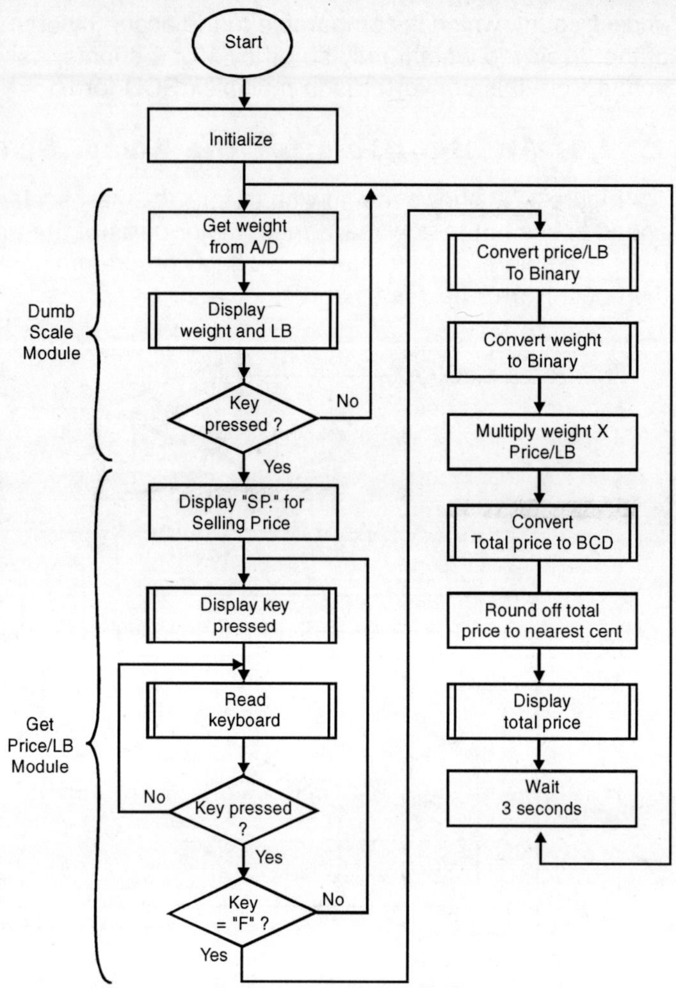

Fig. 12.3. Flow chart for smart scale program.

Computing the price involves multiplying the weight in CD form times the price per pound in BCD form. It is not easy to do a BCD x BCD multiply, so we took an alternate route to get there. We converted both the Wight and the price pound to their binary equivalents and then multiplied the binary numbers. Another procedure converts the binary numbers. Another procedure converts the binary result of the multiplication to BCD. The BCD result is rounded to the nearest cent and displayed in the data field. The letters "Pr" are displayed in the address field to indicate program goes back to reading and displaying weight over and over, until a key is pressed.

12.3. A MICROCOMPUTER-BASED INDUSTRIAL PROCESS-CONTROL SYSTEM

One area in which microprocessors and microcomputers have had a major impact is industrial process control. Process control involves first measuring system variables such as motor speed, temperature, the flow of reactants, the level of a liquid in a tank, the thickness of a material, etc. The output of the controller then adjusts the value of each variable until it is equal to a predetermined value of each variable until it is equal to a predetermined value called a set point. The system controller must maintain and it must compensate as quickly and accurately as possible for any change in the variable caused by, for example, increased load on a motor. A simple example will show the traditional approach to control of a process variable and explain some of the terms used in control systems.

The circuit in Fig. 12.4 shows an analog approach to controlling the speed of a dc motor. Attached to the shaft of the motor is a dc generator, or tachometer, which puts out a voltage is typically to the speed of the motor. The output voltage is typically a few volts per 1000 rpm. A fraction of the output voltage from the tachometer is fed back to the inverting input of the power amplifier driving the motor. A positive as a set point. When the power is turned on, the motor accelerates until the voltage fed back from the tachometer to the inverting input of the amplifier is nearly equal to the set point voltage.

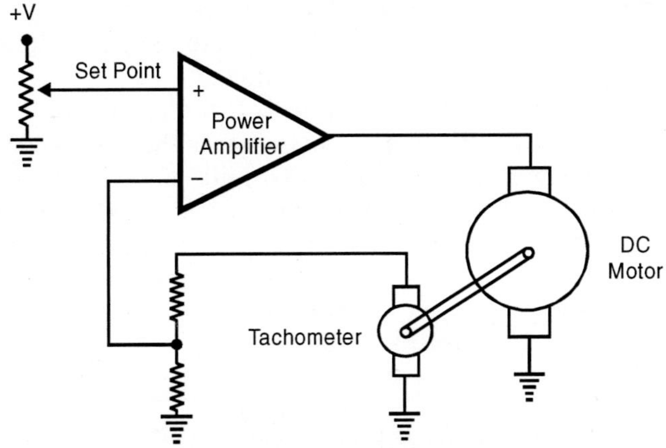

Fig. 12.4. Circuit for controlling the speed of dc motor using feedback from a tachometer.

If the load on the motor is increased, the motor will initially slow down, and the voltage output from the tachometer will decrease. This will increase the difference in voltage between the inputs of the amplifier and cause it to drive more current to the motor. The increased current will increase the speed fo the motor to nearly the speed it had before the increased load was added. A similar reaction takes place if the load on the motor is decreased.

Using negative feedback to control a system such as this is often called servo control. A control loop of this type keeps the motor speed quite constant for applications where the load on the motor does not change much. Somo hard-disk drive motors and high quality phonograph turntables use this method of speed control.

For application in which the load and/or the set point changes drastically, there are several potential problems. The first of these is overshoot when you change the set point. Fig. 12.5(a), shows an example of this. In this case the new set point and bounces up and down for a while. The time it takes the bouncing to settle within a specified error band is called the setting time. This type of response is referred to as underdamped and is similar to the response that a car with bad shock absorbers will make when it hits a bump. The ringing can be prevented by adding damping to the system. However, if too much damping is added, the response to a change in set point may look like that shown in Fig. 12.5(b). This type response is referred to as an overdamped response. The difficulty with this type response is that it takes a long time for the variable to reach the new set point. For best performance, the damping must be custom designed for a particular system.

Another problem in any process control system is residual error. Figure 12.5(c) shows the response a control system such as the motor speed controller in Fig. 12.5 will have when more load is added on the motor. The motor initially slows down, so the voltage out of the tachometer decreases. As we said before, this increases the voltage difference between the amplifier inputs and causes and amplifier output to increase. Increased amplifier output increases the spread of the motor and thereby the output from the tachometer. When the system reaches aquarium, however, there is some noticeable difference between the set point and the voltage fed back from the tachometer. It is this difference or residual error which is amplified by the gain of the amplifier to produce the additional drive for the motor. For stability reasons, the gain of many control systems cannot be too high. Therefore, even if you adjust the speed of a motor, for example, to be exactly at a given speed for one load, when you change the load there will always be some residual error between the set point and the actual output.

To help solve these problems, circuits with more complex feedback are used. Fig. 12.6 shows a circuit which represents the different types of feedback commonly used. First note in this circuit that the output power amplifier is an added with four inputs. The current supplied to the summing point of the adder by the set point input produces the basic output drive current. The other three inputs do not supply any current unless there is a difference between the set point and the feedback voltage from the techometer. Amplifier I is another adder whose function is to compare the set point voltage with the feedback voltage from the tachometer. Let's assume the two input resistor, RI and R2, are voltage from the tachometer is positive, there wil be no net current through the feedback resistor of the amplifier if the two voltages are equal in magnitude. In other words, if the speed of the motor be zero, and amplifier 2, 3 and 4 ill contribute no current to the summing junction of the power amp.

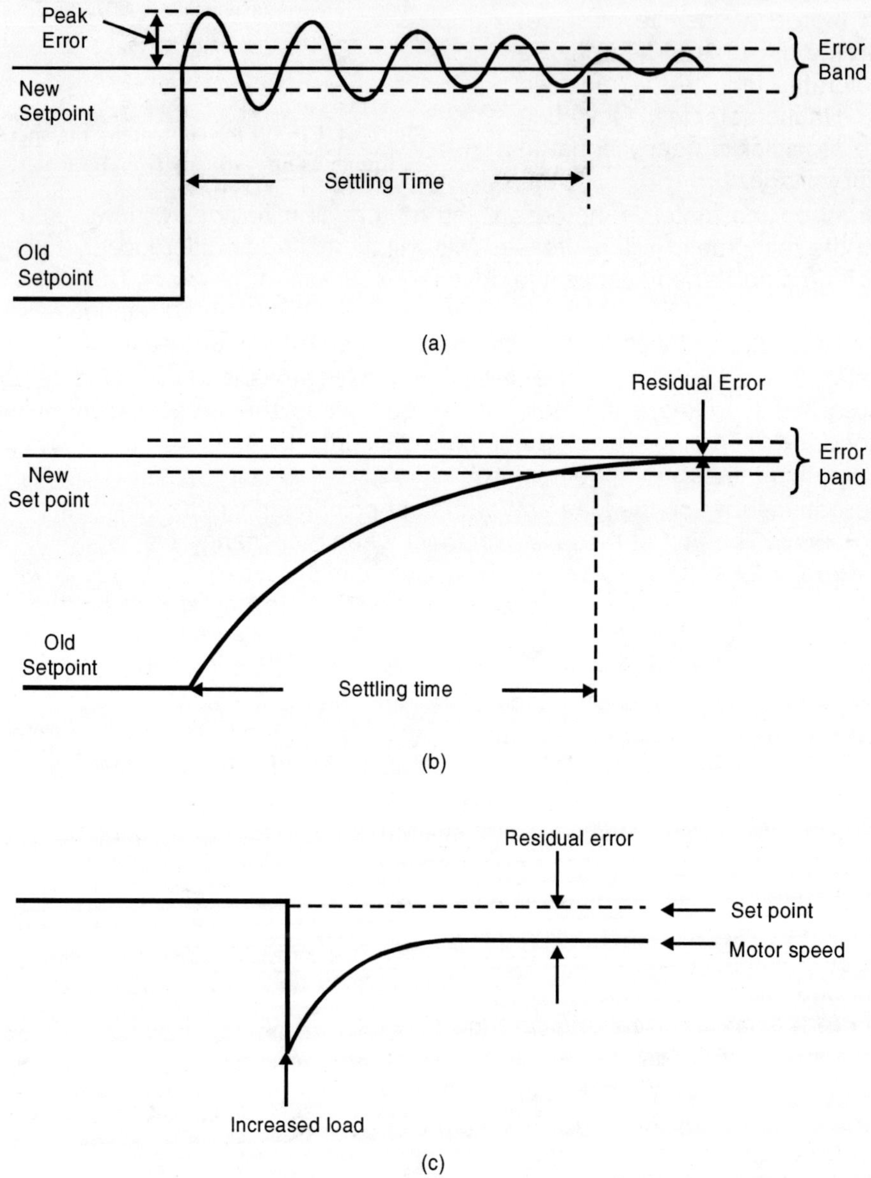

Fig. 12.5. When set point or load of a system is changed. (*a*) Undershoot, (*b*) Overshoot, (*c*) Load Change.

Now suppose that you add more load on the motor, slowing it down. The tachometer voltage is no longer equal to the set-point voltage, so amplifier I now has some output. This error signal on the output produces three types of feedback to the summing junction of the power amp.

Amplifier I produces simple dc deedback proportional to the difference between the set point and the tachometer output. This is exactly the same effect as the voltage divider on the tachometer output in Fig. 12.4. Proportional feedback, as this is called, will correct for most of the effect of the increased load, but, as we discussed before, there will always be some residual errors.

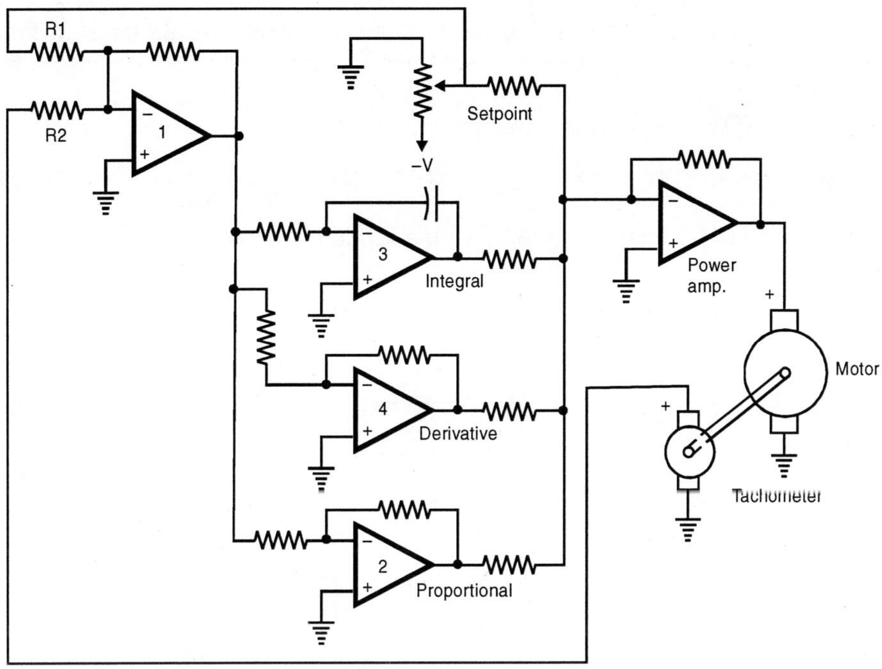

Fig. 12.6. Circuit showing proportional, integral and derivative feedback.

The cure for residual error is to use some integral feedback. Amplifiere 3 in Fig. 12.6 provides this type of feedback. Remember from a previous discussion that this circuit produces a ramp on its output whenever a voltage is applied to its input. For the example here, the integrator will ramp up or ramp down a long as there is any error signal present on Its Input. By rampIng up and down just a tiny bit about the set point. Too much integral feedback, however, will cause the output to oscillate up and down. Also, feedback only slowly affects the output because the error signal must be present for some time before the integrator has much output.

To improve the response time of the system, amplifier 4 in Fig. 12.6 supplies a third type of feedback called derivative feedback. Derivative feedback is a signal proportional to the rate of change of the error signal. If the load on the system is suddenly changed, the derivative amplifier circuit will give a quick shot of feedback to try to correct the error. When the error signal is first applied to the differentiator circuit, the capacitor in series with the input is not charged, so it acts like a short circuit. This initially lets a large current flow, so the amplifier has a sizeable output. As the capacitor charges, the current decreases, so the feedback from the differntiator decreases. The differentiator essentially gives the amplifier a quick pulse of feedback to help correct for the increased load. Too much derivative feedback can cause the system to overshoot and oscillate.

The point here is that by using a combination of some or all of these type of feedback, a given feedback controlled system can be adjusted for optimum response to changes in load or set point. Process control loops that use all three types of feedback are called proportional integral dertvative or PID control loops. Because process variables change much more slowly than the microsecond operation of a microcomputer, a microcomputer with some simple input and output circuitry can perform all the functions of the analog circuitry in Fig. 12.6 for several PID loops.

Figure 12.7 shows a block diagram of a microcomputer based process control system. Data acquisition systems convert the analog signals from various sensors to digital values that can be read in and processed by the microcomputer. A keyboard and display in the system allow the user to enter set-point values, to read the current values of process variables, and to issue commands. Relays, D/A converters, solenoid valves, and other actuators are used to control process variables under program direction. A programmable timer in the system determines the rate at which control loops are serviced.

12.4. MICROCOMPUTER BASED PROCESS-CONTROL SYSTEMS

Microcomputer based process-control systems range from a small programmable controller, which might be used to control a machine on a factory floor, to a large minicomputer refinery. To show you how these microcomputer based control systems work, here's an example system you can build and experiment with.

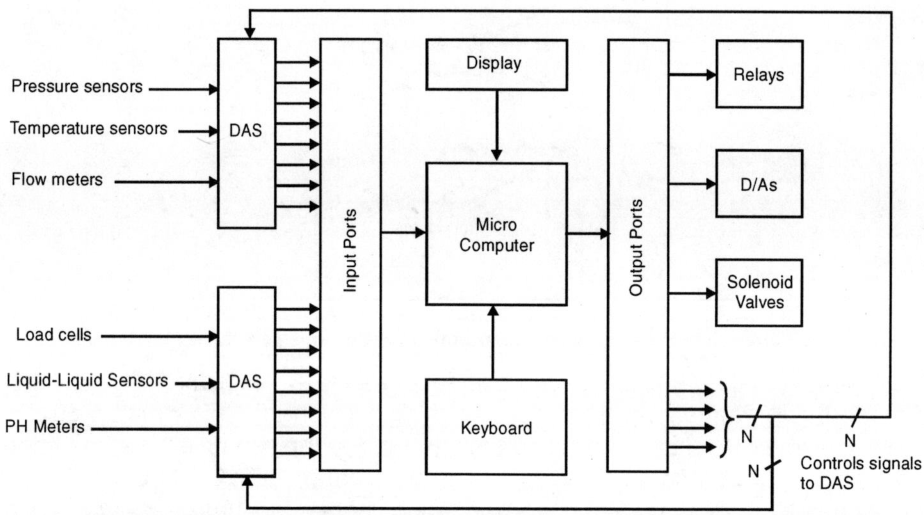

Fig. 12.7. Block Diagram of microprocessor based process control system.

12.5. HARDWARE FOR CONTROL SYSTEMS AND TEMPERATURE CONTROLLER

To build the hardware for this project, we started by adding an 8254 programmable timer and an 8259A priority interrupt controller to an SDK-86 board. The timer is initialized to produce 1-kHz clock ticks. The 8259A provides interrupt inputs for the clock tick interrupts and for keyboard interrupts. We built the actual temperature sensing and detecting circuitry on a separate prototyping board and connected it to some ports on the SDK-86 with a ribbon cable.

The temperature sensing element in the circuit is an LM35 precision Celsius temperature sensor. The voltage between the output pin and the ground pin of this device will be 0 V at 0° C and will increase by 10 mV for each increase of 1° C above that. The 300 kW resistor connecting the output of the LM35 to –15 V allows the output to go negative for temperatures below 0° C. (If you are operating with ±12.V supplies, use a 240 kW resistor.) This makes the circuit able to measure temperatures over the range of –55 to + 150° C. For our application here, we use only the positive part of the output range, but we though you might find this circuit useful for some of your other projects. An LM308 buffers and amplifiers the signal from the sensor by 2 so that the signal uses a greater part of the input range of the A/D converter. This improves the noise immunity and resolution.

The ADC0808 converter used here is an 8-input data acquisition system. You tell the device which input signal you want digitized with a 3-bit address you send to the ADC, ADB and ADA inputs. This 8-input device was chosen so that other control loops could be added later. Some Schmitt trigger inverters in a 74C14 are connected as an oscillator to produce a 300 kHz clock for the DAS. The voltage drop across an LM329 low drift zener is buffered by an LM308 amplifier to produce a V_{CC} and a V_{REF} of 5.12 V for the A/D converter. With this of 20 mV each. Since the temperature sensor signal is amplified by 2, each degree Celsius of temperature change will produce an output change of 20 mV, or one step on the A/D converter. This gives us a resolution of I° C. which is about equal to the typical accuracy of the sensor. The advantage o using V_{REF} as the V_{CC} for the device is that this voltage will not have the switching noise that the digital V_{CC} line has. The control inputs and data outputs of the A/D converter are simply connected to SDK-86 ports as shown.

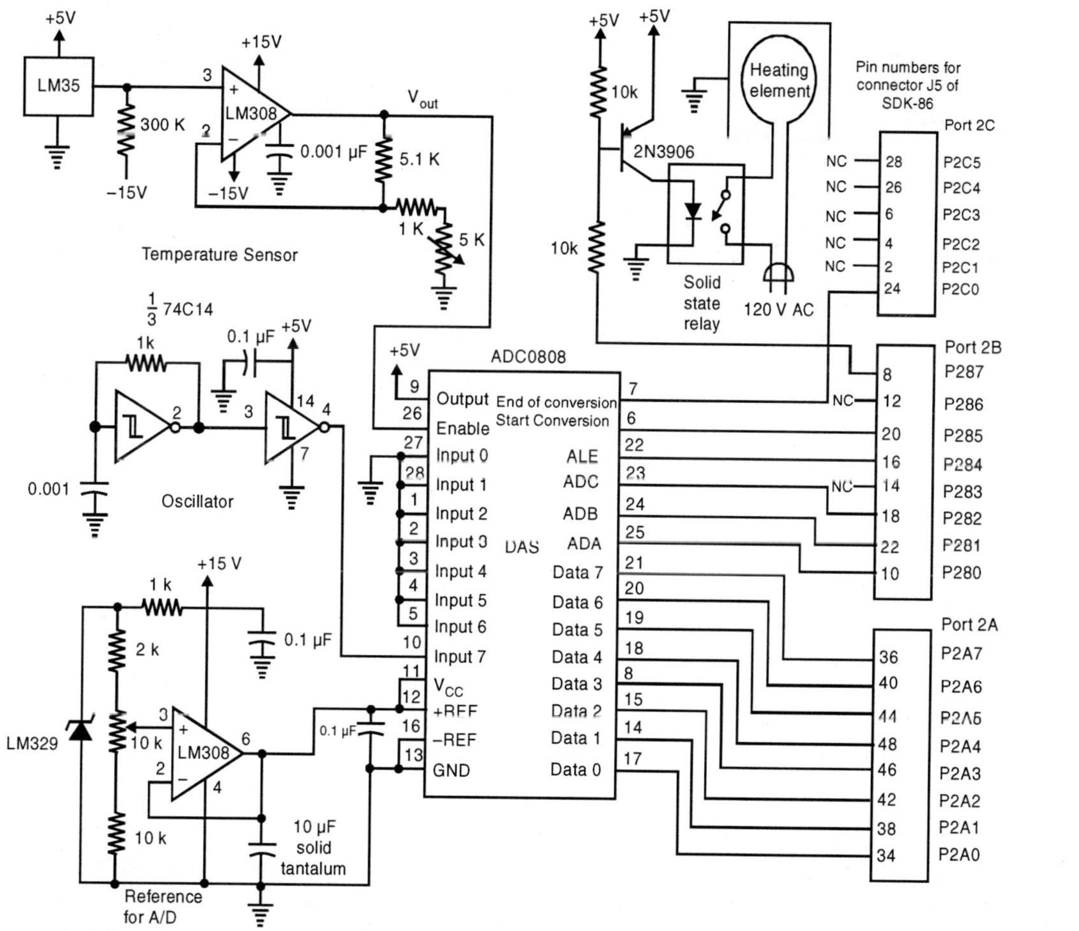

Fig. 12.8. Temperature sensing and Heater Control circuitry for microprocessor based controller.

Figure 12.9 shows the timing waveforms and parameters for the ADCO808. Note the sequence in which control signals must be sent to the device. The 3-bit address of the desired input channels is first sent to the multiplexer inputs. After at least 50 ns. the ALE input is sent high. After another 2.5 ms, the ALE input is sent high. After another 2 ms. The ALE input is sent high. After another 2.5 ms. The START CONVERSION input is sent high and then low. Then the ALE input is brought low again. When the END OF CONVERSION signal from the A/D converter is found to be high the 8- bit data value which represents the temperature can be read in.

To control the power delivered to the heater, we used a 25 A,0 V turn on, solid state relay. With this relay we can control heater. To control the amount of heat put out by the heater. we very the duty cycle of pulses sent to the relay.

For very low power applications, a D/A converter and a power amplifier could be used to drive the heater. However, in high power applications this is not very practical because the power amplifier may dissipate as much or more power than the load. For example, an amplifier intended to control 5000 W. The D/A converter approach has the added disadvantage that it cannot directly use the available ac lines voltage.

The driver transistor on the input of the solid-state relay supplies the drive for the relay, isolates the port pin from the relay, and holds the relay in the off position when the power is first turned on. Port pins, remember, are in floating state after a reset, so some method must be used to hold external circuitry in a known state until the port is initialized and the desired value is output to the port. Now that you know how the hardware is connected, we can explain the operation of the controller program.

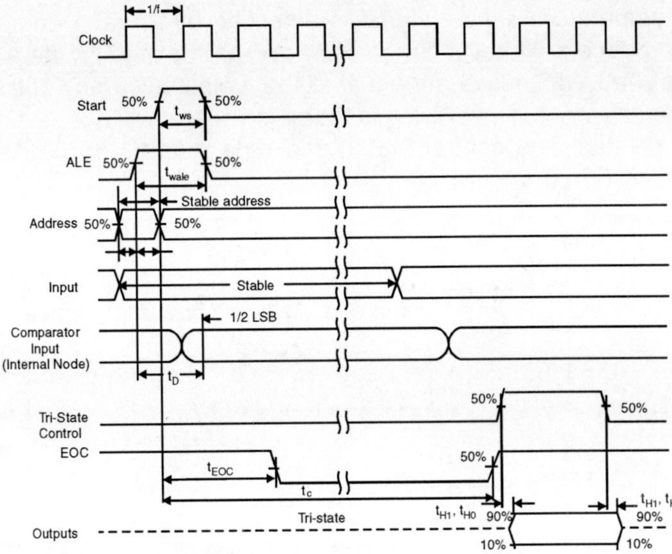

Fig. 12.9. Timing waveform for the ADC 0808 data acquisition system.

REVIEW QUESTIONS

1. Explain with the help of a diagram the smart scale operation using a microprocessor ?
2. Give the algorithm used for the smart scale calibration ?
3. Explain with the help of a Circuit diagram the Load cell interface circuitry and A/D converter for smart scale ?
4. Draw the Flow chart for smart scale program and explain in detail ?
5. Explain with the help of a Circuit diagram the controlling of the speed of dc motor using feedback from a tachometer ?
6. Explain in detail the microprocessor based industrial process control system ?
7. Draw the graph and explain in detail what happens when the set point or load of a system is changed ?
8. With the help of a Circuit diagram show the proportional, integral and derivative feedback ?
9. Draw block diagram of microprocessor based process control system and explain it in detail ?
10. Using a microprocessor draw a circuit for temperature controller and explain it in detail ?
11. Explain in detail the Timing waveform for the ADC 0808 data acquisition system ?

❑❑❑

CHAPTER 13

8051 MICROCONTROLLER

13.1. INTRODUCTION—IT IS ALL FARADAY'S FAULT

INSIDE THIS CHAPTER

In the beginning there was no electricity... People lived in an ignorant bliss, having no idea that energy was everywhere around them and that it could be tamed. It was good... And then came Faraday, and the ball started to roll, picking up the speed slowly but surely...

Eventually, first machines using the new energy source were built, and with these came the first people who could "understand the electricity". Time passed on and just about the time civilized world got used to this new concept and let the self-proclaimed experts toy with it, somebody came up with an idea to put electrons in a glass tube. At the first glance, it seemed like a good idea... but there was no turning back as electronics was born and the ball kept on rolling faster...

New science called for new scientists. White collars took the place of blue collars and there came the people who could "understand the electronics". And while the world stared in awe, the conspirators split up in two factions : those who specialized in software and those who specialized in hardware. Younger and more enthusiastic than their teachers, both groups carried on their efforts, but following the different paths. While the first advanced steadily to this day, those dedicated to hardware got carried away by their success and eventually discovered a transistor.

Up to that crucial moment, the situation could have been controlled to a certain extent - but the public underestimated the importance of this breakthrough and made a grave mistake! In a naive attempt to slow down the technological revolution, world market opened itself for a flood of electronic products, thus closing the circle. Market value of components rapidly decreased, electronics making its way to the younger populace. The ball started to fall freely...

Discovery of the integrated circuits and the processors only sped up the things, making the computers as available as ever. Computers started to upgrade themselves. The resulting deflation brought electronics new wave of devotees, making it omnipresent nowadays. Yet another circle closed - computers fell into the hands of the masses and their rule supreme had begun...

While this drama took place, nameless hobbyists and professionals across the world kept working on their projects, although still divided into two large factions.

And then somebody came up with another great idea : why not create an all purpose component ? A cheap and universal integrated circuit that could be programmed and used in any device, anywhere. Technology was up to the task, there was an interested market... why not ? Soul and body were united into one powerful MICROCONTROLLER.

13.1.1. What are Microcontrollers and What are They for ?

As with everything that is good, this powerful component is actually very simple in its essence. It was built using the tested solutions and ingredients by the following recipe :

1. Processor was removed from the simplest of computers to be used as the "brain" for the upcoming system.
2. Depending on the manufacturers' taste, some memory was added, a few A/D converters, timers, I/O communication lines, etc.
3. It was all placed in a standard casing.
4. Simple software that everybody could learn was developed for controlling the thing.

A variety of microcontrollers has been constructed in this manner, becoming a man's subtle yet indispensable companion in everyday life. Their incredible simplicity and flexibility has earned our trust awhile ago - if you have an idea of utilizing a microcontroller for the most trivial of tasks, know that somebody has already been there...

There are three decisive facts responsible for such a success of microcontrollers :

- Their powerful, cleverly chosen electronics is able to control a variety of processes and devices (industrial automatics, voltage, temperature, engines, etc.) independently or by means of I/O instruments such as switches, buttons, sensors, LCD screens, relays...
- Their low cost makes them suitable for installing in places which attracted no such interest in the past. This is the fact accountable for today's market being swamped with cheap automatons and "intelligent" toys.
- Writing and loading a program into microcontroller requires practically no previous schooling. All that is required is : any PC (software is very friendly and intuitive) and one simple device (programmer) for loading a written program into microcontroller.

So if you are into electronics, you will undoubtedly want to master the great potential of microcontrollers and put it to good use.

13.2. HOW DOES MICROCONTROLLER WORK

Although there are plenty of different microcontrollers and programs available, they all share a lot in common between them. This means that if you master one model, you will be able to handle them all eventually.

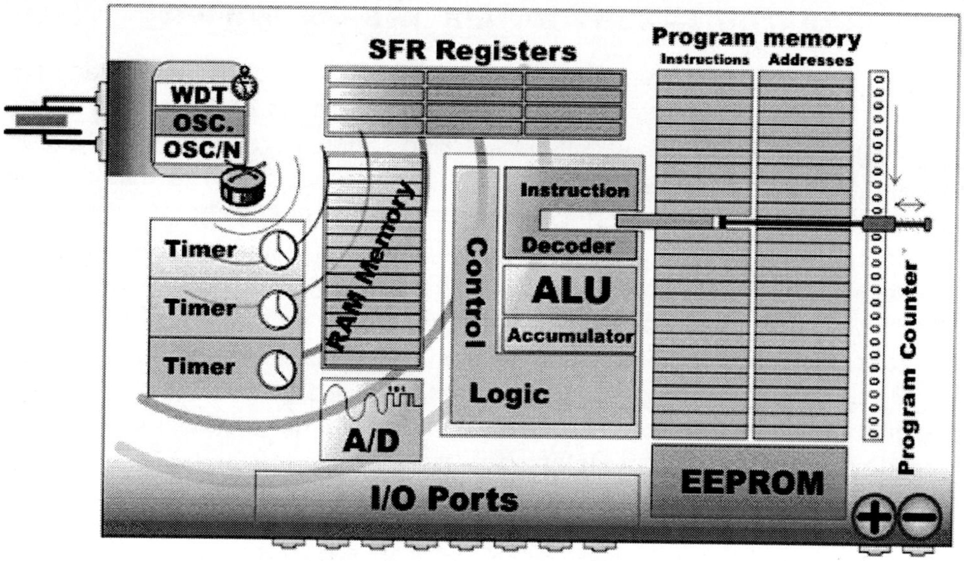

Fig. 13.1. General Microcontroller Block Diagram.

Typical scenario goes something like this :

1. Power is off and all is still... Program is loaded, everything is set, with no indications of what is about to take place...
2. As soon as the power is on, it all happens too quickly to follow! First one to register the change is the control logic. It halts all the circuits except the quartz oscillator. First few milliseconds pass in hurried preparations and loading of parasitic capacitances...
3. As voltage reaches its maximum, oscillator frequency stabilizes. SFR registers are loaded with bits indicating the states of the subsystems, and all the pins are designated as input. Impulse sequence starts dictating the pace and the electronics is now fully operational. From this point on, time is measured in microseconds and nanoseconds.

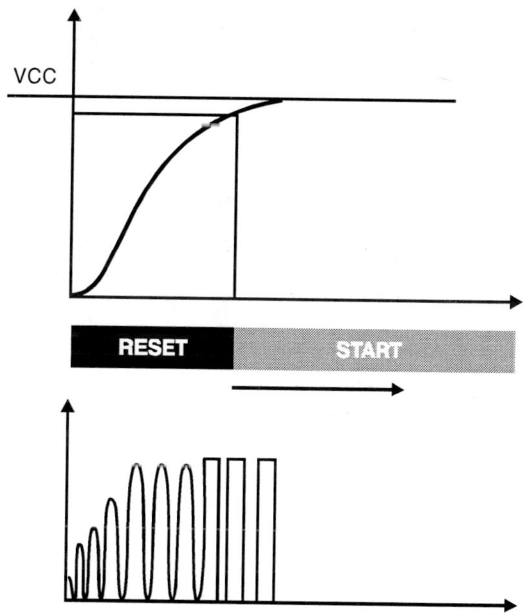

Fig. 13.2. Voltage Graph at Reset operation.

4. Program counter is reset to zero address of the program memory. Instruction is sent from this address to the instruction decoder where it is interpreted and executed promptly.
5. Program counter is incremented by one as the whole process plays out again... (several millions of times per second).

Obviously, it all works at tremendous speeds, and is also pretty simple. But it would not be especially useful if it was not for additional systems which round out the MCU.

13.3. INTRODUCTION—8051 UNDER THE MAGNIFIER

A struggle has been going on between MCU manufacturers for quite a long time, each of them trying to best respond to the ever-increasing demands of the market. Every couple of days there is a brand new chip available, working at higher frequency, with more memory or with better A/D converters. And yet, a closer look to their interior reveals the same or at least very similar structural design referred to as "8051 compatibility". What is it all about ?

The story began in the 80's when Intel introduced their microcontroller family MCS 8051 to the market. Although this family had quite limited capabilities by today's notions, it quickly captivated the world and became the standard for what is today understood as 'microcontroller'. The most significant cause for such a success can be found in the cleverly chosen configuration which can satisfy a diversity of needs, yet allowing for continuous upgrades (in form of new controllers). In a brief period of time, a decent amount of software has been developed for 8051, making further changes of the hardware core simply uneconomical. Consequently, there is a variety of MCUs available today, basically just the upgraded 8051 models. What exactly makes this microcontroller so special and universal that it is still manufactured by all the major companies, just under a different label ?

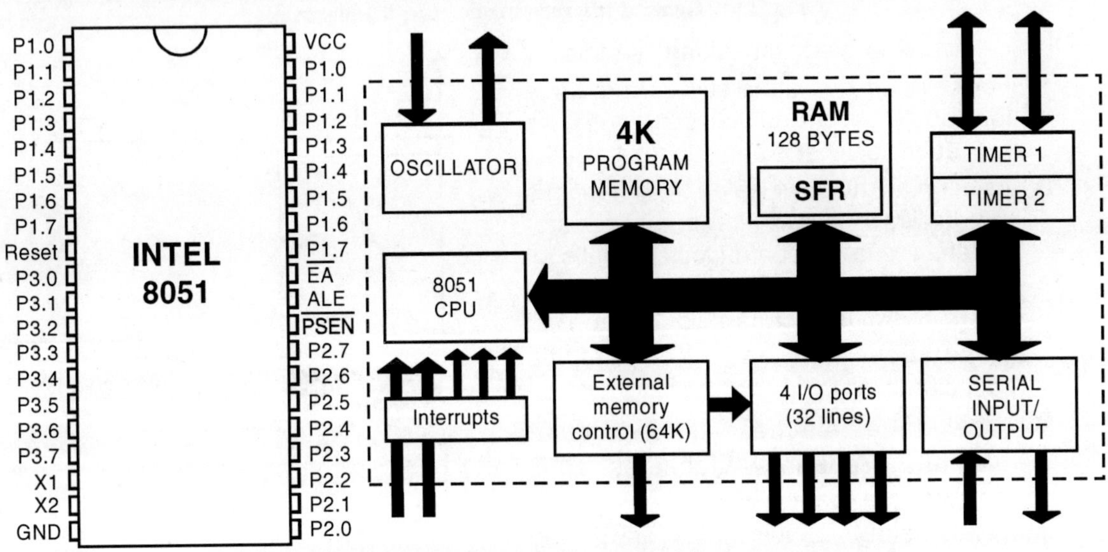

Fig. 13.3. Simplified Bock diagram of 8051 and its Pin Configuration.

As can be seen on the image above, there is nothing particularly remarkable about MCU 8051 :

- ➢ 4 kilobytes of ROM is neither too little nor too much.
- ➢ 128 bytes of RAM (SFR registers included) can satisfy the basic needs, but is not really astounding.
- ➢ 4 ports totaling 32 I/O lines, are usually sufficient for connecting to the environs and are by no means luxury.

Obviously, 8051 configuration is intended to satisfy the needs of programmers developing the controlling devices and instruments. This is one part of its key to success : there is nothing missing, yet there is no lavishness; it is meant for the average user. The other clue can be found in the organization of RAM, Central Processor Unit (CPU), and ports - all of which maximally utilize the available resources and allow further upgrades.

13.3.1. Internal Architecture for 8051

The internal architecture of 8051 is presented in the Fig. 13.4.

The functional description of each block is presented briefly below :

(1) Accumulator (ACC) : The accumulator register acts as an operand register, in case of some instructions. This may either be implicit or specified in the instruction. The ACC register has been allotted an address in the ON chip special register bank.

Block Diagram

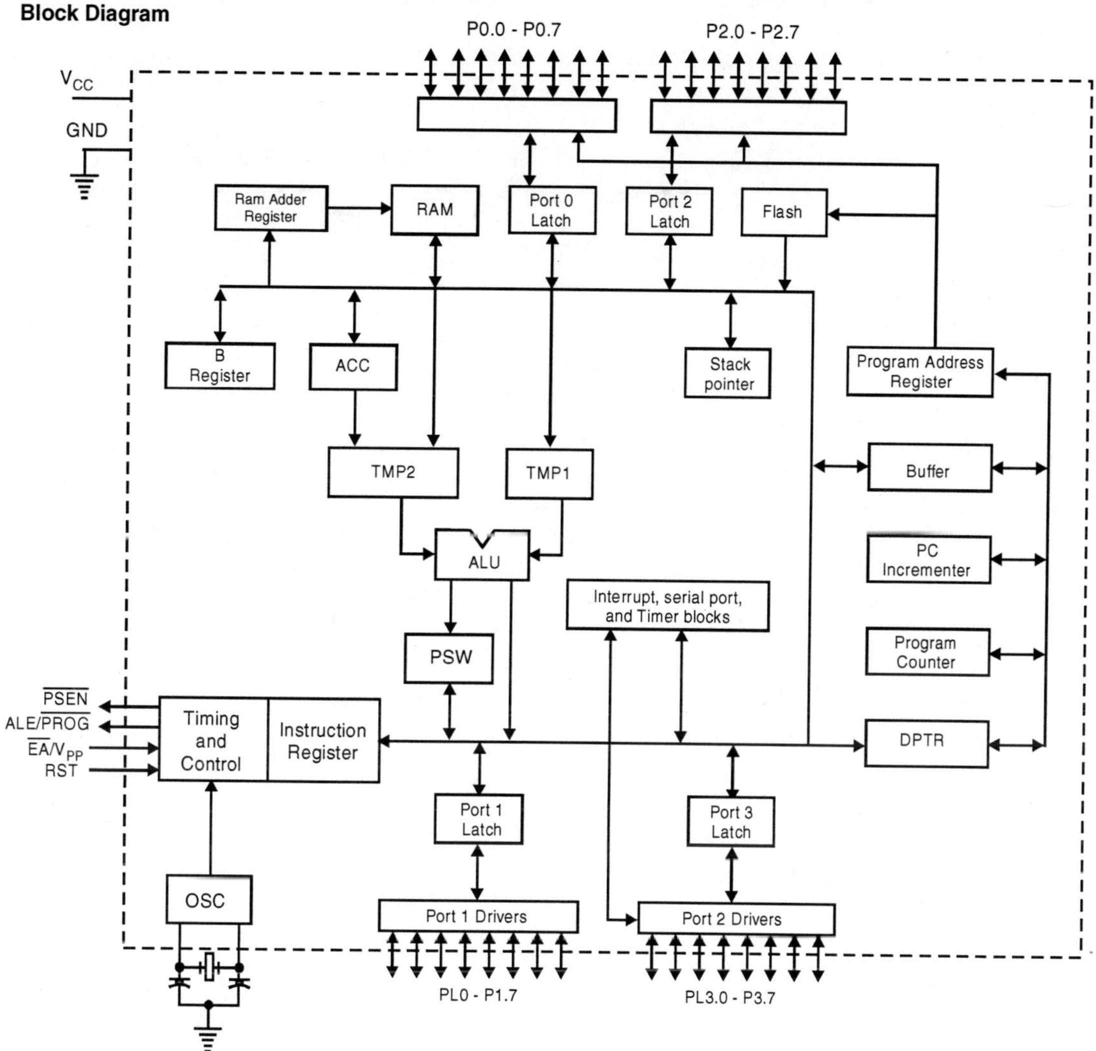

Fig. 13.4. Internal architecture of 8051.

(2) B Register : This register is used to store one of the operands for multiply and divide instructions. In other instructions, it may just be used as a scratch pad. This register is considered as special function register.

(3) Program Status Word (PSW) : This set of flags contains the status information and is considered as one of the special function registers.

(4) Stack Pointer (SP) : This 8-bit wide register is incremented before the data is stored onto the stack using push or call instructions. This register contains 8-bit stack top address.

Stack may be defined anywhere in the on chip 128 byte RAM. After reset the SP register is initialized to 07. After each write to stack operation the 8-bit contents of the operands are stored on to the stack, after incrementing the SP register by one. Thus if SP contains 07H, the forthcoming PUSH operation will store the data at address 08H in the internal RAM. The SP content will be incremented to 08. The 8051 stack is not a top down data structure, like other Intel processors. This register has also been allotted an address in the special function register bank.

(5) **Data Pointer (DTPR) :** This 16-bit register contains a higher byte (DPH) and the lower bye (DPL) of a 16 bit external data RAM address. It is accessed as a 16-bit register or two 8-bit registers as specified above. It has been allotted two addresses in the special function register bank for its two bytes DPH and DPL.

(6) **PORT 0 to 3 Latches and Drivers** : These four latches and drivers pairs are allotted to each of the four on chip I/O ports. These latches have been allotted addresses in the special function register band. Using the allotted addresses the user can communicate with these ports. These are identified as P0, P1, P2, P3.

(7) **Serial Data Buffer** : The serial data buffer internally contains two independent registers, one of them is a transmit buffer which is necessarily a parallel in serial out register. The other is called recessive buffer which is a serial in parallel out register. Loading a byte to the transmit buffer initiates serial transmission of that byte. The serial data buffer is identified as SBUF, and is one of the special function registers. If a byte is written as SBUF it initiates serial transmission and if the SBUF is read, it reads serial data.

(8) **Timer Registers** :- These two 16 bit registers can be accessed as their lower and upper bytes. For example TLO represents the lower byte of the timing register 0, while THO represents the higher byte of the timing register 0. Similarly TL1 and TH1 represents lower and higher bytes of the timing register 1. All these registers can be accessed using the four addresses allotted to them which lie in the special function registers SFR address range, i.e. 80h to FF.

(9) **Control Registers** : The special function registers IP, IE, TMOD, TCON, SCON and PCON contain control and status information for interrupts, timers/counters and serial ports. All these registers have been allotted addresses in the special function register bank of 8051.

(10) **Timing and Control Unit** : This unit derives all the necessary timing and control signals required for the internal operations of the circuit. It also derives the control signals required for controlling the external system bus.

(11) **Oscillator** : This circuit generates the basic timing clock signal for the operation of the circuit using crystal oscillator.

(12) **Instruction Register** : This register decodes the opcode of an instruction to be executed and gives information to the timing and control unit to generate necessary signal for the execution of the instruction.

(13) **EPROM and Program Address Registers** : These blocks provide an on chip EPROM and a mechanism to internally address it.

(14) **RAM and RAM Address Register** : These blocks provide internal 128 bytes of RAM and a mechanism to address it internally.

(15) **ALU :** The arithmetic and logic unit performs 8-byte arithmetic and logical operations over the operands held by the temporary registers TMP1 and TMP2. Users cannot access these temporary registers.

(16) SFR Register Bank : This is a set of special function registers which can be addressed using their respective addresses which lie in the range 80H to FFH. Finally, the interrupt, serial port and timer units control and perform their specific functions under the control of the timing and control unit.

13.3.2. Pins on the Case

- **1–8 : Port 1**; Each of these pins can be used as either input or output according to your needs. Also, pins 1 and 2 (P1.0 and P1.1) have special functions associated with Timer 2.

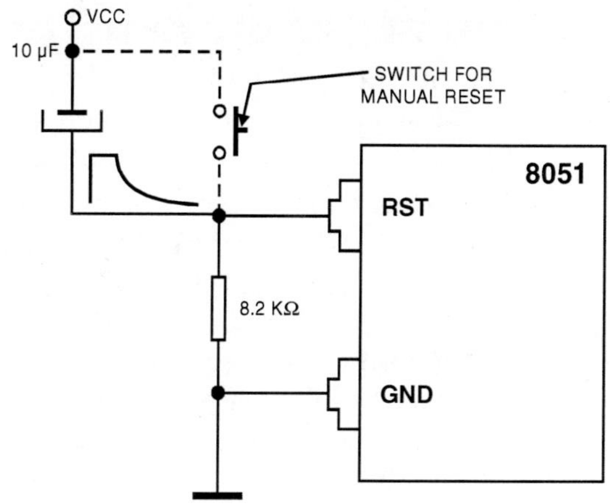

Fig. 13.5. Generation of Reset Signal.

- **9 : Reset Signal**; high logical state on this input halts the MCU and clears all the registers. Bringing this pin back to logical state zero starts the program anew as if the power had just been turned on. In another words, positive voltage impulse on this pin resets the MCU. Depending on the device's purpose and environs, this pin is usually connected to the push-button, reset-upon-start circuit or a brown out reset circuit (covered in the previous chapter). The image shows one simple circuit for safe reset upon starting the controller. It is utilized in situations when power fails to reach its optimal voltage.
- **10-17 : Port 3**; As with Port 1, each of these pins can be used as universal input or output. However, each pin of Port 3 has an alternative function :
  - Pin 10 : RXD - serial input for asynchronous communication or serial output for synchronous communication.
  - Pin 11 : TXD - serial output for asynchronous communication or clock output for synchronous communication
  - Pin 12 : INT0 - input for interrupt 0
  - Pin 13 : INT1 - input for interrupt 1
  - Pin 14 : T0 - clock input of counter 0
  - Pin 15 : T1 - clock input of counter 1
  - Pin 16 : WR - signal for writing to external (add-on) RAM memory
  - Pin 17 : RD - signal for reading from external RAM memory
- **18-19 : X2 and X1**; Input and output of internal oscillator. Quartz crystal controlling the frequency commonly connects to these pins. Capacitances within the oscillator mechanism (see the image) are not critical and are normally about 30 pF. Instead of a quartz crystal, miniature ceramic resonators can be used for dictating the pace. In that case, manufacturers recommend using somewhat higher capacitances (about 47 pF). New MCUs work at frequencies from 0 Hz to 50 MHz+.

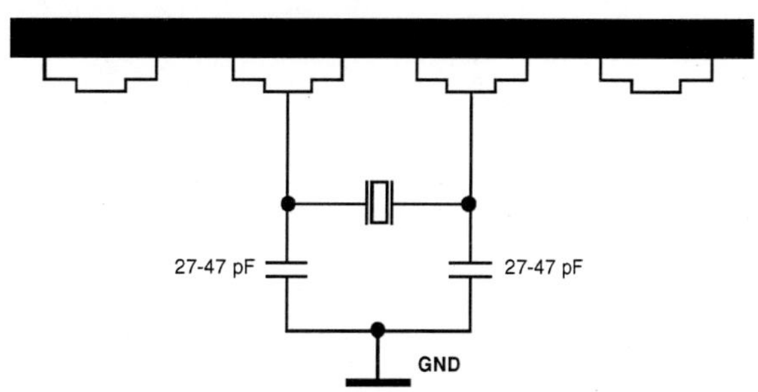

Fig. 13.6. X2 and X1 generation.

- **20 : GND**; Ground

- **21- 28 : Port 2**; If external memory is not present, pins of Port 2 act as universal input/output. If external memory is present, this is the location of the higher address byte, i.e. addresses A8 – A14. It is important to note that in cases when not all the 8 bits are used for addressing the memory (i.e. memory is smaller than 64 kB), the rest of the unused bits are not available as input/output.
- **29 : PSEN**; MCU activates this bit (brings to low state) upon each reading of byte (instruction) from program memory. If external ROM is used for storing the program, PSEN is directly connected to its control pins.
- **30 : ALE**; Before each reading of the external memory, MCU sends the lower byte of the address register (addresses A0 – A7) to port P0 and activates the output ALE. External register (74HCT373 or 74HCT375 circuits are common), memorizes the state of port P0 upon receiving a signal from ALE pin, and uses it as part of the address for memory chip. During the second part of the mechanical MCU cycle, signal on ALE is off, and port P0 is used as *Data Bus.* In this way, by adding only one cheap integrated circuit, data from port can be multiplexed and the port simultaneously used for transferring both addresses and data.
- **31 : EA**; Bringing this pin to the logical state zero (mass) designates the ports P2 and P3 for transferring addresses regardless of the presence of the internal memory. This means that even if there is a program loaded in the MCU it will not be executed, but the one from the external ROM will be used instead. Conversely, bringing the pin to the high logical state causes the controller to use both memories, first the internal, and then the external (if present).
- **32-39 : Port 0**; Similar to Port 2, pins of Port 0 can be used as universal input/ output, if external memory is not used. If external memory is used, P0 behaves as address output (A0 – A7) when ALE pin is at high logical level, or as data output (Data Bus) when ALE pin is at low logical level.
- **40 : VCC**; Power +5 V.

13.4. INPUT–OUTPUT (I/O) PORTS

Every MCU from 8051 family has 4 I/O ports of 8 bits each. This provides the user with 32 I/O lines for connecting MCU to the environs. Unlike the case with other controllers, there is no specific SFR register for designating pins as input or output. Instead, the port itself is in charge : 0=output, 1=input. If particular pin on the case is needed as output, the appropriate bit of I/O port should be cleared. This will generate 0 V on the specified controller pin. Similarly, if particular pin on the case is needed as input, the appropriate bit of I/O port should be set. This will designate the pin as input, generating +5 V as a side effect (as with every TTL input).

Port 0

Port 0 has two fold role : if external memory is used, it contains the lower address byte (addresses A0-A7), otherwise all bits of the port are either input or output. Another feature of this port comes to play when it has been designated as output. Unlike other ports, Port 0 lacks the "pull up" resistor (resistor with +5 V on one end). This seemingly insignificant change has the following consequences :

- When designated as input, pin of Port 0 acts as high impedance offering the infinite input resistance with no "inner" voltage.

➢ When designated as output, pin acts as "open drain". Clearing a port bit grounds the appropriate pin on the case (0 V). Setting a port bit makes the pin act as high impedance. Therefore, to get positive logic (5 V) at output, external "pull up" resistor needs to be added for connecting the pin to the positive pole.

Therefore, to get one (5 V) on the output, external "pull up" resistor needs to be added for connecting the pin to the positive pole.

Port 1

This is "true" I/O port, devoid of dual function characteristic for Port 0. Having the "pull up" resistor, Port 1 is fully compatible with TTL circuits.

Port 2

When using external memory, this port contains the higher address byte (addresses A8–A15), similar to Port 0. Otherwise, it can be used as universal I/O port.

Port 3

Beside its role as universal I/O port, each pin of Port 3 has an alternate function. In order to use one of these functions, the pin in question has to be designated as input, i.e. the appropriate bit of register P3 needs to be set. From a hardware standpoint, Port 3 is similar to Port 0.

As can be seen from the individual descriptions of the ports, they all share highly similar structure. However, you need to consider which task should be assigned to which port. For example : if utilizing port as output with high level (5 V), avoid using Port 0 as its pins cannot produce high logical level without an additional resistor connected to +5 V. If using other port to a same end, bear in mind that built-in resistors have relatively high values, producing the currents limited to few hundreds of amperes as pin output.

13.4.1. Memory Under the Magnifier

During the runtime, microcontroller uses two different types of memory : one for holding the program being executed (ROM memory), and the other for temporary storage of data and auxiliary variables (RAM memory). Depending on the particular model from 8051 family, this is usually few kilobytes of ROM and 128/256 bytes of RAM. This amount is built-in and is sufficient for common tasks performed "independently" by the MCU. However, 8051 can address up to 64KB of external memory. These can be separate memory blocks, (separate RAM chip and ROM chip) totaling 128KB of memory on MCU which is a real programming goody.

ROM memory

First models from 8051 family lacked the internal program memory, but it could be added externally in a form of a separate chip. These MCUs can be recognized by their mark which begins with 803 (e.g. 8031 or 8032). New models have built-in ROM, although there are substantial variations. With some models internal memory cannot be programmed directly by the user. Instead, the user needs to proceed the program to the manufacturer, so that the MCU can be programmed (masked) appropriately in the process of fabrication. Obviously, this option is cost-effective only for large series. Fortunately, there are MCU models ideal for experimentation and small specialized series. Many manufacturers deliver controllers that can be programmed directly by the user. These come in a ceramic case with an opening (EPROM version) or in a plastic case without an opening (EEPROM version). This book deals with one of the latter models that can be programmed via simple programmer, even if the chip has already been mounted to the designated device.

RAM memory

As previously stated, RAM is used for storing temporary data and auxiliary results generated during the runtime. Apart from that, RAM comprises a number of registers : hardware counters and timers, I/O ports, buffer for serial connection, etc. With older versions, RAM spanned 256 locations, while new models feature additional 128 registers. First 256 memory locations form the basis of RAM (addresses 0 – FFh) of every 8051 MCU. Locations that are available to the user span addresses from 0 to 7Fh, i.e. first 128 registers, and this part of RAM is split into several blocks as can be seen in the image below.

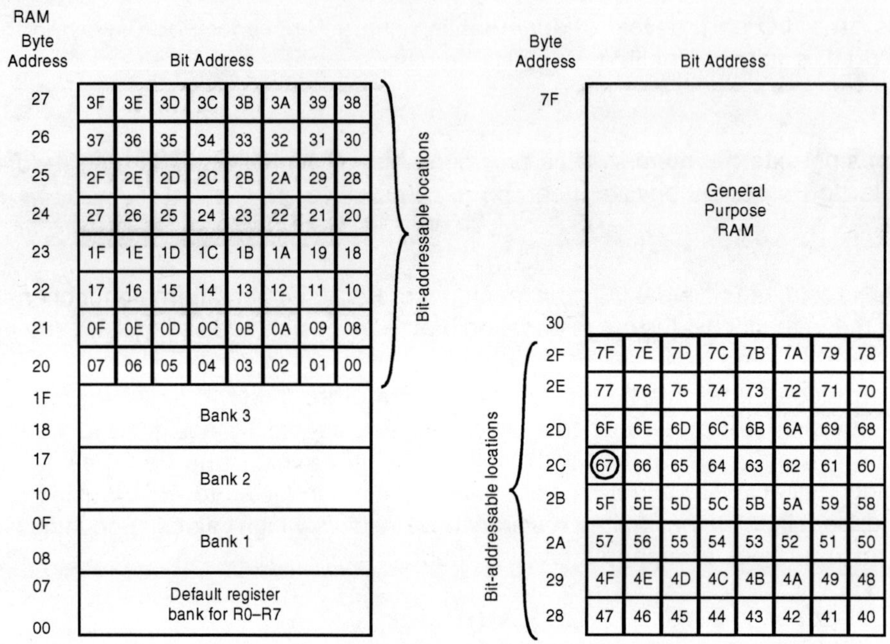

Fig. 13.7. Bit addressable RAM

- First block comprises 4 "banks" of 8 registers each, marked as R0 - R7. To address these, the parent bank has to be selected.
- Second memory block (range 20h – 2Fh) is bit-addressable, meaning that every belonging bit has its own address (0 to 7Fh). Since the block comprises 16 of these registers, there is a total of 128 addressable bits. (Bit 0 of byte 20h has bit address 0, while bit 7 of byte 2Fh has bit address 7Fh).
- Third is the group of available registers at addresses 2Fh – 7Fh (total of 80 locations) without special features or a preset purpose.

Extra Memory Block

To satisfy the programmers' ever-increasing demands for RAM, latest 8051 models were added an extra memory block of 128 locations. But it is not all that simple... The problem lies in the fact that the electronics which addresses RAM employs 1 byte (8 bits), reaching only the first 256 locations. Therefore, a little trick had to be applied in order to keep the existing 8-bit architecture for the sake of compatibility with older models. The idea is to make the additional memory block share the addresses with the existent locations intended for SFR registers (80h - FFh). For distinguishing these two physically separate memory areas, different methods of addressing are used : if SFR registers are in question, direct addressing is used; for extra RAM locations, indirect addressing is used.

Memory Expanding

In case the built-in amount of memory (either RAM or ROM) is not sufficient for your needs, there is always an option of adding two external 64 KB memory chips. When added, they are addressed and accessed via I/O ports P2 and P3. From user's point of view it's all very simple, because if properly connected most of the job is carried out automatically by MCU.

8051 MCU has two separate read signals, RD# (P3.7) and PSEN#. The first one is active when reading byte from the external data memory (RAM), and the second one is active when reading byte from the external program memory (ROM). Both signals are active on low logical level. The following image shows a typical scheme for such expansion using separate chips for RAM and ROM, known as *Harvard architecture.*

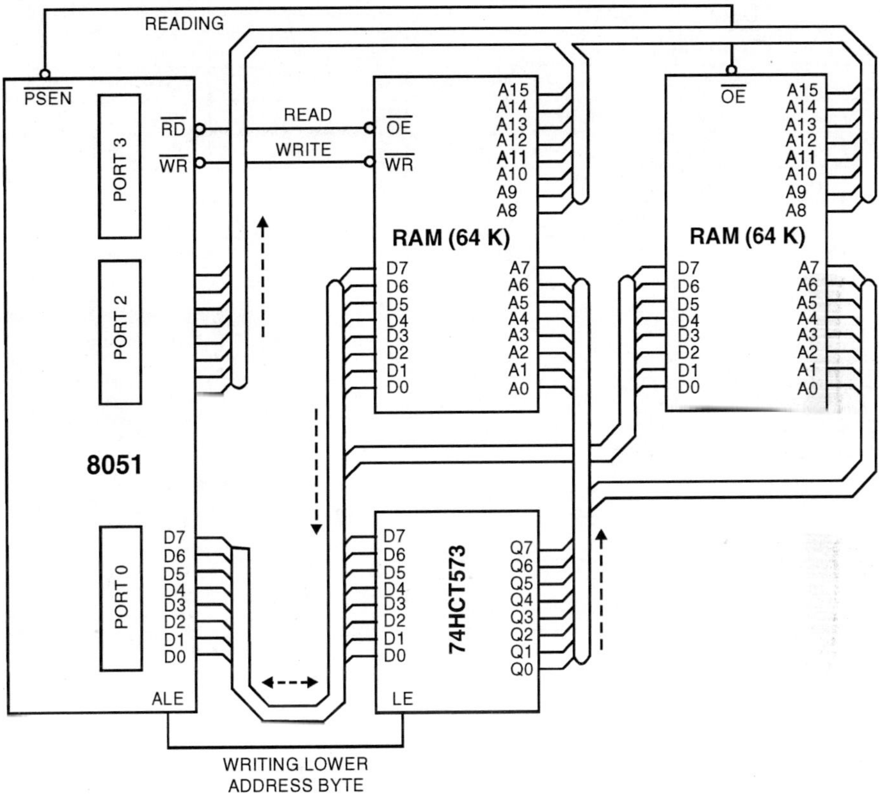

Fig. 13.8. Memory Expansion of 8051.

Memory can be also mapped as a single block, functioning as both data memory and program memory simultaneously (only one memory chip is used). This approach is known as *Von Neumann architecture*. To be able to read the same block using RD# or PSEN#, these two signals were combined via logical AND. In this way, output of AND circuit is low if any of the two inputs is low.

Using the Harvard architecture effectively doubles MCU memory, but that's not the only advantage offered by the method. Keeping the program code separated from the data makes the controller more reliable since there is no writing to the program memory.

13.4.2. SFR Registers (Special Function Registers)

SFR registers can be seen as a sort of control panel for managing and monitoring the microcontroller. Every register and each of the belonging bits has its name, specified address in RAM and strictly defined role (e.g. controlling the timer, interrupt, serial connection, etc.). Although there are 128 available memory slots for allocating SFR registers, the basic core shared by 8051 MCUs has but 22 registers. The rest has been left open intentionally to allow future upgrades while retaining the compatibility with earlier models. This fact makes possible to use programs developed for obsolete models long ago.

Addr. (Hex.)	Mark	Full name	Addr. (Hex.)	Mark	Full name
80	P0	Port 0	8D	TH1	Timer/Counter1 High Byte
81	SP	Stack Pointer	90	P1	Port 1
	DPTR	Data Pointer	98	SCON	Serial Port Control
82	DPL	Data Low Pointer	99	SBUF	Serial Data Port
83	DPH	Data High Pointer	A0	P2	Port 2
87	PCON	Power Control	A8	IE	Interrupt Enable
88	TCON	Timer/Counter Control	B0	P3	Port 3
89	TMOD	Timer/Counter Mode Control	B8	IP	Interrupt Priority Control
8A	TL0	Timer/Counter 0 Low Byte	D0	PSW	Progam Status Word
8B	TL1	Timer/Counter 1 Low Byte	E0	ACC(A)	Accumulator
8C	TH0	Timer/Counter 0 High Byte	F0	B	B Register

Fig. 13.9. SFR Register set.

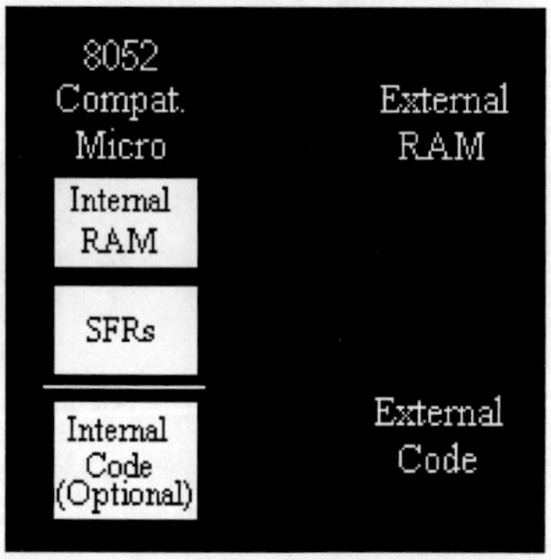

Fig. 13.10. On chip and Off chip memory for 8051.

The 8051 has three very general types of memory. To effectively program the 8051 it is necessary to have a basic understanding of these memory types.

The memory types are illustrated in the following graphic. They are : On-Chip Memory, External Code Memory, and External RAM.

On-Chip Memory refers to any memory (Code, RAM, or other) that physically exists on the microcontroller itself. On-chip memory can be of several types, but we'll get into that shortly.

External Code Memory is code (or program) memory that resides off-chip. This is often in the form of an external EPROM.

External RAM is RAM memory that resides off-chip. This is often in the form of standard static RAM or flash RAM.

Code Memory

Code memory is the memory that holds the actual 8051 program that is to be run. This memory is limited to 64K and comes in many shapes and sizes : Code memory may be found *on-chip*, either burned into the microcontroller as ROM or EPROM. Code may also be stored completely *off-chip* in an external ROM or, more commonly, an external EPROM. Flash RAM is also another popular method of storing a program. Various combinations of these memory types may also be used—that is to say, it is possible to have 4 K of code memory *on-chip* and 64k of code memory *off-chip* in an EPROM.

When the program is stored on-chip the 64 K maximum is often reduced to 4 k, 8 k, or 16 k. This varies depending on the version of the chip that is being used. Each version offers specific capabilities and one of the distinguishing factors from chip to chip is how much ROM/EPROM space the chip has.

However, code memory is most commonly implemented as off-chip EPROM. This is especially true in low-cost development systems and in systems developed by students.

TIPS

Since code memory is restricted to 64 K, 8051 programs are limited to 64 K. Some assemblers and compilers offer ways to get around this limit when used with specially wired hardware. However, without such special compilers and hardware, programs are limited to 64 K.

External RAM

As an obvious opposite of *Internal RAM*, the 8051 also supports what is called *External RAM*.

As the name suggests, External RAM is any random access memory which is found *off-chip*. Since the memory is off-chip it is not as flexible in terms of accessing, and is also slower. For example, to increment an Internal RAM location by 1 requires only 1 instruction and 1 instruction cycle. To increment a 1-byte value stored in External RAM requires 4 instructions and 7 instruction cycles. In this case, external memory is 7 times slower!

What External RAM loses in speed and flexibility it gains in quantity. While Internal RAM is limited to 128 bytes (256 bytes with an 8052), the 8051 supports External RAM up to 64K.

TIPS

The 8051 may only address 64k of RAM. To expand RAM beyond this limit requires programming and hardware tricks. You may have to do this "by hand" since many compilers and assemblers, while providing support for programs in excess of 64k, do not support more than 64k of RAM. This is rather strange since it has been my experience that programs can usually fit in 64k but often RAM is what is lacking. Thus if you need more than 64k of RAM, check to see if your compiler supports it— but if it doesn't, be prepared to do it by hand.

13.4.3. On-Chip Memory

As mentioned at the beginning of this chapter, the 8051 includes a certain amount of on-chip memory. On-chip memory is really one of two types : Internal RAM and Special Function Register (SFR) memory. The layout of the 8051's internal memory is presented in the following memory map shown in Fig. 13.11.

As is illustrated in this map, the 8051 has a bank of 128 bytes of *Internal RAM*. This Internal RAM is found *on-chip* on the 8051 so it is the fastest RAM available, and it is also the most flexible in terms of reading, writing, and modifying its contents. Internal RAM is volatile, so when the 8051 is reset this memory is cleared.

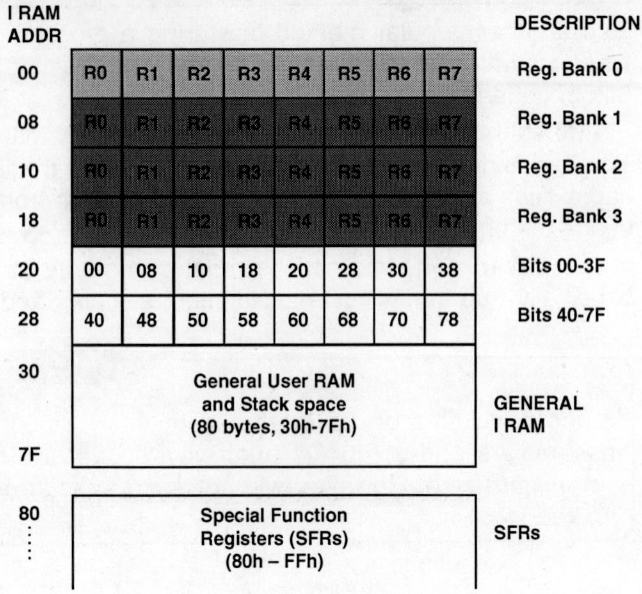

Fig. 13.11. 8051's internal memory.

The 128 bytes of internal ram is subdivided as shown on the memory map. The first 8 bytes (00h - 07h) are "register bank 0". By manipulating certain SFRs, a program may choose to use register banks 1, 2, or 3. These alternative register banks are located in internal RAM in addresses 08h through 1Fh. We'll discuss "register banks" more in a later chapter. For now it is sufficient to know that they "live" and are part of internal RAM.

Bit Memory also lives and is part of internal RAM. We'll talk more about bit memory very shortly, but for now just keep in mind that bit memory actually resides in internal RAM, from addresses 20h through 2Fh.

The 80 bytes remaining of Internal RAM, from addresses 30h through 7Fh, may be used by user variables that need to be accessed frequently or at high-speed. This area is also utilized by the microcontroller as a storage area for the operating *stack*. This fact severely limits the 8051s stack since, as illustrated in the memory map, the area reserved for the stack is only 80 bytes—and usually it is less since this 80 bytes has to be shared between the stack and user variables.

13.4.4. Register Banks

The 8051 uses 8 "R" registers which are used in many of its instructions. These "R" registers are numbered from 0 through 7 (R0, R1, R2, R3, R4, R5, R6, and R7). These registers are generally used to assist in manipulating values and moving data from one memory location to another. For example, to add the value of R4 to the accumulator, we would execute the following instruction :

```
ADD A,R4
```

Thus if the Accumulator (A) contained the value 6 and R4 contained the value 3, the Accumulator would contain the value 9 after this instruction was executed.

However, as the memory map shows, the "R" Register R4 is really part of Internal RAM. Specifically, R4 is address 04h. This can be see in the bright green section of the memory map. Thus the above instruction accomplishes the same thing as the following operation :

```
ADD A,04h
```

This instruction adds the value found in Internal RAM address 04h to the value of the Accumulator, leaving the result in the Accumulator. Since R4 is really Internal RAM 04h, the above instruction effectively accomplished the same thing.

But watch out! As the memory map shows, the 8051 has four distinct register banks. When the 8051 is first booted up, register bank 0 (addresses 00h through 07h) is used by default. However, your program may instruct the 8051 to use one of the alternate register banks; i.e., register banks 1, 2, or 3. In this case, R4 will no longer be the same as Internal RAM address 04h. For example, if your program instructs the 8051 to use register bank 3, "R" register R4 will now be synonomous with Internal RAM address 1Ch.

The concept of register banks adds a great level of flexibility to the 8051, especially when dealing with interrupts (we'll talk about interrupts later). However, always remember that the register banks really reside in the first 32 bytes of Internal RAM.

TIPS

If you only use the first register bank (i.e. bank 0), you may use Internal RAM locations 08h through 1Fh for your own use. But if you plan to use register banks 1, 2, or 3, be very careful about using addresses below 20h as you may end up overwriting the value of your "R" registers!

13.4.5. Bit Memory

The 8051, being a communications-oriented microcontroller, gives the user the ability to access a number of *bit variables*. These variables may be either 1 or 0.

There are 128 bit variables available to the user, numberd 00h through 7Fh. The user may make use of these variables with commands such as SETB and CLR. For example, to set bit number 24 (hex) to 1 you would execute the instruction :

```
SETB 24h
```

It is important to note that Bit Memory is really a part of Internal RAM. In fact, the 128 bit variables occupy the 16 bytes of Internal RAM from 20h through 2Fh. Thus, if you write the value FFh to Internal RAM address 20h youve effectively set bits 00h through 07h. That is to say that :

```
MOV 20h,#0FFh
```

is equivalent to :

```
SETB 00h
SETB 01h
SETB 02h
SETB 03h
SETB 04h
SETB 05h
SETB 06h
SETB 07h
```

As Illustrated above, blt memory isnt really a new type of memory. Its really just a subset of Internal RAM. But since the 8051 provides special instructions to access these 16 bytes of memory on a bit by bit basis it is useful to think of it as a separate type of memory. However, always keep in mind that it is just a subset of Internal RAM—and that operations performed on Internal RAM can change the values of the bit variables.

TIPS

If your program does not use bit variables, you may use Internal RAM locations 20h through 2Fh for your own use. But if you plan to use bit variables, be very careful about using addresses from 20h through 2Fh as you may end up overwriting the value of your bits!

Bit variables 00h through 7Fh are for user-defined functions in their programs. However, bit variables 80h and above are actually used to access certain SFRs on a bit-by-bit basis. For example, if output lines P0.0 through P0.7 are all clear (0) and you want to turn on the P0.0 output line you may either execute :

```
MOV A, #01h
MOV PO, A
```

or you may execute :

```
SET B 80.0h
```

Both these instructions accomplish the same thing. However, using the SETB command will turn on the P0.0 line without effecting the status of any of the other P0 output lines. The MOV command effectively turns off all the other output lines which, in some cases, may not be acceptable.

TIPS

By default, the 8051 initializes the *Stack Pointer* (SP) to 07h when the microcontroller is booted. This means that the stack will start at address 08h and expand upwards. If you will be using the alternate register banks (banks 1, 2 or 3) you must initialize the stack pointer to an address above the highest register bank you will be using, otherwise the stack will overwrite your alternate register banks. Similarly, if you will be using bit variables it is usually a good idea to initialize the stack pointer to some value greater than 2Fh to guarantee that your bit variables are protected from the stack.

13.4.6. Special Function Register (SFR) Memory

Special Function Registers (SFRs) are areas of memory that control specific functionality of the 8051 processor. For example, four SFRs permit access to the 8051s 32 input/output lines. Another SFR allows a program to read or write to the 8051s serial port. Other SFRs allow the user to set the serial baud rate, control and access timers, and configure the 8051s interrupt system.

When programming, SFRs have the illusion of being Internal Memory. For example, if you want to write the value "1" to Internal RAM location 50 hex you would execute the instruction :

```
MOV 50h,#01h
```

Similarly, if you want to write the value "1" to the 8051s serial port you would write this value to the **SBUF** SFR, which has an SFR address of 99 Hex. Thus, to write the value "1" to the serial port you would execute the instruction :

```
MOV 99h,#01h
```

As you can see, it appears that the SFR is part of Internal Memory. This is not the case. When using this method of memory access (its called direct address), any instruction that has an address of 00h through 7Fh refers to an Internal RAM memory address; any instruction with an address of 80h through FFh refers to an SFR control register.

TIPS

SFRs are used to control the way the 8051 functions. Each SFR has a specific purpose and format which will be discussed later. Not all addresses above 80h are assigned to SFRs. However, this area may NOT be used as additional RAM memory even if a given address has not been assigned to an SFR.

What Are SFRs ?

The 8051 is a flexible microcontroller with a relatively large number of modes of operations. Your program may inspect and/or change the operating mode of the 8051 by manipulating the values of the 8051's Special Function Registers (SFRs).

SFRs are accessed as if they were normal Internal RAM. The only difference is that Internal RAM is from address 00h through 7Fh whereas SFR registers exist in the address range of 80h through FFh.

Each SFR has an address (80h through FFh) and a name. The following chart provides a graphical presentation of the 8051's SFRs, their names, and their address.

P0	SP	DPL	DPH				PCON
TCON	TMOD	TL0	TL1	TH0	TH1		
P1							
SCON	SBUF						
P2							
IE							
P3							
IP							
PSW							
ACC							
B							

Fig. 13.12. SFR register details.

As you can see, although the address range of 80h through FFh offer 128 possible addresses, there are only 21 SFRs in a standard 8051. All other addresses in the SFR range (80h through FFh) are considered invalid. Writing to or reading from these registers may produce undefined values or behavior.

TIPS

It is recommended that you not read or write to SFR addresses that have not been assigned to an SFR. Doing so may provoke undefined behavior and may cause your program to be incompatible with other 8051-derivatives that use the given SFR for some other purpose.

13.4.7. SFR Types

As mentioned in the chart itself, the SFRs that have a blue background are SFRs related to the I/O ports. The 8051 has four I/O ports of 8 bits, for a total of 32 I/O lines. Whether a given I/O line is high or low and the value read from the line are controlled by the SFRs in green.

The SFRs with yellow backgrouns are SFRs which in some way control the operation or the configuration of some aspect of the 8051. For example, **TCON** controls the timers, SCON controls the serial port.

Bit Addressable RAM

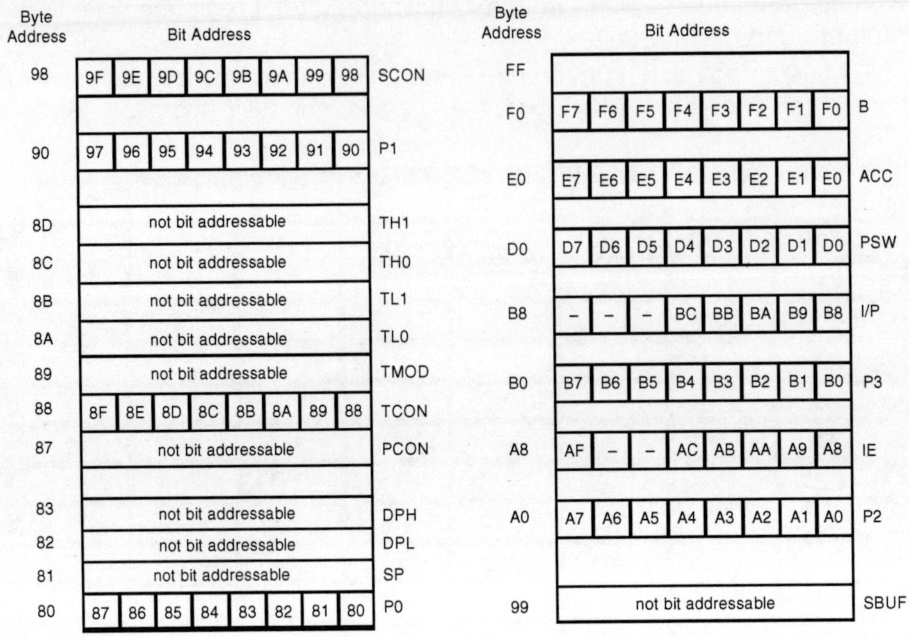

Fig. 13.13. Summary of the 8051 on-chip data memory (Special Function Registers).

The remaining SFRs, with green backgrounds, are "other SFRs." These SFRs can be thought of as auxillary SFRs in the sense that they don't directly configure the 8051 but obviously the 8051 cannot operate without them. For example, once the serial port has been configured using **SCON**, the program may read or write to the serial port using the **SBUF** register.

TIPS

The SFRs whose names appear in red in the chart above are SFRs that may be accessed via bit operations (i.e., using the SETB and CLR instructions). The other SFRs cannot be accessed using bit operations. As you can see, all SFRs that whose addresses are divisible by 8 can be accessed with bit operations.

13.4.8. SFR Descriptions

This section will endeavour to quickly overview each of the standard SFRs found in the above SFR chart map. It is not the intention of this section to fully explain the functionality of each SFR—this information will be covered in separate chapters of the tutorial. This section is to just give you a general idea of what each SFR does.

(1) P0 (Port 0, Address 80h, Bit-Addressable) : This is input/output port 0. Each bit of this SFR corresponds to one of the pins on the microcontroller. For example, bit 0 of port 0 is pin P0.0, bit 7 is pin P0.7. Writing a value of 1 to a bit of this SFR will send a high level on the corresponding I/O pin whereas a value of 0 will bring it to a low level.

TIPS

While the 8051 has four I/O port (P0, P1, P2, and P3), if your hardware uses external RAM or external code memory (i.e., your program is stored in an external ROM or EPROM chip or if you are using external RAM chips) you may not use P0 or P2. This is because the 8051 uses ports P0 and P2 to address the external memory. Thus if you are using external RAM or code memory you may only use ports P1 and P3 for your own use.

(2) **SP (Stack Pointer, Address 81h)** : This is the stack pointer of the microcontroller. This SFR indicates where the next value to be taken from the stack will be read from in Internal RAM. If you push a value onto the stack, the value will be written to the address of SP + 1. That is to say, if SP holds the value 07h, a PUSH instruction will push the value onto the stack at address 08h. This SFR is modified by all instructions which modify the stack, such as PUSH, POP, LCALL, RET, RETI, and whenever interrupts are provoked by the microcontroller.

TIPS

The SP SFR, on startup, is initialized to 07h. This means the stack will start at 08h and start expanding upward in internal RAM. Since alternate register banks 1, 2, and 3 as well as the user bit variables occupy internal RAM from addresses 08h through 2Fh, it is necessary to initialize SP in your program to some other value if you will be using the alternate register banks and/or bit memory. It's not a bad idea to initialize SP to 2Fh as the first instruction of every one of your programs unless you are 100% sure you will not be using the register banks and bit variables.

(3) **DPL/DPH (Data Pointer Low/High, Addresses 82h/83h)** : The SFRs DPL and DPH work together to represent a 16-bit value called the *Data Pointer*. The data pointer is used in operations regarding external RAM and some instructions involving code memory. Since it is an unsigned two-byte integer value, it can represent values from 0000h to FFFFh (0 through 65,535 decimal).

TIPS

DPTR is really DPH and DPL taken together as a 16-bit value. In reality, you almost always have to deal with DPTR one byte at a time. For example, to push DPTR onto the stack you must first push DPL and then DPH. You can't simply plush DPTR onto the stack. Additionally, there is an instruction to "increment DPTR." When you execute this instruction, the two bytes are operated upon as a 16-bit value. However, there is no instruction that decrements DPTR. If you wish to decrement the value of DPTR, you must write your own code to do so.

(4) **PCON (Power Control, Addresses 87h)** : The Power Control SFR is used to control the 8051's power control modes. Certain operation modes of the 8051 allow the 8051 to go into a type of "sleep" mode which requires much less power. These modes of operation are controlled through PCON. Additionally, one of the bits in PCON is used to double the effective baud rate of the 8051's serial port.

(5) **TCON (Timer Control, Addresses 88h, Bit-Addressable)** : The Timer Control SFR is used to configure and modify the way in which the 8051's two timers operate. This SFR controls whether each of the two timers is running or stopped and contains a flag to indicate that each timer has overflowed. Additionally, some non-timer related bits are located in the TCON SFR. These bits are used to configure the way in which the external interrupts are activated and also contain the external interrupt flags which are set when an external interrupt has occured.

(6) **TMOD (Timer Mode, Addresses 89h)** : The Timer Mode SFR is used to configure the mode of operation of each of the two timers. Using this SFR your program may configure each timer to be a 16-bit timer, an 8-bit autoreload timer, a 13-bit timer, or two separate timers. Additionally, you may configure the timers to only count when an external pin is activated or to count "events" that are indicated on an external pin.

(7) **TL0/TH0 (Timer 0 Low/High, Addresses 8Ah/8Ch)** : These two SFRs, taken together, represent timer 0. Their exact behavior depends on how the timer is configured in the TMOD SFR; however, these timers always count up. What is configurable is how and when they increment in value.

(8) **TL1/TH1 (Timer 1 Low/High, Addresses 8Bh/8Dh)** : These two SFRs, taken together, represent timer 1. Their exact behavior depends on how the timer is configured in the TMOD SFR; however, these timers always count up. What is configurable is how and when they increment in value.

(9) **P1 (Port 1, Address 90h, Bit-Addressable)** : This is input/output port 1. Each bit of this SFR corresponds to one of the pins on the microcontroller. For example, bit 0 of port 1 is pin P1.0, bit 7 is pin P1.7. Writing a value of 1 to a bit of this SFR will send a high level on the corresponding I/O pin whereas a value of 0 will bring it to a low level.

(10) **SCON (Serial Control, Addresses 98h, Bit-Addressable)** : The Serial Control SFR is used to configure the behavior of the 8051's on-board serial port. This SFR controls the baud rate of the serial port, whether the serial port is activated to receive data, and also contains flags that are set when a byte is successfully sent or received.

TIPS

To use the 8051's on-board serial port, it is generally necessary to initialize the following SFRs : SCON, TCON, and TMOD. This is because SCON controls the serial port. However, in most cases the program will wish to use one of the timers to establish the serial port's baud rate. In this case, it is necessary to configure timer 1 by initializing TCON and TMOD.

(11) **SBUF (Serial Control, Addresses 99h)** : The Serial Buffer SFR is used to send and receive data via the on-board serial port. Any value written to SBUF will be sent out the serial port's TXD pin. Likewise, any value which the 8051 receives via the serial port's RXD pin will be delivered to the user program via SBUF. In other words, SBUF serves as the output port when written to and as an input port when read from.

(12) **P2 (Port 2, Address A0h, Bit-Addressable)** : This is input/output port 2. Each bit of this SFR corresponds to one of the pins on the microcontroller. For example, bit 0 of port 2 is pin P2.0, bit 7 is pin P2.7. Writing a value of 1 to a bit of this SFR will send a high level on the corresponding I/O pin whereas a value of 0 will bring it to a low level.

TIPS

While the 8051 has four I/O port (P0, P1, P2, and P3), if your hardware uses external RAM or external code memory (i.e., your program is stored in an external ROM or EPROM chip or if you are using external RAM chips) you may not use P0 or P2. This is because the 8051 uses ports P0 and P2 to address the external memory. Thus if you are using external RAM or code memory you may only use ports P1 and P3 for your own use.

(13) **IE (Interrupt Enable, Addresses A8h)** : The Interrupt Enable SFR is used to enable and disable specific interrupts. The low 7 bits of the SFR are used to enable/disable the specific interrupts, where as the highest bit is used to enable or disable ALL interrupts. Thus, if the high bit of IE is 0 all interrupts are disabled regardless of whether an individual interrupt is enabled by setting a lower bit.

(14) **P3 (Port 3, Address B0h, Bit-Addressable)** : This is input/output port 3. Each bit of this SFR corresponds to one of the pins on the microcontroller. For example, bit 0 of port 3 is pin P3.0, bit 7 is pin P3.7. Writing a value of 1 to a bit of this SFR will send a high level on the corresponding I/O pin whereas a value of 0 will bring it to a low level.

(15) **IP (Interrupt Priority, Addresses B8h, Bit-Addressable)** : The Interrupt Priority SFR is used to specify the relative priority of each interrupt. On the 8051, an interrupt may either be of low (0) priority or high (1) priority. An interrupt may only interrupt interrupts of lower priority. For example, if we configure the 8051 so that all interrupts are of low priority except the serial interrupt, the serial interrupt will always be able to interrupt the system, even if another interrupt is currently executing. However, if a serial interrupt is executing no other interrupt will be able to interrupt the serial interrupt routine since the serial interrupt routine has the highest priority.

(16) **PSW (Program Status Word, Addresses D0h, Bit-Addressable)** : The Program Status Word is used to store a number of important bits that are set and cleared by 8051 instructions. The PSW SFR contains the carry flag, the auxiliary carry flag, the overflow flag, and the parity flag. Additionally, the PSW register contains the register bank select flags which are used to select which of the "R" register banks are currently selected.

TIPS

If you write an interrupt handler routine, it is a very good idea to *always* save the PSW SFR on the stack and restore it when your interrupt is complete. Many 8051 instructions modify the bits of PSW. If your interrupt routine does not guarantee that PSW is the same upon exit as it was upon entry, your program is bound to behave rather erradically and unpredictably—and it will be tricky to debug since the behavior will tend not to make any sense.

(17) **ACC (Accumulator, Addresses E0h, Bit-Addressable)** : The Accumulator is one of the most-used SFRs on the 8051 since it is involved in so many instructions. The Accumulator resides as an SFR at E0h, which means the instruction **MOV A,#20h** is really the same as **MOV E0h,#20h**. However, it is a good idea to use the first method since it only requires two bytes whereas the second option requires three bytes.

(18) **B (B Register, Addresses F0h, Bit-Addressable)** : The "B" register is used in two instructions : the multiply and divide operations. The B register is also commonly used by programmers as an auxiliary register to temporarily store values.

13.4.9. Other SFRs

The chart above is a summary of all the SFRs that exist in a standard 8051. All derivative microcontrollers of the 8051 must support these basic SFRs in order to maintain compatability with the underlying MSCS51 standard.

A common practice when semiconductor firms wish to develop a new 8051 derivative is to add additional SFRs to support new functions that exist in the new chip.

For example, the Dallas Semiconductor DS80C320 is upwards compatible with the 8051. This means that any program that runs on a standard 8051 should run without modification on the DS80C320. This means that all the SFRs defined above also apply to the Dallas component.

However, since the DS80C320 provides many new features that the standard 8051 does not, there must be some way to control and configure these new features. This is accomplished by adding additional SFRs to those listed here. For example, since the DS80C320 supports two serial ports (as opposed to just one on the 8051), the SFRs SBUF2 and SCON2 have been added. In addition to all the SFRs listed above, the DS80C320 also recognizes these two new

SFRs as valid and uses their values to determine the mode of operation of the secondary serial port. Obviously, these new SFRs have been assigned to SFR addresses that were unused in the original 8051. In this manner, new 8051 derivative chips may be developed which will run existing 8051 programs.

TIPS

If you write a program that utilizes new SFRs that are specific to a given derivative chip and not included in the above SFR list, your program will not run properly on a standard 8051 where that SFR does not exist. Thus, only use non-standard SFRs if you are sure that your program wil only have to run on that specific microcontroller. Likewise, if you write code that uses non-standard SFRs and subsequently share it with a third-party, be sure to let that party know that your code is using non-standard SFRs to save them the headache of realizing that due to strange behaviour at run-time.

13.4.10. The Accumulator

If youve worked with any other assembly languages you will be familiar with the concept of an *Accumulator* register.

The Accumulator, as its name suggests, is used as a general register to accumulate the results of a large number of instructions. It can hold an 8-bit (1-byte) value and is the most versatile register the 8051 has due to the shear number of instructions that make use of the accumulator. More than half of the 8051s 255 instructions manipulate or use the accumulator in some way.

For example, if you want to add the number 10 and 20, the resulting 30 will be stored in the Accumulator. Once you have a value in the Accumulator you may continue processing the value or you may store it in another register or in memory.

13.4.11. The "R" Registers

The "R" registers are a set of eight registers that are named R0, R1, etc. up to and including R7.

These registers are used as auxillary registers in many operations. To continue with the above example, perhaps you are adding 10 and 20. The original number 10 may be stored in the Accumulator whereas the value 20 may be stored in, say, register R4. To process the addition you would execute the command :

```
ADD A,R4
```

After executing this instruction the Accumulator will contain the value 30.

You may think of the "R" registers as very important auxillary, or "helper", registers. The Accumulator alone would not be very useful if it were not for these "R" registers.

The "R" registers are also used to temporarily store values. For example, lets say you want to add the values in R1 and R2 together and then subtract the values of R3 and R4. One way to do this would be :

```
MOV A,R3   ;Move the value of R3 into the accumulator
ADD A,R4   ;Add the value of R4
MOV R5,A   ;Store the resulting value temporarily in R5
MOV A,R1   ;Move the value of R1 into the accumulator
ADD A,R2   ;Add the value of R2
SUBB A,R5 ;Subtract the value of R5 (which now contains R3 + R4)
```

As you can see, we used R5 to temporarily hold the sum of R3 and R4. Of course, this isnt the most efficient way to calculate (R1+R2) – (R3 +R4) but it does illustrate the use of the "R" registers as a way to store values temporarily.

13.4.12. The "B" Register

The "B" register is very similar to the Accumulator in the sense that it may hold an 8-bit (1-byte) value.

The "B" register is only used by two 8051 instructions : MUL AB and DIV AB. Thus, if you want to quickly and easily multiply or divide A by another number, you may store the other number in "B" and make use of these two instructions.

Aside from the MUL and DIV instructions, the "B" register is often used as yet another temporary storage register much like a ninth "R" register.

13.4.13. The Data Pointer (DPTR)

The Data Pointer (DPTR) is the 8051s only user-accessable 16-bit (2-byte) register. The Accumulator, "R" registers, and "B" register are all 1-byte values.

DPTR, as the name suggests, is used to point to data. It is used by a number of commands which allow the 8051 to access external memory. When the 8051 accesses external memory it will access external memory at the address indicated by DPTR.

While DPTR is most often used to point to data in external memory, many programmers often take advantge of the fact that its the only true 16-bit register available. It is often used to store 2-byte values which have nothing to do with memory locations.

13.4.14. The Program Counter (PC)

The Program Counter (PC) is a 2-byte address which tells the 8051 where the next instruction to execute is found in memory. When the 8051 is initialized PC always starts at 0000h and is incremented each time an instruction is executed. It is important to note that PC isnt always incremented by one. Since some instructions require 2 or 3 bytes the PC will be incremented by 2 or 3 in these cases.

The Program Counter is special in that there is no way to directly modify its value. That is to say, you cant do something like PC=2430h. On the other hand, if you execute LJMP 2430h youve effectively accomplished the same thing.

It is also interesting to note that while you may change the value of PC (by executing a jump instruction, etc.) there is no way to read the value of PC. That is to say, there is no way to ask the 8051 "What address are you about to execute ?" As it turns out, this is not completely true : There is one trick that may be used to determine the current value of PC. This trick will be covered in a later chapter.

13.4.15. The Stack Pointer (SP)

The Stack Pointer, like all registers except DPTR and PC, may hold an 8-bit (1-byte) value. The Stack Pointer is used to indicate where the next value to be removed from the stack should be taken from.

When you push a value onto the stack, the 8051 first increments the value of SP and then stores the value at the resulting memory location.

When you pop a value off the stack, the 8051 returns the value from the memory location indicated by SP, and then decrements the value of SP.

This order of operation is important. When the 8051 is initialized SP will be initialized to 07h. If you immediately push a value onto the stack, the value will be stored in Internal RAM address 08h. This makes sense taking into account what was mentioned two paragraphs above : First the 8051 will increment the value of SP (from 07h to 08h) and then will store the pushed value at that memory address (08h).

SP is modified directly by the 8051 by six instructions : PUSH, POP, ACALL, LCALL, RET, and RETI. It is also used intrinsically whenever an interrupt is triggered (more on interrupts later. Dont worry about them for now!).

13.5. ADDRESSING MODES

An "addressing mode" refers to how you are addressing a given memory location. In summary, the addressing modes are as follows, with an example of each :

```
Immediate Addressing MOV A,#20h
Direct Addressing         MOV A,30h
Indirect Addressing       MOV A,@R0
External Direct      MOVX A,@DPTR
Code Indirect             MOVC A,@A+DPTR
```

Each of these addressing modes provides important flexibility.

(1) Immediate Addressing

Immediate addressing is so-named because the value to be stored in memory immediately follows the operation code in memory. That is to say, the instruction itself dictates what value will be stored in memory.

For example, the instruction :

```
MOV A,#20h
```

This instruction uses Immediate Addressing because the Accumulator will be loaded with the value that immediately follows; in this case 20 (hexidecimal).

Immediate addressing is very fast since the value to be loaded is included in the instruction. However, since the value to be loaded is fixed at compile-time it is not very flexible.

(2) Direct Addressing

Direct addressing is so-named because the value to be stored in memory is obtained by directly retrieving it from another memory location. For example :

```
MOV A,30h
```

This instruction will read the data out of Internal RAM address 30 (hexidecimal) and store it in the Accumulator.

Direct addressing is generally fast since, although the value to be loaded isnt included in the instruction, it is quickly accessable since it is stored in the 8051s Internal RAM. It is also much more flexible than Immediate Addressing since the value to be loaded is whatever is found at the given address—which may be variable.

Also, it is important to note that when using direct addressing any instruction which refers to an address between 00h and 7Fh is referring to Internal Memory. Any instruction which refers to an address between 80h and FFh is referring to the SFR control registers that control the 8051 microcontroller itself.

The obvious question that may arise is, "If direct addressing an address from 80h through FFh refers to SFRs, how can I access the upper 128 bytes of Internal RAM that are available on the 8052 ?" The answer is : You cant access them using direct addressing. As stated, if you directly refer to an address of 80h through FFh you will be referring to an SFR. However, you may access the 8052s upper 128 bytes of RAM by using the next addressing mode, "indirect addressing."

(3) Indirect Addressing

Indirect addressing is a very powerful addressing mode which in many cases provides an exceptional level of flexibility. Indirect addressing is also the only way to access the extra 128 bytes of Internal RAM found on an 8052.

Indirect addressing appears as follows :

```
MOV A,@R0
```

This instruction causes the 8051 to analyze the value of the R0 register. The 8051 will then load the accumulator with the value from Internal RAM which is found at the address indicated by R0.

For example, lets say R0 holds the value 40h and Internal RAM address 40h holds the value 67h. When the above instruction is executed the 8051 will check the value of R0. Since R0 holds 40h the 8051 will get the value out of Internal RAM address 40h (which holds 67h) and store it in the Accumulator. Thus, the Accumulator ends up holding 67h.

Indirect addressing always refers to Internal RAM; it never refers to an SFR. Thus, in a prior example we mentioned that SFR 99h can be used to write a value to the serial port. Thus one may think that the following would be a valid solution to write the value 1 to the serial port :

```
MOV R0,#99h ;Load the address of the serial port
MOV @R0,#01h     ;Send 01 to the serial port - WRONG!!
```

This is not valid. Since indirect addressing always refers to Internal RAM these two instructions would write the value 01h to Internal RAM address 99h on an 8052. On an 8051 these two instructions would produce an undefined result since the 8051 only has 128 bytes of Internal RAM.

(4) External Direct

External Memory is accessed using a suite of instructions which use what I call "External Direct" addressing. I call it this because it appears to be direct addressing, but it is used to access external memory rather than internal memory.

There are only two commands that use External Direct addressing mode :

```
MOVX A,@DPTR
MOVX @DPTR,A
```

As you can see, both commands utilize DPTR. In these instructions, DPTR must first be loaded with the address of external memory that you wish to read or write. Once DPTR holds the correct external memory address, the first command will move the contents of that external memory address into the Accumulator. The second command will do the opposite : it will allow you to write the value of the Accumulator to the external memory address pointed to by DPTR.

(5) External Indirect

External memory can also be accessed using a form of indirect addressing which I call External Indirect addressing. This form of addressing is usually only used in relatively small projects that have a very small amount of external RAM. An example of this addressing mode is :

```
MOVX @R0,A
```

Once again, the value of R0 is first read and the value of the Accumulator is written to that address in External RAM. Since the value of @R0 can only be 00h through FFh the project would effectively be limited to 256 bytes of External RAM. There are relatively simple hardware/ software tricks that can be implemented to access more than 256 bytes of memory using External Indirect addressing; however, it is usually easier to use External Direct addressing if your project has more than 256 bytes of External RAM.

When an 8051 is first initialized, it resets the PC to 0000h. The 8051 then begins to execute instructions sequentially in memory unless a program instruction causes the PC to be otherwise altered. There are various instructions that can modify the value of the PC; specifically, conditional branching instructions, direct jumps and calls, and "returns" from subroutines. Additionally, interrupts, when enabled, can cause the program flow to deviate from its otherwise sequential scheme.

(6) Conditional Branching

The 8051 contains a suite of instructions which, as a group, are referred to as "conditional branching" instructions. These instructions cause program execution to follow a non-sequential path if a certain condition is true.

Take, for example, the JB instruction. This instruction means "Jump if Bit Set." An example of the JB instruction might be :

```
        JB 45h,HELLO
        NOP
    HELLO :  ....
```

In this case, the 8051 will analyze the contents of bit 45h. If the bit is set program execution will jump immediately to the label HELLO, skipping the NOP instruction. If the bit is not set the conditional branch fails and program execution continues, as usual, with the NOP instruction which follows.

Conditional branching is really the fundamental building block of program logic since all "decisions" are accomplished by using conditional branching. Conditional branching can be thought of as the "IF...THEN" structure in 8051 assembly language.

An important note worth mentioning about conditional branching is that the program may only branch to instructions located withim 128 bytes prior to or 127 bytes following the address which follows the conditional branch instruction. This means that in the above example the label HELLO must be within +/- 128 bytes of the memory address which contains the conditional branching instruction.

(7) Direct Jumps

While conditional branching is extremely important, it is often necessary to make a direct branch to a given memory location without basing it on a given logical decision. This is equivalent to saying "Goto" in BASIC. In this case you want the program flow to continue at a given memory address without considering any conditions.

This is accomplished in the 8051 using "Direct Jump and Call" instructions. As illustrated in the last paragraph, this suite of instructions causes program flow to change unconditionally.

Consider the example :

```
        LJMP NEW_ADDRESS
        .
        .
        .
    NEW_ADDRESS : ....
```

The LJMP instruction in this example means "Long Jump." When the 8051 executes this instruction the PC is loaded with the address of NEW_ADDRESS and program execution continues sequentially from there.

The obvious difference between the Direct Jump and Call instructions and the conditional branching is that with Direct Jumps and Calls program flow always changes. With conditional branching program flow only changes if a certain condition is true.

It is worth mentioning that, aside from LJMP, there are two other instructions which cause a direct jump to occur : the SJMP and AJMP commands. Functionally, these two commands perform the exact same function as the LJMP command—that is to say, they always cause program flow to continue at the address indicated by the command. However, SJMP and AJMP differ in the following ways :

- The SJMP command, like the conditional branching instructions, can only jump to an address within +/- 128 bytes of the SJMP command.
- The AJMP command can only jump to an address that is in the same 2k block of memory as the AJMP command. That is to say, if the AJMP command is at code memory location 650h, it can only do a jump to addresses 0000h through 07FFh (0 through 2047, decimal).

You may be asking yourself, "Why would I want to use the SJMP or AJMP command which have restrictions as to how far they can jump if they do the same thing as the LJMP command which can jump anywhere in memory ?" The answer is simple : The LJMP command requires three bytes of code memory whereas both the SJMP and AJMP commands require only two. Thus, if you are developing an application that has memory restrictions you can often save quite a bit of memory using the 2-byte AJMP/SJMP instructions instead of the 3-byte instruction.

Recently, I wrote a program that required 2100 bytes of memory but I had a memory restriction of 2K (2048 bytes). I did a search/replace changing all LJMPs to AJMPs and the program shrunk downto 1950 bytes. Thus, without changing any logic whatsoever in my program I saved 150 bytes and was able to meet my 2048 byte memory restriction.

Note : Some quality assemblers will actually do the above conversion for you automatically. That is, theyll automatically change your LJMPs to SJMPs whenever possible. This is a nifty and very powerful capability that you may want to look for in an assembler if you plan to develop many projects that have relatively tight memory restrictions.

(8) Direct Calls

Another operation that will be familiar to seasoned programmers is the LCALL instruction. This is similar to a "Gosub" command in Basic.

When the 8051 executes an LCALL instruction it immediately pushes the current Program Counter onto the stack and then continues executing code at the address indicated by the LCALL instruction.

(9) Returns from Routines

Another structure that can cause program flow to change is the "Return from Subroutine" instruction, known as RET in 8051 Assembly Language.

The RET instruction, when executed, returns to the address following the instruction that called the given subroutine. More accurately, it returns to the address that is stored on the stack.

The RET command is direct in the sense that it always changes program flow without basing it on a condition, but is variable in the sense that where program flow continues can be different each time the RET instruction is executed depending on from where the subroutine was called originally.

13.6. INTERRUPTS

An interrupt is a special feature which allows the 8051 to provide the illusion of "multi-tasking," although in reality the 8051 is only doing one thing at a time. The word "interrupt" can often be subsituted with the word "event."

An interrupt is triggered whenever a corresponding event occurs. When the event occurs, the 8051 temporarily puts "on hold" the normal execution of the program and executes a special section of code referred to as an interrupt handler. The interrupt handler performs whatever special functions are required to handle the event and then returns control to the 8051 at which point program execution continues as if it had never been interrupted.

The topic of interrupts is somewhat tricky and very important. For that reason, an entire chapter will be dedicated to the topic. For now, suffice it to say that Interrupts can cause program flow to change.

In order to understand—and better make use of—the 8051, it is necessary to understand some underlying information concerning timing.

The 8051 operates based on an external crystal. This is an electrical device which, when energy is applied, emits pulses at a fixed frequency. One can find crystals of virtually any frequency depending on the application requirements. When using an 8051, the most common crystal frequencies are 12 megahertz and 11.059 megahertz—with 11.059 being much more common. Why would anyone pick such an odd-ball frequency ? Theres a real reason for it—it has to do with generating baud rates and well talk more about it in the Serial Communication chapter. For the remainder of this discussion well assume that were using an 11.059 Mhz crystal.

13.6.1. Using Timers to Measure Time

Obviously, one of the primary uses of timers is to measure time. We will discuss this use of timers first and will subsequently discuss the use of timers to count events. When a timer is used to measure time it is also called an "interval timer" since it is measuring the time of the interval between two events.

How long does a timer take to count ?

First, its worth mentioning that when a timer is in interval timer mode (as opposed to event counter mode) and correctly configured, it will increment by 1 every machine cycle. As you will recall from the previous chapter, a single machine cycle consists of 12 crystal pulses. Thus a running timer will be incremented :

```
11,059,000 / 12 = 921,583
```

921,583 times per second. Unlike instructions—some of which require 1 machine cycle, others 2, and others 4—the timers are consistent : They will always be incremented once per machine cycle. Thus if a timer has counted from 0 to 50,000 you may calculate :

```
50,000 / 921,583 = .0542
```

.0542 seconds have passed. In plain English, about half of a tenth of a second, or one-twentieth of a second.

Obviously its not very useful to know .0542 seconds have passed. If you want to execute an event once per second youd have to wait for the timer to count from 0 to 50,000 18.45 times. How can you wait "half of a time ?" You cant. So we come to another important calculation.

Lets say we want to know how many times the timer will be incremented in .05 seconds. We can do simple multiplication :

```
.05 * 921,583 = 46,079.15.
```

This tells us that it will take .05 seconds (1/20th of a second) to count from 0 to 46,079. Actually, it will take it .049999837 seconds—so were off by .000000163 seconds—however, thats close enough for government work. Consider that if you were building a watch based on the 8051 and made the above assumption your watch would only gain about one second every 2 months. Again, I think thats accurate enough for most applications—I wish my watch only gained one second every two months!

Obviously, this is a little more useful. If you know it takes 1/20th of a second to count from 0 to 46,079 and you want to execute some event every second you simply wait for the timer to count from 0 to 46,079 twenty times; then you execute your event, reset the timers, and wait for the timer to count up another 20 times. In this manner you will effectively execute your event once per second, accurate to within thousandths of a second.

Thus, we now have a system with which to measure time. All we need to review is how to control the timers and initialize them to provide us with the information we need.

Luckily, this isnt necessary. Interrupts let us forget about checking for the condition. The microcontroller itself will check for the condition automatically and when the condition is met will jump to a subroutine (called an interrupt handler), execute the code, then return. In this case, our subroutine would be nothing more than :

```
CPL P3.0
RETI
```

First, youll notice the CLR TF0 command has disappeared. Thats because when the 8051 executes our “timer 0 interrupt routine,” it automatically clears the TF0 flag. Youll also notice that instead of a normal **RET** instruction we have a **RETI** instruction. The RETI instruction does the same thing as a RET instruction, but tells the 8051 that an interrupt routine has finished. You must always end your interrupt handlers with RETI.

Thus, every 65536 instruction cycles we execute the CPL instruction and the RETI instruction. Those two instructions together require 3 instruction cycles, and weve accomplished the same goal as the first example that required 1312 instruction cycles. As far as the toggling of P3.0 goes, our code is 437 times more efficient! Not to mention its much easier to read and understand because we dont have to remember to always check for the timer 0 flag in our main program. We just setup the interrupt and forget about it, secure in the knowledge that the 8051 will execute our code whenever its necessary.

The same idea applies to receiving data via the serial port. One way to do it is to continuously check the status of the RI flag in an endless loop. Or we could check the RI flag as part of a larger program loop. However, in the latter case we run the risk of *missing* characters—what happens if a character is received right after we do the check, the rest of our program executes, and before we even check **RI** a second character has come in. We will lose the first character. With interrupts, the 8051 will put the main program “on hold” and call our special routine to handle the reception of a character. Thus, we neither have to put an ugly check in our main code nor will we lose characters.

13.6.2. What Events Can Trigger Interrupts, and Where do they go ?

We can configure the 8051 so that any of the following events will cause an interrupt :

- Timer 0 Overflow.
- Timer 1 Overflow.
- Reception/Transmission of Serial Character.
- External Event 0.

➢ External Event 1.

In other words, we can configure the 8051 so that when Timer 0 Overflows or when a character is sent/received, the appropriate interrupt handler routines are called.

Obviously we need to be able to distinguish between various interrupts and executing different code depending on what interrupt was triggered. This is accomplished by jumping to a fixed address when a given interrupt occurs.

Interrupt	Flag	Interrupt Handler Address
External 0	IE0	0003h
Timer 0	TF0	000Bh
External 1	IE1	0013h
Timer 1	TF1	001Bh
Serial	RI/TI	0023h

By consulting the above chart we see that whenever Timer 0 overflows (i.e., the TF0 bit is set), the main program will be temporarily suspended and control will jump to 000BH. It is assumed that we have code at address 000BH that handles the situation of Timer 0 overflowing.

13.6.3. Setting Up Interrupts

By default at powerup, all interrupts are disabled. This means that even if, for example, the TF0 bit is set, the 8051 will not execute the interrupt. Your program must specifically tell the 8051 that it wishes to enable interrupts and specifically which interrupts it wishes to enable.

Your program may enable and disable interrupts by modifying the IE SFR (A8h) :

Bit	Name	Bit Address	Explanation of Function
7	EA	AFh	Global Interrupt Enable/Disable
6	-	AEh	Undefined
5	-	ADh	Undefined
4	ES	ACh	Enable Serial Interrupt
3	ET1	ABh	Enable Timer 1 Interrupt
2	EX1	AAh	Enable External 1 Interrupt
1	ET0	A9h	Enable Timer 0 Interrupt
0	EX0	A8h	Enable External 0 Interrupt

As you can see, each of the 8051s interrupts has its own bit in the IE SFR. You enable a given interrupt by setting the corresponding bit. For example, if you wish to enable Timer 1 Interrupt, you would execute either :

```
MOV IE,#08h
or
SETB ET1
```

Both of the above instructions set bit 3 of IE, thus enabling Timer 1 Interrupt. Once Timer 1 Interrupt is enabled, whenever the TF1 bit is set, the 8051 will automatically put "on hold" the main program and execute the Timer 1 Interrupt Handler at address 001Bh.

However, before Timer 1 Interrupt (or any other interrupt) is truly enabled, you must also set bit 7 of IE. Bit 7, the Global Interupt Enable/Disable, enables or disables all interrupts simultaneously. That is to say, if bit 7 is cleared then no interrupts will occur, even if all the other bits of IE are set. Setting bit 7 will enable all the interrupts that have been selected by setting other bits in IE. This is useful in program execution if you have time-critical code that needs to execute. In this case, you may need the code to execute from start to finish without any interrupt getting in the way. To accomplish this you can simply clear bit 7 of IE (CLR EA) and then set it after your time-criticial code is done.

So, to sum up what has been stated in this section, to enable the Timer 1 Interrupt the most common approach is to execute the following two instructions :

```
SETB ET1
SETB EA
```

Thereafter, the Timer 1 Interrupt Handler at 01Bh will automatically be called whenever the TF1 bit is set (upon Timer 1 overflow).

13.6.4. Polling Sequence

The 8051 automatically evaluates whether an interrupt should occur after every instruction. When checking for interrupt conditions, it checks them in the following order :

- External 0 Interrupt
- Timer 0 Interrupt
- External 1 Interrupt
- Timer 1 Interrupt
- Serial Interrupt

This means that if a Serial Interrupt occurs at the exact same instant that an External 0 Interrupt occurs, the External 0 Interrupt will be executed first and the Serial Interrupt will be executed once the External 0 Interrupt has completed.

13.6.5. Interrupt Priorities

The 8051 offers two levels of interrupt priority : high and low. By using interrupt priorities you may assign higher priority to certain interrupt conditions.

For example, you may have enabled Timer 1 Interrupt which is automatically called every time Timer 1 overflows. Additionally, you may have enabled the Serial Interrupt which is called every time a character is received via the serial port. However, you may consider that receiving a character is much more important than the timer interrupt. In this case, if Timer 1 Interrupt is already executing you may wish that the serial interrupt itself interrupts the Timer 1 Interrupt. When the serial interrupt is complete, control passes back to Timer 1 Interrupt and finally back to the main program. You may accomplish this by assigning a high priority to the Serial Interrupt and a low priority to the Timer 1 Interrupt.

Interrupt priorities are controlled by the **IP** SFR (B8h). The IP SFR has the following format :

Bit	Name	Bit Address	Explanation of Function
7	–	–	Undefined
6	–	–	Undefined
5	–	–	Undefined
4	PS	BCh	Serial Interrupt Priority
3	PT1	BBh	Timer 1 Interrupt Priority
2	PX1	BAh	External 1 Interrupt Priority
1	PT0	B9h	Timer 0 Interrupt Priority
0	PX0	B8h	External 0 Interrupt Priority

When considering interrupt priorities, the following rules apply :

- Nothing can interrupt a high-priority interrupt—not even another high priority interrupt.
- A high-priority interrupt may interrupt a low-priority interrupt.
- A low-priority interrupt may only occur if no other interrupt is already executing.
- If two interrupts occur at the same time, the interrupt with higher priority will execute first. If both interrupts are of the same priority the interrupt which is serviced first by polling sequence will be executed first.

13.6.6. What Happens When an Interrupt Occurs ?

When an interrupt is triggered, the following actions are taken automatically by the microcontroller :

- The current Program Counter is saved on the stack, low-byte first.
- Interrupts of the same and lower priority are blocked.
- In the case of Timer and External interrupts, the corresponding interrupt flag is cleared.
- Program execution transfers to the corresponding interrupt handler vector address.
- The Interrupt Handler Routine executes.

Take special note of the third step : If the interrupt being handled is a Timer or External interrupt, the microcontroller automatically clears the interrupt flag before passing control to your interrupt handler routine. This means it is not necessary that you clear the bit in your code.

13.6.7. What Happens When an Interrupt Ends ?

An interrupt ends when your program executes the RETI (Return from Interrupt) instruction. When the RETI instruction is executed the following actions are taken by the microcontroller :

- Two bytes are popped off the stack into the Program Counter to restore normal program execution.
- Interrupt status is restored to its pre-interrupt status.

Serial Interrupts

Serial Interrupts are slightly different than the rest of the interrupts. This is due to the fact that there are two interrupt flags : RI and TI. If either flag is set, a serial interrupt is triggered. As you will recall from the section on the serial port, the RI bit is set when a byte is received by the serial port and the TI bit is set when a byte has been sent.

This means that when your serial interrupt is executed, it may have been triggered because the RI flag was set or because the TI flag was set—or because both flags were set. Thus, your routine must check the status of these flags to determine what action is appropriate. Also, since the 8051 does not automatically clear the RI and TI flags you must clear these bits in your interrupt handler.

A brief code example is in order :

```
INT_SERIAL : JNB RI,CHECK_TI ;If the RI flag is not set, we jump to check TI
MOV A,SBUF   ;If we got to this line, its because the RI bit *was* set
CLR RI       ;Clear the RI bit after weve processed it
CHECK_TI :   JNB TI,EXIT_INT ;If the TI flag is not set, we jump to the exit point
CLR TI       ;Clear the TI bit before we send another character
MOV SBUF,#A ;Send another character to the serial port
EXIT_INT :   RETI
```

As you can see, our code checks the status of both interrupts flags. If both flags were set, both sections of code will be executed. Also note that each section of code clears its corresponding interrupt flag. If you forget to clear the interrupt bits, the serial interrupt will be executed over and over until you clear the bit. Thus it is very important that you always clear the interrupt flags in a serial interrupt.

13.7. INSTRUCTION SET FOR 8051

- **ACALL** : Absolute Call
- **ADD, ADDC** : Add Accumulator (With Carry)
- **AJMP** : Absolute Jump
- **ANL** : Bitwise AND
- **CJNE** : Compare and Jump if Not Equal
- **CLR** : Clear Register
- **CPL** : Complement Register
- **DA** : Decimal Adjust
- **DEC** : Decrement Register
- **DIV** : Divide Accumulator by B
- **DJNZ** : Decrement Register and Jump if Not Zero
- **INC** : Increment Register
- **JB** : Jump if Bit Set
- **JBC** : Jump if Bit Set and Clear Bit
- **JC** : Jump if Carry Set
- **JMP** : Jump to Address
- **JNB** : Jump if Bit Not Set
- **JNC** : Jump if Carry Not Set
- **JNZ** : Jump if Accumulator Not Zero
- **JZ** : Jump if Accumulator Zero
- **LCALL** : Long Call
- **LJMP** : Long Jump
- **MOV** : Move Memory
- **MOVC** : Move Code Memory
- **MOVX** : Move Extended Memory
- **MUL** : Multiply Accumulator by B
- **NOP** : No Operation
- **ORL** : Bitwise OR
- **POP** : Pop Value From Stack
- **PUSH** : Push Value Onto Stack
- **RET** : Return From Subroutine
- **RETI** : Return From Interrupt

- **RL** : Rotate Accumulator Left
- **RLC** : Rotate Accumulator Left Through Carry
- **RR** : Rotate Accumulator Right
- **RRC** : Rotate Accumulator Right Through Carry
- **SETB** : Set Bit
- **SJMP** : Short Jump
- **SUBB** : Subtract From Accumulator with Borrow
- **SWAP** : Swap Accumulator Nibbles
- **XCH** : Exchange Bytes
- **XCHD** : Exchange Digits
- **XRL** : Bitwise Exclusive OR
- **Undefined** : Undefined Instruction

13.8. APPLICATION OF MICROCONTROLLERS

Microcontrollers have found application in nearly each and every walk of life.

The advancement in technology is to make the life of the human being easy and comfortable. This can be well supported with the following example.

13.8.1. Microcontroller Based Washing Machine

The use of washing machine in daily use is well known but now the efforts are made to make the system intelligent.

For this the system has to perform certain functions which incorporate very little attention of the user.

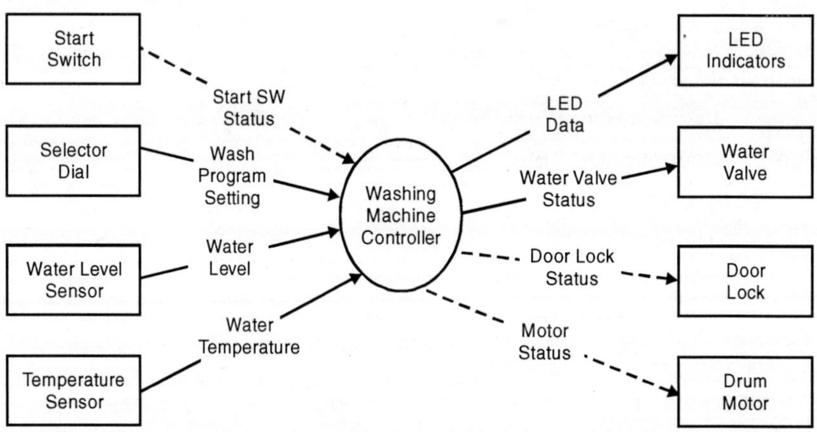

Fig. 13.14. Block Diagram for Microcontroller based Washing Machine.

The block diagram shown in Fig. 13.14 employs the use of water level sensors and temperature indicator and open loop controller. Besides the use of convenventional switches (start switch, selector dial, motor drum control, door lock).

The control scheme involves use of port 3.

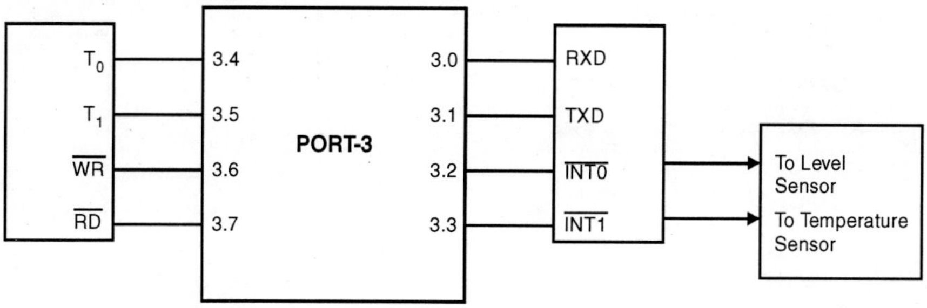

Fig. 13.15. Interfacing diagram for sensor interface.

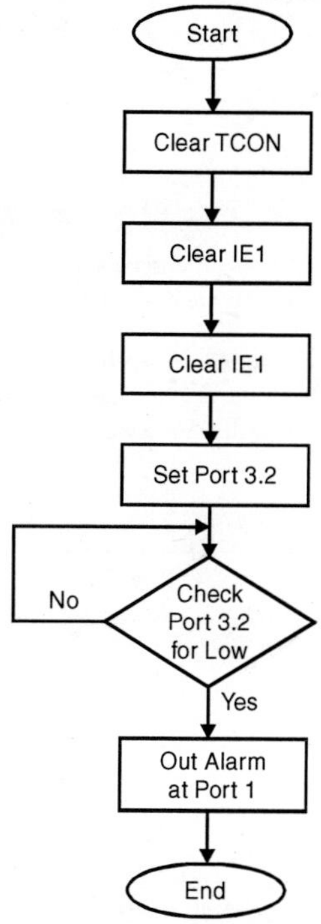

Fig. 13.16. Flow chart for Washing Machine

An appropriate program is written for sensing the level and temperature (pre defined). Once the required set point is attained/ achieved the program executes and an warning signal is sent off to intimate that the whole system is required to be under the control of the user.

The following flow chart is for level indication where a single set point has been set. Once the level is attained at the set point the sensor sends the signal to the microcontroller connected at p. 3.2. At this the high to low transition i.e. $\overline{INT0}$ status is observed and hence the program works.

13.8.2. Microcontroller Based Car Washing System

The car washing system is an assembly line procedure, which involves the following series of steps :

(a) Wet clean

(b) Foam application

(c) Wind screen solution applicator

(d) Wet clean

(e) Drier.

All of the above mentioned steps involves some time interval for proper application. It would be quite clear that microcontroller too finds an application in this field.

As the car enters the washing area the timer is set for each application and at the end of the same the car is made to undergo the next application with some relaxation time and the procedure is repeated for the above mentioned steps.

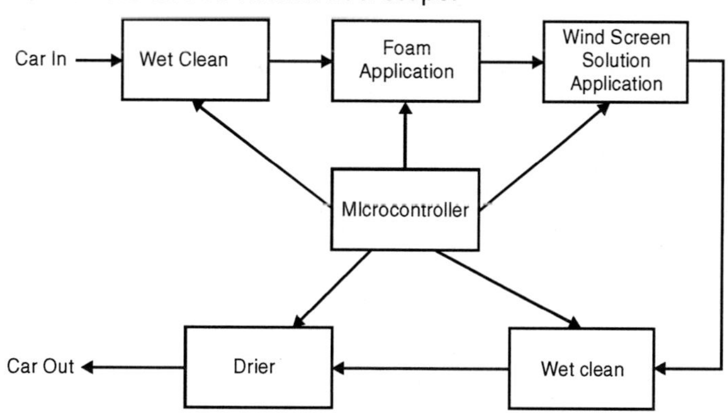

Fig. 13.17. Block Diagram for Microcontroller based car washing system.

Instruction Set for Microprocessor [In alphabetical Order]

(1)

Operation :	ACALL
Function :	Absolute Call Within 2K Block
Syntax :	ACALL *code address*

Instructions	OpCode	Bytes	Cycles	Flags
ACALL *page0*	0x11	2	2	None
ACALL *page1*	0x31	2	2	None
ACALL *page2*	0x51	2	2	None
ACALL *page3*	0x71	2	2	None
ACALL *page4*	0x91	2	2	None
ACALL *page5*	0xB1	2	2	None
ACALL *page6*	0xD1	2	2	None
ACALL *page7*	0xF1	2	2	None

Description : ACALL unconditionally calls a subroutine at the indicated *code address*. ACALL pushes the address of the instruction that follows ACALL onto the stack, least-significant-byte first, most-significant-byte second. The Program Counter is then updated so that program execution continues at the indicated address.

The new value for the Program Counter is calculated by replacing the least-significant-byte of the Program Counter with the second byte of the ACALL instruction, and replacing bits 0-2 of the most-significant-byte of the Program Counter with 3 bits that indicate the page. Bits 3-7 of the most-significant-byte of the Program Counter remain unchanged.

Since only 11 bits of the Program Counter are affected by ACALL, calls may only be made to routines located within the same 2k block as the first byte that follows ACALL.

(2)

Operation :	ADD, ADDC
Function :	Add Accumulator, Add Accumulator With Carry
Syntax :	ADD A,*operand* ADDC A,*operand*

Instructions	OpCode	Bytes	Cycles	Flags
ADD A,#*data*	0x24	2	1	C, AC, OV
ADD A,*iram addr*	0x25	2	1	C, AC, OV
ADD A,@R0	0x26	1	1	C, AC, OV
ADD A,@R1	0x27	1	1	C, AC, OV
ADD A,R0	0x28	1	1	C, AC, OV
ADD A,R1	0x29	1	1	C, AC, OV
ADD A,R2	0x2A	1	1	C, AC, OV
ADD A,R3	0x2B	1	1	C, AC, OV
ADD A,R4	0x2C	1	1	C, AC, OV
ADD A,R5	0x2D	1	1	C, AC, OV
ADD A,R6	0x2E	1	1	C, AC, OV
ADD A,R7	0x2F	1	1	C, AC, OV

Instructions	OpCode	Bytes	Cycles	Flags
ADDC A,#*data*	0x34	2	1	C, AC, OV
ADDC A,*iram addr*	0x35	2	1	C, AC, OV
ADDC A,@R0	0x36	1	1	C, AC, OV
ADDC A,@R1	0x37	1	1	C, AC, OV
ADDC A,R0	0x38	1	1	C, AC, OV
ADDC A,R1	0x39	1	1	C, AC, OV
ADDC A,R2	0x3A	1	1	C, AC, OV
ADDC A,R3	0x3B	1	1	C, AC, OV
ADDC A,R4	0x3C	1	1	C, AC, OV
ADDC A,R5	0x3D	1	1	C, AC, OV
ADDC A,R6	0x3E	1	1	C, AC, OV
ADDC A,R7	0x3F	1	1	C, AC, OV

Description : ADD and ADDC both add the value operand to the value of the Accumulator, leaving the resulting value in the Accumulator. The value operand is not affected. ADD and ADDC function identically except that ADDC adds the value of operand as well as the value of the Carry flag whereas ADD does not add the Carry flag to the result.

The Carry bit (C) is set if there is a carry-out of bit 7. In other words, if the unsigned summed value of the Accumulator, operand and (in the case of ADDC) the Carry flag exceeds 255 Carry is set. Otherwise, the Carry bit is cleared.

The Auxiliary Carry (AC) bit is set if there is a carry-out of bit 3. In other words, if the unsigned summed value of the low nibble of the Accumulator, operand and (in the case of ADDC) the Carry flag exceeds 15 the Auxiliary Carry flag is set. Otherwise, the Auxiliary Carry flag is cleared.

The Overflow (OV) bit is set if there is a carry-out of bit 6 or out of bit 7, but not both. In other words, if the addition of the Accumulator, operand and (in the case of ADDC) the Carry flag treated as signed values results in a value that is out of the range of a signed byte (-128 through +127) the Overflow flag is set. Otherwise, the Overflow flag is cleared.

(3)

Operation :	**AJMP**
Function :	**Absolute Jump Within 2K Block**
Syntax :	**AJMP *code address***

Instructions	OpCode	Bytes	Cycles	Flags
AJMP *page0*	0x01	2	2	None
AJMP *page1*	0x21	2	2	None
AJMP *page2*	0x41	2	2	None
AJMP *page3*	0x61	2	2	None
AJMP *page4*	0x81	2	2	None
AJMP *page5*	0xA1	2	2	None
AJMP *page6*	0xC1	2	2	None
AJMP *page7*	0xE1	2	2	None

Description : AJMP unconditionally jumps to the indicated *code address*. The new value for the Program Counter is calculated by replacing the least-significant-byte of the Program

Counter with the second byte of the AJMP instruction, and replacing bits 0-2 of the most-significant-byte of the Program Counter with 3 bits that indicate the page *of the byte following the AJMP instruction*. Bits 3-7 of the most-significant-byte of the Program Counter remain unchanged.

Since only 11 bits of the Program Counter are affected by AJMP, jumps may only be made to code located within the same 2k block as the first byte that follows AJMP.

(4)

Operation :	ANL
Function :	Bitwise AND
Syntax :	ANL *operand1, operand2*

Instructions	OpCode	Bytes	Cycles	Flags
ANL *iram addr*,A	0x52	2	1	None
ANL *iram addr,#data*	0x53	3	2	None
ANL A,#*data*	0x54	2	1	None
ANL A,*iram addr*	0x55	2	1	None
ANL A,@R0	0x56	1	1	None
ANL A,@R1	0x57	1	1	None
ANL A,R0	0x58	1	1	None
ANL A,R1	0x59	1	1	None
ANL A,R2	0x5A	1	1	None
ANL A,R3	0x5B	1	1	None
ANL A,R4	0x5C	1	1	None
ANL A,R5	0x5D	1	1	None
ANL A,R6	0x5E	1	1	None
ANL A,R7	0x5F	1	1	None
ANL C,*bit addr*	0x82	2	1	C
ANL C,/*bit addr*	0xB0	2	1	C

Description : ANL does a bitwise "AND" operation between *operand1* and *operand2*, leaving the resulting value in *operand1*. The value of operand2 is not affected. A logical "AND" compares the bits of each operand and sets the corresponding bit in the resulting byte only if the bit was set in both of the original operands, otherwise the resulting bit is cleared.

(5)

Operation :	CJNE
Function :	Compare and Jump If Not Equal
Syntax :	CJNE *operand1,operand2,reladdr*

Instructions	OpCode	Bytes	Cycles	Flags
CJNE A,#*data,reladdr*	0xB4	3	2	C
CJNE A,*iram addr,reladdr*	0xB5	3	2	C
CJNE @R0,#*data,reladdr*	0xB6	3	2	C
CJNE @R1,#*data,reladdr*	0xB7	3	2	C
CJNE R0,#*data,reladdr*	0xB8	3	2	C
CJNE R1,#*data,reladdr*	0xB9	3	2	C
CJNE R2,#*data,reladdr*	0xBA	3	2	C
CJNE R3,#*data,reladdr*	0xBB	3	2	C
CJNE R4,#*data,reladdr*	0xBC	3	2	C
CJNE R5,#*data,reladdr*	0xBD	3	2	C
CJNE R6,#*data,reladdr*	0xBE	3	2	C
CJNE R7,#*data,reladdr*	0xBF	3	2	C

Description : CJNE compares the value of *operand1* and *operand2* and branches to the indicated relative address if *operand1* and *operand2* are not equal. If the two operands are equal program flow continues with the instruction following the CJNE instruction.

The Carry bit (C) is set if *operand1* is less than *operand2*, otherwise it is cleared.

(6)

Operation :	CLR
Function :	Clear Register
Syntax :	CLR *register*

Instructions	OpCode	Bytes	Cycles	Flags
CLR *bit addr*	0xC2	2	1	None
CLR C	0xC3	1	1	C
CLR A	0xE4	1	1	None

Description : CLR clears (sets to 0) all the bit(s) of the indicated register. If the register is a bit (including the carry bit), only the specified bit is affected. Clearing the Accumulator sets the Accumulator's value to 0.

(7)

Operation :	CPL
Function :	Complement Register
Syntax :	CPL *operand*

Instructions	OpCode	Bytes	Cycles	Flags
CPL A	0xF4	1	1	None
CPL C	0xB3	1	1	C
CPL *bit addr*	0xB2	2	1	None

Description : CPL complements *operand*, leaving the result in *operand*. If *operand* is a single bit then the state of the bit will be reversed. If *operand* is the Accumulator then all the bits in the Accumulator will be reversed. This can be thought of as "Accumulator Logical Exclusive OR 255" or as "255-Accumulator." If the *operand* refers to a bit of an output Port, the value that will be complemented is based on the last value written to that bit, not the last value read from it.

(8)

Operation :	DA
Function :	Decimal Adjust Accumulator
Syntax :	DA A

Instructions	OpCode	Bytes	Cycles	Flags
DA	0xD4	1	1	C

Description : DA adjusts the contents of the Accumulator to correspond to a BCD (Binary Coded Decimal) number after two BCD numbers have been added by the ADD or ADDC instruction. If the carry bit is set or if the value of bits 0-3 exceed 9, 0x06 is added to the accumulator. If the carry bit was set when the instruction began, or if 0x06 was added to the accumulator in the first step, 0x60 is added to the accumulator.

The Carry bit (C) is set if the resulting value is greater than 0x99, otherwise it is cleared.

(9)

Operation :	**DEC**
Function :	**Decrement Register**
Syntax :	**DEC *register***

Instructions	OpCode	Bytes	Cycles	Flags
DEC A	0x14	1	1	None
DEC *iram addr*	0x15	2	1	None
DEC @R0	0x16	1	1	None
DEC @R1	0x17	1	1	None
DEC R0	0x18	1	1	None
DEC R1	0x19	1	1	None
DEC R2	0x1A	1	1	None
DEC R3	0x1B	1	1	None
DEC R4	0x1C	1	1	None
DEC R5	0x1D	1	1	None
DEC R6	0x1E	1	1	None
DEC R7	0x1F	1	1	None

Description : DEC decrements the value of *register* by 1. If the initial value of *register* is 0, decrementing the value will cause it to reset to 255 (0xFF Hex). Note : The Carry Flag is NOT set when the value "rolls over" from 0 to 255.

(10)

Operation :	**DIV**
Function :	**Divide Accumulator by B**
Syntax :	**DIV AB**

Instructions	OpCode	Bytes	Cycles	Flags
DIV AB	0x84	1	1	C, OV

Description : Divides the unsigned value of the Accumulator by the unsigned value of the "B" register. The resulting quotient is placed in the Accumulator and the remainder is placed in the "B" register.

The Carry flag (C) is always cleared.

The Overflow flag (OV) is set if division by 0 was attempted, otherwise it is cleared.

(11)

Operation :	DJNZ
Function :	Decrement and Jump if Not Zero
Syntax :	DJNZ *register,reladdr*

Instructions	OpCode	Bytes	Cycles	Flags
DJNZ *iram addr,reladdr*	0xD5	3	2	None
DJNZ R0,*reladdr*	0xD8	2	2	None
DJNZ R1,*reladdr*	0xD9	2	2	None
DJNZ R2,*reladdr*	0xDA	2	2	None
DJNZ R3,*reladdr*	0xDB	2	2	None
DJNZ R4,*reladdr*	0xDC	2	2	None
DJNZ R5,*reladdr*	0xDD	2	2	None
DJNZ R6,*reladdr*	0xDE	2	2	None
DJNZ R7,*reladdr*	0xDF	2	2	None

Description : DJNZ decrements the value of *register* by 1. If the initial value of *register* is 0, decrementing the value will cause it to reset to 255 (0xFF Hex). If the new value of *register* is not 0 the program will branch to the address indicated by *relative addr.* If the new value of *register* is 0 program flow continues with the instruction following the DJNZ instruction.

(12)

Operation :	INC
Function :	Increment Register
Syntax :	INC *register*

Instructions	OpCode	Bytes	Cycles	Flags
INC A	0x04	1	1	None
INC *iram addr*	0x05	2	1	None
INC @R0	0x06	1	1	None
INC @R1	0x07	1	1	None
INC R0	0x08	1	1	None
INC R1	0x09	1	1	None
INC R2	0x0A	1	1	None
INC R3	0x0B	1	1	None
INC R4	0x0C	1	1	None
INC R5	0x0D	1	1	None
INC R6	0x0E	1	1	None
INC R7	0x0F	1	1	None
INC DPTR	0xA3	1	2	None

Description : INC increments the value of *register* by 1. If the initial value of *register* is 255 (0xFF Hex), incrementing the value will cause it to reset to 0.

Note : The Carry Flag is NOT set when the value "rolls over" from 255 to 0.

In the case of "INC DPTR", the value two-byte unsigned integer value of DPTR is incremented. If the initial value of DPTR is 65535 (0xFFFF Hex), incrementing the value will cause it to reset to 0. Again, the Carry Flag is NOT set when the value of DPTR "rolls over" from 65535 to 0.

(13)

Operation :	JB
Function :	Jump if Bit Set
Syntax :	JB *bit addr, reladdr*

Instructions	OpCode	Bytes	Cycles	Flags
JB *bit addr,reladdr*	0x20	3	2	None

Description : JB branches to the address indicated by *reladdr* if the bit indicated by *bit addr* is set. If the bit is not set program execution continues with the instruction following the JB instruction.

(14)

Operation :	JBC
Function :	Jump if Bit Set and Clear Bit
Syntax :	JB *bit addr, reladdr*

Instructions	OpCode	Bytes	Cycles	Flags
JBC *bit addr,reladdr*	0x10	3	2	None

Description : JBC will branch to the address indicated by *reladdr* if the bit indicated by *bit addr* is set. Before branching to *reladdr* the instruction will clear the indicated bit. If the bit is not set program execution continues with the instruction following the JBC instruction.

(15)

Operation :	JC
Function :	Jump if Carry Set
Syntax :	JC *reladdr*

Instructions	OpCode	Bytes	Cycles	Flags
JC *reladdr*	0x40	2	2	None

Description : JC will branch to the address indicated by *reladdr* if the Carry Bit is set. If the Carry Bit is not set program execution continues with the instruction following the JC instruction.

(16)

Operation :	JMP
Function :	Jump to Data Pointer + Accumulator
Syntax :	JMP @A+DPTR

Instructions	OpCode	Bytes	Cycles	Flags
JMP @A+DPTR	0x73	1	2	None

Description : JMP jumps unconditionally to the address represented by the sum of the value of DPTR and the value of the Accumulator.

(17)

Operation :	JNB
Function :	Jump if Bit Not Set
Syntax :	JNB *bit addr,reladdr*

Instructions	OpCode	Bytes	Cycles	Flags
JNB *bit addr,reladdr*	0x30	3	2	None

Description : JNB will branch to the address indicated by *reladdress* if the indicated bit is not set. If the bit is set program execution continues with the instruction following the JNB instruction.

(18)

Operation :	JNC
Function :	Jump if Carry Not Set
Syntax :	JNC *reladdr*

Instructions	OpCode	Bytes	Cycles	Flags
JNC *reladdr*	0x50	2	2	None

Description : JNC branches to the address indicated by *reladdr* if the carry bit is not set. If the carry bit is set program execution continues with the instruction following the JNB instruction.

(19)

Operation :	JNZ
Function :	Jump if Accumulator Not Zero
Syntax :	JNZ *reladdr*

Instructions	OpCodc	Bytes	Cycles	Flags
JNZ *reladdr*	0x70	2	2	None

Description : JNZ will branch to the address indicated by *reladdr* if the Accumulator contains any value except 0. If the value of the Accumulator is zero program execution continues with the instruction following the JNZ instruction.

(20)

Operation :	JZ
Function :	Jump if Accumulator Zero
Syntax :	JNZ *reladdr*

Instructions	OpCode	Bytes	Cycles	Flags
JZ *reladdr*	0x60	2	2	None

Description : JZ branches to the address indicated by *reladdr* if the Accumulator contains the value 0. If the value of the Accumulator is non-zero program execution continues with the instruction following the JNZ instruction.

(21)

Operation :	LCALL
Function :	Long Call
Syntax :	LCALL *code addr*

Instructions	OpCode	Bytes	Cycles	Flags
LCALL *code addr*	0x12	3	2	None

Description : LCALL calls a program subroutine. LCALL increments the program counter by 3 (to point to the instruction following LCALL) and pushes that value onto the stack (low byte first, high byte second). The Program Counter is then set to the 16-bit value which follows the LCALL opcode, causing program execution to continue at that address.

(22)

Operation :	LJMP
Function :	Long Jump
Syntax :	LJMP *code addr*

Instructions	OpCode	Bytes	Cycles	Flags
LJMP *code addr*	0x02	3	2	None

Description : LJMP jumps unconditionally to the specified *code addr*.

(23)

Operation :	MOV
Function :	Move Memory
Syntax :	MOV *operand1,operand2*

Instructions	OpCode	Bytes	Cycles	Flags
MOV @R0,*#data*	0x76	2	1	None
MOV @R1,*#data*	0x77	2	1	None
MOV @R0,A	0xF6	1	1	None
MOV @R1,A	0xF7	1	1	None
MOV @R0,*iram addr*	0xA6	2	2	None
MOV @R1,*iram addr*	0xA7	2	2	None
MOV A,*#data*	0x74	2	1	None
MOV A,@R0	0xE6	1	1	None
MOV A,@R1	0xE7	1	1	None
MOV A,R0	0xE8	1	1	None
MOV A,R1	0xE9	1	1	None
MOV A,R2	0xEA	1	1	None
MOV A,R3	0xEB	1	1	None
MOV A,R4	0xEC	1	1	None
MOV A,R5	0xED	1	1	None
MOV A,R6	0xEE	1	1	None
MOV A,R7	0xEF	1	1	None
MOV A,*iram addr*	0xE5	2	1	None
MOV C,*bit addr*	0xA2	2	1	C
MOV DPTR,*#data16*	0x90	3	2	None
MOV R0,*#data*	0x78	2	1	None
MOV R1,*#data*	0x79	2	1	None
MOV R2,*#data*	0x7A	2	1	None
MOV R3,*#data*	0x7B	2	1	None
MOV R4,*#data*	0x7C	2	1	None
MOV R5,*#data*	0x7D	2	1	None
MOV R6,*#data*	0x7E	2	1	None
MOV R7,*#data*	0x7F	2	1	None
MOV R0,A	0xF8	1	1	None
MOV R1,A	0xF9	1	1	None
MOV R2,A	0xFA	1	1	None
MOV R3,A	0xFB	1	1	None
MOV R4,A	0xFC	1	1	None
MOV R5,A	0xFD	1	1	None
MOV R6,A	0xFE	1	1	None
MOV R7,A	0xFF	1	1	None
MOV R0,*iram addr*	0xA8	2	2	None
MOV R1,*iram addr*	0xA9	2	2	None
MOV R2,*iram addr*	0xAA	2	2	None
MOV R3,*iram addr*	0xAB	2	2	None
MOV R4,*iram addr*	0xAC	2	2	None
MOV R5,*iram addr*	0xAD	2	2	None
MOV R6,*iram addr*	0xAE	2	2	None
MOV R7,*iram addr*	0xAF	2	2	None
MOV *bit addr*,C	0x92	2	2	None
MOV *iram addr*,*#data*	0x75	3	2	None
MOV *iram addr*,@R0	0x86	2	2	None
MOV *iram addr*,@R1	0x87	2	2	None
MOV *iram addr*,R0	0x88	2	2	None
MOV *iram addr*,R1	0x89	2	2	None
MOV *iram addr*,R2	0x8A	2	2	None
MOV *iram addr*,R3	0x8B	2	2	None
MOV *iram addr*,R4	0x8C	2	2	None
MOV *iram addr*,R5	0x8D	2	2	None
MOV *iram addr*,R6	0x8E	2	2	None
MOV *iram addr*,R7	0x8F	2	2	None
MOV *iram addr*,A	0xF5	2	1	None
MOV *iram addr*,*iram addr*	0x85	3	2	None

Description : MOV copies the value of *operand2* into *operand1*. The value of *operand2* is not affected. Both *operand1* and *operand2* must be in Internal RAM. No flags are affected unless the instruction is moving the value of a bit into the carry bit in which case the carry bit is affected or unless the instruction is moving a value into the PSW register (which contains all the program flags).

Note : In the case of "MOV iram addr,iram addr", the operand bytes of the instruction are stored in reverse order. That is, the instruction consisting of the bytes 0x85, 0x20, 0x50 means "Move the contents of Internal RAM location 0x20 to Internal RAM location 0x50" whereas the opposite would be generally presumed.

(24)

Operation :	**MOVC**
Function :	**Move Code Byte to Accumulator**
Syntax :	**MOVC A,@A+*register***

Instructions	OpCode	Bytes	Cycles	Flags
MOVC A,@A+DPTR	0x93	1	2	None
MOVC A,@A+PC	0x83	1	1	None

Description : MOVC moves a byte from Code Memory into the Accumulator. The Code Memory address from which the byte will be moved is calculated by summing the value of the Accumulator with either DPTR or the Program Counter (PC). In the case of the Program Counter, PC is first incremented by 1 before being summed with the Accumulator.

(25)

Operation :	**MOVX**
Function :	**Move Data To/From External Memory (XRAM)**
Syntax :	**MOVX *operand1,operand2***

Instructions	OpCode	Bytes	Cycles	Flags
MOVX @DPTR,A	0xF0	1	2	None
MOVX @R0,A	0xF2	1	2	None
MOVX @R1,A	0xF3	1	2	None
MOVX A,@DPTR	0xE0	1	2	None
MOVX A,@R0	0xE2	1	2	None
MOVX A,@R1	0xE3	1	2	None

Description : MOVX moves a byte to or from External Memory into or from the Accumulator.

If *operand1* is @DPTR, the Accumulator is moved to the 16-bit External Memory address indicated by DPTR. This instruction uses both P0 (port 0) and P2 (port 2) to output the 16-bit address and data. If *operand2* is DPTR then the byte is moved from External Memory into the Accumulator.

If *operand1* is @R0 or @R1, the Accumulator is moved to the 8-bit External Memory address indicated by the specified Register. This instruction uses only P0 (port 0) to output the 8-bit address and data. P2 (port 2) is not affected. If *operand2* is @R0 or @R1 then the byte is moved from External Memory into the Accumulator.

(26)

Operation :	MUL
Function :	Multiply Accumulator by B
Syntax :	MUL AB

Instructions	OpCode	Bytes	Cycles	Flags
MUL AB	0xA4	1	4	C, OV

Description : Multiples the unsigned value of the Accumulator by the unsigned value of the "B" register. The least significant byte of the result is placed in the Accumulator and the most-significant-byte is placed in the "B" register.

The Carry Flag (C) is always cleared.

The Overflow Flag (OV) is set if the result is greater than 255 (if the most-significant byte is not zero), otherwise it is cleared.

(27)

Operation :	NOP
Function :	None, waste time
Syntax :	No Operation

Instructions	OpCode	Bytes	Cycles	Flags
NOP	0x00	1	1	None

Description : NOP, as it's name suggests, causes No Operation to take place for one machine cycle. NOP is generally used only for timing purposes. Absolutely no flags or registers are affected.

(28)

Operation :	ORL
Function :	Bitwise OR
Syntax :	ORL *operand1,operand2*

Instructions	OpCode	Bytes	Cycles	Flags
ORL *iram addr*,A	0x42	2	1	None
ORL *iram addr*,#*data*	0x43	3	2	None
ORL A,#*data*	0x44	2	1	None
ORL A,*iram addr*	0x45	2	1	None
ORL A,@R0	0x46	1	1	None
ORL A,@R1	0x47	1	1	None
ORL A,R0	0x48	1	1	None
ORL A,R1	0x49	1	1	None
ORL A,R2	0x4A	1	1	None
ORL A,R3	0x4B	1	1	None
ORL A,R4	0x4C	1	1	None
ORL A,R5	0x4D	1	1	None
ORL A,R6	0x4E	1	1	None
ORL A,R7	0x4F	1	1	None
ORL C,*bit addr*	0x72	2	2	C
ORL C,/*bit addr*	0xA0	2	1	C

Description : ORL does a bitwise "OR" operation between *operand1* and *operand2*, leaving the resulting value in *operand1*. The value of *operand2* is not affected. A logical "OR" compares the bits of each operand and sets the corresponding bit in the resulting byte if the bit was set in either of the original operands, otherwise the resulting bit is cleared.

(29)

Operation :	**POP**
Function :	**Pop Value From Stack**
Syntax :	**POP**

Instructions	OpCode	Bytes	Cycles	Flags
POP *iram addr*	0xD0	2	2	None

Description : POP "pops" the last value placed on the stack into the *iram addr* specified. In other words, POP will load *iram addr* with the value of the Internal RAM address pointed to by the current Stack Pointer. The stack pointer is then decremented by 1.

(30)

Operation :	**PUSH**
Function :	**Push Value Onto Stack**
Syntax :	**PUSH**

Instructions	OpCode	Bytes	Cycles	Flags
PUSH *iram addr*	0xC0	2	2	None

Description : PUSH "pushes" the value of the specified *iram addr* onto the stack. PUSH first increments the value of the Stack Pointer by 1, then takes the value stored in *iram addr* and stores it in Internal RAM at the location pointed to by the incremented Stack Pointer.

(31)

Operation :	**RET**
Function :	**Return From Subroutine**
Syntax :	**RET**

Instructions	OpCode	Bytes	Cycles	Flags
RET	0x22	1	2	None

Description : RET is used to return from a subroutine previously called by LCALL or ACALL. Program execution continues at the address that is calculated by popping the topmost 2 bytes off the stack. The most-significant-byte is popped off the stack first, followed by the least-significant-byte.

(32)

Operation :	**RETI**
Function :	**Return From Interrupt**
Syntax :	**RETI**

Instructions	OpCode	Bytes	Cycles	Flags
RETI	0x32	1	2	None

Description : RETI is used to return from an interrupt service routine. RETI first enables interrupts of equal and lower priorities to the interrupt that is terminating. Program execution continues at the address that is calculated by popping the topmost 2 bytes off the stack. The most-significant-byte is popped off the stack first, followed by the least-significant-byte.

RETI functions identically to RET if it is executed outside of an interrupt service routine.

(33)

Operation :	**RL**
Function :	**Rotate Accumulator Left**
Syntax :	**RL A**

Instructions	OpCode	Bytes	Cycles	Flags
RL A	0x23	1	1	C

Description : Shifts the bits of the Accumulator to the left. The left-most bit (bit 7) of the Accumulator is loaded into bit 0.

(34)

Operation :	**RLC**
Function :	**Rotate Accumulator Left Through Carry**
Syntax :	**RLC A**

Instructions	OpCode	Bytes	Cycles	Flags
RLC A	0x33	1	1	C

Description : Shifts the bits of the Accumulator to the left. The left-most bit (bit 7) of the Accumulator is loaded into the Carry Flag, and the original Carry Flag is loaded into bit 0 of the Accumulator. This function can be used to quickly multiply a byte by 2.

(35)

Operation :	**RR**
Function :	**Rotate Accumulator Right**
Syntax :	**RR A**

Instructions	OpCode	Bytes	Cycles	Flags
RR A	0x03	1	1	None

Description : Shifts the bits of the Accumulator to the right. The right-most bit (bit 0) of the Accumulator is loaded into bit 7.

(36)

Operation :	RRC
Function :	Rotate Accumulator Right Through Carry
Syntax :	RRC A

Instructions	OpCode	Bytes	Cycles	Flags
RRC A	0x13	1	1	C

Description : Shifts the bits of the Accumulator to the right. The right-most bit (bit 0) of the Accumulator is loaded into the Carry Flag, and the original Carry Flag is loaded into bit 7. This function can be used to quickly divide a byte by 2.

(37)

Operation :	SETB
Function :	Set Bit
Syntax :	SETB *bit addr*

Instructions	OpCode	Bytes	Cycles	Flags
SETB C 0xD3	1	1	C	
SETB *bit addr*	0xD2	2	1	None

Description : Sets the specified bit.

(38)

Operation :	SJMP
Function :	Short Jump
Syntax :	SJMP *reladdr*

Instructions	OpCode	Bytes	Cycles	Flags
SJMP *reladdr*	0x80	2	2	None

Description : SJMP jumps unconditionally to the address specified *reladdr*. *Reladdr* must be within –128 or +127 bytes of the instruction that follows the SJMP instruction.

(39)

Operation :	SUBB
Function :	Subtract from Accumulator With Borrow
Syntax :	SUBB A,*operand*

Instructions	OpCode	Bytes	Cycles	Flags
SUBB A,#*data*	0x94	2	1	C, AC, OV
SUBB A,*iram addr*	0x95	2	1	C, AC, OV
SUBB A,@R0	0x96	1	1	C, AC, OV
SUBB A,@R1	0x97	1	1	C, AC, OV
SUBB A,R0	0x98	1	1	C, AC, OV
SUBB A,R1	0x99	1	1	C, AC, OV
SUBB A,R2	0x9A	1	1	C, AC, OV
SUBB A,R3	0x9B	1	1	C, AC, OV
SUBB A,R4	0x9C	1	1	C, AC, OV
SUBB A,R5	0x9D	1	1	C, AC, OV
SUBB A,R6	0x9E	1	1	C, AC, OV
SUBB A,R7	0x9F	1	1	C, AC, OV

Description : SUBB subtract the value of *operand* from the value of the Accumulator, leaving the resulting value in the Accumulator. The value *operand* is not affected.

The Carry Bit (C) is set if a borrow was required for bit 7, otherwise it is cleared. In other words, if the unsigned value being subtracted is greater than the Accumulator the Carry Flag is set.

The Auxiliary Carry (AC) bit is set if a borrow was required for bit 3, otherwise it is cleared. In other words, the bit is set if the low nibble of the value being subtracted was greater than the low nibble of the Accumulator.

The Overflow (OV) bit is set if a borrow was required for bit 6 or for bit 7, but not both. In other words, the subtraction of two signed bytes resulted in a value outside the range of a signed byte (–128 to 127). Otherwise it is cleared.

(40)

Operation :	SWAP
Function :	Swap Accumulator Nibbles
Syntax :	SWAP A

Instructions	OpCode	Bytes	Cycles	Flags
SWAP A	0xC4	1	1	None

Description : SWAP swaps bits 0-3 of the Accumulator with bits 4-7 of the Accumulator. This instruction is identical to executing “RR A” or “RL A” four times.

(41)

Operation :	Undefined instruction
Function :	Undefined
Syntax :	???

Instructions	OpCode	Bytes	Cycles	Flags
???	0xA5	1	1	C

Description : The "Undefined" instruction is, as the name suggests, not a documented instruction. The 8051 supports 255 instructions and OpCode 0xA5 is the single OpCode that is not used by any documented function. Since it is not documented nor defined it is not recommended that it be executed. However, based on my research, executing this undefined instruction takes 1 machine cycle and appears to have no effect on the system except that the Carry Bit always seems to be set.

Note : We received input from an hobbyprojects.com user that the undefined instruction really has a format of *Undefined bit1, bit2* and effectively copies the value of bit2 to bit1. In this case, it would be a three-byte instruction. We haven't had an opportunity to verify or disprove this report, so we present it to the world as "additional information."

(42)

Operation :	XCH
Function :	Exchange Bytes
Syntax :	XCH A,*register*

Instructions	OpCode	Bytes	Cycles	Flags
XCH A,@R0	0xC6	1	1	None
XCH A,@R1	0xC7	1	1	None
XCH A,R0	0xC8	1	1	None
XCH A,R1	0xC9	1	1	None
XCH A,R2	0xCA	1	1	None
XCH A,R3	0xCB	1	1	None
XCH A,R4	0xCC	1	1	None
XCH A,R5	0xCD	1	1	None
XCH A,R6	0xCE	1	1	None
XCH A,R7	0xCF	1	1	None
XCH A,*iram addr*	0xC5	2	1	None

Description : Exchanges the value of the Accumulator with the value contained in *register*.

(43)

Operation :	XCHD
Function :	Exchange Digit
Syntax :	XCHD A,*[@R0/@R1]*

Instructions	OpCode	Bytes	Cycles	Flags
XCHD A,@R0	0xD6	1	1	None
XCHD A,@R1	0xD7	1	1	None

Description : Exchanges bits 0-3 of the Accumulator with bits 0-3 of the Internal RAM address pointed to indirectly by R0 or R1. Bits 4-7 of each register are unaffected.

(44)

Operation :	XRL
Function :	Bitwise Exclusive OR
Syntax :	XRL *operand1,operand2*

Instructions	OpCode	Bytes	Cycles	Flags
XRL *iram addr*,A	0x62	2	1	None
XRL *iram addr*,#*data*	0x63	3	2	None
XRL A,#*data*	0x64	2	1	None
XRL A,*iram addr*	0x65	2	1	None
XRL A,@R0	0x66	1	1	None
XRL A,@R1	0x67	1	1	None
XRL A,R0	0x68	1	1	None
XRL A,R1	0x69	1	1	None
XRL A,R2	0x6A	1	1	None
XRL A,R3	0x6B	1	1	None
XRL A,R4	0x6C	1	1	None
XRL A,R5	0x6D	1	1	None
XRL A,R6	0x6E	1	1	None
XRL A,R7	0x6F	1	1	None

Description : XRL does a bitwise "EXCLUSIVE OR" operation between *operand1* and *operand2*, leaving the resulting value in *operand1*. The value of operand2 is not affected. A logical "EXCLUSIVE OR" compares the bits of each operand and sets the corresponding bit in the resulting byte if the bit was set in either (but not both) of the original operands, otherwise the bit is cleared.

REVIEW QUESTIONS

1. Enlist the salient features of 8051 family of microprocessors ?
2. Discuss the advantages of microcontroller based systems over microprocessor based systems ?
3. Draw and discuss the internal architecture of 8051 ?
4. Discuss the register set of 8051 ?
5. Draw and explain the pin configuration of the 8051 in detail ?
6. Draw and discuss the bit definitions of the PCON, SFR of 8051 ?
7. Draw and discuss the bit definitions of the TCON, SFR of 8051 ?
8. Draw and discuss the bit definitions of the SCON, SFR of 8051 ?
9. Draw and discuss the bit definitions of the TMOD, SFR of 8051 ?
10. Draw and discuss the bit definitions of the IE, SFR of 8051 ?
11. Draw and discuss the bit definitions of the IP, SFR of 8051 ?
12. Draw and discuss the bit definitions of the PSW, SFR of 8051 ?
13. Explain in detail the pin configuration of 8051 ?
14. Explain the concept of Memory Expanding ?
15. Explain in detail the memory structure for 8051 ?
16. Explain the Interrupt Structure of 8051 ?
17. Explain in detail the instruction set of 8051?
18. What are the advantages and disadvantages of using Harvard architecture in 8051 ?
19. How much maximum external program memory can be interfaced?
20. Explain PSW SFR. Give the application differences between Carry and Overflow flags?
21. What are the power consumptions in power down and idle modes?
22. Explain Quasi Bidirectional ports of 8051?
23. What is the status of all registers on reset?
24. What is the maximum delay the Timer0 produces when 8051 is operated at 12MHz?
25. Explain how in Serial communication mode 0 expands I/O lines with the help of shift

26. On what condition JZ quit become true.
27. What are the values of RS0 and RS1 of PSW when 19H location is treated as a register?
28. Explain JBC bit, exit instruction?
29. Does DA A instruction converts binary number to BCD number? Explain under what conditions the BCD numbers gets adjusted after BCD addition
30. Differentiate between RET and RETI instructions
31. What are program branch ranges of SJMP, AJMP and LJMP instructions
32. Write an 8051 assembly program to check a byte is a palindrome. Palindrome is a byte or a word or words when read left or right it will be the same. Like for e.g. C3H
    (a) 11000011 (b) MALAYALAM.
33. Write an 8051 assembly program to produce a software delay of 1 minute.
34. Write an 8051 assembly program to multiply two 16 bit numbers, using shift left and add algorithm.
35. Write an 8051 assembly program to compute the square root of a 16 bit number using shift left and subtract method
36. Write an 8051 assembly program to find LCM of two 16 bit numbers.
37. Write an 8051 assembly program to search a key in a array of 16 bit numbers using Binary search algorithm.
38. Sketch and explain the interface of 5x4 key matrix using 74923.
39. Write an 8051 assembly program segment to input the code of keys.
40. Sketch and explain the interface of 4X7Segment LEDs in multiplexed mode.
41. Write an 8051 assembly program segment to flash CEDT at 5Hz
42. Sketch the interface of a 16ch x 1line LCD to the 8051 microcontroller. Write an 8051 assembly program segment to display any Logo

33\. Sketch the interface of a serial ADC MAX 192 to the 8051 microcontroller. Write an 8051 assembly program segment to read an analog signal through the ADC

34\. Sketch the interface of a dual DAC 7303 to the 8051 microcontroller. Output the DAC outputs are connected to an analog comparator. A LED is connected to the output of the comparator for indication. Write an 8051 assembly program segment to flash if one analog input is greater than the other.

35\. Sketch the interface of a RTC 1302 to the 8051 microcontroller. Write an 8051 assembly program segment to read and write real time into the RTC.

❑❑❑

APPENDIX

08H	READ STANDARD INPUT WITHOUT ECHO
Entry	AH = 08H
Exit	AL = ASCII character
Notes	Performs as function 07H, except that it reads the standard input device. The standard input device can be assigned as either the keyboard or the COM port. This function also responds to a control break, where function 06H and 07H do not. A control-break causes INT 23H to execute. By default, this functions as does function 07H.
09H	DISPLAY A CHARACTER STRING
Entry	AH = 09H
	DS:DX = address of the character string
Notes	The character string must end with an ASCII (24H). The character string can be of any length and may contains control characters such as carriage return (0DH) and line feed (0AH)
0AH	BUFFERED KEYBOARD INPUT
Entry	AH = 0AH
	DS:DX = address of keyboard input buffer
Notes	The first byte of the buffer contains the size of the buffer (up to 255). The second byte is filled with the number of characters typed upon return. The third byte through the end of the buffer contains the character string typed followed by a carriage return (0DH). This function continues to read the keyboard (displaying data as typed) until either the specified number of characters are typed or until the enter key is typed.

DOS Function Calls

In order to use DOS function calls, always place the function number into register AH and load all other pertinent information into registers as described in the table as entry data. Once this is accomplished, follow with.an INT 21H to execute the DOS function.

Example : To display ASCII on CRT at the current cursor position with DOS instruction call.

```
0000  B4  06  MOV AH, 6    ; load function 06H
0002  B2  41  MOV DL, 'A'  ; select letter 'A'
0004  CD  21  INT 21H      ; call DOS function
```

List of DOS Function Calls

00H	TERMINATE A PROGRAM
Entry	AH = 00H
	CS = program segment prefix address
Exit	DOS is entered.
01H	READ THE KEYBOARD
Entry	AH = 01H
Exit	AL = ASCII character
Notes	If Al = 00H the function call must be invoked again to read an extended ASCII character. This function call automatically echoes whatever is typed to the video screen.
02H	WRITE TO STANDARD OUTPUT DEVICE
Entry	AH = 02H
	DL = ASCII character to be displayed
Notes	This function call normally displays data on the video display
03H	READ CHARACTER FROM COM1
Entry	AH = 03H
Exit	AL = ASCII character read from the communications port
Notes	This function call reads data from the serial communications port
04H	WRITE TO COM1
Entry	AH = 04H
	DL = character to be sent, out of COM1
Notes	This function transmits data through the serial communications port. The COM port assignment can be changed to use other COM ports with functions 03H and 04H by using the DOS MODE command to reassign COMI to another COM port.
0BH	TEST STATUS OF THE STANDARD INPUT DEVICE
Entry	AH = 0BH
Exit	AL = status of the input device
Notes	This function tests the standard input device to determine if data are available. If AL = 00, no data are available, if AL = 0FFH then data are available that must. be input using function number 08H.
0CH	CLEAR KEYBOARD BUFFER AND INVOKE KEYBOARD FUNCTION
Entry	AH = 0CH
Exit	AL = 01H, 06H, 07H, or 0AH
Notes	See exit for functions 01H, 06H, 07H or OAH
0DH	The keyboard buffer holds keystrokes while programs execute other tasks.

	This function empties or clears the buffer and then invokes the keyboard function located in register AL.
Entry	FLUSH DISK BUFFERS AH = 0DH
Notes	Erases all file names stored in disk buffers. This function does not close the files specified by the disk buffers, so care must be exercised in its usage.
05H	WRITE TO LPT1
Entry	AH = 05H
	DL = ASCII character to be printed
Notes	Prints DL on the line printer attached to LPT1. Note that the line printer port can be changed with the DOS MODE command.
06H	DIRECT CONSOLE READ/WRITE AH = 06H
Entry	DL = 0FFH or DL = ASCII character
Exit	AL = ASCII character
Notes	IF DL = OFFH on entry, then this function reads the console. If DL = ASCII character, then this function displays the ASCII character on the console (CON) video screen.
	If a character is read from the console keyboard, the zero flag (ZF) indicates whether a character was typed, A zero condition indicates no key was typed, and a not-zero condition indicates that AL contains the ASCII code of the key or a 00H. If AL = 00H, the function must again be invoked to read an extended ASCII character from the keyboard. Note that the key does not echo to the video screen.
07H	DIRECT CONSOLE INPUT WITHOUT ECHO
Entry	AH = 07H
Exit	AL = ASCII character
Notes	This functions exactly as function number 06H with DL = 0FFH, but it will not return from the function until the key is typed.
0EH	SELECT DEFAULT DISK DRIVE
Entry	AH = 0EH
	DL = desired default disk drive number
Exit	AL = the total number of drives present in the system
Notes	Drive A = 00H, drive B = 01H. drive C = 02H and so forth
0FH	@ OPEN FILE WITH FCB
Entry	AH = 0FH
	DS:DX = address of the unopened file control block (FCB)
Exit	AL = 00H if file found
	AL = 0FFH if file not found
Notes	The file control block (FCB) is only used with early DOS software and should never be used with new programs. File control blocks do not allow path names as do the newer file access function codes presented later. To open a file, the file must either be present on the disk or be created with function call 16H.
10H	@ CLOSE FILE WITH FCB
Entry	AH = 10H
	DS:DX = address of the opened file control block (FCB)

Exit	AL = 00H if the closed
	AL = 0FFH if error found
Notes	Errors that occurs usually indicate either that the disk is full or the media is bad.
11H	@SEARCH FOR FIRST MATCH (FCB)
Entry	AH = 11H
	DS:DX = address of the file control block to be searched
Exit	AL = 00H if file found
	AL = 0FFH if file is not found
Notes	Wild card character (? or *) may be used to search for a file name. The wild card character matches any character and the * matches any name or extension.
12H	@ SEARCH FOR NEXT MATCH (FCB)
Entry	AH = 12H
	DS:DX = address of the file control block to be searched
Exit	AL = 00H if file found
	AL = 0FFH if file not found
Notes	This function is used after function 11H finds the first matching file name.
13H	@DELET FILE USING FCB
Entry	AH = 13H
	DS:DX = address of the file control block to be deleted.
Exit	AL = 00H if file deleted
	AL = 0FFH if error occurred
Notes	Errors that most often occur are defective media errors
14H	@SEQUENTIAL READ (FCB)
Entry	AH = 14H
	DS:DX = address of the file control block to be read
Exit	AL = 00H if read successful
	AL = 01H if end of file reached
	AL = 02H if DTA had a segment wrap
	AL = 03H if less than 128 bytes were read
15H	@SEQUENTIAL WRITE (FCB)
Entry	AH = 15H
	DS:DX = address of the file control block to be written
Exit	AL = 00H if write successful
	AL = 01H if disk is full
	AL = 02H if DTA had a segment wrap
16H	@CREATE A FILE (FCB)
Entry	AH = 16H
	DS:DX = address of an unopened file control block
Exit	AL = 00H if file created
	AL = 01H if disk is full

17H	@RENAME A FILE (FCB)
Entry	AH = 17H
	DS:DX - address of a modified file control block
Exit	AL = 00H if file renamed
	AL = 01H if error occurred
19H	RETURN CURRENT DRIVE
Entry	AH = 19H
Exit	AL = current drive
Notes	AL = 00H for drive A, 01H for drive B. and so forth
1AH	SET DISK TRANSFER AREA
Entry	AH = 1AH
	DS:DX = address of new DTA
Notes	The disk trasfer area i» normally located within the program segment prefix at offset address 80H. The DTA is used by DOS for all disk data transfers using file control blocks.
1BH	GET DEFAULT DRIVE FILE ALLOCATION TABLE (FAT)
Entry	AH = IBM
Exit	AL = number of sectors per cluster
	DS:BX = address of the media-descriptor
	CS = size of a sector in bytes
	CX = number of clusters on drive
Notes	The DS register is chained by this function, so make sure to save it before this function.
1CH	GET ANY DRIVE FILE ALLOCATION TABLE (FAT)
Entry	AH = 1CH
	PL = disk drive number
Exit	AL = number of sectors per cluster
	DS:BX = address of the media-descriptor
	CX = size of a sector in bytes
	DX = number of clusters on drive
21H	@RANDOM READ USING FOB
Entry	AH = 21H
	DS:DX = address of opened FOB
Exit	AL = 00H if read successful
	AL = 01H if end of file reached
	AL = 02H if the segment wrapped
	AL = 03H if less than 128 bytes read
22H	@RANDOM WRITE USING FOB
Entry	AH = 22H
	DS:DX = address of opened FOB
Exit	AL = 00H if write successful
	AL = 01H if disk full
	AL = 02H if the segment wrapped

23H	@RETURN NUMBER OF RECORDS (FOB)
Entry	AH = 23H
	DS:DX = address of FOB
	AL = 00H number of records
	AL = 0FFH if file not found
24H	@SET RELATIVE RECORD SIZE (FOB)
Entry	AH = 24H
	DS:DX = address of FCB
Note	Sets the record field to the value contained in the FCB
25H	SET INTERRUPT VECTOR
Entry	AH = 25H
	AL = interrupt vector number
	DS:DX = address of new interrupt procedure
Notes	Before changing the interrupt vector, it is suggested that the current interrupt vector first be saved using DOS function 35H. This allows a back link so the original vector can later be restored.
26H	CREATE NEW PROGRAM SEGMENT PREFIX
Entry	AH = 26 H
	DX = segment, address of new PSP
27H	@RANDOM FILE BLOCK READ (FCB)
Entry	AH = 27H
	CX = the number of records
	DS:DX = address of opened FCB
Exit	AL = 00H if read successful
	AL = 0IH if end of file reached
	AL = 02H if the segment wrapped
	AL = 03H if less than 128 bytes read
	CX = the number of records read
28H	@RANDOM FILE BLOCK WRITE (FCB)
Entry	AH = 28 H
	CX = the number of records
	DS:DX = address of opened FOB
Exit	AL = 00H if write successful
	AL = 01H if disk full
	AL = 0211 if the segment wrapped
	CX = the number of records written
29H	@PARSE COMMAND LINE (FCB)
Entry	AH = 29H
	AL = parse mask
	DS:SI = address of FOB
	DS:DI — address of command line
Exit	AL = 00H if no file name characters found

	AL = 01H if file name characters found
	AL = 0FFH if drive specifier incorrect.
	DS:SI = address of character after name
	DS:DI = address first byte of FOB
2AH	READ SYSTEM DATE
Entry	AH = 2AH
Exit	AL = day of the week
	CX = the year (1980-2099)
	DH = the month
	DL = day of the month
Notes	The day of the week is encoded as Sunday = 00H through Saturday = 06H. The year is a binary number equal to 1980 through 2099.
2BH	SET SYSTEM DATE
Entry	AH = 2BH
	CX = the year (1980-2099)
	DH = the month
	DL = day of the month
2CH	READ SYSTEM TIME
Entry	AH = 20H
Exit	CM = hours (0-23)
	CL = minutes
	DH = seconds
	DL = hundredths of seconds
Notes	All times are returned in binary form, and hundredths of second may not be available.
2DH	SET SYSTEM TIME
Entry	AH = 2DH
	CH = hours
	CL = minutes
	DH = seconds
	DL = hundredths of seconds
2EH	DISK VERIFY WRITE
Entry	AH = 2EH
	AL = 00H to disable verify on write
	AL = 01H to enable verify on write
Notes	By default, disk verify is disabled
2FH	READ DISK TRANSFER AREA ADDRESS
Entry	AH = 2FH
Exit	ES:BX = contains DTA address
30H	READ DOS VERSION NUMBER
Entry	AH = 30H
Exit	AH = fractional version number

	AL = whole number version number
Notes	For example, DOS version number 3.2 is returned as a 3 in AL and 1 14H in AH
31H	TERMINATE AND STAY RESIDENT (TSR)
Entry	AH = 31H
	AL = the DOS return code
	DX = number of paragraphs to reserve for program.
Notes	A paragraph is 16 bytes and the DOS return code is read at the batch file level with ERRORCODE
33H	TEST CONTROL-BREAK
Entry	AM = 33H
	AL = 00H to request current control-break
	AL = 01H to change control-break
	DL = 00H to disable control-break
	DL = 01H to enable control-break
Exit	DL = current control-break state
34H	GET ADDRESS OF In DOS FLAG
Entry	AH = 34H
Exit	ES:BX = address of InDOS flag
Notes	The InDOS flag is available in DOS versions 3.2 or newer and indicates DOS activity. If InDOS = 00H, DOS if inactive or 0FFH if DOS is active and pursuing another operation.
35H	READ INTERRUPT VECTOR
Entry	AH = 35H
	AL = Interrupt vector number
Exit	ES : BX = address stored at vector
Notes	This DOS function is used with function 25H to install/remove interrupt handlers.
36H	DETERMINE FREE DISK SPACE
Entry	AH = 36H
	DL = Drive number
Exit	AX = FFFFH if drive invalid
	AX = Number of sectors per cluster
	BX = Number of free clusters
	CX = bytes per sector
	DX = number of clusters on drive
Notes	The default disk drive is DL = 00H, drive A = 01H, drive B = 02H, and so forth
38H	RETURN COUNTRY CODE
Entry	AH = 38H
	AL = 00H for current country code
	BX = 16-bit country code
	DS:DX = data buffer address

Exit	AX = error code if carry set
	BX = counter code
	DS : DX = data buffer address
39H	CREATE SUB-DIRECTORY
Entry	AH = 39H
	DS : DX = address of ASCII-Z string subdirectory name
Exit	AX = error code if carry set
Notes	The ASCII-Z string is the name of the subdirectory in ASCII code ended with a 00H instead of a carriage return/line feed.
3AH	ERASE SUB-DIRECTORY
Entry	AH = 3 AH
	DS : DX = address of ASCII-Z string subdirectory name
Exit	AX = error code if carry set
3BH	CHANGE SUBDIRECTORY
Entry	AH = 3BH
	DS : DX = address of new ASCII-Z string subdirectory name
Exit	AX = error code if carry set
3CH	CREATE A NEW FILE
Entry	AH = 3CH
	CX = attribute word
	DS : DX = address of ASCII-Z string file name
Exit	AX = error code if carry set
	AX = file handle if carry cleared
Notes	The attribute word can contain any of the following *(added together):* 01H read only access, 02H = hidden file or directory, 04H = system file, 08H = volume label, 10H - subdirectory, and 20H = archive bit. In most cases, a file is created with 0000H.
3DH	OPEN A FILE
	EntryAH= 3DH
	AL = access code
	DS : DX = address of ASCII-Z string file name
	ExitAX = errror code if carry set
	AX = file handle if carry cleared
Notes	The access code in AL = 00H for a read-only access, AL = 01H for a write-only access, and AL = 02H for a read/write access.
	For shared file in a network environment, bit 4 of AL = 1 will deny read/write access, bit 5 of AL = 1 will deny a write access, bits 4 and 5 of AL = 1 will deny read access, bit 6 of AL = 1 denies none, bit 7 of AL = 0 causes the file to be inherited by child, if bit 7 of AL = 1, file is restricted to current process.
3EH	CLOSE A FILE
Entry	AM = 3EH
	BX = file handle
Exit	AX = error code if carry set

3FH	READ A FILE
Entry	AH = 3FH
	BX = file handle
	OX = number of bytes to be read
	DS:DX = address of file buffer to hold data road
Exit	AX = error code if carry set
	AX = number of bytes read if carry cleared
40H	WRITE A FILE
Entry	AH = 4011
	BX = file handle
	CX = number of bytes to write
	DS:DX = address of file buffer that holds write data
Exit	AX = error code if carry set
	AX = number of bytes written if carry cleared
41H	DELETE A FILE
Entry	AM = 41H
	DS:DX = address of ASCII-2 string file name
Exit	AX = error code if carry set
42H	MOVE FILE POINTER
Entry	AH = 42H
	AL = move technique
	BX = file handle
	CX:DX = number of bytes pointer moved
Exit	AX = error code if carry set.
	AX:DX = bytes pointer moved
Notes	The move technique causes the pointer to move from the start of the file if AL = 0011, from the current location if AL = 01H, and from the end of the file if AL = U2H. The count is stored so DX contains the least-significant 16-bits and either 0X or AX contains the most significant, 16-bits.
43H	READ/WRITE FILE ATTRIBUTES
Entry	AH = 43H
	AL = 00H to read attributes
	AL = 01H to write attributes
	CX = attribute word (see function 3CH)
	DS:DX = address of ASCII-Z file name
Exit	AX = error code if carry set
	CX = attribute word of carry cleared
44H	I/O DEVICE CONTROL (IOTCL)
Entry	AH = 44H
	AL = sub-function code (see notes)
Exit	AX = error code (see function 59H) if carry set
Notes	The sub-function codes found in AL are as follows :
	00H = read device status

Entry BX = handle

Exit DX = status

01H = write device status

Entry : BX = file handle, DH = 0, DL = device information

Exit : Ax = error code if carry set

02H = read control data from character device

Entry : BX = file handle, CX = number of bytes, DS:DX = I/O buffer address

Exit : AX = number of bytes read

03H = write control data to character device

Entry : BX = file handle, CX = number of bytes, DS:DX = I/O buffer address

Exit : AX = number of bytes written

04H = read control data from block device

Entry : BL = drive number (0 = default, 1 = A, 2 = B etc.)

CX = number of bytes, DS:DX = I/O buffer address

Exit : AX = number of bytes read

05H = write control data to block device

Entry : BL = drive number, CX = number of bytes

DS:DX = I/O buffer address

Exit : AX = number of bytes written

06H = Check input status

Entry : BX = file handle

Exit : AL = 00H ready or FFH not ready

07H = check output status

Entry : BX = file handle

Exit : AL = 00H ready or FFH not ready

08H = removable media ?

Entry : BL = drive number

Exit : AL = 00H removable, 01H fixed

09 H =netword block device ?

Entry : BL = drive number

Exit : bit 12 of DX set for network block device

0AH = local or network character device ?

Entry : BX = file handle

Exit : bit of 15 of DX set for network character device

OBH = change entry count (must have SHARE.EXE loaded)

Entry : CX = delay loop count, DX = retry count

Exit : AX = error code if carry set

OCH = generic I/O control for character devices

Entry : BX = file handle, CH = category, CL = function

Categories00H = unknown, 01H = COM port, 02H = CON, 05H = LPT ports

Function

Cl = 45H; set iteration count

CL = 4AH; select code page

CL = 4CH; start code page preparation

CL = 4DH; end code page preparation

CL = 5FH; set display information

CL = 65H; get iteration count

CL = 6AH; query preparation

CL = 7FH; get display information

0DH = generic I/O control for block devices

Entry : BL = drive number, CH = category, CL = function,

DS : DX = address of parameter block

Category : 08H = disk drive

Function :

CL = 40H; set device parameters

CL = 41H; write track

CL= 42H; format and verify track

CL = 46H; set media ID code

CL = 47H; set access flag

CL = 60H; get device parameters

CL = 61H; read track

CL = 62H; verify track

CL = 66H; get media ID code

CL = 67H; get access code

0EH = return logical device map

Entry : BL = drive number

Exit : AL = number of last device

0FH = change logical device map

Entry : BL = drive number

Exit : AL = number of last device

45H	DUPLICATE FILE HANDLE
Entry	AH = 45H
	BX = current file handle
Exit	AX = error code if carry set
	AX = duplicate file handle
46H	FORCE DUPLICATE FILE HANDLE
Entry	AH = 46H
	BX = current file handle
	CX = new file handle
Exit	AX = error code if carry set
Notes	This function works like function 45H except that function 45H allows DOS to select the new handle, while this function allows the user to select the new handle.

47H	READ CURRENT DIRECTORY
Entry	AH = 47H
	DL = drive number
	DS:SI = address of 64 byte buffer for directory name
Exit	DS:SI address current directory name if carry cleared
Notes	Drive A = 00, drive B = 01, and so forth
48H	ALLOCATE MEMORY BLOCK
Entry	AH = 48H
	BX = number of paragraphs to allocate
	CX = new file handle
Exit	BX = largest block available if carry cleared
49H	RELEASE ALLOCATED MEMORY BLOCK
Entry	AH = 49H
	ES = segment address of block to be released
	CX = new file handle
Exit	Carry indicates an error if set
4AH	MODIFY ALLOCATED MEMORY BLOCK
Entry	AH = 4AH
	BX = new block size in paragraphs
	BS = segment address of block to be modified
Exit	BX = largest block available if carry clearod
4BH	LOAD OR EXECUTE A PROGRAM
Entry	AH = 4BH
	AL = function code
	ES:BX = address of parameter block
	DS:DX = address ASCII-Z string command
Exit	Carry indicates an error if set

Notes : The function codes are AL = 00H to load and execute a program, AL = 01H to load a program but not execute it, AL = 03H to load a program overlay, and AL = 05H to enter the EX.EC state.

4CH	TERMINATE A PROCESS
Entry	AH = 4CH
	AL = error code
Exit	Returns control to DOS
Notes	This function returns control to DOS with the error code saved so it can be obtained using DOS ERROR LEVEL batch processing system. We normally use this function with an error code of OOH to return to DOS.
4DH	READ RETURN CODE
Entry	AH = 4DH
Exit	AX = return error code
Notes	This function is used to obtain the return status code created by executing a program with DOS function 4BH. The return codes are AX = OOOOH for a normal-no error-termination, AX = 0001H for a control-break termination,

	AX = 0002H for a critical device error, and AX = 0003H for a termination by an INT 31H.
4EH	FIND FIRST MATCHING FILE
Entry	AH = 4EH
	CX = file attributes
	DS:DX = address ASCII-Z string file name
Exit	Carry is set for file not found
Notes	This function searches the current or named directory for the first matching file. Upon exit, the DTA contains the file information.
4FH	FIND NEXT MATCHING FILE
Entry	AH = 4FH
Exit	carry is set for file not found
Notes	This function is used after the first file is found with function 4EH
50H	SET PROGRAM SEGMENT PREFIX (PSP) ADDRESS
Entry	AH = 50H
	BX = offset address of the new PSP
Notes	Extreme care must be used with this function because no error recovery is possible.
51H	GET PSP ADDRESS
Entry	AH = 51H
Exit	BX = current PSP segment address
	54HREAD DISK VERIFY STATUS
Entry	AH = 54H
Exit	AL = 00H if verify off
	AL = 01H if verify on
56H	RENAME FILE
Entry	AH = 56H
	ES:DI = address of ASCII-Z string containing new file name
	DS:DX = address of ASCII-Z string containing file to be renamed
Exit	Carry is set for error condition
57H	READ FILE'S DATE AND TIME STAMP
Entry	AH = 57H
	AL = function code
	BX = file handle
	CX = now time
	DX = now data
Exit	Carry is set for error condition
	CX = time if carry cleared
	HX = time if carry cleared
Notes	AL = 00H to road date and lime or 01H to write date and time
59H	GET EXTENDED ERROR INFORMATION
Entry	AH = 50H
	BX = 0000H for DOS version 3.X

Exit

AX = extended error code
BH = error class
BL = recommended action
CH = locus

Notes

Following are the extended error codes found in AX :

0001H = invalid function number
0002H = file not found
0008H = path not found
0004H = no file handles available
0005H = access denied
000611 = file handle invalid
0007H = memory control block failure
0008H = insufficient memory
0009H = memory block address invalid
0000H = environment, failure
000BH = format invalid
000CH = access code invalid
000DH = data invalid
000EH = unknown unit.
000FH = disk drive invalid
0010H - attempted to remove current directory
0011H = not same device
0012H = no more files
0013H = disk write-protected
0014H = unknown unit
0015H = drive not ready
0016H = unknown command
0017H = data error (CRC check error)
0018H = bad request structure length
0019H = seek error
001AH = unknown media type
001BH = sector not found
001CH = printer out of paper
001DH = write fault
001EH = read fault
001FH = general failure
0020H = sharing violation
0021H = lock violation
0022H = disk change invalid
0023H = FCB unavailable
0024H = sharing buffer exceeded
0025H = code page mismatch

0026H = handle end of file operation not completed
0027H = disk full
0028H-0031H reserved
0032H = unsupported network request
0033H = remote machine not listed
0034H = duplicate name on network
0035H = network name on network
0036H = network busy
0037H = device no longer exists on network
0038H = net BIOS command limit exceeded
0039H = error in network adapter hardware
003AH = incorrect response from network
003BH = unexpected network error
003CH = remote adapter is incompatible.
003DH = print queue is full
003EH = not enough room for print file
003Fh = print file was deleted
0040H = network name deleted
0041H = network access denied
0042H = incorrect network device type
0043H = network name not found
0044H = network name exceeded limit
0045H = netBIOS session limit exceeded
0046H = temporary pause
0047H = network request not accepted
0048H = print or disk redirection pause
0049H-004FH reserved
0050H = file already exists
0051H = duplicate FCB
0052H = cannot make directory
0053H = failure in INT 24H (critical error)
0054H = too many re-directions
0055H = duplicate redirection
0056H = invalid password
0057H = invalid parameter
0058H = network write failure
0059H = function not supported by network
005AH = required system component not installed
0065H = device not selected

Following are the error class codes as found in BH :

01H = no resources available
02H = temporary error

03H	= authorization error
04H	= internal software error
05H	= hardware error
06H	= system failure
07H	= application software error
08H	= item not found
09H	= invalid format
0AH	= item blocked
0BH	= media error
0CH	= item already exists
0DH	= unknown error

Following in the recommended action as found in BL :

01H	= retry operation
02H	= delay and retry operation
03H	= user retry
04H	= abort processing
05H	= immediate exit
06H	= ignore error
07H	= retry with user intervention

Following is a list of locus in CH :

01H	= unknown source
02H	= block device error
03H	= network area
04H	= serial device error
05H	= memory error

5AH CREATE UNIQUE FILE NAME

Entry AH = 5AH

CX = attribute code

DS:DX = address of the appended directory name

Exit Carry is set for error condition

AX = file handle if carry cleared

DS:DX = address of the appended directory name

Notes The ASCI1-Z file directory path must end with a backslash (\). On exit, the directory name is appended with a unique file name.

5BH CREATE A DOS FILE

EntryAH = 5BH

CX = attribute code

DS:DX = address of the ASCII-Z string contain the file name

ExitCarry is set for error condition

AX = file handle if carry cleared

Notes The function only works in DOS version 3.X or higher. It is almost identical to function 3CH, except that function 3CH erases the file; if it already exists, while function 5BH reports that the file exists without erasing it.

5CH	LOCK/UNLOCK FILE CONTENTS
Entry	AH = 5CH
	BX = file handle
	CX:DX = offset address of locked/unlocked area
	SI:DI = number of bytes to lock or unlock beginning at offset
	ExitCarry is set for error condition
5DH	SET EXTENDED ERROR INFORMATION
Entry	AH = 5DH
	AL = 0AH
	DS:DX = address of the extended error data structure
Notes	This function is used by DOS version 3.1 or higher to store extended error information.
5EH	NETWORK/PRINTER
Entry	AH = 5EH
	AL = 00H (get network name)
	DS:DX = address of the ASCII-Z string containing network name
Exit	Carry is set for error condition
	CL = netBIOS number if carry cleared
Entry	AH = 5EH
	AL = 02H (define network printer)
	BX = redirection list
	CX = length of setup string
	DS:DX = address of printer setup buffer
Exit	Carry is set for error condition
Entry	AH = 5EH
	AL = 03H (read network printer setup string)
	BX = redirection list
	DS:DX = address of printer setup buffer
Exit	Carry is set for error condition
	CX = length of setup string if carry cleared
	ES:DI = address of printer setup buffer
62H	GET PSP ADDRESS
Entry	AH = 62H
Exit	BX = segment address of the current program
Notes	The function only works in DOS version 3.0 or higher.
65H	GET EXTENDED COUNTRY INFORMATION
Entry	AH = 65H
	AL = function code
	ES:DI = address of buffer to receive information
	Carry is set for error condition
	CX = length of country information

Notes	The function only works in DOS version 3.3 or higher
66H	GET/SET CODE PAGE
Entry	AH = 65H
	AL = function code
	BX = code page number
Exit	Carry is set for error condition
	BX = active code page number
	DX = default code page number
Notes	A function code in AL of 01H gets the code page number, and a code of 02H sets the code page number
67H	SET HANDLE COUNT
Entry	AH = 67H
	BX = number of handles desired
Exit	Carry is set for error condition
Notes	This function is available for DOS version 3.3 or higher.
68H	COMMIT FILE
Entry	AH = 68H
	BX = handle number
Exit	Carry is set for error condition; otherwise, the date and time stamp is written to directory
Notes	This function is available for DOS version 3.3 or higher
6CH	EXTENDED OPEN FILE
Entry	AH = 6CH
	AL = 00H
	BX = open mode
	CX = attributes
	DX = open flag
	DS:SI = address of ASCII-Z string file name
Exit	AX = error code if carry is set
	AX = handle if carry is cleared
	CX = 0001H file existed and was opened
	CX = 0002H file did not exist and was created
Notes	This function is available for DOS version 4.0 or higher.

BIOS Function Calls

Apart from DOS function call INT 211.1. Some other BIOS function calls are useful in controlling the I/O the computer. Unlike I NT 2111, which exists in the DOS program, the BIOS function calls are found stored in the system and video BIOS ROMs. These BIOS ROM functions directly control the I/O devices with or without DOS loaded into a system.

INT 10H

The INT 10H BIOS interrupt is often called the video services interrupt because it directly controls the video display in a system. The 1NT 10H instruction uses register. All to select the video service provided by this interrupt. *(The video BIOS ROMS is located on the video board and varies from one video card to another).*

TABLE 1. Video Display Modes.

Mode	Type	Columns	Rows	Resolution	Standard	Colours
00H	Text	40	25	320 × 200	CGA	2
00H	Text	10	25	320 × 250	EGA	2
00H	Text	40	25	360 × 400	VGA	2
01H	Text	40	25	320 × 200	CGA	16
02H	Text	10	25	320 × 350	EGA	16
03H	Text	40	25	360 × 640	VGA	16
02H	Text	80	25	640 × 200	CGA	2
02H	Text	80	25	640 × 350	EGA	2
03H	Text	80	25	720 × 400	VGA	2
03H	Text	80	25	640 × 200	CGA	16
03H	Text	80	25	640 × 350	EGA	16
03H	Text	80	25	720 × 400	VGA	16
03H	Graphics	80	25	320 × 200	CGA	4
05H	Graphics	80	25	320 × 350	CGA	2
06H	Graphics	80	25	640 × 200	CGA	2
07H	Text	80	25	720 × 350	EGA	4
07H	Text	80	25	720 × 400	VGA	4
0DH	Graphics	80	25	320 × 200	CGA	16
0EH	Graphics	80	25	640 × 200	CGA	16
0FH	Graphics	80	25	640 × 350	EGA	4
10H	Graphics	80	25	640 × 350	EGA	16
11H	Graphics	80	30	640 x 480	VGA	2
12H	Graphics	80	30	640 × 480	VGA	16
13H	Graphics	40	25	320 × 200	VGA	256

The VGA can use any mode listed, while the other displays are more restrictive in use. Additional higher resolution modes are explained later in this section.

Example : Short sequence of instruction to place video display into mode 03H operation.

(This mode is available on CGA, EGA and VGA displays. This mode allows the display to draw test data with 16 colours at various resolutions dependent upon the display adapter)

```
0000   B4  00  MOV AH 0   ;select mode
0002   B0  03  MOV AL 3   ; mode is 03H
0004   CD  10  INT 10H
```

Cursor Control and Other Standard Features

Appendix II illustrates the function codes (*placed in AH*) used *to control the cursor on the video display. (These cursor control functions will work on any video display from the CGA display to the latent super VGA display. It also lists the functions used to display data and change to a different character set).*

Video BIOS (INT 10H) functions

00H	SELECT VIDEO MODE
Entry	AH = 00H
	AL = mode number
Exit	Most changed and screen cleared
01H	SELECT CURSOR TYPE
Entry	AH = 02H
	CH = starting line number
	CL = ending line number
Exit	Cursor size changed
02H	SELECT CURSOR POSITION
Entry	AH = 02H
	BH = page number (usually 0)
	DH = row number (beginning with 0)
	DL = column number (beginning with 0)
Exit	Changes cursor to new position
03H	READ CURSOR POSITION
Entry	AH = 03H
	BH = page number
Exit	CH = starting line (cursor size)
	CL = ending line (cursor size)
	DH = current row
	DL = current column
04H	READ LIGHT PEN
Entry	AH = 04H (not supported in VGA)
Exit	AH = 0, light pen triggered
	BX = pixel column
	CX = pixel row
	DH = character row
	DL = character column
05H	SELECT DISPLAY PAGE
Entry	AH = 05H
	AL = page number
Exit	Page number selected, Following are the valid page numbers :
	Mode 0 and 1 support pages 0-7
	Mode 2 and 3 support pages 0-7
	Mode 4, 5 and 6 support page 0
	Mode 7 and D support pages 0-7
	Mode E supports pages 0-3
	Mode F and 10 support pages 0-1
	Mode 11, 12 and 13 support page 0
Notes	Most modern displays use page 0 for most operations.

06H	SCROLL PAGE UP
Entry	AH = 06H
	AL = number of lines to scroll (0 clears window)
	BH = character attribute for new lines
	CH = top row of scroll window
	CL = left column of scroll window
	DH = bottom row of scroll window
	DL = right column of scroll window
Exit	Scrolls window from the bottom toward the top of the screen. Black lines fill the bottom using the character attribute in BH.
07H	SCROLL PAGE DOWN
Entry	AH = 07H
	AL = number of lines to scroll (0 clears window)
	BH = character attribute for new lines
	CH = top row of scroll window
	CL = left column of scroll window
	DH = bottom row of scroll window
	DL = right column of scroll window
Exit	Scrolls window from the top toward the bottom of the screen. Black lines fill from the top using the character attribute in BH.
08H	READ ATTRIBUTE/CHARACTER AT CURRENT CURSOR POSITION
Entry	AH = 08H
	BH = page number
Exit	AL = ASCII character code
	AH = character attribute
Notes	This function does not advance the cursor
09H	WRITE ATTRIBUTE/CHARACTER AT CURRENT CURSOR POSITION
Entry	AH = 09H
	AL = ASCII character code
	BH = page number
	BL = character attribute
	CX = number of characters to write
Notes	This function does not advance the cursor.
0AH	WRITE CHARACTER AT CURRENT CURSOR POSITION
Entry	AH = 0Ah
	AL = ASCII character code
	BH = page number
	CX = number of characters to write
Notes	This function does not advance the cursor
0FH	READ VIDEO MODE
Entry	AH = 0FH
Exit	Al = current video mode

	AH = number of character columns
	BH = page number
10H	SET VGA PALETTE REGISTER
Entry	AH = 10H
	AL = 10H
	BX = colour number (0-255)
	CH = green (0-63)
	CL = blue (0-63)
	DH = red (0-63)
Exit	Palette register colour is changed.
Notes	The first 16 colours (0-15) are used in the 16-colour, VGA text mode and other modes
10H	READ VGA PALETTE REGISTER
Entry	AH = 10H
	AL = 15H
	BX = colour number (0-255)
Exit	CH = green
	CL = blue
	DH = red
11H	GET ROM CHARACTER SET
Entry	AH = 11H
	AL = 30H
	BH = 2 = ROM 8 × 14 character set
	BH = 3 × ROM 8 × 8 character set
	BH = 4 × ROM 8 × 8 extended character set
	BH = 5 × ROM 9 × 14 character set
	BH = 6 × ROM 8 × 16 character set
	BH = 7 × ROM 9 × 16 character set
Exit	CX = bytes per character
	DL = rows per character
	ES:BP = address of character set

❑❑❑

QUESTION BANK

ARCHITECTURE OF MICROPROCESSORS

(a) General Definitions of Mini Computers etc.

1. Differentiate between a microprocessor and a micro controller.
2. Differentiate between a microprocessor and digital signal processor.

(b) Overview of 8085 Microprocessor

1. List the internal registers in 8085 microprocessor and their abbreviations and lengths. Describe the primary function of each register.
2. List five levels of interrupts in 8085 microprocessor with priority.
3. Interface a key to SID pin of 8085 Microprocessor.
4. Interface a LED to SOD pin of 8085 Microprocessor.
5. In 8085 microprocessor which has higher the priority NMI or DMA.
6. What are the differences between Memory mapped I/O and I/O mapped I/O.

(c) Overview of 8086 Microprocessor

1. Explain the need of segmentation.
2. List the internal registers in 8086 Microprocessor.
3. Explain the roles of BIU and EU.
4. What are the advantages of pipelining.
5. Explain all the flags in 8086.

(d) Signals and pins of 8086 Microprocessor

1. List the signals in minimum and maximum modes.
2. Explain the roles of pins TEST, LOCK.
3. Which are the pins of 8086 that are to be connected to interface 8284 and explain their functions ?
4. Which are the pins of 8086 that are to be connected to 8087 and explain their functions.

ASSEMBLY LANGUAGE OF 8086

(a) Description of Instructions

1. If AL = –9 and BL = 4710 after IDIV BL what are the values of AL and AH.

2. IF AX = -20010 and CX = 670H after IMUL CX what are the values of AX and DX
3. Explain AAA, AAD, AAM, AAS instructions with examples.
4. Explain DAA, DAS instructions with examples.
5. Explain the instructions related to the fixed and variable ports.
6. Explain the instructions related to arithmetic and logical shift.
7. How REP instruction is used along with string instructions.
8. Explain different types of CALL instructions

(b) Assembly Directives

1. What is the length of bytes reserved for the following directive

```
STORE DW 100
DUP(0)
```

2. What is the difference between ENDS and ENDP directives.
3. Explain PTR directive.

(c) Algorithms with assembly software programs

1. Write an algorithm to convert BCD to Binary numbers. Write 8086 assembly program to convert two digit BCD number to hexadecimal number
2. Write an algorithm to convert Binary number to BCD number. Write 8086 assembly program to convert one byte Binary number to BCD.
3. Write an algorithm to evaluate a factorial of an integer number N. Write an assembly program using recursive procedure.
4. Write an algorithm to find GCD of two numbers. Write an assembly program to find GCD of two words.
5. Write an algorithm and assembly program to sort the numbers in an array in descending order using bubble sort method

INTERFACING WITH 8086

(a) Interfacing with RAM's and ROM's

1. Sketch and explain the interface of 32K x 16 ROMs using a decoder in minimum mode. What is the maximum access time of ROMs such that it does not require wait states when 8086 operates at 8 MHz
2. Sketch and explain the interface of 8K x 16 RAMs using a decoder in minimum mode. What is the maximum access time of RAMs such that it does not require wait states when 8086 operates at 8 MHz
3. Sketch and explain the 8086 bus activities during read machine cycle

(b) Interfacing with peripheral IC's like 8255 etc.

1. Sketch and explain the interface of PPI 8255 to the 8086 microprocessor in minimum mode. Interface 8 LEDs to the port B of 8255. Interface 8 keys to the port A. Write an 8086 assembly program to read the key status and output on to the 8 LEDs
   (i) Interface an 8 bit ADC 808 to port A. Derive control signals from port C. Write an 8086 assembly program segment to read an analog signal.
   (ii) Interface an 8 bit DAC 08 to port A. Write an 8086 assembly program segment to output a ramp.

(iii) Interface 16 ch x 1Line LCD to port A. Derive control signals from port C. Write an 8086 assembly program segment to flash WELCOME TO CEDT.

2. Sketch and explain the interface of PIT 8254 to the 8086 microprocessor in minimum mode. Write an 8086 assembly program to generate a clock of 10 Hz on the OUT 0 pin. Write an 8086 assembly program to generate a hardware triggerable mono-shot of 1 msec pulse width.
3. Sketch and explain the interface of 8279 to the 8086 microprocessor in minimum mode. Interface 8x8 key pad and 16x 7 Seg LED display. Write an 8086 assembly program to read the key codes of keys and display -IISc-BANGALORE-
4. Sketch and explain the interface of PIC 8259 to the 8086 microprocessor in minimum mode. Show the cascading of additional eight 8259s to provide 64 external interrupts. Write an 8086 assembly program to initialize master 8259 and slaves.

ARCHITECTURE OF MICROCONTROLLERS

(a) Overview of 8051 micro controller architecture

1. How Harvard architecture is implemented in 8051
2. List the features of 8051
3. What are the alternate functions of Port 3, port 2, and Port 0
4. List all SRF's
5. How do you differentiate between SFR bits and internal memory bits
6. What are the advantages of register banks
7. What is the maximum program memory and data memory can be interfaced externally
8. List all interrupt vectors.

(b) Overview of 8096 micro controller architecture

1. What is RALU architecture in 8096
2. List all the features of 8096
3. What is reset location of 8096
4. Explain HSI and HSO in 8096
5. How DAC is realized using PWM output of 8096
6. Explain three operands instruction in 8096
7. Explain watch dog timer in 8096
8. Explain ADC in 8096
9. What is the maximum baud rate possible in 8096
10. Bring out the differences between T1 and T2 timers in 8096.

ASSEMBLY LANGUAGE OF 8051

(a) Description of instructions

1. Explain MUL AB and DIV AB instructions.
2. Explain two indirect indexed instructions.
3. Explain all flags in 8051.
4. Explain all Boolean instructions.
5. Explain all compare instructions.
6. Write a program to decrement the value of DPTR.

(b) Assembly Directives

1. Explain 8051 assembler directives EQU, SET and BIT with one example each.
2. Explain 8051 assembler directives DBIT and DS with one example each.

(c) Algorithms with 8051 assembly language programs

1. Write an 8051 assembly program to exchange the data using PUSH and POP instructions.
2. Write an 8051 assembly program to convert two digit BCD number to hexadecimal number
3. Write an 8051 assembly program to convert one byte Binary number to BCD.
4. Write an 8051 assembly program to evaluate the factorial of an integer number N using recursive procedure.
5. Write an 8051 assembly program to find GCD of two numbers.
6. Write an 8051 assembly program to sort the number in an array using bubble sort method.

INTERFACING WITH 8051

(a) Interfacing with peripherals like keyboards, LEDs, 7 segment LEDs, LCDs, ADCs, etc,.

1. Sketch and explain the interface of 8x4 key matrix using 8:1 multiplexer, and a 3:8 decoder to 8051 microcontroller. Write an 8051 assembly program segment to input the code of keys.
2. Switch the interface of a single red LED to 8051 microcontroller. Write an 8051 assembly program segment to blink the LED at 5Hz.
3. Sketch the interface of a 4x7 segment LEDs to the 8051 microcontroller. Write an 8051 assembly program segment to display I.I.Sc.
4. Sketch the interface of a 16ch x 1line LCD to the 8051 microcontroller. Write an 8051 assembly program segment to display NPTEL.
5. Sketch the interface of an ADC 808 to the 8051 microcontroller. Write an 8051 assembly program segment to read an analog signal through the ADC.
6. Sketch the interface of a DAC 08 to the 8051 microcontroller. Write an 8051 assembly program segment to output a ramp signal through the DAC.
7. Sketch the interface of a RTC 1307 to the 8051 microcontroller. Write an 8051 assembly program segment to read and write real time into the RTC.

HIGH END PROCESSORS

(a) Introduction to 80386 and 80486

1. List all the additional features that the 80386 microprocessor has over 8086.
2. What is the main difference between the 80386 DX and 80386 SX microprocessor
3. How much the physical memory can 80386 address in real mode and in protected mode.
4. How are the tasks in 80386 system protected from each other.
5. How is an 80386 switched into virtual 8086 mode during task switch.
6. Describe three major additions or improvements that the 80486 processor has over 80386 processor.

❑❑❑

MODEL TEST PAPERS

SAMPLE PAPER-1

UNIT-1

1. (a) Explain in detail the Execution Unit of 8086 μP, What are the applications of registers present in EU ?
   (b) Discuss in brief that how the evolution of microprocessors took place. What are various applications of microprocessor ?
2. (a) What is the function of the 8086 queue and how docs it speed up process operation ?
   (b) What are main differences between an 8086 operating in minimum mode and operating in maximum mode ?
   (c) Discuss how the 8086 CIR and reset signals are generated using 8284.

UNIT-II

3. (a) Explain the following instructions
   IDIV, ESC, JNA, SCAS, LOCK.
   (b) Explain instruction formats of 8086 μP.
   (c) What do you understand byProcedures and Macros ? Discuss.
4. (a) Write a 8086 program to produce a packet BCD from two ASCII encoded digits. The first ASCII digit (3) is located in AL. The second ASCII digit (9) is located in BL. The result (the packed BCD) is to be left in AL.
   (b) Write the 8086 instruction that will perform the indicated operation :
      (i) Copies AL to BL.
      (ii) Copies DL to a memory location whose offset is in BX.
   (c) What do you understand by the term modular programming ? Discuss in brief.

UNIT-III

5. (a) Describe the memory mapped I/O and direct I/O. Give the main advantage and disadvantage of each.

(b) Describe the timing diagram of an 8086 microprocessor signals for Read operation in minimum mode of operation.

(c) Write a short note on TMS 4500.

6. (a) How address decoding is done in 8086 while interfacing ROM chips ? Discuss with the help of some examples.

(b) Draw and discuss the timing diagram of 8086 μP signals during write operation in the minimum mode.

UNIT-IV

7. (a) Draw the discuss the internal block diagram of 8255A PPI.

(b) What do you understand by microcomputer video displays ? Discuss their operation.

(c) Interface 4 × 4 keymatrix with 8086 through 8255 PPI. Draw and discuss the flow chart to perform the following functions :

(i) To test that all the keys are open.

(ii) To find out the key code for the pressed key.

8. (a) Discuss briefly the D/A converter operation and application with respect to interfacing with μP.

(b) What is intel 8259 chip ? Discuss its use and operation in a 8086 microprocessor based system.

(c) Describe the interrupt response of an 8086 processor.

SAMPLE PAPER-2

UNIT-1

1. (a) If the Data Segment Register DS contains 4000 H, what physical address will the instruction MOV AL. [234 BH] read ?

(b) Discuss the advantages of segmentation of address space in 8086 microprocessor.

(c) Discuss how 8086 operates in maximum mode. Compare minimum mode and maximum mode of operation.

2. (a) Give the pin diagram of 8086 μP chip. Discuss the functions of each pin.

(b) Discuss how 8086 C/K and reset signals are generated using 8284 ?

(c) Discuss the role of 'WAIT STATE' in the operation of 8086 μP. How these are generated ?

UNIT-II

3. (a) Using WHILE-DO structure, draw a flow chart. Write Pseudo code and 8086 programme for the following problems. If the temperature of an oven is less than 100°C, turn the heater ON and wait for the temperature to reach 100°C. If the temperature is at or above 100°C then turn the heater OFF.

(b) What is the difference between recursive and reentrant procedure ? Write the programme for finding the value of n factorial.

4. (a) Write the 8086 program to compute the averages of 4 bytes stored in an array in memory.

(b) Spot the grammatical syntax errors in the following instructions :

(i) MOV, BH, AX

(ii) IN BL, 04H

(iii) ADD AL, 2073 H.

(c) What do you understand by pointers and index registers in 8086 Up ? Discuss in brief.

UNIT-III

5. (a) What do you understand by the term DRAM controller ? Discuss the working of TMS 4500 Controller in brief.

(b) Draw and discuss the timing diagram of 8086 μP during Read and write operations.

6. (a) How address decoding is done in 8086 while interfacing ROM Chips ? Describe with the help of a suitable example.

(b) What is the difference the memory mapped I/O and direct I/O ? Give the main advantages and disadvantages of each.

UNIT-IV

7. (a) Describe the use of CAS 0, CAS 1, and CAS 2 lines in a system with a cascaded 8259 A.

(b) Describe the functions with pin diagram of the following chips :

(i) 8255 (ii) 8251.

8. (a) Describe the role of a DMA chip in Microprocessor Based Systems.

(b) The starting address of the subroutine is 934 B : 1252H. If the interrupting device supplies vector type 41 H, what are the locations where the starting address of the subroutine is stored ?

(c) Write short note on "Microcomputer Video Displays".

SAMPLE PAPER-3

UNIT-1

1. (a) Explain the following 8085 pins :

RST 7-5, SOD, READY, ALE.

(b) Draw timing diagram of OP-Code fetch cycle of 8085.(7)

(c) List the steps 8085 will execute to process : CALL 2070 H instruction.

2. (a) Draw cell structures of ROM, PROM, DRAM, UV PROM memories.

(b) Interface 16K × 8 RAM, 8K × 8 PROM with 8086 Mention memory ICs.

UNIT-II

3. (a) Draw block diagram of 8086 μP and explain its PSW.
   (b) Explain the concept of memory segmentation and its advantages.
4. (a) Explain following 8086 instructions :
   XLAT, LDS, AAM, SCASW.
   (b) Explain following assemble directives :
   EXTRN, PUBLIC, PTR, EQU.
   (c) Write 8086 ALP to generate 10 elements of Fibonacci series.

UNIT-III

5. (a) Explain working of 8255 in BSR mode and "Bidrectional bus" mode. Write control words and status words along with suitable diagrams.
   (b) Interface AD 574 ADC with 8086 using 8255. Use INTR pin of 8255.
   Interface one 7-seg display and one push button key with 8086 using 8255 in isolated I/O mode. Write a program to realize up/down decade counter on the display when key is closed.

UNIT-IV

6. (a) Differentiate between interrupt and DMA.
   (b) Explain in detail, Intel's 8237.
7. (a) Draw interconnection showing 8086 interfaced to three 8259's in master slave configuration innon buffered mode.
   (b) Write initialization instructions for all 8259's working with following features : Level triggered. Automatic EOI and fully nested mode. Assume that master is mapped to addresses 007C and 007E, Slave-1 to 008C and 008E, slave-2 to 009C and 009E.

SAMPLE PAPER-4

UNIT-1

1. (a) Explain following : 8085 instructions CC, JC, RNC SI, XTML.
   (b) Write-8085 program to multiply two 16-bit numbers.
2. (i) Explain and draw Interrupt Control unit of 8085.
   (ii) Interface 16 K RAM and 8 K UVPROM with 8085 starting at address 1234 μ. Also draw address decoding circuit and mention the memory map.

UNIT-II

3. (i) Explain the following 8086 signals $\overline{\text{LOCK}}, \overline{\text{TEST}}$, $\overline{\text{RQ}}/\overline{\text{GT}}$, QSI, QSO.
   (ii) Explain 8086 addressing modes.
   (iii) Write a program for 8086, to compute determinant of 3 × 3 matrix.
4. (i) Draw 8086 Read/Write timing diagrams - when operating in minimum mode.

(ii) Explain following 8086 instructions STOSB, LDS, LOCK, NEG.
(iii) Write 8086 ALP to translate temperature from °F scale to °C scale.

UNIT-III

5. (i) Explain Mode 1 of 8255. Write all possible CWs and SWs.
(ii) Design 16-bit I/O port using 8255 and interface it with 8086 using isolated I/O addressing.
6. (i) Explain the terms-Resolution, accuracy, stability and monotonicity of DACs.
(ii) Interface 8-bit DAC with 8085 to derive 12V D.C. motor. Write ALP for variable rpm of the motor.

UNIT-IV

7. (i) Explain block diagram of 8259.
(ii) Interface three 8259s with 8086 in master slave configuration.
(iii) Write 8086 instructions to initialize 8259s.
8. (i) Explain DMA with the help of a block diagram.
(ii) Draw diagram to interface 8237 with 8086.
(iii) Explain 8253 control word format.

SAMPLE PAPER-5

UNIT-1

1. (i) Explain following 8085 signals
(S1, S0), READY, SID, RST5.5.
(ii) Draw timing dagram for execution of instruction :
MVI A, 32 H.
(iii) List the steps 8085 will execute to process the instruction :
CALL 2075 H.
2. (i) Draw cell structures of RAM, ROM, DRAM, UUPROM.
(ii) Design decoding circuit to select 65 KB of memory divided into 8 modules. Use 74138 IC and mention address range for each module.
(iii) Differentiate between Memory mapped I/O and Isolated I/O.

UNIT-II

3. (i) Draw block diagram of 8086 μP. Explain address and data register and Queue.
(ii) Explain control flags of 8086. Write instructions to set/reset them.
(iii) Explain the concept of memory segmentation.
4. (i) Explain following assembler directives :
EXTRN, PUBLIC, PTR, ASSUME
(ii) Explain following 8086 instructions :
LOCK ESC, TEST, LOOP.
(iii) Write PROC to multiply two 64-bit numbers.

UNIT-III

5. (i) Explain BSR and Bidirectional BUS mode of 8255.

(ii) Interface 8-bit ADC with 8255. Use interrupt driven I/O to read 10 samples. Write ALP.

6. (i) What is sucessive approximation type ADC ? Explain its principle and draw its block diagram.

(ii) Design successive approximation type ADC using 8-bit DAC, 8255 and 8085.

UNIT-IV

7. (i) Differentiate between Interrupt and DMA.

(ii) Explain ICWs of 8259.

(iii) Interface temperature sensor LM35 with 8086 using 8259. Write ALP to read temperature.

8. Write short notes on the following :

(i) 8237

(ii) 8253

SAMPLE PAPER-6

UNIT-1

1. Explain various types of interrupts indetail in 8085 microprocessor and describe the interrupt process for them in detail and what are the major issues in implementing the interrupts ?

2. (a) Explain following instructions in detail with the help of examples :

(i) RAR

(ii) CPI data

(iii) CALL addr (Label)

(iv) SPHL

(v) XRI data.

(b) Describe sequence of events that occurs when 8085 MPU reads from memory with the help of timing diagram.

UNIT-II

3. (a) Explain the following addressing modes with examples :

(i) Base Index Addressing Mode.

(ii) Register Relative Addressing Mode.

(iii) Intra-segment Indirect Mode.

(iv) Implicit Addressing Mode.

(b) Construct the machine codes and give the addressing mode for the following instructions :

(i) ADD C17D [BX] D1], DX; opcode for ADD = 000000.

(ii) MOV AX, [BX] : Opcode for MOV = 100010.

4. (a) Explain the following instructions with examples :
   (i) AAA
   (ii) XLAT
   (iii) RCL
   (iv) RET.
   (b) Write an assembly language program to find any power of any number.

UNIT-III

5. (a) Explain the features of mode '1' and Mode '2' of 8255 in detail.
   (b) Interface 16-bit 8255 ports with 8086. Address of port 'A' is FOH. Draw interface circuitry.
6. (a) Interface digital to analog converter of R-2R ladder type with an assembly language program to generate a sawtooth waveform of period 1 ms with V equals to 5 volts. Take prot 'A' as an output port.
   (b) Explain operation of Successive Approximation Type analog to digital converter with the help of circuitry.

UNIT-IV

7. (a) Explain the architecture and register organisation of DMA controller (8237) in detail with diagram.
   (b) Explain the following modes of DMA operation with 8237 in detail :
   (i) Demand Transfer Mode.
   (ii) Cascade Mode.
8. (a) Give the architecture and signal description of programmable interval timer chip 8253 using diagram.
   (b) Give pin configuration of 8259 i.e. priority Interrupt Controller.

SAMPLE PAPER-7

UNIT-1

1. (a) Draw the internal architecture of 8085 processor. Explain in detail.
   (b) Write a program for multiplication of two 16-bit numbers.
2. Define an interrupt. Describe the interrupt response of an 8085 processor.

UNIT-II

3. Draw block diagram of 8086 μp. Explain address and data registers and queue.
4. Explain following 8086 instructions :
   ESC, LOOP, LOCK, ESC.

UNIT-III

5. What is successive approximation type ADC ? Explain its principle and draw its block diagram.
6. Explain BSR and Bidirectional BUS mode of 8255.

UNIT-IV

7. (i) Explain ICW of 8259.
(ii) Differentiate between interrupt and DMA.

8. Write short note on :
(i) 8237
(ii) 8253

SAMPLE PAPER-8

UNIT-I

1. (a) Explain the following instructions in detail with the help of examples :
(i) CPI data
(ii) SPHL
(iii) RAR
(iv) XRI data
(b) Explain various types of interrupts in detail in 8085 microprocessor.

UNIT-II

2. (a) Explain operation of Dual Slope Type Analog to Digital Converter with help of suitable diagram.
(b) Explain the block diagram of 8255.

3. Interface 8-bit ADC with 8255. Use interrupt driven I/O to read 10 samples. Write ALP.

UNIT-III

4. (a) Differentiate between interrupt and DMA.
(b) Write short notes on 8253.

5. Give the architecture and signal description of programmable interval timer chip 8053 using diagram.

SAMPLE PAPER-9

UNIT-I

1. With the help of block diagram and Pin diagram. Explain 8085 microprocessor.

2. Write an assembly language program to find any power of any number.

3. With the help of pin diagram and block diagram. Explain 8086 microprocessor.

UNIT-II

4. (a) Explain operation of successive approximation type analog to digital converter.
(b) Explain the features of mode 1 and 2 of 8255 in detail.

5. Interface 8-bit ADC with 8255. Use interrupt driven I/O to read 10 samples. Write ALP.

UNIT-III

7. (a) Draw the Pin diagram of 8259. Explain the function of each Pin.
(b) Draw the block diagram of 8253 and explain the function of each block.

8. Discuss the DMA process. Draw and explain the block diagram of 8237 DMA.

SAMPLE PAPER-10

UNIT-I

1. (a) Discuss addressing modes of 8085 micro processor
(b) Describes the machine cycle of 8085.

2. Draw cell structures of RAM, ROM, DRAM, UV PROM.

UNIT-II

3. (a) Explain control flags of 8086. Write instruction to set/reset them.
(b) Draw block diagram of 8086 μp.

4. (a) Explain following 8086 instructions-ESC, LOOP, TEST
(b) Write PROC to multiply two 64-bit numbers.

❑❑❑

INDEX

A

B

C

D

E

F

G

H

I

J

L

M

N

O

P

R

U

V

W

Z

❑❑❑

Win Prizes

ATTENTION READERS

Win Prizes

We request for your frank assessment regarding various aspects of the book as given below :

MICROPROCESSORS AND INTERFACING

By : NIKHIL MARRIWALA

Please Fill up the Sheet in Neat and Clean Writing

(i) What topics of your syllabus, important for your examination point of view, are not covered in this book ?

....

....

....

....

(ii) What are the chapters wherein the treatment of the subject matter is not systematic or organised or updated ?

....

....

....

....

(iii) Have you come across misprints/mistakes/factual inaccuracies in the book ? Please specify the chapters and the page numbers.

....

....

....

....

(iv) What other books on the same subject have you found/heard better than the present book ? Please specify in terms of price and quality.

....

....

....

....

(v) Further suggestions and comments for the improvement of this book.

....

....

....

....

(PS : If need be please attach a separate sheet to write your views)

OTHER DETAILS

(i) Who recommended you this book ? (Please tick (✓) in the box near the option relevant to you.

☐ Teacher ☐ Friends ☐ Bookseller

....

....

....

....

(ii) Name of the recommending teacher, his designation and address:

....

....

....

....

(iii) Name and address of the bookseller you purchased the book from :

....

....

....

....

(iv) Name and address of your institution (Please mention the University or Board as the case may be)

....

....

....

....

(v) Your name and complete address :

....

....

....

....

(vi) Course you are enrolled in :

....

....

....

....

The best assessment will be awarded each month : The award will be in the form of our publications as decided by the Editorial Board.

Please mail the filled up coupon at the vearliest to :

S.K. KATARIA & SONS®

4885/109, Prakash Mahal,

Dr. Subhash Bhargav Lane,

Opposite Delhi Medical Association,

Daryaganj, New Delhi-110002 (INDIA)

☎: +91-11-23243489, 43551243;

Telefax: +91-11-23243489

e-mail: katariabooks@yahoo.com; katariabook@gmail.com

Website: www.skkatariaandsons.com